Foreland Basins and Fold Belts

Edited by

Roger W. Macqueen
and
Dale A. Leckie
Geological Survey of Canada
Institute of Sedimentary and Petroleum Geology
Calgary, Alberta, Canada

AAPG Memoir 55

Chief Editor of World Petroleum Basin Memoirs:
Anny B. Coury

Published by
The American Association of Petroleum Geologists
Tulsa, Oklahoma, U.S.A.

Published 1992

ISBN: 0-89181-334-9

Association Editor: Susan A. Longacre
Science Director: Gary D. Howell
Publications Manager: Cathleen P. Williams
Special Projects Editor: Anne H. Thomas
Production: Custom Editorial Productions, Inc., Cincinnati, Ohio

Basin Memoir Editors:

Active Margin Basins
Kevin T. Biddle

Interior Cratonic Basins
Morris W. Leighton
Dennis R. Kolata
Donald F. Oltz
J. James Eidel

Divergent/Passive Margin Basins
J. D. (Jack) Edwards
P. A. Santogrossi

Interior Rift Basins
John C. Harms
Susan M. Landon

Foreland Basins and Fold Belts
Roger W. Macqueen
Dale A. Leckie

AAPG
Wishes to thank the following
for their generous contributions
to

Foreland Basins
and
Fold Belts

Dr. Robert N. Erlich

Marathon International Petroleum G.B. Ltd.

Contributions are applied against the production costs of the publication, thus directly reducing the book's purchase price and thereby increasing the availability of the volume to a greater audience.

About the Editors

Roger W. Macqueen is Head of the Western Canada Regional Studies group at the Geological Survey of Canada's Calgary office, the Institute of Sedimentary and Petroleum Geology. Born and raised in Toronto, Canada, Dr. Macqueen holds Bachelor's and Master's degrees in Geology from the University of Toronto (1958, 1960), and a Ph.D. in Geology from Princeton University (1965). In addition to work with the Geological Survey of Canada, he has been employed in the oil industry and mining industry, and in geological consulting: in these positions he has conducted research and geological field studies in England, the Persian Gulf, and the United States, as well as many parts of Canada. Earlier in his career Dr. Macqueen was Professor of Earth Sciences at the University of Waterloo, Ontario, Canada for nine years, and Associate Professor of Earth Sciences at the University of Toronto for one year. He is a past president of the Geological Association of Canada, and has been on the executive committees of the Canadian Society of Petroleum Geologists and the Rocky Mountain Section of the Geological Society of America. He has also been Chairman of the North American Committee on Stratigraphic Nomenclature, and has served on the editorial boards of the *Bulletin of Canadian Petroleum Geology* and the *Canadian Journal of Earth Sciences*. He is a member of the Canadian Association, World Petroleum Congresses.

Dr. Macqueen's general research interests and responsibilities include regional geology and geophysics of sedimentary basins and their mineral and energy resources. His specific interests include the origin of Mississippi Valley-type lead-zinc deposits, and the origin and evolution of the oil sands, heavy oils, and bitumens of the Western Canada sedimentary basin. Previously he has served as a co-editor of two volumes: American Geophysical Union Geophysical Monograph 48, *Origin and Evolution of Sedimentary Basins and their Energy and Mineral Resources* (1989; with R. A. Price and R. W. Hutchinson), and *Sedimentology and Geochemistry, 9th Carboniferous Congress* (1984; with E. S. Belt).

About the Editors

Dale A. Leckie was born in Alberta, Canada. He holds a B.Sc. degree in Geography from the University of Alberta (1977), and M.Sc. (Geography, 1979) and Ph.D. (Geology, 1983) degrees from McMaster University, Hamilton, Canada. From 1983 to 1985 he was employed by Petro-Canada Resources Ltd. in Calgary, first as a research geologist and later as staff geologist. In 1985 Dr. Leckie joined the Geological Survey of Canada's Calgary office, the Institute of Sedimentary and Petroleum Geology, as a Research Scientist.

Dr. Leckie's responsibilities with the Geological Survey of Canada center around the regional geology, sedimentology, and origin of Cretaceous sediments of the Western Canada basin, in subsurface and outcrop. His specific research interests include the study of marine and nonmarine sedimentary strata at basin scales and at detailed scales. He is particularly interested in, and is an authority on, relationships among clastic sequences, sea level changes, and regional tectonics. He regularly serves as a team leader in multidisciplinary studies involving paleontologists, geochemists, and other specialists working together to solve applied stratigraphic problems in subsurface and outcrop settings. He served as co-editor of *Sequences, Stratigraphy, Sedimentology: Surface and Subsurface,* Memoir 15 of the Canadian Society of Petroleum Geologists (1988; with D. P. James). He is an active researcher on sequence stratigraphy within the Cretaceous Resources, Events and Rhythms working group, co-sponsored by SEPM and the Global Sedimentary Geology Program.

In 1992, Dr. Leckie was visiting scientist at the Institute of Geological and Nuclear Sciences Ltd., Lower Hutt, New Zealand. This provided him with an opportunity to study foreland basin sedimentation associated with an active orogenic belt, the Southern Alps of New Zealand.

Foreword

The World Petroleum Basins project was conceived in order to aid explorationists in deciding what basins, or parts of basins, show the greatest promise for the most profitable return on investment. One way of achieving this goal is to supply these explorationists with a broad comparable base of data and concepts to improve their forecasts through analog techniques. Several volumes have been compiled to emphasize the geological factors that combine to make a basin productive, and therefore assist in evaluating the geologic risks of a new venture. Ultimate rankings, risk assessments, and business decisions are usually tempered, and often overridden, by nongeologic factors, such as economic climate, geography, and political stability, which lie beyond the scope of this project.

Based on such considerations, the Publications Committee of the American Association of Petroleum Geologists resolved in 1983 to publish a series of volumes on world petroleum basins, with each volume being devoted to one major class of basins. An ad-hoc committee,[1] headed by Anny B. Coury, U.S.G.S. Denver, was charged with the overall planning of the project, with deciding the optimum number of volumes, and with the general organization of each volume. It was further decided that in each volume (or major basin class) a significant portion should be devoted to the detailed description of one maturely explored "type" or "model" basin, while additional summary papers would discuss other basins or provinces of similar type.

Studies comparing frontier provinces, basins, or undrilled prospects with productive counterparts of similar tectonostratigraphic setting have always played an important role in the evaluation of the hydrocarbon potential of new ventures. Such "analogs," when properly documented and of the same general class as the new target, can assist in establishing plausible limits to as-yet-unknown geologic factors, such as reservoir thickness and distribution, trap size and kind, and maturation history, within the framework of the target that is to be assessed. Statistically, such exercises in comparative geology raise the confidence level of the assessment. They also serve to provide concepts to be evaluated and tested in similar areas.

To use analogs in basin assessment without bias or danger of miscomparison or misapplication, a geologist needs to have access to a broad-based worldwide database that must be integrated with innovative geologic concepts. The assemblage of such concepts and a database is, however, a time-consuming and costly task that most smaller organizations cannot afford. The entry of many smaller oil companies and independents into the international exploration theater during recent decades has therefore increased the demand for a readily available catalog of analogs.

The committee in charge of the petroleum basin series wrestled long and hard over the issues of classification, choice of analogs, and number of volumes needed to do justice to the overall objective. Several recently published classification schemes of petroleum basins and provinces (e.g., Bally, 1975; Bally and Snelson, 1980; Kingston et al., 1983a, 1983b; Klemme, 1975, 1980, 1986) were discussed to arrive at a compromise for volume topics and principal analogs. The ad-hoc committee proposed and the full Publications Committee of the Association voted in 1984 to proceed with the preparation and publication of five volumes, one for each of the following classes:

Divergent/Passive Margin Basins
Cratonic Basins
Active Margin Basins
Foreland Basins and Fold Belts
Interior Rift Basins

For each volume, one or more editors were selected, who then proceeded to solicit contributions for that volume. Each editor was given a generalized table of contents and a suggested list of illustrations so that the series would have internal coherence and so that the basin descriptions would be more easily comparable. We hope that these volumes will become a welcome and valuable collection of analogs assisting explorationists for many years to come.

[1]Kaspar Arbenz, Anny B. Coury, Michael A. Fisher, James A. Helwig, David R. Kingston, H. Douglas Klemme

REFERENCES CITED

Bally, A. W., 1975, A geodynamic scenario for hydrocarbon occurrences: Proceedings of the 9th World Petroleum Congress, Tokyo, v. 2, Applied Science Publication, Essex, England, p. 33–34.

Bally, A. W., and S. Snelson, 1980, Realms of subsidence, *in* D. A. Miall, ed., Facts and principles of world petroleum occurrences: Canadian Society of Petroleum Geologists Memoir 6, p. 9–94.

Kingston, D. R., C. P. Dishroon, and P. A. Williams, 1983a, Global basin classification system, AAPG Bulletin, v. 67, p. 2175–2193.

Kingston, D. R., C. P. Dishroon, and P. A. Williams, 1983b, Hydrocarbon plays and global basin classification: AAPG Bulletin, v. 67, p. 2194-2198.

Klemme, H. D., 1975, Giant oil fields related to their geologic setting—a possible guide to exploration: Bulletin of Canadian Petroleum Geology, v. 23, p. 30-66.

Klemme, H. D., 1980, Petroleum basins—classifications and characteristics: Journal of Petroleum Geology, v. 3, p. 187-207.

Klemme, H. D., 1986, Field size distribution related to basin characteristics, *in* D. D. Rice, ed., Oil and gas assessment—methods and applications: AAPG Studies in Geology #21, p. 85-89.

Table of Contents

Introduction

Roger W. Macqueen
Dale A. Leckie

Geological Survey of Canada
Institute of Sedimentary and Petroleum Geology
Calgary, Alberta, Canada

A RATIONALE AND SOME GENERAL COMMENTS

In an engrossing and timely history of oil and the oil industry, Yergin (1991) observed that hydrocarbons are central to the security, the prosperity, and the very nature of civilization. Three themes underlie the story of oil according to Yergin: (1) the rise and development of capitalism and modern business; (2) the interrelationship of oil as a commodity with national strategies and global politics and power; and (3) the evolution of western societies into "Hydrocarbon Societies," populated by "Hydrocarbon Man." Yergin (1991, p. 14) observed of western societies that: "Today, we are so dependent on oil, and oil is so embedded in our daily doings, that we hardly stop to comprehend its pervasive significance. It is oil that makes possible where we live, how we live, how we commute to work, how we travel—even where we conduct our courtships. It is the lifeblood of suburban communities. Oil (and natural gas) are the essential components in the fertilizer on which world agriculture depends; oil makes it possible to transport food to the totally non-self-sufficient megacities of the world. Oil also provides the plastics and chemicals that are the bricks and mortar of contemporary civilization, a civilization that would collapse if the world's oil wells suddenly went dry."

It is in this vein—the global importance of oil to all aspects of modern society—that the American Association of Petroleum Geologists decided to publish a series of volumes summarizing current knowledge on the diverse types of sedimentary basins, thereby offering new insights into the origins of these basins and the hydrocarbon resources they contain. The treatment of the basins in this series uses the analog approach, whereby attempts are made to answer questions centering around the natures of these basins, how they compare with one another, what is known of their hydrocarbon contents and histories within the framework of basin evolution, and what their future hydrocarbon potentials might be. This volume, the fifth in the series, deals with foreland basins and fold belts. Previous volumes cover active margin basins (Biddle, 1991); interior cratonic basins (Leighton et al., 1990); lacustrine basin exploration (Katz, 1990), and divergent/passive margins (Edwards and Santogrossi, 1989).

The economic significance of hydrocarbons occurring within the deposits of foreland basins and fold belts is immense. The six basins described in this volume are estimated to contain in excess of 81.2 billion m^3 (511 billion bbl) of oil, 238 million m^3 (1.5 billion bbl) of natural gas liquids, 23.6 trillion m^3 (833 tcf) of gas, 481 billion m^3 (17 tcf) of coal-bed methane, and 461 billion m^3 (2.9 trillion bbl) of heavy oil.

FORELAND BASINS AND FOLD BELTS: DEFINITION AND ORIGIN

Price and Mountjoy (1971) were among the first to use the term "foreland fold and thrust belt" for the Canadian Rocky Mountain belt, which falls within the foreland fold and thrust belt category as defined in this volume. Subsequently, Dickinson (1974) formally introduced the term "foreland basin" to describe retro-arc and peripheral basins in which the basin-fill was deposited on continental crust or older rifted margin sedimentary prisms; he also discussed foreland basins and their plate tectonic settings. Later classifications (e.g., Bally and Snelson, 1980; St. John et al., 1984) further categorized foreland basins and their associated fold belts as (1) continental, multicycle basins, including cratonic margin composite basins and accreted margin basins; (2) foredeeps and underlying platforms; or (3) back-arc basins, depending on the details of the present or former tectonic setting. In earlier, preplate tectonic terminology, foreland basins are one variety of the exogeosynclines of Marshal Kay (1951). Price (1973), using the Western Canada foreland basin and fold belt as an example, qualitatively suggested that foreland basins owe their origins to the regional isostatic subsidence that results from the supracrustal load of a developing fold-thrust belt. It is this form of subsidence that creates the foredeep in which sediments accumulate. This elegant hypothesis was developed and refined quantitatively by Beaumont (1981) and by others, as discussed herein.

As will be seen in this volume, all foreland basins and fold belts are involved to some extent with both their plate tectonic settings and, particularly, their relationships to underlying earlier sedimentary basins—commonly platformal or continental margin basins.

For the purposes of this volume, a foreland basin is considered to be a succession of sedimentary rocks deposited in a cratonic region adjacent to an active orogenic belt. Sediments are derived mainly from the orogenic belt and thicken toward it. Foreland basin sediments in proximity to the orogenic belt are commonly involved in post-depositional folding and overthrusting toward the craton, resulting in tectonic shortening and cannibalization of previously deposited material (Leckie and Smith, 1992).

PURPOSE OF THE VOLUME

The purpose of this volume is to provide details of six foreland basins and fold belts (Figure 1) and the hydrocarbon resources they contain. One of these, the Western Canada foreland basin and fold belt, serves as the type example and is accordingly presented in considerable detail. The Western Canada foreland basin was chosen as the type example because it is reasonably mature in terms of hydrocarbon exploration, because it shares many attributes normally regarded as characteristic of foreland basins and fold belts, and particularly because there is a wealth of public domain subsurface information (and publications thereof) on this basin. This phenomenal, publicly accessible database, resulting from more than 100 years of hydrocarbon exploration and enlightened governmental policies, has permitted systematic assembly and public release of most of the pertinent exploration data (with the notable exception of seismic reflection lines).

Volume Outline and Approach

This volume has three parts. Part One deals with the type example, the Western Canada foreland basin and fold belt. Part Two consists of syntheses of five other foreland basins (Figure 1): three from North America (the U.S. Rocky Mountain foreland basin, the Ouachita foreland basin, and the North Slope of Alaska foreland basin and fold belt); the Zagros foreland basin of the Persian Gulf region; and the Eastern Venezuela foreland basin. Part Three consists of an overview and synthesis of foreland basins and fold belts, including comparisons of the various examples presented, discussion of unresolved problems, future hydrocarbon potential, and an attempt to identify/evaluate the factors that are responsible for the highly variable hydrocarbon endowments of foreland basins and fold belts.

Western Canada Foreland Basin and Fold Belt

The geologic setting, stratigraphy, and evolution of the Western Canada foreland basin and fold belt and the hydrocarbon resources it contains are presented in five chapters. Chapter 1, by Dale Leckie and David Smith, provides the setting and stratigraphic framework of the basin, which was an elongate, asymmetrical trough that developed between the Rocky Mountain fold and thrust belt and the stable interior platform of the North American craton. At its maximum extent, this foreland basin, including the United States portion, was more than 6000 km long, stretching from the Arctic Ocean to the Gulf of Mexico, and up to 1600 km wide, extending from Ontario to British Columbia. The bounds of the Western Canada foreland basin are, for this volume, from the southern part of Canada's Northwest Territories (60°N) to the Canada–USA political boundary (49°), and eastward into Manitoba where the influence of cordilleran tectonism disappears. In excess of 6 km of sediment was deposited along the western margin of the basin during its evolution from the Middle Jurassic to the middle Tertiary. As is the case for most other foreland basins, the Western Canada foreland basin is closely linked to the underlying Cambrian to Middle Jurassic cratonic platform basin. The beginning of the foreland basin is customarily marked by the first influx of westerly derived detritus from the rising Cordilleran orogene to the west. However, in Chapter 4, Stockmal et al. indicate that the real beginning of the foreland basin should be placed at that level in the sedimentary column where the first effects of tectonically induced subsidence can be recognized; the difficulty is that this is not an easily recognized boundary. Leckie and Smith outline the stratigraphy of the Western Canada foreland basin, recognizing five depositional cycles, each bounded by major unconformities or by regional lithologic changes and each consisting of one or more clastic wedges mostly derived from the Cordilleran orogene to the west. These cycles are the result of the complex interplay between the tectonic development of the cordillera and eustatic sea level changes. This cyclic framework, with minor variations, reappears in later chapters on basinal development and hydrocarbon play descriptions.

In Chapter 2, Macomb Jervey develops computer-based simulations that are aimed at clarifying the relationships between tectonics and sea level changes, particularly as they affect the distribution and geometric form of siliciclastic wedges such as those making up the Western Canada foreland basin. The interaction of tectonic subsidence, sediment supply, and sea level change is a four-dimensional problem in time and space, and thus can be modeled to arrive at a better understanding of how these variables may have related to one another. This exercise is instructive in that it suggests that the rate and magnitude of relative sea level change played a fundamental role in shaping the internal facies distribution, geometry, and nature of bounding surfaces of the siliciclastic sequences that fill the foreland basin. Although at present the approach must be essentially qualitative and theoretical owing to the lack of precise dating of time surfaces in the rock record and the lack of agreement in some instances on the definition of specific

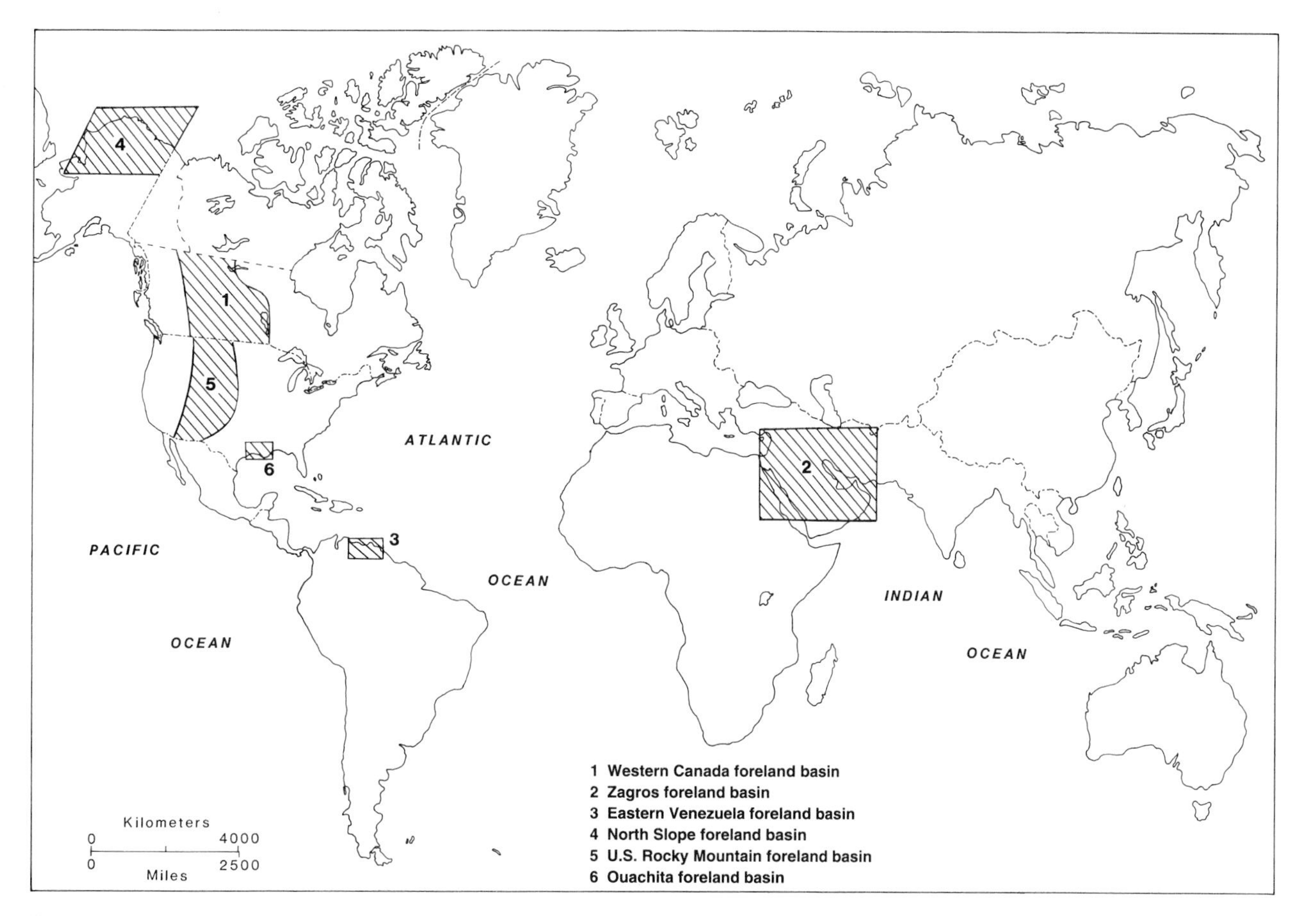

FIGURE 1. Location of foreland fold and thrust belts described in this volume.

stratal sequences, it is clear that the predictive capability of sequence stratigraphy will lead to a greater understanding of the distribution of hydrocarbon source and reservoir facies, and thus to improved methods of petroleum exploration in foreland basins. Jervey then uses two examples from the Western Canada foreland basin to illustrate the general applicability of his modeling.

Peter Fermor and Ian Moffat review the tectonics and structure of the Western Canada foreland basin and fold belt in Chapter 3, emphasizing developments in the cordillera to the west and their possible influence in foreland basin history. They outline the structure of the Rocky Mountain fold and thrust belt, which involves the Mesozoic clastic wedges of the foreland basin and the middle to late Paleozoic sediments of the underlying miogeocline-platform prism. The fold and thrust belt, for much of its length, exhibits the classic thin-skinned structural style. Thrust faulting, with associated folding, was the dominant deformation mechanism in the foreland belt. The style of thrusting and its relationship to folding was influenced by several factors, including the geometry and composition of the deformed sedimentary successions, the magnitude of shortening, and the lateral extent of basal detachments. For example, there is a significant change in structural style between the southern Canadian fold and thrust belt west of Calgary, Alberta and that of the central and northern fold and thrust belt in northeastern British Columbia. Fermor and Moffat review hydrocarbon occurrences in the Foothills belt, which is a significant source of natural gas. The major reservoir units are not foreland basin sediments, however; they are Devonian, Mississippian, and Triassic carbonates, all rocks belonging originally to the underlying miogeocline-platform succession, but deformed during cordilleran tectonism and the development of the foreland basin. The hydrocarbons found in these reservoirs were probably generated around the time of greatest burial, which broadly coincided with the time of maximum thrusting in the Late Cretaceous–Paleocene.

The question of possible relationships between accreting terranes within the developing Canadian Cordillera to the west and the foreland basin clastic wedges is explored by Glen Stockmal, Douglas Cant, and Sebastian Bell in Chapter 4. Geodynamic models have quantified the concept that loading of a cratonic margin by overthrust sheets generates an adjacent foreland basin (e.g., Beaumont, 1981; Stockmal and Beaumont, 1987). This relationship has two important aspects: first, the load flexurally depresses the lithosphere, creating a foredeep into which sediment can be shed; and second, the erosion of the resulting topography of the telescoped overthrust belt provides large volumes of clastic detritus to fill the foredeep. Thus foreland basin stratigraphy and tectonic-loading history are mechanically linked through the lithosphere, and the study of the origin and evolution of the foreland basin has to be treated on a lithospheric scale to be most completely understood. This point raises questions about the mechanical nature and homogeneity of the lithosphere. The tectonic mechanisms involved in foreland basin subsidence include emplacement of overthrust sheets, accretion of exotic terranes, and changes in lithospheric plate boundary dynamics. Geodynamic modeling, although quantitative in nature, at present permits only qualitative conclusions regarding the relative influences of these factors, because of the large number of uncertainties in providing geologic constraints for the models. In theory, the observed foreland basin stratigraphy should be able to be used to estimate the sizes and positions of the inferred tectonic loads. Stockmal et al. observe that the clastic wedges outlined by Leckie and Smith in Chapter 1 can be linked at least generally to the major tectonic events in the cordillera, particularly to the accretion of terranes or superterranes. The central questions here revolve around how far from terrane docking locations such effects will occur, and how closely in time the development of foreland basin clastic wedges matches what is known of the history of terrane accretion events.

In Chapter 5, Jack Porter looks at the discovery and exploration history of the Western Canada basin, specifically presenting information on the foreland basin where possible or appropriate. In the time of discovery from the late 1700s to the mid 1800s, Europeans noted the occurrence of coal, bituminous sands, and salt springs. Serious geologic investigation began about the middle of the 19th century, predating the burgeoning permanent settlement that occurred following the completion of the Canadian Pacific Railway in 1885. In several instances, the first drilling for oil was pioneered by scientists of the Geological Survey of Canada, particularly in what is now known as the Athabasca oil sands deposit. This early work was underlain by the correct interpretation that the bituminous sand represented altered conventional oil, and accordingly was aimed at discovering what were then believed to be vast reserves of unaltered conventional oil below the surface occurrences of the bituminous sands. Although we know this not to be the case, this work heralded later oil sands studies, and began the process of subsurface exploration that led to the oil and gas discoveries of the early years of this century, which underlie the development of the petroleum industry in Western Canada.

Porter reviews the fruits of exploration through 1989 in Chapter 6, on hydrocarbon reserves. Although the emphasis is on conventional oil and gas reserves and their stratigraphic and geographic distributions, Porter observes in passing that reserves of the immense oil sands and bitumens of the foreland basin are estimated at 195 billion m^3, whereas the initial established foreland basin conventional reserves are only 1 billion m^3. Indeed, the area of exposure of the oil sands and bitumens represents about 6% of the total area of the foreland basin, and the Athabasca-Wabasca deposit represents the largest known natural occurrence of hydrocarbons in the world. The problem of the origin of the oil sands and bitumens is dealt with in subsequent chapters. Through the use of carefully curated and publicly available provincial data, Porter has shown that most of the oil and gas discoveries occur in 12 stratigraphic intervals, which relate to the depositional cycles recognized by Leckie and Smith in Chapter 1. Porter's study suggests that the foreland basin originally contained 29% of western Canada's conventional oil, of which 64% had been produced by 1985; the other 71% of the conventional oil, of which 73% had been produced by 1985, was contained by the underlying miogeocline-platform wedge. A further significant fact is that the giant Pembina oil field (97% of reserves), occurring mostly in the Upper Cretaceous Cardium Formation of the foreland basin, originally hosted 34% of the conventional oil reserves of the foreland basin, arguing for particularly favorable source, trap, and seal relationships. With respect to natural gas, the foreland basin contained ~50% of the established reserves through 1985, with about 36% having been produced; the miogeocline-platform wedge contained the other ~50%, with some 41% having been produced by 1985. There were no significant reserve additions from 1985 to 1989. That the Western Canada basin is a reasonably mature basin in terms of oil exploration seems to be borne out by these data and by the fact that, by 1969, 97% of the established reserves of crude oil had been discovered; by 1985, some 61% of the crude oil and natural gas liquids had been produced. In terms of conventional reserves, clearly the foreland basin is a gas-prone basin—although once the immense reserves of oil sands and bitumens are considered, it is also clear that the basin was not always gas-prone. A great deal of oil has been generated. In comparing the Western Canada foreland basin with the Venezuela foreland basin, Porter notes that the Venezuelan reserves of conventional oil are more than three times those of the Western Canada foreland basin. This can be partly accounted for by the normal fault trapping in the Venezuela basin, a feature that is essentially absent from the Canadian scene.

Chapter 7 is concerned with the definition and distribution of the geologic plays that govern oil and gas occurrences in the foreland basin, as presented by Jim Barclay and Dave Smith. A hydrocarbon play is defined (Podruski et al., 1988) as a group of fields, pools, and/or prospects that have common geologic characteristics and a common origin. Fields belonging to a specific play normally share a common trap style, stratigraphic position, depositional and structural setting, diagenetic history, hydrodynamic regime, seal rock, and source rock. Barclay and Smith observe that each of the five depositional cycles or assemblages outlined by Leckie and Smith in Chapter 1 has a proximal, a medial, and a distal component. The building blocks of these cycles or assemblages are coarsening-upward progradational sequences, and the reservoirs are shoreline-related or fluvial sandstones and conglomerates normally occurring at or near the tops of individual cycles. As observed by other contributors, the controls on the nature and distribution of these progradational cycles are tectonic activity, sea level changes, climate, and sediment supply. On the scale of hydrocarbon reservoirs, these factors normally control the variations in trap types. For each depositional cycle or assemblage, Barclay and Smith outline typical traps and reserve figures. This approach is related to, but broader than, the stratigraphic interval treatment used by Porter in Chapter 6. Barclay and Smith observe that assemblage 2, the Lower Cretaceous Mannville assemblage, is the most widespread clastic wedge of the five assemblages and contained 32% of the conventional oil and 53% of the marketable gas. Some 16,000 natural gas pools and 1200 oil pools have been discovered in this interval, which is also famous for its complicated stratigraphy resulting from the interplay of sea level changes, tectonic pulses, and the fact that fluvial sandstones are common reservoirs. Assemblage 3 of the Colorado Group, however, contains the Pembina field—the only giant oil field of the Canadian foreland basin (Carmalt and St. John, 1986; field No. 133 in a worldwide listing of 509 fields grouped according to initial reserves). Barclay and Smith also observe that regional trapping and hydrocarbon filling of the underlying miogeocline-platform wedge succession was fundamentally dependent on the development of the foreland basin, which controlled burial, maturation, hydrocarbon generation, and migration within the miogeocline-platform succession, including the enhancement of updip basin geometry through pronounced westward tilting.

Chapter 8, by Dan Potocki and Ian Hutcheon, deals with the lithology and diagenesis of sandstones in the Western Canada foreland basin. As noted by Barclay and Smith in Chapter 7, and by others, typical foreland basin reservoir facies are shoreline-related or fluvial sandstones or conglomerates. Foreland basin reservoir facies are typically litharenites, rich in lithic fragments and quartz but poor in feldspars. Variations on this theme include the presence at some levels of metamorphic, igneous, and volcanic debris in clastic wedges, and minor volumes of craton-derived quartzose sediment. Because the pattern of diagenesis for many of the sandstones was set by the original framework mineralogy of these rocks, reservoir quality commonly can be predicted by understanding the major framework-dependent diagenetic modifications. These include kaolinite and early carbonate cements (produced by mixing of marine and meteoric waters) and a suite of clay minerals including smectite, illite, and chlorite, produced at greater depths through rock/water interactions. Later modifications of these burial-derived diagenetic assemblages include those reactions attendant on late-stage flushing of foreland basin sandstones by topography-driven meteoric waters that entered clastic wedges in Tertiary uplift caused by the Laramide orogeny. There is also developing evidence that, in certain reservoir facies, relatively fresh water may have interacted with saline (Na-Ca-Cl) waters derived from the underlying miogeocline-platform wedge. This environment appears to have governed the precipitation or solution of carbonate cements. This episode of uplift, introduction of meteoric water, and potential for mixing of highly saline and relatively fresh waters distinguishes the diagenesis of foreland basin settings from those basins in which uplift was not an important factor.

The thermal history of the Western Canada foreland basin is reviewed in Chapter 9 by Kirk Osadetz, Walter Jones, Jacek Majorowicz, David Pearson, and Laverne Stasiuk. These authors argue that, although the geothermal history of a foreland basin can be studied through the use of many indicators, only coalification data exist in sufficient abundance to permit definition of paleogeothermal fields. For the Western Canada foreland basin, coalification patterns result from a geothermal gradient field similar to that observed at present. Geothermal gradients at present decrease toward the Disturbed belt, believed to be the result of advective heat transfer caused by hydrodynamic recharge from regions of elevated topography. This effect should have exerted a similar control on foreland basin sediments from the Middle Jurassic on, wherever topographic relief existed. In the Disturbed belt, coal-rank variations suggest predeformational (normal burial) coalification. For the southern part of the Disturbed belt in southernmost Alberta and British Columbia, anomalous coalification patterns are present. There is much scope for further study of maturation patterns in actual or potential source facies, especially through the use of other maturation indicators.

Chapter 10, by Steve Creaney and Jim Allan, deals with petroleum systems of the foreland basin. By combining available geologic and geochemical databases, these authors derive a history of the source, migration, and trapping of hydrocarbons on a basinal scale. Their work shows that it is impossible to consider the foreland basin alone in terms of petroleum systems. The underlying miogeocline-platform wedge is a fundamental part of the petroleum geology of the foreland basin. Thus the study of petroleum systems has to be on the scale of the Western Canada sedimentary basin, of which the foreland basin is merely the upper part. Two separate petroleum-containing successions of sediments are present in the foreland basin. The younger of these consists of the Upper Cretaceous Viking-Belly River interval, which on compelling geochemical evidence contains oil generated exclusively from the Colorado shale. These oils are virtually unaltered, in strong contrast to oils at lower levels in the foreland basin. The underlying Nikanassin-Mannville interval contains oils generated from a variety of older source rocks that were favorably oriented to facilitate extensive updip and, in some cases, upsection migration. The question of the source(s) of the enormous reserves of oil sands and bitumens remains unresolved, but organic geochemical evidence suggests that these materials were derived from Mississippian or younger source facies: Middle and Upper Devonian oils and source facies have distinctive compositions that are not recognized in any of the oil sands or bitumens.

Examples of Other Foreland Basins

Zagros Foreland Basin, Persian Gulf

The Zagros foreland basin and its supergiant oil fields in the Persian Gulf are described in Chapter 11 by Ziad Beydoun, M. W. Hughes Clarke, and Robert Stoneley. This foreland basin, surely the most important on a global scale in terms of hydrocarbon reserves, contains one-quarter of the oil reserves of the northeast Arabian shelf, which, in turn, contains about 58% and 25%, respectively, of the world's recoverable oil and gas reserves! These reserves dwarf the known reserves in all other foreland basins, including those described in this volume. The Zagros foreland basin and Zagros mountain range were created when oceanic crust of the Arabian plate was subducted northward beneath the Eurasian margin, beginning in the late Eocene. The Zagros foreland basin developed across the Arabian shelf (a divergent or passive margin setting), southwest of the Zagros suture. According to Beydoun et al., on geochemical and geologic evidence, all of the hydrocarbons within the Zagros basin were sourced from the underlying Jurassic to mid-Cretaceous passive margin deposits, which predate the Zagros orogene. Some 12 levels of source facies are known, ranging in age from late Proterozoic to Tertiary. Late Tertiary compressional tectonics associated with the Zagros orogene created vertical migration routes up from breached Mesozoic traps into newly created traps in Tertiary strata. These traps are well-connected, intensely fractured, high-amplitude anticlines. The upper (and highly effective) seal consists of extensive Miocene evaporite deposits. Where the anticlines were breached, hydrocarbons escaped to the surface. Beydoun et al. speculate that as much oil may have been lost as has been trapped! Sedimentary overburden deposited into the Zagros foreland basin increased the maturity of the underlying passive margin source rocks, increasing gas and oil generation from these strata. Reservoirs within the Zagros foreland basin are predominantly carbonates deposited prior to the main continental collision, but deformed during the collision. Thus, the carbonate-dominated Zagros foreland basin differs from all the other basins discussed in this volume in that it was not filled with siliciclastic sediment. The enormous Zagros foreland basin traps are anticlines that can be up to 190 km (118 mi) long with amplitudes of 6 to 10 km (3.7 to 6.2 mi), a fact that, when combined with rich source facies, favorable maturation and tectonic history, and effective evaporite seals, goes some way toward explaining the incredible hydrocarbon richness of this foreland basin setting. The vast horizontal scale of these reservoirs is perhaps a key factor in the richness of this setting. Because of the importance of the pre-foreland basin succession in generating hydrocarbons, Beydoun, Clarke, and Stoneley summarize herein the full history of the Arabian shelf as precursor to the discussion of the Zagros foreland basin.

Eastern Venezuela Foreland Basin

Robert Erlich and Steve Barrett synthesize the Eastern Venezuela foreland basin (actually, two basins) in Chapter 12. The Eastern Venezuela foreland basin is a composite basin, much like the Ouachita foreland basin discussed by Meckel, Smith, and Wells in Chapter 15, consisting of the Guarico and Maturin subbasins separated by the Urica arch. The age of the Venezuela basin is variable; the basin was initiated during the early to middle Eocene in the west and during the late Oligocene to middle Miocene in the east and in Trinidad. In this foreland basin, the style and timing of plate boundary processes were critical to the sedimentary fill. Deformation continues today. Most of the hydrocarbons of the Eastern Venezuela basin are concentrated in the Orinoco tar belt which, along with the oil sands, heavy oils, and bitumen of the Western Canada foreland basin, constitute the world's two major tar sands deposits. The main conventional oil reservoirs occur in Oligocene sandstones with production being largely from normal and reverse fault closures. There is less production from stratigraphic, wrench-fault traps and anticlinal traps associated with overthrusting. This structural style developed as the foreland basin depocenter migrated eastward and southward over time. Much of the trapped conventional oil was derived from underlying Cretaceous strata of a preexisting shelf succession, with migration distances of 150 to 325 km (93 to 202 mi).

North Slope of Alaska Foreland Basin

Sediments filling the North Slope foreland basin, described in Chapter 13 by Ken Bird and C. Molenaar, consist of siliciclastic basinal, basin-slope, shallow marine, and nonmarine deposits. The age of the basin is considered to be Middle Jurassic to late Tertiary. The North Slope foreland basin has a rifted northern margin and a compressional southern margin, a situation that Bird and Molenaar consider to be unique in foreland basins. Within the North Slope region, there appear to be three oil and gas systems, only the upper two of which occur within foreland basin deposits. The most prolific system occurs beneath the foreland basin, and includes North America's largest conventional hydrocarbon field, Prudhoe Bay. Of the 12 plays in the region, however, seven belong to the North Slope foreland basin. Foreland basin reservoir rocks occur in sandstones deposited in environments varying from deep marine to nonmarine. The seven major plays within the foreland basin deposits include structural plays of faulted anticlinal traps, deltaic and turbiditic stratigraphic traps, and structural-stratigraphic traps.

U.S. Rocky Mountain Foreland Basin

The structural and stratigraphic evolution and hydrocarbon distribution of the U.S. Rocky Mountain foreland basin (actually, a series of smaller basins) are described in Chapter 14 by Robbie Gries, John Dolson, and Robert Raynolds. These basins resulted from the collision of the Pacific plates with the North American plate. This composite nature makes the setting of the U.S. Rocky Mountain foreland basin different from that of the Western Canada foreland basin. During the early Tertiary, there were several episodes of uplift within the U.S. Rocky Mountain foreland basin setting, with significant vertical movements of basement-involved compressional blocks that resulted in partitioning of the region into numerous smaller subbasins and highlands. This activity is believed to be related to flat plate subduction, a regional process that was predominant in the deformation of these basins. This form of basement-involved deformation did not take place to such a degree in Canada, perhaps because of the absence of the flat plate subduction associated with the Western Canada foreland basin setting. Future hydrocarbon plays in this region include subvolcanic plays beneath the extensive Eocene–Oligocene volcanic cover. These plays are unique to this setting.

Ouachita Foreland Basin

The Ouachita foreland basin, described by Lawrence Meckel, David Smith, and Leon Wells in Chapter 15, differs significantly from other foreland basins considered in this volume because much of it is covered by younger Mesozoic clastic sediments, and because it is the oldest of the foreland basins considered here, being Pennsylvanian-Permian in age. The Ouachita foreland basin consists of seven present-day structural basins, now mainly located beneath younger sediments. These basins once formed a more extensive late Paleozoic foreland basin but are now separated by a series of arches and/or uplifts. The Ouachita basin apparently resulted from the oblique collision of the Afro-South American plate with North America during the Early Pennsylvanian and the Permian. Basin-fill ranges from alluvial and coastal plain sediments to deep-water turbidites. Because so much of the basin is covered, the question of the source(s) of these sediments is unresolved. Hydrocarbons occur primarily in conventional structural traps and nonconventional "deep-basin traps."

SOME CLOSING REMARKS

The foreland basins examined in this volume include, or are associated with, some of the most hydrocarbon-productive basins in the world. Any attempt to understand and evaluate these basins in terms of their present supplies of and future potentials for hydrocarbons is an exercise in comparison and contrast among a great variety of settings and controls. What are the key factors, and what are those of lesser or little importance? Put another way, why are some foreland basins so incredibly rich in hydrocarbons, whereas others contain only relatively trivial amounts? These are important questions, and they underlie the rationale for this series of volumes. There are other more general questions that occur on a different scale—that of the global arena of hydrocarbon production and consumption in the closing years of the 20th century.

According to sources quoted by Yergin (1991), between 1985 and 1990 the free world's reserves of conventional oil increased from 97.8 billion to 146 billion m^3 (615 billion to 917 billion bbl): nearly all of this increased supply is concentrated in five major producing nations of the Persian Gulf and in Venezuela. By 1990, the Persian Gulf's oil reserves had reached about 70% of the free world's oil.

A complicating factor in the oil equation is the contribution of the burning of fossil fuels to what has become known as the "greenhouse effect"—the increase in global temperatures expected to result from the effects of such gases as carbon dioxide and methane in the upper atmosphere. This factor seems certain to lead to changes in society's patterns of consumption of fossil fuels, but it is difficult to see how these changes can take place over less than a minimum of several decades. The concentration of oil reserves in areas remote from those of high consumption as well as the continuing problem of political instability in the Middle East imply that exploration will continue globally for some time to come: indeed it may be prudent for nations to achieve self-sufficiency where possible. Thus it seems clear in the final years of the 20th century that "Hydrocarbon Man" will persist in his quest for oil and the riches and power that this quest conveys, as so eloquently documented by Yergin (1991). It is also clear that foreland basins will continue to be explored and to provide their share of hydrocarbons for the foreseeable future.

ACKNOWLEDGMENTS

We would like to thank the reviewers of the chapters in this volume, who substantially improved its quality. Reviewers include Sebastian Bell, John Bloch, W. Brosgé, Elliot Burden, Chuck Chapin, Jim Christopher, Jim Dixon, Martin Fowler, K. W. Glennie, H. R. Grunau, Brad Hayes, Dale Issler, Paul Johnstone, Jack Lerbekmo, E. Murany, Margot McMechan, Dave McDonald, Greg Nadon, Peter Nederlof, Jack Porter, Ray Price, Gerry Reinson, Peter Schwans, Glen Stockmal, S. Talukdar, Bob Thompson, and Roger Walker. R. Rahmani provided valuable input into the direction of the volume during its inception. Claudia Thompson assisted with much typing and logistical help.

REFERENCES CITED

Bally, A. W., and S. Snelson, 1980, Realms of subsidence, *in* A. D. Miall, ed., Facts and principles of world petroleum occurrence: Canadian Society of Petroleum Geologists Memoir 6, p. 9-75.

Barclay, J. E., and D. G. Smith, 1992, Western Canada foreland basin oil and gas plays, this volume.

Beaumont, C., 1981, Foreland basins and foldbelts: The Geophysical Journal of the Royal Astronomical Society, v. 65, p. 291-329.

Beydoun, Z. R., M. W. Hughes Clarke, and R. Stoneley, 1992, Petroleum in the Zagros basin: a late Tertiary foreland basin overprinted onto the outer edge of a vast hydrocarbon-rich Paleozoic-Mesozoic passive-margin shelf, this volume.

Biddle, K. T., 1991, Active margin basins: AAPG Memoir 52, 342 p.

Bird, K. J., and C. M. Molenaar, 1992, The North Slope foreland basin, Alaska, this volume.

Carmalt, S. W., and B. St. John, 1986, Giant oil and gas fields, *in* M. T. Halbouty, ed., Future petroleum provinces of the world: AAPG Memoir 40, p. 11-53.

Creaney, S., and J. Allan, 1992, Petroleum systems in the foreland basin of Western Canada, this volume.

Dickinson, W. R., 1974, Plate tectonics and sedimentation, *in* W. R. Dickinson, ed., Tectonics and sedimentation: SEPM Special Publication 22, p. 1-27.

Edwards, J. D., and P. A. Santogrossi, 1989, Divergent/passive margins: AAPG Memoir 48, 252 p.

Erlich, R. N., and S. F. Barrett, 1992, Petroleum geology of the Eastern Venezuela foreland basin, this volume.

Fermor, P. R., and I. W. Moffat, 1992, Tectonics and structure of the Western Canada foreland basin, this volume.

Gries, R., J. C. Dolson, and R. G. H. Raynolds, 1992, Structural and stratigraphic evolution and hydrocarbon distribution, Rocky Mountain foreland, this volume.

Jervey, M., 1992, Siliciclastic sequence development in foreland basins, with examples from the Western Canada foreland basin, this volume.

Katz, B. J., 1990, Lacustrine basin exploration: case studies and modern analogs: AAPG Memoir 50, 340 p.

Kay, M., 1951, North American geosynclines: GSA Memoir 48, 143 p.

Leckie, D., and D. Smith, 1992, Regional setting, evolution, and depositional cycles of the Western Canada foreland basin, this volume.

Leighton, M. W., D. R. Kolata, D. F. Oltz, and J. J. Eidel, 1990, Interior cratonic basins: AAPG Memoir 51, 819 p.

Meckel, L. D., D. Smith, and L. A. Wells, 1992, Ouachita foredeep basins: regional paleogeography and habitat of hydrocarbons, this volume.

Osadetz, K. G., F. W. Jones, J. A. Majorowicz, D. E. Pearson, and L. D. Stasiuk, 1992, Thermal history of the Cordilleran foreland basin in western Canada: a review, this volume.

Podruski, J. A., J. E. Barclay, A. P. Hamblin, P. J. Lee, K. G. Osadetz, R. M. Procter, G. C. Taylor, R. F. Conn, and J. A. Christie, 1988, Conventional oil resources of Western Canada (light and medium): Geological Survey of Canada Paper 87-26, 149 p.

Porter, J., 1992, Conventional hydrocarbon reserves of the Western Canada foreland basin, this volume.

Porter, J., 1992, Early surface and subsurface investigations of the Western Canada sedimentary basin, this volume.

Potocki, D. J., and I. Hutcheon, 1992, Lithology and diagenesis of sandstones in the Western Canada foreland basin, this volume.

Price, R. A., 1973, Large-scale gravitational flow of supracrustal rocks, southern Canadian Rockies, *in* K. A. De Jong and R. A. Scholten, eds., Gravity and tectonics: New York, John Wiley, p. 491-502.

Price, R. A., and E. W. Mountjoy, 1971, Geologic structure of the Canadian Rocky Mountains between Bow and Athabasca Rivers—a progress report, *in* J. O. Wheeler, ed., Structure of the Southern Canadian Cordillera: Geological Association of Canada Special Paper 6, p. 7-26.

St. John, B., A. W. Bally, and H. D. Klemme, 1984, Sedimentary provinces of the world: hydrocarbon productive and nonproductive: AAPG, 35 p.

Stockmal, G. S., and C. Beaumont, 1987, Geodynamic models of convergent margin tectonics: the southern Canadian Cordillera and the Swiss Alps, *in* C. Beaumont and A. J. Tankard, eds., Sedimentary basins and basin-forming mechanisms: Canadian Society of Petroleum Geologists Memoir 12, and Atlantic Geoscience Society Special Publication 5, p. 393-411.

Stockmal, G. S., D. J. Cant, and J. S. Bell, 1992, Relationship of the stratigraphy of the Western Canada foreland basin to cordilleran tectonics: insights from geodynamic models, this volume.

Yergin, D., 1991: The prize—the epic quest for oil, money and power: New York, Simon and Schuster, 877 + xxxii p.

Chapter 1

Regional Setting, Evolution, and Depositional Cycles of the Western Canada Foreland Basin

Dale A. Leckie

Geological Survey of Canada
Institute of Sedimentary and Petroleum Geology
Calgary, Alberta, Canada

David G. Smith

Canadian Hunter Exploration Ltd.
Calgary, Alberta, Canada

INTRODUCTION

The term "foreland basin" was formally introduced by Dickinson (1974) to describe retro-arc and peripheral basins in which the basin-fill was deposited on continental crust or older rifted margin sedimentary prisms. For this volume, a foreland basin is defined as a succession of sedimentary rocks deposited in a cratonic region adjacent to an active orogenic belt. Sediments are derived mainly from the orogenic belt and thicken toward it. Foreland basin sediments in proximity to the orogenic belt are commonly involved in postdepositional folding and overthrusting toward the craton, resulting in tectonic shortening and cannibalization of previously deposited material.

Examples of peripheral foreland basins include the Indo-Gangetic basin south of the Himalayas and the Taiwan basin. In the former, intercontinental collision is occurring between the Indian subcontinent and Asia and there is partial subduction of one continental margin (the B-type subduction of Bally and Snelson, 1980). Peripheral basins will not be discussed further.

The Mesozoic and Cenozoic western North American foreland basin is a retro-arc basin, as depicted in Figure 1. Retro-arc basins form on the cratonic sides of thrust belts, adjacent to magmatic arcs formed by subduction of significant amounts of continental crust (the A-type subduction of Bally and Snelson, 1980). A modern example of a retro-arc foreland basin is that formed adjacent to and east of the Subandean thrust belt of South America.

The western North American foreland basin was an elongate trough that developed between the eastern flanks of the ancestral Rocky Mountains and the stable interior platform represented by the North American craton (Figure 2). At its maximum extent, this foreland basin was more than 6000 km long, stretching from the Arctic Ocean to the Gulf of Mexico and up to 1600 km wide, extending from westernmost Ontario to central British Columbia. In excess of 6 km of sediment was deposited along the western margin of the basin during its evolution from the Middle Jurassic to the middle Tertiary.

The Western Interior Seaway repeatedly occupied this foreland basin, which developed as a direct result of crustal loading along the elongate fold and thrust belt that formed the spine of western North America (Price, 1973; Beaumont, 1981). The major factor controlling basin subsidence and sedimentation was the combination of plutonism, volcanism, and lithospheric loading in the thrust belt caused by subduction that extended from Alaska to Mexico; sediment loading within the foreland played a secondary role (Jordan, 1981). Major fluctuations in sea level, some of which were eustatically controlled, also affected basin sedimentation (Jervey, 1992).

This chapter describes the regional tectonic setting, evolution, stratigraphy, depositional cycles, and paleogeography of the Western Canadian segment of the North American foreland basin and provides the framework for the remainder of this volume. The evolution of the foreland basin succession corresponds to the Zuni sequence

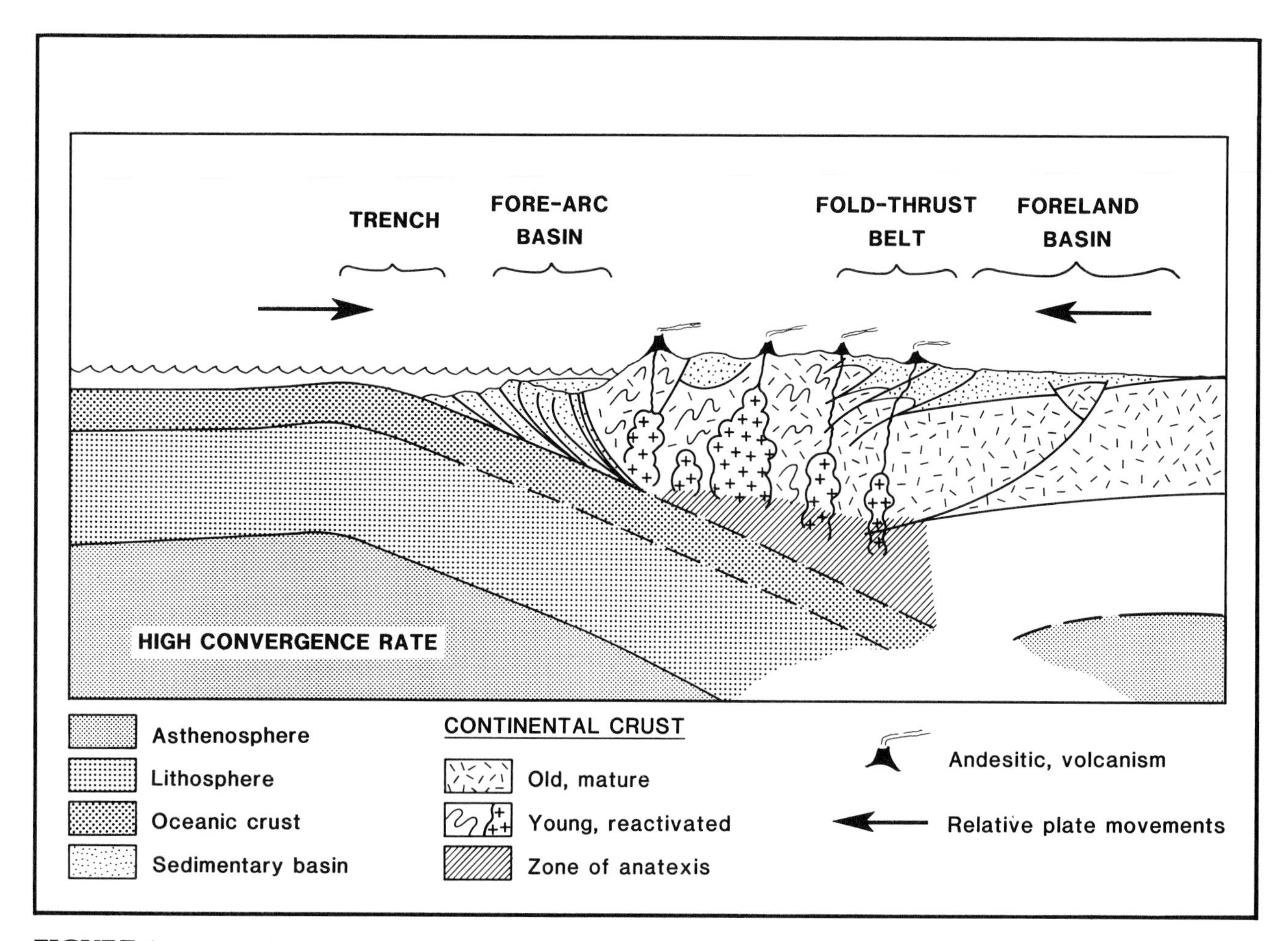

FIGURE 1. **Subduction along an Andean-type continental margin magmatic arc. This model is most commonly used to explain the formation of the Western Canada foreland basin. (Modified from Bally and Snelson, 1980.)**

of Sloss (1963), which extends from the Middle Jurassic to the Paleocene.

EVOLUTION OF THE WESTERN CANADA FORELAND BASIN

Cordilleran Tectonostratigraphic Belts and Complexes

The Western Canada foreland basin is closely related to the evolution of the Western Canada Cordillera, which began in the Early to Middle Jurassic (see Stockmal et al., 1992). Several major collisional tectonic events that occurred from the Middle Jurassic to the early Tertiary resulted in the formation of five tectonostratigraphic zones in western Alberta and British Columbia (Figure 3). These include the Rocky Mountain, Intermontane, and Insular belts, consisting primarily of unmetamorphosed to weakly metamorphosed to low-grade volcanic and sedimentary strata. These belts are separated by two sutural complexes, the Omineca belt and the Coast plutonic complex, comprising high-grade metamorphic and plutonic rocks. The sediments of the foreland basin, located in eastern British Columbia, Alberta, Saskatchewan, and Manitoba, make up a westward-thickening wedge of unmetamorphosed clastics. Older foreland basin sediments have been structurally deformed in the Rocky Mountain fold and thrust belt. A recent summary of the evolution of the Cordilleran orogene has been provided by Monger (1989).

Accretion and Adjacent Foreland Basin Evolution

The distribution and tectonic history of these five accretionary tectonostratigraphic zones suggest that the evolution of the Western Canada Cordillera and adjacent Western Canada basin (including the foreland basin) can be divided into three stages (Porter et al., 1982; Fermor and Moffat, 1992).

Stage 1 represents a generally passive, continental margin terrace wedge that existed throughout the Paleozoic and ended in the Early to Middle Jurassic. This passive continental margin comprised, from west to east, eugeoclinal, miogeoclinal, and platformal sediments that may have been up to 20 km thick in the west (Price, 1981). Events that may have had some bearing on the evolution

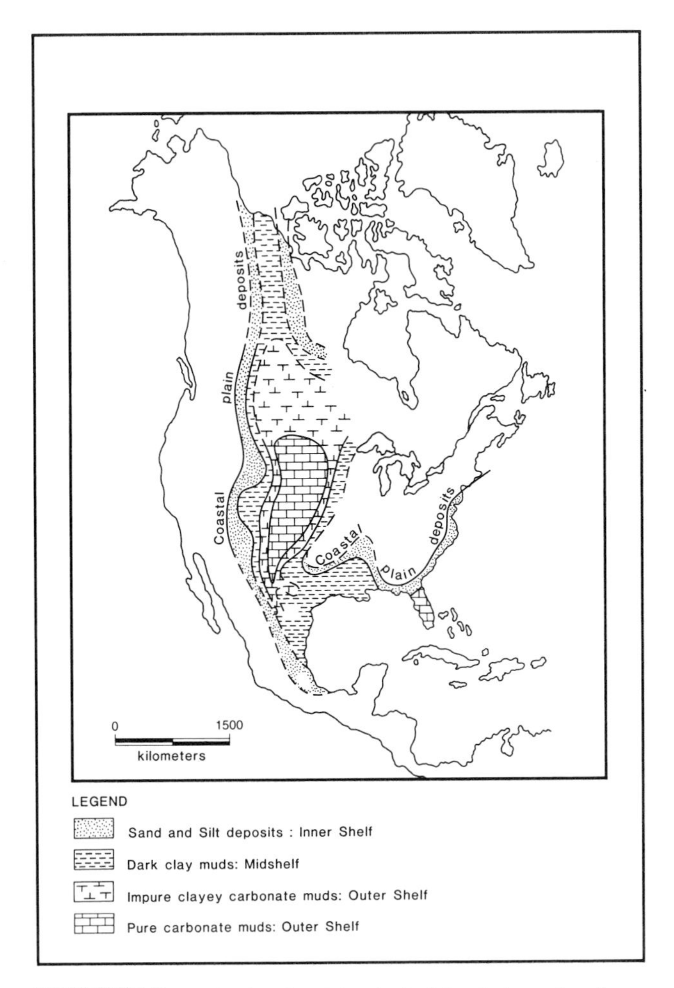

FIGURE 2. The foreland basin in North America during the Turonian (modified from Kauffman, 1984, and McNeil and Caldwell, 1981; paleolatitude from Habicht, 1979).

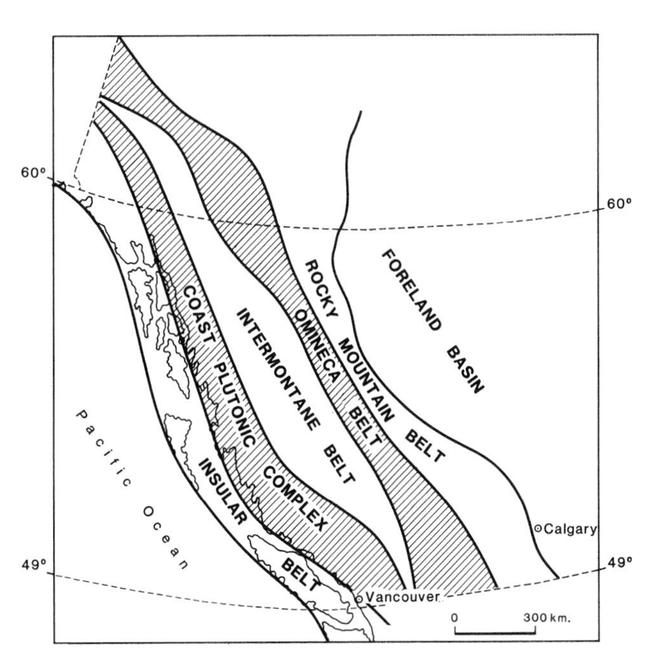

FIGURE 3. The five tectonostratigraphic zones of Western Canada and their relationship to the Western Canada foreland basin.

of the Western Canada sedimentary basin include periods of rapid subsidence, rifting, and possible earlier foreland basin-fill during the late Proterozoic, Late Devonian–Carboniferous, and Triassic (Thompson et al., 1987). During the Late Triassic (230 to 214 Ma), extensive volcanism took place in the Wrangell, Quesnel, and Stikine terranes, now found in the Intermontane belt, prior to their accretion to the western margin of the continent (Armstrong, 1988). Deformation and metamorphism of the Omineca belt began in the Early Jurassic (Price et al., 1985).

Stage 2 records the oblique collision of "foreign" terranes, which now generally correspond to the Intermontane belt (Figure 3), with the westward-moving North American craton (Monger et al., 1982). The Intermontane belt today consists of amalgamated terranes composed of oceanic volcanic arc assemblages on a basement of Triassic and upper Paleozoic rocks (Price et al., 1985). The collision, associated with the Early Jurassic Columbian orogeny (McMechan and Thompson, in press), resulted in the compression of the western part of the passive margin wedge between the Intermontane belt and the North American craton. The suture of the Intermontane belt with the craton is represented by high-grade metamorphic and granitic rocks of the Omineca belt (Figure 3). Metamorphic minerals within the Omineca belt indicate burial depths of 20 to 27 km, probably as a result of westward-dipping subduction (Ghent et al., 1977; Price et al., 1985).

Supracrustal platform rocks were scraped off the westward-subducting continental plate, and then horizontally compressed and translated eastward as stacked thrust sheets now present in the Rocky Mountain fold and thrust belt. Uplift took place where metamorphic rocks of the Omineca belt were wedged under the platform sediments, resulting in a prolific sediment source and the progradation of two early sedimentary cycles into the foreland basin. The tectonic thickening and eastward overthrusting of the continental terrace deposits caused loading and downward isostatic flexure of the underlying lithosphere, resulting in the formation of the foreland basin (Price, 1973). Much of the tectonic thickening may have initially taken place below sea level, thereby limiting the development of an early, extensive, high-relief source area. Upper Jurassic to Lower Cretaceous sediments deposited in the foreland basin consist of quartzose and chert-rich detritus with local occurrences of metamorphically derived muscovite and volcanic pebbles (M. McMechan, cited in Eisbacher, 1985). Price et al. (1985) identified an early and a late component of the Columbian orogeny from the subsidence history of crystalline basement rocks. McMechan and Thompson (in press) indicate that south of 54°N, a major drainage divide was created by sedimentary thrust sheets within the Rocky Mountain belt. North of 54°N, the drainage divide lay farther to the west, within the Omineca belt.

By the Middle Jurassic, most of the foreign terranes had been welded to North America, but considerably farther south than their present locations. Composite right-lateral strike-slip movement from at least the mid-Cretaceous through the Oligocene translocated the accreted terranes northward by up to 900 km (Gabrielse, 1985).

During the latest Jurassic to the late Neocomian Stage, there was generally very little magmatism within the Canadian Cordillera. From the post-Neocomian Stage to the Late Cretaceous, volcanism and associated sedimentation from an Andean-type continental margin magmatic arc prevailed (Armstrong, 1988), resulting in the deposition of a thick coarse-clastic continental succession. During the mid-Cretaceous, convergence of the Intermontane belt with the craton appears to have waned, resulting in a period of tectonic quiescence (Price and Mountjoy, 1970; Stott, 1984b), although changes in plate motion and internal stress regimes may have resulted in large-scale downflexing of the craton (Lambeck et al., 1987). This period also coincided with mid-Cretaceous global sea level rise on the order of 200 to 300 m (Haq et al., 1987). In any event, maximum subsidence and a low coarse-clastic sediment supply during this period combined to produce a thick marine shale succession in the foreland basin, as discussed below.

Stage 3 marks resumption of Late Cretaceous to Paleocene oblique convergence. A second major, foreign exotic terrane, generally corresponding to the Insular belt shown in Figure 3, collided with the North American craton and the previously accreted Intermontane belt, causing the Laramide orogeny. The Coast plutonic complex (Figure 3) represents the suture along which the Insular belt accreted to North America. Renewed thrusting and stacking resulted in eastward expansion of the foreland basin and the deposition of a thick sequence of mostly terrigenous sediments. In the southern Canadian Rocky Mountains, cratonic strata and overlying foreland basin strata were shortened by up to 200 km between the early Campanian and the latest Eocene as a result of this late stage compression (Price, 1981).

A major lull in magmatism in the cordillera from the middle Maastrichtian to the late Paleocene (about 70 to 60 Ma) (Armstrong, 1988) resulted in a massive isostatic uplift of the orogene and foreland basin, and a significant erosion surface developed. Foreland basin–style sedimentation and compressive deformation ceased during the Eocene, with the deposition and subsequent folding of the sediments in the eastern fold and thrust belt (Price, 1981; Wernicke et al., 1987; McMechan and Thompson, in press). During the Eocene, the Omineca belt was subjected to major crustal extension, with concomitant high heat flow and high rates of uplift (Brown and Read, 1983; Archibald et al., 1984; Brown and Journeay, 1987). Within the Western Canada Cordillera, the magmatic lull was followed by intense magmatism from the Paleocene to the middle Miocene (64 to 40 Ma). Right-lateral strike-slip movement along the Tintina–Northern Rocky Mountain trench fault zone continued during this time (McMechan and Thompson, 1989).

The Western Canada Foreland Basin

Foreland basins are asymmetric, being deepest on the fold-thrust belt side because of the combined effects of thrust-plate loading in the orogenic belt and subsidiary sediment loading in the basin itself (Figure 4; Price, 1973; Jordan, 1981). In the Western Canada foreland basin, the amount and rate of subsidence were greatest adjacent to the advancing thrust sheets and it was here that the greatest amount of sediment accumulated, resulting in stacked, westward-thickening clastic wedges. The asymmetric subsidence of the basin also affected drainage patterns, creating a prevailing drainage system that was largely parallel to the basin axis (e.g., Leckie and Walker, 1982; Taylor and Walker, 1984). Recent modeling of foreland basins indicates that basin-axial drainage patterns are characteristic of underfilled basins caused by rapid thrusting (Flemings and Jordan, 1989). By contrast, overfilled foreland basins have a drainage pattern that is orthogonal to the mountain belt. As the fold-thrust belt continued to migrate eastward, pre-foreland basin platformal and miogeoclinal as well as foreland basin strata were progressively incorporated into the deformed belt. The deformed belt was compressively shortened, penecontemporaneous with, and after, basin sedimentation, by up to 50 km in the north (Thompson, 1981) and 200 km in the south (Bally et al., 1966).

Physiographic Components and Stratigraphy

The foreland basin (including foreland basin deformed sediments within the deformed belt) can be divided into five northwest-southeast-trending paleophysiographic components, each distinguished by different water depths, sedimentation rates, facies, subsidence rates, and tectonic stability (Kauffman, 1977; McNeil and Caldwell,

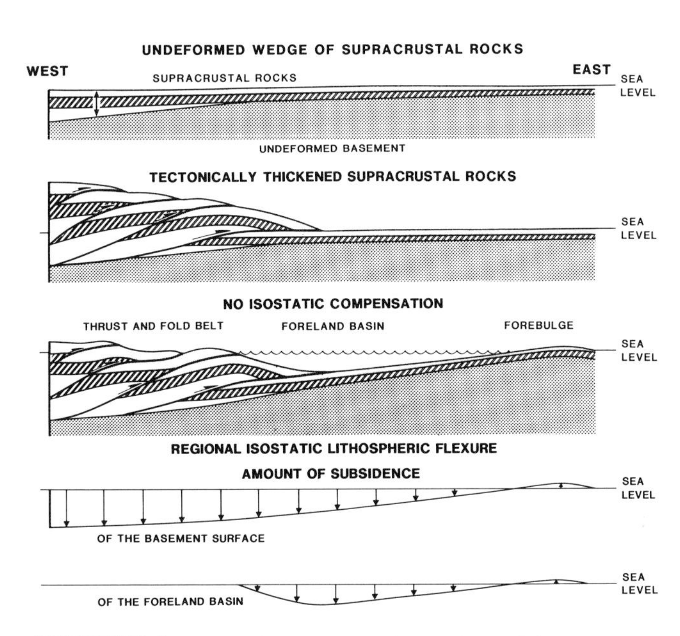

FIGURE 4. Model describing formation of the Western Canada foreland basin. Supracrustal rocks are overthrust as a result of plate accretion and compression. This results in isostatic flexure of the lithosphere and formation of the foreland basin into which synorogenic detritus is deposited. (Modified from Price, 1973.)

1981). The major components are, from west to east (Figure 5):

1. A tectonically active highland.
2. A zone of maximum subsidence.
3. A zone of high subsidence.
4. A broad tectonic hinge zone.
5. A stable eastern platform.

The tectonically active highland is represented by the cordilleran fold-thrust belt and its associated plutonism and volcanism (including the Omineca and Rocky Mountain belts; see Figure 5).

The zone of maximum subsidence is adjacent to the Disturbed belt and bordered on the east by a forebulge. Sediment consists of thick pulses of coarse clastics from the tectonically active highlands. This zone alternated between alluvial, coastal plain and shallow marine conditions. Water depths during marine incursions were probably less than 50 m. This zone ("axis of the foreland basin") shifted progressively eastward in tandem with the shift of cordilleran deformation (Price and Mountjoy, 1970; Stott, 1984b). The forebulge may not necessarily have been a topographic high, depending on the rate of thrusting in the cordillera (Flemings and Jordan, 1989).

The zone of high subsidence is a broad trough that was the deepest-water portion of the basin and received the thickest accumulation of fine- to medium-grained detritus during regressive events; maximum water depths were 200 to 300 m.

The broad tectonic hinge zone migrated in tandem with cordilleran deformation and was characterized by abrupt facies changes and low to moderate rates of sedimentation and subsidence. This zone is dominated by fine-grained clastics and carbonates. The stratigraphic successions are thin and include numerous disconformities. Water depths were 100 to 200 m.

The stable eastern platform is a shelf characterized by low subsidence and low rates of sedimentation. Sediments are fine grained except in coastal areas. Carbonate-rich sediments are also common in the east. Stratigraphic successions are thin with numerous disconformities; water depths were less than 100 m.

In general, sedimentary fill within the foreland basin consists primarily of sandstone, siltstone, and shale with considerable thicknesses of conglomerate along the western margin close to the predominant source area. There were also significant amounts of eastern and northeastern sourced clastics. Thin limestone beds are common in the eastern portions of the basin where the clastic input was significantly reduced during periods of maximum marine transgression.

Figure 6 illustrates a series of representative regional stratigraphic cross sections along and across the foreland basin.

Structural Elements

The Western Canada sedimentary basin is characterized by several major tectonic and physiographic elements

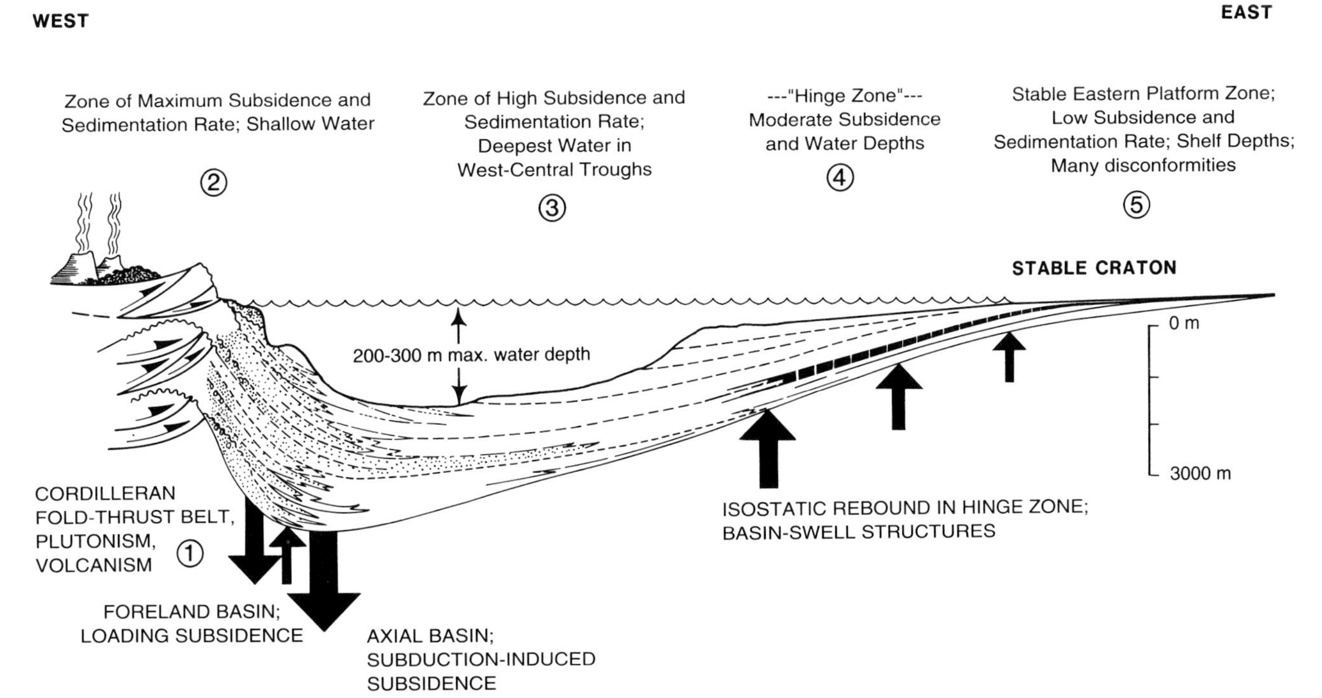

FIGURE 5. Idealized structural and stratigraphic cross section across the Western Canada foreland basin at a time of maximum transgression. The positions, directions, and sizes of arrows indicate relative thrusting, subsidence, and uplift (rebound). The numbers in circles refer to the major components described in the text. (Modified from Kauffman, 1984.)

that were present during the evolution of the foreland basin (Figure 7). The present-day expression of the Rocky Mountains in the Rocky Mountain belt is largely the result of the Late Cretaceous and Paleocene Laramide orogeny.

In northern Alberta, the block-faulted Peace River arch was uplifted during the early Cambrian and remained as a structural high until approximately the Late Devonian, when it began to subside differentially and became a basin. Subsidence continued through the remainder of the Paleozoic and the Mesozoic with as much as 100 to 150 m of subsidence localized in small areas during the Cretaceous (Cant, 1988). During the latest Cretaceous and Tertiary, the Peace River arch again became a positive-relief feature as a result of Laramide deformation farther west (Williams, 1958).

The Sweetgrass arch (Figure 7), located in southern Alberta and northern Montana, separates the Williston basin from the Alberta basin. This arch comprises several subarches, domes, and northwest-trending, north-plunging, faulted folds (Tovell, 1958; Podruski, 1988). The arch was a positive structural feature during the Laramide orogeny and during the Jurassic (Hayes, 1983; Podruski, 1988). During the Neocomian, reactivation of the Sweetgrass arch was responsible for the creation of broad uplands and deeply incised valleys in southwestern Saskatchewan (Christopher, 1964). During the Late Cretaceous, the arch appears to have been inactive (Shepard and Bartow, 1986), but subsequently was uplifted during the Laramide orogeny. The early Eocene (Marvin et al., 1980) plutonic intrusions forming the Sweetgrass Hills in northern Montana occur along the axis of the Sweetgrass arch. The Sweetgrass arch is still expressed in present-day surface geology.

The Punnichy arch, which forms the northern margin of the Williston basin in Saskatchewan, was uplifted during the middle Albian (Christopher, 1984). Much of the Aptian to middle Albian succession in Saskatchewan was eroded over the crest of this feature during a middle Albian lowstand of relative sea level.

The Williston basin, located in Saskatchewan, Montana, and North Dakota, is generally considered to be part of the foreland basin although it is essentially an intracratonic sag that began during the Mississippian and continued through to the Jurassic. Although the expression of a basin is not evident in Lower Cretaceous depositional patterns, isopachs indicate renewed subsidence during the Late Cretaceous (Shurr et al., 1989).

The Swift Current platform, located in southwestern Saskatchewan, dips southeastward within the Williston basin. This feature, located on the eastern flank of the Sweetgrass arch, started to rise during the Oxfordian (Christopher, 1964) but subsided in the middle Albian (Koziol, 1988).

The Alberta syncline, located in western Alberta, is relatively young, having formed during the Laramide orogeny. The eastern limb of the Alberta syncline conforms to the west-dipping Precambrian basement; the

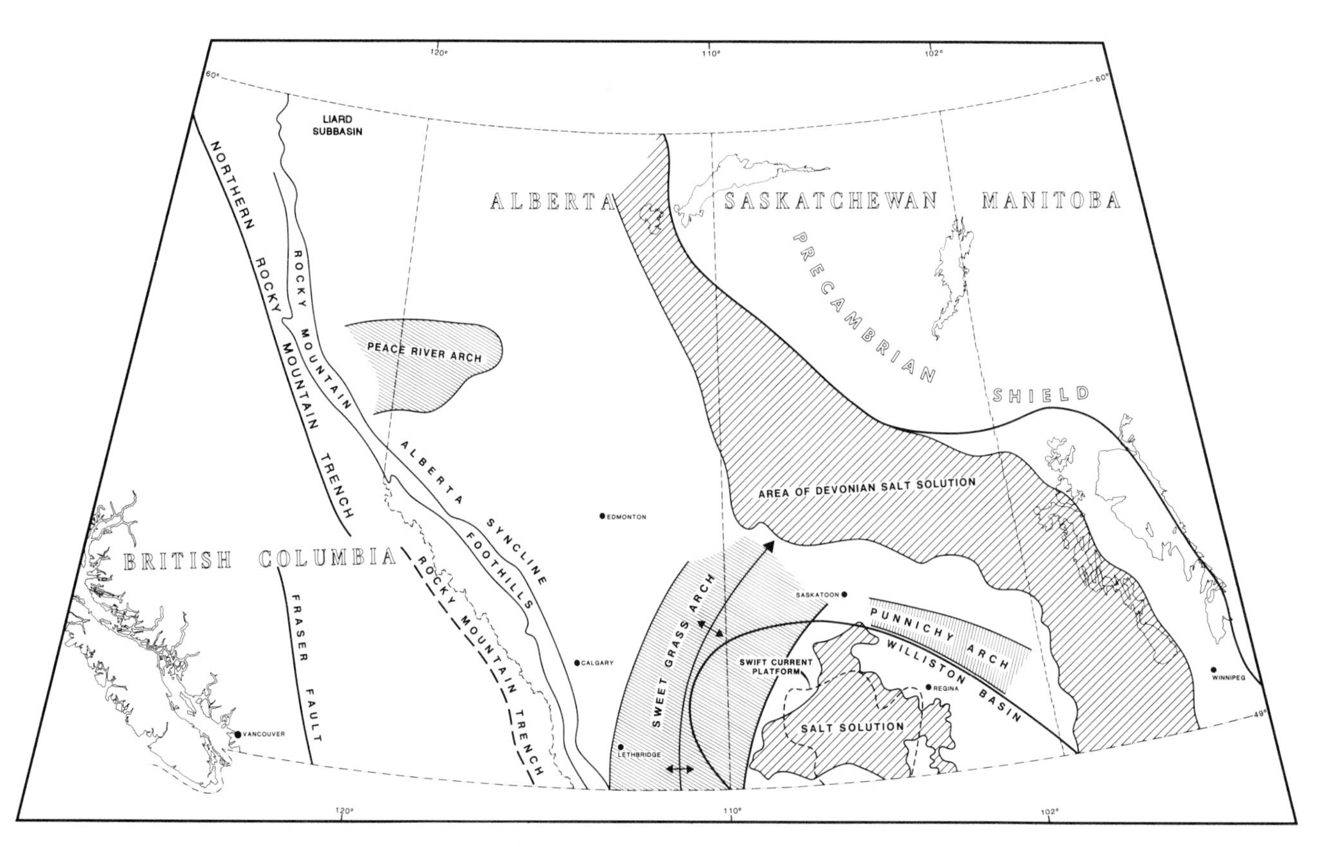

FIGURE 7. Major tectonic and physiographic elements that affected sedimentation within the Western Canada foreland basin.

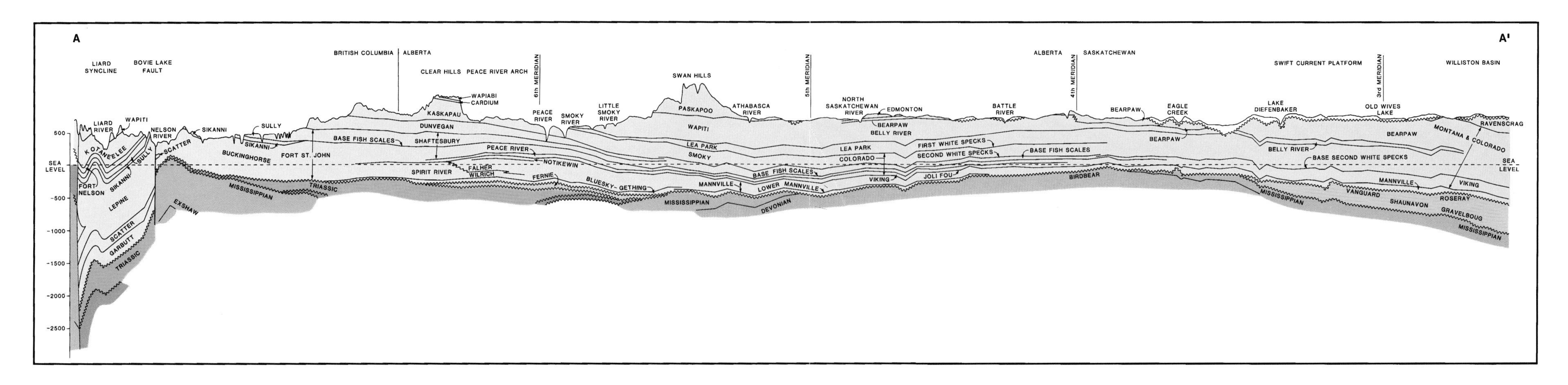

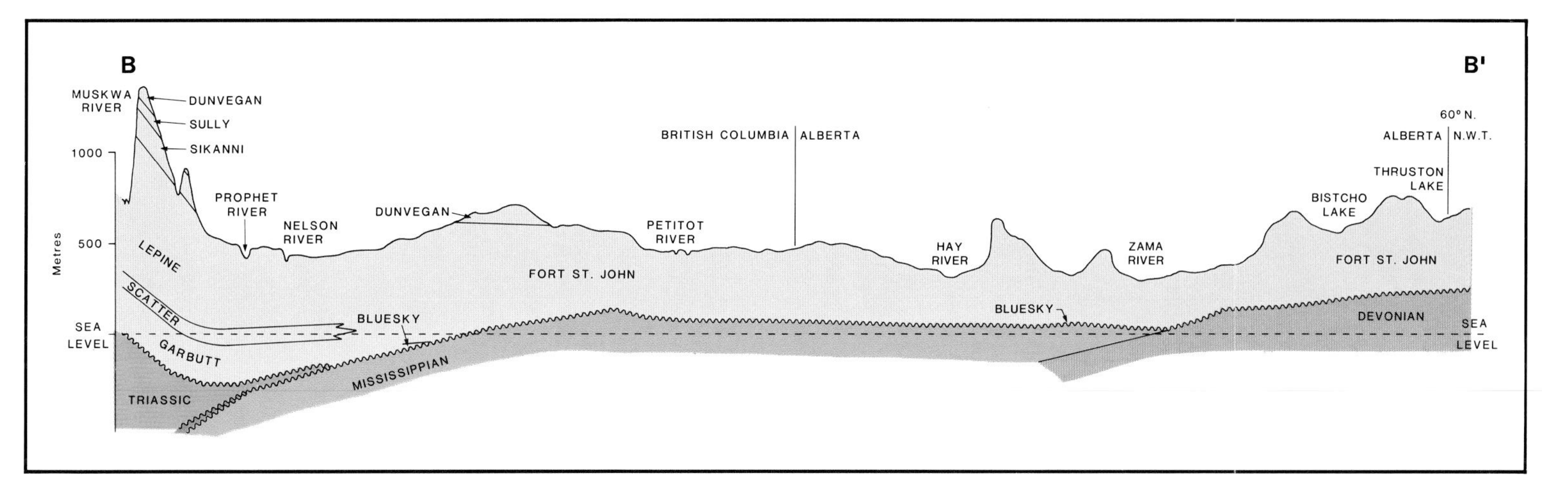

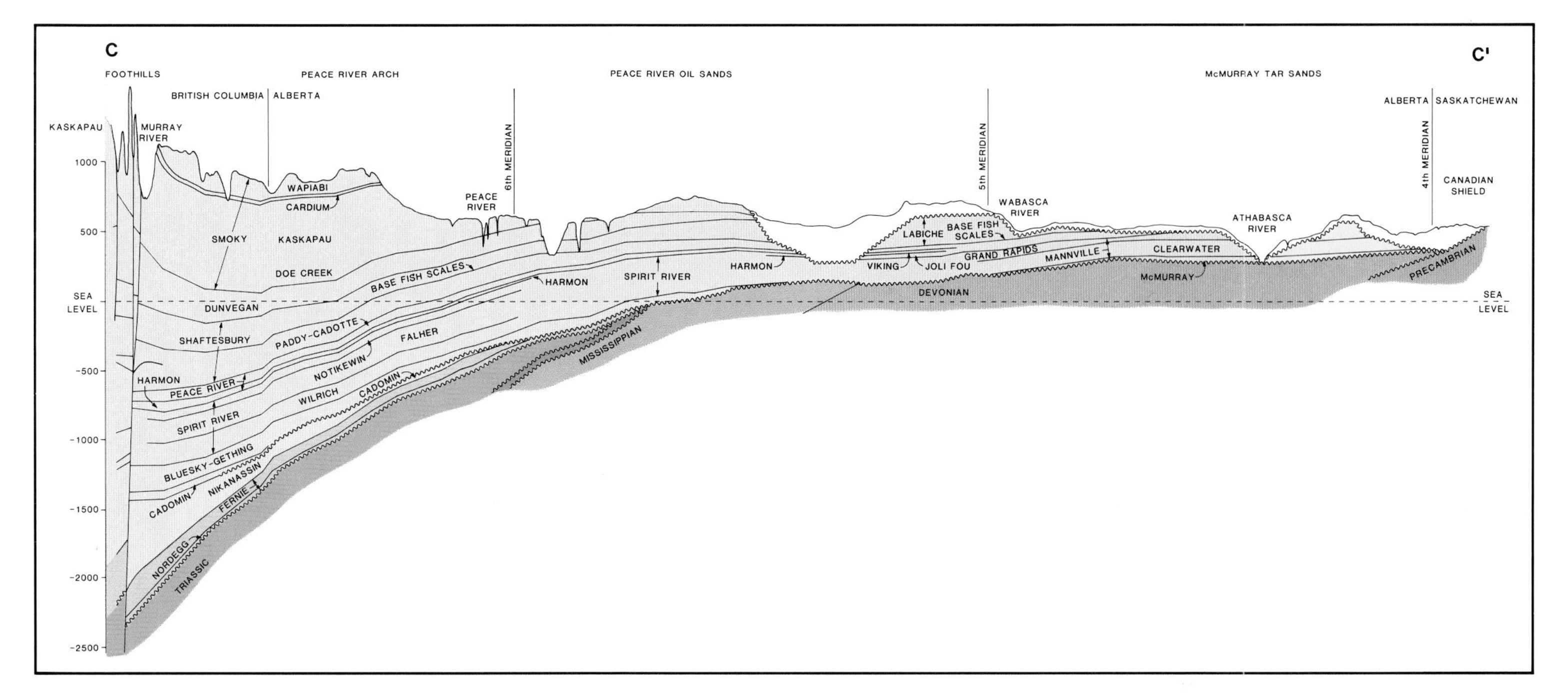

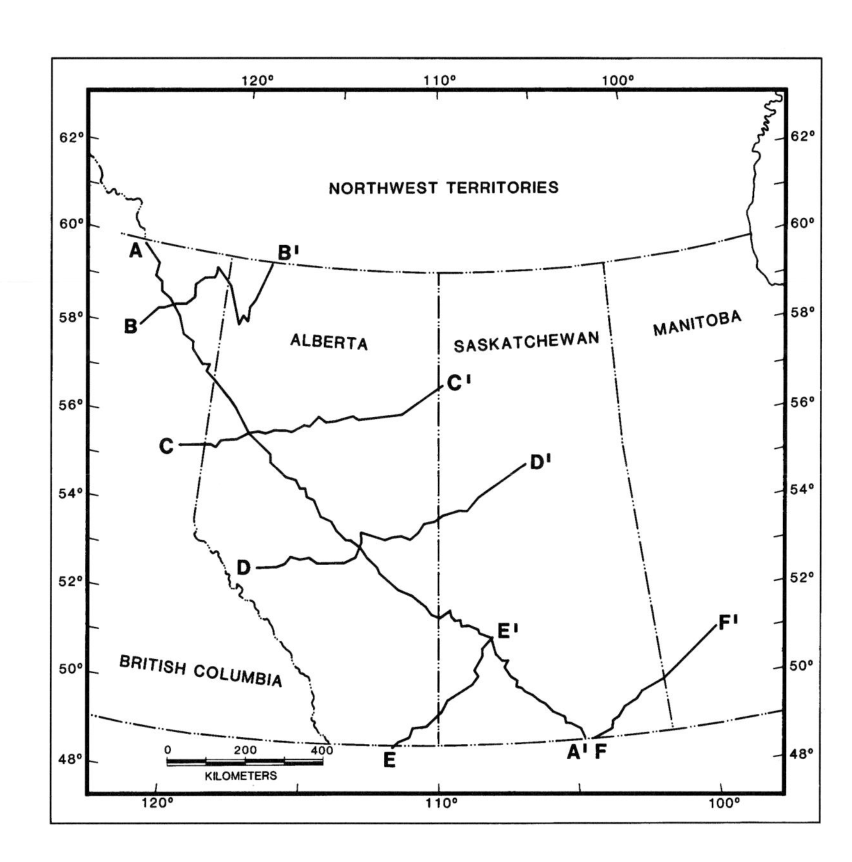

FIGURE 6. **Regional cross sections across the Western Canada foreland basin (modified from Wright, 1984).**

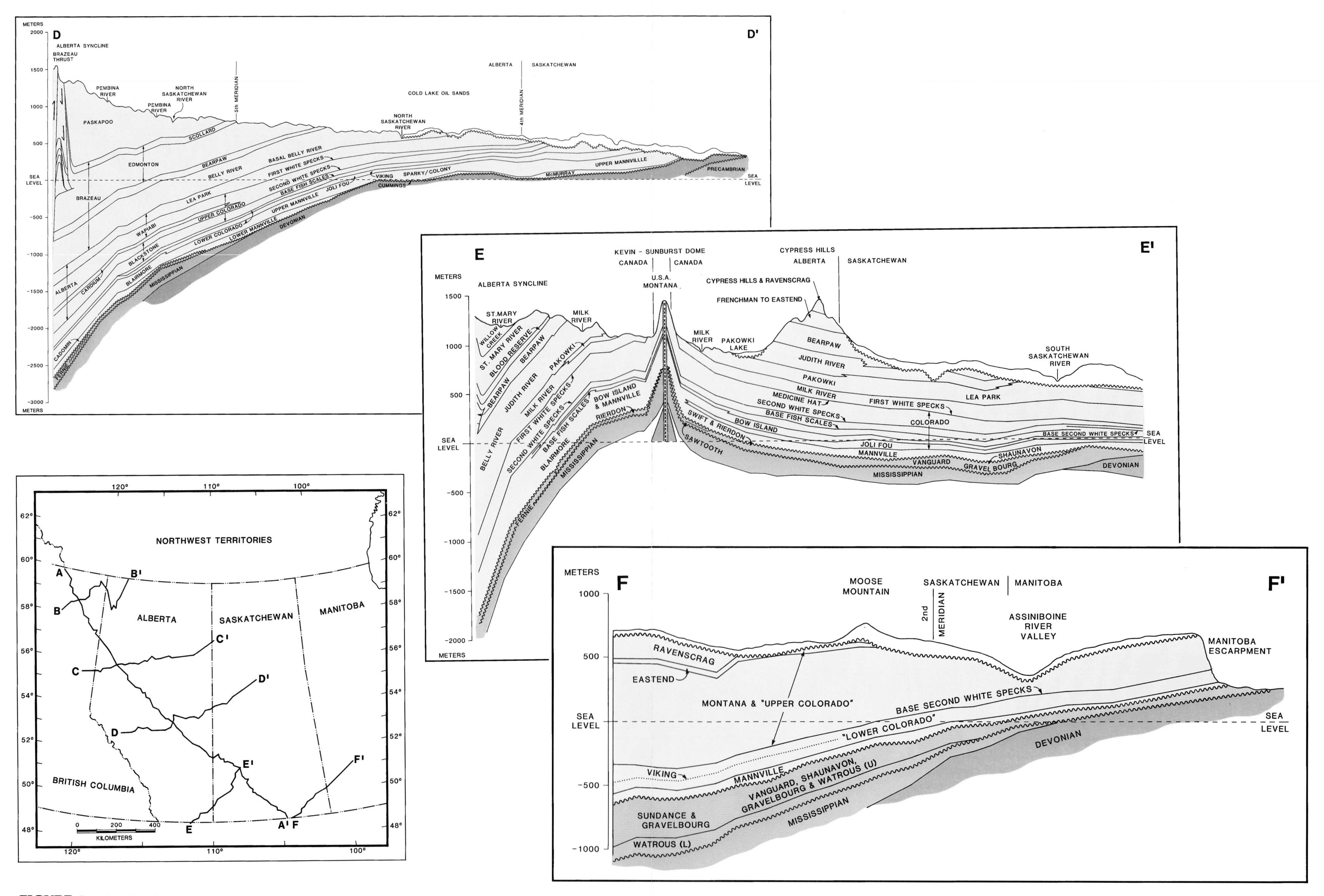

FIGURE 6. (Continued)

western limb is formed by the east-dipping strata at the margin of the Rocky Mountain Foothills.

A series of paleotopographical highs, consisting of Jurassic, Mississippian, and Devonian strata, trended northwest to southeast across the basin during the Neocomian period of uplift and erosion (Rudkin, 1964). These highs subdivided the foreland basin into several subbasins and played a major role in influencing drainage patterns during Aptian and Albian deposition.

Large- and small-scale structures created by the dissolution of Devonian salt beds are common in the eastern part of the basin (Figure 7). Salt-solution structures, which range in age from Late Devonian to Holocene, were generated by the collapse of overlying strata into salt-bed solution cavities that can be up to several hundred meters thick. Devonian salt removal prior to, or penecontemporaneous with, younger sedimentation has also affected sediment thicknesses and depositional patterns in various parts of the basin (Simpson, 1988). The heavy oil and tar sand deposits in Saskatchewan and northeastern Alberta are trapped in part in massive structural traps created by updip salt removal from the Devonian Elk Point Group during the Cretaceous and Tertiary (Edmunds, 1980).

A structure map constructed on the base of the Lower Cretaceous Fish Scale Zone (Figure 8) illustrates the expression of several of the major structural elements including the Peace River arch, Williston basin, and Sweetgrass arch. These structures affected early sedimentation patterns in the basin and are still evident today as is the evidence for extreme subsidence along the Rocky Mountain Foothills.

DEPOSITIONAL CYCLES AND PALEOGEOGRAPHY

Cycles, Bounding Surfaces, and Hiatal Gaps

A regional correlation chart of strata deposited within the Western Canada foreland basin is shown in Enclosure 1(A). For descriptive purposes, the depositional history of the foreland basin has been divided into five cycles of

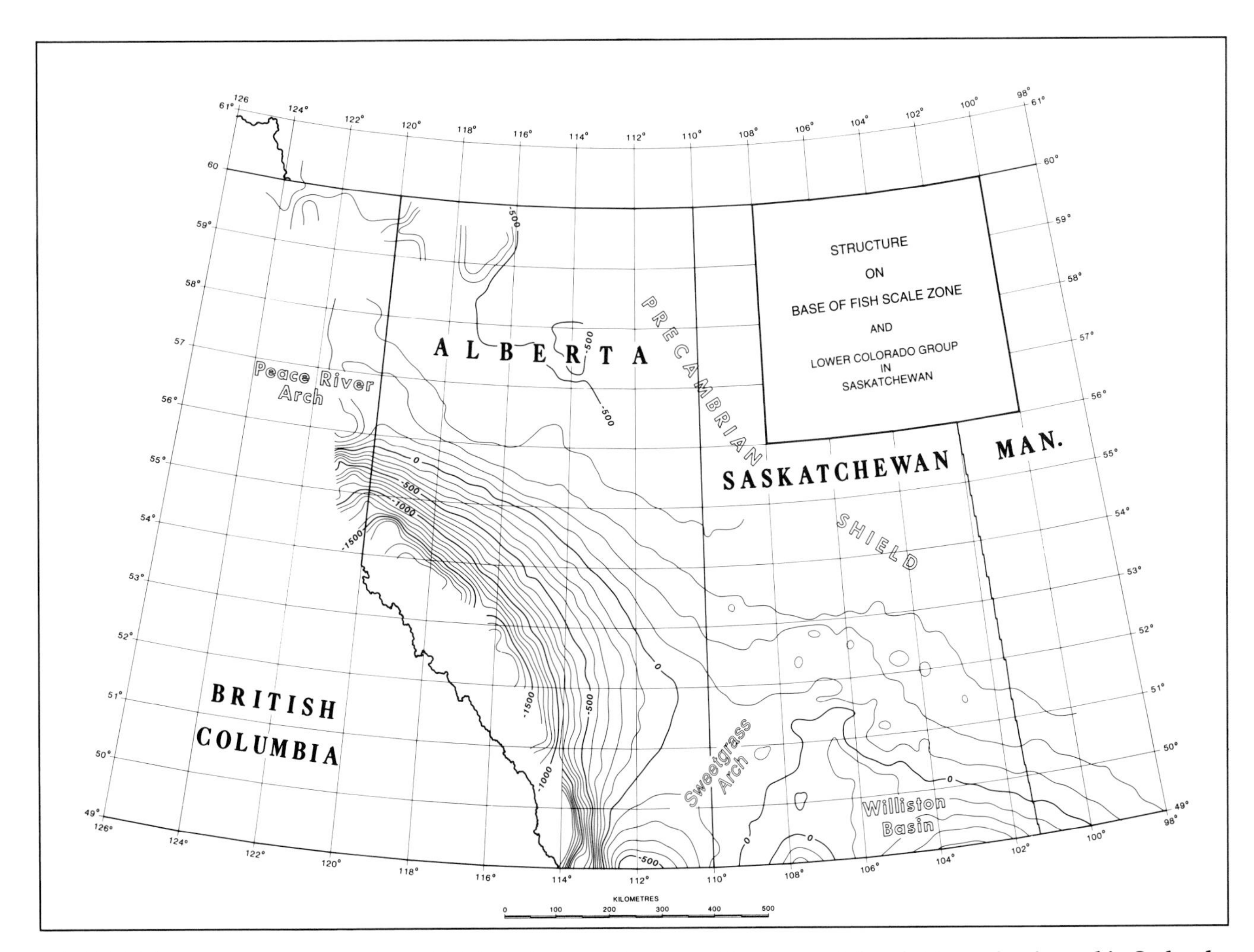

FIGURE 8. **Structure map on base of Fish Scale Zone (the base of the Lower Colorado Formation is used in Saskatchewan). Original data based on preliminary map from the new Atlas of Western Canada. Contour interval, 100 m. Scale, 1:500,000. (G. Mossop, personal communication, 1990.)**

differing ages, each representing strata bounded by major unconformities or lithologic changes. These are, in ascending order:

Cycle 1: Oxfordian to late Valanginian: base of Fernie shales to top Nikanassin Formation/ Kootenay Group

Cycle 2: Hauterivian to Albian: Cadomin Formation/Dina Member to top of the Mannville Group

Cycle 3: Albian to Campanian: base of Joli Fou Formation/Paddy Member to top of Wapiabi Formation/Colorado Group

Cycle 4: Campanian to early Eocene: Saunders Group (base Belly River to top of Porcupine Hills Formation/Paskapoo Formation)

Cycle 5: Eocene to Pliocene: Cypress Hills, Wood Mountain, and Empress formations

The rationale for the subdivision of these cycles is as follows. The base of Cycle 1, in the Green or Gray Beds of the Late Jurassic Fernie Formation, marks the first widespread appearance of sediments that were derived from the rising cordillera to the west. Virtually all older deposits have an eastern, cratonic origin and are considered as "pre-foreland basin" passive margin sediments. The top of Cycle 1 is represented by the unconformity dividing the Lower Cretaceous Cadomin Formation from the underlying Kootenay Group. The hiatus between Cycles 1 and 2 may be in excess of 27 million years (Figure 9). The top of Cycle 2 is marked by the regional flooding event represented by the marine shales of the Colorado/Alberta Group. The top of Cycle 3 is designated as the base of the Belly River/Lea Park/Milk River Formation above which there is a pronounced lithologic change. A major unconformity at the top of the Porcupine Hills/Ravenscrag Formation marks the top of Cycle 4. Cycle 5 is represented by the Tertiary conglomerates that cap the uplands of several areas of Alberta.

Each of the cycles, except the top of Cycle 3, is delineated by major hiatal-bounding surfaces. The contact between Cycles 3 and 4 corresponds to the beginning of a major influx of volcanic sediment into the basin (Stott, 1984b; Mack and Jerzykiewicz, 1989; Potocki and Hutcheon, 1992). Generally, each succeeding cycle records a major reorganization of the paleogeography within the basin and, accordingly, a different regime of petroleum exploration plays as discussed by Barclay and Smith (1992). Although several sequences within the various cycles may approximately coincide with global sea level lowstands or highstands (Haq et al., 1987; Jervey, 1992), each cycle as a whole corresponds to specific tectonic events in the cordillera to the west.

A noticeable aspect of the sedimentary record is the presence of numerous basin-wide short and long periods of nondeposition or erosion. Recent studies have demonstrated that subtle but significant gaps in the sedimentary record are evident at even finer levels of detail. Figure 9 shows the sedimentary record of the middle to upper Albian from three areas in northwestern Alberta and adjacent British Columbia. In the Monkman Pass area of British Columbia, individual paleosols within flood plain deposits of the Boulder Creek Formation may have taken

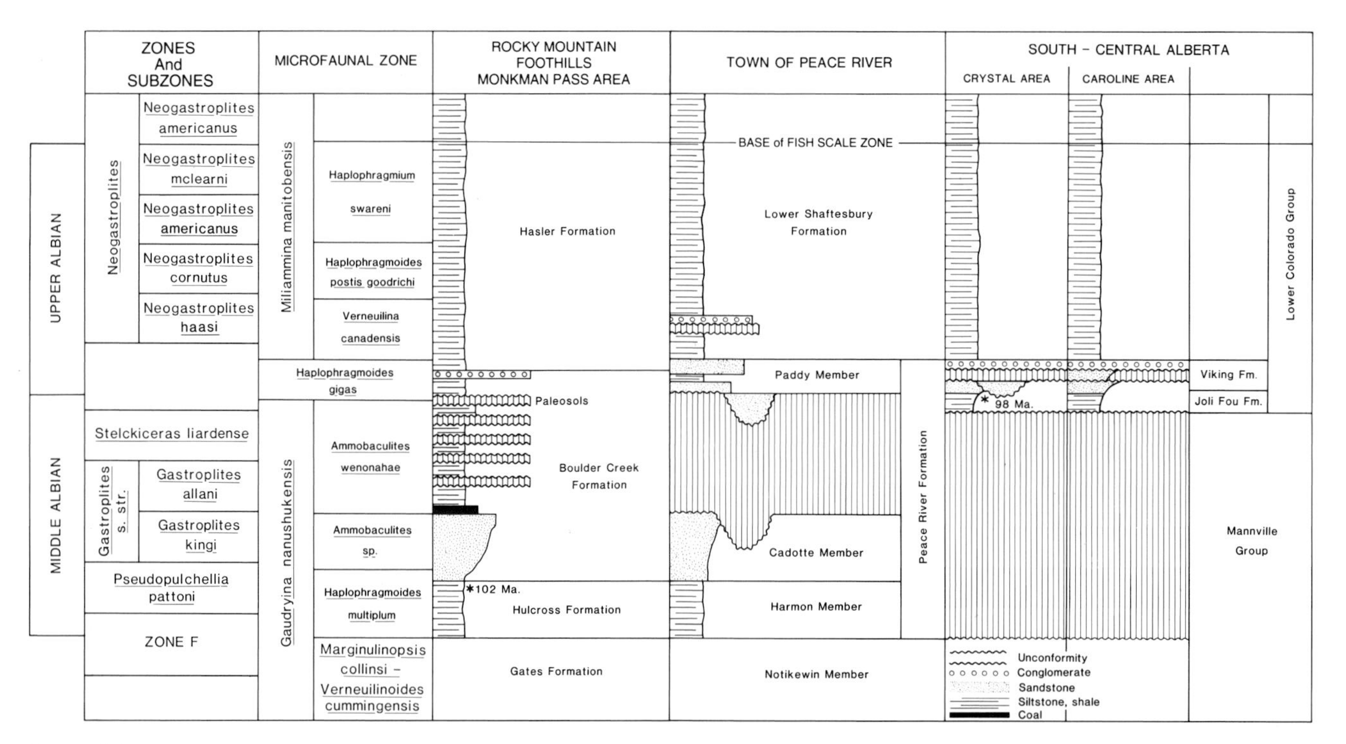

FIGURE 9. Stratigraphy of middle and upper Albian strata from the Foothills of British Columbia (Monkman Pass area), northern Alberta (Peace River), and south-central Alberta. (Modified from Leckie and Reinson, in press.)

as long as 10,000 years to form, indicating a depositional hiatus of at least this duration (Leckie et al., 1989). Cumulatively, 15 paleosols recognized within the 90 m of the Boulder Creek Formation may represent a hiatus of 150,000 years. The development of the paleosols is coincident with periods of valley incisement during sea level lowstand, when the rate of sedimentation on the flood plain was reduced. Near the town of Peace River, 260 km northeast, the relationship of the Paddy Member to the underlying Cadotte Member (Figure 9) is unconformable and is the result of erosion within a large valley that cut into the Cadotte Member during a regional middle Albian lowstand (see Jervey, 1992). The amount of missing section may represent half of middle Albian time. In south-central Alberta, there are hiatal gaps within the Viking Formation and a major hiatus below the Joli Fou Formation where the entire middle Albian section is missing.

A second example illustrating the abundance of time gaps, represented by unconformities, occurs in the Turonian Cardium Formation, which consists of six upward-coarsening mudstone to sandstone sequences capped by erosively based conglomerate bodies (Plint et al., 1986). The conglomerates vary from a thin veneer to 20 m thick. The Cardium Formation was deposited over a period of 1 to 2 m.y., and the unconformities are interpreted to be related to high-frequency changes in relative sea level caused by tectonic loading of the basin (Plint et al., 1988).

Enclosure 1(B) shows the variability of lithology and thickness of the foreland basin succession in different areas of the basin. What is most evident is the large proportion of shale and siltstone which, for the Lower Cretaceous, makes up as much as 65% of the preserved rock record (Parsons, 1973). The largest volumes of sands and gravels were deposited in the southern Alberta and British Columbia Foothills close to the cordilleran source. The greatest sediment thickness occurs in the southern Rocky Mountain Foothills [Section 3, Enclosure 1(B); zone of maximum subsidence, Figure 5] where more than 4800 m was deposited. By comparison, the preserved succession in the Manitoba escarpment [stable shelf, Figure 5; section 7, Enclosure 1(B)] is about 150 m.

Cycle 1: Base of Fernie shales to top of Nikanassin Formation/Kootenay Group (Oxfordian to late Valanginian)

The lowermost cycle of the foreland basin succession contains Oxfordian to late Valanginian sediment deposited as a result of the collisional tectonics associated with the Columbian orogeny (i.e., the collision of the Intermontane belt with western North America). During this time, the foreland basin likely was narrow and elongate, subsiding most rapidly in northeastern British Columbia (Stott, 1984b). The Fernie-Minnes/Kootenay succession [Enclosure 1(A)] reached a thickness of 2.7 km in the Foothills of northeastern British Columbia (Stott, 1984b) and 1.7 km in the Fernie basin of southeastern British Columbia (Gibson, 1977). The end of Cycle 1 is represented by a major basin-wide unconformity. The youngest sediment preserved beneath the unconformity is the late Valanginian Bickford Formation in northeastern British Columbia. Most of the preserved sediment of Cycle 1 is found in the Rocky Mountain Foothills, in the Williston basin, and in the subsurface adjacent to the Foothills of northeastern British Columbia and northwestern Alberta.

Westerly derived sediment from the uplifted Omineca belt and the newly formed Rocky Mountains of the Columbian orogene first occurs in the upper parts of the Oxfordian Fernie Formation [Enclosure 1(A)]. The upper Fernie unconformity (base of Cycle 1) truncates progressively older strata eastward and northeastward. The overlying Green Beds of the Fernie Formation, which consist of glauconitic mudstone and quartzose, craton-derived sandstones, represent the initial flooding of the foreland basin (Poulton, 1984). The younger Passage Beds and Kootenay Group record basin-filling in a northerly direction along the axis of the basin (Figure 10). The Passage beds (siltstone and shale) represent marine deposition during the initial stages of basin-fill. Sandstones of the Morrissey Formation, which is the basal formation of the Kootenay Group, were deposited along a northerly to northwesterly prograding high-energy shoreline. The Morrissey Formation is overlain by the fluviodeltaic quartz and chert-rich sandstones, siltstones, and shales of the Mist Mountain Formation (Gibson, 1977). Extensive peat deposits at the base of the Mist Mountain Formation accumulated on strand plain sediments of the Morrissey Formation. The Kootenay Group grades northward into the increasingly marine sediments of the Nikanassin Formation (central Foothills) and the Minnes Group (northern Foothills) (Stott, 1984b).

In the south-central part of the basin (Figure 11), correlatives to the Oxfordian Green Beds of the Fernie Formation are the glauconitic beds of the lowermost Swift/Success formations (southern Alberta and Saskatchewan). These latter formations consist of noncalcareous shales coarsening upward into lenticular-bedded, glauconitic and quartzose sandstone deposited in a shallow marine shelf to shoreface setting on the northern margin of the seaway. The S-2 or upper Success Formation was deposited by southward-flowing braided and meandering streams (Christopher, 1964). In northern British Columbia, the quartzose littoral sandstones of the Monteith Formation and their counterparts in the Nikanassin Formation are foreland basin sediments but likely were derived from the craton (Stott, 1984b).

Cycle 2: Cadomin Formation to top of Cadotte Member/top of Mannville Group (Hauterivian to Albian)

The contact between Cycles 1 and 2 is represented by a major post-Valanginian unconformity ("pre-Cretaceous unconformity") at the base of the Cadomin Formation and its equivalents. This unconformity extends throughout the Rocky Mountain Foothills, east across the prairies, and onto the Canadian shield in Saskatchewan and Manitoba [Figure 6 and Enclosure 1(A)]. In the westernmost outcrops, little or no hiatus is apparent between the Cadomin and underlying strata (Rapson, 1965; Gibson, 1977; Stott, 1984a; Ricketts and Sweet, 1986). The erosion surface truncates the Minnes and Fernie groups in the

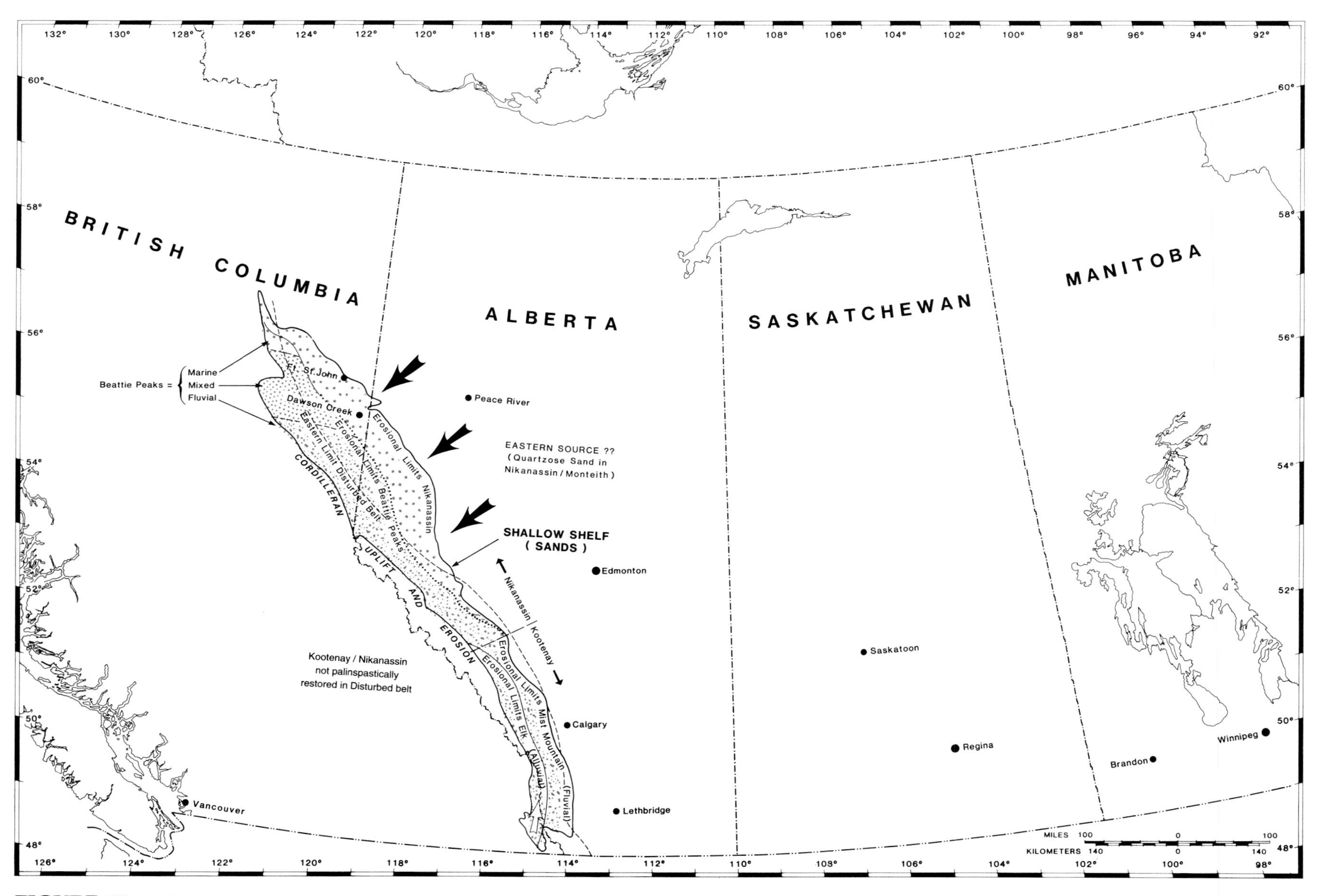

FIGURE 10. Paleogeographic reconstruction of the Kootenay and Nikanassin groups including the Monteith and Beattie Peaks formations deposited during Cycle 1.

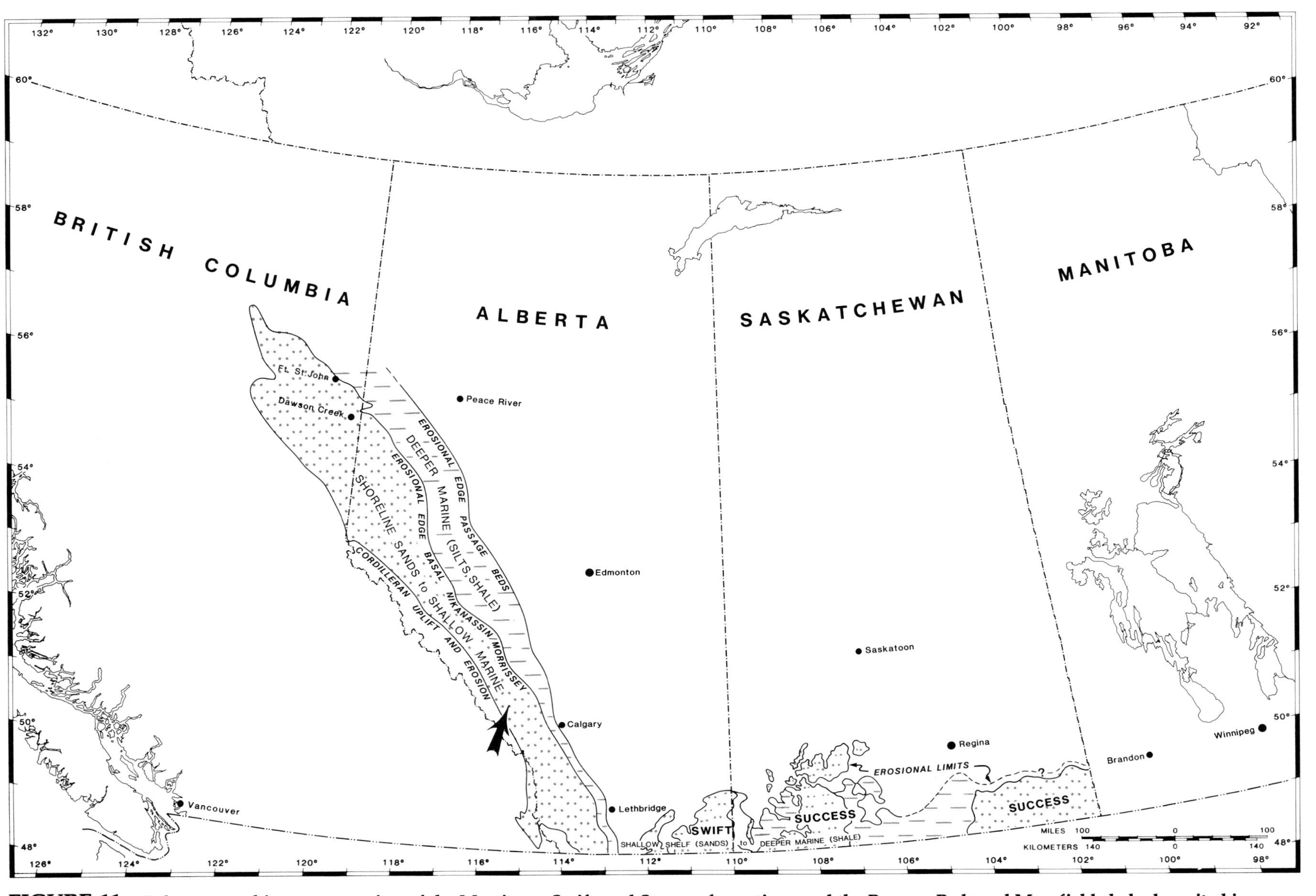

FIGURE 11. Paleogeographic reconstruction of the Morrissey, Swift, and Success formations and the Passage Beds and Masefield shale deposited in Cycle 1.

west and cuts progressively older strata to the east and north (Figures 6 and 12) until, at its easternmost extent in Saskatchewan, it transects rocks of Cambrian age (Christopher, 1984). Erosion beneath the unconformity resulted in very irregular regional topography, consisting of valleys and interfluves that controlled sedimentation patterns during subsequent Early Cretaceous deposition. A paleogeologic map of the post-Valanginian erosional surface (Figure 12) shows the increasing age of subunconformity units eastward across the basin. The thickness of the sediments of Cycle 2 preserved in the interior plains is shown in Figure 13. Maximum preserved sediment thickness occurs in northeastern British Columbia over the then-subsiding Peace River arch, and in southwestern Alberta.

Paleodrainage and Paleotopography

The paleogeography of Cycle 2 sediments is depicted in Figures 14 through 20. At the beginning of Cycle 2 deposition, a series of regional highs trended northwest-southeast across the basin, the largest being the Medicine Hat-Swift Current Highlands in the southern Prairies (Figure 14). In southeastern Alberta and southwestern Saskatchewan, uplift of the Sweetgrass arch and the Swift Current platform (Figure 7) resulted in deep erosion of the Cutbank, Whitlash, and pre-Cantuar valleys (Christopher, 1964, 1980; Hayes, 1986). In northern Alberta, these regional highs are referred to as the Fox Creek escarpment and the Keg River and Red Earth highlands. The Fox Creek escarpment is a series of northwest-trending low-relief hills that separated the Spirit River paleodrainage from more easterly drainage systems. Other major valley systems include the Edmonton and Assiniboia paleovalleys (Figure 14). The Cadomin, Cutbank, Basal Quartz, Ellerslie, and Dina formations across Alberta represent sandstones deposited by northwest-flowing rivers within these valley systems. In southern Saskatchewan, a system of northwest-, north-, and east-trending valleys originating from southwest Saskatchewan can be mapped as the McCloud Member and the Dina Formation. The "Detrital" zone (Deville Formation) of southern Alberta and Saskatchewan, which is a waxy shale to brecciated, quartzose, chert-rich sandstone in a clayey matrix, is a well-developed paleosol capping karstic Mississippian limestone highlands.

Cadomin/Ellerslie/Gething/Dina Sedimentation

The Cadomin Formation, deposited in the Spirit River-Cutbank drainage system, is a chert and quartzite pebble conglomerate locally up to 200 m thick and possibly derived from several point-source alluvial fans along the rising cordillera (Figure 14; Stott, 1968; McLean, 1977; Schultheis and Mountjoy, 1978). The Cadomin gravels are interpreted to have been transported eastward as alluvial fans and braidplain(?) deposits into the axis of the basin where they entered the drainage system of the northwest-flowing Spirit River (McLean, 1977). North of the Peace River arch, Cadomin gravels thin in a northeasterly direction and grade laterally into finer clastic sediments. The Cadomin Formation is absent east of the topographically high Fox Creek escarpment.

The overlying and, in part, laterally equivalent Gething Formation contains deltaic sediments that interfinger northward with nearshore and shallow marine sandstone and shale. These sediments, in turn, grade into marine shale and siltstone of the offshore environment. Shoreline coals are found northward well into northern British Columbia (Stott, 1973). This sandstone/shale/coal complex comprises the Sikanni Chief delta (Figure 14), which was fed by the Spirit River and Edmonton Valley paleodrainage. The Keg River delta, a smaller complex between the Keg River and Red Earth highlands, was probably fed by an arm of the Edmonton paleodrainage. The Fort McMurray delta received its sediment from the Assiniboia paleodrainage and its tributaries in Saskatchewan (Figure 14).

Deposition of the Cadomin Formation occurred during a renewed episode of subsidence that accompanied the emplacement of thrust sheets in the adjacent fold belt. Erosion of the thrust sheets would have resulted in isostatic rebound and flexural uplift of the foreland basin and associated orogene, causing detritus to be shed out into the basin (Heller et al., 1988). With this model the Cadomin sediments appear to represent an interval of tectonic quiescence in the cordillera when much of the Rocky Mountains and Eastern foreland basin were being isostatically uplifted and sediments older than Hauterivian were regionally beveled. The major lull in magmatic activity in the Western Cordillera during the Early Cretaceous (135 to 125 Ma; Armstrong, 1988) relates to the cessation in tectonic activity.

Following deposition of the Cadomin Formation and correlative units, progressive but intermittent flooding of the foreland basin from the north occurred (Figure 15). In northern areas, marine and deltaic deposits are first evident in the Gething Formation at the Sikanni Chief and Keg River deltas (Figure 14). In the northeast, the flooding is represented by the brackish, estuarine, and deltaic deposits of the McMurray Formation (the Fort McMurray delta). In southern and central Alberta, the younger brackish-bay shales, fine-grained sandstones, and argillaceous limestones of the Calcareous/Ostracod Member marked the flooding event (Glaister, 1959). In western Saskatchewan, the flooding event is characterized by brackish shale and fine-grained sandstone of the Cummings Member and in eastern Saskatchewan by the fluvial valley-fill deposits of the Cantuar Formation (Putnam, 1989). In central Alberta, the low-lying drainage network of the Spirit River and Edmonton paleovalleys were the first to be flooded and filled with brackish-water estuarine deposits of the Ellerslie Formation, representing the leading edge of the marine transgression (Banerjee and Davies, 1988). In Saskatchewan, the kaolinite-rich quartz sandstone of the Dina Member within the Assiniboia paleovalley system (Figure 14) was reworked by the advancing sea.

The transgression of the Boreal Sea continued southward across Alberta, Saskatchewan, and the northern United States along the preexisting drainage network. Tidally influenced sedimentation resulted in isolated sands of the Ostracod Member within and near the

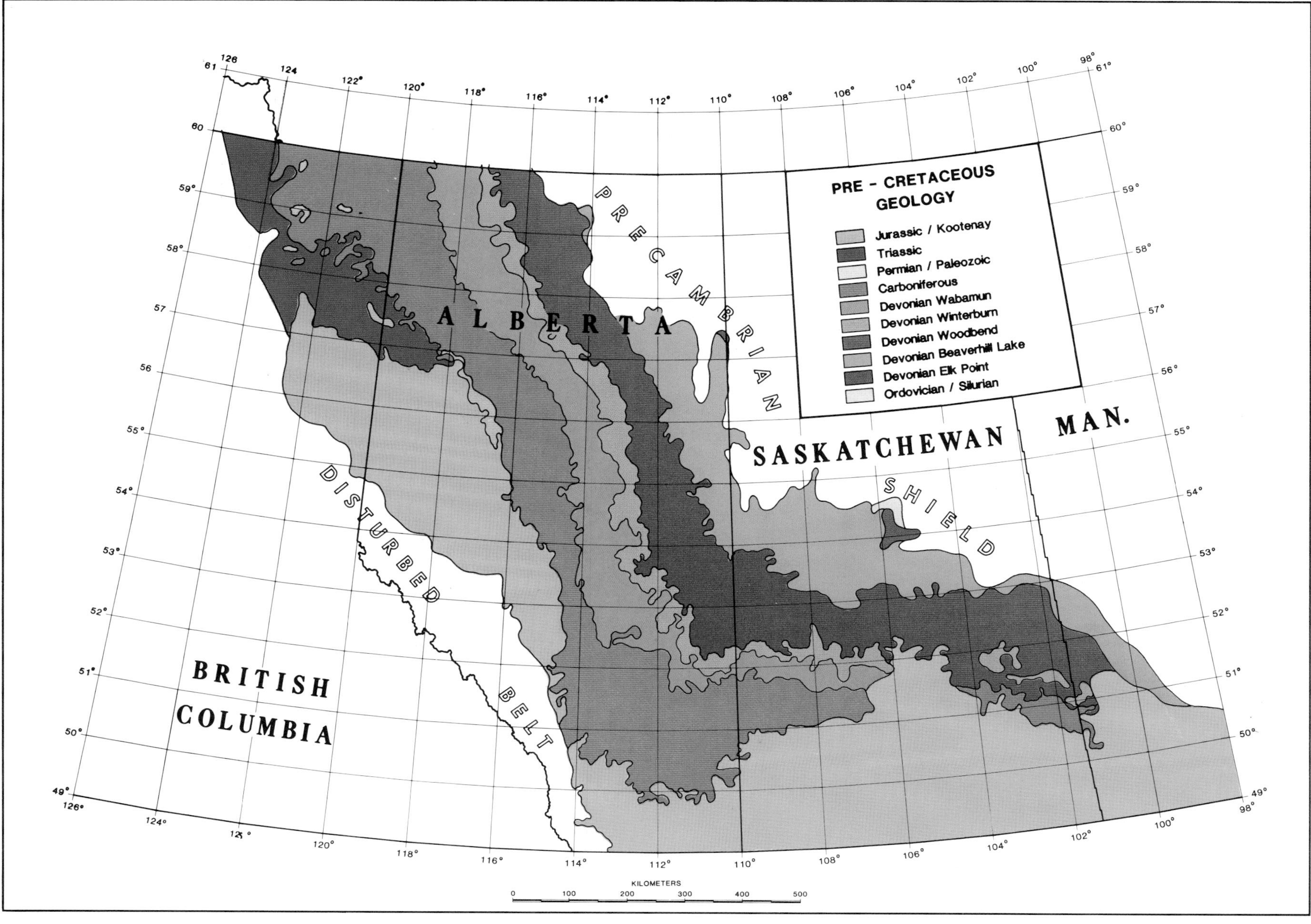

FIGURE 12. Post-Valanginian or "pre-Cretaceous" geology. Original data based on preliminary map from the new Atlas of Western Canada. (G. Mossop, personal communication, 1990.)

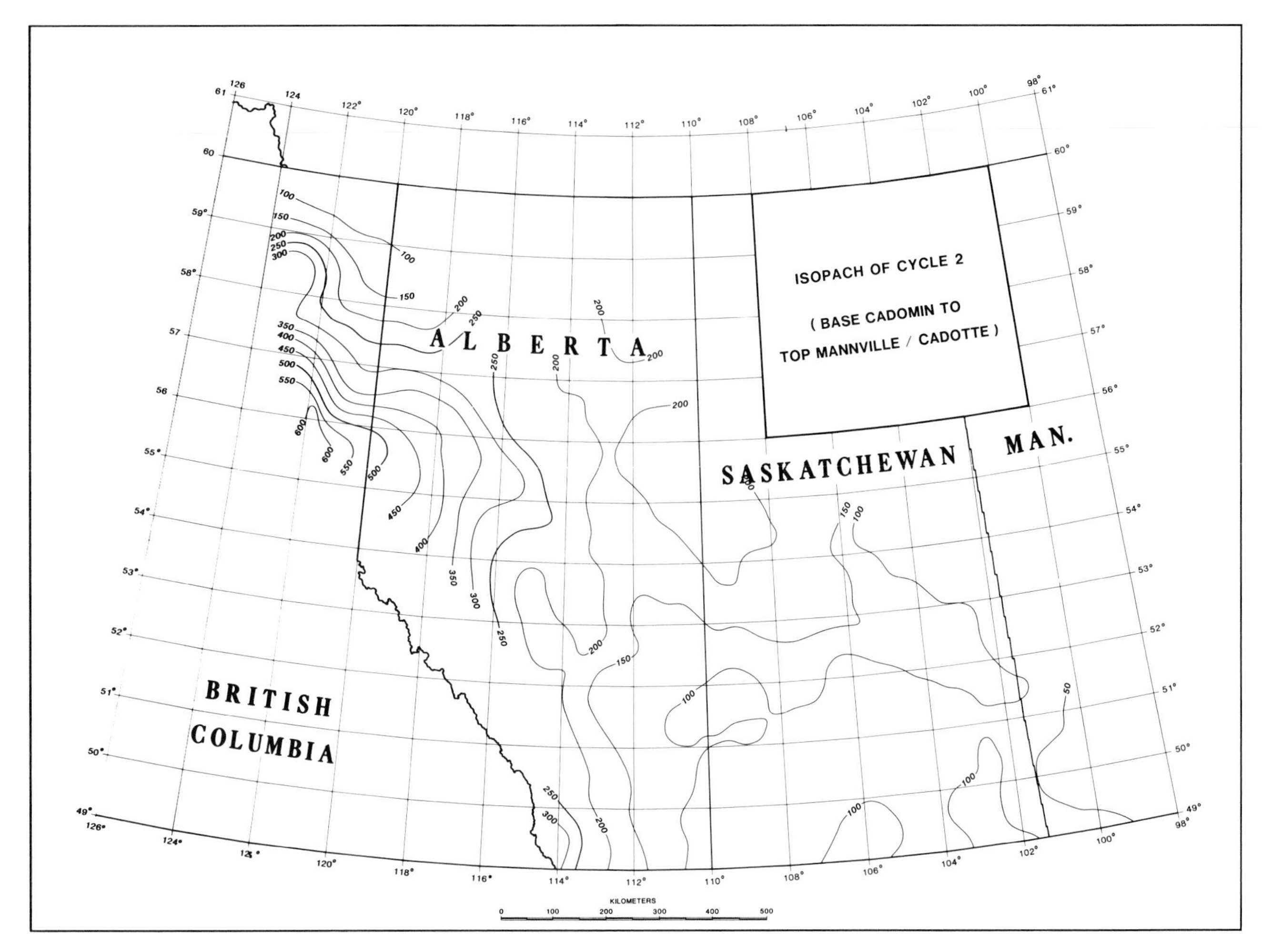

FIGURE 13. Isopach map of Cycle 2 from the base of the Cadomin Formation to the top of the Mannville Group. Original data based on preliminary map from the new Atlas of Western Canada. Contour interval, 50 m. (G. Mossop, personal communication, 1990.)

mouths of estuaries. In southern and central Alberta, prior to inundation by the Moosebar Sea proper, black calcareous muds and minor sandstones of the Calcareous Member (Ostracod limestone/Gladstone equivalent) were deposited in large lakes and swamps that were periodically linked to the sea (Figure 15; McLean and Wall, 1981). Renewal of volcanic activity in the Western Cordillera is indicated by the abundance of volcanic rock fragments and feldspars in the Calcareous Member and the predominance of volcanically derived, smectitic clays of the Bantry shales of the Ostracod Member (Farshori, 1983).

Glauconite/Bluesky/Wabiskaw/Lloydminster Sedimentation

As the transgression progressed, the discontinuous sandstones of the Glauconite, Bluesky, Wabiskaw, and Lloydminster units were deposited as retrogradational shoreline, estuarine, and shallow-shelf deposits (Figure 16). Fluctuations in relative sea level during this time resulted in incised valleys, estuarine-fill sequences, and local progradational shorelines. Paleovalleys infilled with estuarine and brackish-water sediments are found as far south as northern Montana (Burden, 1982). In west-central Alberta, three transgressive-regressive events are recognized in the retrogradational Glauconite and Bluesky formations within the overall larger-scale advance of the Moosebar Sea (Rosenthal, 1988). Maximum transgression occurred below the Glauconite Formation. A paleogeographic reconstruction of the Hoadley-Strachan sandstone complex of the Glauconite Formation is shown in Figure 17. The Hoadley shoreline and offshore sands are dissected by fluvial channels, some of which had headwaters in southern Alberta.

Continued transgression resulted in deeper-water sediments being deposited in the Moosebar Sea (Figure 17)—sediments that are now preserved as shales of the Moosebar and Clearwater formations, and the Lloydminster and Wilrich members. At its maximum transgression, the Moosebar Sea inundated eastern British Columbia, Alberta, and Saskatchewan, leaving only isolated uplands in the Swift Current, Kindersley, and Moosomin areas (Figures 15 and 16). Maximum depth of the Moosebar Sea in the Alberta Foothills was 100 m or less (McLean

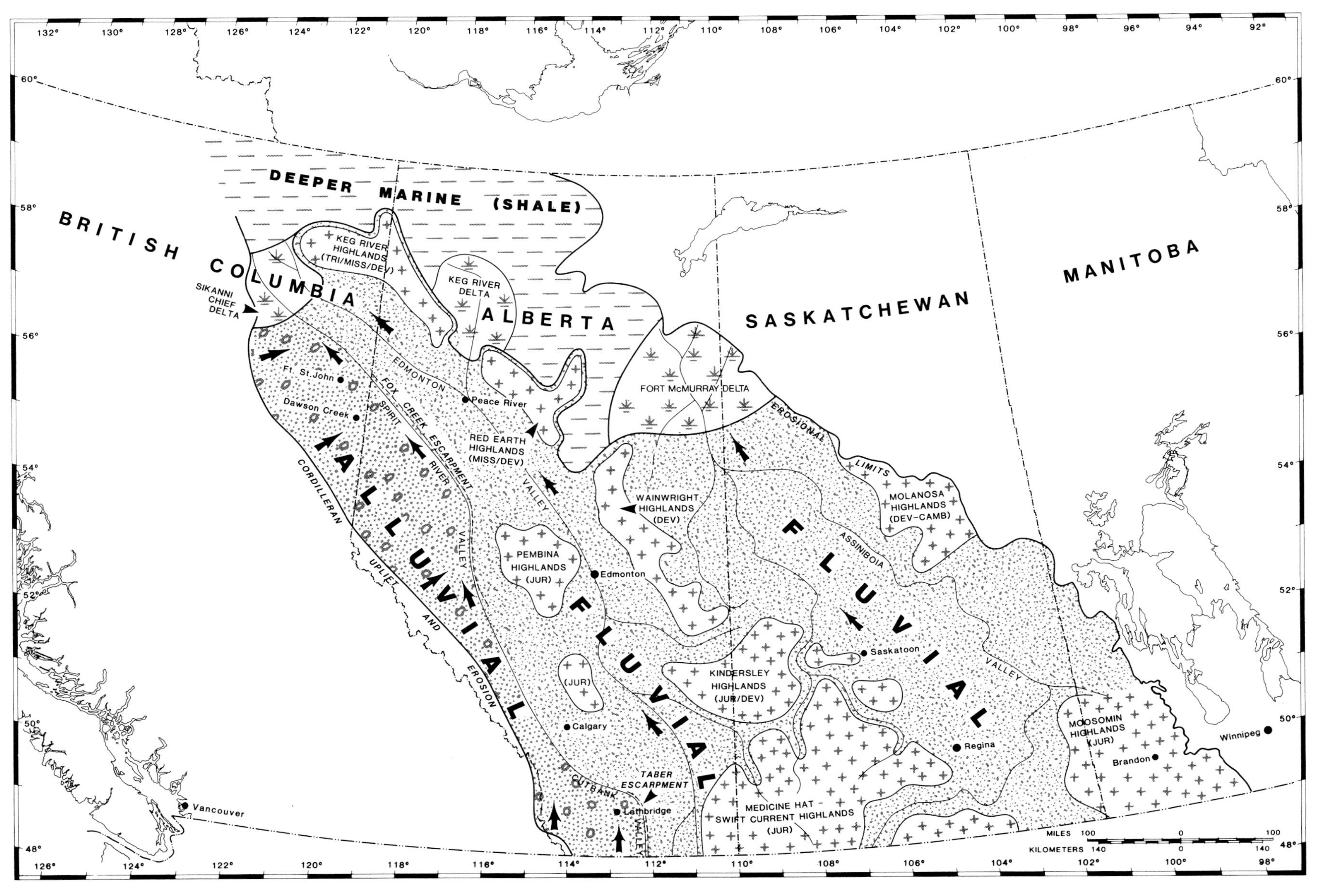

FIGURE 14. Paleogeographic reconstruction of the Cadomin, Ellerslie, Gething, and Dina formations during deposition of Cycle 2.

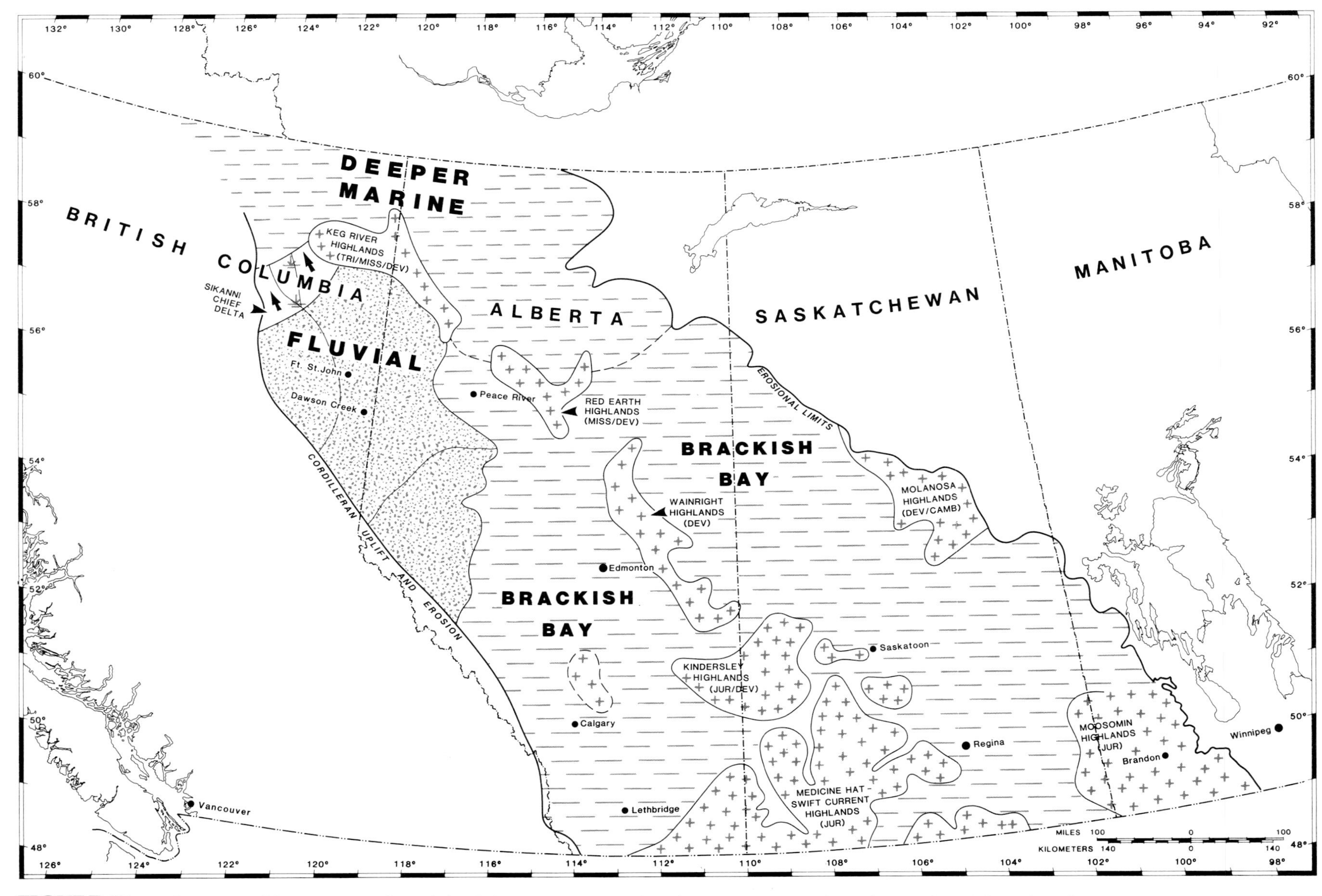

FIGURE 15. Paleogeographic reconstruction of the Calcareous, Ostracod, and Cummings members during deposition of Cycle 2.

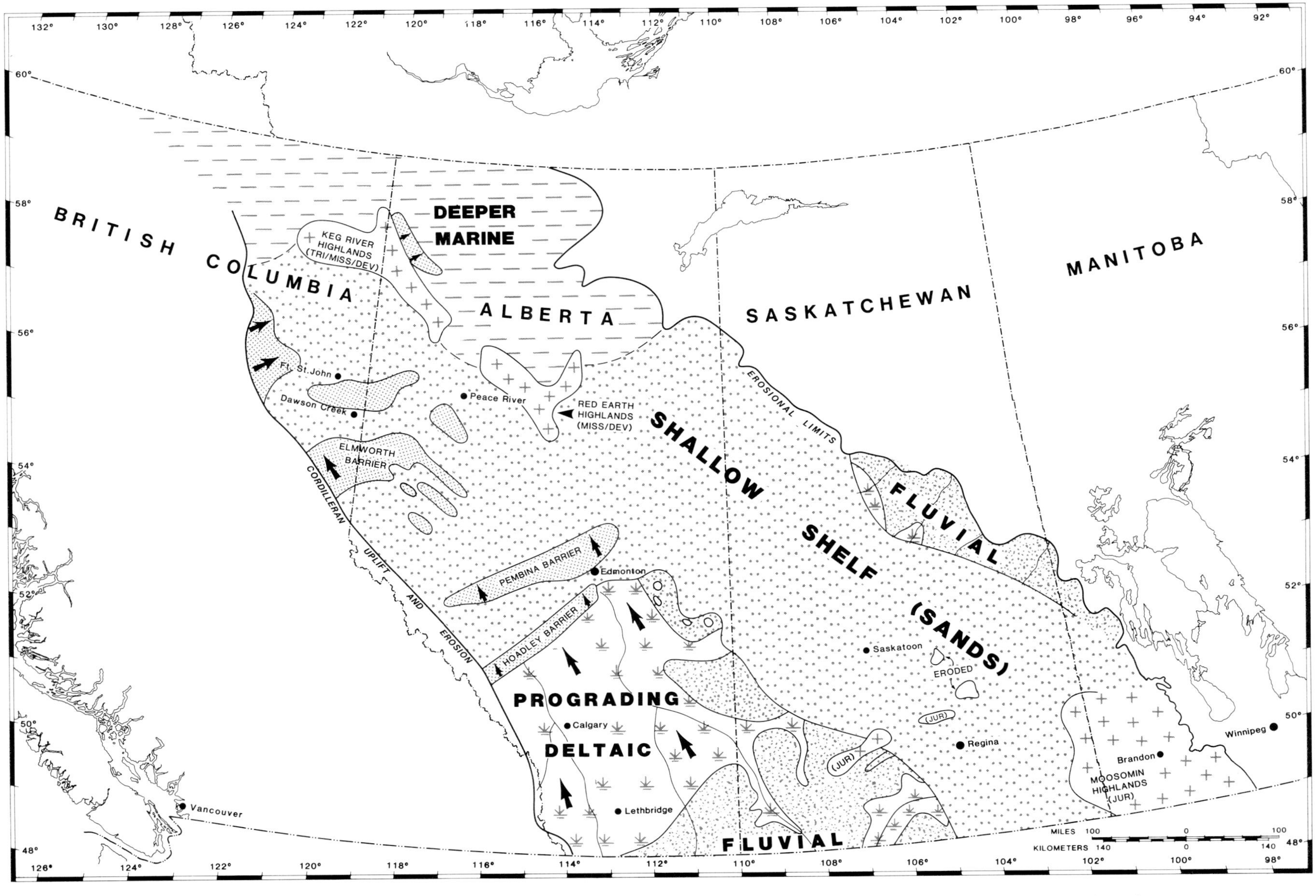

FIGURE 16. Paleogeographic reconstruction of the Glauconite, Bluesky, Wabiskaw, and Lloydminster units during deposition of Cycle 2.

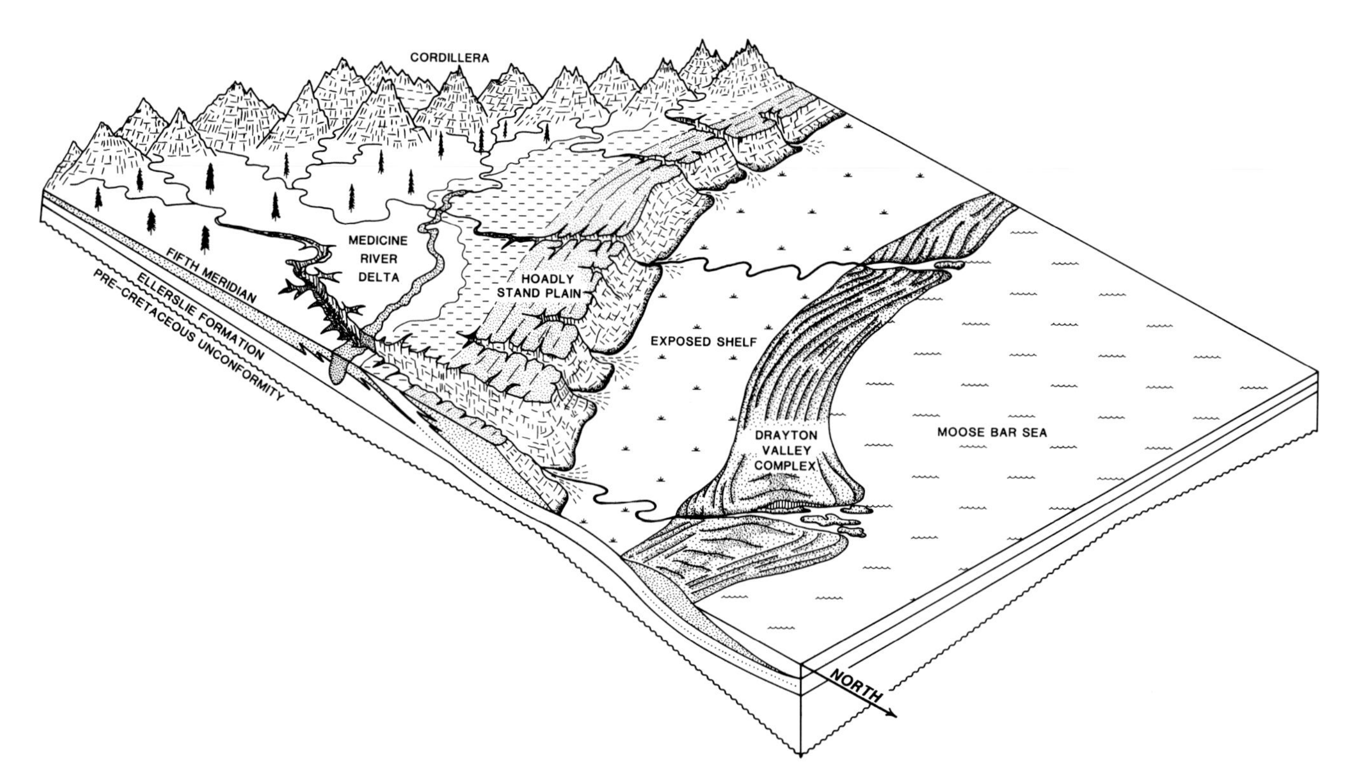

FIGURE 17. Paleogeographic reconstruction of the Glauconite Formation (Cycle 2) during the later stages of its deposition in west-central Alberta. This cartoon depicts a lowstand in sea level that followed the progradation of the Hoadley-Strachan complex. (Modified from Rosenthal, 1988.)

and Wall, 1981) and in Saskatchewan was on the order of about 30 m (Christopher, 1984).

Upper Mannville and Equivalent Strata Sedimentation

Following the incursion of the Moosebar Sea, a thick clastic wedge of the upper part of the Mannville/Blairmore Group prograded northward across the foreland basin, forming extensive flood plain and stacked-shoreline sandstones and conglomerates of the Falher, Notikewin, and Cadotte members (of the Spirit River and Peace River formations) and the Grand Rapids Formation (Figures 18, 19, and 20). The northern limit of this regression appears to have been controlled by the subsiding Peace River arch. Measured sections typical of the Falher, Cadotte, and other wave-dominated coastal successions within this sequence are shown in Figure 21. These paleoshorelines can be traced from Foothills outcrop eastward for several hundred kilometers. Falher Member, Notikewin Member, and Boulder Creek Formation/Cadotte Member fluvial channels were cut by streams that flowed in a northerly direction from sources in the cordillera and entered the sea along west-northwest-trending shorelines. In the eastern portion of the basin, sediment was supplied from the Canadian shield. By the end of Blairmore/Mannville time, most of the preexisting Paleozoic to Jurassic highs had been buried except for the peaks of the uplands in Saskatchewan. This middle Albian period of extensive progradation was also responsible for the highly glauconitic Scatter Formation, which prograded east-southeastward into the Liard basin area in the extreme northeast of British Columbia (Figure 18). In southeastern Saskatchewan and western Manitoba, shallow-water shelf sediments were deposited in the Pense basin (Figure 19). During Cadotte time (Figure 20), shoreline sediments reached their northernmost extent. Sediments of the Cadotte shoreline were westerly derived and are traceable for more than 300 km in the Peace River region of northern Alberta.

The climate during deposition of Cycle 2 is generally considered to have been warm and humid. The extensive coal deposits within the Gething and Gates formations formed in a warm, humid-maritime setting along and inland of the coast in northeastern British Columbia and northwestern Alberta. The Cadomin Formation was also largely deposited in a warm, humid setting. However, in southern Alberta, much farther from the coastline, thick deposits of red beds in the Blairmore Group and in-situ caliche as well as transported caliche clasts in the Cadomin Formation indicate arid to semiarid conditions (Ricketts and Sweet, 1986).

In southwestern Alberta, the middle to late Albian Crowsnest Volcanics of the Mill Creek Formation represent the only in-situ extrusive volcanic rocks preserved within the foreland basin. A palinspastic reconstruction of the present position of the Crowsnest Volcanics places the site of intrusion about 200 km to the west, near Cranbrook, British Columbia (Norris, 1964).

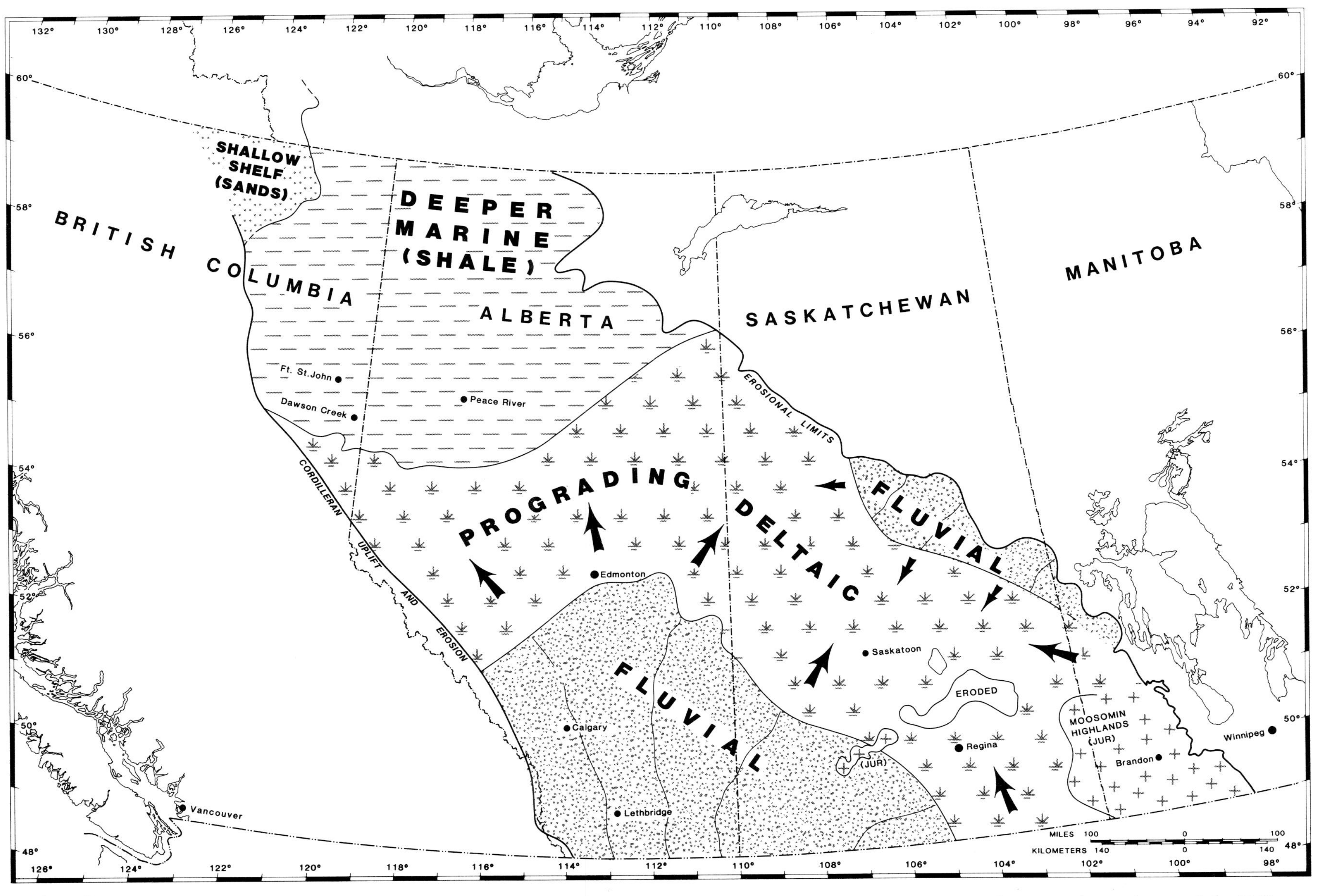

FIGURE 18. Paleogeographic reconstruction of the Upper Mannville, Rex, General Petroleum, Waseca, Gates, and Clearwater formations and the Falher Member of the Spirit River Formation deposited during Cycle 2.

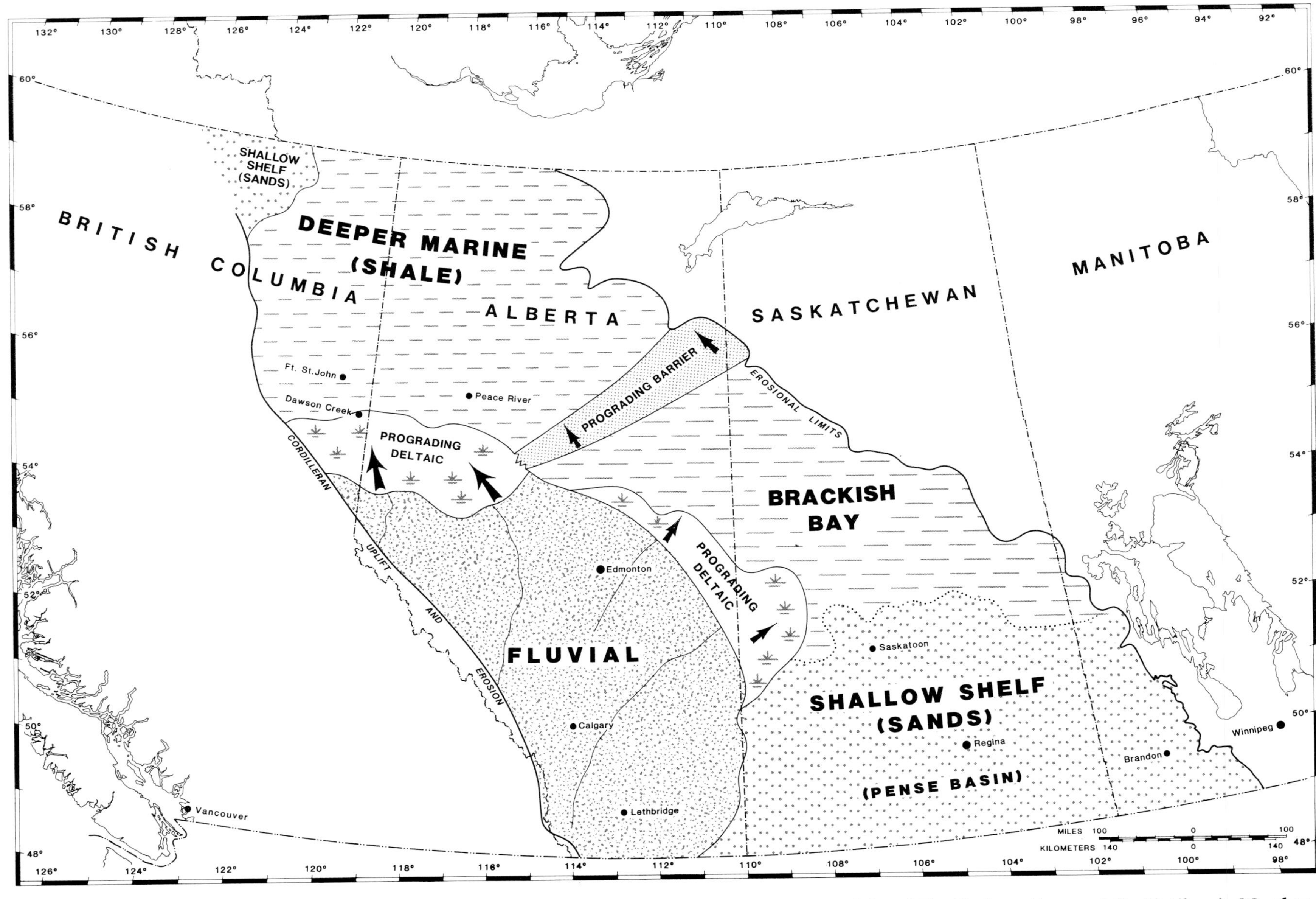

FIGURE 19. Paleogeographic reconstruction of the Upper Mannville, Colony, McLaren, Pense, and Grand Rapids formations and the Notikewin Member deposited during Cycle 2.

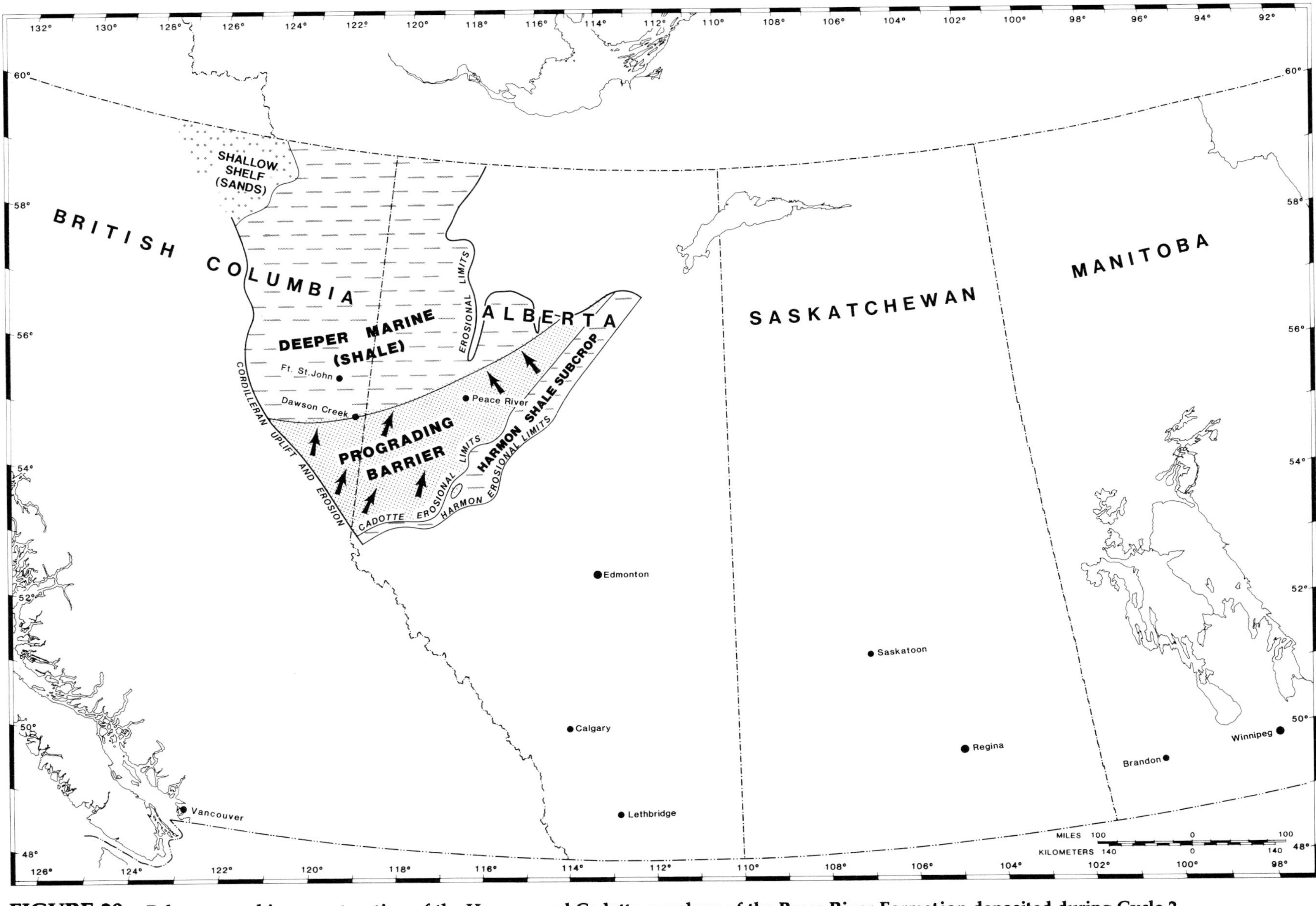

FIGURE 20. Paleogeographic reconstruction of the Harmon and Cadotte members of the Peace River Formation deposited during Cycle 2.

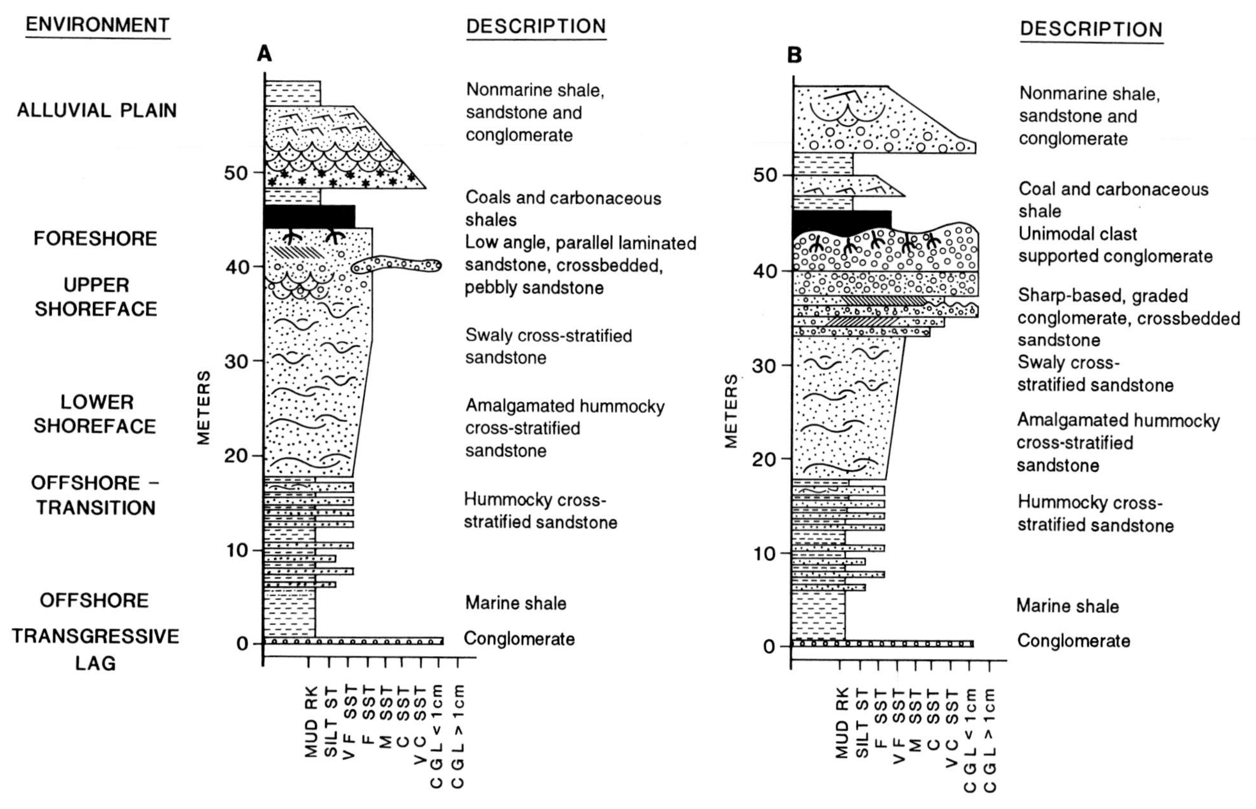

FIGURE 21. Typical measured sections of upward-coarsening successions deposited along wave-dominated, progradational shorelines such as represented by the Falher Member, Cadotte Member, Gates Formation, and Glauconite Formation (Cycle 2). If a sand source is available, vertical succession will resemble A; if a gravel source is available, vertical succession will resemble B.

Structural elements active in the basin during deposition of Cycle 2 include the Peace River arch, the Punnichy arch, and subsidence caused by regional Devonian salt solution (Figure 7). The Peace River arch, which was a topographical low during Cycle 2 time, affected regional and local sedimentation patterns of the Bullhead and Fort St. John groups (Stott, 1968, 1973; Leckie, 1986b; Cant, 1988). The northern depositional limits of the Falher and Cadotte member shorelines appear to have been controlled by the arch. Uplift of the Sweetgrass–North Battleford arch and the Swift Current platform during the Neocomian resulted in incision of deep valleys that were subsequently filled during the Aptian and early Albian with fluvial and estuarine sediments (Christopher, 1984). By the end of the Albian, the Swift Current platform had subsided approximately 50 m (Koziol, 1988). The Punnichy arch of central Saskatchewan, which separates the Williston basin from the Alberta basin, was uplifted in the middle Albian (Christopher, 1980). Salt solution in the Middle Devonian Prairie Formation evaporites in southern and central Saskatchewan and the Lower Devonian Cold Lake and Lotsburg formation evaporites in northern Alberta and Saskatchewan contributed to an irregular erosional topography with relief of 30 to 100 m (Christopher, 1984). In northeastern Alberta and northwestern Saskatchewan, the thickness of the Mannville Group was controlled by relief on the post-Valanginian unconformity with more localized structural control being the result of dissolution of Paleozoic evaporites. The Severn arch of north-central Manitoba likely contributed quartz and kaolinite-rich sediments to the eastern platform of the foreland basin during much of Mannville time (Porter et al., 1982; Christopher, 1984).

Cycle 3: Base of Joli Fou Formation/Paddy Member to top of Wapiabi Formation/ Colorado Group (Albian to Campanian)

The base of Cycle 3 occurs at the contact between the Cadotte and Paddy members in northwestern Alberta. In north-central and southern Alberta, the contact occurs at the base of the Joli Fou Formation; in Manitoba, it is at the base of the Ashville Formation; and in Saskatchewan it is at the top of the Pense Formation [or at the base of the Joli Fou/Spinney Hill formations; Enclosure 1(A)]. An isopach map of Cycle 3 is shown in Figure 22.

The deposition of sediments of Cycle 3 corresponds to a long-term period of global sea level rise from the Cenomanian through the Santonian (Haq et al., 1987; Jervey, 1992). It was apparently coincident with a regional downflexing of the North American craton (Lambeck et al., 1987). Cycle 3 includes three major marine inundations separated by two lowstand, regressive pulses represented by the Peace River–Viking and Cardium formations [En-

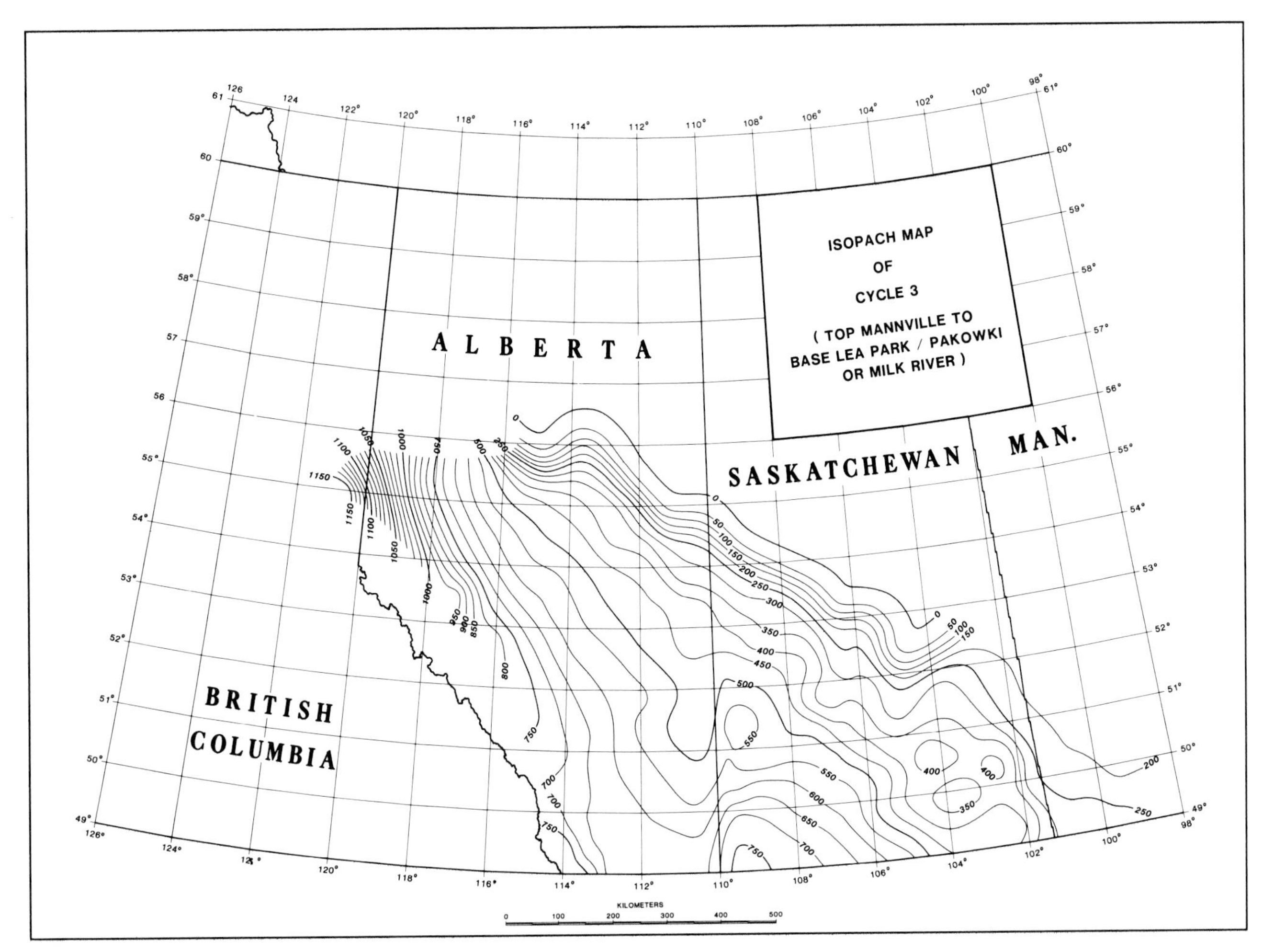

FIGURE 22. Isopachs of Cycle 3. Original data based on preliminary map from the new Atlas of Western Canada. Contour interval, 50 m. (G. Mossop, personal communication, 1990.)

closure 1(A)]. During the highstands, warm marine waters from the Gulf of Mexico mixed with the boreal waters extending south from the Arctic. Lithologically, this cycle consists primarily of marine shale and is in excess of 1100 m thick. Economically significant sandstone and gravel units within Cycle 3 include the Viking and Cardium formations, the Chungo Member, and the discontinuous deltaic sands of the Dunvegan Formation.

Bow Island/Viking/Pelican/Paddy/Newcastle Sedimentation

In northern Alberta, a sea level lowstand resulted in an incised valley system at the base of Cycle 3, several hundreds of kilometers long, cut into the middle Albian Cadotte Member, from the Rocky Mountain Foothills into the interior plains. This incised valley system is coeval with multiple paleosols in the Boulder Creek Formation in northern British Columbia (Leckie et al., 1989) and in the Mill Creek/Bow Island formations in southern Alberta. These paleosols formed when sedimentation rates on the flood plains decreased in the more westerly portions of the basin during lowstands. The subsequent sea level rise deposited the estuarine, shallow-bay, and shoreline deposits of the Paddy Member.

Warm waters that extended from the Gulf of Mexico as far north as central Alberta deposited the marine shales of the upper Albian Joli Fou Formation that unconformably overlie the Mannville Formation (Figure 23). The overlying Viking Formation is 15 to 75 m thick, and consists of marine sandstone and conglomerate throughout much of central Alberta and Saskatchewan, but grades into nonmarine sediments in southwestern Alberta. The marine beds are generally considered to be of shallow-water origin and locally tidally influenced. Several sea level lowstands may be present in the Viking Formation, with the major one being dated at 97 Ma (Beaumont, 1984; Leckie, 1986a, Hein et al., 1986; Reinson et al., 1988). During these sea level lowstands, many incised valleys were cut and subsequently infilled with estuarine sediments during sea level rise, such as the one containing the thick conglomerates of the Crystal oil field (Reinson et al., 1988). The Viking-equivalent Newcastle Formation consists of shelf sands (Figure 23) that extended north from Montana into Saskatchewan and Manitoba.

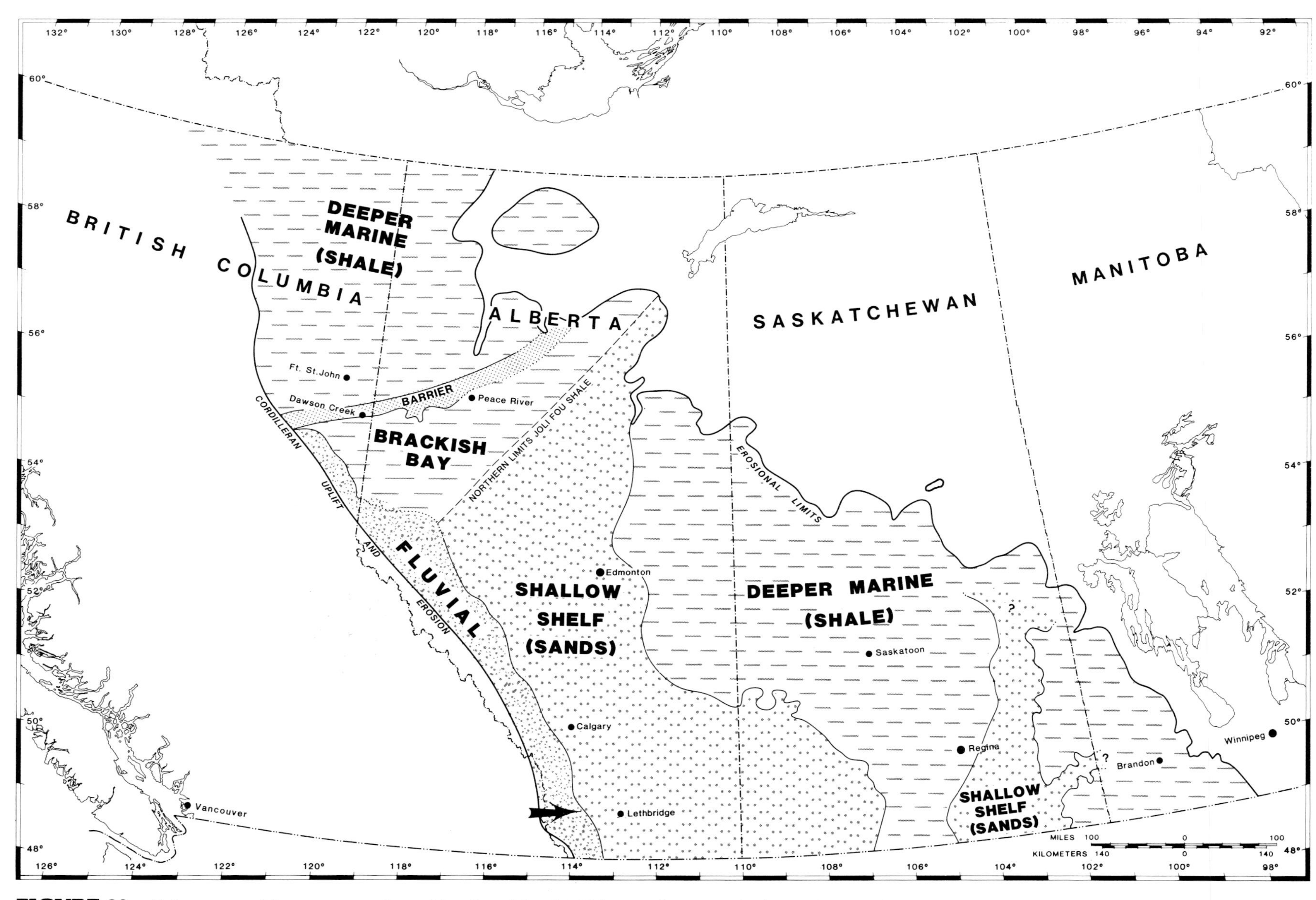

FIGURE 23. Paleogeographic reconstruction of the Bow Island, Viking, and Newcastle formations and the Paddy and Pelican members deposited during Cycle 3.

The shales of the uppermost Albian Shaftesbury Formation in northern Alberta and British Columbia represent a major basin-wide transgression during which the Boreal Sea connected with the northward-advancing seas from the Gulf of Mexico. In southern Alberta, the oldest record of the transgression is the Cenomanian Blackstone Formation where it onlaps the Blairmore Group. The lowermost Cenomanian Fish Scale Zone is a basin-wide marker containing abundant fish remains deposited with finely laminated, unbioturbated sandstones and siltstones. The Fish Scale Zone, which is characterized by high total organic carbon contents (up to 8%) and a low concentration of benthic foraminifera, likely was deposited under poorly oxygenated bottom conditions. In Saskatchewan and Manitoba, the base of the Fish Scale Zone may represent a major hiatus with substages missing below it (Caldwell, 1984), whereas in western Alberta, strata above and below the Fish Scale Zone appear to be more conformable (Stelck and Armstrong, 1981). The Fish Scale Zone is generally considered to contain a condensed section deposited during a peak transgression that occurred during deposition of the Colorado Group.

Dunvegan Sedimentation

The middle Cenomanian Dunvegan Formation of northern British Columbia and northwestern Alberta represents a southeastward thinning and fining, fluviodeltaic wedge deposited above the Shaftesbury Formation (Stott, 1984a). At outcrops in northeastern British Columbia and southern Yukon Territory, gravels and coarse sands were deposited in a braided-river and alluvial-plain setting (Figure 24). In northwestern Alberta, seven sand-rich, progradational cycles separated by regional transgressive surfaces represent the southeastward advance of the Dunvegan delta (Bhattacharya, 1988). The progradation of the Dunvegan delta, although attributed to global lowering of sea level at 94 Ma (Bhattacharya, 1988), also coincides with a major uplift in the Omineca and Intermontane belts (Tater, 1965; Stott, 1984a). South of latitude 54°N, the Dunvegan Formation grades into marine shales. The Simonette Channel is inferred to have been eroded during a sea level lowstand and then infilled with estuarine sands during the subsequent sea level rise. Beyond the southern limit of the Dunvegan delta, sedimentation in the basin was dominated largely by pelagic deposition.

The Dunvegan Formation is overlain by shales of the Cenomanian to Turonian Kaskapau Formation, which contain the retrogradational and transgressive, shallow marine sandstones of the Doe Creek, Pouce Coupe, and Howard Creek members. Wallace-Dudley and Leckie (1988) considered these sands to be similar in origin to the shoals described by Nelson et al. (1982) on the modern transgressive Bering Sea epicontinental shelf.

The sea level rise that began in the late Albian and reached its peak during the early Turonian is inferred to be eustatic (Haq et al., 1987). It resulted in across-the-basin deposition of the coccolithic Second White Speckled Shales (Vimy Member), which delineate the Cenomanian-Turonian boundary (Stelck and Wall, 1954). The Second White Speckled Shale contains up to 11% Type II dispersed total organic carbon and thus may have hydrocarbon-generation potential (Creaney and Allan, 1992). The generalized distribution of sediments within the foreland basin during the early Turonian is shown in Figure 2. The peak of the Turonian transgression is represented by the argillaceous Laurier Limestone Beds in Manitoba (McNeil and Caldwell, 1981). The calcium carbonate content of this part of the section increases eastward away from the cordillera.

Cardium Sedimentation

The upper Turonian is marked by a regressive event in the west that is capped by an erosional unconformity that can be traced across the basin. In the west, this unconformity is overlain by the conglomerates of the Cardium Formation, whereas in Manitoba, it lies between the Morden and Niobrara formations (McNeil and Caldwell, 1981). The unconformity approximately coincides with the 90 Ma eustatic lowstand of Haq et al. (1987), although Plint and Walker (1987) attributed Cardium shoreline progradation and unconformities within the formation to tectonic uplift at the western margin of the basin. The depositional history of the Cardium Formation is complex, with six upward-coarsening cycles capped by erosional surfaces (Plint et al., 1988). The Cardium shoreline trended northwest-southeast and prograded eastward (Figure 25).

The Cardium Formation and its correlative units are disconformably(?) overlain by the Coniacian to Campanian marine shales of the upper Colorado Group, Wapiabi Formation, and Niobrara Formation, which represent a second major marine inundation during Cycle 3. The peak of this marine transgression is represented by the planktonic foraminifera found in the First White Specks Zone. In the First and Second White Specks zones and perhaps the Fish Scale Zone, the presence of biogenic chalk and planktonic foraminifers indicate open marine conditions within the seaway during the peaks of the marine transgressions (Jervey, 1992).

The Santonian Niobrara Formation in Manitoba, which is equivalent to the First White Specks Zone, is erosionally truncated under the Pierre Group (McNeil, 1984). This sub-Pierre unconformity is well documented in the east but is not distinguishable in the western part of the basin.

Milk River/Chungo/Chinook Sedimentation

The final regressive event of Cycle 3 occurred during the early Campanian, and is represented by extensive shoreline sandstones of the Milk River Formation and the Chungo Member, which extend from southeastern Alberta and Montana to the central Alberta Foothills (Figure 26). The shoreline prograded as a wave-dominated sheet of sandstone that extended laterally for at least 350 km. The influence of tides in the foreland basin at this time is recorded in an extensive tidal-inlet sequence preserved in outcrop at Writing-On-Stone Park in southern Alberta (Figure 27; Cheel and Leckie, 1990). During early to late Campanian time, sea level rose again and the marine shales of the Pakowki/Nomad Member were deposited

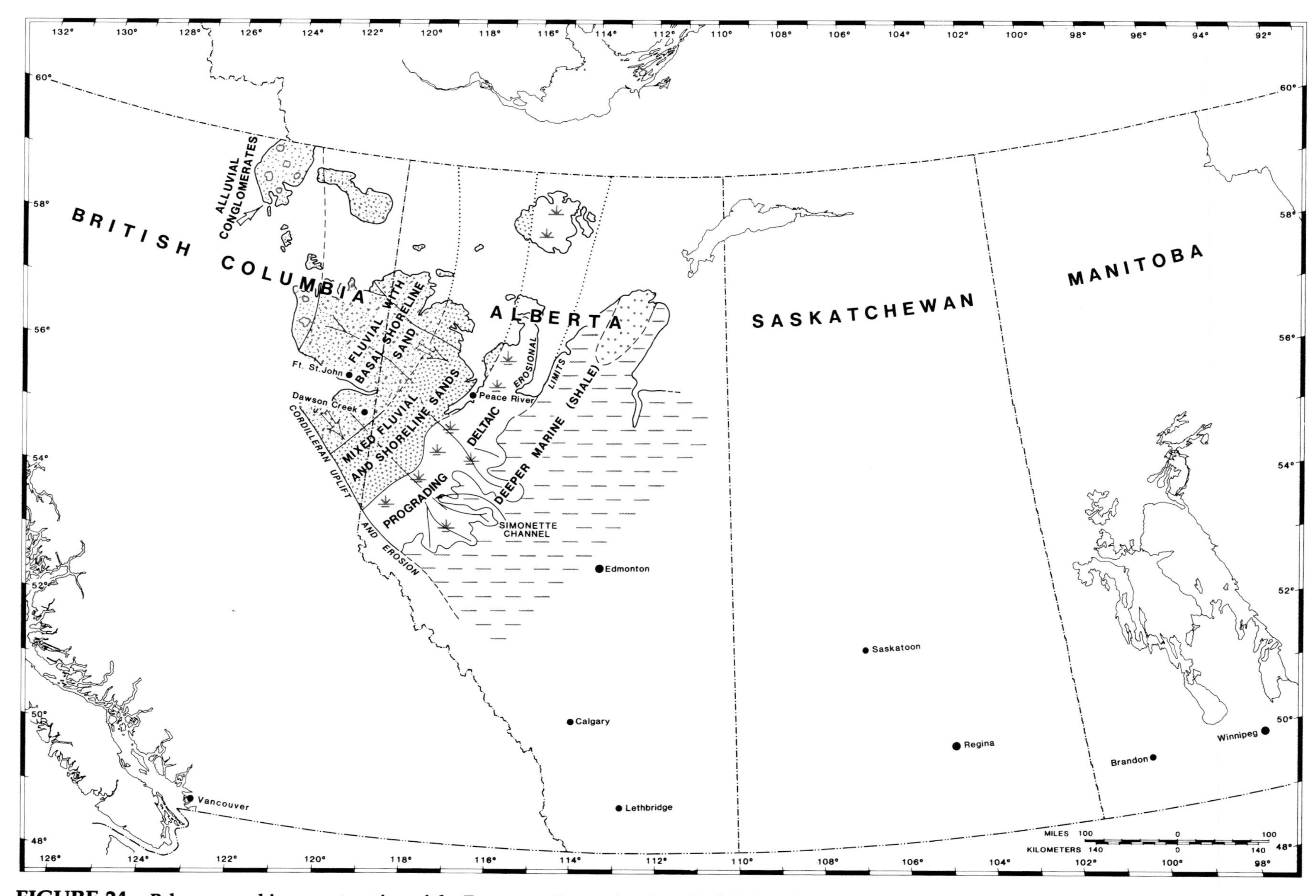

FIGURE 24. Paleogeographic reconstruction of the Dunvegan Formation deposited during Cycle 3.

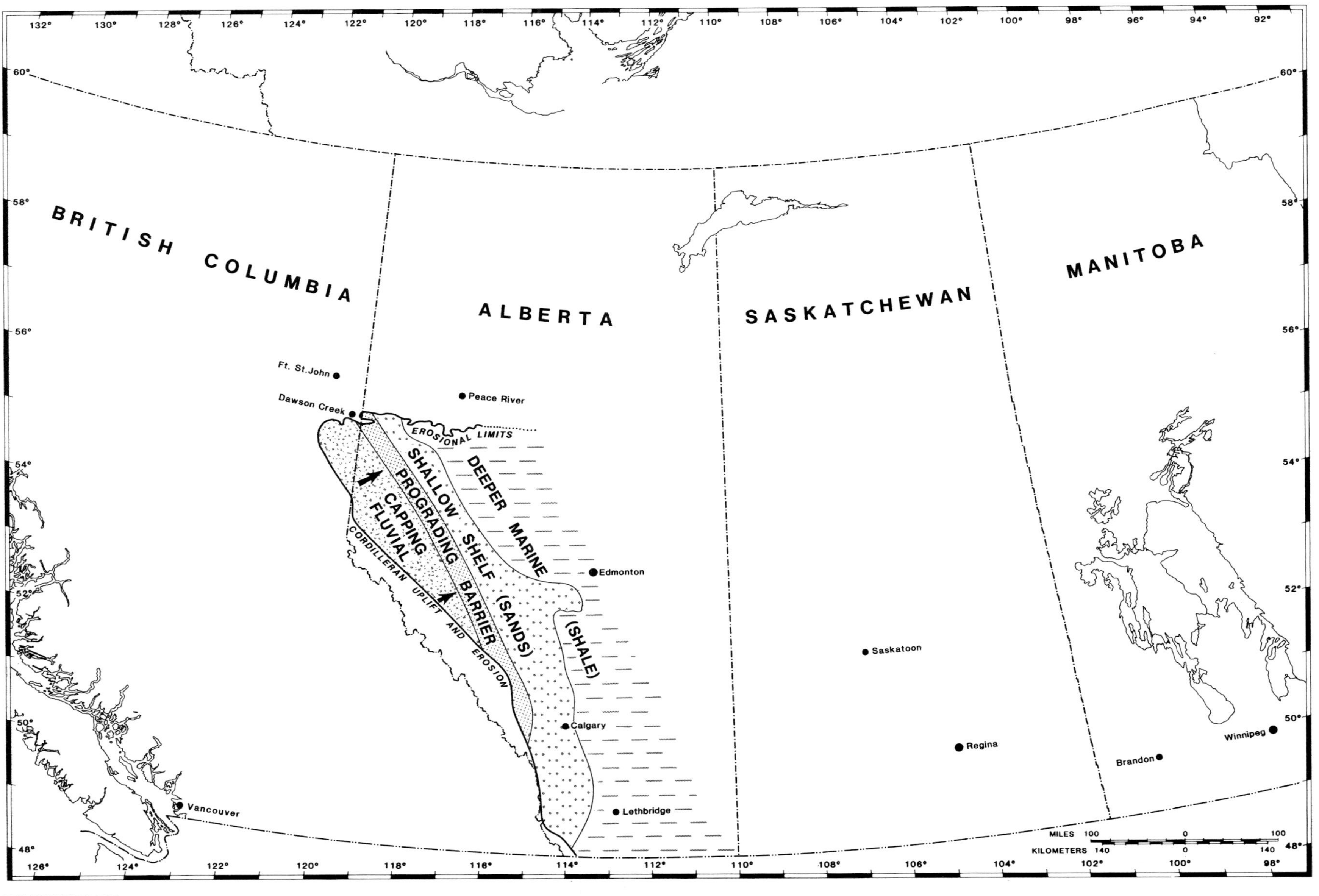

FIGURE 25. Paleogeographic reconstruction of the Cardium Formation deposited during Cycle 3.

FIGURE 26. Paleogeographic reconstruction of the Milk River Formation and the Chungo and Chinook members deposited during Cycle 3.

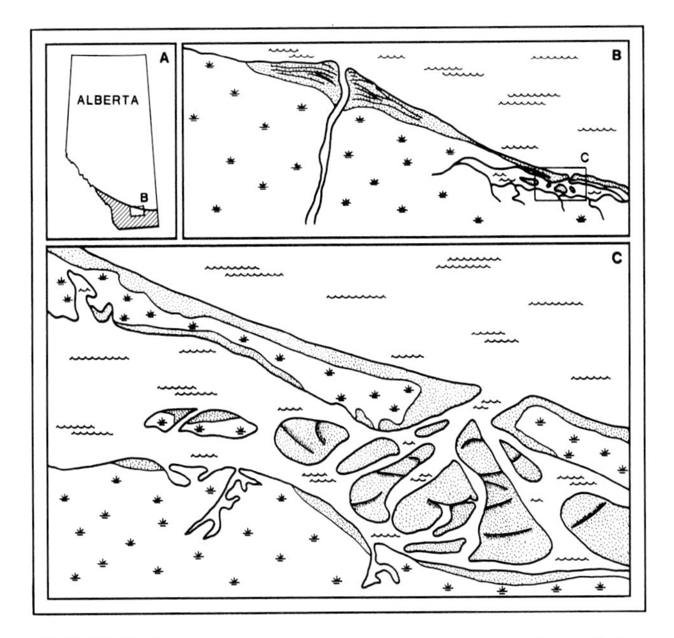

FIGURE 27. **Paleoenvironmental interpretation of the Milk River Formation (Cycle 3) at Writing-on-Stone Provincial Park in southeastern Alberta. (Modified from Cheel and Leckie, 1990.)**

(Clagget Cyclothem of Kauffman, 1977). Isopachs of the Pakowki/Nomad Member thin regularly in a westerly direction due to the advancing deltaic deposits of Cycle 4.

Planktonic faunas indicate a warm temperate climate in at least the eastern part of the basin during the latest Cenomanian, Turonian, and early Santonian (McNeil, 1984). During maximum marine transgressions, warm waters from the Gulf of Mexico may have extended as far north as 54°N latitude, warming water temperatures by up to 5°C to a temperature near 20°C (McNeil and Caldwell, 1981). The winds that dispersed volcanic ash (bentonites) within the Joli Fou and Viking formations across southern Alberta blew toward the northeast (Amajor, 1985).

Cycle 4: Saunders Group/(base Belly River to top of Porcupine Hills) (Campanian to early Eocene)

Cycle 4 represents an interval of primarily nonmarine deposition of clastic detritus making up the Campanian to Maastrichtian Belly River, St. Mary River, Edmonton, Willow Creek, Brazeau, and Wapiti formations and the Paleocene Paskapoo Formation (Figures 28 and 29) with one marine incursion represented by shales of the Bearpaw Formation. The top of Cycle 4 is represented by a basin-wide erosion surface over which lie sediments of Cycle 5. Sediments of Cycle 4 were deposited during an episode of subsidence and thrusting, resulting from collision of the Intermontane belt with North America. The uplifted Omineca belt created a major drainage divide that prevented detritus of the Intermontane belt from entering the foreland basin (Mack and Jerzykiewicz, 1989). The final episode of foreland basin sedimentation is marked by deposition of the Paskapoo Formation during the Paleocene. However, deformation of the easternmost Paleocene strata of the Coalspur, Paskapoo, Willow Creek, and Porcupine Hills formations in the Rocky Mountain Foothills indicates that compressional stresses continued until at least the Paleocene. An isopach map of Cycle 4 is shown in Figure 30.

Subsidence rates and sediment supply within the foreland basin were very high at this time (Stott, 1984b; Stockmal and Beaumont, 1987). During Belly River/Judith River time, a wedge of nonmarine sediments up to 1370 m thick (Wall and Rosene, 1977) was shed eastward. Belly River sediments are primarily fluvial channel and associated flood plain, crevasse splay, and lacustrine in origin. In east-central Alberta, the correlative Judith River Formation was deposited by easterly to southeasterly flowing braided and low-sinuosity meandering streams (Koster, 1984; Wood, 1985; Visser, 1986) flowing into an estuarine, channel-dominated shoreline. Sediments of the Brazeau Formation in the central Alberta Foothills were deposited by northeast-flowing meandering rivers (McLean and Jerzykiewicz, 1978); thick coals accumulated in meandering channel backswamps. The overall upward-coarsening nature of the Brazeau Formation may reflect an easterly migrating, and thus increasingly proximal, thrust zone. Despite the tremendous volume of cordillera-derived detritus in western Alberta [e.g., compare column 3 with columns 6 and 7, Enclosure 1(B)], much of the eastern foreland basin was sediment-starved, resulting in deposition of the chalky Gregory Member in Manitoba (McNeil, 1984). A west-to-east facies change from fluvio-deltaic in western Alberta to predominantly marine in eastern Saskatchewan was demonstrated by Stelck et al. (1972) for sediments of the St. Mary River and Belly River formations.

Abundant caliche, red beds, and a "semiarid flood plain facies" occur in the uppermost portion of the Belly River Formation in southwestern Alberta, indicating a warm and semiarid climate (Jerzykiewicz and Sweet, 1988). In the central Alberta Foothills, peats accumulated in extensive swamps in an alluvial setting. In southeastern Alberta and southwestern Saskatchewan, the Oldman Formation is generally considered to consist of nonmarine coastal plain deposits, whereas the underlying Foremost Formation represents deltaic and strandline deposits. The Judith River Formation thins toward and over the Sweetgrass arch (Figure 30; McLean, 1971), suggesting that it was a structural high at this time. The sedimentation pattern of the Claggett, Judith River, and Bearpaw formations in Saskatchewan was affected by dissolution of Devonian evaporites (McLean, 1971).

The marine shales of the Bearpaw Formation (Odanah Member or Pierre shale in Manitoba) represent the last major marine inundation to have affected the foreland basin. The Bearpaw Sea extended from the Gulf of Mexico through to the Arctic (Wall and Singh, 1975), with water depths on the order of 45 to 60 m (McLean, 1971). Several upward-coarsening marine sandstone units occur within the Bearpaw Formation (Link and Childerhose, 1931). The transition from the Bearpaw Formation to the overlying coal-bearing strata of the Horseshoe

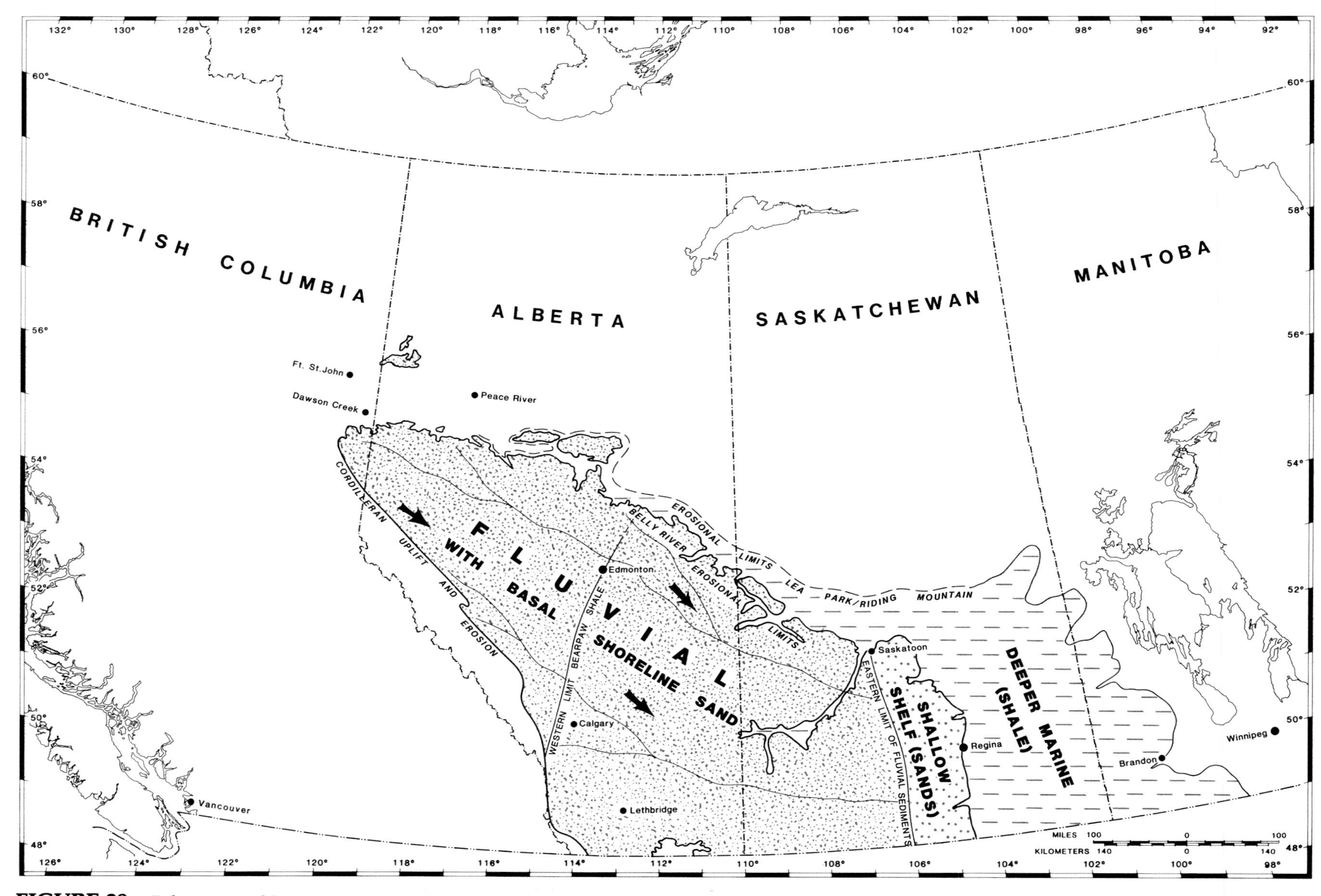

FIGURE 28. Paleogeographic reconstruction of the Belly River Formation and equivalent units deposited during Cycle 4.

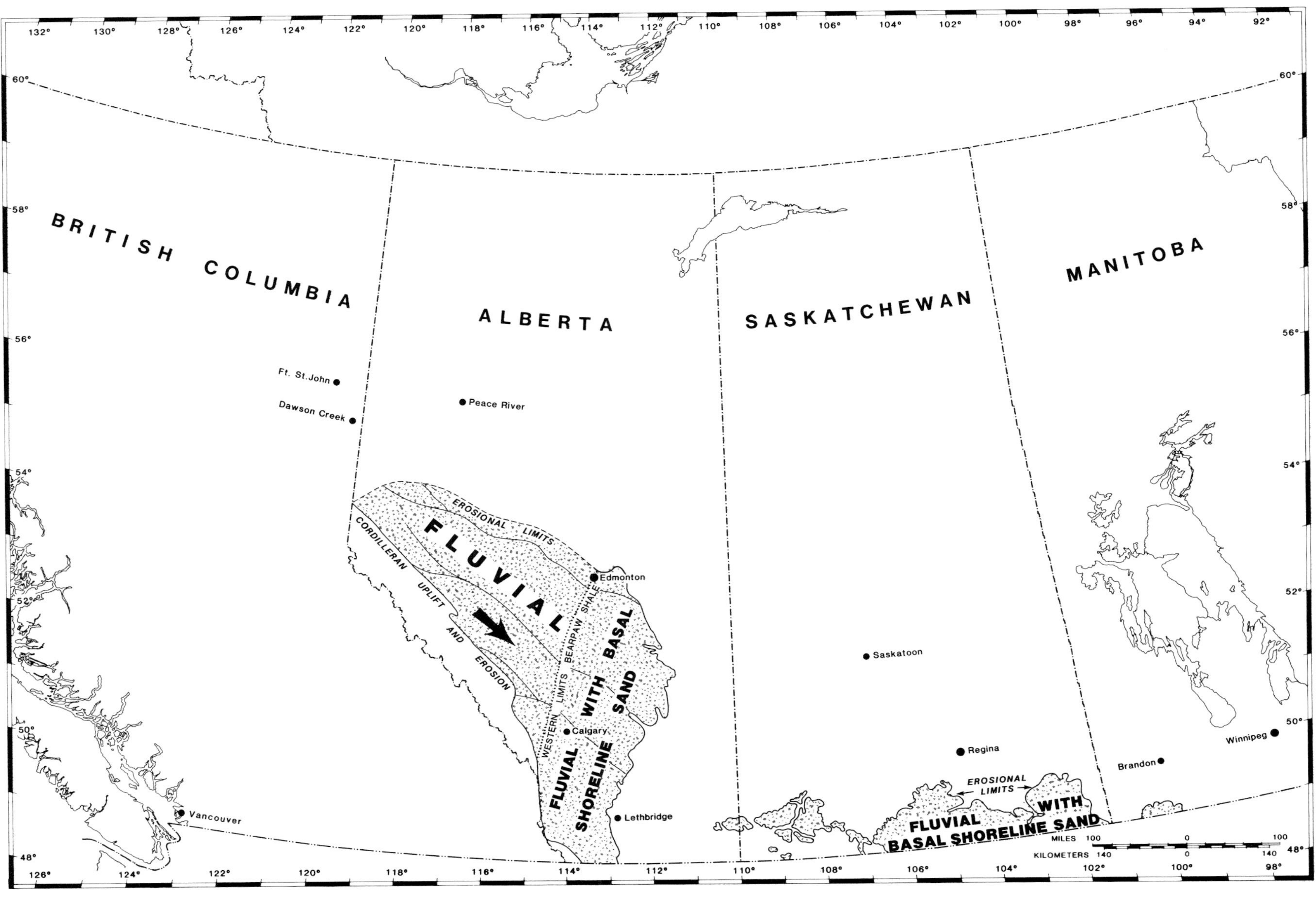

FIGURE 29. Paleogeographic reconstruction of the Edmonton, Willow Creek, Brazeau, Wapiti, and Paskapoo formations deposited during Cycle 4.

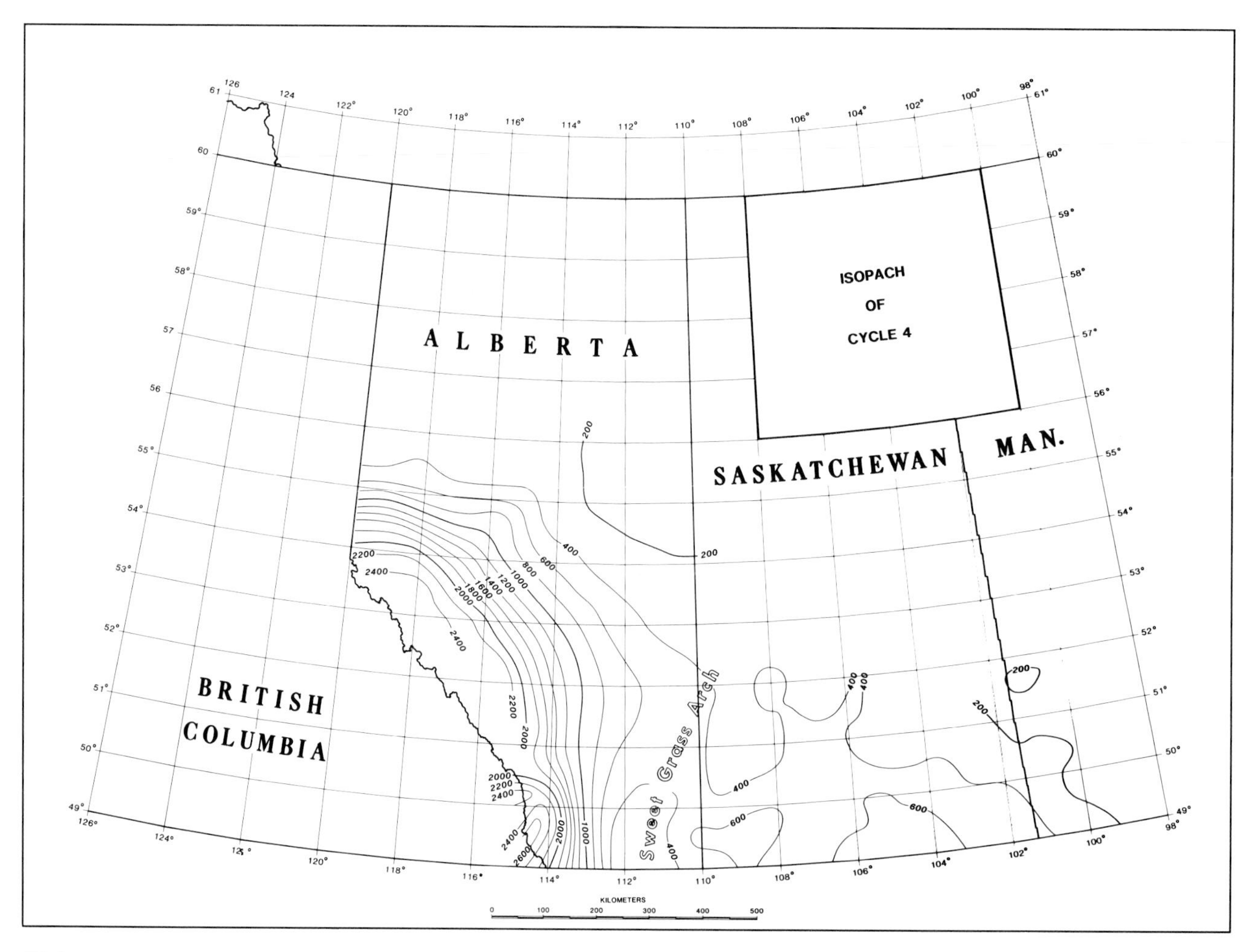

FIGURE 30. Isopach map of sediments deposited during Cycle 4. Original data based on preliminary map from the new Atlas of Western Canada. Contour interval, 200 m. (G. Mossop, personal communication, 1990.)

Canyon Formation in south-central Alberta is represented by an upward-coarsening, fluvially or tide-dominated deltaic sequence scoured by lowstand fluvial activity and backfilled as estuaries (Rahmani, 1988; Saunders and Pemberton, 1988). Significant peat accumulated in the associated coastal plain during the transgression and regression of the Bearpaw Sea (Jerzykiewicz and Sweet, 1988). Its siliceous nature has been attributed to the dissolution of siliceous microfossils, possibly deposited at the peak of the Bearpaw transgression (McNeil and Caldwell, 1981). As the Bearpaw Sea receded, the coastal sediments of the Blood Reserve Formation (southern Alberta), Eastend Formation (Saskatchewan), and Boissevain Formation (Manitoba) were deposited, followed by the nonmarine St. Mary River and Willow Creek formations (Frenchman and lower Ravenscrag formations in Saskatchewan). The shorelines of the Blood Reserve Formation varied from low-energy to wave-dominated, with a significant tidal influence as inferred from the presence of tidal-channel deposits and oyster bioherms (Nadon, 1988).

Cycle 4 sediments were derived primarily from volcanic, low-grade metamorphic and sedimentary sources with minor input from granitic and gneissic sources (Mack and Jerzykiewicz, 1989). Volcanic rocks in the Elkhorn Mountains of southwestern Montana have been cited as a sediment source for the Belly River Formation and equivalent strata in the southern foreland basin (McLean, 1971), although this has been questioned by Mack and Jerzykiewicz (1989), who suggested a westerly source in the Omineca and Rocky Mountain belts. In southwestern Alberta, rivers depositing the St. Mary River Formation flowed north to northeastward out of Montana (Nadon, 1988), whereas in south-central Alberta, the rivers of the correlative Edmonton Formation flowed south to southeastward (Rahmani, 1981).

Mack and Jerzykiewicz (1989) recognized five stages in which thrusting alternated with episodes of tectonic quiescence throughout the deposition of Cycle 4. The thrust episodes (stages I, III, and V) incorporated sedimentary strata in the Rocky Mountain belt, which was eroded and resulted in deposition of the Willow Creek Formation and the Entrance Conglomerate [stage III; Enclosure 1(A)], and the Paskapoo Formation and the High Divide Ridge Conglomerate (stage V). These sediments are characterized by a coarse grain size, thick channel sandstones, and a predominance of carbonate rock fragments and chert. Fluvial channels flowed northeastward out of the mountains. During the tectonically inactive times, the St. Mary River Formation (stage II) and the upper portion of the Brazeau Formation (stage IV) were deposited. These are characterized by finer-grained sediment, thinner channel sandstones, and major coal seams. The sediment generally contains increased proportions of metamorphic detritus derived from the Omineca belt and shales of the western Rocky Mountain belt with a decrease in carbonate and chert. The change in provenance reflects erosion of the easternmost thrust sheets and a westward shift in the source terrain. During the quiescent periods, two paleocurrent trends developed—one perpendicular to the basin and the second parallel to the basin axis.

Beyond the most northerly and westerly limits of the Bearpaw transgression (Figure 28), the term "Brazeau Formation" is used to represent sediments lying above the top of the Wapiabi Formation and below the Coalspur Formation (Jerzykiewicz and Sweet, 1988). Detritus was derived from andesitic/dacitic volcanic and low-grade metamorphic rocks in the Omineca belt and sedimentary rocks in the Rocky Mountain belt (Mack and Jerzykiewicz, 1989).

The climate during deposition of Cycle 4 alternated from warm and humid to semiarid. Major caliche beds at the top of the Belly River Formation and in the Willow Creek and Porcupine Hills formations are restricted to southern Alberta and may have required hundreds of thousands of years of semiarid climate to form (Jerzykiewicz and Sweet, 1988). The climate became increasingly humid northward (Jerzykiewicz and Sweet, 1988). In south-central Alberta, the Judith River Formation was deposited in a humid tropical to subtropical coastal plain setting having a mildly seasonal rainfall (Visser, 1986). Palynoflora and coal in sediments below and above the Bearpaw Formation indicate a warm temperate to subtropical, humid setting in environs proximal to the Bearpaw Sea (Dodson, 1971; Jerzykiewicz and Sweet, 1988). Mean annual rainfall may have been on the order of 120 cm or more (Béland and Russell, 1978). Water temperatures in the Bearpaw Sea were approximately 17 to 27°C based on oxygen and carbon isotopes from ammonites (Forester et al., 1977). The prevailing winds were westerly (Lloyd, 1982).

Cycle 5: Tertiary Conglomerates (Eocene to Pliocene)

Tertiary gravels, ranging in age from late Eocene to Pliocene, form the uplands in at least ten isolated plateaus throughout southern Saskatchewan and Alberta, and constitute much of the sediment of Cycle 5. Cycle 5 deposition postdates the phase of foreland basin compressional deformation. The origin of the gravels is enigmatic, being very coarse-grained (clasts to >40 cm maximum length) and far from the cordillera. At the time of deposition, overthrusting in the cordillera had ceased and the deformed belt and foreland basin were being isostatically uplifted and eroded. Major tectonic uplift and erosion in the Omineca belt are indicated by potassium-argon dates of ~50 Ma (Price et al., 1985). The gravels overlie a major unconformity throughout most of the basin [Enclosure 1(A)]. The unconformity directly follows and may be related to magmatic quiescence in the cordillera during the Paleocene (70 to 60 Ma; Armstrong, 1988). As much as 1.5 to 2.0 km of sediment may have been eroded from the western side of the basin during this period of erosion (Magara, 1976; Hacquebard, 1977). From latest Paleocene to middle Eocene (55 to 45 Ma), the cordillera was again affected by extension-related magmatism in all the terranes (Armstrong, 1988).

Recent studies have indicated that the coarse boulder conglomerate comprising the Cypress Hills Formation in southern Alberta and Saskatchewan was resedimented during the Oligocene by magmatic intrusions originating in the Sweetgrass Hills and Bearpaw Mountains of northern Montana (Figure 31; Leckie and Cheel, 1989).

The western Canada basin was characterized by major geographic variations in climate during the Paleocene portion of Cycle 5 (Jerzykiewicz and Sweet, 1988). Silcretes and vertebrate faunal remains in the Oligocene Cypress Hills Formation of southwest Saskatchewan indicate that semiarid conditions existed at least during early Cycle 5 time. However, northward in west-central Alberta, the climate was characterized by much more humid conditions as indicated by better-preserved coal deposits.

ACKNOWLEDGMENTS

An earlier version of this manuscript benefited from rigorous review by Jim Barclay, Elliot Burden, Jim Christopher, Brad Hayes, Paul Johnstone, Roger Macqueen, and Greg Nadon. Grant Mossop provided the regional isopach and structure maps, which are prototypes for the new Atlas of Western Canada. Drafting was done by Alex Swacha.

REFERENCES CITED

Amajor, L. C., 1985, Biotite grain size distribution and source area of the Lower Cretaceous Viking bentonites, Alberta, Canada: Bulletin of Canadian Petroleum Geology, v. 33, p. 471–478.

Archibald, D. A., T. E. Krogh, R. L. Armstrong, and E. Farrar, 1984, Chronology and tectonic implications of magmatism and metamorphism, southern Kootenay arc and neighboring regions, southwestern British Columbia, part II: mid Cretaceous to Eocene: Canadian Journal of Earth Sciences, v. 21, p. 567–583.

Armstrong, R. L., 1988, Mesozoic and early Cenozoic magmatic evolution of the Canadian Cordillera, *in* S. P. Clark, B. C. Burchfiel, and J. Suppe, eds., Processes in continental lithospheric deformation: GSA Special Paper 218, p. 55–92.

Bally, A. W., P. L. Gordy, and G. A. Stewart, 1966, Structure, seismic data, and orogenic evolution of southern Canadian Rocky Mountains: Bulletin of Canadian Petroleum Geology, v. 14, p. 337–381.

Bally, A. W., and S. Snelson, 1980, Realms of subsidence, *in*

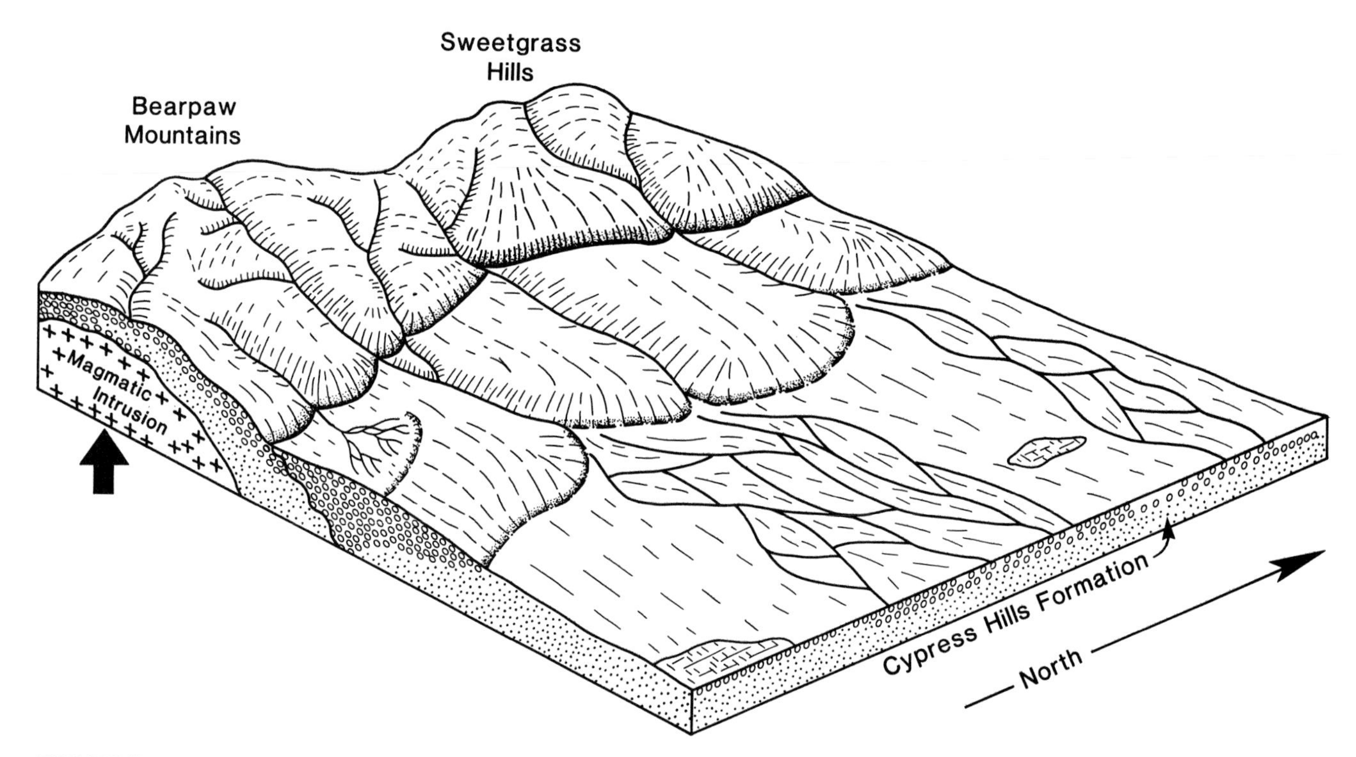

FIGURE 31. Depositional model used to account for deposition of the boulder conglomerates and debris flow deposits of the Cypress Hills Formation (Cycle 5) found in the Cypress Hills. (Modified from Leckie and Cheel, 1989.)

A. D. Miall, ed., Facts and principles of world petroleum occurrence: Canadian Society of Petroleum Geologists Memoir 6, p. 9–75.

Banerjee, I. B., and E. H. Davies, 1988, An integrated lithostratigraphic and palynostratigraphic study of the Ostracode zone and adjacent strata in the Edmonton embayment, central Alberta, *in* D. P. James and D. A. Leckie, eds., Sequences, stratigraphy, sedimentology: surface and subsurface: Canadian Society of Petroleum Geologists Memoir 15, p. 261–274.

Barclay, J. E., and D. G. Smith, 1992, Western Canada foreland basin oil and gas plays, this volume.

Beaumont, C., 1981, Foreland basins: Geophysical Journal of the Royal Astronomical Society, v. 65, p. 291–329.

Beaumont, E. A., 1984, Retrogradational shelf sedimentation; Lower Cretaceous Viking Formation, central Alberta, *in* R. W. Tillman and C. T. Siemers, eds., Siliciclastic shelf sediments: SEPM Special Publication 34, p. 163–177.

Béland, P., and D. A. Russell, 1978, Paleoecology of Dinosaur Provincial Park (Cretaceous), Alberta, interpreted from the distribution of articulated vertebrate remains: Canadian Journal of Earth Sciences, v. 15, p. 1012–1024.

Bhattacharya, J., 1988, Autocyclic and allocyclic sequences in river and wave-dominated deltaic sediments of the Upper Cretaceous Dunvegan Formation, Alberta: core examples, *in* D. P. James and D. A. Leckie, eds., Sequences, stratigraphy, sedimentology: surface and subsurface: Canadian Society of Petroleum Geologists Memoir 15, p. 25–32.

Brown, R. L., and J. M. Journeay, 1987, Tectonic denudation of the Shuswap metamorphic terrane of southeastern British Columbia: Geology, v. 15, p. 142–146.

Brown, R. L., and P. B. Read, 1983, Shuswap terrane of British Columbia: a Mesozoic "core complex": Geology, v. 11, p. 164–168.

Burden, E. T., 1982, Lower Cretaceous terrestrial palynomorph biostratigraphy of the McMurray Formation, northeastern Alberta: Ph.D. thesis, University of Calgary, 59 p.

Caldwell, W. G. E., 1984, The Lower-Upper Cretaceous boundary in the southern interior plains: Geological Association of Canada/Mineralogical Association of Canada, Joint Annual Meeting, London, Ontario, Program with Abstracts, v. 9, p. 50.

Cant, D. J., 1988, Regional structure and development of the Peace River arch, Alberta; a Paleozoic failed-rift system?: Bulletin of Canadian Petroleum Geology, v. 36, p. 284–295.

Cheel, R. J., and D. A. Leckie, 1990, A tidal-inlet complex in the Cretaceous epeiric sea of North America: Virgelle Member, Milk River Formation, southern Alberta: Sedimentology, v. 37, p. 67–82.

Christopher, J. E., 1964, The Middle Jurassic Shaunavon Formation of southwestern Saskatchewan: Saskatchewan Department of Mineral Resources Report 95, 96 p.

Christopher, J. E., 1980, The Lower Cretaceous Mannville Group of Saskatchewan—a tectonic overview, *in* L. S. Beck, J. E. Christopher, and D. M. Kent, eds., Lloydminster and beyond: geology of Mannville hydrocarbon reservoirs: Saskatchewan Geological Society Special Publication No. 5, p. 3–32.

Christopher, J. E., 1984, The Lower Cretaceous Mannville Group, northern Williston basin region, Canada, *in* D. F. Stott and D. J. Glass, eds., The Mesozoic of middle North America: Canadian Society of Petroleum

Geologists Memoir 9, p. 109-126.
Cowie, J. W., and M. G. Bassett, 1989, Global stratigraphic chart: Supplement of Episodes, 12.
Creaney, S., and J. Allan, 1992, Petroleum systems in the foreland basin of Western Canada, this volume.
Dickinson, W. R., 1974, Plate tectonics and sedimentation, *in* W. R. Dickinson, ed., Tectonics and sedimentation: SEPM Special Publication 22, p. 1-27.
Dodson, P., 1971, Sedimentology and taphonomy of the Oldman Formation (Campanian), Dinosaur Provincial Park, Alberta (Canada): Palaeogeography, Palaeoclimatology, Palaeoecology, v. 10, p. 21-74.
Edmunds, R. H., 1980, Salt removal and oil entrapment, *in* A. D. Miall, ed., Facts and principles of world oil occurrence: Canadian Society of Petroleum Geologists Memoir 6, p. 988.
Eisbacher, G. H., 1985, Pericollisional strike-slip faults and synorogenic basins, Canadian Cordillera, *in* K. T. Biddle and N. Christie-Blick, eds., Strike-slip deformation, basin formation and sedimentation: SEPM Special Publication 31, p. 265-282.
Farshori, M. Z., 1983, Glauconite sandstone, Countess field "H" pool, southern Alberta, *in* J. R. McLean and G. E. Reinson, eds., Sedimentology of selected Mesozoic clastic sequence: Calgary, Alberta, Canadian Society of Petroleum Geologists, p. 27-42.
Fermor, P. R., and I. W. Moffat, 1992, Tectonics and structure of the Western Canada foreland basin, this volume.
Flemings, P. B., and T. E. Jordan, 1989, A synthetic stratigraphic model of foreland basin development: Journal of Geophysical Research, B, Solid Earth and Planets, v. 94, p. 3851-3866.
Forester, R. W., W. G. E. Caldwell, and F. H. Oro, 1977, Oxygen and carbon isotopic study of ammonites from the Late Cretaceous Bearpaw Formation in southwestern Saskatchewan: Canadian Journal of Earth Sciences, v. 14, p. 2086-2100.
Gabrielse, H., 1985, Major dextral transcurrent displacements along the northern Rocky Mountain Trench and related lineaments in north-central British Columbia: GSA Bulletin, v. 96, p. 1-14.
Ghent, E. D., J. Nicholls, M. Z. Stout, and B. Rottenfusser, 1977, Clinopyroxene amphibolite boudins from Three Valley Gap, British Columbia: The Canadian Mineralogist, v. 15, p. 269-282.
Gibson, D., 1977, Sedimentary facies in the Jura-Cretaceous Kootenay Formation, Crowsnest Pass area, southwestern Alberta and southeastern British Columbia: Bulletin of Canadian Petroleum Geology, v. 4, p. 767-791.
Glaister, R. P., 1959, Lower Cretaceous of southern Alberta and adjoining areas: AAPG Bulletin, v. 43, p. 590-640.
Habicht, J. K. A., 1979, Paleoclimate, paleomagnetism and continental drift: Tulsa, Oklahoma, AAPG Studies in Geology No. 9, 31 p.
Hacquebard, P. A., 1977, Rank of coal as an index of organic metamorphism for oil and gas in Alberta, *in* G. Deroo, T. G. Powell, B. Tissot, and R. G. McGrossan, eds., The origin and migration of petroleum in the Western Canadian sedimentary basin, Alberta: a geotechnical and thermal maturation study: Geological Survey of Canada Bulletin 262, p. 11-22.
Haq, B. U., J. Hardenbol, and P. R. Vail, 1987, Chronology of fluctuating sea levels since the Triassic: Science, v. 235, p. 1156-1167.
Hayes, B. J. R., 1983, Stratigraphy and petroleum potential of the Swift Formation (Upper Jurassic), southern Alberta and north-central Montana: Bulletin of Canadian Petroleum Geology, v. 31, p. 37-52.
Hayes, B. J. R., 1986, Stratigraphy of the basal Cretaceous Lower Mannville Formation, southern Alberta and north-central Montana: Bulletin of Canadian Petroleum Geology, v. 34, p. 30-48.
Hein, F. J., M. E. Dean, A. M. Delure, S. K. Grant, G. A. Robb, and F. J. Longstaffe, 1986, Regional sedimentology of the Viking Formation, Caroline, Garrington and Harmattan East fields, western south-central Alberta: storm and current-influenced shelf settings: Bulletin of Canadian Petroleum Geology, v. 34, p. 91-110.
Heller, P. L., C. L. Angevine, N. S. Winslow, and C. Paola, 1988, Two-phase stratigraphic model of foreland-basin sequences: Geology, v. 16, p. 501-504.
Jervey, M., 1992, Siliciclastic sequence development in foreland basins, with examples from the Western Canada foreland basin, this volume.
Jerzykiewicz, T., and A. R. Sweet, 1988, Sedimentological and palynological evidence of regional climatic changes in the Campanian to Paleocene sediments of the Rocky Mountain Foothills, Canada: Sedimentary Geology, v. 59, p. 29-76.
Jordan, T. E., 1981, Thrust loads and foreland basin evolution, Cretaceous, western United States: AAPG Bulletin, v. 65, p. 2506-2520.
Kauffman, E. G., 1977, Geological and biological overview: Western Interior Cretaceous basin: Mountain Geologist, v. 14, p. 79-99.
Kauffman, E. G., 1984, Paleobiogeography and evolutionary dynamic response in the Cretaceous Western Interior Seaway of North America, *in* G. E. G. Westermann, ed., Jurassic-Cretaceous biochronology and biogeography of North America: Geological Association of Canada Special Paper 27, p. 273-306.
Koster, E. H., 1984, Paleochannel sedimentology in the upper Judith River Formation (Campanian), Dinosaur Provincial Park, southeastern Alberta, *in* D. F. Stott and D. J. Glass, eds., The Mesozoic of middle North America: Canadian Society of Petroleum Geologists Memoir 9, p. 556.
Koziol, B. L., 1988, Stratigraphy of the Viking Formation (Albian) and Newcastle Member (Albian) of the Ashville Formation in Saskatchewan: unpublished Master's thesis, University of Saskatchewan, 238 p.
Lambeck, K., S. Cloetingh, and H. McQueen, 1987, Intraplate stress and apparent changes in sea level: the basins of northwestern Europe, *in* C. Beamont and A. J. Tankard, eds., Sedimentary basins and basin-forming mechanisms: Canadian Society of Petroleum Geologists Memoir 12, p. 259-268.
Leckie, D. A., 1986a, Tidally influenced, transgressive shelf sediments in the Viking Formation, Caroline, Alberta: Bulletin of Canadian Petroleum Geologists, v. 34, p. 111-125.
Leckie, D. A., 1986b, Rates, controls, and sand-body geometries of transgressive-regressive cycles: Cretaceous Moosebar and Gates formations, British Colum-

bia: AAPG Bulletin, v. 70, p. 516-535.

Leckie, D. A., and R. J. Cheel, 1989, The Cypress Hills Formation (upper Eocene to Miocene): a semi-arid braidplain deposit resulting from intrusive uplift: Canadian Journal of Earth Sciences, v. 26, p. 1918-1931.

Leckie, D. A., C. Fox, and C. Tarnocai, 1989, Multiple paleosols of the late Albian Boulder Creek Formation, British Columbia, Canada: Sedimentology, v. 36, p. 307-323.

Leckie, D. A., and G. E. Reinson, in press, Effects of middle to late Albian sea level fluctuations in the Cretaceous Interior seaway, Western Canada, *in* W. G. E. Caldwell and E. G. Kauffman, eds., Evolution of the Western Interior basin: Geological Association of Canada Special Publication.

Leckie, D. A., and R. G. Walker, 1982, Storm- and tide-dominated shorelines in the Cretaceous Moosebar-Lower Gates interval: outcrop equivalents of the Deep Basin gas trap in Western Canada: AAPG Bulletin, v. 66, p. 138-157.

Link, T. A., and A. J. Childerhose, 1931, Bearpaw shale and contiguous formations in Lethbridge area, Alberta: AAPG Bulletin, v. 15, p. 1227-1242.

Lloyd, C. R., 1982, The mid-Cretaceous Earth; paleogeography, ocean circulation and temperature, atmospheric circulation: Journal of Geology, v. 90, p. 393-413.

Mack, G. H., and T. Jerzykiewicz, 1989, Provenance of post-Wapiabi sandstones and its implications for Campanian to Paleocene tectonic history of the southern Canadian Cordillera: Canadian Journal of Earth Sciences, v. 26, p. 665-676.

Magara, K., 1976, Thickness of removed sedimentary rocks, paleopore pressure and paleotemperature, southwestern part of the Western Canada basin: AAPG Bulletin, v. 60, p. 554-565.

Marvin, R. F., B. C. Hearn, Jr., H. H. Mehnert, C. W. Naeser, R. E. Zartman, and D. A. Lindsey, 1980, Late Cretaceous-Paleocene-Eocene igneous activity in north-central Montana: Isochron/West, No. 29, p. 5-25.

McLean, J. R., 1971, Stratigraphy of the Upper Cretaceous Judith River Formation in the Canadian Great Plains: Saskatchewan Research Council, Geological Division Report 11, 96 p.

McLean, J. R., 1977, The Cadomin Formation: stratigraphy, sedimentology and tectonic implication: Bulletin of Canadian Petroleum Geology, v. 25, p. 792-787.

McLean, J. R., and T. Jerzykiewicz, 1978, Cyclicity, tectonics and coal: some aspects of fluvial sedimentology in the Brazeau-Paskapoo formations, Coal Valley area, Alberta, Canada, *in* A. D. Miall, ed., Fluvial sedimentology: Canadian Society of Petroleum Geologists Memoir 5, p. 441-469.

McLean, J. R., and J. H. Wall, 1981, The Early Cretaceous Moosebar Sea in Alberta: Bulletin of Canadian Petroleum Geology, v. 29, p. 334-377.

McMechan, M. E., and R. I. Thompson, 1989, Structural style and history of the Rocky Mountain fold and thrust belt, *in* B. D. Ricketts, ed., Western Canada sedimentary basin, a case history: Calgary, Alberta, Canadian Society of Petroleum Geologists, p. 47-72.

McMechan, M. E., and R. I. Thompson, in press, The Canadian Cordilleran fold and thrust belt south of 66°N and its influence on the Western Interior basin, *in* W. G. E. Caldwell and E. Kauffman, eds., Evolution of the Western Interior basin: Geological Association of Canada Special Publication.

McNeil, D. H., 1984, The eastern facies of the Cretaceous System in the Canadian Western Interior, *in* D. F. Stott and D. J. Glass, eds., The Mesozoic of middle North America: Canadian Society of Petroleum Geologists Memoir 9, p. 145-171.

McNeil, D. H., and W. G. E. Caldwell, 1981, Cretaceous rocks and their foraminifera in the Manitoba escarpment: Geological Association of Canada Special Paper No. 21, 439 p.

Monger, J. W. H., 1989, Overview of cordilleran geology, *in* B. D. Ricketts, ed., Western Canada sedimentary basin, a case history: Calgary, Alberta, Canadian Society of Petroleum Geologists, p. 9-32.

Monger, J. W. H., R. A. Price, and D. J. Templeman-Kluit, 1982, Tectonic accretion and the origin of the two major metamorphic and plutonic welts in the Canadian Cordillera: Geology, v. 10, p. 70-75.

Nadon, G., 1988, Tectonic controls on sedimentation within a foreland basin: the Bearpaw, Blood Reserve and St. Mary River formations, southwestern Alberta, *in* Field guide to sequences, stratigraphy, sedimentology: Society of Petroleum Geologists Surface and Subsurface Technical Meeting, September 14-16, Calgary, Alberta, 85 p.

Nelson, C. H., W. Dupré, M. Field, and J. D. Howard, 1982, Variation in sand body types on the eastern Bering Sea epicontinental shelf, *in* C. H. Nelson and S. E. Nio, eds., The northeastern Bering Shelf: new perspectives of epicontinental shelf processes and depositional products: Geologie en Mijnbouw, v. 61, p. 37-48.

Norris, D. K., 1964, The Lower Cretaceous of the southeastern Canadian Cordillera: Bulletin of Canadian Petroleum Geology, v. 12, p. 201-237.

Parsons, W. H., 1973, Alberta, *in* R. G. McCrossan, ed., The future petroleum provinces of Canada—their geology and potential: Canadian Society of Petroleum Geologists Memoir 1, p. 73-120.

Plint, A. G., and R. G. Walker, 1987, Cardium Formation 8, facies and environments of the Cardium shoreline and coastal plain in the Kakwa field and adjacent areas, northwestern Alberta: Bulletin of Canadian Petroleum Geology, v. 35, p. 48-64.

Plint, A. G., R. G. Walker, and K. M. Bergman, 1986, Cardium Formation 6, stratigraphic framework of the Cardium in subsurface: Bulletin of Canadian Petroleum Geology, v. 34, p. 213-225.

Plint, A. G., R. G. Walker, and W. M. Duke, 1988, An outcrop to subsurface correlation of the Cardium Formation in Alberta, *in* D. P. James and D. A. Leckie, eds., Sequences, stratigraphy, sedimentology: surface and subsurface: Canadian Society of Petroleum Geologists Memoir 15, p. 167-183.

Podruski, J. A., 1988, Contrasting character of the Peace River and Sweetgrass arches, Western Canada sedimentary basin: Geoscience Canada, v. 15, p. 94-97.

Porter, J. W., R. A. Price, and R. G. McCrossan, 1982, The Western Canada sedimentary basin: Philosophical Transactions of the Royal Society of London, Series A, v. 305, p. 169-192.

Potocki, D. J., and I. Hutcheon, 1992, Lithology and diagenesis of sandstones in the Western Canada foreland basin, this volume.

Poulton, T., 1984, The Jurassic of the Canadian Western Interior, from 49°N latitude to Beaufort Sea, *in* D. F. Stott and D. J. Glass, eds., The Mesozoic of middle North America: Canadian Society of Petroleum Geologists Memoir 9, p. 15-42.

Price, R. A., 1973, Large-scale gravitational flow of supracrustal rocks, southern Canadian Rockies, *in* K. A. De Jong and R. A. Scholten, eds., Gravity and tectonics: New York, John Wiley, p. 491-502.

Price, R. A., 1981, The Cordilleran foreland thrust and fold belt in the southern Canadian Rocky Mountains, *in* K. R. McClay and N. J. Price, eds., Thrust and nappe tectonics: Geological Society of London Special Publication 9, p. 427-448.

Price, R. A., J. W. H. Monger, and J. A. Roddick, 1985, Cordilleran cross-section: Calgary to Vancouver, trip 3, *in* D. Tempelman-Kluit, ed., Field guides to geology and mineral deposits in the southern Canadian Cordillera; GSA Cordilleran Section Meeting: Vancouver, British Columbia, Geological Survey of Canada, p. 3.1-3.85.

Price, R., and E. W. Mountjoy, 1970, Geologic structure of the Canadian Rocky Mountains between Bow and Athabasca rivers—a progress report, *in* J. O. Wheeler, Structure of the southern Canadian Cordillera: Geological Association of Canada Special Paper 6, p. 7-25.

Putnam, P. E., 1989, Geological controls on hydrocarbon entrapment within the Lower Cretaceous Cantuar Formation, Wapella field, southeastern Saskatchewan: Bulletin of Canadian Petroleum Geology, v. 37, p. 389-400.

Rahmani, R. A., 1981, Facies relationships and paleoenvironments of a Late Cretaceous tide-dominated delta, Drumheller, Alberta, *in* R. I. Thompson and D. G. Cook, Field guides to geological and mineralogical deposits; Calgary '81 Annual Meeting, Geological Association of Canada, Mineralogical Association of Canada, and Geological and Geophysical Union: Calgary, Alberta, University of Calgary, Department of Geology and Geophysics, p. 159-176.

Rahmani, R. A., 1988, Estuarine tidal channel and nearshore sedimentation of a Late Cretaceous epicontinental sea, Drumheller, Alberta, Canada, *in* P. L. de Boer, A. van Gelder, and S. D. Nio, eds., Tide-influenced sedimentary environments and facies: Dordrecht, Reidel Publishing Co., p. 433-471.

Rapson, J. E., 1965, Petrography and derivation of Jurassic-Cretaceous clastic rocks, southern Rocky Mountains, Canada: AAPG Bulletin, v. 49, p. 1426-1452.

Reinson, G. E., J. E. Clark, and A. E. Foscolos, 1988, Reservoir geology of the Crystal Viking field, Lower Cretaceous estuarine tidal-channel-bay complex, south-central Alberta: AAPG Bulletin, v. 72, p. 1270-1294.

Ricketts, B. D., and A. R. Sweet, 1986, Stratigraphy, sedimentology and palynology of the Kootenay-Blairmore transition in southwestern Alberta and southeastern British Columbia: Geological Survey of Canada Paper 84-15, 41 p.

Rosenthal, L., 1988, Wave-dominated shorelines and incised channel trends: Lower Cretaceous Glauconite Formation, west-central Alberta, *in* D. P. James and D. A. Leckie, eds., Sequences, stratigraphy, sedimentology: surface and subsurface: Canadian Society of Petroleum Geologists Memoir 15, p. 207-220.

Rudkin, R. A., 1964, Lower Cretaceous, Chapter 11, *in* R. G. McCrossan and R. P. Glaister, eds., Geological history of Western Canada: Calgary, Alberta, Alberta Society of Petroleum Geologists, p. 157-168.

Saunders, T., and S. G. Pemberton, 1988, Trace fossils and sedimentology of a Late Cretaceous progradational barrier island sequence: Bearpaw-Horseshoe Canyon Formation transition, Dorothy, Alberta, *in* Field guide to sequences, stratigraphy, sedimentology: Society of Petroleum Geologists Surface and Subsurface Technical Meeting, September 14-16, Calgary, Alberta, 85 p.

Schultheis, N. H., and E. W. Mountjoy, 1978, Cadomin conglomerate of western Alberta—a result of Early Cretaceous uplift of the Main Ranges: Bulletin of Canadian Petroleum Geology, v. 26, p. 297-342.

Shepard, W., and B. Bartow, 1986, Tectonic history of the Sweetgrass arch, a key to finding new hydrocarbons, Montana and Alberta, *in* J. H. Noll and K. M. Doyle, eds., Rocky Mountain oil and gas fields; Wyoming Geological Association Symposium-1986: Casper, Wyoming, Wyoming Geological Association, p. 9-19.

Shurr, G. W., L. O. Anna, and J. A. Peterson, 1989, Zuni sequence in Williston basin—evidence for Mesozoic paleotectonism: AAPG Bulletin, v. 73, p. 68-87.

Simpson, F., 1988, Solution-generated collapse (SGC) structures associated with bedded evaporites: significance to base-metal and hydrocarbon localization: Geoscience Canada, v. 15, p. 89-93.

Sloss, L. L., 1963, Sequences in the cratonic interior of North America: Bulletin of the GSA, v. 74, p. 93-113.

Stelck, C. R., and J. Armstrong, 1981, Neogastroplites from southern Alberta: Bulletin of Canadian Petroleum Geology, v. 29, p. 399-407.

Stelck, C. R., and J. H. Wall, 1954, Kaskapau foraminifera from Peace River area of Western Canada: Research Council of Alberta Report 70, p. 1-62.

Stelck, C. R., J. H. Wall, G. D. Williams, and G. B. Mellon, 1972, Cretaceous and Jurassic of the Rocky Mountain Foothills of Alberta, *in* D. J. Glass, ed., International Geological Congress Guidebook No. 24, Part A20, 51 p.

Stockmal, G. S., and C. Beaumont, 1987, Geodynamic models of convergent margin tectonics: the southern Canadian Cordillera and the Swiss Alps, *in* C. Beaumont and A. J. Tankard, eds., Sedimentary basins and basin-forming mechanisms: Canadian Society of Petroleum Geologists Memoir 12, p. 393-411.

Stockmal, G. S., D. J. Cant, and J. S. Bell, 1992, Relationship of the stratigraphy of the Western Canada foreland basin to cordilleran tectonics: insights from geodynamic models, this volume.

Stott, D. F., 1968, Lower Cretaceous Bullhead and Fort St. John groups between Smoky and Peace rivers, Rocky Mountain Foothills, Alberta and British Columbia: Geological Survey of Canada Bulletin 152, 279 p.

Stott, D. F., 1973, Lower Cretaceous Bullhead Group between Bullmoose Mountain and Tetsa River, Rocky Mountain Foothills, northeastern British Columbia: Geological Survey of Canada Bulletin 219, 228 p.

Stott, D. F., 1984a, Lower Cretaceous Fort St. John Group and Upper Cretaceous Dunvegan Formation of the foothills and plains of Alberta, British Columbia, District of MacKenzie and Yukon Territory: Geological Survey of Canada Bulletin 328, 124 p.

Stott, D. F., 1984b, Cretaceous sequences of the foothills of the Canadian Rocky Mountains, *in* D. F. Stott and D. J. Glass, The Mesozoic of middle North America: Canadian Society of Petroleum Geologists Memoir 9, p. 85–107.

Tater, J. M., 1965, Potassium-argon dates as provenance indicators for the Dunvegan Formation: Bulletin of Canadian Petroleum Geology, v. 13, p. 503–508.

Taylor, D. R., and R. G. Walker, 1984, Depositional environments and paleogeography in the Albian Moosebar Formation and adjacent fluvial Gladstone and Beaver Mines formations, Alberta: Canadian Journal of Earth Sciences, v. 21, p. 698–714.

Thompson, B., E. Mercier, and C. Roots, 1987, Extension and its influence on Canadian Cordilleran passive-margin evolution, *in* M. P. Coward, J. F. Dewey, and P. L. Hancock, Continental extensional tectonics: Geological Society of London Special Publication No. 28, p. 409–417.

Thompson, R. I., 1981, The nature and significance of large "blind" thrusts within the northern Rocky Mountains of Canada, *in* K. R. McClay and N. J. Price, eds., Thrust and nappe tectonics: Geological Society of London Special Publication No. 9, p. 449-462.

Tovell, W. M., 1958, The development of the Sweetgrass arch, southern Alberta: Geological Association of Canada Proceedings, v. 10, p. 19–30.

Visser, J., 1986, Sedimentology and taphonomy of a *Styracosaurus* bone bed in the Late Cretaceous Judith River Formation, Dinosaur Provincial Park, Alberta: unpublished M.Sc. thesis, University of Calgary, 150 p.

Wall, J. H., and K. K. Rosene, 1977, Upper Cretaceous stratigraphy and micropaleontology of the Crowsnest Pass–Waterton area, southern Alberta Foothills: Bulletin of Canadian Petroleum Geology, v. 25, p. 842–867.

Wall, J. H., and C. Singh, 1975, A Late Cretaceous microfossil assemblage from the Buffalo Head Hills, north-central Alberta: Canadian Journal of Earth Sciences, v. 12, p. 1157–1174.

Wallace-Dudley, K. E., and D. A. Leckie, 1988, Preliminary observations on the sedimentology of the Doe Creek Member, Kaskapau Formation, in the Valhalla field, northwestern Alberta, *in* D. P. James and D. A. Leckie, eds., Sequences, stratigraphy, sedimentology: surface and subsurface: Canadian Society of Petroleum Geologists Memoir 15, p. 485–496.

Wernicke, B. P., R. L. Christiansen, P. C. England, and L. J. Sonder, 1987, Tectonomagmatic evolution of Cenozoic extension in the North American Cordillera, *in* M. P. Coward, J. F. Dewey, and P. L. Hancock, Continental extensional tectonics: Geological Society of London Special Publication No. 28, p. 203–221.

Williams, G. K., 1958, Influence of the Peace River arch on Mesozoic strata: Journal of Alberta Society of Petroleum Geologists, v. 6, p. 74–81.

Wood, J., 1985, Sedimentology of the Late Cretaceous Judith River Formation, "Cathedral" area, Dinosaur Provincial Park, Alberta: unpublished M.Sc. thesis, University of Calgary, 215 p.

Wright, G. N., ed., 1984, The Western Canada sedimentary basin illustrating basin stratigraphy and structure: Canadian Society of Petroleum Geologists/Geological Association of Canada (poster).

Chapter 2

Siliciclastic Sequence Development in Foreland Basins, with Examples from the Western Canada Foreland Basin

Macomb T. Jervey

Jervey Geological Consulting
Mosier, Oregon, U.S.A.

INTRODUCTION

Unconformity-bounded depositional sequences are the fundamental building blocks of the stratigraphic record. The post-Middle Jurassic succession in the Western Canada foreland basin comprises siliciclastic sequences derived primarily from the active thrust belt to the west. Tectonic elements of the developing Cretaceous foreland basin, such as arches, foredeeps, and the bordering mountainous source terrain, were of great importance in their influence on rates of sediment supply, subsidence, and accommodation for sediment accumulation (Leckie and Smith, 1992; Potocki and Hutcheon, 1992).

Concurrent with structural deformation of the basin, eustatic sea level fluctuation may have imposed additional controls on the nature of the evolving sedimentary sequences. Although considerable evidence has been presented in the literature regarding the importance of eustasy in sequence stratigraphy (e.g., Vail et al., 1977; Haq et al., 1988; Olsson, 1988), the relative importance of sea level change and tectonics in the process of sequence development in the Western Canada foreland basin is unclear. Nevertheless, it is apparent that changes in relative sea level, whether tectonic or eustatic in origin, are the driving mechanism producing many of the observed characteristics of sedimentary sequences, both from the large-scale seismic perspective and at the scale of core and outcrop studies.

Figure 1 diagrammatically illustrates post-Middle Jurassic stratigraphy along the western margin of the Western Canada foreland basin. Thick wedges of coarse-grained siliciclastics, derived ultimately from the west, prograded at different times in an easterly, northerly, or southerly direction. Changes in sediment supply and relative sea level produced a cyclic pattern of regression and transgression which, in turn, resulted in the large-scale patterns of facies change depicted in the diagram. The stratigraphic nomenclature shown in Figure 1 is the system of formal and informal names by which these cycles are generally recognized within the Cretaceous strata of the foreland basin of Western Canada. The purpose of this paper is to examine depositional cyclicity in foreland basins and its relationship to changes in eustatic and relative sea level, using the Western Canada foreland basin as an example.

The interaction of tectonic deformation, sediment supply, and sea level change is a four-dimensional problem of time and space. Sequence development occurs in a dynamic environment in which these major variables control the distribution of depositional facies, the geometry of the resulting sedimentary units, and the nature of the sequence-bounding surfaces. A numerical model of basin-filling was created to clarify these relationships. The following discussion will focus on the implications of this model with respect to (1) sequence development in asymmetrical basins, (2) depositional facies distribution, and (3) seismic expression of sequence stratigraphy.

Although the modeling process, as described below, uses the combined effects of subsidence and eustasy to drive changes in relative sea level, it should be understood that uplift and depression of the basin caused by tectonic events alone could give a similar response pattern. Differentiation of the tectonic and eustatic signals in the stratigraphy of the Western Canada foreland basin must await a more complete understanding of the timing of tectonic events and the relationship of global sea level change to observed depositional cyclicity.

THE SEQUENCE MODEL

The models discussed in this paper are generic and are not designed to represent a specific set of Cretaceous

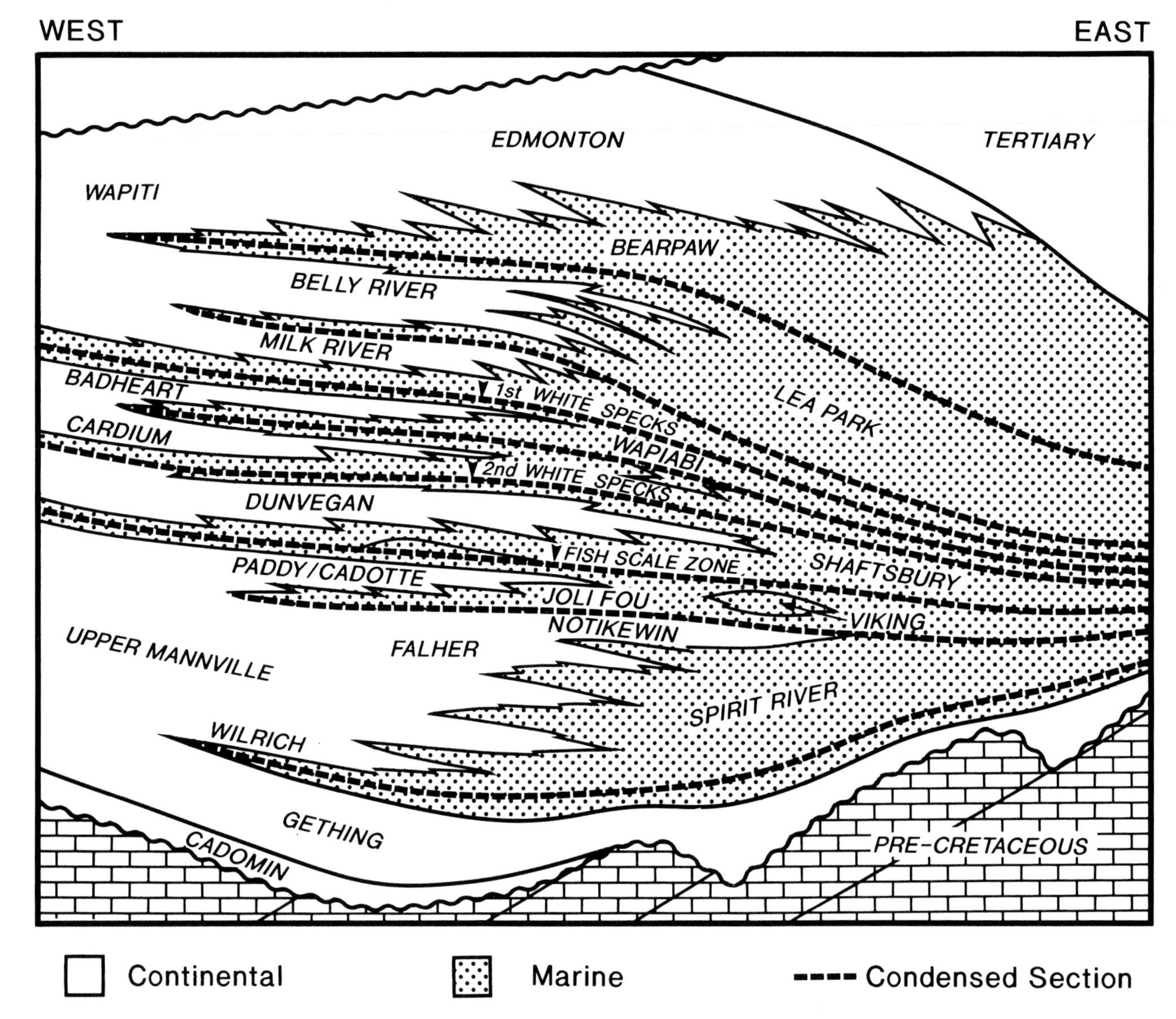

FIGURE 1. **Post-Middle Jurassic stratigraphy, western margin, Western Canada foreland basin. Diagrammatic representation of the major siliciclastic transgressive/regressive cycles of sedimentation. Stratigraphic names are formal and informal nomenclature in common use.**

sequences in Western Canada. Rather, they are presented to demonstrate general principles of sequence development.

Basin Geometry

The modeled foreland basin includes a thrusted mountain belt to the west, a craton to the east, and a depositional basin between (Figure 2). The basin collapses through model time to achieve a strongly asymmetrical profile as the mountains rise to the west. The configuration of the model basin, as shown in Figure 2, is based on a simple, mathematical, time-varying function that is assumed to represent the combined effect of all tectonic processes active during model time. The mechanisms involved in the evolution of foreland basin geometry are discussed elsewhere in this volume (Stockmal et al., 1992).

Model distance and depth are measured relative to the mathematical origin of model space at a hinge of no subsidence at the western margin of the depositional basin (i.e., at the coordinate origin of Figure 2). Distances referred to in subsequent discussion are measured relative to this hinge position at the "mountain front."

Modeling Procedure

The modeling process is two-dimensional in nature and produces a model profile oriented perpendicular to the "mountain front." Key model variables are calculated through iterative integration of the area under the functions that represent the model depositional surfaces

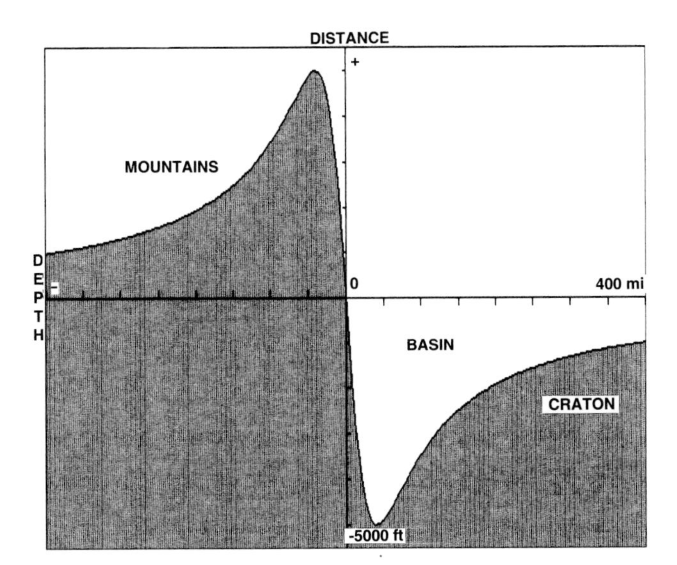

FIGURE 2. Model of basin geometry in profile. The hinge at the coordinate origin is a model position of zero subsidence. The basin is a depocenter for siliciclastics derived from the mountains to the west. The mountains and the western basin margin simulate a thrusted terrain of premodel rocks.

described below. The iteration converges to a solution at model time when the model profile area is equal to the integrated rate of model area increase—a proxy for sediment supply. This calculation is performed at small time increments to obtain the model data array.

The models reported on here are constructed as if the model profiles were located in the valleys of streams arising in the mountains. The stream profile and its response to changes in base level are therefore important considerations in defining model sequence geometry.

A three-dimensional representation of the model basin can be obtained by aligning a number of model profiles side-by-side down the axis of the basin. A geologically meaningful distribution of model profiles is achieved by careful selection of input parameters relative to the intended position of the model in the basin. The models presented in this paper have been calibrated to represent a gradient in subsidence rate from a position on the flank of an arch, with a relatively low subsidence rate, to a position in a rapidly subsiding foredeep.

Model Parameters

Because the intent of this paper is to clarify the role of sea level change in sequence development, variation in other parameters, such as model sediment supply and subsidence, has been controlled to allow the effects of sea level change to be observed.

Sea Level

Modeling is initiated at a lowstand of eustatic sea level and proceeds forward in geologic time from time zero. Eustatic sea level changes in an irregular but smoothly fluctuating pattern described by the convolution of two sinusoids with amplitudes and periods determined at model initiation. The sea level curve used for the models presented here was selected to represent a series of major third-order fluctuations, in the sense of Haq et al. (1988), with a maximum amplitude of 400 feet (122 m) and a periodicity of about 5 million years.

Sediment Supply

Sediment influx rate is modeled as the rate of increase of the model cross-sectional area. The rate of supply is held constant through model time for the models discussed here. Sediment from the mountainous source terrain is assumed to be uniformly dispersed along the axis of the basin. Progradational/transgressive cycles of sedimentation fill the basin from west to east in response to variations in accommodation. Although the craton was a significant source of sediment during the Cretaceous in the Western Canada foreland basin, these sediments are not considered in the modeling process.

Subsidence

The model subsidence rate is a function of distance from the "mountain front" and is assumed to be most rapid at the basin axis (40 mi, or 64 km, from the hinge). When multiple models are considered in a three-dimensional basin model, subsidence may also vary parallel to the basin margin or "mountain front."

For the models presented in this paper, the subsidence rate measured at the basin axis varies from 1.8 cm/1000 yr (0.7 in./1000 yr) on the model arch to 4.3 cm/1000 yr (1.7 in./1000 yr) in the foredeep. These rates are within the average range calculated from present-day Cretaceous sediment thicknesses in the Western Canada foreland basin.

Model Depositional Surfaces

Model depositional surfaces consist of: (1) a profile of exponentially declining gradient that simulates a stream profile extending from the "mountain front" to the coastal plain, (2) a coastal plain, modeled as a flat surface at sea level extending from the base of the stream profile to the coast, and (3) a marine clinoform slope extending from the coastline to the basin floor.

The stream profile is assumed to maintain its gradient by aggradation of fluvial sediments wherever accommodation beneath the profile is increasing, or by erosion wherever accommodation is decreasing. It is further assumed that, relative to the time scale of the model, gradient adjustment is instantaneous.

When relative sea level is rising at the model coast, the stream profile is assumed to be depositional as subsidence creates accommodation. Sediment is ponded near sea level to form a broad, low-lying, coastal plain.

When relative sea level falls, model stream profiles first respond by lengthening and extending across the coastal plain. As a result, the coastal plain depositional surface becomes narrower. If relative sea level falls across the entire model land surface, comprising the stream profile and the coastal plain, erosion and stream valley entrenchment are assumed as the stream profile is depressed into preexisting model sediments. In this case, the stream profile extends to the coast, and a coastal plain is lacking. See Posamentier et al. (1988) and Posamentier and Vail (1988) for a thorough discussion of fluvial response to sea level change.

A clinoform is a depositional response to a gradient in sedimentation rate from high near the coast to low offshore. In the modeling process, a clinoform surface is declared to be a plane with a specified seaward slope. Model slopes are 0.5 degrees consistent with observed magnitudes of slope on clinoforms in the Western Canada foreland basin. The coastline is assumed to lie at the top of the clinoform either at the basinward margin of the coastal plain or the simulated stream profile, depending on the depositional scenario. The extent to which each of the model depositional surfaces is developed varies throughout model time as model geometry changes.

Facies Zones

Each of the model depositional surfaces described above controls the distribution of associated model environmental facies. These general facies associations are termed "facies zones," denoting their distribution in both space and time. The following paragraphs summarize basic model assumptions concerning the relationship of environmental facies and the mathematical functions that describe the depositional surfaces.

Alluvial

The alluvial facies zone is an area of high-energy fluvial deposition, including environments such as braided streams, braidplains, and alluvial fans, associated with the model stream profile. On basin maps, such as Figure 3, entrenched streams may be embedded in the map area designated as the alluvial facies zone.

Coastal Plain

The coastal plain facies zone is assumed to comprise sediments derived from low-gradient fluvial systems (meandering streams) and associated interfluvial and coastal facies. Estuarine valley-fill is also included in the coastal plain facies zone.

Clinoform

During progradation, marine sediments of the clinoform facies zone extend from the coastline down the depositional slope to intersect the basin floor or a preexisting depositional surface. These clinoform sediments are assumed to be "shale-prone" when relative sea level is rising at the coast and "sand-prone" when sea level is falling and erosion is underway. In the latter case, it is assumed that streams reaching the coast provide a source of coarse sediment that is reworked along the coast and redistributed to the shelf.

Transgressive Shelf

During transgression, two model scenarios are possible: (1) transgressive truncation and (2) coastal retrogradation. Transgressive truncation is assumed to occur when all of the model cross-sectional area must be contributed to subaerial deposition to keep pace with increasing accommodation during sea level rise. The model coastline is assumed to retreat by the process of ravinement and shoreface retreat described by Nummedal and Swift (1987). The transgressive-shelf facies zone comprises a sediment-starved environment in which a condensed section of reworked coastal and nearshore facies and offshore sediments is deposited on the underlying, beveled coastal plain.

In contrast, coastal retrogradation is assumed when the coast-to-basin geometry permits continued development of a clinoform during transgression. In this case, the coastline is assumed to backstep with partial or complete preservation of coastal parasequences in the sense of Van Wagoner et al. (1988).

Basin Floor

The basin floor is assumed to be a nondepositional environment that extends from the seaward limit of clinoform deposition to the cratonic shoreline. The model does not account for any craton-sourced sedimentation.

Basement

The term "basement" refers to premodel rocks that subside to form the model basin. In time-stratigraphic diagrams and basin map views, "basement" refers to premodel rock exposed subaerially above the coastline on the cratonic basin margin.

SEQUENCE DEVELOPMENT

Figure 3 illustrates, in plan view, the modeled basin configuration used in the following discussion. At the time of model initiation (time 0), a narrow seaway exists between the mountains and the cratonic basement. During the 20 million years of model time, the basin subsides more rapidly to the north (left) to create a foredeep. In the south (right), a structural arch is maintained on which slow subsidence prevails.

At the end of model time (time = 20 million years), a sedimentary prism extends along the "mountain front" and partially fills the basin. The prism consists of: (1) a wedge of alluvial facies near the mountains that is broadest on the arch where the basin is filled from side to side with continental sediments; (2) a belt of coastal plain facies that borders the sea and laps against the arch; (3) a marine clinoform that extends to the basin floor in the north (left), but that overlaps a transgressive shelf in the center of the basin; and (4) a transgressive shelf that is the drowned coastal plain of a previous progradation. The shelf is bounded on the northern flank by a drowned, inactive clinoform and on the southern flank by the submerged arch. The basin floor is shown with dashed water

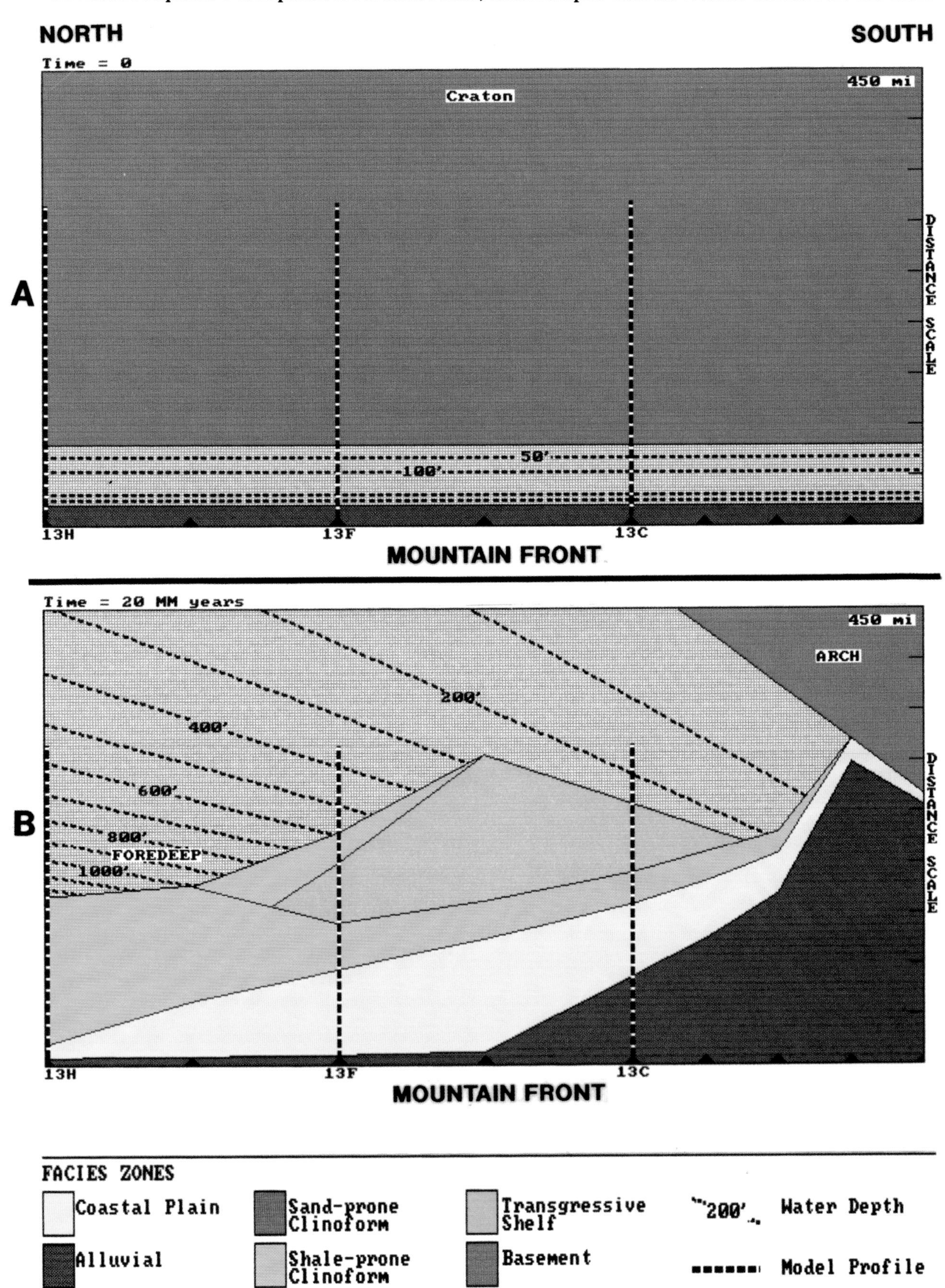

FIGURE 3. Paleogeography of modeled basin. (A) Time = 0. Water depth of linear seaway is 125 ft (38 m). (B) Time = 20 m.y. Rapid subsidence in the north has created a foredeep; slower subsidence in the south has resulted in an arch. Sedimentation has filled the basin on the arch, and progradation across a broad, drowned shelf is underway. In the foredeep, progradation into deep water prevails. Black triangles show distribution of models used in the simulation.

depth contours. It extends to the cratonic shoreline where the basement is exposed on the arch.

Three depositional models are examined and contrasted to illustrate sequence development on the flank of an arch (profile 13C), at an intermediate position (profile 13F), and in a foredeep position (profile 13H). The geographic positions of these model profiles are shown in Figure 3 by the model number annotations along the lower figure margin. The dashed lines at the profile locations indicate the west-east orientations and geographic extents of the profiles.

Chronostratigraphy

The evolution of model sequences in time and space is shown in the chronostratigraphic diagrams of Figures 4, 5, and 6 for the slow, moderate, and rapid subsidence cases, respectively. Model time is shown on the vertical axis and increases from 0 at model initiation to 20 million years at model termination. Distance from the hinge or "mountain front" is plotted along the horizontal axis, and a maximum distance of 328 miles (528 km) is represented.

The eustatic sea level curve employed in the modeling process is illustrated vertically on the right-hand side of each diagram, and the curve is a key reference in the following discussion. Four cycles of eustatic sea level change (C1 to C4) are represented in Figures 4 through 6. Sea level fluctuates in an irregular sinusoidal pattern with a periodicity of about 5 million years and a maximum amplitude of 400 feet (122 m).

The column next to the eustatic curve plots the time distribution of the systems tracts (ST) of Posamentier and Vail (1988). Systems tracts partition sequences into facies groups related to position on the relative sea level curve.

The remainder of each diagram plots the distribution of model sequences and facies zones in time and space. Four depositional sequences are labeled S1 through S4 in the models of Figures 4, 5, and 6. Shifts of the boundaries of the facies zones reflect the following:

1. *Encroachment and coastal onlap.* The coastal plain expands to the left at the expense of the alluvial facies zone.

2. *Alluvial expansion.* The alluvial facies zone expands to the right at the expense of the coastal plain.

3. *Progradation.* The basinward margin of the coastal plain or the hingeward margin of the clinoform facies zone extends to the right or seaward.

4. *Transgression.* The basinward margin of the coastal plain or the hingeward margin of the clinoform facies zone extends to the left. The transgressive-shelf facies zone is shown where previous progradational cycles of sedimentation have been submerged. The process of transgression is assumed to be truncational (see previous discussion). This facies zone is bounded in model time and space by the beveled surface of the underlying coastal plain and the bases of the clinoforms of the succeeding progradation. The facies zone is limited on the right by either the drowned clinoform of the preceding progradation or the submerged basement, if the basin was filled prior to transgression. In reality, the condensed section of the transgressive-shelf facies zone would extend beyond the shelf to drape the drowned clinoform and the basin floor.

5. *Erosion.* The white areas of the diagrams represent the occurrence of subaerial erosion during which deposition is restricted to the coastal and marine environment. A filled basin is indicated when deposition or erosion impinges on the basement to the right.

6. *Basement.* Excursions of the basement merely reflect the changing position of the shoreline on the cratonic margin as sea level rises and falls.

Also shown on the diagrams are envelopes (enclosed in light dashed lines) in which relative sea level is rising or falling, as indicated. Because sediment influx is constant among the models, differences in sequence development reflect only differences in subsidence rate or relative sea level change.

Sea Level Fall

In areas of slow subsidence, eustatic sea level fluctuations cause correspondingly large fluctuations in relative sea level. In these areas, modeling has shown that eustatic sea level fall is often accompanied by erosion. In contrast, as the rate of subsidence increases, only the most profound drops in eustatic sea level cause relative sea level to fall. In the latter case, eustatic sea level change may be reflected merely in perturbations on a continuously rising curve of relative sea level.

Alluvial Expansion

The initial responses of model sequences to falling eustatic sea level are narrowing of the coastal plain, rapid progradation, and expansion of alluvial deposition near the hingeward basin margin. This phase of alluvial expansion is shown in Figures 4, 5, and 6 at the beginning of sea level fall of eustatic cycles C1 to C4. The coastal plain narrows because relative sea level falls near the hinge as eustatic sea level falls. Accommodation for coastal plain deposition, therefore, decreases near the mountains.

The modeling assumption is that high-gradient streams will begin to extend across the area of potential coastal plain erosion to intersect the hingeward margin of the coastal plain. The point of intersection is the equilibrium point or bayline of Posamentier et al. (1988), where the rate of subsidence is balanced by the rate of eustatic sea level fall. Basinward of the equilibrium point, relative sea level and base level continue to rise, and the coastal plain facies zone continues to develop. This depositional scenario was also used in earlier sequence models by Jervey (1988).

In areas of rapid subsidence where erosion is lacking, relative sea level continues to rise at the coast throughout

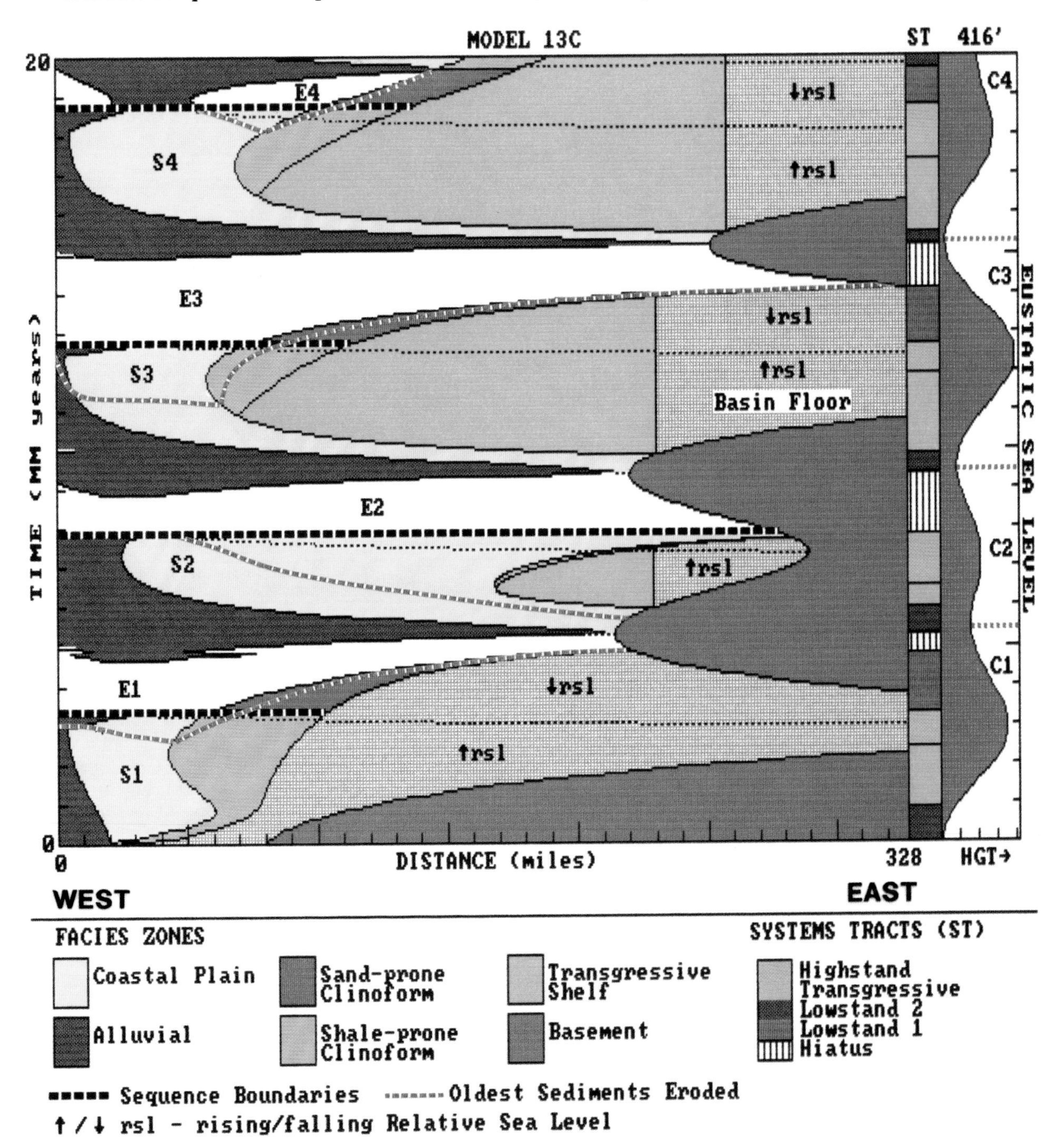

FIGURE 4. Chronostratigraphy of slow subsidence case (Profile 13C). Eustatic sea level cycles are labeled C1 to C4. Sequences and erosional phases are labeled S1 to S4 and E1 to E4, respectively. Colors represent the distribution of model associations of environmental facies and the systems tracts in which these facies zones occur. Heavy dashed lines indicate sequence boundaries and the ages of the oldest sediments eroded at unconformities, as indicated in the legend. Lighter dashed lines enclose envelopes in which relative sea level is rising or falling, as shown.

eustatic sea level fall. In this case, alluvial expansion continues to the inflection point on the curve of falling eustatic sea level. At this point, the rate of eustatic fall begins to decline, and relative sea level begins to rise at the hingeward margin of the coastal plain. Modeling suggests that the stream profile is depositional as it aggrades and adjusts to intersect the inner margin of the coastal plain.

This scenario of alluvial expansion and continuous coastal plain deposition is illustrated in sea level cycles C2 and C4 in Figure 5 and C1, C2, and C4 in Figure 6. Alluvial expansion continues to the indicated boundaries of the sequences associated with these eustatic cycles.

During alluvial expansion, relative sea level continues to rise at the basin axis because of the rapid rate of subsidence there. However, previous basin-filling may extend

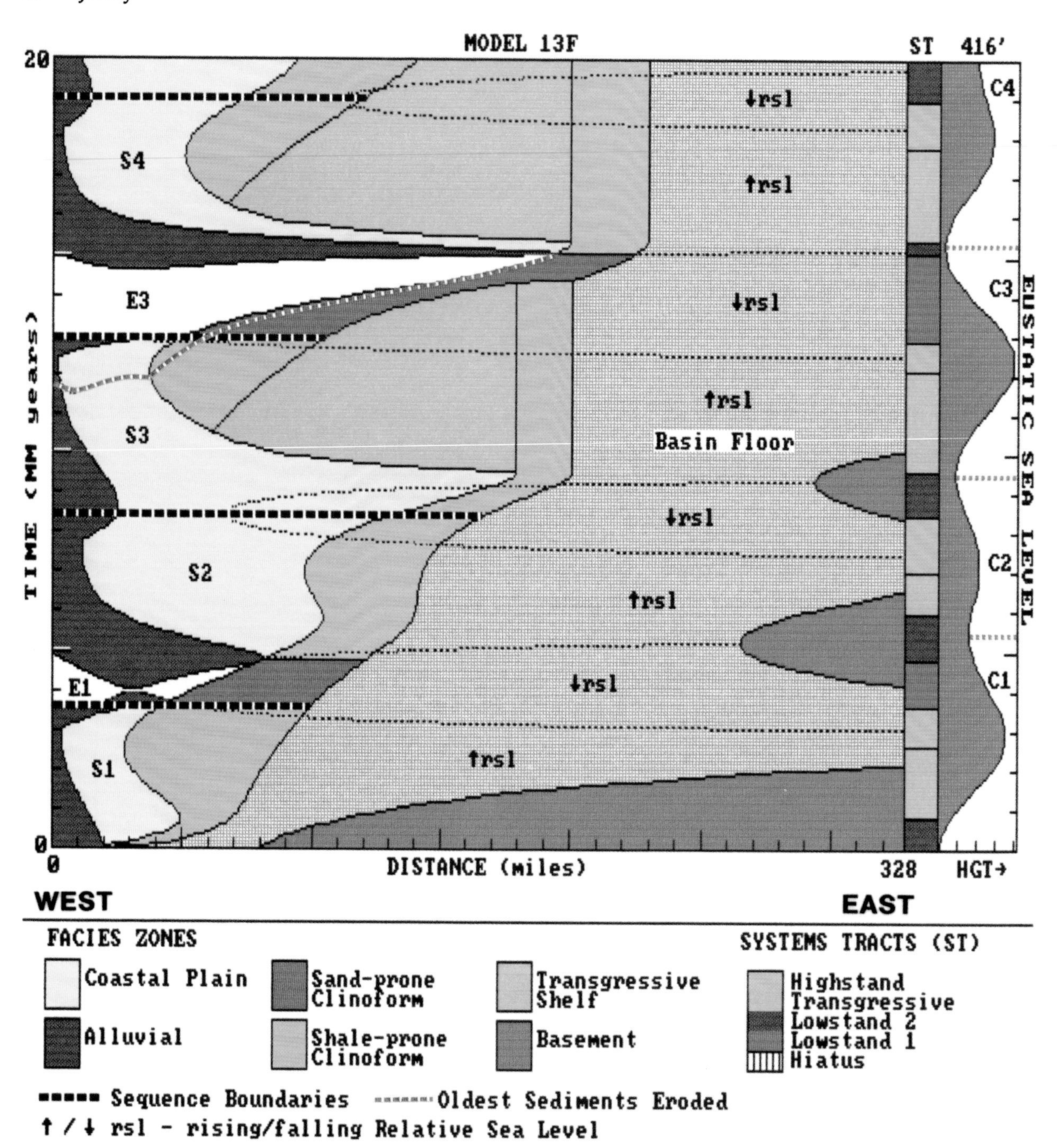

FIGURE 5. Chronostratigraphy of moderate subsidence case (Profile 13F). Eustatic sea level cycles are labeled C1 to C4. Sequences and erosional phases are labeled S1 to S4 and E1 to E4, respectively. Colors represent the distribution of model associations of environmental facies and the systems tracts in which these facies zones occur. Heavy dashed lines indicate sequence boundaries and the ages of the oldest sediments eroded at unconformities, as indicated in the legend. Lighter dashed lines enclose envelopes in which relative sea level is rising or falling, as shown.

the coastal plain beyond the basin axis into areas of low subsidence where relative sea level will fall. In this case, erosion may occur on the coastal plain in these areas. This situation is illustrated in Figure 5 at 8 million years. The envelope of falling relative sea level (light dashed line) intersects the coastal plain, indicating a time of erosion.

Erosion

Following periods of alluvial expansion, continued eustatic sea level fall results in erosion and stream incision in areas of slow subsidence. Relative sea level falls at the coast, and the stream profile is depressed into pre-

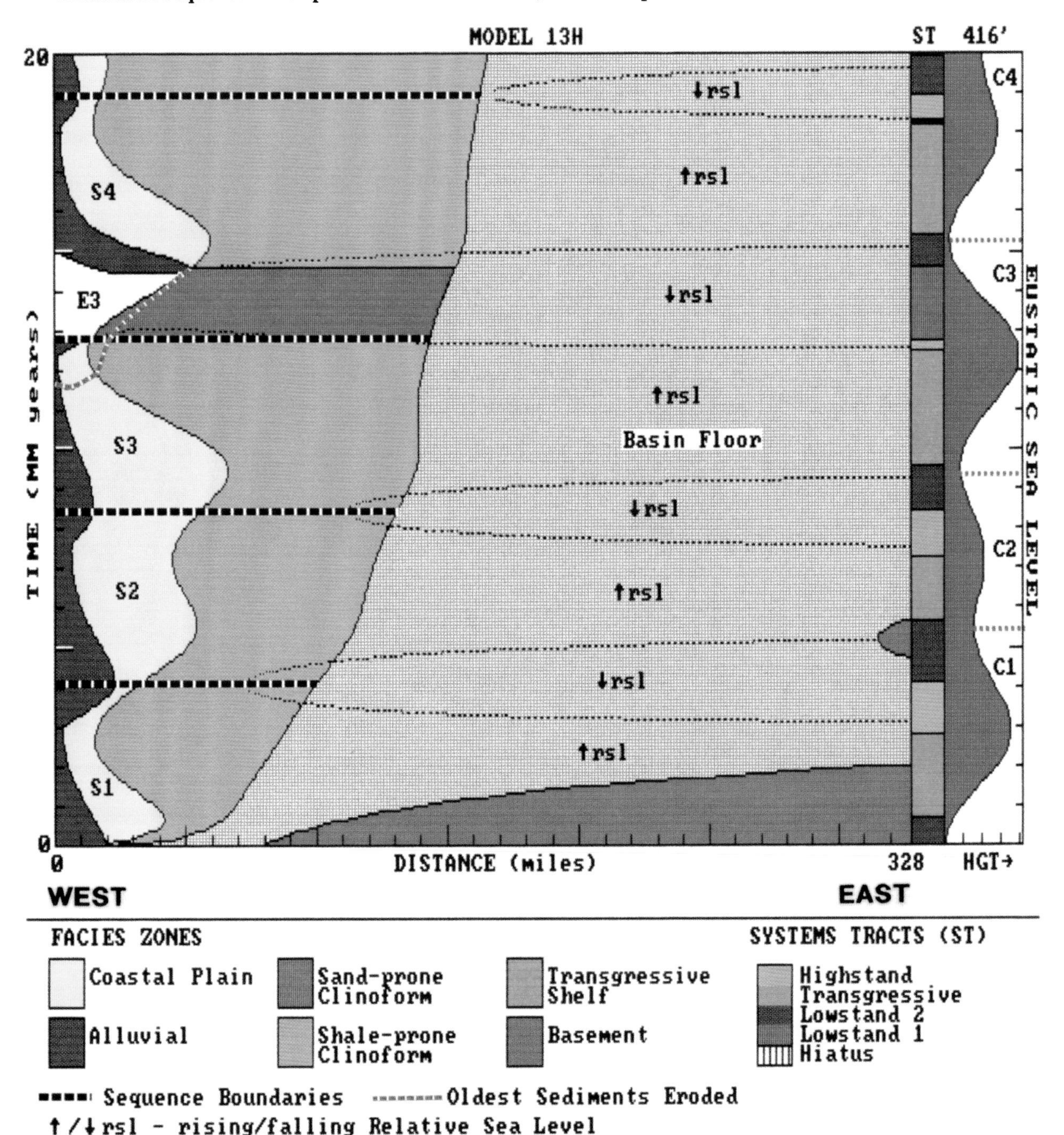

FIGURE 6. Chronostratigraphy of rapid subsidence case (Profile 13H). Eustatic sea level cycles are labeled C1 to C4. Sequences and erosional phases are labeled S1 to S4 and E1 to E4, respectively. Colors represent the distribution of model associations of environmental facies and the systems tracts in which these facies zones occur. Heavy dashed lines indicate sequence boundaries and the ages of the oldest sediments eroded at unconformities, as indicated in the legend. Lighter dashed lines enclose envelopes in which relative sea level is rising or falling, as shown.

existing alluvial and coastal plain sediments. The sequences in Figure 4 are separated by significant periods of subaerial erosion (labeled E1 to E4) driven by falling eustatic and relative sea level. During erosional periods E1 to E3, the rate of fall is sufficiently rapid to cause erosion across the entire land surface.

In contrast, during erosional period E4, deposition of alluvial sediments is maintained near the basin axis (40 mi, or 64 km, from the hinge), where the rate of subsidence is most rapid. Hingeward and toward the craton, erosion prevails because relative sea level is falling there. Thus, modeling suggests that continuous deposition of fluvial sediments may occur in the axis of the basin while erosion is underway toward the basin margins.

As the rate of subsidence increases toward the model foredeep (Figures 5 and 6), subaerial erosion declines in importance, and only the more profound drops in eustatic sea level result in relative sea level fall across the model land surface. For example, in the case of moderate subsidence (Figure 5), only erosional periods E1 and E3 are represented as times of subaerial erosion. For other cycles, falling eustatic sea level is accompanied by expansion of the alluvial facies zone across the coastal plain, and coastal plain deposition is continuous across the sequence boundaries.

The ages of the oldest sediments eroded in the model alluvial valleys of Figures 4, 5, and 6 are shown by the red dashed lines. In the valleys, younger sediments are stripped away and contribute to coastal and marine deposition. Away from the valleys (i.e., out of the plane of the model profiles), a more complete sedimentary record is preserved. Examination of the three figures shows that alluvial valley incision is most profound where subsidence is slow (Figure 4). Rapid falls in relative sea level depress the stream profile deeply into the preexisting sediments. Note that the transgressive event shown in Figure 4 at 6 million years has been completely removed by valley incision. Valley-filling alluvial facies of sequence S3 are deeply nested in the coastal plain of sequence S2 although they are significantly younger. Figures 4 and 5 show that rapid coastal progradation accompanies periods of subaerial erosion. Because coastal progradation and clinoform deposition are synchronous with erosion, these facies are cannibalized as the erosional streams advance behind the coastline. The streams are entrenched within valley walls composed of marine sediments.

It is quite likely that valley entrenchment produces an increase in coarse sediment supplied to the coast as most sediment is bypassed through the basin drainage system. Model clinoform sediments are therefore termed "sand-prone" when relative sea level is falling at the coast. Coarse sand and gravel, brought to the coast, may be transported laterally by longshore currents and accumulate on beach ridges.

Concurrently, marine base level falls in shallow shelfal areas, and coarse sediment may be transported basinward across the shelves. In shallow-water basins, such as the Western Canada foreland basin, the occurrence of submarine erosional discontinuities and shelf sands may be correlated with periods of relative sea level fall with respect to the marine depositional surface. Davis and Byers (1989) appealed to this process to explain the distribution of shelf sands in the Mancos Formation of the U.S. Cretaceous foreland basin.

Figures 4 and 5 show that broad platforms are developed as a result of rapid progradation associated with hingeward erosion. In the case of Figure 4, each of the first three episodes of sea level fall is accompanied by filling of the basin, and the erosional land surface extends from the mountains to the craton. Erosional period E3 in Figure 5 extends the coastline far into the basin, overstepping previous progradational cycles.

In areas of rapid subsidence where erosion is lacking, the rate of coastal progradation is slower. Because accommodation is increasing on the coastal plain and in the alluvial environment, the assumption is that coarse sediment will be stored in these zones. Therefore, periods of eustatic fall may not result in anomalously coarse coastal or marine deposits. Marine base level continually rises, and aggradation of shale-prone shelf and clinoform sediments prevails. This scenario of progradation is illustrated in sea level cycles C2 and C4 in Figure 5 and C1, C2, and C4 in Figure 6, where the progradational clinoforms are shale-prone ramps extending to the basin floor.

Sea Level Rise

During sea level rise, each sequence displays a pattern of encroachment, progradation, and transgression that reflects the control of eustatic and relative sea level on the geometry of the sequence.

Encroachment

As the rate of eustatic fall declines toward lowstand, relative sea level begins to rise in axial positions in the basin where subsidence is maximal. In response, alluvial valleys begin to accumulate fluvial sediments as the profile of equilibrium adjusts accordingly. This is illustrated in Figures 4 and 5 by the earliest appearance of sediments of the alluvial facies zone near the basin axis following periods of subaerial erosion.

Interestingly, because of the synchroneity of coastal deposition and subaerial erosion, sediments deposited above the unconformity may be older than sediments preserved below the unconformity at a different location. This effect is best observed following erosional period E3 in Figure 5. The oldest alluvial sediments above the erosional hiatus are older than the youngest sand-prone clinoform sediments below the hiatus. This modeling result may be of theoretical interest only, because the age determinations required to demonstrate this relationship in the rock record may not be accurate enough to do so.

Following periods of erosion, the commencement of relative sea level rise at the coast is assumed to allow coastal plain deposition to resume. This is illustrated in Figures 4, 5, and 6 as the expansion of coastal plain facies zones near eustatic sea level lowstands. In areas of rapid subsidence, where alluvial expansion continues without a significant erosional hiatus, expansion of the coastal plain begins at the inflection point of falling sea level, as mentioned above. For example, in Figure 6 the expansion of the alluvial facies zone toward the coast continues during the sea level falls of eustatic cycles C1, C2, and C4 until the inflection points, at which the rates of fall are maximum, are reached. Subsequently, the rates of eustatic fall begin to decline, relative sea level begins to rise at the inner margin of the coastal plain, and encroachment of the coastal plain upon the alluvial surface begins.

The chronostratigraphic diagrams show that rises in relative and eustatic sea level result in aggradation of coastal plain and alluvial sediments. As relative sea level rises, the coastal plain encroaches upon the area of alluvial deposition. Where incised valleys have been formed, this encroachment may take the form of estuarine onlap within the valley walls as well as a more general widening of the zone of coastal sedimentation. The area of coastal plain deposition expands both in a hingeward direction, as base level rises and accommodation increases, and seaward as progradation occurs.

Transgression

Following lowstand, Figures 4, 5, and 6 show that as eustatic sea level rises rapidly the coastline recedes hingeward. When broad sedimentary platforms have developed prior to transgression, as in Figure 4 (sequences S2 to S4) or Figure 5 (sequences S3 and S4), the coastal plain is rapidly transgressed. In order to accommodate rising base level, all sediment must be stored in the coastal plain and alluvial facies zones. As a consequence, these broad platforms are rapidly inundated by the rising sea. The scenario is one of transgressive truncation, as discussed previously. If significant erosion has preceded sea level rise, estuarine conditions are likely near the coast as the incised valleys are drowned.

The model transgressive shelves are nondepositional. The illustrated areas of the transgressive shelf zone simulate depositional settings in which condensed sections of pebble lag and associated facies, such as glauconitic sands, shell hash, phosphorites, and black shales, accumulate on starved shelves and basinward in deeper water.

The rate of relative sea level rise begins to decline following the inflection point on the eustatic sea level curve, where the rate of rise is maximized. Figures 4 and 5 show that progradation of shale-prone clinoforms begins near the time of this inflection point, and the area of marine shale deposition expands across the drowned shelf. The resumption of clinoform deposition is assumed when the slope of the beveled transgressive surface is equal to the depositional slope of the model clinoforms. This circumstance simulates the slowing of transgression and the aggradation and preservation of coastal facies. The coastline continues to recede, now in a retrogradational manner, until early highstand.

In contrast, when the coastal plain is narrow and relative sea level rise is rapid, transgression is retrogradational (see previous discussion), and the coastline backsteps, accompanied by continued marine deposition on the clinoform slope. For example, the transgressive episodes of Figure 6 (rapid subsidence) are retrogradational because the sedimentary wedge is narrow. Although the coastal plain narrows during transgression, rapid aggradation continues as a result of increasing accommodation. The slope of the retreating coastline is such that the model assumption of coastal preservation applies. As the rate of rise declines toward highstand, the rate of transgression diminishes. Regression begins again when the rate of relative sea level rise is balanced by sediment supply.

Progradation

Coastal progradation accompanies eustatic and relative sea level rise when the rate of sediment supply is more than sufficient to fill the volumes required by alluvial and coastal plain accommodation. The modeling assumption is that most of the coarse-grained sediment is trapped in continental depositional environments. Marine deposits are, therefore, termed "shale-prone" during sea level rise.

In areas of slow subsidence, represented by Figure 4, progradation is likely to result in filling of the basin. Much of this basin-filling is accomplished during erosional phases, as discussed above. Thereafter, rising base level causes encroachment upon both the alluvial surface and the craton.

As the rate of subsidence increases toward the model foredeep (Figures 5 and 6), progradation only partially fills the basin. The aggradational component of the sedimentary prism increases as required to keep pace with accommodation development. For instance, as a consequence of rapid subsidence in Figure 6, the sediment prism is confined near the source area, coastal plains are narrow, and shale-prone clinoforms are marine aggradational ramps that extend many miles to intersect the basin floor.

Sequence Boundaries

The model sequences (S1 to S4) are bounded by unconformities or their correlative conformities. These sequence boundaries are indicated on the chronostratigraphic diagrams of Figures 4, 5, and 6. The erosional hiatuses, labeled E1 to E4, are periods of time during which sequence-bounding unconformities are formed. These boundaries are characterized by significant erosional truncation and are classified as Type 1 boundaries in the sense of Posamentier and Vail (1988). They are contiguous, within the basin, with conformable sequence boundaries in areas of more rapid subsidence that are classified as Type 2 boundaries.

The gradation from one type of boundary to another within the model basin is illustrated by the upper boundary of sequence S2. The S2 sequence boundary in Figure 4 is strongly erosional and would be termed Type 1 in the terminology of Posamentier and Vail (1988). The upper S2 boundary becomes partially conformable in Figure 5, because coastal plain deposition is continuous across the boundary at the basin axis. In Figure 6, the S2 boundary is entirely conformable and would be termed Type 2. It is therefore evident that the nature of sequence boundaries is entirely dependent on the history of relative sea level change across the basin.

Note that the erosional sequence boundaries are older than the wedges of sand-prone clinoform deposition and lie beneath them. These wedges constitute, by definition, lowstand units of the succeeding sequences (Baum and Vail, 1988; Posamentier and Vail, 1988). Although these units may be conformable along the line of the model profile (i.e., in the model stream valleys), away from the stream valleys they would exhibit unconformable relationships with the underlying shelf and adjacent incised land surface of the preceding sequence.

Figures 4 through 6 show that erosion is centered about the inflection point of falling eustatic sea level in cycles C1 to C4, where the rate of fall is maximum. In the case of rapid falls in sea level, erosion may begin near highstand and continue almost to lowstand. The lowstand wedge of the sand-prone clinoform facies zone may occupy most of the falling sea level time period, as shown for erosional period E3 (Figures 5 and 6).

Interestingly, the erosion surfaces associated with model Type 1 boundaries extend across the lowstand units as erosion proceeds throughout the hiatal period. The lowstand units are, therefore, bounded both below and above by unconformities. The actual sequence boundary at the base of the lowstand unit may be the more subtle of the two, because the top of the lowstand wedge is likely to be deeply incised.

Figures 4 through 6 illustrate age relationships at sequence boundaries. The ages of sequence boundaries are determined where they become conformable toward the basin (Baum and Vail, 1988). In the case of the model basin, conformity is achieved in areas of high subsidence rate, represented by Figures 5 and 6, where Type 2 boundaries are developed. Figures 5 and 6 show that conformable sequence boundaries are located, in time, at inflection points of eustatic sea level fall. Following the inflection point, relative sea level begins to rise in hingeward areas and coastal onlap of the next sequence ensues. The ages of Type 2 boundaries are, therefore, determined by the age of eustatic inflection.

A comparison of the sequence boundaries in Figures 4, 5, and 6 shows that Type 1 boundaries, which underlie the zones of lowstand sand-prone clinoforms, are significantly older than the inflection points of sea level fall. Their ages are determined by the age of the lowstand wedge. This modeling result suggests that sequence boundaries may be diachronous in basins with pronounced geographic variation in subsidence.

Figure 7 summarizes the model basin configurations, in map view, at the ends of sequences S1 through S4 when the sequence boundaries are fully developed. In each of the maps in Figure 7, the arch is to the south (right) and the foredeep is to the north (left). The model profiles of Figures 4, 5, and 6 extend west-east across the basin. The envelopes of small crosses enclose areas in which relative sea level is rising. Outside these envelopes, relative sea level is falling across the entire basin and erosion is underway onshore. An envelope is not shown for sequence S3 because relative sea level is falling everywhere within the map area.

The gradation from Type 1 boundaries on the arch to Type 2 boundaries in the foredeep is evident for sequences S1, S2, and S4. On the flanks of the arch, erosion is underway, and incised valleys of the alluvial facies zone extend across the entire basin. The erosion surface in this area would be described as a Type 1 sequence boundary. Toward the foredeep, where relative sea level is rising, deposition is continuous, and the sequence boundary in this area is Type 2. In the case of model sequence S3, the sequence boundary is erosional and Type 1 everywhere within the map area.

The map coastline is found at the basinward margin of the coastal plain, where one is developed, or at the margin of the alluvial facies zone. Note that the coastline becomes oriented in a northwest-southeast direction because of rapid basin-filling toward the arch. Streams flow down the axis of the basin to the coast. A similar reorientation of the coastline is observed in the Western Canada foreland basin where the Sweetgrass arch limits accommodation in southern Alberta.

Wedges of sand-prone clinoform sediments are deposited along the coastlines of the alluvial facies zones in Figure 7. As discussed previously, these lowstand units overlie the Type 1 sequence boundaries. However, the maps show that these units are deposited at the same time as are the shale-prone clinoforms and coastal plains that underlie the sequence boundaries in map areas of Type 2 boundaries. This relationship again demonstrates the diachronous nature of model sequence boundaries.

SYSTEMS TRACTS

A systems tract is defined by Van Wagoner et al. (1988) as "a linkage of contemporaneous depositional systems" associated with a specific segment of the eustatic and relative sea level curve. The following discussion uses the terminology of Van Wagoner et al. (1988) and Posamentier and Vail (1988) to describe the occurrence of model systems tracts. The time distribution of model systems tracts is shown in Figures 4, 5, and 6.

Three systems tracts—lowstand, transgressive, and highstand—are recognized in the modeling process.

Lowstand Systems Tract

The lowstand systems tract refers to deposition during falling relative sea level and the subsequent initial rise. The lowstand systems tract of the Western Canada foreland basin is analogous to the lowstand deposits of ramp-style continental margins described by Posamentier and Vail (1988). In the western Canadian setting, the post-Middle Jurassic foreland basin remained shallow throughout its history. The lowstand systems tract comprises coastal and shelf facies that are equivalent in their time of deposition to lowstand fans and lowstand wedges of deep-water continental margins. Two subdivisions of the lowstand systems tract are defined in this paper: lowstand 1 and lowstand 2.

Lowstand 1 deposits are the wedges of sand-prone clinoform facies associated with subaerial erosion and valley incision. The lowstand 1 systems tract begins when relative sea level begins to fall at the coast and ends when relative sea level begins to rise at the coast. This systems tract is, therefore, coincident with the formation of Type 1 sequence boundaries. This systems tract corresponds to the first part of the lowstand wedge as defined by Van Wagoner et al. (1988) for a ramp-style basin margin.

The lowstand 1 systems tract includes the sediment bypass zone of the entrenched land surface and the coastal and marine deposits. Near the mouths of streams form lowstand deltas that are progressively eroded as sea level falls. In high-energy coastal environments, the sediments of the lowstand deltas are redistributed by coastal waves and currents to accumulate as terraced beach ridges. In lower-energy shelfal settings, the deltaic morphology may be preserved. Because no space is available for landward deposition, coastal deposits of the lowstand 1 systems

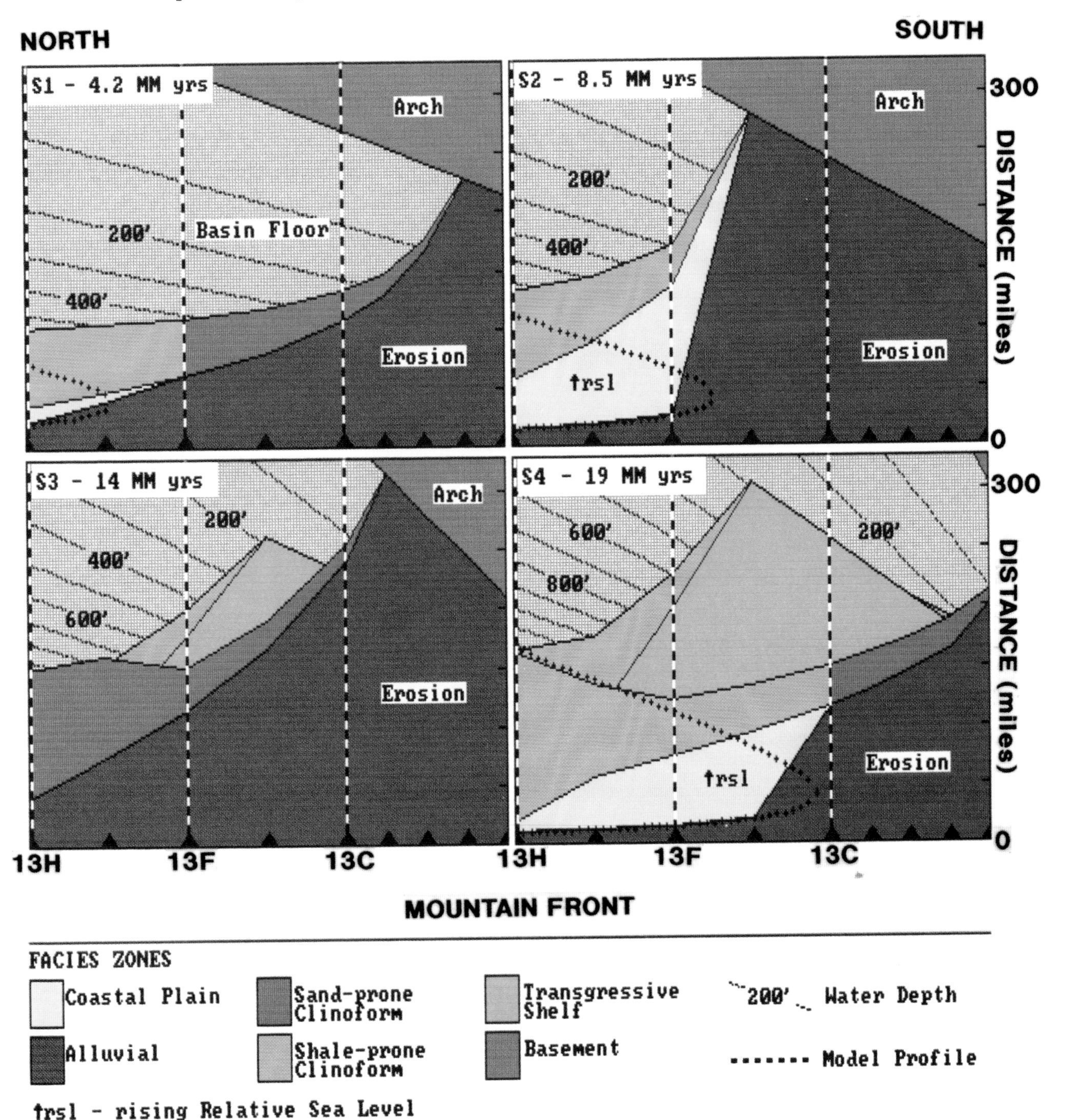

FIGURE 7. Four paleogeographic maps of the modeled foreland basin. These maps represent basin configurations and facies distributions at the terminations of sequences S1 through S4 when coastal onlap of their respective sequence boundaries begins. The foredeep is to the north (left), and the arch is to the south (right). Envelopes of heavy dots define areas of erosion (labeled) and deposition at the end of sequence time. Contour interval for water depth in the basin, shown only outside the area of deposition, is 100 ft (30.5 m). The distribution of models used in the paleogeographic simulation is shown by black triangles. The locations of model profiles 13C, 13F, and 13H, discussed in the text, are highlighted.

tract are characterized by depositional and erosional toplap, and have limited preservation potential.

Offshore, marine base level falls with falling sea level, and storms more frequently rework the shelf surface. Shelf and coastal sands with sharp, scoured bases may be prevalent in the model sand-prone clinoform facies zone. Plint (1988) employed this process to explain coarse-grained, sharp-based units of the Turonian Cardium Formation in the Cretaceous strata of the Western Canada foreland basin.

The lowstand 1 systems tract is an unconformity-bounded unit. It overlies and onlaps a Type 1 sequence boundary below and underlies and is onlapped by the lowstand 2 systems tract above.

The lowstand 2 systems tract develops during the initial rise of relative sea level. The lowstand 2 systems tract begins when relative sea level begins to rise following alluvial expansion or erosion and ends when either transgression begins or the inflection point of sea level rise is reached. This systems tract may follow the lowstand 1 systems tract or may, in the case of a Type 2 sequence boundary, directly follow the highstand systems tract of the previous sequence. This systems tract is analogous to the second part of the lowstand wedge defined by Van Wagoner et al. (1988) for a ramp-style basin margin.

The lowstand 2 systems tract includes the initial deposits of alluvial aggradation in the incised valleys concurrent with and following subaerial erosion. Modeling suggests that rapid subsidence in the basin axis may promote early aggradation of alluvial sediments there while coeval stream-valley erosion continues in more slowly subsiding areas toward the basin margins. This depositional situation is illustrated in Figure 4 during erosional periods E1 to E4.

As relative sea level rises more rapidly, accommodation is developed at the coast and within the alluvial facies zone. The lowstand 2 systems tract includes progradational and aggradational facies of the expanding coastal plain facies zone. If valley entrenchment precedes sea level rise, coastal plain facies initially encroach within the valleys at the coast and onlap the upland erosion surface along the coast between valleys. At the same time, alluvial facies fill upper valley reaches and interfinger with estuarine facies toward the coast. With continued rise in sea level, the erosional land surface may be buried beneath the onlapping units of the lowstand 2 systems tract.

If, on the other hand, the lowstand 2 systems tract is preceded by a Type 2 sequence boundary, the scenario is one of a more general aggradation and expansion of coastal plain facies against the alluvial plain near the mountains. Rising base level may promote estuarine sedimentation along the entire coastline in areas of coastal embayment.

Expansion of the coastal plain facies zone is accompanied by progradation of coastal parasequence sets (Van Wagoner et al., 1988) comprising beach, shoreface, and backbeach associations separated by minor drowning surfaces. As the rate of sea level rise increases, the parasequences aggrade and progradation declines.

In the models presented in this paper, the lowstand 2 systems tract is always superseded by the transgressive systems tract. However, other models have shown that the lowstand 2 systems tract may pass directly into the highstand systems tract, at the rising inflection point, if an appropriate balance between sea level rise and sediment supply is achieved.

Transgressive Systems Tract

The transgressive systems tract is deposited above the lowstand 2 systems tract. The base of the transgressive systems tract is recognized as the first backstepping parasequence set prior to regional transgression or, in the sense of the models of Figures 4, 5, and 6, the beginning of the hingeward shift of the coastline. This systems tract ends, in the definition of this paper, when progradation resumes.

The transgressive systems tract is characterized, according to Van Wagoner et al. (1988), by retrogradational coastal parasequence sets. However, a distinction is made in this paper between retrogradation and transgressive truncation (see previous discussion). The coasts of the transgressive systems tract of Figures 4 and 5 are assumed to be predominantly truncational where the transgressive-shelf facies zone is indicated. The model coastline simulates backstepping in which parasequences are deeply eroded during shoreface retreat. Modeling suggests that although coastal retreat may be accomplished by parasequence backstepping, significant preservation of parasequence sets may be confined to aggradational portions of the transgressive cycle.

The model transgressive coastline is termed "retrogradational" where the transgressive systems tract includes the shale-prone clinoform facies zone. This situation indicates that significant aggradation of parasequences is occurring at the coast, accompanied by deposition of normally graded shelf sediments. In the model diagrams, the clinoform facies zone continues to expand across the basin floor during retrogradation. The preservation of coastal facies improves during retrogradation.

The depositional facies of the transgressive systems tract include a complex assemblage of shelf, coastal, and estuarine environments (see Nummedal and Swift, 1987). The transgressive systems tract may encroach upon the previous sequence boundary depending on the nature and extent of the underlying lowstand 2 systems tract. Where deeply incised topography of a Type 1 boundary exists, true estuarine conditions may ensue as the transgression is driven far up the valleys. This situation occurs following erosional period E3 in Figures 4, 5, and 6, in which sedimentation associated with the short-lived lowstand 2 progradation and subsequent transgression is contained within the incised valley of the model profile (see model profile discussion below). The transgressive systems tract is characterized by rapid aggradation of the coastal plain facies zone and onlap of the alluvial wedge adjacent to the hinge.

Highstand Systems Tract

The base of the highstand systems tract is the preceding transgressive systems tract, when transgression has occurred, or is otherwise taken to be the inflection point of rising sea level. This systems tract ends at either a Type 1 or a Type 2 sequence boundary.

The highstand systems tract is characterized first by aggradation and onlap of the alluvial facies zone as sea level continues to rise and later by expansion of alluvial deposition and progradation as sea level falls. The reduction in accommodation development associated with falling sea level may give rise to a highstand horizon of sand-prone fluvial deposition of enhanced sand continuity.

At the coast, initial phases of highstand coastal sedimentation consist of aggradational parasequence sets as sea level rises rapidly. However, as the rate of rise declines and sea level fall begins, the coast progrades rapidly. Clinoforms of the highstand systems tract downlap against the preceding transgressive systems tract, the lowstand systems tract, or the basin floor.

Highstand deposition ends either with the commencement of subaerial erosion or at the inflection point of falling sea level, which is the base of the lowstand 2 systems tract. Figures 4, 5, and 6 show that erosion may remove the entire nonmarine portion of the highstand systems tract if erosion proceeds over a long period of time. More likely, highstand sediments will be preserved on the interfluves and removed in the incised valleys.

A careful comparison of model systems tracts (Figures 4, 5, and 6) shows that geographic variation in subsidence systematically affects the time of initiation and termination of systems tract development. Because the lowstand 1 systems tract is associated with Type 1 sequence boundaries, this systems tract is more prevalent in areas of slow subsidence, represented by Figure 4. As subsidence increases toward the model foredeep, Figures 5 and 6 show that Type 1 boundaries pass into Type 2 boundaries (except for E3), and the lowstand 1 wedge is replaced by an expanded highstand and lowstand 2 systems tract. In addition, the time interval of the transgressive systems tract increases in Figures 5 and 6, reflecting the longer time of relative sea level rise in areas of higher subsidence rate. Clearly, then, changes in relative sea level history across the basin will control the distribution of systems tracts and their associated depositional facies.

MODEL PROFILES

Model data can be displayed in the form of basin profiles that demonstrate features of sequence geometry, seismic stratigraphy, and facies distribution in cross section. The profiles are constructed to lie exactly along stream profiles, so that erosion, when it occurs, is maximized along the line of profile. The profiles are assumed to be oriented in a west-east direction perpendicular to the north-south-trending mountain front of the model basin.

Simulated Seismic Profiles

Figures 8, 9, and 10 are model cross sections for the arch position (profile 13C), the intermediate position (profile 13F), and the foredeep position (profile 13H), respectively. Sequences S1 to S4 are labeled and sequence boundaries are shown. The model eustatic sea level curve is annotated to show the sequence boundaries in relationship to eustatic cycles C1 to C4.

The sediment prism is subdivided into systems tracts corresponding to those of Figures 4, 5, and 6. Isotime lines that simulate seismic reflection horizons are superimposed. These lines are drawn at model depositional surfaces of the stream profile, the coastal plain, and the clinoform. The model eustatic sea level curve is annotated to show the sequence boundaries in relationship to eustatic cycles C1 to C4.

The model coastline is found on the profiles at the basinward margin of the coastal plain, indicated by the heavier dashed lines. Where alluvial time horizons rest on those of the clinoform, erosion has removed the coastal facies along the line of the model profile. In this case, model stream valleys incise into marine sediments in Figures 8, 9, and 10, and subsequently fill with alluvial and coastal plain sediments.

The spacing of the coastal plain and alluvial time horizons demonstrates the rate of aggradation in these environments and is, therefore, a measure of the rate of accommodation development or relative sea level rise. Where these time lines are widely spaced, rapid rise of relative sea level is indicated at the time of deposition. Conversely, closely spaced time horizons indicate slow relative sea level rise. In the marine environment, the clinoform time horizons indicate rapid progradation where they are widely spaced, and slow progradation or condensation where they are closely spaced.

Sequence Geometry

Model sequences are wedge-shaped sedimentary prisms that fill the basin from west to east on model profiles. The profiles illustrate a general change in sequence geometry from the flank of the arch (Figure 8) to the foredeep (Figure 10), which reflects the gradient in subsidence rate.

Figure 8 shows that sequence preservation is incomplete on the flanks of the model arch, and that the Type 1 sequence-bounding unconformities are strongly erosional. Coastal plain time horizons are truncated below each of the sequence boundaries, as indicated by termination against the unconformity. Likewise, coastal and marine sediments of the clinoform are truncated, and clinoform time horizons toplap beneath succeeding sequences. For example, most of sequence S2 has been removed in the plane of the model profile by deep erosion on the eastern side of the basin.

As the rate of subsidence increases toward the foredeep (Figures 9 and 10), the degree of sequence preservation increases with increasing accommodation, and sequence boundaries are conformable. In the foredeep (Figure 10), only the upper boundary of sequence S3 is truncational. The pattern of time horizons at the other boundaries shows either depositional toplap or conformity of coastal plain sedimentation.

In Figure 8, rapid progradation of lowstand clinoform units fills the basin in sequences S1 to S3. In areas of increased subsidence rate (Figures 9 and 10), rapid accommodation increase at the basin axis (40 mi, or 64 km, from the hinge) requires greater sediment volumes in aggradational facies zones. Figure 9 shows the effect of increasing subsidence in that progradation only partially fills the basin as greater volumes of sediment are stored in the coastal plain. This trend is continued in the foredeep (Figure 10), where the sedimentary prism is confined to the basin axis and a narrow coastal plain is maintained near the hinge.

On the flanks of the arch, clinoform deposition may be interrupted by transgression and the formation of broad, transgressive ravinement surfaces. Highstand and lowstand units prograde across these shelves. Transgressive surfaces within sequences S3 and S4 (Figures 8 and 9)

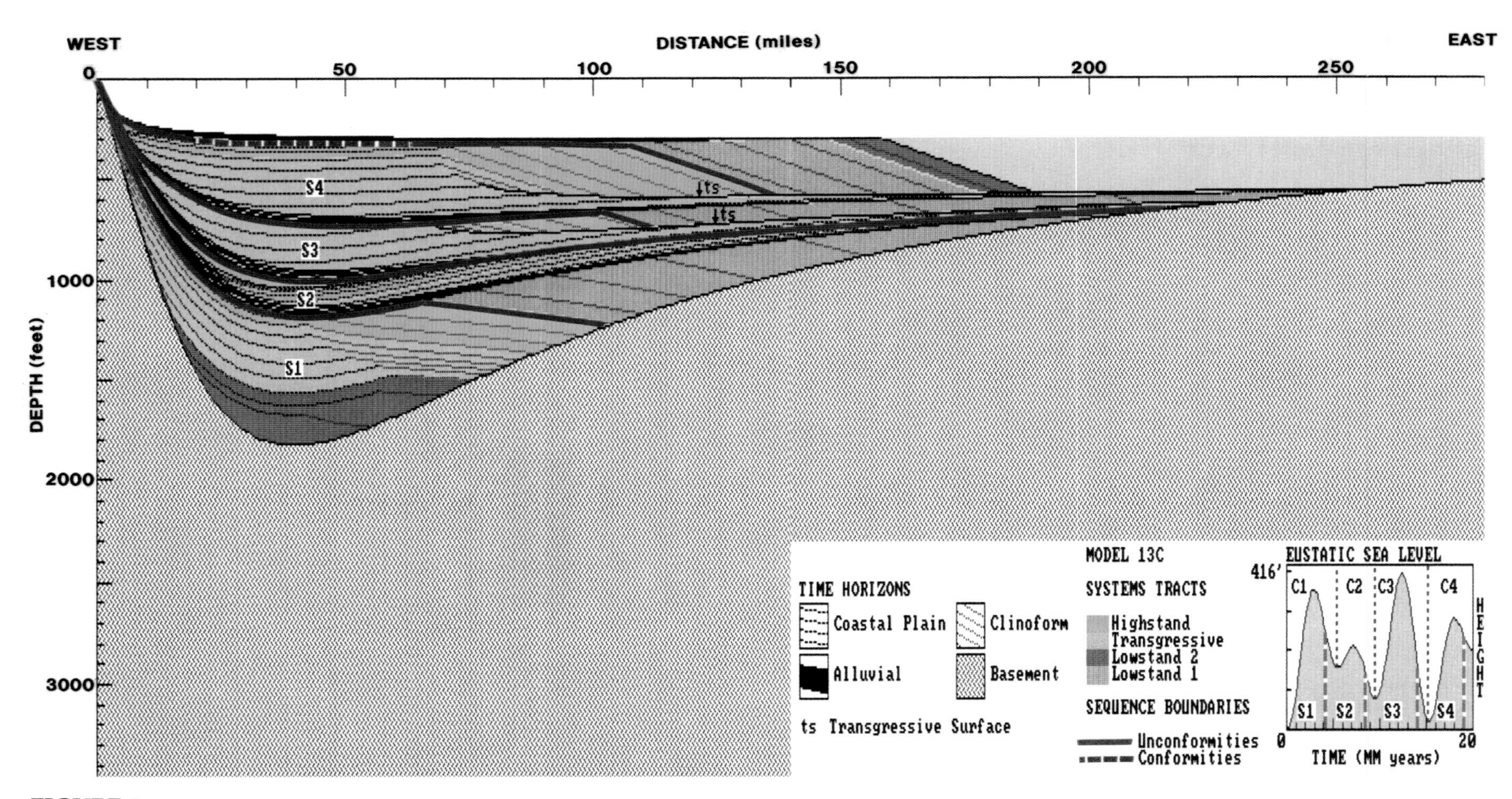

FIGURE 8. Simulated seismic profile with systems tract overlay: slow subsidence—arch position (profile 13C). Black lines are model depositional surfaces drawn at time intervals of 300,000 years. They simulate seismic reflections or time lines. They achieve their curved form through subsidence of the depositional surfaces. Terminations of these surfaces occur through truncation, onlap, downlap, and toplap in a representation of seismic stratigraphic geometry. Type 1 and Type 2 sequence boundaries are indicated by unconformable and conformable boundaries, respectively. The sequences are partitioned into systems tracts or facies groups controlled by relative sea level change.

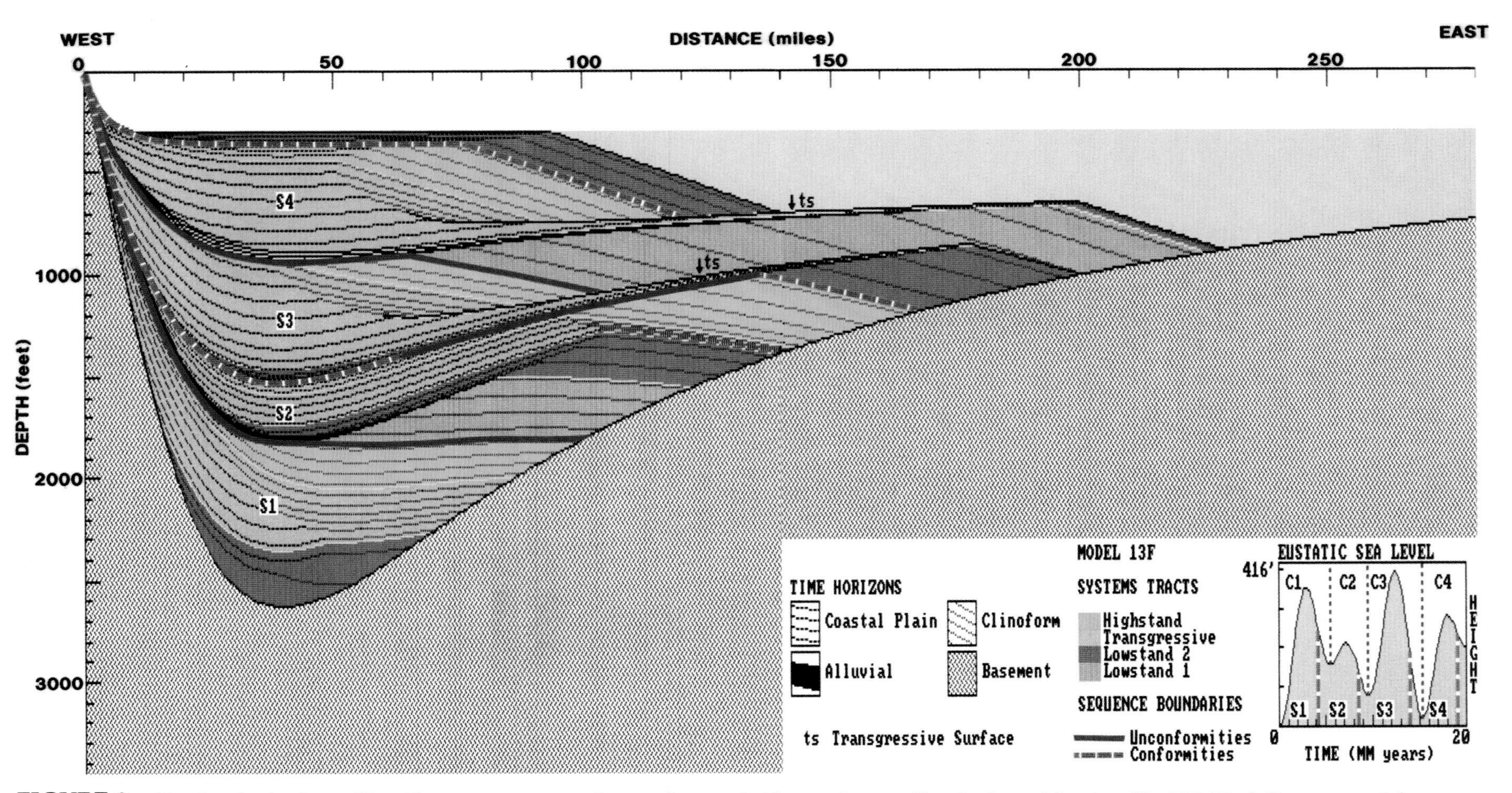

FIGURE 9. Simulated seismic profile with systems tract overlay: moderate subsidence—intermediate basin position (profile 13F). Black lines are model depositional surfaces drawn at time intervals of 300,000 years. They simulate seismic reflections or time lines. They achieve their curved form through subsidence of the depositional surfaces. Terminations of these surfaces occur through truncation, onlap, downlap, and toplap in a representation of seismic stratigraphic geometry. Type 1 and Type 2 sequence boundaries are indicated by unconformable and conformable boundaries, respectively. The sequences are partitioned into systems tracts or facies groups controlled by relative sea level change.

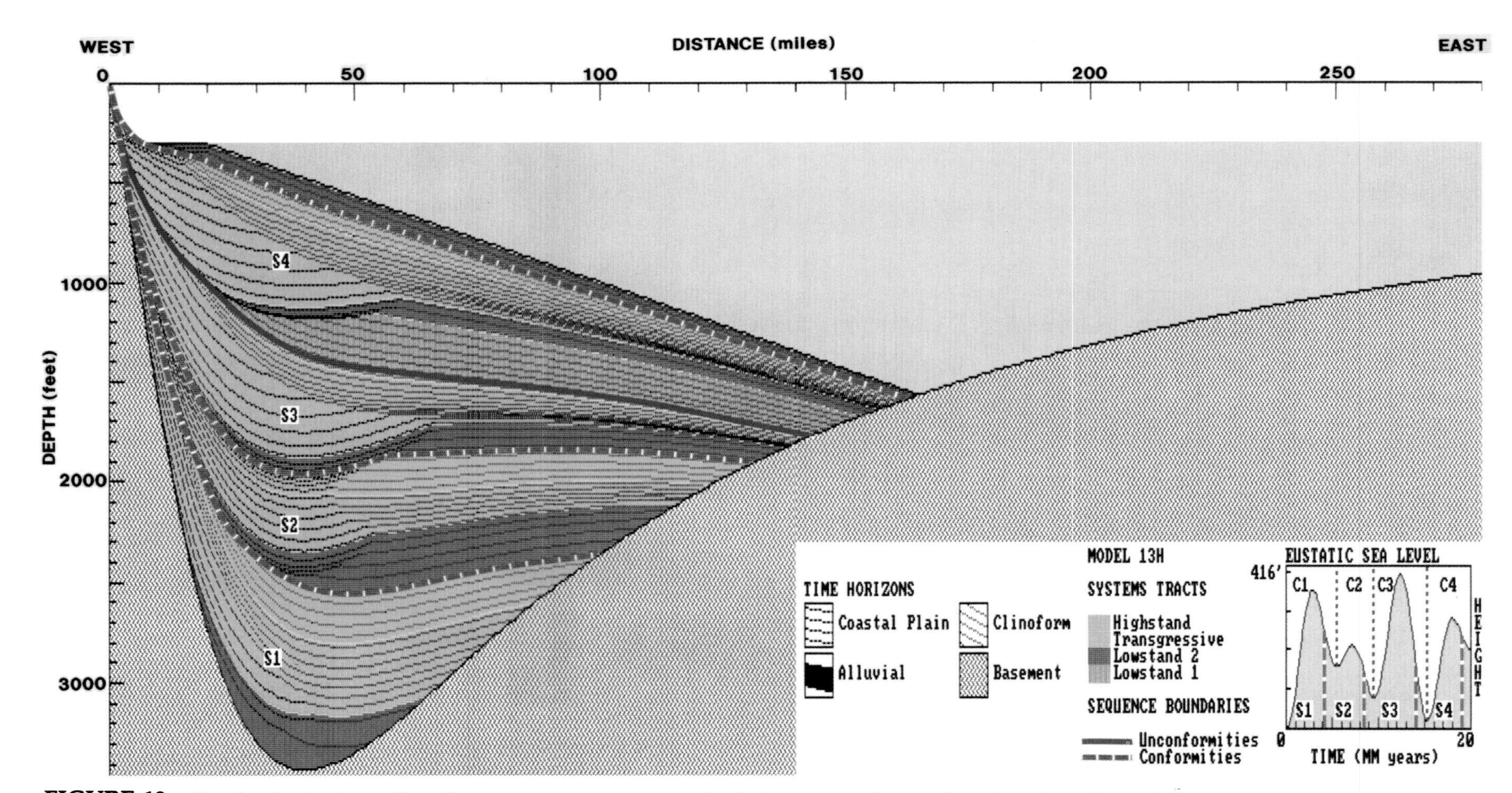

FIGURE 10. Simulated seismic profile with systems tract overlay: rapid subsidence—foredeep position (profile 13H). Black lines are model depositional surfaces drawn at time intervals of 300,000 years. They simulate seismic reflections or time lines. They achieve their curved form through subsidence of the depositional surfaces. Terminations of these surfaces occur through truncation, onlap, downlap, and toplap in a representation of seismic stratigraphic geometry. Type 1 and Type 2 sequence boundaries are indicated by unconformable and conformable boundaries, respectively. The sequences are partitioned into systems tracts or facies groups controlled by relative sea level change.

represent condensed sections of reworked coastal and marine deposition. Toward the foredeep, transgressions are retrogradational, and transgressive surfaces are not developed. Here, continuous clinoform deposition produces a broad ramp of shaly, basinal sediments that extends to the basin floor. Condensation of section in this setting is indicated by the closely spaced clinoform time horizons. In Figure 10, condensed deposition on the clinoform corresponds to times of maximum coastal retrogradation, when most sediment is stored on the coastal plain and marine sedimentation is reduced.

Time-Horizon Patterns

The configuration of model time horizons reflects both the pattern of deposition (or erosion) and subsidence. For instance, coastal plain and clinoform surfaces, originally planar, have been warped by subsidence into their curved profile shapes. Three patterns of seismic reflection termination, simulated by termination of model time horizons, are most useful in describing basin behavior. These are onlap, downlap, and toplap. Each sequence displays patterns of line terminations that reflect relative sea level change. The reader is referred to Mitchum et al. (1977) for a review of seismic stratigraphic concepts.

Onlap. Rising eustatic and relative sea level results in progressive onlap of the coastal plain on the alluvial facies zone in all sequences of Figures 8, 9, and 10. This is shown by the termination of coastal plain time horizons against alluvial horizons. Onlap is characteristic of the lowstand 2, transgressive, and highstand systems tracts.

The position of onlap is an important criterion in recognizing the falls in relative sea level that result in sequence-bounding unconformities. Figures 8, 9, and 10 show that the fall and subsequent rise of relative sea level result in a downward or basinward shift in the positions of onlapping time-horizon terminations. Below the sequence-bounding unconformities, the simulated reflectors onlap near the hinge in the transgressive and highstand systems tracts. Above the boundary, model coastal plain horizons terminate in the lower reaches of the valleys in the lowstand 2 systems tract. This effect is most easily observed above the upper S2 and S3 boundaries in Figures 9 and 10.

The evidence of valley incision and the basinward shift of onlapping reflection terminations are critical in assessing the importance of relative sea level change in sequence evolution. Although transgressive/regressive cycles can be produced by variations in sediment supply or stillstands in long-term patterns of sea level rise or fall, sequence-bounding unconformities require falls in depositional base level. These changes in base level may be induced by uplift or eustatic sea level fall.

Downlap. The model profiles show that clinoform horizons downlap on the basin floor or on transgressive surfaces. The expansion of clinoform deposition across transgressive shelves produces major seismic discontinuities known as downlap surfaces. These surfaces are recognized in Figures 8 and 9 by the downlapping terminations of simulated clinoform reflectors in sequences S3 and S4. These surfaces and the condensed sections associated with them are important geologic features and form regional subsurface correlation markers. Major examples from the Western Canada foreland basin are known informally as the First and Second White Specks and the Fish Scale Zone. Downlap surfaces are contained within sequences as features of the transgressive systems tract. The downlap surface is younger than the lower sequence boundary but may be stratigraphically equivalent to the boundary when transgressive erosion truncates the previous section. Downlap surfaces may be overlain by systems tracts of succeeding sequences, as they are in Figures 8 and 9.

In the model foredeep (Figure 10), downlap surfaces are not well developed, but a subtle downlapping pattern is recognizable above the intervals of condensed clinoform deposition. However, the predominant pattern is ramp aggradation and downlap on the basin floor.

Toplap. As sea level rises toward highstand, accommodation for coastal plain and alluvial deposition is reduced. The later stages of the highstand systems tract are characterized by depositional toplap reflecting the loss of accommodation. This pattern of toplap can be seen best in Figure 10, sequences S1 and S2.

As relative sea level falls, the reduction in accommodation results in erosional toplap. The lowstand 1 systems tract is truncated beneath the subaerial erosion surface. Previous highstand units are also truncated, so that erosional toplap is observed in the lines of the model profiles. Highstand onlapping coastal plain time horizons are preserved on the interfluves (i.e., between model profiles).

Systems Tracts. Each sequence of the model profiles includes the lowstand, transgressive, and highstand systems tracts. In areas of slow subsidence (Figure 8), long periods of relative sea level fall are represented by extensive lowstand coastal and marine deposits of the lowstand 1 systems tract. These form a major basin-filling component of sedimentation on the flank of the arch. The lowstand 1 systems tract is shown to rest on model Type 1 sequence boundaries, and sedimentation is confined to the marine basin. Erosion extends across the top of the lowstand units as sea level continues to fall. As accommodation increases toward the foredeep, the lowstand 1 systems tract is not developed, and the highstand systems tract is onlapped by the lowstand 2 systems tract of the next sequence. In Figure 10, only the rapid sea level fall of cycle C3 produces a lowstand 1 unit.

Figures 8, 9, and 10 show that the lowstand 2 systems tract onlaps either the lowstand 1 wedge, where it is present, or the highstand systems tract. On the arch (Figure 8), the lowstand 2 systems tract is primarily progradational. Aggradation is poorly developed because relative sea level rise occurs late in the eustatic cycle, and the coastal plain is then rapidly transgressed. The deeply incised alluvial valleys in this environment may entirely contain the lowstand 2 systems tract. Toward the foredeep (Figures 9 and 10), the lowstand 2 systems tract becomes a more significant component of the sequences. Lowstand onlap begins early at the inflection point on the falling eustatic sea level curve, and the coastal plain facies zone is not confined by entrenched valleys.

Figures 8, 9, and 10 indicate that the transgressive systems tract contains the bulk of coastal plain aggradation during relative sea level rise. The highstand systems tract is relatively less important—particularly in the foredeep, where the narrow highstand coastal plain limits significant aggradation. Here, the highstand systems tract is primarily progradational or poorly developed.

FACIES PROFILES

Figures 11, 12, and 13 illustrate the distribution of model facies zones in cross section for the arch position (profile 13C), the intermediate position (profile 13F), and the foredeep position (profile 13H), respectively. Sequences S1 to S4 are labeled, and sequence boundaries are shown. The model eustatic sea level curve is annotated to show the sequence boundaries in relationship to eustatic cycles C1 to C4. These diagrams can be compared with previous profiles to place the facies zones in the context of systems tracts and simulated seismic reflection patterns.

The discussion of facies distribution incorporates the modeling assumption of constant sediment supply along the length of the basin. Obviously, other distributions are possible if supply varies as a result of either cyclic tectonics or point sources of sediment. However, the principle that variations in subsidence can control general facies patterns is believed to be valid and most easily demonstrated when sediment supply is held constant. The facies relationships described below should be regarded as end members of a continuum in which sea level, sediment supply, and subsidence are all variable in time and space.

Alluvial

High-energy alluvial facies are thickest near the arch, where they fill incised stream valleys at Type 1 sequence boundaries (Figures 11 and 12). They are the first deposits of the lowstand 2 systems tract when relative sea level rise initiates aggradation within the valleys. With continued sea level rise, the alluvial facies zone thickens and interfingers with encroaching low-energy coastal plain facies in the lower reaches of the valleys. Alluvial sedimentation continues during rapid aggradation of coastal plain facies of the transgressive systems tract, but is progressively restricted to hingeward areas by the expanding coastal plain. As the incised valleys are filled, facies change from braided stream valley-fill to braidplain as the area of deposition becomes a generalized apron along the mountain front. In the foredeep (Figure 13), rapid aggradation of coastal plain facies restricts alluvial deposits to thin veneers near the mountains.

Relative sea level fall in the later stages of the highstand systems tract causes the alluvial facies zone to thicken and extend across the coastal plain. For most sequences of the model profiles, alluvial sediments in the area of expansion are too thin to observe or are removed by erosion. However, in sequence S2 in Figure 11, the wedge of alluvial facies above the coastal plain, at its hingeward termination, represents deposits of alluvial expansion.

Coastal Plain

The bulk of model aggradation involves low-energy fluvial, interfluvial, and paralic facies of the coastal plain facies zone. Following Type 1 sequence boundaries, such as those in Figures 11 and 12, these facies first appear as estuarine valley-fill in the lower reaches of the drowned valleys of the lowstand 2 systems tract. In the case of Type 2 boundaries, the coastal plain encroaches upon the apron of alluvial sediments near the mountains as the area of accommodation expands. For the model profiles presented here, most coastal plain aggradation is accomplished in the transgressive systems tract.

The coastal plain is subdivided into sand-prone and shale-prone coastal plain facies zones. Sand-prone coastal plain facies are assumed to be associated with slow relative sea level rise, usually near a sequence boundary. Slow rates of accommodation development on the coastal plain may favor the formation of multistory fluvial sand bodies as the surface is continually reworked over long time periods. In this setting, point-bar sands of high-sinuosity streams merge across broad expanses of the coastal plain as laterally continuous and vertically stacked sands. In contrast, when relative sea level rises rapidly, drainage systems aggrade, and the sand bodies are separated laterally and vertically by thick shale lenses of the interfluvial environment.

In areas of slow subsidence toward the arch, sand-prone coastal plain occupies a greater proportion of coastal plain deposition. Figures 11 and 12 show that the coastal plain of sequence S2 is entirely composed of sand-prone facies because relative sea level rises only slightly in response to the moderate rise in eustatic sea level of cycle C2. Figure 13 indicates that sequence S2 is predominantly shale-prone—a result of the more rapid subsidence in the foredeep. Note that the sand-prone coastal plain zone occurs almost exclusively within the lowstand 2 and highstand systems tracts. The transgressive systems tract includes shale-prone coastal plain because of rapid accommodation development in transgressive phases of sequence evolution.

On the facies profiles, sand-prone coastal plain is shown at both Type 1 and Type 2 sequence boundaries. At the upper boundary of sequence S3, the facies zone has been removed by erosion along the line of profile. A criterion for the recognition of Type 2 sequence boundaries, within an otherwise continuous record of continental deposition, may be the presence of zones of high fluvial sand continuity.

Beach/Shoreface

A beach/shoreface unit is indicated on model facies profiles at the coast where the coast is progradational or retrogradational. The beach and shoreface are assumed to comprise a sand body 30 ft (9 m) thick except during deposition of the sand-prone clinoform facies zone. Con-

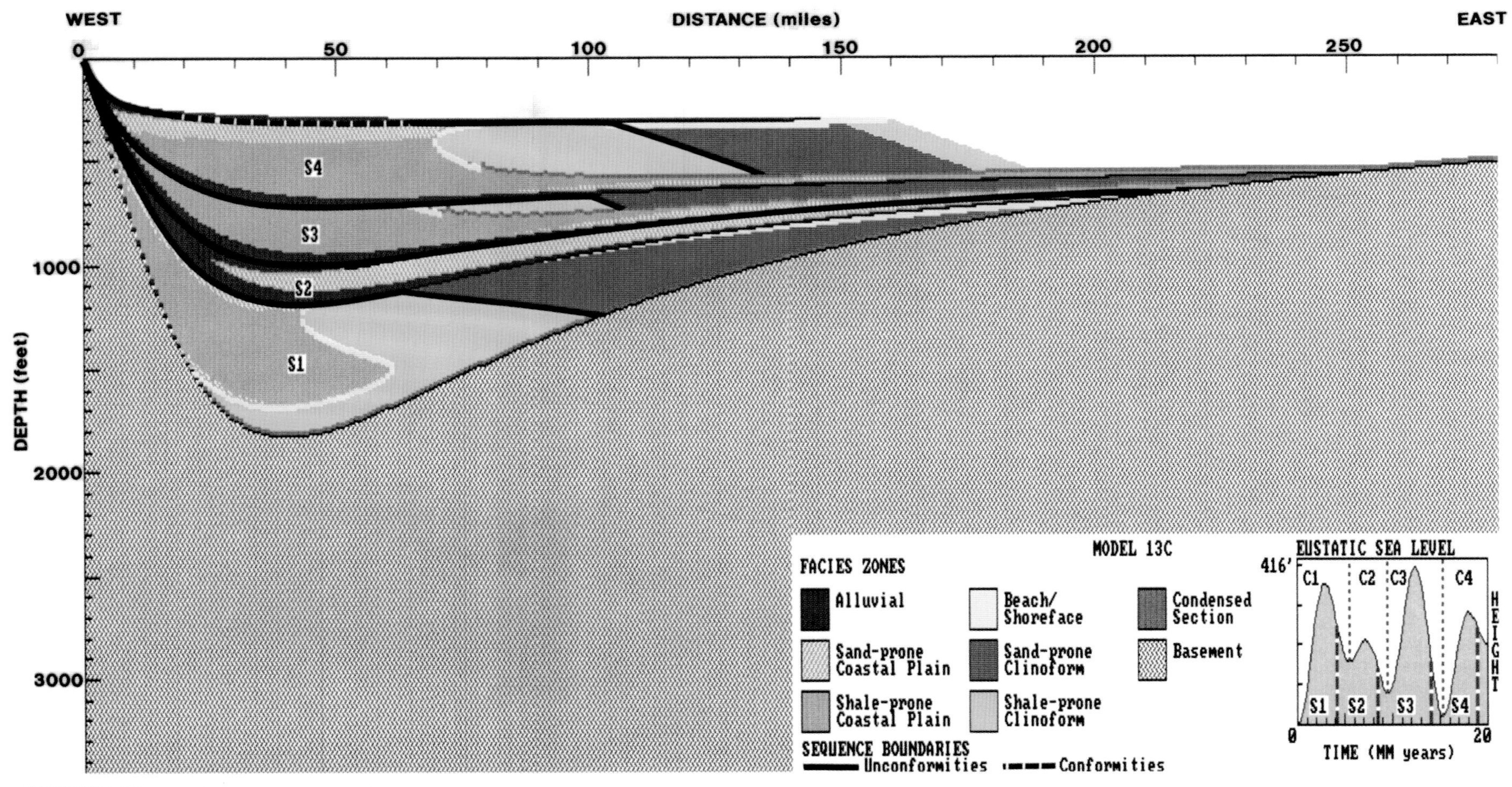

FIGURE 11. Facies cross section: slow subsidence—arch position (profile 13C). Model facies zones are represented in west to east cross section. Lithofacies and environmental facies are those assumed to be associated with each of the model depositional surfaces during sequence evolution. Type 1 and Type 2 sequence boundaries are indicated by unconformable and conformable boundaries, respectively.

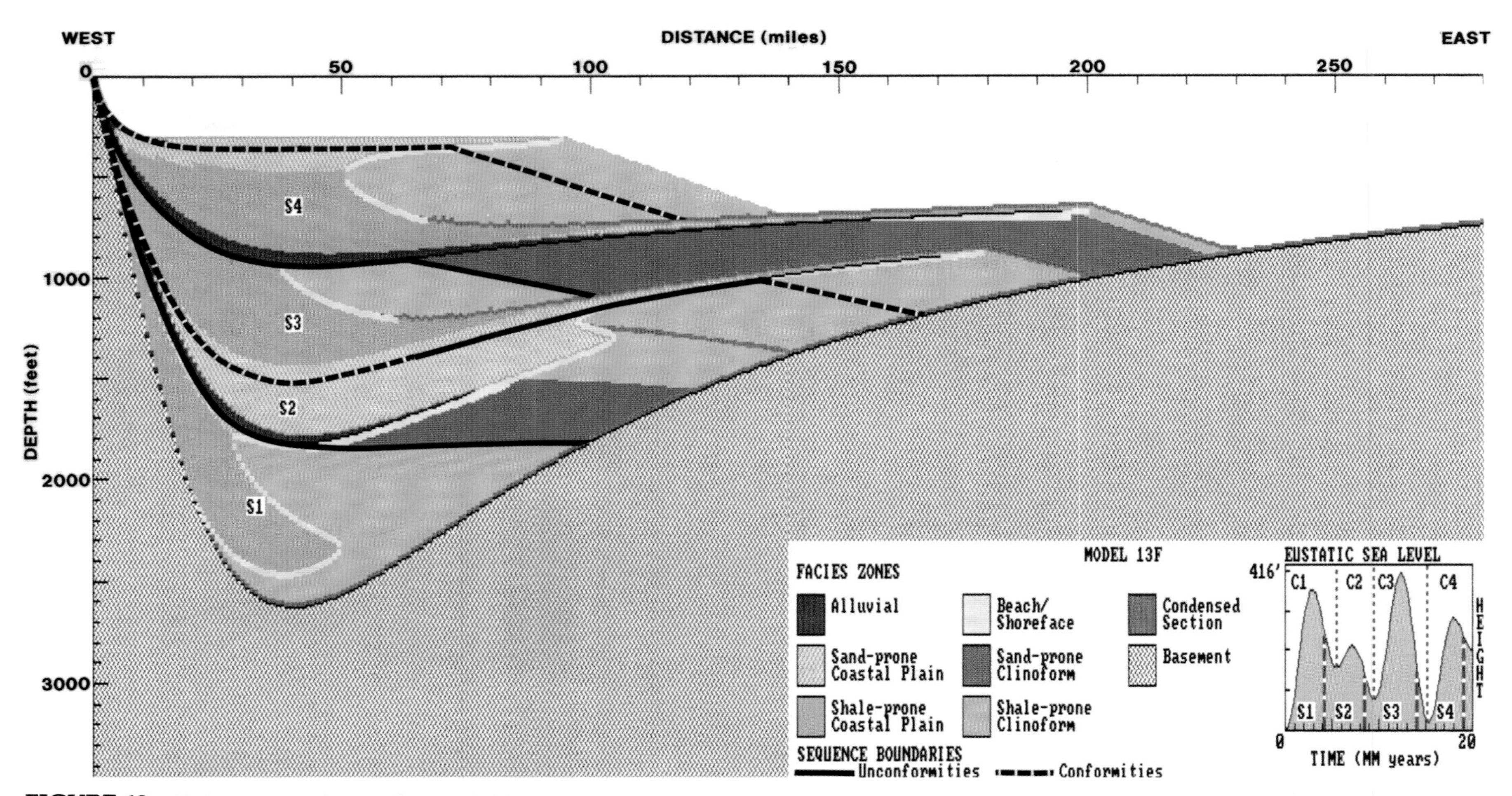

FIGURE 12. Facies cross section: moderate subsidence—intermediate basin position (profile 13F). Model facies zones are represented in west to east cross section. Lithofacies and environmental facies are those assumed to be associated with each of the model depositional surfaces during sequence evolution. Type 1 and Type 2 sequence boundaries are indicated by unconformable and conformable boundaries, respectively.

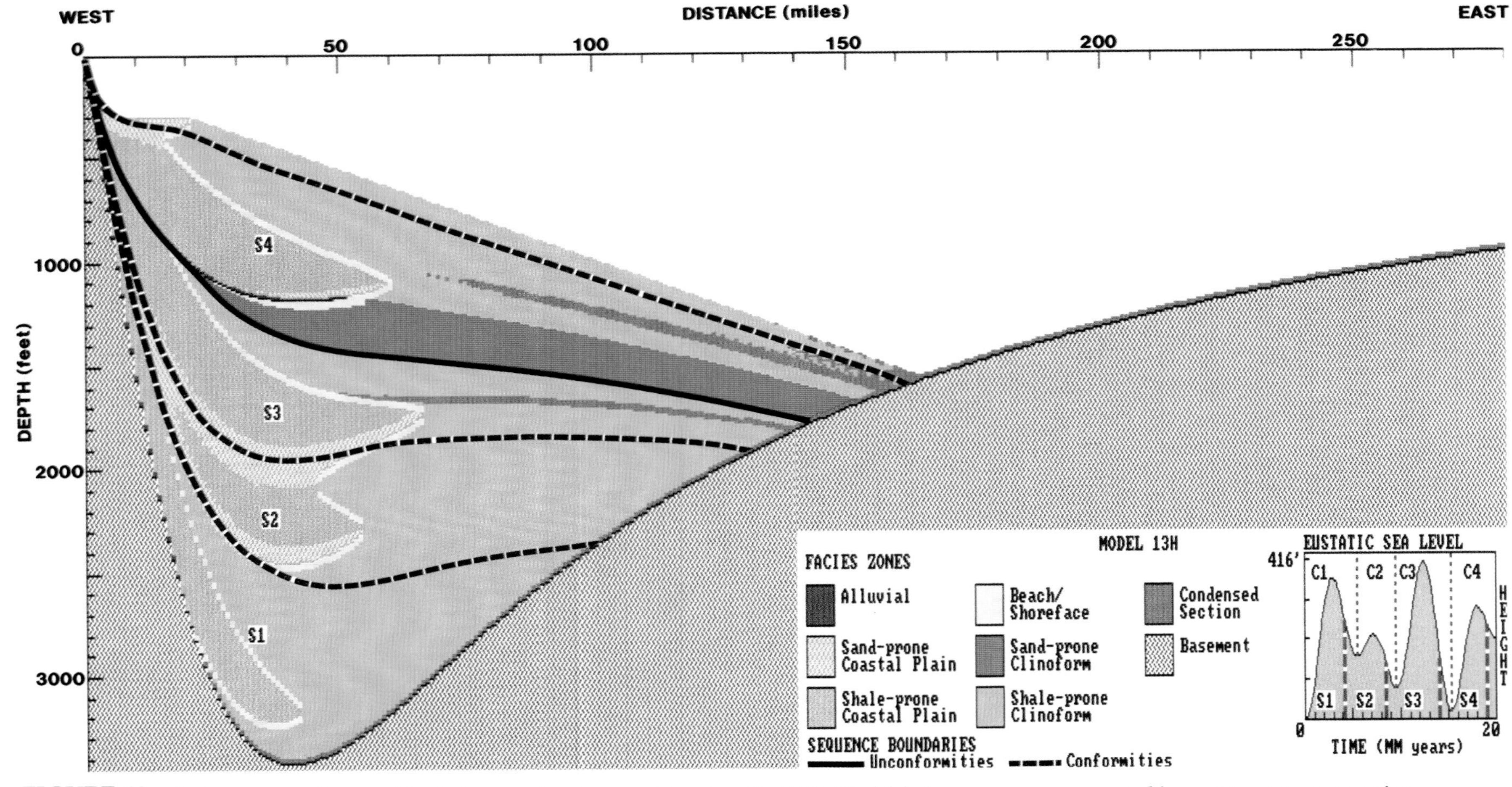

FIGURE 13. Facies cross section: rapid subsidence—foredeep position (profile 13H). Model facies zones are represented in west to east cross section. Lithofacies and environmental facies are those assumed to be associated with each of the model depositional surfaces during sequence evolution. Type 1 and Type 2 sequence boundaries are indicated by unconformable and conformable boundaries, respectively.

ditions then are assumed to favor a sandier shoreface, and a beach/shoreface thickness of 50 ft (15 m) is used.

The absence of the beach/shoreface along transgressive surfaces illustrates the modeling assumption of transgressive truncation. In areas of Figures 11 and 12 where the condensed section rests on coastal plain, the beach is assumed to be beveled during transgression. The eroded sediments are redistributed along the coast and eventually incorporated into the condensed section as relict shelf deposits. Where beach/shoreface is shown along the transgressive coastline, transgression is retrogradational, and the coast backsteps. In the latter case, partial or complete preservation of coastal deposits is expected.

On the model profiles, the beach/shoreface is truncated beneath Type 1 sequence boundaries and the erosion surfaces that cap the lowstand 1 systems tract. Truncation is illustrated in Figures 11 and 12 in sequences S1 to S3. Away from model profiles (i.e., between entrenched valleys), lowstand beaches may be preserved as a basinward-sloping strand plain, if longshore transport is efficient, or as incised lowstand deltas in lower-energy coastal settings. An example of the formation of lowstand strand plains is provided by the present-day Brazilian coast (Dominguez et al., 1987).

Clinoform

Sand-prone lowstand, and shale-prone highstand marine facies form the model clinoform. The low slope (< 0.5 degrees) of the clinoform allows large areas of the marine depositional surface to fall within the zone of storm-wave base. Sea level fall is likely to produce numerous discontinuities of local and regional extent on these broad ramps. Sands of the lowstand 1 systems tract may extend down the clinoform and rest disconformably on the intercalated shales.

Condensed Section

One of the more complex and variable facies zones is the condensed section. In this paper, the model condensed section represents reworked coastal deposits as well as marine facies that reflect sediment starvation on the transgressive shelf and in the adjacent basin. These facies include pebble beds, shell hash, glauconitic sands, phosphorites, and black shales. On model profiles, the condensed section is indicated where the rate of marine sedimentation is very slow, (i.e., on the basin floor, on transgressive shelves, and on the clinoform when deposition is reduced during coastal retrogradation). Where coastal aggradation is minimal, condensed marine facies may rest directly on beveled estuarine and coastal plain units. This is the case in the scenario of transgressive truncation shown in Figures 11 and 12. When transgression is accomplished by backstepping and aggradation, the condensed section may lie within the transgressive marine shale above the shoreline. This is the situation in Figure 13, where condensed deposition is indicated on the clinoform during retrogradation.

SUBSURFACE EXAMPLES IN THE WESTERN CANADA FORELAND BASIN

Recent subsurface and outcrop studies indicate that the major progradational sedimentary prisms of the post-Middle Jurassic section of the Western Canada foreland basin formed in an environment of fluctuating relative sea level similar to the modeled scenarios. However, numerical modeling can provide only a first-order approximation of stratigraphic complexity. The large-scale models of this paper generalize the detail revealed in subsurface studies. Note the difference in scale between the model profile figures cited in the text and the subsurface examples.

The following discussion will focus on several examples from the subsurface of the Western Canada foreland basin that illustrate model stratigraphic relationships. Figure 14 is a location map for the subsurface studies cited below. Figure 15 illustrates Cretaceous stratigraphy of the Alberta plains for reference.

Dunvegan/Doe Creek

The Cenomanian Dunvegan Formation occurs in western Alberta and provides a subsurface example of prograda-

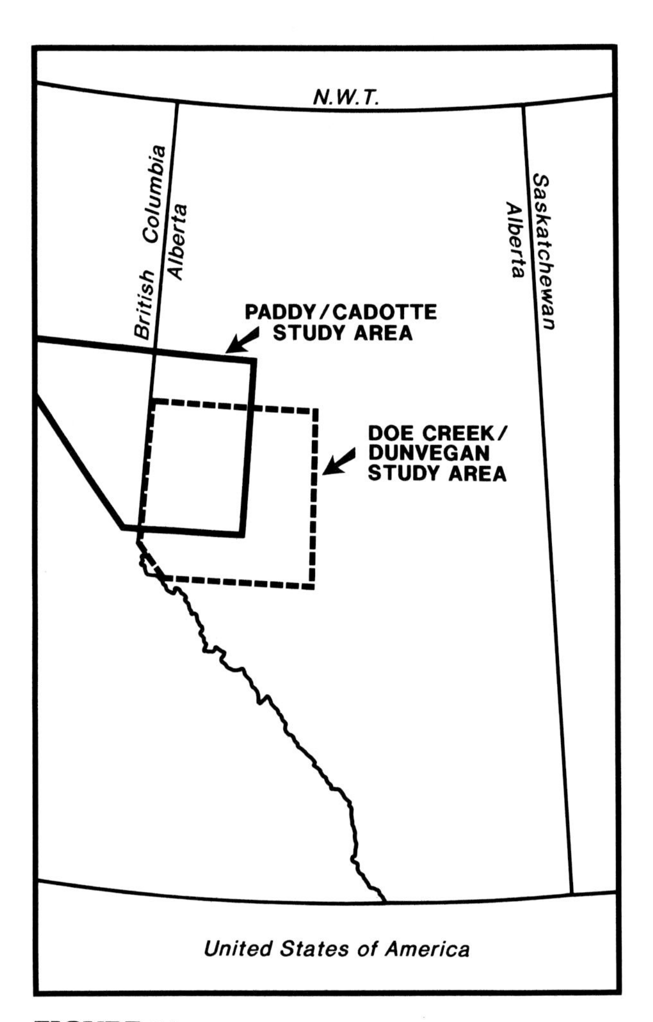

FIGURE 14. **Location map for subsurface examples.**

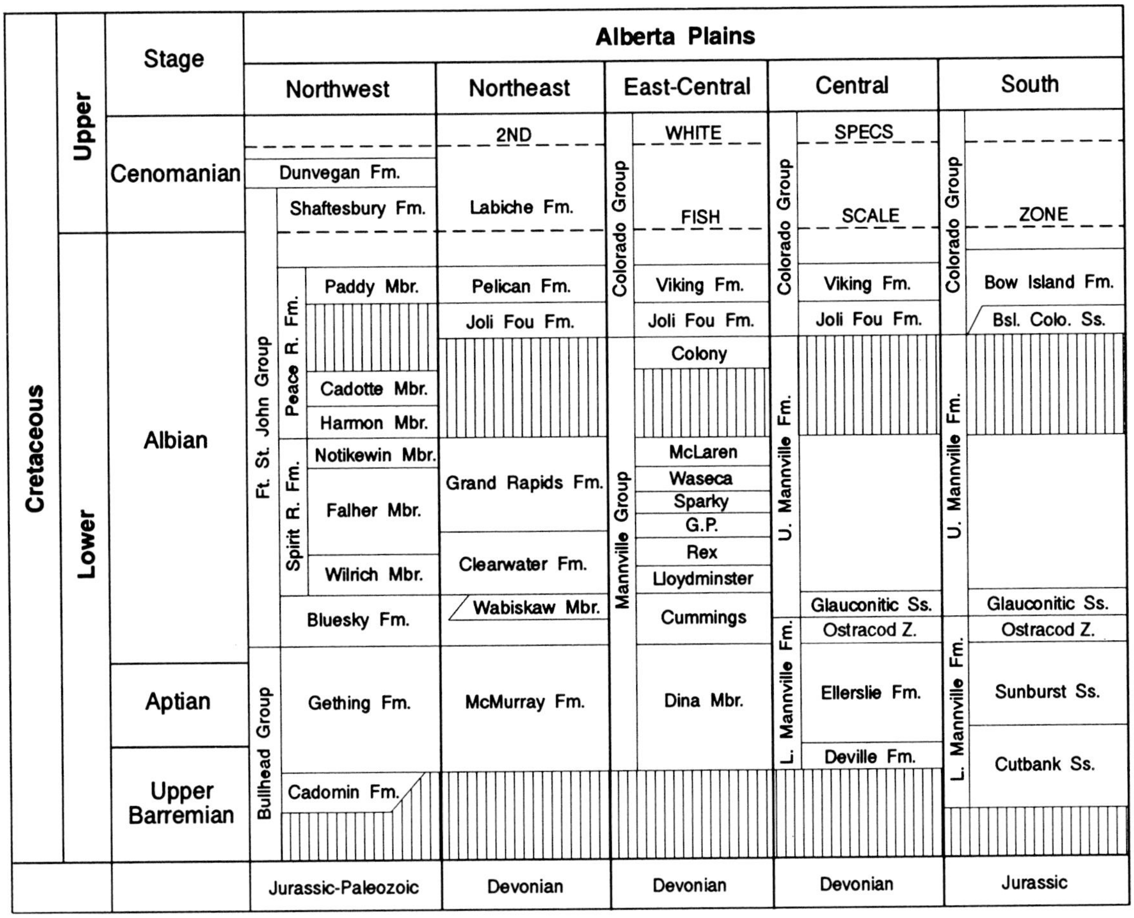

			Alberta Plains				
		Stage	Northwest	Northeast	East-Central	Central	South
Cretaceous	Upper	Cenomanian		2ND	WHITE (Colorado Group)	SPECS (Colorado Group)	(Colorado Group)
			Dunvegan Fm.				
			Shaftesbury Fm. (Ft. St. John Group)	Labiche Fm.	FISH	SCALE	ZONE
	Lower	Albian	Paddy Mbr. (Peace R. Fm.)	Pelican Fm.	Viking Fm.	Viking Fm.	Bow Island Fm.
				Joli Fou Fm.	Joli Fou Fm.	Joli Fou Fm.	Bsl. Colo. Ss.
			Cadotte Mbr. (Peace R. Fm.)		Colony	(U. Mannville Fm.)	(U. Mannville Fm.)
			Harmon Mbr. (Peace R. Fm.)				
			Notikewin Mbr. (Spirit R. Fm.)	Grand Rapids Fm.	McLaren (Mannville Group)		
			Falher Mbr. (Spirit R. Fm.)		Waseca		
					Sparky		
					G.P.		
			Wilrich Mbr. (Spirit R. Fm.)	Clearwater Fm.	Rex		
					Lloydminster		
			Bluesky Fm.	Wabiskaw Mbr.	Cummings	Glauconitic Ss.	Glauconitic Ss.
						Ostracod Z. (L. Mannville Fm.)	Ostracod Z. (L. Mannville Fm.)
		Aptian	Gething Fm. (Bullhead Group)	McMurray Fm.	Dina Mbr.	Ellerslie Fm.	Sunburst Ss.
		Upper Barremian	Cadomin Fm.			Deville Fm.	Cutbank Ss.
			Jurassic-Paleozoic	Devonian	Devonian	Devonian	Jurassic

FIGURE 15. A partial Cretaceous stratigraphy of the Alberta plains (from Strobl, 1988).

tional/retrogradational cycles that characterize models presented in this paper. Features of the Dunvegan Formation, as described by Bhattacharya (1988), can be compared with the modeled sequence stratigraphy of Figures 9 and 12 (profile 13F).

Dunvegan stratigraphy is illustrated in Figure 16. Members of the Dunvegan Formation were interpreted by Bhattacharya (1988) as regressive units separated by regionally persistent transgressive surfaces. The members can be grouped into an older progradational phase comprising members E, F, and G, and a younger aggradational/retrogradational phase comprising members A through D. The Dunvegan wedge downlaps on the condensed section of the Fish Scale Unit (FSU) and is bounded above by the condensed section of the Second White Speckled Shale, a coccolith-rich interval in the Kaskapau Formation.

Bhattacharya (1988) recognized a sequence boundary at the top of member E. The boundary is coincident with the overlying transgressive surface northwest of the channel at Simonette (Figure 16) but underlies a lowstand deltaic wedge at Bigstone. In this interpretation, the Simonette "channel" was an incised valley that supplied sediment to the "Bigstone Delta" during a lowstand of relative sea level. With subsequent rise in sea level, the delta onlapped the sequence boundary as the Simonette valley filled with estuarine sands. The lowstand delta and member E were then transgressed. Subsequently, continued sea level rise was accompanied by deposition of members A to D.

Dunvegan stratigraphy can be compared to sequences S3 and S4 of Figures 9 and 12. The condensed intervals that bound the Dunvegan Formation are represented by the condensed-section facies zones that lie above and below the sequence S3 highstand and sequence S4 lowstand progradation. Progradational units of sequences S3 and S4 downlap against the condensed section of the transgressive systems tract—a model representation of Dunvegan progradational downlap of Figure 16.

The progradational phase of the Dunvegan Formation, northwest of Bigstone, is composed of coastal and clinoform units of the highstand systems tract. Dunvegan regressive cycles of members E, F, and G combine to form the shale-prone clinoform highstand unit of sequence S3 at the scale of Figures 9 and 12. Dunvegan and model highstand units are characterized by depositional toplap.

The lowstand Bigstone Delta demonstrates a significant basinward shift in coastal onlap from a position far to the west of the study area to a downslope position on the last highstand clinoform of the Dunvegan progradation (Figure 16). The downslope position of onlap and the incised valley at Simonette indicate a Type 1 sequence

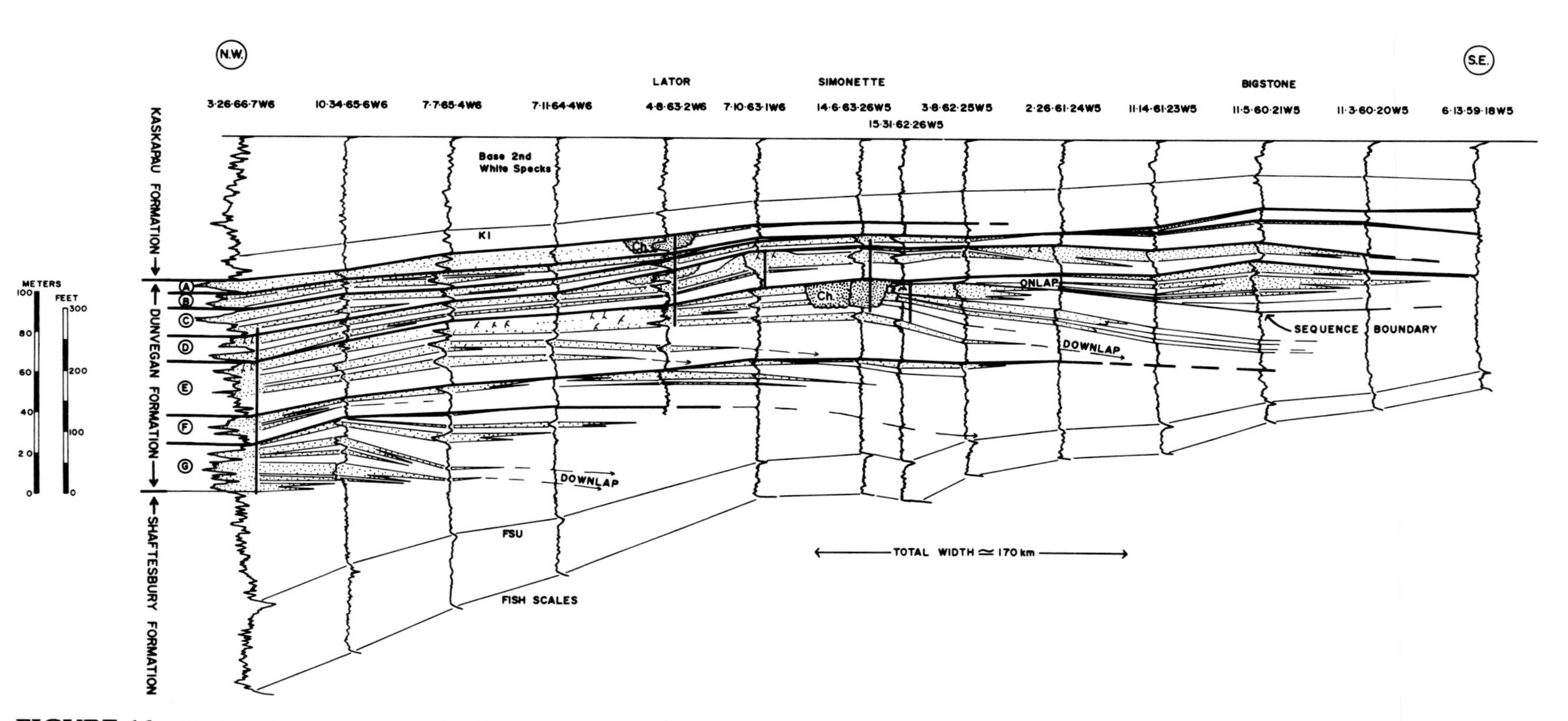

FIGURE 16. Regional cross section of the Dunvegan Formation in western Alberta. Nonmarine facies are indicated by root symbols. Log traces are gamma ray. (From Bhattacharya, 1988.)

boundary. The incised valley of the model profile (Figure 12) simulates the Simonette valley because they both truncate well into the underlying marine section.

The Bigstone Delta is a unit of the lowstand 1 and lowstand 2 systems tracts of this paper and is represented in Figure 12 by the sand-prone clinoform and distal shale-prone clinoform facies of sequence S4. The computer model differs significantly from the lowstand geometry of the Dunvegan Formation in the volume of sediment included in the lowstand unit. The Bigstone Delta apparently represents a rapid but short-lived sea level drop compared with the assumed model sea level curve.

The onlapping pattern of deposition of the Bigstone Delta suggests that the bulk of deposition occurred during sea level rise in the lowstand 2 systems tract. The estuarine valley-fill of the Simonette "channel" is analogous to model valley-filling facies of the coastal plain facies zone, characteristic of lowstand 2 deposition.

Members A to D in Figure 16 are represented by coastal aggradation at the distal end of sequence S4 lowstand progradation in Figure 12 (at 200 mi, or 322 km, from the hinge). The transgressive surface that caps member E is the base of the transgressive systems tract. The retrogradational members that comprise the transgressive systems tract of the Dunvegan Formation are generalized in the model simulation of transgression in sequence S4.

Retrogradation continued into the Doe Creek Member (Figure 17), which was described by Wallace-Dudley and Leckie (1988). Five stacked, backstepping sand units were interpreted to have been deposited as shoals during the transgression of the Dunvegan delta. These units are coeval with coastal plain deposition to the west. The ultimate limit of Doe Creek transgression is lost in the mountains or at the surface to the north. At the scale of Figure 12, the Dunvegan/Doe Creek backstepping cycles are simulated by the modeled retrogradation of the sequence S4 coastline between the distances of 50 and 80 miles (80 to 130 km) on the model profile.

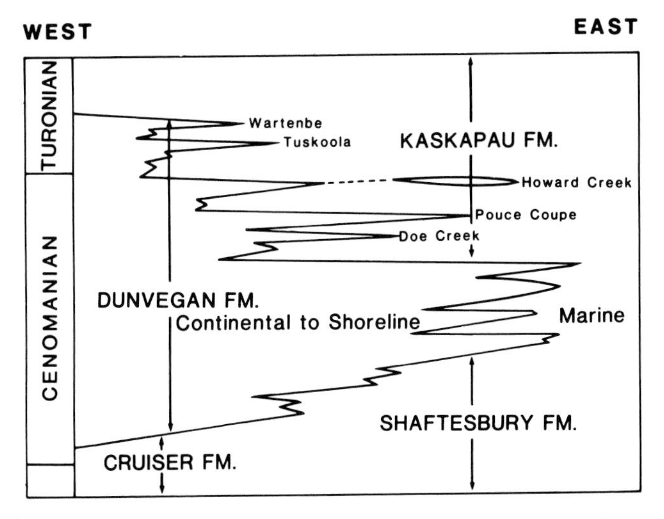

FIGURE 17. Diagrammatic cross section through the Peace River plains into northeastern British Columbia, showing the interfingering relationship of the lower Kaskapau Formation and the underlying Dunvegan Formation. Note the Doe Creek Member referenced in this paper. (From Wallace-Dudley and Leckie, 1988.)

Paddy/Cadotte

The Cadotte and Paddy members of the Peace River Formation, of middle to late Albian age, comprise a regressive wedge that covers a large area of west-central Alberta and northeastern British Columbia. Figure 18 illustrates stratigraphy of the Paddy and Cadotte members in north-south cross section.

The Cadotte Member is composed of coastal sands and conglomerates overlain by coastal plain and bay facies of the Paddy Member. Offshore facies of the Cadotte progradation downlap in the underlying Harmon shale, a marine unit representing distal aggradation on the Albian shelf. At the terminus of the Cadotte progradation, the Paddy and Cadotte members merge to form a thick, aggradational barrier behind which a complex of estuarine and bay sediments accumulated (Rahmani and Smith, 1988). The top of the Paddy Member is a regional transgressive surface that may bevel the underlying coastal deposits. The Paddy Member is overlain by the marine Shaftsbury shale, which contains the condensed section of the Fish Scale Zone about 150 ft (50 m) above the top of the Paddy Member.

Hayes (1988) recognized a paleodrainage system in the Noel area of British Columbia that truncates the coastal facies of the Cadotte well into the distal shoreface (Figure 19). In cross section (Figure 20), the incised valleys are nested in the normal coastal succession and are filled with fluvial and estuarine sands and conglomerates. According to Hayes (1988), valley incision in the Cadotte Member was caused by a fall in relative sea level followed by infilling during the subsequent sea level rise.

The interpretation can be made that the Cadotte Member is a strand plain that accumulated during declining relative sea level with concurrent or subsequent incision of entrenched valleys. In this view, the valleys described by Hayes (1988) are elements of an erosion surface composed of interfluves, where a normal Cadotte succession is preserved, and valleys. Cadotte lowstand deposition terminates in the thick coastal facies at the northern limit of Cadotte beach/shoreface deposition (Figure 18).

The fluvial/estuarine valley-fill described by Hayes (1988) represents the initial encroachment of the Paddy sequence upon the incised strand plain of the Cadotte Member. With continued relative sea level rise, the Paddy coastal plain facies completely onlapped the Cadotte Member as the Paddy shoreline aggraded in the north (see Figure 18). This interpretation is supported by the age of the Paddy Member, which is significantly younger than the Cadotte Member (Figure 15, northwest Alberta plains). The hiatus between the two members may represent the period of erosion and coastal onlap.

Paddy/Cadotte stratigraphy can be compared to model elements of Figures 9 and 12. Falling relative sea level during Cadotte deposition is indicated not only by the entrenched valleys described by Hayes (1988), but also by depositional and erosional toplap beneath the

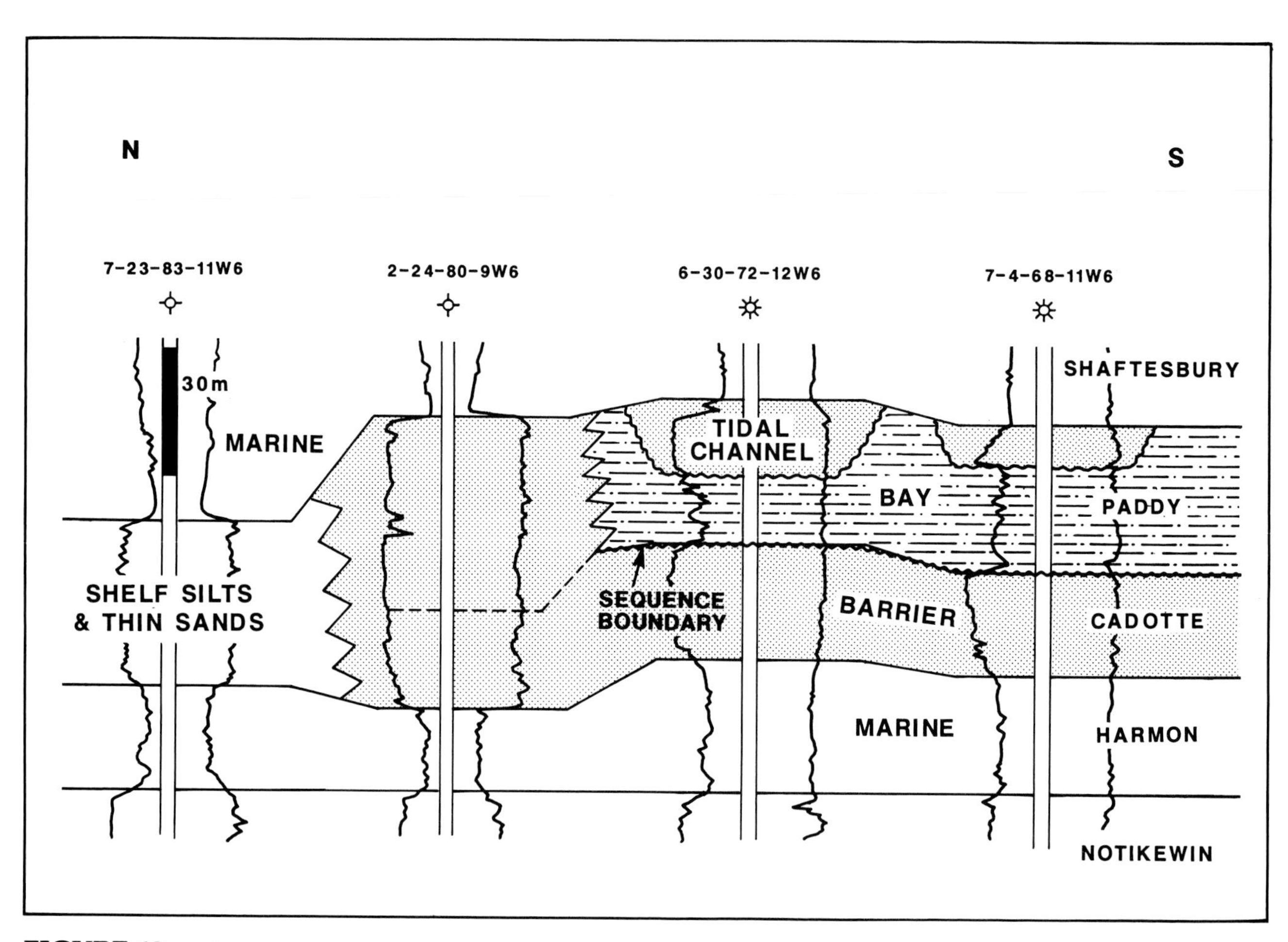

FIGURE 18. Diagrammatic north-south cross section on the south flank of the Peace River arch in western Alberta, showing the stratigraphic context of the Paddy and Cadotte members of the Peace River Formation. (Courtesy of D. G. Smith.)

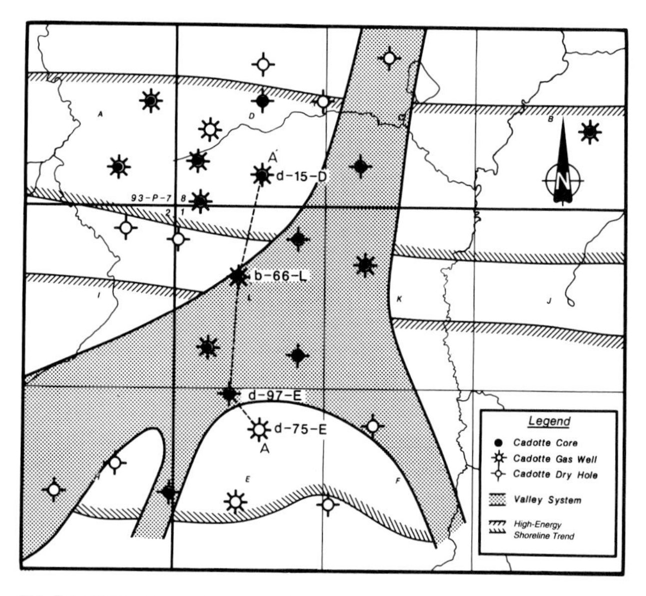

FIGURE 19. Cadotte facies map in the Noel area of British Columbia (from Hayes, 1988).

younger Paddy Member. The Cadotte strand plain is, therefore, analogous to the sand-prone clinoform wedge of the lowstand 1 systems tract of sequence S4. In this interpretation, the lowstand Cadotte strand plain should overlie a Type 1 sequence boundary developed either at the top of the underlying Notekewin or at the end of Cadotte highstand progradation. On the model profile of Figure 12, the valley that truncates sequence S3 and the lowstand 1 unit of S4 simulate the Cadotte valley system.

The lowstand 2 systems tract includes (1) the initial onlap of Paddy fluvial/estuarine and coastal plain facies in the entrenched valleys, and (2) the aggradation of the Paddy coastal barrier and expansion of the coastal plain and bay environments across the Cadotte strand plain. Encroachment and aggradation within the lowstand 2 systems tract of sequence S4 (Figure 9 at 200 mi) simulate these events.

Continued Paddy onlap is accomplished in the transgressive systems tract. The sequence S4 transgression of Figures 9 and 12 represents the regional beveling and transgression of the Paddy coastal plain preceding deposition of the Shaftsbury shale. The marine shale unit

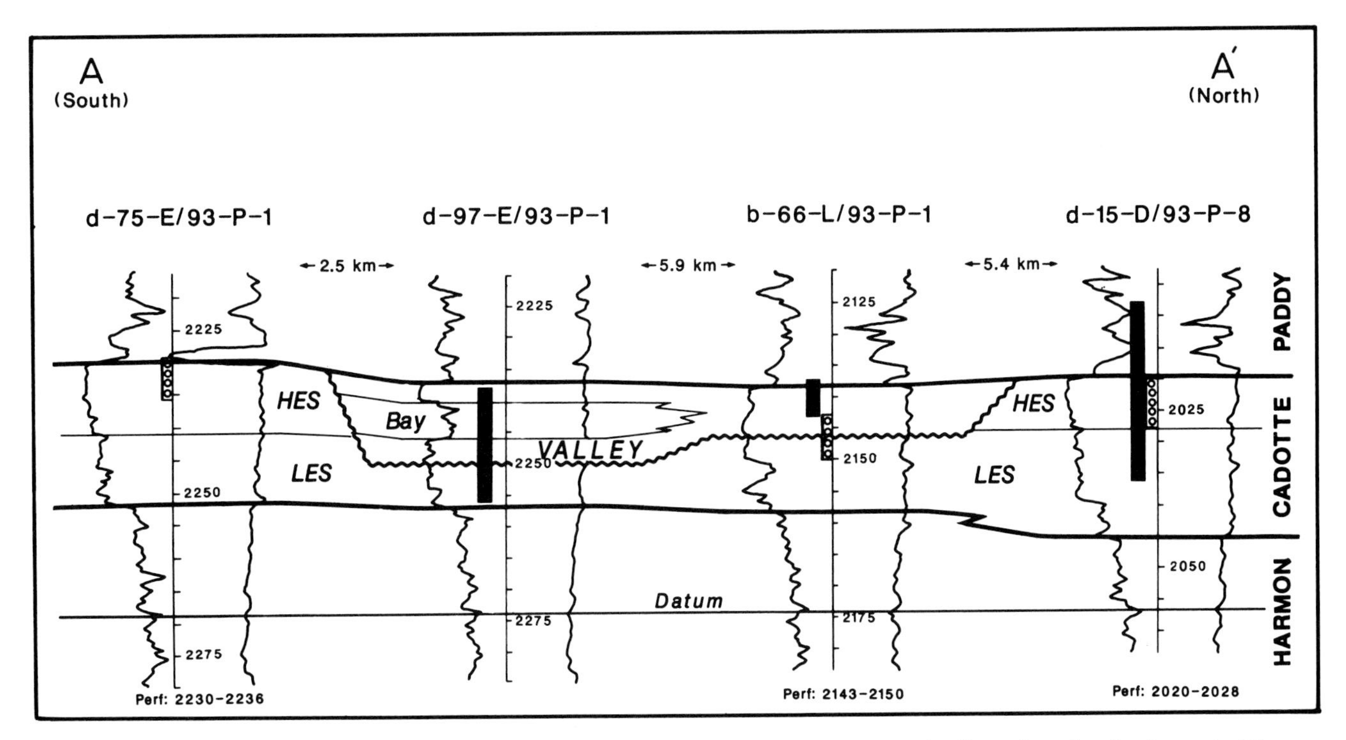

FIGURE 20. Stratigraphic cross section A-A′ in the Paddy/Cadotte succession. The line of section is shown in Figure 19. Black bars indicate cored intervals; circles indicate perforated (gas-bearing) intervals. HES stands for high-energy shoreface and foreshore; LES stands for low-energy (distal) shoreface. (From Hayes, 1988.)

below the Fish Scale Zone condensed interval probably indicates coastal retrogradation southwest of the area of subsurface control prior to the maximum drowning represented by the Fish Scale Zone. In Figure 12, the condensed section above the sequence S4 transgressive surface simulates the Fish Scale Zone.

CRETACEOUS CHRONOSTRATIGRAPHY

Stratigraphic principles derived from sequence models discussed in this paper can be applied to the description of Cretaceous chronostratigraphy in the Western Canada foreland basin, diagrammatically illustrated in Figure 21. Cyclic patterns of regression and transgression, erosional sequence boundaries, and transgressive hiatal periods of the western Canadian post-Middle Jurassic succession are correlated with the global changes in relative and eustatic sea level published by Haq et al. (1988). In the foreland thrust belt, Cretaceous stratigraphic relationships have been obscured through deformation or recent erosion (west of the line shown in Figure 21).

The sequences displayed in Figure 21 are themselves composed of smaller sequences and generalize the complexity revealed in subsurface studies. For example, Leckie (1986) and Carmichael (1988) described multiple sequences in the Gates Formation (Falher equivalent) that are merely represented in Figure 21 as the Falher progradation. Likewise, Strobl (1988) and Rosenthal (1988) described several different ages of valley incision in the Glauconite interval that are represented in Figure 21 as a single erosional event. However, it is at the scale of Figure 21 that global sequence correlations can best be achieved.

Figure 21 shows that known and inferred periods of Cretaceous subaerial erosion in Western Canada can be readily correlated to significant global sea level falls. For instance, the angular unconformity at the base of the Cretaceous may have been formed, in part, during the profound global Valanginian lowstand (126 Ma) although, in Western Canada, structural deformation was likely the dominant cause of relative sea level fall.

Documented episodes of stream valley incision and/or lowstand deposition in the Ostracod (early Albian), Glauconite (middle Albian), Cadotte (middle Albian), Dunvegan (Cenomanian), Cardium (Turonian), and Badheart (Santonian) formations appear to correspond closely to global Type 1 sequence boundaries at 108, 103, 98, 94, 90, and 85 Ma, respectively (see Strobl, 1988; Hayes, 1988; Rosenthal, 1988; Bhattacharya, 1988; Plint et al., 1988; and Plint and Walker, 1987). Other Type 1 boundaries shown in the chronostratigraphy are conjectural and are based on the time-position of the sequence with respect to the global coastal onlap curve and known facies distributions.

Figure 21 describes the Cretaceous sequences in terms of generalized facies zones used in sequence models. The timing of large-scale regressive/transgressive cyclicity is compatible with the pattern of major global eustatic lowstands discussed above. Condensed sections, such as the First and Second White Specks and the Fish Scale Zone (Figure 21), represent time intervals of reduced sedimentation in the Cretaceous marine basin corresponding to maximum transgression. These transgressive periods are positioned between major global lowstands. Subsequent periods of rapid coastal progradation and marine deposition (clinoform and bottom set

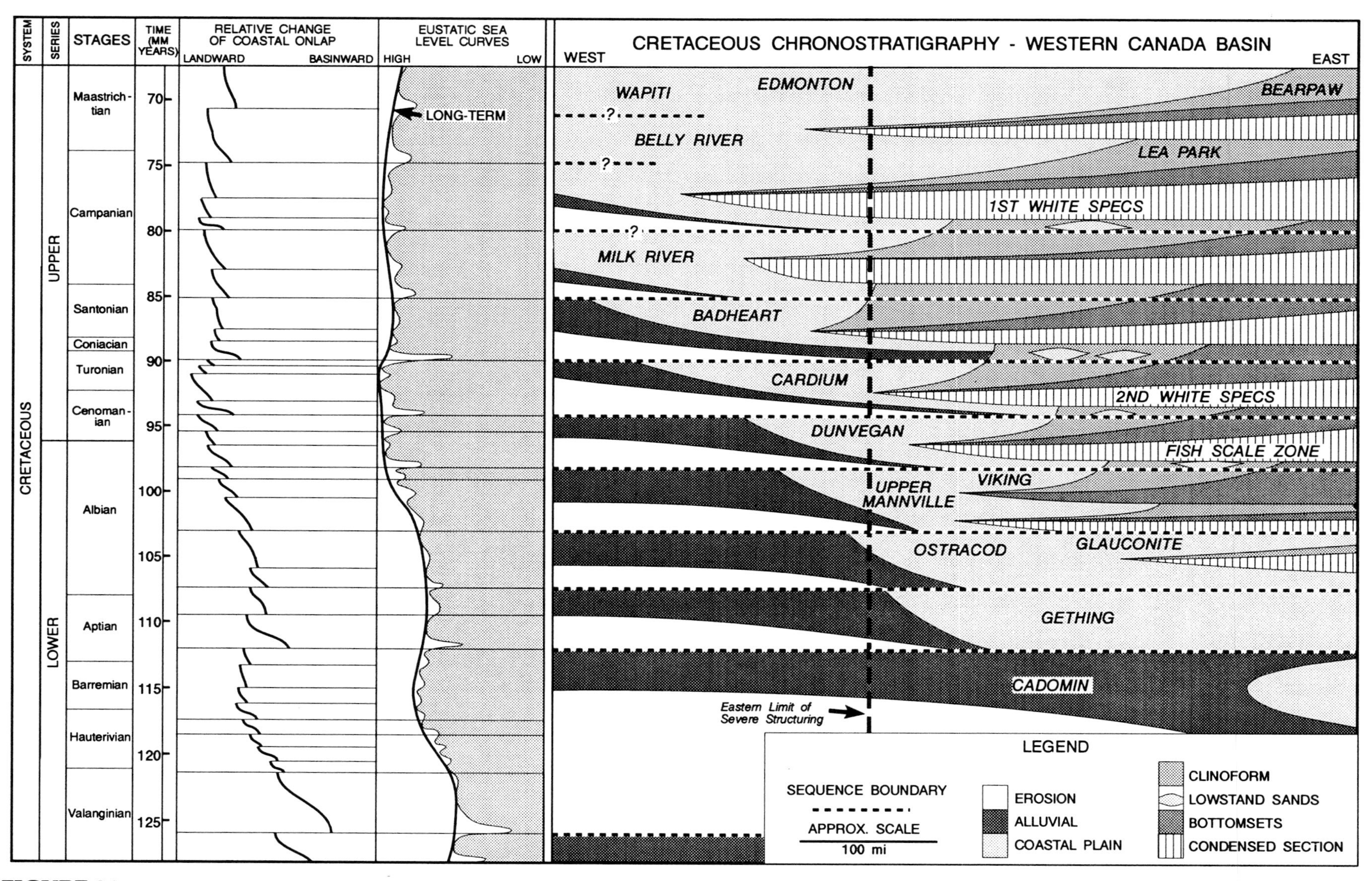

FIGURE 21. Cretaceous chronostratigraphy of the Western Canada foreland basin. Diagrammatic representation of Cretaceous sequence stratigraphy using concepts derived from numerical models of sequence development in foreland basins. Type 1 erosional sequence boundaries are associated with many of the Cretaceous sequences and are correlative to major global sea level lowstands. Cretaceous strata west of the eastern limit of severe structuring and recent erosion are deformed, and stratigraphic relationships are speculative. Curves of coastal onlap position and eustatic sea level are from Haq et al. (1988).

units of Figure 21) are capped by the Type 1 sequence boundaries correlatable to the major global lowstands of Haq et al. (1988).

Long-term Cretaceous sea level change also appears to be reflected in the general distribution of marine and nonmarine facies. Rising sea level, following the Valanginian, was accompanied by a gradual expansion of marine facies. The initial basin-fill of the Gething Formation was inundated by the Ostracod transgression (Leckie and Smith, 1992). Thereafter, the regressive prisms were progressively reduced in areal extent. During the long-term highstand, from late Albian to late Campanian, the basin was dominated by marine shale, with coastal and nonmarine sediments largely confined to the western (and eastern) basin margin. Long-term sea level fall, beginning in the late Campanian, may have played a role in the basin-wide regressions of the Belly River and Edmonton formations, which eventually filled the basin.

SUMMARY AND CONCLUSIONS

The modeling process is useful in establishing basin-scale relationships involving fundamental geologic parameters. Models identify basic constraints to stratigraphic conjecture and form working hypotheses to be subjected to the test of outcrop and subsurface data. The models presented in this paper suggest differences in sequence architecture that may be related to the positions of structural elements within foreland basins. The confirmation of these relationships is the subject of future work.

Modeling demonstrates that the rate and magnitude of eustatic and relative sea level change play a primary role in the evolution of siliciclastic sequences. Rapid falls of eustatic sea level produce sequence-bounding subaerial unconformities characterized by stream valley incision and general erosion of the land surface. Erosion is most pronounced when relative sea level falls at the coast, and the shelf is exposed.

When eustatic sea level falls slowly, relative sea level may continue to rise. In this case, erosion is minimal, and coastal plain deposition is continuous across the sequence boundary. The position of the boundary may be marked by a basinward shift in coastal onlap and an increase in sandy facies of the coastal plain facies zone, reflecting a reduction in accommodation.

Sequence boundaries are centered, in time, about the inflection point of falling sea level when the rate of fall is maximum. Declining rates of eustatic fall, subsequent to the inflection point, result in relative sea level rise, encroachment of coastal and estuarine facies upon the erosion surface, and alluviation of entrenched valleys.

Rising relative sea level favors continual expansion of coastal facies zones as base level is raised. Rapid eustatic rises result in widespread transgression. Transgression may take the form of a beveling of the coast, through the process of shoreface retreat, or retrogradation or backstepping of coastal deposition. The nature of transgression is a function of the width of the coastal plain and the rate of rise.

Transgression is maximum at or near the inflection point on the rising sea level curve when the rate of rise is maximum. Condensed deposition on the shelf and in the basin accompanies transgression, as the rate of marine sedimentation is reduced. Subsequent decline in the rate of rise toward highstand allows progradation of the coastline to commence and increase in magnitude as relative sea level begins to fall in a continuation of cyclic sea level change.

The strong asymmetry in subsidence rate across the Western Canada foreland basin has important consequences for sequence development. Relative sea level may be rising in axial basin settings at the same time as relative sea level is falling on structural arches and along the basin margins. Modeling has shown that these geographic variations of relative sea level change affect the stratal patterns, the environmental facies, and the nature of sequence boundaries forming in different parts of the basin. Table 1 summarizes contrasting features of sequence geometry and facies on an arch (slow subsidence) and in the foredeep (rapid subsidence).

Stratigraphic studies in the subsurface of Western Canada suggest that principles derived from modeling can be applied to the interpretation of the Cretaceous record. The Cretaceous section clearly needs more precise biostratigraphic definition before the speculative chronostratigraphic relationships presented in Figure 21 can be fully demonstrated. Undoubtedly, tectonic events and mechanisms of uplift and crustal deformation as described by Stockmal et al. (1992) have been important causes of relative sea level change. However, the correlation of events in Figure 21 suggests that global sea level change has been a major influence on sequence development in the Cretaceous strata of the Western Canada foreland basin. As sequence stratigraphic concepts become more widely applied, important relationships will be discovered that will advance our understanding of the stratigraphy of the Western Canada sedimentary basin.

ACKNOWLEDGMENTS

I would particularly like to acknowledge the many helpful suggestions that Peter Schwans, Dale Leckie, and Roger Macqueen contributed in reviewing this paper. I am also grateful for the editing and critical reading of my wife, Gay.

REFERENCES CITED

Baum, G. R., and P. R. Vail, 1988, Sequence stratigraphic concepts applied to Paleogene outcrops, Gulf and Atlantic basins, *in* C. K. Wilgus, B. S. Hastings, C. A. Ross, H. W. Posamentier, J. C. Van Wagoner, and C. G. St. C. Kendall, eds., Sea-level changes: an integrated approach: SEPM Special Publication No. 42, p. 309–327.

Bhattacharya, J., 1988, Autocyclic and allocyclic sequences in river- and wave-dominated deltaic sediments of the Upper Cretaceous, Dunvegan Formation, Alberta: core examples, *in* D. P. James and D. A. Leckie, eds., Sequences, stratigraphy, sedimentology: surface and subsurface: Canadian Society of Petroleum Geologists Memoir 15, p. 25–32.

TABLE 1. Comparison of Modeled Sequence Characteristics in Areas of Slow and Rapid Subsidence.

	Slow Subsidence	Rapid Subsidence
	Sequence Geometry	
Sequences	Thin, areally extensive, basin-filling prisms.	Thick prisms confined near sediment source.
Boundaries	Numerous Type 1 erosional boundaries. Deeply incised stream valleys during eustatic falls. Boundaries represent significant hiatus.	Dominantly Type 2 conformable boundaries. Minimal valley entrenchment during eustatic falls. Boundaries represent little to no hiatus.
	Facies Zones	
Alluvial	Widespread braidplains, thick valley-fill.	Thin braidplains, areally restricted near the mountains.
Coastal plain	Thin, but areally extensive. Dominantly sand-prone; laterally coalesced fluvial sands.	Thick, but narrow adjacent to source Dominantly shale-prone; aggradational architecture.
Sand-prone clinoform	Broad strand plains during eustatic falls. Potential for numerous marine sand bodies.	Narrow strand plains during most rapid eustatic falls. Limited coastal zone of potential marine sand deposition.
Shale-prone clinoform	Restricted to rapid eustatic rises.	Broad, aggradational ramps of marine shale deposition.
Transgressive shelf	Broad, drowned shelves; ravinement surfaces. Condensed sections include reworked coastal sands and conglomerates.	Retrogradational coasts; parasequence backstepping. Condensed sections mainly marine facies.
	Systems Tracts	
Lowstand 1	Areally extensive, numerous unconformity-bounded wedges.	Minor component of basin-filling limited to most rapid eustatic falls.
Lowstand 2	Onlap initiated at the coast and within entrenched valleys. Eustuarine valley-fill significant.	Onlap initiated at the inner margin of the coastal plain. Paralic facies along embayed coasts.
Transgressive	Truncational coastlines. Estuarine valley-fill within entrenched valleys.	Retrogradational coastlines. Aggradational coastal plain.
Highstand	Deeply truncated near incised valleys during eustatic fall.	Conformable with superseding lowstand units.
	Seismic Reflections	
Truncation	Deep truncation near incised valleys.	Mainly conformable.
Onlap	Pronounced basinward shift; broad area of coastal onlap.	Onlap confined near hinge.
Toplap	Extensive area associated with lowstand regression.	Toplap only during most rapid eustatic falls.
Downlap	Extensive downlap surfaces within sequences.	Internal downlap not well developed.

Carmichael, S. M. M., 1988, Linear estuarine conglomerate bodies formed during a mid-Albian marine transgression: "Upper Gates" Formation, Rocky Mountain foothills of northeastern British Columbia, *in* D. P. James and D. A. Leckie, eds., Sequences, stratigraphy, sedimentology: surface and subsurface: Canadian Society of Petroleum Geologists Memoir 15, p. 49-61.

Davis, H. R., and C. W. Byers, 1989, Shelf sandstones in the Mowry Shale: evidence for deposition during Cretaceous sea level falls: Journal of Sedimentary Petrology, v. 59, p. 548-560.

Dominguez, J. M. L., L. Martin, and A. C. S. P. Bittencourt, 1987, Sea-level history and Quaternary evolution of river mouth-associated beach-ridge plains along the east-southeast Brazilian coast: a summary, *in* D. Nummedal, O. H. Pilkey, and J. D. Howard, eds., Sea-level fluctuation and coastal evolution: SEPM Special Publication No. 41, p. 115-127.

Haq, B. U., J. Hardenbol, and P. R. Vail, 1988, Mesozoic and Cenozoic chronostratigraphy and cycles of sea-level change, *in* C. K. Wilgus, B. S. Hastings, C. A. Ross, H. W. Posamentier, J. C. Van Wagoner, and C. G. St. C. Kendall, eds., Sea-level changes: an integrated approach: SEPM Special Publication No. 42, p. 71-108.

Hayes, B. J. R., 1988, Incision of a Cadotte Member paleovalley-system at Noel, British Columbia—evidence of late Albian sea-level fall, *in* D. P. James and D. A. Leckie, eds., Sequences, stratigraphy, sedimentology: surface and subsurface: Canadian Society of Petroleum Geologists Memoir 15, p. 97-105.

Jervey, M. T., 1988, Quantitative geological modeling of siliciclastic rock sequences and their seismic expression, *in* C. K. Wilgus, B. S. Hastings, C. A. Ross, H. W. Posamentier, J. C. Van Wagoner, and C. G. St. C. Kendall, eds., Sea-level changes: an integrated approach: SEPM Special Publication No. 42, p. 47-69.

Leckie, D. A., 1986, Rates, controls and sand body geometries of transgressive-regressive cycles: Cretaceous Moosebar and Gates formations, British Columbia: AAPG Bulletin, v. 70, p. 516-535.

Leckie, D., and D. Smith, 1992, Regional setting, evolution, and depositional cycles of the Western Canada foreland basin, this volume.

Mitchum, R. M., Jr., P. R. Vail, and J. B. Sangree, 1977, Seismic stratigraphy and global changes of sea level: part 6, stratigraphic interpretation of seismic reflection patterns in depositional sequences, *in* C. E. Payton, ed., Seismic stratigraphy—applications to hydrocarbon exploration: AAPG Memoir 26, p. 117-133.

Nummedal, D., and D. J. P. Swift, 1987, Transgressive stratigraphy at sequence-bounding unconformities: some principles derived from Holocene and Cretaceous examples, *in* D. Nummedal, O. H. Pilkey, and J. D. Howard, eds., Sea-level fluctuation and coastal evolution: SEPM Special Publication No. 41, p. 241-260.

Olsson, R. K., 1988, Foraminiferal modeling of sea-level change in the Late Cretaceous of New Jersey, *in* C. K. Wilgus, B. S. Hastings, C. A. Ross, H. W. Posamentier, J. C. Van Wagoner, and C. G. St. C. Kendall, eds., Sea-level changes: an integrated approach: SEPM Special Publication No. 42, p. 289-297.

Plint, A. G., 1988, Sharp-based shoreface sequences and "offshore bars" in the Cardium Formation of Alberta: their relationship to relative changes in sea level, *in* C. K. Wilgus, B. S. Hastings, C. A. Ross, H. W. Posamentier, J. C. Van Wagoner, and C. G. St. C. Kendall, eds., Sea-level changes: an integrated approach: SEPM Special Publication No. 42, p. 357-370.

Plint, A. G., and R. G. Walker, 1987, Morphology and origin of an erosion surface cut into the Bad Heart Formation during major sea-level change, Santonian of west-central Alberta: Journal of Sedimentary Petrology, v. 57, p. 639-650.

Plint, A. G., R. G. Walker, and W. L. Duke, 1988, An outcrop to subsurface correlation of the Cardium Formation in Alberta, *in* D. P. James and D. A. Leckie, eds., Sequences, stratigraphy, sedimentology: surface and subsurface: Canadian Society of Petroleum Geologists Memoir 15, p. 161-183.

Posamentier, H. W., and P. R. Vail, 1988, Eustatic controls on clastic deposition II—sequence and systems tract models, *in* C. K. Wilgus, B. S. Hastings, C. A. Ross, H. W. Posamentier, J. C. Van Wagoner, and C. G. St. C. Kendall, eds., Sea-level changes: an integrated approach: SEPM Special Publication No. 42, p. 125-154.

Posamentier, H. W., M. T. Jervey, and P. R. Vail, 1988, Eustatic controls on clastic deposition I—conceptual framework, *in* C. K. Wilgus, B. S. Hastings, C. A. Ross, H. W. Posamentier, J. C. Van Wagoner, and C. G. St. C. Kendall, eds., Sea-level changes: an integrated approach: SEPM Special Publication No. 42, p. 109-124.

Potocki, D. J., and I. Hutcheon, 1992, Lithology and diagenesis of sandstones in the Western Canada foreland basin, this volume.

Rahmani, R. A., and D. G. Smith, 1988, The Cadotte Member of northwestern Alberta: a high energy barred shoreline, *in* D. P. James and D. A. Leckie, eds., Sequences, stratigraphy, sedimentology: surface and subsurface: Canadian Society of Petroleum Geologists Memoir 15, p. 431-437.

Rosenthal, L., 1988, Wave-dominated shorelines and incised channel trends: Lower Cretaceous Glauconite Formation, west-central Alberta, *in* D. P. James and D. A. Leckie, eds., Sequences, stratigraphy, sedimentology: surface and subsurface: Canadian Society of Petroleum Geologists Memoir 15, p. 207-219.

Stockmal, G. S., D. J. Cant, and J. S. Bell, 1992, Relationship of the stratigraphy of the Western Canada foreland basin to cordilleran tectonics: insights from geodynamic models, this volume.

Strobl, R. S., 1988, The effects of sea-level fluctuations on prograding shorelines and estuarine valley-fill sequences in the Glauconitic Member, Medicine River field and adjacent areas, *in* D. P. James and D. A. Leckie, eds., Sequences, stratigraphy, sedimentology: surface and subsurface: Canadian Society of Petroleum Geologists Memoir 15, p. 221-236.

Vail, P. R., R. M. Mitchum, Jr., and S. Thompson III, 1977, Seismic stratigraphy and global changes in sea level: part 3, relative changes of sea level from coastal onlap, *in* C. E. Payton, ed., Seismic stratigraphy—applications to hydrocarbon exploration: AAPG Memoir 26, p. 63-81.

Van Wagoner, J. C., H. W. Posamentier, R. M. Mitchum, Jr.,

P. R. Vail, J. F. Sarg, T. S. Loutit, and J. Hardenbol, 1988, An overview of the fundamentals of sequence stratigraphy and key definitions, *in* C. K. Wilgus, B. S. Hastings, C. A. Ross, H. W. Posamentier, J. C. Van Wagoner, and C. G. St. C. Kendall, eds., Sea-level changes: an integrated approach: SEPM Special Publication No. 42, p. 39–45.

Wallace-Dudley, K. E., and D. A. Leckie, 1988, Preliminary observations on the sedimentology of the Doe Creek Member, Kaskapau Formation, in the Valhalla field, northwestern Alberta, *in* D. P. James and D. A. Leckie, eds., Sequences, stratigraphy, sedimentology: surface and subsurface: Canadian Society of Petroleum Geologists Memoir 15, p. 485–496.

Chapter 3

Tectonics and Structure of the Western Canada Foreland Basin

Peter R. Fermor

Shell Canada, Calgary, Alberta, Canada

Ian W. Moffat

Gulf Canada Resources, Calgary, Alberta, Canada

ABSTRACT

In Western Canada, a series of deformational events, probably the result of the accretion of a series of allochthonous terranes and accompanying subduction, have deformed the miogeoclinal sediment prism that previously formed the western margin of North America. The miogeoclinal prism was composed of a series of Proterozoic, Paleozoic, and early Mesozoic sedimentary successions, apparently deposited in response to an extended series of episodes of rifting and subsidence. Tectonism leading to the development of the Western Canada foreland basin commenced in the Middle Jurassic, apparently the result of the initial collision between North America and the allochthonous terranes to the west. Various clastic wedge successions were shed eastward into the foreland basin through Late Jurassic to Paleocene, but at present they can only be loosely linked to identified tectonic episodes in the interior of the Canadian Cordillera. Transpressional tectonics, and deposition of foreland basin sediments, terminated at about the Paleocene–Eocene boundary as a change in tectonic regime produced large-scale extension in the interior of the Cordillera that lasted through the mid-Tertiary.

The foreland fold and thrust belt, developed along the eastern margin of the cordillera, overlaps and deforms sediments of the foreland basin. For much of its length through Canada, the foreland belt exhibits a thin-skinned structural style. The North American craton, essentially undeformed, extends westward beneath detached, horizontally compressed, and tectonically thickened strata of the miogeocline, the cratonic platform cover, and the foreland basin. Thrust faulting, with associated folding, was the dominant deformation mechanism in the foreland belt. The style of thrusting, and its relationship to folding, were influenced by several factors such as the geometry and composition of the deformed sedimentary sequence, the magnitude of shortening, the lateral extent of the basal detachment, and possibly pore-fluid overpressures.

The Foothills belt is a significant source of hydrocarbons. The major reservoir units are Devonian, Mississippian, and Triassic dolostones, all rocks belonging originally to the miogeoclinal succession. Hydrocarbon gases and oils filling the reservoirs were probably produced around the time of greatest burial, which broadly coincided with the time of maximum thrusting (Late Cretaceous–Paleocene).

THE CORDILLERAN COLLAGE AND THE FORELAND FOLD AND THRUST BELT

In Western Canada, a series of Mesozoic and early Tertiary deformational episodes have deformed, thickened, and transported eastward the westward-thickening miogeoclinal sediment wedge that previously formed the western margin of the North American continent (Monger et al., 1982). These events, probably the consequence of the accretion of a series of allochthonous terranes and accompanying subduction and transcurrent faulting, created the Canadian portion of the North American Cordillera. The Canadian Cordillera is divisible into five morphotectonic belts (Figure 1; Wheeler and Gabrielse, 1972; Monger et al., 1982; Monger, 1989). Two of these belts, the Omineca crystalline belt and the Coast plutonic complex, represent major regional tectonic welts that expose the metamorphic/plutonic cordilleran infrastructure and physically separate the other three belts (Monger et al., 1982). The Rocky Mountain, Intermontane, and Insular belts are, on the other hand, composed of unmetamorphosed or low-grade metamorphic rocks that represent remnants of the cordilleran stratigraphic record. The Canadian Cordilleran fold and thrust belt (equivalent to the Rocky Mountain belt of Wheeler and Gabrielse, 1972) can be further divided into the Foothills and the Rocky Mountains (e.g., Bally et al., 1966).

The Foothills in the southern part of the Canadian Cordillera are characterized by closely spaced, easterly verging imbricate thrust slices of the Mesozoic-Cenozoic clastic foreland basin succession. These thrust slices are bounded by listric thrust faults that merge at depth into relatively flat-lying thrusts carrying Paleozoic platformal carbonates in their hanging walls. In turn, these thrusts all root into a major basal detachment above the passive Precambrian crystalline basement, producing a structural style referred to as "thin-skinned." In the northern Foothills, near the Peace River, the Inner Foothills form a topographically high fold terrain with surface exposures of box-style, kink-folded Mesozoic sedimentary rocks inferred to be underlain by "blind" thrust sheets of Paleozoic and Mesozoic strata (Fitzgerald, 1968; Thompson, 1979, 1981; McMechan, 1985; McMechan and Thompson, 1989). These folds merge eastward into low-amplitude, long-wavelength folds in the Outer Foothills that die out gradually toward the Plains.

The Rocky Mountains in the southern part of the cordillera are subdivided into the Front Ranges, Main Ranges, and Western Ranges subprovinces (North and Henderson, 1954; Bally et al., 1966), which are distinguished by characteristic topography, stratigraphy, and structural style and are bounded by major en echelon thrust fault systems. The southern Rocky Mountains expose successively deeper parts of the miogeoclinal sedimentary wedge from east to west.

The Front Ranges subprovince consists of imbricate thrust sheets involving Paleozoic carbonates and, in the south, Proterozoic sediments of the Belt-Purcell Supergroup, separated from underlying Precambrian crystalline basement by a major detachment. In the southern Rocky Mountains, the Front Ranges are bounded on their eastern margin by two faults of major displacement, the Lewis thrust south of Crowsnest Pass and the McConnell-Livingstone thrust system between Crowsnest Pass and the Athabasca River.

In turn, the Main Ranges are typically, although not entirely, bounded on their eastern flank by major thrust faults. The Main Ranges expose lower Paleozoic strata that are tightly folded and faulted in the Western Ranges subprovince.

In the northern Rocky Mountains, the linear fault systems that bound subdivisions to the south gradually die out, structural styles change, the belt is considerably narrower, and tectonic shortening appears to decrease (Thompson, 1979; McMechan and Thompson, 1989). In contrast to the southern Rockies, no large thrust separates the Foothills and Rocky Mountains subprovinces. In the northern Rocky Mountains north of the Peace River (lat. 56°N), the boundary is represented by a transition zone on the eastern flanks of a series of large anticlinoria, presumably formed by the structural repetition of middle and lower Paleozoic carbonates beneath the Upper Devonian to Lower Mississippian Besa River shales (Thompson, 1979). Still farther north (about lat. 60°N), folding is the dominant structural style in the Rocky Mountains as well as in the Foothills.

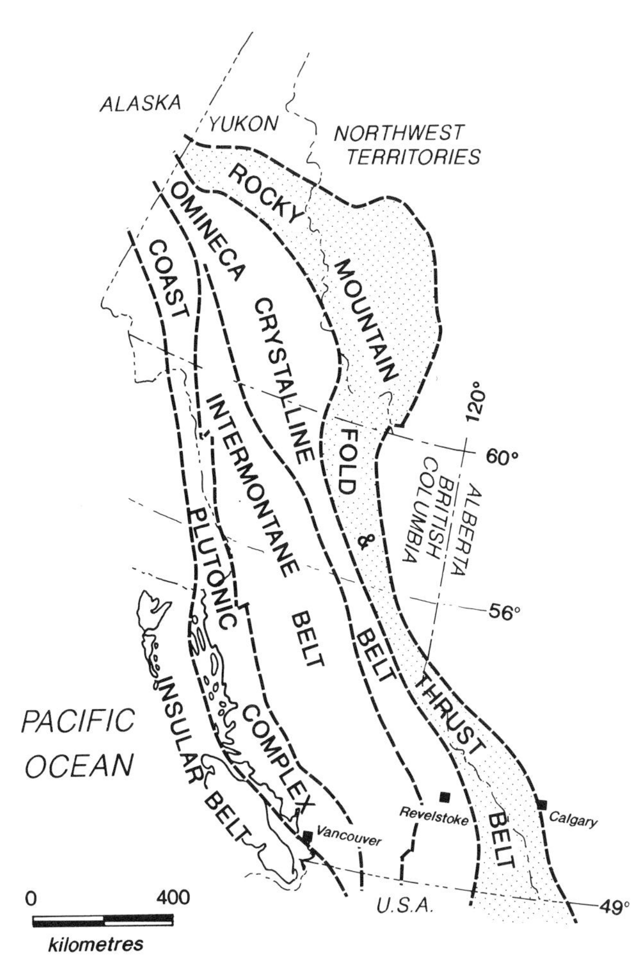

FIGURE 1. Morphotectonic belts of the Canadian Cordillera. The Rocky Mountain fold and thrust belt is stippled. Summary descriptions of belts are as follows. (1) Rocky Mountain fold and thrust belt: Northeasterly tapering wedge of mid-Proterozoic to Upper Jurassic (1500 to 150 Ma) miogeosynclinal and platform carbonates and craton-derived clastics, and overlying Upper Jurassic to Paleogene cordillera-derived clastics; horizontally compressed and displaced up to 200 km northeastward onto craton in Late Jurassic to Paleogene. (2) Omineca crystalline belt: Mid-Proterozoic to mid-Paleozoic miogeoclinal rock, Paleozoic and lower Mesozoic volcanogenic and pelitic rock, local Precambrian crystalline basement, highly deformed and variably metamorphosed up to high grades in mid-Mesozoic to early Tertiary time, and intruded by Jurassic and Cretaceous plutons. (3) Intermontane belt: Upper Paleozoic to mid-Mesozoic marine volcanic and sedimentary rock, mid-Mesozoic to upper Tertiary marine and nonmarine sedimentary and volcanic rock; granitic intrusions comagmatic with the volcanics; deformed at various times from early Mesozoic to Neogene. (4) Coast plutonic complex: Sedimentary and volcanic strata of known late Paleozoic to Tertiary age and probable early Paleozoic and Precambrian age, variably metamorphosed up to high grades, and dominant, mainly Cretaceous and Tertiary granitic rock. (5) Insular belt: Upper Cambrian to Neogene volcanic and sedimentary strata, granitic rocks in part comagmatic with the volcanics; deformed at various times from Paleozoic to Neogene. Adapted from Monger et al. (1982) and Monger (1984).

Cordilleran Deformational History

In this review we discuss the geometry, chronology, kinematics, and mechanics of structures forming the western deformed boundary of the Western Canada foreland basin, and the relationships of these to depositional events within the basin. However, any such discussion would be incomplete without first addressing the nature of tectonism across the entire cordillera, because deformation processes and events are dynamically linked over its full width. For more detailed analyses of the tectonic history of the Western Canadian Cordillera, the reader is referred to Monger and Price (1979), Price (1981), Monger et al. (1982), Monger (1984, 1989), and Brown et al. (1986).

As palinspastically restored, the sediments composing the Rocky Mountain belt form a wedge of miogeoclinal and exogeoclinal (foreland basin) strata ranging in age from Proterozoic through Tertiary. The wedge tapers eastward to a relatively thin succession of undeformed platformal and foreland basin sediments (Price, 1981). The stratigraphy in the Rocky Mountain belt records, in most general terms, three principal tectonic regimes. The oldest, which is characterized by a period of rifting, possible transtension, and subsidence from Proterozoic to Late Triassic, was followed by a period of convergence and transpression from Jurassic to Tertiary, which in turn was followed by an extensional event in Eocene to Oligocene time.

Pre-Foreland Basin Tectonic History, Cordilleran Orogene

Thickness and facies variations in the Proterozoic Belt-Purcell Supergroup exposed in the southernmost Canadian Rocky Mountains suggest accumulation in a basin formed by rifting of a preexisting Precambrian continental landmass (McMechan, M. E., 1981; Price, 1981). A similar tectonic setting is suggested for the late Precambrian Windermere and Miette supergroups exposed in the central portion of the Canadian Rockies. Based on the analysis of tectonic subsidence curves, Bond and Kominz (1984) suggested that the latest phase of continental break-up and onset of seafloor spreading off the west coast of what is now North America did not occur until latest Precambrian and earliest Cambrian. These authors further suggested that the older Precambrian rifting event could represent either (1) an early phase of a prolonged rifting sequence or (2) the major rifting sequence. If (2) is the case, the late Precambrian to Early Cambrian event merely modified the existent proto-Pacific margin. In various parts of the cordillera of the Yukon and British Columbia, the presence of Devonian alkalic and tholeiitic rift-related volcanics, and synsedimentary faults that control sedimentary facies and thicknesses, suggests periods of extension and/or strike-slip faulting during the Late Devonian. This faulting produced block uplift and subsequent subsidence of outer miogeoclinal sediments in the late Paleozoic (Abbot, 1987; Gordey, 1987; Gordey et al., 1987; Struik, 1987). It is possible that extensional structures in the northern Cordillera are coeval with compressional structures in southeastern British Columbia that are thought to be related to the Antler orogeny (Gehrels and Smith, 1987). Minor northeast-southwest and southeast-northwest extensional offsets that appear to have influenced sedimentary facies and thickness variations in the otherwise undeformed Plains region south of Calgary (Kanasewich et al., 1969) and north of 55°N latitude (Sikabonyi and Rodgers, 1959; Gosselin, 1987; Viau, 1987; Cant, 1988) may be a result of Paleozoic extension along the Western North American cratonic margin. The multiphase late Paleozoic uplift-subsidence history suggested by thickness variations and contact relationships in the Peace River arch area (Figure 2) of northern Alberta and northeastern British Columbia (Sikabonyi and Rodgers, 1959; Lavoie, 1958; deMille, 1958) also support the existence of several stages of extension, subsidence, and graben-edge uplift in the British Columbia Cordillera at this time (Struik, 1987).

West of the Rocky Mountain belt, rocks exposed in the various morphotectonic belts outlined above comprise a series of tectonostratigraphic terranes also known as allochthonous terranes (Figure 3). Each terrane is characterized by a distinctive tectonostratigraphic record that differs from the coeval records in adjacent terranes. Each is separated from adjacent terranes by profound structural discontinuities (Monger, 1977; Monger and Price, 1979; Monger et al., 1982). Reasoning based on comparisons of paleomagnetic (Irving and Yole, 1972; Van der Voo et al., 1980; Irving et al., 1980), paleontologic (Monger and Ross, 1971; Tipper, 1981), and structural information (Tempelman-Kluit, 1979; Gabrielse, 1985) suggests that these terranes are allochthonous to North American miogeoclinal rocks. Sedimentologic information and dating of crosscutting plutons indicates that the various terranes were amalgamated into two composite terranes prior to accretion to the North American cratonic margin (e.g., Monger et al., 1982). The Stikine, Cache Creek, Quesnel, and Slide Mountain terranes were linked into a composite terrane (Terrane I or Intermontane Superterrane), possibly as early as the Late Triassic in the northern cordillera but not until the Middle Jurassic in the southern cordillera, prior to accretion to the cratonic margin (Figure 4) (Cordey et al., 1987). These terranes may have originated as portions of a volcanic arc and associated back-arc basins that formed adjacent to a continental rise and slope sequence along the western margin of North America; in this sense, these terranes may all be native to North America (e.g., Wernicke and Klepacki, 1988). The Wrangell and Alexander terranes were amalgamated into a second composite terrane (Terrane II, or Insular Superterrane; Figure 3) probably by the Middle Pennsylvanian (approximately 309 Ma), based on U-Pb Zircon and ^{40}K-^{40}Ar hornblende data (Gardner et al., 1988), and almost certainly by the Oxfordian, prior to accretion to the new continental margin in the Cretaceous (Monger et al., 1982; Monger, 1984, 1989; Armstrong, 1988).

By as early as the Late Triassic, rocks of both oceanic and island-arc affinities were being underthrust along an east-dipping subduction zone beneath the ancestral western margin of the Quesnel terrane, forming the Cache Creek Group accretionary complex (Monger, 1984; Mortimer, 1986; Gordey et al., 1987), but the ancient cratonic margin of Western Canada was not deformed and metamorphosed until the initial convergence of the amalgamated allochthonous terranes during the Jurassic and Cretaceous. In the Yukon, obduction (overthrusting) of displaced terranes comprising ophiolitic cataclasites, metasedimentary rocks, and arc-related granites occurred prior to the mid-Cretaceous, possibly in the Late Jurassic or Early Cretaceous (Tempelman-Kluit, 1979). In northern British Columbia, Stikine terrane rocks were underthrust beneath Cache Creek Group rocks along the Nahlin fault beginning in the Middle Jurassic. In southern British Columbia, easterly thrusting of an ancestral oceanic basin began in the Early Jurassic, resulting in obduction of Slide Mountain terrane rocks onto North American rocks (Struik, 1982; Archibald et al., 1983; Klepacki and Wheeler,

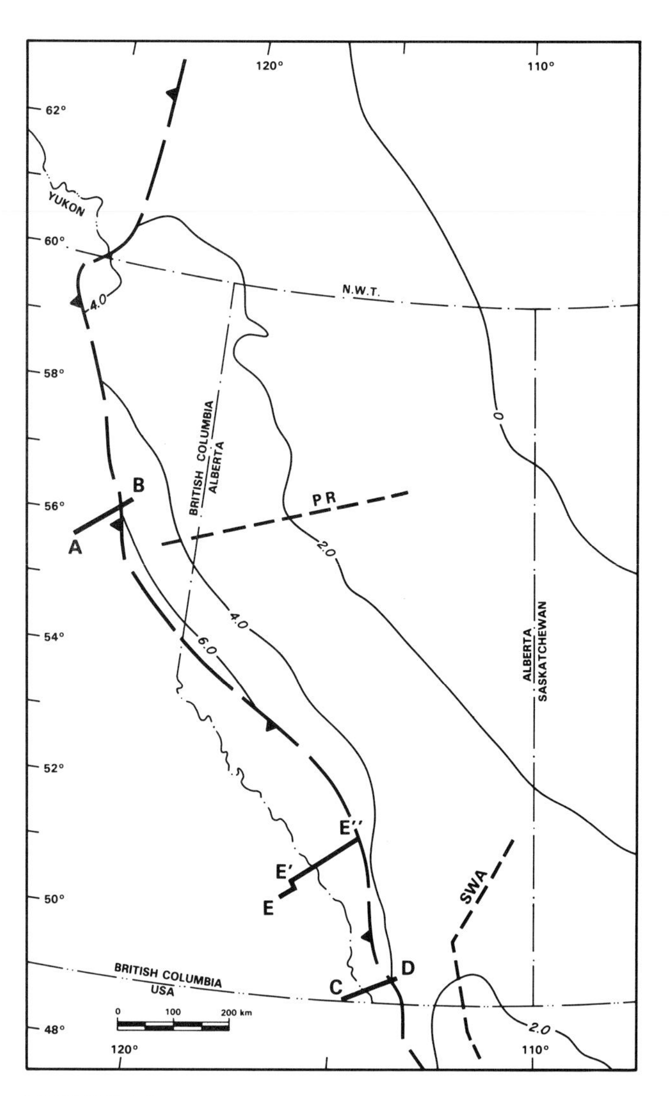

FIGURE 2. Location map. Contours represent total thickness of sedimentary strata in Western Canada basin, in kilometers. Heavy dashed lines represent locations of Peace River arch (PR) and Sweetgrass arch (SWA). Dashed line with triangles represents eastern boundary of Deformed belt. Heavy solid lines represent locations of regional structural cross sections. A-B is section in Figure 8, E-E′-E″ is section in Figure 7, and C-D is section in Figure 6. Map after Porter et al. (1981).

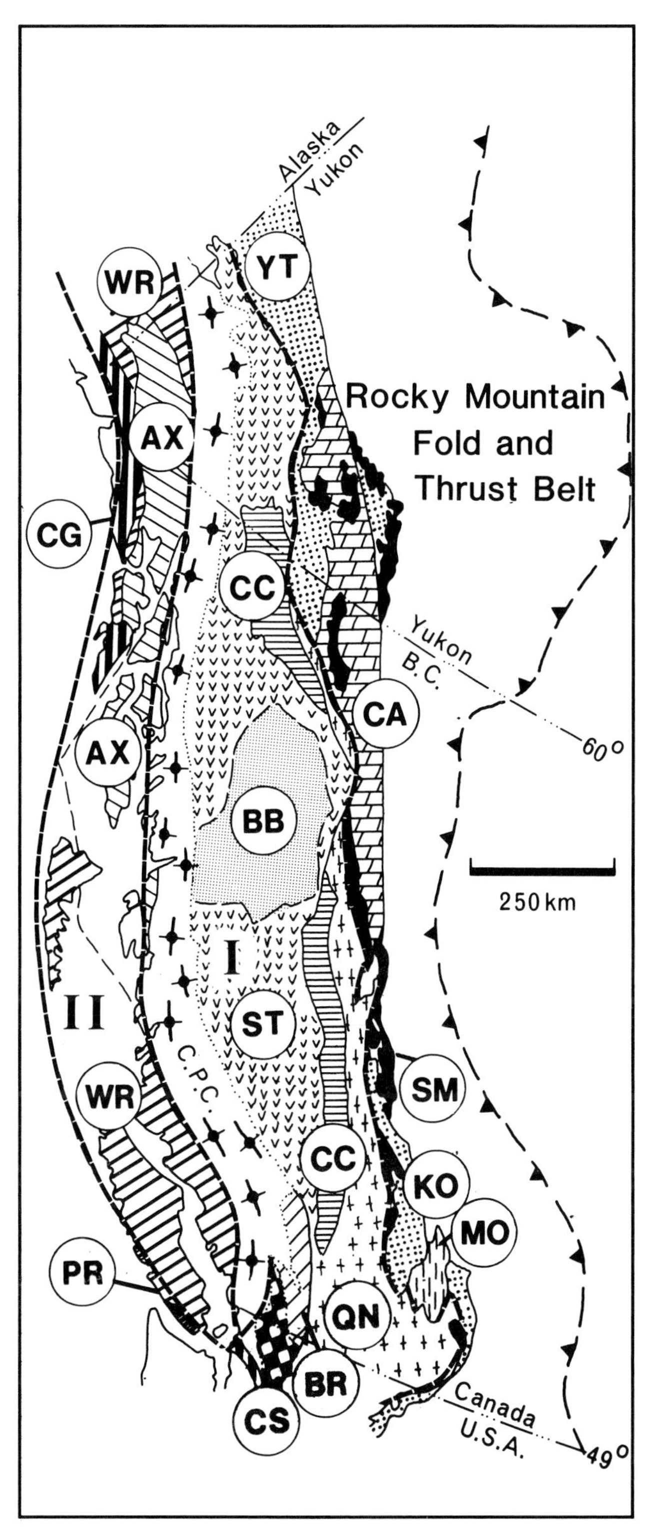

FIGURE 3. Map illustrating the major tectonostratigraphic terranes of the Canadian Cordillera. AX: Alexander terrane; BR: Bridge River terrane; CA: Cassiar terrane; CC: Cache Creek terrane; CG: Chugach terrane; CS: Cascade terrane; KO: Kootenay terrane; MO: Monashee terrane; PR: Pacific Rim terrane; QN: Quesnel terrane; SM: Slide Mountain terrane; ST: Stikine terrane; WR: Wrangell terrane; YT: Yukon-Tanana terrane. The largest of the successor basins, the Bowser basin, is indicated by BB. Adapted from Monger (1984).

1985). The convergence of terranes in the southern cordillera is complex, with initial Jurassic convergence apparently resulting in east-verging folds followed by underthrusting or "tectonic wedging" of allochthonous terrane rocks beneath the cordilleran miogeocline (Price, 1981; Archibald et al., 1983; Price, 1986; Brown et al., 1986). Tectonic wedging here has resulted in extensive crustal thickening and regional metamorphism. Crustal thickening in this tectonic "hinterland" apparently produced the eastward-dipping topographic slope that controlled sediment transport directions along the western margin of the foreland basin. It has been proposed that formation of thrust slices involving basement rocks transmitted displacements toward the Rocky Mountain foreland (Read and Brown, 1981; Brown and Read, 1983; Brown et al., 1986), resulting in initiation of the foreland fold and thrust belt in the Late Jurassic.

Lateral Displacements and Their Influences in the Eastern Cordillera

The occurrence of major dextral displacement along a series of north- to northwest-trending faults in the Tintina trench-northern Rocky Mountain trench (TT-NRMT) (Figure 5) in the mid-Cretaceous to Eocene resulted in a cumulative displacement of approximately 450 to 500 km (Tempelman-Kluit, 1979; Gabrielse, 1985; Butler et al., 1988). However, it is difficult to reconcile this 500 km of post mid-Cretaceous strike slip with the geology of southern British Columbia. Price (1979) and Price and Carmichael (1986) have shown that the TT-NRMT fault follows a small circle north of 54°N latitude but bends toward the craton south of 54°N to form a restraining bend across which there must have been convergence (Figure 5). They suggested that most of the 450 to 500 km of net right-hand slip on the TT-NRMT fault must have been transformed into Late Cretaceous and Paleocene oblique compression south of 54°N latitude, but that some slip was transferred during the early and middle Eocene to the en echelon Fraser River–Straight Creek fault system, which also follows a small-circle path about the same pole as the TT-NRMT fault. This theory provides an explanation for the recognized southward increase in Late Cretaceous to early Tertiary horizontal compression across the Rocky Mountain fold and thrust belt, resulting in distinctly different deformational geometries in the foreland fold and thrust belt along its strike. It also accounts for the early and middle Eocene east-west crustal extension in southern British Columbia, where right-hand en echelon offset between the two interacting fault systems may have produced the low-angle normal faults along the flanks of the uplifted portions of the cordilleran metamorphic core complexes (Tempelman-Kluit and Parkinson, 1986; Brown and Journeay, 1987). However, these mechanisms can account for only a portion of the postulated strike-slip displacement of up to 500 km of Late Cretaceous–Paleocene age in the northern cordillera.

South of about 55°30′N, the McLeod Lake fault zone, indicated as having strike-slip offset (Struik, 1989) as well as extensional movement, more closely follows the small circle through the TT-NRMT fault system. The McLeod

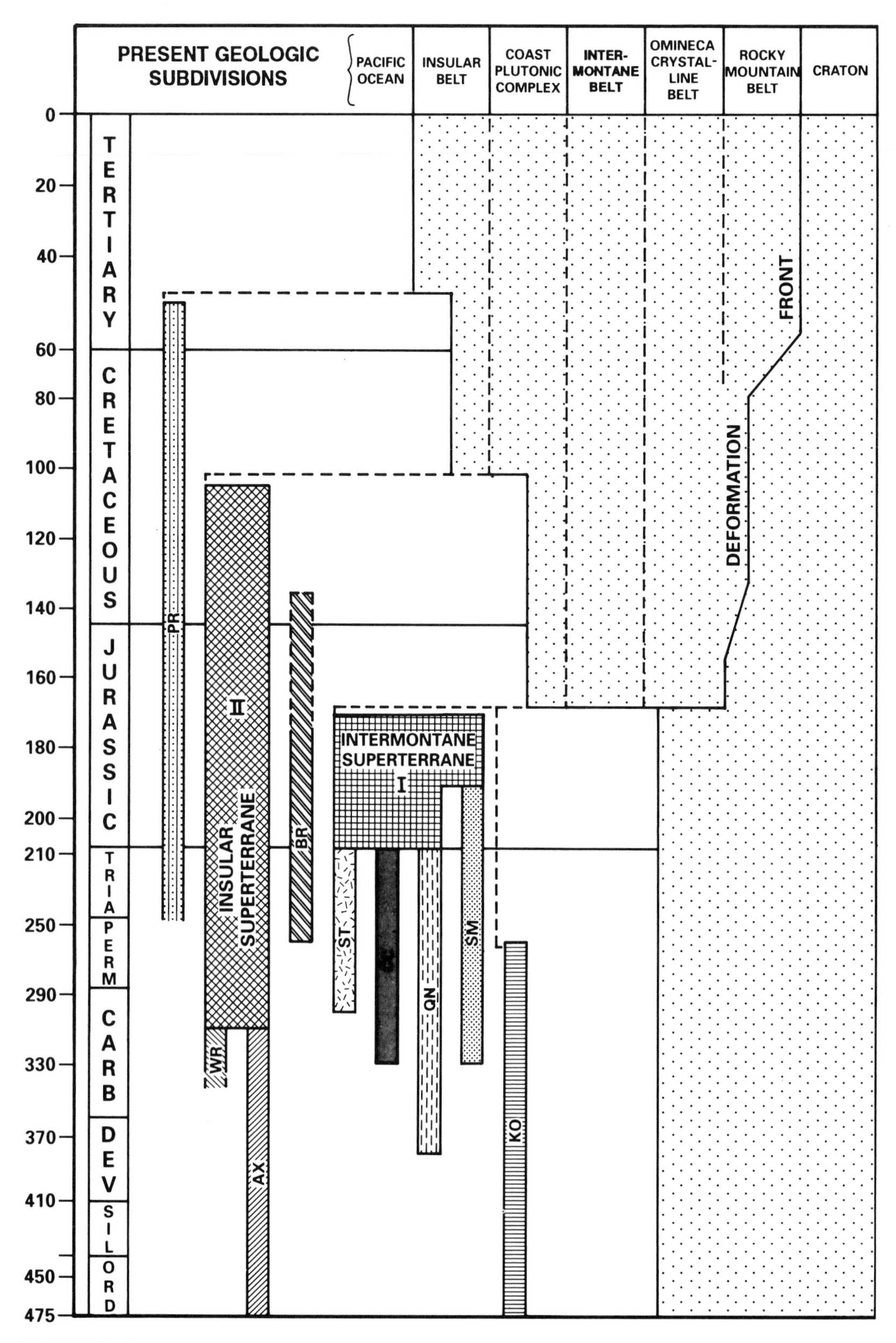

FIGURE 4. (Left) Time-space diagram of allochthonous terranes in the Canadian Cordillera. Diagram in part after Monger et al. (1982). Shown are depositional intervals spanned by each terrane, and overlap assemblages in the Insular and Intermontane superterranes, indicating coalescence prior to accretion to western North America. Unknown spatial relationships during deposition of allochthonous terranes are indicated by vertical spaces separating them. Times of accretion to western North America are indicated by horizontal dashed lines. Approximate boundaries of present morphotectonic belts are indicated by vertical dotted lines. Stippled area

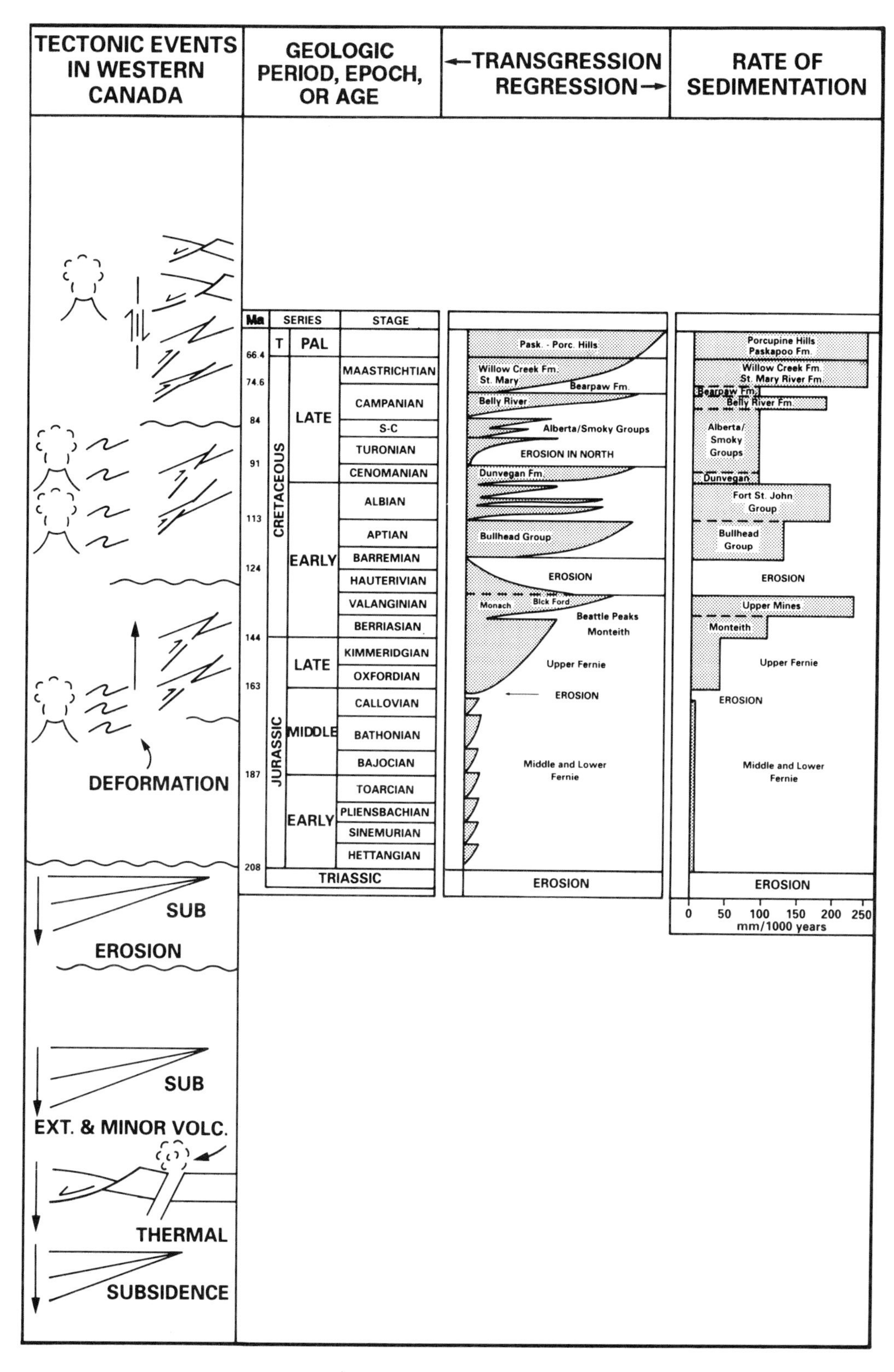

represents rocks belonging to the North American plate. See Figure 3 for identification of abbreviations. Boundaries between terranes may be thrust faults, strike-slip faults, or other complex relationships. (Right) Tectonic events in Western Canada, and depositional events in the foreland basin. Depositional events in the foreland basin are derived from Stott (1984), and the diagram tends to represent events in the more northern portions of the basin. The DNAG time scale is used.

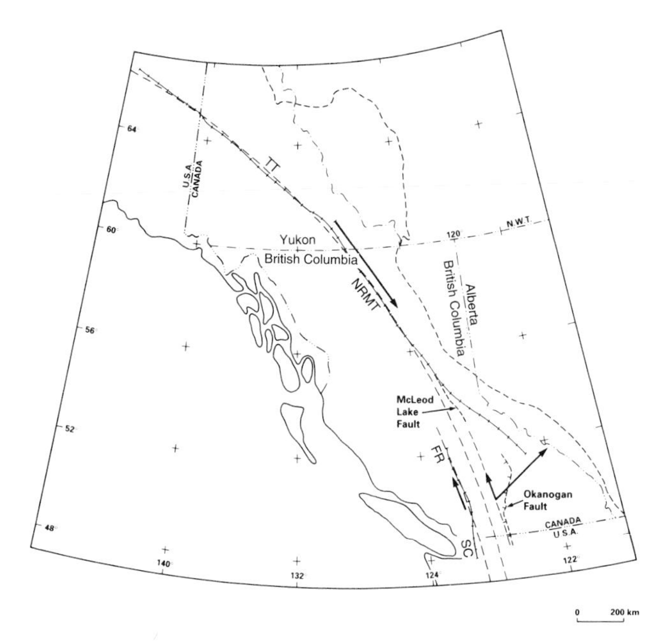

FIGURE 5. A stereographic projection of the Canadian Cordillera centered on the axis of the best-fitting, small-circle curve for the Tintina trench (TT)-northern Rocky Mountain trench (NRMT) fault zone and Fraser River (FR)-Straight Creek (SC) fault zone (after Price and Carmichael, 1986). Extensions of small circles through the FR-SC and TT-NRMT fault zones are shown by dashed lines, as well as the extension of a similar small circle through the end of the McLeod Lake fault zone. Heavy arrows show right-hand offset on the TT-NRMT fault zone (450 km) and FR-SC fault zone (110 km). The diverging pair of heavy arrows shows the magnitude and direction of the likely maximum amount of compression across the southeastern portion of the Canadian Cordillera (300 km); and the magnitude of the component of this compression, which lies on the extension of the small circle extending from the NRMT via the McLeod Lake fault. Offset on the TT-NRMT and FR-SC fault zones is post mid-Cretaceous. Compression across the southeastern Cordillera may be as early as Jurassic and therefore the 300-km figure represents a maximum for post mid-Cretaceous compression.

Lake fault zone may carry the principal strike-slip component of displacement south of about 55°30′N. Struik (1989) has suggested that it is linked with the Fraser fault system by an en echelon system of strike-slip and extensional faults.

Crustal Shortening

As shown in Figure 5, cumulative shortening within the thrust belt and Omineca belt at about 51°N is estimated to be, at most, between about 200 and 300 km (Price and Mountjoy, 1970; Price and Fermor, 1985)—the maximum shortening recognized along the length of the belt. The net displacement direction is probably approximately perpendicular to the strike of the thrust belt, N50° to N55°E, because of the absence within the thrust belt of features interpreted as being indicative of transpression. This is consistent with the dominant displacement direction derived from slip indicators in coal seams (Norris, 1971a) and the matching of fault cutoffs in the Alberta Foothills (P. Fermor, unpublished data). At 50 to 51°N, the small circle fitting the TT-NRMT has a NNE-SSW orientation, and the component of shortening along this direction would be up to about 120 km (Figure 5), depending on the magnitude of shortening. However, a significant although ill-defined portion of this displacement is pre-Late Cretaceous and probably predates most displacement on the TT-NRMT fault zone. If the 80- to 110-km offset on the en echelon FR-SC (Fraser River-Straight Creek) fault is added, the sum of these displacements is 200 to 230 km, leaving at least 220 km of strike-slip offset unaccounted for. Where this displacement is carried is unknown.

Gabrielse (1985) concluded that, if no major strike-slip displacement occurred along the southern Rocky Mountain trench, major dislocations must have occurred within, or flanking, the metamorphic complexes of the Omineca crystalline belt. An extension of the small circle from the McLeod Lake fault zone aligns approximately with the Okanogan Valley fault (Figure 5), a major throughgoing low-angle extensional fault that forms the western boundary of the metamorphic complexes in southern British Columbia (Parrish et al., 1988). The alignment may be purely coincidental, however.

TIMING OF DEFORMATION

Timing of deformation in an orogenic terrane may be assessed through two principal approaches:

1. Dating, by various radiometric methods, of the ages of cooling through various temperatures of plutonic and metamorphic rocks within the "core zone" of the orogenic terrane.

2. Paleontological dating of clastic sediments believed to have been derived from the rising orogenic terrane.

The timing must be derived from known or inferred relationships of the dated material to structural features (e.g., minerals forming a cleavage or a pluton cutting a fault), subsidence of sedimentary basins, or uplift of a source terrane. Stratigraphic overlaps on orogenic structures and the age of deformed or structurally overlapped strata may also provide important constraints, as may fission track data. It is the nature of these indicators that their relationships to the time or period of deformation is often indirect, and that they normally only place limits or brackets on the timing.

With these limitations in mind, the following is a brief summary of the timing of events known or inferred to be related to deformation of the eastern portion of the Canadian Cordillera and adjacent portions of the Western Canada foreland basin, and a synopsis of the relationships of these events to the sedimentary sequences deposited along the western border of the basin during the Mesozoic and early Cenozoic.

Jurassic and Early Cretaceous

Along the western margin of the foreland basin, Jurassic strata exposed in the Rocky Mountains and Foothills and preserved below adjacent portions of the Plains unconformably (or disconformably) overlie Triassic and older strata. These Triassic and older rocks consist either of platformal deposits or of part of a miogeoclinal wedge presumably formed along an Atlantic-type passive continental margin. Strata of Early and Middle Jurassic age consist of a relatively thin, condensed marine sequence of black shale, limestone, and thin sandstones principally derived from the east with some possibly westerly derived sandstone present in Middle Jurassic strata (Stott, 1984; Poulton, 1984; Hall, 1984; Stronach, 1984).

The first clear stratigraphic record of a western source, and by inference deformation, uplift, and erosion in what is now the interior of the Canadian Cordillera, occurs in Late Jurassic sediments (Fernie Formation, Minnes and Kootenay groups; Cycle 1 of Leckie and Smith, 1992). Late Jurassic (broadly, during the Oxfordian and Kimmeridgian, 163 to 152 Ma) marine shales of the Fernie Formation are interbedded with large quantities of siltstone and sandstone along their western preserved edge in the Rocky Mountains (see Figure 4). Sandstone begins to dominate in the latest Jurassic, beginning in the south in the upper part of the Fernie Formation, up through the overlying nonmarine, coal-bearing Kootenay Group (spanning much of Kimmeridgian through Tithonian time, 156 to 144 Ma), and, in the north, through the nearshore marine Monteith Formation (Tithonian, 152 to 144 Ma). In northeastern British Columbia, the Monteith grades westward to nonmarine facies.

The 1.2 km of preserved Kootenay Group sediments in southeastern British Columbia and over 2.1 km of Monteith Formation sediments in northeastern British Columbia, much of them deposited nearly at or below sea level, indicate significant crustal depression. This appears to have resulted from loading by thrust sheets formed in what are now interior portions of the cordillera (Price, 1973; Beaumont, 1981). The Kootenay Group sediments are considered to have been derived from thrust sheets now forming the westernmost portions of the southern Rocky Mountains and/or from areas farther to the west such as the Purcell Mountains, and the Kootenay Arc and Selkirk Mountains in the southern Omineca belt (e.g., Norris, 1964, 1971b; Gibson, 1977, 1985).

In contrast, conglomerates in the nonmarine facies of the Monteith Formation near 54°N contain volcanic and foliated greenstone clasts (M. McMechan and D. F. Stott, in Poulton, 1984). Presumably these clasts were derived from the allochthonous terranes to the west, indicating that the drainage divide lay within the terranes at this latitude, and, conversely, the lack of such clasts in the Kootenay Group suggests that the divide may have been to the east of the terranes in the south.

The initial metamorphic and deformational episode in the metamorphic complexes (at the levels now exposed at the surface) occurred at the same time as, or somewhat earlier than, the onset of coarse clastic sedimentation in the Western Canada foreland basin during the Middle Jurassic. Archibald et al. (1983) studied plutonic and metamorphic rocks in the so-called Kootenay Arc of southeastern British Columbia. This is a zone of intense, penetrative deformation and, locally, amphibolite facies regional metamorphism that is part of a regional metamorphic culmination in the southern part of the Omineca belt. On the basis of the work of these authors (K-Ar and Rb-Sr mineral dates, U-Pb zircon dates, and ^{40}Ar-^{39}Ar step-heating experiments on micas) and previous studies by others, it appears that in neighboring, lower-grade areas, regional metamorphism had ceased by the Middle Jurassic (175 to 165 Ma, Callovian; see Figure 4). In the higher-grade area, Archibald et al. (1983) concluded that penetrative deformation and regional metamorphism occurred at or shortly before 166 Ma (mid-Callovian). The sedimentary rocks (of North American origin) that formed the protolith for the metamorphic rocks were carried rapidly to crustal depths of more than 20 km in the Middle Jurassic (180 to about 162 Ma, early Bajocian through the Callovian-Oxfordian boundary), while undergoing deformation and metamorphism. Granitic rocks were emplaced during and after deformation in the Middle to Late Jurassic (170 to 155 Ma, early Bathonian through the Oxfordian-Kimmeridgian boundary). The previously deformed and metamorphosed rocks were uplifted rapidly (165 to 150 Ma, mid-Callovian through Tithonian) as the allochthonous terrane Quesnellia (part of the composite Terrane I, the Intermontane Superterrane) was driven into and beneath them (Figure 4). The plutons cut both Paleozoic miogeoclinal rocks of North American origin and strata of the allochthonous terranes containing Toarcian and Bajocian ammonites (193 to 176 Ma), marking the closure of the basin between North America and Terrane I (Intermontane Superterrane) in the Middle Jurassic. The age of the tail of the Nelson batholith, which is deformed in the zone surrounding the fault that thrust miogeoclinal North American strata over Quesnellia, is 124 to 150 m.y.—i.e., Tithonian (latest Jurassic) through the Hauterivian-Barremian boundary (Early Cretaceous) (Archibald et al., 1983). The Kuskanax batholith, which intrudes the northern part of the Kootenay Arc and apparently is syntectonic, is dated at $173 \pm \sim 5$ Ma (Bathonian-Bajocian) by U-Pb dating of zircons (Parrish and Wheeler, 1983).

North of the Kootenay Arc in the high-grade metamorphic terrain of the Revelstoke region, Rb-Sr dating of intrusives postdating initial phases of compressional tectonics indicates that the primary phase of deformation (at the levels now exposed at the surface) had ended by the Late Jurassic (Duncan et al., 1979). Wanless and Reesor (1975) inferred Middle Jurassic high-grade metamorphism in the Thor-Odin dome area south of Revelstoke.

Still farther north, in another metamorphic culmination called the Wolverine complex at about 56°N in north-central British Columbia, Parrish (1979) found that the minimum age of the initial period of metamorphism and deformation is 154 to 166 m.y. (Callovian to Kimmeridgian—i.e., Middle to Late Jurassic) and that the metamorphosed rocks were subsequently uplifted over an extended period of time and probably had completed more than half their cooling by 80 Ma (Late Cretaceous) (see Figure 4 for time scale).

The time lag between deformation and the observed onset of coarse clastic sedimentation may have been the result of a delay between the initiation of collision and the development of the significant topographic relief required before coarse clastics could reach the westernmost part of the Western Canada foreland basin now preserved from erosion. However, some portions of the metamorphic culmination—particularly the more northerly portions such as the Wolverine complex—may have been displaced northward a significant distance on strike-slip faults whose activity postdates the time of initial metamorphism. Thus their present positions relative to the foreland basin may differ significantly from their positions during the Mesozoic.

The south-to-north change in Late Jurassic facies from nonmarine through nearshore marine to offshore marine, and paleocurrent directions, suggest dominant input of sediment into the southern end of a south-to-north-sloping foreland basin trough (Gibson, 1977, 1985; Hamblin and Walker, 1979). This implies that deformation and uplift in the metamorphic complexes began slightly earlier in the south, possibly as a result of slightly earlier collision of the allochthonous terranes in this area. Radiometric dating studies of intrusives just to the west of the southernmost portion of the Canadian Rocky Mountains, in the Kootenay Arc, suggest that compression and uplift continued in the metamorphic culmination into the Early Cretaceous (Archibald et al., 1983) with uplift of as much as 8 to 10 km.

Thrust faults in the western portion of the southern Canadian Rocky Mountains appear to be linked to northeast-trending, right-hand reverse faults cut by Early to mid-Cretaceous intrusives (122 and 94 Ma), suggesting a mid-Cretaceous upper limit for the time of last displacement on these thrust faults (Hoy and van der Heyden, 1988).

The development of the southernmost Alberta portion of the foreland basin was affected by the Sweetgrass arch, which extends from the Little Belt Mountains of west-central Montana northward into south-central Alberta (SWA in Figure 2). The arch began its major development as a positive feature in the Middle Jurassic, and strong uplift followed by peneplanation occurred in the Late Jurassic and Early Cretaceous. To the south, in Montana, crystalline basement rocks were probably exposed along the crest of the arch in Oxfordian time (Shepard and Bartow, 1986). Succeeding Cretaceous strata appear rather uniform in thickness and facies over the arch (Shepard and Bartow, 1986), indicating quiescence through the remainder of the Cretaceous and possibly into the Paleocene.

Early Cretaceous

In late Valanginian to Barremian time (130 to 120 Ma), uplift in the metamorphic belt appears to have slowed considerably (e.g., Archibald et al., 1983), there was little or no igneous activity (R. L. Armstrong, in Eisbacher, 1985; Armstrong, 1988), and there was considerable erosion of the older foredeep sediments, producing a widespread unconformity. This unconformity may be absent only in the westernmost exposures of the succession (Gibson, 1977, 1985; Stott, 1984). The unconformity appears to represent significant isostatic adjustment of the craton (Stott, 1984), possibly related to a decrease in the rate of thrust loading.

Renewed deposition of Cycle 2 (Leckie and Smith, 1992) commenced with a very widespread, but usually thin, chert/pebble-rich conglomeratic unit, the Cadomin Formation, which mantled the erosion surface. A wide variety of lithologies occurs in the pebbles, including rare volcanic pebbles. However, Schultheis and Mountjoy (1978) found that the majority of pebbles could have been derived from rocks of late Paleozoic age, indicating a source area that could have been as far west as the Selkirk Mountains (in the Omineca belt) and as far east as the strata now exposed in the Main Ranges. Presumably the distance to and the nature of the most westerly source varied considerably along the length of the mountains (McLean, 1977). Above the Cadomin Formation, the sand-rich deposits of the second clastic wedge (Stott, 1984) (or Cycle II of Leckie and Smith, 1992) continue; their ages range from Barremian through Cenomanian (120 to 95 Ma). These deposits coincide with renewed intrusion of plutons into a dormant superstructure in the metamorphic culminations (Figure 4), accompanying renewed heating, deformation, and metamorphism at deeper levels, during the approximate interval from 115 to 90 Ma (Aptian, Albian, and Cenomanian) (Archibald et al., 1984). The volcanic carapace to these intrusives may have yielded the feldspathic detritus contained in the second clastic wedge. Some K-Ar dates from low-grade metamorphic rocks in the thrust belt also fall into this time interval and may indicate uplift (Eisbacher, 1985). Late Early Cretaceous strata occur in a fault zone, surrounded by Proterozoic and lower Paleozoic strata, in the Western Ranges of the Rocky Mountains, at about 50°N latitude (Leech, 1966). Whether in place or out of place relative to surrounding strata, these outcrops indicate the deposition of Early Cretaceous sedimentary strata far to the west of the presently preserved western edge of the foreland basin.

An unconformity that developed above the second clastic wedge in the Foothills may represent another reduction in tectonic activity and thrust loading, lasting as much as 10 million years during the early Late Cretaceous (95 to 85 Ma). In the central Foothills (the portion astride the Alberta–British Columbia boundary and the Peace River arch; Figure 2), the succession may be continuous. This could indicate uninterrupted tectonic activity or alternatively may be explained as a reflection of the collapse of the arch during the Mesozoic (Cant, 1988).

Late Cretaceous and Early Tertiary

The third clastic wedge commenced with a marine flooding phase that persisted through the early Campanian (approximately 80 Ma) and was dominated by shale deposition. A general westward coarsening of the sedimentary succession (Stott, 1963) and continued subsidence suggest that erosion and tectonic activity persisted to the west during this period. Tectonic activity increased in intensity toward the mid-Campanian (approximately 79 Ma) as alluvial, coarse-grained sandstone and shale were deposited in the foredeep. This major Late Cretaceous to Paleocene episode of crustal shortening, thrust faulting, uplift, and accompanying sedimentation produced the Foothills, the Front Ranges, and parts of, and possibly nearly all of, the Main Ranges. It was accompanied by a distinct lull in magmatism. The cause is uncertain but has been attributed to an oblique collision of the second composite terrane, Terrane II, forming much of what is now coastal British Columbia and the Alaska Panhandle, with the previously accreted Terrane I (Monger et al., 1982) (see Figure 4).

The coarse clastic sequence or third clastic wedge is best preserved in the southern part of the foreland basin in Alberta. Where it is not preserved in some portions of the Foothills in northeastern British Columbia, its original thickness and extent cannot be determined. Deposition of this third clastic wedge persisted in Alberta to at least the Paleocene, which is the age of the youngest preserved deposits. The Bearpaw Formation marine shale tongue, of upper Campanian age (approximately 76 Ma), is sometimes taken as representing the divide between the third and fourth clastic wedges, although coarse clastic sedimentation was continuous through this time. The entire succession will be referred to here as the third clastic wedge.

Major thrust faults in the Alberta Front Ranges, such as the Lewis and McConnell thrusts, overthrusted and crosscut Cenomanian to Santonian (97 to 84 Ma) marine strata and early Campanian (84 to 80 Ma) nonmarine deposits of the Western Canada foreland basin, thus establishing a lower limit for the time of displacement on these faults. The youngest deposits preserved in the foredeep—of early (to middle?) Paleocene age (about 62 Ma)—are deformed by folds produced by subsurface thrusting at the eastern edge of the Foothills, and thus the termination of thrusting must postdate the mid-Paleocene (Bally et al., 1966; Jones, 1971). The syntectonically deposited Kishenehn Formation dates motion on the Flathead normal fault, which merges with the Lewis thrust, as earliest to late Oligocene (McMechan, R. D., 1981). Therefore we can conclude that thrusting in the southernmost Canadian Rockies had terminated by this time.

Although evidence of large-scale crustal shortening during this time is best developed along the southern margin of the basin, very large-scale right-lateral strike-slip faulting, such as along the Tintina trench-northern Rocky Mountain trench (TT-NRMT) fault zone, developed in the Northern Rockies from about Late Cretaceous onward. Cumulative displacements on this and other fault systems farther west were possibly up to about 900 km (Gabrielse, 1985), although half of this displacement could have occurred as early as the Late Devonian. Paleocene and/or Eocene nonmarine clastics occur along many of the strike-slip fault zones, and Eocene volcanic rocks are present along some of the faults, suggesting an early Tertiary age for this strike-slip displacement, which transported rocks west of the faults northward with respect to the North American craton.

During the latest Paleocene and/or earliest Eocene (at about 58 Ma), a large portion of the Canadian Cordillera underwent a major change in tectonic regime from transpressional deformation to extension along faults oriented north to northeast. Crosscutting relationships among intrusives, thrust zones, and normal fault zones in the metamorphic core complexes in the southern Omineca belt date the change at about 58 to 59 Ma ± 3 Ma (Carr, 1989). Crustal scale extension persisted through the mid-Eocene (45 Ma; Parrish et al., 1988), producing downdropped blocks of the suprastructure and intervening uplifted domes that expose portions of the former infrastructure within the interior portions of the orogenic belt.

The change from transpression to transtension in the southern Canadian Cordillera in latest Paleocene or earliest Eocene could be a result of a small change in plate motions and/or the late initiation of the Fraser River-Strait Creek fault zone (Price and Carmichael, 1986). Alternatively, gravitationally driven crustal extension of the uplifting core complex, a consequence of stacking of crystalline basement rocks (Coney and Harms, 1984), may be the cause. Tertiary normal faults that occur in the southern Rocky Mountain trench and along preexisting thrusts (e.g., Flathead fault) in the southeastern Canadian Rocky Mountains (Bally et al., 1966; Dahlstrom, 1970) may be a later expression of this widespread extensional event, but probably are unrelated.

Tilting of Eocene (54 to 50 Ma) plutons on the eastern flanks of the Sweetgrass arch (Figure 2), in southernmost Alberta and northern Montana, indicates a probable post-Eocene uplift of the arch. An Oligocene-Miocene age is favored for this event by Shepard and Bartow (1986), in accordance with the age of uplift of the Little Belt Mountains, coinciding approximately with the age of the extension faulting in the Front Ranges of the Rocky Mountains along the international boundary.

STYLES OF DEFORMATION

Owing to the series of Mesozoic and early Tertiary deformational episodes, strata forming much of the miogeocline, part of the cover of the cratonic platform, and the western margin of the foreland basin were horizontally compressed and tectonically thickened. In the southern Canadian part of the fold and thrust belt, these strata were detached from underlying rocks and displaced northeastward by easterly directed shallow thrust faulting and detachment folding. The rocks forming the craton remained undeformed, except for a gentle westward tilt imposed as a consequence of the additional loading of the lithosphere imposed by the stacking of the thrust sheets (e.g., Bally et al., 1966; Dahlstrom, 1970; Price, 1973, 1981). The resulting style of deformation is termed "thin-skinned."

Thin-skinned tectonics are characteristic of the fold and thrust belt from south of the Canada-U.S. border to about 56°N. Petroleum industry seismic data indicate that north of 56°N, tilted reflective sequences, possibly representing folded and faulted strata of late Precambrian and/or early Paleozoic age, lie beneath late Paleozoic and Mesozoic sedimentary strata of the western portion of the plains and at least the eastern portion of the Foothills. These rocks are, at least in certain areas, involved in deformation in what is termed a "thick-skinned" style, where no basal detachment is evident.

Structures in the Cordilleran foreland thrust and fold belt developed at temperatures insufficient to produce major recrystallization, hence this discussion involves structural styles developed in essentially nonmetamorphic rocks, with only minor development of pressure-solution cleavage.

Geometry

The Basal Detachment

The concept of a thin-skinned style derives from recognition of a *basal detachment surface* (or sole thrust). This is a throughgoing low-angle fault surface that underlies the length and breadth of the deformed belt, and is the surface along which the strata of the miogeocline, platform, and foreland basin have been moved northeastward relative to the underlying craton. Folding and thrusting within the deformed belt are the mechanisms by which the wedge of sediments above the basal detachment has been internally thickened and shortened. Faults within this deformed wedge ultimately merge at depth with the basal detachment. Displacement on each fault within this wedge represents but a small component of the principal northeasterly displacement carried on the basal detachment.

A "thin-skinned" style of deformation for the Rocky Mountain foreland fold and thrust belt was demonstrated by Bally et al. (1966). They recognized the presence of planar reflections from near-horizontal, undeformed Cambrian strata beneath the plains extending westward below tilted and discontinuous reflections produced by strata in the deformed belt. Drilling at the eastern margin of the deformed belt demonstrates that the undeformed Cambrian strata rest upon high-grade metamorphic rocks, which also must extend westward beneath the deformed belt. A throughgoing fault surface must separate the undeformed Cambrian strata and the craton beneath from the deformed strata above (Figures 6, 7, and 8).

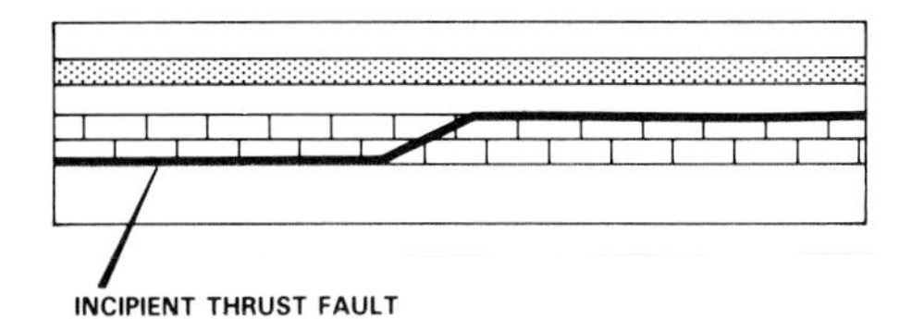

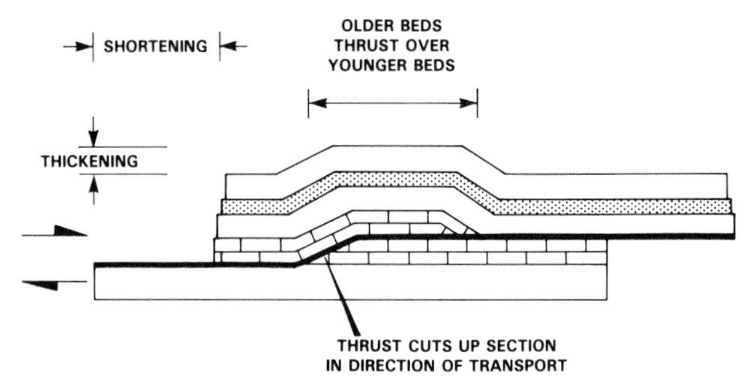

FIGURE 9. Model of "ideal" thrust fault. Upper panel shows trajectory of incipient thrust fault, parallel to layering at base and top of carbonate unit (brick pattern), and crosscutting carbonate at fairly steep angle along "ramp" linking the two-layer parallel portions of the fault. Lower panel illustrates geometric consequences of fault displacement over ramp. Offset terminations of layers in the hanging wall of the thrust indicate accommodation of folding by interstratal slip. Diagram designed by R. D. McMechan.

Thrust Fault Geometry

Thrust faults are defined as reverse faults that cut stratigraphic layering at a low angle relative to bedding and repeat and shorten the strata. The mass of rock above the thrust fault (in the hanging wall) is termed the "thrust sheet." In sedimentary strata, thrust faults step up through the stratigraphy in the direction of displacement. In cross section, thrust faults parallel layering for considerable distances, abruptly crosscut or ramp across layering at angles as steep as 30°, then once again parallel layering (Figure 9). This "stair-step" geometry was first described by Rich (1934) in the Appalachians and was applied to the Alberta Foothills by Douglas (1950). Cross sections and seismic data show that the thrust faults (or "thrusts") commonly parallel bedding in thinly bedded and less competent (weaker) units and crosscut bedding in more competent (stronger) units. However, it is not uncommon for thrust faults to parallel bedding for distances of many kilometers in more competent units such as carbonate rocks.

Because thrust faults cut up-section through stratigraphic layering in the direction of motion, they produce stratigraphic repetition (Figure 9), thickening the succession. The succession will retain its original thickness only if the thrust is parallel to layering in both hanging wall and footwall and if motion is insufficient to place older layers onto younger ones (upper part of Figure 9). Thrust faults do not cause bed omission, do not thin the stratigraphic succession, and do not, except in unusual circumstances, thrust younger beds over older ones. Major thrust faults are characterized by stratigraphic repetitions of several kilometers and lateral continuity of tens of kilometers. The largest known thrust in the Western Canada foreland belt, the Lewis thrust, extends over a length of greater than 400 km, has a maximum displacement of approximately 90 km, and produces a stratigraphic repetition of up to 10 km or more.

Even the largest thrusts eventually lose their identities along strike. They either die out or merge with other thrusts. Displacement carried on a thrust that dies out is transferred to conical folds, which commonly propagate laterally at the tip of a thrust (Elliott, 1977; Stockmal, 1979), and/or to en echelon thrust faults (Dahlstrom, 1970; Gardner and Spang, 1973). As a result of the transfer, displacement across the belt does not change significantly in the vicinity of a thrust tip.

Fold Geometry

Folds within the Western Canada foreland belt are both (1) concentric—i.e., folds around which the thickness of beds remains constant and that die out toward a center of curvature where the fold is cut off against a detachment horizon and/or faults—and (2) chevron-type or kink folds in which crumpling, flowage, and faulting in the co[illegible] the fold produce folds with relatively straight limbs an[illegible] narrow hinge zones, and in which the geometry can persist through a considerable stratigraphic thickness

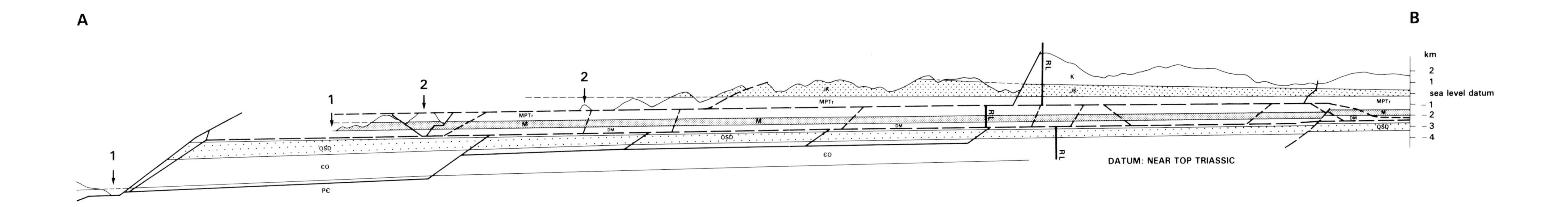

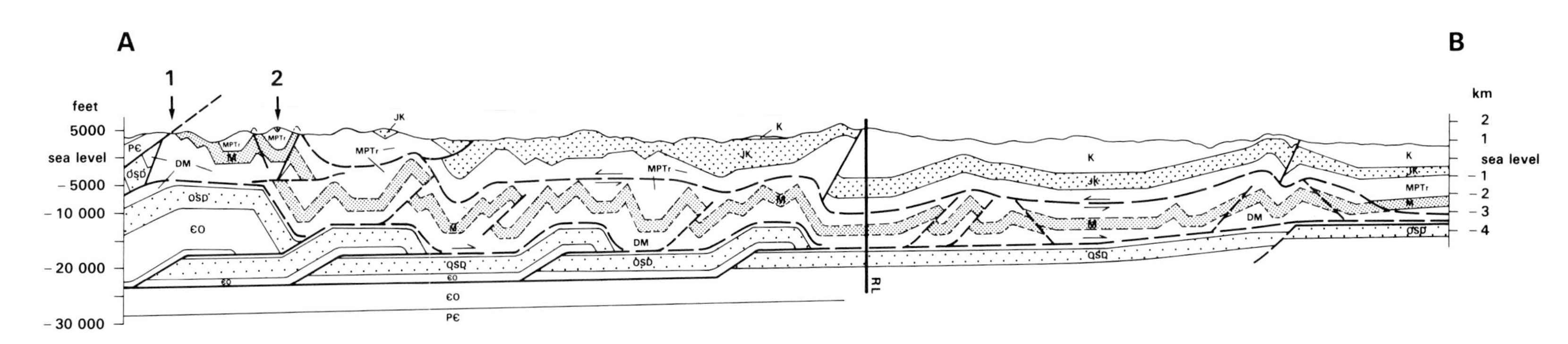

FIGURE 8. Schematic structural cross section and accompanying palinspastic restoration of the Foothills belt in the area south of the Peace River, reproduced with modifications from McMechan (1985). Reproduced with permission of the Canadian Society of Petroleum Geologists. Location of cross section (A-B) is shown in Figure 2. RL is a reference line joined in deformed section. Numbered arrows indicate the palinspastically restored positions of two points joined in the deformed section. Stratigraphic abbreviations are as follows. Pɛ: Proterozoic sediments; ɛO: Cambrian-Ordovician; OSD: Ordovician, Silurian, and Devonian (carbonates); DM: Devonian-Mississippian (shales); M: Mississippian (carbonates); MPTr: Mississippian, Permian, and Triassic (shales), JK: Jurassic and Cretaceous (sand and shale); K: Cretaceous (sand and shale). Recent industry seismic data indicate that the structure differs significantly in certain aspects from McMechan's concept. The section is a conceptual solution to the apparent problem of greater shortening being present at depth than is evident in the surface exposures.

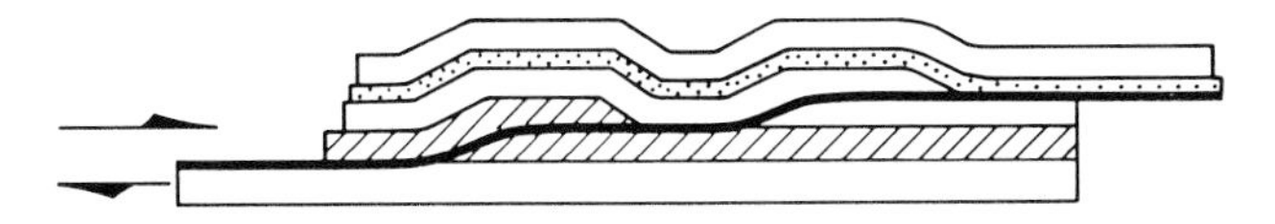

FIGURE 10. Model of thrust fault crosscutting bedding in stair-step geometry, producing anticlines and synclines in its hanging wall as a result of displacement over the ramps and flats forming the stair steps. Note that both anticlines and synclines terminate downward against the thrust fault surface. Folding is accommodated by interstratal slip, as shown by the offset terminations of the layers in the hanging wall of the thrust. Diagram designed by R. D. McMechan.

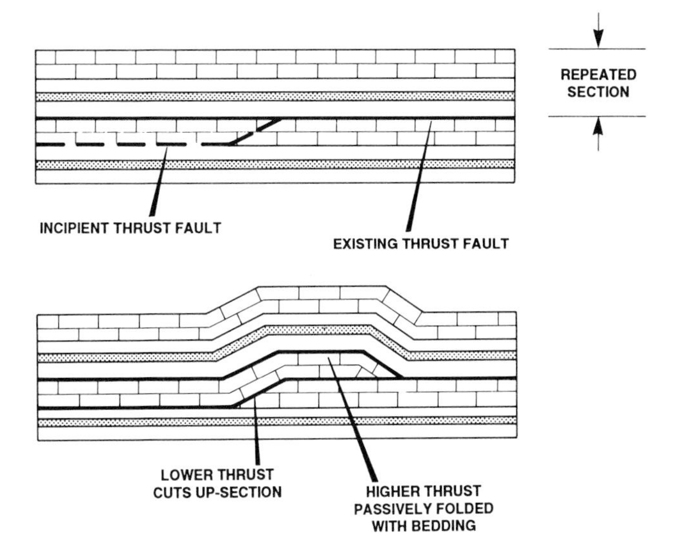

FIGURE 11. Diagram illustrating model for the development of a folded thrust fault. Upper panel illustrates stratigraphic section repeated on flat thrust fault, which for simplicity of illustration is shown as parallel to the bedding in both footwall and hanging wall. The path of a younger, incipient thrust fault is shown joining the footwall of the existing thrust fault. Lower panel illustrates configuration after displacement over ramp on younger, lower fault, which produces folding of both bedding and older thrust, which is passively folded along with surrounding bedding.

(Dahlstrom, 1970; Gardner and Spang, 1973; Elliott, 1977; Stockmal, 1979).

Folds are often intimately associated with thrust faults. Anticlines may have thrust faults extending upward into their cores. Occasionally, minor thrust faults are observed to root in the cores of synclines, the thrust fault originating along bedding surfaces across which displacement occurs during flexural-slip folding. Typically, folds terminate upward or downward against detachment surfaces that are thrust faults. Folding above a thrust fault is produced as a direct consequence of the stair-step geometry typical of thrust faults (Figure 10).

Although kink-style folds in the southern portion of the foreland fold and thrust belt are generally overturned in the direction of transport, those in the north tend to be box-style folds with east and west limbs both steeply dipping (Fitzgerald, 1968).

Folded thrusts, in which the fault is folded along with layering, are common in the foreland fold and thrust belt (e.g., Bally et al., 1966; Dahlstrom, 1970). Fold amplitudes may be as much as several kilometers. Folded thrusts always cut upward through the layering in the direction of displacement, irrespective of their present geometry. They are formed by motion on a lower thrust fault located in the footwall of an existing fault (Figure 11). The upper thrust will then be passively folded along with bedding as motion on the younger, lower fault produces a fold. Therefore, the present attitude of a fault is not necessarily the same as or even similar to its original attitude, nor a key to the nature of the fault.

The explanation for folded thrusts has important implications regarding the development of the thrust belt as a whole. It requires that the structurally lower (and therefore commonly more easterly; Figures 6 and 7) faults are the younger, and it implies that in general the thrust and fold belt developed progressively from west to east.

Where thrust faults in the foreland belt converge, they rarely crosscut, but almost always merge. This is not the case in more interior parts of the cordillera, where there were several episodes of deformation. Where a flat-lying thrust is present, underlying thrust faults cutting up-section will merge with it. These faults split off ("splay") from some lower thrust (the floor thrust), steepen as they climb upward through the intervening layering, then turn parallel with layering, flattening and merging upward with the upper thrust (the roof thrust). Displacement on these sigmoidally shaped link thrusts commonly produces sigmoidally shaped bedding within the fault slices (Figure 12). The structure is termed a "duplex fault structure" or "duplex" (Dahlstrom, 1970; Elliott and Johnson, 1980).

Where each link thrust splits off, the floor thrust displacement is reduced by the link thrust displacement, and where a link thrust merges with the roof thrust, the roof thrust displacement is similarly increased.

Several superposed duplex structures are illustrated in Figure 6 in the central portion of the cross section, in the vicinity of the Clarke Range. The lowest duplex involves the Paleozoic carbonate section in a system of faults that splay from the basal detachment in the Cambrian and merge upward into a roof thrust in the base of the Jurassic shale section. Both faults and layering are sigmoidal. The fault slices within the lowest duplex are the targets of wells drilled for dry sour gas and CO_2. The next-higher duplex thickens the Jurassic marine shale section. This presence of the latter duplex is inferred because the Shell Westcastle 5-7-4-3W5M well penetrated a Jurassic shale section approximately 1400 m thick, about 4 times its normal thickness. Above this inferred duplex is a third, poorly developed duplex, which thickens Cretaceous strata beneath the Lewis thrust. A duplex structure is also present above the Lewis thrust within Precambrian sediments at the base of the Lewis thrust sheet (Fermor and Price, 1987).

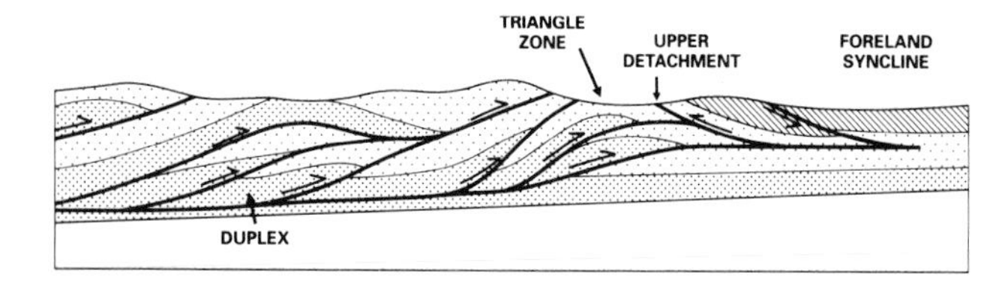

FIGURE 12. Conceptual diagram illustrating a duplex fault structure and a triangle zone. Arrows indicate sense of motion on faults. After Jones (1987).

Duplex-type structures are a common feature throughout the length of the foreland fold and thrust belt. The Waterton, Moose Mountain, Limestone, and Clearwater gas pools are all trapped in stacked thrust sheets of Paleozoic strata that form duplex-like structures beneath overlying folded thrust sheets.

Triangle Zones

Much of the eastern boundary of the Alberta portion of the foreland fold and thrust belt is formed by a structure that resembles a duplex, but with the significant difference that strata above the roof thrust are displaced southwestward with respect to the sediments below the thrust, opposite to the sense of displacement on the floor thrust, link thrusts, and the thrust belt as a whole.

This structure, the "triangle zone," is shown in one conceptual view in Figure 12, and along the right-hand margins of the regional cross sections in Figures 6 and 7. A seismic section illustrating the structure is shown in Figure 13 (from Teal, 1983). The name derives from the east dips on the east side above the roof thrust, the west dips in thrust fault slices within the structure, and near-horizontal layering below the floor thrust in undisturbed sediments.

The thrust slices form a wedge, with the roof and floor thrusts converging at its eastern tip. Each of the thrust slices in succession formed the leading edge of a wedge that was driven eastward into the sediments of the foreland basin sequence. The wedge grew by accreting newly formed fault slices at its tip. Strata above the roof thrust, tilted and uplifted as the wedge was driven beneath them, remained attached to the undeformed sediments of the foreland basin.

It appears, on the basis of industry seismic data, that other triangle zones may exist within more interior parts of the Alberta Foothills. Such structures are difficult to distinguish from duplex structures because their overall geometries are similar. Presumably, these triangle zones were formed at or near the eastern boundary of the thrust belt, and were incorporated within it as new thrust faults developed below and to the east.

Recently, the concept of the triangle zone has been extended by Jones (1982) and McMechan (1985). Jones proposed that prior to erosion the roof thrust of the triangle zone originally extended tens of kilometers west of its present erosional edge, so that much of the Alberta Foothills and Front Ranges were part of a single giant triangular zone-type structure. This implies that all of the thrusts in the Foothills and some of those in the Front Ranges were formed as great link thrusts within this structure. McMechan (1985) proposed that the British Columbia Foothills in the vicinity of the Peace River are composed of a similar structure, but with the present level of erosion well above the roof thrust. Jones (1982) has applied his interpretation to the Foothills west of Calgary, in the area of the cross section shown in Figure 7, and McMechan's interpretation is shown by the cross section in Figure 8.

Such interpretations are difficult to test, because critical evidence has been removed by erosion or is deeply buried. However, for both to be correct, there must be some area along the Foothills belt where erosion is at an intermediate level, where the eroded edge of the roof thrust swings across from the eastern boundary of the foothills toward the western border. No such feature has yet been recognized. Recent industry seismic data of moderate quality in the vicinity of McMechan's (1985) section strongly suggest that the structure of this portion of the northern Foothills differs significantly in certain aspects from McMechan's interpretation. Nonetheless, McMechan's section schematically illustrates a conceptual solution to a significant problem in understanding the structure of the British Columbia Foothills—that is, the apparent presence of greater structural shortening at depth in Paleozoic and Triassic strata than is evident in the overlying Jurassic-Cretaceous sediments of the foreland basin succession. These Jurassic-Cretaceous strata are evidently tightly folded, but few thrust faults are mapped in them and the shortening does not appear to be great. However, they typically are poorly exposed, and it may be that several faults and detachments within this succession have been overlooked.

A concept similar to that of the triangle zone is the *blind thrust* (Thompson, 1979). In the portion of the foreland belt north of the Peace River (56°N), major thrust faults cutting through the Paleozoic shelf carbonates disappear to the east and along strike into a thick Devonian-Mississippian marine shale sequence. No corresponding major thrusts are mapped emerging to the east in the tightly folded strata above the Devonian-Mississippian shales. This geometry may be explained if the thrust sheets of Paleozoic carbonates are in effect the fault slices in a giant triangle-zone type of structure, with the floor thrust deep in the Paleozoic succession and the roof thrust embedded in the poorly exposed Devonian-Mississippian shales, which may also act as a detachment horizon for the tightly folded Mesozoic strata above.

Transverse Faults

Faults that intersect regional structural trends at high angles have been termed "tear," "wrench," "cross," or "transverse" faults by various authors. The term "transverse fault" is preferred because it is strictly geometric. Douglas (1956, 1958) was one of the first to recognize transverse faults in the Alberta Foothills. Transverse faults may be classified as (1) tear faults that crosscut strata within one thrust sheet or crosscut strata and thrust faults within a packet of thrust sheets, and (2) transverse ramps that are an integral part of a thrust fault surface.

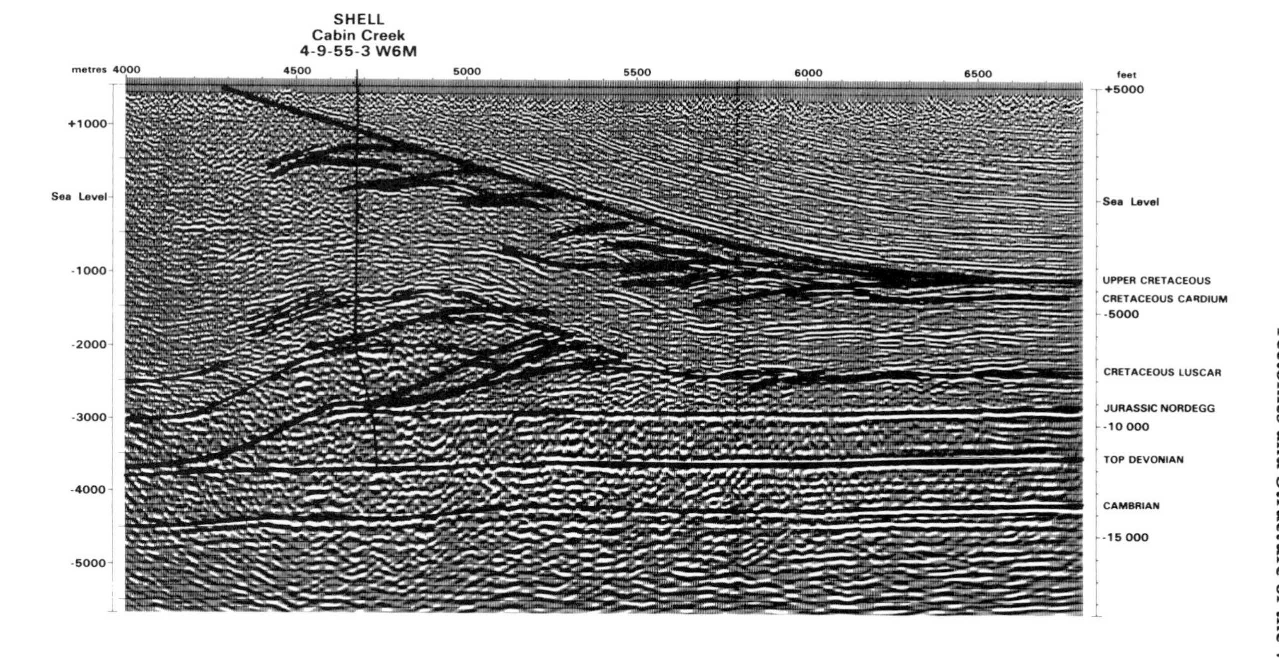

FIGURE 13. Seismic depth section illustrating geometry of the "triangle zone" at Cabin Creek, Alberta, along eastern margin of Foothills belt. Location of seismic section is shown on accompanying map. Section illustrates seismic signature of multiple fault slices forming faulted wedge that splits apart Cretaceous and Tertiary strata of foreland basin. Reproduced with modifications from Teal (1983).

Tear faults may be perpendicular or oblique to the thrust faults bounding the thrust sheets. They separate panels within thrust sheets that have undergone differing amounts of displacement. Transverse ramps can produce folds in the overlying thrust sheet because slip along the thrust will juxtapose the transverse ramps with flats in adjacent parts of the hanging wall and footwall of the thrust.

Transverse ramps may also form as lateral ramps through previously folded strata, as has been interpreted for a portion of the Rundle thrust in the Front Ranges near Canmore, Alberta (Moffat, 1981; Spang et al., 1981; Moffat and Spang, 1984).

Transverse ramps produce the principal zones of abrupt structural plunge in the Alberta portion of the foreland fold and thrust belt, such as in the Limestone Mountain area in the central Alberta Foothills, where a tight grid of seismic lines and wells documents a transverse ramp that abruptly cuts out the entire 2 km thick Paleozoic carbonate section in the Brazeau thrust sheet.

Factors Influencing Structural Styles and Geometry

The nature and distribution of stresses at the western margin of the thrust and fold belt were important factors influencing structural style. Transformation of strike-slip faulting into thrust faulting (Price and Carmichael, 1986) and/or some other mechanism(s) are responsible for the large increase in shortening from north to south within the foreland fold and thrust belt. Estimated shortening varies from about 70 km at about 58°N (Thompson, 1981) to a maximum of 200 km at about 52°N (Price and Mountjoy, 1970) before decreasing to about 150 to 170 km at the international boundary. The increased strain in the southern part of the fold and thrust belt at least in part explains the persistence of large thrust structures of Paleozoic strata in the subsurface from 52°N south into Montana, whereas large thrust structures are less common north of 52°N.

A factor seldom addressed is the extent of the basal detachment and thus the geographical limits of the thin-skinned structural style. In the vicinity of the international boundary, industry seismic data (Bally et al., 1966) indicate that west-dipping, near-basement Cambrian strata continue westward as far as the Rocky Mountain trench, the western boundary of the Rocky Mountains (Figure 6). More recent unpublished industry data from west of the Rocky Mountain trench indicate that equivalent reflectors are absent there; and unidentified dipping reflectors can be observed extending to the base of the seismic records, representing depths of 10 to 12 km or more. This evidence suggests that at this latitude there is no well-defined basal detachment west of the trench.

The presence of dipping reflectors extending to depths of 12 km and more beneath Mesozoic and Paleozoic strata of the northern British Columbia Foothills (north of 56°N) suggests that the older Precambrian high-grade metamorphic craton may be absent there. If present, it is relatively thin and perhaps broken by faulting. This implies that the "basement" may be weaker than the highly metamorphosed and greatly deformed (and thus nonreflecting) rocks that form the basement to the south, east of the Rocky Mountain trench. Indeed, seismic reflection data indicate that some folding and faulting of Cretaceous or Tertiary age do affect this "soft" basement north of 56°N latitude, with the overlying upper Paleozoic and Mesozoic strata passively deformed along with it in a "thick-skinned" style. Shortening accommodated by this mechanism is relatively minor. A much greater amount is taken up in detachment folding and faulting involving upper Precambrian, Paleozoic, and Mesozoic strata in the Foothills and mountains, in a thin-skinned style similar but not identical to that found farther south. Still farther north, close to the northern margin of the Rocky Mountains (60°N), the eastern margin of the deformed belt is composed of large anticlines, with subsidiary thrust faults, involving strata of Mesozoic, Paleozoic, and probably Precambrian ages. The style here appears to be thick-skinned.

Among the most important influences on structural style are the compositions, thicknesses, and lateral extents of the stratigraphic units. Because structural strike does not correspond to earlier established depositional facies trends, the fold and thrust belt, developed in sediments deposited along a continental margin, will exhibit broad changes in structural style along its length and breadth as it involves changing depositional facies. Deformational style can also be influenced by variations in depositional facies on a local scale. These topics will be discussed with reference to the regional cross sections and palinspastic reconstructions in Figures 6, 7, and 8, the locations of which are shown in Figure 2.

The southernmost part of the Canadian portion of the foreland fold and thrust belt is dominated by the Lewis thrust sheet, a thick, broadly folded sheet of mid-Proterozoic to Cretaceous sedimentary rock that was probably 10 km or more thick when emplaced. This thick succession of well-indurated carbonate, sandstone, and argillite maintained a relative structural integrity during thrusting. A thin remnant of this formerly extensive mass of rock remains preserved from erosion east of the Flathead normal fault along the line of cross section in Figure 6; a much more complete, although broken up, portion remains preserved west of the Flathead fault. In contrast, the relatively thin sequence of Paleozoic shelf carbonates overridden by the Lewis thrust sheet is broken by a considerable number of thrust faults. The overlying, less-competent Mesozoic clastic strata of the foreland basin succession are even more intensely faulted and folded, and are detached from the Paleozoic strata by thrust faults in the Fernie shales at the base of the Mesozoic sequence.

Farther north, in the vicinity of the Bow River Valley (Figure 7), the structural style of the Foothills is similar to that in the southern cross section. There are no Proterozoic strata in the Front Ranges, which are cut into several westward-dipping thrust sheets formed largely of upper Paleozoic shelf carbonates, with the base of the Mesozoic clastic sequence preserved as a series of small erosional wedges beneath overlying thrust sheets. The Eastern Main Ranges are composed of broadly folded thrust sheets containing a rapidly westward-thickening lower

Paleozoic miogeoclinal carbonate sequence beneath the more gradually thickening upper Paleozoic shelf carbonate sequence. In contrast, to the west of an abrupt facies change across which most of the lower Paleozoic carbonates change to shale and calcareous shale, tight folding and associated cleavage development are the predominant structural style in the Western Main Ranges (Balkwill, 1972; Cook, 1975; Gardner, 1977). Proterozoic strata exposed in the Main Ranges and farther west are cleaved and folded, but are also cut by major thrust faults.

At the latitude of the northernmost cross section (McMechan, 1985; Figure 8, location on Figure 2), the structural style in the upper Paleozoic and Mesozoic strata of the Foothills is dominated by large kink folds that often have boxlike geometries (Fitzgerald, 1968). The predominance of folding versus thrusting may be a consequence of the large proportion of shale present in the Devonian-Mississippian Besa River shale (Fitzgerald, 1968; Thompson, 1979; McMechan, 1985). In addition, the Mesozoic foreland basin succession contains much more shale than does this succession in the southern Foothills. Although not illustrated in McMechan's (1985) cross section, Jurassic shales at the base of the Mesozoic foreland basin sequence are an important detachment horizon, as shown in the other cross sections. In contrast, the lower Paleozoic succession exposed along the eastern margin of the mountains is relatively thick and competent, having been deformed into several major thrust sheets that are relatively undeformed internally (Fitzgerald, 1968; Thompson, 1979; McMechan, 1985). Farther west, however, the lower Paleozoic carbonates shale out, and the structural style changes abruptly to tight folding with associated cleavage (Thompson, 1979, 1981).

An important deformation mechanism is pore-fluid overpressuring. High pore-fluid pressure, resulting from rapid loading, at the toe of the advancing thrust belt can facilitate sliding along preexisting planes of failure if they have the appropriate orientation and aid in the progradation (growth) of the fold and thrust belt wedge (Gretener, 1972, 1977). Recent results from the Ocean Drilling Program across the deformation front of the Barbados ridge accretionary prism indicate that areas of active faulting have high associated fluid pressures, the chemistry of which is consistent with dewatering of smectites and long-range transport of hydrocarbon-bearing solutions (Moore et al., 1988). Deformation driven by fluid overpressuring may occur in areas that are (1) subject to rapid loading of thick thrust sheets, (2) characterized by thick sequences of shale at depths at which shale dewatering is important, or (3) characterized by thick, organic-rich sequences at depths at which generation of oil and/or gas is important. Such conditions may have existed in the foreland fold and thrust belt during deformation.

In the central Alberta Foothills, shales of the Upper Cretaceous (Cenomanian-Turonian) Blackstone Formation are less dense than would be predicted from the normal compaction trend. This anomaly has been interpreted as a relic of overpressures that persisted in this zone until after the time of maximum burial, thus preserving the shales in their undercompacted state until considerable erosion reduced lithostatic pressures to a level too low to compact the shales further (D. Woodland, personal communication, 1989). Jones (1987) has suggested that overpressure regimes may favor the formation of triangle zones and blind thrusts. If this is the case, the extensive triangle zone along the eastern margin of the Foothills suggests that overpressures affected the structural style of the foreland belt.

In summary, several significant factors influencing structural style are broadly understood. However, the controls determining particular aspects or details of structural style, such as the locations, dimensions, and magnitudes of faults, remain poorly understood and a subject for future research.

HYDROCARBON ACCUMULATIONS

The oil and gas industry in Alberta received its first "boost" from discoveries made at Turner Valley (Porter, 1992), where oil is contained in porous Mississippian carbonates, and in Cretaceous sandstones trapped in a thrusted anticlinal structural closure in the "triangle zone" of southern Alberta. These discoveries resulted from drilling during the period of 1913 to 1936, which gradually delineated the nature of the oil/gas/condensate reservoir, the only known significant oil accumulation in the Foothills belt. The original estimated in-place reserves are 43×10^9 m^3 (1.5 tcf) of gas and 160×10^6 m^3 (1.01 billion barrels) of 39° API oil. One of the recent major gas/condensate discoveries in the Foothills zone is located in the Millarville-North Turner Valley area (Figure 14). The thrust sheets penetrated here form lateral en echelon imbricates of the Turner Valley thrust sheet. The discovery of this separate trap 70 years after initial drilling in the Turner Valley area attests to the "frontier" nature of exploration in the Foothills.

Figure 14 illustrates the locations of major hydrocarbon pools in the Foothills belt of Alberta and British Columbia (see also Barclay and Smith, 1992). The relative paucity of pools in the northern Foothills of Alberta and the Foothills of northeastern British Columbia is attributed to a decrease in porous Paleozoic carbonate reservoirs northwestward within the Alberta Foothills and the more remote location and rugged topography of the British Columbia part of the belt, where exploration is much less mature. A series of significant discoveries within the British Columbia portion of the belt has been recently announced.

Reservoirs

Figure 15 outlines the major stratigraphic units of economic interest in the Foothills belt. The most productive level is the cyclical dolomitized skeletal limestones of the Mississippian Rundle Group. These were deposited within shoal complexes on a shallow shelf margin that deepened gradually westward (Bamber et al., 1981). Porosity development and preservation were complex and appear to be related to a multistage history involving dolomitization of preexisting grains and calcite and anhydrite precipitation in pore spaces (Illing, 1959).

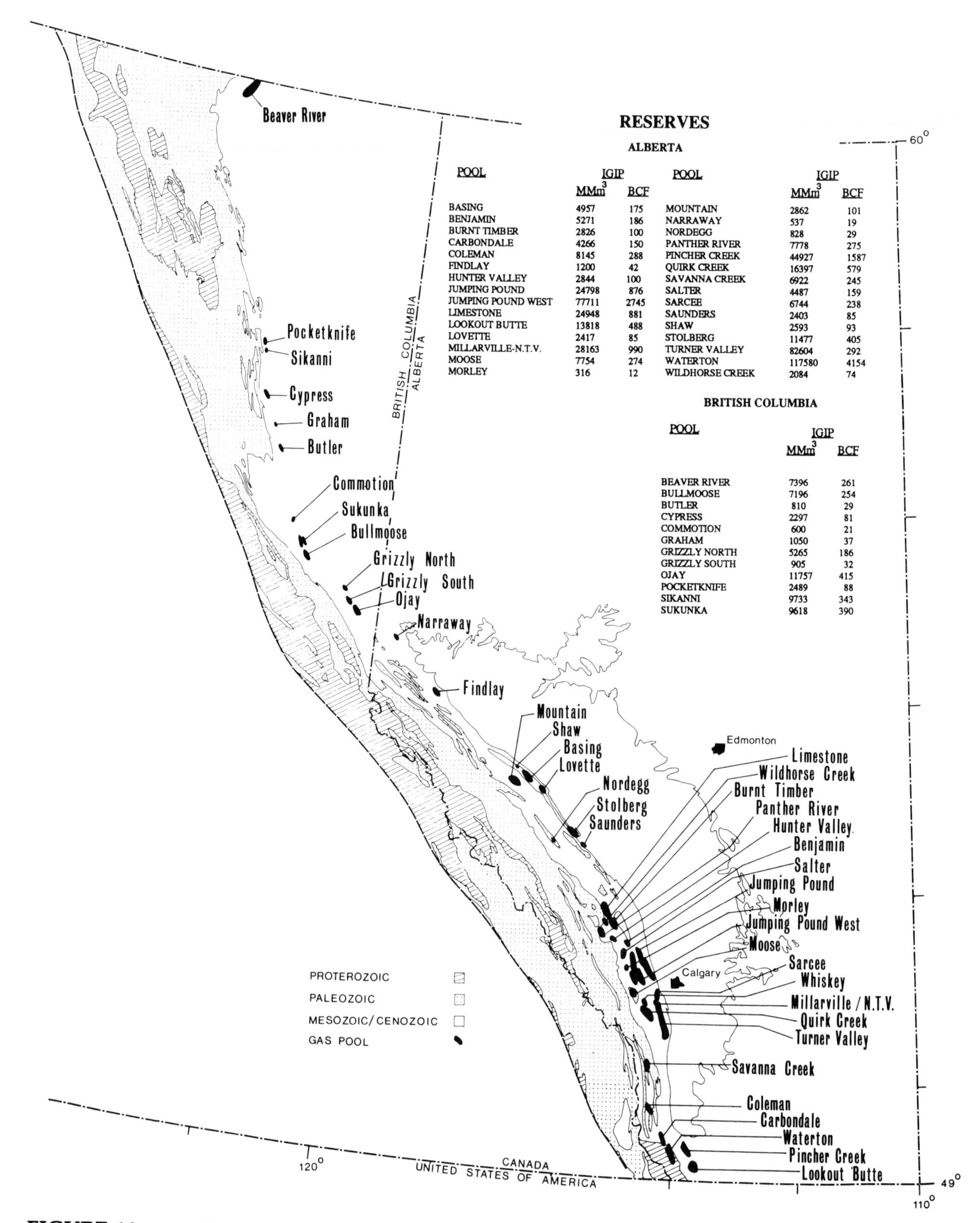

RESERVES

ALBERTA

POOL	IGIP MMm³	BCF	POOL	IGIP MMm³	BCF
BASING	4957	175	MOUNTAIN	2862	101
BENJAMIN	5271	186	NARRAWAY	537	19
BURNT TIMBER	2826	100	NORDEGG	828	29
CARBONDALE	4266	150	PANTHER RIVER	7778	275
COLEMAN	8145	288	PINCHER CREEK	44927	1587
FINDLAY	1200	42	QUIRK CREEK	16397	579
HUNTER VALLEY	2844	100	SAVANNA CREEK	6922	245
JUMPING POUND	24798	876	SALTER	4487	159
JUMPING POUND WEST	77711	2745	SARCEE	6744	238
LIMESTONE	24948	881	SAUNDERS	2403	85
LOOKOUT BUTTE	13818	488	SHAW	2593	93
LOVETTE	2417	85	STOLBERG	11477	405
MILLARVILLE-N.T.V.	28163	990	TURNER VALLEY	82604	292
MOOSE	7754	274	WATERTON	117580	4154
MORLEY	316	12	WILDHORSE CREEK	2084	74

BRITISH COLUMBIA

POOL	IGIP MMm³	BCF
BEAVER RIVER	7396	261
BULLMOOSE	7196	254
BUTLER	810	29
CYPRESS	2297	81
COMMOTION	600	21
GRAHAM	1050	37
GRIZZLY NORTH	5265	186
GRIZZLY SOUTH	905	32
OJAY	11757	415
POCKETKNIFE	2489	88
SIKANNI	9733	343
SUKUNKA	9618	390

FIGURE 14. Map illustrating the locations of significant gas pools present in the Foothills zone of the fold and thrust belt. Reserves for individual pools are derived from Energy Resources Conservation Board (1986).

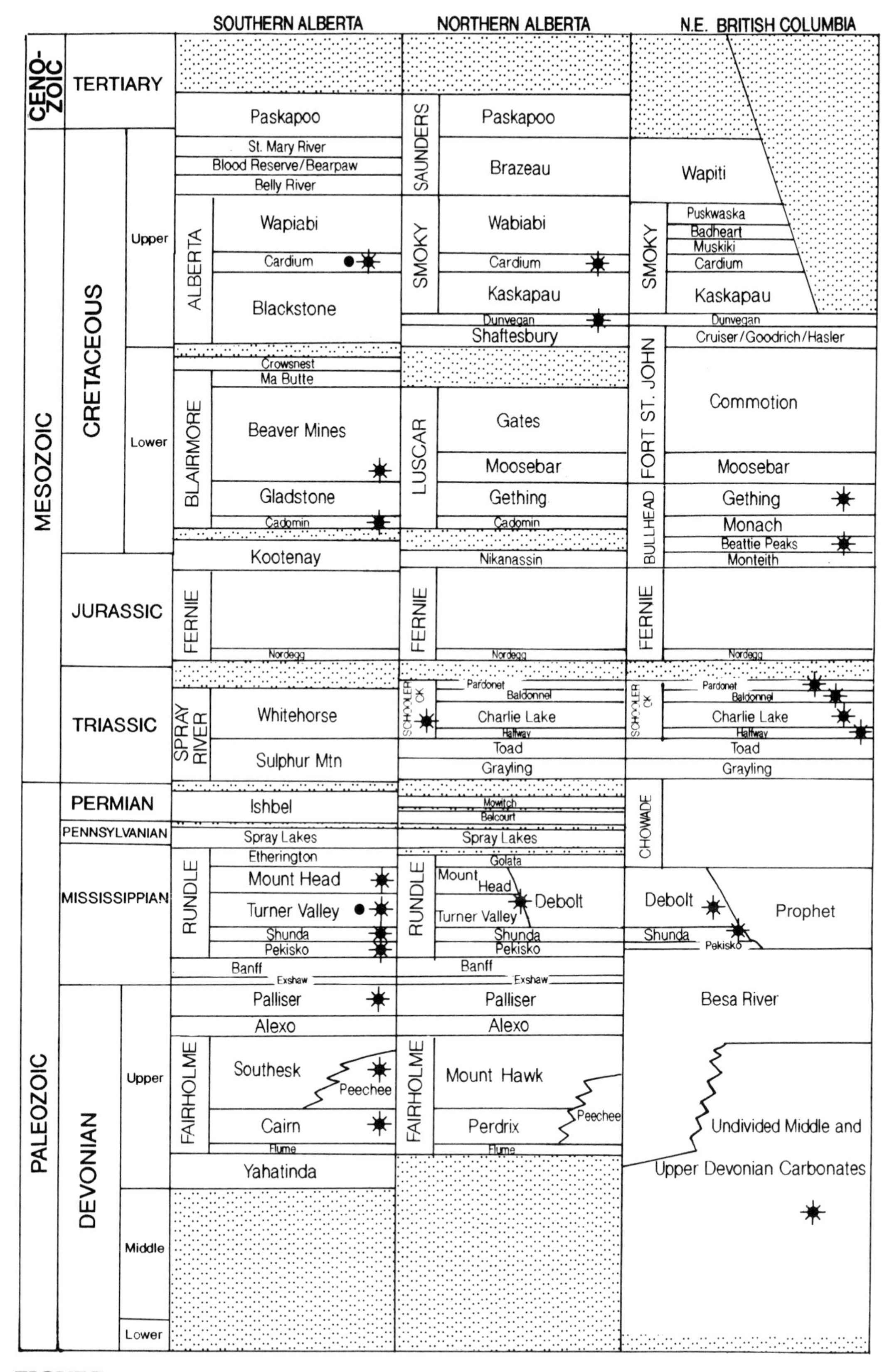

FIGURE 15. Stratigraphic correlation chart for the Foothills of southern Alberta, northern Alberta, and northeastern British Columbia. Formation and group nomenclature is adapted from the Energy Resources Conservation Board Table of Formations, Alberta and British Columbia, December, 1985, with adjustments to reflect recent changes recognized by the Geological Survey of Canada. Note that local terminology may differ. Zones that have produced gas and/or oil are indicated with the appropriate symbol.

Gas fields in the Devonian Wabamun Formation are concentrated in the southernmost Foothills at Limestone, Burnt Timber, Panther, Moose, Whiskey, Coleman, and Waterton. The Wabamun Formation was deposited on a broad carbonate ramp that prograded west and north into deeper water. Facies patterns suggest a gradual change from "updip" evaporitic dolostones in the east and south through reservoir-bearing dolostones in the Foothills and Front Ranges of southern Alberta to open-marine limestones in the Foothills and Front Ranges of central Alberta. Carbonate facies pass laterally into basinal marine shales in the Foothills of northeastern British Columbia (Basset and Stout, 1967). The porous component of the Wabamun Formation in the Foothills belt, the Crossfield Member, is a dolomite lithosome contained between tight, anhydrite-rich dolomite beds, and was deposited during a marine transgressive event. The presence of secondary anhydrites coupled with elevated temperatures associated with thrust belt burial aided the thermal sulfate reduction process, resulting in the high values of H_2S in Wabamun reservoirs (Eliuk, 1984). Particularly high values of H_2S at Panther River, Salter, and Limestone have rendered the Wabamun gas unretrievable with present economics (Eliuk and Hunter, 1987).

The Devonian Peechee/Leduc equivalent reefal buildups (Figure 15) produce gas at the Limestone and Quirk Creek gas fields. Porous Nisku reservoirs are also present at Limestone and Panther River. Isolated Devonian biostromal and biohermal buildups are secondary targets because seismic quality is too poor to delineate stratigraphic variations in areas where structures with complex geometries are present.

In the northern Alberta and British Columbia Foothills, the primary target is Triassic carbonates and clastics that comprise a westward-thickening, shoaling-upward succession of marine siltstone, sandstone, limestone, and dolostone. The predominant Triassic reservoir is Baldonnel Formation bioclastic dolostone, which is productive in several gas fields in British Columbia. Carbonates in the Charlie Lake Formation and clastics in the Halfway Formation (both Triassic units) are also localized potential reservoirs. The development of a relatively high-amplitude duplex zone, formed by numerous imbricate thrusts, at the Bullmoose and Sukunka fields has probably produced the vertical fracturing that contributes significantly to permeability in these fields. The presence of anhydrite within the Schooler Creek Group evaporitic section, coupled with high paleotemperatures, has contributed to the localized occurrence of relatively high H_2S gas and, hence, marketable sulfur reserves.

The upper dolomitic unit of the Mississippian Debolt Formation forms a hydrocarbon trap in the more northerly part of the British Columbia Foothills at the Graham, Kobes, Sikanni, and Pocketknife fields (Figure 14). In this area, a combination of post-Mississippian normal faulting, concomitant erosion, and later Laramide compressional uplift have resulted in pop-up structures that bound zones of laterally variable lithologic facies.

Gas has also been produced from a low-porosity, highly fractured Middle Devonian dolomite reservoir at the Beaver River field, along the boundary with the Yukon Territory.

Hydrocarbon Maturation and Migration

The timing of hydrocarbon generation and migration relative to deformation is of obvious importance to exploration in the Foothills belt. Information concerning hydrocarbon maturation is generally obtained indirectly by the measurement and subsequent modeling of coal rank. Deroo et al. (1977) used burial history models indicating that source rocks in the Western Canada sedimentary basin did not reach peak hydrocarbon genesis until the Late Cretaceous and early Tertiary. The overall pattern of increasing Cretaceous coal rank from east to west in the Alberta Foothills and adjacent Plains region is generally assumed to be a result of increasing depth of preorogenic burial in the late Mesozoic–Early Cretaceous foreland basin (Hacquebard and Donaldson, 1974; Hacquebard, 1975; Deroo et al., 1977; also see Osadetz et al., 1992). However, investigations of vitrinite reflectance variations in the Foothills of Alberta and British Columbia have indicated that a portion of the coalification history has been locally both preorogenic and postorogenic and hence synorogenic on the regional scale (Kalkreuth and McMechan, 1984; Kalkreuth and Langenberg, 1986; Pearson and Grieve, 1985; England and Bustin, 1986). More direct evidence for this timing relationship is evident in the central Alberta Foothills at the Nordegg field, where sour dry gas is trapped in Mississippian Turner Valley carbonates within the relatively shallow (1600 m) Brazeau thrust sheet (Bland, 1967), whereas to the south oil was recovered from porous Cambrian dolostone at a similar depth in the same thrust sheet (Henderson, 1943). If this oil was sourced from adjacent footwall Cretaceous shales, maturation in the area has been both pre- and postdeformation whereas migration has been postdeformation.

ACKNOWLEDGMENTS

This review paper is based on the work of a great number of earth scientists who have worked over many years to bring our knowledge of the Canadian Cordillera to its present level. The authors' own original contributions are but a small part of our accumulated knowledge. We thank Shell Canada and Gulf Canada Resources for supporting our work. Shell Canada contributed the services of Doreen Neilson, who typed the manuscript, and Eva Necas and Randy Stroobant, who prepared many of the illustrations. We thank Steven Johnston, Dale Leckie, Ray Price, Bob Thompson, and others who reviewed earlier versions of the manuscript. Their criticisms and contributions have substantially improved this chapter.

REFERENCES CITED

Abbot, J. G., 1987, Devonian extension and wrench(?) tectonics near Macmillan Pass, Yukon Territory Canada (abstract), *in* Second international symposium on the Devonian programs and abstracts: Calgary, Alberta, Canadian Society of Petroleum Geologists, unpaginated.

Archibald, D. A., J. K. Glover, R. A. Price, E. Farrar, and D. M. Carmichael, 1983, Geochronology and tectonic implications of magmatism and metamorphism, southern Kootenay Arc and neighboring regions, southeastern British Columbia: Part I, Jurassic to mid-Cretaceous: Canadian Journal of Earth Sciences, v. 20, p. 1891-1913.

Archibald, D. A., T. E. Krogh, R. L. Armstrong, and E. Farrar, 1984, Geochronology and tectonic implications of magmatism and metamorphism, southern Kootenay Arc and neighboring regions, southeastern British Columbia: Part II, mid-Cretaceous to Eocene: Canadian Journal of Earth Sciences, v. 21, p. 567-583.

Armstrong, R. L., 1988, Mesozoic and early Cenozoic magmatic evolution of the Canadian Cordillera: GSA Special Paper 218, p. 55-91.

Balkwill, H. R., 1972, Structural geology, lower Kicking Horse River region, Rocky Mountains, British Columbia: Bulletin of Canadian Petroleum Geology, v. 20, p. 608-633.

Bally, A. W., P. L. Gordy, and G. A. Stewart, 1966, Structure, seismic data, and orogenic evolution of southern Canadian Rocky Mountains: Bulletin of Canadian Petroleum Geology, v. 14, p. 337-381.

Bamber, E. W., R. W. Macqueen, and N. C. Ollerenshaw, 1981, Mississippian stratigraphy and sedimentology, Canyon Creek (Moose Mountain), Alberta, *in* R. I. Thompson and D. G. Cook, eds., Field guides to geology and mineral deposits, Geological Association of Canada/Mineralogical Association of Canada Calgary '81 annual meeting: Calgary, Alberta, University of Calgary Department of Geology and Geophysics, p. 177-194.

Barclay, J. E., and D. G. Smith, 1992, Western Canada foreland basin oil and gas plays, this volume.

Bassett, H. G., and J. G. Stout, 1967, Devonian of western Canada, *in* D. H. Oswald, ed., International symposium on the Devonian System: Calgary, Alberta, Canadian Society of Petroleum Geologists, v. 1, p. 717-752.

Beaumont, C., 1981, Foreland basins: The Geophysical Journal of the Royal Astronomical Society, v. 65, p. 291-329.

Bland, R. C. J., 1967, Nordegg field, gas fields of Alberta: Journal of the Alberta Society of Petroleum Geologists, v. 15, p. 244-245.

Bond, G. C., and M. A. Kominz, 1984, Construction of tectonic subsidence curves for the early Paleozoic miogeocline, southern Canadian Rocky Mountains: implications for subsidence mechanisms, age of breakup and crustal thinning: GSA Bulletin, v. 95, p. 155-173.

Brown, R. L., and M. J. Journeay, 1987, Tectonic denudation of the Shuswap metamorphic terrane of southeastern British Columbia: Geology, v. 15, p. 142-146.

Brown, R. L., and P. B. Read, 1983, Shuswap terrane of British Columbia: a Mesozoic "core complex": Geology, v. 11, p. 164-168.

Brown, R. L., M. J. Journeay, L. S. Lane, D. C. Murphy, and C. J. Rees, 1986, Obduction, backfolding and piggyback thrusting in the metamorphic hinterland of the southeastern Canadian Cordillera: Journal of Structural Geology, v. 8, p. 255-268.

Butler, R. F., T. A. Harms, and H. Gabrielse, 1988, Cretaceous remagnetization in Sylvester Allochthon: limits to post-105 Ma northward displacement of north-central British Columbia: Canadian Journal of Earth Sciences, v. 25, p. 1316-1322.

Cant, D. J., 1988, Regional structure and development of the Peace River arch, Alberta: a Paleozoic failed-rift system?: Bulletin of Canadian Petroleum Geology, v. 36, p. 284-295.

Carr, S. D., 1989, Implications of early Eocene Ladybird granite in the Thor-Odin-Pinnacles area, southern British Columbia, *in* Current research Part E: Geological Survey of Canada Paper 89-1E, p. 69-77.

Coney, P. J., and T. A. Harms, 1984, Cordilleran metamorphic core complexes: Cenozoic extensional relics of Mesozoic compression: Geology, v. 12, p. 550-554.

Cook, D. G., 1975, Structural style influenced by lithofacies, Rocky Mountain main ranges, Alberta-British Columbia: Geological Survey of Canada Bulletin 233, 73 p.

Cordey, F., N. Mortimer, P. DeWever, and J. W. H. Monger, 1987, Significance of Jurassic radiolarians from the Cache Creek terrane, British Columbia: Geology, v. 15, p. 1151-1154.

Dahlstrom, C. D., 1970, Structural geology in the eastern margin of the Canadian Rocky Mountains: Bulletin of Canadian Petroleum Geology, v. 14, p. 332-406.

deMille, G., 1958, Pre-Mississippian history of the Peace River arch, *in* J. C. Scott, ed., Symposium on the Peace River arch: Journal of the Alberta Society of Petroleum Geologists, v. 6, p. 61-69.

Deroo, G., T. G. Powell, B. Tissott, and R. G. McCrossan, 1977, The origin and migration of petroleum in the western Canadian sedimentary basin, Alberta: Geological Survey of Canada Bulletin 262, 200 p.

Douglas, R. J. W., 1950, Callum Creek, Langford Creek, and Gap map-areas, Alberta: Geological Survey of Canada Memoir 255, 124 p.

Douglas, R. J. W., 1956, Nordegg, Alberta (geologic map and report): Geological Survey of Canada Paper 55-34, 31 p.

Douglas, R. J. W., 1958, Mount Head map-area, Alberta: Geological Survey of Canada Memoir 291, 241 p.

Duncan, J. J., R. R. Parrish, and R. L. Armstrong, 1979, Rb/Sr geochronology of post-tectonic intrusive events in the Omineca crystalline belt, southeastern British Columbia (abstract), *in* Evolution of the cratonic margin and related mineral deposits: Geological Association of Canada Cordilleran Section Program and Abstracts, p. 15.

Eisbacher, G. H., 1985, Pericollisional strike-slip faults and synorogenic basins, Canadian Cordillera, *in* K. T. Biddle and N. Christie-Blick, eds., Strike-slip deformation, basin formation and sedimentation: SEPM Special Publication 37, p. 265-282.

Eliuk, L., 1984, A hypothesis for the origin of hydrogen sulphide in Devonian Crossfield Member dolomite, Wabamun Formation, Alberta, Canada, *in* L. Eliuk, J. Kaldi, N. Watts, and G. Harrison, eds., Carbonates in subsurface and outcrop, Canadian Society of Petroleum Geologists Core Conference proceedings: Calgary, Alberta, Canadian Society of Petroleum Geologists, p. 245-289.

Eliuk, L., and D. F. Hunter, 1987, Wabamun Group structural thrust fault fields: the Limestone-Burnt Timber example, *in* F. Krause and G. Burrowes, eds., Second international symposium on the Devonian System: Calgary, Alberta, Canadian Society of Petroleum Geologists, p. 39-62.

Elliott, D., 1977, Some aspects of the geometry and mechanics of thrust belts: Canadian Society of Petroleum Geologists, 8th Annual Seminar Pub. Notes, Continuing Education Department, University of Calgary, v. 1, 2.

Elliott, D., and M. R. W. Johnson, 1980, Structural evolution in the northern part of the Moine thrust belt, NW Scotland: Transactions of the Royal Society of Edinburgh, v. 71 (pt. 2), p. 69-96.

Energy Resources Conservation Board, 1986, Alberta's reserves of crude oil, oil sands, gas, natural gas liquids and sulphur: Reserve Report Series ERCB-18, Chapters 3 and 4, irregularly paginated.

England, T. D. J., and R. M. Bustin, 1986, Thermal maturation of the western Canadian sedimentary basin south of the Red Deer River: (I) Alberta Plains: Bulletin of Canadian Petroleum Geology, v. 34, p. 71-90.

Fermor, P. R., and R. A. Price, 1987, Multiduplex structure along the base of the Lewis thrust sheet in the southern Canadian Rockies: Bulletin of Canadian Petroleum Geology, v. 35, p. 159-185.

Fitzgerald, E. L., 1968, Structure of British Columbia Foothills, Canada: AAPG Bulletin, v. 52, p. 641-664.

Gabrielse, H., 1985, Major dextral transcurrent displacements along the northern Rocky Mountain trench and related lineaments in north-central British Columbia: GSA Bulletin, v. 96, p. 1-14.

Gardner, D. A. C., 1977, Structural geology and metamorphism of calcareous lower Paleozoic slates, Blaeberry River-Redburn Creek area, near Golden, British Columbia: unpublished Ph.D. thesis, Queen's University, Kingston, Ontario, 224 p.

Gardner, D. A. C., and J. H. Spang, 1973, Model studies of the displacement transfer associated with overthrust faulting: Bulletin of Canadian Petroleum Geology, v. 21, p. 534-552.

Gardner, M. C., S. C. Bergman, G. W. Cushing, E. M. Mackevett, G. Plafker, R. B. Campbell, C. J. Dodds, W. C. McClelland, and P. A. Mueller, 1988, Pennsylvanian pluton stitching of Wrangellia and the Alexander terrane, Wrangell Mountains, Alaska: Geology, v. 16, p. 967-971.

Gehrels, G. E., and M. T. Smith, 1987, "Antler" Allochthon in the Kootenay Arc?: Geology, v. 15, p. 769-770.

Gibson, D. W., 1977, Sedimentary facies in the Jura-Cretaceous Kootenay Formation, Crowsnest Pass area, southwestern Alberta and southeastern British Columbia: Bulletin of Canadian Petroleum Geology, v. 25, p. 767-791.

Gibson, D. W., 1985, Stratigraphy, sedimentology and depositional environments of the coal-bearing Jurassic-Cretaceous Kootenay Group, Alberta and British Columbia: Geological Survey of Canada Bulletin 357, 108 p.

Gordey, S. P., 1987, Devono-Mississippian sedimentation and tectonism in the Canadian cordilleran miogeocline (abstract), *in* Second international symposium on the Devonian System program and abstracts: Calgary, Alberta, Canadian Society of Petroleum Geologists, p. 98.

Gordey, S. P., S. G. Abbott, D. G. Tempelman-Kluit, and H. Gabrielse, 1987, "Antler" clastics in the Canadian Cordillera: Geology, v. 15, p. 103-107.

Gosselin, E. G., 1987, Petroleum geology of the Slave Point Formation (Middle Devonian) in the Peace River arch region of northwestern Alberta (abstract), *in* Second international symposium on the Devonian System program and abstracts: Calgary, Alberta, Canadian Society of Petroleum Geologists, p. 100.

Gretener, P. E., 1972, Thoughts on overthrust faulting in a layered sequence: Bulletin of Canadian Petroleum Geology, v. 20, p. 583-607.

Gretener, P. E., 1977, On the character of thrust faults with particular reference to the basal tongues: Bulletin of Canadian Petroleum Geology, v. 25, p. 110-122.

Hacquebard, P. A., 1975, Correlation between coal rank, paleotemperature, and petroleum occurrence in Alberta, *in* Report of activities: Geological Survey of Canada Paper 75-1, pt. B, p. 5-9.

Hacquebard, P. A., and J. R. Donaldson, 1974, Studies of coals in the Rocky Mountains and Inner Foothills belt, Canada: GSA Special Paper 153, p. 75-94.

Hall, R. L., 1984, Lithostratigraphy and biostratigraphy of the Fernie Formation (Jurassic) in the southern Canadian Rocky Mountains, *in* D. F. Stott and D. J. Glass, The Mesozoic of middle North America: Canadian Society of Petroleum Geologists Memoir 9, p. 233-247.

Hamblin, A. P., and R. G. Walker, 1979, Storm-dominated shallow marine deposits: the Fernie-Kootenay (Jurassic) transition, southern Rocky Mountains: Canadian Journal of Earth Sciences, v. 16, p. 1673-1690.

Henderson, J. F., 1943, Tay River map sheet: Canadian Department of Mines and Resources, Map 840A.

Hoy, T., and P. van der Heyden, 1988, Geochemistry, geochronology and tectonic implications of two quartz monzonite intrusions, Purcell Mountains, southeastern British Columbia: Canadian Journal of Earth Sciences, v. 25, p. 106-115.

Illing, L. V., 1959, Cyclic carbonate sedimentation in the Mississippian at Moose Dome, southwest Alberta, *in* Alberta Society of Petroleum Geologists Guidebook, 9th annual field conference: Calgary, Alberta, Alberta Society of Petroleum Geologists, p. 36-52.

Irving, E., and R. W. Yole, 1972, Paleomagnetism and the kinematic history of mafic and ultramafic rocks in fold mountain belts: Canadian Earth Physics Branch Publication, v. 42, p. 87-95.

Irving, E., J. W. H. Monger, and R. W. Yole, 1980, New paleomagnetic evidence for displaced terranes in British Columbia, *in* D. W. Strangway, ed., The continental crust and its mineral deposits: Geological Association of Canada Special Paper 20, p. 441-456.

Jones, P. B., 1971, Folded faults and sequence of thrusting in Alberta Foothills: AAPG Bulletin, v. 55, p. 292-306.

Jones, P. B., 1982, Oil and gas beneath east-dipping underthrust faults in the Alberta Foothills, *in* R. B. Powers, ed., Geologic studies of the cordilleran thrust belt: Denver, Colorado, Rocky Mountain Association of Petroleum Geologists, p. 61-74.

Jones, P. B., 1987, Quantitative geometry of thrust and fold belt structures: Tulsa, Oklahoma, AAPG, 26 p.

Kalkreuth, W. D., and W. C. Langenberg, 1986, The timing of coalification in relation to structural events in the Grande Cache area, Alberta, Canada: Canadian Journal of Earth Sciences, v. 23, p. 1103-1116.

Kalkreuth, W. D., and M. E. McMechan, 1984, Regional pattern of thermal maturation as determined from coal rank studies, Rocky Mountain Foothills and Front Ranges north of Grande Cache, Alberta—implications for petroleum exploration: Bulletin of Canadian Petroleum Geology, v. 32, p. 249-271.

Kanasewich, E. R., R. M. Clowes, and C. H. McCloughan, 1969, A buried Precambrian rift in western Canada, *in* The world rift system—international mantle committee, upper mantle science report number 19: Tectonophysics, v. 8, p. 513-527.

Klepacki, D. W., and J. O. Wheeler, 1985, Stratigraphic and structural relations of the Milford, Kaslo and Slocan Groups, Goat Range, Lardeau and Nelson map areas, British Columbia, *in* Current research Part A: Geological Survey of Canada Paper 85-1A, p. 277-286.

Lavoie, D. H., 1958, The Peace River arch during Mississippian and Permo-Pennsylvanian time: Journal of the Alberta Society of Petroleum Geologists, v. 6, p. 69-74.

Leckie, D., and D. Smith, 1992, Regional setting, evolution, and depositional cycles of the Western Canada foreland basin, this volume.

Leech, G. B., 1966, Cretaceous strata in the west face of the Rocky Mountains: Geological Survey of Canada Paper 66-1, p. 72-73.

McLean, J. R., 1977, The Cadomin Formation: stratigraphy, sedimentology and tectonic implications: Bulletin of Canadian Petroleum Geology, v. 25, p. 792-827.

McMechan, M. E., 1981, The middle Proterozoic Purcell Supergroup in the southeastern Rocky and southeastern Purcell Mountains, British Columbia, and the initiation of the cordilleran miogeocline, southern Canada and adjacent United States: Bulletin of Canadian Petroleum Geology, v. 29, p. 583-621.

McMechan, M. E., 1985, Low taper triangle zone geometry: an interpretation for the Rocky Mountain Foothills, Pine Pass-Peace River area, British Columbia: Bulletin of Canadian Petroleum Geology, v. 33, p. 31-38.

McMechan, M. E., and R. I. Thompson, 1989, Structural style and history of the Rocky Mountain fold and thrust belt, *in* B. D. Ricketts, ed., Western Canada sedimentary basin: Calgary, Alberta, Canadian Society of Petroleum Geologists, p. 47-72.

McMechan, R. D., 1981, Stratigraphy, sedimentology, structure and tectonic implications of the Oligocene Kishenehn Formation, Flathead Valley graben, southeastern British Columbia: unpublished Ph.D. thesis, Queen's University, Kingston, Ontario, 579 p.

Moffat, I. W., 1981, Geometry and mechanisms of transverse faulting, Rocky Mountain Front Ranges, Canmore, Alberta: unpublished M.Sc. thesis, University of Calgary, Calgary, Alberta, 193 p.

Moffat, I. W., and J. H. Spang, 1984, Origin of transverse faulting, Rocky Mountain Front Ranges, Canmore, Alberta: Bulletin of Canadian Petroleum Geology, v. 32, p. 147-161.

Monger, J. W. H., 1977, Upper Paleozoic rocks of the Canadian Cordillera and their bearing on cordilleran evolution: Canadian Journal of Earth Sciences, v. 14, p. 1832-1859.

Monger, J. W. H., 1984, Cordilleran tectonics, a Canadian perspective: Bulletin Societié Geologic France, v. 2, p. 255-278.

Monger, J. W. H., 1989, Overview of cordilleran geology, *in* B. D. Ricketts, ed., Western Canada sedimentary basin: Calgary, Alberta, Canadian Society of Petroleum Geologists, p. 9-32.

Monger, J. W. H., and R. A. Price, 1979, Geodynamic evolution of the Canadian Cordillera: progress and problems: Canadian Journal of Earth Sciences, v. 16, p. 771-791.

Monger, J. W. H., and C. A. Ross, 1971, Distribution of fusulinaceans in the western Canadian Cordillera: Canadian Journal of Earth Sciences, v. 8, p. 259-278.

Monger, J. W. H., R. A. Price, and D. J. Tempelman-Kluit, 1982, Tectonic accretion and the origin of the two major metamorphic and plutonic welts in the Canadian Cordillera: Geology, v. 10, p. 75-79.

Moore, J. C., A. Mascle, E. Taylor, P. Andreieff, F. Alvarez, R. Barnes, C. Beck, J. Behrmann, G. Blanc, K. Brown, M. Clark, J. Dolan, A. Fisher, J. Gieskes, M. Hounslow, P. McClellan, K. Moran, Y. Ogawa, T. Sakai, J. Schoonmaker, P. Vrolijk, R. Wilkens, and C. Williams, 1988, Tectonics and hydrogeology of the northern Barbados ridge: results from Ocean Drilling Program leg 110: GSA Bulletin, v. 100, p. 1578-1593.

Mortimer, N., 1986, Late Triassic, arc-related potassic igneous rocks in the North American Cordillera: Geology, v. 14, p. 1035-1038.

Norris, D. K., 1964, The Lower Cretaceous of the southeastern Canadian Cordillera: Bulletin of Canadian Petroleum Geology, v. 12, p. 512-535.

Norris, D. K., 1971a, Comparative study of the Castle River and other folds in the eastern cordillera of Canada: Geological Survey of Canada Bulletin 205, 58 p.

Norris, D. K., 1971b, The geology and coal potential of the Cascade coal basin, *in* I. A. R. Halladay and D. K. Mathewson, eds., A guide to the geology of the eastern cordillera along the Trans-Canada Highway between Calgary, Alberta and Revelstoke, British Columbia: Calgary, Alberta, Alberta Society of Petroleum Geologists, p. 25-39.

North, F. K., and G. G. L. Henderson, 1954, Summary of the geology of the southern Rocky Mountains of Canada: *in* Alberta Society of Petroleum Geologists Guidebook; 4th annual field conference: Calgary, Alberta, Alberta Society of Petroleum Geologists, p. 15-81.

Osadetz, K. G., F. W. Jones, J. A. Majorowicz, D. E. Pearson, and L. D. Stasiuk, 1992, Thermal history of the cordilleran foreland basin in western Canada: a review, this volume.

Parrish, R. P., 1979, Geochronology and tectonics of the northern Wolverine complex, British Columbia: Canadian Journal of Earth Sciences, v. 16, p. 1428-1438.

Parrish, R. P., and J. O. Wheeler, 1983, A U-Pb zircon age from the Kuskanax batholith, southeastern British Columbia: Canadian Journal of Earth Sciences, v. 20, p. 1751-1756.

Parrish, R. P., S. D. Carr, and D. L. Parkinson, 1988, Eocene extensional tectonics and geochronology of the southern Omineca belt, British Columbia and Washington:

Tectonics, v. 7, p. 181–212.

Pearson, D. E., and D. A. Grieve, 1985, Rank variation, coalification pattern and coal quality in the Crowsnest coalfield, British Columbia: Canadian Insitute of Mining and Metallurgy Bulletin, v. 78, p. 39–46.

Porter, J. W., 1992, Early surface and subsurface investigations of the Western Canada sedimentary basin, this volume.

Porter, J. W., R. A. Price, and R. G. McGrossan, 1981, The Western Canada sedimentary basin: Philosophical Transactions of the Royal Society of London, v. A305, p. 169–192.

Poulton, T. P., 1984, The Jurassic of the Canadian Western Interior, from 49 degrees N latitude to Beaufort Sea, *in* D. F. Stott and D. J. Glass, eds., The Mesozoic of middle North America: Canadian Society of Petroleum Geologists Memoir 9, p. 15–41.

Price, R. A., 1973, Large-scale gravitational flow of supracrustal rocks, southern Canadian Rockies, *in* K. A. DeJong and R. Schotten, eds., Gravity and tectonics: New York, Wiley and Sons, p. 491–502.

Price, R. A., 1979, Intracontinental ductile crustal spreading linking the Fraser River and northern Rocky Mountain trench transform fault zones, south-central British Columbia and northeast Washington (abstract): GSA Abstract Programs, v. 11, p. 499.

Price, R. A., 1981, The cordilleran foreland thrust and fold belt in the southern Canadian Rocky Mountains, *in* K. R. McClay and N. J. Price, eds., Thrust and nappe tectonics: Geological Society of London Special Publication 9, p. 427–448.

Price, R. A., 1986, The southeastern Canadian Cordillera: thrust faulting, tectonic wedging and delamination of the lithosphere: Journal of Structural Geology, v. 8, p. 239–254.

Price, R. A., and D. M. Carmichael, 1986, Geometric test for Late Cretaceous–Paleogene intracontinental transform faulting in the Canadian Cordillera: Geology v. 14, p. 468–471.

Price, R. A., and P. R. Fermor, 1985, Structure section of the cordilleran foreland thrust and fold belt west of Calgary, Alberta: Geological Survey of Canada Paper 84-14, 1 sheet.

Price, R. A., and E. W. Mountjoy, 1970, Geologic structure of the Canadian Rocky Mountains between Bow and Athabasca Rivers—a progress report, *in* J. O. Wheeler, ed., Structure of the southern Canadian Cordillera: Geological Association of Canada Special Paper No. 6, p. 7–25.

Read, P. B., and R. L. Brown, 1981, Columbia River fault zone: southeastern margin of the Shuswap and Monashee complexes, southern British Columbia: Canadian Journal of Earth Sciences, v. 18, p. 1127–1145.

Rich, J. L., 1934, Mechanics of low-angle overthrust faulting illustrated by Cumberland thrust block, Virginia, Kentucky and Tennessee: AAPG Bulletin, v. 18, p. 1584–1596.

Schultheis, N., and E. W. Mountjoy, 1978, Cadomin conglomerate of western Alberta, a result of Early Cretaceous uplift of the main ranges: Bulletin of Canadian Petroleum Geology, v. 26, p. 297–342.

Shepard, W., and B. Bartow, 1986, Tectonic history of the Sweetgrass arch, a key to finding new hydrocarbons, Montana and Alberta, *in* J. H. Noll and K. M. Doyle, Rocky Mountain oil and gas fields, Wyoming Geological Association 1986 symposium: Casper, Wyoming, Wyoming Geological Association, p. 9–19.

Sikabonyi, L. A., and W. J. Rodgers, 1959, Paleozoic tectonics and sedimentation in the northern half of the west Canadian basin: Alberta Society of Petroleum Geologists Journal, v. 7, p. 193–216.

Spang, J. H., S. B. Brown, G. S. Stockmal, and I. W. Moffat, 1981, Structural geology of the Front Ranges, *in* R. I. Thompson and D. G. Cook, eds., Field guides to geology and mineral deposits, Geological Association of Canada/Mineralogical Association of Canada Calgary '81 annual meeting: Calgary, Alberta, University of Calgary Department of Geology and Geophysics, p. 101–120.

Stockmal, G. S., 1979, Structural geology of the northern termination of the Lewis thrust, Front Ranges, southern Canadian Rocky Mountains: unpublished M.Sc. thesis, University of Calgary, Calgary, Alberta, p. 174.

Stott, D. F., 1963, The Cretaceous Alberta Group and equivalent rocks, Rocky Mountain Foothills, Alberta: Geological Survey of Canada Memoir 317, 306 p.

Stott, D. F., 1984, Cretaceous sequences of the Foothills of the Canadian Rocky Mountains, *in* D. F. Stott and D. J. Glass, eds., The Mesozoic of middle North America: Canadian Society of Petroleum Geologists Memoir 9, p. 85–107.

Stronach, N. J., 1984, Depositional environments and cycles in the Jurassic Fernie Formation, southern Canadian Rocky Mountains, *in* D. F. Stott and D. J. Glass, eds., The Mesozoic of middle North America: Canadian Society of Petroleum Geologists Memoir 9, p. 43–67.

Struik, L. C., 1982, A re-examination of the type area of the Devono-Mississippian Cariboo Orogeny, central British Columbia: Canadian Journal of Earth Sciences, v. 18, p. 1767–1775.

Struik, L. C., 1987, Devonian breakup of North America (abstract), *in* Second international symposium on the Devonian System programs and abstracts: Calgary, Alberta, Canadian Society of Petroleum Geologists, p. 215.

Struik, L. C., 1989, Regional geology of the McLeod Lake map area, British Columbia, *in* Current research Part E: Geological Survey of Canada Paper 89-1E, p. 109–114.

Teal, P. R., 1983, The triangle zone at Cabin Creek, Alberta, *in* A. W. Bally, ed., Seismic expression of structural styles: a picture and work atlas: AAPG Studies in Geology Series 15, v. 3, p. 3.4.1.48–3.4.1.53.

Tempelman-Kluit, D. J., 1979, Transported cataclasite, ophiolite and granodiorite in Yukon: evidence of arc-continent collision: Geological Survey of Canada Paper 79-14, 27 p.

Tempelman-Kluit, D. J., and D. Parkinson, 1986, Extension across the Eocene Okanagan crustal shear in southern British Columbia: Geology, v. 14, p. 318–321.

Thompson, R. I., 1979, A structural interpretation across part of the northern Rocky Mountains, British Columbia, Canada: Canadian Journal of Earth Sciences, v. 16, p. 1228–1241.

Thompson, R. I., 1981, The nature and significance of large

"blind thrusts" within the northern Rocky Mountains of Canada, *in* R. B. Powers, ed., Geologic studies of the cordilleran thrust belt: Denver, Colorado, Rocky Mountain Association of Geologists, p. 47-60.

Tipper, H. W., 1981, Offset of an upper Pliensbachian geographic zonation in the North American Cordillera by transcurrent movement: Canadian Journal of Earth Sciences, v. 18, p. 1788-1792.

Van der Voo, R., M. Jones, C. S. Gromme, G. D. Eberlein, and M. Churchkin, Jr., 1980, Paleozoic paleomagnetism and northward drift of the Alexander terrane, southeastern Alaska: Journal of Geophysical Research, v. 85, p. 5281-5296.

Viau, C. A., 1987, Structural control on the sedimentological development of the Swan Hills Formation, Swan Hills field, Alberta, Canada (abstract), *in* Second international symposium on the Devonian System programs and abstracts: Calgary, Alberta, Canadian Society of Petroleum Geologists, p. 235.

Wanless, R. K., and J. E. Reesor, 1975, Precambrian zircon age of orthogneiss in the Shuswap Metamorphic complex, British Columbia: Canadian Journal of Earth Sciences, v. 12, p. 326-332.

Wernicke, B., and D. W. Klepacki, 1988, Escape hypothesis for the Stikine block: Geology, v. 16, p. 461-464.

Wheeler, J. O., and H. Gabrielse, 1972, The cordilleran structural province, *in* R. A. Price and R. J. W. Douglas, eds., Variations in tectonic styles in Canada: Geological Association of Canada Special Paper No. 11, p. 1-81.

Chapter 4

Relationship of the Stratigraphy of the Western Canada Foreland Basin to Cordilleran Tectonics: Insights from Geodynamic Models

Glen S. Stockmal
Douglas J. Cant
J. Sebastian Bell

Geological Survey of Canada
Institute of Sedimentary and Petroleum Geology
Calgary, Alberta, Canada

ABSTRACT

Geodynamic models of foreland basin development have enabled researchers to quantify the mechanical relationships between foreland basin subsidence and tectonic loads that cause down-flexing of the lithosphere. These models also provide a qualitative framework for construction of foreland basin stratigraphic sequences that are anticipated as having resulted from loading events caused by emplacement of overthrust sheets, terrane accretions, and changes in plate-boundary dynamics.

Initial terrane accretion and associated overthrust emplacement at a preexisting passive margin are anticipated to result in an unconformity-bounded foreland succession that will exhibit a shallowing-upward pattern similar to the classic Flysch to Molasse succession of the European Alpine basins. A basal unconformity develops as a result of cratonward migration of the peripheral bulge associated with lithospheric flexure caused by the tectonically advancing terrane. The shallowing of sedimentary facies occurs because initially low sediment supply, a consequence of little or no subaerial expression across the terrane at early stages, is succeeded by a progressively coarser and more voluminous supply as significant tectonic uplift occurs across the terrane. The upper unconformity may develop primarily as a result of (1) relaxation of lithospheric bending stresses through time following overthrusting, resulting in migration of the peripheral bulge toward the orogene, and/or (2) reduction of the flexural load on the lithosphere through erosion or tectonic denudation of the overthrust belt, causing regional uplift or basin "rebound."

Later clastic wedges, associated with subsequent overthrust events sufficient to significantly modify the size or position of the tectonic load depressing the foreland basin, may depart from this idealized sequence in that (1) basal unconformities may not develop if eustatic sea level is high, or if there is little or no time lag between events; and (2) a clear shallowing-upward trend may be precluded if sediment supply can keep pace with relative subsidence, impeding the normal nonmarine to marine transition between stacked clastic wedges. In the case of alternative (1), however, the transition from one clastic wedge to another may be marked by a transition from predominantly nonmarine to marine conditions. Some clastic units appear to correlate in time with known eustatic changes; others are considered indicative of changes in tectonic loading.

The stratigraphy of the Alberta basin has been subdivided by comparing it with the idealized sequence resulting from an individual tectonic loading (overthrust and/or accretion) event, modified to account for conditions prevalent after initial accretion. The ages of the six clastic wedges recognized (Fernie and Kootenay groups; Mannville Group; upper Fort St. John Group and Dunvegan Formation; Smoky Group and Belly River Formation; Edmonton Group; Paskapoo Formation—and their lateral equivalents) are compared with the times of accretion of Cordilleran terranes (Intermontane superterrane, terranes of the North Cascades and the Coast belt, Insular superterrane, and "outer" terranes of the latest Cretaceous–early Tertiary thrust stack). Some mechanical implications of a cause-and-effect relationship between terrane collisions and clastic wedges, if one exists, are discussed; it is also shown that most accreted terranes are too distant from the foreland basin to have influenced subsidence directly through tectonic loading of the lithosphere. General discrepancies in timing between collisions and clastic wedges may result from many factors, but the relatively well-understood temporal relations in the foreland basin provide an opportunity to more thoroughly investigate the use of foreland stratigraphy to constrain models of deformation and crustal thickening in the Cordillera.

INTRODUCTION

The evolution of foreland sedimentary basins can be simulated using quantitative geodynamic models that account for the flexural response of the Earth's lithosphere to tectonic loads. If the principal geologic events that contributed to development of a specific basin can be modeled accurately, then an understanding of the sensitivity of the model output to the input parameters should lead to an improved understanding of the tectonic subsidence, sediment supply, and gross stratigraphy of that basin (e.g., Beaumont, 1981; Jordan, 1981; Quinlan and Beaumont, 1984; Stockmal et al., 1986; Stockmal and Beaumont, 1987; Beaumont et al., 1987, 1988, in press-a). A synthetic model cannot explain everything at all scales, but it can indicate which processes are physically capable of leading to particular events and, conversely, it can suggest that certain interpretations of tectonic behavior are physically implausible.

One purpose of this volume is to compile information about one of the world's best-known foreland basins, the Western Canada foreland basin (Figure 1), with a view toward encouraging refinement of basin models—partly by identifying those areas on which more information is needed for progress. Geodynamic models have quantified the notion that loading of a cratonic margin by overthrust sheets generates an adjacent foreland basin. This relationship exists because (1) the load flexurally depresses the lithosphere, thereby creating a "foredeep" into which sediment can be shed, and (2) the erosion of topography produced across the telescoped overthrust belt provides large volumes of clastic detritus to fill the foredeep. Therefore, foreland basin stratigraphy and the tectonic loading history of the adjacent overthrust belt are not independent; there is a mechanical linkage through the lithosphere (Price, 1973; Beaumont, 1981). In theory, the observed stratigraphy can be used to estimate the sizes and positions of the inferred tectonic loads.

The approach of estimating tectonic loads through time was applied to the Western Canada foreland basin by Beaumont (1981), Stockmal and Beaumont (1987), and Beaumont et al. (in press-a). In these models, tectonic loads were adjusted iteratively such that the distributed thickness of the computed foreland basin-fill for a given time interval closely matched observations. At present, most foreland basin models are two-dimensional only,

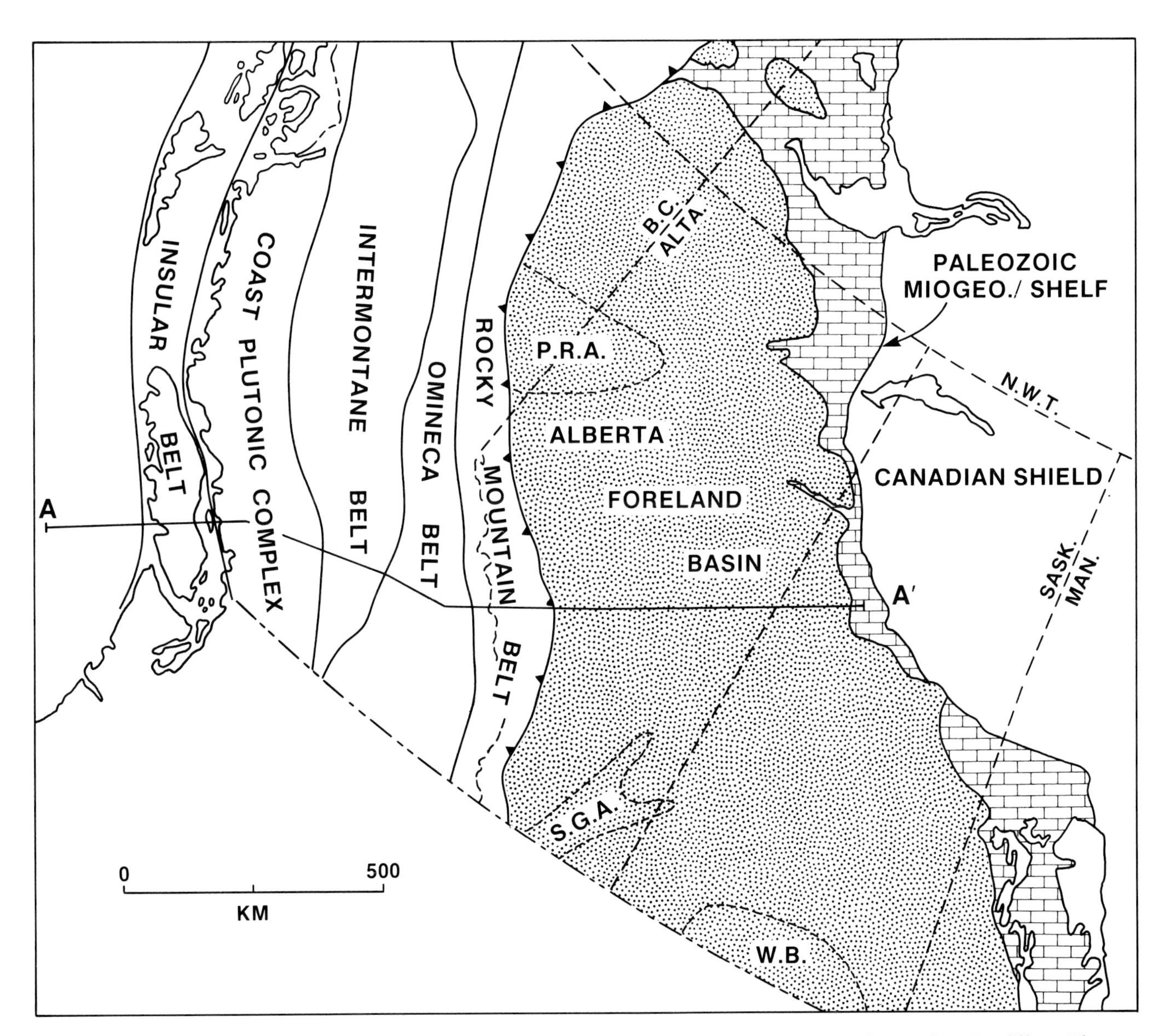

FIGURE 1. Tectonic subdivisions and features of the Western Canada foreland basin and Canadian Cordillera. The basin widens from 500 to 600 km in central Alberta to almost 1000 km in southern Saskatchewan, because of flexural yoking with the cratonic Williston basin (Beaumont et al., in press-a). A-A′ = line of section in Figure 12. P.R.A. = Peace River arch; S.G.A. = Sweetgrass arch; W.B. = Williston basin. From Cant and Stockmal (1989).

constructed to reproduce structural and stratigraphic cross sections (however, see Quinlan and Beaumont, 1984; Beaumont et al., 1987, 1988, in press-a).

In this chapter, we pursue four objectives. First, we discuss the mechanical linkage between a generalized foreland basin and its adjacent overthrust belt, emphasizing the anticipated internal stratigraphy of the basin. Second, we derive an "idealized" stratigraphic succession, based in part on predictions of geodynamic models. Third, we compare the idealized succession with sedimentary sequences within the Western Canada foreland basin, remembering to account qualitatively for variations in eustatic sea level and the relative rates of subsidence and sediment supply; we distinguish six discrete clastic wedges by this comparison. And fourth, we compare the ages of these six wedges with the times of collision of accreted terranes in the Canadian Cordillera.

This exercise tests the general concept that the Jurassic through Paleocene sediments in the Western Canada basin reflect crustal-scale and thin-skinned overthrusting driven by Cordilleran terrane accretion and associated plate boundary processes to the west (see Fermor and Moffat, 1992). Geodynamic considerations and the observed stratigraphy suggest that successive cycles of load-driven subsidence and erosion-related uplift are regional in nature because of flexure of the lithosphere; these took place against a backdrop of changing sea level. We do not dispute that the internal sedimentology and stratigraphy of some clastic wedges, and the deposition of some widespread units, were probably controlled by eustatic sea level variations (see Jervey, 1992), but previous geodynamic models (e.g., Beaumont, 1981; Jordan, 1981) coupled with basin-wide observations (e.g., Stott, 1984) strongly suggest that eustasy was not the primary mechanism that orchestrated the development of the largest-scale stratigraphic units.

GEODYNAMIC MODELS

The goal of geodynamic modeling of foreland basins is to understand quantitatively the mechanics of basin evolution. In detail, these models depend critically on the rheology of the lithosphere because it controls the flexural response to loads. To a first approximation, however, the Earth's lithosphere can be viewed as a simple elastic plate (e.g., Turcotte and Schubert, 1982). A load on the plate—resulting from stacking of overthrust sheets, for example—causes regional isostatic adjustment that is expressed as flexure (Figure 2). In response to an isolated load, the plate is downflexed over a specific wavelength, the dimensions of which depend on the strength of the plate (its flexural rigidity). The resultant lithospheric bending gives rise to depressions (foredeeps) into which sediment is shed. At the distal edge of a foredeep, an uplifted area develops (the peripheral bulge or forebulge) in response to the downflexure. The downflexed shape of a basin developed on an elastic lithosphere will not change with time, provided the magnitude of the load does not change, because the elastic bending stresses cannot relax (Figure 2a). Beaumont (1981) recognized that such behavior was not compatible with important details of the stratigraphic record in foreland basins, and postulated an alternative model plate rheology to accommodate relaxation of lithospheric bending stresses.

Such alternative models include uniform viscoelastic (Beaumont, 1981) and temperature-dependent viscoelastic rheologies (viscosity in the latter decreases exponentially with increasing temperature; Quinlan and Beaumont, 1984). The bending stresses relax through time following loading, decreasing the ability of the lithosphere to support flexurally the initial load applied (Figure 2b). If the magnitude of the load remains constant, the peripheral bulge migrates back toward the load, resulting in narrowing and *deepening* of the foredeep. This response can account for stratigraphic features such as regional erosion surfaces in the foreland basin that pass into conformable surfaces toward the orogene (see figure 19 in Quinlan and Beaumont, 1984).

The shapes of the deflected beams in Figure 2 represent flexure resulting from an isolated "point load" (or a "line load" acting on a plate in three dimensions, extending in and out of the plane of the figure). In the case of loading by an orogene-wide overthrust wedge, the flexural shape is the cumulative response to a great number of point loads, spaced very close together. The deflected plate beneath a very broad overthrust load may be almost horizontal beneath the load, but will have an associated foredeep beside the load.

The flexed shape of the lithosphere depends not only on its flexural rigidity (strength), but also on its continuity. In the simplest terms and for a two-dimensional case, the lithosphere can be modeled either as a continuous plate, extending to infinity in both directions, or as a "broken" plate, where the plate has a discrete edge and continues to infinity in only one direction (Figure 3). The proximity of the overthrust load and its associated foreland basin to the edge of the lithospheric plate (the plate tectonic boundary) determines whether a continuous plate or a broken plate is a more appropriate model. In two dimensions, a continuous plate is loosely analogous to a leaf spring in a car's suspension, whereas a broken plate acts somewhat like a diving board. If the load is positioned within one flexural half-wavelength (as defined in Figure 3) of the edge of the plate, it is better to use a broken plate model to determine the flexural response. If the load is positioned farther away, the continuous plate model is more appropriate. By consideration of the actual physical dimension of the flexural half-wavelengths for continuous and broken plates, certain bounds can be placed on the influence of loads that result in subsidence and infilling.

These flexural models indicate that loads cannot directly contribute to downflexing of a foreland basin unless they are applied at a distance of less than one lithospheric flexural half-wavelength from that basin. Overthrusting directly adjacent to the basin provides such direct loads, but terrane accretion need not, unless emplacement onto the lithosphere occurs within one flexural half-wavelength of the basin. Accreted terranes emplaced farther than one half-wavelength from the basin can, however, indirectly influence basin stratigraphy if they displace and deform earlier-accreted terranes lying closer to the basin and/or generate renewed thrusting and

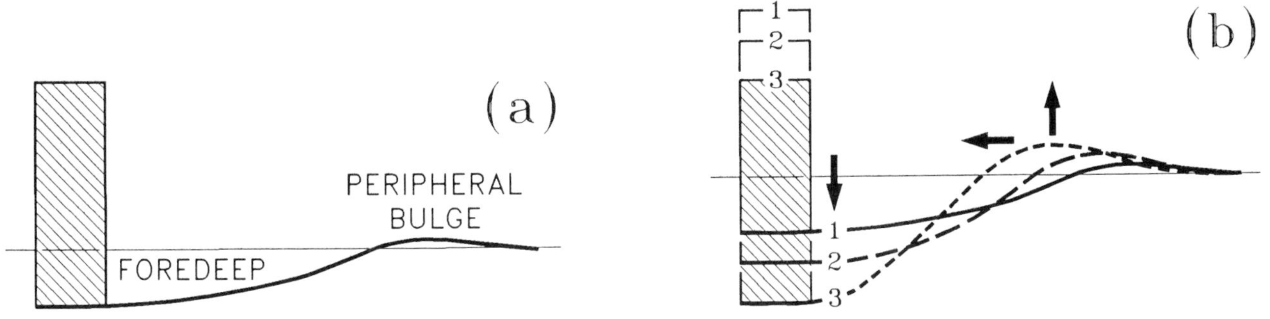

FIGURE 2. Flexural response of elastic (a) and viscoelastic (b) plates overlying an inviscid fluid. The load (shaded column) causes a deflection in the elastic plate manifested by a depression next to the load (the foredeep) and a distant flexural uplift (the peripheral bulge). For the elastic case (a), the shape of the flexed plate will not change with time if the magnitude of the load and the strength of the plate are also time-invariant. In a viscoelastic plate (b), however, the bending stresses induced by loading can relax with time, effectively weakening the flexural rigidity of the plate. Configuration 1 in (b), which is identical to the elastic response in (a), is the shape assumed initially by the viscoelastic plate if the load is applied instantaneously. With time, bending stresses are relaxed, and the shape of the flexed plate passes progressively through configurations 2 and 3; the peripheral bulge migrates toward the load and increases in amplitude, while the basin adjacent to the load deepens and narrows (arrows). In the important case of a temperature-dependent viscoelastic plate, such as that used by Quinlan and Beaumont (1984), the thermal structure of the lithosphere controls the effective viscosity at various depths. In realistic cases, viscous relaxation occurs relatively quickly immediately after loading as the hottest (most deep and weak) portions of the lithosphere deform. The rate of relaxation decreases exponentially, however, because the cooler and stronger portions of the lithosphere relax more slowly. In fact, the strongest regions of such a viscoelastic lithosphere will not relax even on time scales many times the age of the Earth.

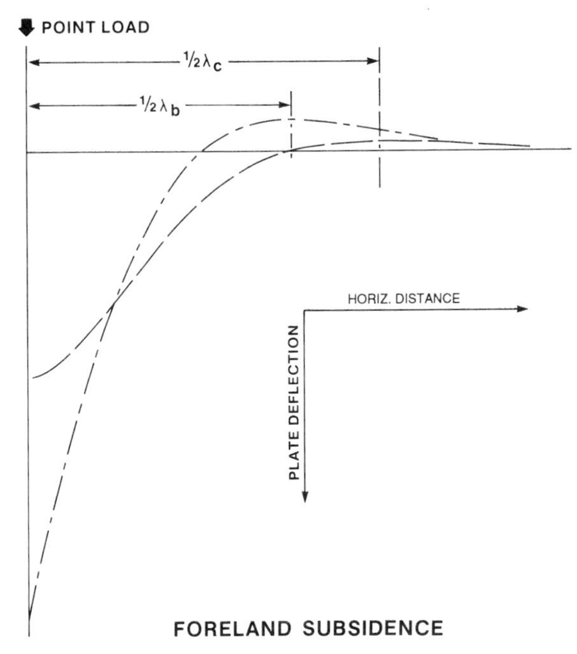

FIGURE 3. A comparison of shapes and magnitudes of deflections of continuous and broken elastic plates (definitions in text). The broken plate (long-short dashed line) is loaded at its edge. One-half of the continuous plate (long dashed line) is shown; the remainder would be a mirror image to the left. For a given flexural rigidity and point load, the broken plate will be depressed twice as much as the continuous plate. At the same time, the forebulge produced on the broken plate will be significantly larger and also closer to the load than its continuous-plate counterpart. The half-wavelength of flexure ($1/2\,\lambda$) is the distance between the point load and the forebulge. From Cant and Stockmal (1989).

telescoping within the miogeoclinal wedge. Flexure resulting from vertical loading of a laterally uniform lithospheric plate is the simplest case. In practice, most basins develop on lithosphere that is not uniform, but which has geometry and strength that reflect its previous geologic history. Prior to the evolution of the Western Canada foreland basin, the western Canadian continental margin was a westward-facing passive margin and the site of miogeoclinal accumulation of carbonates and clastics throughout most of the Paleozoic (McCrossan and Glaister, 1964). Much of the miogeocline and shelf/platform was structurally telescoped and overthrust eastward between the Jurassic and the earliest Tertiary (Porter et al., 1982; Fermor and Moffat, 1992). If this tectonic telescoping was largely the result of one or more collisions at the North American plate boundary, as suggested by present understanding of plate tectonics, the initial convergence is likely to have involved crustal-scale overthrusting whereby an eastward-moving terrane and associated overthrust sheets attempted to override the western Canadian continental margin.

Appropriate consideration of the passive margin has important implications for the evolution of the Western Canada foreland basin, as demonstrated by the geodynamic models of Stockmal et al. (1986) and Stockmal and Beaumont (1987). These models explicitly incorporate a preexisting passive margin (Figure 4, step I), providing a "hole" into which a very thick overthrust wedge (approximately 20+ km) can be emplaced, even though the wedge will have little to no topographic expression above sea level (Figure 4, steps II and III). By explicitly including the passive-margin bathymetry in the model it is possible to explain how a large flexural load can be applied at a plate margin (causing a substantial flexural foredeep), and yet not be associated initially with a major influx of clastic detritus (owing to low or negligible topog-

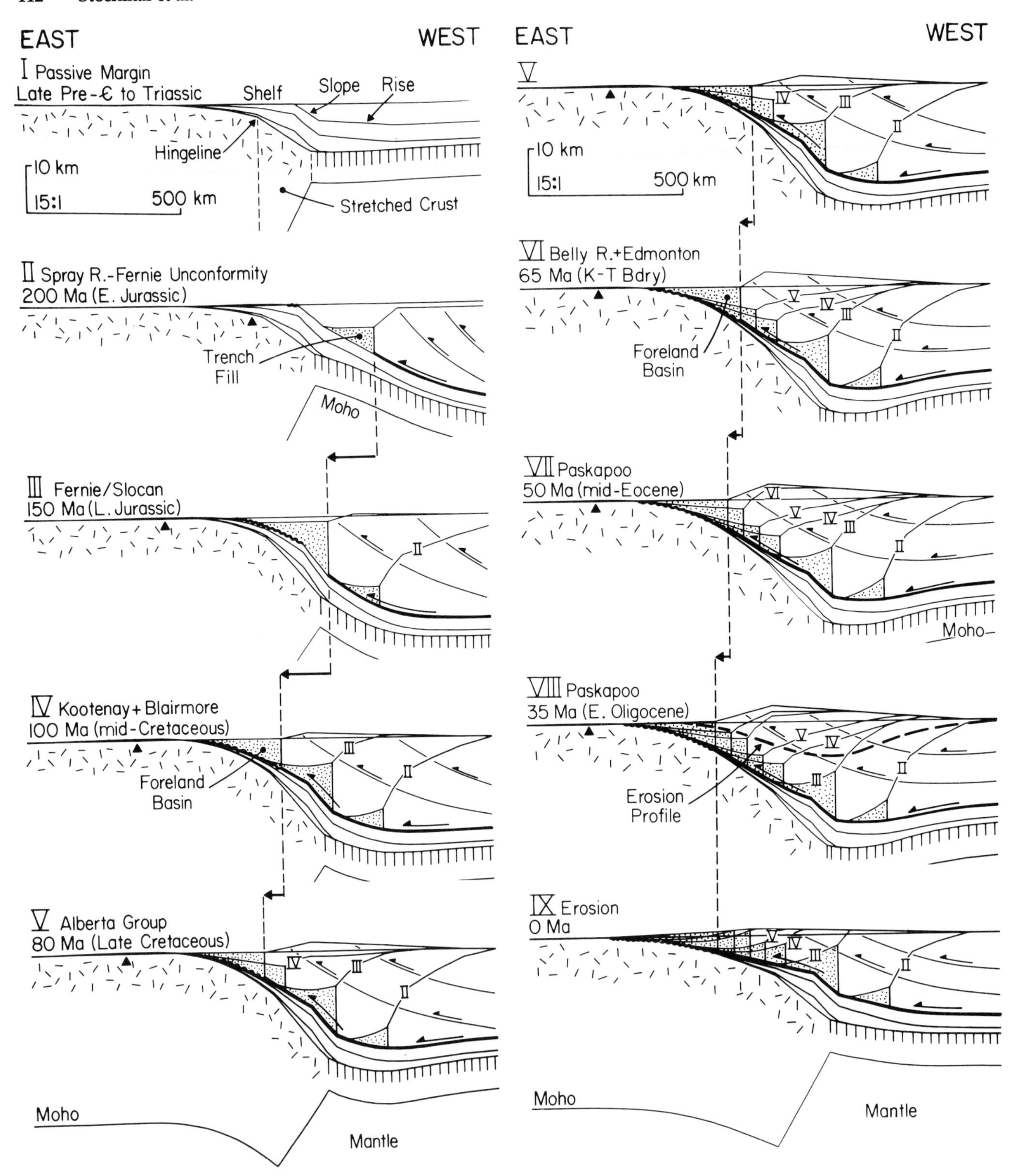

FIGURE 4. Geodynamic model of the Canadian Cordillera and the Western Canada foreland basin, from Stockmal and Beaumont (1987; refer to text for details). The cross sections are shown with a vertical exaggeration of 15:1 to highlight basin stratigraphy. Following previous conventions of placing the continent to the left, the sections are viewed from the north. The time steps were chosen to group major stratigraphic intervals, rounded to the nearest 5 million years, coupled with a logistical requirement to minimize computations. Step I shows the passive margin just prior to emplacement of the first overthrust load. Steps II and III demonstrate that a significant thickness of overthrusts, producing considerable flexure of the lithosphere, can be emplaced onto a passive margin with little or no topographic expression. Time steps IV through VIII illustrate the progressive emplacement of the growing overthrust wedge. Rapid growth of the wedge is required between steps V and VI to downflex the foreland sufficiently to accommodate the great thickness of Belly River and Edmonton. Time step IX shows the present-day cross section, following prolonged erosion. Random dashed pattern: continental crust. Vertical ruled pattern: oceanic crust. Stippled pattern: foreland basin or trench fill sediments. Solid triangle: position of flexural bulge.

raphy). However, as the overthrust wedge rides up the flank of the craton, significant topography is produced, and the basin fills to sea level and beyond (Figure 4, steps IV through XIII). This deep to shallow pattern of infilling implied by these models is supported by the stratigraphy of the Western Canada foreland basin, as noted below.

IDEALIZED STRATIGRAPHY

Geodynamic models demonstrate how emplacement of an overthrust wedge onto a preexisting passive margin produces a foredeep above the miogeocline-shelf succession, into which clastic material from the overthrust wedge is shed. A recognizable, "idealized" foreland stratigraphic succession may be expected, as constructed qualitatively by Cant and Stockmal (1989) with reference to a continuous episode of overthrusting, illustrated in four steps in Figure 5.

As the overthrust wedge approaches the passive cratonic margin, the associated foredeep and peripheral bulge migrate ahead of it, similar to a standing wave (Figure 5a). As the peripheral bulge migrates across the shelf and platform, uplift may be sufficient to cause subaerial exposure and erosion (cf. Jacobi, 1981). With continued bulge migration into shallower water, subaerial exposure becomes more certain. At this stage, the overthrust wedge occupies a position above the slope or rise of the old passive margin, in deep water (Figure 5b). Therefore, sediment shed from the overthrusts will be deposited in deep-water conditions (the Flysch stage of Bertrand, 1897; cf. Hsu, 1970; Crook, 1989).

As the overthrust wedge advances up the cratonic margin, the foredeep and peripheral bulge continue to migrate cratonward as well. The portions of the shelf and platform previously uplifted and exposed to erosion across the flexural bulge then subside, producing a "basal unconformity" beneath the clastic wedge now being shed from the overthrusts. As the overthrust wedge increases in topographic expression as a result of its movement up the flank of the craton, it sheds larger volumes of material into the foreland basin (Figure 5c). Deep-water conditions in the foreland basin are eventually replaced by shallow-water to nonmarine conditions, and the classic transition from flysch to molasse occurs (cf. Bertrand, 1897; Crook, 1989).

As convergence ceases or slows, bending stresses in the plate may relax at a rate higher than can be produced by emplacement of additional overthrust loads. This results in the migration of the peripheral bulge back across the basin toward the load, which may generate an unconformity on the top of the clastic wedge. Reducing the tectonic load on the lithosphere through erosion or tectonic denudation results in regional epeirogenic uplift that may also produce an unconformity (Figure 5d). However, the regional extents of these two types of unconformity differ, as do the magnitudes and distributions of erosion.

An unconformity that is caused by migration of the peripheral bulge back toward the load will die out toward the orogene, where the basin is deepening as a result of relaxation of bending stresses (Figures 6a and 2b). The greatest erosion in this case will occur within the basin, away from the deformation front (see figure 18 in Quinlan and Beaumont, 1984). In contrast, an unconformity caused by regional uplift resulting from removal of the tectonic load may be present across most of the basin, and the magnitude of erosion in the basin will increase toward the orogene to a maximum at the deformation front (Figure 6b). In practice, both processes are operative and will interfere constructively and destructively.

Therefore, the anticipated or "idealized" foreland basin stratigraphic succession, associated with the initial emplacement of an overthrust wedge onto a passive mar-

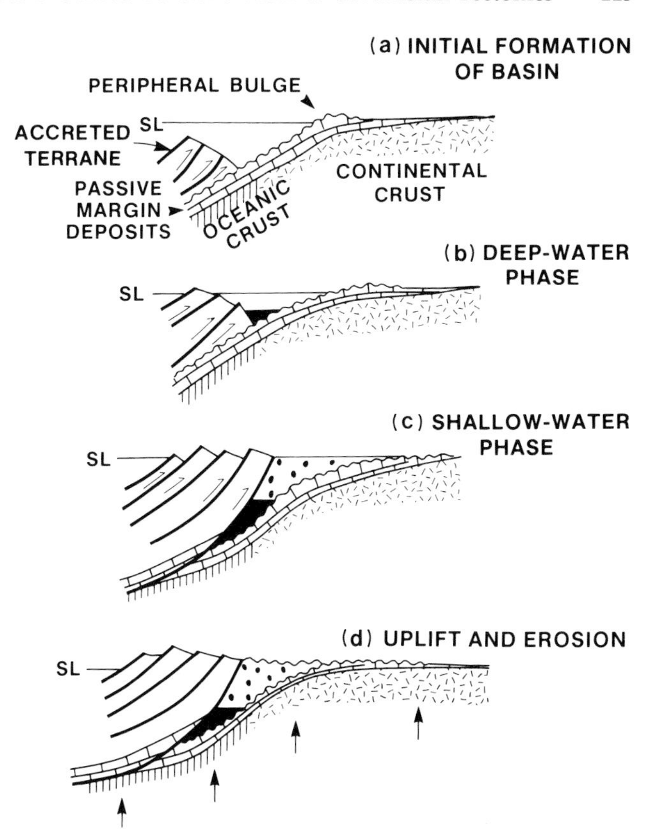

FIGURE 5. Idealized model of terrane accretion and development of the foreland basin. (a) Initial stage of accretion causes depression of the shelf and slope and uplift across the peripheral bulge. A basal unconformity is cut as the shelf and platform rocks are uplifted across the bulge. Little sediment is shed into the basin because of the absence of significant uplift above sea level. (b) Low rates of sedimentation plus rapid flexurally induced subsidence result in initial deep-water conditions in the foreland basin. Sediment shed from limited uplifted areas is deposited as deep-water "flysch." (c) Continued convergence results in uplift across the overthrust wedge and high sedimentation rates; the foreland basin fills to or above sea level with shallow-water to nonmarine "molasse." (d) Following the period of overthrusting, erosion across the orogene reduces the load and the basin undergoes epeirogenic uplift resulting in an upper regional unconformity. From Cant and Stockmal (1989).

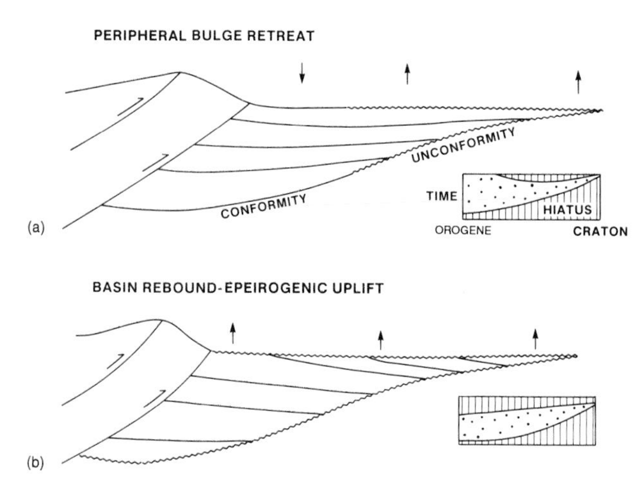

FIGURE 6. Stratigraphy produced by relaxation of bending stresses and erosion of the tectonic load. (a) An unconformity produced by a peripheral bulge migrating back toward the orogene should die out into a conformity as the orogene is approached, because relaxation of bending stresses results in deepening of the foredeep adjacent to the load (Figure 2; cf. figure 19 in Quinlan and Beaumont, 1984). (b) An unconformity resulting from erosion of the orogenic belt and partial removal of the tectonic load. If the associated uplift is sufficient to mask the effects of stress relaxation illustrated in (a), the unconformity will transect the entire basin and the greatest erosion in the basin will be immediately adjacent to the orogenic belt. From Cant and Stockmal (1989).

gin and deposited above the preexisting shelf and platform, is unconformity bounded and coarsens upward from deep- to shallow-water facies (Figure 7; cf. Covey, 1986; Crook, 1989). Subsequent clastic wedges, produced by renewed overthrusting (perhaps resulting from accretion of terranes outboard of the initial overthrust pile), may deviate from this ideal succession, because of one or more of the following. (1) The situation of initial accretion of a terrane onto a preexisting passive margin is predisposed toward deep-water conditions at the base of the clastic wedge; later overthrusting events would downflex a basin whose depositional base level will be much closer to or even above sea level, and therefore predisposed to shallow-water to nonmarine conditions. (2) Basin sedimentation rates may be high relative to tectonic subsidence rates, resulting in the maintenance of sedimentary base level at or above sea level. (3) Eustatic increases in sea level, not explicitly included here, alter both the accommodation space available and the magnitude of uplift required to create an unconformity. (4) A brief time period between wedges (e.g., as between the Belly River and Edmonton wedges, discussed below) may be insufficient for significant relaxation of the flexural bulge and for regional uplift resulting from erosion.

Outboard of the passive margin shelf edge, where initial water depths exceed 300 to 400 m, peripheral bulge uplift will be insufficient to produce a basal erosional unconformity. Identifying the transition from passive margin to foreland deposition becomes problematical, and is partly a matter of perspective. Conceptually, we would prefer to place the boundary not where craton-derived material gives way to overthrust-derived material, but where the driving mechanism of subsidence changes from thermal subsidence and miogeoclinal sediment loading (the passive margin phase) to flexural loading by overthrusts (the foreland basin phase). This is particularly relevant where the first "foreland" sediments deposited, following passage of the flexural bulge across a carbonate platform, are platform carbonates themselves (as in the Appalachians of western Newfoundland; Jacobi, 1981). However, in the absence of a clear "basal unconformity," the transition from shelf to foreland in relatively deep-water clastics is difficult to identify.

If the foreland basin is already filled to sedimentary base level at a time of introduction of additional detritus, this clastic material will pass through the basin rather than be deposited (the "steady state" of Jamieson and Beaumont, 1988; cf. Covey, 1986). Therefore, the presence of a discrete clastic wedge, bounded by unconformities or by abrupt transitions from nonmarine to marine or from shallower- to deeper-water conditions, implies not only a supply of detritus but also a tectonic loading event sufficient to depress flexurally the foreland and create accommodation space. The ultimate source of the driving forces behind the tectonic loading events is presumably plate tectonics operating along the active plate boundary.

Although plate boundary processes, including terrane accretion, are envisaged as causing widespread deformation within the orogene, which in turn results in tectonic loading, the physical position of such loading must be within one lithospheric flexural half-wavelength of the basin if the loading is to have a flexural effect. As discussed below, this implies that the tectonic loads holding down the Western Canada foreland basin must be thrust sheets within the most inboard accreted terranes and the telescoped miogeocline; the in-filling foreland sediments themselves also contribute to basin flexure.

CLASTIC WEDGES IN THE WESTERN CANADA FORELAND BASIN

The gross stratigraphy of the Western Canada foreland basin was interpreted by Porter et al. (1982) in terms of superterrane collisions in the Cordilleran orogene. We suggest that a somewhat finer subdivision of the foreland succession can be made, which may correlate with tectonic loading events in the orogene. We consider the stratigraphic succession found in western to central Alberta and northeastern British Columbia, shown in simplified form in Figure 8. Subdivision of Alberta basin strata into groups and formations is dependent on geographic position and whether information was gained principally through outcrops or boreholes. We use subdivisions corresponding to usage in the Central and Northwest Plains (for details, see Leckie and Smith, 1992).

A comparison of the idealized foreland basin sequence with this generalized stratigraphy of the Western Canada foreland basin suggests a subdivision into six discrete clastic wedges (Figure 9; Cant and Stockmal, 1989). These six wedges correspond to the five cycles

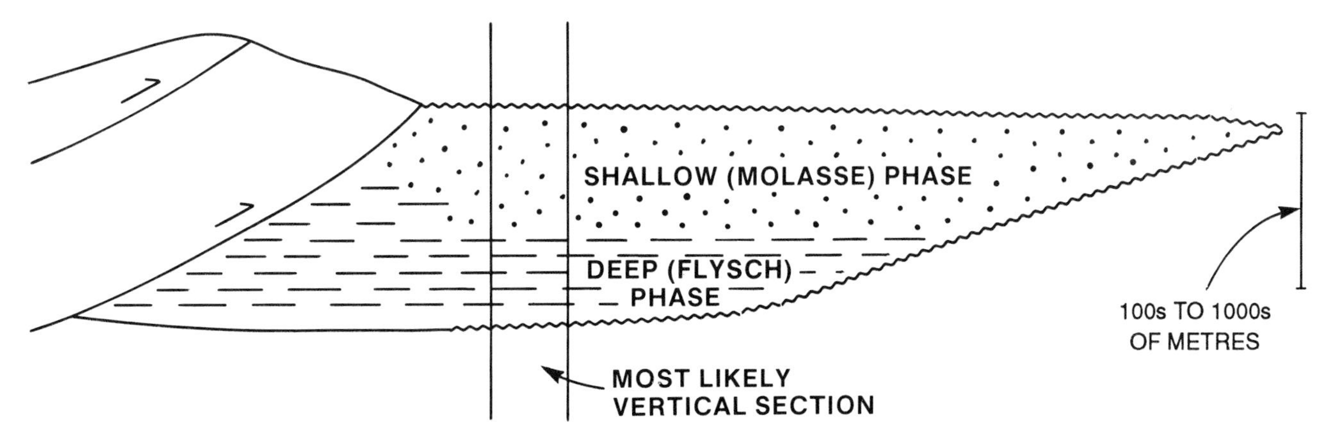

FIGURE 7. **Idealized foreland basin sequence, unconformity bounded with a coarsening and shallowing-upward progression. From Cant and Stockmal (1989).**

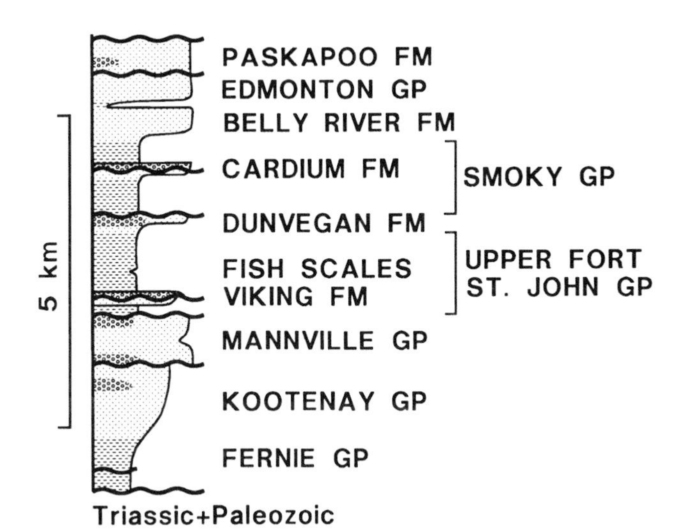

FIGURE 8. **Generalized stratigraphy of the Western Canada foreland basin; thicknesses are approximate because of basin-wide variability. The Viking and Cardium sandstones and conglomerates are considered products of eustatic sea level drops, whereas other discrete, first-order clastic wedges (Figure 9) are considered products of tectonic loading and downflexure of the foreland basin.**

identified by Leckie and Smith (1992), except that we emphasize the separation of the Belly River Formation from the Edmonton Group (resulting in the additional unit). As anticipated, some wedges deviate somewhat from the "ideal" succession, but they are distinct packages bounded either by widespread unconformities or by significant and rapid changes in sedimentary facies (e.g., nonmarine to marine conditions).

The generalized stratigraphy used here corresponds closely to that of Stott (1984), with the addition of the Tertiary Paskapoo Formation and the deletion of some thin sandstone units in northeastern British Columbia (Goodrich and Sikanni formations). Details of many units have been simplified or neglected, owing largely to the regional nature of our investigation. The clastic wedges (numbered 1 to 6) are described below.

The classic study of Eisbacher et al. (1974) subdivided the Alberta foreland basin into four lithological "assemblages," which were grouped into pairs to form two major coarsening-upward molasse successions or "megacycles." However, this grouping now appears to be an oversimplification. For example, their lower molasse succession comprises the "Upper Fernie" and "Kootenay-Blairmore" assemblages, although a major unconformity separates the Kootenay Formation and the Blairmore Group, thereby disrupting the coarsening-upward pattern.

(1) Fernie and Kootenay Groups

The first clastic-wedge succession begins within the Jurassic Fernie Group shales, which rest unconformably on truncated Paleozoic and Triassic platformal and miogeoclinal rocks of the former passive margin. The Fernie Group is marine and of deep-shelf to perhaps upper-slope origin, but the point of transition from passive margin subsidence to foreland subsidence is unclear. The first westerly derived detritus occurs above a sub-Oxfordian disconformity (Poulton, 1984), but the beveled basal unconformity (e.g., Stronach, 1984) may represent initial peripheral bulge migration (Figure 9). The overlying and conformable Upper Jurassic through early Neocomian nonmarine Kootenay Group (Nikanassin equivalent) is truncated above by a major unconformity. This upper unconformity represents a period of approximately 12 million years, encompassing most of the Hauterivian and Barremian (Stott, 1984), although Ricketts and Sweet (1986) have suggested that the contact in the westernmost exposures is essentially conformable. The unconformity-bounded and shallowing-upward nature of the Fernie-Kootenay succession matches the idealized sequence very well. One deviation from the classic Alpine flysch is the

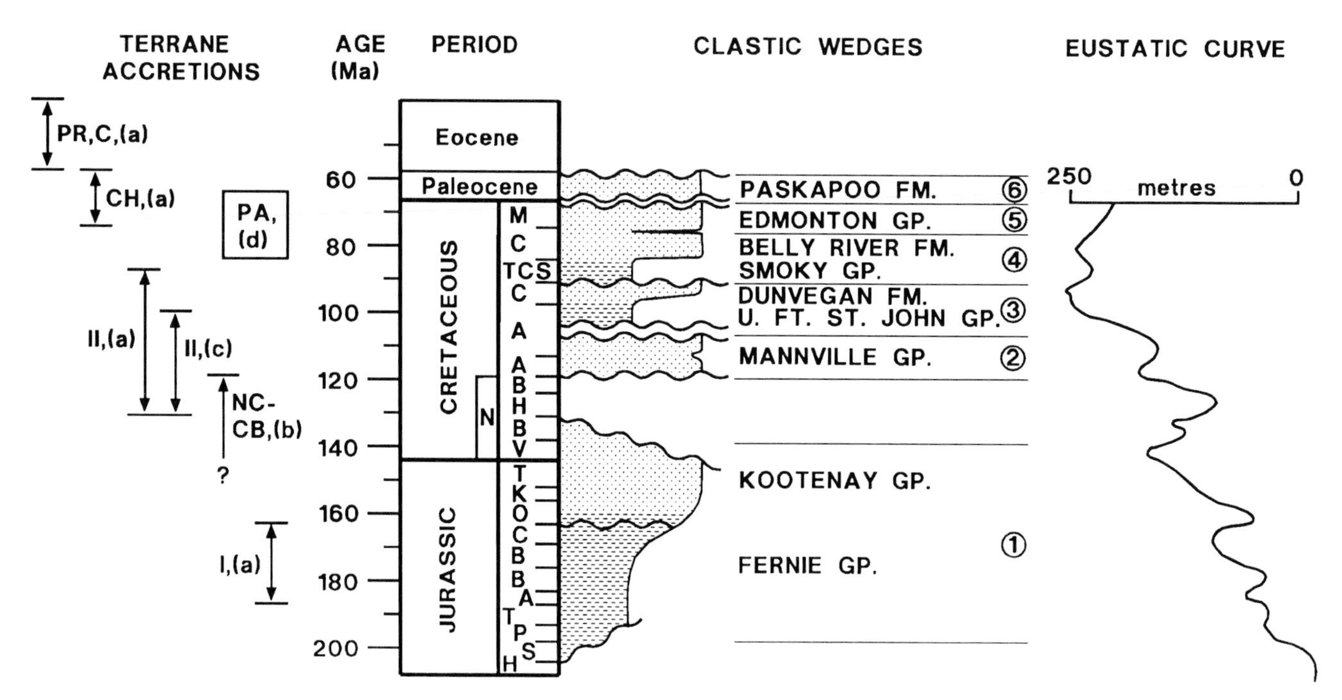

FIGURE 9. The six first-order clastic wedges of the Western Canada foreland basin shown as a function of time (absolute ages and Jurassic and Cretaceous age names, abbreviated by single letters, from the DNAG timescale; Palmer, 1983), with a eustatic sea level curve from Haq et al. (1987), and the times of accretion of allochthonous terranes (see Figure 10) taken from: (a) Gabrielse and Yorath (1989); (b) Monger et al. (in press); and (c) Thorkelson and Smith (1989). I = Intermontane superterrane; NC-CB = North Cascades and Coastal belt terranes; II = Insular superterrane; CH = Chugach terrane; PR = Pacific Rim terrane; and C = Crescent terrane. Eustatic drops are seen not to correspond in any general way with the stratigraphy, with a possible exception at approximately 125 Ma, where the brief relative lowstand may have enhanced tectonic uplift effects. Shown also is the period of development of the Purcell anticlinorium (PA), as given by (d) Price (1981).

apparent absence of true deep-water turbidites. However, this may simply reflect lack of preservation.

(2) Mannville Group

The Fernie-Kootenay sequence is overlain unconformably by the shallow marine to nonmarine, Aptian to middle Albian Mannville Group (Blairmore and Bullhead-lower Fort St. John equivalent). In southern and central Alberta and Saskatchewan, the Mannville Group lacks a deep to shallow progression of infill, but this may be a consequence of eustatic fluctuations, as suggested by large-scale regressions and transgressions in its internal stratigraphy (see Cant and Stockmal, in press). The top of the Mannville is marked by a middle Albian unconformity (Stelck and Kramers, 1980). The time gap represented by this unconformity is only a few million years, constrained by a continuous section overlying the Peace River arch (Cant, 1988). The lower part of the Mannville is largely absent along the axis of the Western Canada foreland basin, where the unit onlaps a chain of paleogeographic "islands" formed from Paleozoic rocks (Rudkin, 1964; see Leckie and Smith, 1992). This unusual "axial swell" may be a consequence of the peripheral bulge.

(3) Upper Fort St. John Group and Dunvegan Formation

Shale deposition predominated during late Albian to Cenomanian time, as an interval approximately 500 m thick accumulated. This period was a time of generally rising sea level (Vail et al., 1977; Haq et al., 1987) and the establishment of water depths of up to approximately 300 m in the Western Canada foreland basin. The well-preserved organic matter of the Albian/Cenomanian Fish Scale Zone may reflect anoxic conditions in relatively deep water. The Albian Viking Formation sandstone, however, which represents the only major influx of coarse clastics in this shale interval (Figure 8), can be interpreted as the depositional consequence of a eustatic sea level drop on the basis of its stratigraphy and sedimentology, comparison with global sea level curves, and its similarities to equivalent units in the western U.S.A. (Hein et al., 1986; Cant and Hein, 1986; Leckie, 1986). The base of the nonmarine Cenomanian Dunvegan Formation is gradational with the underlying shallow marine shales. The conglomerates and sandstones of the Dunvegan form a wedge up to 300 m thick (Stott, 1984) that extends from northeastern British Columbia into central Alberta. The upper section of the Dunvegan is truncated by an unconformity in the north, whereas to the south its facies equivalent passes upward into marine shales (Stott, 1982). The relative importance of eustasy to tectonism is presently unclear, but eustatic fluctuations may have caused channeling and delta progradation (Bhattacharya, 1988).

(4) Smoky Group and Belly River Formation

Global sea levels were generally high during Turonian to Campanian time (Vail et al., 1977; Haq et al., 1987; Figure 9), possibly promoting the deposition of what were predominantly shales in a few hundred meters of water. Deep water is reflected in the deposition of pelagic carbonates (Second White Speckled Shale) and the preserved marine organic debris. The upper Turonian Cardium (Figure 8) and Santonian Bad Heart formations are the only major sandstone units in this interval. As with the Viking Formation, they are also now believed to have resulted from eustatic sea level drops (Cant and Hein, 1986; Bergman and Walker, 1987; Plint and Walker, 1987). Therefore, we do not ascribe the same level of tectonic significance to the Viking, Cardium, and Bad Heart formations that we do to the Dunvegan Formation. The Campanian Belly River Formation is up to 800 m thick, grades upward from marine shale to nonmarine sediments, and represents a large supply of coarse detritus to the basin. The marine Bearpaw shale, conformably overlying the Belly River Formation, forms the base of the next clastic wedge, represented by the Edmonton Group. In northwest Alberta, the Belly River Formation and Edmonton Group are not stratigraphically separable; Leckie and Smith (1992) place them in the same depositional cycle.

(5) Edmonton Group

The Maastrichtian Edmonton Group shows a shallowing-upward trend in south-central Alberta, from the marine Bearpaw shale to nonmarine coal-bearing clastics. The time interval between the tectonic loading events responsible for the Belly River and Edmonton wedges may have been too brief for formation of an unconformity resulting from erosional unloading or lithospheric relaxation (as noted above). Over much of the plains, the top of the Edmonton Group is truncated by an unconformity, perhaps reflecting erosional unloading in the orogene prior to loading, resulting in deposition of the next clastic wedge.

(6) Paskapoo Formation

The Paleocene Paskapoo Formation is a nonmarine wedge of sandstone, shale, and minor conglomerate, bounded below by an unconformity (Lerbekmo et al., 1990) and above by the present-day erosion surface. A considerable thickness of Tertiary sediments is believed to have formerly rested above the Paskapoo Formation; estimates of the overburden removed by erosion (e.g., Magara, 1976; Hacquebard, 1977; Nurkowski, 1984; Osadetz et al., 1992) range up to 3 km along the western edge of the basin. This erosion probably reflects regional flexural uplift across the basin resulting from reduction of the tectonic load in the adjacent orogene (through erosion, and possibly through Eocene extension in British Columbia).

The generalized stratigraphy of the Alberta portion of the Western Canada foreland basin is not everywhere exactly as described here. For example, the unconformity between the Edmonton and Paskapoo wedges is in some places partly occupied by lacustrine and other nonmarine sediments.

CORDILLERAN ACCRETION AND FORELAND WEDGES

The recognition of discrete, allochthonous terranes in the North American Cordillera has greatly aided the unraveling of complex histories of deformation and accretion across orogenes in general and the Canadian Cordillera in particular (e.g., Monger and Price, 1979; Coney et al., 1980; Price et al., 1981, 1985; Monger et al., 1982, in press; Monger, 1989). The collage of terranes in the Canadian Cordillera is dominated by two large composite terranes: the Intermontane superterrane and the Insular superterrane (Figure 10; see also Fermor and Moffat, 1992). In addition, numerous smaller terranes, some perhaps composite, are recognized (Monger, 1989, states that over 200 North American Cordillera terranes have been delineated). The timing of terrane accretion is constrained by overlap assemblages, stitching intrusions, and deformational-metamorphic history (but not through correlation with the foreland succession). With regard to intrusive and metamorphic dates, however, care must be taken in relating these to times of tectonic loading and hence foreland subsidence. Jamieson and Beaumont (1988) discussed pressure-temperature-time paths in developing orogenes, and clearly illustrated that high-grade metamorphism and granites generated by partial melting can significantly precede emplacement of a terrane onto a preexisting margin (see their figure 6).

Figure 11 schematically illustrates the initial accretion of an allochthonous terrane to the flank of a craton occupied by a passive margin sequence (Figures 11a and b). This situation is similar to that in Figure 5, and is applied in the models of Stockmal et al. (1986) and Stockmal and Beaumont (1987; Figure 4, steps I through IV). Figure 11c illustrates a possible configuration for accretion of other terranes outboard of the first accreted terrane (see also the discussion by Monger et al., in press, regarding the requisite switch in the sense of subduction). Such later accretion events can result in cratonward translation and tectonic thickening of already accreted and more inboard terranes, given the present understanding of "critical taper" concepts (Chapple, 1978; Dahlen and Suppe, 1988) and the implications for internal deformation of an orogenic wedge as a result of structural addition of material in a compressional environment. A similar scenario is illustrated by Jamieson and Beaumont (1988; see their figure 2).

In this manner, even small terranes, accreted farther than one flexural half-wavelength away from the basin, may influence the stratigraphy of the foreland by deforming inboard terranes and the miogeoclinal sequence. Ter-

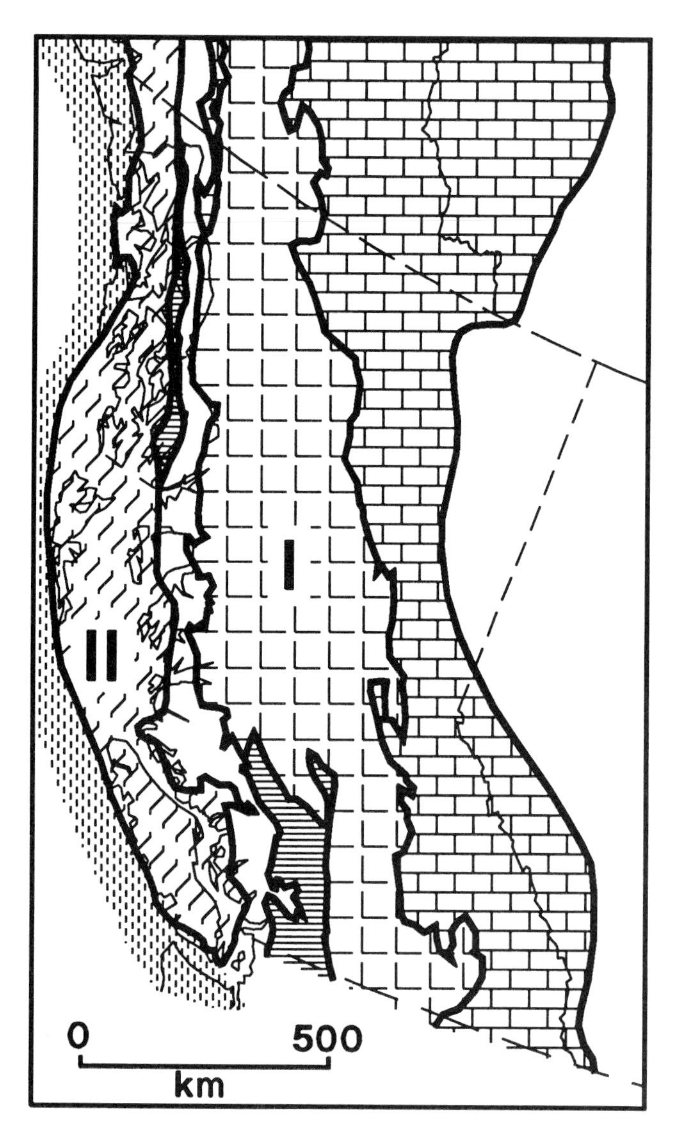

FIGURE 10. Allochthonous terranes of the Canadian Cordillera; simplified after Gabrielse and Yorath (1989). I = composite Intermontane superterrane (Cache Creek, Dorsey, Quesnellia, Slide Mountain, Stikinia, and Windy-McKinley terranes); II = composite Insular superterrane (Alexander and Wrangellia terranes); brick pattern = deformed North American miogeocline and shelf sequence, plus pericratonic terranes (Kootenay, Nisling, and Pelly Gneiss terranes); horizontal ruled pattern = North Cascades and Coast belt terranes (Bridge River, Cadwallader, Chilliwack, Harrison, Methow, Shuksan, and Taku terranes); vertical ruled pattern = "outer terranes" of the latest Cretaceous–early Tertiary thrust stack (Chugach, Crescent, Hoh, Olympic Core, Ozette, Pacific Rim, and Yakutat terranes); unpatterned area between I and II = undivided metamorphics.

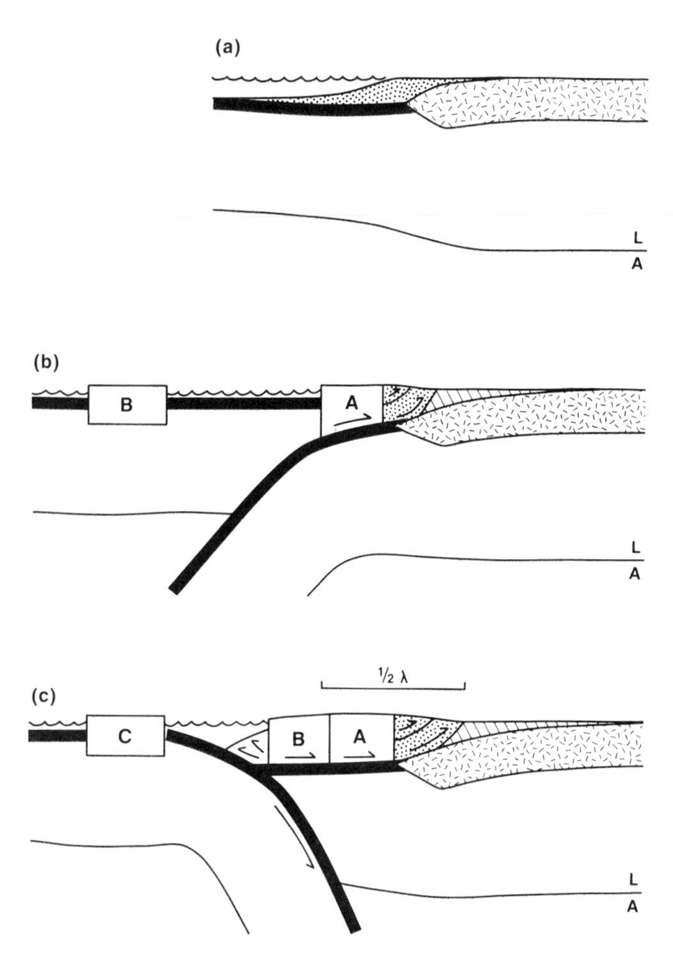

FIGURE 11. Schematic illustration of terrane accretion and consequences for the foreland basin. Extreme vertical exaggeration. (a) Passive margin draped by miogeoclinal wedge (stippled) above continental (random dashed pattern) and oceanic (black) crust. L/A = lithosphere/asthenosphere boundary. (b) Accretion of terrane A above oceanward-dipping subduction zone. Passive margin wedge is deformed and thrust cratonward. Foreland basin (diagonal lines) develops in foredeep created by load of terrane A and thickened and displaced miogeoclinal rocks. (c) Geometry following accretion of terrane B but prior to accretion of terrane C. In this hypothetical case, the sense of subduction reversed following accretion of A and "clogging" of the subduction zone with buoyant continental lithosphere. Terrane B is more than one half-wavelength from the foreland basin; therefore, its influence is felt by displacement and thickening of terrane A and the miogeoclinal wedge (compare b and c). When terrane C collides with B, it will be stripped from the downgoing plate and will displace B, A, and the miogeoclinal wedge cratonward. Modified after Stockmal and Beaumont (1987).

ranes accreted in a strongly transpressional environment, as in the case of the North American Cordillera (e.g., Price and Carmichael, 1986; Wernicke and Klepacki, 1988; Monger et al., in press), would influence the Western Canada foreland basin proportional to their perpendicular component of convergence. Subsequent strike-slip removal or replacement of terrane B or C in Figure 11c, for example, will not necessarily affect the basin if it does not cause redistribution of mass either in terrane A or in the deformed miogeoclinal wedge, because terranes B and C are too far removed from the basin to directly influence it through flexure.

In the case of the western Canadian Cordillera, the accretion of the Intermontane superterrane loosely coincides in time with initial stages of deposition of the Fernie-Kootenay wedge (Figure 9). A similar temporal relationship between early foreland basin sedimentation and deformation within and west of the Omineca belt was pointed out by Eisbacher et al. (1974). The published time of accretion (Middle Jurassic; Gabrielse and Yorath, 1989) somewhat precedes the age of westerly derived detritus (latest Middle Jurassic to early Neocomian). This accretion date, based principally on K-Ar and U-Pb dates of granitic intrusions (Monger et al., in press), can be reconciled with the foreland basin record because significant high-grade metamorphism and even granite production through partial melting within a large allochthonous terrane can be expected as the terrane structurally thickens in its attempt to override the toe of the cratonic margin (Jamieson and Beaumont, 1988). In addition, the portion of the pre-Oxfordian Fernie Group deposited under conditions of foreland (flexural) subsidence is uncertain.

Deposition of the Mannville and upper Fort St. John–Dunvegan clastic wedges loosely overlapped in time with accretion of terranes in the North Cascades, the Coastal belt, and the Insular superterrane (Figure 9). In terms of their present surface expression (Figure 10), the North Cascades and Coastal belt terranes are considerably smaller in size than either superterrane. Nevertheless, their accretion could have resulted in thickening and cratonward displacement of the Intermontane terrane and parts of the miogeocline, perhaps as suggested schematically in Figure 11c, sufficient to depress the foreland. The explanation for the discrepancy between the areally restricted nonmarine Dunvegan wedge (see figure 23 in Stott, 1984) and the temporally correlative but areally extensive Insular superterrane is uncertain. Perhaps accretion of terranes of the Coastal belt, such as the Taku terrane (Gabrielse and Yorath, 1989), provided the tectonic "push" that resulted in loading adjacent to the region of Dunvegan deposition. Other such terranes, in addition to Taku, could have been either structurally removed by intense suturing between the two superterranes or so heavily intruded by and metamorphosed with rocks of the Coast plutonic complex as to be unrecognizable (Monger et al., in press; i.e., the unpatterned area of undivided metamorphics in Figure 10).

The tectonic setting of post-Kootenay deposition is complicated by the required switch in the sense of subduction following collision of the Intermontane superterrane. The switch to eastward-directed subduction may have been accompanied by a shallowing of the subducting slab, resulting in entrained flow in the upper asthenosphere and dynamic downflexure of the lithosphere beneath the foreland basin (Mitrovica et al., 1989; Beaumont et al., in press-a, see their figure 17). This effect may account in part for the discrete Mannville wedge, for which Beaumont (1981) was forced to propose a large eustatic sea level rise in order to model accurately its cross-sectional shape. Such an imposed relative sea level rise could be produced by the dynamic effects of shallow subduction, which would also broaden the basin considerably, as seen for the Mannville wedge. Interestingly, the Albian Crowsnest volcanics, which cap the Blairmore Group (Mannville equivalent) in southern Alberta and which occupy such an anomalous position so far into the foreland (cf. figure 4 in Armstrong, 1988), could be related to such a shallowly subducting slab.

The protracted accretion of the Insular superterrane, perhaps stretching from the Hauterivian through to the Coniacian, followed closely in time by accretion of the Chugach and Pacific Rim terranes from Maastrichtian through Eocene time (Gabrielse and Yorath, 1989), correlate loosely with near-continuous deposition of the Smoky–Belly River, Edmonton, and Paskapoo wedges (Figure 9). Eisbacher et al. (1974) had also grouped the Alberta Group through Paskapoo Formation succession into an upper "megacycle," correlating its deposition with structural shortening in the eastern cordillera. Eocene accretion of the Crescent and Olympic Core terranes (Figure 9; Gabrielse and Yorath, 1989) has no known correlative in the foreland basin, perhaps because of intervening extensional tectonics in southeastern British Columbia.

Although "accretion" of the Insular superterrane appears to have been completed by Santonian time (Gabrielse and Yorath, 1989), the compressive tectonics associated with the accretion event could have persisted at least through Campanian time to result in tectonic loading and lithospheric flexure sufficient to accommodate the Turonian through Campanian Smoky–Belly River wedge. This period also overlaps in time with the Campanian through early Paleocene development of the Purcell anticlinorium (Figure 9; Price, 1981). The Purcell anticlinorium is a major structural feature of the southern Rocky Mountain belt, developed as a result of wholesale thickening and detachment of Proterozoic belt-Purcell miogeoclinal clastics. These rocks, deposited outboard of the ancient crustal hinge line as a northeasterly tapering prism, were structurally displaced inboard of the hinge line to produce the anticlinorium (see figure 5 in Price, 1981). The timing of this major structure is apparently coincident with a rapid period of overthrust loading required to flexurally accommodate the Belly River and Edmonton wedges (Figure 4, step VI; see also figure 8 in Stockmal and Beaumont, 1987). The late stages of Insular superterrane accretion, coupled with plate boundary forces associated with Chugach–Pacific Rim accretion, may have driven tectonic development of the anticlinorium (cf. figure 9b in Stockmal and Beaumont, 1987).

Our discussion here of timing relationships differs in detail from that of Cant and Stockmal (1989). The clear one-to-one temporal correlations that Cant and Stockmal believed to exist were based on accretion ages and terrane definitions published in the early to mid-1980s, and these ages and definitions have recently been revised. This degree of volatility in the current understanding of the timing of accretion and the nature and number of discrete terranes suggests that the loose correlations we have drawn here will also undergo revision. However, this does not alter our basic hypothesis that tectonic loading of the edge of the North American craton by cordilleran overthrust sheets, causing flexural subsidence of the Alberta basin, should be related in some manner (through both "direct" and "indirect" flexural influences) to terrane

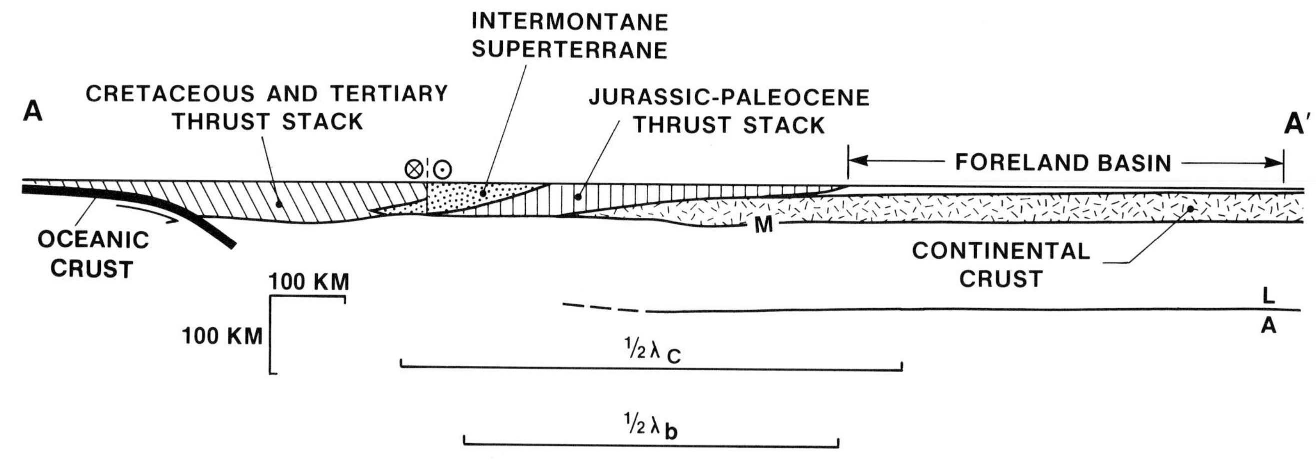

FIGURE 12. True-scale simplified cross section across the Western Canada foreland basin and the Canadian Cordillera; modified after DNAG Transect B2 (Monger et al., 1985). Line of section shown in Figure 1. Jurassic–Paleocene thrust stack includes telescoped miogeoclinal wedge and cannibalized foreland sediments; Cretaceous and Tertiary thrust stack includes Insular superterrane and other outboard terranes. M = Moho; L/A = lithosphere/asthenosphere boundary; λ_c and λ_b are the half-wavelengths of continuous and broken uniform elastic plates of strength appropriate for the North American lithosphere.

accretion events and/or plate boundary dynamics. Differences in timing may also reflect the fact that "accretion" is not a sharply defined process mechanically, and is therefore difficult to relate to compressional deformation of miogeoclinal/shelf rocks and already accreted terranes.

Figure 12 is a true-scale cross section of the Canadian Cordillera and the Western Canada foreland basin (modified from DNAG Transect B2, Monger et al., 1985), showing the positions of accreted terranes relative to the basin, and the flexural half-wavelengths of continuous and broken elastic plates (approximately 700 and 500 km, respectively, with sediment-filled foredeeps) of appropriate strength for the North American craton (Stockmal and Beaumont, 1987). The transported miogeoclinal wedge (Rocky Mountain belt and eastern Omineca belt, Figure 1) is seen, by virtue of its proximity to the foreland basin (Figure 12), to provide the principal load downflexing the lithosphere to produce the basin. The Intermontane superterrane (Intermontane belt and western Omineca belt, Figure 1) is sufficiently close to the western portion of the foreland basin to have affected it directly to a small degree (Figure 12). The outboard terranes, however, are clearly positioned too far from the basin to have had substantial direct effects, even if the continuous plate model is assumed (Figure 12). Regardless of how restricted they are areally, accretion of these outboard terranes, or their equivalents now displaced northward by strike-slip movement, may have tectonically thickened and/or displaced much of the overthrust package cratonward, as shown in Figure 11c, to produce a stratigraphic signature in the basin. In addition, major strike-slip faults (shown schematically in Figure 12) may cut the lithosphere, producing a broken plate geometry, requiring that overthrust loads be transferred onto the end of the plate to provide a flexural load (cf. figure 11 in Stockmal and Beaumont, 1987).

Note that we are comparing computed flexural half-wavelengths with the present-day distances between accreted terranes and the foreland basin. These observed distances could be considered minimum bounds because structural telescoping occurred simultaneously with foreland basin deposition. An additional complicating factor is the potentially large Eocene extension in the southern Cordillera, which partly counterbalanced the earlier structural shortening. In general, however, because palinspastic position should be taken into account, the accreted terranes were farther removed from the foreland basin in the past than at present. This enhances the argument that terranes outboard of the Intermontane superterrane could only indirectly (in the sense above) influence foreland basin subsidence and sedimentation by thickening and displacing intervening terranes and the miogeocline/shelf succession.

In detail, the linkage between foreland basin deposition and discrete loading events is undoubtedly more complex than discussed here. Present evidence, however, suggests a loose temporal correlation between the docking of accretionary terranes on the western edge of North America and the development of clastic wedges in the Western Canada foreland basin. Discrepancies are probably indicative of the complex chain of processes through which distant accreted terranes may influence foreland basin subsidence; they should be resolvable with an improved understanding of the events and processes involved.

SUMMARY

Geodynamic models of the role of lithospheric flexure in foreland basin development have increased our understanding of the configurations and mechanical causes of regional unconformities and first-order changes in sedimentary facies that bound discrete foreland clastic wedges. Similar models suggest that initial terrane accretion at passive continental margins, and subsequent inboard deformation associated with accretion of more outboard terranes and changes in plate boundary processes, provide the tectonic loads that deflect the lithosphere to produce foreland basins.

An idealized foreland basin sequence, derived in a qualitative fashion from the predictions of these geodynamic models, can be compared with observed foreland basin stratigraphy to identify discrete wedges associated with discrete tectonic loading events. Such wedges, identified in the Western Canada foreland basin, correlate loosely in time with the accretion of allochthonous terranes in the Canadian Cordillera.

It is broadly accepted that the observed stratigraphy in the foreland basin provides an excellent and sensitive (although incomplete) record of tectonics in the adjacent orogene. It follows, therefore, that we may be able to use the ages of clastic wedges to make inferences regarding the timing of tectonic loading events. In the simplest terms, we would anticipate these loading events to be caused by significant terrane accretions (if there is a convergent component to their motion, rather than purely strike slip) or by changes in plate boundary dynamics (e.g., changes in angle of subduction, age of subducted oceanic lithosphere, velocity of convergence, etc.). A marriage of research in the foreland basin and the Cordillera, especially west of the Omineca belt, should prove to be very profitable.

Considerable work must be done before we are able to map accurately the form, distribution, bounding surfaces, temporal relations, and longitudinal variations of and within the various clastic wedges in the Western Canada foreland basin. We hope that this paper, although speculative, emphasizes the need for interdisciplinary studies and demonstrates that our understanding of the evolution of the Western Canada foreland basin is likely to advance fastest in a research environment that includes construction of synthetic geodynamic and stratigraphic models (e.g., Kleinspehn and Paola, 1988; Cross, 1990), collection of new geologic and geophysical data (e.g., Cook et al., 1988), and the gleaning of new insights from data already gathered.

ADDENDUM

Since this paper was submitted and critically reviewed, several important papers have appeared that bear upon the details of relating discrete clastic wedges to specific events. Perhaps most notable are papers dealing with basin-filling processes (e.g., Angevine et al., 1990; Flemings and Jordan, 1990; Jordan and Flemings, 1991) and the influence of climatic variations (e.g., Beaumont et al., in press-b). Evidence has also accumulated suggesting that the two superterranes were already linked as far back as Middle Jurassic time, calling into question the major mid-Cretaceous collision event (van der Heyden, 1992). We have not attempted to rewrite our present paper to accommodate the views expressed in these new contributions, but rather draw the reader's attention to them.

ACKNOWLEDGMENTS

We wish to thank Margot McMechan and John Bloch for careful and highly constructive critical reviews, Chris Beaumont for numerous general discussions, Tony Edwards and Ross Boutilier for internal reviews of an earlier version of this paper, and Roger Macqueen and Dale Leckie for editorial comments.

REFERENCES CITED

Angevine, C. L., P. L. Heller, and C. Paola, 1990, Quantitative sedimentary basin modeling: AAPG, Continuing Education Course Note Series #32, 133 p.

Armstrong, R. L., 1988, Mesozoic and early Cenozoic magmatic evolution of the Canadian Cordillera, *in* S. P. Clark, Jr., B. C. Burchfiel, and J. Suppe, eds., Processes in continental lithospheric deformation: GSA Special Paper 218, p. 55–91.

Beaumont, C., 1981, Foreland basins: The Geophysical Journal of the Royal Astronomical Society, v. 65, p. 291–329.

Beaumont, C., G. M. Quinlan, and J. Hamilton, 1987, The Alleghanian orogeny and its relationship to the evolution of the Eastern Interior, North America, *in* C. Beaumont and A. J. Tankard, eds., Sedimentary basins and basin-forming mechanisms: Canadian Society of Petroleum Geologists, Memoir 12, and Atlantic Geoscience Society, Special Publication 5, p. 425–445.

Beaumont, C., G. Quinlan, and J. Hamilton, 1988, Orogeny and stratigraphy: numerical models of the Paleozoic in the Eastern Interior of North America: Tectonics, v. 7, p. 389–416.

Beaumont, C., G. M. Quinlan, and G. S. Stockmal, in press-a, The evolution of the Western Interior basin: causes, consequences, and unsolved problems: Geological Association of Canada, Special Paper Series, "Geodynamic Evolution of the Western Interior Basin."

Beaumont, C., P. Fullsack, and J. Hamilton, in press-b, Erosional control of active compressional orogens, *in* K. McClay, ed., Proceedings of thrust tectonics 1990.

Bergman, K. M., and R. G. Walker, 1987, The importance of sea level fluctuations in the formation of linear conglomerate bodies: Carrot Creek Member of Cardium Formation, Cretaceous Western Interior seaway, Alberta, Canada: Journal of Sedimentary Petrology, v. 57, p. 651–665.

Bertrand, M., 1897, Structure des Alpes Francaises et recurrence de certaines facies sedimentaires: VIe International Congress (Zurich), p. 161–177.

Bhattacharya, J., 1988, Autocyclic and allocyclic sequences in river- and wave-dominated deltaic sediments of the Upper Cretaceous, Dunvegan Formation, Alberta: core examples, *in* D. P. James and D. A. Leckie, eds., Sequences, stratigraphy, sedimentology: surface and subsurface: Canadian Society of Petroleum Geologists Memoir 15, p. 25–32.

Cant, D. J., 1988, Regional structure and development of the Peace River arch, Alberta: a Paleozoic failed-rift system?: Bulletin of Canadian Petroleum Geology, v. 36, p. 284–295.

Cant, D. J., and F. J. Hein, 1986, Depositional sequences in ancient shelf sediments: some contrasts in style, *in* R. J. Knight and J. R. McLean, eds., Shelf sands and sandstones: Canadian Society of Petroleum Geologists, Memoir 11, p. 303–312.

Cant, D. J., and G. S. Stockmal, 1989, Stratigraphy of the Alberta foreland basin: an interpretation in terms of cordilleran tectonics: Canadian Journal of Earth Sciences, v. 26, p. 1964–1975.

Cant, D. J., and G. S. Stockmal, in press, Some controls on sedimentary sequences in foreland basins: examples from the Alberta basin: Special Publication of the International Association of Sedimentologists, edited by R. Steel.

Chapple, W. M., 1978, Mechanics of thin-skinned fold-and-thrust belts: GSA Bulletin, v. 89, p. 1189–1198.

Coney, P. J., D. L. Jones, and J. W. H. Monger, 1980, Cordilleran suspect terranes: Nature, v. 288, p. 329–333.

Cook, F. A., A. G. Green, P. S. Simony, R. A. Price, R. Parrish, B. Milkereit, P. L. Gordy, R. L. Brown, K. C. Coflin, and C. Patenaude, 1988, Lithoprobe seismic reflection structure of the southeastern Canadian Cordillera: initial results: Tectonics, v. 7, p. 157–180.

Covey, M., 1986, The evolution of foreland basins to steady state: evidence from the western Taiwan foreland basin, *in* P. A. Allen and P. Homewood, eds., Foreland basins: International Association of Sedimentologists, Special Publication 8, p. 77–90.

Crook, K. A. W., 1989, Suturing history of an allochthonous terrane at a modern plate boundary traced by flysch-to-molasse facies transitions: Sedimentary Geology, v. 61, p. 49–79.

Cross, T. A., 1990, Quantitative dynamic stratigraphy: Prentice Hall.

Dahlen, F. A., and J. Suppe, 1988, Mechanics, growth, and erosion of mountain belts, *in* S. P. Clark Jr., B. C. Burchfiel, and J. Suppe, eds., Processes in continental lithospheric deformation: GSA Special Paper 218, p. 161–178.

Eisbacher, G. H., M. A. Carrigy, and R. B. Campbell, 1974, Paleodrainage patterns and late-orogenic basins of the Canadian Cordillera, *in* W. R. Dickinson, ed., Tectonics and sedimentation: SEPM Special Publication No. 22, p. 143–166.

Fermor P. R., and I. W. Moffat, 1992, Tectonics and structure of the Western Canada foreland basin, this volume.

Flemings, P. B., and T. E. Jordan, 1990, Stratigraphic modeling of foreland basins: interpreting thrust deformation and lithosphere rheology: Geology, v. 18, p. 430–434.

Gabrielse, H., and C. J. Yorath, 1989, DNAG #4: The Cordilleran orogen in Canada: Geoscience Canada, v. 16, p. 67–83.

Hacquebard, P. A., 1977, Rank of coal as an index of organic metamorphism for oil and gas in Alberta, *in* The origin and migration of petroleum in the Western Canada sedimentary basin, Alberta: Geological Survey of Canada, Bulletin 262, p. 11–22.

Haq, B. U., J. Hardenbol, and P. R. Vail, 1987, Chronology of fluctuating sea levels since the Triassic: Science, v. 235, p. 1156–1167.

Hein, F. J., M. E. Dean, A. M. Delure, S. K. Grant, G. A. Robb, and F. J. Longstaffe, 1986, The Viking Formation in the Caroline, Garrington, and Harmattan East fields, western south-central Alberta: sedimentology and paleogeography: Bulletin of Canadian Petroleum Geology, v. 34, p. 91–110.

Hsu, K. J., 1970, The meaning of the word flysch—a short historic search, *in* Flysch sedimentology in North America: Geological Association of Canada, Special Paper 7, p. 1–11.

Jacobi, R. D., 1981, Peripheral bulge—a causal mechanism for the lower/middle Ordovician unconformity along the western margin of the Northern Appalachians: Earth and Planetary Science Letters, v. 56, p. 245–251.

Jamieson, R. A., and C. Beaumont, 1988, Orogeny and metamorphism: a model for deformation and pressure-temperature-time paths with applications to the central and southern Appalachians: Tectonics, v. 7, p. 417–445.

Jervey, M. T., 1992, Siliciclastic sequence development in foreland basins, with examples from the Western Canada foreland basin, this volume.

Jordan, T. E., 1981, Thrust loads and foreland basin evolution, Cretaceous, western United States: AAPG Bulletin, v. 65, p. 2506–2520.

Jordan, T. E., and P. B. Flemings, 1991, Large-scale stratigraphic architecture, eustatic variation, and unsteady tectonism: a theoretical evaluation: Journal of Geophysical Research, v. 96, n. B4, p. 6681–6699.

Kleinspehn, K. L., and C. Paola, 1988, New perspectives in basin analysis: New York, Springer-Verlag, 453 p.

Leckie, D., 1986, Tidally influenced, transgressive shelf sediments in the Viking Formation, Caroline, Alberta: Bulletin of Canadian Petroleum Geology, v. 34, p. 111–125.

Leckie, D., and D. Smith, 1992, Regional setting, evolution, and depositional cycles of the Western Canada foreland basin, this volume.

Lerbekmo, J. F., M. E. Evans, and G. S. Hoye, 1990, Magnetostratigraphic evidence bearing on the magnitude of the sub-Paskapoo disconformity in the Scollard Canyon–Ardley area of the Red Deer Valley, Alberta: Bulletin of Canadian Petroleum Geology, v. 38, p. 197–202.

Magara, K., 1976, Thickness of removed sedimentary rocks, paleopore pressure, and paleotemperature, southwestern part of the western Canadian basin: AAPG Bulletin, v. 60, p. 554–565.

McCrossan, R. G., and R. P. Glaister, 1964, Geological history of western Canada: Calgary, Alberta, Alberta Society of Petroleum Geologists, 232 p.

Mitrovica, J. X., C. Beaumont, and G. T. Jarvis, 1989, Tilting of continental interiors by the dynamic effects of subduction: Tectonics, v. 8, p. 1079–1094.

Monger, J. W. H., 1989, Overview of cordilleran geology, Chapter 2, *in* B. D. Ricketts, ed., Western Canada sedimentary basin: a case history: Canadian Society of Petroleum Geologists, p. 9–32.

Monger, J. W. H., and R. A. Price, 1979, Geodynamic evolution of the Canadian Cordillera—progress and problems: Canadian Journal of Earth Sciences, v. 16, p. 771–791.

Monger, J. W. H., R. A. Price, and D. J. Tempelman-Kluit, 1982, Tectonic accretion and the origin of the two major metamorphic and plutonic welts in the Canadian Cordillera: Geology, v. 10, p. 70–75.

Monger, J. W. H., R. M. Clowes, R. A. Price, P. S. Simony, R. P. Riddihough, and G. J. Woodsworth, 1985, Continent-ocean transect B2, Juan de Fuca plate to Alberta plains: GSA Transects Program, v. 2, 21 p.

Monger, J. W. H., R. M. Clowes, D. S. Cowan, C. J. Potter, R. A. Price, and C. J. Yorath, in press, Continent-ocean transitions in western North America between latitudes 46 and 56 degrees: transects B1, B2, B3, *in* Decade of North American Geology Volume: GSA.

Nurkowski, J. R., 1984, Coal quality, coal rank variation and its relation to reconstructed overburden, Upper Cretaceous and Tertiary plains coals, Alberta, Canada: AAPG Bulletin, v. 68, p. 285–295.

Osadetz, K. G., F. W. Jones, J. A. Majorowicz, D. E. Pearson, and L. D. Stasiuk, 1992, Thermal history of the Cordilleran foreland basin in Western Canada, this volume.

Palmer, A. R., 1983, The Decade of North American Geology: 1983 Geologic time scale: Geology, v. 11, p. 503–504.

Plint, A. G., and R. G. Walker, 1987, Morphology and origin of an erosion surface cut into the Bad Heart Formation during major sea-level change, Santonian of west-central Alberta, Canada: Journal of Sedimentary Petrology, v. 57, p. 639–650.

Porter, J. W., R. A. Price, and R. G. McCrossan, 1982, The Western Canada sedimentary basin: Philosophical Transactions of the Royal Society of London, Series A, v. 305, p. 169–192.

Poulton, T. P., 1984, The Jurassic of the Canadian Western Interior, from 49°N latitude to Beaufort Sea, *in* D. F. Stott and D. J. Glass, eds., The Mesozoic of middle North America: Canadian Society of Petroleum Geologists Memoir 9, p. 15–41.

Price, R. A., 1973, Large-scale gravitational flow of supracrustal rocks, southern Canadian Rocky Mountains, *in* K. A. De Jong and R. A. Scholten, eds., Gravity and tectonics: New York, Wiley-Interscience, p. 491–502.

Price, R. A., 1981, The Cordilleran foreland thrust and fold belt in the southern Canadian Rocky Mountains, *in* K. R. McClay and N. J. Price, eds., Thrust and nappe tectonics: Geological Society of London Special Publication No. 9, p. 427–448.

Price, R. A., and D. M. Carmichael, 1986, Geometric test for Late Cretaceous-Paleogene intracontinental transform faulting in the Canadian Cordillera: Geology, v. 14, p. 468–471.

Price, R. A., J. W. H. Monger, and J. E. Muller, 1981, Cordilleran cross-section, Calgary to Victoria, *in* R. I. Thompson and D. G. Cook, eds., Field guides to geological and mineralogical deposits, Calgary '81 Annual Meeting, Geological Association of Canada, Mineralogical Association of Canada, and Geological and Geophysical Union: Calgary, Alberta, University of Calgary, Department of Geology and Geophysics, p. 261–334.

Price, R. A., J. W. H. Monger, and J. A. Roddick, 1985, Cordilleran cross-section; Calgary to Vancouver, trip 3, *in* D. Tempelman-Kluit, ed., Field guides to geology and mineral deposits in the southern Canadian Cordillera, Geological Association of America, Cordilleran Section annual meeting: Vancouver, British Columbia, Geological Survey of Canada, p. 3.1–3.85.

Quinlan, G. M., and C. Beaumont, 1984, Appalachian thrusting, lithospheric flexure and the Paleozoic stratigraphy of the Eastern Interior of North America: Canadian Journal of Earth Sciences, v. 21, p. 973–996.

Ricketts, B. D., and A. R. Sweet, 1986, Stratigraphy, sedimentology and palynology of the Kootenay-Blairmore transition in southwestern Alberta and southeastern British Columbia: Geological Survey of Canada, Paper 84-15, 41 p.

Rudkin, R. A., 1964, Lower Cretaceous, Chapter 11, *in* R. G. McCrossan and R. P. Glaister, eds., Geological history of western Canada: Calgary, Alberta, Alberta Society of Petroleum Geologists, p. 156–168.

Stelck, C. R., and J. W. Kramers, 1980, *Freboldiceras* from the Grand Rapids Formation of north-central Alberta: Bulletin of Canadian Petroleum Geology, v. 28, p. 509–521.

Stockmal, G. S., and C. Beaumont, 1987, Geodynamic models of convergent margin tectonics: the southern Canadian Cordillera and the Swiss Alps, *in* C. Beaumont and A. J. Tankard, eds., Sedimentary basins and basin-forming mechanisms: Canadian Society of Petroleum Geologists, Memoir 12, and Atlantic Geoscience Society, Special Publication 5, p. 393–411.

Stockmal, G. S., C. Beaumont, and R. Boutilier, 1986, Geodynamic models of convergent margin tectonics; transition from rifted margin to overthrust belt and consequences for foreland-basin development: AAPG Bulletin, v. 70, p. 181–190.

Stott, D. F., 1982, Lower Cretaceous Fort St. John Group and Upper Cretaceous Dunvegan Formation of the Foothills and Plains of Alberta, British Columbia, District of Mackenzie and Yukon Territory: Geological Survey of Canada, Bulletin 328, 124 p.

Stott, D. F., 1984, Cretaceous sequences of the Foothills of the Canadian Rocky Mountains, *in* D. F. Stott and D. J. Glass, eds., The Mesozoic of middle North America: Canadian Society of Petroleum Geologists, Memoir 9, p. 85–107.

Stronach, N. J., 1984, Depositional environments and cycles in the Jurassic Fernie Formation, southern Canadian Rocky Mountains, *in* D. F. Stott and D. J. Glass, eds., The Mesozoic of middle North America: Canadian Society of Petroleum Geologists, Memoir 9, p. 43–67.

Thorkelson, D. J., and A. D. Smith, 1989, Arc and intraplate volcanism in the Spences Bridge Group: implications for Cretaceous tectonics in the Canadian Cordillera: Geology, v. 17, p. 1093–1096.

Turcotte, D. L., and G. Schubert, 1982, Geodynamics; applications of continuum physics to geological problems: New York, John Wiley and Sons, 450 p.

Vail, P. R., R. M. Mitchum Jr., and S. Thompson III, 1977, Seismic stratigraphy and global changes of sea level, Part 4, global cycles of relative changes of sea level, *in* C. E. Payton, ed., Seismic stratigraphy—applications to

hydrocarbon exploration: AAPG Memoir 26, p. 83–97.
van der Heyden, P., 1992, A Middle Jurassic to early Tertiary Andean-Sierran arc model for the Coast belt of British Columbia: Tectonics, v. 11, p. 82–97.
Wernicke, B., and D. W. Klepacki, 1988, Escape hypothesis for the Stikine block: Geology, v. 16, p. 461–464.

Chapter 5

Early Surface and Subsurface Investigations of the Western Canada Sedimentary Basin

J. W. Porter

Calgary, Alberta, Canada

INTRODUCTION

As noted elsewhere throughout this volume, the Western Canada foreland basin makes up only the younger part of the Western Canada sedimentary basin, consisting of those generally westerly derived sediments of Middle Jurassic and younger age; the underlying miogeocline-platform wedge makes up the remainder of the Western Canada sedimentary basin. In terms of both the history of the region and of oil and gas exploration, it is neither feasible nor desirable to treat the foreland basin separately. Accordingly, this account outlines the early history of the Western Canada sedimentary basin, with foreland basin developments highlighted as appropriate.

Exploratory drilling for oil or gas in the Western Canada sedimentary basin did not commence until the 1880s. However, the occurrence of pitch or tar deposits along the banks of the Athabasca River in northeastern Alberta had been known for over 100 years. Alfred R. C. Selwyn, the second director of the Geological Survey of Canada, stated in his Report of Progress for 1873–74 (Selwyn, 1874) that "there seems but little doubt that Canada has her [sic] a salt and oil bearing region surpassing in extent and productive capacity any hitherto developed on the American continent." He was alluding to Western Canada, from the Mackenzie River to the Arctic Ocean. Selwyn's optimism was influenced by his awareness of the exposures of the Athabasca bituminous sands and their conjectured extension beneath the surface.

The documentation, by explorers and fur traders of the late 18th century and early 19th century, of mineral occurrences as well as physiographic and geologic phenomena throughout Western Canada, attests to their "eye for the unusual" and uncanny interest in natural science. It was primarily the surveyors of the Hudson's Bay Company, notably Philip Turnor and Peter Fidler, as well as Edward Umfreville, David Thompson, Alexander Mackenzie and Alexander Henry of the North West Company, who best documented the principal geographic features as well as observations bearing on the geology of the regions they traversed. Besides an inherent interest in natural science, each possessed a rudimentary knowledge of the then nascent science of geology. Additional information concerning the locations of unusual natural phenomena was no doubt provided by the native people.

TIME OF DISCOVERY: LATE 1700s TO MID-1800s

Coal Occurrences

The naturally occurring substance that was most commonly reported by these early intrepid travelers was coal, which was invariably exposed along the banks of the rivers they traversed in Western Canada. The occurrence of coal attracted interest because of its use as a fuel for the burgeoning industrial revolution. Coke had become a substitute for charcoal in the process of iron smelting. Float coal was first reported by Edward Umfreville (Umfreville, 1954) on the North Saskatchewan River in 1786. Alexander Mackenzie had observed exposures of smoldering coal beds at the confluence of the Great Bear River with the Mackenzie, some 46 mi (74 km) upstream from Norman Wells, during his return in 1789 from his epic river voyage to the Arctic Ocean (Mackenzie, 1970). Alexander McKay, Mackenzie's second-in-command on his subsequent voyage to the Pacific Ocean, reported to Mackenzie his sighting in the Peace River Canyon of smoldering coal seams in May 1793 (Mackenzie, 1970). Peter Fidler had also viewed exposures of coal during the same year in southeastern Alberta at the confluence of Rosebud Creek and the Red Deer River while returning from his overland trip to the Rocky Mountains (Burpee, 1908). The significance of Fidler's discovery is attested by its incorporation in Alexander Mackenzie's map of 1801 on which he noted "Edge Coal Creek—great quantity of Coal in this Creek." This same notation appears on William Clark's map of 1809 as well as in Arrowsmith's 1811 map of North America.

David Thompson (Warkentin, 1964) reported coal beds along the bank of the North Saskatchewan River near Rocky Mountain House when he first visited this post in the autumn of 1800 (Morton, 1939), one year after the post's establishment. Alexander Henry observed the same exposure while residing at this post in 1811 (Henry and Thompson, 1897), stating in his journal that "This coal abounds along the Saskatchewan, in some places forming solid beds several feet thick for several acres, which are washed by the river. It is always overlaid with the thick beds of soil, and sometimes mixed with earth, clay and stones, running in horizontal veins. In its pure state our smiths use it for the forge with equal proportions of charcoal made of birch or aspen, which answers every purpose for making and repairing our axes and other tools."

Samuel Black (Rich, 1955), while exploring the Peace River and its tributaries in 1824, recorded smoldering coal beds along the Smoky River and near Fort Dunvegan on the Peace River. It was not until August 1857 that James Hector of the Palliser Expedition (Spry, 1963) reported the finding of float lignite coal in the valley of the Souris River near its confluence with the Assiniboine, as well as in situ lignite farther up the Souris in the vicinity of Roche Percée in southeastern Saskatchewan. The first report of the occurrence of coal in the Cypress Hills (Bell, 1874) was made by Isaac Cowie of the Hudson's Bay Company to Robert Bell of the Geological Survey of Canada. Cowie spent the winter of 1871–1872 at his post in the valley of Frenchman Creek near present-day Eastend, Saskatchewan and had observed lignitic coal beds exposed in the Paleocene Ravenscrag Formation along the upper flanks of the valley.

The coal reported by these early explorers of Western Canada was both geographically and stratigraphically widespread. The age of the in situ coal as well as the float fragments ranges from the Jurassic–Cretaceous transition (Kootenay Formation) to Lower Cretaceous (Mannville Group) to Upper Cretaceous (Dunvegan, Belly River, and Edmonton formations) to Paleocene (Paskapoo and Ravenscrag formations) and Eocene (Great Bear River location) (see Leckie and Smith, 1992, for stratigraphy). In terms of rank (Douglas, 1970), the occurrences in the Foothills and Rocky Mountains are bituminous, whereas those in the Central Plains are predominantly subbituminous with lignitic occurrences in the Southern Plains.

Bituminous Sands

Connecticut-born Peter Pond was the first nonnative to observe the deposits of bituminous sands located in what is now northeastern Alberta. Outfitted with trading goods, men, and canoes through his newly formed partnerships with "Pedlars" who were headquartered on the North Saskatchewan River, Pond arrived at the Athabasca River in 1778 (Coues, 1897; Morton, 1939). He wintered on the Athabasca River (post No. 18 as recorded on his earliest map) some 40 mi (64 km) above Lake Athabasca.

The "tar"- or "pitch"-impregnated sandstone and its occurrence as "bituminous springs" that sporadically cropped out along the Athabasca River for some 100 miles (161 km) was commercially inconsequential to the late 18th and early 19th century observers. Its presence provoked a curiosity as one of nature's oddities. Alexander Mackenzie, on viewing this phenomenon in 1787, was first to describe some of the deposits' physical properties as well as their prevailing use. In his journal he states that "At about twenty-four miles [39 km] from the Fork [Clearwater and Athabasca confluence], are some bitumenous [sic] fountains, into which a pole of twenty feet [6.1 m] long may be inserted without the least resistance. The bitumen is in a fluid state, and when mixed with gum, or the resinous substance collected from the spruce fir, serves to gum the canoes. In its heated state it emits a smell like that of sea-coal. The banks of the river, which are there very elevated, discover veins of the same bitumenous quality" (Mackenzie, 1970).

Dr. John Richardson, surgeon and naturalist with Captain John Franklin's Arctic Expedition of 1819 to 1822, was the first scientist to describe the stratigraphic relationship and properties of the petroleum occurrences along the Athabasca River. He studied the deposits in July 1820 and recorded the results in his "Geognostical Observations" (Richardson, 1828), which report was incorporated in Franklin's "Narrative of a Journey to the Shores of the Polar Sea, 1819–22." In reference to its habitat, Richardson stated "This mineral exists in great abundance in this district. We never observe it flowing from the limestone, but always above it, and generally agglutinating the beds of sand into a kind of pitchy sandstone."

Salt Springs

The discovery of salt springs was noted by the fur traders inasmuch as these springs were potential sources of salt for the preservation and flavoring of fish and meat. Alexander Henry of the North West Company ascended the Saskatchewan River in August 1808 on his way to winter at Fort Vermilion. He wrote in his journal (Coues, 1897) that on arriving at the confluence of the Carrot River with the Saskatchewan, he encountered a "freeman" (independent trader) tented by the river. Henry stated that "He [freeman] had passed part of the summer up this river [Carrot] where there were several salt springs, and had made a considerable quantity of salt, which he had brought to dispose of to our men on their way to the interior, where this article is not found." Some 100 years later, W. McInnes, of the Geological Survey of Canada, observed these same springs and noted them on his Map 58A that was issued in 1914.

MID-19TH CENTURY GEOLOGIC INVESTIGATIONS

Prior to the British-sponsored Palliser Expedition of 1857–1860 and the contemporaneous Canadian sponsored Assinniboine [sic] and Saskatchewan Exploring Expedition of 1858, no attempt had been made to map the distribution and succession of surface strata now known to comprise the Western Canada sedimentary basin. The first half of the 19th century had witnessed several British and American expeditions that traversed portions of Western Canada. Naturalists accompanied these parties and, apart from collecting fossils (mainly Paleozoic species) and describing physiographic regions, made no attempt to differentiate the country on the basis of sedimentary rocks. Notwithstanding, Alexander Mackenzie earlier had recognized the geologic demarcation between what is now known as the Precambrian shield and the Western Canada sedimentary basin. In respect to Lake Winnipeg (Mackenzie, 1970), he noted that "Along the West banks of the former [Lake Winnipeg] is to be seen, at intervals, and traced in the line of the direction of the plains, a soft rock of lime-stone [sic], in thin and nearly horizontal stratas [sic], particularly on the Beaver [Amisk], Cedar, Winipic [Winnipeg] and Superior lakes, as also in the beds of rivers crossing that line. It is also remarkable that, at the narrowest part of Lake Winipic [Winnipeg], where it is not more than two miles in breadth, the West side is faced with rocks of this stone thirty feet [9.1 m] perpendicular; while, on the East side, the rocks are more elevated, and of a dark-grey granite." Considering the infancy of geologic science in the late 18th century, one must conclude that Alexander Mackenzie had an extraordinary propensity to interpret natural phenomena.

In 1853, F. B. Meek and F. V. Hayden undertook a reconnaissance geologic survey of the Upper Missouri River and adjoining Badlands of what is now northwestern North Dakota and northeastern Montana. At that time, this area was part of the Nebraska Territory. Their expedition had been sponsored by James Hall, who was New York's state geologist. As a result of that season's work and Hayden's independent surveys the following two seasons, an informal subdivision of the Cretaceous rocks in the northern Great Plains of the United States was achieved. A five-fold numerical terminology was applied to those beds so identified by their fossil contents in respect to the earlier biostratigraphic units resolved from the Cretaceous exposures in what is now Nebraska. Meek and Hayden subsequently replaced their numerical terminology with corresponding formal stratigraphic names, in ascending order, as follows: Dakota Group (No. 1), Fort Benton Group (No. 2), Niobrara Group (No. 3), Fort Pierre Group (No. 4), and Fox Hills Beds (No. 5).

Dr. James Hector of Edinburgh, Scotland was the geologist and physician attached to the Palliser Expedition, which covered the western part of British North America between 1857 and 1860. He was aware of the biostratigraphic differentiation of the Cretaceous strata resolved by Meek and Hayden in the Western Interior Plains of the United States and applied this knowledge in an attempt to recognize coeval beds in what is now Western Canada. However, instead of utilizing Meek and Hayden's five-fold numerical terminology, he substituted an alphabetical terminology and reversed the sequence (in descending order A to D), the latter being in contradiction to descriptive chronology as expressed from oldest to youngest (Kupsch and Caldwell, 1982). Hector introduced the term "Long River shale" for the deposit that cropped out near the river that bears this name in southern Manitoba near the village of Clearwater. This term, which was ephemeral, nevertheless has the distinction of being the first published stratigraphic name coined in the Western Canada sedimentary basin. Stratigraphically this unit is incorporated in the Riding Mountain Formation (Campanian Stage) of today's terminology. Probably Hector's greatest contribution was his recognition of the three major physiographic "steps" or "levels" in Western Canada. Their bedrock surfaces, as it was later learned, were essentially expressed by lower Paleozoic strata in the east, Cretaceous strata in the Central Plains, and Tertiary strata in the southwestern Canadian Plains. Hector was somewhat skeptical of the role of glaciation in respect to the ubiquitous surficial deposits, stating that "The whole country traversed by the Expedition during the last year had been overspread by surficial deposits of great thickness. Although these might be all included under the group of Northern Drifts in the ordinary acceptation, still it is probable that they consist of deposits of very different ages, and circumstances of deposition." The results of Hector's three field seasons were published in the Quarterly Journal of the Geological Society of London and included a colored regional map (Figure 1). He also published, in 1859, two geologic cross sections, running from Hudson's Bay to Vancouver Island.

Henry Youle Hind was Hector's contemporary in the first attempt to synthesize the geology of what was later designated Western Canada. As a naturalist and geologist, he was appointed leader of the Assinniboine [sic] and Saskatchewan Exploring Expedition of 1858. Both the Canadian-sponsored expedition and Palliser's British-backed effort were launched at a time when the Hudson's Bay Company's fur empire was waning as a result of the declining interest in beaver fur. At the same time, the British government was contemplating revoking the Company's charter in respect to its Rupert's Land interests.

Hind's reconnaissance surveys involved the eastern borders of the Western Canada sedimentary basin. He examined in detail the lower Paleozoics of Lakes Winnipeg and Winnipegosis as well as the Cretaceous along the Manitoba escarpment. At Deer Island on Lake Winnipeg, Hind applied a numerical terminology (in ascending order; strata No. 1 to No. 4) to what is now recognized as Middle Ordovician (Winnipeg and Lower Red River) (Hind, 1860). Hind equated the rocks on the west side of Lake Winnipeg from Grand Rapids to Big Black Island (Black Island) as belonging to the Silurian System as represented in the state of New York.

Hind postulated that, although no Jurassic or Carboniferous strata (Hind, 1860) could be found along the drift-covered base of the Manitoba escarpment where Cretaceous rocks appeared to overlie Devonian rocks, the elevation differential between the two steps was sufficient to suggest the possibility that "rocks of Carboniferous, Permian, Triassic or Jurassic may yet be found." Jurassic rocks were later discovered during subsequent exploratory drilling.

Probably Hind's greatest observation with respect to the drift cover (Hind, 1860) was the evidence that "glacial or stranded ice" as well as the action of "water and floating ice [icebergs and flows]" were responsible for the present land forms. On the basis of his regional field studies, Hind published a geologic map (Figure 2) of the

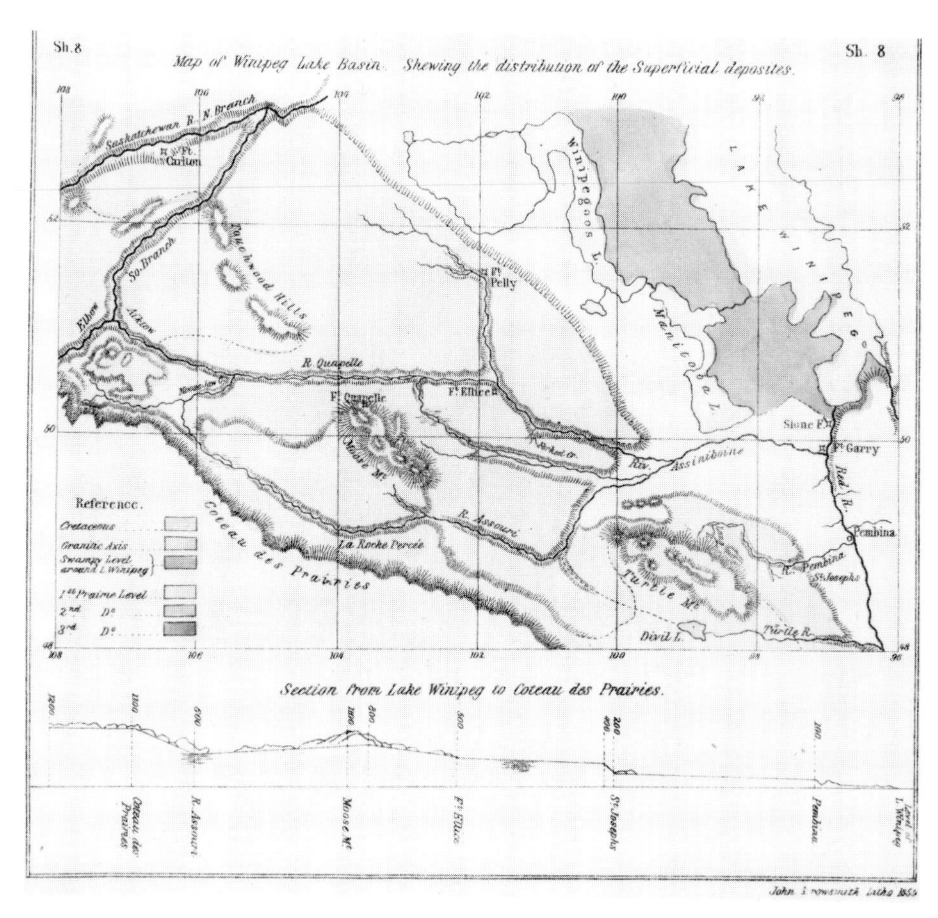

FIGURE 1. Dr. James Hector's geologic map and regional cross section of eastern Rupert's Land, published in 1859 and showing his three prairie steps or levels based on the regional expression of the predrift lower Paleozoics, Cretaceous, and Tertiary (Coteau des Prairies). He spent three seasons (1857–1859) with the Palliser expedition in Western Canada. Glenbow Archives NA-789-149.

eastern portion of Rupert's Land, with accompanying cross sections, in 1860.

EARLY EXPLORATION AND DISCOVERY: MID TO LATE 1800s

Historic Résumé

The surrender by the Hudson's Bay Company of its Rupert's Land territory to the newly formed Dominion of Canada was agreed upon in principle in November 1869. The pending transfer of the same year led to the first insurrection by the Red River Métis (mixed blood) led by Louis Riel. The tenuous "provisional government" established by Riel lasted less than nine months. The passing of the Manitoba Act in May 1870, just one month prior to the time that royal assent was given to the Rupert's Land transfer, resulted in the creation of the Province of Manitoba. Subsequently Colonel Wolseley's military expedition was sent to Fort Garry via the Lake Superior–Dawson Route to affirm Canada's sovereignty over the newly created province. Louis Riel escaped capture by Wolseley troops by a self-imposed banishment that was reinforced by a reward for his capture.

The Northwest Territories (Figure 3) at this period consisted of several Hudson's Bay Company trading depots or forts along the major rivers whose routes encroached on the parkland belt lying between the open prairie and heavily forested areas. Principal among these were Fort Garry, at the confluence of the Assiniboine and Red rivers; Fort Ellice, on the Assiniboine River a few miles below its confluence with the Qu'Appelle; Fort Pelly, on the upper Assiniboine; and Fort Carlton, below the elbow of the North Saskatchewan River. Fort Pitt and Edmonton House occupied positions to the northwest of Fort Carlton along the North Saskatchewan. Missions ministering to the various Indian tribes of the Northwest Territories had been established at Grand Rapids, The Pas, Prince Albert, Lac Ste. Anne, and Victoria, all located on the North Saskatchewan River. Between 1874 and 1875, the newly created Northwest Mounted Police established posts at Fort Calgary, Fort Macleod, Fort Edmonton, Fort Walsh, Swan River (Livingston), and Fort Saskatchewan. In 1874 the Livingston post (Turner, 1950) on Swan River was designated temporary capital of the Northwest Terri-

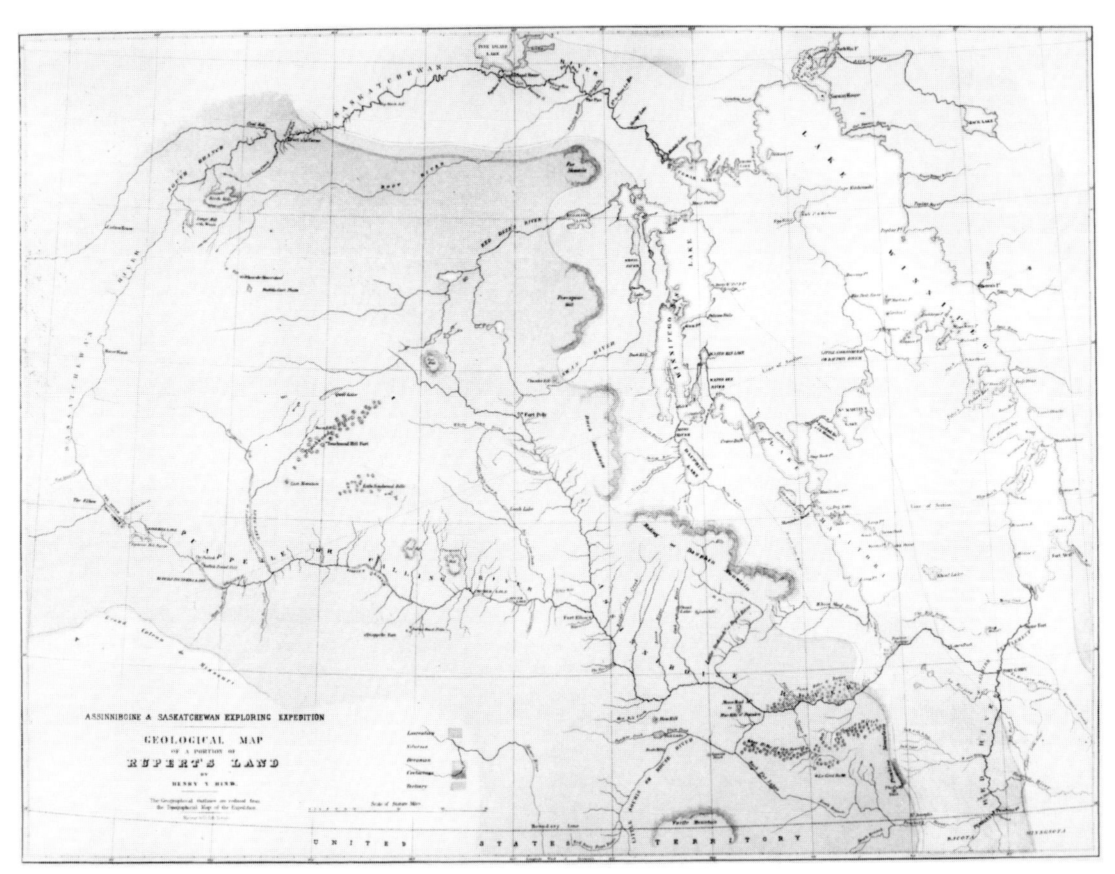

FIGURE 2. Henry Youle Hind's geologic map of Rupert's Land, prepared in 1860. As a geologist and naturalist, he was the leader of the Canadian-sponsored Assinniboine [sic] and Saskatchewan Exploring Expedition of 1858. Apart from his stratigraphic differentiation of the pre-Devonian sequence exposed at Lake Winnipeg, he recognized the geomorphological role of the glacial cover in influencing the present topography of the prairies of Western Canada. Glenbow Archives NA-789-150.

tories, but two years later the seat of government was transferred to the Northwest Mounted Police's newly established post at Battleford.

Between 1871 and 1877, seven treaties were signed in which the Indians of the Northwest Territories abrogated their entire rights to the territory comprising their ancestral hunting grounds (Wilson, 1939). During this period, the native population of this vast area numbered approximately 27,000 (Turner, 1950), with the Blackfoot proper and Cree being the dominant groups. The Métis and whites, for the most part occupying the Red River Valley, numbered about 10,000 and 3000, respectively.

Transportation

During the 1870s, the main trail interconnecting the fur trading establishments extended from Fort Garry (Winnipeg) to Edmonton House (Figure 3). It was known as the Carlton Trail, and had principal secondary trails branching off to Fort Pelly, Fort Qu'Appelle, and Wood Mountain country (Russell, 1959). These trails, which originally were the pathways of the itinerant tribes of the Plains, were later used by Hudson's Bay Company freighters, Métis buffalo hunters, missionaries, Northwest Mounted Police, scientific parties, adventurous travelers, and the first settlers. Although Red River carts were the main purveyors for the transport of freight, the horse-drawn buckboard became popular in the 1870s.

Prior to the advent of the railways, sternwheeler and sidewheeler steamboats were introduced to the major navigable rivers of the Northwest Territories (McFadden, 1953). In 1859 the first sternwheeler or paddlewheeler ascended the Red River. This was the *Anson Northup*, an American vessel named after its owner and captain. The S.S. *Dakota* ascended the Assiniboine River to Portage la Prairie in 1874, and by 1879 the S.S. *Marquette* reached Fort Ellice. Two years later, the *Marquette* ascended this same river to Fort Pelly, located near its upper reaches. The Hudson's Bay Company constructed the first steamboat on the Saskatchewan above Grand Rapids in the winter of 1872–1873 (Peel, 1964). This sternwheeler was not christened prior to its launching and ominously was wrecked in the Demi-Charge Rapids some 10 mi (16 km) upstream. The following year, the second Hudson's Bay Company sternwheeler was built—also above Grand Rapids—and was christened S.S. *Northcote*. The *Northcote* reached Fort

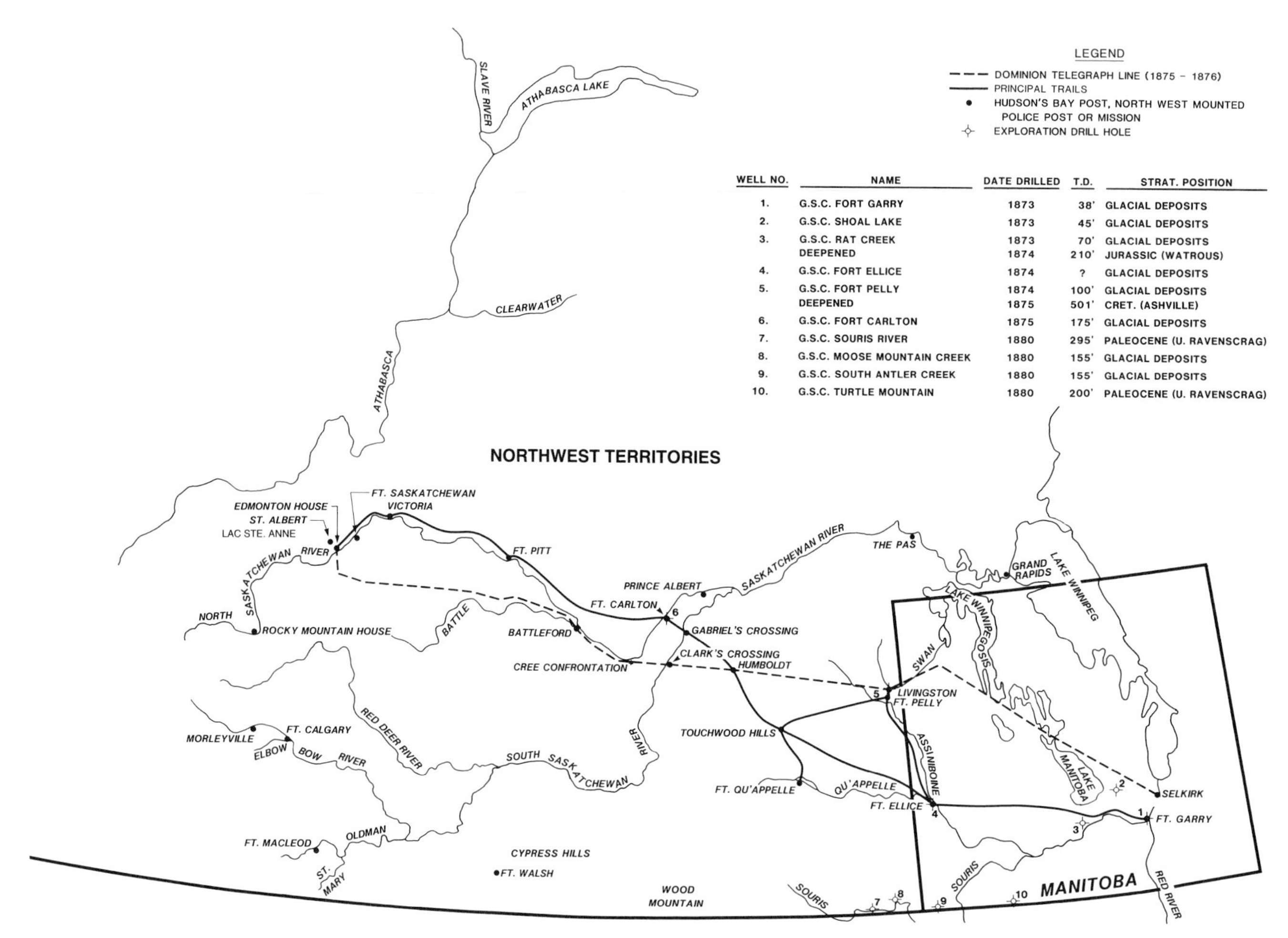

WELL NO.	NAME	DATE DRILLED	T.D.	STRAT. POSITION
1.	G.S.C. FORT GARRY	1873	38'	GLACIAL DEPOSITS
2.	G.S.C. SHOAL LAKE	1873	45'	GLACIAL DEPOSITS
3.	G.S.C. RAT CREEK	1873	70'	GLACIAL DEPOSITS
	DEEPENED	1874	210'	JURASSIC (WATROUS)
4.	G.S.C. FORT ELLICE	1874	?	GLACIAL DEPOSITS
5.	G.S.C. FORT PELLY	1874	100'	GLACIAL DEPOSITS
	DEEPENED	1875	501'	CRET. (ASHVILLE)
6.	G.S.C. FORT CARLTON	1875	175'	GLACIAL DEPOSITS
7.	G.S.C. SOURIS RIVER	1880	295'	PALEOCENE (U. RAVENSCRAG)
8.	G.S.C. MOOSE MOUNTAIN CREEK	1880	155'	GLACIAL DEPOSITS
9.	G.S.C. SOUTH ANTLER CREEK	1880	155'	GLACIAL DEPOSITS
10.	G.S.C. TURTLE MOUNTAIN	1880	200'	PALEOCENE (U. RAVENSCRAG)

FIGURE 3. Northwest Territories and Manitoba, 1873–1880, showing exploration drill sites.

Carlton, on the North Saskatchewan, the same year. A year later, in 1875, the *Northcote* visited Edmonton House for the first time. Although utilitarian by design, these very shallow-drafted vessels held an unbridled fascination for all riverside inhabitants of the Northwest Territories, with their wood-fired boilers belching black smoke and with the piercing pitch of their whistles as they ceremoniously approached river settlements.

The inclusion of British Columbia as a province within the Confederation in July 1871 was contingent on the fulfillment of an agreement with the Dominion government to construct a transcontinental railway linking this new province with Eastern Canada. Sanford Fleming (Grant, 1873), chief engineer of the proposed project, surveyed the region from Selkirk, Manitoba to the Yellow Head Pass during August and September of 1872. Fleming had selected a northern route within Hind's "fertile belt" (1860), which at the time contained the more populated area of the Northwest Territories. The preference for this northern route was also reinforced by Palliser's arid "triangle" represented by the area of present-day southwestern Saskatchewan and southeastern Alberta. This northern surveyed route was initially to be the route of the proposed Canadian Pacific Railway, but by 1881 it had been scrapped in favor of the present southern route.

Two years following the completion of Fleming's northern route survey, the Dominion government awarded contracts for the building of a telegraph line (Macdonald, 1930) along its designated right of way. Winnipeg, formerly known as Fort Garry, already served by the American telegraph system, was to be tied into the proposed line commencing at Selkirk, Manitoba. By July 1876 it had been completed (Figure 3) to Livingston on Swan River, and by November of the same year to Fort Saskatchewan via Humbolt, Clarke's Crossing on the South Saskatchewan River, and Battleford on the North Saskatchewan.

Events Leading to First Exploratory Drilling

The acquisition of Rupert's Land by the Dominion of Canada in 1870 set into motion a series of events that were directly related to the first exploratory drilling in Western Canada. First, the Geological Survey of Canada (Zaslow, 1975), established in 1842, was given a mandate to investigate the geology and assess the mineral potential of some 500,000 mi^2 (1,294,900 km^2) of this newly acquired territory. Second, the 49th parallel of latitude, establishing the boundary between Western Canada and the United States, was to be surveyed. Third, the route of the proposed transcontinental (Canadian Pacific) railway was also to be surveyed. At this juncture, the most reliable accounts of the geology of this vast country were those generated during the contemporaneous expeditions of Palliser and Hind some dozen years earlier.

Alfred R. C. Selwyn, who became the second director of the Geological Survey of Canada in 1869, spent the field season of 1871 investigating the geology and mineral potential of the proposed Canadian Pacific Railway route through the mountains of British Columbia. In 1873, he conducted a reconnaissance survey of the geology between Fort Garry and Rocky Mountain House, the latter post located on the North Saskatchewan River. His journey took him along the Carlton Trail (Figure 3) to Edmonton House. He traveled by buckboard, utilizing Red River carts for freighting his equipment. From Edmonton House, he left the rest of his party and, with a guide, made a trip southwestward to Rocky Mountain House by buckboard. Selwyn reported the numerous salt (sodium sulfate) lakes associated with depressions on the drift-covered plains. At Edmonton House, he examined Cretaceous coal seams on the banks of the North Saskatchewan River (Selwyn, 1874). He examined the layer of brown greasy clay (bentonite) above the coal seam, and noted that Hector (1861) had previously referred to it as soap clay because "it was used by the women of the fort for washing blankets."

Selwyn was without doubt disappointed by the lack of bedrock exposures, particularly on the plains. The ubiquitous drift cover (Selwyn, 1874) led him to remark, "From Fort Garry westward, on the route which we followed, no exposures of solid unmoved rocks were seen till within a few miles of Edmonton. For the whole of this distance—885.52 miles [1425.1 km] as measured by odometer, an universal mantle of drift and superficial deposits, sand, clay and gravel is spread over the face of the country." This observation unquestionably led him to recognize that it would require exploratory drilling not only to aid in the interpretation of the stratigraphic succession of the plains but also to access the potential for coal, water, and petroleum resources.

Selwyn departed from Rocky Mountain House in mid-September and returned to Fort Garry via the North Saskatchewan River water route to Lake Winnipeg. He traveled not by canoe but by the larger, more serviceable York boat.

George M. Dawson of the Geological Survey of Canada was appointed geologist and botanist to the North American Boundary Commission Survey of 1873–1874. Dawson's most important contribution from his assessment of potential resources was his mapping of coal (lignite) beds exposed in the Souris River valley and its tributaries. The Roche Percée coal beds of the Souris River valley had been examined previously by James Hector (1861) in June 1857. Dawson believed that the "Lignite Tertiary" beds were a stratigraphic component of Hector's third prairie level (Warkentin, 1964). By identifying the coal beds as Tertiary (Paleocene) in age, Dawson concluded that the third prairie level was a physiographic expression of a Tertiary plateau (Dawson, 1875) of regional magnitude.

Mechanized steam-powered drilling equipment was first put to use in the Western Canada sedimentary basin in 1873, a mere 15 years after the sinking of commercial oil wells at Oil Springs (Petrolia), Ontario, and 14 years after the drilling of a well at Watson's Flats near Titusville, Pennsylvania. The drilling project, sponsored by the Dominion of Canada, was coordinated by the Geological Survey of Canada. Financial assistance was given by the Canadian Pacific Railway's service fund in the wake of the construction of its transcontinental track.

Alfred R. C. Selwyn stated the objectives (Selwyn, 1874) for the proposed boring operations in his Report of Progress for 1873–74 "with a view of hastening and facilitating the geological exploration of the North-West territory especially in connection with the determination of facts relating to water supply and the occurrence of available beds of coal and other useful minerals in proximity to the proposed course of the Canadian Pacific railway, it was thought advisable to commence a series of borings to be extended across the plains from Fort Garry westward." In respect to what questions (Selwyn, 1874) may be answered, Selwyn further mentions that "Amongst these may be the formation of artesian wells, the existence of springs of petroleum and brine and also deposits of rock salt and other valuable minerals, as well as the thickness and character of coal seams."

Selwyn's choice of boring locations was dictated primarily by existing trails in proximity to Fleming's northern expedition route of 1872 for a transcontinental railway. The search for oil or gas was preempted by his preoccupation with the discovery of coal in the subsurface. This priority was understandable at a time when coal held an advantage over wood for locomotive fuel and fresh water was needed for the creation of steam. Water from the alkali (sodium sulfate) lakes of the prairies was totally unsatisfactory. Selwyn's opinion of the future hydrocarbon potential as well as the abundance of salt in the Western Canada sedimentary basin was predicated on his awareness of the numerous sodium chloride salt springs on the west side of Lake Winnipegosis as well as the bituminous sands of the Athabasca River valley. Henry Hind, in early October 1858, visited with a James Monkman who had been manufacturing salt for some 40 years at "Salt Springs" located near the present town of Winnipegosis, Manitoba (Hind, 1860). Hand-dug shallow wells were located over the springs to collect the brine. Iron evaporation pans and crude stone furnaces were used in the process. The bulk of the salt produced from this primitive operation was packed in birch-bark boxes and transported to the Red River Colony for sale. Monkman was probably the first entrepreneur involved in the exploitation of a mineral resource in the Western Canada sedimentary basin.

Initial Drilling

Between August 14 and October 27, 1873, three boreholes were drilled in southern Manitoba under the direction of

Alfred Selwyn of the Geological Survey of Canada (Waud, 1874). Alexander McDonald, a "practical" engineer, arrived in Fort Garry (Winnipeg) by a Red River steamboat the first week of August to take charge of the operations. According to W. B. Waud, his supervisor, McDonald brought with him "the Diamond Drill, a boiler with force pump and fittings, gearing for working the drill by horse-power, 400 feet [121.9 m] of 2½ in. [6.4 cm] diameter, tubular drill rods, 150 feet [45.7 m] of 3" [7.6 cm] tubing (wrought iron), annular and hollow boring heads with diamonds, and an independent steam pump with hose and other fittings." This equipment, weighing more than six tons, was purchased in New York and was conveyed by the Northern Pacific Railroad to its terminus at Moorhead, Minnesota on the Red River. The drilling equipment was probably not carried aboard the steamboat that conveyed McDonald down the Red River to Fort Garry, but was placed on a barge and towed by the steamboat.

The first borehole was spudded on August 14, 1873 at Fort Garry, "on the left bank of the Assineboine [sic] near the military barracks" (Selwyn, 1874). The second borehole, which commenced on September 8, was located approximately 38 mi (61 km) northwest of the first hole and was located within a short distance of the southeast corner of Shoal Lake. The third and final borehole of the season was commenced on October 16 near present-day Burnside, Manitoba on Rat Creek, where pioneer Kenneth McKenzie's farm was located. Operations were suspended on October 27 because of inclement weather. The drilling equipment, including the boiler, was freighted by oxen to Portage la Prairie where it was stored for the winter.

Of these three wells (Figure 3), drilled to total depths of 38, 45, and 70 ft (11.6, 13.7, and 21.3 m), respectively, none was able to reach bedrock. The entire season's operations were, to say the least, disappointing and plagued with problems. Principally, it was the inability to penetrate the drift and encounter bedrock despite attempts made with the various chisel, auger, and diamond bits employed. The gravel and boulder components of the glacial drift resulted in continual caving of the boreholes, which necessitated the driving of 3-in. (7.6-cm) casing. Caving within the open hole, below the base of maximum casing point, determined the depth to which the 2½-in. (6.4-cm) bit could penetrate. The logistics of transporting the cumbersome and heavy drilling equipment, the lack of spare parts to facilitate repairs, and the shortness of the season had compounded the difficulty.

Two drilling parties were assigned to the Northwest Territories in 1874 (Selwyn, 1876). According to Selwyn, they were to be used "to ascertain where the eastern limits of the Cretaceous coal bearing rocks is [sic], and at the same time whether artesian wells affording good water can be made upon the prairies where surface water is either very scarce, or for the most part too saline for domestic purposes." Sodium sulfate (Glauber's salt) brined sloughs and lakes of the prairies were well known to travelers of the time.

The borehole (Figure 4) at McKenzie's farm (Rat Creek) was reentered in July 1874. The well was deepened to 210 ft (64 m), of which the lower 122 ft (37.2 m) was drilled through bedrock. This lower section consisted of "white limestone" underlain by "grey rock." Selwyn (1876) postulated that the "cream colored limestone shewn [sic] in the section is certainly either of Devonian or Silurian age" and further stated that "The dark grey

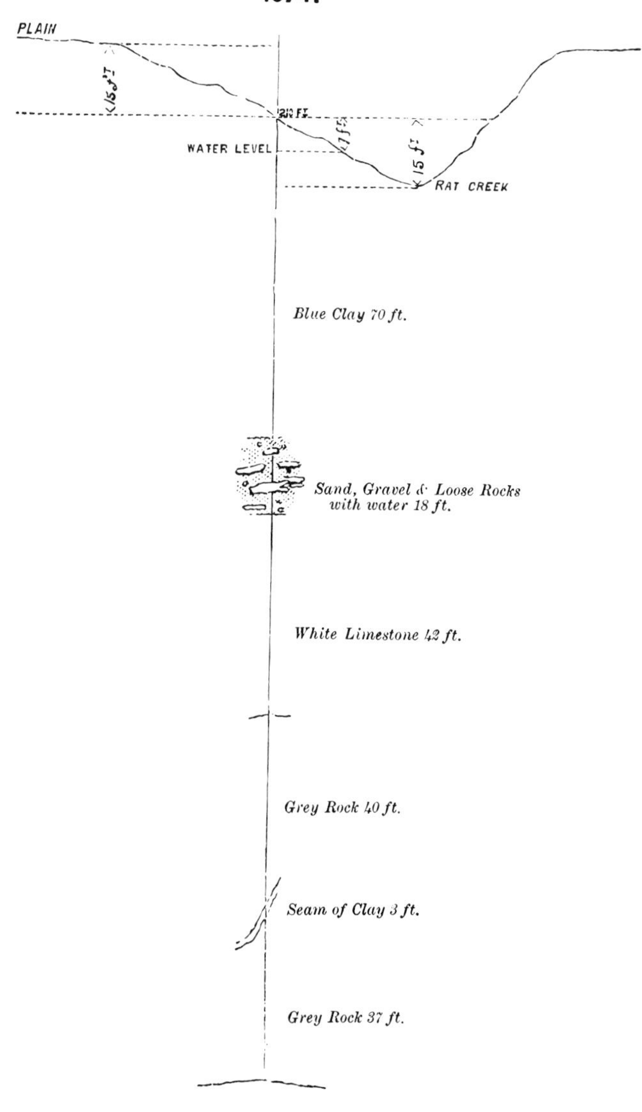

FIGURE 4. The first stratigraphic log interpreted from subsurface rock cuttings in the Western Canada sedimentary basin. The borehole, known as McKenzie's Farm (Rat Creek) and located in W1/2, Sec. 35, Twp. 11, Rge. 8WPM, was drilled during 1873–1874 by the Geological Survey of Canada. The log was prepared by R. C. Selwyn. The borehole reached a total depth of 210 ft (64 m), having encountered a Middle Jurassic section beneath glacial drift.

fine-grained rock beneath the limestone is unlike any rock, that has so far as I am aware, been observed cropping out in this region." It now appears that the entire section of bedrock penetrated is probably of Middle Jurassic (Watrous) in age and that the underlying "grey rock" is actually anhydrite. No coal was encountered, and Selwyn correctly concluded that coal-associated sediments could not be found in sediments underlying the "first prairie steppe." This well appears to have been the first to have documented the penetration of pre-Pleistocene sediments associated with the Western Canada sedimentary basin.

Following the completion of the McKenzie's farm well, the drilling equipment and boiler were transported along the Carlton Trail by oxen-drawn wagons to Fort Ellice. Selwyn alludes in his 1874–75 report that a fourth borehole (Figure 3) was drilled near the fort and presumably at the base of the Assiniboine River valley. He implies that this boring did not encounter bedrock.

Under an agreement made with a Mr. Fairbank of Petrolia, Ontario, a borehole was scheduled to be drilled near Fort Pelly to ascertain, according to Selwyn, the "reported occurrence of lignite or coal in this region; and also, if possible, endeavor to establish and define the eastern limits of Cretaceous rocks and their relation to the Devonian, Silurian or older rocks which succeed them in that direction" (Bell, 1876). Accordingly, the crew, under the direction of Joseph Ward, left Petrolia with their drilling equipment on June 30, 1874 and traveled the Lake Superior–Dawson Route to Fort Garry.

Robert Bell of the Geological Survey of Canada, along with his assistant, J. W. Spencer, had been assigned that same season to engage in the reconnaissance mapping of an area encompassing Lake Winnipeg, Lake Winnipegosis, and the Manitoba escarpment (Riding, Duck, and Porcupine mountains as well as the Pasquia Hills). Fort Pelly was Bell's first working base, and from here, according to Selwyn's letter of instruction (Bell, 1876), he was to "select a sight for the bore-hole in such a position as to avoid, if possible, the having to penetrate a great thickness of drift before reaching the rock; also as conveniently situated as practicable as regards timber suitable for derrick, walking-beam and other purposes, as well as obtaining the requisite supply of water and fuel for the engine."

The site proposed by Bell appears to have been located at Swan River Crossing, being the ford on the trail from Fort Pelly to Swan Lake. Presumably it was here that Bell located the only exposure of bedrock (Bell, 1876) in the general area of Fort Pelly and reported that "a soft bluish-grey shale occurs in the south bank of the river." However, this original selection was abandoned for a drill site (Figure 3) located at the confluence of Snake (Spruce) Creek and the Swan River some 9 mi (14.5 km) north by northeast of Fort Pelly and 4 mi (6.4 km) upstream from Swan River Crossing.

Joseph Ward of Petrolia, Ontario and his party, including the drilling equipment, arrived at Fort Pelly on August 24, 1874. From there they proceeded north along the west bank of Snake Creek, a trail used by early travelers when portaging between the Assiniboine and Swan rivers. Ward's decision to select this particular location for the fifth borehole may have been based on the proximity to the barracks erected nearby for the future stationing of a detachment of Northwest Mounted Police. Commissioner French had arrived on October 21 to establish Swan River Barracks (Turner, 1950). The location, originally known as Livingstone, was designated as a site for a railway station along Sandford Fleming's proposed northern route. It had also been selected as the temporary capital of the Northwest Territories, and it would seem probable that Commissioner French and his "E" troops under Inspector Carvell had arrived in time to observe the drilling operation prior to its suspension. Cold weather commenced on October 25, and operations ceased after reaching a depth of only 100 ft (30.5 m). Swan River Barracks was to remain headquarters for the Northwest Mounted Police until 1876 (Turner, 1950), when it was moved to Fort Macleod. This was necessitated following the massacre of Custer and his 7th Cavalry by the Sioux and Cheyenne on the Little Big Horn River in Montana on June 26, 1876.

The following summer, the drilling party reentered the borehole on July 6, 1875, and by October 9 the contracted total depth of 500 ft (152 m) was surpassed by 1 ft (0.3 m). The section penetrated consisted of Upper Cretaceous marine shale (Vermilion River–Favel–Ashville formations). No coal beds were penetrated, because the Lower Cretaceous (Swan River Formation) was not reached. No report of any biogenic gas (in the Favel Formation of Kamsack field) or immature oil (Ashville shales of the Pasquia Hills) was made. Notwithstanding, this Fort Pelly borehole (Figure 3) can be officially considered to have been the first stratigraphic and/or exploratory well to have encountered Cretaceous strata in the Western Canada sedimentary basin.

R. W. Ells, geologist with the Geological Survey of Canada, had been instructed by Selwyn in April 1875 to proceed with drilling equipment to the elbow of the North Saskatchewan near its confluence with Eagle Creek and there locate a suitable drill site (Ells, 1877). Ells' assistant was John Highman, an engineer who had worked under Alexander McDonald during the deepening of the McKenzie's farm (Rat Creek) borehole in 1874.

After purchasing "drift-boring tools and casing" at Petrolia that spring, Ells proceeded to Fort Ellice. There he obtained the boiler and engine that had been stored there following the drilling of the Fort Ellice borehole in 1874. Highman, who had arrived earlier, had made arrangements to engage freighters to transport the heavy equipment by oxen-drawn wagons to the proposed drill site. Some time after June 15, the party departed for the South Saskatchewan River. They reached Gabriel's Crossing (Figure 3) on July 13 by following the Carlton Trail, a distance of some 275 mi (443 km), having averaged 10 mi (16 km) per day. The following two days were spent in moving the wagons and equipment across the swiftly flowing South Saskatchewan River to its west bank.

From Gabriel's Crossing, Ells, Highman, and an Indian guide left the party and equipment and proceeded by buckboard to Eagle Creek. There Ells was unable to locate a suitable drill site because of the steepness of the boulder-strewn river bank. Finally a site was found some 10 mi (16 km) above the elbow of the North Saskatchewan

River where timber was available for construction of the framework and walking beam.

On returning to Gabriel's Crossing, the three reported that they had been confronted at the elbow by a band of 25 Cree Indians led by their chief, Mistiwassis ("The Little Child") (Ells, 1877). He requested that a council be held with Ells, Highman, and their guide concerning their intent to drill in the area. According to Ells, "the Indians stated they were a deputation sent by the great body of Plains Cree to put in a protest against any Government party carrying on work in their country before a treaty had been made with them for their lands"; and further, "if we chose to go on in opposition to their expressed wishes, we do so at our own risk." Treaty No. 6 with the Plains Cree was not to be signed until the next year. Discreetly, Ells and his two men retraced the trail to Gabriel's Crossing and, with the rest of the party and equipment, proceeded to Fort Carlton.

Mistiwassis and some members of his band had previously been involved in a similar confrontation at Fort Carlton in July 1875 (Macdonald, 1930). He, along with a fellow Cree chief named Altacouppe, had protested the construction of the telegraph line through their country lying north of the South Saskatchewan River. George Wright, who was in charge of the construction crew, was aware that a treaty had yet to be made with the Plains Cree. As a consequence, the wire and insulators were stockpiled at Clark's Crossing (Figure 3). The Cree were overtly suspicious of the telegraph line or "speaking iron," believing it to be a medium that possessed magic inasmuch as soldiers could somehow be quickly transported through the wire. They also believed the telegraph line would scare their game away.

Between August 12 and September 8, 1875, the sixth borehole was completed at Fort Carlton. In order to minimize the amount of drift to be penetrated, the borehole was located over a preexisting water well some 35 ft (10.7 m) deep. The subsequent drilling failed to reach bedrock and bottomed at 175 ft (53.3 m) in the drift.

Although only two of the first six wells drilled (Figure 3) were able to encounter bedrock, they were, in the context of time and place, a very significant achievement. Their locations ranged over a distance of some 450 mi (724 km) of prairie and parkland terrain. The heavy and cumbersome drilling equipment, including a boiler, was freighted by oxen-powered wagons traveling at an incredulously low speed of 1½ mph (2.4 km/hr). Apart from the logistics of transportation and "on the spot repairs," the drill sites were dictated by the accessibility of timber of sufficient girth to build a sturdy framework (derrick) and walking beam as well as a source of fresh water for the generation of steam.

Before returning east, Ells, as part of his mandate for his season's work in the Northwest Territories, was to collect mineral specimens of potential economic interest for the pending "Centennial Exhibition" in 1876 to celebrate America's first 100 years of existence. He collected, among other specimens, placer gold and coal samples from the North Saskatchewan River. Before departing from Fort Carlton, Ells made a request to Chief Commissioner Graham (Ells, 1877) of the Hudson's Bay Company "to have forward from the Athabasca country two bottles of petroleum." Graham had offered use of the Hudson's Bay Company's steamboat S.S. *Northcote* for transporting the drilling equipment and boiler up the North Saskatchewan River. Ells had speculated on drilling in the Victoria or Edmonton area. This did not materialize until a drilling program predicated on the subsurface extent and nature of the Athabasca tar sands was conducted by the Geological Survey of Canada from 1894 to 1899 in the region between Edmonton and McMurray.

In the spring of 1880, Selwyn made plans to drill a series of boreholes in the Souris River valley (Selwyn, 1881a, b). He reported that the purpose of the drilling was "to obtain more precise information respecting the Tertiary lignite-coal seams of the Souris River, more especially as regards their eastern extension from the known outcrops in the vicinity of Roche Percée." This decision undoubtedly resulted from George Dawson's field work and his report on the geology and resources of the 49th parallel (Dawson, 1875) in connection with the International Boundary Survey of 1873.

Subsequently four boreholes (numbered 7 through 10 in Figure 3) were drilled between July 12 and October 6, 1880. They were spaced 115 miles (185 km) apart along the international border and confined within a corridor 14 mi (22.5 km) wide. The drilling contractors were McGurvey and Highman of Petrolia, Ontario. The entire month of June was spent transporting the engine and drilling works by horse-drawn wagons from West Lynn, where the Red River crosses the international border, to Roche Percée. The principal obstacles impeding their efforts were the crossings of Souris River and its adjoining boulder-strewn coulees. The four boreholes were drilled in the following order: No. 7, Souris Valley; No. 8, Moose Mountain Creek; No. 9, South Antler Creek; and No. 10, Turtle Mountain. Their respective total depths were 295, 155, 155, and 200 ft (89.9, 47.2, 47.2, and 61 m). The only borehole to encounter lignite was No. 7 (Souris Valley). At a depth of 68 ft (20.7 m), some traces of lignite were observed in the cuttings, with a 6-ft (1.8-m) seam of lignite being penetrated after reaching a depth of 273 ft (83.2 m).

Selwyn commented that, after the first drill site was selected, "men were employed getting out the timber required for the derrick, engine-bed, etc., much hunting up and down the river being required before suitable trees could be found" (Selwyn, 1881b). He further stated that "The site for the first bore was fixed, photographs taken [Figure 5], and a section measured from the left bank of the Souris south to where the east bank of Short Creek intersects the Line [49th parallel], a distance of about four and a half miles [7.2 km]."

In the same text the extraction of coal from a 5-ft (1.5-m) seam exposed near the junction of Short Creek and the Souris River is described. In the spring of 1880, W. D. Sutherland of Winnipeg dug a 78-ft (23.8-m) adit into the seam and extracted a quantity of lignite (Selwyn, 1881b). Several tons of lignite were subsequently barged down the Souris and Assiniboine rivers to Winnipeg. This Sutherland "mine" site may well constitute the first attempt at commercial exploitation of the vast lignite deposits of the Estevan-Bienfait area of present southern Saskatchewan.

The construction of the rail line of the Canadian Pacific Railway in 1882 and 1883 across the southern part of the Canadian Prairies led to exploratory drilling by the

FIGURE 5. Canadian-type pole rig drilling for Tertiary coal seams on the Souris River (NW1/4, Sec. 31, Twp. 1, Rge. 5W2M) near Roche Percée. This was the first of four exploratory wells drilled by the Geological Survey of Canada from July to October 1880 to evaluate coal reserves in what is now southeastern Saskatchewan. Total depth, 295 ft (89.9 m). Glenbow Archives NA-302-10.

company for water and to a lesser extent for coal. The results of this drilling not only aided in the initial unraveling of the stratigraphy of the Western Canada sedimentary basin, but also resulted in the serendipitous discovery and subsequent development of a giant gas field.

As an incentive for the construction of the "colonization" rail lines, the backers were given land grants by the Dominion government that included not only surface rights but also ownership of the subsurface mineral rights. The pattern of the land grants was in the nature of a checkerboard, and was usually based on a formula of 6400 ac (2590 ha) for each line mile (1.6 km) constructed. The borings invariably were made near railway stations along the right of way and were well distanced.

Between 1883 and 1885, wells were drilled in the towns of Rosenfeld and Solsgirth located on a branch line of the Canadian Pacific Railway in the province of Manitoba. During the same period, wells were also drilled near the Grenfell, McLean, and Belle Plaine stations, along the main line of this railway in the district of Assiniboia in what is now southern Saskatchewan, as well as near the Langevin, Cassils, and Gleichen stations in the district of Alberta.

George M. Dawson of the Geological Survey of Canada had interpreted the drillers' descriptions of the lithology encountered in the aforementioned eight wells (Dawson, 1886) and presented his findings to the Royal Society of Canada on May 26, 1886. The Rosenfeld station well has the distinction of having been the first exploratory well in the Western Canada sedimentary basin to have penetrated the entire Phanerozoic section and bottomed in Precambrian rock at a total depth of 1037 ft (316 m). After being drilled through some 143 ft (43.6 m) of glacial drift cover, it encountered 92 ft (28 m) of Middle Jurassic (Watrous Formation) strata overlying 810 ft (247 m) of Middle Ordovician (Stony Mountain, Red River, and Winnipeg formations). Dawson was correct in his identification of a lower Paleozoic sequence (Maquoketa, Trenton, Galena, and St. Peter formations). In the absence of exposure of Jurassic strata between the Cretaceous of the Second Prairie Step and the exposed lower Paleozoics of the First Prairie Step, he understandably omitted their possible presence.

EARLY OIL AND GAS DISCOVERIES

First Discovery of Natural Gas

The first of two wells put down at the Langevin (Carlstadt and/or Alderson) station by the Canadian Pacific Railway in 1883 encountered a heavy flow of gas in the Upper Cretaceous Milk River sand at a depth of 1060 ft (323 m), and the gas ignited when a depth of 1155 ft (352 m) had been reached (Dawson, 1886). This resulted in the destruction of the derrick. Gas from this well was used to fire the boilers for the second attempt made the following year. This skidded well attained a total depth of 1426 ft (434.6 m), and gas obtained from both wells was used for many years by the Canadian Pacific Railway. This initial discovery and the subsequent development of the Alderson field resulted in proven initial established reserves of 474.2×10^9 ft^3 (13.43×10^9 m^3) of gas. When combined with subsequently discovered gas reserves in contiguous Milk River sandstone reservoirs, they represented a grand total of initial established reserves of 4.54×10^{12} ft^3 (128.6×10^9 m^3).

Dawson (1886) speculated that "It is unfortunate, for several reasons, that the boring at Langevin was not carried still deeper," and further, "There is, also, probably on the line of the railway no better place in which, by penetrating the Cretaceous series [Dakota sandstones], to ascertain whether it is underlain by Devonian rocks like those of the Athabasca region, and whether these maintain their petroleum-bearing character so far south." He apparently was alluding to a Devonian source for the Athabasca (McMurray Formation) bituminous sands.

Gas was also discovered by chance in the Milk River Sand reservoir at a shallow depth of 650 ft (198 m) at the town of Medicine Hat in 1890 (Clapp, 1915). The objective of this exploratory well had been for coal. A drilling rig for a follow-up well was supplied by the Canadian Pacific Railway with the town financing the drilling costs. It was anticipated that deeper prospective sands could be reached. The Milk River sandstone reservoir at Medicine Hat was found through subsequent development drilling to contain initial established reserves of 1.087×10^{12} ft^3 (30.78×10^9 m^3) of gas. In 1904 a deeper well, drilled by the town, discovered gas in a sandstone reservoir at 980 ft (298.7 m), which became known as the Medicine Hat gas sandstone. The well penetrated the sandstone reservoir and bottomed in the Colorado shale at a depth of 1010 ft (307.8 m). This reservoir, through the coalescing of subsequently discovered Medicine Hat gas pools, ultimately developed into an area in excess of 9300 mi^2 (24,085 km^2), with contained initial established reserves of 4.46×10^{12} ft^3 (126.3×10^9 m^3) of gas. Together, the superposed Medicine Hat and Milk River sand reservoirs contained initially a total of 9.0×10^{12} ft^3 (254.9×10^9 m^3). This supergiant reserve of shallow immature gas constitutes the largest of the Canadian gas fields found to date.

Petroleum Indications: Oil Shales

In the summer and early autumn of 1887, J. B. Tyrrell of the Geological Survey of Canada and his assistant, D. B. Dowling, were engaged in a geologic reconnaissance survey of the Riding Mountain-Duck Mountain area of southwestern Manitoba. The survey traverse was dictated for the most part by existing cart trails, some of which followed the courses of tributaries of the Assiniboine River as well as those emptying into Dauphin Lake. Two horse-drawn Red River carts were used to carry their equipment and food supplies, with their drivers doubling as guides. A buckboard buggy served to transport the two geologists (Figure 6). An odometer attached near the axle of the left rear wheel of the buggy measured the distance between observation stations or stops along the route of their traverse.

Prior to the commencement of their field work, Tyrrell and Dowling were "aware of the occurrence in it of salt springs, gypsum deposits and strong indications of petroleum" in this general area of Manitoba (Selwyn, 1889). The first of these mineral occurrences was based on the long-known salt springs of the Dawson Bay region of Lake Winnipegosis. Gypsum exposures on the north side of Lake St. Martin had been documented previously. As to the petroleum indications, Selwyn (1889) remarked on his activities of 1887 that "In some places, as on Vermilion and Ochre rivers, the dark shales [Vermilion River Formation] were found to contain a small quantity of mineral oil, and it is quite possible that borings may reveal reservoirs holding this oil in considerable quantities." He did not refer to the petroleum occurrences as "seeps" or "springs," and one must conclude that his inference related to the heavy or immature oil characteristic of oil shales. In 1907, W. McInnes of the Geological Survey of Canada observed oil shales associated with the same formation (Vermilion River) in the Pasquia Hills some 200 mi (322 km) to the northwest. He indicated the location of these shales on his map entitled "Churchill and Nelson Rivers," issued in 1914 as Map 58A.

The occurrence of petroleum associated with shales cropping out in the Vermilion River valley may have been a factor in the formation in Western Canada of the first entrepreneurial exploration company, known as the Manitoba Oil Company. In the spring of 1887, this company spudded a well on the west bank of the Vermilion River where the old cart trail crossed the river (SW1/4, Sec. 22, Twp. 23, Rge. 20WPM). This places the well site on the north boundary of the present Riding Mountain National Park.

Apparently, Tyrrell and Dowling visited the drill site and conferred with the drillers the same summer. After leaving Strathclair, they followed the trail north toward Dauphin Lake via the Little Saskatchewan (Minnedosa) River to the height of land and then down the trail on the west side of the Vermilion River to the drill site located where the trail crosses to the opposite bank. Selwyn made the following reference to their visit (Selwyn, 1889) in his summary report of 1887-88: "During the season of 1887, a number of enterprising gentlemen of Manitoba sank an eight-inch borehole on the banks of the Vermilion River to a depth of 300 feet [91. 4 m] in the hope of obtaining a supply of petroleum. Mr. Tyrrell was then consulted, and they were told that they would have to bore through Cretaceous shales to at least 420 feet [128 m] below the surface, below which they would in all probability shortly strike the Devonian limestones. If petroleum was to be found, as far as there was any evidence to show, it would be in beds immediately overlying this limestone." Tyrrell deduced this on the basis of the Dakota (McMurray) sand hosting the bitumen along the Athabasca River. On the basis of this advice, albeit gratis, Tyrrell has the distinction of being the first consulting petroleum geologist in Western Canada.

The drilling rig used was a percussion or cable tool type that had been freighted north a distance of some 50 mi (80.5 km) from Strathclair (Tyrrell, 1891), a station on the Manitoba and Northwestern Railway line, to the drill site. After reaching a depth of 292 ft (89 m), which stratigraphically was within the Vermilion River Formation, the drilling machinery broke down. This would appear to be the approximate time of Tyrrell and Dowling's arrival. Operations were halted that year, and, presumably on the advice of the two geologists, plans were made for deeper drilling the following year.

In the spring of 1888, the rig was skidded to a new location a short distance down the valley of the river, and drilling was resumed. This second attempt attained a depth of 743 ft (226.5 m) before it was abandoned. Selwyn (1889) further commented on the information obtained from the section encountered in his summary report by stating that "During the past summer the bore was sunk to a depth of 743 feet [226.5 m], the Devonian limestone having been struck at a depth of 422 feet [128.6 m], two feet [0.6 m] below the depth stated by Mr. Tyrrell. Flows of salt water are reported from the well, but no petroleum or gas."

On the basis of the drillers' log and cutting specimens described by Tyrrell, the well appears to have encountered at least 160 ft (48.8 m) of Middle Jurassic arenaceous limestone, white (anhydrite) gypsum, and red shale. Whether the lowermost 68 ft (20.7 m), consisting of pink and cream-colored limestone including red shale, represent Devonian beds younger than those exposed in the Lake Winnipegosis area is problematic. The seeming absence of the typical basal Middle Jurassic fine-grained red sandstone interspersed with larger, frosted sand grains

FIGURE 6. Geological Survey of Canada field party resting after a lunch stop near the Riding Mountain-Duck Mountain area of southwestern Manitoba during the summer of 1887. J. B. Tyrrell, the party chief, is probably the person seated in front of the buckboard. The other three men are freighters and/or guides. The Red River supply carts were pony drawn. Note odometer attached near axle of left rear wheel of buckboard for measuring distances between stops. Tyrrell's assistant, D. B. Dowling, may have taken this picture. Glenbow Archives NA-302-8.

makes a Devonian age probable. In all fairness, Tyrrell had no cause to suspect the presence of rocks of Jurassic age encountered in the well, because no exposures of beds of this age are present between the Cretaceous strata of the Manitoba escarpment and the lower Paleozoics of the Interlake area. The evaporites of Gypsumville are considered to be an outlier and probably of Middle Jurassic age.

Tyrrell (1891) concluded that the "palaeozoic beds passed through in the Vermilion River boring [420 to 743 feet; 128 to 226.5 m] represent an upward continuation of the Point Wilkins limestone, and therefore in the main overlie the highest Devonian beds seen on the shore of Swan Lake or Lake Winnipegosis."

The Vermilion River boring appears to represent the first privately financed exploratory venture drilled solely for the discovery of oil or gas in the Western Canada sedimentary basin. Its location, like that of the Geological Survey of Canada's Fort Pelly borehole, is located on Tyrrell and Dowling's geologic map of northwestern Manitoba dated 1891.

Oil Seepages

Surface seepages of petroleum or natural gas have understandably led prospectors to conclude that their source reservoirs in the subsurface could be discovered by drilling in the vicinity. The majority of oil fields discovered in the last half of the 19th century and the early part of the 20th century were confirmed as a result of surface seepages. The Western Canada sedimentary basin, apart from the bituminous sands of the Athabasca River and the oil shales of the Pasquia Hills, is essentially devoid of oil seepages. Many of the early wells drilled prior to the appreciation of advanced exploration techniques were predicated on the presence of "oil" scum from prairie water wells. The delicate iridescent coating of the water's surface invariably proved to be iron oxide residue. However, seepages of light (39° API; 830 kg/m^3) paraffinic oil do occur along Cameron (Oil) Creek, which drains into Waterton Lakes in extreme southwestern Alberta, and on Kishinena and Sage creeks, both of which are tributaries of the Flathead River in the adjoining area of British Columbia on the west side of the South Kootenay pass. Geologically, the Waterton area is structurally complicated by the Lewis overthrust fault, which involved the detachment of Proterozoic strata and their eastward juxtaposition over Cretaceous rocks. The oil from the seeps appears to have been hosted by the older rocks but may have been sourced by the Cretaceous rocks.

William Rodney (1969) gives credit to Kootenai Brown for having been made aware of the oil seeps on Cameron Creek by local Indians in the mid-1880s. A. R. C. Selwyn (1892) of the Geological Survey of Canada observed the same seeps during his investigations of the southern

Rocky Mountains during the field season of 1891. Selwyn quoted his colleague G. M. Dawson as referring to the oil's habitat "as a somewhat anomalous occurrence of petroleum."

Rodney (1969) gave an account, taken from the Fort Macleod Gazette dated October 3, 1889, of a Mr. Osborne of the Standard Oil Company visiting the site of the seeps on Cameron Creek and extolling the potential of the area for the finding of commercial quantities of oil. This may have been the American geologist C. D. Osborne, the first professional petroleum geologist, acting on behalf of a large American oil company, to visit Western Canada specifically to investigate a potential drilling prospect. Rodney's narrative reveals that the Alberta Petroleum and Prospecting Company was formed and local capital for drilling was raised, and further that the Fort Macleod Gazette had reported in late 1891 that "two holes had been sunk a short distance." Neither the Geological Survey of Canada report nor Alberta's Schedule of Oil and Gas Wells report compiled through December 31, 1930 makes mention of these wells.

It was not until 1902 that drilling activity was resumed, and some 16 wells were drilled from then until the end of 1908. The original discovery well was drilled by the Rocky Mountain Development Company and has been referred to as the John Lineham No. 1 (Figure 7), located on Cameron (Oil) Creek (Lsd. 16, Sec. 30, Twp. 1, Rge. 30W4M). By early autumn of the following year, the well encountered an oil reservoir at a depth of 1020 ft (311 m). An unsubstantiated report claimed that the well yielded a flush production of some 300 bbl (47.7 m^3) per day. Lineham and his partner, A. P. Patrick, anticipating an oil boom, legally subdivided the area into lots and christened the place "Oil City." Following its flush production, the Lineham discovery never produced more than 20 bbl (3.2 m^3) per day, it being the most prolific of all the wells that were to follow. A refinery of sorts was built at the site. Most "successful" wells produced at best a few barrels per day before expiring, and by the commencement of World War I interest in the area had dissipated. The discovery at Turner Valley (Sheep Creek) of wet gas in the Lower Cretaceous Home sand in May 1914 served to divert attention away from the Waterton area. The last unsuccessful drilling attempt at Oil City was made in 1932.

FIGURE 7. **Canadian-type pole-tool rig drilling at Oil City, Alberta in 1902. This well, the Rocky Mountain Development Company (John Lineham) No. 1, was the discovery well. Note pole rods stacked against the derrick support and large barrels for storing bailed crude in the background. Glenbow Archives NA-1585-3.**

Oil Sands: Early History

Geologic Assessment of Bituminous Sands

Robert Bell of the Geological Survey of Canada was the first professional geologist to investigate thoroughly the regional distribution of the surface expression of the Athabasca bituminous sands. His mandate from Director Alfred Selwyn (Bell, 1885) for the field season of 1882 was "to investigate the geology of the heretofore unexplored portion of the Athabasca River between the junction of the Lac la Biche River and the Clearwater." The instruction included a stipulation that stated "especially with reference to the mode of occurrence of petroleum and asphalt." Previous travelers, who had observed the black sands along a 100-mi (161-km) segment of the river's course, had treated them as a curious natural phenomenon. Bell can be given additional credit for first recognizing their commercial potential and contemplating the problem involved in separating the sand from the bitumen.

Bell's report of his field work, published in 1885, included some very interesting observations and conclusions in respect to the nature of the bituminous sands of which the following are taken from his text (Bell, 1885):

1. "There is little doubt but that the vast quantity of somewhat altered petroleum contained in the soft Cretaceous sandstones of the Athabasca region have been derived from the Devonian limestones, immediately underlying them, which are probably very thick."

2. "Where the contact of the sandy petroleum-bearing strata [McMurray Formation] with the higher Cretaceous rocks [Clearwater Formation] was seen at the Drowned Rapid, it was observed that the oil was prevented from passing upward by tenacious clayey strata."

3. "The drift resting on the black petroleum-bearing strata was nowhere observed to be impregnated with the oil, showing that it had saturated the Cretaceous strata, probably as a thin liquid, and became altered to its present state long before the glacial period."

4. "The thickened and blackened residue, which now saturates the sand and renders it plastic, has resulted from the escape of more volatile hydrocarbons and

the simultaneous oxidation of those remaining. This itself may have, in the course of time, prevented any further escape of the petroleum from the limestone below."

5. "The subterranean accumulations of oil may be expected to be found on the principal anticlinals or domes in the limestones [Devonian] without reference to the attitude of the unconformably overlying Cretaceous sands."

6. "A furnace or stove might be contrived so as to burn this material (bituminous sand). Perhaps a grate constructed on the plan adopted for burning sawdust, with additional contrivance for removing the sand, would be found to succeed, and, if so, the banks of the Athabasca would furnish an inexhaustible supply of fuel . . . yet the material [bituminous sand] occurs in such enormous quantities that a profitable means of extracting the oil and paraffin which it contains may be found."

7. "The principal obstacle in the way of a speedy development of the oil-fields of the Athabasca is their distance from a sufficient market."

Robert Bell was a visionary in recognizing the potential of the bituminous sands and the necessity to separate the sand from the bitumen in order to refine successfully the residue. His belief (Bell, 1885) in a Devonian sourced and hosted oil was based on his having observed that "The walls of the traverse joints and other spaces in the limestone were frequently observed to be blackened with petroleum," and that "some irregular cavities contain inspissated pitch." He assumed any subsurface oil reservoirs found in the Devonian would not have been altered (rendered asphaltic) by atmospheric exposure and therefore would be lighter and less viscous. Bell suggested (Bell, 1885) that the bitumen residue "may have, in the course of time, prevented any further escape of the petroleum from the limestone below." This reasoning is understandable because the concept of long-range migration of petroleum was at that juncture yet to be seriously considered. Moreover, the regional phenomenon of meteoric water washing and attendant biodegradation of paraffinic petroleum to asphaltic petroleum was yet to be postulated (see Creaney and Allan, 1992).

In 1984, a map entitled "Map of part of the Athabasca River To Illustrate Dr. R. Bell's Exploration" was published by the Geological and Natural History Survey of Canada to compliment Bell's report of 1882–1884. This map details the location and description of "petroleum bearing strata" along the Athabasca River with 30 annotated observations along the margins. Bell indicates intermittent occurrences of "petroleum bearing strata" from Drowned Rapid in the south to Pointe aux Trembles in the north, a distance of some 160 mi (257.5 km) along the course of the Athabasca River.

The first mention of exploratory drilling in the Athabasca bituminous sands area was made by Malcolm McLeod (McDonald, 1872) in 1872. McLeod had edited the journal of Chief Factor Archibald McDonald, who had accompanied Sir George Simpson, Governor of the Hudson's Bay Company, on a canoe trip from Hudson Bay to the Pacific Ocean. McDonald wrote that the party "passed the Bituminous Springs [Athabasca bituminous sands area]" on August 10, 1828. McLeod, in editing McDonald's notes, quoted Sir John Richardson as remarking on his visit at the same place in July 1820 that "it [bitumen] seems rather to increase than impair the fertility of the soil." McLeod further made his own comment that " 'Striking oil' would seem to be rather an easy process in such grounds. It would be worth trying."

Drilling for Subsurface Extent of Bituminous Sands

The Geological Survey of Canada commenced a three-well drilling program in 1894, 12 years after Bell's investigation. The prevailing question to be answered by the drilling was stated succinctly by Wyatt Malcolm (1913) in Memoir No. 29-E: "Since the bitumen has probably been derived from petroleum by evaporation and chemical changes, it was thought that at some distance from their outcrop the sands would be saturated with liquid petroleum."

Two of the wells were drilled on the banks of the Athabasca River, and the third on the North Saskatchewan River. The reason for these locations, apart from being downdip from the bituminous sand exposures, was the proximity to settlements and accessibility by existing trails and riverboats.

The first well was drilled at Athabasca Landing between 1894 and 1896. The location was approximately 134 mi (215.6 km) south by southwest of the most southerly exposure of the bituminous sand on the banks of the Athabasca River. Continuous caving problems prevented the well from reaching the Dakota (McMurray) sand, and it was abandoned while in the Clearwater shales at a total depth of 1770 ft (539.5 m).

The second well was spudded in 1897 at Victoria (Pakan) (Figure 8), located on the north bank of the North Saskatchewan River and 177 mi (285 km) south of the most southerly exposure of the bituminous sands. This attempt, like that at Athabasca Landing, was plagued with caving hole problems. It was unable to reach its objective in the basal Cretaceous sand, falling short by an estimated 250 ft (76 km). The Victoria well was finally abandoned in 1899 at a total depth of 1840 ft (561 m). Some minor gas shows were encountered.

The third well was drilled on the banks of the Athabasca River 2 mi (3 km) above the mouth of Pelican River at Pelican Portage. This well, spudded in 1897, encountered a strong gas flow at the top of the Dakota (McMurray) sand at 750 ft (228.6 m). On reaching 820 ft (250 m), an uncontrollable flow of gas necessitated suspension of operations. Casing could not be run nor the well capped. The gas flow subsequently subsided as a result of asphalt plugging at the base of the casing. Drilling was resumed the following year, and, after the hole had been deepened an additional 17 ft (5.2 m) to 837 ft (255 m), flowing gas once again forced suspension of operations. The flowing well remained uncontrolled for 15 years.

FIGURE 8. **An early cable-tool rig drilling on the bank of the North Saskatchewan River in 1898. This well, known as the Victoria (Pakan) well, was the second of three drilled between 1894 and 1899 by the Geological Survey of Canada in an attempt to trace the subsurface extent of the exposed McMurray bituminous sands. The other two were drilled at Athabasca Landing and Pelican Portage. The Victoria well, plagued with caving problems, was abandoned in 1899 at a depth of 1840 ft (561 m), some 250 ft (76 m) above the basal Cretaceous bituminous sands objective. The river location provided accessibility of drilling equipment and boiler by a sternwheeler riverboat. Glenbow Archives NA-302-11.**

Instead of the Pelican Portage well encountering lower-gravity conventional oil in the Dakota (McMurray) sand, heavy nonconventional bituminous oil and gas were found. This no doubt was disappointing but did indicate the probable existence of continuous heavy oil saturated sand strata of the same age some 56 mi (90 km) south by southwest of its surface expression on the Athabasca River. Because the Devonian limestone was considered to be the source of this oil, expectation of lighter conventional crude oil hosted by this system was yet to be ruled out.

The gas reservoir discovered at Pelican Portage, later to be designated the Portage gas field in 1972, contained initial established reserves of marketable gas of 54.3×10^9 ft^3 (1.54×10^9 m^3). Thus the Portage field represents the first discovered basal Mannville (McMurray) sand gas field in Western Canada. It was the third gas field following the shallow Milk River gas sand reservoirs discovered at Langevin and Medicine Hat in 1883 and 1890, respectively.

Initial Exploratory Drilling by Private Oil Companies in the Athabasca Bituminous Sands Area

The first exploratory drilling in the general environs of the bituminous sands occurred between 1907 and the commencement of World War I. Drilling was confined to a 50-mi (80.5-km) corridor along a stretch of the Athabasca River between Fort McMurray and a point some 10 mi (16.1 km) down river from Fort MacKay near the present site of Bitumount. The areas of the drilling operations were dictated both by the proximity to the bituminous sand exposures and the river access for transportation of drilling equipment and personnel.

During this period at least six private interest groups had obtained drilling leases, the most notable of which were Alfred von Hammerstein and Associates, Athabasca Oil Company, and Athabasca Asphalt Company. The hopes of "gushers" had been predicated on Robert Bell's observations in 1882 that the bitumen that agglutinated the sands at the surface was inspissated petroleum altered to its present state by evaporation and oxidation. He inferred that pools of less-viscous petroleum may have been trapped in the immediate area in Devonian reservoirs that could be structurally expressed as anticlines or domes. However, the majority of the wells drilled failed to reach the Devonian (Waterways Formation). For those wells that penetrated the basal Lower Cretaceous bituminous sand (McMurray Formation), no liquid petroleum reached the surface or even entered any of the boreholes. It became obvious that the asphaltic nature of the McMurray sand, where encountered as a result of drilling, was similar in character to that of its surface expression. All of the wells had been drilled with cable tools, which had necessitated alternate drilling and bailing, and such a procedure would have provided a natural escape route for any fluid petroleum encountered. Bailing attempts to stimulate production ended in failure.

In 1910, the Northern Alberta Exploration Company spudded a well known as North-West McMurray No. 1 (NE1/4, Sec. 17, Twp. 89, Rge. 9W4M) located near the town of McMurray. By the late summer of 1912 it had reached a total depth of 1475 ft (449.6 m), having bottomed 336 ft (102.4 m) into the Precambrian. This well encountered the McMurray bituminous sands, but, like the other exploratory wells drilled to that date along the Athabasca River, it was unproductive. However, the significance of this venture was the fact that it was the first well drilled in the Western Canada sedimentary basin to have encountered salt in penetrating the Middle Devonian Elk Point Group.

Early Investigations and Research Leading to the First Commercial Attempts at Separation and Refining of Bitumen

Commercial interest in the bituminous sands for the extraction or separation of the bitumen and its subsequent refining was not seriously entertained until 1913, when Sidney C. Ells of the Federal Mines Branch commenced a regional assessment of the McMurray sand outcropping along the Athabasca River. He obtained, by the use of a hand auger, some 200 core samples of bituminous sand representative of exposures along a 185-mi (297.7-km) course of the river (Ferguson, 1985). Apart from recognizing the sand's potential use as a road-surfacing material, Ells concluded from his early research that its economic importance lay in a method of extracting the bitumen from the sand through the use of solvents, heat distillation, or hot-water washing (Ferguson, 1985).

Following the end of World War I, research was transferred from the Federal Mines Branch to the newly formed Research Council of Alberta, which was created in 1921. Karl Clark, who had previously been Ells' overseer with the Mines Branch, joined the Research Council and continued the extraction studies that Ells had formerly undertaken. By conducting initial experiments with a process of hot-water treatment of the bituminous sand, Clark by 1923 had concluded that this method was the most practical. The following year a pilot separation plant was established at Edmonton's Dunvegan railway yards (Ferguson, 1985). Although subsequent modification of the Edmonton plant led to marginal success, it was dismantled in 1929 and rebuilt on the bank of the Clearwater River near Waterways (Ferguson, 1985), which had become in 1925 the northern terminus of the Northern Alberta Railways line. Some bitumen was produced at the Clearwater plant in the fall of 1929, but greater success was achieved in the summer of 1930.

During the 1930s, two entrepreneurs were to capitalize on the technological achievement of Clark and his colleagues at the Research Council of Alberta. The first was Robert Fitzsimmons, whose International Bitumen Company plant (Figure 9) commenced operations in 1930 at Bitumount, located some 48 mi (77 km) down the Athabasca River from the Research Council's Clearwater plant. Although Fitzsimmons' operation at Bitumount extracted some bitumen from the McMurray sands in late 1930, the plant was plagued with financial problems and was forced to close between 1932 and 1937 (Ferguson, 1985). During 1937 and 1938, following its rebuilding, it successfully produced and refined crude bitumen. However, owing to financial obligations to its employees, it went into insolvency and was subsequently sold in 1942 and renamed Oil Sands Limited (Ferguson, 1985).

The second private interest operation of a commercial extraction plant and refinery was pioneered by Max Ball, an American, whose company was known as Abasand Oil. This plant was constructed in 1936 at the Horn River reserve near the town of Fort McMurray. In 1941, after a series of setbacks, the Abasand plant went into successful operation (Ferguson, 1985), utilizing a combination of hot-water treatment and solvent distillation. Unfortunately, the plant was destroyed by fire the same year.

The initial research and pioneering of the extraction and refining of the bitumen obtained from the McMurray "oil sands" by visionaries such as Ells, Clark, Fitzsimmons, and Ball led to the construction of two massive mining operations. Great Canadian Oil Sands, later to be renamed Suncor, commenced mining operations in 1967, followed in 1978 by Syncrude Limited. Both plants are located on the west side of the Athabasca River in the Ruth-Mildred Lakes area, some 20 and 26 mi (32 and 42 km), respectively, north of Fort McMurray. In 1989, the combined daily production of the Syncrude and Suncor plants was in excess of 200×10^3 bbl (31.8×10^3 m^3) of synthetic crude oil, which represents approximately 14% of the total daily production of conventional crude plus natural gas liquids from the Western Canada sedimentary basin.

FIGURE 9. **Robert Fitzsimmons' extraction plant and refinery, known as International Bitumen, located at Bitumount some 48 mi (77 km) down the Athabasca River from Fort McMurray (circa 1936). Some bitumen was extracted from bituminous sands exposed in this area by Fitzsimmons in a primitive experimental operation in late 1930. Glenbow Archives NA-3394-57.**

FOUNDATIONS OF THE MODERN WESTERN CANADIAN PETROLEUM INDUSTRY

Bow Island and Viking Gas Fields

During the first two decades of the 20th century, over 200 exploratory wells for oil and gas were drilled in the Western Canada sedimentary basin. The great majority of these wells were located in the province of Alberta. Companies involved in exploration during this period were small and locally financed. Exceptions were the railway companies—in particular, the Canadian Pacific Railway. The serendipitous discovery of gas by the Canadian Pacific Railway at Langevin station in 1883, during drilling for a fresh water reservoir, gave this railway company an acute awareness of the added benefits to be derived from the discovery of oil or gas on their extensive grant lands.

The first significant discovery during this period was Bow Island No. 1, which "blew in" during February 1909 with an unrestricted flow of dry gas at 8.5×10^6 ft^3 (0.24×10^6 m^3) per day. The producing reservoir, which was encountered at a depth of 1900 ft (579 m), was named Bow Island sand after the nearby station. It became affectionately known as "Old Glory." This name is inextricably linked with its discoverer, Eugene Coste. He had formerly worked as a geologist with the Geological Survey of Canada, but left the government to devote his efforts to the search for oil and gas. In respect to the initial drilling at Bow Island, Coste worked as a consultant for the Canadian Pacific Railway, with his group receiving financial backing for this well from the railway. By April 1911, Coste's company, known as Prairie Fuel, had drilled seven successful development wells and had begun searching for a market in order to exploit their reserves. In August of the same year, Prairie Fuel merged with Calgary Natural Gas and Calgary Gas to form a common exploration, producing, transmission, and marketing entity. This new company was called Canadian Western Natural Gas, Light, Heat and Power Company. By the summer of 1912, a 16-in. (41-cm) gas pipeline (Figure 10) had been completed to Medicine Hat, Lethbridge, and Calgary, and was supplying natural gas to those cities as well as to the intervening towns. The Bow Island gas field, when drilled out, had original initial established reserves of marketable gas amounting to 102.6×10^9 ft^3 (2.9×10^9 m^3).

Five years later, in 1914, gas was discovered 75 mi (120.7 km) southeast of the city of Edmonton. The discovery well, Northwestern Utilities No. 1, found dry gas in a Lower Cretaceous sand reservoir at a depth of 2180 ft (664.5 m). It was named Viking sand after the nearby town. This productive sand occupied a stratigraphic position within the Cretaceous sequence that was somewhat similar to that of the Bow Island sand of the field of the same name. By 1924, 11 successful development wells had been completed by Northwestern Utilities, with open flows of gas ranging from 1.35 to 9.0×10^6 ft^3 (0.04 to 0.25×10^6 m^3) per day. Exploitation of the Viking field commenced the same year when the company completed a pipeline to Edmonton. This field, through extensive development drilling, merged with the Kinsella pool. The Viking-Kinsella field, as it is now known, had original initial established reserves of marketable gas amounting to 1.029×10^{12} ft^3 (29.1×10^9 m^3).

Turner Valley Gas Condensate Discovery in Lower Cretaceous

The investigation of gas seepages by a local rancher named W. S. Herron along Sheep Creek, located 20 mi (32.2 km) southwest of Calgary, led to the formation of Calgary Petroleum Products and the discovery of the Turner Valley oil field. William Elder and A. W. Dingman were Herron's principal partners in the drilling syndicate. The "Dingman" discovery well, drilled by Calgary Petroleum Products, was located on the yet-undetected crest of an elongated, faulted, and overthrusted anticline. It was spudded on January 25, 1913, and by June had encountered, at a depth of 1560 ft (475.5 m) within the Upper Cretaceous, a sand that produced an open flow of 2.0×10^6 ft^3 (0.06×10^6 m^3) per day, with associated light amber

FIGURE 10. Steam-powered tractor hauling 12 wagons of 16-in. (41-cm) casing at Claresholm, Alberta for Canadian Western Natural Gas, Light, Heat and Power Company's gas transmission line from Bow Island field to Calgary in 1912. Construction was completed the same year. Glenbow Archives NA-44-14.

condensate amounting to 10 to 15 bbl (1.59 to 2.38 m^3). Excitement mounted in the ensuing months, and by December 1913 an unprecedented oil boom occurred with the formation of more than 500 promotional companies. In May of 1914, the well, after encountering wet gas shows in intervening sands below the first occurrence, "blew in" at a depth of 2718 ft (828.4 m) with an unrestricted flow of 4.0×10^6 ft^3 (0.113×10^6 m^3) of wet gas and associated condensate (Figure 11). A small gas absorption plant was subsequently built, and 48,000 bbl (7631 m^3) of natural gas liquids were produced at the expense of the flared gas prior to the plant's destruction by fire in 1920. By 1915, Turner Valley's first oil boom had deflated as a result of meager successes exacerbated by a preoccupation with World War I. The total initial established reserves of natural gas liquids from the productive Lower Cretaceous (Home) sand amounted to a rather insignificant 68×10^3 bbl (10.8×10^3 m^3). The ramifications of the "Dingman" discovery led in 1921 to Imperial Oil's investigation of the Turner Valley structure through their newly created subsidiary known as Royalite Oil Company. Royalite acquired Calgary Petroleum Products the same year. In October 1924, Royalite No. 4 was drilled into the Mississippian (Turner Valley Formation) "lime," encountering a structural gas cap at a depth of 3507 to 3740 ft (1068.9 to 1140 m). This porous dolomite reservoir flowed 20×10^6 ft^3 (566×10^3 m^3) of gas and 500 bbl (79.5 m^3) of natural gas liquids per day and set into motion Turner Valley's second oil boom.

Norman Wells Discovery

Imperial Oil, through its subsidiary exploration company known as the Northwest Company, had been drilling exploration wells in Western Canada since 1917. Northwest Company McMurray No. 1 appears to have been the first of 25 unsuccessful exploratory wells for oil that were located in Alberta and Saskatchewan and drilled between 1917 and 1930. Of these 25 wells, 23 were drilled in Alberta and two in Saskatchewan. Their locations ranged from the Foothills to extreme southern Alberta and southwestern Saskatchewan as well as east-central Alberta and west-central Saskatchewan. Two wells were drilled in the McMurray bituminous sands area.

Imperial Oil, however, did make a very significant discovery in the remote Mackenzie River valley, to the north of the Western Canada sedimentary basin, during

FIGURE 11. Sampling of condensate from Turner Valley's Calgary Petroleum Products (Dingman) No. 1 discovery well in July 1914. Four notable oil pioneers are present: W. S. Herron, extreme left; A. W. Dingman, beside tap on end of pipe; R. A. Brown, holding white coat; and driller T. A. P. Frost, holding mug. Glenbow Archives NA-2119-4.

the aforementioned period. In 1914, T. O. Bosworth, a geologist who had previously worked for Shell Oil Company, was hired by a Calgary syndicate to stake claims along an area of Oil (Bosworth) Creek located some 50 mi (80.5 km) downstream from Fort Norman (Owen, 1975). Seepages along this creek had been previously investigated and sampled by J. K. Cornwall in 1911. An Indian he had employed to search for seeps in the general area had made the original discovery. The Northwest Company, an exploration subsidiary of Imperial Oil, acquired the syndicate's claims during World War I. T. A. (Ted) Link (Figure 12) visited the site for Imperial Oil in 1919 to investigate the nature of the seepages. By constructing seepage pits, he was able to localize an area where the seepages were the most prolific and thereby selected a drill site. Fort Norman became the staging area (Figure 13) to which drilling equipment and fuel were transported. The Discovery No. 1 well (Stewart, 1944) was drilled in 1920 and found light oil (38° API; 835 kg/m^3) in an Upper Devonian (Kee Scarp Member) stromatoporoid and coralline reef reservoir. The ultimate size of this major oil field was not fully realized until additional development wells fully delineated the Norman Wells field in the late 1970s.

FIGURE 12. Dr. T. A. "Ted" Link of Imperial Oil serving as an "instrument man" in the operation of a plane table traverse up Vermilion Creek in the Fort Norman area of the Northwest Territories in the summer of 1920. Glenbow Archives NA-5056-13.

Recognition of Western Canada Sedimentary Basin

The first mentions of the regional structural components that serve to define the Western Canada sedimentary basin were made by George Dawson and J. B. Tyrrell of the Geological Survey of Canada in the late 19th century. Tyrrell's observation was made in respect to the Precambrian surface on the eastern side of Lake Winnipeg (Tyrell, 1891); he stated that "The undulating surface of these crystalline rocks declines gently to the west beneath paleozoic beds." In Dawson's publication, "Report on the Geology and Resources of the 49th Parallel," in a section entitled "On the Lignite Tertiary Formation from the Souris River to the 108th Meridian," he referred to the Pierre Group (Dawson, 1881) of Upper Cretaceous shales: "Though usually in appearance quite horizontal, these beds must have a general light westerly dip, which carries them beneath the lignite groups of the Souris River." Later in this report he referred to "the eastern margin of the basin" in respect to his Upper Cretaceous Pembina Mountain Group of southwestern Manitoba. His discussion entitled "On Certain Borings in Manitoba and the Northwest Territory," published in 1886, makes reference to the "interior basin" (Dawson, 1886).

By 1913, Wyatt Malcolm, in Geological Survey of Canada Memoir 29-E entitled "Oil and Gas Prospects of the Northwest Provinces of Canada," recognized the role of the sub-Cretaceous unconformity (Malcolm, 1913). He stated that "In the eastern part of the plains a great unconformity exists between the Palaeozoic systems, consisting of limestones, dolomites, and shales, and the Cretaceous system, consisting of shales and sandstones, so that we find the Dakota sandstones of the Cretaceous system resting directly upon limestones of the Devonian system." He further stated that "In the west, deposition during Carboniferous, Triassic and Jurassic times has to a great extent bridged over the unconformity seen in the east." Malcolm referred to the presently recognized pre-Bajocian northeast-plunging Sweetgrass arch (Malcolm, 1913), which bifurcates the Western Canada basin into a western Alberta component (Alberta basin) and its southeastern counterpart (Williston basin). No geographic names were applied to differentiate these major features. Instead, he noted that "A very broad anticline crosses the boundary in the vicinity of the Sweet Grass hills," and "West of these is a broad syncline occupying a large part of Alberta." The expression of the aforementioned regional anticlinal feature in Alberta was first named "Bow Island anticline" (Slipper, 1919; Dowling, 1919).

In 1919, the term "Alberta syncline" was first used by D. B. Dowling (1919) of the Geological Survey of Canada in respect to the foredeep Tertiary sandstone deposits of western Alberta. His term "Central Basin" (Dowling, 1920) was used at the same time in reference to the Western Canada sedimentary basin, and his term "Central Plains terrace" (Dowling, 1919) was used for the eastern side of

FIGURE 13. Fort Norman, located on the Mackenzie River, showing Imperial Oil Company's Junkers monoplane and riverboat. Note improvised wharf supported by small craft. Casing and oil drums along the shore indicate staging area for drilling operations. Photo taken in summer of 1921. Glenbow Archives NA-781-13.

this basin. Dowling's regional structure contour maps of various Cretaceous stratigraphic horizons and his accompanying stratigraphic cross sections are truly remarkable, considering the limited subsurface information that had been derived from exploratory wells to that date. Nine colored maps and four cross sections are contained in Dowling's Part 1 of Memoir 116, in which his contribution (Dowling, 1919) is entitled "The Structure and Correlation of the Formations underlying Alberta, Saskatchewan and Manitoba." His accompanying regional structure contour map "showing tops of Lower Pierre shale [Bearpaw Formation]" clearly differentiates the two structural components or subbasins within the Western Canada sedimentary basin. In his figure 1, entitled "Well sections, arranged in west-east order, showing correlation of the geological formations," which stratigraphic datum is the base of the Milk River sand, Dowling correctly showed the presence of Jurassic strata in the Moose Jaw well, which was spudded in 1913. Although Jurassic strata, which included anhydrite and red clastics, were encountered in three wells drilled in Manitoba during the late 19th century, the apparent absence of fauna precluded the identification of these rocks as Jurassic. They were assumed to have a Devonian affinity based on lithology and the proximity to exposed Devonian rocks in the Interlake area of south-central Manitoba.

The term "Dakota Basin" (McCoy and Keyte, 1934) was in use in the very early 1930s. The geographic expression of this basin was defined by the distribution of the Lower Cretaceous Dakota sandstone contained by the Central Montana uplift on the west, the Black Hills on the southwest, the Sioux arch on the southeast, and the Precambrian shield on the east. The name "Williston Basin" (Dobbin and Erdmann, 1934) was used contemporaneously with the term "Dakota Basin." The derivation of the former can be attributed to W. T. Thom, Jr. of Princeton University and the United States Geological Survey, who, in a press release in 1923 (Denison, 1953), had reported "The great structural depression centering near Williston." Thom had based this on the evidence of a few widespread North Dakota exploratory wells that had encountered the Dakota sandstone. As late as 1943, both basin names were used interchangeably (Hennen, 1943), but the name "Dakota" was eventually superseded by "Williston." Initially the latter basin's northwestern expression included not only south-central Saskatchewan and southwestern Manitoba but also northern Alberta and northeastern British Columbia (Ballard, 1942).

The name "Moose Jaw syncline," representing the northern extension of the Williston basin, was in usage as early as 1941 (Webb et al., 1941) and persisted throughout the 1940s (Layer, 1949). By the mid-1950s it had fallen into disuse at the expense of the all-encompassing term "Williston Basin."

By 1931, the designation "Alberta syncline" was also known by the less-popular and ephemeral name "Edmonton syncline" (Link, 1931). The outline of the Alberta syncline was roughly defined by the distribution of the sediments associated with the Paleocene Paskapoo Formation, which expressed the foredeep component of the Western Canada sedimentary basin. The term "Alberta Basin" (Layer, 1958) was first introduced in 1955. It not only incorporated the foredeep component (Alberta syncline) but also included the more stable shelf component flanking the Precambrian shield. The current usage of the terms "Williston Basin" and "Alberta Basin" are recognized in the context that they are bipartite elements of the Western Canada sedimentary basin, being separated by the Sweetgrass-North Battleford arch (Stebinger, 1917; McLearn and Hume, 1927; Layer, 1958).

Establishment of Industrial and Professional Geologists

By 1930, in excess of 400 combined exploratory and development wells had been drilled in the Western Canada sedimentary basin. Rotary drilling, as opposed to the "standard" or cable-tool method, was beginning to gain acceptance, and the first commercial "heavy oil" field had been discovered. As well, an experimental refraction seismograph survey had been conducted. By 1929, a total of 365 companies had been registered to conduct exploration for oil and gas in the western provinces. The vast majority were small, locally formed and financed companies. Among the larger were Imperial Oil, including its subsidiaries Northwest Oil and Royalite Oil; British American Oil; the Hudson Bay-Marland of Texas partnership; Home Oil; and the Calgary and Edmonton Corporation. Both the Canadian Pacific and Canadian National railways were indirectly involved in exploration because of their very large grant or "fee" land holdings. Prior to this juncture, all Crown petroleum and natural gas mineral rights lands were administered by federal rather than provincial laws. It was not until 1930 that the ownership of these rights was transferred to the provinces.

Approximately 40 professional geologists and engineers were directly involved with the private sector in the search for oil and gas in Western Canada by 1930. Of this number, half were Americans originating for the most part from the mid-continent area and to a lesser extent from California. The industry was aided by geologists associated with the Geological Survey of Canada's Boring Division in the gathering of statistical and lithological data obtained from the various exploration wells. The interpreted results, including descriptive lithologic logs and age determinations, were published each year in their summary reports. Of paramount importance during this period of the 1920s and early 1930s was the contribution made by George S. Hume of the Geological Survey of Canada. He was greatly respected, not only by his geologic colleagues but by the industry at large, for his contribution to the interpretation of the stratigraphy and structure of Turner Valley and related foothill features as well as the "heavy oil" fields discovered in the Lower Cretaceous sands of east-central Alberta.

The Alberta Society of Petroleum Geologists was founded in Calgary on December 17, 1927 in the board of directors room of the Canadian Western Natural Gas, Light, Heat and Power Company. Among the 12 founding members present were such distinguished geologists as T. A. Link of Imperial Oil, J. A. Allan of the University of Alberta, and S. E. Slipper of the Canadian Western Natural Gas, Light, Heat and Power Company. S. E. Slipper was the Society's first president. An affiliation of the Society's 35 members was made with the American Association of Petroleum Geologists in 1928 (Allan, 1953). By 1990, the Society's total membership had risen to nearly 4000 members. On January 18, 1973, the Society's name was changed to the Canadian Society of Petroleum Geologists.

Evolution of Stratigraphic Terminology

The terminology applied to the Cretaceous rocks of Western Canada prior to 1920 was, to all intents and purposes, borrowed from the prevailing nomenclature associated with the Cretaceous of the north-central plains of the United States. As a result of the drilling activities of the 1920s and attendant surface and subsurface studies, new Cretaceous terminology evolved. Such terms as Dakota, Benton, Niobrara, Pierre, and Foxhills had been introduced for the most part by Meek and Hayden by 1861. Some of these formational names were originally prefixed by "Fort" after their type localities near military posts on the upper Missouri River. The name Laramie (Fort Union), designating the Tertiary Paleocene, also found its place in early Geological Survey of Canada reports. These American names gradually became obsolete and were replaced by names of Canadian derivation as a greater understanding of the surface stratigraphy evolved. In particular, the recognition of paralic and nonmarine sand beds associated with the predominantly marine Upper Cretaceous of southern and western Alberta necessitated this refinement. Terms such as Cardium (D. D. Cairnes-1907), Milk River (G. M. Dawson-1875), Belly River (G. M. Dawson-1883), Edmonton (J. B. Tyrrell-1887), and Paskapoo (J. B. Tyrrell-1887) met these requirements. Paskapoo is a Cree Indian name meaning "Blind Man." It appears to be the first of many Indian names applied to both surface and subsurface units in the Western Canada sedimentary basin in the ensuing years.

Evolution of Mineral Rights Ownership

The present diverse ownership of mineral rights, including hydrocarbons within the Western Canada sedimentary basin, is the result of the historic events that created Canada. This long evolution, spanning a period of more than 300 years, initially involved the Hudson's Bay Company, and much later the Dominion government, the aboriginal natives, as well as the Métis, the railway companies, the original settlers, and finally the western provinces.

The Hudson's Bay Company, or "The Governor and Company of Adventurers of England Trading into Hudson's Bay," had its inception in the year 1670 as a result of Charles II, King of England, granting his cousin Prince Rupert and his business colleagues a charter for the "sole trade and commerce" of lands drained by Hudson Bay. This exclusivity of ownership involved not only the fur and other surface resources but the subsurface resources as well. The grant encompassed a vast region of 1.5 million mi^2 (3.9 million km^2), which included most of the present area of the prairie provinces. Rupert's Land, as it was known, was governed by the Hudson's Bay Company for 200 years. In 1870 the Deed of Surrender came into effect and the transfer of Rupert's Land to the new Dominion of Canada was a *fait accompli.* Canada made a compensatory payment of 300,000 pounds and agreed to assign to the Company one-twentieth of the area to be opened for settlement in the "fertile belt" of Western Canada. The subsurface mineral rights were included in these lands, which in total amounted to more than 108,000 mi^2 (280,000 km^2). Specifically, a checkerboard representation amounting to 5% of the total arable lands was decided in which section 8, and three-quarters of section 26, of each township were allotted to the Company. Where such lands were under water, alternative lands could be selected. Undoubtedly the agriculture potential of the selection was uppermost in the minds of the company officials.

Further assignments of mineral rights were applied to those lands designated as Indian reservation lands located in Western Canada following the negotiation of the seven major treaties made between 1871 and 1877. The aggregate areal extent of Indian reservation lands in the western provinces amounts to some 4500 mi^2 (11,654 km^2).

Probably the most important factors in the inducement for settlement in Western Canada were the construction of the "colonization railways" and the offer of free land for the settlers by the Dominion government. As an incentive to create a network of rail lines throughout the "fertile belt," the early railway companies were offered land grants based on the formula of 10 sections (25.9 km^2) for each line mile (1.6 km) of track constructed. The subsurface mineral rights were included in these land grants. Commencing with the Canadian Pacific Railway and followed by the Calgary and Edmonton Railway, the Great Northern Railway, and the Grand Trunk Pacific Railway, some 50,000 mi^2 (129,490 km^2) of land was allotted them as a result of the construction of their respective rail lines. The lands were selected in a checkerboard fashion or in a massive "en bloc" pattern, the latter in the case of the Calgary and Edmonton Railway, where at the time ranching and irrigation rather than grain farming were believed to be best suited for this area of Alberta. The configurations of these grants usually consisted of narrow corridors straddling the railroad right of way for a distance of 20 mi (32.2 km) on either side.

The aforementioned railway companies all became involved as explorationists for oil and gas as a result of their acquisition of this "fee" or grant mineral land. Initially their prime concern in drilling was to create good supplies of fresh water and potentially exploitable coal deposits. The early discovery of gas in southeastern Alberta, followed by the discovery in 1914 of gas condensate at the Turner Valley (Sheep Creek) field, resulted in the Canadian Pacific Railway and later the Calgary and Edmonton Land Company becoming involved in oil and gas exploration. Likewise, the Hudson's Bay Company became indirectly involved in exploration as early as 1926, when it granted an exclusive option to Marland Oil Company of Ponca City, Oklahoma to explore its lands. The Marland Oil Company subsequently was acquired by the Continental Oil Company, which later became an exploration partner of Hudson's Bay Oil and Gas Company.

Beginning in 1870 and continuing until late 1887, a homesteader earned title to both the surface and subsurface rights by homesteading on a designated quarter section (0.25 mi^2, or 0.65 km^2). This entitlement became known as a "Queen Victoria Grant." Subsequently, with the knowledge of widespread coal resources, the Dominion government reserved all "mines and minerals" in the name of the Crown on all lands lying west of the third meridian. As a consequence, with the later movement of settlers westward, the areas of southwestern Manitoba and southeastern Saskatchewan contain a much greater representation of "freehold" mineral rights on homestead lands than western Saskatchewan and Alberta. In excess of 60,000 mi^2 (155,388 km^2) of "freehold" mineral rights were acquired by homesteaders from the Dominion government prior to 1888.

The Dominion government was not an exception in reserving unto themselves the mineral rights of lands of which they disposed. After 1902, the Canadian Pacific Railway, through sales of their lands, first reserved the coal rights and later petroleum followed by all mines and minerals. The reservation of coal rights was understandable at a time when coal-fired steam locomotives were beginning to replace the wood burners. The Calgary and Edmonton Railway incorporated the Calgary and Edmonton Land Company to solely administer the sale to settlers of their 2900 mi^2 (7510 km^2) of land grants received from the Dominion government for fulfilling their agreement in the construction of their rail line. The mineral or subsurface rights were reserved in the subsequent sale of this land. Thus, this company and other railway companies or their land subsidiaries became potential candidates in the early search for oil and gas in Western Canada.

In 1930, each of the western provinces received a transfer from the government of Canada of those "Dominion Crown Lands" contained within their respective borders. This transfer of natural resources, including the mineral rights, resulted in these lands being designated "Provincial Crown Lands." The exception involved those lands contained within the boundaries of the national parks of Western Canada, representing a combined area of some 26,000 mi^2 (67,335 km^2). The Indian reservation lands continued to be administrated by the government of Canada. The northern portion of the Western Canada sedimentary basin includes the southern portion of the Northwest Territories, and the mineral rights under these lands were not relinquished. They are designated as "Canada Lands" as opposed to "Provincial Lands."

The potpourri of subsurface mineral ownership within the 500,000-mi^2 (1,294,900-km^2) confines of the Western Canada sedimentary basin has in retrospect enabled not only the multinational and large Canadian independent companies but also the small independent companies to compete for land on which to explore. A consequence has been a more rapid discovery of the potential reserves of oil and gas in the Western Canada sedimentary basin.

Surface-Expressed Structures

In the early 1920s, prior to the advent of the seismograph as an exploration tool, the location of an exploratory well involving geologic resolution was predicated on surface-expressed structure and especially the anticline feature. The search by geologic field parties (Figure 14) for structures similar to Turner Valley resulted in a concentration of reconnaissance surface mapping along the Foothills and Front Ranges structural belt of southwestern Alberta.

FIGURE 14. Rio Bravo (Canadian Superior) Oil Company's field camp "snowed in" at Porcupine River, Alberta in October 1945. The inaccessibility of Foothills terrain necessitated the reliance on horses for transportation. Party chief N. W. ("Nick") Nicols was assisted by J. M. (John) Andrichuk and L. R. (John) Baxendale in mapping Foothills structure by plane table survey. Glenbow Archives NA-5377-1.

By the late 1940s, this investigation had been extended to the northwest by several companies, notably Shell Oil, along the entire structural belt of western Alberta and into northeast British Columbia (Figure 15).

The presence of glacial drift cover over the greater part of the Western Canada sedimentary basin prevented an examination of the underlying bedrock. This was in sharp contrast to the mid-continent and western plains areas of the United States, where surface mapping was *de rigueur* and ultimately responsible for the discovery of many major fields. However, a fault-type structure was mapped and drilled by the Northwest (Imperial Oil) Company between 1920 and 1923 in a partially unglaciated area of extreme southwestern Saskatchewan. The well, Northwest Boundary No. 1, was located on the south or upturned side of an east-west surface-expressed fault. The well was spudded in Belly River sands, with the country rock being the younger Bearpaw shale. Many hundreds of feet of throw was believed to extend to the Precambrian. No oil or gas was discovered, and the well bottomed at a depth of 3940 ft (1200.9 m) in Jurassic shale. Apparently the fault, of a listric variety, was emplaced during the Tertiary, postdating the generation and migration of oil in the Western Canada sedimentary basin. This fault structure was related to the inception, during Eocene to middle Miocene time, of the Bearpaw Mountains located to the south in north-central Montana. It, like similar faults fronting the deep-seated volcanic intrusions of western Montana, is believed to sole out in the Colorado shale.

Wainwright: First Commercial Discovery of Heavy Oil

At Wainwright, Alberta in June 1925, discovery of commercial heavy oil from a Lower Cretaceous sand reservoir heralded the future discoveries of many large fields of heavy oil in reservoirs of this age. The initial discovery, British Petroleum 3B, found 22° API (922 kg/m^3) oil at 2233 ft (680.6 m). When completely developed, its original initial established reserves amounted to 78.0 $\times$ 10^6 bbl (12.4 $\times$ 10^6 m^3). To date, some 31 heavy oil fields hosted by Lower Cretaceous sand reservoirs and each in excess of 10.0 $\times$ 10^6 bbl (1.6 $\times$ 10^6 m^3) have been discovered in east-central Alberta, west-central Saskatchewan, and southern Alberta. Their total initial established reserves of recoverable oil are 983.6 $\times$ 10^6 bbl (156.3 $\times$ 10^6 m^3). This represents about 20% of the total proven conventional oil (not oil sands or bitumen) found to date in the Western Canada sedimentary basin.

British Petroleum No. 3, spudded in 1924 and abandoned in 1925, was drilled at the same location as its successor, British Petroleum No. 3B, which was a skidded location that was drilled as a completion well. The British Petroleum No. 3 well appears to have been one of the earliest rotary rigs utilized for oil and gas exploration in the Western Canada sedimentary basin. This well did continuous wireline coring from 1300 to 2086 ft (396 to 636 m). Some 54 cores were recovered out of 58 attempts, with total recovery of 320 ft (98 m). This represents a 41% recovery of the section cored. From a Geological Survey of

FIGURE 15. Geological Survey of Canada field party under the leadership of Dr. C. O. "Con" Hage (center), loading canoes for departure from a camp on Fort Nelson River, British Columbia, in the summer of 1944. Note portable stove on smaller "catamaran" attached canoe. Glenbow Archives NA-4450-69.

Canada description of the cores by D. C. Maddox in 1925, the coring appears to have commenced in the Second Speckled Shale Formation and bottomed between the base of the Viking Formation and the Mannville Group.

Subsequent Discoveries

By the end of 1930, three other commercial heavy oil discoveries had been made in the Western Canada sedimentary basin in addition to Wainwright. The second find was made in 1927 at Skiff, located in southern Alberta. Here the Devenish No. 1 well found 19° API (940 kg/m^3) oil in the Jurassic Sawtooth sand. Although the original initial established recoverable reserves of this first Jurassic discovery were less than 1 million bbl (159 × 10^3 m^3), it nevertheless revealed the potential of reservoirs of Jurassic age to host prolific reserves of oil. To date, 15 fields of heavy oil, with initial established reserves of recoverable oil in excess of 10 × 10^6 bbl (1.59 × 10^6 m^3) are hosted by both Upper and Middle Jurassic reservoirs. Their total reserves amount to 477.1 × 10^6 bbl (75.8 × 10^6 m^3). These principal Jurassic fields are all located in southwestern Saskatchewan.

In 1928, heavy oil was discovered at Dina, located northeast of Wainwright near the Saskatchewan border. The discovery well, Meridian No. 1, found 14.5° API (969 kg/m^3) oil in a sand reservoir (Sparky) near the top of the Mannville Group.

A third Cretaceous discovery of 20° API (934 kg/m^3) oil was made in 1929 at Red Coulee (Figure 16) in extreme southern Alberta near the international border. The discovery well, Vanalta Oils Red Coulee No. 1, obtained production from the "basal quartz" sand of the Mannville Group. This small field, when fully developed, straddled the international border and was subsequently named Border-Red Coulee.

Oil production from the Western Canada sedimentary basin through December 31, 1930 was restricted to five fields, all located in Alberta—namely, Turner Valley, Wainwright, Skiff, Dina, and Red Coulee. Total cumulative production of the four heavy oil fields was a mere 104 × 10^3 bbl (16.5 × 10^3 m^3). However, the vast majority of production, totaling 3.7 × 10^6 bbl (588 × 10^3 m^3), consisted of condensate oil from wet gas production at Turner Valley.

The commencement of the depression in 1930 significantly affected the level of exploration in Western Canada. The momentum that had been established as a result of the discovery of Mississippian-reservoired wet gas at Turner Valley in 1924 slowly diminished until June 1936, when Turner Valley Royalties No. 1 was drilled and major crude oil reserves were discovered. The well, located on the west flank of the Turner Valley structure, had encountered the oil "leg" beneath the Mississippian wet gas "cap." As a result, the Turner Valley field (Figure 17) became Canada's first major oil and gas field, hosting some 140 × 10^6 bbl (22.2 × 10^6 m^3) of initial established oil reserves and 2.5 × 10^{12} ft^3 (70.8 × 10^9 m^3) of marketable

FIGURE 16. Preparation for "shooting" a well in the Red Coulee field, straddling the Montana-Alberta border, with nitroglycerine, October 31, 1930. The "nitro" was used to stimulate production or in some cases to "bring the well in." It was apparently transported in the metal box occupying the "rumble" seat of this early convertible. No doubt the driver was unable to obtain insurance! Glenbow Archives NA-711-169.

gas. Unfortunately, much of the gas was flared; a rough estimate suggests that 1.5×10^{12} ft^3 (42.5×10^9 m^3) of the field's gas reserves were flared (Figure 18) in the acquisition of associated oil prior to the Alberta government's establishment of a Conservation Board in 1938 (currently the Alberta Energy Resources Conservation Board). In consequence of this third Turner Valley oil boom, a resurgence of drilling activity took place, not only in the development of the field but across the entire Western Canada sedimentary basin. By the end of 1939, in excess of 600 exploratory wells had been drilled since the beginning of the century. Discoveries of heavy crude were made, notably at Taber (1937), Princess (1939), Vermilion (1939), and Lloydminster (1938). Of these four, the last was the most significant. Lloydminster Royalties No. 1 was completed in November 1938 after discovering heavy crude (10 to 18° API; 1000 to 946 kg/m^3) in the Upper Mannville Sparky sand reservoir. The field, straddling the Alberta-Saskatchewan border, had original initial established recoverable reserves of 67.9×10^6 bbl (10.8×10^6 m^3). Development drilling was particularly active in Saskatchewan, commencing in 1945 and continuing into the mid-1950s. The geologist who was intimately associated with this development phase was Professor F. H. (Harry) Edmunds of the Geology Department of the University of Saskatchewan. During the 1930s he had visited the sites of many of the exploratory wells drilled in Saskatchewan on behalf of the operators. He was tireless in his efforts to record descriptions of drill cuttings and cores so as to gain a more comprehensive understanding of the stratigraphy and oil and gas potential of the basin. In his consulting capacity, his unassuming yet scholarly approach gained him respect and admiration among not only his colleagues but also the industry at large. Apart from his university duties, Harry Edmunds was introduced to the Lloydminster area when the first commercial gas was discovered in 1934 with the completion of Lloydminster No. 1.

In 1939, shallow biogenic methane gas was encountered in a sandstone member of the Upper Cretaceous Second Speckled Shale in the drilling of California Standard Princess C.P.R. No. 1 near Medicine Hat, Alberta. Subsequently, the coalescing of adjoining common reservoired pools at Alderson, Verger, Medicine Hat, and Suffield resulted in the proving of initial established reserves of marketable gas amounting to 1.36×10^{12} ft^3 (38.51×10^9 m^3). Gas from this same stratigraphic level had been discovered in 1937 in a fractured calcareous shale (Upper Cretaceous Boyne Member) several miles immediately south of the town of Kamsack, Saskatchewan, located near the Manitoba border. This biogenic gas was reservoired at a very shallow depth of less than 200 ft (61 m). Between the aforementioned date and 1949, some 37 wells were drilled by the cable-tool method into this gas reservoir. The reserves were exceedingly small, as was the reservoir pressure. A pipeline to Kamsack was constructed, however, and a few businesses were supplied with gas for heating purposes.

FIGURE 17. View of Turner Valley field during the mid-1930s, showing mostly wooden derricks still remaining at drill sites following completions. Sheep Creek is in foreground. Glenbow Archives NA-67-73.

ADVANCES IN GEOPHYSICS AND DRILLING, AND SOME SIDE ISSUES

Quacks and Bogus Oil-Finding Devices

The history of the development of Kamsack's inconsequential gas field is a colorful one. The selection of some of the drilling sites was reportedly based on dowsing and, in some cases, downright chicanery. The glaciated terrain flanking the Assiniboine River provided the necessary "anticline structures" on which to locate wells. One would-be oil finder is reported to have constructed a device that contained a bullfrog concealed in a cigar box along with a copper plate and battery. Under the box was an attached doorbell button wired to the battery and copper plate. As the operator slowly carried his "biophysical instrument" across the "anticline," he would press the concealed button, wherewith the bullfrog would receive an electric shock sufficient to cause the cigar box to vibrate. A stick was stuck in the ground at this location, marking the site for drilling operations.

More sophisticated devices were displayed by seemingly sincere and dedicated oil finders who visited the offices of geologists to exhibit their brass-trimmed oak boxes and tripods. They would be prepared to demonstrate the capabilities of their instruments either in the field or over a map in the office for a fee. During the late 1940s, a Lloydminster operator who claimed to be able to locate oil deposits was put to the test when he was asked to locate in a field some buried drums filled with Lloydminster crude. He was reported to have had a measure of success, but this success was believed to be a result of his acute perception in detecting evidence of freshly disturbed terrain rather than his avowed psychic propensity for locating hitherto undiscovered oil fields.

A "Patent Automatic Oil Finder" known as the "Mansfield" was advertised for use in the late 1930s when drilling activity at Turner Valley was at its height. A series of "Directions for taking an Observation" was noted in the advertisement. This particular instrument measured the conductivity of air currents. The time of day and type of weather condition were essential for its optimum performance. Point 1 of the operating directions stated in part that "These currents naturally seek the path of greatest conductivity and are therefore strongest in the immediate vicinity of the water blanket which, in the form of shale, slate, clay etc. covers oil in its natural state."

In the period immediately following the Leduc discovery, a more creditable but nevertheless questionable geophysical method of detecting the underground presence of hydrocarbons was introduced in both Alberta and Saskatchewan. This method involved the differential emissivity of radioactive substances as applied to natural crude and its host rock. When applied to known fields, this detection method was invariably successful in terms of the operators' implicit biases. The dowsers and psychics involved in oil finding have always elicited in onlookers a bemused skepticism about, and yet a fascination for, the geologist or geophysicist. The state of the art of today's exploration techniques has all but spelled their demise. Notwithstanding, these charlatans, as well as those who believed in their unorthodox scientific methods, undoubtedly made some small, albeit questionable, contribution to the overall exploratory drilling during the early history of the Western Canada sedimentary basin.

Early Geophysical Surveys

The important role played by geophysics in selection of locations for potential oil and gas fields has evolved from

FIGURE 18. Controlled flare from Turner Valley field well with oil storage tanks to right of stationary rotary rig. Photo taken during the mid-1930s. The "oil leg" of the reservoir was not found until 1936, and probably as much as 1.5 × 10^{12} ft^3 (42.5 × 10^9 m^3) of natural gas was flared until commencement of World War I, when the Alberta government instituted conservation regulations to prevent such waste. The combined field flaring could be seen at night in Calgary. Turner Valley was humorously referred to as "Hell's Half Acres." Glenbow Archives NA-67-143.

one eliciting skepticism and avoidance to one instilling belief and acceptance. The early stigma of the charlatan posing as an "oil finder" with his "black box" was not easily erased when the first geophysicists appeared in the field. Their amusing sobriquet "doodlebugger" was coined at that time.

The beginnings of the applications of the various geophysical methods occurred in the Gulf Coast area of the United States during the early 1920s. These methods included electrical, gravity (including torsion-balance), magnetometer, and seismic surveys.

Torsion-balance and seismic instruments were used experimentally in 1920 and 1923, respectively, over known piercement salt dome structures in the Gulf Coast oil and gas region. By 1924 they were proven and accepted tools in the exploration for new fields. These two methods were first applied in the search for new salt dome structures and then to the areas beyond the Gulf Coast in different structural settings.

One of the earliest documented drillings of a rank exploration well in Western Canada, predicated on a seismically resolved "structure," appears to have been the Pine Hills Petroleum Avonlea No. 1 well located in the Dirt Hills or Claybank area south of Moose Jaw, Saskatchewan. Some questionable partial closure expressed in the Upper Cretaceous Whitemud Formation had been recognized as early as 1926 by W. S. Dyer of the Geological Survey of Canada. By 1932, according to G. S. Hume (1933), "Geophysical research first by electrical surveys and later by reflection seismograph methods are reported to have outlined an anticline in the Dirt Hill area south of Moose Jaw." The Avonlea well was spudded in July 1933 and abandoned in October of the following year after having reached a total depth of 4363 ft (1329.8 m) in sediments of Middle Jurassic age.

In early 1932, the Parco Twin River No. 1 exploratory well, located on the west flank of the Sweetgrass arch, was completed as a potential Mississippian Rundle limestone producer. However, the completion was unsuccessful, and, following a nitroglycerine fracturing of the limestone reservoir, the well gave water at the expense of the oil initially bailed from the hole. This well was drilled with a standard or cable-tool system. The same summer a reflection seismic survey (Clarke, 1982) by Geophysical Service Corporation was conducted for Nordon Corporation across a township area that included the site of the aforementioned suspended well. Based on the results of the survey, which deemed the suspended well to be an "edge" well, a step-out known as Nordon (Roney) Twin River No. 2 was drilled 2 mi (3.2 km) to the northeast. It proved to be a marginal oil producer from the Rundle limestone when completed in 1933 but was finally abandoned in 1945 after a series of unsuccessful completion attempts. Interestingly, this step-out well was drilled by a combination rig having a steel erected derrick. It commenced with a rotary system through the Mesozoic clastic section and converted to a cable-tool system on encountering the Mississippian Rundle limestone (Clarke, 1982). The resolution by a reflection seismic survey of a structurally improved location for a step-out well appears, in the case of the Twin River wells, to have been a first in Canada. Notwithstanding, refraction seismic work was conducted in the Turner Valley field as early as 1928.

Heiland Research appears to have been the first seismic company to conduct a reconnaissance survey in the Western Canada sedimentary basin. Under contract by Canadian Western Natural Gas, Light, Heat and Power Company, they operated during the summer of 1934 in the High River (Figure 19) area of southeastern Alberta. The following year (Norman J. Christie, personal communication, 1990), the crew was employed by Imperial Oil

FIGURE 19. **Observer Randy Langlois standing at the back of Heiland Research Corporation's recording car near High River, Alberta in the summer of 1934. This reconnaissance survey was for Canadian Western Natural Gas, Light, Heat and Power Company and was the first undertaken in Western Canada. Glenbow Archives NA-5103-14.**

in the Lethbridge area. E. J. P. (Emile) van der Linden was the party chief. A four-point pattern of spot shooting, with drill holes spaced at intervals of two miles in each direction, gave nine control points per township. A charge of 50 lb (22.7 kg) was used in each hole. The surface of the Paleozoic was anticipated to be a recognizable reflector for mapping purposes, but results proved it to be indistinct and unreliable.

Following the discovery in 1936 of the Del Bonita field located on the west flank of the Sweetgrass arch, some reflection seismic was reported to have been shot (Hume, 1944). The oil reservoir of this abandoned field is hosted by the Mississippian Rundle Group.

Commencing in 1940 and continuing until the autumn of 1945, a regional seismic survey was conducted by Carter Oil crews for Imperial-Norcanols Oil across southern and west-central Saskatchewan. This reflection seismic survey, including reconnaissance gravity as well as an extensive core-drill biostratigraphic program, was coordinated by J. C. (Cam) Sproule from headquarters in Moose Jaw, Saskatchewan. Fourteen unsuccessful exploratory wells were drilled in conjunction with the geophysical and geologic surveys. After leaving Saskatchewan, the two Carter crews continued their reconnaissance seismic surveys northwestward to the Edmonton, Alberta area, which efforts ultimately culminated in the discovery of the Devonian Leduc field in February 1947. Heiland Exploration Company (Layer, 1949) is credited with detailing the Leduc anomaly.

During 1942, as a military support strategy, the Canol project was created as a joint Canadian and American government operation to construct a pipeline from the Norman Wells field, Northwest Territories, to Whitehorse, Yukon Territory, a distance of some 600 mi (966 km). Imperial Oil's mandate in the overall operation was to develop its Norman Wells field as well as conduct exploration surveys in the general area that would lead to possible exploratory drill sites. Production at this time from four completed wells in the field supplied an on-site small refinery with some 840 bbl (133 m^3) per day (Stewart, 1944). Between 1942 and 1944, an additional 56 development wells were drilled, proving the field's expanded reserves to be categorized as major in size. Production averaged 3368 bbl (535 m^3) per day during 1944, part of which was moved through the completed pipeline to a refinery constructed at Whitehorse the same year.

During the period 1944 to 1946, Imperial conducted seismic surveys northwestward from Norman Wells to a distance of 80 miles (129 km) in the environs of the Mackenzie River valley. This was the first seismograph survey work to have been conducted in the search for petroleum in Northern Canada. Twelve exploratory wells, predicated on seismic and surface geologic surveys, were drilled, but all were abandoned.

The Jumping Pound field, located 20 miles (32 km) northwest of Calgary, was discovered by Shell Oil of Canada in 1944. The trapping mechanism of this wet-gas field is a complex folded and faulted overthrust structure involving Mississippian Rundle limestone that contains the reservoir. Heiland Exploration did seismic work in the area as early as 1942, followed by United Geophysical in 1944. This discovery was very significant because it not only can be credited as the first tangible seismic discovery in the Western Canada sedimentary basin, but also demonstrated the ability to map successfully by seismic means folded and faulted traps along the entire Foothills belt of western Alberta and northeastern British Columbia. Following the discovery, United Geophysical in 1945 ran a continuous seismic profile (Norman J. Christie, personal communication, 1990) from the northwest edge of the city of Calgary (Nose Hill) to Cochrane in the vicinity of Jumping Pound, a distance of 20 miles (32 km).

The first discovery in Alberta of an oil pool associated with Devonian rocks was at Princess in the southeastern part of the province, made by California Standard in 1944. Subsequent follow-up drilling proved the reserves to be inconsequential. Some seismic, probably by Heiland Exploration, is believed to have been shot in the general area as early as 1940. This was followed by a gravity survey (J. D. Weir, personal communication, 1990), which revealed a density anomaly over which the Devonian pool was found. A deep California Standard exploratory well drilled in 1939 and 1940 (Princess C.P.R. No. 1), and contiguous with the five completed wells, encountered the Precambrian at an anomalously high elevation, which appears to confirm the density anomaly.

Evolution of Drilling Rigs in Western Canada

The history of oil and gas exploration in the Western Canada sedimentary basin cannot be portrayed satisfactorily without reference to the type and evolution of the drilling rigs that served as instruments in the discovery of the present conventional oil and gas reserves. They also provided the rock cuttings and cores that enabled the geologic fraternity to synthesize the basin's history.

People have always been fascinated by drilling rigs. Their sudden introduction, whether into a rustic rural community or a hinterland setting, served to attract the local inhabitants. They were invariably amazed at the sheer size of the derricks as well as the operation of the drilling equipment. The prevailing sense of a "discovery" beneath the surface gave them an awareness of a new dimension of their environment.

The first type of percussion drilling assembly used in Western Canada specifically for oil exploration contained a simple derrick constructed of three timbers forming a tripod that supported a pulley. It was known as a "pole-tool rig," because a series of poles or rods served to connect the bit to the surface. Because of its extensive use in the early development of the Petrolia field in Ontario during the 1860s and 1870s, it was designated a Canadian-type rig. This same type of rig was employed in the drilling of the Rocky Mountain Development (John Lineham) No. 1 oil discovery (Figure 7) in 1902 at Oil City, located just west of Waterton Lakes, Alberta. The Geological Survey of Canada also utilized a Canadian-type rig (Figure 5) in their subsurface assessment of Tertiary lignite deposits in southeastern Saskatchewan in the summer and autumn of 1880.

The American standard cable-tool drilling rig, in which a cable (hemp rope and later steel cable) was used for lifting or lowering the drilling tools, was also introduced in Western Canada during the latter part of the 19th century. The derrick consisted of a four-sided tapered structure framed by horizontal and diagonal supports. Timber was invariably used in the construction of the cable-tool derrick, although a derrick constructed of bolted steel rods and incorporating a cable-tool system was employed in the drilling of Sullivan Creek No. 1, located near Flat Creek, Alberta, in 1941. Steam-powered cable-tool rigs were used by the Canadian Pacific Railway in their search for water at various station sites in Western Canada during the 1880s. Likewise, the Geological Survey of Canada utilized cable-tool rigs (Figure 8) in their assessment of the subsurface expression of the bituminous sands during the late 1890s.

Portable truck-mounted drilling machines (spudders) with single or twin-braced wooden masts were utilized in Western Canada prior to 1950 for shallow exploratory drilling—principally in the Medicine Hat, Alberta and Kamsack, Saskatchewan (Figure 20) areas. They were gasoline powered and characteristically utilized a rocker and floating sheave device that replaced the walking beam of the conventional cable-tool drilling rig.

The first reported drilling of an exploratory well with a rotary system in Western Canada occurred in 1913. This well, known as Fairholm Oil and Gas No. 1, was located at the southeast extremity of Edmonton, Alberta. This was

FIGURE 20. Kamsack Coutts No. 1 drilling in the Kamsack gas field of eastern Saskatchewan in August 1947. This truck-mounted, gasoline-powered cable-tool spudder had a twin-braced wooden mast. A rocker and floating sheave device replaced the reciprocating walking beam so characteristic of the conventional cable-tool rig. Photo credit, J. W. Porter.

only 12 years after the first introduction of this drilling method in the Spindletop field in Texas. By 1925, rotary rigs began to supplant some of the slower-drilling cable-tool rigs in the Turner Valley and Wainwright fields. The late 1920s witnessed a few combination rigs employing both cable-tool and rotary systems. One such rig drilled the Unity Valley No. 1 well in 1927, located in west-central Saskatchewan. The phasing out of the large American standard cable-tool rigs did not occur until after World War II. The last massive wooden-derrick cable-tool system was used at the site of the Alliance (Trans-Alta.) No. 2 exploratory well drilled in the summer of 1947 along the banks of the South Saskatchewan River a few miles west of the Alberta-Saskatchewan border. Steam-powered rotary rigs, utilizing steel-constructed stationary derricks, were used in the Princess field (Figure 21), located in southeastern Alberta, by the California Standard Company in the late 1940s.

Use of stationary derricks, whether of steel or timber, began to wane in the early 1940s with the advent of rolled tubular steel derricks that could be partly dismantled, transported, and reassembled at a new drilling site. At the same time, steam-powered engines were replaced by more efficient diesel engines. Imperial Oil was one of the leaders in this transition. Their Northwest (Imperial) Tilley No. 1, spudded in late 1941 in southeastern Alberta, was drilled with a diesel-powered Franks rotary skid rig. This well, in all probability, was the first in Western Canada to be drilled in such a manner. Imperial subsequently employed three diesel-powered portable rigs to drill 14 deep exploratory wells in Saskatchewan between 1942 and 1946.

Over 100 years have passed since the first exploratory well was drilled in the search for oil beneath the glacial mantle of the Western Canada sedimentary basin. Subsequently, a seemingly inexhaustible number of reports and papers, prepared by both industry and government earth scientists, have documented aspects of the basin's geology and the habitats of the hydrocarbons it hosts. The Western Canada sedimentary basin, and in particular the Western Canada foreland basin, could be considered a model for the less-explored and less-understood foreland basins of the world.

FIGURE 21. California Standard Princess C.P.R. No. 14-22-A (Lsd. 8, Sec. 22, Twp. 20, Rge. 12W4M) in 1947. This rotary rig, with its large steel-girded stationary derrick, required the power generated by three steam boilers for the drill bits to reach the lower Paleozoics. Glenbow Archives NA-5103-13.

ACKNOWLEDGMENTS

I wish to thank Drs. Roger W. Macqueen, Dale A. Leckie, and J. Sebastian Bell of the Institute of Sedimentary and Petroleum Geology, Calgary, for their critical reading of the manuscript as well as their helpful suggestions. Mrs. Claudia Thompson, also of the institute, typed the numerous drafts, and to her I also extend thanks. I am grateful for the help of Mrs. Lynette Walton, Assistant Chief Archivist, Glenbow Museum, Calgary, for her cooperation in expediting the selection of photographs used in this paper. Credit is given to the Glenbow Archives for permission to use these photographs in this publication. The author wishes to thank Mobil Oil Canada for its generous help in the use of its library services and especially Brian Scott of Mobil's Drafting and Reproduction Department for drafting Figure 1.

REFERENCES CITED

Allan, A. J., 1953, Guest editorial: Alberta Society of Petroleum Geologists Bulletin, v. 1, n. 6, p. 1.

Ballard, W. N., 1942, Regional geology of the Dakota basin: AAPG Bulletin, v. 26, p. 1557–1584.

Bell, R., 1874, Report on the country between Red River and the South Saskatchewan, with notes on the geology of the region between Lake Superior and Red River: Geological Survey of Canada Report of Progress for 1873–74, p. 66–90.

Bell, R., 1876, Report on the country west of Lakes Manitoba and Winnipegosis, with notes on the geology of Lake Winnipeg: Geological Survey of Canada Report of Progress for 1874–75, p. 25–56.

Bell, R., 1885, Report on part of the basin of the Athabasca River, North West Territory: Geological and Natural History Survey and Museum of Canada Report of Progress for 1882–1884, p. CC1–CC37.

Burpee, L. J., 1908, The search for the western sea: the story of the exploration of northwestern America: Toronto, Ontario, The Musson Book Company Ltd., 651 p.

Clapp, F. G., et al., 1915, Petroleum and natural gas resources of Canada: Canada Department of Mines, Mines Branch Report 291, v. 2, 404 p.

Clarke, P. J., 1982, Fifty years ago: October 11, 1932, from the archives of the Glenbow-Alberta Institute: Oilweek, v. 32, p. 9.

Coues, E., 1897, Henry and Thompson Journals, v. 11, reprinted 1965: Minneapolis, Ross and Haines, p. 470-471.

Creaney, S., and J. Allan, 1992, Petroleum systems in the foreland basin of Western Canada, this volume.

Dawson, G. M., 1875, Report on the geology and resources of the region in the vicinity of the forty-ninth parallel from Lake of the Woods to the Rocky Mountains: Ottawa, Ontario, British North America Boundary Commission, 379 p.

Dawson, G. M., 1881, On the Lignite Tertiary Formation from the Souris River to the 108th meridian: Geological and Natural History Survey of Canada Report of Progress for 1879-80, Report A, p. 12A-49A.

Dawson, G. M., 1886, On certain borings in Manitoba and the Northwest Territory: Proceedings and Transactions of the Royal Society of Canada, v. 4, sec. IV, p. 85-99 .

Denison, A. R., 1953, Structure and stratigraphy in relation to recent oil and gas discoveries: AAPG Bulletin, v. 1, n. 1, 2, 3, and 4, p. 3.

Dobbin, C. E., and C. E. Erdmann, 1934, Geologic occurrence of oil and gas in Montana, *in* W. E. Wrather and F. H. Lahee, eds., Problems of petroleum geology: AAPG, Sidney Powers Memorial Volume, p. 695-718.

Douglas, R. J. W., 1970, Geology and economic minerals of Canada: Geological Survey of Canada Economic Geology Report No. 1, 838 p.

Dowling, D. B., 1919, The structure and correlation of the formations underlying Alberta, Saskatchewan and Manitoba, *in* D. B. Dowling, S. E. Slipper, and F. H. McLearn, eds., Investigations in the gas and oil fields of Alberta, Saskatchewan, and Manitoba: Canada Department of Mines Geological Survey Memoir 116, p. 1-9.

Dowling, D. B., 1920, Oil possibilities and developments in the great plains: Canada Department of Mines Geological Survey Summary Report 1919, Part C, p. 20C-24C.

Ells, R. W., 1877, Report on the boring operation in the North-West Territory, Summer of 1875: Geological and Natural History Survey of Canada Report of Progress for 1875-76, p. 281-291.

Ferguson, B. G., 1985, Athabasca oil sands: northern resource exploration, 1875-1951: Alberta Culture/ Canadian Plains Research Center, 283 p.

Grant, G. M., 1873, Ocean to ocean: Sandford Fleming's expedition through Canada in 1872: reprinted 1925, Toronto, Ontario, Radisson Society of Canada Ltd., 412 p.

Hector, J., 1861, On the geology of the country between Lake Superior and the Pacific Ocean (between the 48th and 54th parallels of latitude), visited by the Government Exploring Expedition under the command of Captain J. Palliser (1857-60): Quarterly Journal of the Geological Society of London, v. 17 , p. 388-445.

Hennen, R. V., 1943, Tertiary geology and oil and gas prospects in Dakota basin of North Dakota: AAPG Bulletin, v. 27, p. 1567-1594.

Henry, A., and D. Thompson, 1897, New light on the early history of the greater northwest: the manuscript journals of Alexander Henry, fur trader of the Northwest Company, 1799-1814; exploration and adventure among the Indians on the Red, Saskatchewan, Missouri, and Columbia rivers, edited with copious critical commentary by Elliott Coues (volume 2): New York, Francis P. Harper, 702 p.

Hind, H. Y., 1860, Narrative of the Canadian Red River Exploring Expedition of 1857 and of the Assinniboine and Saskatchewan Exploring Expedition of 1858: reprinted 1971, Edmonton, Alberta, M. G. Hurtig Ltd., 472 p.

Hume, G. S., 1933, Oil and gas in Western Canada: Canada Department of Mines Geological Survey Economic Geology Series No. 5 (second edition), 359 p.

Hume, G. S., 1944, Petroleum developments in Canada in 1943: AAPG Bulletin, v. 28, p. 864-872.

Kupsch, W. O., and W. G. E. Caldwell, 1982, Some nineteenth-century studies of the Cretaceous System in the Williston basin—their contribution to modern stratigraphy, *in* J. E. Christopher, J. Kaldi, C. E. Dunn, D. M. Kent, and J. A. Lorsong, eds., Fourth International Williston Basin Symposium: Saskatchewan Geological Society Special Publication No. 6, p. 277-293.

Layer, D. B., 1949, Leduc oil field, Alberta, a Devonian coral-reef discovery: AAPG Bulletin, v. 33, p. 572-602.

Layer, D. B., 1958, Characteristics of major oil and gas accumulations in the Alberta basin, *in* L. G. Weeks, ed., Habitat of oil—a symposium: Tulsa, Oklahoma, AAPG, p. 113-128.

Leckie, D. A., and D. G. Smith, 1992, Regional setting, evolution, and depositional cycles of the Western Canada foreland basin, this volume.

Link, T. A., 1931, Alberta syncline, Canada: AAPG Bulletin, v. 15, p. 491-507.

Macdonald, J. S., 1930, The Dominion telegraph: Canadian North-West Historical Society Publications, v. 1, n. 6, 64 p.

Mackenzie, Sir Alexander, 1970, The journal and letters of Sir Alexander Mackenzie; edited by W. Kaye Lamb: Toronto, Ontario, MacMillan of Canada, 551 p.

Malcolm, W., 1913, Oil and gas prospects of the northwest provinces of Canada: Canada Department of Mines Geological Survey Memoir 29-E, 99 p.

McCoy, A. W., and W. R. Keyte, 1934, Present interpretations of the structural theory for oil and gas migration and accumulations, *in* W. E. Wrather and F. H. Lahee, eds., Problems of petroleum geology: AAPG, Sidney Powers Memorial Volume, p. 253-307.

McDonald, A., 1872, Peace River a canoe voyage from Hudson's Bay to Pacific by Sir George Simpson in 1828: journal of the late Chief Factor Archibald McDonald (Hon. Hudson's Bay Company), who accompanied him; edited with notes by Malcom McLeod: reprinted 1971, Edmonton, Alberta, M. G. Hurtig Ltd., 119 p.

McFadden, M., 1953, Assiniboine steamboats: The Beaver a

Magazine of the North, Outfit 284, June 1953, p. 38-42.
McLearn, F. H., and G. S. Hume, 1927, The stratigraphy and oil prospects of Alberta, Canada: AAPG Bulletin, v. 11, p. 237-260.
Morton, A. S., 1939, A history of the Canadian West to 1870-71, being a history of Rupert's Land (The Hudson's Bay Company's territory) and of the North-West Territory (including the Pacific slope): London, Thomas Nelson and Sons Ltd., 987 p.
Owen, E. W., 1975, Trek of the oil finders: a history of exploration for petroleum: AAPG Memoir 6, 1647 p.
Peel, B., 1964, First steamboats on the Saskatchewan: The Beaver a Magazine of the North, Outfit 295, Autumn 1964, p. 16-21.
Rich, E. E., ed., 1955, Black's Rocky Mountain Journal 1824: The Publications of the Hudson's Bay Record Society, v. 18, 260 p.
Richardson, J., 1828, Geognostical observations, *in* Sir John Franklin, Narrative of a journey to the shores of the Polar Sea, in the years 1819-22: London, John Murray; reprinted 1969, Edmonton, Alberta, M. G. Hurtig Ltd. 768 p.
Rodney, W., 1969, Kootenai Brown his life and times: Sidney, British Columbia, Gray's Publishing Ltd., 252 p.
Russell, R. C., 1959, A minister takes the Carlton Trail: The Beaver a Magazine of the North, Outfit 290, Winter 1959, p. 4-11.
Selwyn, A. R. C., 1874, Introductory report: Geological Survey of Canada Report of Progress for 1873-74, p. 1-9.
Selwyn, A. R. C., 1876, Summary report of the operations of the Geological Corps to 31st December, 1875: Geological Survey of Canada Report of Progress for 1874-75, p. 1-23.
Selwyn, A. R. C., 1881a, Introductory report: Geological and Natural History Survey of Canada Report of Progress for 1879-80, p. 1-9.
Selwyn, A. R. C., 1881b, Report on boring operations in the Souris River Valley 1880: Geological and Natural History Survey Report of Progress for 1879-80, p. A1-A11.
Selwyn, A. R. C., 1889, Summary reports of the operations of the Geological Survey for the years 1887 and 1888: Geological and Natural History Survey of Canada Annual Report 1887-88, v. 3, report A, p. 1A-117A.
Selwyn, A. R. C., 1892, Summary reports on the operations of the Geological Survey for the year 1891: Geological Survey of Canada Annual Report, v. 5, pt. 1, report A (1891), p. 3A-93A.
Slipper, S. E., 1919, Sketch of the geology of southern and central Alberta, *in* D. B. Dowling, S. E. Slipper, and F. H. McLearn, eds., Investigations in the gas and oil fields of Alberta, Saskatchewan and Manitoba: Canada Department of Mines Geological Survey Memoir 116, p. 13-24.
Spry, I. M., 1963, The Palliser Expedition—an account of John Palliser's British North American Expedition 1857-1860: Toronto, Ontario, The MacMillan Company of Canada Ltd., 310 p.
Stebinger, E., 1917, Possibilities of oil and gas in north-central Montana: U.S. Geological Survey Bulletin 641-C, p. 49-91.
Stewart, J. S., 1944, Petroleum possibilities in Mackenzie River Valley, Northwest Territories: Canadian Institute of Mining and Metallurgy Transactions, v. 47, p. 152-171.
Turner, J. P., 1950, The North-West Mounted Police 1873-1893: Ottawa, Ontario, E. Cloutier King's Printer, v. 1, 340 p.
Tyrrell, J. B., 1891, Three deep wells in Manitoba: Transactions of the Royal Society of Canada, v. 9, sec. IV, p. 91-104.
Umfreville, E., 1954, The present state of Hudson's Bay, containing a full description of that settlement, and the adjacent country, and likewise of the fur trade, with hints for its improvement; edited with an introduction and notes by W. S. Wallace: The Canadian Historical Studies, v. 5, 122 p.
Warkentin, J., 1964, The Western Interior of Canada: a record of geographical discovery, 1612-1917; The Carleton Library No. 15: Toronto, Ontario, McClelland and Stewart, 304 p.
Waud, W. B., 1874, Report on operations in Manitoba with the diamond-pointed steam drill, 1873: Report of Progress for 1873-74, p. 12-16.
Webb, J. B., J. S. Irvin, E. J. Hunt, W. D. C. Mackenzie, S. E. Slipper, R. V. Johnson, J. O. G. Sanderson, J. O. Galloway, and M. W. Ball, 1941, Possible future oil provinces in Western Canada: AAPG Bulletin, v. 25, p. 1450.
Wilson, C., 1939, Indian treaties: The Beaver Outfit 269, Hudson's Bay Company, Winnipeg, p. 38-41.
Zaslow, M., 1975, Reading the rocks, the story of the Geological Survey of Canada 1842-1972: Ottawa, Ontario, The Macmillan Company of Canada Ltd. in association with the Department of Energy, Mines and Resources and Information Canada, Ottawa, Ontario, 599 p.

Chapter 6

Conventional Hydrocarbon Reserves of the Western Canada Foreland Basin

J. W. Porter

Calgary, Alberta, Canada

INTRODUCTION

Before any discussion is presented of the collated statistical data of conventional hydrocarbon reserves contained within the Western Canada foreland basin, some prefatory remarks will be addressed to the associated presence of nonconventional heavy oil, oil sands ("tar sands"), and bitumen. A general understanding of the natures and settings of these deposits is a prerequisite to a better appreciation of the maps and charts relating to conventional crude in the body of this text.

In this study, no attempt has been made to include a detailed synthesis of the reserves of the heavy oil, oil sands, and bitumen deposits. Those interested are referred to Govier (1974), Outtrim and Evans (1978), Wilson and Bennet (1985), Energy Resources Conservation Board (1986), and Meyer and De Witt (1990).

CONVENTIONAL AND NONCONVENTIONAL OIL

The habitat of the immense deposits of heavy oil and bitumen within the foreland basin succession is within Lower Cretaceous sand reservoirs. Extensive amounts of bitumen also occur in underlying Paleozoic carbonate rocks of the miogeocline platform wedge (the "carbonate triangle"; e.g., Mossop et al., 1981): these deposits are very similar geochemically to the Cretaceous deposits (Brooks et al., 1989). Geologically, the reservoirs are positioned along the northeast flank of the basin (Figure 1) and extend in part to the updip limit of Lower Cretaceous strata. Geographically, the main deposits of heavy oil, oil sands, and bitumen occur in three regions of north-central Alberta—Peace River, Athabasca-Wabasca, and Cold Lake (Figure 1)—encompassing 29,000 mi^2 (75,104 km^2), which represents 6% of the areal extent of the Western Canada foreland basin. Recent reserve data provided by the Energy Resources Conservation Board of Alberta indicate that Cretaceous oil sands and bitumen reserves are slightly less than 1.227×10^{12} bbl (195×10^9 m^3), with Paleozoic carbonate triangle bitumen reserves of 449.4×10^9 bbl (71.4×10^9 m^3)(Energy Resources Conservation Board, 1989).

For comparison, the initial established reserves of conventional crude oil plus natural gas liquids, based on waterflood projects and other enhanced recovery schemes, contained within the Western Canada foreland basin amount to approximately 6.4×10^9 bbl (1.0×10^9 m^3). Paradoxically, these conventional crude oil reserves represent less than 2% of the combined conventional and nonconventional ultimate recoverable reserves of the Western Canada foreland basin. Several conclusions are apparent from this very disproportionate relationship. First, the Western Canada foreland basin and its precursor, the Western Canada cratonic platform basin (pre-Middle Jurassic succession), must contain immense and widespread volumes of source rocks. Second, optimum conditions were present in the Western Canada sedimentary basin for the prolific generation of hydrocarbons from these source rocks. Third, in-situ entrapment of this generated conventional crude was incidental to its escape updip along northeasterly migration routes toward the Precambrian shield. Fourth, as we will expand on later in the text, there was a progressive conversion of much of the original mobile crude during its migration to an eventual static state, mainly through biodegradation, with accompanying water washing and oxidation (e.g., Brooks et al., 1988). Both the great extent and enormous size of these heavy oil, oil sand, and bitumen deposits attest to the failure of the host rocks to trap locally much of the migrating hydrocarbons. In essence, the conventional and nonconventional crude oil deposits are part of a continuum, the latter crude type ultimately manifesting its own seal through progressive stages of biodegradation. This megaseal of heavy oil, oil sands, and bitumen, and not the host strata, appears to have been responsible for the

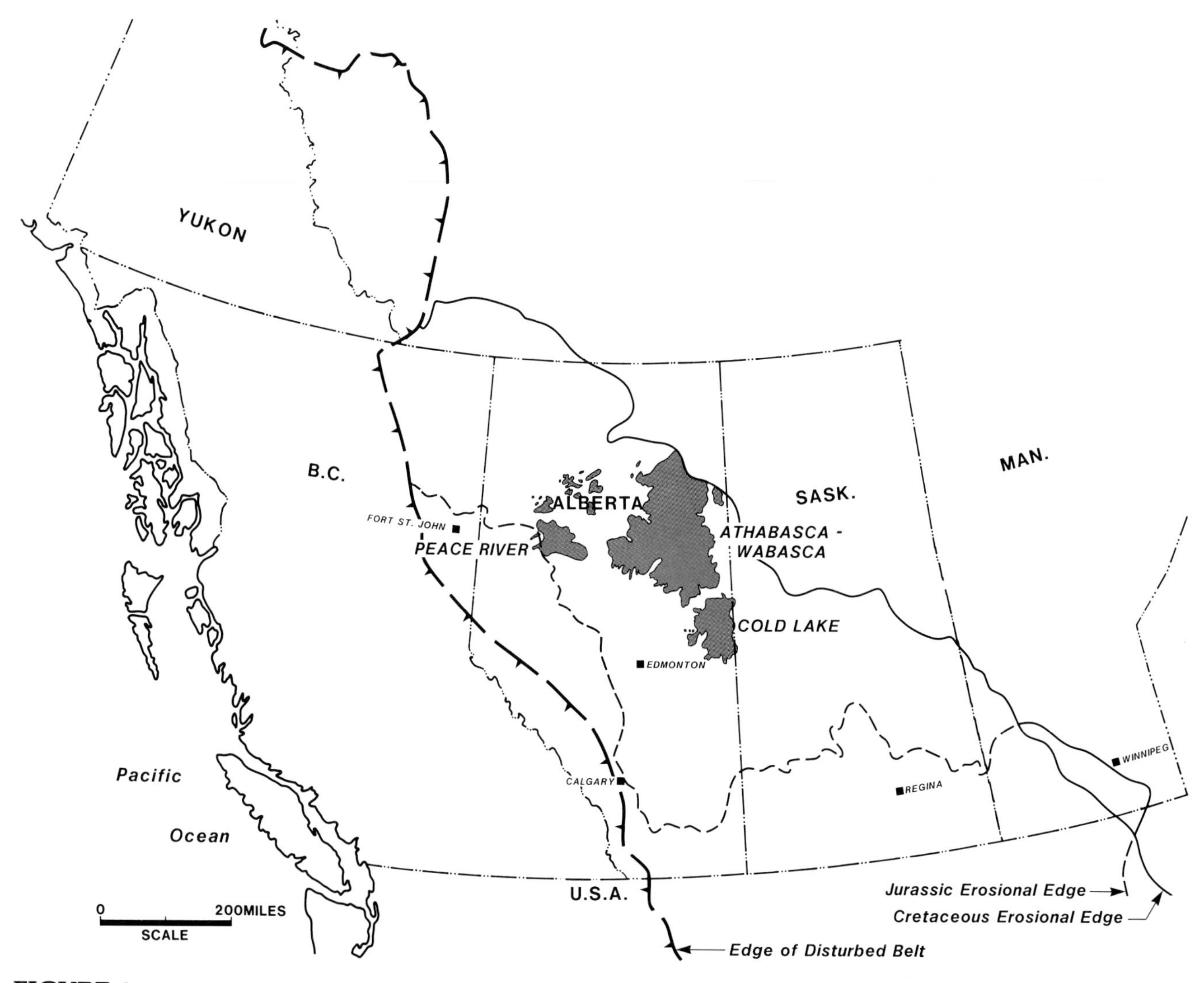

FIGURE 1. Distribution of nonconventional heavy oil, oil sands ("tar sands"), and bitumen in the Western Canada foreland basin.

retention of the vast reserves of nonconventional oil flanking the northeast rim of the basin.

The updip limit of the bulk of these vast deposits hosted by Cretaceous sands coincides with the dissolution edge of the underlying Middle Devonian salt beds contained within the Elk Point Group (Leckie and Smith, 1992). This relationship appears to be no coincidence, because the regional dip flattens abruptly, which probably served to suppress the transmissibility of the migrating oil, thus abetting its gathering *en masse* at its present position. The invading meteoric water provided microbes that progressively biodegraded the migrating oil and concomitantly dissolved the entire northeastern extension of these salt beds fronting the Precambrian shield (Porter et al., 1982).

SOURCES OF NONCONVENTIONAL OIL (HEAVY OIL, BITUMEN)

The question arises as to where all this crude oil originated. The Western Canada foreland basin is not only the repository for indigenous hydrocarbons but also includes those hydrocarbons that were generated in the antecedent Western Canada cratonic platform basin (see Creaney and Allan, 1992). The sub-Cretaceous unconformity reveals evidence of post-Jurassic removal of preexisting Jurassic sedimentary strata across the entire northeast flank of the basin, exposing progressively older Paleozoic strata in the direction of the Precambrian shield (Leckie and Smith, 1992). It was on this erosional surface that the Lower Cretaceous basal sands were deposited. These diachronous sands may have been the conduit not only for Jurassic-generated hydrocarbons but also for cratonic, platform-sourced hydrocarbons originating from strata of Devonian–Mississippian age. Because of the characteristic fluviatile and deltaic sedimentation, the Lower Cretaceous Mannville Group appears to have been the principal conduit for the regional migration of multisourced hydrocarbons generated in the Western Canada sedimentary basin.

The Late Cretaceous westward tilt and attendant beveling of the Mannville Group flanking the Precambrian shield resulted in a reduction of its original sedimentary expression. The exposure of porous Mannville sands at the erosional edge permitted large-scale introduction of meteoric water. Biodegradation processes appear to have occurred where hydrocarbons moving up the regional dip encountered the basinward encroachment of invading meteoric water. The degree of encroachment of meteoric water is expressed basinward by the gravity spectrum of Mannville-hosted crude oil from low-gravity nonconventional (biodegraded) crude on the basin's upper flank to high-gravity conventional nonbiodegraded crude as the foredeep is approached in a westerly and southwesterly direction. This situation is well shown by comparing the degree of biodegradation of Mannville-hosted oils with formation-water salinities, as first presented by Deroo et al. (1974, figure 6, p. 156). These data verify that formation-water salinities are inversely proportional to the degree of biodegradation shown by crude oils in this succession.

The Peace River, Athabasca-Wabasca, and Cold Lake deposits are believed for the most part to have undergone long-range migration. The near-surface presence of this heavy oil and crude bitumen might at first sight tend to suggest in-situ generated immature crude lacking an adequate burial history. However, the Late Cretaceous synorogenic loading of the basin's foredeep during the initiation of the Laramide orogeny is believed to have triggered much of the progressive generation of crude throughout the Western Canada sedimentary basin (see Creaney and Allan, 1992).

Recent organic geochemical work on nonconventional heavy oil, oil sands, and bitumen from the Western Canada sedimentary basin (including foreland basin materials) indicates (a) that the geochemical characteristics of the nonconventional materials strongly support the concept that these materials originally were conventional oils that were altered subsequently, mainly by biodegradation (Evans et al., 1971); and (b) that all Cretaceous-hosted heavy oils, oil sands, and bitumens are very similar in organic geochemical composition, the differences being most closely related to the degree of biodegradation (Brooks et al., 1988, 1989). Comparative organic geochemical work on conventional crude oils and their actual prospective source rocks from the Western Canada sedimentary basin suggests that sources for the nonconventional materials are likely to have been Devonian–Mississippian or Mesozoic, because the geochemical characteristics of Upper and Middle Devonian conventional oils and demonstrated source rocks are very different from those of the heavy oils, oil sands, and bitumens (Brooks et al., 1988, 1989, 1990; also see Creaney and Allan, 1992).

CONVENTIONAL CRUDE OIL AND NATURAL GAS RESERVES

Database and Methodology

The database utilized for the preparation of the maps and charts accompanying this text was obtained from statistical reports published by the provinces of British Columbia (Province of British Columbia, 1985), Alberta (Energy Resources Conservation Board, 1986, 1989), and Saskatchewan (Province of Saskatchewan, 1985). The respective reports include reserves and related data through December 31, 1985. The locations and extents of the oil and gas fields shown on the maps were transferred from maps published by the Geological Survey of Canada (Wallace-Dudley, 1981a, b).

The methodology pursued in the synthesis of the data was to subdivide the total reserves of a multipool field on the basis of where the reserves occur in the stratigraphic column. Figure 2 displays the principal oil and gas reservoirs to which the hydrocarbon reserves were assigned. This simplified approach was taken in order to demonstrate the comparative roles of the various designated stratigraphic units in harboring hydrocarbons. The sum total of vertically stacked pools within a designated reservoir unit is indicated as a "field" in Figures 3 to 22, inclusive. Only "fields" with minimum initial established reserves of 10×10^6 bbl (1.59×10^6 m^3) of oil, or its equivalent 60×10^9 ft^3 (1.69×10^9 m^3) of gas, are shown for utilitarian sake. Isogravity contours, utilizing a contour interval of 5° API (1036 kg/m^3), are shown on each of the oil "field" maps. A common vertical scale is used on both the oil and gas "field" maps to demonstrate the latter's comparative size in terms of oil equivalent of gas (1 bbl of oil equivalent to 6000 ft^3 of gas).

Stratigraphic Intervals of Pooled Hydrocarbons

Twelve stratigraphic intervals represent the principal oil- and gas-producing reservoirs in the Western Canada foreland basin (Figure 2). The Jurassic succession contains two reservoir units: the intra-Jurassic, as represented by the Middle Jurassic Shaunavon–Sawtooth–Rock Creek reservoirs; and the Upper Jurassic Swift-Vanguard reservoirs, which subcrop at the pre-Cretaceous unconformity. The Lower Cretaceous Mannville Group is subdivided into two reservoir units, the Lower Mannville or pre-"Glauconitic" sandstone and the Upper Mannville or post-Ostracod zone. The remaining eight reservoir units represent widespread regressive sand wedges. The Basal Colorado Sand and the Cadotte–Viking–Bow Island reservoirs are assigned to the Lower Cretaceous. Upper Cretaceous reservoirs are designated in ascending order: Doe Creek–Dunvegan, Second White Speckled Shale, Cardium, Medicine Hat, Milk River, and Belly River. Each reservoir unit, with the exception of the Second White Speckled

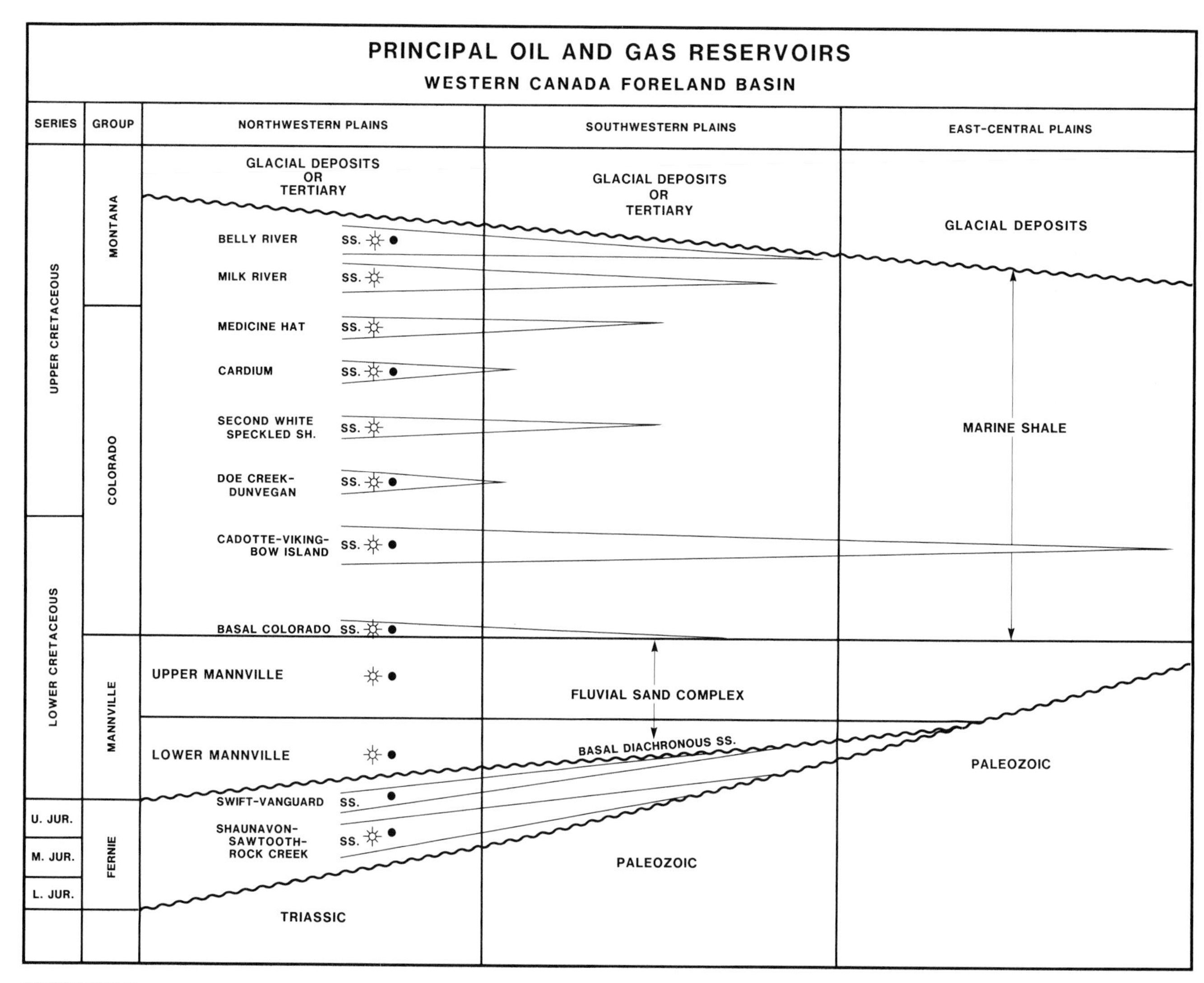

FIGURE 2. Generalized Western Canada foreland basin wedge, exhibiting principal oil- and gas-producing reservoir units.

Shale, Medicine Hat, and Milk River, contains at least one "field" with initial established oil reserves in excess of 10 $\times$ 10^6 bbl (1.59 $\times$ 10^6 m^3); all contain at least one gas "field" in excess of 60 $\times$ 10^9 ft^3 (1.69 $\times$ 10^9 m^3) with the single exception of the Swift-Vanguard reservoir unit. Some 148 gas "fields" and 84 conventional oil "fields" including natural gas liquid "fields," meeting the aforementioned minimum reserves requirements, have been identified and assigned to their respective habitats within one or more of the 12 ascribed reservoir units.

The dominant lithology of all of the aforementioned reservoir units is sandstone. In general, the loss of effective porosity on the west or foredeep portion of the foreland basin is a result of excessive depth of burial and attendant diagenetic processes. To the east and northeast side of the basin, the loss of porosity within the regressive sand wedges is attributable to the reduction of grain size, being a function of the sand's distance from its westward provenance.

The converse is evident with respect to the Mannville reservoirs, where there is an enhancement of total porosity in the direction of the exposed Precambrian shield. Pre-Cretaceous recycled sand sequences flanking the shield afforded a source for much of the Mannville sedimentation. This phenomenon, coupled with the lack of lateral shale seals, reinforces the concept of the ability of surface-derived water to induce a megaseal, thus inhibiting the eventual surface dissipation of the migrating crude oil.

HYDROCARBON ANALYSIS OF RESERVOIR LIMITS

Jurassic and Lower Cretaceous Conventional Crude Oil

Figures 3, 5, 6, and 8 represent the conventional crude oil reserves contained within the intra-Jurassic, subcrop

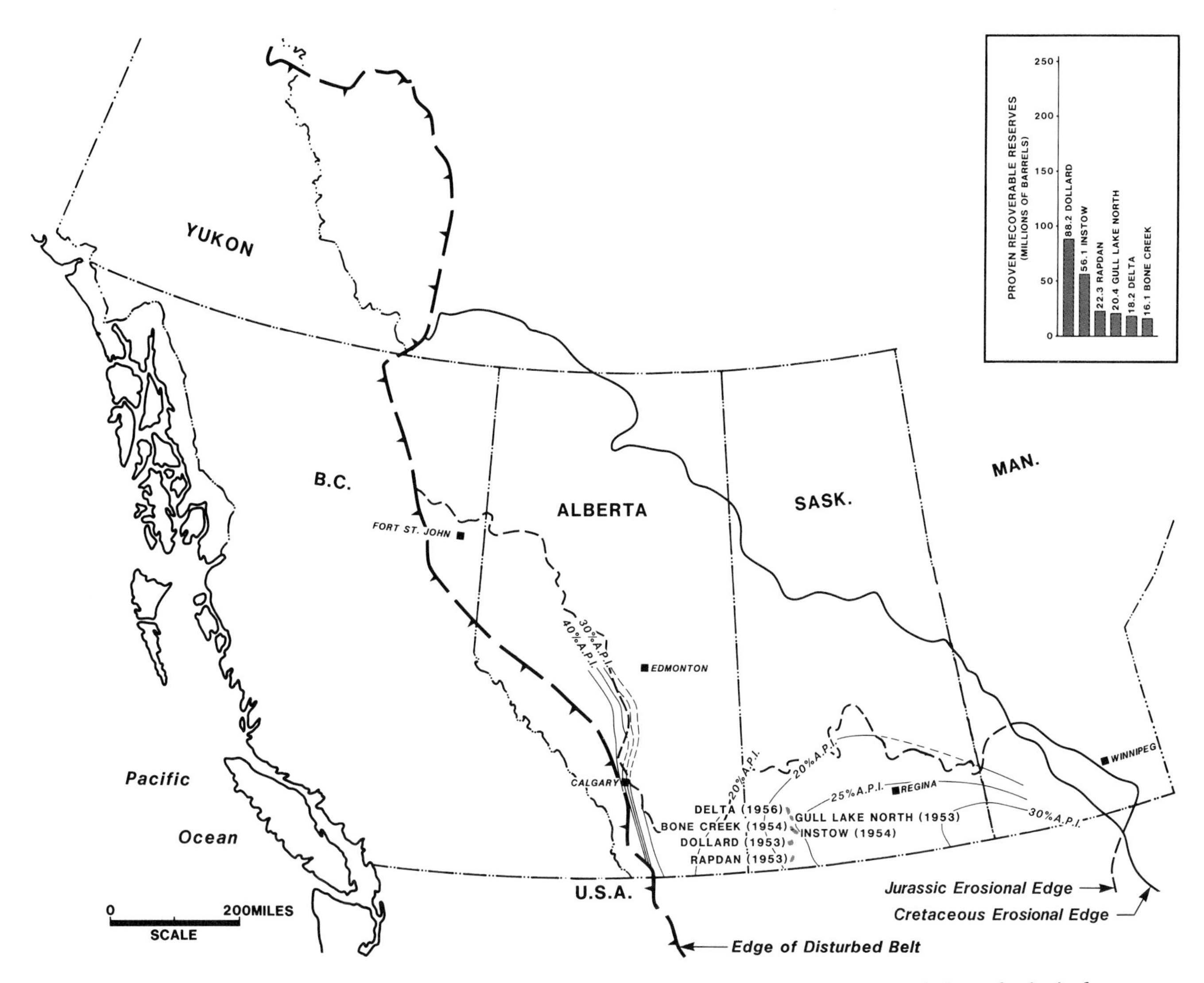

FIGURE 3. Intra-Jurassic Shaunavon-Sawtooth-Rock Creek reservoir unit: distribution and sizes of principal conventional oil fields associated with this unit, including corresponding isogravity pattern.

Jurassic, Lower Cretaceous Lower Mannville, and Lower Cretaceous Upper Mannville reservoir units, respectively. The most obvious relationship is the increment in both the number of fields (Figures 5, 6, and 8) and their cumulative reserves from the oldest reservoir unit to the youngest. There is also a corresponding eastward and northeastward migration through time of the respective field patterns (Figures 5, 6, and 8), with the youngest fields occurring in west-central Saskatchewan. These patterns are attributed to the lack of effective sealing mechanisms to prevent both long-range lateral and vertical migration of hydrocarbons within the Mannville Group.

The isogravity patterns of three of the four reservoir units (Figures 5, 6, and 8) exhibit a similar spectrum of decreased API gravity to the east and north, representing progressive biodegradation toward the basin's flank. The distribution of the heavy oil and bitumen crude ("tar sands") represents the end of this spectrum as part of a total hydrocarbon continuum. Topography-driven meteoric waters also originated from north-central Montana, aiding in biodegradation of the oil contained in the Jurassic reservoir units (Figures 3 and 5) of southwestern Saskatchewan. Jurassic-Cretaceous strata were exhumed in north-central Montana during the Laramide orogeny, as expressed in the Bearpaw and Little Rockies Mountains.

Jurassic and Lower Cretaceous Natural Gas

The gas field maps, Figures 4, 7, and 9, which represent the intra-Jurassic, Lower Cretaceous Lower Mannville, and Lower Cretaceous Upper Mannville reservoir units, respectively, compliment the corresponding oil field maps inasmuch as a similar time-space relationship is evident. The intra-Jurassic gas fields (Figure 4) of central Alberta are located near the updip erosional edge of this system, whereas several of the larger Upper Mannville gas fields (Figure 9) occur well up toward the basin's flank. However, the Elmworth gas field, which is the largest field of

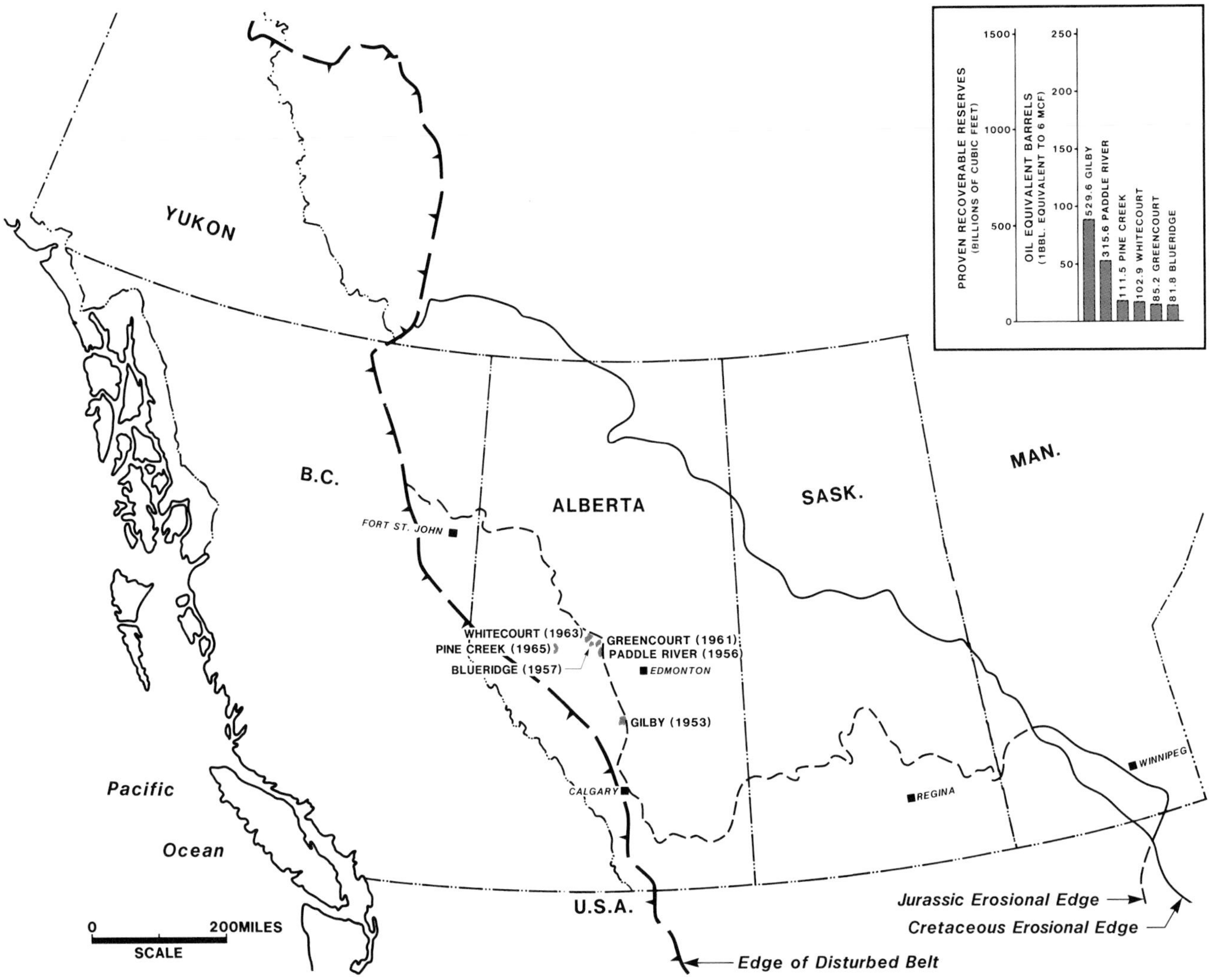

FIGURE 4. Intra-Jurassic Shaunavon-Sawtooth-Rock Creek reservoir unit: distribution and sizes of principal gas fields associated with this unit.

this reservoir unit, occupies a foredeep position. The location of the Elmworth gas field is attributed to the very low-permeability character of the reservoir induced through excessive depth of burial with attendant diagenetic implications.

Uppermost Lower Cretaceous and Upper Cretaceous Conventional Crude Oil

Figures 10, 12, 14, 17, and 21 represent the oil fields associated with the Basal Colorado Sand and Cadotte-Viking-Bow Island units of the Lower Cretaceous and the Doe Creek-Dunvegan, Cardium, and Belly River units of the Upper Cretaceous. These reservoir units, with the possible exception of the Basal Colorado Sand, are isolated vertically and laterally by marine shale throughout the basin and are unique in their respective stratigraphic realms in harboring hydrocarbons. The limit to which entrapment occurs shelfward is a function of the eastward expression of the reservoir unit. The intensity of the orogenic pulses initiating these regressive sand cycles and the effects of changes in relative sea level varied, as revealed by the distribution patterns of the corresponding reservoir units. For example, the Cadotte-Viking-Bow Island unit has the greatest shelfward expression of all the Cretaceous sandstone cycles with fields shown in west-central Saskatchewan (Figure 12).

In terms of size of reserves, the giant Cardium accumulation associated with the Pembina field dwarfs all other Western Canada foreland basin conventional oil

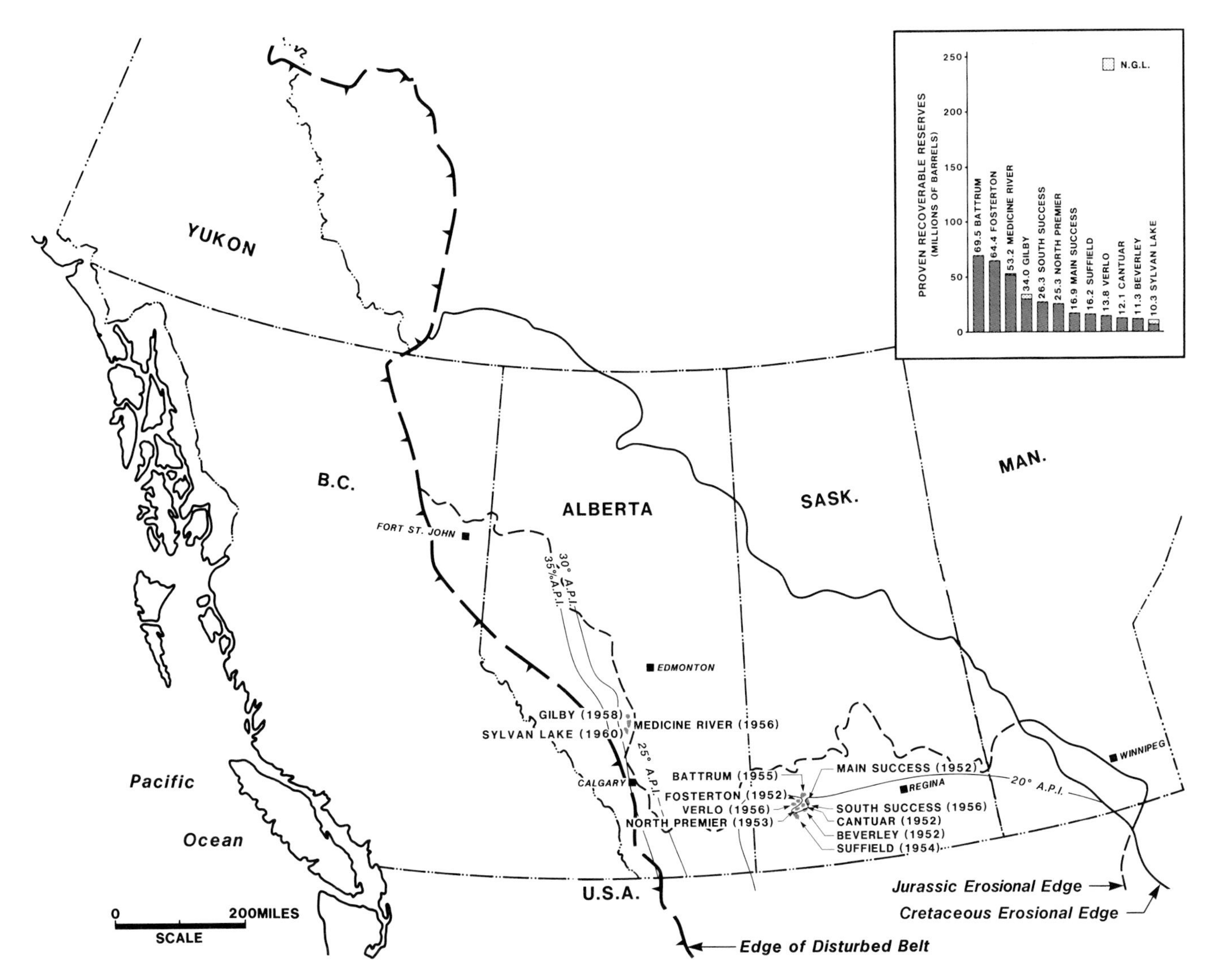

FIGURE 5. **Subcrop Jurassic Swift-Vanguard reservoir unit: distribution and sizes of principal conventional oil plus natural gas liquid fields associated with this unit, including corresponding isogravity pattern.**

fields. This anomaly, in terms of sheer size and geographic extent, appears to be a function of its basin position with respect to the synsedimentary loading hinge demarcating the foreland shelf from its foredeep counterpart. A diagenetic seal limits the basinward extent of the Cardium reservoir (Nielsen and Porter, 1984), whereas the updip limits on the shelf are controlled by the sand size and attendant "shale-out."

The API gravity values of the aforementioned five regressive sand reservoirs fall in the category of light-gravity oil. The low sulfur-paraffinic nature of the crude associated with these reservoirs demonstrates the lack of contamination by meteoric waters and attendant biodegradation. The Basal Colorado Sand reservoir unit appears to be an exception inasmuch as communication with the subjacent Upper Mannville reservoirs is suspected. The medium-gravity, low-sulfur quality of the Cessford field (Figure 10) supports this conclusion.

Uppermost Lower Cretaceous and Upper Cretaceous Natural Gas

Gas fields are associated with each of the eight regressive sand reservoirs noted above. Figures 11 and 13 represent productive gas reservoir units in the Lower Cretaceous Basal Colorado Sand and Cadotte-Viking-Bow Island. The remaining six productive gas reservoirs (Figures 15, 16, 18, 19, 20, and 22) occur within the Upper Cretaceous and are in ascending order: Doe Creek-Dunvegan, Second White Speckled Shale, Cardium, Medicine Hat, Milk River, and Belly River.

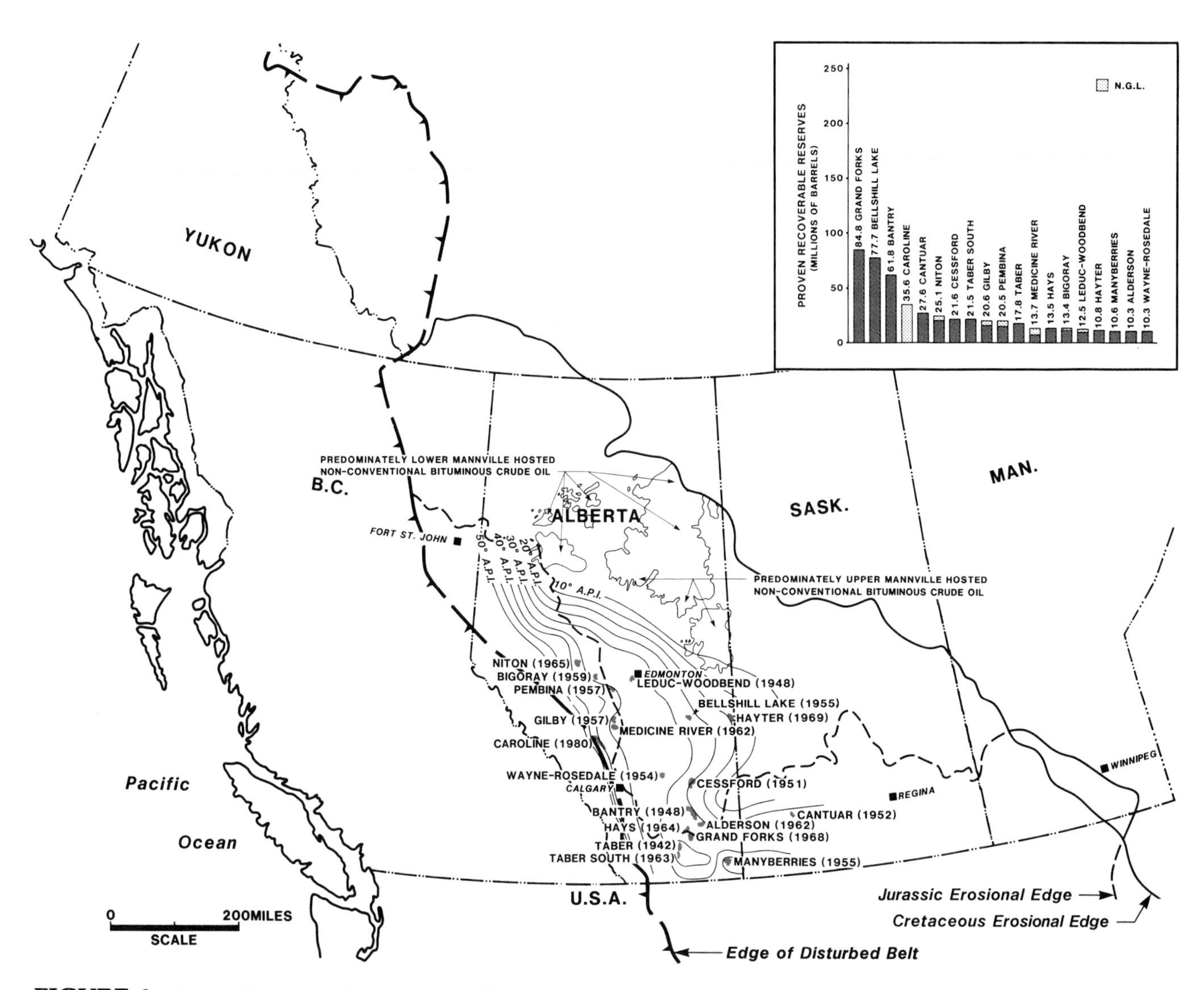

FIGURE 6. Lower Cretaceous Lower Mannville reservoir unit: distribution and sizes of principal conventional oil plus natural gas liquid fields associated with this unit, including corresponding isogravity pattern.

The Cadotte-Viking-Bow Island reservoir unit (Figure 13) harbors the greatest number of gas fields as well as the largest total marketable gas reserves of all the regressive sand units. Much of this gas, as well as the oil (Figure 12) associated with this unit, is believed to have undergone long-range migration. The positioning of the major reserves of both the gas and oil fields well up on the basin's flank supports this interpretation. Furthermore, the shallow depth of burial, approximately 2700 ft (823 m), is not geothermally compatible with the degree of maturity of the crude oil contained in the Viking pools of east-central Alberta and west-central Saskatchewan (Deroo et al., 1977). Much of the gas is trapped at the updip limit of the regional reservoir sand sheets.

Shallow accumulation of very large reserves of biogenic immature gas occurs in the Second White Speckled Shale, Medicine Hat, and Milk River reservoir units, as shown in Figures 16, 19, and 20. The gas accumulations associated with these three reservoir units have common geographic and structural affinities. They straddle the regional northeast-trending Sweetgrass-Battleford arch, which separates Alberta's portion of the foreland basin's shelf component from its Saskatchewan counterpart.

The Medicine Hat and Milk River low-permeability reservoir gas accumulations consist of stratigraphic meldings of 13 and 12 "fields," respectively. The very nature of their geographic concentration places their aggregate reserves in the giant field category, both with reserves on the order of 4500×10^9 ft^3 (126.8×10^9 m^3).

OVERVIEW OF FORELAND BASIN HYDROCARBON RESERVES

Figure 23 illustrates the chronology of discovery and size distribution of major oil fields associated with the West-

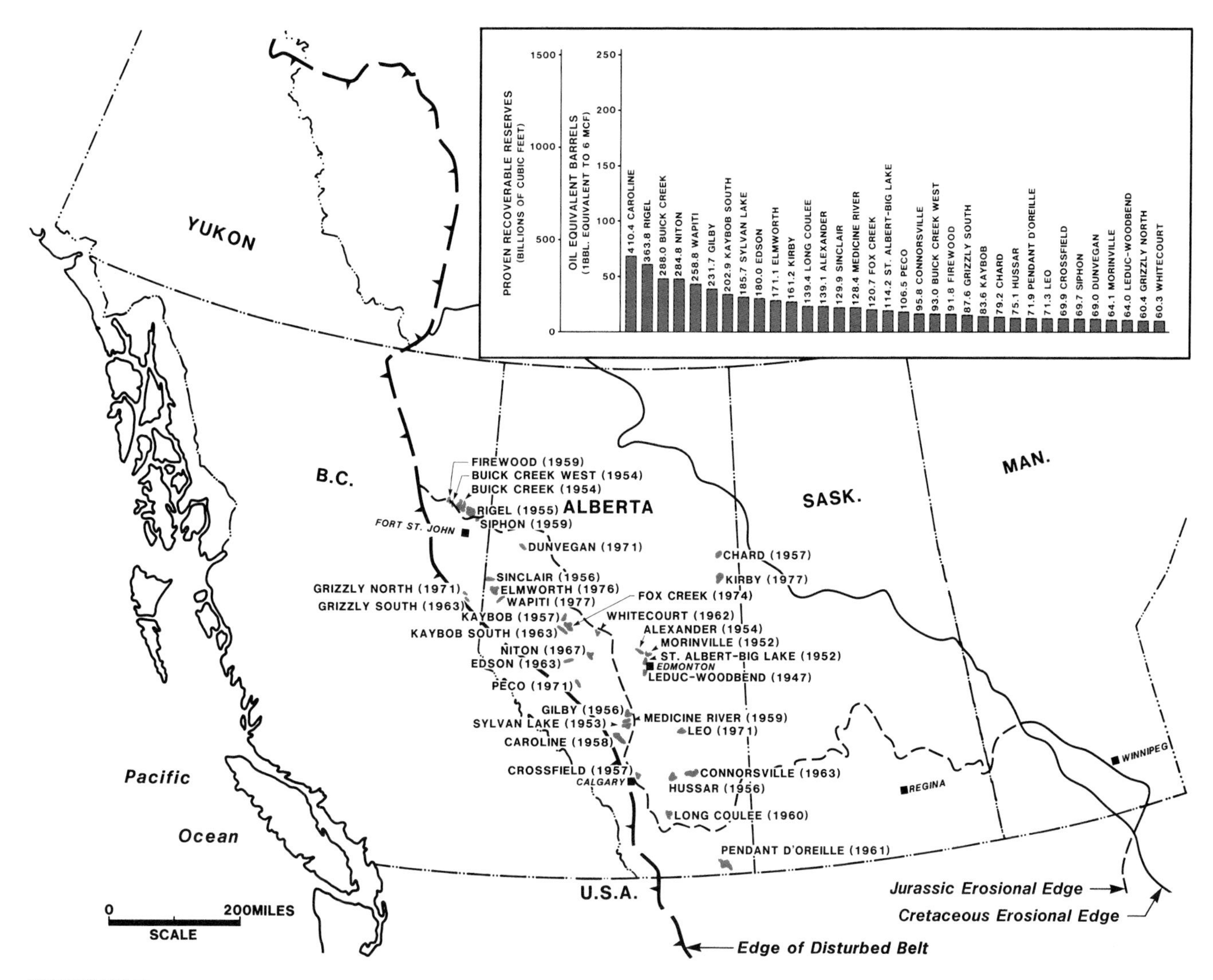

FIGURE 7. Lower Cretaceous Lower Mannville reservoir unit: distribution and sizes of principal gas fields associated with this unit.

ern Canada foreland basin. A designated field is represented by the total number of stratigraphically differentiated pools contained within a common geographic area. The perimeter of a defined field may encompass one or more hydrocarbon-bearing reservoir units. Figure 23 also includes pie diagrams that display the proportions of produced (cumulative) and remaining reserves for both conventional oil and conventional oil plus natural gas liquids. A size classification subdivision, including numbers of "fields" and their corresponding aggregate reserves, is also displayed.

The plot of the five fields shown in Figure 23 does not reveal a log-normal distribution pattern that might be expected of a basin that has undergone a protracted period of exploratory drilling. The anomalously large reserves of the Pembina field, which include not only the principal Cardium reservoir unit but five other units as well, represent 34% of the total established reserves of conventional oil in the Western Canada foreland basin. Of the six reservoir units associated with the Pembina field, the Cardium reservoir hosts 92% of their combined reserves.

As of the end of December 1985, the cumulative initial established reserves of conventional crude oil amounted to 4873.3×10^6 bbl (774.4×10^6 m^3), of which 3131.8×10^6 bbl (497.7×10^6 m^3) had been produced. In addition, 1477.9×10^6 bbl (234.9×10^6 m^3) of natural gas liquids had been found in the Western Canada foreland basin, for a combined total of 6351.2×10^6 bbl (1009.3×10^6 m^3). Of this amount, 61%, or 3854.4×10^6 bbl (612.5×10^6 m^3), had been produced as of December 31, 1985.

Commencing in 1948, the reserves of conventional crude oil discovered in the Western Canada foreland basin increased by significant annual increments until the end of 1969. By that time, 97% of the established crude reserves had been discovered. The 16 years from the end

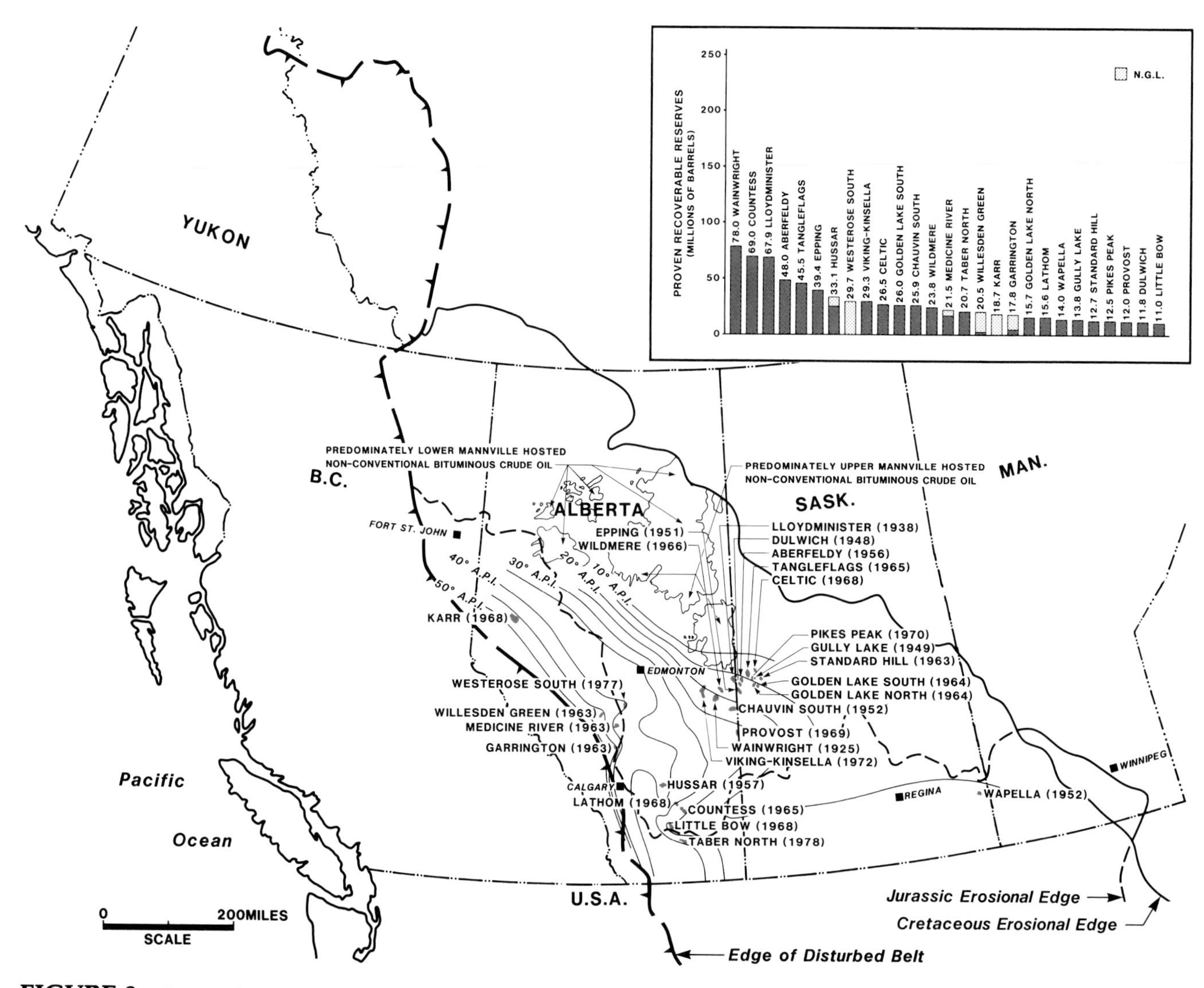

FIGURE 8. Lower Cretaceous Upper Mannville reservoir unit: distribution and sizes of principal conventional oil plus natural gas liquid fields associated with this unit, including corresponding isogravity pattern.

of 1969 to the end of 1985 witnessed a sharp decline in the discovery rate, with an average yearly increment of only about 10.0×10^6 bbl (1.59×10^6 m^3), which attests to the exploration maturity of the Western Canada foreland basin. This takes cognizance of annual reserve revisions based on in-field development drilling and on enhanced recovery methods applied to existing fields.

Figure 24 illustrates the chronology of discovery and size distribution of major gas fields associated with the Western Canada foreland basin. These gas fields, like their oil counterparts displayed in Figure 23, include the combined reserves of pools associated with one or more reservoir units. A pie diagram also illustrates the relationship of cumulative production to the remaining reserves of marketable gas.

A log-normal distribution pattern of the 16 major gas fields is somewhat apparent but is skewed by the very large Elmworth field discovered in 1976. The reason for this anomaly is that the field is located in the foredeep ("deep basin") part of the foreland basin, where excessive depth to, and attendant compaction diagenesis of, potential sand reservoirs, coupled with economic constraints, had previously precluded extensive exploration.

The cumulative reserves of marketable gas amount to $62,657.2 \times 10^9$ ft^3 (1765.3×10^9 m^3), of which 35.9%, or $22,485.4 \times 10^9$ ft^3 (633.5×10^9 m^3), had been produced as of December 31, 1985. When we establish the oil equivalent (1 bbl equivalent to 6000 ft^3) of the total gas reserves as $10,442.9 \times 10^6$ bbl (1659.5×10^6 m^3) and compare this figure with the total initial estimated reserves of 4873.3×10^6 bbl (774.4×10^6 m^3) of conventional crude oil, it is obvious that the Western Canada foreland basin is more gas prone than oil prone. This is also confirmed by comparing the total numbers of giant, major, and minor oil

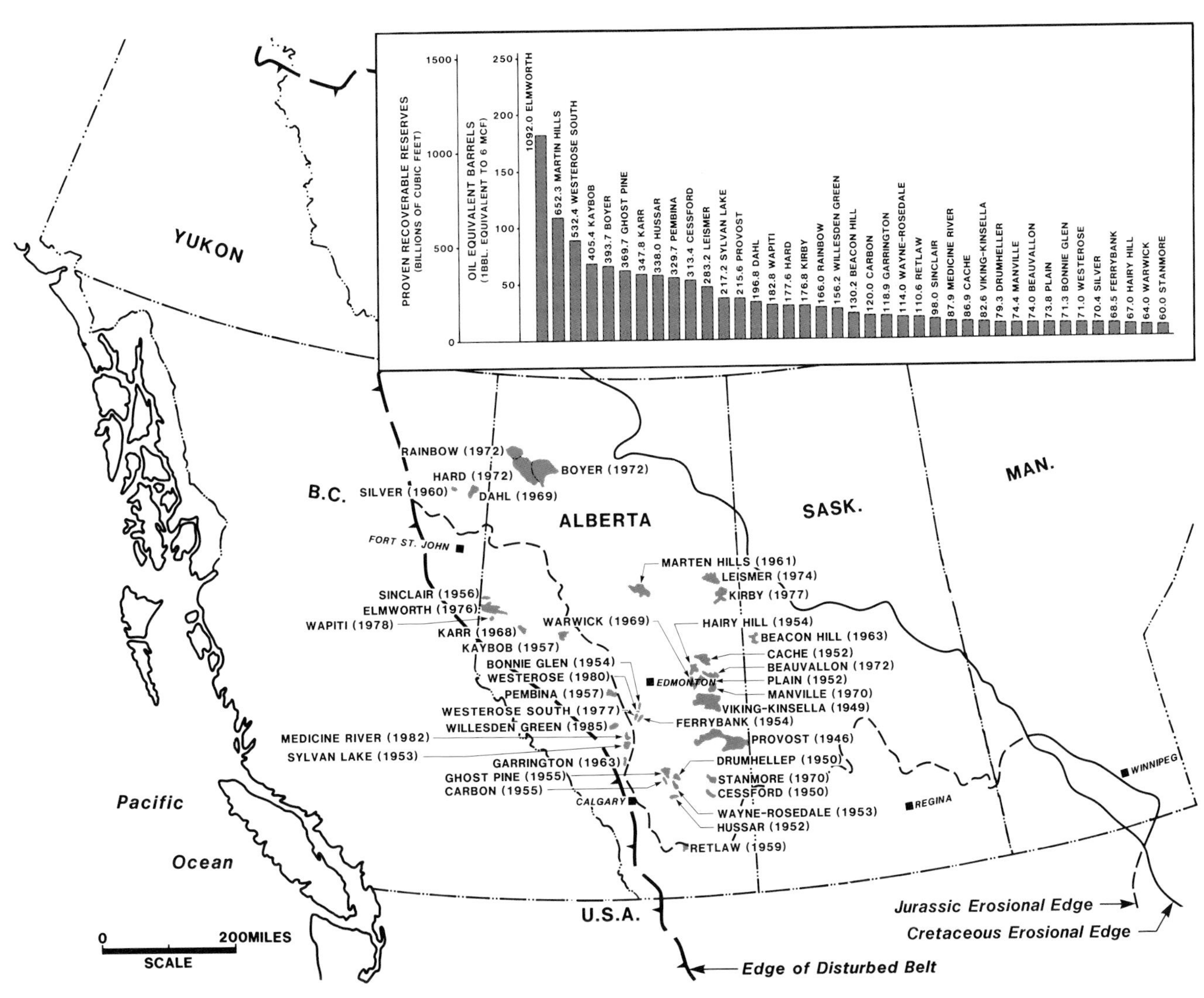

FIGURE 9. Lower Cretaceous Upper Mannville reservoir unit: distribution and sizes of principal gas fields associated with this unit.

versus gas fields as listed in the summary columns in Figures 23 and 24.

The exploration maturity of the foreland basin is likewise demonstrable in respect not only to the size-frequency relationship of the crude oil pools but to marketable gas as well, inasmuch as 97% of the total initial established reserves had been discovered by the end of 1977. Gas reserves, in terms of oil equivalent, continued to climb beyond the level of the discovered oil reserves, commencing in 1970. This apparently can be attributed to the emphasis on oil exploration during the pre-1975 period of depressed gas prices and attendant market constraints.

Figures 25 and 26 illustrate, respectively, the proportions of cumulative initial established reserves of conventional crude oil and marketable gas represented by the various reservoir units. These units have been grouped to indicate the percentage proportions represented by the Upper and Middle Jurassic, Lower Cretaceous, and Upper Cretaceous intervals.

In reference to conventional oil, as displayed in Figure 25, the Cardium reservoir unit hosts the largest reserves, comprising 39.5% of the total. This can be attributed almost entirely to the reserves of this unit in the Pembina oil field. The Upper Mannville reservoir unit follows, comprising 21.4% of the total reserves of conventional crude oil.

In contrast, the marketable gas reserves assigned to the Cardium reservoir unit in Figure 26 constitute a mere 4.3% of the total reserves, whereas the Upper Mannville reservoir unit is the dominant host with 34.1% of the total reserves. The Milk River, Medicine Hat, and Second White Speckled Shale reservoir units are generally considered barren of conventional oil but are shown to contain a

FIGURE 10. Lower Cretaceous Basal Colorado Sand reservoir unit: distribution and sizes of principal conventional oil fields associated with this unit, including corresponding isogravity pattern.

combined 18.4% share of the total initial established marketable gas reserves. These reserves are characteristically immature or biogenic in composition. Recently, considerable effort has been made in exploring for fractured oil-bearing reservoirs in the Second White Speckled Shale in the westernmost Interior Plains.

Figures 27 and 28 represent the proportions, expressed as percentages, of the initial established reserves of conventional oil and marketable gas, respectively, that have been assigned to specific stratigraphic habitats. The sandstone reservoirs comprising the Basal Colorado, Cadotte-Viking-Bow Island, Doe Creek-Dunvegan, Cardium, and Belly River reservoir units contain 54.4%, or 2651.1 $\times$ 10^6 bbl (421.3 $\times$ 10^6 m^3), of the total oil reserves. The reserves associated with the regressive sandstone cycles contained in the Cardium and Belly River reservoir units of the Pembina oil field account for 62.7% of the aforementioned reserves.

The quality of the crude associated with the regressive sand cycles is characteristically "sweet" and has not been subjected to water washing and attendant biodegradation. The crude oils associated with the intra-Jurassic marine sands, the Jurassic-Cretaceous unconformity-related sands, and the Lower Cretaceous fluvial and delta-related sands have all been subjected to varying degrees of biodegradation and water washing as expressed by the gravity spectrum patterns of the isogravity maps in Figures 3, 5, 6, and 8. There is an upward progressive increase of reserves from the intra-Jurassic marine sands to the Lower Cretaceous fluvial and delta

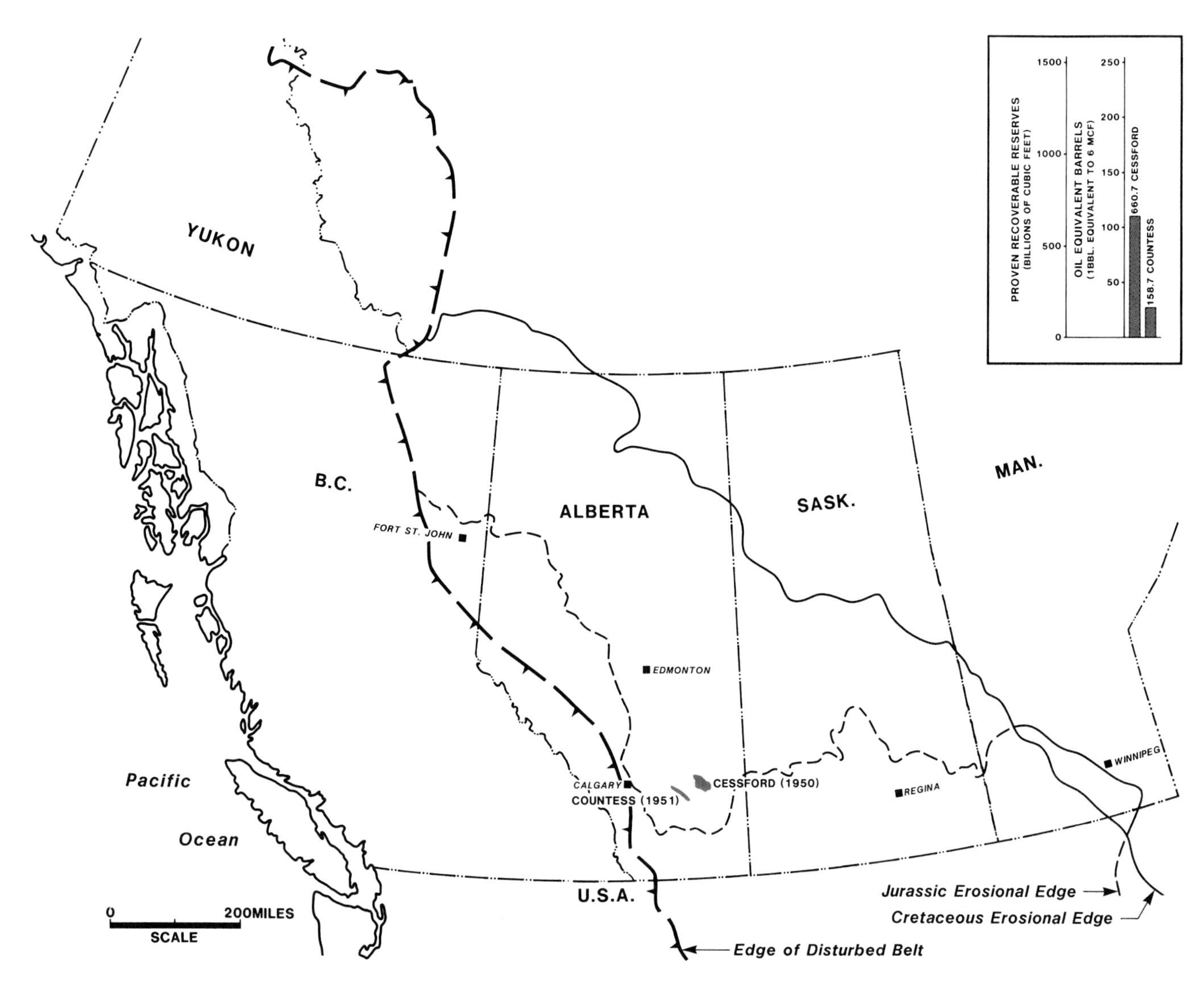

FIGURE 11. Lower Cretaceous Basal Colorado Sand reservoir unit: distribution and sizes of principal gas fields associated with this unit.

sand reservoirs. The ultimate stratigraphic ascension of the crude was terminated by the regionally related marine shales of the Colorado Group. This impervious regional cap rock was instrumental in preventing further ascension and ultimate dispersion of migrating hydrocarbons. The nonconventional heavy oil, bitumen, and oil sands ("tar sands") owe their preservation not only to biodegradation processes but also to the Colorado shale "roof," which served to contain the meteoric waters within the Lower Cretaceous Mannville group.

The stratigraphic habitats of marketable gas (Figure 28) can be compared to those of the conventional crude oil (Figure 27). Like the oil, the greatest share of gas reserves is contained in the regressive sandstone cycles, representing 44.2% of the total reserves, or $27{,}694.4 \times 10^9$ ft^3 (780.3×10^9 m^3). Likewise, the progressive increase of gas reserves upward from Jurassic-Cretaceous unconformity-related sandstone to Lower Cretaceous fluvial and deltaic sandstone is similar to that displayed by the conventional crude (Figure 27). The fact that a greater proportion of gas (34.1%) than of conventional oil (21.4%) is contained within the Mannville fluvial and deltaic sandstone may be explained by the gas-prone nature of the basin's foredeep, coupled with the preferential mobility of gas compared with that of oil.

Figures 29 and 30 illustrate the proportions of the cumulative initial established reserves of conventional crude oil and marketable gas, respectively, by geographic allocation. The absence of significant hydrocarbons associated with the Western Canada foreland basin's exten-

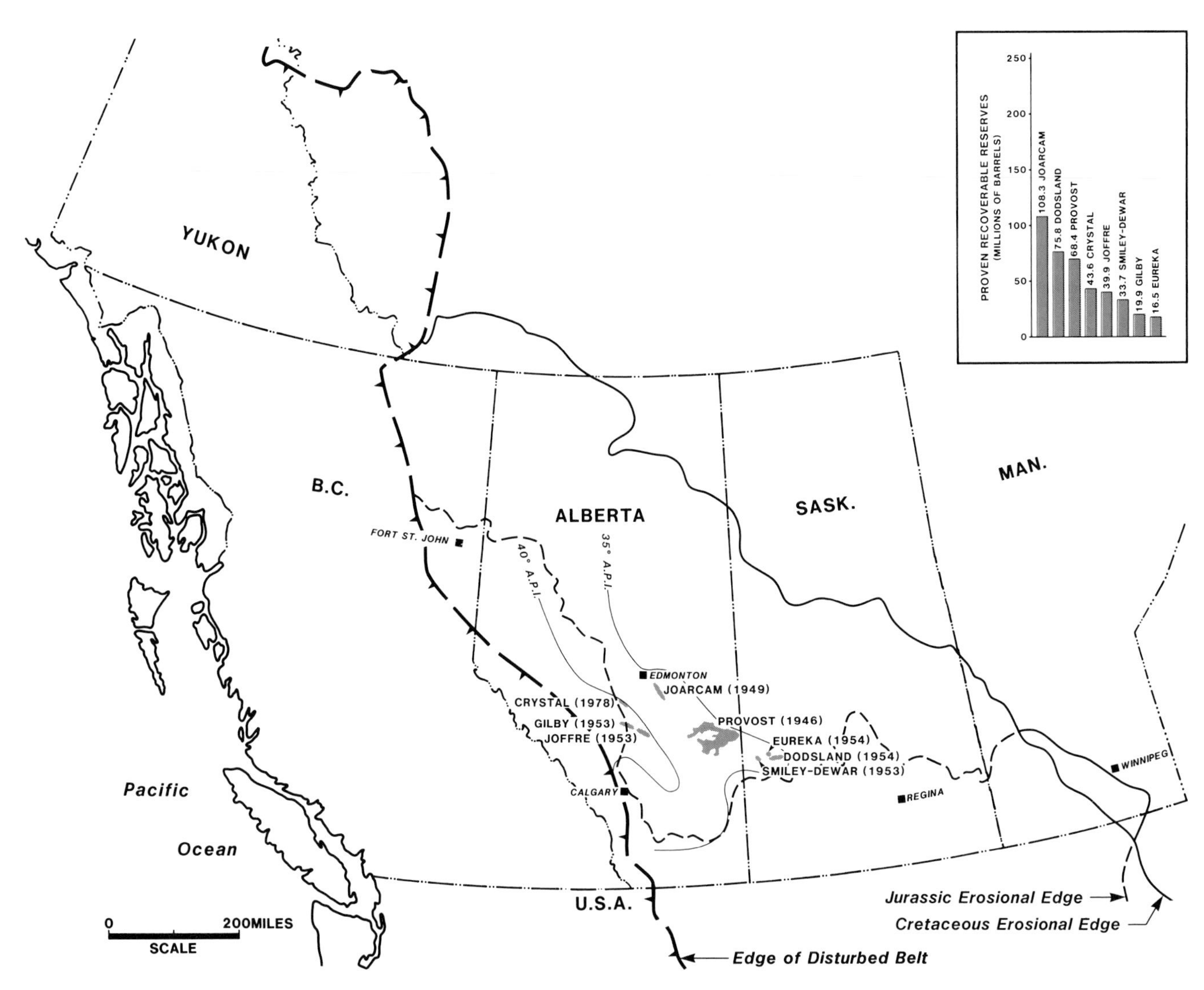

FIGURE 12. **Lower Cretaceous Cadotte-Viking-Bow Island reservoir unit: distribution and sizes of principal conventional oil fields associated with this unit, including corresponding isogravity pattern.**

sion into south-central Saskatchewan, southeastern Saskatchewan, and southwestern Manitoba has long been enigmatic. Probably the most obvious answer is indicated by the absence of deposits of heavy oil, oil sands, and bitumen up the regional dip from the aforementioned areas and along the outcrop of Lower Cretaceous (Swan River Group) sands; oils generated in Alberta generally did not migrate into Saskatchewan, and the local Saskatchewan successions are immature. The ability of the Williston basin's basal Jurassic evaporites (Watrous Formation) to prevent the escape of oil generated and hosted in reservoirs associated with the cratonic platform (Paleozoic systems) appears to have some significant bearing on this problem.

The significance of the western Alberta and northeastern British Columbia portion of the Western Canada foreland basin as a generative environment for hydrocarbons cannot be overestimated. In contrast, the Western Canada foreland basin's sedimentary expression within the Williston basin is represented by relatively thin shelf deposits that were not subjected to deep burial. Consequently, Williston basin would appear to have been a relatively inadequate environment for indigenously sourced crude to have been generated. The isolated and small Wapella oil field, discovered in 1952, produced from a basal Cretaceous sand and to a much lesser degree from the subcrop Jurassic sands. This isolated occurrence of foreland basin crude oil has long provided a rationale for

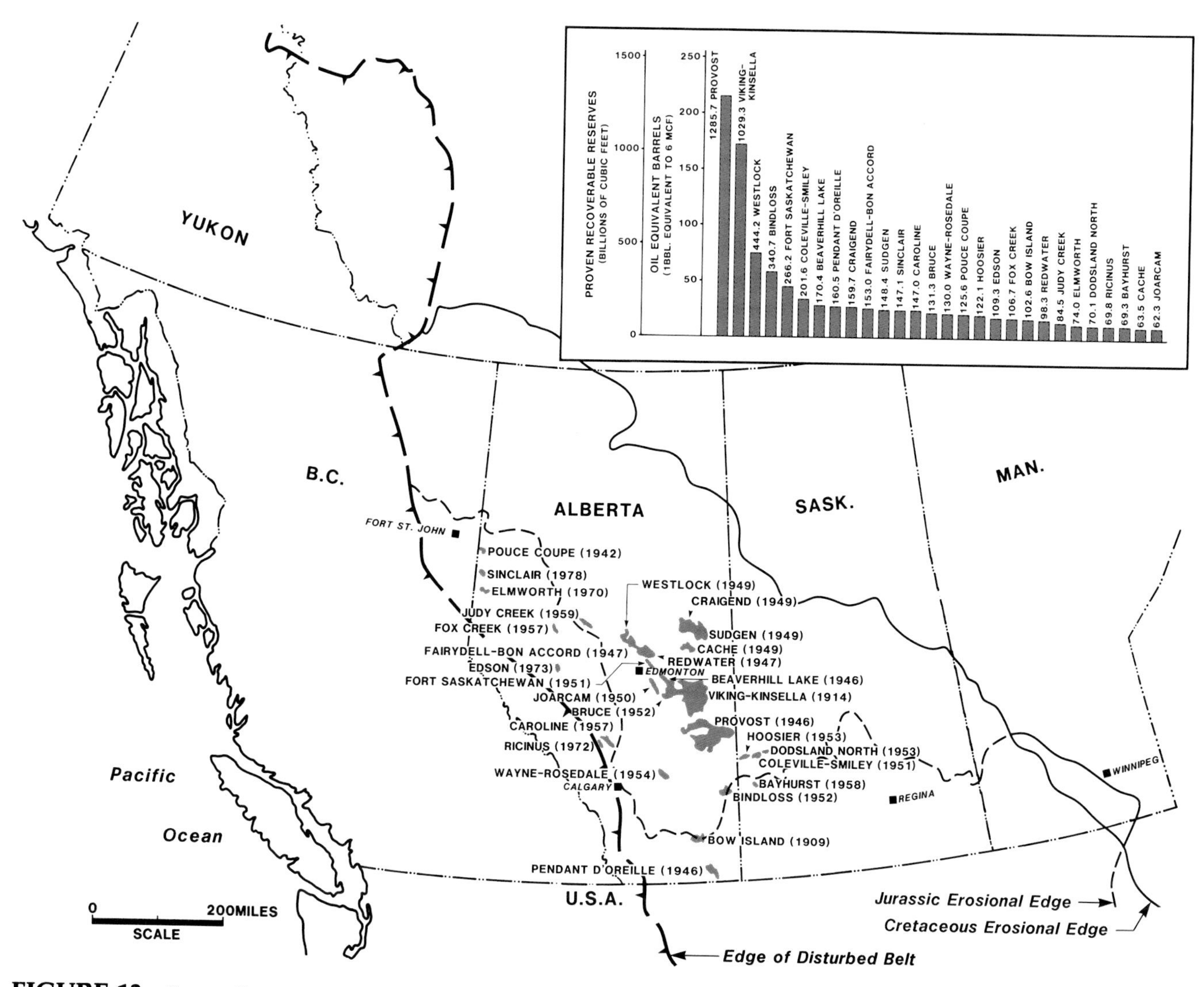

FIGURE 13. Lower Cretaceous Cadotte-Viking-Bow Island reservoir unit: distribution and sizes of principal gas fields associated with this unit.

obtaining prolific production from reservoirs of similar age in southeastern and south-central Saskatchewan. Extensive exploratory drilling for similar-sized fields has been disappointing. The crude oil at Wapella is believed to have been generated in Jurassic shales and to have undergone long-range migration.

Figures 31 and 32 compare the total conventional crude oil and marketable gas reserves, respectively, of the Western Canada foreland basin with its precursor or antecedent Western Canada cratonic platform basin. The foreland basin's reservoir rock type is essentially sandstone, whereas the cratonic platform rock is for the most part carbonate.

In Figure 31, the conventional crude oil reserves of the Western Canada foreland basin comprise 29% of the total $16{,}815.3 \times 10^6$ bbl (2672.1×10^6 m^3) found through December 31, 1985 in the Western Canada sedimentary basin. Of the initial established crude oil reserves associated with the cratonic platform reservoirs, 73% has been produced, compared with 64.3% for the foreland basin.

Figure 32 shows that the Western Canada foreland basin harbors 50.4% of the $124{,}295.8 \times 10^9$ ft^3 (3501.9×10^9 m^3) of the initial established marketable gas reserves discovered through December 31, 1985 in the Western Canada sedimentary basin. Of the initial established reserves of marketable gas associated with the cratonic

FIGURE 14. Upper Cretaceous Doe Creek–Dunvegan reservoir unit: distribution and sizes of principal conventional oil fields associated with this unit, including corresponding isogravity pattern.

platform reservoirs, 47% has been produced, whereas only 35.9% of the reserves of the foreland basin reservoirs has been produced. This disparity in the proportions of both crude oil and gas reserves produced in the respective basin types may be attributable to the greater productive capacity of the cratonic platform basin's reservoirs, which is a function of better porosity, better permeability, and attendant higher reservoir pressures.

SUMMARY OF HYDROCARBONS IN WESTERN CANADA FORELAND BASIN

The Western Canada foreland basin was unique as a crucible for the generation of vast quantities of hydrocarbons. Throughout its geologic history, this basin not only generated hydrocarbons from indigenous source rocks but also became a repository for escaping hydrocarbons generated within the subjacent Western Canada cratonic platform basin. The synorogenic development of the foredeep component during latest Cretaceous time was accompanied by the generation of hydrocarbons within the foredeep and their subsequent migration eastward. The major conduits, in the form of basal diachronous sandstones, unconformity-related sandstones, fluvial and deltaic complexes, as well as regressive sand sheets, permitted the transport of much of the generated hydrocarbons to the basin's flank. Long-range migration of hydrocarbons contained by the Lower Cretaceous Mannville Group appears for the most part to have been

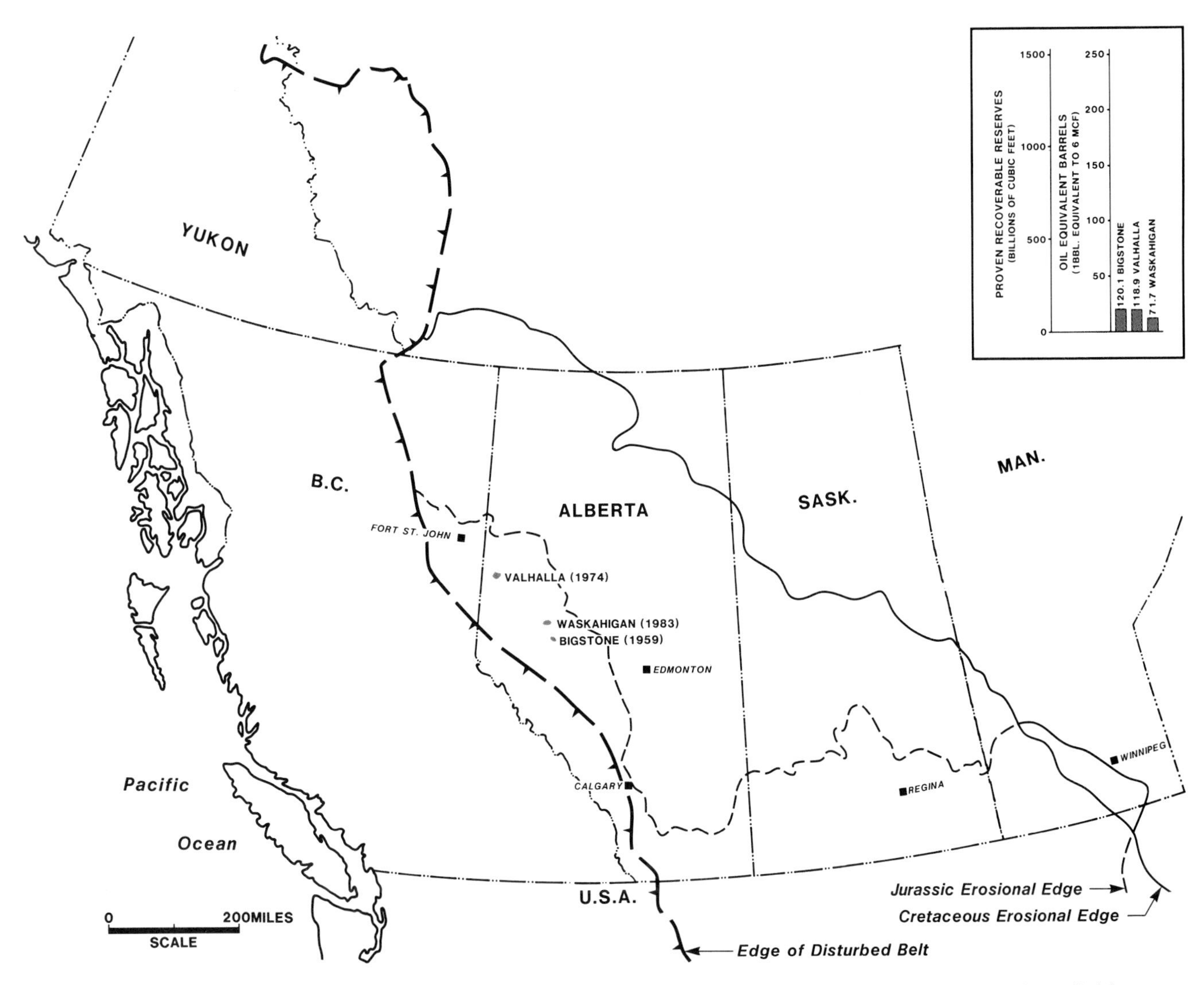

FIGURE 15. **Upper Cretaceous Doe Creek–Dunvegan reservoir unit: distribution and sizes of principal gas fields associated with this unit.**

unrestricted, and was hampered only by its encounter with meteoric water and its subsequent rendering to an immobile state. The Peace River, Athabasca-Wabasca, and Cold Lake deposits of nonconventional heavy oil, oil sands, and bitumens bear this out.

COMPARISON OF WESTERN CANADA AND EASTERN VENEZUELA FORELAND BASINS

The dominance of porous sandstones within the Lower Cretaceous Mannville Group precluded the localized entrapment of much of the migrating crude. If trapping mechanisms in the form of compressional, tensional, or geostatic structures had been present or had been developing on the flank of the Western Canada foreland basin at the time that migration was occurring, a very much greater amount of conventional crude could have been preserved. To substantiate this hypothesis, a comparison of the Eastern Venezuela foreland basin can be made with the Western Canada foreland basin. The Eastern Venezuela basin is bounded on the north by the Serrania del Interior and on the south by the Guayana shield. The sediments contained by it are of Cretaceous and Tertiary age. They are represented by thick sand-shale cyclic sequences ranging from marine to paralic to continental. Production is primarily obtained from Oligocene-Miocene sandstone reservoirs. Two giant fields (Quiriquire

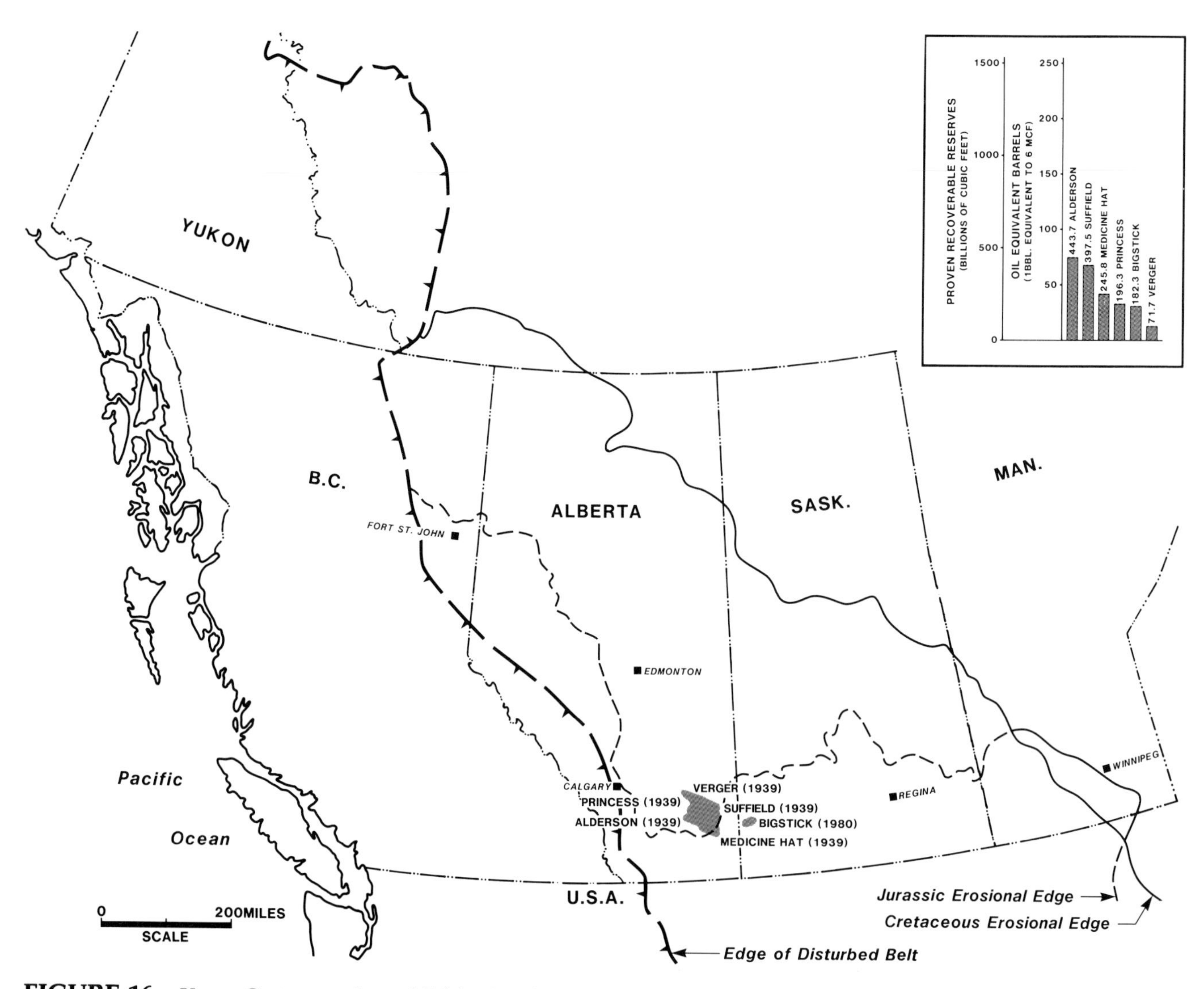

FIGURE 16. **Upper Cretaceous Second White Speckled Shale reservoir unit: distribution and sizes of principal gas fields associated with this unit.**

and Oficina) and 27 major fields account for most of the $16{,}000 \times 10^6$ bbl (2542.5×10^6 m^3) of ultimately recoverable reserves of oil (Roadifer, 1986). These conventional reserves are more than three times the size of the conventional reserves attributable to the Western Canada foreland basin. The oil fields, for the most part, are located on the shelf and upper slope of the foredeep (Renz et al., 1958) and are all located downdip from the Orinoco heavy oil and bitumen oil ("tar sands") belt. The in-place reserves of these nonconventional crude oil deposits are $1{,}200{,}000 \times 10^6$ bbl ($190{,}691.1 \times 10^6$ m^3) (Roadifer, 1986) and approximate the combined reserves of the in-place nonconventional crude oil deposits in the Peace River, Athabasca-Wabasca, and Cold Lake regions of Western Canada. These gargantuan reserves, like those of the Western Canada foreland basin, have undergone long-range migration from the thermally mature foredeep to the uppermost flank of the basin (Roadifer, 1986). Likewise, meteoric water originating from the Guayana shield has progressively biodegraded the migrating crude and subsequently rendered most of it immobile.

The trapping mechanism attributable to the conventional oil fields of the Eastern Venezuela foreland basin, like that of the Western Canada foreland basin, is essentially stratigraphic in nature. In the Eastern Venezuela foreland basin, however, trapping has been significantly enhanced by tensional faulting that either predated or was concomitant with hydrocarbon migration into Oligo-

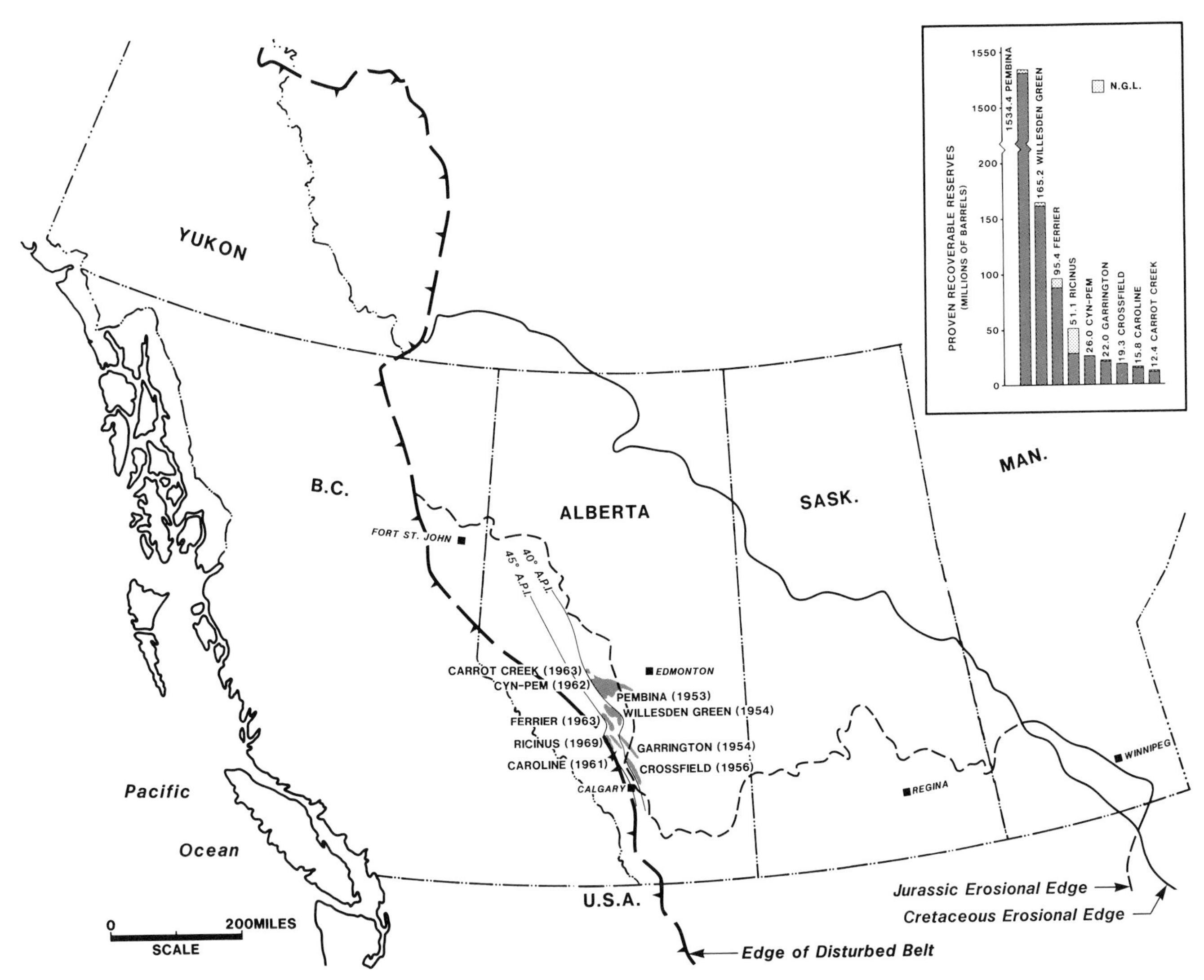

FIGURE 17. Upper Cretaceous Cardium reservoir unit: distribution and sizes of principal conventional oil plus natural gas liquid fields associated with this unit, including corresponding isogravity pattern.

cene sand reservoirs during late Miocene and Pliocene time (Renz et al., 1958). It appears that this normal faulting, originating within the shelf and extending to its interface with the foredeep, was responsible for the significantly larger reserves of conventional crude oil compared with those of the Western Canada foreland basin. Here, in Canada, with the absence of normal faulting, the problem of localized entrapment of conventional crude is compounded by the progressive replacement of shale by sand as the Precambrian shield is approached, resulting in a massive coalescing of reservoir rock. In contrast, the "sweet" conventional crude associated with the regressive sand cycles of the post-Mannville Colorado and Montana Group of the Western Canada foreland basin was destined to be trapped updip at the upper limit of the cyclic sand reservoirs within a massive marine shale section.

In conclusion, the exploration of the Western Canada foreland basin has historically played a secondary role to the exploration of its precursor, the Western Canada cratonic platform basin. The reason for this was the preoccupation with the search for prolific reef reservoirs following the Devonian Leduc reef discovery in February 1947. Much of the exploration since that time has been predicated on the results of seismograph surveys. Whereas Mississippian erosional topography and Devonian reef expressions were identifiable by seismic interpretation, the stratigraphic traps associated with Cretaceous oil and

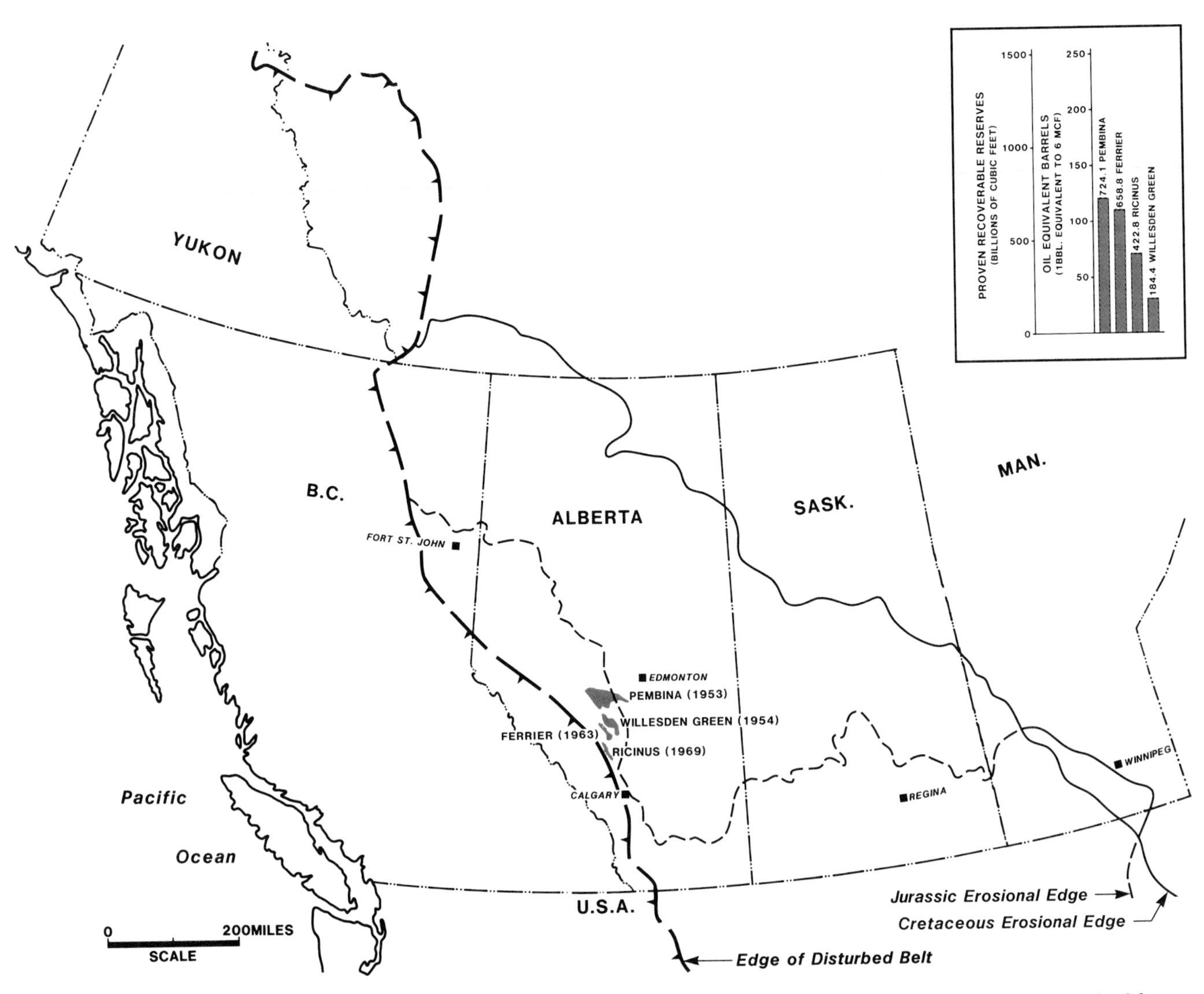

FIGURE 18. Upper Cretaceous Cardium reservoir unit: distribution and sizes of principal gas fields associated with this unit.

gas reservoirs were not. As a consequence, much of the reserves attributable to many of the large Western Canada foreland basin fields—namely, Pembina, Joarcam, and Joffre (see Barclay and Smith, 1992)—were discovered serendipitously in the quest for deeper Paleozoic objectives. It was not until the potential of Pembina was recognized in the mid-1950s that exploration became increasingly more directed at specific Cretaceous objectives.

ADDENDUM

Since this paper was completed, additional statistical reserves data through December 31, 1989 have been tabulated for conventional oil, natural gas liquids, and marketable gas (see Table 1). Also incorporated for comparison purposes are the reserves data for the aforementioned categories through December 31, 1985. The sources of these updated data were the same provincial government annual reports used in preparing this paper and listed in the reference section.

Comparison of the initial established reserves of conventional oil, natural gas liquids, and marketable gas associated with the Western Canada foreland basin through December 31, 1985 and through December 31, 1989 shows significant increases in all three categories. However, these increments were not caused primarily by the discovery of new pools but for the most part were the result of a reevaluation of existing reserves based on

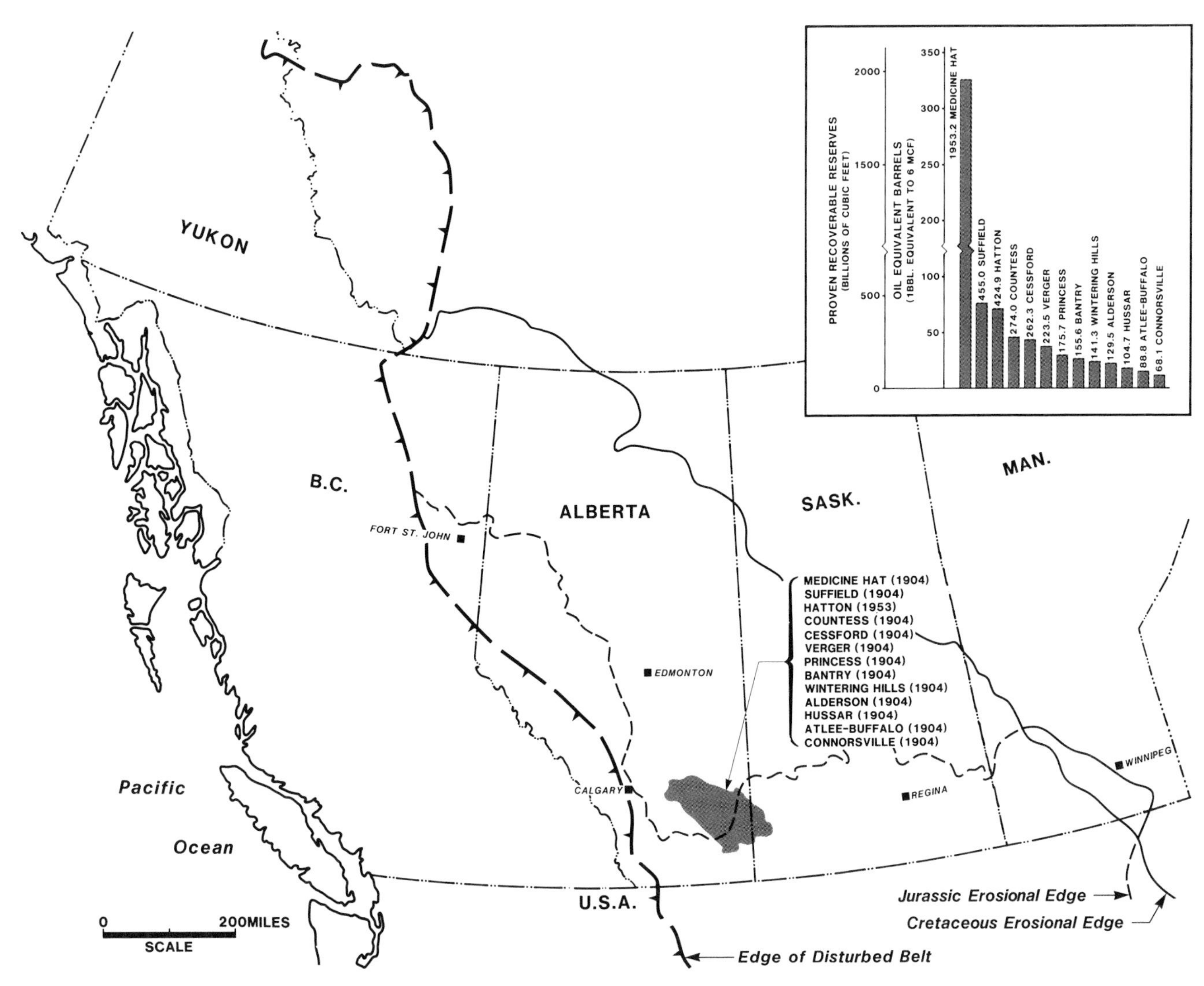

FIGURE 19. **Upper Cretaceous Medicine Hat reservoir unit: distribution and sizes of principal gas fields associated with this unit.**

enhanced recovery technology involving existing and new waterflood projects. The extensional and in-field drilling of existing pools was also an important factor contributing to the reassessment of the existing reserves.

The remaining reserves of conventional oil and natural gas liquids as of December 31, 1989 show increases of 3 and 13%, respectively, over those listed through December 31, 1985. For the same period, however, remaining reserves of marketable gas show a 4% decrease, reflecting an increased production capacity that has offset a 7% increase in the initial established reserves of marketable gas.

The application of improved technology in an effort to maximize the ultimate recovery of existing reserves of conventional oil, natural gas liquids, and marketable gas is an ongoing process. Coupled with escalating prices for these commodities, upward readjustments of existing remaining reserves will continue. In the historical context of oil and gas exploration, not only in the Western Canada foreland basin but in other basins as well, the initial appraisals of reserves associated with new field discoveries have been understandably conservative. This bodes well for the ultimate recovery of hydrocarbons in existing fields as well as in those yet undiscovered in the extensively explored Western Canada foreland basin.

ACKNOWLEDGMENTS

I wish to thank Drs. Roger W. Macqueen and Dale A. Leckie of the Institute of Sedimentary and Petroleum

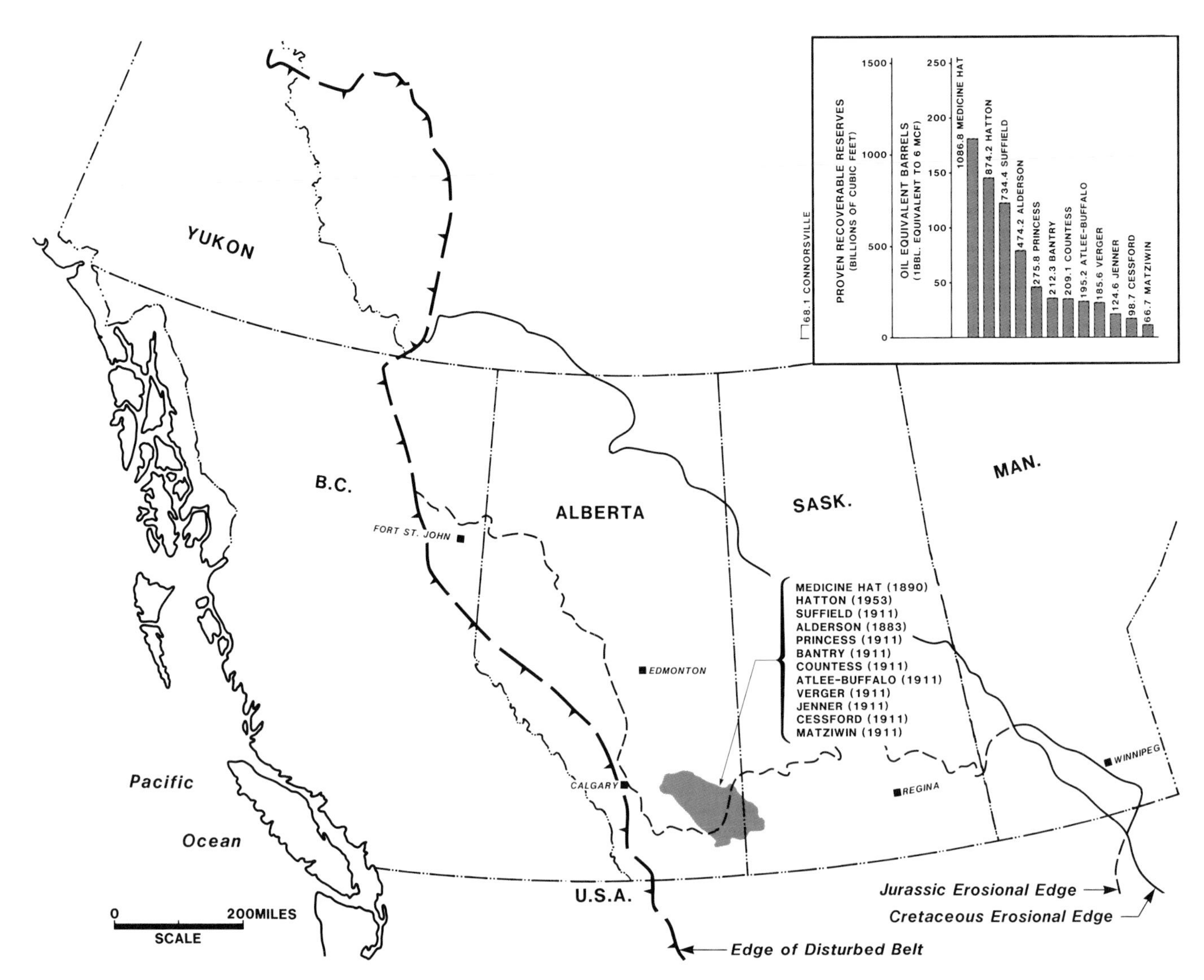

FIGURE 20. **Upper Cretaceous Milk River reservoir unit: distribution and sizes of principal gas fields associated with this unit.**

Geology for their critical reading of the manuscript as well as their helpful suggestions. Mrs. Claudia Thompson, also of the institute, typed the numerous drafts, and to her I also extend thanks. I also wish to thank Mobil Oil Canada for its generous help in the use of its library services and especially Brian Scott of Mobil's Drafting and Reproduction Department for drafting the illustrations.

REFERENCES CITED

Barclay, J. E., and D. G. Smith, 1992, Western Canada foreland basin oil and gas plays, this volume.

Brooks, P. W., M. G. Fowler, and R. W. Macqueen, 1988, Biological marker and conventional organic geochemistry of oil sands/heavy oils, Western Canada basin: Organic Geochemistry, v. 12, p. 519–538.

Brooks, P. W., M. G. Fowler, and R. W. Macqueen, 1989, Biomarker geochemistry of Cretaceous oil sands, heavy oils and Paleozoic carbonate trend bitumens, Western Canada basin, *in* R. G. Meyer and E. J. Wiggins, eds., Proceedings, Fourth UNITAR-UNDP International Conference on Heavy Crudes and Tar Sands: Edmonton, Alberta, AOSTRA, v. 2, Geology, Chemistry, p. 594–606; discussion, p. 629–631.

Brooks, P. W., M. G. Fowler, and R. W. Macqueen, 1990, Use of biomarkers, including aromatic steroids, to indicate

FIGURE 21. Upper Cretaceous Belly River reservoir unit: distribution and sizes of principal conventional oil plus natural gas liquid fields associated with this unit, including corresponding isogravity pattern.

relationships between oil sands/heavy oils and Paleozoic carbonate trend bitumens, Western Canada basin; *in* D. J. Lazar, ed., Proceedings of the 1989 Eastern Oil Shale Symposium, November 15–17, 1989: Institute for Mining and Minerals Research, University of Kentucky, p. 104–111.

Creaney, S., and J. Allan, 1992, Petroleum systems in the foreland basin of Western Canada, this volume.

Deroo, G., T. G. Powell, B. Tissot, and R. G. McCrossan, 1977, The origin and migration of petroleum in the Western Canadian sedimentary basin, Alberta: Geological Survey of Canada Bulletin 262, 136 p.

Deroo, G., B. Tissot, R. G. McCrossan, and F. Der, 1974, Geochemistry of heavy oils of Alberta, *in* L. V. Hills, ed., Oil sands, fuel of the future: Canadian Society of Petroleum Geologists Memoir 3, p. 148–167.

Energy Resources Conservation Board, 1986, Alberta's reserves of crude oil, oil sands, gas, natural gas liquids, and sulphur: Calgary, Alberta, Province of Alberta Energy Resources Conservation Board, Report ST86-18.

Energy Resources Conservation Board, 1989, Alberta's reserves of crude oil, oil sands, gas, natural gas liquids, and sulphur at December 31, 1989: Calgary, Alberta, Province of Alberta Energy Resources Conservation Board, Report ST90-18.

Evans, C. R., M. A. Rogers, and N. J. L. Bailey, 1971, Evolution and alteration of petroleum in Western Canada: Chemical Geology, v. 8, p. 147–170.

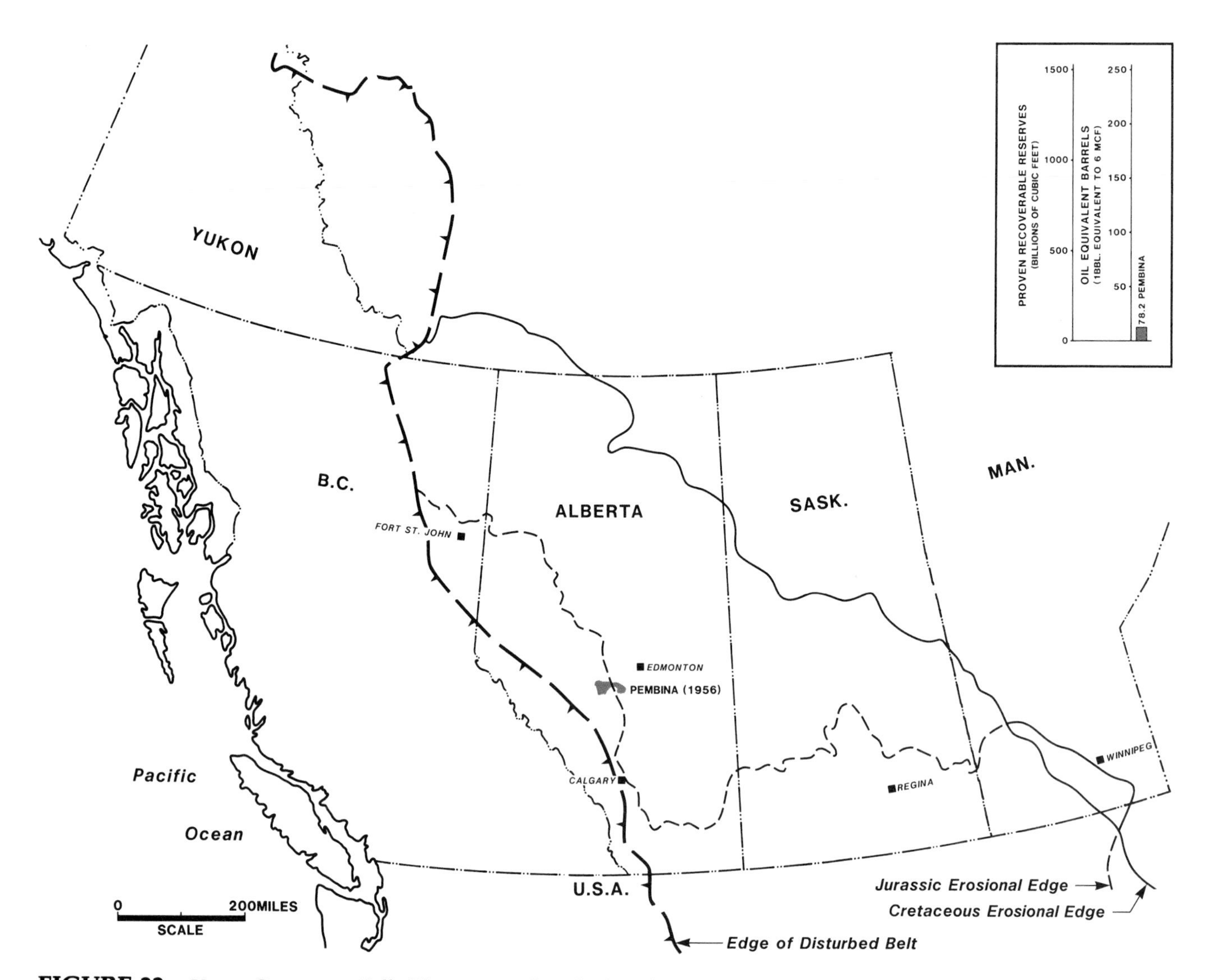

FIGURE 22. Upper Cretaceous Belly River reservoir unit: distribution and sizes of principal gas fields associated with this unit.

Govier, G. W., 1974, Alberta's oil sands in the energy supply picture, *in* L. V. Hills, ed., Oil sands, fuel of the future: Canadian Society of Petroleum Geologists Memoir 3, p. 35-49.

Leckie, D., and D. Smith, 1992, Regional setting, evolution, and depositional cycles of the Western Canada foreland basin, this volume.

Meyer, R. F., and W. D. De Witt, Jr., 1990, Definition and world resources of natural bitumens: United States Geological Survey Bulletin 1944, 14 p.

Mossop, G. D., J. W. Kramers, P. D. Flach, and B. A. Rottenfusser, 1981, Geology of Alberta's oil sands and heavy oil deposits, *in* R. G. Meyer and C. T. Steele, eds., The future of heavy crude oil and tar sands: New York, United Nations Institute for Training and Research, McGraw-Hill, p. 197-207.

Nielsen, A. R., and J. W. Porter, 1984, Pembina oil field—in retrospect, *in* D. F. Stott and D. J. Glass, eds., The Mesozoic of middle North America: Canadian Society of Petroleum Geologists Memoir 9, p. 1-13.

Outtrim, C. P., and R. G. Evans, 1978, Alberta's oil sands reserves and their evaluation, *in* D. A. Redford and A. G. Winestock, eds., The oil sands of Canada-Venezuela: Canadian Institute of Mining and Metallurgy Special Volume No. 17, p. 36-66.

Porter, J. W., R. A. Price, and R. G. McCrossan, 1982, The Western Canada sedimentary basin, *in* P. A. Kent, ed., The evolution of sedimentary basins: The Royal Society

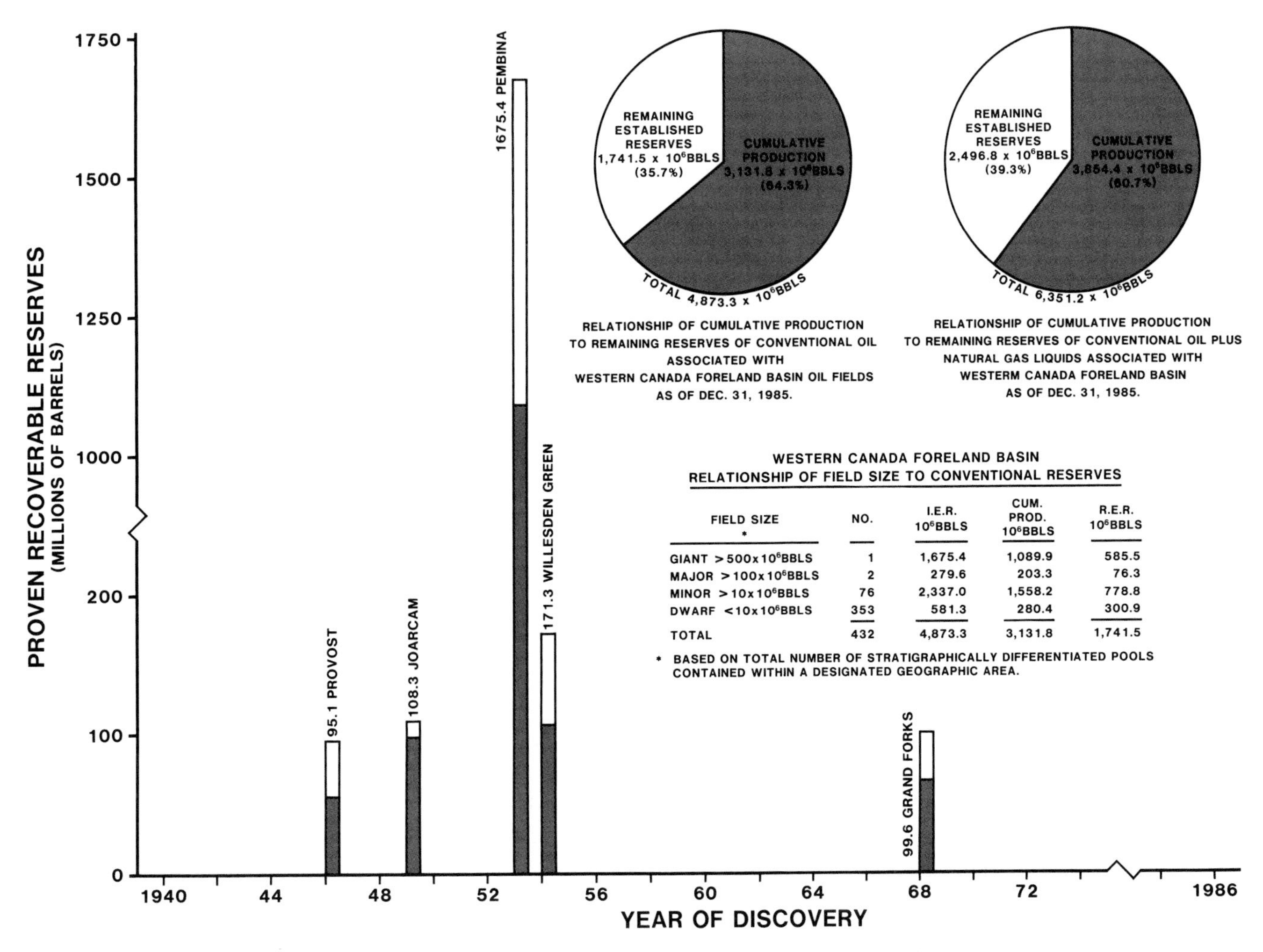

FIELD SIZE *	NO.	I.E.R. 10^6BBLS	CUM. PROD. 10^6BBLS	R.E.R. 10^6BBLS
GIANT >500x10^6BBLS	1	1,675.4	1,089.9	585.5
MAJOR >100x10^6BBLS	2	279.6	203.3	76.3
MINOR >10x10^6BBLS	76	2,337.0	1,558.2	778.8
DWARF <10x10^6BBLS	353	581.3	280.4	300.9
TOTAL	432	4,873.3	3,131.8	1,741.5

* BASED ON TOTAL NUMBER OF STRATIGRAPHICALLY DIFFERENTIATED POOLS CONTAINED WITHIN A DESIGNATED GEOGRAPHIC AREA.

FIGURE 23. Chronology of discovery and size distribution of major oil fields discovered in the Western Canada foreland basin. Total reserves of conventional oil and conventional oil plus natural gas liquids are indicated in the accompanying pie diagrams.

of London, p. 181.

Province of British Columbia, 1985, Hydrocarbon and by-product reserves in British Columbia: Victoria, British Columbia, Petroleum Resources Division, Ministry of Energy, Mines and Petroleum Resources, Publication 1985-12-31.

Province of Saskatchewan, 1985, Reservoir annual: Regina, Saskatchewan, Province of Saskatchewan Department of Energy and Mines.

Renz, H. H., H. Alberding, K. F. Dallmus, J. M. Patterson, R. H. Robie, N. E. Weisbord, and J. MasVall, 1958, The Eastern Venezuelan basin, *in* L. G. Weeks, ed., Habitat of oil: Tulsa, Oklahoma, AAPG, p. 551–600.

Roadifer, R. E., 1986, How heavy oil occurs worldwide: giant fields—2: Oil and Gas Journal, March 3, 1986, v. 84, n. 9, p. 111–115.

Wallace-Dudley, K. E., 1981a, Gas pools of Western Canada: Geological Survey of Canada Map 1558A.

Wallace-Dudley, K. E., 1981b, Oil pools of Western Canada: Geological Survey of Canada Map 1559A.

Wilson, M., and R. W. Bennet, 1985, Evaluation of Saskatchewan's heavy oil reserves: Province of Saskatchewan Department of Energy and Mines, Saskatchewan Geological Survey open file report, 25 p.

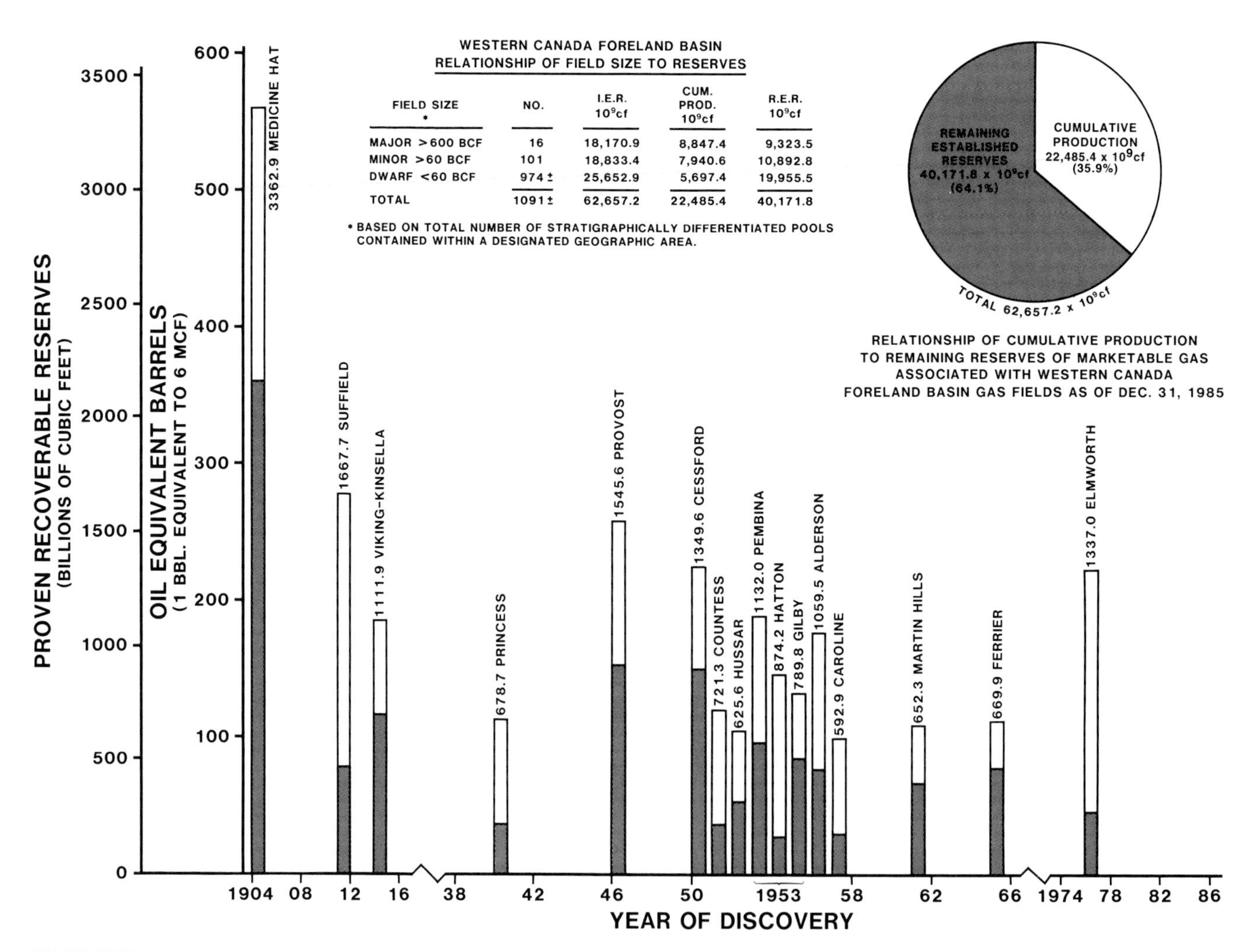

WESTERN CANADA FORELAND BASIN
RELATIONSHIP OF FIELD SIZE TO RESERVES

FIELD SIZE *	NO.	I.E.R. 10^9cf	CUM. PROD. 10^9cf	R.E.R. 10^9cf
MAJOR >600 BCF	16	18,170.9	8,847.4	9,323.5
MINOR >60 BCF	101	18,833.4	7,940.6	10,892.8
DWARF <60 BCF	974±	25,652.9	5,697.4	19,955.5
TOTAL	1091±	62,657.2	22,485.4	40,171.8

* BASED ON TOTAL NUMBER OF STRATIGRAPHICALLY DIFFERENTIATED POOLS CONTAINED WITHIN A DESIGNATED GEOGRAPHIC AREA.

FIGURE 24. Chronology of discovery and size distribution of major gas fields discovered in the Western Canada foreland basin. Total reserves of gas are indicated in the accompanying pie diagram.

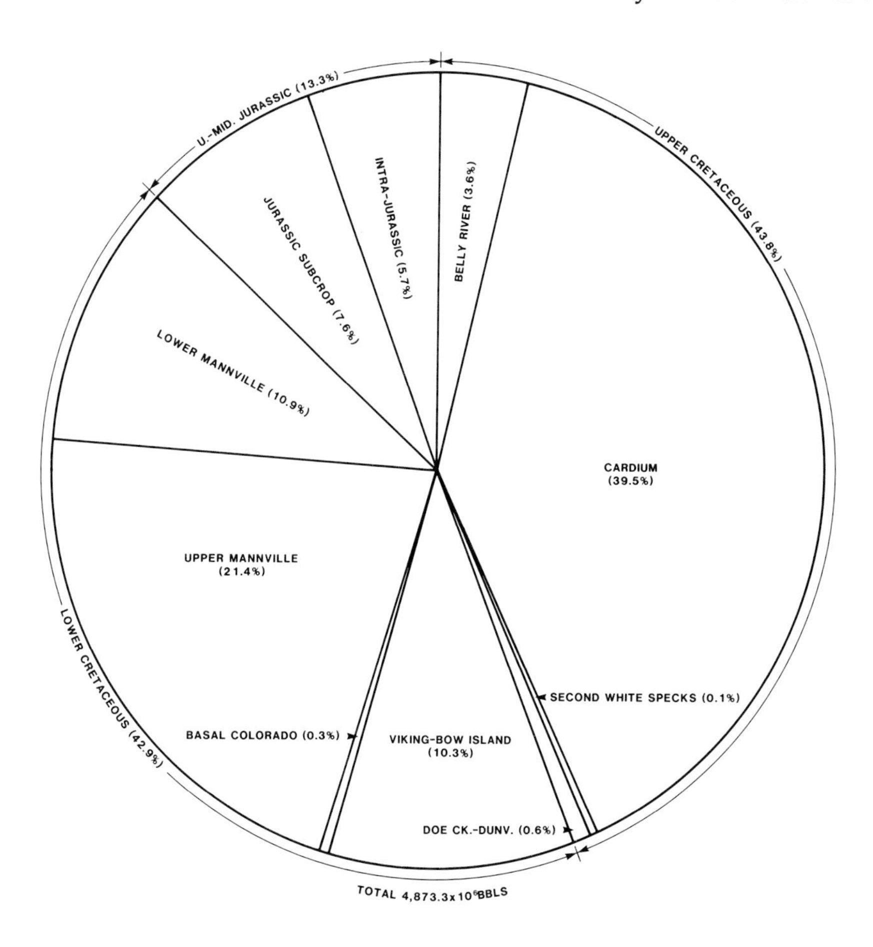

FIGURE 25. Western Canada foreland basin conventional initial established crude oil reserves by geologic age, based on principal reservoir units. Note: Figure for total is in million (10^6) barrels.

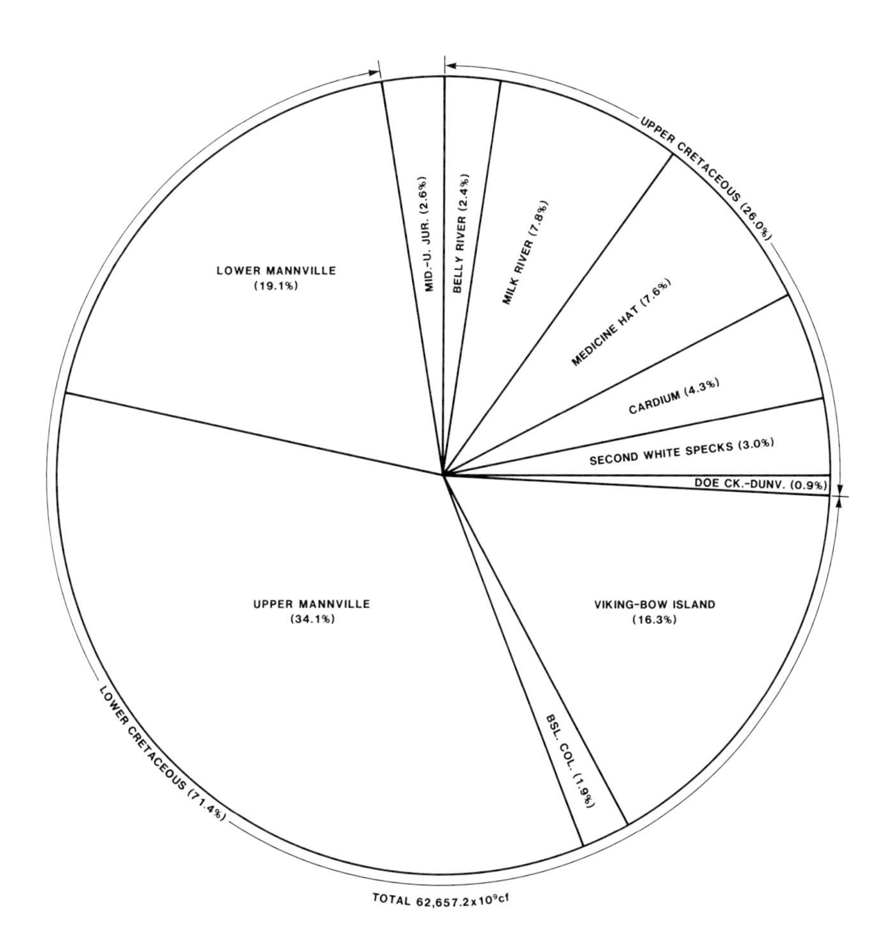

FIGURE 26. Western Canada foreland basin initial established marketable gas reserves by geologic age, based on principal reservoir units. Note: Figure for total is in billion (10^9) cubic feet.

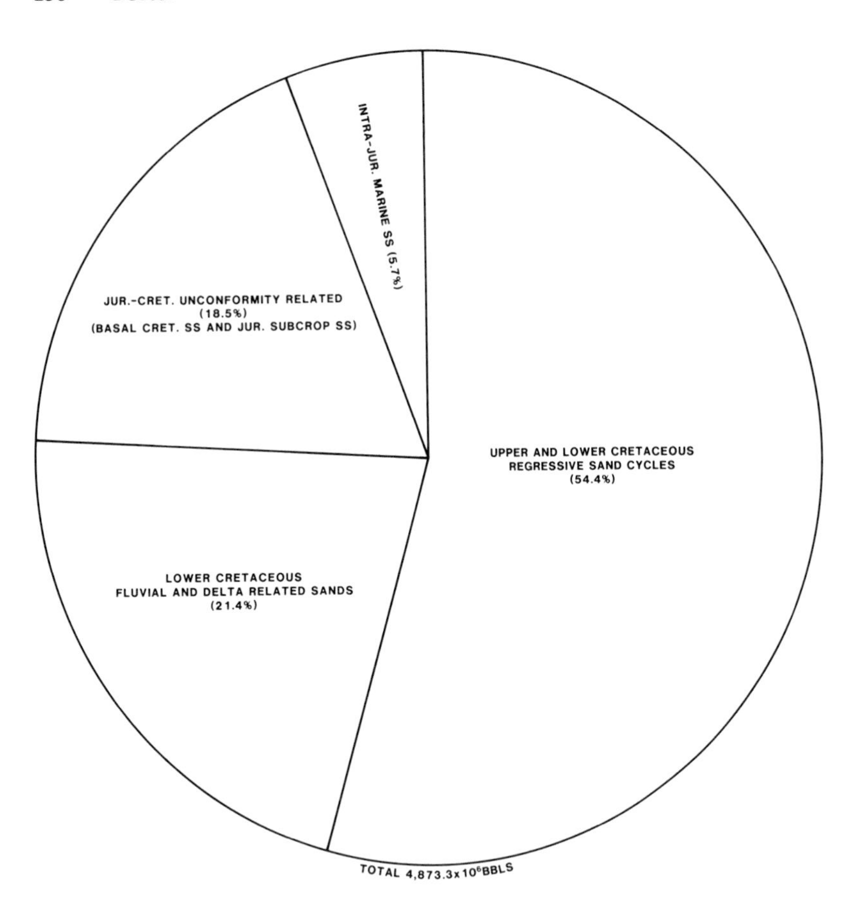

FIGURE 27. Western Canada foreland basin conventional initial established crude oil reserves by stratigraphic habitat. Note: Figure for total is in million (10^6) barrels.

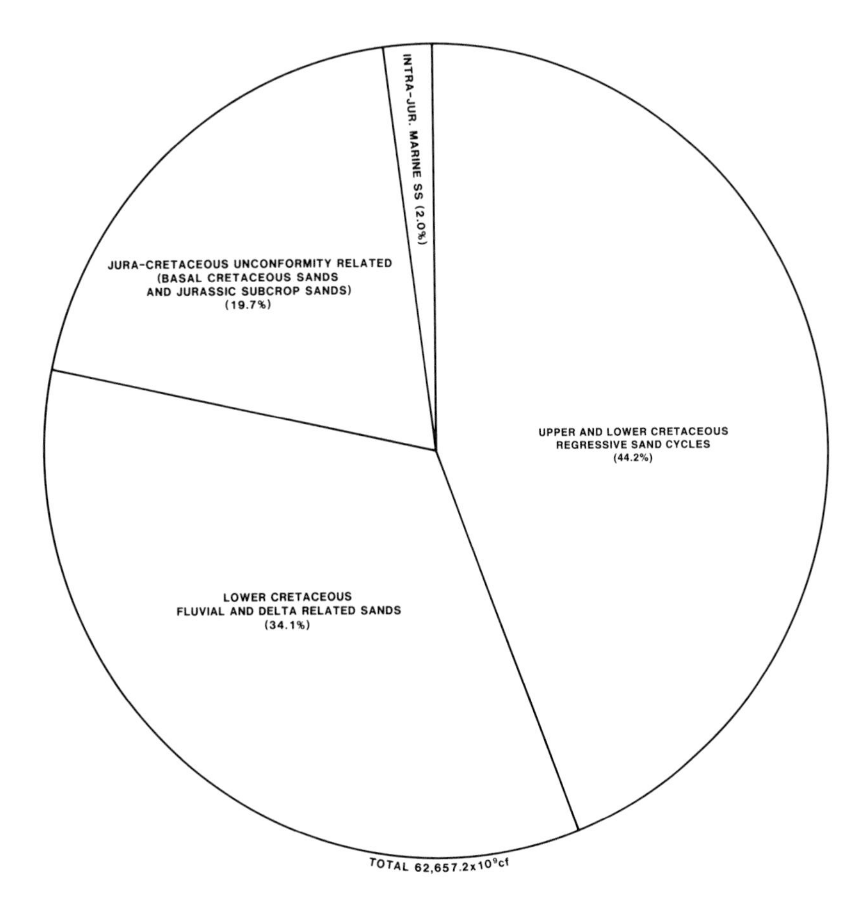

FIGURE 28. Western Canada foreland basin initial established marketable gas reserves by stratigraphic habitat. Note: Figure for total is in billion (10^9) cubic feet.

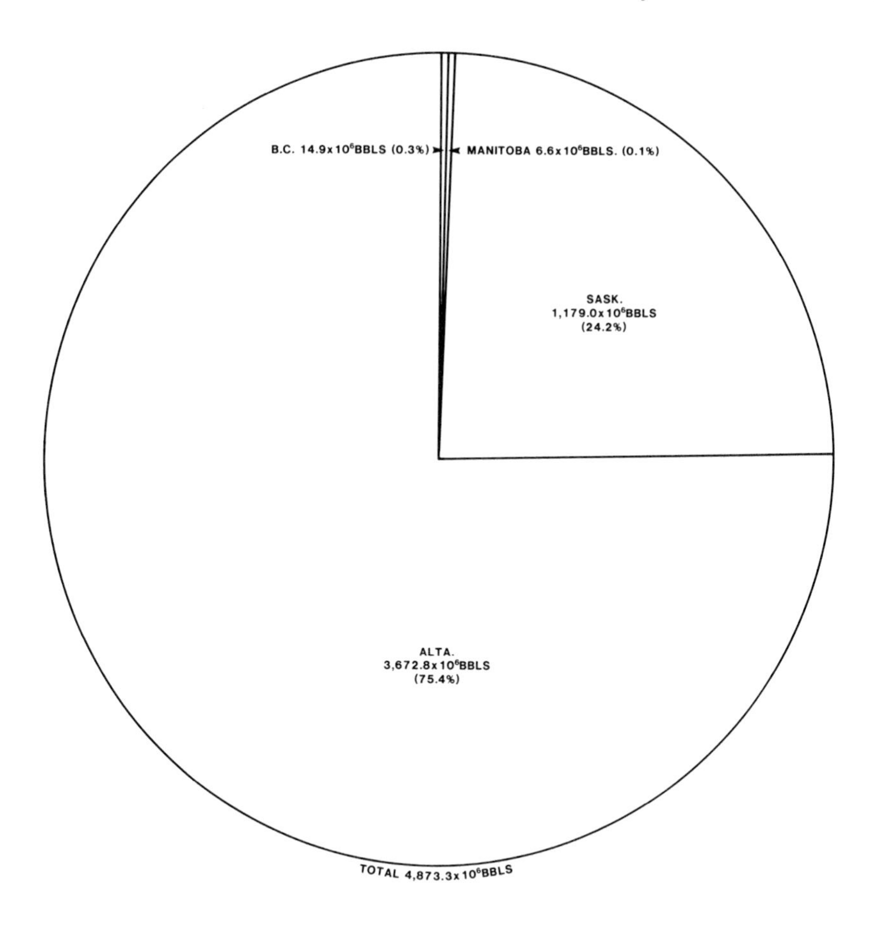

FIGURE 29. Western Canada foreland basin initial established marketable crude oil reserves by geographic allocation. Note: Figures given are in million (10^6) barrels.

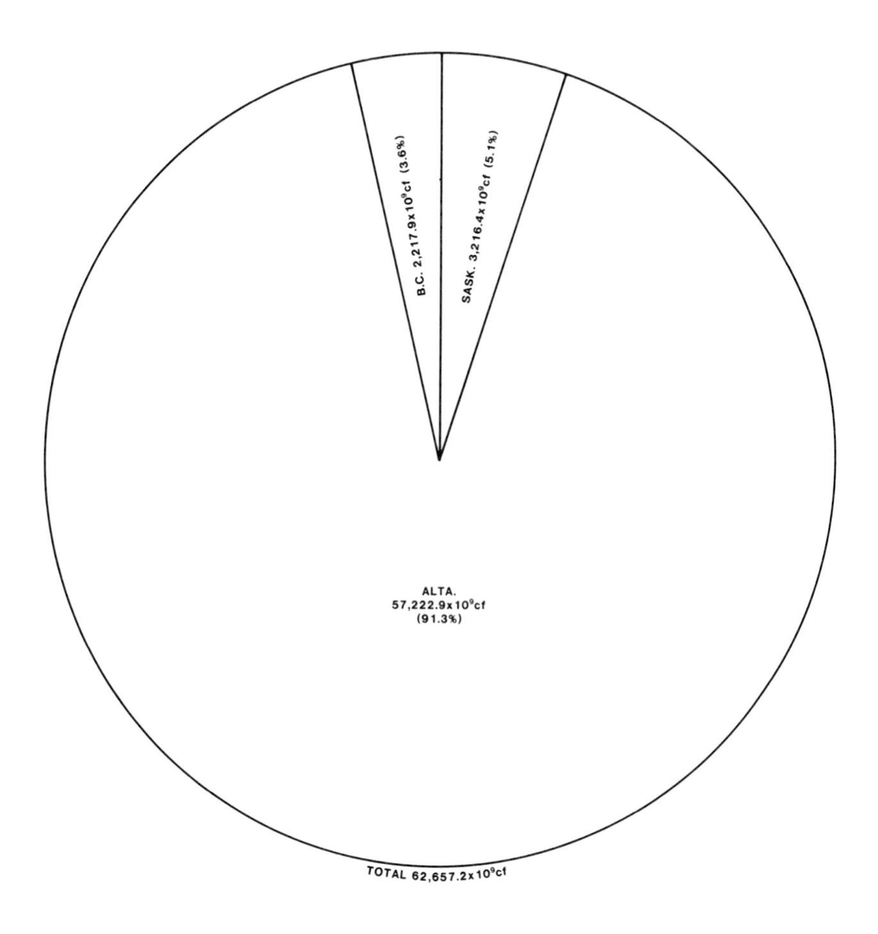

FIGURE 30. Western Canada foreland basin initial established marketable gas reserves by geographic allocation. Note: Figures given are in billion (10^9) cubic feet.

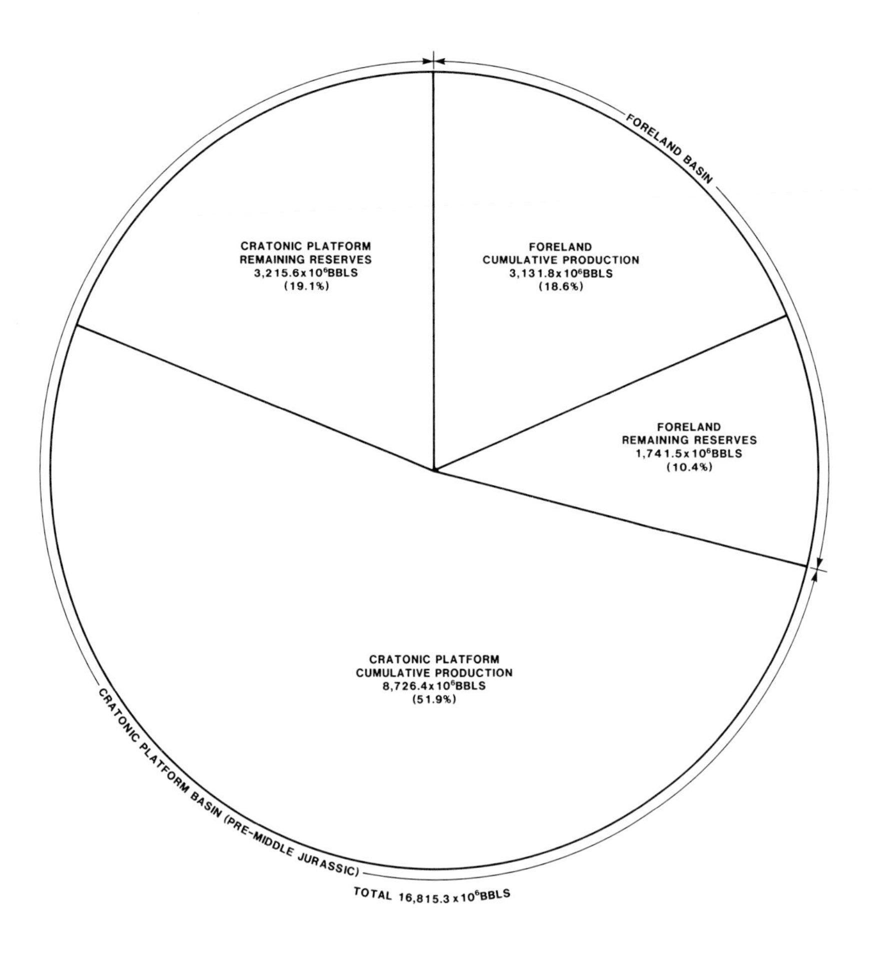

FIGURE 31. Comparison of conventional initial crude oil reserves of the Western Canada foreland basin with those of its precursor, the Western Canada cratonic platform basin. Note: Figures given are in million (10^6) barrels.

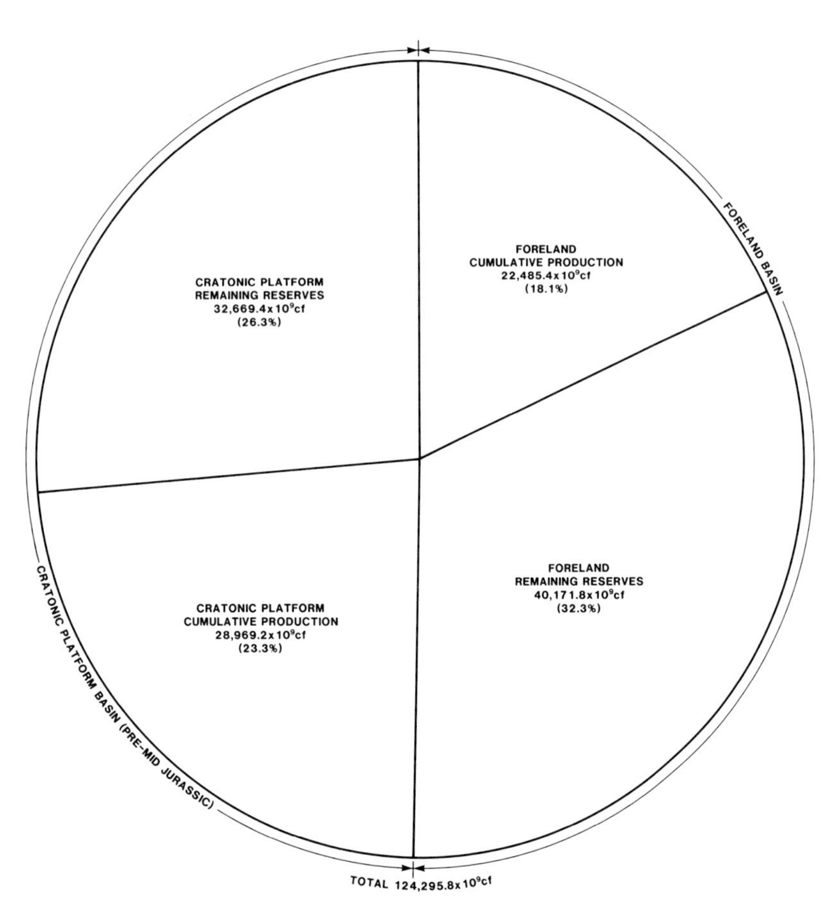

FIGURE 32. Comparison of initial established marketable gas reserves of the Western Canada foreland basin with those of its precursor, the Western Canada cratonic platform basin. Note: Figures given are in billion (10^9) cubic feet.

TABLE 1. Summary of Hydrocarbon Reserves in the Western Canada Foreland, Cratonic Platform, and Sedimentary Basins.

	Conventional Oil and Natural Gas Liquids					
	Initial Established Reserves		Cumulative Produced Reserves		Remaining Established Reserves	
Basin	10^6 bbl	10^6 m^3	10^6 bbl	10^6 m^3	10^6 bbl	10^6 m^3
	Conventional Oil Through Dec. 31, 1985					
Foreland	4,873.3	774.4	3,131.8	497.7	1,741.5	276.7
Cratonic platform	11,942.0	1897.7	8,726.4	1386.7	3,215.6	511.1
Sedimentary	16,815.3	2672.1	11,858.2	1884.4	4,957.1	787.8
	Conventional Oil Through Dec. 31, 1989					
Foreland	5,579.7	886.6	3,786.1	601.6	1,793.6	285.0
Cratonic platform	12,407.8	1971.7	9,762.3	1551.3	2,645.5	420.4
Sedimentary	17,987.5	2858.3	13,548.4	2152.9	4,439.1	705.4
	Natural Gas Liquids Through Dec. 31, 1985					
Foreland	1,477.9	234.8	722.5	114.8	755.4	120.0
Cratonic platform	2,522.7	400.9	1,235.0	196.3	1,287.7	204.6
Sedimentary	4,000.6	635.7	1,957.5	311.1	2,043.1	324.6
	Natural Gas Liquids Through Dec. 31, 1989					
Foreland	1,702.7	270.5	849.8	135.0	852.9	135.5
Cratonic platform	2,911.8	462.7	1,453.3	230.9	1,458.5	231.8
Sedimentary	4,614.5	733.2	2,303.1	365.9	2,311.4	367.3

	Marketable Gas					
	Initial Established Reserves		Cumulative Produced Reserves		Remaining Established Reserves	
Basin	10^9 ft^3	10^9 m^3	10^9 ft^3	10^9 m^3	10^9 ft^3	10^9 m^3
	Marketable Gas Through Dec. 31, 1985					
Foreland	62,657.2	1765.3	22,485.4	633.5	40,171.8	1131.8
Cratonic platform	61,638.6	1736.6	28,939.2	815.3	32,699.4	921.3
Sedimentary	124,295.8	3501.9	51,424.6	1448.8	72,871.2	2053.1
	Marketable Gas Through Dec. 31, 1989					
Foreland	67,025.5	1888.4	28,578.6	805.2	38,446.9	1083.2
Cratonic platform	65,366.1	1841.6	35,136.4	989.9	30,229.7	851.7
Sedimentary	132,391.6	3730.0	63,715.0	1795.1	68,676.6	1934.9

Chapter 7

Western Canada Foreland Basin Oil and Gas Plays

J. E. Barclay

Geological Survey of Canada
Calgary, Alberta, Canada

David G. Smith

Canadian Hunter Exploration Ltd.
Calgary, Alberta, Canada

ABSTRACT

The stages of Western Canada foreland basin tectonic evolution are recorded as five stratigraphically distinct depositional assemblages. Each assemblage can be characterized by a series of proximal, medial, and distal facies belts that migrated with and recorded foreland basin regressive or transgressive events. Each facies belt and each assemblage has a unique set of traps and plays controlled by the types of rocks available as reservoirs and seals. Deposits within the facies belts that make up the assemblages are composed typically of stacked progradational sequences that are commonly punctuated by either transgressive or erosional events.

The fundamental building blocks of the five assemblages are coarsening-upward progradational sequences that are meters to tens of meters in thickness. The sequences consist of offshore deposits overlain successively by shoreface to shoreline to continental sediments. Most foreland basin oil and gas fields are found in the higher-energy, porous, shoreline-related or fluvial sandstones and conglomerates near the tops of individual sequences.

Tectonic subsidence induced by orogenic activity, sea level, climate, and sediment supply controlled the sequence character and sequence distribution and also controlled the geometric relationships among reservoirs, source rocks, and seal rocks. Thus, variations in these controlling factors created differences in trap style within the sequences. For example, distinctive traps originated through a sea level fall that initiated erosion and valley incisement of marine sediments in a progradational sequence, followed by a sea level rise that caused sediment infill of these incised valleys by reservoir sandstones and conglomerates.

Because the five assemblages are stratigraphically distinct and represent unique geologic histories, the oil and gas plays of the foreland basin can be grouped within their hosting depositional assemblages as follows.

In the **Fernie/Nikanassin Assemblage** (Upper Jurassic/basal Cretaceous shallow marine to continental deposits), the main plays comprise intra-Jurassic shallow marine, shoreline to fluvial sandstone reservoirs and uppermost Jurassic sandstone erosional remnants isolated by the sub-Cretaceous unconformity. Erosional remnant, channel, and valley-fill traps produce oil and gas in southern Alberta and Saskatchewan (Swift and Success formations), and fluvial channel traps produce gas in the northern Alberta and northeastern British Columbia Deep Basin and Foothills (Nikanassin Formation).

In the **Mannville Assemblage** (Lower Cretaceous fluviodeltaic-littoral deposits), the main plays occur within continental basal sandstones deposited transgressively on the sub-Cretaceous unconformity and within deltaic to estuarine sandy deposits overlying the basal sandstones. Complex geologic relationships created a variety of stratigraphic traps; major petroleum accumulations occur in, for example, Basal Quartz fluvial and estuarine sandstones, Athabasca and Peace River deltaic oil sands, and Dunlevy/Buick Creek fluvioestuarine channel complexes. This assemblage dominates foreland basin gas reserves and unconventional heavy oil reserves.

In the **Colorado Assemblage** (middle Cretaceous marine deposits), the main plays occur in thin, coarsening-upward sandstone sequences commonly overlain unconformably by conglomerates with the sequences pinching out into marine shales. Major oil and gas producers are conglomerates and sandstones of the Viking, Cardium, and Medicine Hat formations: these units dominate foreland basin conventional (light to medium) oil reserves.

In the **Saunders/Edmonton Assemblage** (Upper Cretaceous marine units transitional to lower Tertiary continental deposits), the main plays occur within the basal part of the assemblage in deltaic and fluvial sandstones of the Milk River and Belly River formations, which formations represent the final episode of marine deposition in the foreland basin. The upper part of the assemblage is mainly continental deposits that are essentially barren of petroleum because of a lack of source and seal rocks.

The **Cypress Hills/Hand Hills Assemblage** (upper Tertiary continental deposits) consists of thick, gravelly, continental strata that represent proximal molassic deposits and is currently considered poorly prospective for hydrocarbons. Although channel sandstone and conglomerate reservoir facies are abundant, petroleum potential is limited by the near-surface and surface location of the assemblage, by a lack of petroleum source rock and seal rock facies, and by low thermal maturities.

INTRODUCTION

The foreland basin portion of the Western Canada sedimentary basin contains 29% of the sedimentary basin's conventional oil (light to medium gravity) reserves and 50% of its marketable gas reserves (Porter, 1992; see also Parsons, 1973; Podruski et al., 1988). Foreland basin conventional oil reserves are 1×10^9 m^3 and marketable gas reserves are 1765×10^9 m^3 (ibid.). The conventional oil reserves, however, comprise less than 1% of the total foreland basin oil reserves, paling beside the 195×10^9 m^3 of unconventional heavy oil reserves also hosted in foreland basin strata (ibid.). This paper describes the major oil and gas plays hosting the conventional oil and gas reserves and thus provides a summary not previously available in the literature.

The play descriptions were derived by reviewing the available stratigraphic and petroleum-geologic literature on the foreland basin and combining the information from that review with our knowledge of the plays. The description of plays emphasizes the control on petroleum occurrences by five foreland basin depositional assemblages (see Leckie and Smith, 1992) that were, in turn, controlled mainly by tectonism of the orogenic belt. Because the five depositional assemblages control the geologic characteristics of the plays, the plays are described here within the assemblages (Figures 1 and 2).

The five assemblages can be used to portray the migration patterns of proximal to distal facies belts (i.e., the transgression–regression history; see later). Also, depositional patterns within each assemblage control the trapping mechanisms of the plays, and the assemblage descriptions provide the geologic context of major plays (Figures 2 and 3).

The term "play" is used here to refer to a group of fields, pools, and/or prospects that have common geologic characteristics and a common origin (see Podruski et al., 1988, p. 7–13). Fields belonging to a specific play share the following characteristics: trap, stratigraphic position, depositional and structural setting, diagenetic history, hydrodynamic regime, seal rock, and source rock. Most major fields mentioned in the text are plotted on the paleogeographic maps presented as figures in this paper; those fields not on these maps can be located on the maps of Wallace-Dudley (1981a, b). The reader also can find further information on Western Canada petroleum geology and stratigraphy in McCrossan and Glaister (1964), McCrossan (1973), Masters (1984a), Stott and Glass (1984), James and Leckie (1988), Podruski et al. (1988), and Glass (1990).

Foreland Basin Progradational Sequences and Relation to Petroleum Trapping

The basic sedimentation patterns or building blocks of the foreland basin consist of progradational siliciclastic sequences that coarsen upward and are meters to tens of meters in thickness (Figure 4). These sequences are also called shallowing/shoaling-upward or sandier-upward sequences and have been termed "parasequences" by Van Wagoner et al. (1988). The bases of the sequences consist of transgressive deposits that overlie a marine flooding surface (sometimes exhibiting minor erosion). The basal transgressive deposits are overlain by a coarsening-upward, progradational shale to sandstone package that is capped by the next marine flooding surface. The lateral and vertical extents of the sequences are controlled by their bounding surfaces and also by internal erosional breaks. Stacks of these sequences are arranged in progradational or transgressive vertical patterns ("parasequence sets"; Van Wagoner et al., 1988) and form the formations, members, or informal units of the five assemblages that host the plays described in this paper.

These sequences are important because they controlled the deposition of reservoirs, seals, and source rocks as well as their geometric interrelationships (Figures 4 and 5). Typically, the best petroleum reservoirs are the coastal and fluvial sandstones and conglomerates developed at the tops of the progradational sequences. Reservoir sandstones and conglomerates may also have been formed by transgressive deposition and reworking on internal erosional breaks within the progradational sequences. Seal and source rock facies were provided by the underlying marine shales and overlying coastal plain shales found, respectively, near the bases and tops of the progradational sequences (Figures 6 and 7).

Depositional Assemblages and Internal Facies Belts and Their Relationships to Petroleum Trapping

Each of the five foreland basin depositional assemblages has a characteristic paleogeographic and geologic expression within the major geographic areas of the basin (Figures 8 to 20). For example, part of an assemblage deposited close to an active orogenic source area will have different depositional units than those of the distal regions: thus the resultant plays are different. Each assemblage can be characterized by a series of proximal to

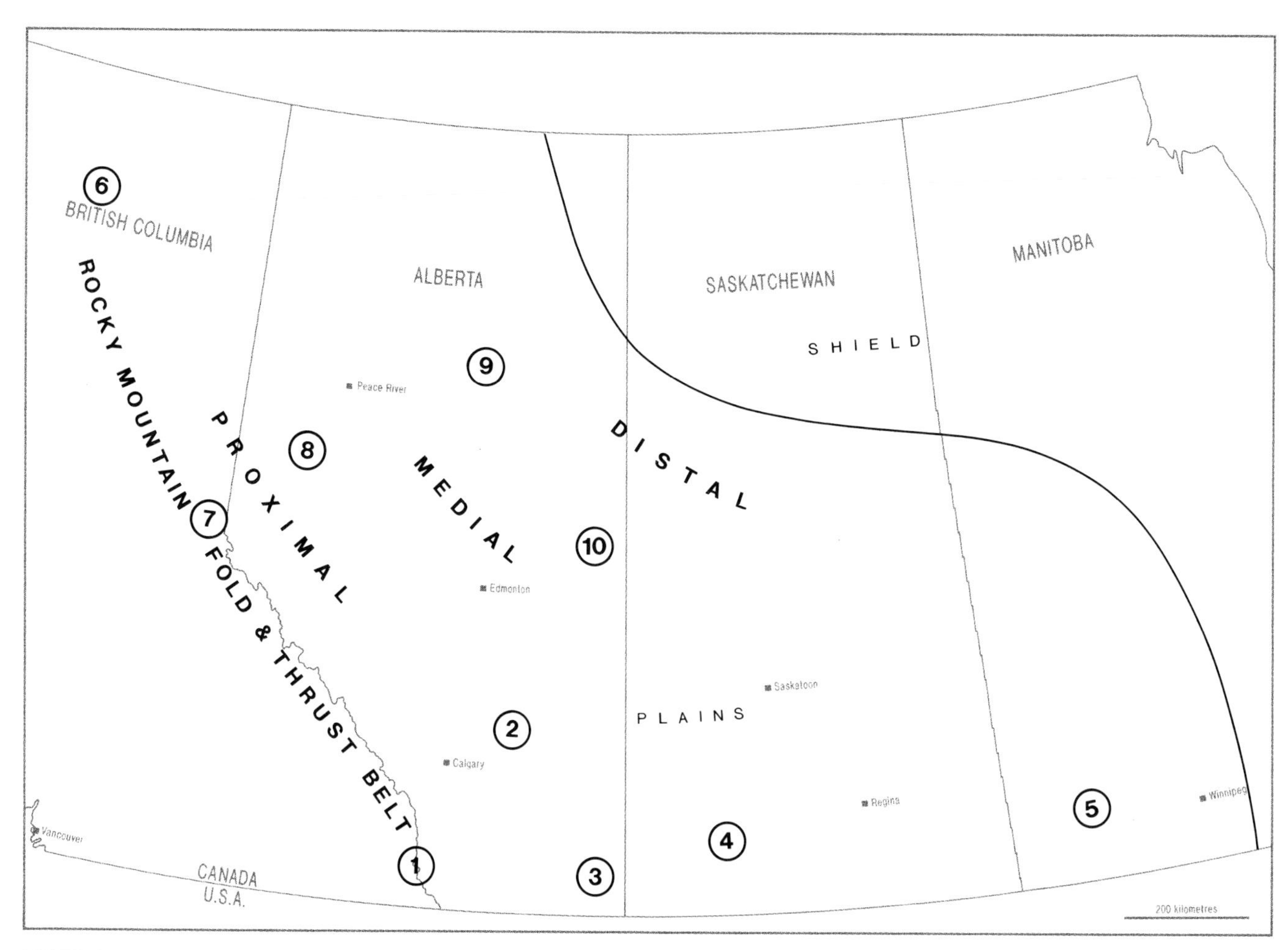

FIGURE 1. Locations of stratigraphic columns numbered 1 through 10 in Figure 2 and of the foreland basin depositional regimes (proximal, medial, and distal) in terms of distance from the fold and thrust belt orogene. Columns 1 to 5 represent the southern foreland basin, and columns 6 to 10 represent the northern foreland basin. This figure can be compared with the continental setting shown in Figures 3 and 7 of Leckie and Smith (1992).

distal facies belts that migrated with and recorded foreland basin regressive or transgressive events. Each facies belt has a unique set of traps or plays controlled by the types of rocks available as reservoirs and seals (Figures 2 and 3). The proximal facies belt (Figure 1) has fluvial to shoreline sandstone reservoirs whereas the distal facies belt is characterized by offshore shelf sandstone reservoirs encased in marine shale. The medial facies belt contains reservoirs representing transitional marine to continental deposits, such as shoreface or shoreline sandstones.

Plays in the Western Canada foreland basin involve mainly stratigraphic traps because (1) the Western Canada foreland basin is a large, simple, asymmetric syncline with a steep limb bordering the orogene and a huge monoclinal limb extending east to the Canadian shield, (2) the basin is affected only locally by subtle internal structures, and (3) the basin lacks significant basement-involved structuring. Thus, this basin and its plays contrast with the common structural trapping in the numerous smaller and structurally complex foreland basins in the western United States.

Petroleum Source Rocks

Limited literature is available on source rocks for specific foreland basin plays; we provide the following summary from Creaney and Allan (1992). The source rocks for foreland basin oils form two source rock groups: a preorogenic, cratonic platform Devonian to Middle Jurassic group and a synorogenic, mainly Upper Cretaceous group. The preorogenic sources have filled the reservoirs of the Fernie/Nikanassin and Mannville assemblages (assemblages 1 and 2, Late Jurassic–Early Cretaceous) and the synorogenic sources have filled the reservoirs of the Colorado and Saunders/Edmonton assemblages (assemblages 3 and 4, mainly Late Cretaceous). These "Nikanassin-Mannville" and "Viking-Belly River" oil systems (Creaney and Allan, 1992) are separated by the Albian Joli Fou Formation, an extensive, marine shale unit. Deeper marine, organic-rich, condensed section shales are the main type of source rock in both groups. The main examples of such source beds in the preorogenic group include the Devonian Duvernay Formation, the Devonian–Mississippian Exshaw Formation, the Triassic Doig Formation, and the Jurassic Nordegg Formation. The synorogenic source beds include the First and Second White Speckled Shales and the Fish Scale Zone (Figure 2).

PLAYS OF THE FERNIE/NIKANASSIN ASSEMBLAGE (ASSEMBLAGE 1, OXFORDIAN-NEOCOMIAN)

The Fernie/Nikanassin Assemblage is a clastic wedge that represents the earliest orogenic detritus in the foreland basin. The assemblage contains 13% of the foreland basin's conventional oil reserves and 3% of its marketable gas reserves (Porter, 1992). Deposits are transitional upward from the underlying cratonic platform sediments. The Passage Beds, Transition Beds, Swift Formation, and equivalent units display this upward cratonic platform to foreland basin transition (Hayes, 1983; Poulton, 1984). Thus, within assemblage 1, orogene-sourced, westerly derived strata interfinger with craton-sourced, easterly derived deposits. The lower part of the clastic wedge consists of marine deposits that filled the foredeep—e.g., Green Beds, Fernie, and Passage Beds strata (Figure 2). The marine strata pass upward into continental clastic deposits of the Kootenay Group and the Nikanassin Formation (Figure 2).

Assemblage 1 is truncated by a regional Neocomian unconformity that has removed much of the formerly extensive Jurassic strata (Figure 2). The resultant deeply dissected erosional surface controlled the distribution of assemblage 1 strata and also partly controlled depositional patterns of the succeeding Neocomian-Albian Mannville Assemblage. The unconformity therefore also controlled the occurrence of productive reservoir facies, the main examples of which are discussed below (Tables 1 and 2).

Petroleum plays consist of two main types: offshore bar plays in distal facies belts of lower parts of the assemblage, and fluvial channel plays in proximal facies belts in upper parts of the assemblage. Some plays have shoreline reservoirs of medial facies belts occurring between the lower distal facies belts (marine) and upper proximal facies belts (continental).

Assemblage 1 also has one play in the regional trap referred to as the "Deep Basin" of northwestern Alberta and northeastern British Columbia on the deeply buried west flank of the foreland basin (Masters, 1979). In this unconventional trap (discussed later with assemblage 2 plays that contain most of the Deep Basin gas), gas-saturated strata occur downdip (southwest) of water-saturated porous strata with local structural and stratigraphic gas traps (Masters, 1984b).

PLAYS OF THE MANNVILLE ASSEMBLAGE (ASSEMBLAGE 2, NEOCOMIAN-ALBIAN)

The Mannville Assemblage is the second, and most widespread, clastic wedge of the foreland basin. This assemblage comprises sediments of the Mannville (Blairmore) Group and equivalent strata (Figure 2). The Mannville Assemblage contains 32% of the foreland basin's conventional oil reserves and 53% of its marketable gas reserves (Porter, 1992).

The Mannville Assemblage formed by deposition, in part transgressively, on the "sub-Cretaceous" unconformity surface separating assemblages 1 and 2. Depositional environments are mainly continental and shoreline, representing medial to proximal facies belt deposits (Figures 10 to 14), and stratigraphy is very complicated because of the complex interplay of tectonism, erosion, sediment influx, and sea level changes. This complexity resulted in the development of numerous stratigraphic trap configurations within a variety of depositional facies. As a result, the Mannville Assemblage contains about 16,000 individual gas pools and about 1200 oil pools, which represent most of the foreland basin pools (Saskatchewan Energy and Mines, 1987; Energy Resources Conservation Board, 1988; Energy Resources Division, 1988; D. J. Cant, P. J. Lee, personal communications, 1991). Several typical plays are discussed below and in Tables 3 and 4.

The Lower Mannville "Formation," or lower part of the assemblage, consists mainly of coarse-grained clastics that filled channels eroded on the unconformity developed on the underlying sediments of assemblage 1. Lower Mannville facies change upward from orogene-proximal, alluvial plain deposits to marginal marine deposits of medial facies belts. Lower Mannville plays include Cadomin Formation conglomeratic alluvial plain sheets, Dunlevy Formation fluvial channels, and Basal Mannville marginal marine sandstones (Table 3).

Legend for Figure 2. **Figure 2 appears on pages 196 and 197.**

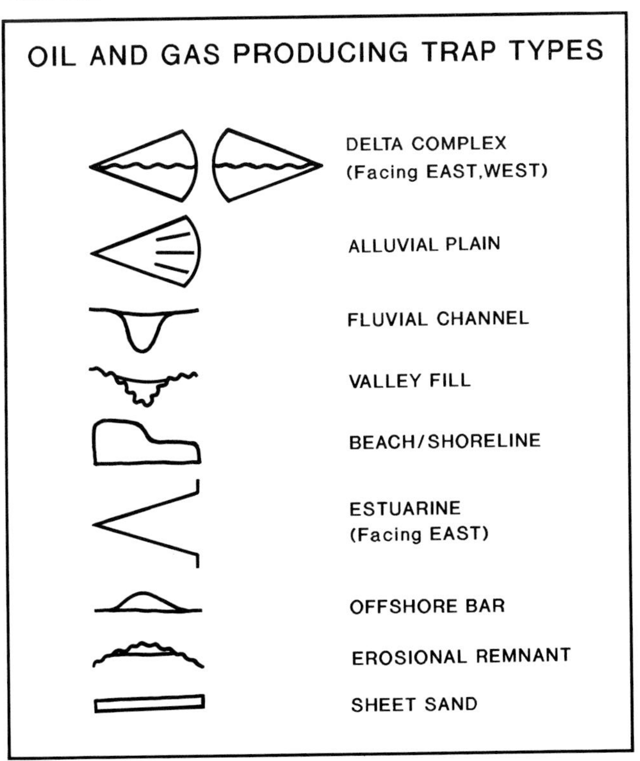

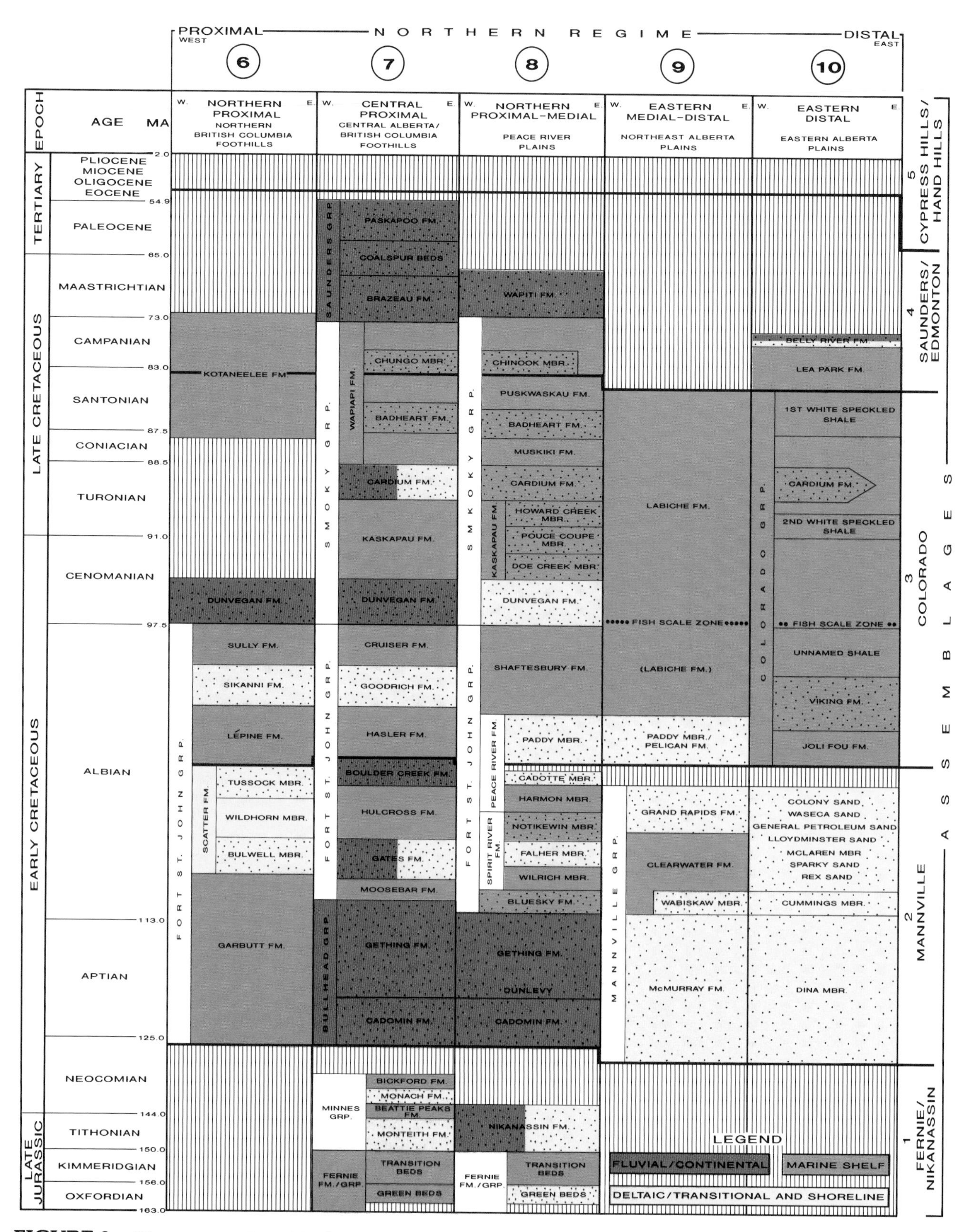

FIGURE 2. West-east stratigraphic charts showing the five major foreland basin depositional assemblages, potential and producing reservoir units, and the trapping mechanisms associated with petroleum-producing units. Producing or potential reservoir sandstones and conglomerates are highlighted in stippling; shales in white; limy units in brick pattern. These charts differ from the Enclosure 1(A) chart of Leckie and Smith (1992) in that petroleum reservoir units are highlighted along with their depositional origins, reservoir unit thicknesses are exaggerated for clarity, and the vertical

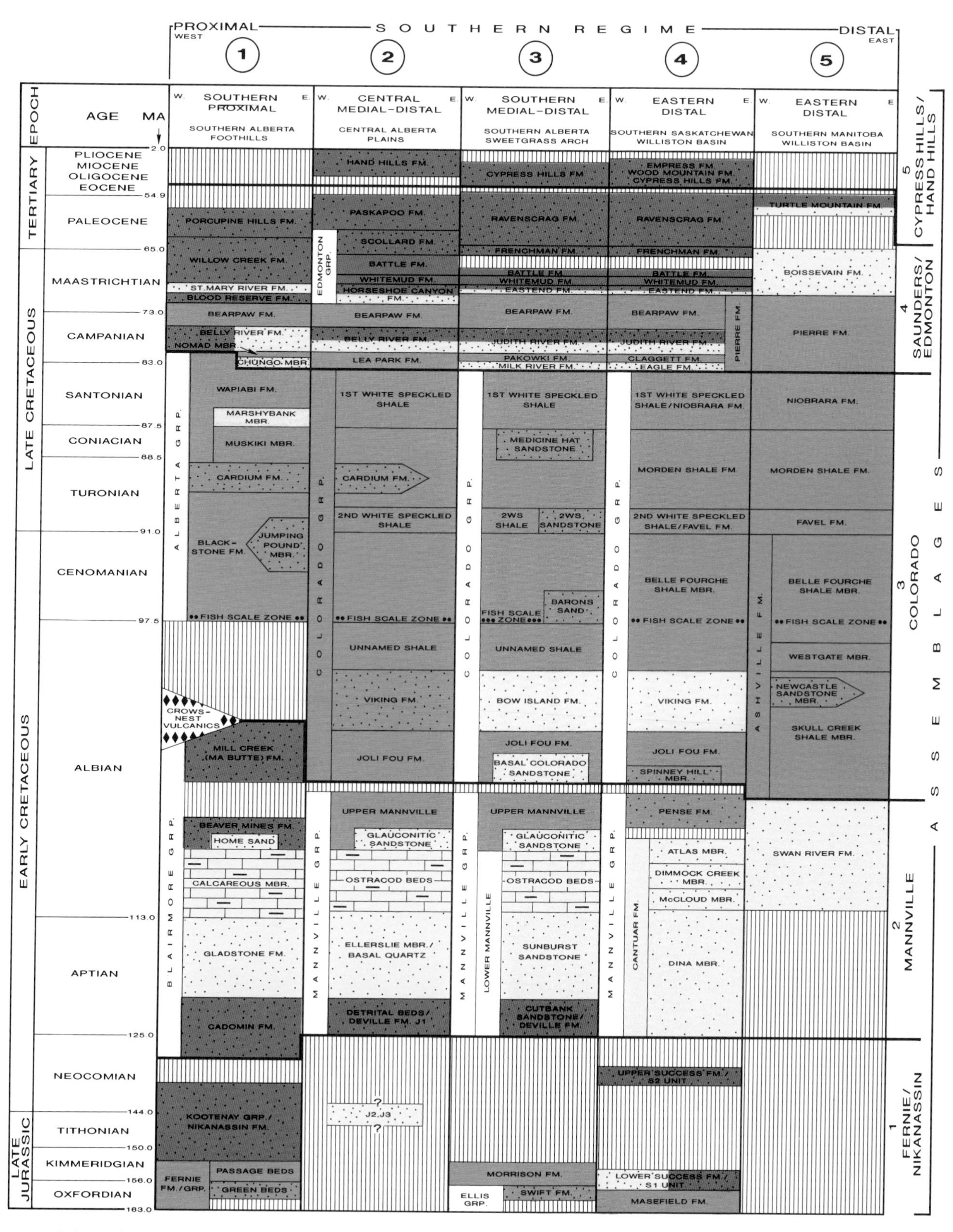

axes of these charts are not directly proportional to time (as are Leckie and Smith's). Post-Paleocene strata are reduced because of a lack of petroleum production. Stratigraphic relationships shown in these charts are illustrated in cross-sectional format in Figure 6 of Leckie and Smith (1992). Formal units—groups, formations, and members—are noted as GRP., FM., and MBR., respectively. Informal units are noted with their most common usages (see Glass, 1990). For column locations, see Figure 1.

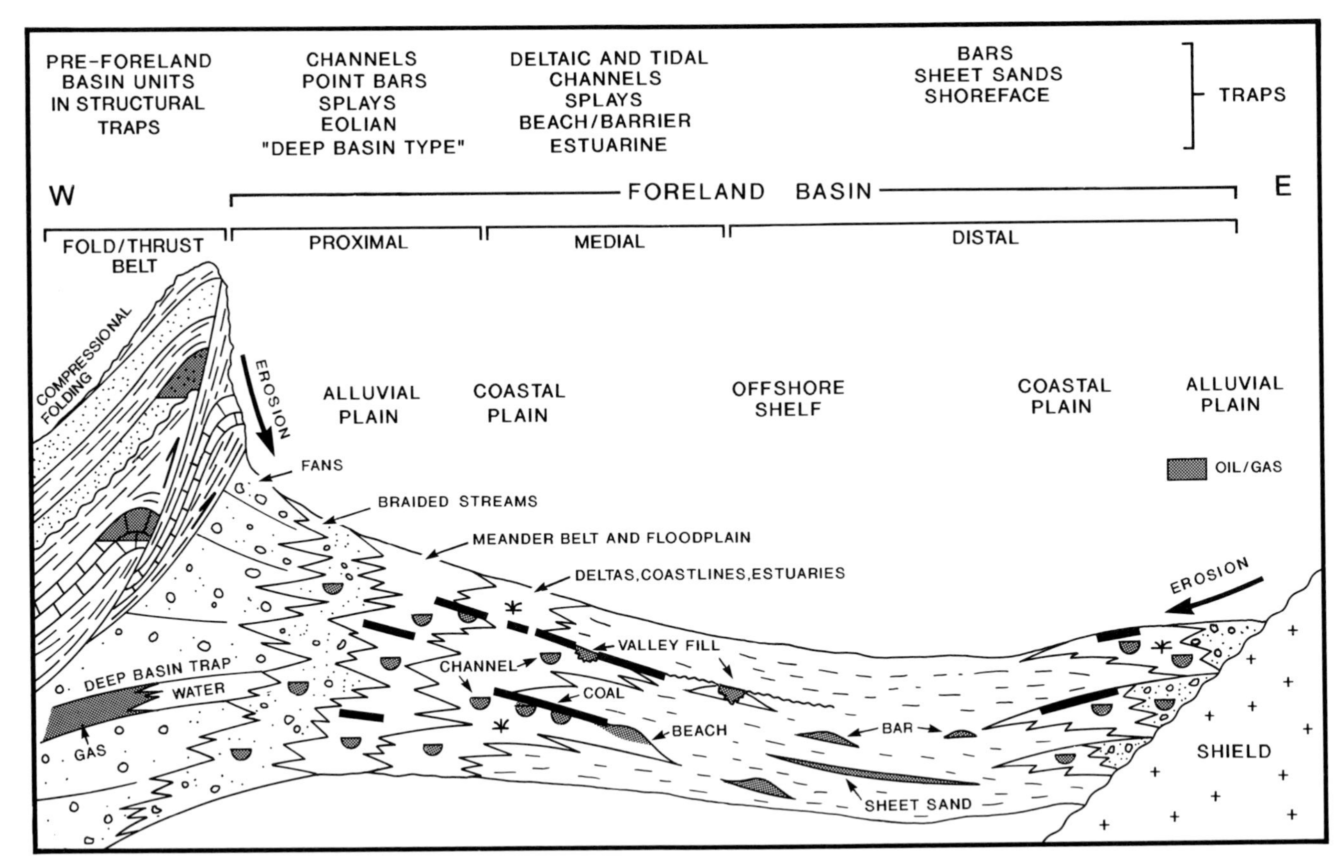

FIGURE 3. Foreland basin petroleum trapping model based on depositional facies distribution and structural styles (sketch developed from the basin model shown in figure 5 of Leckie and Smith, 1992). This sketch represents the typical basin pattern comprising an assemblage that varies through time as transgression pushed facies belts to basin margins or progradation moved shoreline belts to basin center.

Lower Mannville traps typically include reservoir sandstones isolated by deposition on the "sub-Cretaceous" unconformity or isolated by depositional pinch-out (Table 3). Two other common trap types are Deep Basin-type traps with gas-saturated regions below an updip water zone. Drape traps over horst structures related to the Peace River arch are also present.

The Upper Mannville "Formation," or upper part of assemblage 2, comprises deltaic and marginal marine fine-grained clastics deposited in medial facies belts during the overall continued Clearwater-Boreal Sea incursion (Figures 13 and 14; Table 4). The Upper Mannville is typified by progradational fluviodeltaic complexes reflecting abundant orogenic detritus shed into the sea (see Leckie and Smith, 1992).

Upper Mannville reservoir sandstones and conglomerates are commonly developed in shoreline-related facies at the leading edges of these prograding complexes, such as in the Falher Member and Cadotte Member plays (Table 4; Figures 13 and 14). Other sandstone and conglomerate reservoirs are developed in fluvial channels deposited in the coastal plain behind these leading edges. As in the Lower Mannville, traps are typically stratigraphic. Deep Basin-type traps and drape on underlying structures are also common.

The Deep Basin unconventional trap consists of deeply buried, underpressured, gas-saturated (water-poor) strata that occur in the foredeep downdip (southwest) of normally pressured, water-saturated porous strata with local structural and stratigraphic gas traps (Masters, 1984a, b). The gas-saturated region is characterized by stacked "tight sand" sandstone sequences with reduced porosities and permeabilities because of increased cementation, high clay contents, and high compaction. The gas-saturated region is stratigraphically unconfined, and gas is produced from Permian, Triassic, Jurassic, and Cretaceous units. Assemblages 2 and 3 host most of the Deep Basin gas, and the main producers are Lower Cretaceous (assemblage 2) sandstones and conglomerates with the most prolific being those of the Falher Member (Spirit River Formation) at the Elmworth gas field (Masters, 1979, 1984b; McLean, 1979; McMaster, 1981; Cant, 1983; Gies, 1984; Varley, 1984a, b). In the Deep Basin, unconventional "tight-gas sand" reservoirs produce gas after extensive fracture treatment. Conventional reservoirs consisting of porous sandstone and conglomerate account, however, for most of the present gas production.

Deep Basin gas is interpreted as having been generated from coaly beds downdip that, in turn, displaced

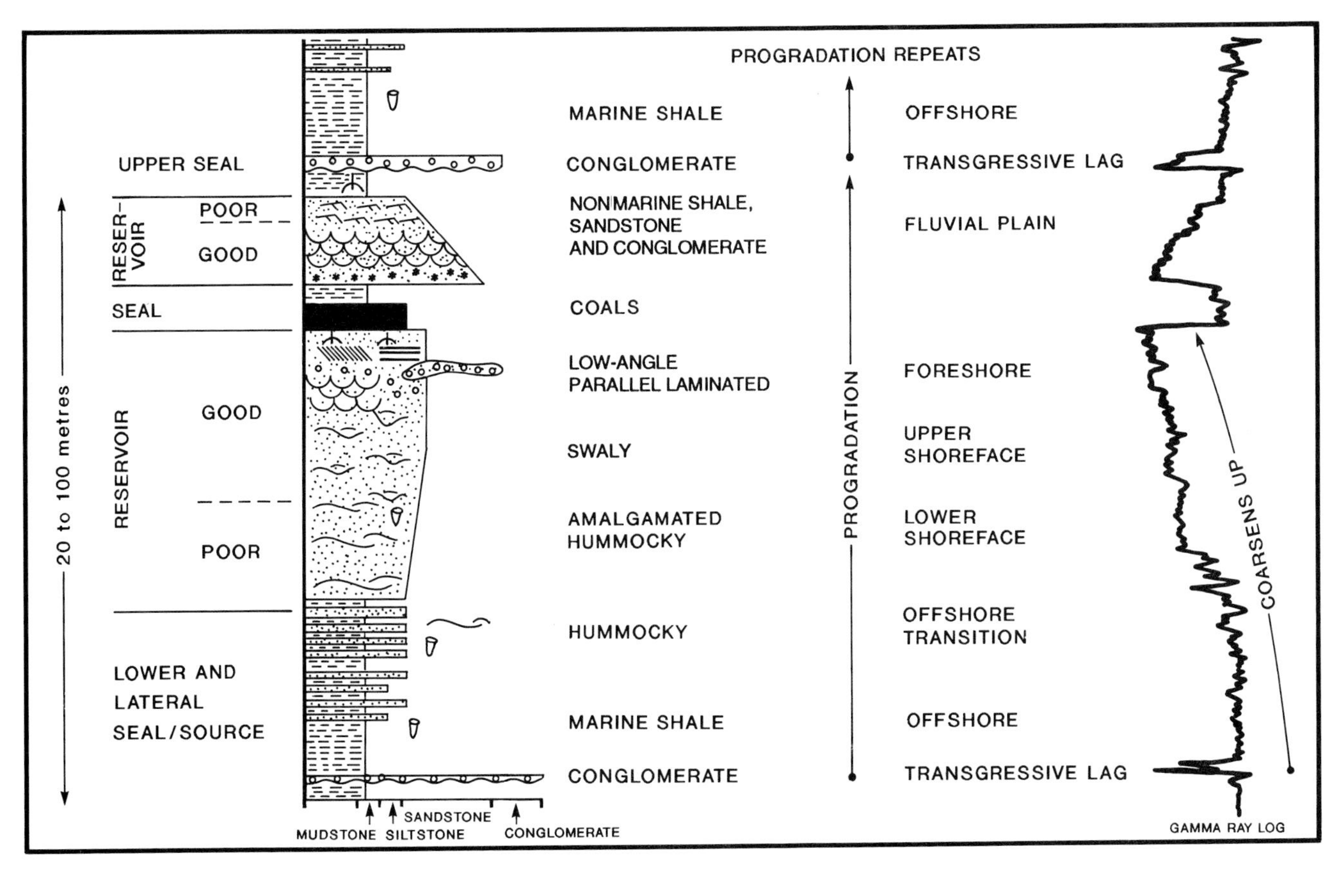

FIGURE 4. Building blocks of the foreland basin. Coarsening-upward, progradational sequence as seen in drill core or outcrop and on well logs. Best reservoir facies occurs in fluvial channels and shoreline-shoreface units (modified from Leckie and Smith, 1992). Seal and source rock units consist of marine or continental shales.

formation water in an updip direction. Although no consensus exists on trap origin, this "upside-down" trap with updip water has been interpreted in several ways, including (1) a dynamic trap in which gas generation exceeds loss through a "leaky" updip seal (Gies, 1984; Masters, 1984b; Welte et al., 1984); (2) a dynamic trap in which downdip water flow from outcrop recharge blocks updip gas migration or balances updip gas leakage (Masters, 1979; Varley, 1984b); (3) a static trap with lowered relative permeability to gas near the updip water zone (Masters, 1979); and (4) a static trap caused by updip permeability barrier(s) of depositional or diagenetic origin (Cant, 1983; Gies, 1984; Rahmani, 1984). Combinations of the above mechanisms have also been proposed for the trap origin by some of these authors.

PLAYS OF THE COLORADO ASSEMBLAGE (ASSEMBLAGE 3, UPPER ALBIAN–UPPER CAMPANIAN)

This assemblage is a thick marine shale package punctuated with thin sandstone and conglomerate reservoir units that were deposited at the tops of progradational sequences (Figure 2). The shale-dominated Colorado Assemblage with mainly distal facies belt deposits contrasts with the coarse-grained clastics deposited in medial and proximal facies belts of the overlying and underlying assemblages 4 and 2. Colorado Assemblage reservoirs contain 51% of the foreland basin's conventional oil reserves and 26% of its marketable gas reserves, and these reserves comprise 15% of the total conventional oil reserves and 13% of the total marketable gas reserves of the Western Canada sedimentary basin (Porter, 1992; see also Parsons, 1973; Podruski et al., 1988).

The sandstones and conglomerates occur typically as thin beds capping the progradational sequences that coarsen upward from underlying marine shale. The conglomerate units form the main reservoir facies, and the sandstones also form important reservoirs (Table 5). The Colorado Assemblage shales contain good petroleum source rocks and are excellent seal rocks (Creaney and Allan, 1992).

The conglomerate and sandstone reservoirs, although thin, contain a disproportionately large volume of hydrocarbons in the foreland basin, commonly in large pools (e.g., Pembina field; see Porter, 1992) because of their favorable source-seal settings.

Plays in this assemblage occur mainly in the southern and central plains because of the concentration of marine shelf sand and conglomerate deposition there. Plays are

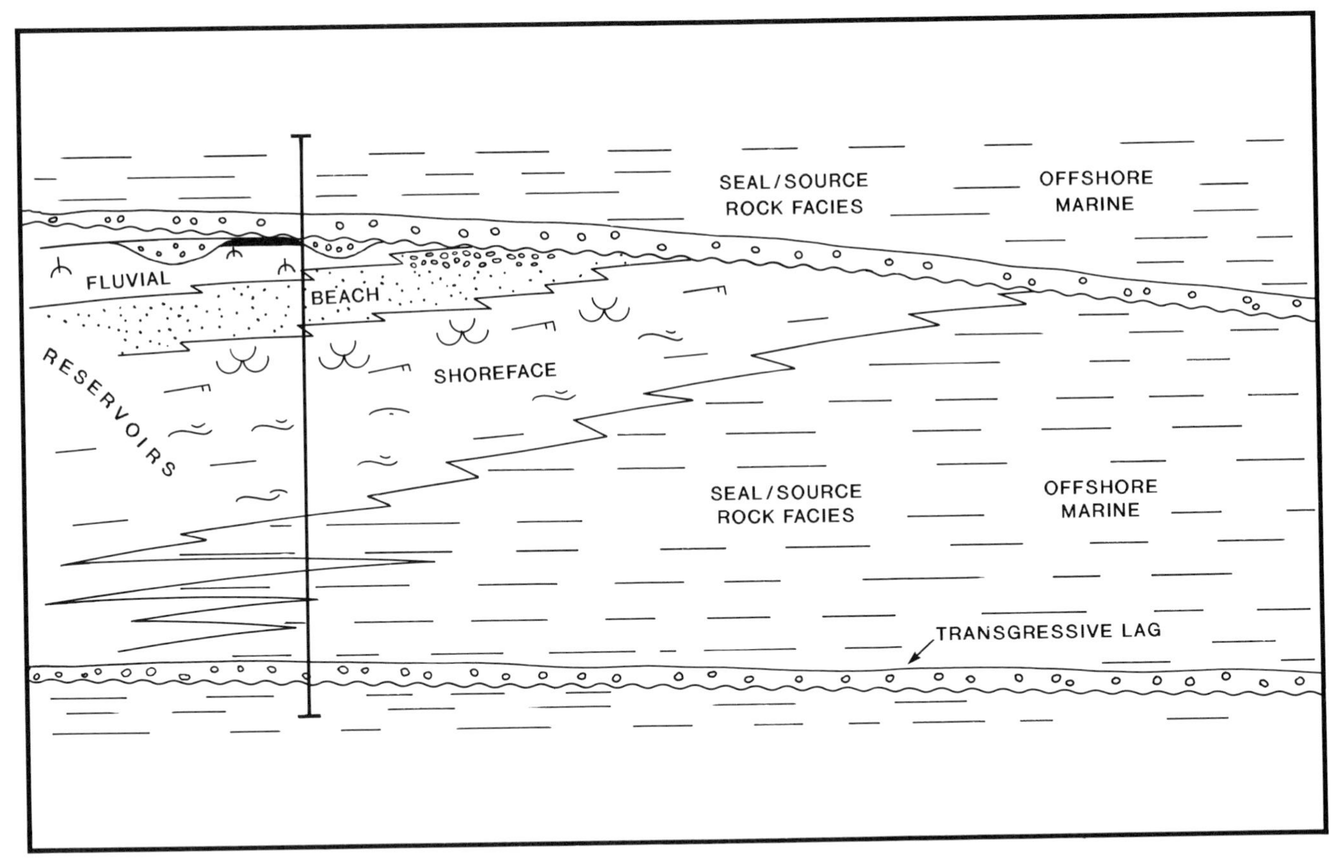

FIGURE 5. Coarsening-upward, progradational sequence in a schematic cross section. Vertical bar corresponds to a blowup of environments at vertical bar in Figure 7.

not developed in the northern or northeastern plains because of erosional edges and sandstone depositional limits.

Major producers from this assemblage are sandstones and conglomerates of the Cardium, Viking, and Medicine Hat formations, and these units dominate conventional oil and gas reserves of the Colorado Assemblage and of the entire foreland basin (Table 5). The largest foreland basin oil fields, for example, occur in Cardium Formation conglomerates and sandstones in west-central Alberta. The Pembina oil field, discovered in 1953 and containing 0.29×10^9 m^3 of oil, is well known as a giant oil field that aided oil industry recognition of the importance of stratigraphic traps (Nielsen, 1957; Nielsen and Porter, 1984). Numerous oil and gas pools also occur in Viking Formation shelf and coastal sandstones and conglomerates in central and eastern Alberta and western Saskatchewan. Significant shallow gas reserves occur in Medicine Hat Formation outer shelf sandstones of southeastern Alberta.

The Cardium Formation hosts 40% of the foreland basin's conventional oil reserves (Porter, 1992) and deserves extra attention in this introduction to Colorado Assemblage plays. The Pembina field was discovered during drilling for Upper Devonian reefs; the oil discovery in a Cardium sandstone pinch-out was one of three secondary targets in the well (Figure 17; Nielsen and Porter, 1984). In most of the major Cardium Formation fields—i.e., Pembina, Willesden Green (Figure 6), Ferrier, Crossfield, and Garrington—oil is associated with solution gas. Deeper, western fields close to the Rocky Mountain Foothills grade with increasing gravity to gas condensate pools (Nielsen and Porter, 1984; Podruski et al., 1988; Porter, 1992).

The main and most thoroughly explored Cardium Formation play involves sandstone and conglomerate reservoirs having a sheetlike geometry. The rock bodies consist of a stacked series of coarsening-upward sandstones and conglomerates (Figures 6 and 7) that are replaced laterally by shale to the northeast, creating an updip stratigraphic seal. Because of the economic importance of the conglomerate reservoirs, much study and discussion have been focused on the problematic origin of the Cardium gravel-sized detritus encased in marine shelf shale (e.g., Beach, 1955; Walker, 1983; Krause and Nelson, 1984). Normal marine processes do not easily account for gravel transport offshore onto the shelf, and interpretations have been proposed that involve turbidity currents, storm currents, river floods, shoreline processes, tidal currents, and lowstand deposition leaving marine-reworked, relict shoreline gravels.

A second and developing play involves Cardium marine sandbodies that fill broad shallow scours and form stratigraphic traps (Walker, 1985; Podruski et al., 1988). These sandbodies are tens of meters thick and several

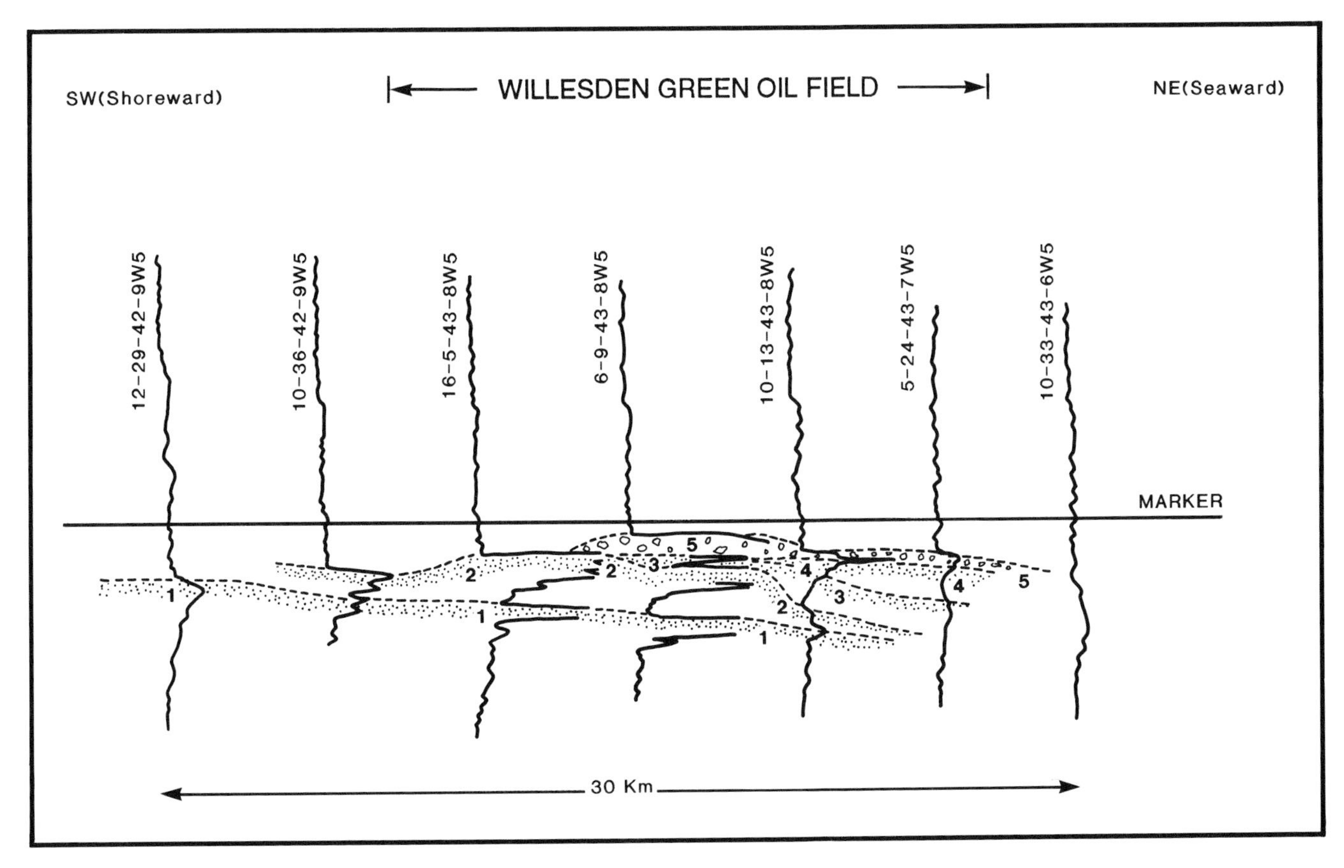

FIGURE 6. Well log example of progradational sequence from Cardium Formation at Willesden Green field, west-central Alberta; see Figure 17 for section location. Numbers 1 to 5 represent progressively younger shingles of reservoir sandstone (stippled) and reservoir conglomerate (open circles) encased within marine shale and display a northeast (seaward) prograding pattern. Electrical resistivity well logs are shown with well locations near the top of the log trace (modified from Keith, 1985).

kilometers long and wide. The sandbodies fill scours cut into Cardium sheet sands and underlying Blackstone Formation shales. The largest scour sandstone fields occur at Ricinus, Cyn-Pem, and Carrot Creek, and they are an order of magnitude smaller than the sheet sand fields (Figure 17). A third Cardium play contains oil trapped in shoreline deposits of proximal to medial facies belts in the northern plains (Nielsen and Porter, 1984; Plint and Walker, 1987; Podruski et al., 1988; Deutsch and Krause, 1990).

PLAYS OF THE SAUNDERS/EDMONTON ASSEMBLAGE (ASSEMBLAGE 4, UPPER CAMPANIAN-PALEOCENE)

The Saunders/Edmonton Assemblage forms a transition from the underlying shale-dominated, distal facies belt marine sediments of the Colorado Assemblage to the conglomerate-dominated, continental proximal facies belts of the overlying Cypress Hills/Hand Hills Assemblage (Figures 18 to 20). The Saunders/Edmonton Assemblage contains 4% of the foreland basin's conventional oil reserves, exclusively in the Belly River Formation, and 18% of its marketable gas reserves, mainly in the Milk River Formation and Medicine Hat sandstone (upper part of assemblage 3) "shallow gas sands" (Porter, 1992).

The Saunders/Edmonton Assemblage comprises two synorogenic molasse packages deposited during the Laramide orogeny: a lower package with progradational, marine to shoreline sequences (distal to medial facies belts) deposited during a Late Cretaceous orogenic phase (Figures 18 and 19); and an upper package of mainly continental conglomerates (proximal facies belts) deposited during an early Tertiary orogenic phase (Figure 20).

The lower Laramide molasse package is the Milk River-Pakowki-Belly River-Bearpaw interval and represents an overall regressive transition from marine units of the underlying Colorado Assemblage (Figures 2, 18, and 19). The transition comprises two major marine to shoreline-continental eastward-prograding sequences. This molasse package consists of Milk River Formation shallow marine to shelf sandstones that prograded eastward over the uppermost Colorado marine shales. The Milk River Formation is capped by Pakowki Formation marine shale (Figure 18), which is, in turn, overlain by the deltaic and continental deposits of the Belly River Forma-

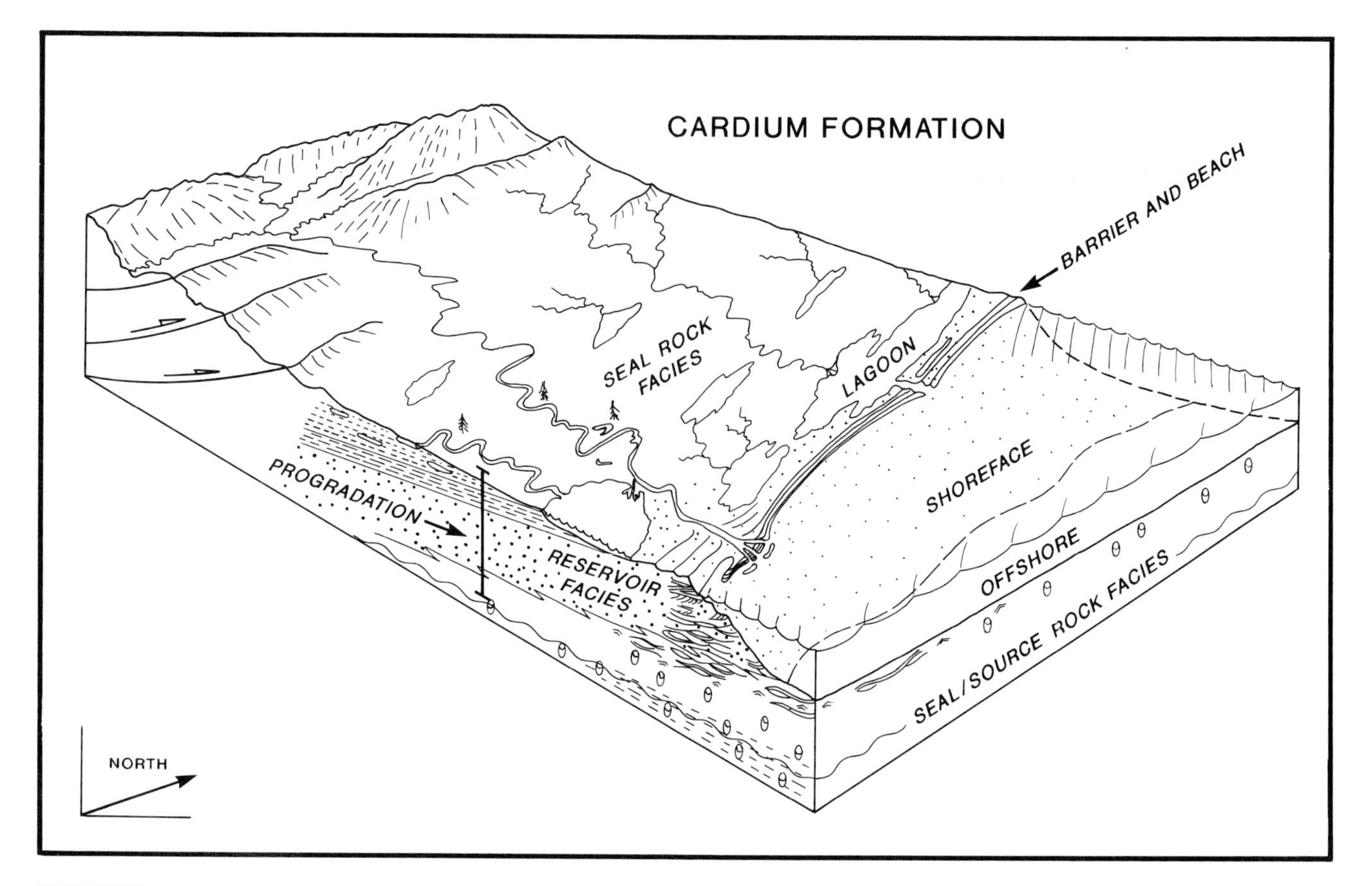

FIGURE 7. Three-dimensional view of progradational sequence emphasizing the bounding of shoreface reservoir-quality sandstones by offshore and coastal plain mudstones (modified from Plint and Walker, 1987). Vertical bar on left side would have a vertical facies sequence as at the vertical bar in Figures 4 and 5.

tion (Figure 19). The Belly River Formation is overlain by marine shale of the Bearpaw Formation, representing the last widespread marine transgression in the foreland basin.

Assemblage 4 oil and gas production occurs only within the marine to continental transitions of progradational sequences in the lower molasse package of assemblage 4. Stratigraphic traps are developed within the large progradational wedges of the Milk River and Belly River formations (Figures 18 and 19). The traps are depositional pinch-outs consisting of sandstones encased in and grading eastward to marine shales—e.g., the Alderson Member, which makes up the Milk River gas field in southern Alberta, is a distal facies of the Milk River Formation sandstones (Figure 18; see also Meijer-Drees and Myhr, 1981).

The upper Laramide molasse package overlies the Bearpaw Formation marine shales and begins with the Edmonton Group sandstone, shale, and coal, which are overlain by coarser-grained continental alluvial deposits of the Paleocene Paskapoo, Porcupine Hills, and Ravenscrag formations (Figures 2 and 20). Areally restricted marine shale tongues (distal facies belts) occur in eastern parts of the foreland basin, representing a final retreat of marine waters from the foreland basin, and are developed at the bases of marine to continental progradational sequences (e.g., "Cannonball Shale" in the Turtle Mountain Formation of Manitoba; Fort Union Formation of North Dakota).

The second molasse package presently lacks significant hydrocarbon reserves. Included in this presently unproductive package are the Edmonton and Saunders groups and the St. Mary River, Blood Reserve, and Willow Creek formations (Figure 20). Important coal deposits are hosted in the Edmonton Group (Scollard and Horseshoe Canyon formations) and the Saunders Group (Coalspur Formation), however. The overlying Paskapoo Formation contains small amounts of gas trapped in fluvial channels (Figure 20). Hydrocarbon potential may exist in these strata because reservoir facies of porous fluvial sandstone and conglomerate are abundant, shale seal rock is present, and source rock facies of shale and coal are also present.

PETROLEUM POTENTIAL OF THE CYPRESS HILLS/HAND HILLS ASSEMBLAGE (ASSEMBLAGE 5, EOCENE–PLIOCENE)

Deposits of assemblage 5 represent the final filling of the foreland basin with a continental, gravelly clastic wedge. Deposits are dominated by proximal facies of braidplain and meander belt alluvial gravels. Eocene to Pliocene

Legend for Figures 8 to 20: **Paleogeographic maps of stages within foreland basin depositional assemblages 1 to 5. Maps display principal depositional facies and significant oil and gas fields (fields shown as black dots and irregular black blobs). Geologic base maps from Leckie and Smith (1992) and modified with data from references in figure captions and with field locations added from Wallace-Dudley (1981a, b), Saskatchewan Energy and Mines (1987), Energy Resources Conservation Board (1988), and Energy Resources Division (1988).**

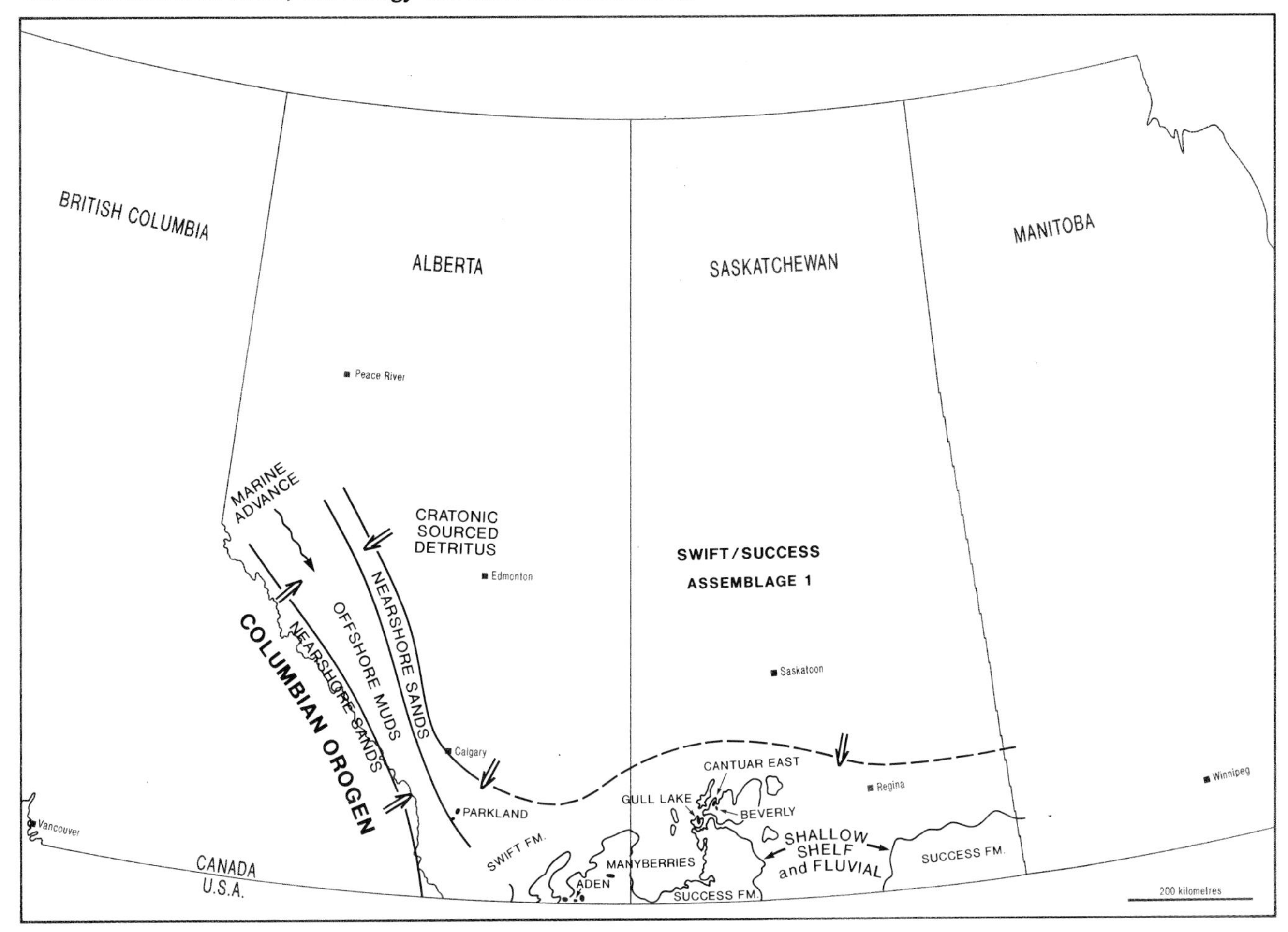

FIGURE 8. **Kootenay/Nikanassin Assemblage 1, lower units: Swift and Success formations paleogeography and oil/gas fields. Main traps are fluvial channel sandstones, sandstone erosional remnants, and shelf sandstones (modified from Leckie and Smith, 1992; Hayes, 1983; Christopher, 1974).**

uplift and erosion have left only thin, scattered erosional remnants of these strata that occur as cap rocks of flat-topped hills in southern Alberta and Saskatchewan. These strata have no known petroleum production and have limited petroleum potential because of their surface and near-surface locations, restricted distribution, and lack of seal rock and source rock facies (Parsons, 1973; Podruski et al., 1988).

PLAYS WITH PRE-FORELAND BASIN RESERVOIRS

Paleozoic and Early Mesozoic Traps Filled Because of Foreland Basin Tectonism

The numerous prolific plays of the Western Canada sedimentary basin with Paleozoic and early Mesozoic reservoirs owe their regional trapping geometry and hydrocarbon filling to Upper Cretaceous–Tertiary (Laramide orogeny) foreland basin tectonics. These pre-foreland basin plays contain 61% of the Western Canada basin's conventional oil reserves and 50% of its marketable gas reserves (Porter, 1992). This orogeny was of fundamental importance to Western Canadian petroleum accumulations, because the foredeep subsidence caused generation of the majority of oils and gases and also enhanced the eastward updip basin geometry, allowing migration of these petroleums into preexisting, mainly stratigraphic traps.

The Paleozoic and early Mesozoic plays differ from the foreland basin plays in that traps were developed solely because of preorogenic sedimentation, but filling of the preorogenic traps by petroleum was caused by the westward tilt imposed by later foredeep subsidence. The lack of Paleozoic tectonism within the Western Canada sedimentary basin allowed the preorogenic traps to

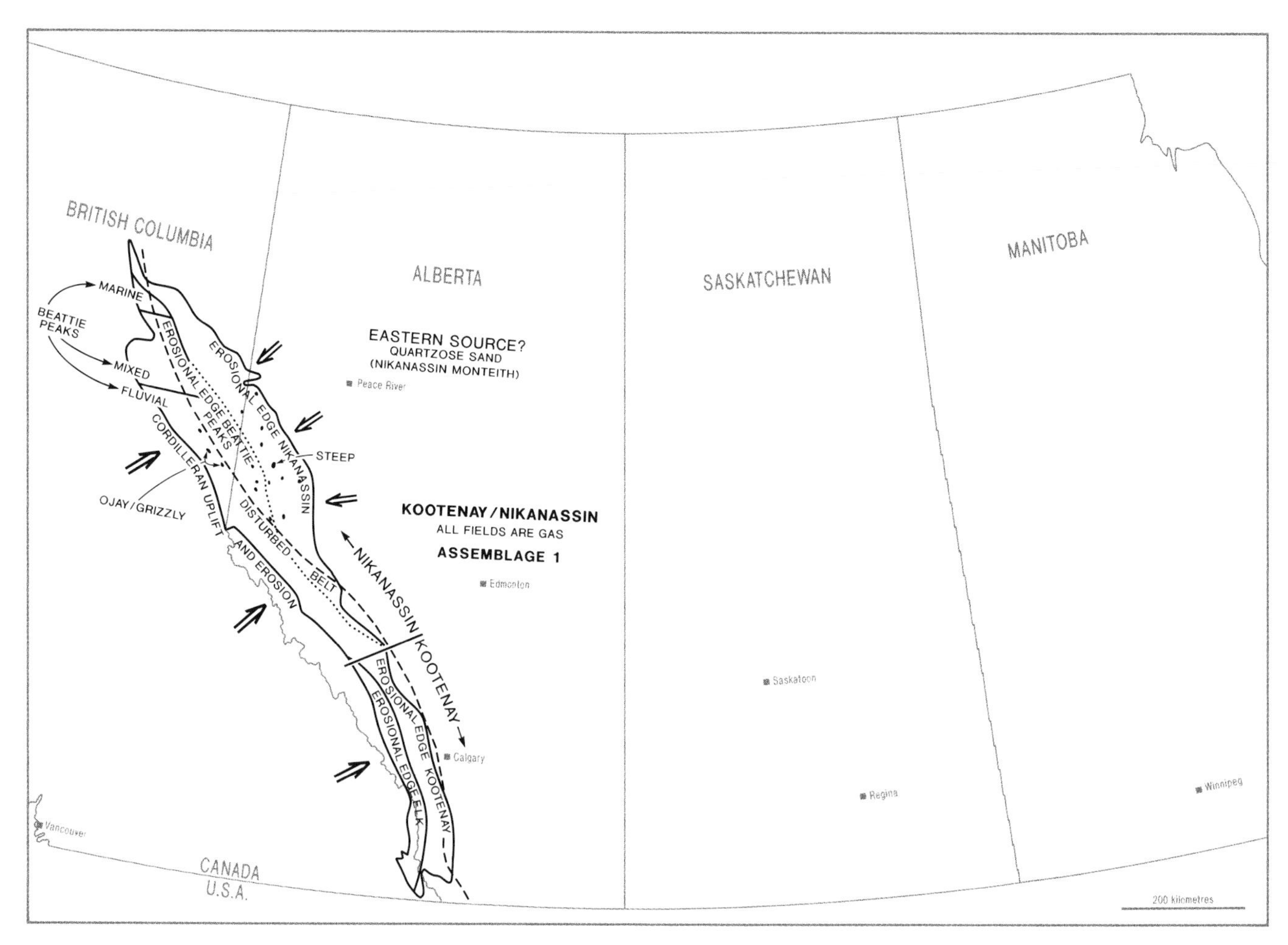

FIGURE 9. **Kootenay/Nikanassin Assemblage 1, upper units: paleogeography and oil/gas fields. Main traps are sandstones in anticlines on thrust faults (west), and shoreline and fluvial sandstones (east). (See Figure 8 for legend.)**

remain undisturbed and preserved the traps for filling by Laramide-generated hydrocarbons. In contrast, for the synorogenic foreland basin plays (i.e., post-Middle Jurassic), both the trap creation and petroleum filling were consequences of both foreland basin sedimentation and foreland basin tectonics. This aspect of the Western Canada sedimentary basin plays reflects the two major tectonic regimes of the sedimentary basin, and although addressed only briefly here, has added to the understanding of the foreland basin's role in petroleum distribution. For thorough descriptions of these relationships, see Creaney and Allan (1992), Porter (1992), and Deroo et al. (1977).

The preorogenic reservoirs dominate Western Canada sedimentary basin hydrocarbon reserves and include traps in Devonian reef chains, Mississippian shelf carbonates, and Permian–Triassic shallow marine sandstones. Most of these reservoirs contain petroleum in preorogenic stratigraphic traps filled by eastward updip petroleum migration. Discussed below, however, are two important preorogenic plays that differ from the above stratigraphic traps because they have a strong structural trapping component that was caused by orogenic tectonism in addition to preorogenic stratigraphic trap development. Traps of the first play were developed by structural folding, and the second play has subcrop edge traps developed by structural tilting.

Foothills Fold and Thrust Belt: Folded and Faulted Reservoirs

Structural traps in the fold and thrust belt traps consist mainly of Mississippian and Triassic carbonate reservoirs involved in thrust faults and folds. The structural traps occur as rollover anticlines perched on the hanging-wall leading edges of thrusts (Figure 3). The main production is from the Mississippian Turner Valley field in southwestern Alberta (Gallup, 1954), which consists of porous crinoidal dolomites, of the Turner Valley Formation (Rundle Group), deposited in a bank margin setting. Permeability is enhanced by fracturing related to folding. The large Turner Valley oil and gas field is approximately 25 km long by 4 km wide within a rollover anticline carried in the NNW-trending Sheep River anticline, which is one of a series of structures parallel to the folded belt hosting

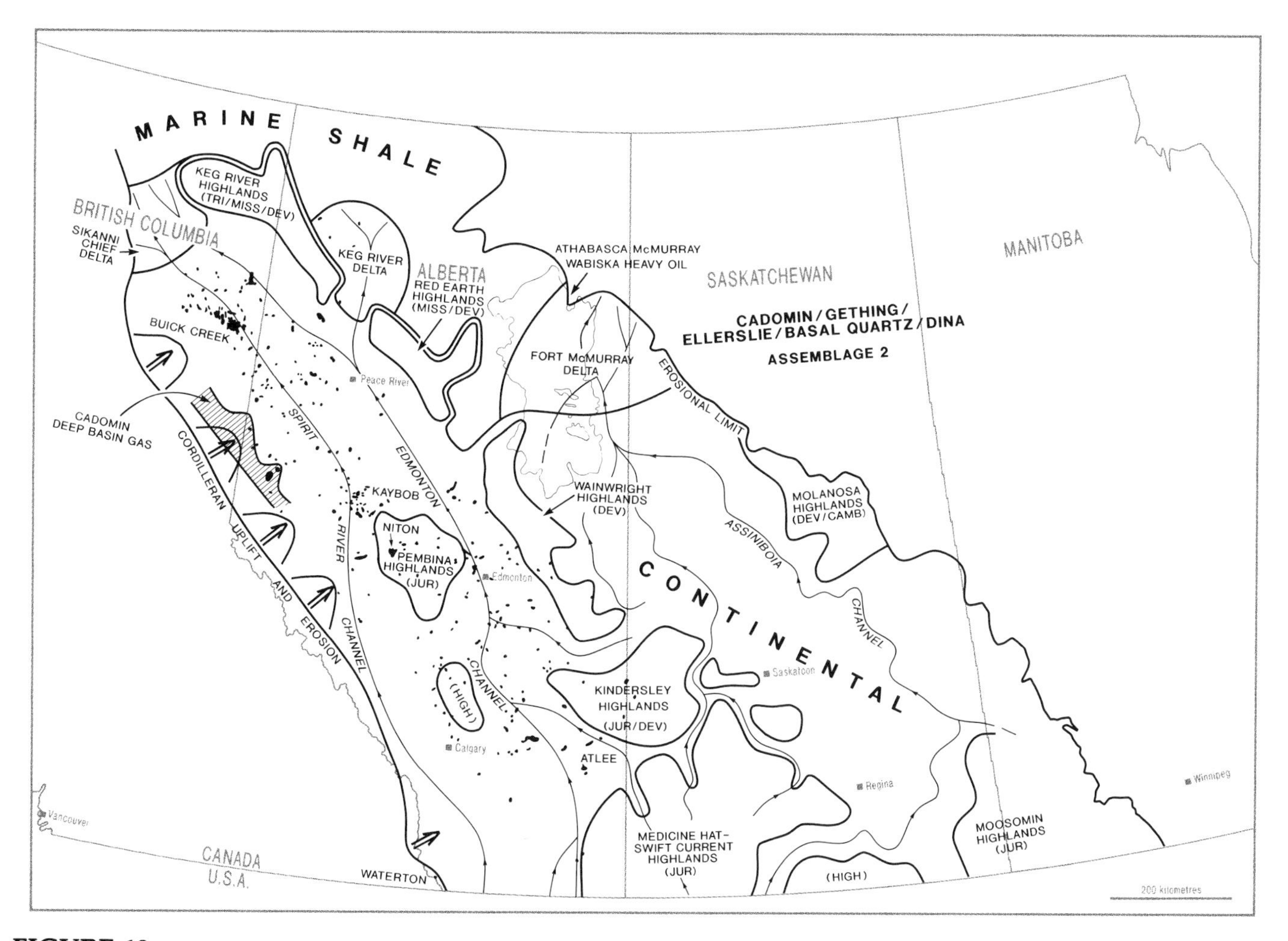

FIGURE 10. Mannville Assemblage 2, lowermost units: "Basal Mannville" paleogeography and oil/gas fields. Main traps are in sandstones of fluvial and estuarine channels, deltas, valley-fills, and alluvial plain sheets (in part modified from Jackson, 1984; see Figure 8 for legend).

similar Mississippian fields (e.g., Quirk Creek, Sarcee, and Savanna Creek; see Tippett, 1987). Other important accumulations in southern Alberta occur to the northwest in the Limestone–Burnt Timber fields and to the southwest in the Waterton–Pincher Creek fields (see maps of Wallace-Dudley, 1981a, b, for field locations). These last fields contrast with the Turner Valley oil leg in that they contain sour gas (Norris and Bally, 1972), possibly reflecting a Paleozoic (preorogenic) basinal carbonate source.

In the northern Foothills fold and thrust belt of northeastern British Columbia, similar accumulations involve fracture-enhanced, low conventional porosity, gas-filled carbonate reservoirs. Fields are developed in the Triassic Pardonet and Baldonnel formations (Sukunka and Bullmoose fields; Barss and Montandon, 1981) and in the Mississippian Debolt Formation (Pocketknife and Sikanni field areas). Less-productive sandstone reservoirs also occur, including the Triassic Halfway and Charlie Lake (Inga Member) formations in the Jedney-Highway-Inga areas and the Jura-Cretaceous Nikanassin Formation in the Grizzly-Ojay fields (see maps of Wallace-Dudley, 1981a, b, for field locations).

Paleozoic and Early Mesozoic (Pre-Foreland Basin) Reservoirs Isolated at the Pre-Mesozoic Unconformity

The widespread pre-Mesozoic unconformity is a cratonward merging of several unconformities that isolate successively older units to the east (Figure 2). Important among these unconformities is that produced by pronounced westward tilting of the basin during earliest Cretaceous and latest Jurassic (approximately Tithonian and Neocomian). Also partly controlled by the pre-Mesozoic unconformity are synorogenic plays of the Fernie/Nikanassin and Mannville assemblages (1 and 2) in which deposition of Upper Jurassic units and basal Mannville and equivalents was controlled by the irregular erosional surfaces of older, mainly Paleozoic, formations.

The unconformity governed the development of several important preorogenic Paleozoic and Mesozoic plays: the Upper Devonian Wabamun Formation subcrop edge play in central Alberta, the Mississippian carbonate subcrop edge plays throughout Alberta, the Mississippian

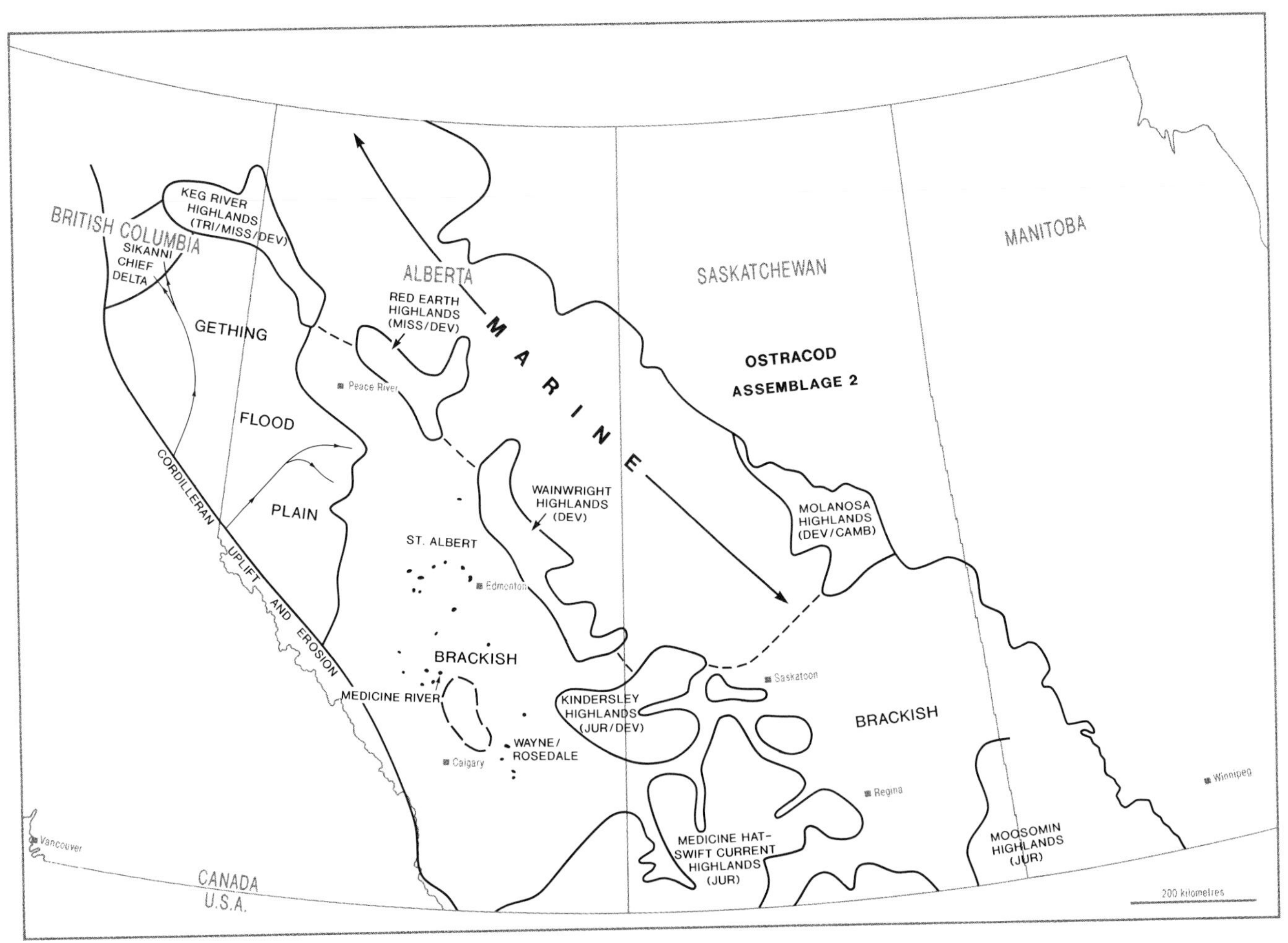

FIGURE 11. Mannville Assemblage 2, lower units: Ostracod (Lower Mannville) paleogeography and oil/gas fields. Main traps are in sandstones of fluvial and estuarine channels and valley-fills (see Figure 8 for legend).

Souris Valley-Tilston, Frobisher-Alida, and Midale subcrop edge plays of Saskatchewan, and some of the pre-orogenic Jurassic units in the Gilby-Medicine River trend of Central Alberta (see Podruski et al., 1988). In many of these plays, porosity has been enhanced by freshwater leaching during exposure events. Also, conversely, diagenetic trapping was developed by occlusion of porosity during cementation. In addition, trapping configurations were controlled by synorogenic deposition of overlying shale facies that provide excellent seal rocks.

CONCLUSIONS

The tectonic evolution of the Western Canada foreland basin is recorded as five stratigraphically distinct depositional assemblages, and each assemblage has a distinct set of petroleum plays. Each assemblage is characterized by a series of proximal, medial, and distal facies belts that migrated with and recorded foreland basin regressive or transgressive events. Each facies belt and each assemblage has a unique set of traps and plays controlled by the types of rocks available as reservoirs and seals. The differences between assemblages are controlled by changes in basin subsidence (orogenically induced), sediment supply, sea level, and climate. The foreland basin depositional assemblages host 29% of the conventional oil reserves, 50% of the marketable gas reserves, and 73% of the enormous reserves of unconventional heavy oil in the Western Canada sedimentary basin.

The stratigraphy and hence the petroleum reservoir-seal relationships of each distinct depositional assemblage are related by a common sedimentation pattern that is the main building block for all of the assemblages. Each assemblage is composed of stacked progradational sequences consisting of marine shale to shoreline and continental sandstone and conglomerate. The best reservoirs are located in the porous, shoreline-related or fluvial sandstones and conglomerates that commonly form the tops of the sequences.

The oil and gas plays of the five depositional assemblages can be summarized as follows.

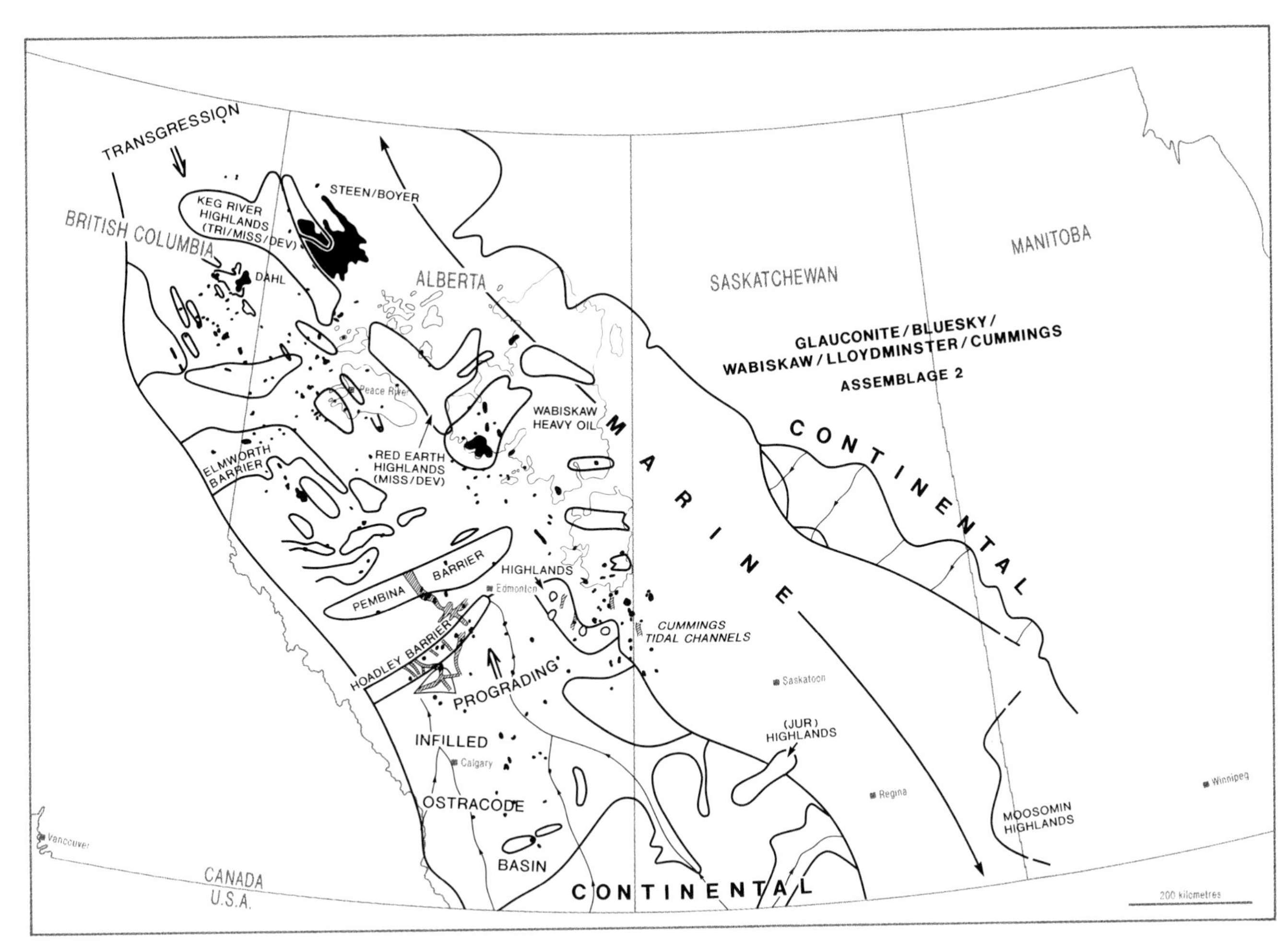

FIGURE 12. **Mannville Assemblage 2, upper units: Lower Mannville paleogeography and oil/gas fields. Main traps are in sandstones of barriers; fluvial, estuarine, tidal, and valley-fill channels; and shelf settings (in part modified from Jackson, 1984; Rosenthal, 1988b; see Figure 8 for legend).**

Fernie/Nikanassin Assemblage (Assemblage 1): Upper Jurassic/Basal Cretaceous

Assemblage 1 comprises distal and proximal facies belt clastics that represent the earliest orogenic detritus and contains shallow marine, shoreline, fluvial channel sandstones in stratigraphic traps and in erosional remnant traps.

Mannville Assemblage (Assemblage 2): Lower Cretaceous Fluviodeltaic-Littoral Clastics

Assemblage 2 represents a supra-unconformity package, and complex geologic relationships have created numerous stratigraphic traps in fluvial, estuarine, shoreline, deltaic, and shallow marine sandstones deposited in medial facies belts. This assemblage dominates gas reserves and unconventional heavy oil reserves of the Western Canada foreland basin.

Colorado Assemblage (Assemblage 3): Mid-Cretaceous Marine Shales

Assemblage 3 contains thin, but prolific, marine-reworked sandstone and conglomerate (in part fluvially derived) reservoirs at the tops of progradational sequences. Plays occur as depositional pinch-out stratigraphic traps within the thick, distal facies belt marine shale assemblage. This assemblage dominates conventional oil reserves of the Western Canada foreland basin.

Saunders/Edmonton Assemblage (Assemblage 4): Upper Cretaceous/Lower Tertiary Marine to Continental Clastic Wedges

Assemblage 4 prograded into the marine foreland basin, eventually filling the basin with proximal facies belt continental deposits. Stratigraphic traps occur within deltaic and shoreline sandstone reservoirs deposited in medial

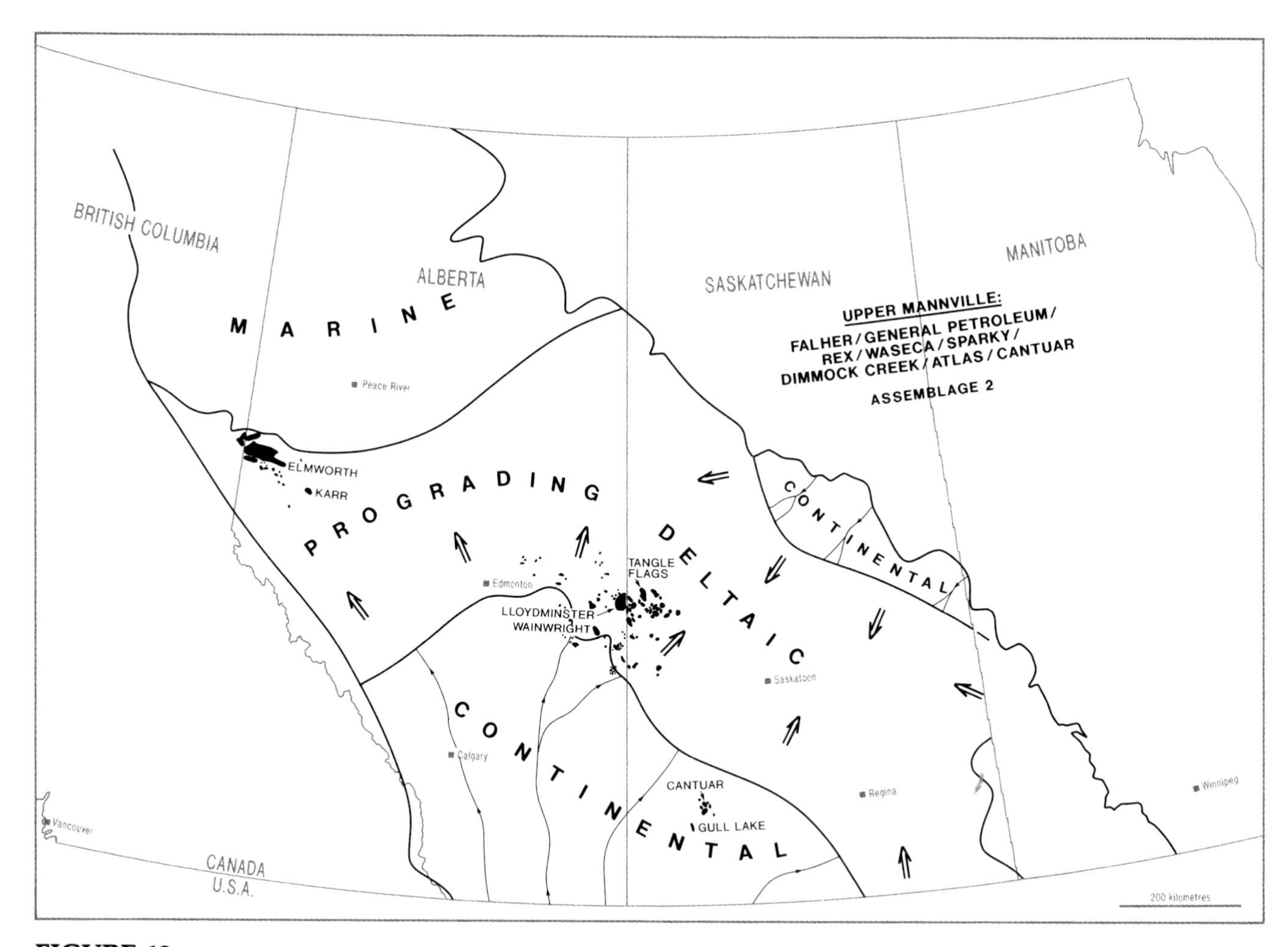

FIGURE 13. Mannville Assemblage 2, upper units: Upper Mannville paleogeography and oil/gas fields. Main traps are in shoreface and deltaic conglomerates and sandstones at the leading edges of progradational fluviodeltaic complexes (see Figure 8 for legend).

facies belts at the bases of progradational sequences in basal parts of the assemblage.

Cypress Hills/Hand Hills Assemblage (Assemblage 5): Upper Tertiary Continental Coarse-Grained Clastics

Assemblage 5 is essentially barren of petroleum because it occurs near the present-day surface and it also lacks significant source and seal rocks. In addition, the areal extent of this assemblage is limited by erosion.

Other Plays

Another important set of petroleum plays includes the oil and gas hosted in preorogenic (pre-foreland basin), cratonic platform deposits (Devonian to Lower Jurassic) that were filled by petroleum generated during foreland basin burial of source rocks and trapped during eastward updip migration. These plays include Devonian reef chains, Mississippian shelf carbonates, and Permian–Triassic shoreline sandstones filled during foredeep tilting. Other preorogenic traps owe both their structural geometries and their petroleum filling to basin tilting and burial, such as folded and faulted traps in Foothills anticlines and reservoirs along erosional edges below the sub-Mesozoic unconformity.

ACKNOWLEDGMENTS

Canadian (and Alberta) Society of Petroleum Geologists publications were particularly helpful in this compilation, especially McCrossan and Glaister (1964), McCrossan (1973), Stott and Glass (1984), James and Leckie (1988), and numerous CSPG Core Conference volumes. American Association of Petroleum Geologists Memoir 38 was also very useful (Masters, 1984a). The following people assisted the authors: A. P. Hamblin, K. G. Osadetz, J. A. Podruski, D. J. Cant, P. White, and I. Banerjee (discussions and/or reviews of early manuscripts); G. E. Fullmer and R. I. Campbell (data compilation); P. Gubitz (drafting); and

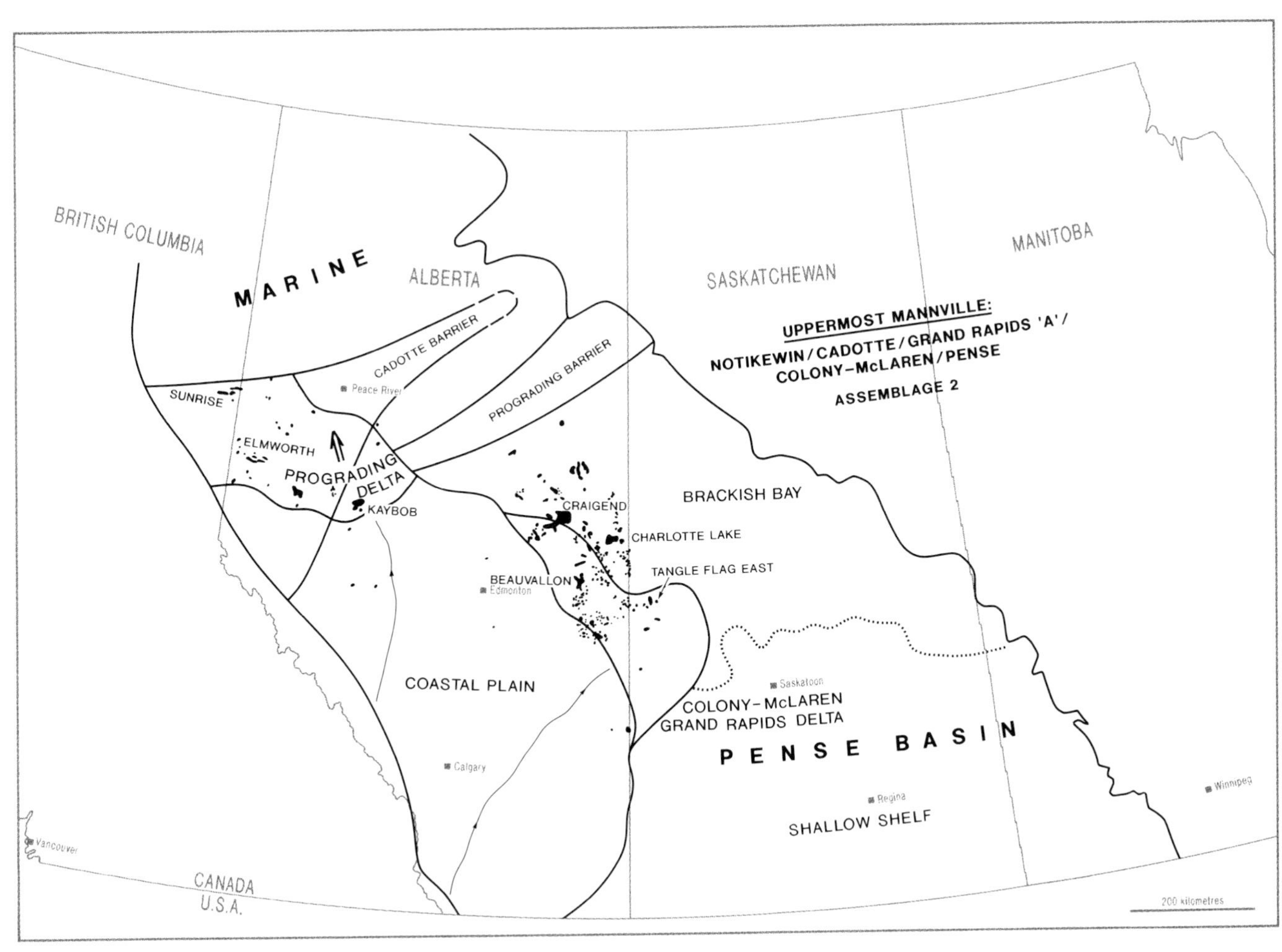

FIGURE 14. Mannville Assemblage 2, uppermost units: "Uppermost" Mannville paleogeography and oil/gas fields. Main traps are in barrier, shoreface, and deltaic channel sandstones (see Figure 8 for legend).

volume editors R. W. Macqueen and D. A. Leckie (guidance, critical reviewing, and arm-twisting). Critical reviews by G. E. Reinson and J. W. Porter assisted greatly in improving the manuscript. Without J. S. Bell's volunteered assistance, this paper would not have been completed.

REFERENCES CITED

Banerjee, I., 1989, Tidal sand sheet origin of the transgressive Basal Colorado Sandstone (Albian): a subsurface study of the Cessford field, southern Alberta: Bulletin of Canadian Petroleum Geology, v. 37, p. 1–17.

Banerjee, I., and E. H. Davies, 1988, An integrated lithostratigraphic and palynostratigraphic study of the Ostracode zone and adjacent strata in the Edmonton embayment, central Alberta, *in* D. P. James and D. A. Leckie, eds., Sequences, stratigraphy, sedimentology: surface and subsurface: Canadian Society of Petroleum Geologists Memoir 15, p. 261–274.

Barss, D. L., and F. A. Montandon, 1981, Sukunka-Bullmoose gas fields: model for a developing trend in the southern Foothills of northeast British Columbia: Bulletin of Canadian Petroleum Geology, v. 29, p. 293–333.

Beach, F. K., 1955, Cardium, a turbidity current deposit: Journal of the Alberta Society of Petroleum Geologists, v. 3, p. 123–125.

Bhattacharya, J., 1988, Autocyclic and allocyclic sequences in river- and wave-dominated deltaic sediments of the Upper Cretaceous, Dunvegan Formation, Alberta: core examples, *in* D. P. James and D. A. Leckie, eds., Sequences, stratigraphy, sedimentology: surface and subsurface: Canadian Society of Petroleum Geologists Memoir 15, p. 25–32.

Bhattacharya, J., 1989, Estuarine channel fills in the Upper Cretaceous Dunvegan Formation: core examples, *in* G. E. Reinson, ed., Modern and ancient examples of clastic tidal deposits—a core and peel workshop, Canadian Society of Petroleum Geologists and Society of Economic Paleontologists and Mineralogists Second International Research Symposium on Clastic Tidal Deposits: Calgary, Alberta, Canadian Society of Petroleum Geologists, p. 37–49.

Cant, D. J., 1983, Spirit River Formation—a stratigraphic-diagenetic gas trap in the Deep Basin of Alberta: AAPG Bulletin, v. 67, p. 577–587.

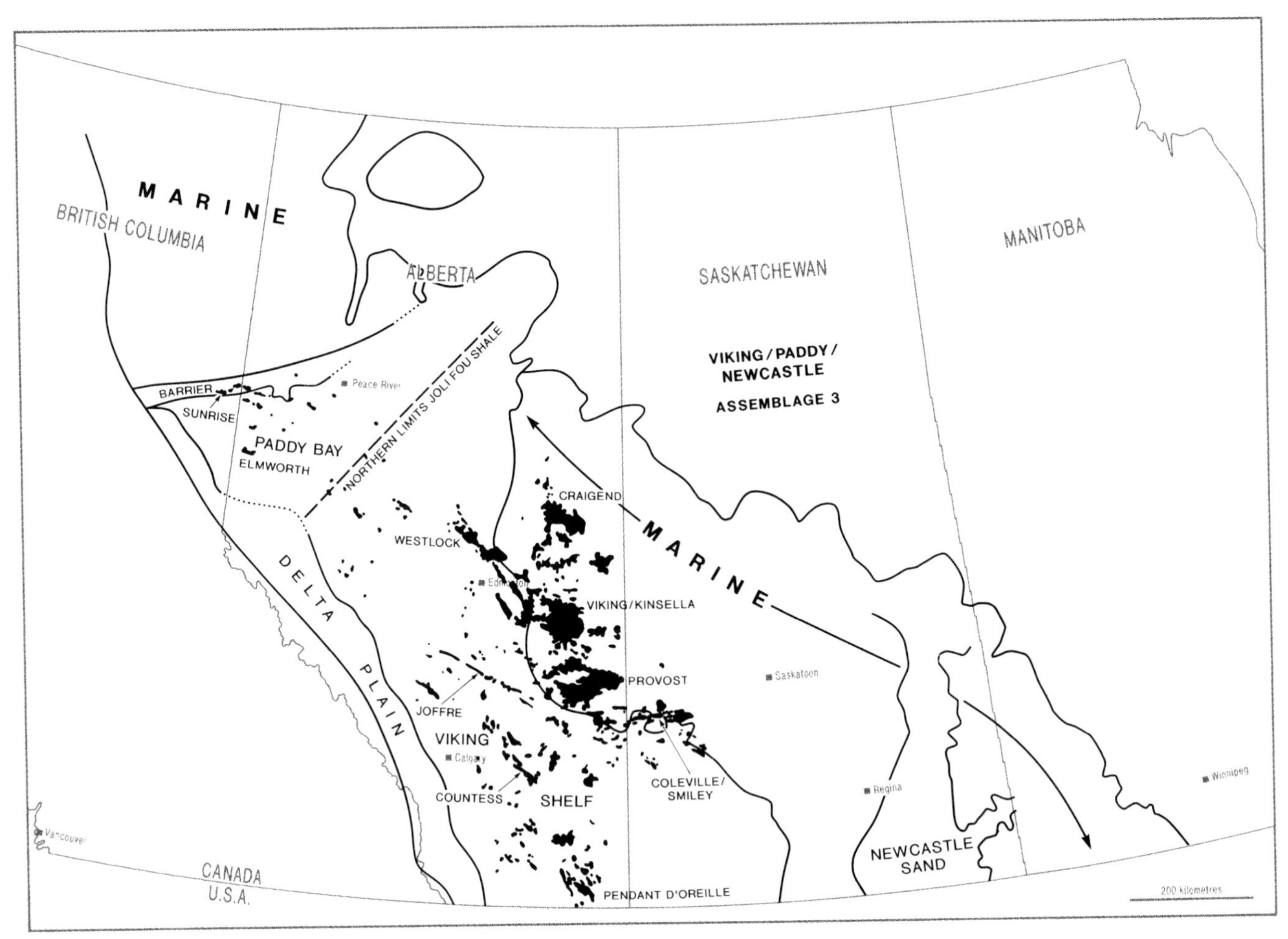

FIGURE 15. **Colorado Assemblage 3, lower units: "Lower" Colorado-Viking paleogeography and oil/gas fields. Main traps are in shelf conglomerates and sandstones and in fluvial, estuarine and valley-fill channel conglomerates and sandstones (see Figure 8 for legend).**

Chiang, K. K., 1984, The giant Hoadley gas field, south-central Alberta, *in* J. A. Masters, ed., Elmworth—case study of a Deep Basin gas field: AAPG Memoir 38, p. 297-313.

Christopher, J. E., 1974, The Upper Jurassic Vanguard and Lower Cretaceous Mannville groups: Saskatchewan Geological Survey Department of Mineral Resources Report No. 151, 349 p.

Christopher, J. E., 1984, The Lower Cretaceous Mannville Group, northern Williston basin region, Canada, *in* D. F. Stott and D. J. Glass, eds., The Mesozoic of middle North America: Canadian Society of Petroleum Geologists Memoir 9, p. 109-126.

Clark, J. E., and G. E. Reinson, 1990, Continuity and performance of an estuarine reservoir, Crystal field, Alberta, Canada, *in* J. H. Barwis, J. G. McPherson, and J. R. Studlick, eds., Sandstone petroleum reservoirs: New York, Springer Verlag, p. 343-361.

Conybeare, C. E. B., 1976, Geomorphology of oil and gas fields in sandstone bodies, *in* Developments in petroleum science 4: Amsterdam, Netherlands, Elsevier Scientific Publishing Company, 341 p.

Creaney, S., and J. Allan, 1992, Petroleum systems in the foreland basin of Western Canada, this volume.

Dawson, G., 1982, Depositional environments and diagenesis of the Hoadley barrier-bar complex: Glauconitic Sandstone (Upper Mannville Formation), central Alberta, *in* J. C. Hopkins, ed., Depositional environments and reservoir facies in some western Canadian oil and gas fields, University of Calgary Core Conference: Calgary, Alberta, University of Calgary, p. 61-73.

Dearborn, D. W., D. A. Leckie, and D. Potocki, 1985, Sedimentary and reservoir geology of Upper Cretaceous Doe Creek Sandstone, Alberta, Canada (abstract): AAPG Bulletin, v. 69, p. 248-249.

de Reuver, F., 1987, The Glauconitic Sandstone in the Hoadley area, internal structure of a progradational shoreline: Unpublished report presented at 1987 Canadian Society of Petroleum Geologists Core Conference, Calgary, Alberta, 14 p.

Deroo, G., T. G. Powell, B. Tissot, R. G. McCrossan, and P. Hacquebard, 1977, The origin and migration of petroleum in the western Canadian sedimentary basin—a geochemical and thermal maturation study: Geological Survey of Canada Bulletin 262, 136 p.

Deutsch, K. B., and F. F. Krause, 1990, A marine to terrestrial succession in the Cardium Formation, Kakwa region, west-central Alberta: implications for relative

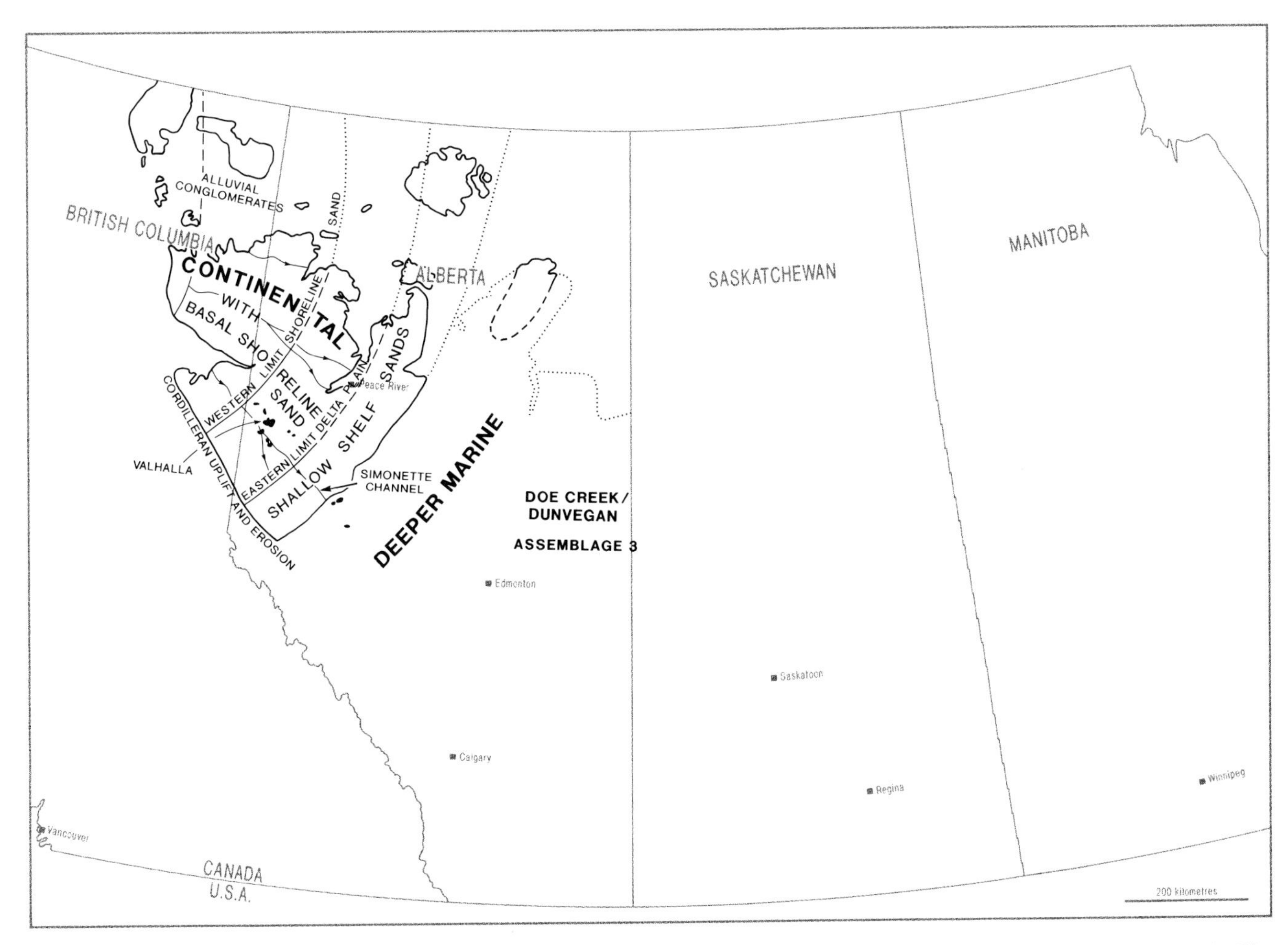

FIGURE 16. Colorado Assemblage 3, middle units: "Middle" Colorado-Dunvegan/Doe Creek paleogeography and oil/gas fields. Main traps are in distributary channels and shelf sandbodies (see Figure 8 for legend).

sea level movements (abstract), *in* Basin perspectives, Canadian Society of Petroleum Geologists 1990 convention program and abstracts: Calgary, Alberta, Canadian Society of Petroleum Geologists, p. 44.

Energy Resources Conservation Board, 1988, ERCB ST89-18: Alberta's reserves of crude oil, oil sands, gas, natural gas liquids, and sulphur: Energy Resources Conservation Board Reserve Report Series ERCB ST89-18, 28th Ed., irregularly paginated (data also taken from related digital tapes).

Energy Resources Division, 1988, Hydrocarbon and by-product reserves in British Columbia 1988: Victoria, British Columbia, Province of British Columbia Ministry of Energy, Mines and Petroleum Resources, 17 p. plus appendices (data also taken from related digital tapes).

Ethier, V., 1982, Channel and shoreline sequences in the Paddy and Cadotte members of the Peace River Formation, Deep Basin of Alberta, *in* J. C. Hopkins, ed., Depositional environments and reservoir facies in some western Canadian oil and gas fields, University of Calgary Core Conference: Calgary, Alberta, University of Calgary, p. 29-41.

Farshori, M. Z., 1983, Glauconitic Sandstone, Countess field "H" pool, southern Alberta, *in* J. R. McLean and G. E. Reinson, eds., Sedimentology of selected Mesozoic clastic sequences, Canadian Society of Petroleum Geologists Corexpo '83: Calgary, Alberta, Canadian Society of Petroleum Geologists, p. 27-41.

Farshori, Z., and J. C. Hopkins, 1987, Fluvial and lacustrine bodies of the Middle Mannville; exploration significance of the Ostracode Beds and Sunburst (Basal Quartz) Beds, Alberta and Montana: Canadian Society of Petroleum Geologists Reservoir, v. 14, n. 6, p. 1-3.

Gallup, W. B., 1954, Geology of Turner Valley oil and gas field, Alberta, Canada, *in* L. M. Clark, ed., Western Canada sedimentary basin symposium: AAPG, Ralph Leslie Rutherford Memorial Volume, p. 397-414.

Gies, R. M., 1984, Case history for a major Alberta Deep Basin gas trap: the Cadomin Formation, *in* J. A. Masters, ed., Elmworth—case study of a Deep Basin gas field: AAPG Memoir 38, p. 115-140.

Glass, D. J., ed., 1990, Lexicon of Canadian stratigraphy, volume 4, Western Canada, including eastern British Columbia, Alberta, Saskatchewan and southern Manitoba: Calgary, Alberta, Canadian Society of Petroleum Geologists, 772 p.

Hankel, R. C., G. R. Davies, and H. R. Krouse, 1989, Eastern Medicine Hat gas field: a shallow, Upper Cretaceous, bacteriogenic gas reservoir of southeastern Alberta:

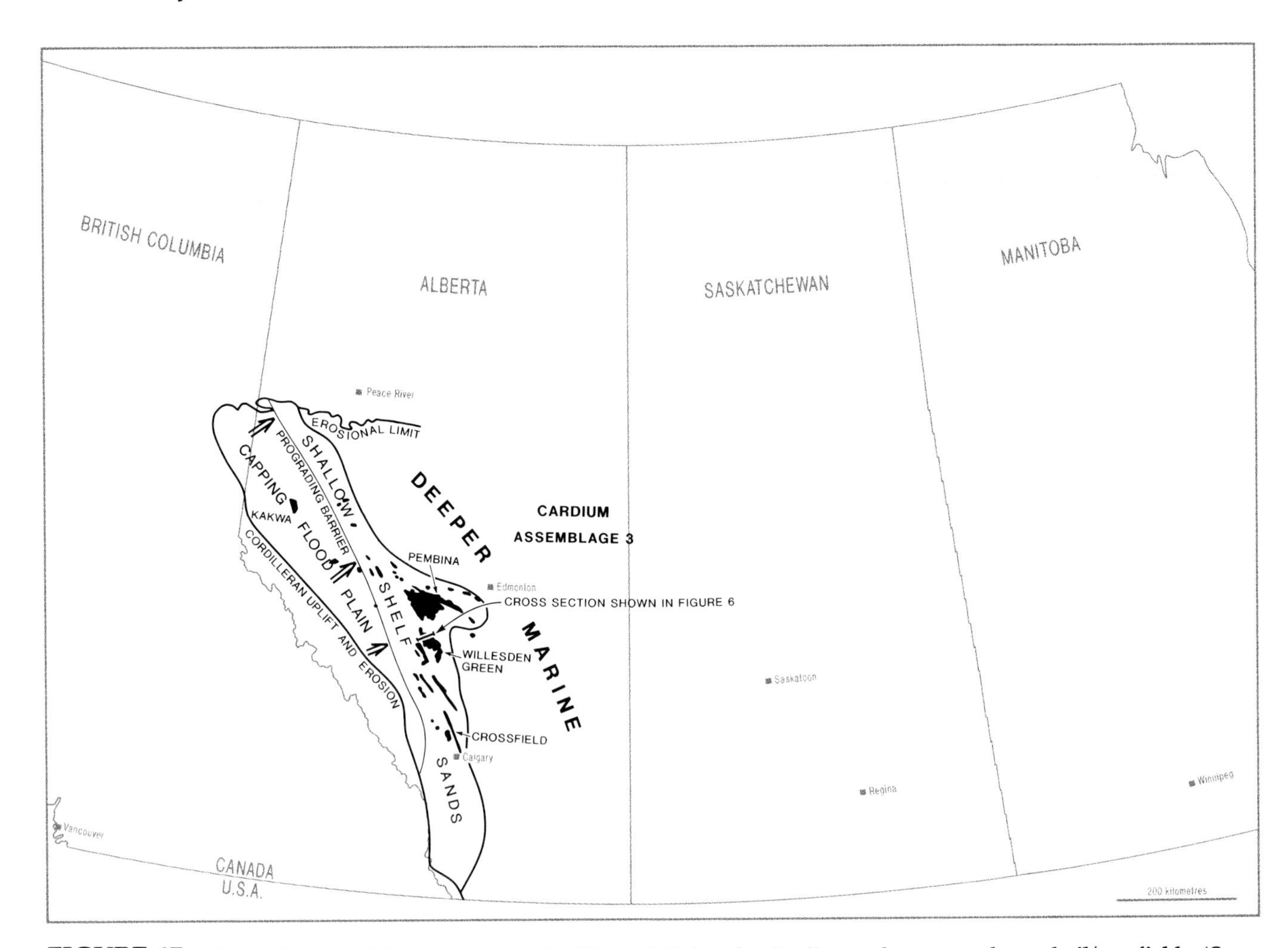

FIGURE 17. Colorado Assemblage 3, upper units: "Upper" Colorado-Cardium paleogeography and oil/gas fields. (See Figure 8 for legend). Traps are in shelf sandstones. Also shoreface sandstones at Kakwa. Bar represents Figure 6 Willesden Green cross-section location.

Bulletin of Canadian Petroleum Geology, v. 37, p. 98-112.

Hayes, B. J. R., 1983, Stratigraphy and petroleum potential of the Swift Formation (Upper Jurassic), southern Alberta and north-central Montana: Bulletin of Canadian Petroleum Geology, v. 31, p. 37-52.

Hayes, B. J. R., 1988, Incision of a Cadotte Member paleovalley-system at Noel, British Columbia—evidence of a late Albian sea-level fall, *in* D. P. James and D. A. Leckie, eds., Sequences, stratigraphy, sedimentology: surface and subsurface: Canadian Society of Petroleum Geologists Memoir 15, p. 97-105.

Hein, F. J., M. E. Dean, A. M. Delure, S. K. Grant, G. A. Robb, and F. J. Longstaffe, 1986, The Viking Formation in the Caroline, Garrington and Harmattan East fields, western south-central Alberta: sedimentology and paleogeography: Bulletin of Canadian Petroleum Geology, v. 34, p. 91-110.

Hopkins, J. C., 1981, Sedimentology of quartzose sandstones of Lower Mannville and associated units, Medicine River area, central Alberta: Bulletin of Canadian Petroleum Geology, v. 29, p. 12-27.

Hopkins, J. C., S. W. Hermanson, and D. C. Lawton, 1982, Morphology of channels and channel-sand bodies in the Glauconitic Sandstone Member (Upper Mannville), Little Bow area, Alberta: Bulletin of Canadian Petroleum Geology, v. 30, p. 274-285.

Horne, J. C., M. O. Hayes, and P. J. Reinhart, 1982, Niton field: an estuarine sandstone reservoir (abstract): AAPG Bulletin, v. 66, p. 583.

Hradsky, M., and M. Griffin, 1984, Sandstone body geometry, reservoir quality and hydrocarbon trapping mechanisms in Lower Cretaceous Mannville Group, Taber/Turin area, southern Alberta, *in* D. F. Stott and D. J. Glass, eds., The Mesozoic of middle North America: Canadian Society of Petroleum Geologists Memoir 9, p. 401-411.

Humphreys, J. T., 1960, Cessford Sand, *in* G. E. Hargreaves, A. D. Hunt, R. De Wit, and L. E. Workman, eds., Lexicon of geologic names in the Western Canada sedimentary basin and Arctic Archipelago: Calgary, Alberta, Alberta Society of Petroleum Geologists, p. 70-71.

Hunt, C. W., 1950, Preliminary report on Whitemud oil field, Alberta, Canada: AAPG Bulletin, v. 34, p. 1795-1801.

Iwuagwu, C. J., and J. F. Lerbekmo, 1984, Application of outcrop information to subsurface exploration for sandstone reservoirs: basal Belly River Formation

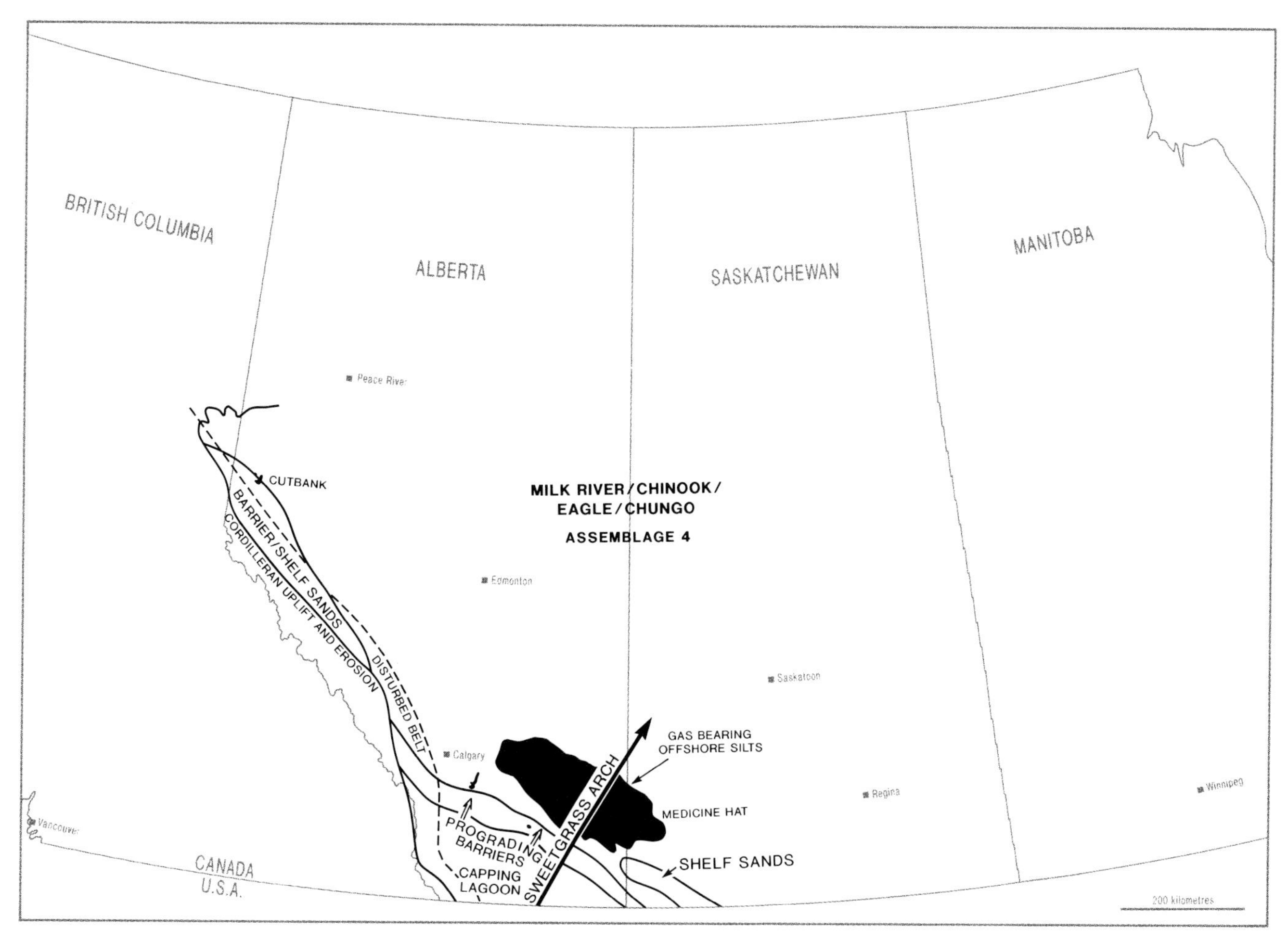

FIGURE 18. Saunders/Edmonton Assemblage 4, lowermost units: "Basal" Saunders/Edmonton–Milk River paleogeography and oil/gas fields. Main traps are in shelf sandstones (see Figure 8 for legend).

(Upper Cretaceous), Alberta Foothills, *in* D. F. Stott and D. J. Glass, eds., The Mesozoic of middle North America: Canadian Society of Petroleum Geologists Memoir 9, p. 387–400.

Jackson, P. C., 1984, Paleogeography of the Lower Cretaceous Mannville Group of Western Canada, *in* J. A. Masters, ed., Elmworth—case study of a Deep Basin gas field: AAPG Memoir 38, p. 49–77.

James, D. P., and D. A. Leckie, eds., 1988, Sequences, stratigraphy, sedimentology: surface and subsurface: Canadian Society of Petroleum Geologists Memoir 15, 586 p.

Keith, D. A. W., 1985, Sedimentology of the Cardium Formation (Upper Cretaceous) Willesden Green field, Alberta: unpublished M.Sc. Thesis, McMaster University, Hamilton, Ontario, 241 p.

Krause, F. F., and D. A. Nelson, 1984, Storm event sedimentation: lithofacies association in the Cardium Formation, Pembina area, west-central Alberta, Canada, *in* D. F. Stott and D. J. Glass, eds., The Mesozoic of middle North America: Canadian Society of Petroleum Geologists Memoir 9, p. 485–511.

Krause, F. F., H. N. Collins, D. A. Nelson, S. D. Machemer, and P. R. French, 1987, Multiscale anatomy of a reservoir: geological characterization of Pembina-Cardium Pool, west-central Alberta, Canada: AAPG Bulletin, v. 71, n. 10, p. 1233–1260.

Kryczka, A. A. W., 1959, The Nikanassin Formation of the type area, near Cadomin, Alberta: unpublished M.Sc. Thesis, University of Alberta, Edmonton, Alberta, 135 p.

Leckie, D. A., 1985, The Lower Cretaceous Notikewin Member (Fort St. John Group), northeastern British Columbia: a progradational barrier island system: Bulletin of Canadian Petroleum Geology, v. 33, p. 39–51.

Leckie, D. A., 1986a, Tidally influenced, transgressive shelf sediments in the Viking Formation, Caroline, Alberta: Bulletin of Canadian Petroleum Geology, v. 34, p. 111–125.

Leckie, D. A., 1986b, Rates, controls, and sand-body geometries of transgressive-regressive cycles: Cretaceous Moosebar and Gates formations, British Columbia: AAPG Bulletin, v. 70, p. 516–535.

Leckie, D. A., 1988, Sedimentology and sequences of the Paddy and Cadotte members along the Peace River, *in* Canadian Society of Petroleum Geologists field guide to sequences, stratigraphy, sedimentology: surface and subsurface technical meeting: Calgary, Alberta, Canadian Society of Petroleum Geologists, 78 p.

Leckie, D. A., and R. J. Cheel, 1986, Tidal channel facies of

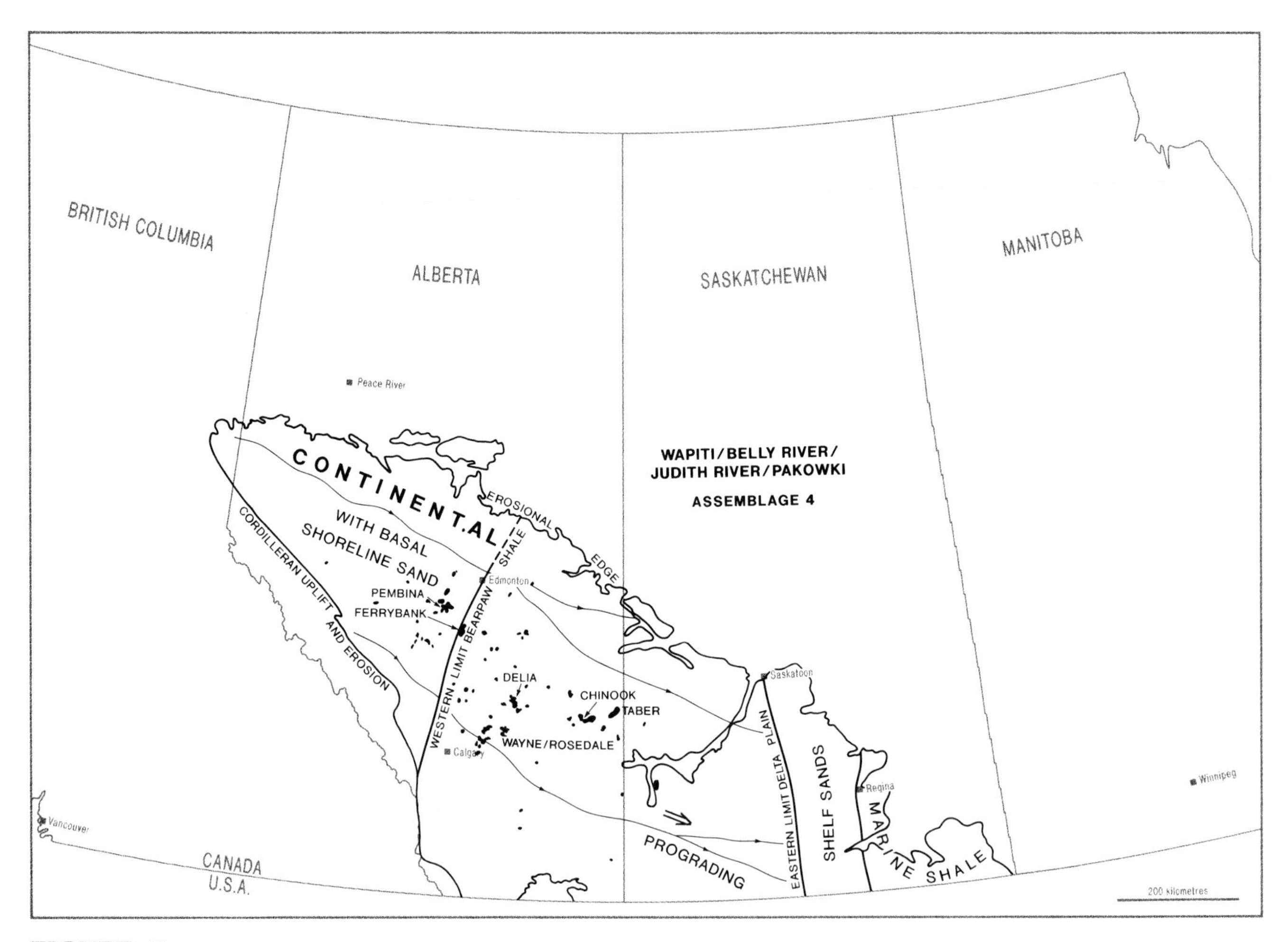

FIGURE 19. **Saunders/Edmonton Assemblage 4, middle units: "Middle" Saunders/Edmonton-Belly River paleogeography and oil/gas fields. Main traps are in basal shoreline and overlying fluvial sandstones (see Figure 8 for legend).**

the Virgelle Member (Cretaceous Milk River Formation), southern Alberta, *in* Current research, Part B: Geological Survey of Canada Paper 86-1B, p. 637-645.

Leckie, D. A., and C. Singh, 1991, Estuarine deposits of the Albian Paddy Member (Peace River Formation) and Lowermost Shaftesbury Formation, Alberta, Canada: Journal of Sedimentary Petrology, v. 61, p. 825-849.

Leckie, D. A., and D. G. Smith, 1992, Regional setting, evolution, and depositional cycles of the Western Canada foreland basin, this volume.

Leckie, D. A., and R. G. Walker, 1982, Storm- and tide-dominated shorelines in Cretaceous Moosebar-Lower Gates interval—outcrop equivalents of Deep Basin gas trap in Western Canada: AAPG Bulletin, v. 66, p. 138-157.

Leckie, D. A., M. R. Staniland, and B. J. Hayes, 1990, Regional maps of the Albian Peace River and Lower Shaftesbury formations on the Peace River arch, northwestern Alberta and northeastern British Columbia, *in* S. C. O'Connell and J. S. Bell, eds., Geology of the Peace River Arch: Bulletin of Canadian Petroleum Geology, v. 38A, p. 176-189.

Masters, J. A., 1979, Deep Basin gas trap, Western Canada: AAPG Bulletin, v. 63, p. 152-181.

Masters, J. A., ed., 1984a, Elmworth—case study of a Deep Basin gas field: AAPG Memoir 38, 316 p.

Masters, J. A., 1984b, Lower Cretaceous oil and gas in Western Canada, *in* J. A. Masters, ed., Elmworth—case study of a Deep Basin gas field: AAPG Memoir 38, p. 1-33.

McCrory, V. L. C., and R. G. Walker, 1986, A storm and tidally-influenced prograding shoreline—Upper Cretaceous Milk River Formation of southern Alberta, Canada: Sedimentology, v. 33, p. 47-60.

McCrossan, R. G., ed., 1973, The future petroleum provinces of Canada—their geology and potential: Canadian Society of Petroleum Geologists Memoir 1, 720 p.

McCrossan, R. G., and R. P. Glaister, eds., 1964, Geological history of Western Canada: Calgary, Alberta, Alberta Society of Petroleum Geologists, 232 p.

McLean, J. R., 1977, The Cadomin Formation: stratigraphy, sedimentology, and tectonic implications: Bulletin of Canadian Petroleum Geology, v. 25, p. 792-827.

McLean, J. R., 1979, Regional considerations of the Elmworth field and the Deep Basin: Bulletin of Canadian Petroleum Geology, v. 27, p. 53-62.

McLean, J. R., 1985, Sedimentological modelling for enhanced oil recovery prospects: an example for the

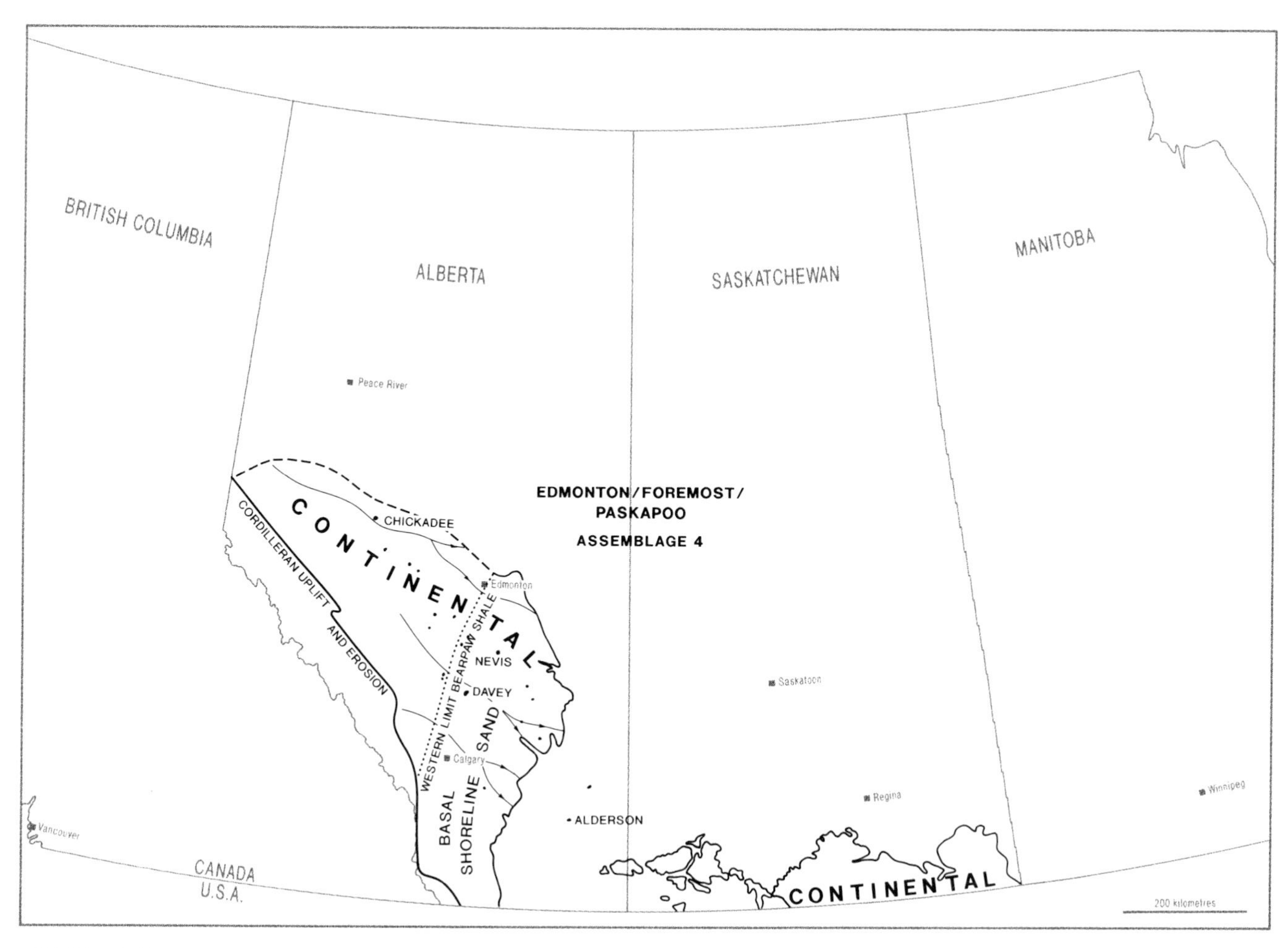

FIGURE 20. Saunders/Edmonton Assemblage 4, upper units: "Upper" Saunders/Edmonton-Paskapoo paleogeography and oil/gas fields. Rare traps are in Paskapoo fluvial channel sandstones and conglomerates (see Figure 8 for legend).

Manyberries Q pool, southeastern Alberta (abstract), *in* Energy—challenges and opportunities, The Petroleum Society of C.I.M. and Canadian Society of Petroleum Geologists Annual Meeting, Edmonton, Alberta, general programme: Canadian Society of Petroleum Geologists, p. 37.

McMaster, G. E., 1981, Gas reservoirs, Deep Basin, Western Canada: Journal of Canadian Petroleum Technology, v. 20, p. 62-66.

Meijer-Drees, N. C., and D. W. Myhr, 1981, The Upper Cretaceous Milk River and Lea Park formations in southeastern Alberta: Bulletin of Canadian Petroleum Geology, v. 29, p. 42-71.

Nielsen, A. R., 1957, Cardium stratigraphy of the Pembina field: Journal of the Alberta Society of Petroleum Geologists, v. 5, p. 64-72.

Nielsen, A. R., and J. W. Porter, 1984, Pembina oil field—in retrospect, *in* D. F. Stott and D. J. Glass, eds., The Mesozoic of middle North America: Canadian Society of Petroleum Geologists Memoir 9, p. 1-13.

Norris, D. K., and A. W. Bally, 1972, Coal, oil, gas and industrial mineral deposits of the Interior Plains, Foothills and Rocky Mountains of Alberta and British Columbia, Excursions A25-C25, *in* D. J. Glass, ed., Guidebook 24th International Geological Congress, Montreal, Quebec: Ottawa, Ontario, Twenty-Fourth International Geological Congress, 108 p.

O'Connell, S. C., 1988, The distribution of Bluesky facies in the region overlying the Peace River arch, northwestern Alberta, *in* D. P. James and D. A. Leckie, eds., Sequences, stratigraphy, sedimentology: surface and subsurface: Canadian Society of Petroleum Geologists Memoir 15, p. 387-399.

Parsons, W. H., 1973, Alberta, *in* R. G. McCrossan, ed., Future petroleum provinces of Canada—their geology and potential: Canadian Society of Petroleum Geologists Memoir 1, p. 73-120.

Plint, A. G., and R. G. Walker, 1987, Cardium Formation 8, facies and environments of the Cardium shoreline and coastal plain in the Kakwa field and adjacent areas, northwestern Alberta: Bulletin of Canadian Petroleum Geology, v. 35, p. 48-64.

Podruski, J. A., J. E. Barclay, A. P. Hamblin, P. J. Lee, K. G. Osadetz, R. M. Procter, and G. C. Taylor, 1988, Conventional oil resources of Western Canada (light and medium): Geological Survey of Canada Paper 87-26, 149 p.

Porter, J. W., 1992, Conventional hydrocarbon reserves of the Western Canada foreland basin, this volume.

Poulton, T. P., 1984, The Jurassic of the Canadian Western Interior, from 49 degrees N latitude to Beaufort Sea, *in*

TABLE 1. Fernie-Nikanassin Assemblage 1. Lower Units—Marine Play Developed During the Early Marine Filling of the Foredeep (see Figure 8).*

Swift "Ribbon Sand" Play

Loc.	Southern and central Alta. plains (Figure 8).
Prox.	Eastern medial to distal (Figure 1). East-derived cratonic detritus mixed with first orogenic detritus (west-derived).
Depo.	Shallow marine bar and interbar.
Fms.	"Ribbon Sand" (upper Swift Fm.), Oxfordian.
Fluids	Low-gravity oils.
Reservoirs	Lenticular-bedded sandstone and siltstone.
Seals	Intercalated mudstone.
Traps	1. **Stratigraphic:** erosional remnants at Swift Fm. updip edge isolated by Lower Cretaceous channels. 2. **Structural:** Laramide-aged, basement-rooted block faulting.
Fields	Butte, Aden.
Refs.	Christopher, 1974; Hayes, 1983; Poulton, 1984; McLean, 1985.

*Plays are described in approximate stratigraphic order from oldest to youngest. Major oil and gas fields mentioned are plotted in Figures 8–20; others are on the maps of Wallace-Dudley (1981a, b).

Heading abbreviations:

Loc. = location in foreland basin
Prox. = proximity to orogene (main source area)
Depo. = depositional setting
Fms. = formations and ages
Refs. = references.

Text abbreviations:

Alta. = Alberta
B.C. = British Columbia
Fm. = Formation
Mbr. = Member
Saskatchewan = Saskatchewan.

D. F. Stott and D. J. Glass, eds., The Mesozoic of middle North America: Canadian Society of Petroleum Geologists Memoir 9, p. 15–41.

Rahmani, R. A., 1984, Facies control of gas trapping, Lower Cretaceous Falher A cycle, Elmworth area, northwestern Alberta, *in* J. A. Masters, ed., Elmworth—case study of a Deep Basin gas field: AAPG Memoir 38, p. 141–152.

Rahmani, R. A., and D. G. Smith, 1988, The Cadotte Member of northwestern Alberta: a high-energy barred shoreline, *in* D. P. James and D. A. Leckie, eds., Sequences, stratigraphy, sedimentology: surface and subsurface: Canadian Society of Petroleum Geologists Memoir 15, p. 431–438.

Rall, R. D., 1980, Stratigraphy, sedimentology and paleotopography of the Lower Jurassic in the Gilby-Medicine River fields, Alberta: unpublished M.Sc. Thesis, University of Calgary, Calgary, Alberta, 142 p.

Reel, C. L., and C. V. Campbell, 1985, Chin Coulee field—a stratigraphic trap in Lower Cretaceous valley-fill sandstones (abstract): AAPG Bulletin, v. 69, p. 300.

Reichenbach, M. E., 1982, Reservoir lithofacies in the Medicine River delta and Hoadley Barrier complex, *in* J. C. Hopkins, ed., Depositional environments and reservoir facies in some western Canadian oil and gas fields, University of Calgary Core Conference: Calgary, Alberta, University of Calgary, p. 52–58.

Reinson, G. E., J. E. Clark, and A. E. Foscolos, 1988, Reservoir geology of the Crystal Viking field, Lower Cretaceous estuarine tidal channel-bay complex, south-central Alberta: AAPG Bulletin, v. 72, p. 1270–1294.

Rosenthal, L. R. P., 1988a, Wave-dominated shorelines and incised channel trends: Lower Cretaceous Glauconite Formation, west-central Alberta, *in* D. P. James and D. A. Leckie, eds., Sequences, stratigraphy, sedimentology: surface and subsurface: Canadian Society of Petroleum Geologists Memoir 15, p. 207–219.

Rosenthal, L. R. P., 1988b, Stratigraphy and depositional facies, Lower Cretaceous Blairmore-Luscar groups, central Alberta Foothills, *in* Canadian Society of Petroleum Geologists field guide to sequences, stratigraphy, sedimentology: surface and subsurface technical meeting: Calgary, Alberta, Canadian Society of Petroleum Geologists, 56 p.

Rosenthal, L. R. P., 1989, Stratigraphic framework and reservoir potential of Jurassic and Lower Cretaceous (Mannville) sandstones in western Alberta: implications for deep gas exploration (abstract): Canadian Society of Petroleum Geologists Reservoir, v. 16, n. 6, p. 1–3.

Ross, D. A., 1986, Paleogeography of the Doe Creek Member, northwest Alberta (abstract), *in* Canada's hydrocarbon reserves for the 21st century, Canadian Society of Petroleum Geologists 1986 Convention program and abstracts: Calgary, Alberta, Canadian Society of Petroleum Geologists, p. 78.

Rudkin, R. A., 1964, Chapter 11, Lower Cretaceous, *in* R. G.

TABLE 2. Fernie/Nikanassin Assemblage 1. Upper Units—Fluvial Channel Plays Developed During Continental Foredeep Filling (see Figures 8, 9).*

Lower Success (S-1 Unit) Fluvial Sandstone Play

Loc.	Southern and central plains (Figure 8).
Prox.	Medial to distal eastern (Figure 1).
Depo.	Fluviodeltaic, lacustrine, shallow marine. Fills topographic lows on underlying Masefield Shale.
Fms.	Lower Success S-1 Unit, Kimmeridgian.
Fluids	Oil and gas.
Reservoirs	Fluvial: channel and valley-fill sandstone.
Seals	Overlying Cantuar Fm. shale(s).
Traps	**Stratigraphic:** erosional remnants overlain by shale(s).
Fields	Beverly East, Cantuar East, Gull Lake.
Refs.	Christopher, 1974, 1984.

Upper Success (S-2 Unit) Fluvial Sandstone Play

Loc.	Southern and central plains (Figure 8).
Prox.	Medial to distal eastern (Figure 1). Williston basin also has east-derived cratonic detritus.
Depo.	Continental.
Fms.	Upper Success S-2 Unit. Late Jurassic to Neocomian, Early Cretaceous.
Fluids	Oil.
Reservoirs	Fluvial channel sandstone.
Seals	Cantuar Fm.(?) shale.
Traps	**Stratigraphic:** mesa and butte-like erosional remnants isolated by shale-filled, deeply incised valleys.
Fields	Cantuar, Beverly, Gull Lake, Success.
Refs.	Christopher, 1984; Podruski et al., 1988.

Nikanassin Sandstone Foothills Structural Play

Loc.	Northern fold and thrust belt (Figure 9).
Prox.	Proximal (Figure 1).
Depo.	Continental.
Fms.	Nikanassin Fm., Kimmeridgian-Neocomian.
Fluids	Dry gas.
Reservoirs	Fluvial channel sandstone.
Seals	Intercalated and overlying shale and tight sandstone.
Traps	**Structural:** rollover anticlines perched on thrust sheet leading edges.
Fields	Ojay, Grizzly, Sukunka.
Refs.	Kryczka, 1959; Masters, 1979.

Nikanassin Sandstone Deep Basin Play

Loc.	Northern plains.
Prox.	Proximal (Figures 1, 9).
Depo.	Continental, some shoreline-shallow marine.
Fms.	Nikanassin Fm., Kimmeridgian-Neocomian.
Fluids	Dry sweet gas.
Reservoirs	Sandstone: fluvial channel, shoreline, and shallow marine bar.

TABLE 2. (Continued)

Seals	Intercalated siltstone and shale.
Traps	**Deep Basin-type:** tight gas sands downdip of regional water aquifer.
Fields	Steep Creek, Sinclair, Bilbo, Albright.
Refs.	Kryczka, 1959; Masters, 1979; McMaster, 1981.

*Plays are described in approximate stratigraphic order from oldest to youngest. Major oil and gas fields mentioned are plotted in Figures 8-20; others are on the maps of Wallace-Dudley (1981a, b).

Heading abbreviations:

Loc. = location in foreland basin
Prox. = proximity to orogene (main source area)
Depo. = depositional setting
Fms. = formations and ages
Refs. = references.

Text abbreviations:

Alta. = Alberta
B.C. = British Columbia
Fm. = Formation
Mbr. = Member
Sask. = Saskatchewan.

McCrossan and R. P. Glaister, eds., Geological history of Western Canada: Calgary, Alberta, Alberta Society of Petroleum Geologists, p. 156-168.

Ryer, T. A., J. C. Horne, and M. O. Hayes, 1984, Aggradational valley-fill model for strata of Cretaceous Lower Mannville Formation, south-central Alberta (abstract), *in* D. F. Stott and D. J. Glass, eds., The Mesozoic of middle North America: Canadian Society of Petroleum Geologists Memoir 9, p. 562.

Saskatchewan Energy and Mines, 1987, Reservoir annual 1987, miscellaneous report 88-1: Regina, Saskatchewan, Saskatchewan Energy and Mines Petroleum and Natural Gas Division, irregularly paginated.

Schultheis, N. H., and E. W. Mountjoy, 1978, Cadomin conglomerate of western Alberta—a result of Early Cretaceous uplift of the Main Ranges: Bulletin of Canadian Petroleum Geology, v. 26, p. 297-342.

Shouldice, J. R., 1979, Nature and potential of Belly River gas sand traps and reservoirs in Western Canada: Bulletin of Canadian Petroleum Geology, v. 27, p. 229-241.

Smith, D. G., C. E. Zorn, and R. M. Sneider, 1984, The paleogeography of the Lower Cretaceous of western Alberta and northeastern British Columbia in and adjacent to the Deep Basin of the Elmworth area, *in* J. A. Masters, ed., Elmworth—case study of a Deep Basin gas field: AAPG Memoir 38, p. 79-114.

Stelck, C. R., and D. A. Leckie, 1990, Biostratigraphy of the Albian Paddy Member (Lower Cretaceous Peace River Formation), Goodfare, Alberta: Canadian Journal of Earth Sciences, v. 27, p. 1159-1169.

Storey, S. R., 1982, Optimum reservoir facies in an immature, shallow-lobate delta system: basal Belly River Formation, Keystone-Pembina area, *in* J. C. Hopkins, ed., Depositional environments and reservoir facies in some western Canadian oil and gas fields, University of Calgary Core Conference: Calgary, Alberta, University of Calgary, p. 3-13.

Stott, D. F., 1972, The Cretaceous Gething delta, northeast British Columbia, *in* Proceedings of the first geological conference on Western Canada coal: Research Council of Alberta, Information Serial No. 60, p. 151-163.

Stott, D. F., 1973, Lower Cretaceous Bullhead Group between Bullmoose Mountain and Tetsa River, Rocky Mountain Foothills, northeastern British Columbia: Geological Survey of Canada Bulletin 219, 228 p.

Stott, D. F., 1975, The Cretaceous System in northeastern British Columbia, *in* W. G. E. Caldwell, ed., The Cretaceous System in the Western Interior of North America: Geological Association of Canada Special Paper No. 13, p. 441-467.

Stott, D. F., 1982, Lower Cretaceous Fort St. John Group and Upper Cretaceous Dunvegan Formation of the Foothills and Plains of Alberta, British Columbia, District of Mackenzie and Yukon Territory: Geological Survey of Canada Bulletin 328, 124 p.

Stott, D. F., and D. J. Glass, eds., 1984, The Mesozoic of middle North America: Canadian Society of Petroleum Geologists Memoir 9, 573 p.

Stott, D. F., W. G. E. Caldwell, D. J. Cant, J. E. Christopher, J. Dixon, E. H. Koster, D. McNeil, and F. Simpson, in press, Cretaceous, *in* D. F. Stott and J. D. Aitken, eds., Sedimentary cover of the North American craton, Canada: The Geology of North America: Tulsa, Oklahoma, Geological Society of America.

Tilley, B. J., 1983, An exceptionally thick barrier-island deposit: Glauconitic Sandstone, Suffield area, southeastern Alberta, *in* J. R. McLean and G. E. Reinson, eds., Sedimentology of selected Mesozoic clastic sequences, Corexpo '83: Calgary, Alberta, Canadian Society of Petroleum Geologists, p. 119-131.

Tippett, C. R., 1987, The Foothills belt and its structurally-controlled gas and oil fields, *in* L. E. Jackson, Jr. and M. C. Wilson, eds., Geology of the Calgary area: Calgary, Alberta, Canadian Society of Petroleum Geologists, p. 65-79.

Van Wagoner, J. C., H. W. Posamentier, R. M. Mitchum, Jr., P. R. Vail, J. F. Sarg, T. S. Loutit, and J. Hardenbol, 1988, An overview of the fundamentals of sequence stratigraphy and key definitions, *in* C. K. Wilgus, B. S. Hastings, C. A. Ross, H. W. Posamentier, J. C. Van Wagoner, and C. G. St. C. Kendall, eds., Sea-level changes: an integrated approach: SEPM Special Publication 42, p. 39–45.

Varley, C. J., 1984a, Sedimentology and hydrocarbon distribution of the Lower Cretaceous Cadomin Formation, northwest Alberta, *in* E. H. Koster and R. J. Steel, eds., Sedimentology of gravels and conglomerates: Canadian Society of Petroleum Geologists Memoir 10, p. 175–187.

Varley, C. J., 1984b, The Cadomin Formation: a model for the Deep Basin type gas trapping mechanism, *in* D. F. Stott and D. J. Glass, eds., The Mesozoic of middle North America: Canadian Society of Petroleum Geologists Memoir 9, p. 471–484.

Walker, R. G., 1983, Cardium Formation 1, "Cardium, a turbidity current deposit" (Beach, 1955): a brief history of ideas: Bulletin of Canadian Petroleum Geology, v. 31, p. 205–212.

Walker, R. G., 1985, Cardium Formation at Ricinus field, Alberta: a channel cut and filled by turbidity currents in Cretaceous Western Interior seaway: AAPG Bulletin, v. 69, p. 1963–1981.

Walker, R. G., and C. H. Eyles, 1988, Geometry and facies of stacked shallow-marine sandier upward sequences dissected by erosion surface, Cardium Formation, Willesden Green, Alberta: AAPG Bulletin, v. 72, p. 1469–1494.

Wallace-Dudley, K. E., 1981a, Gas pools of Western Canada: Geological Survey of Canada Map 1558A.

Wallace-Dudley, K. E., 1981b, Oil pools of Western Canada: Geological Survey of Canada Map 1559A.

Wallace-Dudley, K. E., and D. A. Leckie, 1988, Preliminary observations on the sedimentology of the Doe Creek Member, Kaskapau Formation, in the Valhalla field, northwestern Alberta, *in* D. P. James and D. A. Leckie, eds., Sequences, stratigraphy, sedimentology: surface and subsurface: Canadian Society of Petroleum Geologists Memoir 15, p. 485–496.

Wallace-Dudley, K. E., and D. A. Leckie, in press, Sedimentology and source-rock potential of the Lower Kaskapau Formation (Cenomanian to lowermost Turonian), northwestern Alberta, Canada: Geological Survey of Canada Bulletin.

Welte, D. H., R. G. Schaefer, W. Stoessinger, and M. Radke, 1984, Gas generation and migration in the Deep Basin of Western Canada, *in* J. A. Masters, ed., Elmworth—case study of a Deep Basin gas field: AAPG Memoir 38, p. 35–47.

Williams, G. D., 1963, The Mannville Group (Lower Cretaceous) of central Alberta: Bulletin of Canadian Petroleum Geology, v. 11, p. 350–368.

Williams, G. D., and C. F. Burk, Jr., 1964, Upper Cretaceous, *in* R. G. McCrossan and R. P. Glaister, eds., Geological history of Western Canada: Calgary, Alberta, Alberta Society of Petroleum Geologists, p. 169–189.

Williams, G. D., and C. R. Stelck, 1975, Speculations on the Cretaceous paleogeography of North America, *in* W. G. E. Caldwell, ed., The Cretaceous System in the Western Interior of North America: The Geological Association of Canada Special Paper No. 13, p. 1–20.

Workman, L. E., 1958, Glauconite sandstone in southern Alberta: Journal of the Alberta Society of Petroleum Geologists, v. 6, p. 237–245.

Zaitlin, B. A., and B. C. Schultz, 1984, An estuarine embayment fill model from the Lower Cretaceous Mannville Group, west-central Saskatchewan, *in* D. F. Stott and D. J. Glass, eds., The Mesozoic of middle North America: Canadian Society of Petroleum Geologists Memoir 9, p. 455–469.

TABLE 3. Mannville Assemblage 2. Lower Mannville Continental to Marginal Marine Plays (see Figures 10, 11).*

Transgressive Basal Mannville Sandstone Play

Loc.	Southern and central plains (Figures 2, 10).
Prox.	Medial to distal (Figure 1).
Depo.	Continental to marginal marine strata deposited mainly transgressively and unconformably on dissected Mesozoic and Paleozoic surface. Later reworked by marine incursion.
Fms.	Lower Mannville units: Detrital, Deville, J1, Cutbank, Sunburst, Taber. Aptian.
Fluids	Oils, some low-gravity; sweet gas.
Reservoirs	Diverse sandstones: fluvial, estuarine, coastal plain, deltaic, tidal, and shallow marine.
Seals	Intercalated shale and siltstone units.
Traps	**Stratigraphic:** diverse pinch-outs—e.g., tidal channels, paleovalleys, erosional remnants. Differential compaction of shale enhances traps.
Fields	Gilby, Highvale, Carrot Creek, Turin, Wintering Hills.
Refs.	Williams, 1963; Rudkin, 1964; Williams and Burk, 1964; Hradsky and Griffin, 1984; Podruski et al., 1988.

J2 and J3 Valley-Fill Sandstone Play

Loc.	Central Alta. plains (Figure 10).

TABLE 3. (Continued)

Prox.	Medial eastern (Figure 1).
Depo.	Valley-fill complex on Mississippian carbonates.
Fms.	J2 and J3 units (Basal Mannville?). Late Jurassic and earliest Cretaceous.
Fluids	Oil.
Reservoirs	Valley-fill fluvial channel sandstone.
Seals	Overlying basal Ellerslie shale and confining valley walls.
Traps	**Stratigraphic:** sandstone confined by valley walls; depositional pinch-outs of sandstone within valleys.
Fields	Gilby, Medicine River.
Refs.	Rall, 1980; Hopkins, 1981.

Cadomin Conglomerate Deep Basin Play

Loc.	Northern B.C. and Alta. plains (Figure 10).
Prox.	Proximal (Figure 1).
Depo.	Alluvial fan/braidplain: alluvial plain conglomerate in west, changes facies east to sandstone.
Fms.	Cadomin Fm., Neocomian-Aptian.
Fluids	Sweet gas.
Reservoirs	Chert pebble conglomerate, fluvial channels.
Seals	Intraformational updip fluvial sandstone.
Traps	**Deep Basin trap:** gas zone downdip of water aquifer, plus updip tight sandstone permeability barriers.
Fields	Wapiti, Elmworth, Sinclair, Steep Creek, Red Rock, Bighorn, Windsor.
Refs.	McLean, 1977; Schultheis and Mountjoy, 1978; Gies, 1984; Jackson, 1984; Varley, 1984a, b.

Cadomin Sandstone Structural Play

Loc.	Northern Alta. plains. In water aquifer updip of Deep Basin gas play (Figure 10).
Prox.	Proximal to medial (Figure 1).
Depo.	Braidplain sandstone in "Spirit River Channel" trunk stream system (distal equivalent of Deep Basin play alluvial conglomerate).
Fms.	Cadomin Fm., Neocomian-Aptian.
Fluids	Dry sweet gas.
Reservoirs	Fluvial channel sandstone, lesser conglomerate.
Seals	Overlying Gething Fm. shale, updip tight Cadomin Fm. sandstone.
Traps	**Structural/Stratigraphic?:** drape on basement horsts related to earlier Peace River arch faulting, plus updip pinch-out?
Fields	Kaybob South, Gold Creek, Valhalla, Hamelin Creek, Saddle Hills, Whitelaw.
Refs.	McLean, 1977; Schultheis and Mountjoy, 1978; Gies, 1984; Varley, 1984a, b.

Dunlevy/Buick Creek Channel Play

Loc.	Northern B.C. plains (Figure 10).
Prox.	Northern proximal (Figure 1).
Depo.	Channels lapping onto "Fox Creek Escarpment."
Fms.	"Buick Creek" and "Dunlevy" units, maybe below or equivalent to basal Gething Fm., Aptian.
Fluids	Oil, sweet gas.
Reservoirs	Conglomeratic sandstone channels: fluvial, deltaic, and tidal.
Seals	Top: overlying Gething Fm.(?) shale. Updip: Fernie Gp. shale of confining Fox Creek Escarpment.

TABLE 3. (Continued)

Traps	1. **Stratigraphic:** depositional pinch-out of channels against Fox Creek Escarpment. 2. **Structural:** faulted anticlines on thrust sheet leading edges.
Fields	**Stratigraphic:** Rigel, Buick Creek, Siphon, Beavertail, Fireweed. **Structural:** Inga, Graham, Kobes, Grizzly, Wolverine, Ojay.
Refs.	Stott, 1973; Jackson, 1984.

Gething Fluvial Channel Sandstone Play

Loc.	Northern plains, Alta. and B.C. (Figures 10, 11).
Prox.	Proximal (Figure 1).
Depo.	West-derived fluviodeltaic complex.
Fms.	Gething Fm., Aptian.
Fluids	Mainly gas, some oil.
Reservoirs	Fluvial channel sandstone, local conglomerate.
Seals	Intraformational shale and siltstone.
Traps	1. **Deep Basin trap:** west area in gas zone downdip of water aquifer. 2. **Stratigraphic/Structural:** east area has depositional pinch-outs and drape over Peace River arch horsts.
Fields	**Gas:** Edson, Fox Creek, Peco, Velma, Beavertail, Simonette, and many small fields. **Oil:** Aitken Creek, Sakwatamau, Willow, Mike.
Refs.	Stott, 1972, 1973, 1975; Jackson, 1984; Smith et al., 1984.

Lower Mannville Transgressive Clastics Play

Loc.	Southern and central plains; Alta., southwestern Sask. (Figures 10, 11).
Prox.	Medial to distal (Figure 1).
Depo.	Marginal marine deposits on margins of Boreal Sea, representing reworking of Basal Mannville continental units.
Fms.	Basal Quartz, Ellerslie, Sunburst, Cutbank, Taber. Aptian.
Fluids	Oil, gas, condensate. Most of Western Canada's vast heavy oil reserves in Lower Mannville (e.g., McMurray Fm.).
Reservoirs	Diverse sandstones: fluvial, deltaic, estuarine, valley-fill, shoreline, beach, tidal, barrier, shallow marine.
Seals	Mainly intraformational shale and siltstone.
Traps	1. **Diverse, mainly stratigraphic:** e.g., depositional pinch-outs, valley-fills, erosional remnants. 2. **Structural:** drape over or infill of salt (Devonian) dissolution/collapse structures; structural control on channels and incised valleys.
Fields	Highvale, Bigoray, Wayne-Rosedale, Whitemud, Bellshill, Niton, Taber, Ricinus, Medicine River, Grand Forks, Turin.
Refs.	Hunt, 1950; Williams and Stelck, 1975; Conybeare, 1976; Hopkins, 1981; Horne et al., 1982; Hradsky and Griffin, 1984; Jackson, 1984; Ryer et al., 1984; Zaitlin and Schultz, 1984; Reel and Campbell, 1985; Farshori and Hopkins, 1987; Porter, 1992.

*Plays are described in approximate stratigraphic order from oldest to youngest. Major oil and gas fields mentioned are plotted in Figures 8–20; others are on the maps of Wallace-Dudley (1981a, b).

Heading abbreviations:

Loc. = location in foreland basin
Prox. = proximity to orogene (main source area)
Depo. = depositional setting
Fms. = formations and ages
Refs. = references.

Text abbreviations:

Alta. = Alberta
B.C. = British Columbia
Fm. = Formation
Mbr. = Member
Sask. = Saskatchewan.

TABLE 4. Mannville Assemblage 2. Upper Mannville Plays in Progradational Fluviodeltaic Wedges (see Figures 11, 12, 13, 14).*

Ostracod Lagoonal Sandstone Play

Loc.	Southern and central plains, Alta. (Figure 11).
Prox.	Medial to distal (Figure 1).
Depo.	In brackish "Edmonton Embayment" at south end of Boreal Sea transgressive deposits.
Fms.	Ostracod Mbr. (Lower Mannville "Fm."). Early Albian.
Fluids	Oil.
Reservoirs	Brackish/lagoonal and shallow marine calcareous sandstone.
Seals	Intraformational shale.
Traps	**Stratigraphic:** thin sandstone lenses in shale.
Fields	Pembina, Bigoray, Medicine, Taber.
Refs.	Hradsky and Griffin, 1984; Banerjee and Davies, 1988; Podruski et al., 1988.

Glauconitic Barrier-Deltaic Sandstone Play

Loc.	Southern and central Alta. plains (Figure 12).
Prox.	Medial to distal (Figure 1).
Depo.	Progradational fluviodeltaic complexes.
Fms.	"Glauconitic Sandstone" of Upper Mannville "Fm." Early Albian.
Fluids	Oil, locally low-gravity; gas.
Reservoirs	Barrier/beach, eolian, shallow marine sandstone at tops of progradational sequences. Local sandstone and conglomerate in incised valleys.
Seals	Intraformational and overlying shale.
Traps	**Stratigraphic:** large barrier or wave-dominated coastal-deltaic complexes form large pinch-out traps. Depositional sandstone pinch-outs within channels and valleys and against confining valley wall shale. Local drape over salt collapses or basement faults.
Fields	Bantry, Carbon, Countess, Hussar, Hoadley, Suffield.
Refs.	Workman, 1958; Dawson, 1982; Hopkins et al., 1982; Reichenbach, 1982; Farshori, 1983; Tilley, 1983; Chiang, 1984; Jackson, 1984; de Reuver, 1987; Rosenthal, 1988a, 1989.

Bluesky-Wabiskaw-Cummings Play

Loc.	Northern Alta. and B.C. plains (Figure 12).
Prox.	Medial to distal (Figure 1).
Depo.	Shelf to deltaic complexes prograded during Boreal Sea transgression.
Fms.	Bluesky, Wabiskaw, and Cummings fms. Early Albian.
Fluids	Oil, heavy oil, gas.
Reservoirs	Shoreline/barrier and shallow marine offshore bar sandstone.
Seals	Overlying Clearwater and Wilrich fms. shale.
Traps	1. **Stratigraphic:** sandstone sealed by deltaic and marine shale. 2. **Deep Basin trap:** gas zone downdip of water aquifer. 3. **Structural:** sheet sandstone draped over Peace River arch horsts.
Fields	**Gas:** Keg River, Elmworth, Edson. **Heavy oil:** Wabasca, Athabasca, Peace River oil sands deposits.
Refs.	Stott, 1973, 1975; Jackson, 1984; Smith et al., 1984; O'Connell, 1988.

Falher Conglomerate Deep Basin Play

Loc.	Northern Alta. and B.C. plains (Figure 13).
Prox.	Proximal (Figure 1).

TABLE 4. (Continued)

Depo.	North edge of progradational fluviodeltaic wedge at south margin of Boreal Sea. Shoreface conglomerate near tops of seven stacked marine to shoreface to continental progradational sequences.
Fms.	Falher Mbr. (Spirit River Fm.), Albian.
Fluids	Dry, sweet gas.
Reservoirs	Porous conglomerate and unconventional "tight-gas" sandstone: shoreface, beach, fluvial channel.
Seals	Intraformational shale, siltstone, and coal.
Traps	Deep Basin gas zone downdip of water aquifer.
Fields	Elmworth, Bilbo, Karr, Sinclair, Wapiti, Windsor, Kelly.
Refs.	Masters, 1979, 1984b; McLean, 1979; McMaster, 1981; Leckie and Walker, 1982; Cant, 1983; Jackson, 1984; Rahmani, 1984; Leckie, 1986b.

Notikewin Deep Basin and Peace River Structural Play

Loc.	Northern Alta. and B.C. plains (Figure 14).
Prox.	Proximal to medial (Figure 1).
Depo.	Prograded barriered coastal plain wedge.
Fms.	Notikewin Mbr. (Spirit River Fm.), Albian.
Fluids	Dry, sweet gas.
Reservoirs	Beach, barrier, shallow marine sandstone, conglomerate.
Seals	Overlying Harmon Mbr. shale
Traps	1. **West areas:** Deep Basin gas zone. 2. **East areas: Stratigraphic:** depositional pinch-outs. **Structural:** drape over Peace River arch horsts.
Fields	1. **West areas:** Karr, Elmworth, Wolf S. 2. **East areas:** Kaybob S., Belloy, Heart River.
Refs.	Jackson, 1984; Leckie, 1985.

Cadotte Shoreface/Barrier Sandstone Play

Loc.	Northern Alta. and B.C. plains (Figure 14).
Prox.	Medial to proximal (Figure 1).
Depo.	Sandstone and conglomerate in shoreface/barrier bars at leading edge of prograded fluviodeltaic wedge.
Fms.	Cadotte Mbr. (Peace River Fm.), Albian.
Fluids	Dry, sweet gas.
Reservoirs	Sandstone: barred and nonbarred shoreline, deltaic.
Seals	Overlying coastal plain Paddy Mbr. shale and siltstone, local infill of incised valleys on Cadotte.
Traps	1. **Deep Basin** gas zone in west areas, with local stratigraphic pinch-out traps—e.g., shoreface and valley-fill sandstone. 2. **Structural:** drape over Peace River arch horsts in east areas.
Fields	**Deep Basin:** Elmworth, Sunrise, Doe, Dawson Creek, Bilbo, Noel. **Structural:** Belloy, Saddle Hills, Lator, Webster, Woking.
Refs.	Ethier, 1982; Smith et al., 1984; Hayes, 1988; Leckie, 1988; Rahmani and Smith, 1988; Leckie et al., 1990.

*Plays are described in approximate stratigraphic order from oldest to youngest. Major oil and gas fields mentioned are plotted in Figures 8–20; others are on the maps of Wallace-Dudley (1981a, b).

Heading abbreviations:

Loc. = location in foreland basin
Prox. = proximity to orogene (main source area)
Depo. = depositional setting
Fms. = formations and ages
Refs. = references.

Text abbreviations:

Alta. = Alberta
B.C. = British Columbia
Fm. = Formation
Mbr. = Member
Sask. = Saskatchewan.

TABLE 5 Colorado Assemblage 3. Stratigraphic Traps of Thin Sandstone and Conglomerate Units Encased in Marine Shale (see Figures 2, 15, 16, 17).*

Basal Colorado Transgressive Sandstone Play

Loc.	Southern and central plains, Alta., southwestern Sask.
Prox.	Medial to distal (Figures 1, 2).
Depo.	Tidal sandstone transgressed on eroded Mannville and Jurassic surface.
Fms.	Basal Colorado Sandstone = upper Cessford Sand, Albian.
Fluids	Oil, gas.
Reservoirs	Tidal sand-sheet sandstone in Cessford area.
Seals	Overlying Joli Fou Fm. marine shale.
Traps	1. **Stratigraphic:** depositional pinch-out. 2. **Structural/Stratigraphic:** sands (on unconformity) in lows overlying Devonian salt collapses.
Fields	Cessford, Countess, Enchant, Verger, Hussar, Provost, Wayne-Rosedale.
Refs.	Humphreys, 1960; Banerjee, 1989; J. C. Hopkins, pers. comm., 1989; Stott et al., in press.

Viking Bar, Channel and Sheet Sandstone Play

Loc.	Southern, central plains, southern Alta., southwestern Sask.
Prox.	Medial to distal (Figures 1, 15).
Depo.	Clastic wedge with progradational, coarsening-up sequences, locally truncated and marine reworked by transgression.
Fms.	Viking Fm., Albian.
Fluids	Oil, gas.
Reservoirs	Sandstone and conglomerate: marine shelf bar and sheet, shoreface, estuarine channel.
Seals	Intraformational and overlying Colorado shale.
Traps	**Stratigraphic:** bar and sheet sandstone depositional pinch-outs into shale. Incised valley-fill channel sandstone within valley-wall shale. Sandstone pinch-outs in valleys.
Fields	Joarcam, Joffre, Garrington (bars), Provost (sheet), Crystal (channel).
Refs.	Hein et al., 1986; Leckie, 1986a; Podruski et al., 1988; Reinson et al., 1988; Clark and Reinson, 1990.

Paddy Member—Basal Transgressive Estuarine and Barrier Sandstone Play

Loc.	Northern Alta. and B.C. plains (Figure 15).
Prox.	Proximal to medial (Figure 1).
Depo.	Transgressive Paddy Mbr. coastal plain and estuarine deposits on incised Cadotte Mbr. prograded clastic wedge.
Fms.	Paddy Mbr. (Peace River Fm.), Albian.
Fluids	Gas, minor condensate.
Reservoirs	Tidal channel, estuarine valley-fill and barrier sandstone
Seals	Overlying Shaftesbury Fm. marine shale and intraformational estuarine–coastal plain shale.
Traps	1. **Stratigraphic:** sandstone depositional pinch-outs. 2. **Deep Basin** gas zone in west areas.
Fields	1. **Stratigraphic:** Heart River, Saddle Hills, Dimsdale, Woking, Wembley. 2. **Deep Basin:** Sinclair, Knopcik, Kelly, Valhalla.
Refs.	Smith et al., 1984; Hayes, 1988; Leckie, 1988; Stelck and Leckie, 1990; Leckie and Singh, 1991.

TABLE 5. (Continued)

Dunvegan Delta and Doe Creek Offshore Sandstone Play

Loc. Northern Alta. and B.C. plains (Figure 16).

Prox. Proximal (Figure 1).

Depo. Dunvegan is a progradational fluviodeltaic conglomerate and sandstone clastic wedge. Doe Creek is five stacked shallow marine sandstone progradational sequences equivalent to, east of, and isolated from uppermost Dunvegan. Doe Creek sandstone encased in marine shale.

Fms. Dunvegan Fm., Doe Creek Mbr. (Kaskapau Fm.), Cenomanian.

Fluids Oil, gas.

Reservoirs **Dunvegan:** distributary mouth bar, distributary channel, estuarine channel, and strandplain sandstone **Doe Creek:** shallow marine offshore bar sandstone of shoreface to inner shelf origin. Minor cross-cutting channels or deltaic lobes.

Seals Intraformational and overlying shale and siltstone.

Traps
1. **Stratigraphic:** depositional pinch-outs of progradational sandstone below transgressive shale.
2. **Deep Basin** gas zone in west areas.
3. **Structural:** faulted anticlines on leading edge of thrust sheets may have gas in Foothills.

Fields 1. **Dunv. Stratigraphic:** Simonette, Waskahigan (distributary and estuarine channels), Jayar, Lator. 2. **Doe Creek Stratigraphic:** Valhalla, Knopcik, Pouce Coupe S., Progress, Sinclair, Elmworth.
3. **Dunv. Deep Basin and Structural:** Bigstone, Hinton, Fir, Lynx, Sexsmith.

Refs. Stott, 1982; Dearborn et al., 1985; Ross, 1986; Bhattacharya, 1988, 1989; Wallace-Dudley and Leckie, 1988, in press.

Cardium Sheet and Scour-Fill Sandstone Play

Loc. Southern and central Alta. plains (Figures 6, 7, 17).

Prox. Proximal to medial (Figure 1).

Depo. Sandstone and conglomerate in stacked progradational sequences of a clastic wedge pinching out into shelfal marine shale. Wedge may be fluviodeltaic and transgressively reworked by marine processes.

Fms. Cardium Fm. Turonian.

Fluids Oil, dry and wet gas, condensate.

Reservoirs Stacked coarsening-up sheet sandstone and conglomerate deposited or reworked into broad sheets or elongate bars. Scour-fill sandstone in broad shallow scours cut in shale or sheet sandstone.

Seals Marine shale.

Traps **Stratigraphic:** depositional pinch-outs of sheet and scour-fill sandstone and conglomerate.

Fields
1. **Sheet:** Pembina (giant oil field), Willesden Green, Ferrier, Crossfield, Garrington.
2. **Scour-fill:** Ricinus, Cyn-Pem, Carrot Creek.

Refs. Beach, 1955; Nielsen, 1957; Walker, 1983, 1985; Krause and Nelson, 1984; Nielsen and Porter, 1984; Keith, 1985; Krause et al., 1987; Podruski et al., 1988; Walker and Eyles, 1988 and many citations in above.

Cardium-Kakwa Shoreline Sandstone and Conglomerate Play

Loc. Northern plains, Alta., B.C. (Figures 7, 17).

Prox. Proximal (Figure 1).

Depo. Shorelines of northeastward-prograding, wave-dominated, barriered, coastal plain complex.

Fms. Kakwa Mbr. of Cardium Fm. Turonian.

Fluids Low gravity oil, gas.

Reservoirs Coastal plain-shoreline sandstone and conglomerate of shoreface, beach, river-mouth bar, and fluvial channel environments. Conglomerate interpreted as upper shoreface.

TABLE 5. (Continued)

Seals	Intraformational and overlying (Muskiki Fm.) shale and siltstone.
Traps	**Stratigraphic:** depositional pinch-outs of progradational shoreline sandstone and conglomerate sealed below by marine shale and above by lagoonal-coastal plain shale.
Fields	Kakwa, Wapiti.
Refs.	Nielsen and Porter, 1984; Plint and Walker, 1987; Deutsch and Krause, 1990.

*Plays are described in approximate stratigraphic order from oldest to youngest. Major oil and gas fields mentioned are plotted in Figures 8–20; others are on the maps of Wallace-Dudley (1981a, b).

Heading abbreviations:

Loc. = location in foreland basin
Prox. = proximity to orogene (main source area)
Depo. = depositional setting
Fms. = formations and ages
Refs. = references.

Text abbreviations:

Alta. = Alberta
B.C. = British Columbia
Fm. = Formation
Mbr. = Member
Sask. = Saskatchewan.

TABLE 6. Saunders/Edmonton Assemblage 4. Fluvial Channel and Shoreline Reservoirs at Leading Edges of Prograding Fluviodeltaic Wedges (see Figures 18, 19, 20).*

Milk River Sheet Sandstone Play

Loc.	Southeastern Alta. and southwestern Sask. plains (Figure 18).
Prox.	Distal (Figure 1).
Depo.	Milk River Fm.: broad shallow shelf and shoreface shaly sandstone to siltstone (gas also in Medicine Hat Sandstone, similar facies in progradational clastic wedge near top of Colorado Assemblage).
Fms.	Milk River and Lea Park fms., Alderson Mbr., Upper Santonian–Lower Campanian (Medicine Hat Sandstone = Coniacian).
Fluids	Gas.
Reservoirs	Shallow shelf–shoreface shaly sandstone and siltstone.
Seals	Marine shale: overlying Pakowki Fm. and updip Lea Park Fm.
Traps	**Stratigraphic/Structural/Hydrodynamic:** Depositional pinch-out of sandstone to updip shale. Field straddles Bow Island–Sweetgrass arch. **"Deep Basin"-type trap** with tight gas sand downdip of porous wet sand.
Fields	Medicine Hat (giant shallow gas field), Milk River, Liebenthal, Hatton.
Refs.	Williams and Burk, 1964; Masters, 1979; Meijer-Drees and Myhr, 1981; Leckie and Cheel, 1986; McCrory and Walker, 1986; Hankel et al., 1989.

Belly River Shoreline-Fluvial Sandstone Play

Loc.	Southern and central Alta. plains (Figure 19).
Prox.	Proximal, medial and distal (Figure 1).
Depo.	Eastward-prograded fluviodeltaic clastic wedge.
Fms.	Belly River Fm., "Basal Belly River" unit. Campanian.
Fluids	Oil, gas.
Reservoirs	Main reservoirs (Basal Belly R.) are shoreline-related sandstone: shoreline, estuarine, distributary channel and distributary mouth bar. Secondary reservoirs in overlying (Belly R.) fluvial channel and crevasse-splay sandstone
Seals	Updip marine shale, overlying Bearpaw Fm. marine shale and locally overlying coastal plain muds.

TABLE 6. (Continued)

Traps	1. **Stratigraphic:** depositional sandstone pinch-outs. 2. **Structural influence:** in southeastern Alta., sheet-like basal sandstone drapes over highs on eroded Paleozoic surface. 3. **Diagenetic:** clays and calcitic-sideritic cements occlude porosity.
Fields	1. **Shoreline:** Pembina, Keystone, Herronton, Rowley. **Fluvial:** Ferrier, Davey, Willesden Green, Minnehik-Buck Lake, Gilby. 2. **Drape:** Atlee-Buffalo, Bindloss, Medicine Hat.
Refs.	Shouldice, 1979; Storey, 1982; Iwuagwu and Lerbekmo, 1984; Podruski et al., 1988.

Paskapoo/Edmonton Group Continental Gas Sands Play

Loc.	Southern and central Alta. plains (Figure 20).
Prox.	Proximal to medial (Figure 1).
Depo.	Thick continental molasse clastics.
Fms.	Paskapoo Fm., Paleocene. Edmonton Group, late Campanian to early Paleocene. Foremost Fm., Santonian to Campanian.
Fluids	Gas.
Reservoirs	Probably fluvial channel and crevasse splay conglomerate and sandstone.
Seals	Limited because of porous, coarse-grained clastic units.
Traps	1. **Stratigraphic (?):** possibly depositional pinch-outs of fluvial units. 2. **Structural (?):** Laramide synsedimentary tectonics could create structural traps.
Fields	Bigoray, Sylvan Lake, Willesden Green (all Paskapoo Fm.), Alderson, Davey, Taber, Nevis, Chickadee, Atlee-Buffalo (Edmonton Group), Atlee-Buffalo (Foremost Fm.).
Refs.	Parsons, 1973.

*Plays are described in approximate stratigraphic order from oldest to youngest. Major oil and gas fields mentioned are plotted in Figures 8–20; others are on the maps of Wallace-Dudley (1981a, b).

Heading abbreviations:

Loc. = location in foreland basin
Prox. = proximity to orogene (main source area)
Depo. = depositional setting
Fms. = formations and ages
Refs. = references.

Text abbreviations:

Alta. = Alberta
B.C. = British Columbia
Fm. = Formation
Mbr. = Member
Sask. = Saskatchewan.

Chapter 8

Lithology and Diagenesis of Sandstones in the Western Canada Foreland Basin

Dan Potocki

Shell Canada Limited
Calgary Research Centre
Calgary, Alberta, Canada

Ian Hutcheon

The Department of Geology and Geophysics
The University of Calgary
Calgary, Alberta, Canada

INTRODUCTION

From Middle Jurassic to early Tertiary time, the Western Canada foreland basin filled with mud and silt, and smaller quantities of sand and gravel, derived primarily from the rising cordillera to the west. Most sandstones in the basin are lithic arenites and consist of recycled sedimentary debris, but quartzose and volcano-feldspathic-rich sandstones also are abundant. This paper reviews the lithology, diagenesis, and hydrologic history of sandstones in the basin and is based primarily on published studies, government reports, and unpublished university theses. Because hydrocarbon exploration and associated petrologic and hydrologic studies utilizing subsurface drill core and water analyses have focused upon the Alberta portion of the Western Canada foreland basin, this paper also deals primarily with sandstones in the Alberta foreland basin. In spite of the excellent knowledge of the geology and history of this basin, and the abundant, publicly available core, comparatively few diagenetic studies have been carried out.

Within the Alberta foreland basin, each eastward-tapering wedge of sandstone is interpreted to reflect a period of extensive tectonism within the cordillera (Bally et al., 1966). Orogenic events in the cordillera are thought to be the result of two major periods of deformation resulting from the collision of foreign crustal fragments with the North American plate during Early to Middle Jurassic time (Columbian orogeny) and mid-Cretaceous to early Tertiary time (Laramide orogeny) (Monger et al., 1982; Stockmal et al., 1992).

This paper is divided into three segments. First, the provenance and character of the three major sandstone lithofacies in the Western Canada foreland basin are described. This is followed by a discussion of commonly observed framework-dependent diagenetic pathways for each of the three lithofacies. Patterns of diagenesis locally have little or no relation to original sand composition, but have been unpredictably controlled by fluid hydrology. The third segment discusses the complex hydrologic evolution of the Alberta foreland basin and relates observed diagenetic assemblages to the compositions of formation waters. The water compositions and resulting diagenetic events reflect mixing of underlying saline waters from Paleozoic rocks with the less saline waters contained in the Mesozoic sediments that were deposited during the development of the foreland basin in Alberta.

LITHOLOGY

Introduction

The framework composition of a sandstone represents the end product of complex interactions among factors such as source rock, climate, durability and size of detritus, distance and type of transport, and depositional environment. However, sediment provenance, controlled in turn by the tectonic setting, exerts the greatest influence on sandstone framework composition. Dickinson and Suczek (1979) and Dickinson et al. (1983) demonstrated that the framework mineralogy of terrigenous sandstones from different tectonic regimes commonly plot within discrete and separate fields on ternary quartz-feldspar-lithic (QFL) diagrams. Three general provenance types can be recognized using ternary plots: continental blocks, active mag-

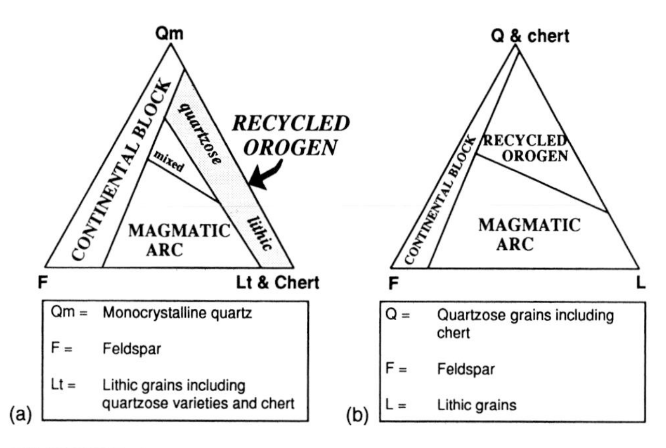

FIGURE 1. Triangular QmFLt (a) and QFL (b) diagrams showing compositional fields defined by Dickinson and Suczek (1979) and modified by Dickinson et al. (1983), reflecting sand derivation from different provenance terranes.

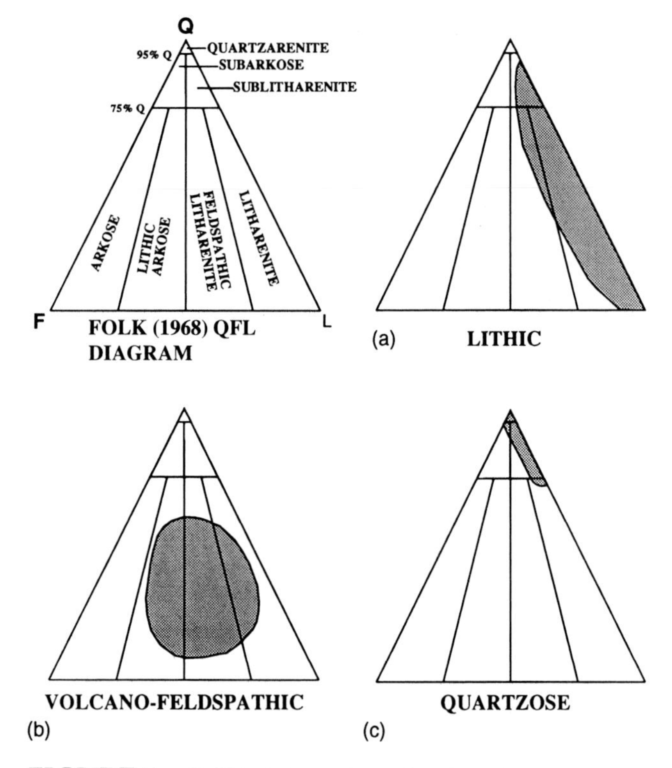

FIGURE 2. Folk's (1968) triangular QFL diagram, where Q represents monocrystalline quartz and metaquartzite grains, F is feldspar, and L is lithic fragments including chert, showing the compositional fields for the lithic (a), volcano-feldspathic (b), and quartzose (c) sandstones defined in the text. The compositional fields were determined by plotting available QFL point-count data (Table 1) for each lithofacies, and then drawing fields by hand to surround the majority of the data points in each lithofacies.

matic arcs, and recycled orogenic terranes, the last including foreland fold and thrust belts (Figure 1).

Within a foreland basin, sediments are primarily derived from preexisting sedimentary and metasedimentary rocks uplifted in the fold and thrust belt. Generally, the fold-thrust belt forms a topographic barrier or drainage divide that protects the basin from sediment in arc or collision orogenes behind the fold-thrust front. Minor volumes of sediment may be derived from the craton on the opposite side of the basin. As a result, foredeep sands typically contain moderate to high amounts of recycled sedimentary debris but little metamorphic and igneous material and feldspar (Dickinson and Suczek, 1979). When chert grains are plotted with other lithic grains at the L pole of a ternary diagram, a continuum of orogenic compositional types ranging from quartzose recycled to lithic recycled is observed (Figure 1a).

The majority of published petrographic studies for sandstones in the Western Canada foreland basin have used the Folk (1968) classification scheme (Figure 2), which differs slightly from the Dickinson and Suczek (1979) QmFLt ternary plot (Figure 1a). On the Dickinson and Suczek QmFLt diagram, only monocrystalline quartz is plotted at the Qm pole, whereas in the Folk scheme both monocrystalline quartz and metaquartzite grains are plotted at the Q pole. Nevertheless, comparing published ternary diagrams for sandstones in the Alberta foreland basin with the provenance fields defined by Dickinson and Suczek (1979) shows that the lithic and quartz-rich, feldspar-poor composition of most sandstones in the basin is consistent with derivation from a recycled orogenic terrane. However, some sandstones in the Western Canada foreland basin have compositions atypical of a recycled orogenic provenance. These include sandstones rich in feldspar and/or volcanic rock fragments and sandstones anomalously rich in monocrystalline quartz. The framework modes of these sandstones reflect derivation, or dilution, from active magmatic arc terranes and stable craton interiors/quartzose recycled orogenes, respectively. Based on these distinct compositional differences, sandstones in the Western Canada foreland basin can be grouped into three major lithofacies (Figures 2 and 3). These are:

1. Lithic lithofacies (Figure 2a)

2. Volcano-feldspathic lithofacies (Figure 2b)

3. Quartzose lithofacies (Figure 2c)

Data sources from the various sandstone units used to define the compositional fields for the three lithofacies are referenced in Table 1. Because framework mineralogy is perhaps the prime variable governing the course of sandstone diagenesis (Dickinson and Suczek, 1979; Schmidt and McDonald, 1979), a certain level of reservoir quality prediction can be achieved by determining framework-dependent diagenetic features characteristic of each of the three lithofacies. Prior to discussing probable compositionally controlled diagenetic pathways, the lithologic properties and provenance of the three lithofacies are discussed.

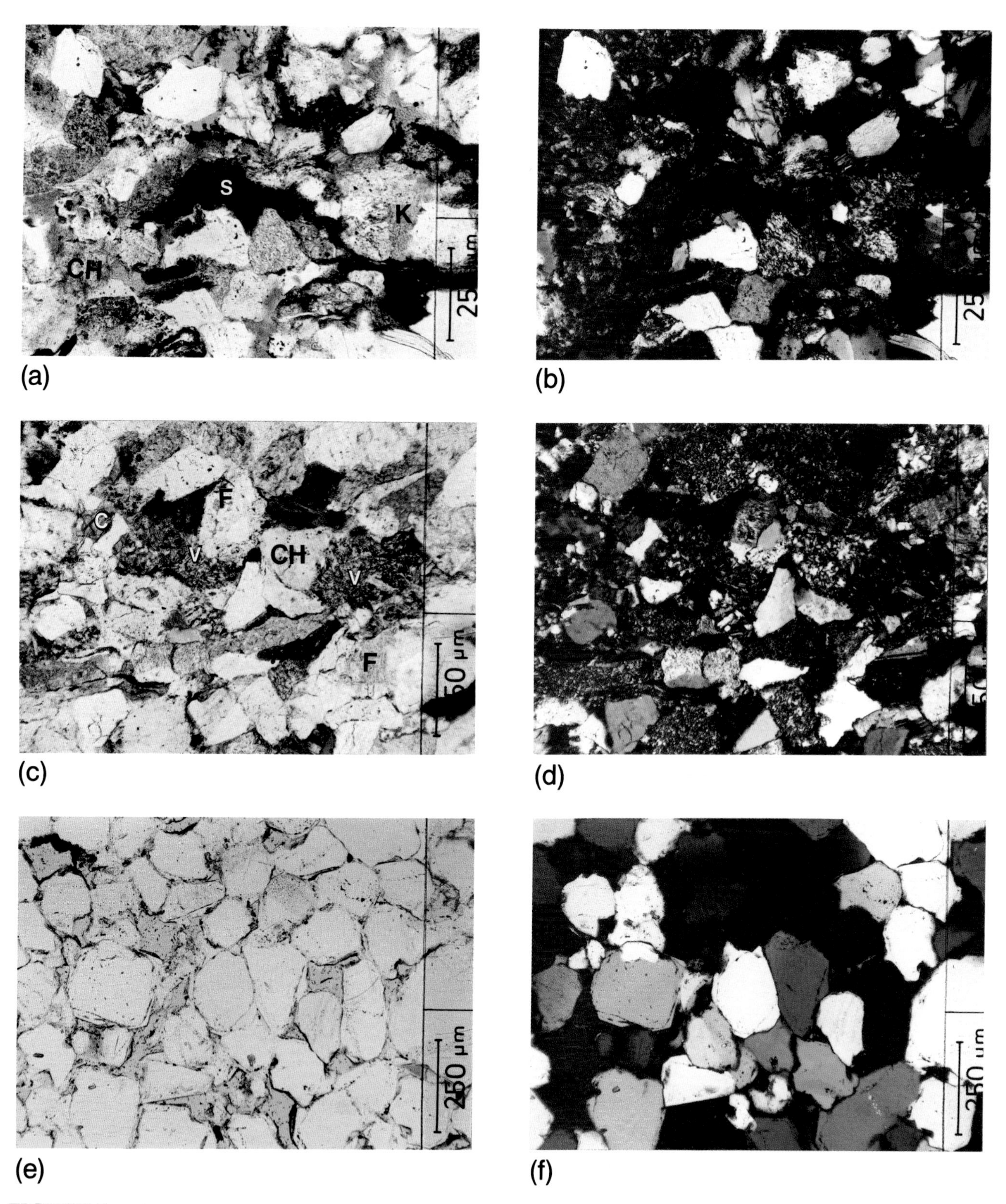

FIGURE 3. Thin-section photographs, all at the same magnification, showing plain-light and crossed-polar views of representative members of the lithic (a, b), volcano-feldspathic (c, d) and quartzose (e, f) lithofacies. Lithic sample from the Dunvegan Formation, Simonette field, west central Alberta. Volcano-feldspathic sample from the Belly River Formation, Highwood River outcrop, southern Alberta Foothills. Quartzose sample from the Basal Quartz sandstone, Niton field, west central Alberta. Legend: (C) calcite; (CH) chert; (F) feldspar; (K) kaolinite; (S) sedimentary rock fragment; (V) volcanic rock fragment.

TABLE 1. Sources of QFL Data and Petrographic Descriptions Used To Define Compositional Fields for the Lithofacies in Figure 2.

Formation	Location	Reference
	Lithic Lithofacies	
Alberta Group	South-central Foothills	Stott (1968, 1984)
Belly River (basal)	Pembina	Iwuagwu and Lerbekmo (1982)
Blairmore (lower)	Southwestern Foothills	Rapson (1965)
Cardium	West-central Plains	Potocki (unpublished)
Dunvegan	West-central Plains	Tater (1964)
Dunvegan	Simonette field	Potocki (unpublished)
Gates	Eastern British Columbia	Leckie (1986)
Glauconitic	Central Plains	Young and Doig (1986)
Glauconitic	South-central Plains	Meshri (1981)
Glauconitic	Suffield field	Tilley and Longstaffe (1984)
Kootenay	South-central Foothills	Gibson (1979, 1985)
Mannville (L and U)	Southern Alberta	Glaister (1959)
Notikewin	West-central Plains	Gardiner (1982)
Porcupine Hills	Southwestern Foothills	Carrigy (1971)
St. Mary River	Southern Plains	Nelson (1968)
Viking	Southeastern Saskatchewan	O'Connell (1981)
Viking (lower)	Provost	Reinson and Foscolos (1986)
Viking	Caroline	Reinson and Foscolos (1986)
Viking	Joffre	Reinson and Foscolos (1986)
	Volcano-Feldspathic Lithofacies	
Foremost and Oldman	Milk River	Ogunyomi and Hills (1977)
Belly River	Southern Plains	Lerbekmo (1963)
Belly River	Ghost Dam	Haywick (1982)
Belly River (basal)	Trap Creek	Nelson and Glaister (1975)
Blackleaf (upper)	Montana	Dyman et al. (1988)
Blood Reserve	Southern Plains	Lerand (1983)
Clearwater	Central Plains	Williams (1964)
Clearwater	Cold Lake	McLean and Putnam (1983)
Clearwater	Cold Lake	Waywanko (1984)
Clearwater	Cold Lake	Putnam and Pedskalny (1983)
Colony	Lloydminster	Prentice and Kramers (1987)
Frenchman	Southwestern Saskatchewan	Misko and Hendry (1979)
Grand Rapids	Central Plains	Williams (1964)
Horsethief	Montana	Bibler and Schmitt (1986)
Mannville (L and U)	Southern Alberta	Glaister (1959)
Paskapoo	Central Plains	Carrigy (1971)
Ravenscrag	Southeastern Plains	Carrigy (1971)
St. Mary River	Southern Foothills	Nelson (1968)
Waseca	Lloydminster	Prentice and Kramers (1987)

TABLE 1. (Continued)

Formation	Location	Reference
Waseca	Lloydminster	Putnam (1982)
Willow Creek	Southwestern Foothills	Carrigy (1971)
	Quartzose Lithofacies	
Ellerslie	Central Plains	Williams (1964)
Ellerslie	Medicine River field	Hopkins (1981)
Ellerslie	Niton field	Potocki (1981)
Ellerslie/Rock Creek	Central Plains	Marion (1982)
Lloydminster	Lloydminster	McLean and Putnam (1983)
Mannville (lower)	Southern Alberta	Glaister (1959)
McLaren	Lloydminster	Prentice and Kramers (1987)
McMurray	Northeastern Plains	Carrigy (1963)
McMurray	Northeastern Plains	Nelson and Glaister (1978)

Lithic Sandstones

Lithic sandstones, which comprise the bulk of the sandstones in the Western Canada foreland basin [Enclosure 1(A)], are characterized by small volumes of feldspar and moderate to large amounts of quartz and sedimentary lithic grains, especially chert (Figure 2a), and have been derived in large part from the Rocky Mountain fold-thrust belt to the west. However, metamorphic and igneous rock fragments and feldspar are locally abundant in many lithic sandstones, indicating sediment contribution from cordilleran terranes west of the fold-thrust belt (i.e., Omineca and Intermontane belts). The mixture of light-colored quartz with dark lithic grains creates a diagnostic "salt and pepper" appearance for these rocks. According to the Folk (1968) classification scheme, lithic sandstones in the Western Canada foreland basin are classified primarily as sublitharenites and litharenites, but feldspar-rich varieties fall into the feldspathic-litharenite field (Figure 2a).

The framework mineralogy of lithic sandstones reflects a mixed provenance dominated by sedimentary rocks, but which also includes metasedimentary, low- to high-grade metamorphic, and volcanic to plutonic igneous source terranes. Quartz, chert, limestone, and dolomite fragments, and arenaceous clastic grains, have been derived primarily from the erosion of uplifted miogeoclinal, late Proterozoic and Paleozoic carbonates and clastics contained within the Rocky Mountain fold and thrust belt. Argillaceous or pelitic sedimentary rock fragments have been derived from deeper-water, miogeoclinal to platformal, calcareous mudstones and argillaceous limestones now exposed in the western portion of the fold and thrust belt (Price et al., 1985). Metamorphic rock fragments and igneous plutonic and volcanic debris, including feldspar, were derived from west of the fold-thrust belt from what are now the Omineca and Intermontane belts (Rapson, 1965; Mellon, 1967; Jansa, 1972; Stott, 1984; Leckie, 1986). The Omineca belt is separated from the fold-thrust belt by the Rocky Mountain Trench, a right-lateral strike-slip fault considered to have been active in Late Cretaceous time (Price et al., 1985). Estimates of displacement along the fault vary from 450 to 900 km (280 to 560 mi) (Templeman-Kluit, 1979; Gabrielse, 1985), indicating that source terranes for Early Cretaceous sandstones, situated west of the fault, may now be displaced hundreds of kilometers north of their original position.

Because of the incorporation of debris from several terranes within the cordillera, it appears that the major control on regional and stratigraphic compositional variations of lithic sandstones in the Alberta foreland basin has been the position of the drainage divide throughout time (Dickinson and Suczek, 1979; McMechan and Thompson, in press). Lithic sandstones deposited during periods of tectonism, when the leading edge of the fold and thrust belt formed a proximal drainage divide, contain abundant coarse-grained sedimentary debris (i.e., quartz, chert, and carbonate) and comparatively little metamorphic and igneous material. Lithic sandstones deposited during periods of tectonic quiescence contain increased metamorphic and igneous debris, including feldspar, relative to clastic and carbonate grains, indicating a westerly shift in the drainage divide owing to erosion of the thrust front and integration into the drainage basin of more westerly source terranes. During the Laramide orogeny, the deformation front of the fold-thrust belt migrated eastward, and foredeep sandstones deposited previously during the Late Jurassic and Early Cretaceous (i.e., cycles 1 and 2 of Leckie and Smith, 1992) were themselves thrust faulted, eroded, and incorporated with older sedimentary debris into Late Cretaceous and younger sandstones (i.e., cycle 3 and 4 sandstones).

The position of the drainage divide is related in turn to the structural style and tectonic history of the fold and thrust belt. McMechan and Thompson (in press) show that, at present, the fold and thrust belt changes northward along its strike from predominantly thick thrust sheets in the south to detached fold complexes with few broad, low-angle thrust sheets north of approximately 54° latitude. It is likely that structural differences between the southern and northern fold-thrust belt have been manifest throughout foredeep sedimentation and have controlled spatial and temporal differences in provenance and sandstone composition.

Provenance studies of sandstones in the southern foreland basin (south of 54° N lat.) indicate several alternations of more and less proximal drainage divides during foredeep sedimentation (Figure 4). The Passage Beds of the upper Fernie Formation, deposited during the early stages of the Columbian orogeny, are believed to represent sediment first derived from the rising cordillera to the west, but the occurrence of abundant mica suggests derivation from uplifted metamorphic terranes in what is now the southern Omineca belt (Poulton, 1984). Overlying lithic sandstones of the Kootenay and Minnes groups are rich in clastic and carbonate sedimentary debris and record the initial development of the Rocky Mountain fold and thrust belt and the establishment of a proximal drainage divide (Gibson, 1985; Poulton, 1984). Within younger Campanian to Paleocene sandstones deposited during the Laramide orogeny, Mack and Jerzykiewicz (1989) demonstrated that relative differences in the amounts of carbonate and chert grains (from the Front Ranges and Eastern Main Ranges of the fold-thrust belt), pelitic sedimentary grains (from the Western Main Ranges), and metamorphic and volcanic debris (from the Omineca belt) reflect three tectonically influenced shifts in the position of the drainage divide (Figure 5). In contrast to lithic sandstones in the southern foreland basin, many lithic sandstones in the northern segment of the basin contain a greater proportion of metamorphic and igneous material from the Omineca and Intermontane belts (e.g., Stott, 1968; Leckie, 1986). Throughout most of the sedimentation history of the northern foredeep, the shallow fold complexes and low thrust sheets were incapable of forming a proximal drainage divide. Only during deposition of the Cadomin Formation did the leading edge of the northern thrust belt form an effective divide, thereby preventing dilution by debris from western terranes (McMechan and Thompson, in press).

The feldspar content of lithic sandstones is commonly less than 10%, but may locally be slightly more than 20%. Generally, feldspar is most abundant in lithic sandstones deposited during periods associated with a distal drainage divide. Certain sandstones deposited during early Albian and Campanian–Maastrichtian time contain anomalously large amounts of feldspar in association with volcanic rock fragments. These volcano-feldspathic sandstones comprise a separate lithofacies and are discussed in detail in the next section. In lithic sandstones, plagioclase feldspar is generally more abundant than orthoclase feldspar. Generally, a high proportion of plagioclase to orthoclase is indicative of derivation from a volcanic source (Folk, 1968). However, interpretations of provenance based on feldspar are commonly obscured because of preburial weathering and postburial diagenesis. For example, owing to dissolution and replacement reactions that are interpreted to be depth-related, Upper Mannville sediments in the western portion of the Western Canada foreland basin have less potassium feldspar and increased albitized feldspar relative to equivalent, but less deeply buried, sandstones farther east (Mellon, 1967; Ghent and Miller, 1974; James, 1985). Nevertheless, in a study of glauconitic sandstones in central Alberta, Young and Doig (1986) suggested that the increased feldspar content of channel sands relative to older, feldspar-poor barrier sands demonstrates a change of source and paleoslope between the two facies.

Although the main control on the composition of detritus shed into the basin has been the position of the drainage divide, compositional variations have occurred in some instances because of weathering in the source area, mechanical and chemical breakdown during transport, breakdown during depositional reworking, or diagenetic removal of soluble framework components (Figure 6). These processes may account for local intraformational, or possibly interformational, compositional variations commonly observed in lithic sands in the Western Canada foreland basin. Byers (1969) has suggested that the volcano-feldspathic Eastend Formation and the overlying lithic Whitemud Formation may have been derived from the same source but that greater weathering in the source area, and during sediment transport, has increased the mineralogic maturity of the Whitemud.

Blatt et al. (1972) have shown that compositional fractionation may also occur because of grain-size differences. Coarser-grained lithic sandstones and conglomerates in the Western Canada foreland basin are generally rich in chert and polymineralic rock fragments, whereas finer lithic sandstones are more quartzose. Some conglomerates consist almost entirely of chert and are classified as chert-arenites (e.g., Cardium, Cadomin, and Spirit River formations). Because grain size is strongly controlled by hydraulic sorting during deposition, distinct compositional groups often are associated with different depositional facies. Fluvial environments commonly trap the coarsest lithic-rich sediment fraction, and pass the finer, quartz-rich sediment to the coast. In high-energy coastal environments, remaining mechanically unstable lithic grains may be destroyed because of increased wave agitation (Mack, 1978). For example, in a study of Belly River sandstones in central Alberta, Storey (1979) found that beach sandstones are more quartzose than channel sandstones because of increased sorting and sediment abrasion in the higher-energy beach environment. However, this pattern is not always observed. For instance, Spirit River chert conglomerates in the Elmworth field are more prevalent in the marine shoreface environment than in associated nonmarine deposits (Cant and Ethier, 1984).

Volcano-Feldspathic Lithofacies

Sandstones in the volcano-feldspathic lithofacies characteristically contain large amounts of feldspar and volcanic

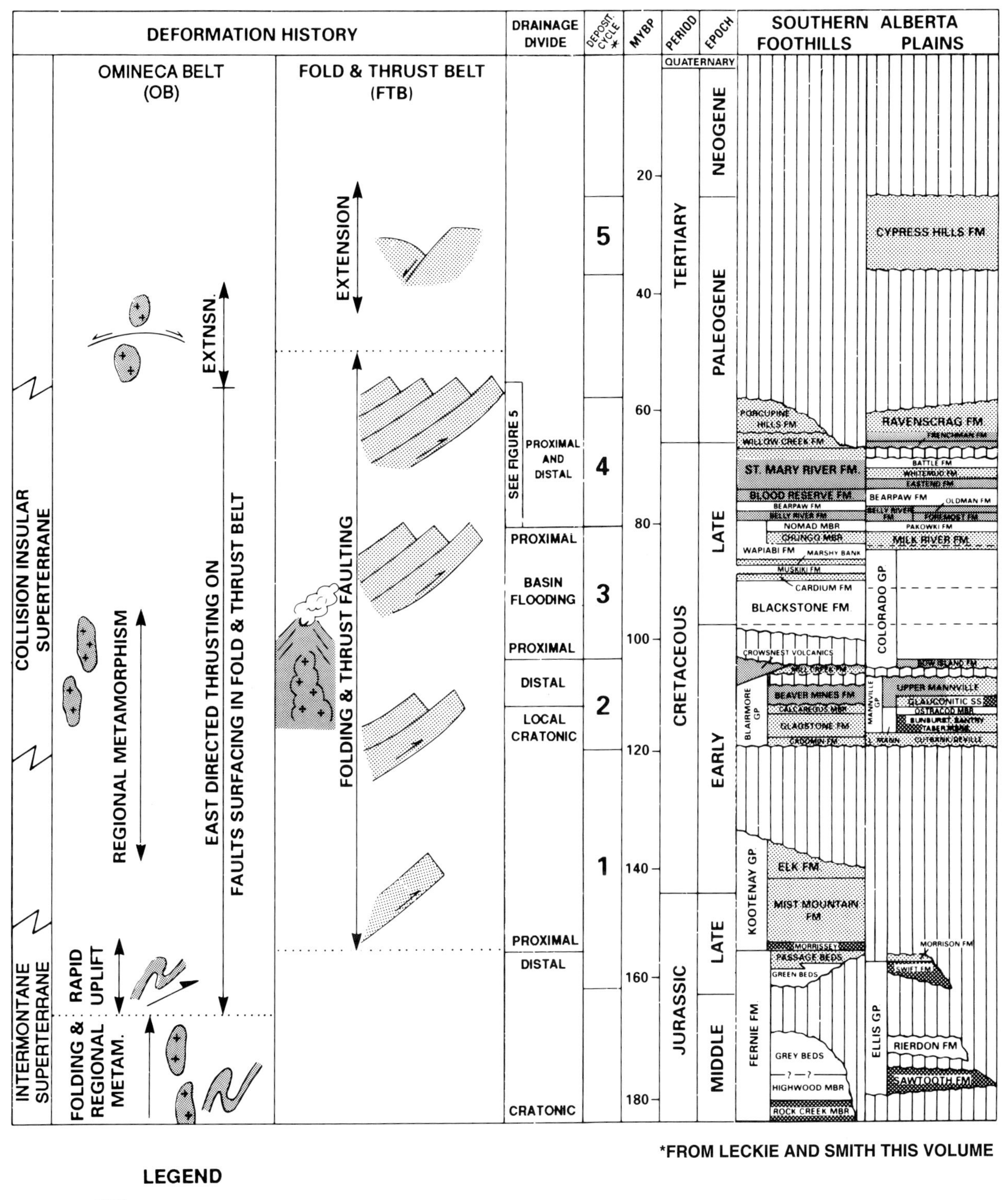

FIGURE 4. Stratigraphic section of the Foothills and Plains of the southern Alberta foreland basin, showing how the relative position of the drainage divide, controlled by deformation and plutonism/volcanism in the fold and thrust belt and Omineca belt, has influenced sandstone composition through time. Sands sourced during periods associated with a distal drainage divide generally contain increased amounts of pelitic debris, igneous and metamorphic rock fragments, and feldspar in relation to sands deposited during more proximal drainage regimes that contain relatively greater amounts of clastic and carbonate rock fragments. Sands derived from the craton to the east are generally quartzose. (Diagram modified from McMechan and Thompson, in press.)

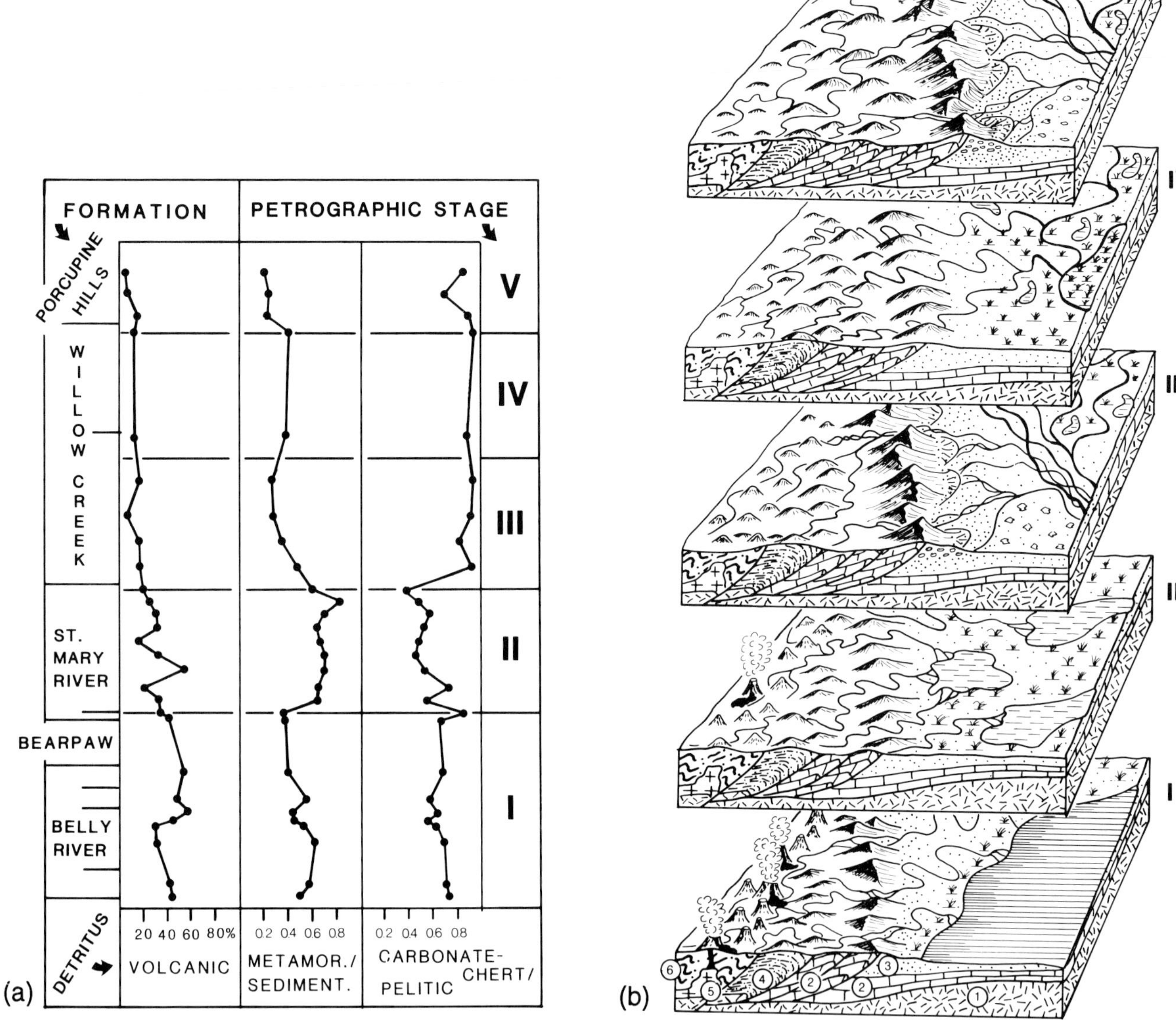

FIGURE 5. Provenance studies of Belly River to Porcupine Hills sandstones in the southern Alberta Foothills indicate that they were derived from several sources, including volcanic and metamorphic rocks of the Omineca belt, pelitic sedimentary rock fragments of the western Main Ranges of the Rocky Mountain fold and thrust belt, and carbonate and chertarenite rock fragments of the eastern Main and Front Ranges. Vertical trends in the relative proportions of framework constituents reveal five petrographic stages (part a), which can be interpreted to represent an initial volcanic (possibly volcanic-tectonic) event (stage I) followed by two thrust events in the fold and thrust belt (stages III and V) separated by periods of tectonic quiescence (stages II and IV, part b). Legend for part b: (I) Precambrian crystalline basement; (II) Paleozoic carbonates and Mesozoic chertarenites of the Front Ranges and eastern Main Ranges; (III) Cretaceous to Paleocene sediments shed into the foreland basin; (IV) Precambrian and lower Paleozoic pelitic rocks of the western Main Ranges; (V) Mesozoic intrusive igneous rocks and associated volcanics of the Omineca belt; (6) Precambrian metasedimentary rocks of the Omineca belt. (From Mack and Jerzykiewicz, 1989; used by permission.)

rock fragments and are generally quartz-poor and matrix-rich. Feldspar and volcanic material may comprise upwards of 70% of the sandstone framework, indicating major sediment contribution from a magmatic arc terrane (Dickinson and Suczek, 1979). Feldspar generally occurs in association with volcanic grains; however, some sandstones in this lithofacies contain abundant volcanic material but negligible feldspar (i.e., volcaniclastic sandstones) and others contain abundant feldspar but little volcanic debris (i.e., lithofeldspathic sandstones). Dickinson and Suczek (1979) show that volcaniclastic sandstones generally are derived from undissected volcanic arcs whereas

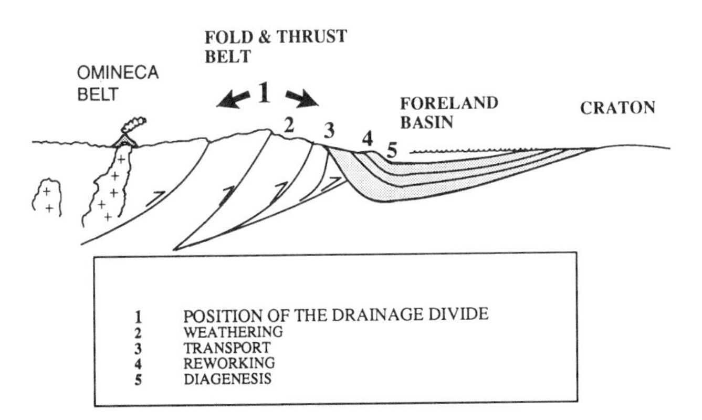

FIGURE 6. Controls on the composition of sandstones in the Western Canada foreland basin.

lithofeldspathic sandstones commonly are derived from the underlying granitic plutons exposed by erosion of the volcanic carapace. To simplify this discussion, sandstones containing significantly greater amounts of volcanic debris and/or feldspar than are present in a lithic sandstone are defined here as volcano-feldspathic sandstones.

Because of their highly variable composition, sandstones in the volcano-feldspathic lithofacies may be classified as litharenites, feldspathic litharenites, or lithic arkoses (Folk, 1968; Figures 2c and 3). The mineralogical and textural immaturity of the majority of the constituents in volcano-feldspathic sandstones suggests a first-cycle origin, rather than derivation from a recycled orogene. Volcano-feldspathic sandstones commonly have abundant amounts of "matrix," some of which may be detrital, but much of which commonly consists of severely compacted or altered framework grains (i.e., pseudomatrix or diagenetic matrix). Laths of detrital chlorite commonly are present in the matrix, and when associated with abundant chloritized volcanic grains and chlorite clay cement, lend a greenish color to volcano-feldspathic sandstones (Glaister, 1959).

Within volcano-feldspathic sandstones the most easily recognizable volcanic grains, but not necessarily the most abundant, consist of relatively large lath-shaped feldspar crystals set in a finer-grained, partially altered groundmass (Figure 3). Devitrified volcanic grains are also abundant but they have a microcrystalline fabric that is not diagnostic and resembles chert, causing incorrect identification. There is a wide range in the type and abundance of feldspar in this lithofacies, but plagioclase is generally dominant, as in the lithic lithofacies. The dominance of plagioclase over orthoclase suggests derivation from a volcanic terrane (Folk, 1968), whereas subequal amounts of the two feldspar types suggest a plutonic source (Dickinson and Suczek, 1979). As with the lithic sandstones, interpretations of provenance based on feldspar content may be of little relevance because of pre- and postburial alteration and dissolution.

Using the depositional megacycles of Leckie and Smith (1992), it is interesting to note that volcano-feldspathic sandstones are essentially confined to megacycles 2 and 4 [i.e., early Albian and Campanian-Maastrichtian time; Enclosure 1(A)] and generally coincide with distal drainage regimes (i.e., periods when the drainage divide was situated within the Omineca or Intermontane belts) and with periods of volcanism. The geographic distribution of volcano-feldspathic sandstones within cycles 2 and 4 is linked to regional sediment dispersal patterns controlled primarily by basin tectonics, but possibly influenced by eustatic sea level changes as well (James, 1985; Cant and Stockmal, 1989).

Volcano-Feldspathic Sandstones in Cycle 2

In the southern portion of the Western Canada foreland basin, the transition from Lower to Upper Mannville and equivalent strata is marked by a significant and abrupt increase in feldspar content, accompanied by the appearance of notable quantities of volcanic rock fragments, slate, and phyllite [Enclosure 1(A)] (Glaister, 1959; Carrigy and Mellon, 1964; Williams, 1964; Mellon, 1967; Ghent and Miller, 1974; Putnam, 1982; Putnam and Pedskalny, 1983). A similar change is observed in coeval rocks in Montana (Dyman et al., 1988). Upper Mannville sandstones also contain increased proportions of unstrained quartz, detrital chlorite, biotite, sodic igneous debris, smectite, and a more diverse heavy mineral assemblage (including apatite, garnet, hornblende, and pyroxene), and commonly are less texturally mature than underlying Lower Mannville sandstones (Glaister, 1959; Williams, 1964; Cameron, 1965; Nelson and Glaister, 1978).

There are various opinions about the origin of the feldspar and volcanic debris. Some believe that this material was derived solely from the erosion of plutons and preexisting volcanic rocks in the Western Cordillera (Rudkin, 1964; Rapson, 1965; Mellon, 1967; Putnam and Pedskalny, 1983). If this was the case, sedimentation of the cycle 2 sandstones would have been triggered by a tectonic event. However, the widespread occurrence of bentonite throughout Upper Mannville strata, in combination with the sudden appearance of fresh feldspar (including high-temperature sanidine), volcanic rock fragments, and sodic igneous material strongly suggests sediment contribution from then-active volcanos (Glaister, 1959; Williams, 1964; Cameron, 1965; Putnam, 1982; James, 1985; Dyman et al., 1988). In this case, sedimentation would have been the result of a combined volcanic and tectonic event.

During Upper Mannville deposition, volcanos are known to have been active in parts of Idaho and Washington (e.g., Idaho Batholith, McGookey et al., 1972) but also are believed to have been present in southeastern British Columbia (e.g., the Crowsnest Volcanics). Although few volcanic rocks are observed today in southeast British Columbia, there are numerous unroofed plutons that likely were the sites of active volcanism when emplaced during Early to mid-Cretaceous time (Figure 7A) (Gabrielse and Reesor, 1974). Because of asymmetric subsidence, major rivers in a foreland basin generally exhibit a basin-axial drainage pattern (Eisbacher et al., 1974). During most of Mannville time, areas of known or potentially active volcanism were drained by streams that initially flowed northeastward but fed major northwest-trending rivers (Figure 7A) (Eisbacher et al., 1974; McGookey et al., 1972; Putnam, 1982; Jackson, 1984; Leckie and Smith, 1992). As a result, volcano-feldspathic sandstones are distributed

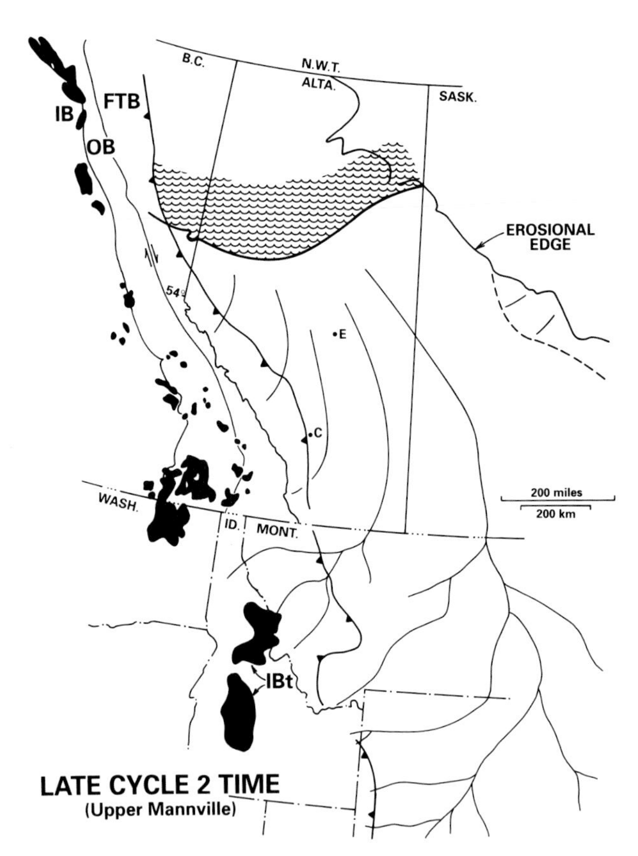

FIGURE 7A. Paleogeographic map during late cycle 2 time of Leckie and Smith, this volume (Upper Mannville), showing probable paleoflow directions (McGookey et al., 1972; Putnam, 1982; Leckie and Smith, 1992) and the locations of Lower and mid-Cretaceous plutons in British Columbia (Gabrielse and Reesor, 1974) and the Idaho Batholith (IBt). During this time the drainage divide had moved westward in the cordillera, allowing contribution of eroded volcanic debris from the western fold-thrust belt (FTB) and Omineca belt (OB). (IB) Intermontane belt.

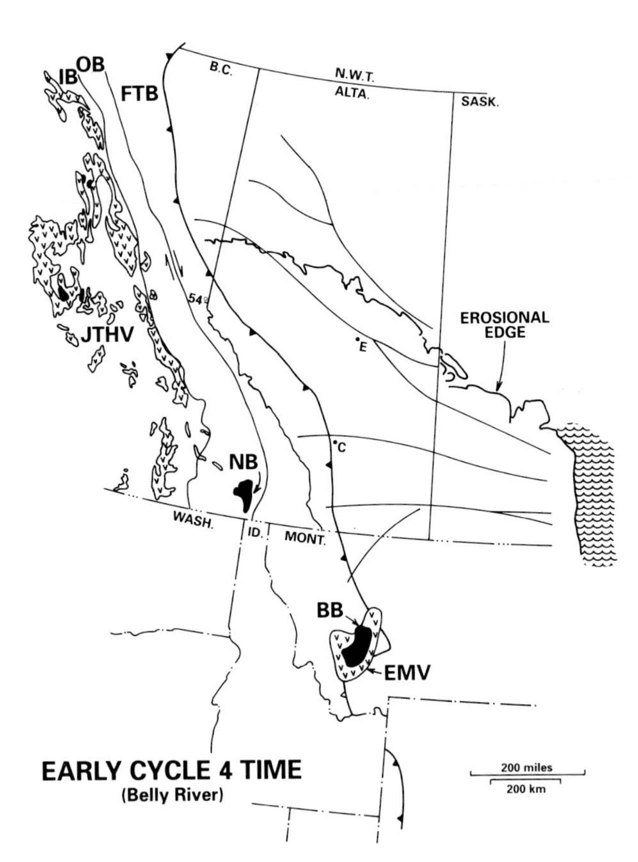

FIGURE 7B. Paleogeographic map during early cycle 4 time (Belly River), showing probable paleoflow directions (Ogunyomi and Hills, 1977; Eisbacher et al., 1974; Haywick, 1982; Rahmani, 1982; Taylor and Walker, 1984; Leckie and Smith, 1992; Mack and Jerzykiewicz, 1989) and the locations of Jurassic Takla-Hazelton volcanics (JTHV), Elkhorn Mountain Volcanics (EMV), Boulder Batholith (BB), Nelson Batholith (NB), and several small Upper Cretaceous plutons (solid fill). (Sources include Gabrielse and Reesor, 1974; McGookey et al., 1972; Tipper et al., 1981.)

throughout most of southern Alberta and, in places, within southwestern Saskatchewan and Montana. Volcano-feldspathic sandstones in cycle 2 sediments are less common in the northern foreland basin, reflecting a westward swing in the locus of mid-Cretaceous intrusions and associated volcanics relative to the fold-thrust belt (McMechan and Thompson, in press).

Volcano-Feldspathic Sandstones in Cycle 4

In places within the Western Canada foreland basin, the transition from cycle 3 to cycle 4 is coincident with a sudden and abrupt increase in volcano-feldspathic sediments. The range of feldspar and rock fragment content of uppermost Cretaceous to lower Tertiary sandstones is variable (Lerbekmo, 1963; Nelson, 1968; Carrigy, 1971; Ogunyomi and Hills, 1977; Iwuagwu and Lerbekmo, 1982; Haywick, 1982; Byers, 1969; Rahmani and Lerbekmo, 1975; Elliot, 1977; Misko and Hendry, 1979; Mack and Jerzykiewicz, 1989).

By Belly River time, alluvial flow within the basin had changed from a northwesterly axial drainage pattern to a southeasterly direction. In northern Alberta, major rivers are interpreted by Eisbacher et al. (1974) to have issued from the Liard and Peace River reentrants and drained southeastward toward the Bearpaw Sea (Figure 7B). Southeast of the Peace River reentrant, uppermost Cretaceous to Paleocene sandstones have southeasterly trends of framework and heavy mineral grain composition, indicating sediment dispersal in that direction (Carrigy, 1971; Eisbacher et al., 1974; Rahmani and Lerbekmo, 1975). Feldspar and volcanogenic material in these sandstones may have been derived from older volcanic terranes such as the Jurassic Takla-Hazelton volcanic complex in British Columbia (Carrigy, 1971; Eisbacher et al., 1974). The dominance of orthoclase relative to plagioclase feldspar in sandstones of the Edmonton Group suggests derivation from older volcanic sources (Chi, 1966).

In contrast to the southeasterly regional drainage pattern, sediment dispersal over southern Alberta, southern Saskatchewan, and Montana was predominantly east and northeast away from the rising cordillera (Mack and Jerzykiewicz, 1989; Taylor and Walker, 1984; McGookey et al., 1972). In these areas, sandstones in cycle 4 are enriched to varying degrees in feldspar and volcanic debris. In the Foothills of southern Alberta and Montana, volcano-feldspathic sandstones occur in the Belly River, Blood Reserve, and St. Mary River formations and in coeval units (Lerbekmo, 1963; Nelson, 1968; Carrigy, 1971; Ogunyomi and Hills, 1977; Lerand, 1983; Bibler and Schmitt, 1986; Mack and Jerzykiewicz, 1989). In southern Saskatchewan, volcano-feldspathic sandstones occur in equivalent strata, but also in younger Paleocene sandstones whose time equivalent strata in the Foothills are lithic [Enclosure 1(A)], suggesting derivation from a different volcano-feldspathic source at that time.

The common occurrence of bentonite in Upper Cretaceous strata throughout these areas is evidence of active local volcanism during Late Cretaceous and early Tertiary time. Paleoflow directions from volcano-feldspathic sandstones in the Plains of southeastern Alberta and Saskatchewan suggest that much of the feldspar and volcanic debris has been derived from the Elkhorn Mountain volcanics to the southwest (Figure 7B) (Byers, 1969; Ogunyomi and Hills, 1977; Misko and Hendry, 1979; Bibler and Schmitt, 1986). The Adel Mountain volcanics in Montana may have contributed volcano-feldspathic debris to the younger Paleocene Ravenscrag sandstones (Misko and Hendry, 1979). In contrast, easterly paleoflow directions from volcano-feldspathic sandstones in the southern Foothills of Alberta suggest derivation from the cordillera, directly to the west. As was the case for cycle 2 sandstones, the absence of volcanic sources in the cordillera has led some authors to suggest derivation from older plutons and associated volcanics such as the Nelson batholith in British Columbia (Lerbekmo, 1963; Nelson, 1968). However, several unroofed plutons in this area likely were the site of active volcanism when they were emplaced during later Cretaceous time (Figure 7B) (Gabrielse and Reesor, 1974). The disappearance of volcano-feldspathic sandstones in the early Tertiary is likely related both to dilution of volcano-feldspathic debris by sedimentary material shed from the uplifted Front Ranges of the Rocky Mountains (Misko and Hendry, 1979) and to erosion of the volcanic source terranes (Nelson, 1968; Carrigy, 1971).

Quartzose Lithofacies

Sandstones in the quartzose lithofacies consist predominantly of well-rounded monocrystalline quartz (Figure 3), reflecting derivation from a stable cratonic interior and/or a quartzose recycled orogenic provenance (Figure 1a; Dickinson and Suczek, 1979), and are classified mostly as quartz arenites and sublitharenites following Folk (1968) (Figure 2c). Although dominated by quartz, smaller quantities of lithic fragments and feldspar may be present. Lithic grains are predominantly quartzose sedimentary rock fragments such as chert and polycrystalline quartz, and any feldspar present is commonly orthoclase. Typically, the heavy mineral suite is dominated by the resistant assemblage of zircon, rutile, and tourmaline. In addition, detrital kaolinite commonly occurs in association with quartzose sandstones.

At no other time in the history of the Western Canada foreland basin are quartzose sandstones as prevalent and widespread as within the Aptian (i.e., lower half of megacycle 2 of Leckie and Smith, 1992) [Enclosure 1(A)]. Quartzose sandstones occur locally and at various stratigraphic levels within older sediments in cycle 1, but within cycle 2 they are essentially confined to Lower Mannville and coeval strata throughout central and eastern Alberta and western Saskatchewan.

Middle Jurassic, pre-cycle 1, quartzose sandstones (i.e., Rock Creek and Sawtooth) were derived from the east (Marion, 1982; Poulton, 1984) and are interpreted to represent the final stages of miogeoclinal sedimentation (Bally et al., 1966). Although the foredeep was developing during deposition of the Rock Creek and Sawtooth sandstones, loading occurred far to the west and was unlikely to have influenced sedimentation of these sands in the east (McMechan and Thompson, in press). Quartzose sandstones within the Late Jurassic to Early Cretaceous Kootenay and Minnes formations (cycle 1) were deposited along the western side of the narrow incipient foredeep, and likely were derived from older quartzose sediments uplifted in the fold and thrust belt (Gibson, 1985). Quartzose sandstones in the Monteith Formation may also have been derived from the craton to the east (Poulton, 1984).

Deposition of quartzose sandstones in the Aptian (cycle 2) occurred after significant expansion of the basin. Most workers agree that the quartzose sandstones within cycle 2 have been derived primarily from the craton to the east (Glaister, 1959; Williams, 1964; Carrigy, 1963, 1971; Mellon, 1967; Jardine, 1974; Bayliss and Levinson, 1976; Hopkins, 1981; Putnam, 1982; Davies, 1983; James, 1985). The widespread occurrence and temporal position of quartzose sandstones above the pre-Cadomin unconformity suggest some relation with cratonic tectonism. McMechan and Thompson (in press) suggest that the pre-Cadomin unconformity reflects cratonic uplift resulting from isostatic adjustment during Late Jurassic and Early Cretaceous time. Such an event could account for the significant flood of craton-derived quartzose debris into the basin in Aptian time. During Lower Mannville deposition, the climate in Alberta was believed to be warm to subtropical. The presence of potassium feldspar, mica, and detrital kaolinite in many cycle 2 quartzose sandstones suggests direct contribution of quartzose debris from the weathered crystalline rocks of the craton itself. However, the high textural maturity of many quartzose sandstones containing accessory chert and a resistant heavy mineral assemblage suggests derivation from the multicycled Precambrian sandstones and Paleozoic carbonates and clastics flanking the craton. It is also possible that the textural and mineralogic maturity of some quartzose sandstones has been enhanced by eolian transport (Hopkins, 1981) or by depositional reworking within high-energy environments (Putnam, 1982).

Lower Blairmore sandstones in the southern Alberta Plains and Foothills are equivalent to Lower Mannville quartzose sandstones, but consist of lithic sandstones as they were derived from the immediately adjacent cordillera. In places within central Alberta, quartz-rich Lower Mannville strata contain increased amounts of sedimentary lithic debris, possibly reflecting sediment mixing from both eastern and western terranes. However, in Aptian time, paleotopography on the Paleozoic surface (see figure 17 in Leckie and Smith, 1992) acted as a barrier that prevented regional mixing of eastern and western derived sands (Glaister, 1959). Hopkins (1981) states that it is sometimes difficult to distinguish between the two sources because Paleozoic carbonates and clastics are present in both eastern and western source areas.

FRAMEWORK-DEPENDENT PATTERNS OF DIAGENESIS

Introduction

In many sandstones in the Western Canada foreland basin, patterns of diagenesis and resultant trends in reservoir quality are closely related to original composition. As a result, a certain level of reservoir prediction can be achieved by evaluating key framework-dependent diagenetic features for each of the three lithofacies. However, in many instances, porosity-modifying diagenetic reactions have been governed by the hydrologic history of the sand. In the following section, framework-dependent patterns of diagenesis for the three major lithofacies are examined.

The extent to which diagenesis proceeds is controlled by factors such as depth and duration of burial, volume and chemistry of fluid flux, and temperature and pressure. In general, the degree of diagenetic modification in the Western Canada foreland basin increases with depth. Because of greater uplift and erosion in early Tertiary time (Hacquebard, 1977; Nurkowski, 1984; Osadetz et al., 1992), sediments in the western portion of the basin show diagenetic features typical of burial depths roughly 1 to 3 km greater than the present depths.

Diagenesis of the Lithic Lithofacies

Important framework-dependent diagenetic features within the lithic lithofacies include ductile grain compaction, quartz cementation, and authigenic clay development. A generalized paragenetic sequence comparing the key framework-dependent diagenetic events for the three lithofacies is shown in Figure 8.

From deposition to burial depths of about 1000 to 1500 m, reservoir quality reduction in most lithic sands occurs through burial compaction (Fuchtbauer, 1967). During compaction, intergranular porosity is irreversibly lost and can be regained only by dissolution of soluble framework constituents. The framework mineralogy of sands, specifically the ductile grain content, is known to influence strongly the rate of porosity loss during burial diagenesis (Nagtegaal, 1978; Scherer, 1987). Many lithic sandstones in the Alberta basin contain appreciable amounts of easily compressible argillaceous, micaceous, and schistose grains. Hence, in the absence of early pore-cementing phases, lithic sandstones commonly suffer significant porosity loss through ductile grain deformation.

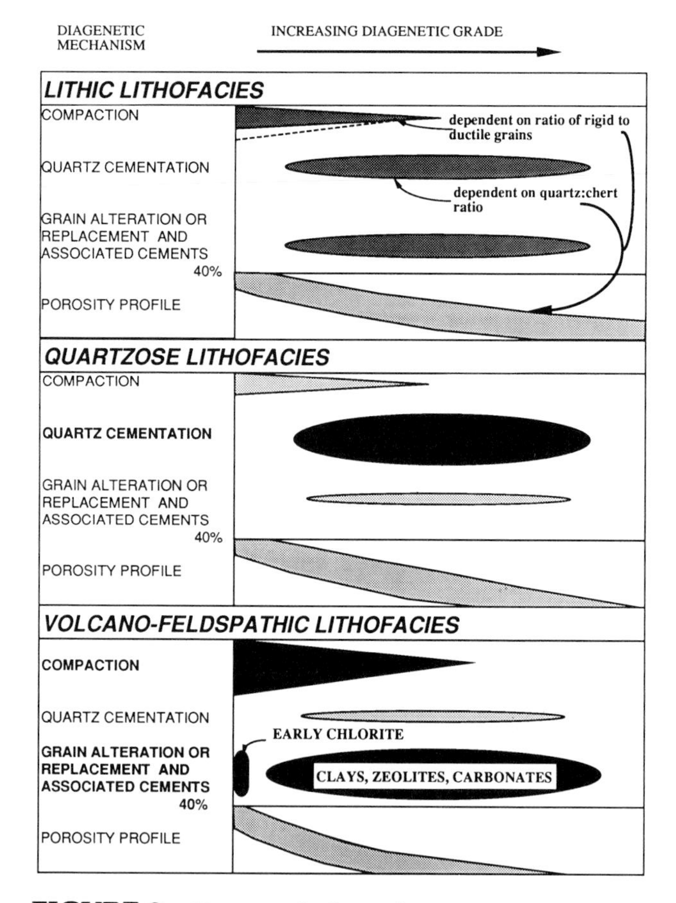

FIGURE 8. Framework-dependent paragenetic sequences for typical, fine-grained sandstones in the lithic, quartzose, and volcano-feldspathic lithofacies.

Houseknecht (1987) has shown that a graph comparing intergranular volume (minus cement porosity) versus volume of cement can be used to illustrate the relative effects of compaction and cementation on intergranular porosity reduction. However, because the relative importance of compactional processes will be exaggerated in ductile-rich sandstones, Houseknecht (1987) cautions about using this technique for ductile-rich sandstones. Bearing this in mind, a compaction/cementation crossplot of representative sandstones from the three lithofacies was constructed to give a crude comparison of the relative importance of compaction and cementation on porosity reduction (Figure 9). As might be anticipated, compactional processes account for the bulk of porosity reduction in volcano-feldspathic sandstones (which are rich in ductile and chemically unstable grains) but play a subordinate role in quartzose sandstones (which are rich in mechanically stable quartz). Because lithic sandstones contain variable proportions of stable quartzose grains, metastable chert, and ductile lithics, it is not surprising

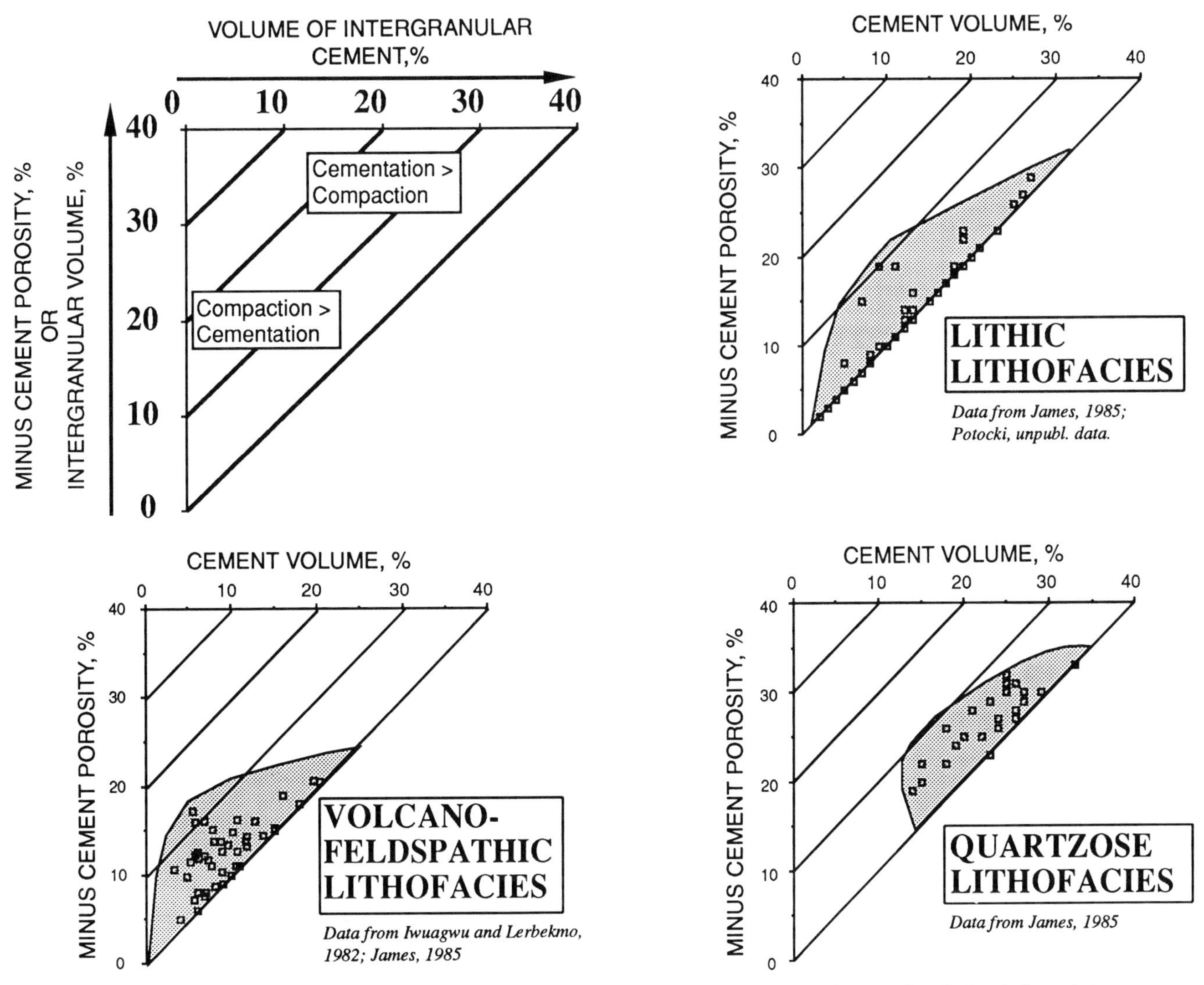

FIGURE 9. Crossplots comparing the relative effects of cementation and compaction on deeply buried sandstones (present burial depth, >2000 m) from the lithic, quartzose and volcano-feldspathic lithofacies. Porosity in all cases has been reduced to less than 10% bulk volume.

that there is wide variation in the degree to which compaction and cementation processes influence porosity loss.

Porosity-depth plots of several representative sandstones from the lithic, volcano-feldspathic, and quartzose lithofacies are illustrated in Figure 10. Although data are limited, especially in the quartzose lithofacies, the plots do reveal that porosity decrease with depth is greatest in the volcano-feldspathic lithofacies. Figure 10 also shows that, at equivalent burial depths (i.e., at approximately 2200 m in this instance), quartzose sandstones may have notably higher porosities than lithic sandstones. However, the wide scatter in porosity values at any given depth, evident for each of the lithofacies, indicates that porosity is also influenced by variables other than depth and ductile grain content.

In deeply buried lithic sandstones, after much porosity has already been lost through compaction, cementation begins to fill the remaining intergranular porosity. Quartz overgrowths generally dominate the framework-dependent cements in the lithic lithofacies, especially in quartz-rich varieties that contain a larger number of potential quartz nucleation sites. Locally, early carbonate cements may be dominant, but carbonate cementation commonly is related to early depositional conditions rather than framework mineralogy. However, because detrital carbonate provides a suitable substrate for carbonate nucleation, carbonate cementation may have been facilitated in lithic sandstones locally containing large amounts of detrital Paleozoic clastic carbonate grains.

Various authors (Cant and Ethier, 1984; James, 1985; Krause et al., 1987; Rosenthal, 1989) have noted in lithic sandstones and conglomerates that monocrystalline quartz grains commonly develop large syntaxial overgrowths whereas chert and polycrystalline quartz develop a myriad of smaller, randomly oriented euhedral crystals. Porosity may be totally obliterated by quartz cement in rocks with a high proportion of monocrystalline quartz, whereas chert-rich sandstones and conglomerates, subjected to the same burial conditions, may preserve more

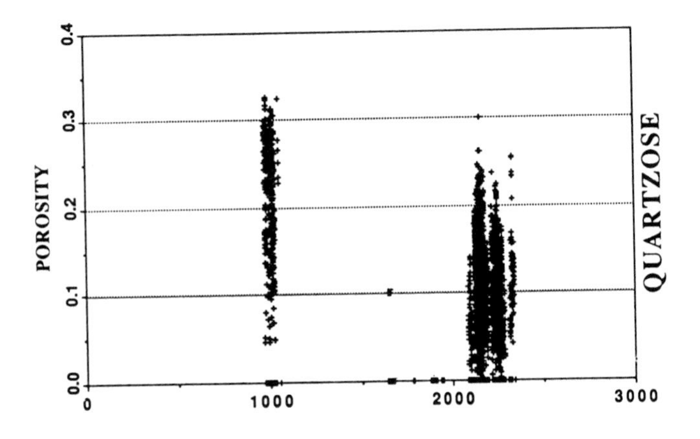

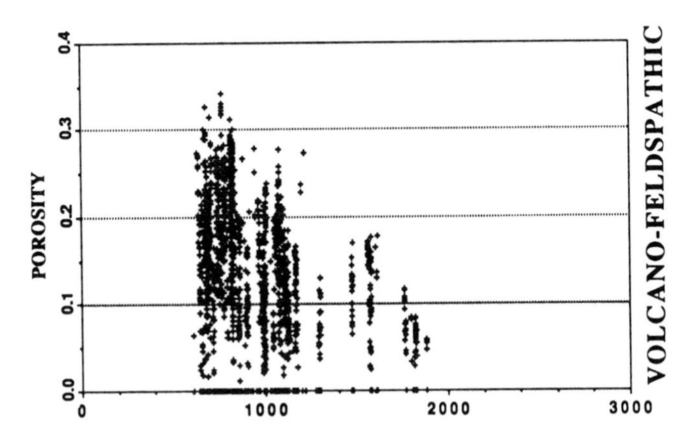

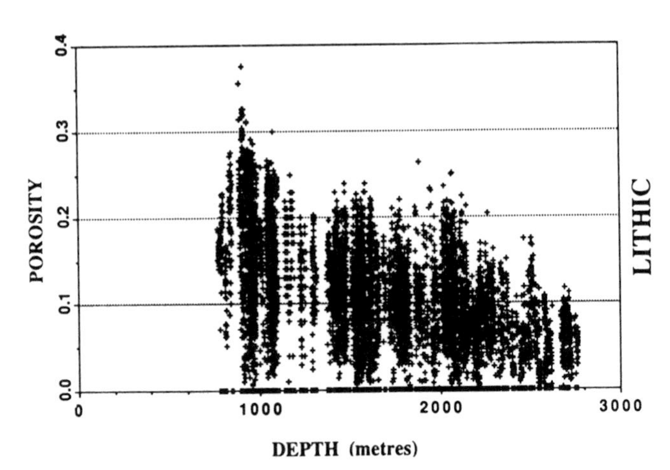

FIGURE 10. Plots of core analysis porosity vs. present-day depth for several representative sandstones from the lithic, volcano-feldspathic, and quartzose lithofacies. Data sources: lithic lithofacies, Cardium sandstone in Pembina and Crossfield fields, Viking sandstone in Joffre and Caroline fields, and Glauconitic sandstone from several locations between Calgary and Edmonton; volcano-feldspathic lithofacies, Belly River and St. Mary River sandstones from several locations in southern Alberta; quartzose lithofacies, Lower Mannville Basal Quartz sandstones in Medicine River, Gilby, and Bantry fields.

intergranular porosity (e.g., Cadomin conglomerate, Ellerslie sandstones). Some of these authors suggest that the mechanism of crystal growth causes the difference in cementation patterns because smaller, randomly oriented crystals common in chert interfere with each other during growth and thus develop overgrowth cement at a lower rate than the larger syntaxial overgrowths on quartz grains.

Rimstidt and Barnes (1980) have studied silica precipitation kinetics, and their experiments show that the rate of quartz precipitation increases with increasing surface area. Finer grain size will result in increased surface area, and this would imply that microcrystalline quartz should show the greatest rate of silica precipitation and development of quartz overgrowths, the reverse of the suggestion that interference by small crystals during growth inhibits overgrowth formation. The silica minerals in chert generally are not documented, but it is likely that chert contains more soluble forms of silica such as chalcedony ($SiO_2 \cdot nH_2O$), as shown in Figure 11. In a rock containing quartz and chalcedony, which is present in chert, the pore water can be stable only with respect to one silica mineral. The greater solubility of chert could cause local dissolution or slow precipitation of quartz on chert and precipitation of quartz on larger, monocrystalline quartz grains. This suggests that differences in cementation in chert- and quartz-rich sandstones may not be related to crystal growth interference.

Regardless of the mechanism that causes quartz overgrowths to form preferentially on monocrystalline quartz, rather than on chert, intergranular porosity is affected differently in quartz-rich rocks than in chert-rich rocks. Chert content generally is greater in relatively coarse-grained sands, and therefore depositionally controlled textural properties can be used to assist in predicting occurrences where chert is more abundant. Sandstone dominated by monocrystalline quartz will be more likely to be cemented than adjacent chert-rich rocks, and thus understanding of depositional environments can provide clues to explore for trends in porosity.

With the increasing temperatures that accompany deeper burial, previously unreactive lithic grains may become susceptible to alteration and dissolution reactions leading to authigenic clay precipitation. Authigenic clay minerals, although seemingly volumetrically insignificant relative to other authigenic cements, may exert substantial control on the reservoir quality and sensitivity of lithic reservoirs in the Alberta basin (e.g., Thomas and Miller, 1980). Most lithic sandstones contain abundant aluminosilicate grains but are poor in feldspar, volcanogenic, and ferromagnesian minerals, and thus kaolinite and illite (usually in lesser amounts) are the two most abundant authigenic clay minerals that have formed from chemical grain breakdown (Carrigy and Mellon, 1964). Many lithic sandstones contain higher amounts of unstable sodium-, calcium-, magnesium-, and iron-bearing minerals, and framework-dependent clay authigenesis has locally favored development of chlorite and/or smectite at the expense of kaolinite or illite. Anomalous occurrences of chlorite and smectite in lithic sandstones may be related to hydrologic influences that are less predictable.

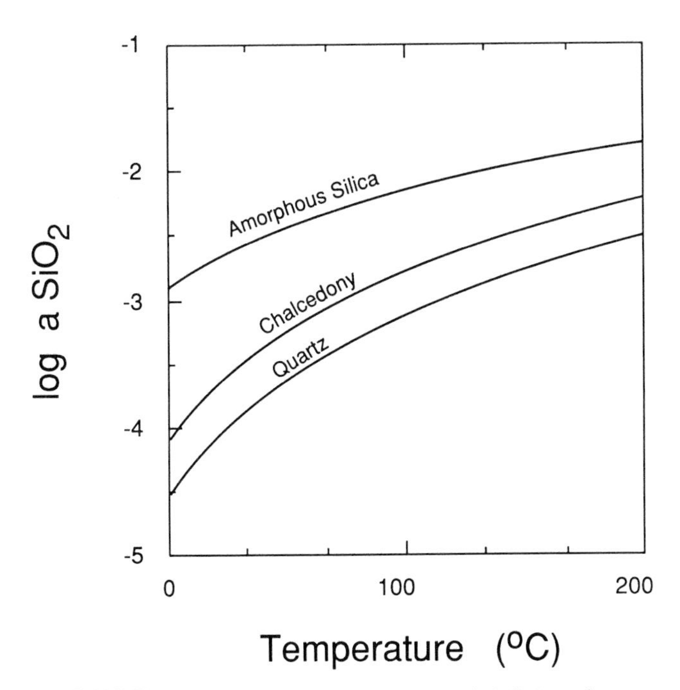

FIGURE 11. Plot showing that the solubilities of different silica minerals (Berman et al., 1985) are different. Relatively high-solubility chert, as the mineral chalcedony, could be dissolved and precipitate as quartz overgrowths.

Diagenesis of the Volcano-Feldspathic Lithofacies

Reservoir quality in the volcano-feldspathic lithofacies is strongly dependent on the mechanical and chemical instability of the feldspar and volcanic constituents. Porosity is reduced primarily by ductile framework compaction and by clay, zeolite, and carbonate cementation but may be regained in later stages of diagenesis through cement or grain dissolution. Figure 8 illustrates a generalized paragenetic sequence highlighting framework-dependent diagenetic features common to the volcano-feldspathic lithofacies.

In addition to volcanogenic grains, volcano-feldspathic sandstones generally contain large amounts of other ductile constituents such as argillaceous sedimentary rock fragments, phyllitic grains, and mica. These components compact easily, resulting in significant reduction of intergranular porosity during early burial. The chemical instability of feldspar and volcanic constituents further facilitates porosity loss by chemical grain breakdown during compaction. Volcanic grains, especially those containing glass, generally are more reactive than feldspar. The degree of porosity loss during burial of volcano-feldspathic sandstones is strongly related to the abundance of feldspar and volcanic rock fragments, and reservoir destruction with depth is greatest in the volcano-feldspathic lithofacies relative to the quartzose and lithic lithofacies (Figures 8 and 10).

Much of the clay, carbonate, and zeolite cement present in volcano-feldspathic sandstones has formed either by direct alteration of feldspar and volcanic debris, or by precipitation from pore waters whose composition was controlled by the local breakdown of these constituents. Authigenic clays are the most common alteration product within volcano-feldspathic sandstones, although chemical conditions have favored zeolite precipitation in some instances. Carrigy and Mellon (1964) and Ghent and Miller (1974) document the occurrence of laumontite and heulandite in the Beaver Mines Formation within the Foothills of southern Alberta.

Because volcano-feldspathic sandstones contain framework minerals rich in silicon, aluminum, calcium, sodium, iron, and magnesium, the authigenic clay minerals that form are a complex group including chlorite, berthierine, smectite, mixed-layer smectite, and kaolinite (Carrigy and Mellon, 1964; Sedimentology Research Group, 1981; Putnam and Pedskalny, 1983; Hutcheon et al., 1989). To illustrate, pore-lining smectitic clays and the iron mineral berthierine (a 7-Å mineral easily confused with kaolinite by x-ray diffraction) frequently dominate the volcano-feldspathic Clearwater oil sands in the Cold Lake region of Alberta (Sedimentology Research Group, 1981; Dean and Nahnybida, 1985; Abercrombie, 1988), whereas kaolinite and illite appear to be the most abundant authigenic clay minerals in volcano-feldspathic-poor sandstones (Carrigy and Mellon, 1964; Glaister, 1959).

Clay minerals commonly are more reactive than many other authigenic constituents, and we might expect that, in addition to the framework mineralogy, they would show some dependence on the hydrologic history of the units in which they are found. Pore-lining clays, including some or all of chlorite, chlorite/smectite, and berthierine, commonly occur in greater abundance in sandstones interpreted to have been deposited in estuarine environments, suggesting that their formation is favored in the presence of fresh to brackish waters (Ayalon and Longstaffe, 1988; Longstaffe, 1986).

The common occurrence of large volumes of calcite grain replacement and calcite intergranular cement within volcano-feldspathic sandstones such as the Belly River Formation (Lerbekmo, 1963; Storey, 1979; Iwuagwu and Lerbekmo, 1982) also suggests some dependency upon framework mineralogy. Surdam and Boles (1979) show that the dissolution and albitization of feldspar grains and volcanic material provide a ready source of Ca^{2+} cations, which, in association with sufficient HCO_3^-, result in carbonate cementation or replacement. In some sandstones containing detrital carbonate grains, carbonate precipitation may be stimulated by the existing carbonate nucleation sites and detrital carbonate may represent a local source of carbonate cement. The degree to which the occurrence of carbonate cement is related to fluid hydrology compared with framework mineralogy is difficult to assess. In the Upper Cretaceous Blood Reserve sandstones of southern Alberta, calcite cement crosscuts primary textural features, suggesting that precipitation may have been related to the former position of the water table or to a mixing zone between early depositional waters (Lerand, 1983). Although framework mineralogy may provide the source of constituents for carbonate cementation, the necessary components also may be transported with moving waters.

Volcano-feldspathic sandstones have limited amounts of quartz overgrowth cement, a feature that may reflect the low percentage of quartz grain substrates in these rocks. In addition, the complex reactions of other minerals in volcano-feldspathic rocks may increase competition for silica, preventing quartz cementation. Furthermore, authigenic pore-lining phases such as berthierine commonly develop during early diagenesis in these sandstones and may hinder subsequent quartz development by covering available nucleation sites. Opaline silica is present in volcano-feldspathic rocks of the Clearwater Formation (Sedimentology Research Group, 1981; Abercrombie, 1988). Dissolved silica levels in these rocks would be anticipated to be high with respect to quartz saturation, and quartz overgrowths should be a common authigenic constituent. However, quartz overgrowths have not developed, suggesting that restrictions on the number of nucleation sites and competition of other mineral reactions for silica must have prevented their development.

Volcano-feldspathic sandstones generally are characterized by lower reservoir quality relative to the other lithofacies. In some instances, though, the chemical instability of the framework allows porosity to be regained at various stages during diagenesis through the dissolution of grains, replacive cements, or intergranular cements. For overall porosity to increase during dissolution, pore waters must remove reaction products from the system rather than allowing them to be redistributed by local reprecipitation of some pore-obstructing authigenic mineral. The secondary porosity observed in many volcano-feldspathic sandstones has not necessarily resulted in a net gain of porosity, but may only represent the exchange of intergranular porosity for porosity produced by dissolution of framework grains.

Diagenesis of the Quartzose Lithofacies

Less porosity is lost by framework compaction in this lithofacies than in the lithic and volcano-feldspathic lithofacies because of the abundance of mechanically stable quartz. Quartz-rich sandstones also tend to have abundant quartz cement and authigenic kaolinite. The widespread occurrence of quartz overgrowths and authigenic kaolinite in quartzose sandstones suggests that diagenetic mineral assemblages have an element of dependence on framework mineralogy. A generalized paragenetic sequence of framework-dependent diagenetic features for this lithofacies is presented in Figure 8.

Without early cementation by soluble cements such as carbonate, or termination of quartz cementation by the emplacement of hydrocarbons or early grain-rimming authigenic phases, intergranular porosity lost in quartz-rich sandstones during compaction and/or quartz cementation has little likelihood of being regained later in diagenesis. In most subsurface environments, the relatively low solubility of silica makes quartz grains and cement poor precursors for secondary intragranular and intergranular porosity (Schmidt and McDonald, 1979). As a result, mineralogically and texturally mature quartzose sandstones may, in some cases, be less porous and permeable than less mature lithic sandstones that have undergone a similar burial history (James, 1985).

Textural relations among authigenic phases in quartzose sandstones in the Western Canada foreland basin commonly imply that quartz cementation commences relatively early during diagenesis and continues throughout burial. Minus cement porosities in some quartz cemented sandstones may be as high as 25 to 30%, indicating pervasive quartz cementation at comparatively shallow burial depths (James, 1985). In pervasively quartz cemented sandstones (i.e., $>10\%$ quartz cement), it is difficult to demonstrate material balance between the volume of silica present as cement and the estimated volume of silica either derived internally or transported from external sources to the sand (McBride, 1989; Potocki, 1989). One possible internal silica source includes the dissolution of more soluble forms of silica present in chert and other silicate grains. However, most quartzose sandstones in the Western Canada foreland basin do not contain (or exhibit evidence of having contained) significant amounts of chemically unstable silicate grains that may have liberated silica upon dissolution. Therefore, pressure solution of quartz grains has often been suggested as a major internal silica source for quartz cementation. Cathodoluminescence work in quartz-rich sandstones elsewhere has shown that sutured boundaries, used as evidence for internal silica generation, actually occur between already existing overgrowths (Sippel, 1968; Zinkernagel, 1978). This does suggest, however, that pressure solution in more deeply buried sandstones and finer-grained sediments may provide an external silica source.

To resolve the apparent silica shortage in some quartz-rich sandstones, various external silica sources and fluid transport mechanisms have been suggested. Grain-dissolution reactions in adjacent or distant shales and sandstones, and the smectite-to-illite transformation, commonly have been cited as external silica sources (Hower et al., 1976; Boles and Franks, 1979). Ascending compactional waters, waters modified by illite-smectite reactions, and meteoric waters recharging into basin margins have been suggested as possible transporting agents of silica (Powers, 1967; Burst, 1969; Perry and Hower, 1970; James, 1985; Longstaffe and Ayalon, 1987; Ayalon and Longstaffe, 1988; McBride, 1989; Potocki, 1989).

At shallow depths, comparatively large volumes of pore fluid are available for silica transport. However, the amount of silica inherent within marine and meteoric pore waters is inadequate for significant silica cementation. If additional silica is somehow added to the waters, low silica solubilities at shallow burial temperatures preclude extensive mass transport of silica. At greater depths, sufficient silica may be generated by various diagenetic reactions in sandstones and shales. Although silica solubilities are higher under increased burial temperatures, estimated volumes of compactional fluid are generally inadequate to transport the large volumes of silica required for cementation (Bjorlykke, 1984; Bloch, 1989). Using oxygen isotope analyses of quartz overgrowths, Longstaffe and Ayalon (1987) have attributed the formation of quartz overgrowths in the Viking Formation to deeply circulating meteoric waters entering the basin during the Laramide uplift. Quartz cements in the Muddy sandstone of the Powder River basin in the United States have also been interpreted as precipitating from meteoric waters (Burrows et al., 1984). Quartz precipitation may have taken place as deeply recharging/circulating meteoric waters mixed with, and cooled, silica-rich compactional waters. Further precipitation may have occurred as silica-laden mixed meteoric-compactional waters cooled during updip migration toward discharge zones.

Shales are variously described as exporters of silica (Boles and Franks, 1979; Siever, 1962; Towe, 1962; Hower, 1983; McBride, 1989) or importers of silica (Hower et al., 1976; Yeh and Savin, 1977). The arguments of these authors are based on studies of the Gulf Coast, a basin in which the smectite-illite reaction has been well documented and may produce silica. Studies of Cretaceous shales in the Western Canada foreland basin are few, but Bloch (1989) shows that the illite in shales of the lower Cretaceous Harmon Member of the Fort St. John Group likely originated from the dissolution of micas and feldspars, rather than from a reaction involving smectite, and that smectite was probably never present in substantial quantities in these shales. James (1985) suggests that the illite-smectite reaction in some Lower Cretaceous shales may have supplied silica, but does not present detailed clay mineral analyses to show evidence of smectite having been present in substantial quantities and having been a precursor for illite. Studies show the Lower Cretaceous shales in Alberta to be poor in smectite (Foscolos and Kodama, 1974; Foscolos et al., 1976; Connolly, 1989). Smectite is observed locally in ashfall deposits (bentonites) in Upper Cretaceous shales in southern Alberta and southwestern Saskatchewan and may be generally more common in Upper Cretaceous than in Lower Cretaceous sediments (Bloch, personal communication). In summary, the silica/water material balance shortage evident in many quartz-cemented quartzose sandstones in the Western Canada foreland basin is a problem for which there are more questions than answers.

The common occurrence of authigenic kaolinite within quartzose sandstones of the Alberta foreland basin has been documented by several workers (Carrigy and Mellon, 1964; Ghent and Miller, 1974; Carrigy, 1963; Hopkins, 1981; Potocki, 1981; Lefebvre and Hutcheon, 1986; Hutcheon and Lefebvre, 1988). Kaolinite that can be interpreted to be authigenic on the basis of textural criteria can be observed to form by recrystallization of detrital kaolinite or as a product of the incongruent dissolution of feldspars, micas, aluminous silicates, or shale rock fragments. Lefebvre and Hutcheon (1986) suggest that kaolinite in quartz-rich sands in the Lloydminster area may have formed from feldspars or detrital micas when meteoric water entered the rocks. Authigenic kaolinite can be observed in some quartzose sandstones in which there is no evidence for a detrital precursor, suggesting that kaolinite has been deposited from solution (i.e., neoformation) (Potocki, unpublished data). The low solubility of aluminum as an inorganic species in most subsurface waters (Curtis, 1985; Boles, 1984) makes aluminum transport over long distances a difficult problem that complicates the understanding of formation of authigenic kaolinite and other aluminosilicates. Surdam and Crossey (1987) have suggested that organic acids, particularly difunctional acids, may allow aluminum to be transported over significant distances, but published compositional data for formation waters in the Western Canada foreland basin do not generally show aluminum or organic acid concentrations, and hence we cannot evaluate this proposed mechanism. Recent data presented by Connolly et al. (1990a, b) should help to resolve this question.

PATTERNS OF DIAGENESIS AND THE HYDROLOGIC REGIME

From the moment of deposition, sand bodies may be subjected to flow of a succession of waters of varying chemistry, the transition from one fluid regime to another being controlled by the subsidence and uplift history of the basin (Galloway, 1984). In a normally subsiding basin, depositional waters are displaced by compactional waters emanating from adjacent fine-grained sediments. If relative sea level changes during early burial, waters of a composition different from the original connate waters may enter the sand body (Machemer and Hutcheon, 1988). During deeper burial, at very low rates of fluid flow, waters produced from mineral reactions may be added to the hydrologic system. Foreland basins such as the Western Canada foreland basin are different from "normal" downgoing basins because they are brought back into a regime of meteoric water flow by epeirogenic uplift. Sediments within a foreland basin generally undergo a more complex hydrologic history that leads to increased diagenetic complexity and decreased predictability of reservoir quality.

The evolution of fluid regimes during burial outlined above allows some general inferences to be made about the flow of waters and possible effects on diagenetic modifications of sediments. During the first tens of meters of burial, the composition of waters is largely controlled by depositional environment. At marine marginal settings, mixing of marine and meteoric waters may take place, and this may cause precipitation of carbonate cements and/or dissolution of cements and clasts. Porosity and permeability are relatively high, allowing for relatively high fluid flow rates. At the low temperatures typical of this stage, mineral reaction rates are relatively low and water-rock interactions are less likely to play a major role in diagenetic modification. Biological processes, such as sulfate reduction and methanogenesis, significantly alter fluid composition and are major contributors to cementation, particularly by carbonates and pyrite in early diagenesis.

At somewhat greater burial depths (on the order of kilometers), fluid flow rates are much lower and movement of fluids is generally upward and outward, driven by compactional processes. Transport of materials in solution to produce authigenic phases is probably on the order of kilometers, with the most soluble constituents probably being transported farthest (carbonate cements, for example). Water-rock interactions become increasingly important in the dissolution of framework grains and the precipitation of authigenic minerals. The organic matter in fine-grained sediments matures and produces water-soluble organic acids that may play a role in dissolution and cementation. Migration and entrapment of oil and gas may cause a slowdown, or cessation, of diagenetic processes. This stage ends with virtual stagnation of the waters and very little transport of material. The much higher solubility of silica and aluminum at high temperatures may allow diagenetic processes to proceed more rapidly than at shallower depths, even though fluid flow is more restricted. During very deep burial, CO_2 and H_2S may be produced from carbonates and sulfates, and these acidic gases may play a role in further dissolution.

Foreland basins are uplifted and introduced into a realm of higher fluid flow and participation of relatively low-salinity meteoric water in diagenetic processes. In the Western Canada foreland basin, the underlying Paleozoic section is dominated by carbonates and evaporites, and fluid salinities typically are high, in the range of 1 to 10%. In the clastic-dominated Mesozoic section, salinities are lower, on the order of 0.01 to 0.1%. These two major divisions of rock and fluid composition are juxtaposed over each other during thrusting. Deformation, combined with the increase in hydrologic potential and availability of meteoric water, makes very late diagenesis by mixing of waters during uplift an important diagenetic process.

In general, large-scale patterns of diagenesis in sandstone reservoirs in the Western Canada foreland basin are not well understood. In many Alberta sandstone reservoirs, diagenesis may have been unpredictably controlled by the hydrologic history. Patterns of diagenesis locally may have little or no relation to original sand composition. In these cases, knowledge of the lithofacies is not sufficient for reliable prediction of reservoir quality.

Depositional and Early Compactional Waters

The chemistry of waters deposited with a sediment reflects the environment in which the sediment was deposited. Sands typically are associated with mud interbeds in the same depositional setting, and, upon compaction, waters in the sand generally are displaced by waters of similar parentage emanating from the adjacent muds. Temperatures during early burial are too low, and the water:rock ratio too high, for the sand framework mineralogy to influence pore water composition appreciably, but low-temperature biological reactions affect the chemistry of compactional waters squeezed from organic-rich muds within the first few meters of burial (Gautier and Claypool, 1984). Marine pore waters are chemically and isotopically distinct from fresh waters, and early diagenetic mineral phases may be distinctive for different depositional settings. Relations between fresh to slightly brackish depositional waters and the occurrence of early diagenetic pore-lining chlorite and/or berthierine, siderite, and kaolinite have been documented in various sandstone lithofacies in the Western Canada foreland basin (Storey, 1979; Meshri, 1981; Longstaffe, 1986; Lefebvre and Hutcheon, 1986; Hutcheon and Lefebvre, 1988; Ayalon and Longstaffe, 1988; Abercrombie, 1988). Although volumetrically negligible, early authigenic pore-lining phases such as berthierine, which may be a precursor to chlorite, and chlorite have been shown to hinder later development of syntaxial quartz overgrowths, thereby preserving reservoir quality.

Carbonates, principally calcite, may form early cements that occupy up to 40% of the rock. Their high solubility compared with other authigenic minerals makes them important as potential precursors for cement dissolution during later burial. Many calcite cements that form early can be related to mixing of marine and meteoric waters (Runnells, 1969; Back and Hanshaw, 1970, 1971; Machemer and Hutcheon, 1988). Calcite cements can also result from bacterial reduction of sulfate and the corresponding production of aqueous carbonate. Isotopic signatures of calcite cements frequently indicate that both processes were active agents of carbonate cementation. Reduction of sulfate produces sulfide, which will react rapidly with available iron to precipitate pyrite cement. Most clastic units in the Western Canada foreland basin contain sufficient iron to have produced at least some early authigenic pyrite. Early silicate cements are more difficult to document except by textural criteria.

During Late Cretaceous and Tertiary time, continental depositional conditions prevailed in western portions of the Alberta foreland basin in response to eastward migration of the fold-thrust belt. This suggests that early diagenetic mineral parageneses in these rocks should reflect fresh to brackish waters (Hitchon and Friedman, 1969). Within gently dipping confined aquifers, typical of many sands deposited on the broad shelf of shallow foreland epeiric seas, small fluctuations in relative sea level (on the order of meters) may have resulted in flushing and mixing of connate reservoir waters over large lateral distances (Machemer and Hutcheon, 1988). Early diagenetic minerals, considered diagnostic of a particular depositional setting, may therefore have precipitated in waters strikingly different from those in which the sediment was originally deposited. Influx of meteoric water may result in leaching and porosity creation. Ayalon and Longstaffe (1988), in a detailed isotopic and mineralogical study of units in the Belly River Formation, show the continuous evolution of diagenetic phases from early meteoric to brackish conditions, deeper diagenetic events, and the eventual return of meteoric influence during uplift in the Tertiary.

Deeper Waters

Prior to the Laramide uplift in the early Tertiary, sediments in the Western Canada foreland basin were buried to progressively greater depths and drainage likely was dominated by compactional flow (Hitchon, 1984). Comparison of porosity lost to compaction in sandstones and shales shows that between 1 and 2 km of burial, the compaction curves cross and the porosity of shales typically is lower than that of sands. The volume of shales is much greater than that of sands, typically comprising 80% of the sediment in most of the Western Canada foreland basin, and flow of fluids from shales to sands would be expected. This flow of fluids from shales is often cited as a major factor in sandstone diagenesis, but studies of the changes in shale mineralogy with depth, particularly in the Western Canada foreland basin, are few. During this stage of diagenesis, organic matter in shales matures and soluble organic acids, oil, and gas are produced. Surdam and coworkers (Surdam et al., 1984; Surdam and Crossey, 1987; Surdam et al., 1989) have proposed the hypothesis that these organic acids are a major influence on dissolution of minerals, particularly feldspars, and precipitation of authigenic kaolinite. However, there are no published data for organic acid content of formation waters in the Western Canada foreland basin. The migration and trapping of hydrocarbons are thought to curtail diagenetic processes. Certainly the presence of oil and gas accumulations suggests that fluid flow must be slow after trapping, and this, combined with the generally low mutual solubility of hydrocarbon and aqueous phases, reinforces the hypothesis. During very late stages of burial, CO_2 can be generated by mineral reactions involving carbonates (Hutcheon et al., 1980; Smith and Ehrenberg, 1989; Hutcheon and Abercrombie, 1990) and H_2S may be generated by reactions of light hydrocarbon gases with anhydrite (Krouse et al., 1988). H_2S may cause dissolution, particularly of carbonates, and CO_2 may cause precipitation or dissolution of carbonates, depending on how the pH is controlled.

Processes characteristic of the compactional regime include formation of illite and chlorite, dissolution of K-feldspar, albitization of plagioclase (including dissolution of feldspar and precipitation of kaolinite), and the precipitation of iron-rich carbonate cements including ankerite. Late-stage kaolinite, commonly of the dickite variety, may be produced by reactions involving K-feldspar or albitization of plagioclase. During this stage of diagenesis, porosity may be regained by dissolution of relatively less stable grains, such as feldspars or lithic fragments, and early-formed calcite or anhydrite cements. The major control on fluid composition is the local buffering of waters by reactions with the minerals in the rocks. At this stage of diagenesis, fluid flow rates are low and temperatures are high, resulting in more rapid reaction rates and allowing the rock to dominate water compositions by water-rock interaction. Hutcheon et al. (1980) have shown the formation of late-stage chlorite and proposed this as a mechanism for production of CO_2 during late stages of diagenesis. The rise of CO_2-charged fluids through the sedimentary section could produce carbonate cements, presuming that the pH is buffered by the mineralogy of the rocks and not the carbonate system itself. Interactions between organic materials, particularly via the soluble organic acids, has been suggested as an influence on inorganic reactions in shales and sandstones. The most recent example is the hypothesis for the development of porosity proposed by Surdam et al. (1989). However, there are no published data on organic acid content of formation waters in the Western Canada foreland basin. Although many of the features in the hypothesis suggested by Surdam et al. (1989) can be observed in the Western Canada foreland basin, the available data are not sufficient to ascribe the generation of porosity and authigenic parageneses to the presence (or absence) of organic acids.

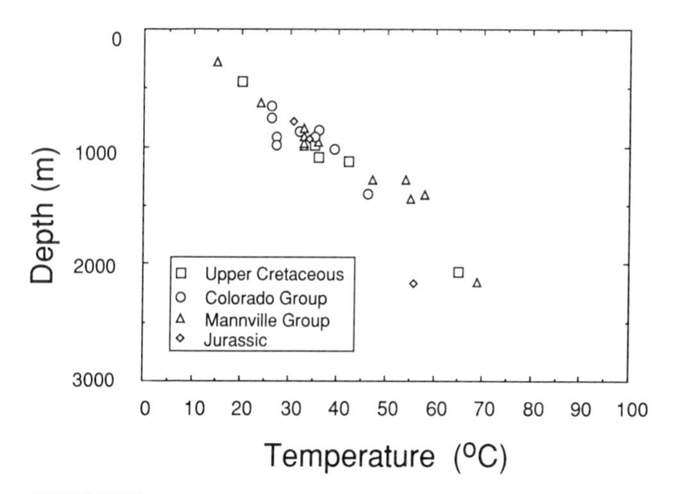

FIGURE 12. Present-day temperatures (bottom hole) from Hitchon and Friedman (1969) for wells that were used for consideration of water compositions in this study.

Later Meteoric Waters

Sediments within the Western Canada foreland basin have been deposited within five major tectonically triggered cycles, each terminated by a period of nondeposition, or uplift and erosion. Topographic highs created during uplift may have provided sufficient hydraulic potential to cause significant recharge of meteoric waters. However, it is unlikely that, prior to the Laramide orogeny, the cordillera was uplifted as high as it was during Eocene time. Since the Laramide orogeny, fluid flow within the basin has been dominated by gravity-driven meteoric recharge (Toth, 1978; Garven and Freeze, 1984). Hitchon and Friedman (1969) and Hitchon et al. (1971) interpret the chemical and isotopic signature of present-day formation waters in most Alberta hydrocarbon reservoirs to reflect the mixing of meteoric water and diagenetically modified seawater.

Temperatures in the past have been higher than they are at present. Figure 12 shows present-day temperature vs. depth for formation water samples to be discussed in this paper. The data are from Hitchon et al. (1971) and Hitchon and Friedman (1969). Waters are mobile within sedimentary sections and, in broad terms, the lithologies and waters of the Paleozoic and Mesozoic parts of the Western Canada foreland basin are quite different. The data of Hitchon et al. (1971) show that the Paleozoic is dominated by carbonates and evaporites with high-salinity waters. Typical total dissolved solids (TDS) contents are 100,000 mg/L or higher, and the waters are rich in Na^+, Ca^{2+} and Cl^- (Figure 13). The Mesozoic rocks are dominantly siliciclastic; their contained waters have TDS contents in the range of 30,000 mg/L, and the composition of these waters is dominated by Na^+, Cl^- and HCO_3^- (Figure 14). Studies of fluid flow regimes by Toth (1978) and Garven and Freeze (1984) suggest that mixing between these waters should take place. Figure 15 shows the concentration of Br^- and Cl^- from Hitchon et al. (1971) for waters from Paleozoic rocks (closed squares), Triassic and

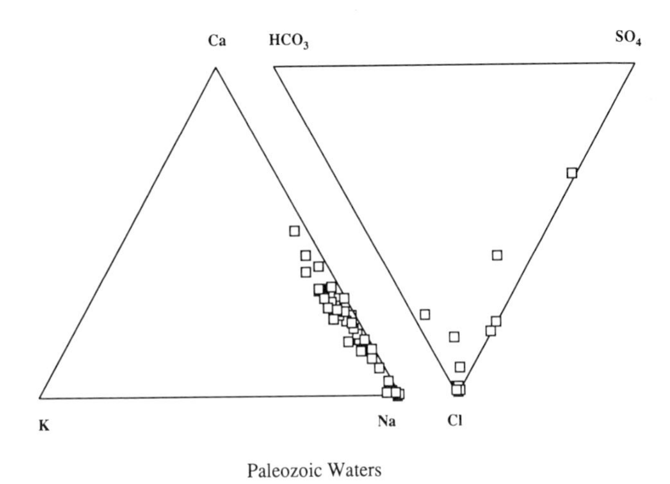

FIGURE 13. Compositions of waters in Paleozoic rocks (data from Hitchon et al., 1971).

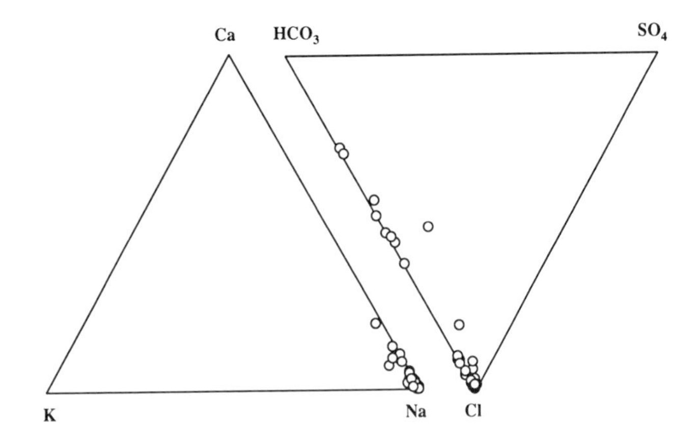

FIGURE 14. Compositions of waters in Mesozoic rocks (data from Hitchon et al., 1971).

Carboniferous rocks (shaded triangles), and Cretaceous and Jurassic rocks (open circles). At low salinities, Br^- and Cl^- are conservative—that is, they do not participate in water-rock reactions—and thus these elements are a measure of the degree to which waters from various sources are mixed. The waters from Paleozoic rocks show a scatter of data, reflecting their high salinity and interaction with evaporitic minerals. However, the formation waters from Triassic-Carboniferous and Cretaceous-Jurassic units show a comparatively linear array, which may be interpreted as representing mixing of fresher waters with more saline waters associated with older rocks. This is evidence that waters in the basin are mixing with each other during topographically driven flow caused by uplift. The mixing of these waters may have important consequences for diagenesis.

Runnells (1969) recognized that the mixing of waters is a possible geologic mechanism for the precipitation and dissolution of carbonates. The evidence that waters in Paleozoic rocks have mixed (and may be mixing at the present day) with waters in Mesozoic rocks suggests that the compositions of these waters should be examined to determine if the mixtures might be agents of precipitation or dissolution of authigenic phases. Because the waters from Paleozoic rocks are relatively rich in calcium and the waters from Jurassic-Cretaceous units are enriched in bicarbonate, their mixtures can be expected to be supersaturated with most carbonate species. In addition, because the interface between the Mesozoic and Paleozoic rocks, and their associated waters, is in the subsurface over much of the Western Canada foreland basin, mixing of the waters during cross-formational fluid flow may be a significant factor in reservoir quality assessment. To examine the effects of mixing Paleozoic and Mesozoic waters, two water analyses from Hitchon et al. (1971) were mixed in various proportions using the geochemical modeling program SOLMNEQF (Aggarwal and Gunter, 1986). The waters chosen were from the Viking Formation (analysis #36, Chigwell field, Lower Colorado Group, from Hitchon et al., 1971) and the Woodbend Group (analysis #25, Worsley field, Upper Devonian, from Hitchon et al., 1971). These specific analyses were chosen because they were closest to saturation with calcite and had a reasonable cation-anion charge balance, although similar results would be obtained for most combinations of waters from Paleozoic rocks with waters from Mesozoic rocks. Figure 16 shows the result of adding incremental amounts of very saline Na-Ca-Cl formation waters from Paleozoic rocks to the less saline Na-HCO_3^- waters typical of Mesozoic rocks. The saturation index (S.I.), a measure of the degree of over- or undersaturation, is plotted on the vertical axis and is defined as the logarithm of the ion activity product divided by the equilibrium constant. For calcite, the ion activity product (IAP) is the product of calcium and bicarbonate ion activities divided by the activity of the hydrogen ion, as described by the equation

$$CaCO_3 + H^+ \leftrightarrow Ca^{2+} + HCO_3^-$$

As small proportions of saline water from Paleozoic rocks are added to the less saline water from Mesozoic rocks, the mixture of the two waters rapidly becomes oversaturated with respect to both dolomite and calcite. The result of mixing in this range of compositions would be expected to be precipitation of calcite and dolomite as cements. One might predict that Mesozoic units that unconformably overlie the Paleozoic, particularly in the more eastern regions of the basin, would be carbonate cemented in zones where mixing has occurred. The saturation diagram (Figure 16) may also have some implications for diagenesis of Paleozoic rocks, as shown in the following section.

For waters with high proportions of saline water, the mixtures are undersaturated with respect to dolomite and calcite, but there is a range of mixtures, between about 10 and 15%, that are oversaturated with calcite and undersaturated with dolomite. Mixtures in this range favor dissolution of dolomite and replacement by calcite. Calcitized dolomites have been observed in Paleozoic rocks of the Grosmont Formation (Theriault and Hutcheon, 1987) and

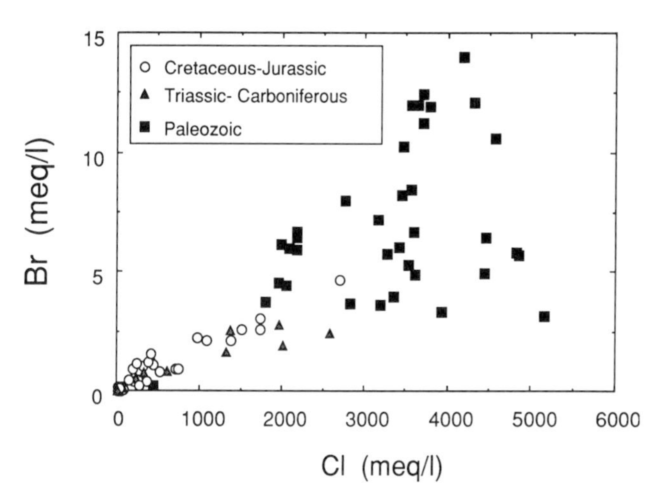

FIGURE 15. The concentration of Br^- and Cl^- (meq/l) from Hitchon et al. (1971) for waters from Paleozoic rocks (closed squares), Triassic and Carboniferous rocks (shaded triangles), and Cretaceous and Jurassic rocks (open circles), suggesting that there is mixing between the saline waters in Paleozoic rocks and the less saline waters in Mesozoic rocks.

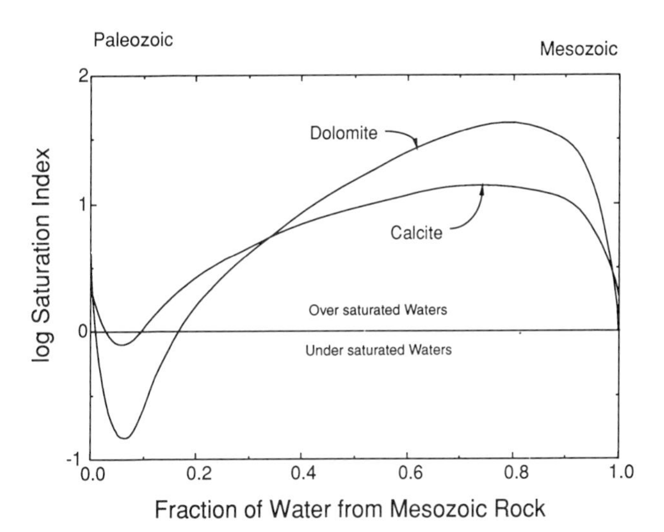

FIGURE 16. Mixtures of very saline Na-Ca-Cl formation waters from Paleozoic rocks and less saline Na-HCO_3^- waters typical of Mesozoic rocks, showing varying degrees of under- or oversaturation with respect to dolomite or calcite. Fractions of relatively low-salinity waters from Mesozoic rocks are shown on the horizontal axis (i.e., 1.0 is "pure" Mesozoic water). If low fractions of waters typical of Mesozoic rocks are mixed with waters typical of Paleozoic rocks, there should be dissolution of carbonates. If high fractions of waters typical of Mesozoic rocks are mixed with waters typical of Paleozoic rocks, there should be precipitation of carbonates.

may have been produced by such a process, supporting the concept that formation water mixing between Mesozoic and Paleozoic rocks can affect diagenesis and reservoir quality and the possibility that cross-formational flow can occur, at least at unconformities.

Dissolution and precipitation of carbonate cements in clastic rocks produced by mixing should be relatively predictable where regional ground water flow trends and the compositions of waters are known. Hydrocarbon emplacement probably is accompanied by flow of water, a process in which mixing could take place. For hydrocarbon accumulations that are trapped by diagenetic cements, it is possible that mixing of in-situ waters with waters accompanying moving hydrocarbons could cause cementation and result in hydrocarbon trapping. The local distribution of such processes in the past would be difficult to predict. Mixing as a mechanism for precipitation of carbonates has implications for hydrocarbon recovery, particularly in waterflooding. Most surface or ground waters are Na^+-HCO_3^- waters. Allowing waters with these compositions to mix during water injection, particularly in units with very saline Na-Ca-Cl waters, could cause cementation in a formation subject to secondary recovery.

The authigenic minerals observed in clastic rocks in the Western Canada foreland basin include kaolinite, smectite, illite, and chlorite. The composition of produced waters reflects the nature of the minerals formed at higher temperatures during thermal enhanced recovery (Abercrombie and Hutcheon, 1986; Hutcheon et al., 1988), and, given the longer times available during diagenesis, one might expect that formation waters would reflect the present-day authigenic mineral assemblage. Hutcheon et al. (1988) observed that waters produced during thermal recovery from sandstones in the quartzose lithofacies tended to indicate reaction between kaolinite and smectite, whereas waters from volcaniclastic lithofacies sandstones tended to indicate reaction between smectite and analcime. Abercrombie (1988), in a detailed study of produced waters from a volcaniclastic sandstone, showed reactions involving smectite to be a major influence on water compositions.

Water analyses from Hitchon et al. (1971) were processed using SOLMNEQF (Aggarwal and Gunter, 1986) to determine the activity (a) ratios [e.g., $aMg^{2+}/(aH^+)^2$, aNa^+/aH^+, aK^+/aH^+, and so on] of various species in solution. The pH values of the solutions were determined using the values listed in Hitchon et al. (1971) and calculated to the temperatures for the same sample intervals as given in Hitchon and Friedman (1969). The activity ratios of various species for the waters from Cretaceous and Jurassic rocks can then be compared to the ratios that would be expected if the authigenic minerals in the rocks are determining the activity ratios in the fluid. The stability fields for authigenic minerals have been calculated using the computer program PTA (Brown et al., 1988a; Brown et al., 1988b; Berman, 1988; Berman et al., 1985). This program contains data for most minerals observed as authigenic phases in sedimentary rocks. Abercrombie (1988) obtained data for Ca-, Mg-, K-, and Na-beidellite, and these data were also included in the calculations.

Figure 17 shows the aNa^+/aH^+ vs. aK^+/aH^+ activity ratios of waters from Hitchon et al. (1971). Filled symbols are waters from lithic or volcaniclastic rocks and open symbols are waters from quartz-rich rocks. It is clear that the aNa^+/aH^+ ratios for waters from volcaniclastic and

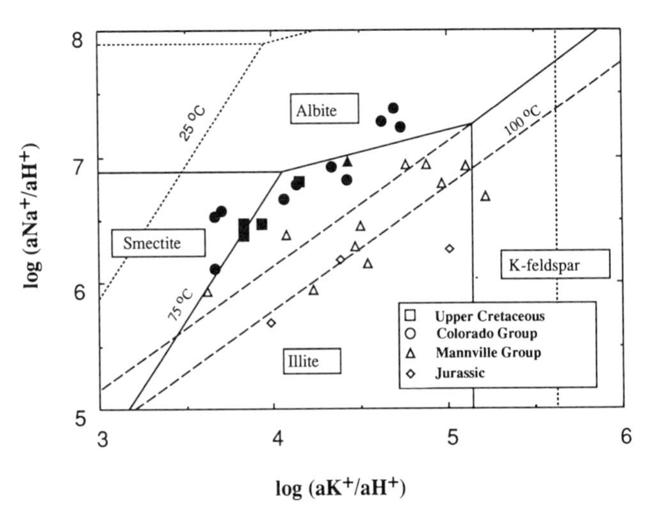

FIGURE 17. The aNa^+/aH^+ vs. aK^+/aH^+ activity ratios of waters from Mesozoic rocks calculated using SOLMNEQF (Aggarwal and Gunter, 1986). Filled symbols are waters from lithic or volcanic-lithic rocks and open symbols are waters from quartz-rich rocks. The phase boundaries between smectite (Na-beidellite), illite, albite, and K-feldspar are shown at 75°C (solid lines) and 25°C (dotted lines). The metastable extension of the phase boundary between albite and K-feldspar is shown at 100°C and 75°C by the dashed lines. The generally good agreement between the phase boundaries and the plotted water compositions can be taken as strong evidence that the activity ratios in solution are being set by the authigenic mineralogy of the host rock.

lithic lithofacies are higher than aNa^+/aH^+ ratios for waters from quartzose lithofacies. The phase boundaries between smectite (Na-beidellite), illite, albite, and K-feldspar are shown at 75°C (solid lines) and 25°C (dotted lines). The metastable extension of the phase boundary between albite and K-feldspar is shown at 100 and 75°C by the dashed lines. The activity ratios in the fluid, represented by the symbols plotted on the diagram, and the phase boundaries between minerals are independent. Correspondence between the phase boundaries and the plotted water compositions can be taken as strong evidence that the activity ratios in solution are being set by the authigenic mineralogy of the host rock. Many of the data points fall on or near phase boundaries. For the waters from lithic or volcanic lithofacies, the data suggest that reactions among smectite, illite, and albite are influencing water compositions, and it might be expected that these minerals would be observed as authigenic phases present in the pore spaces in these rocks. The waters from quartzose lithofacies tend to lie on the phase boundary (or the metastable extension of this boundary) between K-feldspar and albite, suggesting that these minerals, rather than smectite and illite, will be observed as authigenic phases in quartzose rocks. Kaolinite is not shown explicitly in these diagrams, and to consider the effect of kaolinite on water compositions the aNa^+/aH^+ activity ratios are shown as functions of reservoir temperatures taken from Hitchon and Friedman (1969).

The reactions that define the phase boundaries on this figure are generally hydrolysis reactions—that is, they involve the reaction of a cation such as Na^+ with an aluminous silicate such as kaolinite to produce a Na-aluminous silicate and a hydrogen ion (H^+). The observations of Hutcheon et al. (1989) and Abercrombie (1988) indicate that the ratio of aNa^+/aH^+ is controlled by hydrolysis reactions similar to that described above. The Na^+ concentration of waters in Paleozoic rocks generally is much higher than in waters in Mesozoic rocks, and the total Na^+ concentration of mixtures of the two waters is most likely to be controlled by mass transport during fluid flow. In addition, the total Na concentration is on the order of 10^3 or 10^4 and is generally higher for waters in Paleozoic rocks. The aNa^+/aH^+ ratio of waters appears to be mediated by silicate reactions (Figure 17), and thus the inflow of very saline, Na-rich waters from Paleozoic rocks into less saline waters in Mesozoic rocks must be balanced by increases in the activity of H^+, a number on the order of 10^{-6} or 10^{-7}. This corresponds to decreasing the pH, and this process could cause dissolution of calcite or of other minerals whose stability is pH dependent. For each mole of Na^+ added via fluid mixing, depending on the actual stoichiometry of the silicate hydrolysis reactions, a mole of H^+ is produced. If we consider the dissolution of calcite by the reaction

$$CaCO_3 + H^+ \leftrightarrow HCO_3^- + CaS^{2+},$$

one mole of H^+ will dissolve one mole of calcite.

The consequences of the calcite dissolution equation can be estimated using some approximate mass balance considerations. The two waters used in the mixing example for calcite precipitation can be used as an example. A mixture of 10% water from the Paleozoic unit has a molality of Na^+ of 0.71 compared with 0.47 for the Mesozoic water, an increase of 0.24 mole of Na^+, which would require a release of 0.24 mole of H^+ and dissolution of 0.24 mole of calcite. To place this in terms of volume of calcite dissolved, we will use 1 L of water-filled porosity in a rock with 20% porosity, or 5 L (10 kg) of rock plus pore space. Exchanging a single pore volume of fluid in the proportions noted would result in dissolution of 0.24 mole of calcite, or approximately 25 g or 10 cm^3. This is approximately a 1% incremental increase in porosity. Higher proportions of water from Paleozoic formations would result in more dissolution.

The reaction rate of carbonates typically is higher than that of silicates. Accordingly, we might expect that formation water compositions with respect to activity ratios of $aCa^{2+}/(aH^+)^2$ and $aMg^{2+}/(aH^+)^2$ would be set by equilibrium with carbonates. Figure 18 shows the stability of dolomite and calcite versus an exchange reaction between Ca- and Mg-smectite at 75°C (solid lines) and 25°C (dashed lines). The activity ratios of waters do tend to follow the exchange reaction for the smectites, rather than the carbonates, suggesting that, although the activity product that includes HCO_3^- for carbonate dissolution may be set by carbonates, the Ca/Mg ratio seems to be controlled by the silicates. The activity ratios of the water

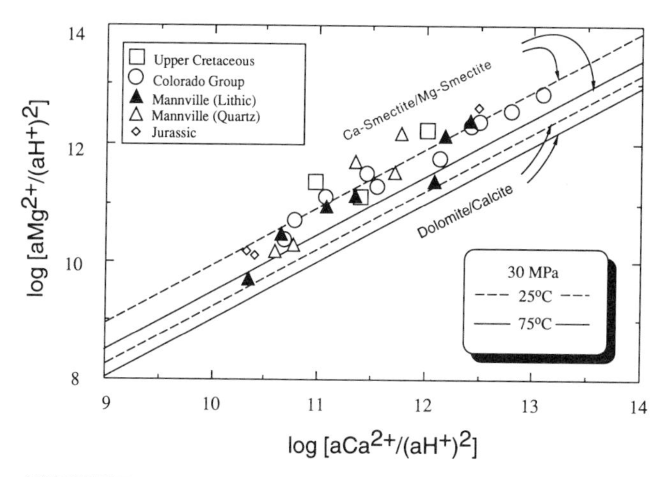

FIGURE 18. The stability of dolomite and calcite and an exchange reaction between Ca- and Mg-smectite at 75° (solid lines) and 25°C (dashed lines). The activity ratios of waters do tend to follow the exchange reaction for the smectites, rather than the carbonates, suggesting that the Ca/Mg ratio seems to be controlled by the silicates. The activity ratios of the water lie within the dolomite stability field, suggesting that water-rock interactions with silicates could be interpreted to be a potential mechanism producing dolomite cementation in clastics.

lie within the dolomite stability field, suggesting that water-rock interactions with silicates could be interpreted to be a potential mechanism producing dolomite cementation in clastics.

SUMMARY

The evidence from the compositions of formation waters strongly suggests that water-rock interaction is a process that is ongoing at the present time. Presumably these interactions also occurred during burial and diagenesis in the past, indicating that the history of diagenesis is a continuum of physical events (sedimentation, burial, fluid compaction, and heat flow) interacting with chemical properties (bulk rock and fluid compositions). Changes in fluid and rock properties therefore are incremental over time and space. The waters responsible for past authigenic events have long passed from the rock, and thus information on past hydrologic regimes can be gained only through utilization of detailed petrographic and geochemical techniques such as petrologic determination of paragenetic sequences, stable isotope geochemistry, and fluid inclusion analysis. There is a tendency for later diagenetic events to be superimposed on, or even to totally obliterate, earlier events. Thus the recognition of diagenetic evolution is difficult. An important goal of applying analytical techniques to the study of diagenesis is to be able to associate specific diagenetic features with a particular hydrologic regime in order to allow extrapolation to unknown portions of the reservoir. The work of Ayalon and Longstaffe (1988) has shown that the trends in isotopic compositions of authigenic minerals are consistent with the general burial history of the Western Canada foreland basin. Early in the burial history, fresh, brackish, or marine waters may have been displaced by waters modified by water-rock interaction during deepest burial. Uplift related to the Laramide orogeny introduced meteoric waters driven by topographic relief, which progressively displaced the original waters that were already in the process of being modified by water-rock interaction. The evidence that waters and rocks interact begs the important question as to how influential is changing water chemistry by flow of meteoric or other waters on the bulk rock mineralogy. Boles (1984) suggests that flow rates in an active sedimentary basin are probably less than a few meters per year. The chemical evidence (Hutcheon et al., 1988) would suggest that metastable chemical equilibrium is established over a time frame too short to be measured in years. The concentration of elements in solution is orders of magnitude lower than the concentration of these elements in rocks, with the general exception of hydrogen and oxygen. These two factors, the approach to equilibrium during water-rock reactions and the higher concentrations of elements in rocks than in water, suggest that high volumes and/or rates of fluid flow are required to cause significant changes in the bulk chemistry of rocks. Oxygen isotopic compositions reflect the possible origin and evolution of waters because oxygen is significantly concentrated in water in comparison with most rocks. For example, the fluid/rock concentration ratio of oxygen is about 1.75, whereas the same ratio for potassium is on the order of 0.0025 to 0.05. Quite clearly, oxygen isotopic values of waters can be reflected in rocks, but to affect the potassium content (and thus the mineralogy) of a rock is much more difficult.

General patterns of flow could be predicted, at least crudely, by combining the sedimentological patterns of basin filling through time, as reported by Jervey (1992), with numerical hydrological models such as those developed by Garven and Freeze (1984) or Hitchon (1984). A realistic model would include the distribution of lithofacies, estimates of the rate of change of horizontal and vertical permeabilities in various units as a function of total compaction, and the general change in relative hydrostatic heads in the rising cordillera. Prediction of diagenetic changes by water-rock interactions using such a model would be very complex. A chemical model for mineral reactions and the approximate compositions of depositional and formation waters would have to be integrated with the flow model, and realistic assumptions about flow rates as a function of time and temperature would have to be produced. Even assuming that the variables of the model could be accurately constrained, the model itself would be extremely cumbersome for numerical calculations. Although general patterns could be predicted with such a model, detailed reservoir-scale predictions would be much more problematic. The empirical approach of understanding lithology, depositional facies, patterns of fluid flow, and burial history, combined with petrographic observations, is probably a more practical and efficient method of determining reservoir quality and how it has been affected by diagenesis.

CONCLUSIONS

The initial composition, and rock-water interactions, of sediments within the Western Canada foreland basin ultimately have been controlled by the tectonic forces responsible for the development, filling, burial, and uplift of the basin itself. The tectonic history, in particular the uplift corresponding to the Laramide orogeny, has had a profound influence on fluid flow, causing late infiltration of relatively fresh meteoric water and mixing of saline waters in underlying Paleozoic units with the less saline waters of the Mesozoic units that comprise the foreland basin sediments. Because it has influenced drainage divides and exposed particular rock types in the source areas at particular times, the tectonic setting has exerted significant control on the framework grain composition of the sediments that were deposited. This in turn determines the course of subsequent diagenesis, providing an opportunity to make crude predictions of reservoir quality as it has been affected by burial and diagenesis.

The lithic lithofacies are recycled orogenic sandstones with diagnostic "salt and pepper" appearance. They are affected variably by compaction and quartz cementation depending on the ductile grain content and the relative proportion of lithic grains to monocrystalline quartz. Typical diagenetic effects are derivation of clay (primarily kaolinite) from feldspar and dissolution of unstable lithic fragments. Because of the chemical and physical instability of components in volcano-feldspathic sandstones, porosity loss is dominated by compaction. Typical diagenetic effects are the alteration of framework grains to produce chlorite, chlorite-smectite, illite, illite-smectite, and kaolinite clay minerals, carbonate cements, and possibly zeolites. The extreme variety and intensity of diagenesis in this lithofacies is a reflection of the complex chemical composition of the original sediments and provides the possibility of the development of secondary porosity by dissolution of framework grains and/or early-formed cements. Potential reservoir quality, although generally low, is extremely variable because of the unpredictable pathways of diagenesis. In contrast, the quartzose lithofacies is chemically and mechanically stable, and quartz cementation is a more important factor and compaction relatively less important in determining porosity loss in this lithofacies. Unpredictable factors that affect reservoir quality are determined by the waters that attend diagenetic reactions. These may produce extensive carbonate cements and/or kaolinite. Quartz cements are chemically stable, leaving little opportunity for porosity to be recovered by dissolution processes in this lithofacies.

The complex hydrologic evolution of the Western Canada foreland basin during uplift and burial has produced local variations that limit the ability to predict reservoir quality in each lithofacies. Generally the first 1000 m of burial has been accompanied by subsurface mixing of marine and meteoric waters. This process has in some rocks produced kaolinite, siderite, and early dissolution of unstable framework grains. Later burial produces different diagenetic minerals, depending on the chemical composition of the rocks in each lithofacies. The intrusion of meteoric water, driven by rising topography, has caused mixing of saline waters from Paleozoic rocks with less saline waters in Mesozoic rocks. Where these units are in contact, there is the potential for dissolution and/or precipitation of authigenic phases, further complicating the prediction of potential reservoir quality.

The composition and diagenetic modification of sediments in the Western Canada foreland basin are inextricably linked with the tectonic forces responsible for its development. Tectonics have influenced sedimentation patterns, provenance, sediment composition, hydrodynamics, and the composition of waters that were present at various stages of burial and uplift. These complex variations of rock and water chemistry have caused equally complicated patterns of rock-water interactions that produce a bewildering array of diagenetic effects. The major period of uplift during the Laramide orogeny distinguishes the diagenetic evolution of this basin from that of other basins that have not undergone significant uplift.

ACKNOWLEDGMENTS

We wish to thank Drs. H. J. Abercrombie, D. J. Cant, R. P. Glaister, J. F. Lerbekmo, F. J. Longstaffe, and D. A. McDonald for reviewing various drafts of this manuscript. Dan Potocki would especially like to thank D. A. McDonald for discussions during the early stages in the formulation of this paper. Gratitude is also expressed to Shell Canada Limited for the use of company resources in the preparation of this work. Ian Hutcheon is grateful to Shell Canada, the Natural Sciences and Engineering Research Council of Canada, and the University of Calgary for their support of a Senior Industrial Fellowship held during the time this paper was in progress.

REFERENCES CITED

Abercrombie, H. J., 1988, Water-rock interaction during diagenesis and thermal recovery, Cold Lake, Alberta: Ph.D. thesis, Department of Geology and Geophysics, The University of Calgary, Calgary, Alberta, 183 p.

Abercrombie, H. J., and I. E. Hutcheon, 1986, Remote monitoring of water-rock-bitumen interactions during steam-assisted heavy oil recovery: Fifth International Symposium on Water-Rock Interactions, Reykjavik, Iceland, August 1986, p. 1–4.

Aggarwal, P. K., and W. D. Gunter, 1986, SOLMNEQF: a FORTRAN-77 version of SOLMNEQ: ARC/AOSTRA Joint Agreement Report 8687-4.

Ayalon, A., and F. J. Longstaffe, 1988, Oxygen isotope studies of diagenesis and pore-water evolution in the Western Canada sedimentary basin: evidence from the Upper Cretaceous basal Belly River Sandstone, Alberta: Journal of Sedimentary Petrology, v. 58, p. 489–505.

Back, W., and B. B. Hanshaw, 1970, Comparison of chemical hydrogeology of the carbonate peninsulas of Florida and Yucatan: Journal of Hydrology, v. 10, p. 330–368.

Back, W., and B. B. Hanshaw, 1971, Rates of physical and chemical processes in a carbonate aquifer, *in* Nonequilibrium systems in natural water chemistry: American Chemical Society Advances in Chemistry Series No. 106, p. 77–93.

Bally, A. W., P. L. Gordy, and G. A. Stewart, 1966, Structure, seismic data, and orogenic evolution of southern Canadian Rocky Mountains: Bulletin of Canadian Petroleum Geology, v. 14, p. 337-381.

Bayliss, P., and A. A. Levinson, 1976, Mineralogical review of the Alberta oil sand deposits (Lower Cretaceous, Mannville Group): Bulletin of Canadian Petroleum Geology, v. 24, p. 211-224.

Berman, R. G., 1988, Internally-consistent thermodynamic data for minerals in the system Na_2O-K_2O-CaO-MgO-FeO-Fe_2O_3-Al_2O_3-SiO_2-TiO_2-H_2O-CO_2: Journal of Petrology, v. 29, p. 445-522.

Berman, R. G., T. H. Brown, and H. J. Greenwood, 1985, An internally consistent thermodynamic data base for minerals in the system Na_2O-K_2O-CaO-MgO-FeO-Fe_2O_3-Al_2O_3-SiO_2-TiO_2-H_2O-CO_2: Atomic Energy of Canada Limited Technical Report 377, 62 p.

Bibler, C. J., and J. G. Schmitt, 1986, Barrier-island coastline deposition and paleogeographic implications of the Upper Cretaceous Horsethief Formation, northern Disturbed belt, Montana: The Mountain Geologist, v. 23, p. 113-127

Bjorlykke, K., 1984, Formation of secondary porosity: how important is it?, *in* D. A. McDonald and R. C. Surdam, eds., Clastic diagenesis: AAPG Memoir 37, p. 277-286.

Blatt, H., G. V. Middleton, and R. C. Murray, 1972, Origin of sedimentary rocks: Englewood Cliffs, NJ, Prentice Hall, 634 p.

Bloch, J. D., 1989, Diagenesis and rock-fluid interaction of Cretaceous Harmon Member (Fort St. John Group) mudstones, Alberta and British Columbia: Ph.D. thesis, Department of Geology and Geophysics, The University of Calgary, Calgary, Alberta, 165 p.

Boles, J. R., 1984, Secondary porosity reactions in the Stevens sandstone, San Joaquin Valley, California, *in* D. A. McDonald and R. C. Surdam, eds., Clastic diagenesis: AAPG Memoir 37, p. 217-224.

Boles, J. R., and S. G. Franks, 1979, Clay diagenesis in Wilcox sandstones of southwest Texas: implications of smectite diagenesis on sandstone cementation: Journal of Sedimentary Petrology, v. 49, p. 55-70.

Brown, T. H., R. G. Berman, and E. H. Perkins, 1988a, GEO-CALC: a software package for rapid calculation of stable pressure-temperature-activity phase diagrams, *in* W. R. Dickinson, ed., GSA 1987 Annual Meeting and Exposition: GSA Abstracts with Programs, v. 19, p. 603.

Brown, T. H., R. G. Berman, and E. H. Perkins, 1988b, GEO-CALC II: PTA-SYSTEM software for calculation and display of pressure-temperature-activity phase diagrams: Vancouver, British Columbia, The University of British Columbia (Manual), 35 p.

Burrows, S. M., S. M. Savin, and J. L. Aronson, 1984, Oxygen isotope evidence for the conditions of diagenesis of the Muddy sandstone, east flank of the Powder River basin, Montana and Wyoming: GSA Abstracts with Programs, v. 16, p. 460.

Burst, J. F., 1969, Diagenesis of Gulf Coast clayey sediments and its possible relation to petroleum migration: AAPG Bulletin, v. 53, p. 73-93.

Byers, P. N., 1969, Mineralogy and origin of the upper Eastend and Whitemud formations of south-central and southwestern Saskatchewan and southeastern Alberta: Canadian Journal of Earth Sciences, v. 6, p. 317-334.

Cameron, E. M., 1965, Application of geochemistry to stratigraphic problems in lower Cretaceous of Western Canada: AAPG Bulletin, v. 49, p. 62-80.

Cant, D. J., and V. G. Ethier, 1984, Lithology-dependent diagenetic control of reservoir properties of conglomerates, Falher Member, Elmworth field, Alberta: AAPG Bulletin, v. 68, p. 1044-1054.

Cant, D. J., and G. S. Stockmal, 1989, The Alberta foreland basin: relationship between stratigraphy and cordilleran terrane-accretion events: Canadian Journal of Earth Sciences, v. 26, p. 1964-1975.

Carrigy, M. A., 1963, Petrology of coarse-grained sands in the lower part of the McMurray Formation, *in* M. A. Carrigy, ed., The K. A. Clark volume, a collection of papers on the Athabasca oil sands: Research Council of Alberta Information Series No. 45, p. 43-54.

Carrigy, M. A., 1971, Lithostratigraphy of the uppermost Cretaceous (Lance) and Paleocene strata of the Alberta Plains: Research Council of Alberta Bulletin No. 27, 161 p.

Carrigy, M. A., and G. B. Mellon, 1964, Authigenic clay mineral cements in Cretaceous and Tertiary sandstones of Alberta: Journal of Sedimentary Petrology, v. 34, p. 461-472.

Chi, B. I., 1966, A petrologic comparison of the Frenchman and upper Edmonton formations: unpublished M.Sc. thesis, University of Alberta, Edmonton, Alberta, 124 p.

Connolly, C. A., 1989, Thermal history and diagenesis of the Wilrich Member shale, Spirit River Formation, northwest Alberta: Bulletin of Canadian Petroleum Geology, v. 37, p. 182-197.

Connolly, C. A., L. M. Walter, H. Baadsgaard, and F. J. Longstaffe, 1990a, Origin and evolution of formation waters, Alberta basin, Western Canada sedimentary basin: I, Chemistry: Applied Geochemistry, v. 5, p. 375-395.

Connolly, C. A., L. M. Walter, H. Baadsgaard, and F. J. Longstaffe, 1990b, Origin and evolution of formation waters, Alberta basin, Western Canada sedimentary basin: II, Isotope systematics and water mixing: Applied Geochemistry, v. 5, p. 397-413.

Curtis, C. D., 1985, Clay mineral precipitation and transformation during burial diagenesis: Philosophical Transactions of the Royal Society of London, Series A, v. 315, p. 91-105.

Davies, G. R., 1983, Sedimentology of the Middle Jurassic Sawtooth Formation of southern Alberta, *in* J. R. McLean and G. E. Reinson, eds., Sedimentology of selected Mesozoic clastic sequences, Corexpo '83: Calgary, Alberta, Canadian Society of Petroleum Geologists, p. 11-25.

Dean, R. S., and C. Nahnybida, 1985, Authigenic trioctahedral clay minerals coating Clearwater Formation sand grains in Cold Lake, Alberta—extended abstract: Applied Clay Science, v. 1, p. 237-238.

Dickinson, W. R., and C. A. Suczek, 1979, Plate tectonics and sandstone compositions: AAPG Bulletin, v. 63, p. 2164-2182.

Dickinson, W. R., L. S. Beard, G. R. Brakenridge, J. L. Erjavec, R. C. Ferguson, K. F. Inman, R. A. Knepp, F. A.

Lindberg, and P. T. Ryberg, 1983, Provenance of North American Phanerozoic sandstones in relation to tectonic setting: GSA Bulletin, v. 94, p. 222-235.

Dyman, T. S., W. J. Perry, Jr., and D. J. Nichols, 1988, Stratigraphy, petrology and provenance of the Albian Blackleaf Formation and the Cenomanian to Turonian lower part of the Frontier Formation in part of Beaverhead and Madison Counties, Montana: The Mountain Geologist, v. 25, p. 113-128.

Eisbacher, G. H., M. A. Carrigy, and R. B. Campbell, 1974, Paleodrainage pattern and late orogenic basins of the Canadian Cordillera, *in* W. R. Dickinson, ed., Tectonics and sedimentation: SEPM Special Publication 22, p. 143-166.

Elliot, T. J., 1977, Mineralogy and grain size analysis of the central section of the Horseshoe Canyon Formation, Drumheller, Alberta: unpublished B.Sc. thesis, University of Waterloo, 75 p.

Folk, R. L., 1968, Petrology of sedimentary rocks—University of Texas: Austin, Texas, Hemphill's, 170 p.

Foscolos, A. E., and H. Kodama, 1974, Diagenesis of clay minerals from Lower Cretaceous shales of northeastern British Columbia: Clays and Clay Minerals, v. 22, p. 319-335.

Foscolos, A. E., T. G. Powell, and P. R. Gunther, 1976, The use of clay minerals and organic geochemical indicators for evaluating the degree of diagenesis and oil generating potential of shales: Geochimica et Cosmochimica Acta, v. 40, p. 953-966.

Fuchtbauer, H., 1967, Influence of different types of sandstone diagenesis on sandstone porosity: Proceedings of the Seventh World Petroleum Congress, p. 354-369.

Gabrielse, H., 1985, Major dextral transcurrent displacements along the northern Rocky Mountain trench and related lineaments in north-central British Columbia: GSA Bulletin, v. 96, p. 1-14.

Gabrielse, H., and J. E. Reesor, 1974, The nature and setting of granitic plutons in the central and eastern parts of the Canadian Cordillera: Pacific Geology, v. 8, p. 109-138.

Galloway, W. E., 1984, Hydrogeologic regimes of sandstone diagenesis, *in* D. A. McDonald and R. C. Surdam, eds., Clastic diagenesis: AAPG Memoir 37, p. 3-14.

Gardiner, S., 1982, Depositional environment, petrology, diagenesis and reservoir aspects of the Notikewin Member, Fort St. John Group, west-central Alberta: unpublished B.Sc. thesis, McMaster University, Hamilton, Ontario, 128 p.

Garven, G., and R. A. Freeze, 1984, Theoretical analysis of the role of groundwater flow in the genesis of stratabound ore deposits: 1, mathematical and numerical model: American Journal of Science, v. 284, p. 1085-1174.

Gautier, D. L., and G. E. Claypool, 1984, Interpretation of methanic diagenesis in ancient sediments by analogy with processes in modern diagenetic environments, *in* D. A. McDonald and R. C. Surdam, eds., Clastic diagenesis: AAPG Memoir 37, p. 111-123.

Ghent, E. D., and B. E. Miller, 1974, Zeolite and clay-carbonate assemblages in the Blairmore Group (Cretaceous), southern Alberta Foothills, Canada: Contributions to Mineralogy and Petrology, v. 44, p. 313-329.

Gibson, D. W., 1979, The Morrissey and Mist Mountain formations—newly defined lithostratigraphic units of the Jura-Cretaceous Kootenay Group, Alberta and British Columbia: Bulletin of Canadian Petroleum Geology, v. 27, p. 183-208.

Gibson, D. W., 1985, Stratigraphy, sedimentology and depositional environments of the coal-bearing Jurassic-Cretaceous Kootenay Group, Alberta and British Columbia: Geological Survey of Canada Bulletin 357, 108 p.

Glaister, R. P., 1959, Lower Cretaceous of southern Alberta and adjoining areas: AAPG Bulletin, v. 43, p. 590-640.

Hacquebard, P. A., 1977, Rank of coal as an index of organic metamorphism for oil and gas in Alberta, *in* G. Deroo, T. G. Powell, B. Tissot, and R. G. McCrossan, eds., The origin and migration of petroleum in the Western Canadian sedimentary basin, Alberta—a geochemical and thermal maturation study: Geological Survey of Canada Bulletin 262, p. 11-22.

Haywick, D. W., 1982, Sedimentology of the Wapiabi-Belly River transition and the Belly River Formation (Upper Cretaceous) near Ghost Dam, Alberta: unpublished B.Sc. thesis, McMaster University, Hamilton, Ontario, 150 p.

Hitchon, B., 1984, Geothermal gradients, hydrodynamics, and hydrocarbon occurrences, Alberta, Canada: AAPG Bulletin, v. 68, p. 713-743.

Hitchon, B., and I. Friedman, 1969, Geochemistry and origin of formation waters in the Western Canada sedimentary basin—I, stable isotopes of hydrogen and oxygen: Geochimica et Cosmochimica Acta, v. 33, p. 1321-1349.

Hitchon, B., G. K. Billings, and J. E. Klovan, 1971, Geochemistry and origin of formation waters in the Western Canada sedimentary basin—III, factors controlling chemical composition: Geochimica et Cosmochimica Acta, v. 35, p. 567-598.

Hopkins, J. C., 1981, Sedimentology of quartzose sandstones of Lower Mannville and associated units, Medicine River area, central Alberta: Bulletin of Canadian Petroleum Geology, v. 29, p. 12-41.

Houseknecht, D. W., 1987, Assessing the relative importance of compaction processes and cementation to reduction of porosity in sandstones: AAPG Bulletin, v. 71, p. 633-642.

Hower, J., 1983, Clay minerals reactions in clastic diagenesis: AAPG Bulletin, v. 67, p. 486 (abstract).

Hower, J., E. V. Eslinger, M. E. Hower, and E. A. Perry, 1976, Mechanisms of burial metamorphism of argillaceous sediment: 1, mineralogical and chemical evidence: GSA Bulletin, v. 87, p. 725-737.

Hutcheon, I., and H. J. Abercrombie, 1990, Carbon dioxide in clastic rocks and silicate hydrolysis: Geology, v. 18, p. 542-544.

Hutcheon, I., and R. Lefebvre, 1988, Sedimentology, diagenesis and thermal effects on petrophysical properties of the Aberfeldy field, Saskatchewan: Bulletin of Canadian Petroleum Geology, v. 36, p. 70-85.

Hutcheon, I., H. J. Abercrombie, P. Putnam, H. R. Krouse, and R. Gardner, 1989, Diagenesis and sedimentology of the Clearwater Formation at Tucker Lake: Bulletin of Canadian Petroleum Geology, v. 37, p. 83-97.

Hutcheon, I., H. J. Abercrombie, M. Shevalier, and C. Nahnybida, 1988, A comparison of formation reactivity in quartz-rich and quartz-poor reservoirs during steam assisted recovery: UNITAR/UNDP Fourth International Conference, Edmonton, v. 3, paper 235, p. 1-12.

Hutcheon, I., A. Oldershaw, and E. Ghent, 1980, Diagenesis of sandstones of the Kootenay Formation at Elk Valley (southeastern British Columbia) and Mount Allan (southwestern Alberta): Geochimica et Cosmochimica Acta, v. 44, p. 1425-1435.

Iwuagwu, C. J., and J. F. Lerbekmo, 1982, The petrology of the basal Belly River sandstone reservoir, Pembina field, Alberta: Bulletin of Canadian Petroleum Geology, v. 30, p. 187-207.

Jackson, P. C., 1984, Paleogeography of the Lower Cretaceous Mannville Group of Western Canada, *in* J. A. Masters, ed., Elmworth—case study of a deep basin gas field: AAPG Memoir 38, p. 49-77.

James, D. P., 1985, Stratigraphy, sedimentology and diagenesis of Upper Jurassic and Lower Cretaceous (Mannville) strata, southwestern Alberta: unpublished Ph.D. thesis, Oxford University, U.K., 225 p.

Jansa, L., 1972, Depositional history of the coal bearing Upper Jurassic-Lower Cretaceous Kootenay Formation, southern Rocky Mountains, Canada: GSA Bulletin, v. 83, p. 3199-3222.

Jardine, D., 1974, Cretaceous oil sands of Western Canada, *in* L. V. Hills, ed., Oil sands, fuel of the future: Canadian Society of Petroleum Geologists Memoir 3, p. 50-67.

Jervey, M. T., 1992, Siliciclastic sequence development in foreland basins, with examples from the Western Canada foreland basin, this volume.

Krause, F. F., H. N. Collins, D. A. Nelson, S. D. Machemer, and P. R. French, 1987, Multiscale anatomy of a reservoir: geological characterization of the Pembina-Cardium pool, west-central Alberta, Canada: AAPG Bulletin, v. 71, p. 1233-1260.

Krouse, H. R., C. A. Viau, L. S. Eliuk, A. Ueda, and S. Halas, 1988, Chemical and isotopic evidence of thermochemical sulfate reduction by light hydrocarbon gases in deep carbonate reservoirs: Nature, v. 333, p. 415-419.

Leckie, D. A., 1986, Petrology and tectonic significance of Gates Formation (Early Cretaceous) sediments in northeast British Columbia: Canadian Journal of Earth Sciences, v. 23, p. 129-141.

Leckie, D., and D. Smith, 1992, Regional setting, evolution, and depositional cycles of the Western Canada foreland basin, this volume.

Lefebvre, R., and I. Hutcheon, 1986, Mineral reactions in quartzose rocks during thermal recovery of heavy oil, Lloydminster, Saskatchewan, Canada: Applied Geochemistry, v. 1, p. 395-405.

Lerand, M. M., 1983, Sedimentology of the Blood Reserve sandstone in southern Alberta, Canadian Society of Petroleum Geologists field trip guide book: Calgary, Alberta, Canadian Society of Petroleum Geologists, 55 p.

Lerbekmo, J. F., 1963, Petrology of the Belly River Formation, southern Alberta Foothills: Sedimentology, v. 2, p. 54-86.

Longstaffe, F. J., 1986, Oxygen isotope studies of diagenesis in the basal Belly River sandstone, Pembina I-Pool, Alberta: Journal of Sedimentary Petrology, v. 56, p. 78-88.

Longstaffe, F. J., and A. Ayalon, 1987, Oxygen-isotope studies of clastic diagenesis in the Lower Cretaceous Viking Formation, Alberta: implications for the role of meteoric water, *in* J. D. Marshall, ed., Diagenesis of sedimentary sequences: (London) Geological Society Special Publication 36, p. 277-296.

Machemer, S. D., and I. Hutcheon, 1988, Geochemistry of early carbonate cements in the Cardium Formation, central Alberta: Journal of Sedimentary Petrology, v. 58, p. 136-147.

Mack, G. H., 1978, The survivability of labile light-mineral grains in fluvial, aeolian and littoral marine environments: the Permian Cutler and Cedar Mesa formations, Moab, Utah: Sedimentology, v. 25, p. 587-604.

Mack G. H., and T. Jerzykiewicz, 1989, Provenance of post-Wapiabi sandstones and its implications for Campanian to Paleocene tectonic history of the southern Canadian Cordillera: Canadian Journal of Earth Sciences, v. 26, p. 665-676.

Marion, D. J., 1982, Sedimentology of the Middle Jurassic Rock Creek Member in the subsurface of west-central Alberta: unpublished B.Sc. thesis, University of Calgary, Calgary, Alberta, 129 p.

McBride, E. F., 1989, Quartz cement in sandstones: a review: Earth-Science reviews, v. 26, p. 69-112.

McGookey, D. P., J. D. Haun, L. A. Hale, H. G. Goodell, D. G. McCubbin, R. J. Weimer, and G. R. Wulf, 1972, Cretaceous system, *in* Geologic atlas of the Rocky Mountain region, USA: Denver, Colorado, Rocky Mountain Association of Geologists, p. 190-228.

McLean, J. R., and P. E. Putnam, 1983, Comparison of heavy oil reservoirs; the Lloydminster Formation, Lloydminster area, and the Clearwater Formation, Cold Lake area, *in* J. R. McLean and G. E. Reinson, eds., Sedimentology of selected Mesozoic clastic sequences, Corexpo '83: Calgary, Alberta, Canadian Society of Petroleum Geologists, p. 81-93.

McMechan M. E., and R. I. Thompson, in press, The Canadian Cordilleran fold and thrust belt south of 66 degrees N and its influence on the Western Interior basin, *in* W. G. E. Caldwell and E. Kaufman, eds., Evolution of the Western Interior foreland basin: Geological Association of Canada Special Paper No. 39.

Mellon, G. B., 1967, Stratigraphy and petrology of the Lower Cretaceous Blairmore and Mannville groups, Alberta Foothills and Plains: Alberta Research Council Bulletin 21, 270 p.

Meshri, I. D., 1981, Deposition and diagenesis of Glauconite sandstone, Berrymore-Lobstick-Bigoray area, south central Alberta: a study of physical chemistry: unpublished Ph.D. thesis, The University of Tulsa, Tulsa, Oklahoma, 130 p.

Misko, R. M., and H. E. Hendry, 1979, The petrology of sands in the uppermost Cretaceous and Paleocene of southern Saskatchewan: a study of composition influenced by grain size, source area and tectonics: Canadian Journal of Earth Sciences, v. 16, p. 38-49.

Monger, J. W. H., R. A. Price, and D. J. Templeman-Kluit, 1982, Tectonic accretion and the origin of the two major metamorphic and plutonic welts in the Canadian Cordillera: Geology, v. 10, p. 75-79.

Nagtegaal, P. J. C., 1978, Sandstone-framework instability as a function of burial diagenesis, *in* Sandstone diagenesis: Journal of the Geological Society of London, v. 135, p. 101-105.

Nelson, H. W., 1968, Petrography of sandstone interbeds in the Upper Cretaceous-Paleocene of Gulf Spring Point: Bulletin of Canadian Petroleum Geology, v. 16, p. 425-430.

Nelson, H. W., and R. P. Glaister, 1975, Trap Creek Belly River section, a deltaic progradational sequence (locality E), *in* M. S. Shawa, ed., Guidebook to selected sedimentary environments in southwestern Alberta, Canada: Calgary, Alberta, Canadian Society of Petroleum Geologists, p. 41-53.

Nelson, H. W., and R. P. Glaister, 1978, Subsurface environmental facies and reservoir relationships of the McMurray oil sands, northeastern Alberta: Bulletin of Canadian Petroleum Geology, v. 26, p. 117-207.

Nurkowski, J. R., 1984, Coal quality, coal rank variation and its relation to reconstructed overburden, Upper Cretaceous and Tertiary Plains coals, Alberta, Canada: AAPG Bulletin, v. 68, p. 285-295.

O'Connell, S. C., 1981, The Viking Formation (Lower Cretaceous) of southeastern Saskatchewan, unpublished M.Sc. thesis, University of Windsor, Windsor, Ontario, 191 p.

Ogunyomi, O., and L. V. Hills, 1977, Depositional environments, Foremost Formation (Late Cretaceous), Milk River area, southern Alberta: Bulletin of Canadian Petroleum Geology, v. 25, p. 929-968.

Osadetz, K. G., F. W. Jones, J. A. Majorowicz, D. E. Pearson, and L. D. Stasiuk, 1992, Thermal history of the cordilleran foreland basin in Western Canada: a review, this volume.

Perry, E. A., and J. Hower, 1970, Burial diagenesis in Gulf Coast pelitic sediments: Clays and Clay Minerals, v. 18, p. 165-177.

Potocki, D. J., 1981, A study of the diagenesis of the Lower Cretaceous Niton Basal Quartz sandstones of western Alberta with specific reference to porosity and permeability modification: unpublished B.Sc. thesis, McMaster University, Hamilton, Ontario, 83 p.

Potocki, D. J., 1989, The influence of deep meteoric invasion on the reservoir quality of Cretaceous sandstones in the Mackenzie delta, Canada, *in* D. L. Miles, ed., Water-rock interaction: WRI6: Proceedings of the Sixth International Symposium on Water-Rock Interaction, Malvern, U.K., p. 561-564.

Poulton, T. P., 1984, The Jurassic of the Canadian Western Interior, from 49 degrees N latitude to Beaufort Sea, *in* D. F. Stott and D. J. Glass, eds., The Mesozoic of middle North America: Canadian Society of Petroleum Geologists Memoir 9, p. 15-41.

Powers, M. C., 1967, Fluid-release mechanisms in compacting marine mudrocks and their importance in oil exploration: AAPG Bulletin, v. 51, p. 1240-1254.

Prentice, M. E., and J. W. Kramers, 1987, Petrology and diagenesis of the McLaren Formation in the Bodo area of east central Alberta: Alberta Geological Survey, Alberta Research Council Report No. 8788-7.

Price, R. A., J. W. H. Monger, and J. A. Roddick, 1985, Cordilleran cross-section Calgary to Vancouver, *in* D. Templeman-Kluit, ed., Field guides to geology and mineral deposits in the southern Canadian Cordillera: GSA cordilleran section meeting: Vancouver, British Columbia, Geological Survey of Canada, p. 3.1-3.85.

Putnam P. E., 1982, Aspects of the petroleum geology of the Lloydminster heavy oil fields, Alberta and Saskatchewan: Bulletin of Canadian Petroleum Geology, v. 30, p. 81-111.

Putnam, P. E., and M. A. Pedskalny, 1983, Provenance of Clearwater Formation reservoir sandstones, Cold Lake, Alberta, with comments on feldspar composition: Bulletin of Canadian Petroleum Geology, v. 31, p. 148-160.

Rahmani, R. A., 1982, Facies relationships and paleoenvironments of a Late Cretaceous tide-dominated delta, Drumheller, Alberta, *in* R. G. Walker, ed., Clastic units of the Front Ranges, Foothills and Plains in the area between Field, B.C. and Drumheller, Alberta: International Association of Sedimentologists, 11th Congress on Sedimentology, Hamilton, Ontario, Guidebook to Excursion 21A, p. 31-60.

Rahmani, R. A., and J. F. Lerbekmo, 1975, Heavy mineral analysis of Upper Cretaceous and Paleocene sandstones in Alberta and adjacent area of Saskatchewan, *in* Cretaceous System in the Western Interior of North America: Geological Association of Canada Special Paper 13, p. 607-632.

Rapson J. E., 1965, Petrography and derivation of Jurassic-Cretaceous clastic rocks, southern Rocky Mountains, Canada: AAPG Bulletin, v. 49, p. 1426-1452.

Reinson, G. E., and A. E. Foscolos, 1986, Trends in sandstone diagenesis with depth of burial, Viking Formation, southern Alberta: Bulletin of Canadian Petroleum Geology, v. 34, p. 126-152.

Rimstidt, J. D., and H. L. Barnes, 1980, The kinetics of silica-water reactions: Geochimica et Cosmochimica Acta, v. 44, p. 1683-1700.

Rosenthal, L. R. P., 1989, The stratigraphy, sedimentology and petrography of the Jurassic to Early Cretaceous clastic wedge in western Alberta: unpublished Ph.D. thesis, University of Manitoba, 500 p.

Rudkin, R. A., 1964, Lower Cretaceous, *in* R. G. McCrossan and R. P. Glaister, eds., Geological history of Western Canada: Calgary, Alberta, Alberta Society of Petroleum Geologists, p. 156-168.

Runnells, D. D., 1969, Diagenesis, chemical sediments and the mixing of natural waters: Journal of Sedimentary Petrology, v. 39, p. 1188-1202.

Scherer, M., 1987, Parameters influencing porosity in sandstones: a model for sandstone porosity prediction: AAPG Bulletin, v. 71, p. 485-491.

Schmidt, V., and D. A. McDonald, 1979, The role of secondary porosity in the course of sandstone diagenesis, *in* P. A. Scholle and P. R. Schluger, eds., Aspects of diagenesis: SEPM Special Publication 26, p. 175-207.

Sedimentology Research Group, A. Oldershaw, I. Hutcheon, and C. Nahnybida, 1981, The effect of in-situ steam injection on Cold Lake oil sands: Part I, mineralogic and petrographic considerations: Bulletin of Canadian Petroleum Geology, v. 29, p. 447-465.

Siever, R., 1962, Silica solubility, 0 degrees-200 degrees C, and the diagenesis of siliceous sediments: Journal of Geology, v. 70, p. 127-151.

Sippel, R. F., 1968, Sandstone petrology, evidence from luminescence petrography: Journal of Sedimentary Petrology, v. 38, p. 530-554.

Smith, J. T., and S. N. Ehrenberg, 1989, Correlation of carbon dioxide abundance with temperature in clastic hydrocarbon reservoirs: relationship to inorganic chemical equilibrium: Marine and Petroleum Geology, v. 6, p. 129-135.

Stockmal, G. S., D. J. Cant, and J. S. Bell, 1992, Relationship of the stratigraphy of the Western Canada foreland basin to cordilleran tectonics: insights from geodynamic models, this volume.

Storey, S. R., 1979, Clay-carbonate diagenesis of deltaic sandstones, basal Belly River Formation (Upper Cretaceous), central Alberta, Canada (abstract): AAPG Bulletin, v. 63, p. 534.

Stott, D. F., 1968, Lower Cretaceous Bullhead and Fort St. John groups, between Smoky and Peace rivers, Rocky Mountain Foothills, Alberta and British Columbia: Geological Survey of Canada Bulletin 152, 279 p.

Stott, D. F., 1984, Cretaceous sequences of the Foothills of the Canadian Rocky Mountains, *in* D. F. Stott and D. J. Glass, eds., The Mesozoic of middle North America: Canadian Society of Petroleum Geologists Memoir 9, p. 85-107.

Surdam, R. C., and J. R. Boles, 1979, Diagenesis of volcanic sandstones, *in* P. A. Scholle and P. R. Schluger, eds., Aspects of diagenesis: SEPM Special Publication 26, p. 227-242.

Surdam, R. C., and L. J. Crossey, 1987, Integrated diagenetic modeling: a process-oriented approach for clastic systems: Annual Review of Earth and Planetary Sciences, v. 15, p. 141-170.

Surdam, R. C., S. W. Boese, and L. G. Crossey, 1984, The chemistry of secondary porosity, *in* D. A. McDonald and R. C. Surdam, eds., Clastic diagenesis: AAPG Memoir 37, p. 127-149.

Surdam, R. C., L. G. Crossey, E. S. Hagen, and H. P. Heasler, 1989, Organic-inorganic interactions and sandstone diagenesis: AAPG Bulletin, v. 73, p. 1-23.

Tater, J. M., 1964, The Dunvegan sandstone of the type area: unpublished M.Sc. thesis, University of Alberta, Edmonton, Alberta, 69 p.

Taylor, D. R., and R. G. Walker, 1984, Depositional environments and paleogeography in the Albian Moosebar Formation and adjacent fluvial Gladstone and Beaver Mines formations, Alberta: Canadian Journal of Earth Sciences, v. 21, p. 698-714.

Templeman-Kluit, D. J., 1979, Transported cataclasite, ophiolite and granodiorite in Yukon: evidence of arc-continent collision: Geological Survey of Canada Paper 79-14, 27 p.

Theriault, F., and I. Hutcheon, 1987, Dolomitization and calcitization of the Devonian Grosmont Formation, northern Alberta: Journal of Sedimentary Petrology, v. 57, p. 955-966.

Thomas, M., and B. Miller, 1980, Diagenesis and rock-fluid interactions in the Cadotte Member from a well in northeastern British Columbia: Bulletin of Canadian Petroleum Geology, v. 28, p. 173-199.

Tilley, B. J., and F. J. Longstaffe, 1984, Controls on hydrocarbon accumulation in Glauconitic sandstone, Suffield heavy oil sands, southern Alberta: AAPG Bulletin, v. 68, p. 1004-1023.

Tipper, H. W., G. J. Woodsworth, and H. Gabrielse, 1981, Tectonic assemblage map of the Canadian Cordillera and adjacent parts of the United States of America: Geological Survey of Canada Map 1505A.

Toth, J., 1978, Gravity-induced cross-formational flow of formation fluids, Red Earth region, Alberta, Canada: analysis, patterns, evolution: Water Resources Research, v. 14, p. 805-843.

Towe, K. M., 1962, Clay mineral diagenesis as a possible source of silica cement in sedimentary rocks: Journal of Sedimentary Petrology, v. 32, p. 26-28.

Waywanko, A. O., 1984, Sedimentary and geophysical well log analysis of the Clearwater Formation, Lower Cretaceous, Cold Lake, Alberta: unpublished M.Sc. thesis, University of Alberta, Edmonton, Alberta, 163 p.

Williams, G. D., 1964, The Mannville Group (Lower Cretaceous) of central Alberta: Bulletin of Canadian Petroleum Geology, v. 11, p. 350-368.

Yeh, H., and S. M. Savin, 1977, Mechanism of burial metamorphism of argillaceous sediments, 3. O-isotope evidence: GSA Bulletin, v. 88, p. 1321-1330.

Young, H. R., and D. J. Doig, 1986, Petrography and provenance of the Glauconitic sandstone, south-central Alberta, with comments on the occurrence of detrital dolomite: Bulletin of Canadian Petroleum Geology, v. 34, p. 408-425.

Zinkernagel, U., 1978, Cathodoluminescence of quartz and its application to sandstone petrology: Contributions to Sedimentology, No. 8, 66 p.

Chapter 9

Thermal History of the Cordilleran Foreland Basin in Western Canada: A Review

Kirk G. Osadetz

Geological Survey of Canada
Institute of Sedimentary and Petroleum Geology
Calgary, Alberta, Canada

F. Walter Jones

Institute of Earth and Planetary Physics, and Department of Physics
University of Alberta
Edmonton, Alberta, Canada

Jacek A. Majorowicz

Institute of Earth and Planetary Physics, and Department of Physics
University of Alberta
Edmonton, Alberta, Canada
Northern Geothermal Consultants
Edmonton, Alberta, Canada

David E. Pearson

David E. Pearson and Associates Ltd.
Golden Head, Victoria, British Columbia, Canada

Laverne D. Stasiuk

Department of Geology
University of Regina
Regina, Saskatchewan, Canada

ABSTRACT

Cordilleran foreland basin geothermal history can be studied with numerous indicators, but only coalification data exist in sufficient abundance, regionally and stratigraphically distributed, to permit definition of a late Laramide (middle Eocene) paleogeothermal gradient field. Interior platform coalification patterns result from sedimentary burial in a geothermal field similar to that observed at the present day. Geothermal gradients decrease toward the mountain front and are believed to be controlled by advective heat transfer caused by hydrodynamic recharge in regions of elevated topography. From the beginning of foreland basin sedimentation (Late Jurassic), the presence of an elevated western source area should have exerted a similar control on the geothermal gradient field. In the Disturbed belt, coal rank variations suggest predeformational coalification. This is consistent with models of coalification in the Northern Foothills, but coalification timing and mechanism in the southern Front Range and Foothills remain unresolved because of anomalous coalification gradients and implausible geologic histories, suggested by present predeformational coalification models. Coalification is a cumulative thermal effect indicator, and it provides information only about peak temperatures. More regional application of other indicators could help to characterize pre-Laramide geothermal environments, the general character of which is unknown.

INTRODUCTION

Cordilleran foreland basin thermal history reflects the most energetic physical Earth process during Cordilleran orogenies, and controls organic thermal maturities that are important for coalification and oil and gas generation. Thermal history also controls metamorphism and diagenesis. Variations in basal heat flux from the lithosphere distinguish classes of tectonic processes, and advective heat transfer accompanying formation fluid motion forms an unrivaled control on chemical reactions that are responsible for the concentration of the Earth's petroleum and mineral resources.

GEOLOGIC AND TECTONIC SETTING

The Western Canada foreland basin lies unconformably on the western North American miogeoclinal succession, a predominantly Paleozoic carbonate and shale succession (Porter et al., 1982). The miogeocline overlies either Canadian shield Precambrian rocks (beneath the undeformed Plains and eastern portions of the Disturbed belt) or thick Proterozoic successions (within the Rocky Mountains) (Porter et al., 1982; Price, 1981). Several westerly derived clastic wedges compose the foreland succession and reflect individual orogenic stages separated by unconformities (Stott, 1984; Leckie and Smith, 1992). Much of the foreland basin lies within the overthrust belt where eastwardly migrating structures have progressively deformed earlier orogenic clastic wedges and preorogenic successions (Price, 1981).

The stratigraphic and tectonic history of the foreland basin and underlying successions has strongly controlled their thermal histories. Evidence from the present geothermal gradient field indicates departures from a thermally conductive state (Majorowicz et al., 1985a; Majorowicz and Jessop, 1981), consistent with large-scale fluid migrations evidenced by oil and gas pool distributions. This circulation controls advective heat transfer from recharge areas in the deformed uplands to discharge areas in the lowlands at the edge of the Plains (Majorowicz and Jessop, 1981; Majorowicz et al., 1985c; Majorowicz et al., 1986). The significant tectonic and geomorphological controls on basin-wide hydrodynamic recharge have been present since the Columbian orogeny began (Gibson, 1985; Archibald et al., 1983). Therefore, hydrodynamic controls on earlier orogenic geothermal environments were probably similar to those characterizing both the late Laramide and the present.

METHODS OF THERMAL HISTORY ANALYSIS

Many techniques are available to study and constrain sedimentary basin thermal histories (Naeser and McCullough, 1989). Some techniques have been extensively applied to the Western Canada foreland basin, whereas others have been applied to limited areas. Techniques can be divided into three types: time and temperature specific, cumulative thermal effect, and temperature specific–time unspecified.

One time- and temperature-specific method is description of the present thermal field. Observation wells, hydrocarbon production, and wireline logging provide current temperature data (Majorowicz et al., 1985a; Majorowicz et al., 1985c; Majorowicz et al., 1986; Majorowicz and Jessop, 1981). A second method determines closure and fission track annealing ages within radiometric systems. Throughout the Plains, Precambrian basement biotite and muscovite K/Ar radiometric ages are older than the basin-fill (Burwash et al., 1962), indicating that temperatures in the sedimentary succession never exceeded the closure temperatures for those minerals—280 and 350°C, respectively (see references in Archibald et al., 1984, p. 575). Fission track annealing age analysis represents a powerful technique, and apatites from Williston basin Precambrian basement (Crowley and

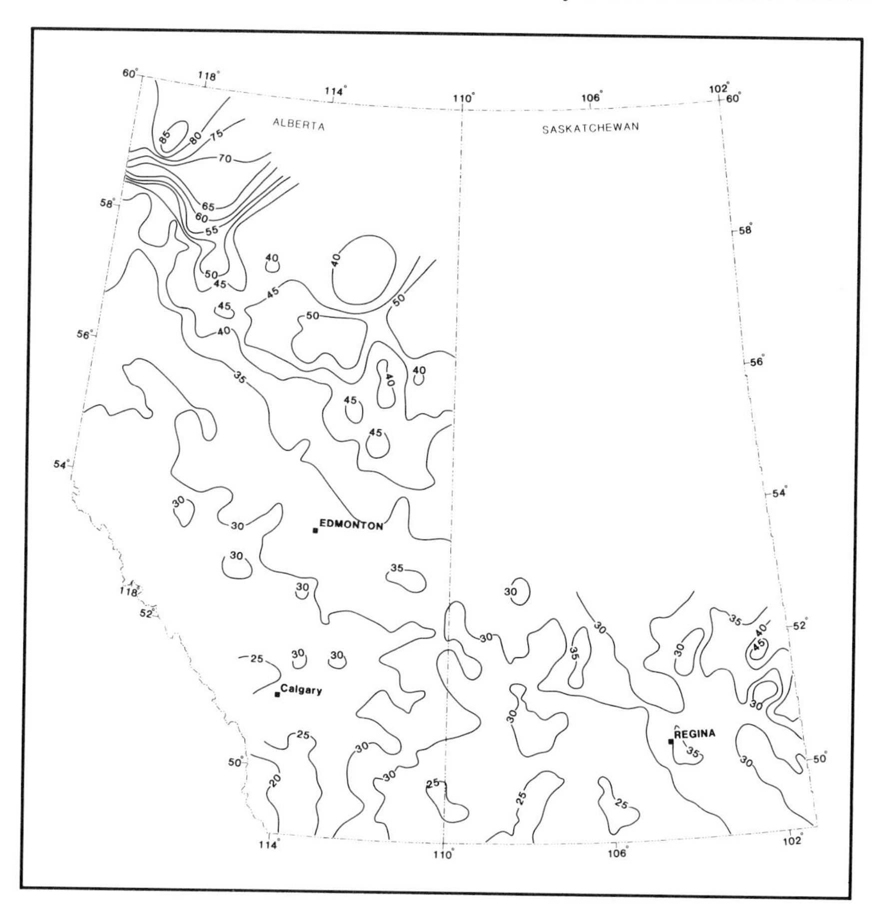

FIGURE 1. Present geothermal gradient field (mK/m) within the foreland basin succession (after Majorowicz et al., 1985b; Majorowicz et al., 1986).

Kulhman, 1988; Crowley et al., 1985) provide important data regarding both the uplift and erosion of Precambrian basement and maximum temperatures reached during the Phanerozoic. Issler et al. (1991) have published the preliminary results from a comprehensive study of apatite fission track data in the Peace River arch region that is consistent with large-scale heat transfer by gravity-driven paleogroundwater similar to that inferred for the end of the Laramide orogeny from regional coalification patterns (Osadetz et al., 1990b) and characteristic of the present geothermal field (Majorowicz et al., 1985a, b).

Organic matter transformation accompanying coalification and hydrocarbon generation measures cumulative thermal effect, and these techniques may be subdivided into those for which a kinetic reaction model exists and those that employ empirical correlations to thermal maturity and thermal history. There is no coherent organic geochemical data set describing either thermal maturity or thermal maturity gradient in the Western Canada sedimentary basin.

Fluid inclusions, mineral phase equilibrium, and mineral isotopic composition are temperature-dependent features that can provide constraints on foreland basin thermal history. None of these techniques provides a specific time at which reactions occur. However, judicious reasoning commonly allows a relative sequence of paragenetic events to be defined (Longstaffe, 1986). Unfortunately, the reactions, or features to be analyzed, commonly occur in the presence of a fluid phase whose composition is not known. This forces investigators to consider independent constraints on temperature and pressure to examine the limits these constraints place on the variation of fluid phase composition. The identification of changes in the composition of formation waters (Longstaffe, 1986) is a useful contribution itself, because a relationship exists between meteoric water influx and advective heat transfer that controls the geothermal gradients (Majorowicz et al., 1985a; Hitchon, 1984).

PRESENT GEOTHERMAL FIELD

A map of the present geothermal gradient in Mesozoic and Cenozoic rocks of the Plains (the foreland succession) derived from wireline logging runs from petroleum exploration wells (Majorowicz et al. 1985a; Majorowicz et al., 1985c; Majorowicz et al., 1986) is given in Figure 1. Figure 1 employs data averaged over 93-km^2 (36-mi^2) areas.

The map's main feature is the correlation among higher elevations, hydraulic recharge areas, and low geothermal gradients. Lower elevations, which are regions of hydraulic discharge, correspond to areas with high geothermal gradients. Isogradient contours increase progressively from 20 mK/m in the extreme southwest to greater than 35 mK/m in the north and east. The calculated heat flow values associated with the observed gradients are much higher than expected for a conductive thermal environment (Beach et al., 1987; Majorowicz et al., 1985b; Majorowicz and Jessop, 1981). A map of present geothermal gradients in the Paleozoic succession is different, but consistent with patterns expected for significant advective heat transfer by formation waters (Beach et al., 1987; Majorowicz et al., 1985a). The inferred departure from a thermally conductive state is not universally accepted. Debate centers on the poor agreement between hydrologically determined fluid velocities and those required to produce the magnitude of observed thermal effects (Majorowicz, 1989).

The technique described above gives an estimate of the average geothermal gradient within a lithostratigraphic succession and cannot identify the roles that individual formations play in lowering or raising gradients in specific areas. The magnitude of the advective heat transfer has not been determined because the basal heat flux distribution has only recently been investigated and there remains some uncertainty in the relationship between heat flux and heat generation.

Two attempts have been made to estimate the basal heat flux into the sedimentary succession for eventually estimating the advective heat transfers. One attempt (Luheshi and Jackson, 1986) compared model heat fluxes with heat fluxes inferred from geothermal gradient field maps. An assumed 60 mW/m^2 basal heat flux resulted in a poor model that could not be substantially improved by altering the hydrological characteristics. A study of the relationship between heat flux and heat generation in Precambrian basement rocks estimates the reduced heat flow and scale depth parameter of radiogenic element distribution (Beach et al., 1987). Both heat generation and heat flow inferred for basement below the sedimentary basin are commonly higher than values assigned to the exposed shield (Figure 2). The reduced heat flow is 43 mW/m^2 with a slope of 7.4 km. If two data points are ignored, then the reduced heat flow is 36 mW/m^2 with a slope of 14.6 km, similar to the exposed shield of the basement tectonic province on which the Phanerozoic sedimentary basin lies (Beach et al., 1987). Three high basement heat generation zones have been identified (Figure 3) (Jones and Majorowicz, 1987) roughly corresponding to long-lived tectonic features.

INTERIOR PLATFORM PALEOGEOTHERMAL FIELD

There have been numerous Interior Platform paleogeothermal studies using various indicators, but only coalification data are sufficiently abundant to attempt a comprehensive geothermal field description using inferred coalification gradients. Quantitative coalification studies were begun by Hacquebard, who described coalification gradient patterns in the Plains (Hacquebard, 1977) and the Disturbed belt (Hacquebard and Donaldson, 1974). In the Plains, Hacquebard recognized pre-epeirogenic coalification accompanying increasing coalification gradients away from the cordillera while providing erosion estimates since the end of the Laramide orogeny. Further studies have confirmed his initial observations and provided regionally distributed data (Cameron, 1991; Kalkreuth and McMechan, 1984, 1988; Kalkreuth and Langenberg, 1986; England, 1984; England and Bustin, 1986; Nurkowski, 1984; Stasiuk, 1988).

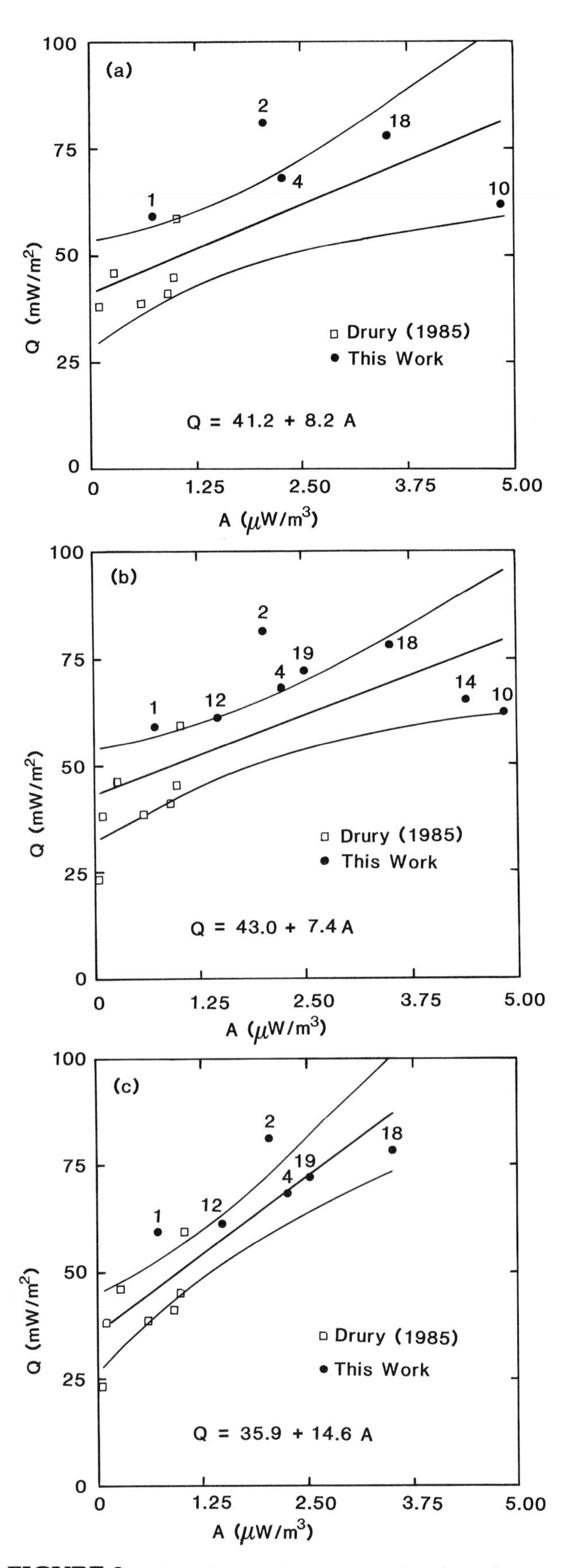

FIGURE 2. Heat flow vs. heat generation for values as found in the 3 × 3 township/range areas together with data collected from the exposed portion of the Canadian shield (from Beach et al., 1987).

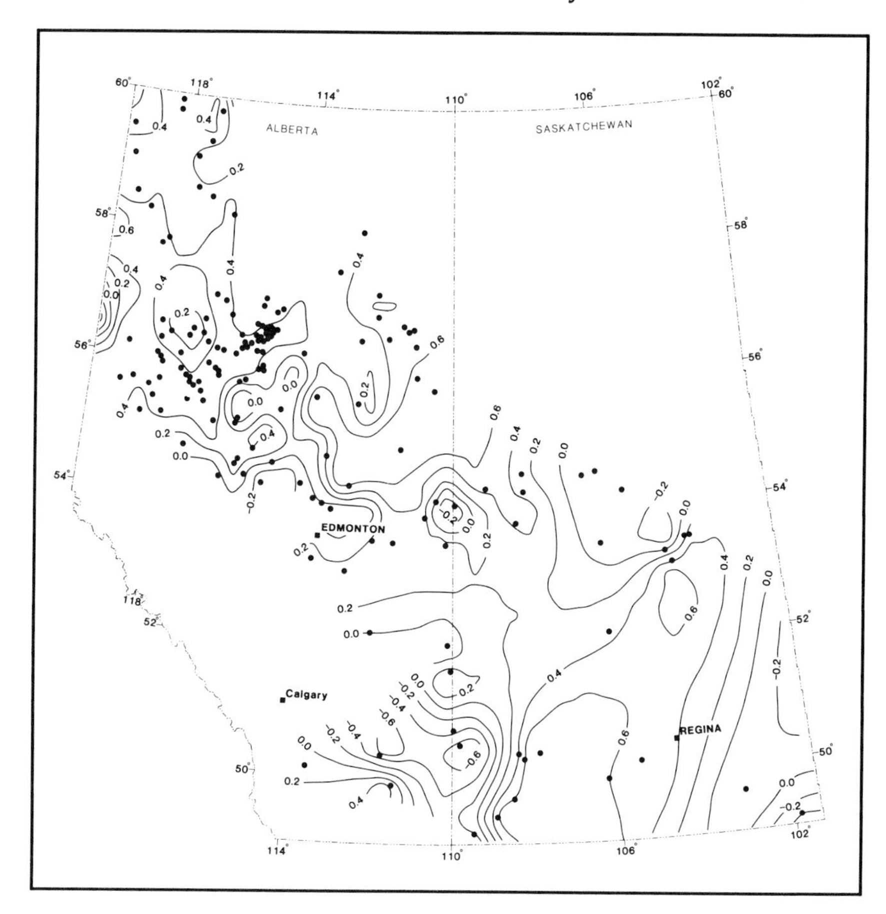

FIGURE 3. Map of $\log_{10}$ (logarithm of heat generation, $\mu W/m^3$) for the Precambrian basement underlying the Plains region of the Western Canada sedimentary basin (from Jones and Majorowicz, 1987).

Coalification Pattern

Interior Platform surface coal rank increases progressively westward across an erosion surface where outcropping strata are progressively younger westward, and near-surface coal equilibrium moisture content decreases westward (England and Bustin, 1986; Nurkowski, 1984; Hacquebard, 1977). Reflectance, measured on a stratigraphic marker, generally increases toward the west (Hacquebard, 1977), whereas coal rank progressively increases with depth (England and Bustin, 1986). Therefore, surface coalification patterns reflect eastward-dipping isorank surfaces (Figure 8) (Nurkowski, 1984). These surfaces indicate coalification accompanying maximum burial beneath the westward-thickening Late Cretaceous and early Tertiary Laramide molasse, followed by post-Laramide epeirogenic uplift and erosion that increases in severity westward (Hacquebard, 1977).

Paleogeothermal Gradient Inference

Interior Platform coalification gradient maps permit paleogeothermal gradient inference at the time of maximum burial (middle Eocene; cf. Kalkreuth and McMechan, 1984). Coalification gradients can be determined directly using reflectance well profiles and considerable gradient data are available (Stasiuk, 1988; England and Bustin, 1986; England, 1984; Kalkreuth and McMechan, 1984, 1988; Cameron, 1991; Hacquebard, 1977) (Figure 9). Unfortunately, individual well coalification gradients are typically imprecise and their extrapolation to zero coalification commonly fails to predict consistent maximum burial patterns or to match the erosion estimates predicted by surface equlibrium moisture data (Osadetz et al., 1990b; Majorowicz et al., 1991a). The Parkland well (Figure 10; Lundell, 1991) uses core samples to compare coalification profile erosion estimates with a cutting sample profile in the nearby 6-2-16-29W4 well (England and Bustin, 1986). Both are compared with the independent erosion estimate using surface equilibrium moisture content (Hacquebard, 1977) (Figure 11). The Parkland well estimate, although imprecise, is consistent with the equilibrium surface moisture content erosion estimate, but compares poorly with the 6-2-16-29W4 well profile erosion estimate (Figure 11). Lack of consistent coalification gradients or erosion estimates is attributed to technical problems not further discussed (Osadetz et al., 1990b; Majorowicz et al., 1991a; Lundell, 1991; Majorowicz et al., 1991b).

An alternative approach uses near-surface coal equilibrium moisture content maximum burial estimates (Osadetz et al., 1990b). Using equilibrium moisture isocontours (Figure 12) to estimate erosion (Figure 5) combined with the remaining Cretaceous and Tertiary sediment thickness (Masters, 1984) provides a depth estimate to the base of the Cretaceous succession during coalification (Figure 13). Knowing Mannville Group reflectance variations (Figure 14) and the foreland succession thickness (Figure 13), and using 0.25% R_0 to represent zero coalification, it is possible to calculate a mid-Eocene paleogeothermal gradient field.

Like an earlier Mannville Group reflectance map (figure 18 in Masters, 1984, p. 17), Figure 14 is characterized by a general increase toward the west. Reflectances range from less than 0.3% R_0, on the northeastern margin of the basin, to approximately 2.0% R_0 in its deepest portions. Unlike the earlier map (Masters, 1984), isoreflectance contours do not conform to burial depth contours, suggesting a more complex paleogeothermal field. Figure 14 is characterized by prominent intrusions of the 0.70% R_0 contour into the Plains. The more northern of these features is poorly delineated. The southern feature forms bifurcating ridges that penetrate the Plains, one looping from Calgary toward Brooks, and the other trending northeast toward Edmonton, where it bifurcates toward Lloydminster and Lac La Biche. A major extension of the 0.50% R_0 contour lies in the keel of the Shaunavon syncline in southwestern Saskatchewan at approximately 108°W.

Paleogeothermal Gradients and Variations

The resulting paleogeothermal gradient map (Figure 15) exhibits both similarities and differences in relation to the present geothermal gradient map. The most striking similarity is the large region where geothermal gradients have not significantly changed. The similarity between the two maps is particularly evident in southern Alberta, where 30-mK/m isogradient contours are generally coincident. A ridge extending from Canmore to Calgary occurs on both maps, as do lower geothermal gradients in the southwestern Plains, adjacent to the Foothills. The pronounced ridge of high geothermal gradients northeast of Edmonton remains a feature of the present gradient field, although the gradients are now higher than in the past.

Hitchon (1984) mapped the maximum burial paleogeothermal gradient field in Alberta and suggested very low paleogeothermal gradients, between 21 and 27 mK/m, with the isogradient lines striking north-south and higher values to the east. Advective heat transfer controlled by hydrodynamics is common to both analyses, despite the marked differences in the pattern and magnitude suggested by Hitchon (1984) compared with that of Osadetz et al. (1990b).

Several regions exist where significant increases in the geothermal gradient have occurred (Figure 16). Higher present gradients occur in the north and northeast, coinciding with present hydrodynamic discharge regions. These are controlled by outcrop geometry, a post-middle Eocene feature of the basin. Inferred paleogeothermal gradients (Figure 15) compare well with paleogeothermal gradients calculated by Beaumont et al. (1985) along the western three-quarters of their isomerization-aromatization profile (Figure 7). However, inferred paleogeothermal gradients (Figure 15) are generally less than 30 mK/m along the eastern part of their profile, where present geothermal gradients are between 35 and 40 mK/m. In contrast, aromatization and isomerization data suggest that geothermal gradients may have exceeded 60 mK/m (Beaumont et al., 1985).

Some gradient increase is observed along the southwestern margin of the Plains. This can be attributed to topographic reduction and accompanying loss of efficiency of the hydraulic recharge system since the Laramide orogeny. The geothermal gradient has decayed elsewhere in the foreland succession. North of the Nesson anticline (103°W), geothermal gradients in the foreland succession follow regional patterns, whereas geothermal gradients in underlying Paleozoic rocks are high (Majorowicz et al., 1986). Lignites in Tertiary strata overlying high geothermal gradients in Paleozoic rocks have higher reflectances than those at similar stratigraphic levels away from anomalous gradients in Paleozoic rocks (Cameron, 1991). Tertiary coal ranks along this feature suggest that the high geothermal gradients associated with this structure must have extended into the foreland basin succession and have been subsequently suppressed. Organic thermal maturities in Paleozoic source rocks are anomalously high in this region (Osadetz et al., 1989; Osadetz et al., 1990a; Price et al., 1984).

Other significant changes (Figure 16) occur in southwestern Saskatchewan parallel to the interprovincial boundary. These follow another meteoric water circulation pathway commonly attributed to hydraulic recharge from the Bearpaw and Little Rocky mountains of eastern Montana. These topographic uplands are related to a Miocene volcanic center, suggesting that associated gradient changes may have resulted from geologic features developed after the end of the Laramide orogeny. The greatest change in geothermal gradients associated with this feature coincides with the Carboniferous Madison Group subcrop below the Lower Cretaceous Mannville Group. Formation water salinity studies have previously suggested that the Madison and Three Forks groups act as conduits for meteoric waters from the uplifts in eastern Montana. These waters are discharged into Cretaceous strata along the subcrop edge of the Madison Group.

Inferred geothermal gradient pattern changes follow oil quality variations, particularly density. Denser biodegraded oils generally occur where the greatest increases in geothermal gradients have occurred, the region along the Alberta-Saskatchewan interprovincial boundary. Porter (1992) shows that Cretaceous oil densities increase progressively toward the outcrop, where hydraulic discharge and significant geothermal gradient changes are inferred. The reason for this association is an unexplained but interesting paradox, because the progressive biodegradation pattern in oil pools is against the inferred water flow pattern. Limits on reflectance and equilibrium moisture content data prevent calculation of paleogeothermal gradients in northwesternmost Alberta. This is an area of very high present geothermal gradients that coincides with local high geothermal gradient anomalies identified for the Devonian (Aulstead and Spencer, 1985).

Other Paleogeothermal Field Indicators

It is possible that the systematic differences in geothermal gradient, as calculated above, arise from systematic errors associated with the mapping and calculation of the coalification gradient, and the inferred relationship between coalification and geothermal gradients. However, this is

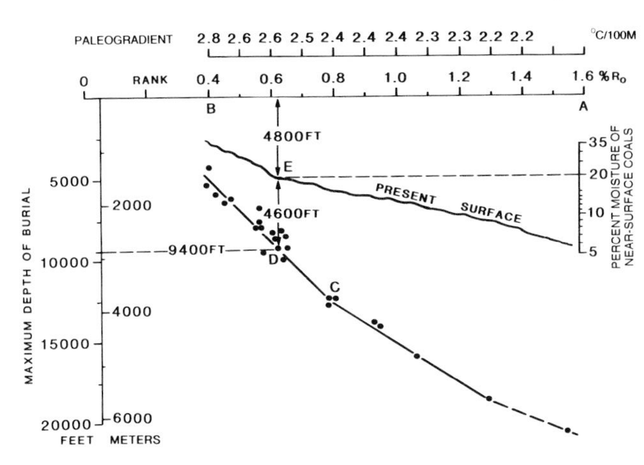

FIGURE 4. Profile across south central Alberta showing Mannville Group coal ranks (reflectance, % R_o) as a function of maximum burial depth at the end of the Laramide orogeny (from Hacquebard, 1977). Garrington oil pool sample is indicated.

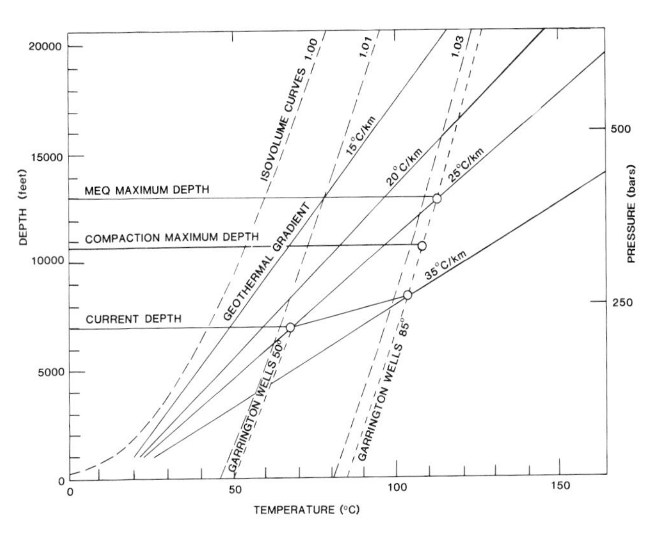

FIGURE 6. Fluid inclusion paleotemperatures from Cardium Formation samples in the Garrington oil pool (from Currie and Nwachukwu, 1974).

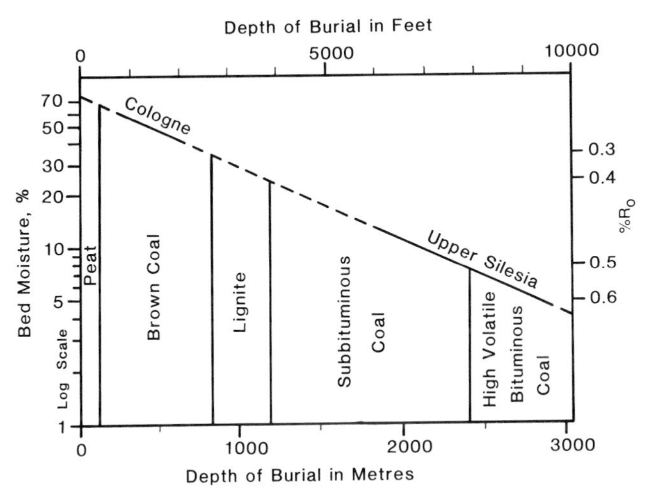

FIGURE 5. Nurkowski's (1984) expression of Hacquebard's (1977) relationship between maximum burial depth and equilibrium moisture content of coals.

unlikely, because in some areas the magnitudes of the geothermal gradients can be checked using other thermal indicators, and there is general agreement within a few millikelvins per meter.

Several studies near the Garrington oil field allow comparison of the present geothermal gradient with the geothermal gradient at the end of the Laramide orogeny. Hacquebard (1977) (Figure 4) indicated 0.96% R_o in Albian Mannville coals underlying the Garrington oil field in a region of 12% surface equilibrium moisture content. This suggests approximately 1890 m (6201 ft) of erosion (Figure 5). The estimated coalification gradient from the surface of maximum burial to the Mannville Group is approximately 0.13 log % R_o/km, suggesting a paleogeothermal gradient of approximately 26 mK/m.

Using the surface equilibrium moisture erosion estimate overlying the Garrington Cardium (Turonian) oil pool, it is possible to reinterpret fluid inclusion data (Currie and Nwachukwu, 1974). Magara (1976) performed such a recalculation using shale compaction trends to estimate the maximum burial depth, but Nurkowski (1984) found that shale compaction estimates are systematically low. The surface equilibrium moisture content erosion estimate makes fluid inclusion paleotemperature estimates consistent with the assumptions used to construct coalification gradient and aromatization-isomerization calculations. Fluid inclusions in the Cardium Formation, currently at a depth of 2121 m (6958 ft), having maximum homogenization temperatures of 112°C, are inferred to have formed at maximum burial depth (Figure 6). This is consistent with a geothermal gradient of approximately 25.4 mK/m using a 10°C surface temperature (Beaumont et al., 1985). "Near present" fluid inclusion data from the same wells suggested geothermal gradients of approximately 25 mK/m, which are consistent with the present geothermal gradients of 24 to 28 mK/m. Beaumont et al. (1985) calculated effective paleogeothermal gradients from aromatization and isomerization reactions in hydrocarbon fractions extracted from Cretaceous shales along a profile slightly north of the Garrington field and the profile studied by Hacquebard. Their estimates were in close agreement with the present geothermal gradient field in this area (Figure 7).

Longstaffe (1986) studied oxygen isotopes of diagenetic minerals in the Campanian basal Belly River Formation sandstone, Pembina I pool, approximately 100 km (62.1 mi) north of the Garrington pool. He determined that early grain-coating chlorites were probably deposited at low temperatures in the presence of brackish waters early in the diagenetic history of the rock, whereas later calcite cement and pore-filling kaolinite and calcite were probably crystallized at higher temperatures in the presence of formation fluids of more meteoric character. His results were consistent with a 24-mK/m paleogeothermal gradient and erosion estimates of 1.3 to 1.4 km (0.81 to 0.87 mi). This agrees with local coalification data assuming that calcite-kaolinite mineral assemblages were precipitated at

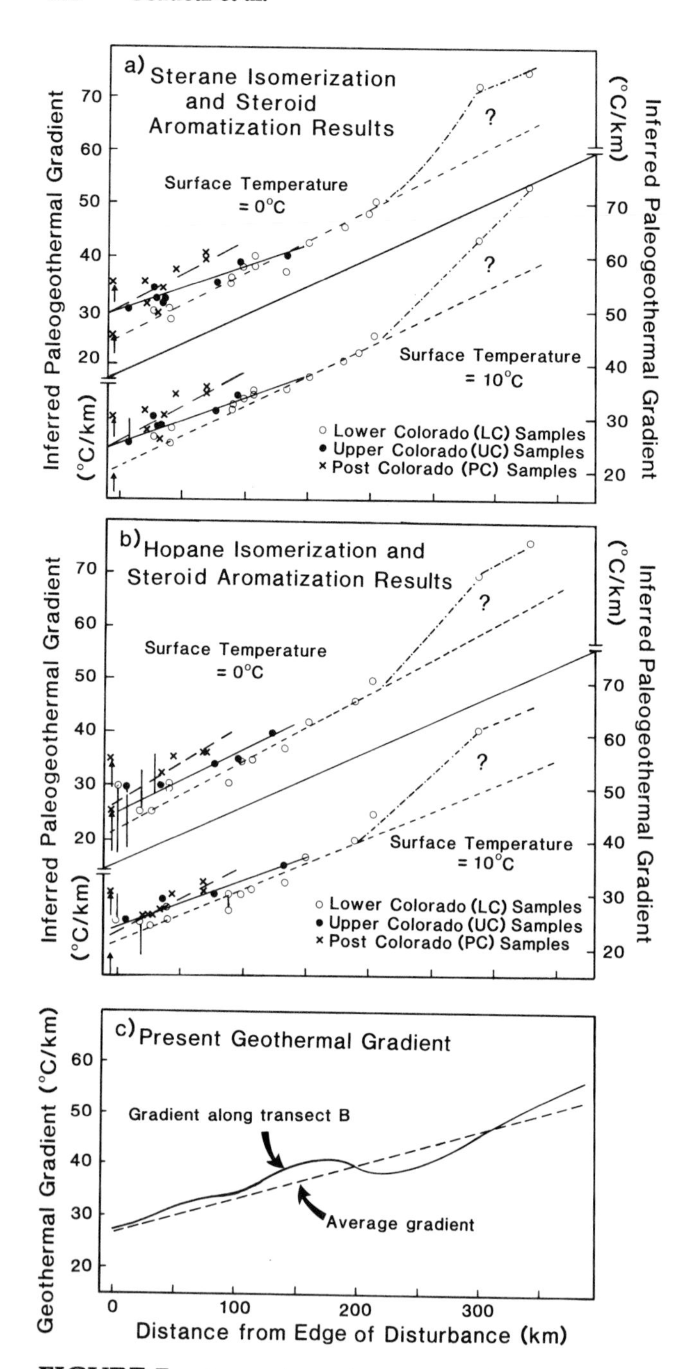

FIGURE 7. Comparison of present geothermal gradient to inferred paleogeothermal gradient at the end of the Laramide orogeny derived from the study of isomerization-aromatization reactions (from Beaumont et al., 1985).

or near the time of maximum burial depth. The present foreland succession geothermal gradient there is 26 mK/m.

These studies suggest that there has been little change in the geothermal gradient from the time of maximum burial to the present, an observation consistent with their position in the basin as suggested by inferred changes in the geothermal gradient pattern (Figure 16), confirming that regional coalification and erosion patterns provide insight into geothermal history.

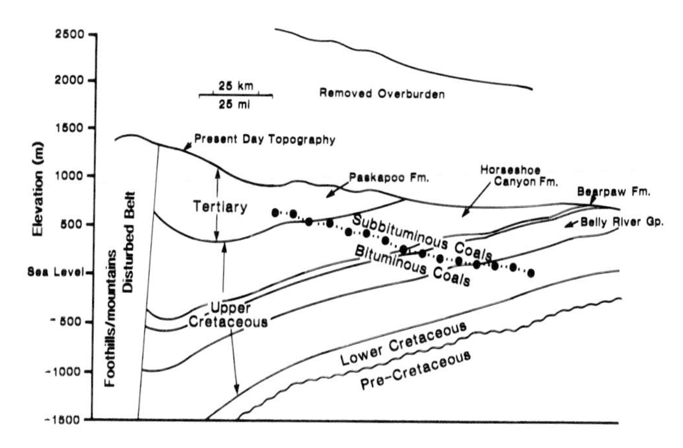

FIGURE 8. Eastward dip of coal isorank contours (from Nurkowski, 1984).

Stoakes and Creaney (1985) formulated a relationship between Upper Devonian Duvernay Formation source rock thermal maturity and present burial depth, showing that maximum burial controlled thermal maturity patterns in Upper Devonian source rocks. In the American Williston basin, Price et al. (1984) described present Devonian and Mississippian Bakken Formation thermal maturities using Rock-Eval pyrolysis. Other local studies are available (Macauley, 1984; Osadetz and Snowdon, 1986a, b). Leckie et al. (1988) examined the pyrolysis results and organic petrography of a 1200-m (3937-ft) thick composite core section through the Lower Cretaceous succession of the northern Foothills. Although the coalification gradient is well constrained, the relationship between reflectance and T_{max} resulted in a positive yet poorly constrained correlation. (T_{max} is a Rock-Eval parameter that is the maximum temperature, in °C, of the S_2 pyrolysis peak.)

Beaumont et al. (1985) examined the isomerization and aromatization of steroidal alkanes and terpanes in hydrocarbon fractions extracted from Cretaceous shales in a line of section across the foreland basin. Using a simplified burial model, constrained by amounts of eroded section consistent with the work of Hacquebard (1977), they calculated that the effective paleogeothermal gradient increased systematically to the east, away from the edge of the Disturbed belt. This pattern follows the general form of the present geothermal gradient field. They also determined that the magnitudes of inferred paleogeothermal gradients along the profile were similar to the present geothermal gradients, except in the easternmost parts.

Preorogenic Geothermal Environment

The thick Tertiary succession overprints previous thermal history in the Plains. Stratigraphic and tectonic analyses show a complex tectonic history that includes passive continental margin formation (Bond and Kominz, 1984) and multiple orogenic phases (Stott, 1984). The foreland basin succession is divided into distinctive orogenic sequences separated by regional unconformities (Stott, 1984; Leckie and Smith, 1992). How these erosional events

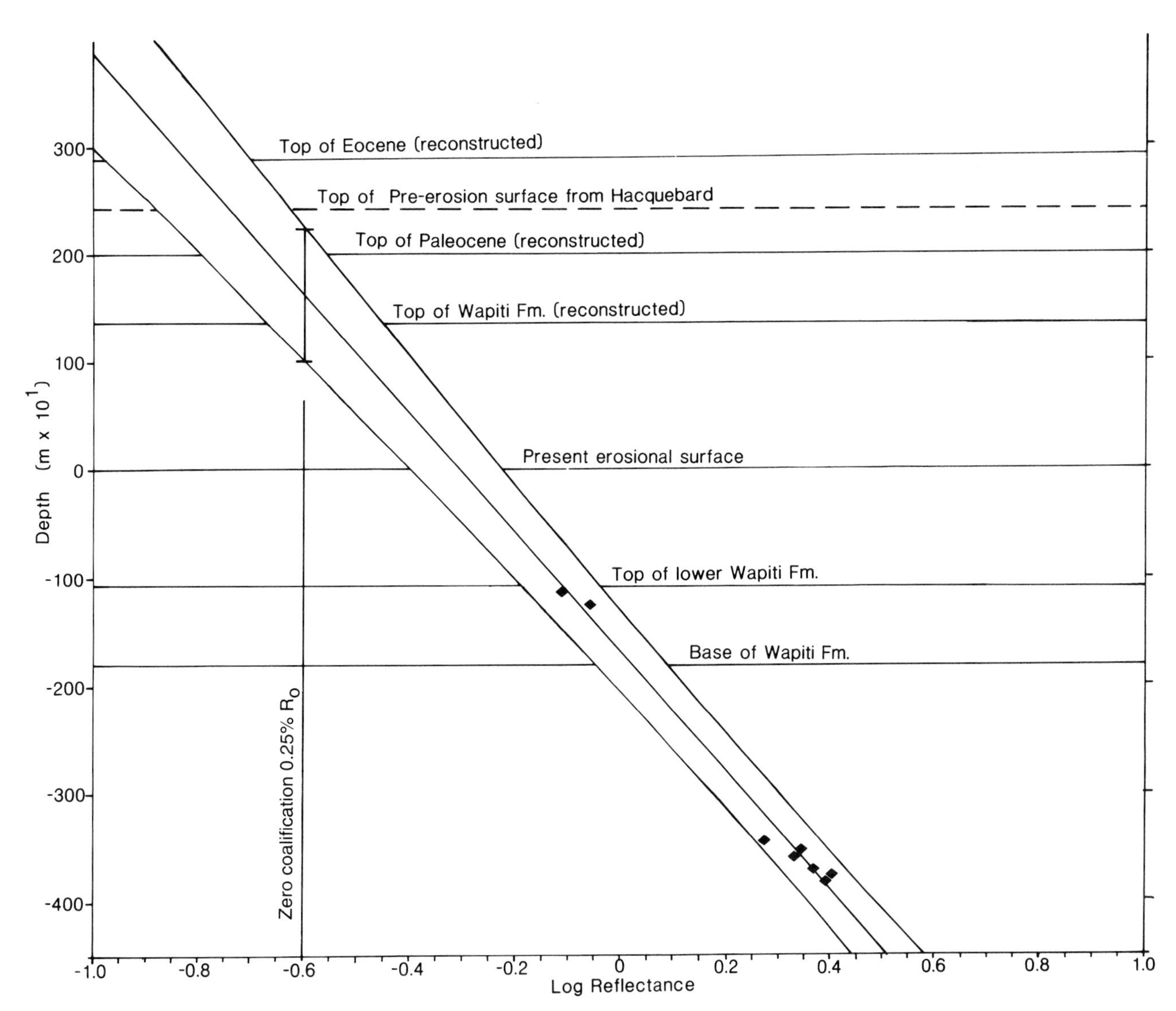

FIGURE 9. **Logarithmic reflectance well profile of Husky and partners Cutbank 16-22 (data from Kalkreuth and McMechan, 1984). Tops of known and inferred stratigraphic layers are indicated. The regressed coalification gradient (log % R_o/km) and 2σ error envelope are employed to infer the effective paleogeothermal gradient (Middleton, 1982) and consistency of the reflectance gradient with the stratigraphic model when the former is extrapolated to a level of zero coalification (0.25 % R_o).**

affected the geothermal pattern and the gradient magnitude is as unknown as the character of all but the last orogenic phase.

There are techniques that constrain pre-middle Eocene geothermal environments in selected areas. Crowley et al. (1985) and Crowley and Kuhlman (1988) studied apatite annealing ages in samples from Precambrian basement beneath the Williston basin. Their results helped to constrain the magnitude and age of the pre-Paleozoic erosional event. Very little is known of Paleozoic geothermal environments. Carbonate lithologies are not as amenable to thermal history analysis. Microfossil coloration indices can be applied (Kalkreuth and McMechan, 1984), but these techniques exhibit little sensitivity within the hydrocarbon generation window and are commonly overprinted by later events. There are several fluid inclusion studies in Devonian rocks (Morrow et al., 1986; Aulstead and Spencer, 1985; Roedder, 1968). Aulstead and Spencer (1985) deduced that pre-Famennian cements in the Middle Devonian Keg River Formation showed both very high maximum homogenization temperatures (235°C) and increasing homogenization temperature with depth. Dolomite cements indicated geothermal gradients as high as 80 mK/m, indicating local geothermal anomalies. Such anomalies may also occur in similar rocks in the Pine Point area (Roedder, 1968), although the timing is less certain there. The general Paleozoic geothermal field remains unknown.

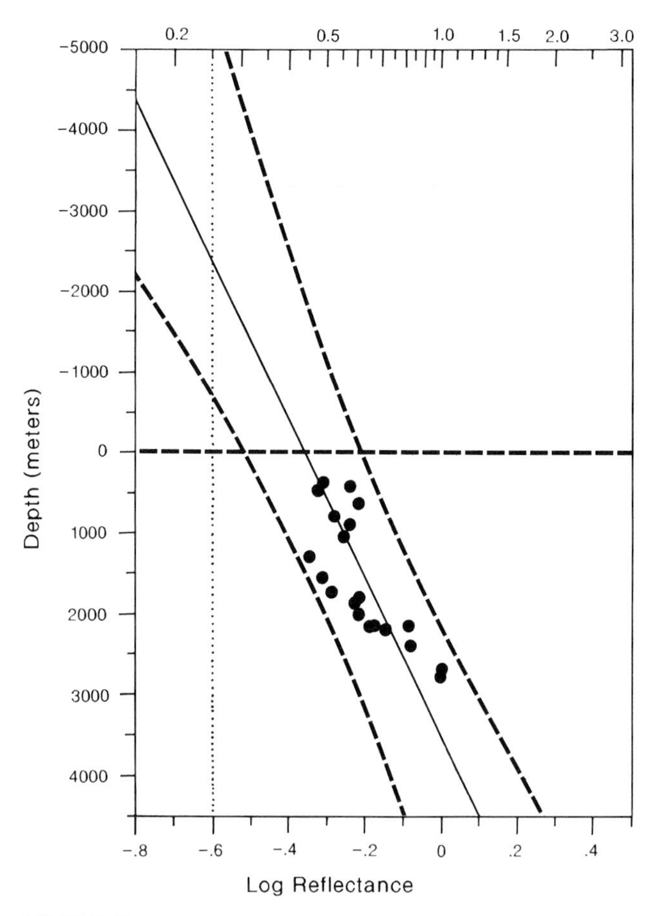

FIGURE 10. Log % R_o/km reflectance gradient in the vicinity of Parkland, Alberta (after Lundell, 1991).

CORDILLERAN GEOTHERMAL ENVIRONMENT

Much of the foreland basin succession lies within the overthrust belt, including successive Columbian foredeeps (Stott, 1984; Gibson, 1985). Currently available Interior Platform thermal indicators provide few constraints on thermal environments older than the late Laramide orogenic phase. Therefore, constraints on the pre-middle Eocene foreland geothermal environment are sought in the Disturbed belt. As for the Plains, coalification is the most commonly used thermal indicator, but other thermal indicators become more useful because of the approach to regional metamorphic grades. However, the cordilleran geothermal environments are difficult to characterize because of uncertainties in the coalification mechanism, the age of structures, and the timing of temperature specific-time unspecified thermal indicators.

In contrast to the Plains, where a simple relationship between coal rank and maximum burial depth exists, calculation of maximum burial depth and the timing of deformation is difficult in the overthrust belt. Few stratigraphic and structural constraints remain, because of pervasive erosion. Structures could be as old as the youngest strata preserved below overthrust plates. However, it is possible that structures are substantially younger and that significant strata important to coalification and thermal reconstructions are not preserved.

The birth of the exogeocline with the deposition of Upper Jurassic sediments in the initial Columbian foredeep indicates uplift and topography in the west (Gibson, 1985; Stott, 1984; Price, 1981), with the possibility that hydraulic recharge and advective heat flow dominated orogenic geothermal environments. Price and Mountjoy (1970) associated the Bourgeau thrust with the western shoreline of the early Columbian foredeep. This suggests that hydraulic recharge was in effect by Late Jurassic time and that the Front Ranges were already evolving during the Columbian orogeny, sparing them and probably much of the Foothills from deep sedimentary burial. Recent geochronological studies of the Purcell anticlinorium (Archibald et al., 1984) suggest that deformation occurred no sooner than latest Cretaceous-earliest Tertiary time. The structural style (Price, 1981) constrains all the structures lying to its east to having evolved later, during a much shorter period than previously considered. Inherent in these models is the possibility that thick uppermost Cretaceous and even Tertiary successions were deposited on the future site of Front Range and Foothills thrust plates, implying quite a different coalification and geothermal history in the deformed foreland basin.

Deep sedimentary burial is consistent with preorogenic coalification and higher paleogeothermal gradients possibly associated with a thermally conductive state. Stratigraphic reconstructions and some coalification gradients suggest high paleogeothermal gradients (Hughes and Cameron, 1985; Hacquebard and Donaldson, 1974). Synorogenic coalification controlled by burial beneath overthrust sheets suggests moderate or low paleogeothermal gradients affected by advective heat transfer long before the mid-Eocene. Some coalification gradient magnitudes, but no coalification patterns, are currently believed to be consistent with this alternative.

Coalification Pattern in the Deformed Belt

In the northern Rocky Mountain Foothills, where structures are dominated by folds and faulted folds (Kalkreuth and Langenberg, 1986), coal ranks in stratigraphic units increase from the inner Foothills to the outer Foothills and reach maximum values near the eastern limit of the deformation and decrease again into the Plains (Kalkreuth and McMechan, 1984, 1988) (Figure 14). The coalification of specific formations decreases to the north along the eastern limit of the deformation. Coalification patterns have been folded in this region (Kalkreuth and Langenberg, 1986). Overturned sections follow Hilt's Law with respect to stratigraphic position (Kalkreuth and McMechan, 1984).

In the southern Foothills, the present complex coalification pattern outcrops along an erosion surface of high topographic and stratigraphic relief. However, higher ranks are generally restricted to older formations. On undeformed restored sections, ranks increase with stratigraphic age and position according to Hilt's Law (Figure 17). This pattern is interpreted to indicate predeforma-

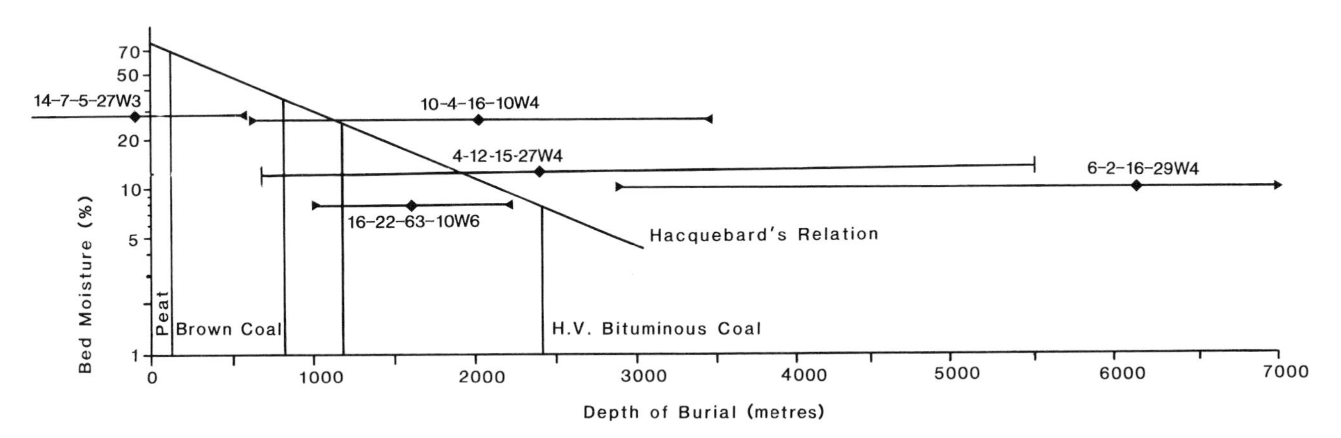

FIGURE 11. **Plot comparing erosion estimates derived from both surface equilibrium moisture content of coal and extrapolation of several coalification gradients to zero coalification. Errors on the extrapolated reflectance gradients are defined by the intersection of the 2σ error envelopes with 0.25% R_o.**

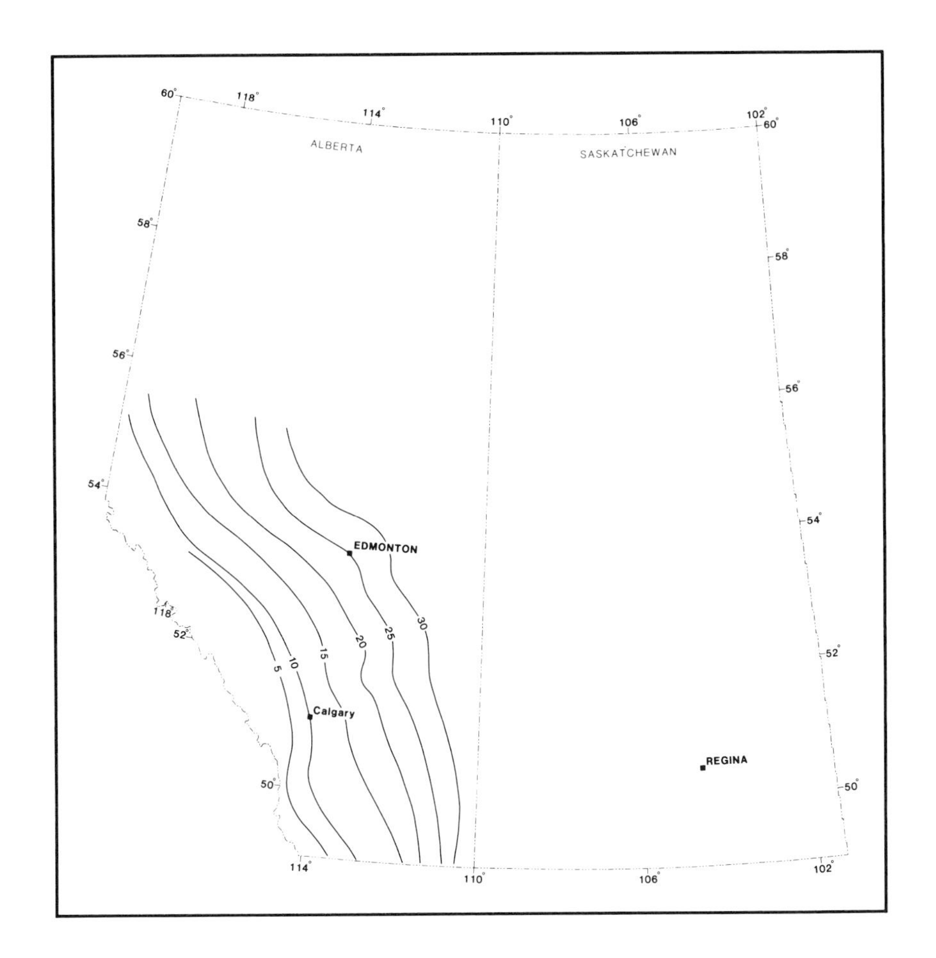

FIGURE 12. **Equilibrium moisture content (isocontours) of near-surface coals (from Hacquebard, 1977).**

tional coalification in the southern Foothills. Isoreflectance lines in restricted areas of the Fernie basin are superimposed on orogenic structure and are offset by later extensional features, indicating at least local postorogenic coalification (Pearson and Grieve, 1985) (Figure 18), but strata in the footwall of the Lewis thrust plate have reflectances similar to those in the same stratigraphic horizon in the overthrust plate (Grieve, 1987), despite 7 km (4.3 mi) of structural separation. Relationships across the Lewis thrust fault are commonly interpeted as the disruption of predeformational isocoalification contours.

Coalification Mechanism and Coalification Gradients

Apparently disrupted coalification patterns that follow Hilt's Law with respect to undeformed stratigraphic

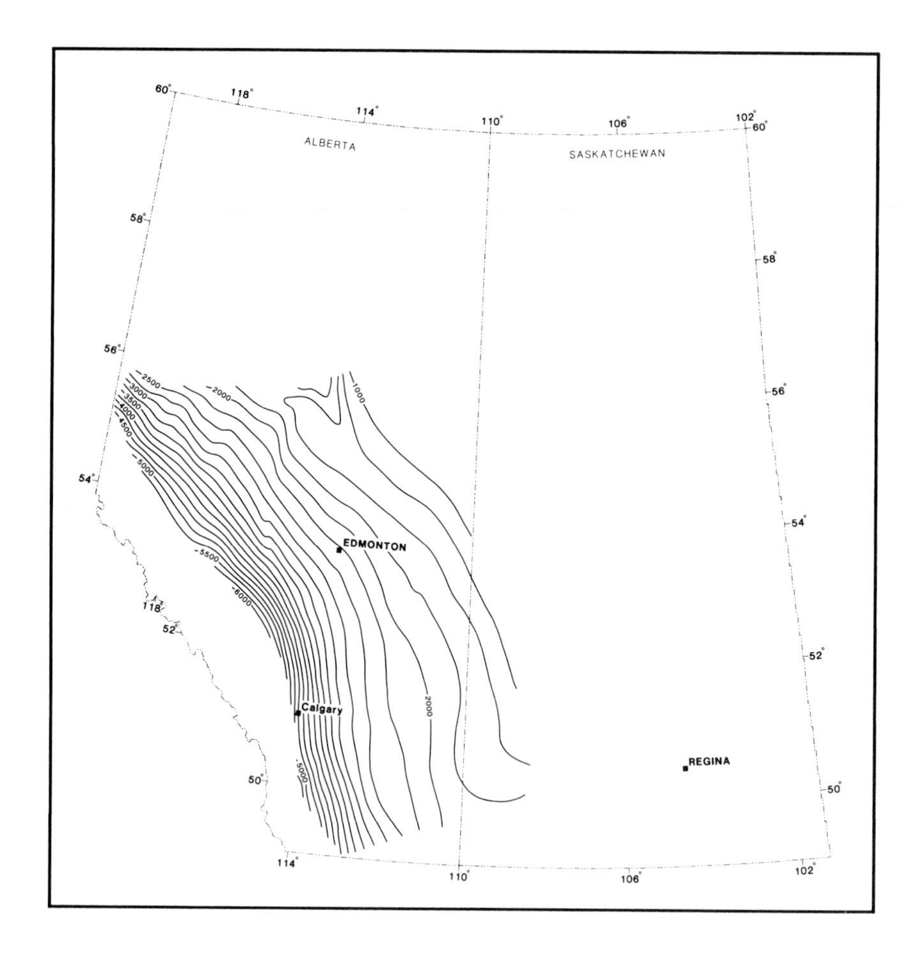

FIGURE 13. Depth to the base of the Cretaceous succession, commonly Mannville Group, in the Plains at the end of the Laramide orogeny in the middle Eocene (from Osadetz et al., 1990b). See text for construction technique.

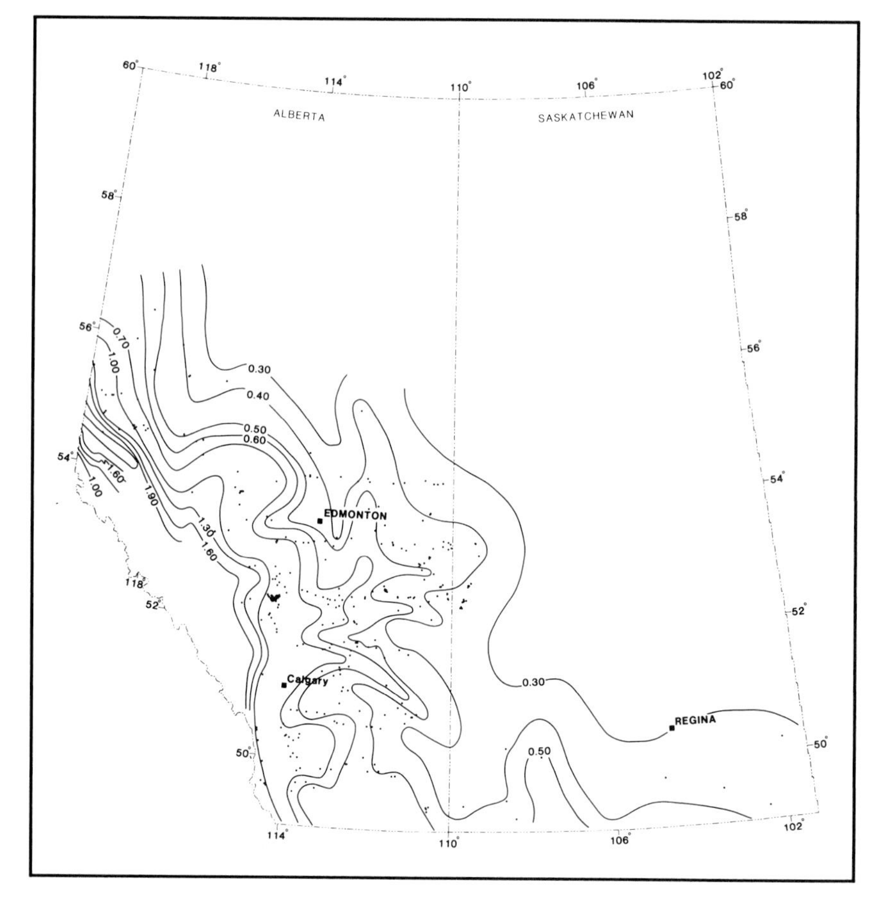

FIGURE 14. Reflectance of Mannville Group sediments from Osadetz et al. (1990b). Plains data from various sources mentioned in Osadetz et al. (1990b). Inset region on the northwestern portion of the map of isoreflectance contours on the Gething Formation, approximately equivalent to the Mannville Group, from Kalkreuth and McMechan (1988).

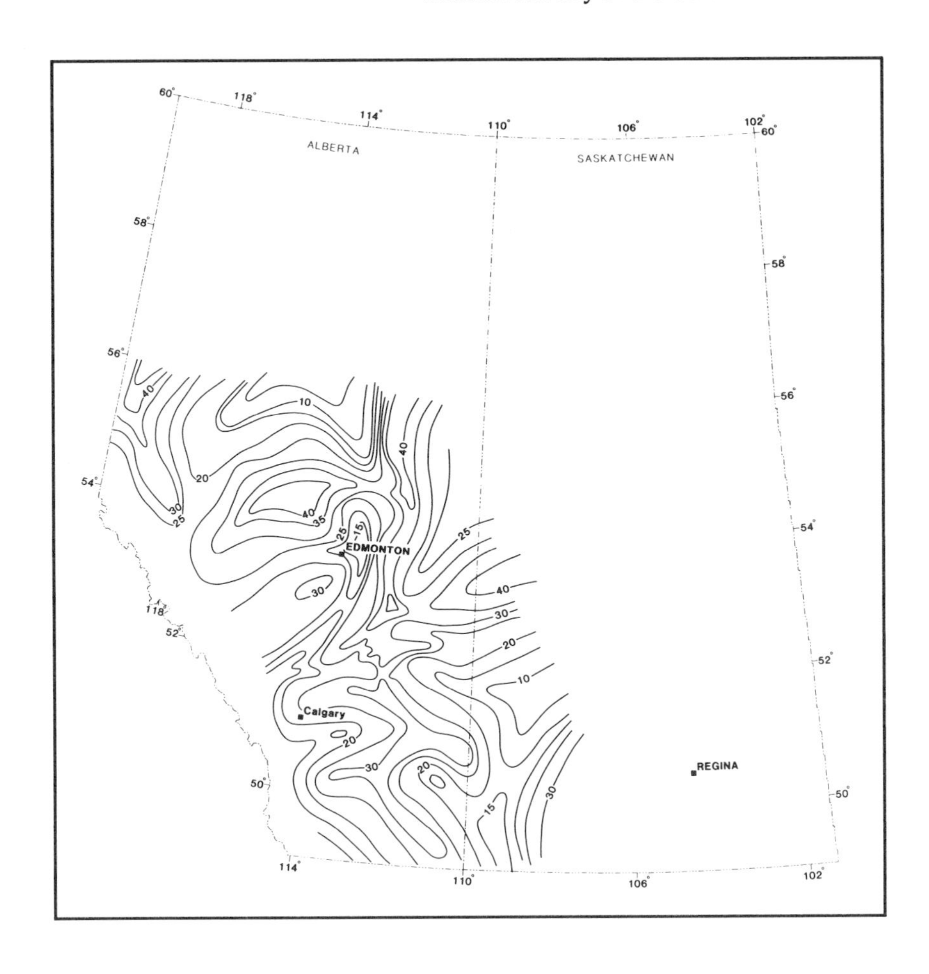

FIGURE 15. Inferred middle Eocene geothermal gradient field based on analysis of Mannville Group reflectance and maximum burial depth, using the technique of Middleton (1982). (From Osadetz et al., 1990b.)

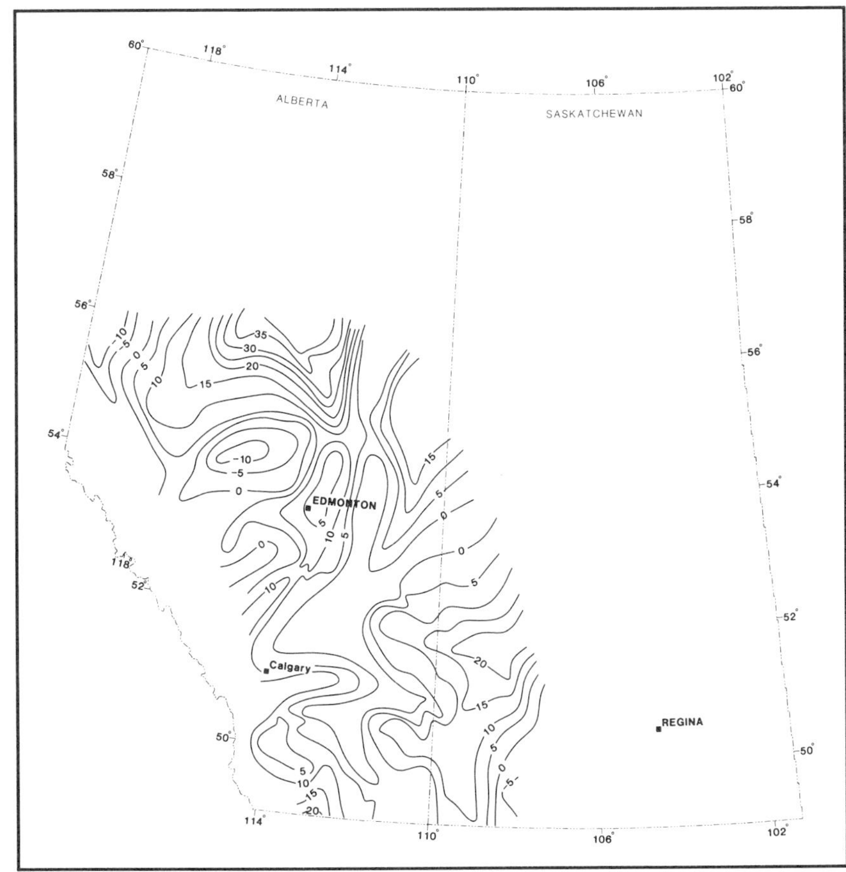

FIGURE 16. Map of the differences between the middle Eocene paleogeothermal gradients and the present geothermal gradients from Osadetz et al. (1990b). See text for discussion of features on this map.

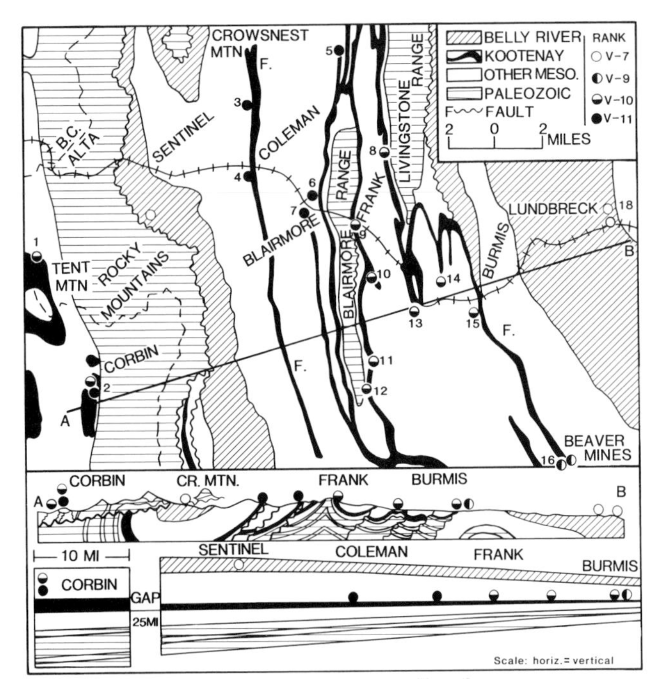

FIGURE 17. Palinspastic profile through the southern Foothills and Front Range used by Hacquebard and Donaldson (1974) to illustrate their inference of predeformational coalification. Notice the high (0.7% R_0) rank of coals in the Campanian Belly River Formation at Sentinel, Alberta. These coals now lie in the immediate footwall of the Lewis thrust.

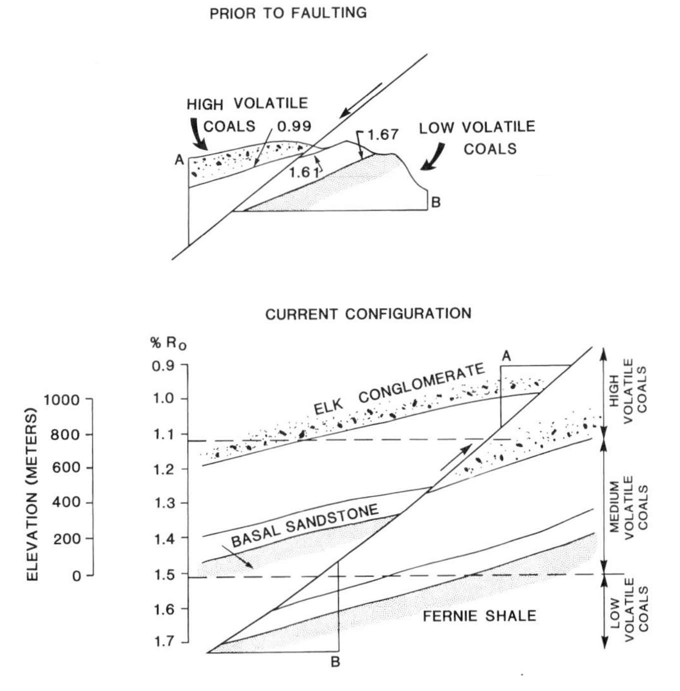

FIGURE 18. Figure from Pearson and Grieve (1985) illustrating the postdeformational nature of coalification in the Front Range. Isoreflectance lines cross the East Crop overthrust.

position appear to be strong evidence for preorogenic coalification. However, depending on the coalification mechanism, continuity of coalification patterns on undeformed sections may be fortuitous. Coalification is a cumulative process. Although significant postdeformational coalification can be discounted, significant syntectonic coalification beneath a cooling thrust plate cannot. Such situations may be characterized by coalification gradients that fail to predict geologically constrained maximum burial depths or unrealistic stratigraphic burials.

In the northern Foothills, the coalification pattern has been combined with time-temperature index (TTI) coalification models to demonstrate the plausibility of predeformational coalification controlled by depth and duration of burial beneath Campanian-Eocene Laramide molasse. Preorogenic coalification in the southern Disturbed belt requires deep burial in post-Campanian time to coalify the Belly River Formation. Coal petrographers and hydrogeologists generally ascribe to such models, but these are generally rejected by stratigraphers and metamorphic petrologists who appeal to syntectonic maximum temperatures after concluding that required sedimentary burial is unreasonable.

Coals in the Cascade, Elk, and Fernie basins of southwestern Alberta and southeastern British Columbia form economically significant coal fields in the Kootenay and Blairmore groups. Hitchon (1984) presented a palinspastic profile extending from Grand Centre to Canmore (Figure 19). Using coalification data (Hacquebard, 1977; Hacquebard and Donaldson, 1974), he suggested that the Cascade basin Kootenay Group was buried to approximately 7.0 km (4.3 mi) and that a Tertiary succession 3.0 km (1.9 mi) thick, greater than that preserved in the Alberta syncline, was deposited on Front Range thrust plates. He inferred this to indicate that late Paleocene geothermal gradients increased westward from approximately 23 mK/m on the Plains to 30 mK/m in the Cascade basin. Similarly, Majorowicz and Jessop (1981) used coalification data to infer that geothermal gradients were higher in the west prior to late Laramide orogeny and effective hydrodynamic recharge.

Observed coalification gradients elsewhere in the southern Front Range intervals are steep (Hughes and Cameron, 1985). If paleogeothermal gradients in the Cascade basin are constrained solely by observations from the Kootenay and Blairmore groups, then coalification gradients are 0.290 log % R_o/km (Hughes and Cameron, 1985) and 0.410 log % R_o/km (Hacquebard and Donaldson, 1974) (Figure 20) and suggest that paleogeothermal gradients of 56 and 80 mK/m existed in the Wind Ridge and Mount Allan sections, respectively. Unfortunately, these gradients fail to predict a reasonable zero coalification surface (Dow, 1977). Not even the Campanian Belly River Formation thickness preserved in the footwall of the McConnell thrust fault is predicted. If the same coalification gradients are constrained by Kootenay Group (latest Jurassic to earliest Cretaceous) coals and

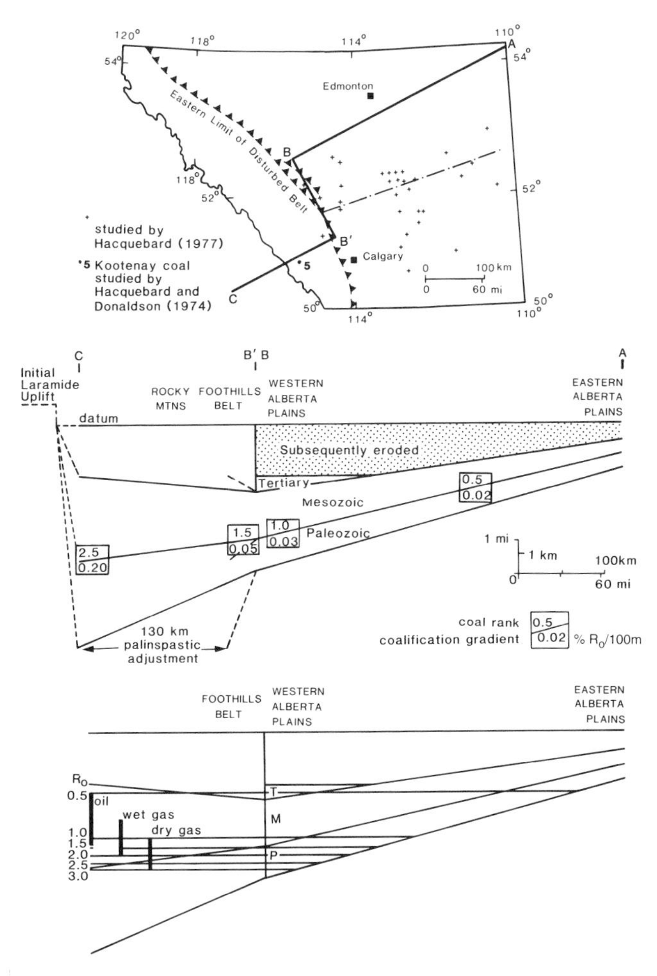

FIGURE 19. Palinspastic profile from Canmore, Alberta across the Plains presented by Hitchon (1984) to illustrate the conditions required for predeformational coalification of the Kootenay Group in the Cascade basin.

maximum burial depths suggested by Hitchon (1984), then inferred coalification gradients are 0.143 and 0.146 log % R_o/km, suggesting paleogeothermal gradients of 27.8 and 28.5 mK/m for the Mount Allan (Hughes and Cameron, 1985) and Wind Ridge (Hacquebard and Donaldson, 1974) regions, respectively. There is, however, no direct evidence for depths of sedimentary burial as suggested by Hitchon (1984).

The Rundle thrust plate is approximately 3.5 km (2.2 mi) thick, and it is equally possible that coal ranks at Canmore reflect tectonic and not sedimentary burial. Hughes and Cameron (1985) suggested that observed coal ranks could be modeled successfully using 43 mK/m orogenic geothermal gradients and only 0.85-km (0.53-mi) tectonic burial beneath the Rundle thrust plate. Deeper tectonic burial would allow lower orogenic geothermal gradients to be considered. Burial beneath the full thickness of the Rundle thrust plate would result in a geothermal gradient of approximately 25 mK/m.

Coal rank in the Campanian Belly River Formation, the youngest preserved beneath the Lewis thrust in the Crowsnest Pass, is approximately 0.7% R_o. Constraining the local coalification gradient with reflectance values from Kootenay Group and nearby coalification gradients (England, 1984; England and Bustin, 1986) requires the Belly River Formation in the footwall of the Lewis thrust, at Sentinel, Alberta, to have been buried to a depth of more than 4.0 km (2.5 mi) to achieve observed coal ranks. It is also possible that this coalification reflects tectonic burial beneath the Lewis thrust plate, which has comparable thickness. The inferred coalification gradients lie between 0.086 and 0.117 log % R_o/km and imply geothermal gradients of 17 and 23 mK/m. Both tectonic and predeformational sedimentary burial would produce similar coalification gradients in this region, although at different times.

No evidence of the thick succession necessary for predeformational coalification is preserved in the southern Front Range, but there is ample evidence for both deep tectonic burial below the Lewis thrust sheet until Oligocene time (McMechan, 1981) and minor postdeformational coalification in the Fernie basin (Pearson and Grieve, 1985) (Figure 18). Coalification gradients in the Front Range are all high (Hacquebard and Donaldson, 1974) and commonly fail to extrapolate to zero coalification at geologically reasonable burial depths. This suggests that they may record a component of postdeformational coalification, particularly in the southern Foothills. However, Kootenay Group coals in the footwall of the Lewis thrust plate remained deeply buried until the early Oligocene (McMechan, 1981). Such long, deep burials produce significantly different coal ranks across the Lewis thrust in model studies. Without evidence for a mechanism to stop syntectonic coalification reactions, such as geothermal gradient decay, the syntectonic coalification model remains dubious.

Other Paleothermal Indicators

Studies of oxygen isotope geothermometry of Belt-Purcell rocks from the Lewis thrust plate in Glacier National Park, Montana indicate that Precambrian sediments in the overthrust plate reached temperatures of 225 to 310°C (Eslinger and Savin, 1973). The age of isotopic equilibrium is uncertain. Eslinger and Savin infer a geothermal gradient of 36 mK/m (Figure 21) if temperatures represent isotopic equilibrium during a Precambrian metamorphic event (McMechan and Price, 1982). Should isotopic equilibrium reflect a later event, such as maximum burial beneath the Phanerozoic succession, then geothermal gradients would be approximately 23 mK/m, roughly comparable to gradients suggested by coalification data from the Crowsnest Pass and Cascade basin. Alternatively, these data may indicate that relatively low geothermal gradients during Late Cretaceous maximum burial neither exceeded middle Precambrian maximum temperatures nor reset middle Precambrian mineral isotopic equilibriums.

Hoffman and Hower (1979), Hutcheon et al. (1980), and Ghent and Miller (1974) studied very low-grade regional metamorphic reactions in the Foothills and Front Ranges and found temperatures in Mesozoic strata consistent with geothermal environments similar to those

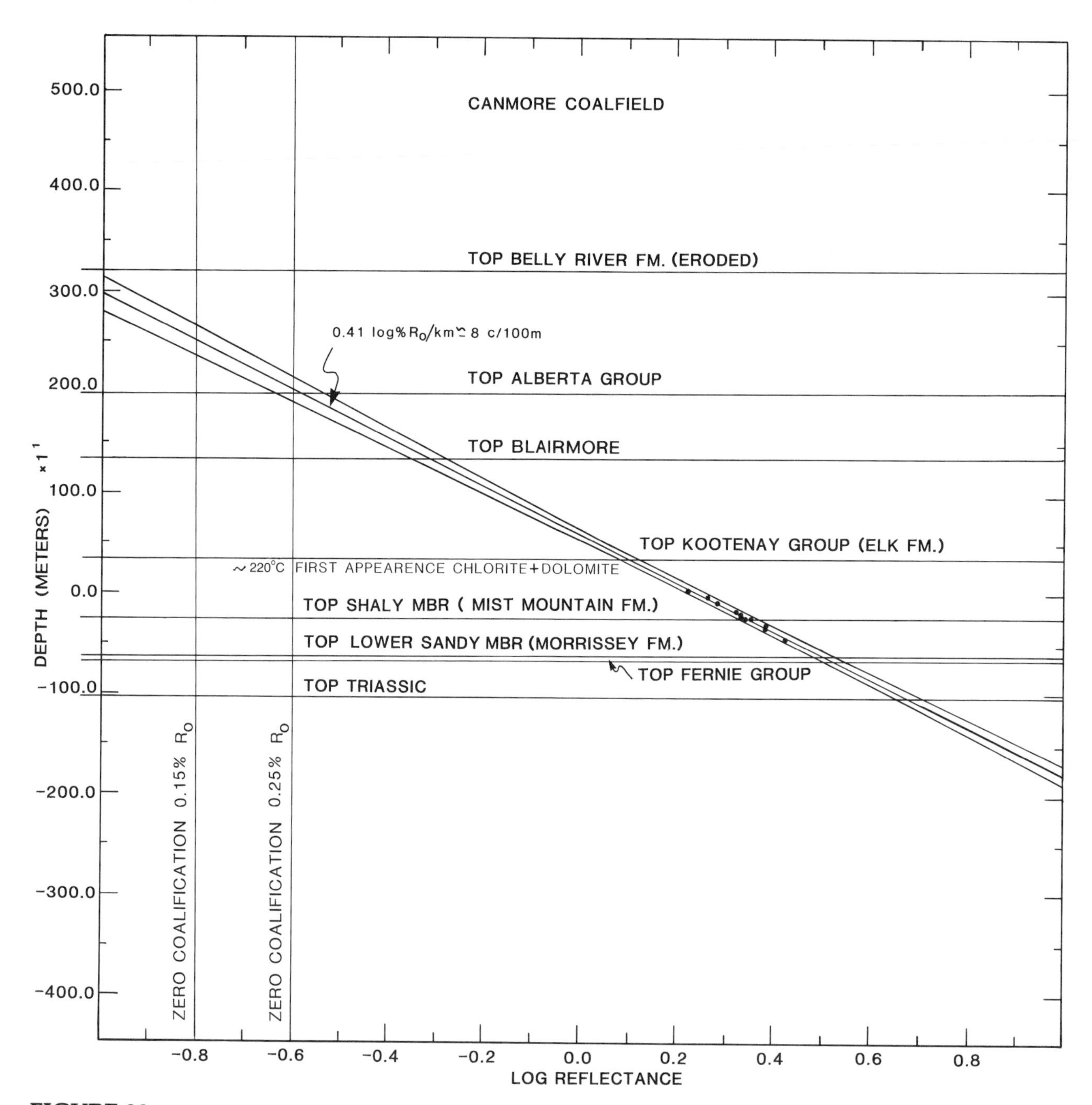

FIGURE 20. Reflectance gradients in the foreland succession of the Cascade basin. Sources of data are discussed in the text. Notice that the log % R_o/km coalification gradients fail to predict burial by stratigraphic thicknesses preserved in the footwall of the McConnell thrust plate, of which the basin is a part. Inferred reflectance gradients consistent with coalification attending burial by the Rundle thrust sheet are indicated.

constrained by coalification and oxygen isotope data. Although Hoffman and Hower (1979) appealed to tectonic loading as a mechanism for achieving elevated temperatures, the mechanism controlling Phanerozoic metamorphism remains as uncertain as the coalification mechanism.

STATE OF GEOTHERMAL HISTORY ANALYSIS

Cordilleran foreland basin thermal environments are probably dominated by advective heat transfer accompanying regional formation water flow (Majorowicz et al.,

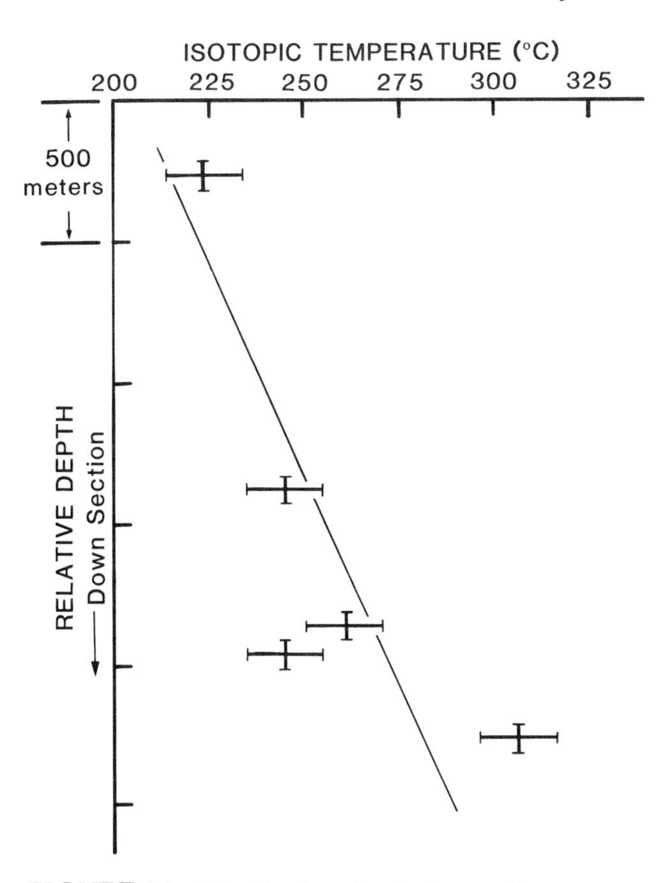

FIGURE 21. Distribution of paleotemperatures with stratigraphic position in the Belt Supergroup, Glacier Park, Montana, as determined by Eslinger and Savin (1973) using the quartz-illite oxygen isotope geothermometer.

1985a; Hitchon, 1984). This is still debated, but the present geothermal gradient field pattern, in which low gradients coincide with areas of higher elevation and hydraulic recharge whereas high gradients generally coincide with low-lying hydraulic discharge regions, is strong evidence for significant advective heat transfer.

In the Plains, coalification records the paleogeothermal field at maximum burial. It is difficult to look beyond the strong overprint that late Laramide successions imposed on thermal indicators to determine Columbian and preorogenic geothermal fields. Nevertheless, the essential topographic elements that drive current advective heat transfer were present from the beginning of the Columbian orogeny. This provides reason to believe that similar geothermal gradient field patterns have characterized each orogenic episode. Foreland successions of the classical defined orogenic phases (Stott, 1984) are separated by erosional unconformities of considerable relief and duration (Leckie and Smith, 1992). The geothermal environment of these epeirogenic or interorogenic events is completely unconstrained and unknown.

In the Plains, coalification was terminated by temperature reduction accompanying epeirogenesis. Topography reduction in mountainous regions, erosion of the Plains, and Tertiary tectonics and volcanism produced identifiable changes in the geothermal gradient field, yet the general pattern remains the same. Within the cordillera, foreland basin coalification mechanisms may have varied, depending on stratigraphic and tectonic burial history, which makes analysis uncertain. Tectonic and stratigraphic analysis suggests a dynamic geothermal history prior to cordilleran orogenies, yet preorogenic geothermal environments, particularly their general features, are largely unknown. Further analysis might indicate regions where preorogenic geothermal history may be determined.

REFERENCES CITED

Archibald, D. A., J. K. Glover, R. A. Price, E. Farrar, and D. M. Carmichael, 1983, Geochronology and tectonic implications of magmatism and metamorphism, southern Kootenay Arc and neighbouring regions, southeastern British Columbia: Part I, Jurassic to mid-Cretaceous: Canadian Journal of Earth Sciences, v. 20, p. 1891-1913.

Archibald, D. A., T. E. Krogh, R. L. Armstrong, and E. Farrar, 1984, Chronology and tectonic implications of magmatism and metamorphism, southern Kootenay Arc and neighbouring regions, southeastern British Columbia: Part II, Mid-Cretaceous to Eocene: Canadian Journal of Earth Sciences, v. 21, p. 567-583.

Aulstead, K. L., and R. J. Spencer, 1985, Diagenesis of the Keg River Formation, northwestern Alberta: fluid inclusion evidence: Bulletin of Canadian Petroleum Geology, v. 33, p. 167-183.

Beach, R. D. W., F. W. Jones, and J. A. Majorowicz, 1987, Heat flow and heat generation estimates for the Churchill Basement of the western Canadian basin in Alberta, Canada: Geothermics, v. 16, p. 1-16.

Beaumont, C., R. Boutilier, A. S. Mackenzie, and J. Rullkoetter, 1985, Isomerization and aromatization of hydrocarbons and the paleothermometry and burial history of the Alberta foreland basin: AAPG Bulletin, v. 69, p. 546-566.

Bond, G. C., and M. A. Kominz, 1984, Construction of tectonic subsidence curves for the early Paleozoic miogeocline, southern Canadian Rocky Mountains: implications for subsidence mechanisms, age of breakup, and crustal thinning: GSA Bulletin, v. 95, p. 155-173.

Burwash, R. A., H. Baadsgaard, and Z. E. Peterman, 1962, Precambrian K-Ar dates from the Western Canada sedimentary basin: Journal of Geophysical Research, v. 67, p. 1617-1625.

Cameron, A. R., 1991, Regional patterns of reflectance distribution in lignites of the Ravenscrag Formation, Saskatchewan, Canada: Organic Geochemistry, v. 17, p. 223-242.

Crowley, K. D., J. L. Ahern, and C. W. Naeser, 1985, Origin and epeirogenic history of the Williston basin; evidence from fission-track analysis of apatite: Geology, v. 13, p. 620-623.

Crowley, K. D., and S. L. Kuhlman, 1988, Apatite thermochronometry of Western Canadian shield; implications

for origin of the Williston basin: Geophysical Research Letters, v. 15, p. 221-224.
Currie, J. B., and S. O. Nwachukwu, 1974, Evidence on incipient fracture porosity in reservoir rocks at depth: Bulletin of Canadian Petroleum Geology, v. 22, p. 42-58.
Dow, W. G., 1977, Kerogen studies and geological interpretations: Journal of Geochemical Exploration, v. 7, p. 79-99.
Drury, M. J., 1985, Heat flow and heat generation in the Churchill Province of the Canadian Shield and their paleotectonic significance: Tectonophysics, v. 115, p. 25-44.
England, T. D. J., 1984, Thermal maturation of the Western Canadian sedimentary basin in the Rocky Mountain Foothills and Plains of Alberta south of the Red Deer River: M.Sc. thesis, University of British Columbia, Vancouver, 171 p.
England, T. D. J., and R. M. Bustin, 1986, Thermal maturation of the Western Canadian sedimentary basin south of the Red Deer River: I, Alberta Plains: Bulletin of Canadian Petroleum Geology, v. 34, p. 71-90.
Eslinger, E. V., and S. M. Savin, 1973, Oxygen isotope geothermometry of the burial metamorphic rocks of the Precambrian Belt Supergroup, Glacier National Park, Montana: GSA Bulletin, v. 84, p. 2549-2560.
Ghent, E. D., and B. E. Miller, 1974, Zeolite and clay-carbonate assemblages in the Blairmore Group (Cretaceous), southern Alberta Foothills, Canada: Contributions to Mineralogy and Petrology, v. 44, p. 313-329.
Gibson, D. W., 1985, Stratigraphy, sedimentology and depositional environments of the coal-bearing Jurassic-Cretaceous Kootenay Group, Alberta and British Columbia: Geological Survey of Canada Bulletin 357, 108 p.
Grieve, D. A., 1987, Coal rank distribution, Flathead coalfield, southeastern British Columbia (82G/2, 82G/7), *in* Geological fieldwork, 1986: British Columbia Ministry of Energy, Mines and Petroleum Resources Paper 1987-1, p. 361-364.
Hacquebard, P. A., 1977, Rank of coal as an index of organic metamorphism for oil and gas in Alberta, *in* G. Deroo, T. G. Powell, B. Tissot, and R. G. McCrossan, eds., The origin and migration of petroleum in the Western Canadian sedimentary basin, Alberta—a geochemical and thermal maturation study: Geological Survey of Canada Bulletin 262, p. 11-22.
Hacquebard, P. A., and J. R. Donaldson, 1974, Rank studies of coals in the Rocky Mountains and inner Foothills belt, Canada, *in* R. R. Dutcher, P. A. Hacquebard, J. M. Schopf, and J. A. Simon, eds., Carbonaceous materials as indicators of metamorphism: GSA Special Paper 153, p. 75-94.
Hitchon, B., 1984, Geothermal gradients, hydrodynamics and hydrocarbon occurrences, Alberta, Canada: AAPG Bulletin, v. 68, p. 713-743.
Hoffman, J., and J. Hower, 1979, Clay mineral assemblages as low grade metamorphic geothermometers: application to the thrust faulted Disturbed belt of Montana, U.S.A., *in* P. A. Scholle and P. R. Schluger, eds., Aspects of diagenesis: SEPM Special Publication No. 26, p. 55-79.
Hughes, J. D., and A. R. Cameron, 1985, Lithology, depositional setting and coal rank-depth relationships in the Jurassic-Cretaceous Kootenay Group at Mount Allan, Cascade coal basin, Alberta: Geological Survey of Canada Paper 81-11, 41 p.
Hutcheon, I., A. Oldershaw, and E. D. Ghent, 1980, Diagenesis of Cretaceous sandstones of the Kootenay Formation at Elk Valley (southeastern British Columbia) and Mt. Allan (southwestern Alberta): Geochimica et Cosmochimica Acta, v. 44, p. 1425-1436.
Issler, D. R., C. Beaumont, S. D. Willett, R. A. Donelick, J. Mooers, and A. Grist, 1991, Preliminary evidence from apatite fission-track data concerning the thermal history of the Peace River arch region, Western Canada sedimentary basin: Bulletin of Canadian Petroleum Geology, v. 38A (1990), p. 250-269.
Jones, F. W., and J. A. Majorowicz, 1987, Regional trends in radiogenic heat generation in the Precambrian basement of the Western Canadian basin: Geophysical Research Letters, v. 14, p. 268-271.
Kalkreuth, W., and C. W. Langenberg, 1986, The timing of coalification in relation to structural events in the Grande Cache area, Alberta, Canada: Canadian Journal of Earth Sciences, v. 23, p. 1103-1116.
Kalkreuth, W., and M. E. McMechan, 1984, Regional pattern of thermal maturation as determined from coal-rank studies, Rocky Mountain Foothills and Front Ranges north of Grande Cache, Alberta—implications for petroleum exploration: Bulletin of Canadian Petroleum Geology, v. 32, p. 249-271.
Kalkreuth, W., and M. E. McMechan, 1988, Burial history and thermal maturity, Rocky Mountain Front Ranges, Foothills and foreland, east-central British Columbia and adjacent Alberta, Canada: AAPG Bulletin, v. 72, p. 1395-1410.
Leckie, D., and D. Smith, 1992, Regional setting, evolution, and depositional cycles of the Western Canada foreland basin, this volume.
Leckie, D., W. Kalkreuth, and L. R. Snowdon, 1988, Source rock potential and thermal maturity of Lower Cretaceous strata, Monkman Pass area, British Columbia: AAPG Bulletin, v. 72, p. 820-838.
Longstaffe, F. J., 1986, Oxygen isotope studies of diagenesis in the basal Belly River sandstone, Pembina I-Pool, Alberta: Journal of Sedimentary Petrology, v. 56, p. 78-88.
Luheshi, M. N., and D. Jackson, 1986, Conductive and convective heat transfer in sedimentary basins, *in* J. Burrus, ed., Thermal modeling in sedimentary basins: Proceedings of the 1er L'Institut Français du Pétrole Exploration Research Conference, Carcans, June 3-7, 1985, p. 219-234.
Lundell, L. L., 1991, Relationship between thermal maturation gradients, geothermal gradients and estimates of the thickness of the eroded foreland section, southern Alberta Plains, Canada: Discussion: Marine and Petroleum Geology, v. 8, p. 241-242.
Macauley, G., 1984, Cretaceous oil shale potential of the Prairie Provinces, Canada: Geological Survey of Canada Open File Report OF-977, 61 p.
Magara, K., 1976, Thickness of removed sedimentary rocks, paleopore pressure and paleotemperature,

southwestern part of Western Canada basin: AAPG Bulletin, v. 60, p. 554-565.

Majorowicz, J. A., 1989, The controversy over the significance of the hydrodynamic effect on heat flow in the Prairies basin: International Union of Geodesy and Geophysics Geophysical Monograph 47, v. 2., p. 101-103.

Majorowicz, J. A., and A. M. Jessop, 1981, Regional heat flow patterns in the Western Canadian sedimentary basin: Tectonophysics, v. 74, p. 209-238.

Majorowicz, J. A., F. W. Jones, M. E. Ertman, K. G. Osadetz, and L. D. Stasiuk, 1991a, Relationship between thermal maturation gradients, geothermal gradients and estimates of the thickness of the eroded foreland section, southern Alberta Plains, Canada: Marine and Petroleum Geology, v. 7, p. 138-152.

Majorowicz, J. A., F. W. Jones, M. E. Ertman, K. G. Osadetz, and L. D. Stasiuk, 1991b, Relationship between thermal maturation gradients, geothermal gradients and estimates of the thickness of the eroded foreland section, southern Alberta Plains, Canada: Reply: Marine and Petroleum Geology, v. 8, p. 242-243.

Majorowicz, J. A., F. W. Jones, and A. M. Jessop, 1986, Geothermics of the Williston basin in Canada in relation to hydrodynamics and hydrocarbon occurrences: Geophysics, v. 51, p. 767-779.

Majorowicz, J. A., F. W. Jones, H. L. Lam, and A. M. Jessop, 1985a, Terrestrial heat flow and geothermal gradients in relation to hydrodynamics in the Alberta basin, Canada: Journal of Geodynamics, v. 4, p. 265-283.

Majorowicz, J. A., F. W. Jones, H. L. Lam, and A. M. Jessop, 1985b, Regional variations of heat flow differences with depth in Alberta, Canada: The Geophysical Journal of the Royal Astronomical Society, v. 81, p. 479-487.

Majorowicz, J. A., M. Rahman, F. W. Jones, and N. J. McMillan, 1985c, The paleogeothermal and present thermal regimes of the Alberta basin and their significance for petroleum occurrences: Bulletin of Canadian Petroleum Geology, v. 33, p. 12-21.

Masters, J. A. 1984, Lower Cretaceous oil and gas in Western Canada, *in* J. A. Masters, ed., Elmworth—case study of a Deep Basin gas field: AAPG Memoir 38, p. 1-33.

McMechan, R. D., 1981, Stratigraphy, sedimentology, structure and tectonic implications of the Oligocene Kishenehn Formation, Flathead Valley graben, southeastern British Columbia: Ph.D. thesis, Queens University, Kingston, 327 p.

McMechan, M. E., and R. A. Price, 1982, Superimposed low-grade metamorphism in the Mount Fisher area, southeastern British Columbia—implications for the East Kootenay orogeny: Canadian Journal of Earth Sciences, v. 19, p. 476-489.

Middleton, M. F., 1982, Tectonic history from vitrinite reflectance: The Geophysical Journal of the Royal Astronomical Society, v. 68, p. 121-132.

Morrow, D. W., G. L. Cumming, and R. B. Koepnick, 1986, Manetoe facies—a gas-bearing, megacrystalline, Devonian dolomite, Yukon and Northwest Territories, Canada: AAPG Bulletin, v. 70, p. 702-720.

Naeser, N. D., and T. H. McCullough, 1989, Thermal history of sedimentary basins; methods and case histories: New York, Springer-Verlag, 319 p.

Nurkowski, J. R., 1984, Coal quality, coal rank variation and its relation to reconstructed overburden, Upper Cretaceous and Tertiary Plains coals, Alberta, Canada: AAPG Bulletin, v. 68, p. 285-295.

Osadetz, K. G., and L. R. Snowdon, 1986a, Petroleum source rock reconnaissance of southern Saskatchewan, *in* Current Research Part A: Geological Survey of Canada Paper 86-1A, p. 609-617.

Osadetz, K. G., and L. R. Snowdon, 1986b, Speculation on the petroleum source rock potential of portions of the Lodgepole Formation (Mississippian) of southern Saskatchewan, *in* Current Research Part B: Geological Survey of Canada Paper 86-1B, p. 647-651.

Osadetz, K. G., F. Goodarzi, L. R. Snowdon, P. W. Brooks, and S. Fayerman, 1990a, Winnipegosis pinnacle reef play in Williston basin, Saskatchewan and North Dakota: oil compositions and effects of oil-based drilling muds on exploration geochemistry, *in* Current Research Part D: Geological Survey of Canada Paper 90-1D, p. 153-163.

Osadetz, K. G., D. E. Pearson, and L. D. Stasiuk, 1990b, Paleogeothermal gradients and changes in the geothermal gradient field of the Alberta Plains, *in* Current Research Part D: Geological Survey of Canada Paper 90-1D, p. 165-178.

Osadetz, K. G., L. R. Snowdon, and L. D. Stasiuk, 1989, Association of enhanced hydrocarbon generation and crustal structure in the Canadian Williston basin, *in* Current Research Part D: Geological Survey of Canada Paper 89-1D, p. 35-47.

Pearson, D. E., and D. A. Grieve, 1985, Rank variation, coalification pattern and coal quality in the Crowsnest coalfield, British Columbia: Canadian Institute of Mining and Metallurgy Bulletin, v. 78, p. 39-46.

Porter, J. W., 1992, Conventional hydrocarbon reserves of the Western Canada foreland basin, this volume.

Porter, J. W., R. A. Price, and R. G. McCrossan, 1982, The Western Canada sedimentary basin, *in* P. Kent, M. H. P. Bott, D. P. McKenzie, and C. A. Williams, eds., The evolution of sedimentary basins: Philosophical Transactions of the Royal Society of London, Series A, v. 305, p. 169-192.

Price, L. C., T. Ging, T. Daws, A. Love, M. Pawlewicz, and D. Anders, 1984, Organic metamorphism in the Mississippian-Devonian Bakken shale, North Dakota portion of the Williston basin, *in* J. Woodward, F. F. Meissner, and J. L. Clayton, eds., Hydrocarbon source rocks of the greater Rocky Mountain region: Denver, Colorado, Rocky Mountain Association of Geologists, p. 83-134.

Price, R. A., 1981, The cordilleran foreland thrust and fold belt in the southern Canadian Rocky Mountains, *in* K. R. McClay and N. J. Price, eds., Thrust and nappe tectonics, International Conference: The Geological Society of London Special Publication No. 9, p. 427-448.

Price, R. A., and E. W. Mountjoy, 1970, Geologic structure of the Canadian Rocky Mountains between Bow and Athabasca rivers—a progress report, *in* J. O. Wheeler, ed., Structure of the southern Canadian Cordillera: Geological Association of Canada Special Paper No. 6, p. 7-39.

Roedder, E., 1968, Temperature, salinity and origin of the ore-forming fluids at Pine Point, Northwest Territories,

Canada, from fluid inclusion studies: Economic Geology, v. 63, p. 439–450.

Stasiuk, L. D., 1988, Thermal maturation and organic petrology of Mesozoic sediments of southern Saskatchewan: M.Sc. thesis, University of Regina, Regina, Saskatchewan, 178 p.

Stoakes, F. A., and S. Creaney, 1985, Sedimentology of a carbonate source rock: the Duvernay Formation of Alberta, Canada, *in* M. W. Longman, K. W. Shanley, R. F. Lindsay, and D. E. Eby, eds., Rocky Mountain carbonate reservoirs: a core workshop: SEPM Core Workshop No. 7, p. 343–375.

Stott, D. F., 1984, Cretaceous sequences of the Foothills of the Canadian Rocky Mountains, *in* D. F. Stott and D. J. Glass, eds., The Mesozoic of middle North America: Canadian Society of Petroleum Geologists Memoir 9, p. 85–107.

Chapter 10

Petroleum Systems in the Foreland Basin of Western Canada

S. Creaney and J. Allan

ESSO Resources Canada Limited
Calgary, Alberta, Canada

INTRODUCTION

Exploration has been intense in the Western Canada basin for over 70 years following the discovery of oil at Turner Valley and with considerable success occurring prior to any routine study of source rocks. The earliest reference to organic-rich mudstones was by J. B. Tyrrell (1892), who described "possible oil shales" along the Manitoba escarpment, and McInnes (1913), who generated a seven gallon/ton oil yield from these shales. Ells (1923) described these rocks as having been deposited in a "muddy sea," and considered them to be age-equivalent to the Niobrara Formation of the United States. These descriptions represent the first attention paid to some of the more significant source rocks of the Western Canada basin.

Gussow (1954) discussed hydrocarbon migration in the Upper Devonian Rimbey-Meadowbrook reef chain of Alberta and speculated that this oil may have contributed to the Athabasca Cretaceous heavy oil deposit. In 1955, Gussow suggested a Jurassic source (undefined) for the Athabasca deposits. Hodgson (1954) used trace element analyses of several crude oil samples to conclude that Devonian- and Cretaceous-reservoired oils were basically very similar. So began a series of research studies that dealt almost exclusively with oil and/or gas samples and sought to understand the source of petroleum in the Western Canada basin. Hitchon (1963a, b, c) described the variations in gas composition in the Western Canada sedimentary basin and postulated migration pathways based on subsurface fluid flow directions. Source rocks were not discussed. Vigrass (1968) summarized the postulated hypotheses for the origin of the tar sands, which at that time were many and varied. Bailey et al. (1974) and Milner et al. (1977) discussed maturity and mechanisms of hydrocarbon alteration in the basin, but provided no source rock data. It was at this time that the vast tar belt began to be understood as a biodegraded remnant of a former supergiant conventional oil field, although some still felt that a local origin from low-maturity sediments could be supported (Montgomery et al., 1974; George et al., 1977).

Deroo et al. (1977) were the first authors to publish combined geochemical rock and reservoired hydrocarbon data. Using a variety of then technically advanced analyses, they divided 105 oil samples into three genetic groups based on a variety of geochemical characteristics. They found that their Group 1 oils (Belly River, Cardium, and Viking) were difficult to distinguish geochemically from their Group 3 oils (Upper Devonian, excluding Wabamun). Their Group 2 oils (Lower Mannville, Jurassic, Mississippian, Wabamun, and the heavy oils at Athabasca) were found to be geochemically quite distinct, having higher sulfur and aromatic hydrocarbon contents and the least amounts of saturated hydrocarbons. Hitchon and Filby (1984) supported this grouping with trace element data but also experienced difficulties distinguishing Groups 1 and 3 oils. Deroo et al. (1977) speculated that the Group 1 oils were sourced from the adjacent Colorado Group shales, the Group 2 oils were sourced from the Mannville itself, and the Group 3 oils were sourced from one or all of the Devonian Beaverhill Lake, Cooking Lake, and Duvernay formations. The oil-source correlations were tentative because the vast majority of the 800 rocks that were assayed for source quality were not actually from oil-source facies. Later, Powell and Snowdon (1980) reiterated the conclusions of Deroo et al. (1977) and suggested that the Mannville heavy oils of the tar belt were generated from the Mannville itself at low levels of maturity.

Stoakes and Creaney (1984) carried out a detailed geochemical and sedimentological study of the Upper Devonian Woodbend Group shales. These authors showed that the organic-rich Duvernay/Majeau Lake basinal laminites were the source of reservoired hydrocarbons in the Leduc reefs of west-central Alberta and could, on geologic grounds, have contributed to the heavy oils in the Mannville. Geochemical support for the heavy oil contribution is lacking, however.

Following an earlier description of extensive gas deposits in low-porosity Mesozoic clastics in the Elmworth area of northwestern Alberta and northeastern British Columbia (Masters, 1979), Welte et al. (1984) postulated that they were sourced from Jurassic–Lower Cretaceous coals and coaly shales at advanced levels of maturity. Reflected-light organic petrography, organic carbon contents, and Rock-Eval pyrolysis data were used to confirm the gas-prone nature of this part of the section. With respect to oil-generative potential, Moshier and Waples (1985) provided a mass-balance calculation, based on the low average TOC and hydrogen indices of Mannville shales, which showed this section to be incapable of generating the required 2.7×10^{11} m^3 (1.7×10^{12} bbl) of oil currently in the tar belt and thus refuting the earlier suggestion of Deroo et al. (1977). Moshier and Waples (1985) suggested the Mississippian Exshaw and unnamed "Jurassic Units" as possible contributing sources.

Du Rouchet (1984, 1985), while discussing migration mechanisms for the tar sand deposits, suggested a Triassic contribution to these deposits. However, this conclusion relied on oil-oil correlation using aromatic hydrocarbon distributions (Triassic-reservoired oil at Inga to Mannville oil at Peace River), with no source rock data. As recently as 1986, Hoffman and Strausz (1986) concluded that "a detailed correlation of suitable mature, organic-rich source rocks, especially those in the Jurassic and Triassic, with the major bitumen deposits are needed to clarify the origin of the bitumens of Northern Alberta." Subsequently, Fowler and Brooks (1987) and Brooks et al. (1988) have used asphaltene pyrolysis and biomarker analyses on heavy oil samples from the major deposits to infer that all belong to a "single" family from an unidentified source rock. Using aromatic steranes (triaromatic steroids), Brooks et al. (1989) supported their earlier conclusion that all the heavy oils are from very similar sources or the same source. They showed further that biomarker distributions from both saturated and aromatic fractions are very similar to conventional Lower Cretaceous–Mississippian conventional oils, but are very different from conventional Middle and Upper Devonian oils. On this basis, they suggested that the Upper Devonian Duvernay source facies has had little if any input to the oil sands/heavy oils/bitumens of the Western Canada sedimentary basin. Several previous authors have also suggested the "single source" idea, including Rubinstein et al. (1977), whereas Leenheer (1984) suggested that the Bakken and its equivalents are the principal source rocks.

Macauley (1984a, b) and Macauley et al. (1985) provided the first stratigraphically constrained geochemical data on organic-rich rocks within the marine Colorado Shale section. These authors analyzed 450 samples from 29 wells and potash shafts across Manitoba, Saskatchewan, and eastern Alberta. Although all their samples were from the immature portion of the Western Canada basin, they showed, for the first time, that the First and Second White Speckled Shales (the Boyne and Favel formations, respectively, in Saskatchewan) were organically enriched (TOC contents up to 11%) and dominated by marine, type II (oil-prone) organic matter (Rock-Eval pyrolysis hydrogen indices up to 600).

Creaney and Allan (1990) and Allan and Creaney (1991) provided the first syntheses of geology and geochemistry to produce an overview of petroleum generation and migration based on actual source rock analyses. These authors identified at least 14 source rocks and 11 related oil families in the Western Canada basin and considered the heavy oils to have been generated principally from the Jurassic Nordegg Member of the Fernie Group and the Mississippian Exshaw Formation. They also gave geologic reasons why Triassic and Devonian sources may have contributed to the heavy oils, but any such contributions were probably relatively small volumetrically. Recently and partly in contrast to the above, Fowler et al. (1989) analyzed two cores of the Nordegg Member of the Fernie Group and used biomarker criteria [principally diasterane content, pristane/phytane ratio, lack of 28,30 bisnorhopanes, and C_{23} tricyclic terpane/17a (H)-hopane ratio] to exclude the Nordegg Member as a contributor to Western Canada's heavy oils.

The present contribution seeks to provide some of the information required to elucidate hydrocarbon sourcing and pooling in sediments of the foreland deep of the Western Canada basin and utilizes a combination of previously unpublished data and material contained in Creaney and Allan (1990) and Allan and Creaney (1991). The nature of the present volume (a review of the Western Canada foreland basin) has precluded the inclusion of the large number of oil-oil and oil-source correlations required to perform a study of this magnitude. This material has been published elsewhere (Allan and Creaney, 1991). Furthermore, on a basinal scale, the perspective presented here is that of the oil explorationist as opposed to that of a detailed analytical oil and source rock geochemistry thesis.

ANALYTICAL METHODOLOGY

The principal resource behind this study is an extensive collection of crude oils, source rock cores, gas analyses, and geophysical well logs from the Western Canada basin. To date, more than 300 oil samples, 200 source rock cores, and 400 gas analyses have been interpreted in the basin model. The classification of the hydrocarbons into families is complex and is based on numerous factors including age and location of reservoir, reservoir type, oil gravity, sulfur content, oil geochemistry, and source rock geochemistry.

Gas chromatography (GC) provides the initial geochemical screen of oils whereby *n*-paraffin and isoprenoid distributions can be examined, approximate maturities assessed, and absence or presence, or extent, of biodegradation determined. The final analysis is the evaluation of naphthenic and aromatic biomarker distributions using computerized gas chromatography–mass spectrometry (C-GC-MS).

Biomarker analysis provides the main evidence for oil-source rock correlations, and all oil/oil-source correlations reported here are based exclusively on the geochemical signatures of core samples rather than cuttings samples. Source rock samples were extracted (dichloromethane as solvent), and GC and C-GC-MS analyses were performed on unfractionated extract or separated saturated and aromatic hydrocarbon fractions. Prior to this step, however, Rock-Eval pyrolysis provided base data on organic richness (TOC), organic matter type via hydrogen indices (HI), and maturity via temperature of maximum rate of evolution of kerogen pyrolysis products

(T_{max}). Where possible, maturity data are complemented by vitrinite reflectance data on adjacent carbonaceous sediments.

In addition to the organic richness and quality determined through core assays, data on source rock thickness and distribution are required for calculation of hydrocarbon yields. To surmount the limitations imposed by insufficient core (in a basinal sense), source quality and quantity are also evaluated from geophysical well logs (Passey et al., 1989, 1990), of which an enormous database exists.

Natural gas analyses provide gas compositional and isotopic data from which maturities can be determined (James, 1983). In combination with wetness indices and H_2S and CO_2 contents, these data complement the oil analyses in compiling evidence for the likely sources and hydrocarbon migration pathways that have been active. The key control on this modeling is detailed knowledge of the geologic structure and stratigraphy of the basin and the timing of significant events that impacted hydrocarbon generation, migration, and entrapment.

FORELAND BASIN SOURCE ROCKS

Late Jurassic–Early Cretaceous Coal-Bearing Section

The earliest deposits of the foreland basin (Figure 1) are dominated by deltaic/coastal plain sedimentation (Leckie and Smith, 1992) that characteristically is coal rich. Welte et al. (1984) provided an excellent geochemical description of this portion of the section in the Elmworth area. Furthermore, these coals are mined at outcrops in the Disturbed belt, and their petrographic composition is well described (Cameron, 1972; Kalkreuth, 1982). The coals are notably low in liptinite and commonly rich in inertinite. Pyrolysis analyses (Figure 2) show these sediments to contain primarily type III kerogen, which led Welte et al. (1984) to conclude that they are very potent gas sources, particularly at the advanced levels of maturity found in the Deep basin (Kalkreuth and McMechan, 1988). Welte et al. (1984) further suggested that the 4.8×10^{11} m^3 (17 tcf) of proved and probable natural gas reservoired in low-porosity Lower Cretaceous sands of Western Canada's Deep basin was largely generated from local high-maturity coals and migrated into adjacent low-porosity, low-permeability sands. Masters (1979) had earlier proposed that these gases are hydrodynamically trapped because the sands are water filled updip from the "tight gas sands" (Barclay and Smith, 1992).

TOC values in associated shales are commonly below 2%, with very low hydrogen indices (Moshier and Waples, 1985). However, the presence of locally restricted, delta plain lacustrine shales containing concentrated amounts of liptinites or sapropelic material cannot be discounted in this sedimentary environment (Stach, 1975). Hence, localized oil or condensate with commercial potential may exist.

Colorado Group

In the foreland basin section of the Western Canada basin, the interaction of sedimentation rate and anoxia can be illustrated schematically by the marine source rock model shown in Figure 3. Sedimentation rate generally decreased from west to east (shelf to basin). In the relatively shallow epicontinental seas of the Western Interior seaway, the rate at which plankton settled to the sea floor from the photic zone is assumed to have been relatively constant across the basin. Thus, in the presence of reduced oxygen conditions (Demaison and Moore, 1980), organic matter accumulated on the seabed, and its concentration, as a percentage of the host sediment (weight percent TOC), was dictated by sedimentation rate. Under these circumstances, when accommodation (Jervey, 1988) was high, distal sedimentation rate would be low and TOC would be high. The model can be used to explain fluctuations in mean organic matter type. The principal organic components of marine source rocks are planktonic/bacterial remains (oil prone) and allochthonous terrigenous organic materials (both gas-prone and inert carbonaceous fragments). The relative supply of these components was influenced by shelfal accommodation. During times of low accommodation, large amounts of terrigenous organics and inorganics would have been supplied to the sediments in basinal locations, whereas, during times of high accommodation, sediments of the distal basin would have been enriched in marine planktonic/bacterial debris. Figure 4 is a crossplot of TOC vs. hydrogen index from a single continuous rock core that contains source and nonsource facies. At TOC values above 2%, the hydrogen index is relatively constant (at a value dictated by maturity) and reflects the compositional characteristics of the marine organic matter. At about 2% TOC, an inflection point occurs in the relationship, indicating the onset of significant dilution by terrigenous inorganics and organics.

The hydrogen index then falls rapidly with decreasing TOC below 2%. This progressive change in richness and organic matter type corresponds to changing sedimentary facies whereby a petroleum source facies can grade gradually or rapidly to a nonsource.

The Colorado Group of upper Albian to Santonian age is a thick, marine shale and siltstone sequence that contains several oil-source horizons. In the southern part of the basin, two principal effective source zones are the Second White Speckled Shale (Cenomanian/Turonian) and the Fish Scale Zone (Albian/Cenomanian). These condensed sections lie close together on the eastern limb of the basin (distal facies) and become increasingly separated stratigraphically toward the Alberta syncline (more proximal facies) (Figure 5). Both contain marine, type II organic matter, with TOC values up to 7% and hydrogen indices up to 450 where immature (Allan and Creaney, 1988). Between the condensed sections are other effective marine, although less rich, source intervals where TOC values range from 2 to 3% and hydrogen indices range up to 300. In the central part of the basin, a younger condensed section (First White Speckled Shale-Santonian) becomes more organic rich and develops source potential. This section is usually less prominent on natural gamma logs than the older condensed sections (Figure 5) but has source quality characteristics similar to those of the older Second White Speckled Shale.

The Colorado Shale Group is mature in the most westerly part of its zone of occurrence, adjacent to the

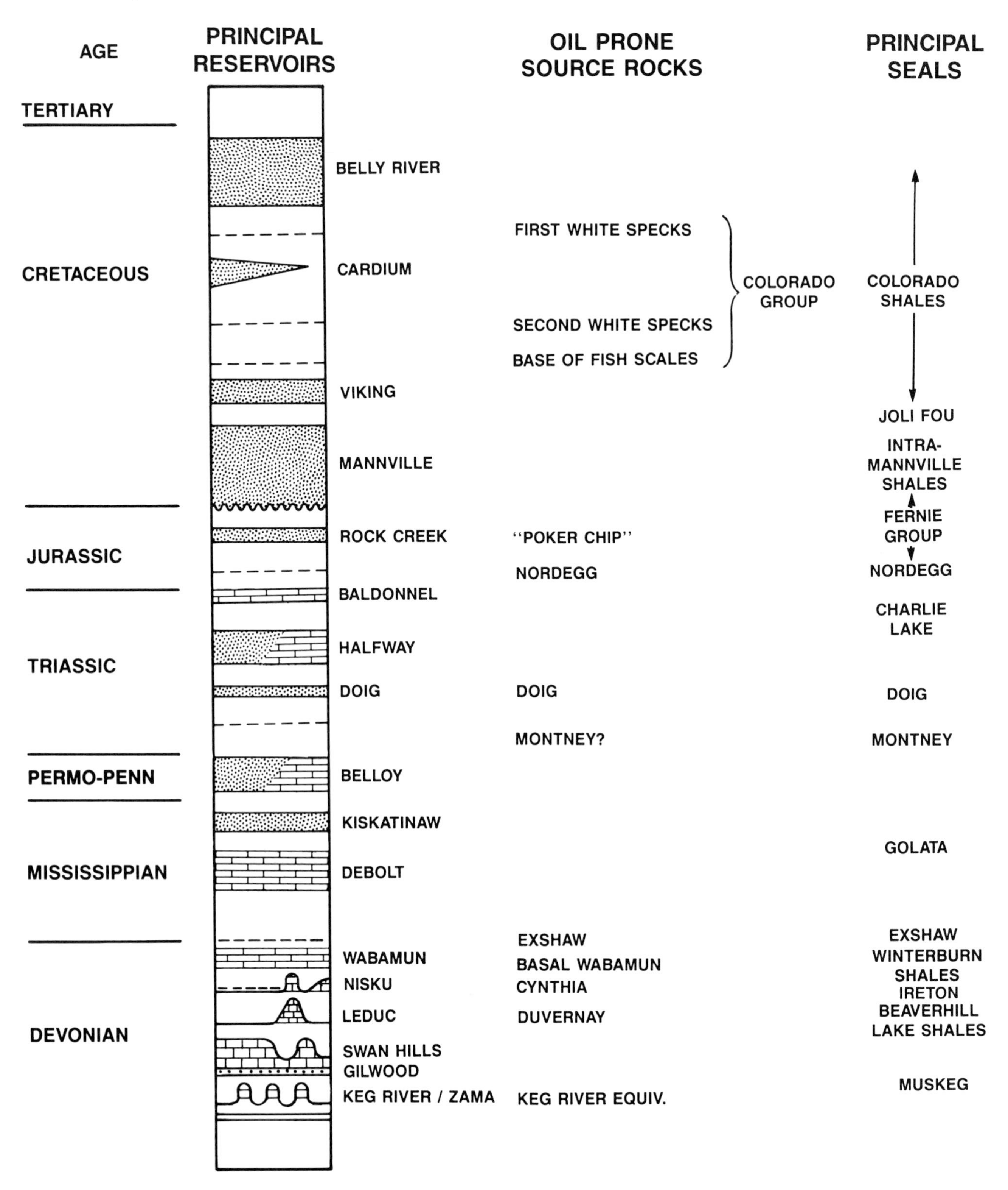

FIGURE 1. Generalized stratigraphy of the Western Canada foreland basin, showing principal hydrocarbon source, seal, and reservoir horizons. (Modified after Allan and Creaney, 1991.)

Disturbed belt (Figure 6). Peak maturity of the Second White Speckled Shale and the Fish Scale Zone lies close to the Disturbed belt in the southern part of the basin. Passing northward, deeper maximum burial causes these two source zones to trend toward overmaturity, and the First White Speckled Shale then becomes mature along the basinal axis. In central Alberta, the maturity contours swing east-west and run into the Disturbed belt. Thus, there is a broad area from central Alberta north to outcrop where the Colorado Shale Group is immature. The out-

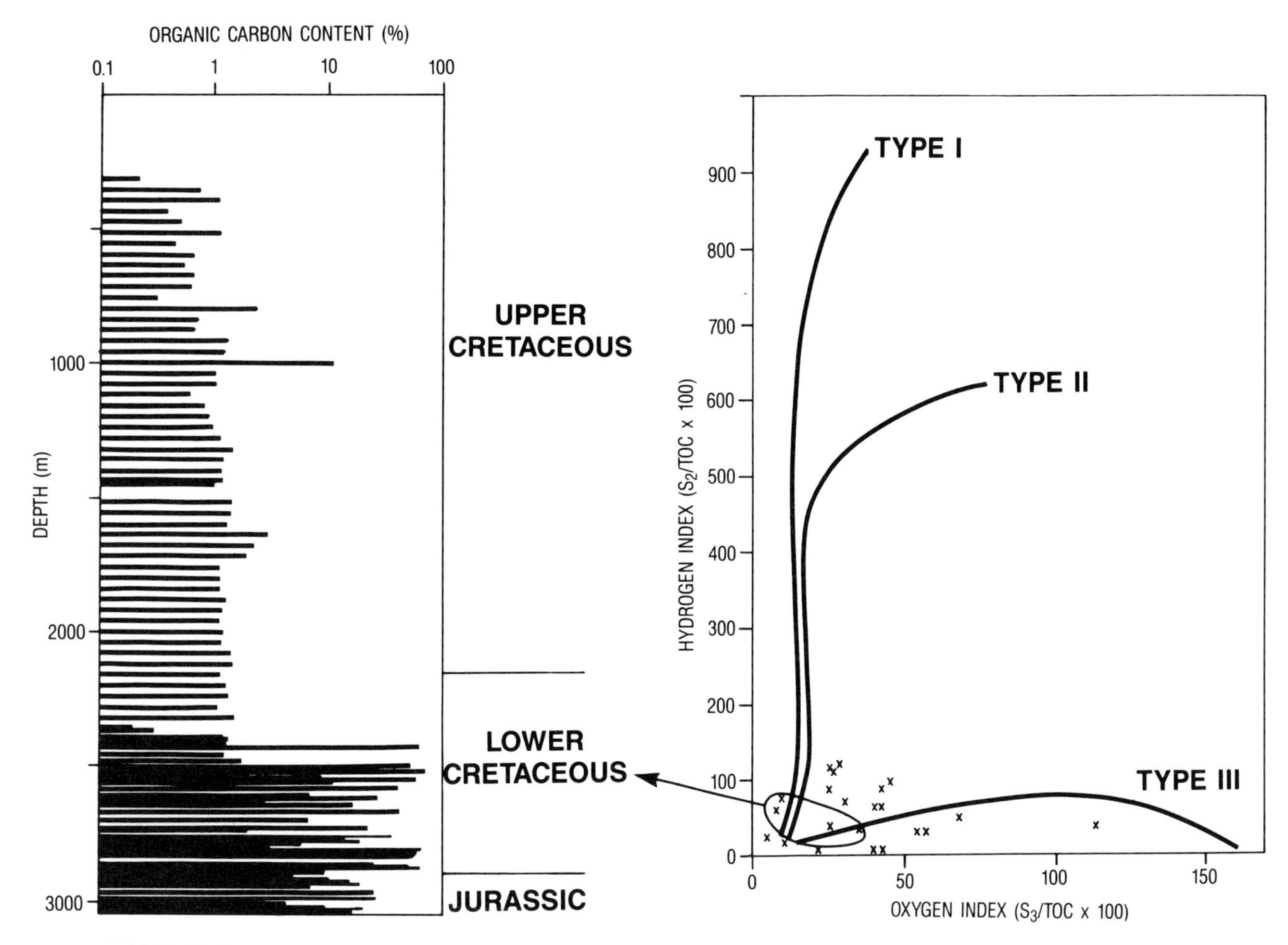

FIGURE 2. Rock-Eval pyrolysis and total organic carbon data for Lower Cretaceous sediments, Deep basin, northwestern Alberta. (Modified after Welte et al., 1984.)

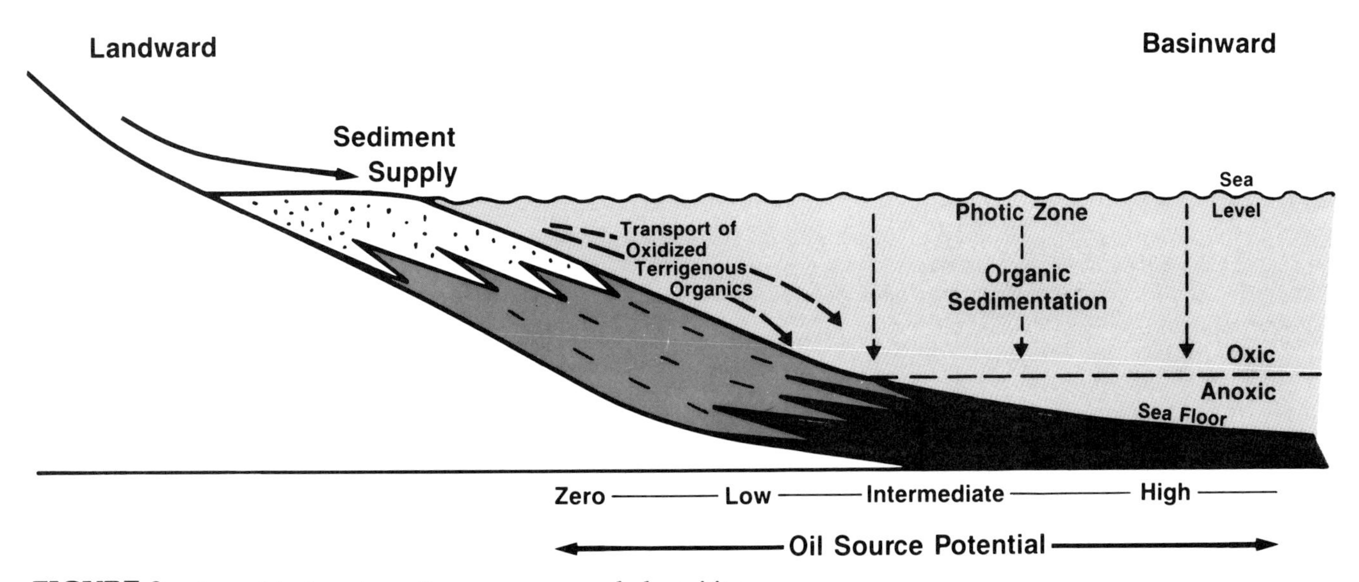

FIGURE 3. A model of marine, oil-prone source rock deposition.

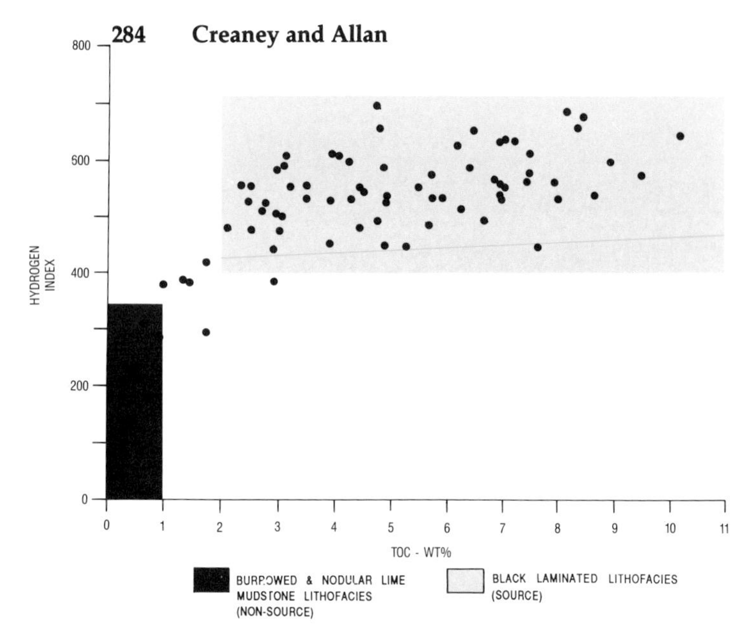

FIGURE 4. Relationship between organic richness (TOC) and organic matter type (HI) for a marine, oil-prone source rock. Data from the Devonian Duvernay Formation at Esso Redwater 10-27-57-21W4.

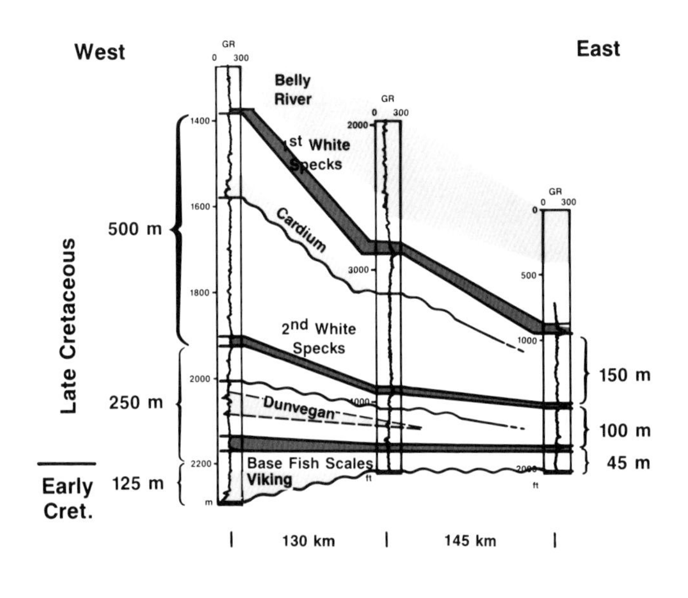

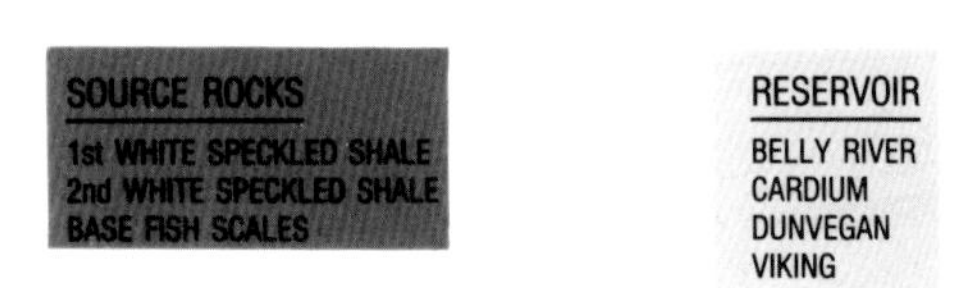

FIGURE 5. Geophysical log signatures of the Colorado Shale, showing stratigraphic relationships in an east-west section.

crop trends southeast across Alberta into central and southern Saskatchewan. Very low maturities are present throughout Saskatchewan, and the condensed sections discussed above have been locally classified as oil shales (Macauley, 1984a, b; Macauley et al., 1985).

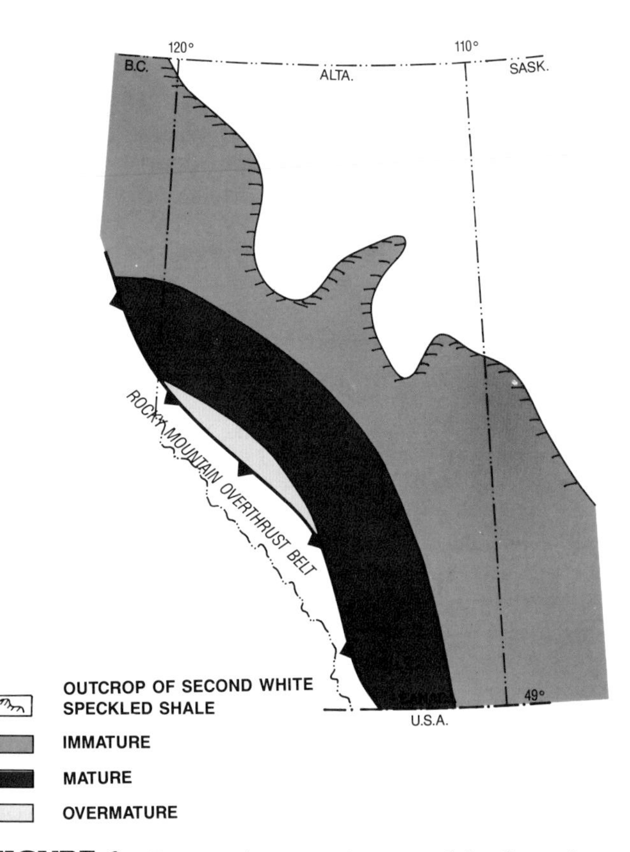

FIGURE 6. Present-day maturity map of the Second White Speckled Shale condensed section.

Figure 7 is modified from Jervey (1992) to highlight zones of low sedimentation rate (condensed sections) in the post–Joli Fou (middle Albian) section in Western Canada. These zones are the most favorable for the occurrence of organic-rich, marine source rocks. To date, this organic enrichment has been observed to occur at the base of the Shaftesbury Formation in northern Alberta (Leckie et al., 1990) and at the Base of Fish Scales and the Second and First White Speckled Shales. Figure 8 shows a well profile at Imperial Union Musreau 11-27-63-5W6, showing a gamma-ray log, stratigraphy, and total organic carbon data through the Base of Fish Scales and Second White Speckled Shale. From this figure, TOC is highest where natural gamma intensity is at its highest within each condensed section. It is also apparent that background TOC through the low-gamma-intensity shales of the Colorado Group is still moderately high (2%). This is a consequence of generally low background sedimentation rates produced by the mid-Cretaceous sea level high generating high shelfal accommodation. Thus it appears that low-oxygen conditions must have first developed in this portion of the Western Interior seaway during the mid-Albian and persisted intermittently until the early Campanian, providing the basin with a substantial volume of potential source rock. It is speculated that the sea floor of the Western Interior seaway may have been protected from polar, oxygenated waters by virtue of its elevation relative to abyssal sea floor, a condition compounded by the relatively warm global conditions of Late Cretaceous time.

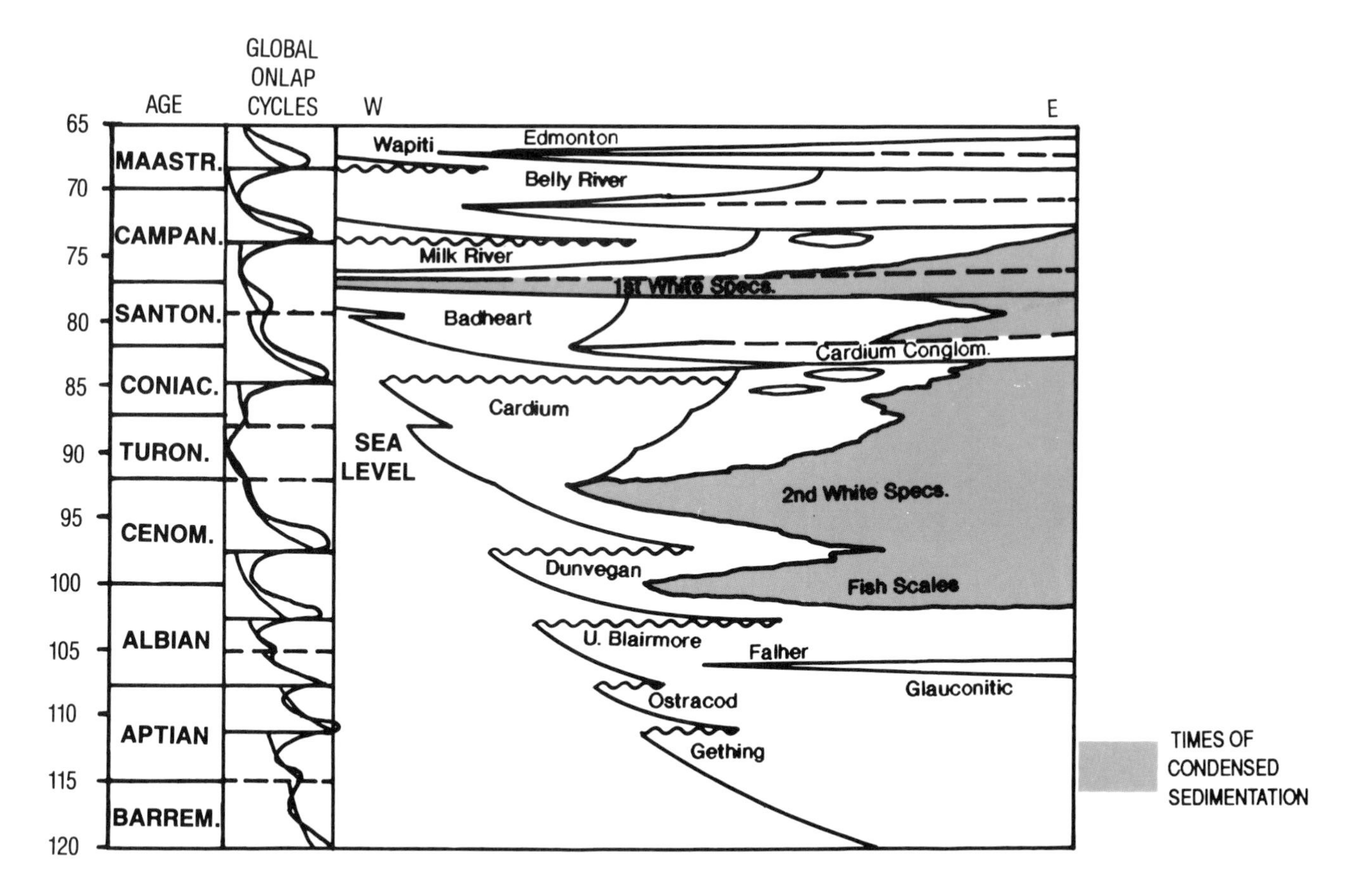

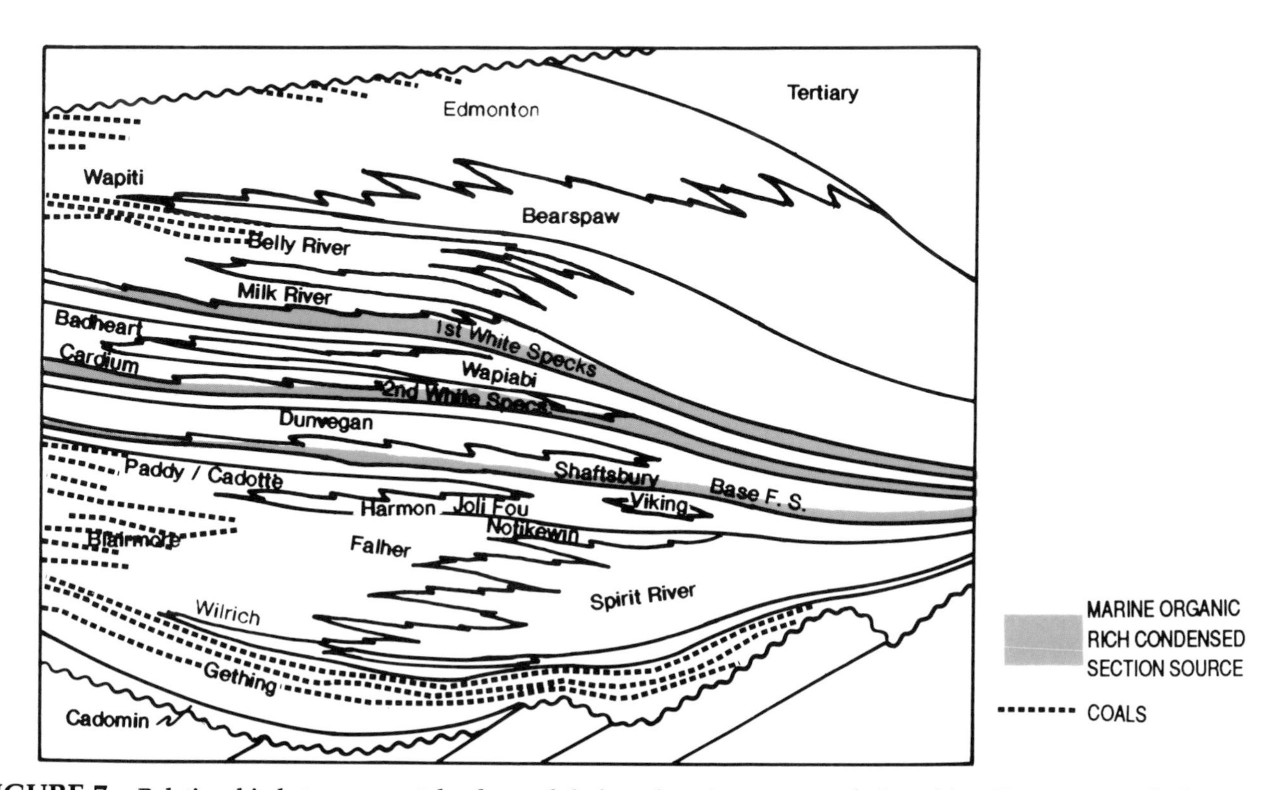

FIGURE 7. Relationship between coastal onlap and timing of marine source rock deposition. Top: source rock time stratigraphy of the Western Canada foreland basin. Bottom: schematic basinal cross section. (Based on Jervey, 1992.)

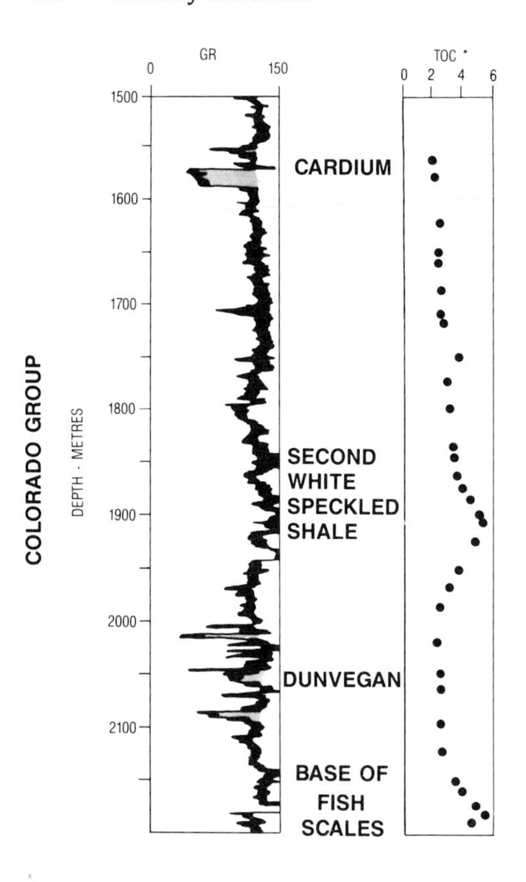

FIGURE 8. Natural gamma-ray log- and cuttings-based TOC profile of the Colorado Shale (Imperial Union Musreau 11-27-63-5W6).

PRE-FORELAND BASIN SOURCE ROCKS

The practical purpose of any regional geochemical study is to explain the known distribution of petroleum in the section under analysis. In the case of the foreland basin section, the total oil-in-place reserves are heavily skewed to heavy oils reservoired in the Mannville (Porter, 1992). In fact, approximately 98% of the oil reserves of the foreland basin occur in the Mannville and, as shown by Creaney and Allan (1990), were not all sourced from within the foreland section. The Mannville has acted as an efficient basin-wide collector for oils leaking from the subcropping pre-foreland succession and therefore for completeness the following discussion of the pre-foreland source rocks is included. Creaney and Allan (1990) believe that the source rocks that have provided oils to the Mannville "collector" are (in order of relative contribution) the Nordegg Member of the Fernie Formation (Jurassic), the Exshaw Formation (Mississippian), and possibly the Duvernay Formation (Devonian) and the Doig Formation (Triassic).

Fernie Group

The basal mudstones of the Lower Jurassic (Sinemurian) Nordegg Member of the Fernie Group have TOC contents up to 33% and hydrogen indices up to 600 (when immature). This Nordegg facies, which is typically less than 20 m thick, was deposited following a major but short-lived marine transgression in Early Jurassic time. The characteristic gamma-log response of the Nordegg (Figure 9) in northwest Alberta reveals a doublet character, suggesting two cycles of relative deepening and concomitant organic enrichment. The basinal facies of the Nordegg contains a sulfur-rich kerogen formed during accumulation of organic detritus in a strongly reducing, sulfur-rich and iron-poor environment. Thus, the Nordegg generates migratable, sulfur-rich oils at very low maturity levels (LOM, 6–7; R_o, 0.40–0.45%). Examples include subcropping oils at Cherhill and Tomahawk.

Across west-central to southern Alberta, the Nordegg Member undergoes a facies change to a more proximal chert- and sand-rich lithofacies, and the oil-source character is progressively diminished. However, deposition of organic-rich rocks in southern Alberta occurred somewhat later in Jurassic time (Figure 9), with preliminary palynological evidence suggesting oil source-rock deposition in the Early Jurassic (Toarcian) (J. Jansonius, 1990, unpublished results). This may be the "Paper" or "Poker Chip" shale, also of the Fernie Group, described by Hall (1984), Poulton (1984), and Stronach (1984). These Lower Jurassic, organic-rich, high-gamma facies occupy fortuitous positions in the stratigraphy of the Western Canada basin. The Jurassic section overlies the Triassic and older formations unconformably, and the source rocks have the opportunity to generate and seal oil directly into older, subcropping sections. Such oils, therefore, have properties that reflect the maturities and organic compositions of the immediately overlying sources. Toward the eastern part of the Jurassic occurrence, the maturity is low (Figure 10) and subcropping oils are commonly of low API gravity and have high sulfur contents (>2 wt.%).

The total thickness of Jurassic sediments is commonly limited. However, the shales of the Fernie Group overlying the organic-rich facies effectively seal off much of the directly overlying Cretaceous sediments from significant amounts of Jurassic oil. The consequence of this is that Jurassic oil is probably channeled updip (e.g., in sandier interbeds of the Fernie Group; see Figure 9) to Mannville subcrop and thence has made its way to the eastern limb of the basin.

Doig Formation

Several organic-rich sediments of Triassic age occur within the Western Canada basin, although with limited present-day extent. The most prolific source is the basal high-gamma facies of the Doig Formation (Anisian age), with measured TOC contents up to 23% and hydrogen indices up to 380 where mature (Figure 11). Organic deposition continued sporadically through the distal facies of the Doig and overlying lower Halfway formations, but did not again attain the prolific source richness of early Doig time. The Doig Formation ranges from mature to overmature, with the onset-of-maturity line approximately coinciding with its erosionally truncated eastern limit (Figure 12).

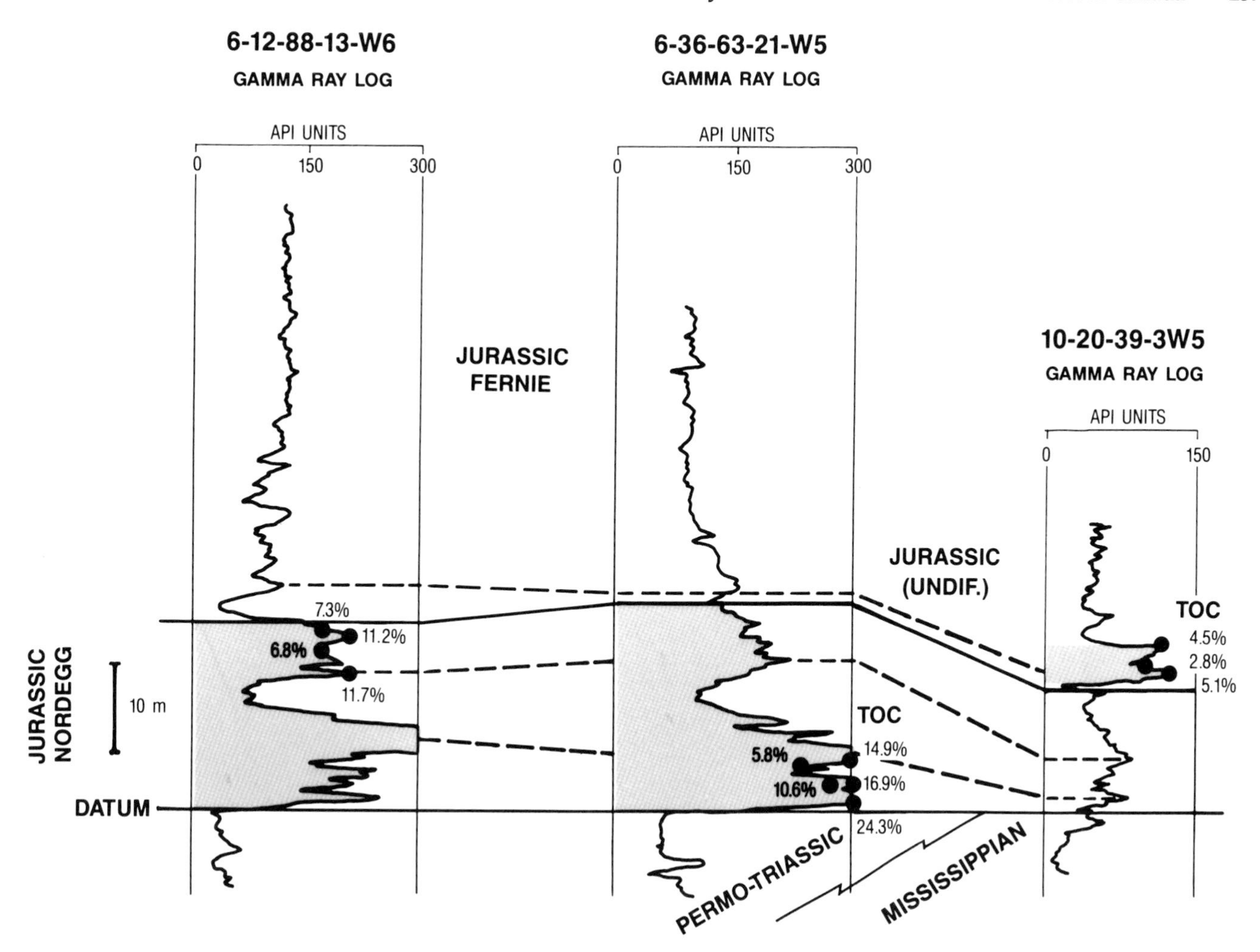

FIGURE 9. **Characteristic natural gamma-ray log signatures of Jurassic source rocks across Alberta. Core-measured total organic carbon values are annotated.**

The Doig Formation is underlain by the thick (up to 300 m, or 984 ft) Montney Shale (Figure 1), which has an organic geochemical signature distinctively different from those of the Doig and younger shales (Allan and Creaney, unpublished data). TOC contents up to 3% and hydrogen indices up to 500 have been found in cored Montney, and its log character suggests local organic enrichment. However, the source character and distribution of source facies within this formation are not yet fully understood.

Most oils currently reservoired in the Halfway and Doig formations, and some oils in the Charlie Lake, show biomarker signatures that correlate them to the Doig (Allan and Creaney, 1991). Du Rouchet (1984, 1985) considered this portion of the Triassic section to be the principal source of hydrocarbons for the massive tar sand accumulations of eastern Alberta. However, over most of their occurrence, the Triassic source rocks are isolated from the Cretaceous Mannville by Triassic evaporites and Jurassic shales, and direct access for Triassic oils to the Mannville is therefore difficult to visualize in any regional sense. Passage stratigraphically downward past the thick Montney Shale is known to have occurred locally, because Doig oil has been identified in some Upper Belloy (Permian) reservoirs (Allan and Creaney, unpublished data). However, these are only limited occurrences, and the Montney acts, for the most part, as a bottom seal for Doig oil.

Exshaw Formation

The Exshaw Shale was deposited following a major drowning event that terminated growth of the carbonate reef systems of the Devonian. The sedimentological relationship between the Exshaw and its underlying Devonian carbonates can be examined at outcrop in the type section at Jura Creek in southern Alberta. The nearly time-equivalent Lodgepole and Bakken shales are major hydrocarbon sources in the Williston basin (Brooks et al., 1987), and the Bakken and its equivalents have been proposed by Leenheer (1984) as the principal sources for the massive heavy-oil accumulations of Western Canada. Brooks et al. (1988) presented an argument that Leenheer's (1984) data did not support this conclusion. The source is thin (10 m), but is commonly extremely organic rich (TOC $\leq$ 20%) and oil prone, with hydrogen indices up to 600. These high values approach those of a type I organic matter, and the algae *Tasmanites* occurs in the shales. The

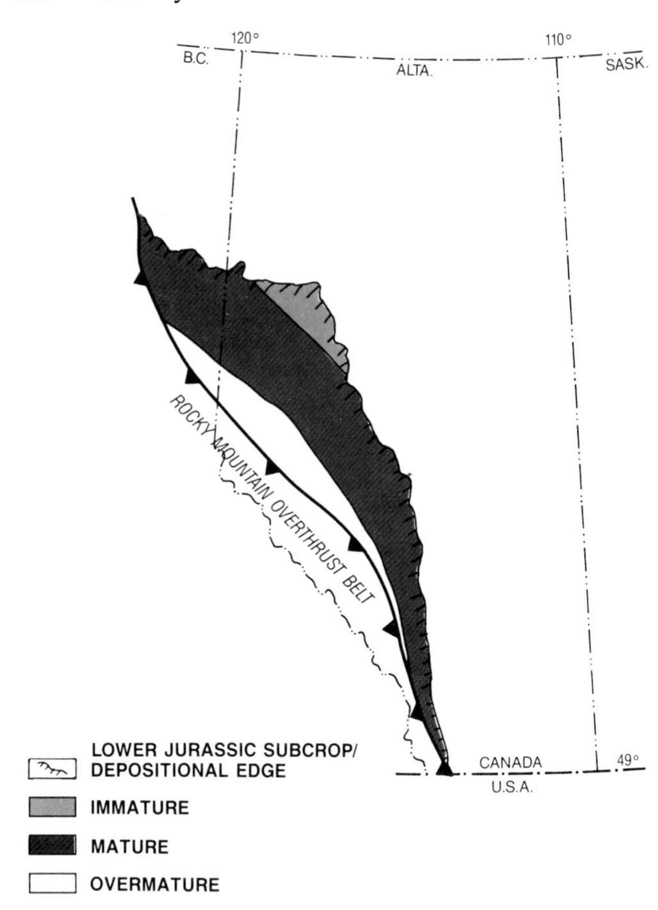

FIGURE 10. **Present-day maturity map of the Lower Jurassic.**

Exshaw has a characteristic log signature, an example of which is shown in Figure 13. It reaches maturity over much of its area (Figure 14). Exshaw-sourced hydrocarbons have accessed both the underlying Wabamun and overlying Banff formations. In the Peace River arch area, the Wabamun is a low-porosity, low-permeability ramp carbonate, and reservoir is developed only along dolomitized fault planes at fields such as Tangent. To the south (farther up the Wabamun shelf), the porosity and permeability improve, thus providing an updip migration conduit. Thus, in the southern portion of Alberta, Exshaw-sourced hydrocarbons can migrate into the Mannville via the top of the subcropping Wabamun.

A similar situation exists for the overlying Banff, with Exshaw-sourced oils trapped in the Banff at fields such as Twining, but with significant volumes expected to have leaked to the Mannville at the Banff subcrop.

Duvernay Formation

A single source rock occurs in the Beaverhill Lake and Woodbend groups of the Upper Devonian. Stoakes and Creaney (1984) provided a detailed description of the Duvernay Formation. This source is time-equivalent to Leduc-age reef buildups and occurs throughout the basin between these buildups (Figure 15). TOC contents range from 2 to 17%, and hydrogen indices from 500 to 600 (typical marine, type II kerogen). This source is immature over much of its occurrence, with a band of progressive "oil window" maturities and, ultimately, overmaturity in the deep, western part of the basin adjacent to the Disturbed belt.

Stoakes and Creaney (1984) provided a comprehensive description of hydrocarbon migration in the Woodbend Group, principally along the Rimbey-Meadowbrook reef chain. Oils generated from mature Duvernay source facies have migrated locally into porous and permeable dolomitized Cooking Lake platform margins and Leduc-reef buildups. This oil then migrated updip along the platform margin, some accessing overlying pinnacle reefs such as Leduc, Redwater, Acheson, etc., with the possibility of further migration updip until it could leak to the overlying Grosmont Formation and then possibly to subcrop beneath the Mannville. However, subsequent biomarker studies (Brooks et al., 1987, 1988) have as yet failed to confirm the presence of purely Duvernay-sourced oil in the Grosmont.

In the area around the Bashaw complex, oil from mature Duvernay has migrated into adjacent Leduc pinnacles, but the proximity of the overlying Camrose Member and Nisku Formation (platform facies) has allowed significant leakage to the Nisku. This oil subsequently accumulated updip in Nisku drape structures over underlying Leduc buildups, as well as in stratigraphic traps. Excess oil migrated to Nisku subcrop beneath the Mannville and thence into the Mannville in the area of Bellshill Lake to Lloydminster. Additionally, certain Mannville oils in the Woodbend-Redwater area may be sourced from the Duvernay, suggesting some possible leakage from the Late Devonian hydrocarbon system. Northwest of the Rimbey-Meadowbrook trend, the Nisku shale basin is developed, effectively top sealing the Duvernay system in that direction. Thus, although the Duvernay source facies is known to extend through northern Alberta, foreland basin sediments are effectively protected from oil ingress from the underlying Devonian source.

Overmaturity of the Duvernay source facies in the deep western portion of the basin has produced significant quantities of gas, which is now found preferentially in reservoirs close to the Disturbed belt. Many Late Devonian gas pools are sour, some with H_2S contents greater than 80% (ERCB, 1988).

OIL GEOCHEMISTRY

Petroleum Geology of Mannville Reservoired Oils

The vast majority (>99%) of the oil contained in Mannville sediments lies in the heavy oil pools which form a broad arcuate belt stretching from the Peace River deposits east and south to the Lloydminster area (Figure 16). The northern limit of the heavy oils (Peace River, Buffalo Head Hills) is constrained by the northerly limit of reservoir facies in Mannville sediments. The eastern limit (Athabasca, Cold

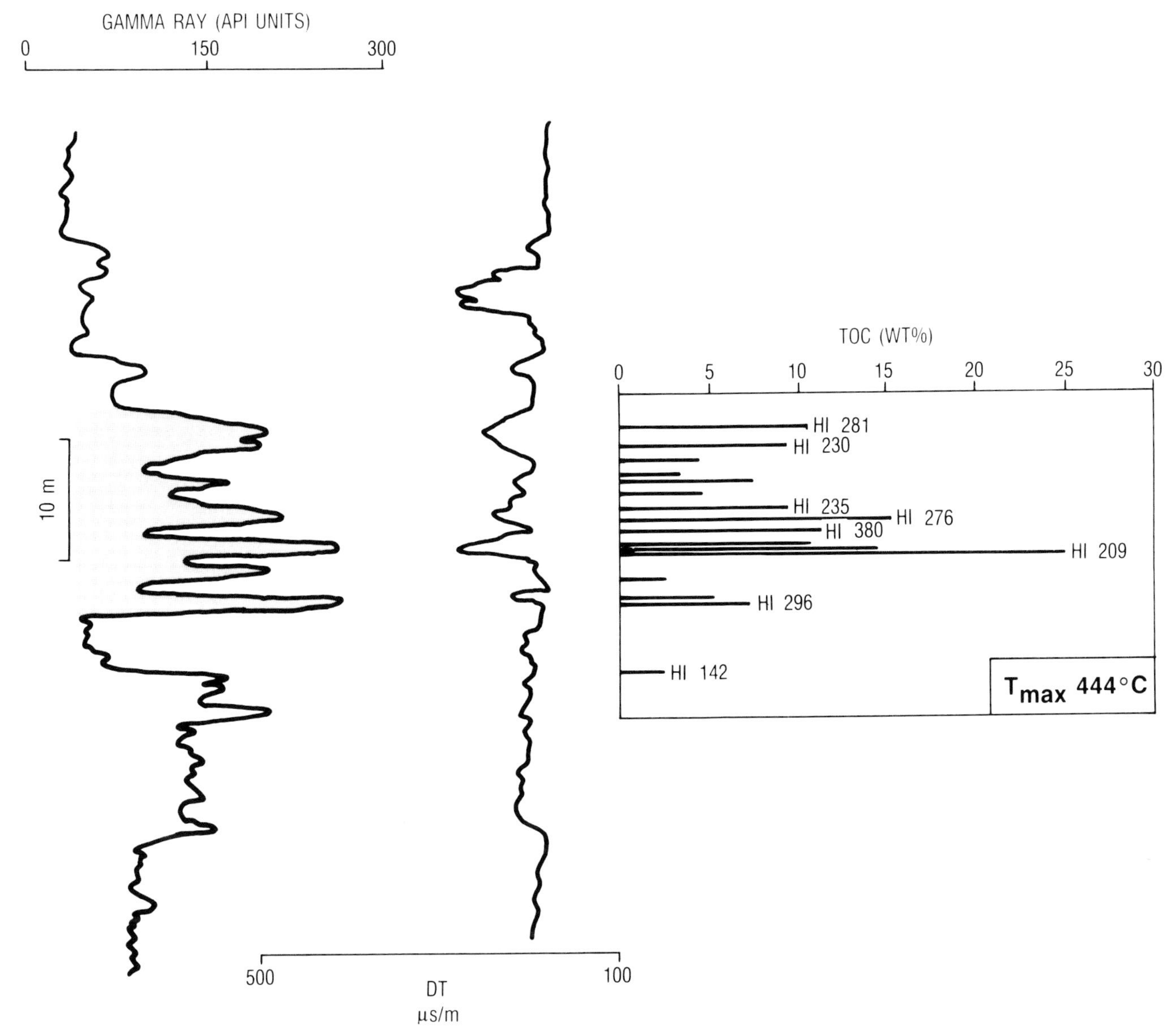

FIGURE 11. Geophysical and geochemical character of the oil-source facies of the Doig Formation. The well is located at 3-22-78-10W6. Core-based total organic carbon, hydrogen index, and average T_{max} data are annotated.

Lake) is a gentle structural feature, probably controlled by salt dissolution. The southern limit of heavy oil occurrences (Lloydminster) is constrained by a combination of reservoir, trap, and source rock drainage factors. To the west and south of the arc of heavy oil fields, reservoir-prone Mannville extends to the edge of the Disturbed belt and the international boundary, respectively.

The structural form of the basin defines the drainage area for source rocks that have contributed to the heavy oils (Figure 17). Geochemical analyses (see below) absolutely rule out post-Mannville-age sources as possible contributors, and the Mannville itself does not contain oil-prone sources (Moshier and Waples, 1985). Of the pre-Mannville section source rocks discussed above, the Jurassic and Mississippian are the most likely candidates from consideration of petroleum geology. It is not coincidental that the subcrops of these two prolific, marine, oil-source-containing sections are located close to the western, downdip edges of the heavy oil belt (Figure 16). These subcrops are postulated to be the major access points for northeasterly, updip-migrating fluids from these sources into the porous sands of the lower Mannville, particularly across the northern half of the drainage area. There, cross-formational vertical discharge of oils from these sources at more westerly locations is limited by extensive thicknesses of low-permeability lithologies (Triassic section above the Mississippian, followed by low-permeability Jurassic sediments above the Nordegg Member).

In the southern part of the basin, the Mannville either sits directly on the Mississippian or is separated from it by

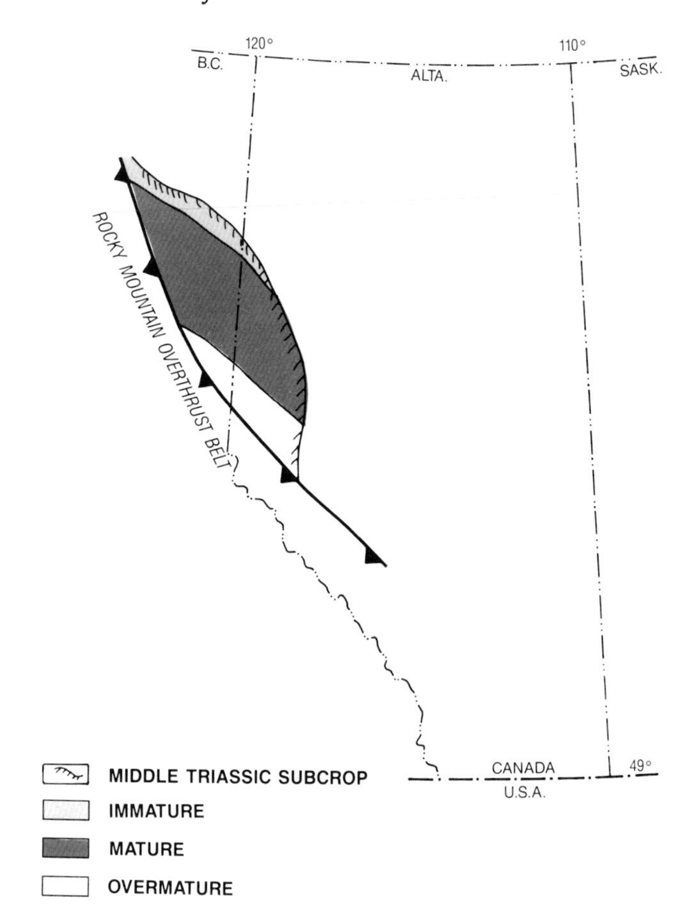

FIGURE 12. Present-day maturity map of the Doig Formation.

fairly thin and sand-rich Jurassic sediments of the Ellis Group. In this area, the Nordegg Member is absent through nondeposition whereas the source facies of the Exshaw Shale is known to extend westward into the Disturbed belt and eastward across the high separating the foreland basin from the Williston basin, where nearly time-equivalent, organic-rich, oil-prone facies are present in the Bakken and Lodgepole formations. In this southern area, a third potential oil source of Toarcian age is present, but its effectiveness is still under investigation. No conventional oil pools have yet been definitively tied to this source rock by geochemical oil-source correlations.

Petroleum Geochemistry of Mannville Conventional Oils

To investigate the hypothesis that the heavy oils are sourced principally from Jurassic and Mississippian sediments, it is clearly advantageous to define type oils from the Nordegg Member and Exshaw Shale. This cannot be done in the Mannville with certainty, because the problem of commingled oils is always a risk. It is in fact done most effectively where the source rock also acts as the seal on the trap. For the Exshaw Shale, a type oil is found in the Peace River area in a Wabamun drill-stem test taken just below the Exshaw contact. The geochemical characteristics of this oil (32° API; sulfur, 0.7 wt. %) are shown in Figure 18. In southern Alberta, a type oil has not yet been defined, but partial geochemical characterization of Exshaw-sourced fluids is available from a stain in the Wabamun near Claresholm (Figure 19). Comparison of the sterane, terpane, and aryl isoprenoid biomarkers suggests that certain minor organic facies changes are evident between these two locations. These changes continue into the Williston basin, as evidenced by the geochemical character of imputed Bakken- and Lodgepole-sourced oils in Saskatchewan pools (Leenheer, 1984; Brooks et al., 1987).

The Nordegg Member source rock is found only in the northern part of the basin (Figure 9). Type oils for this source are in the Baldonnel (Triassic) at Rycroft and in the Belloy (Permian) at Virginia Hills. The geochemical character of these two typically sulfurous oils is shown in Figures 20 and 21. The high sulfur contents (up to 4 wt. %) are typical of conventional oils sourced out of the Nordegg Member, but are usually associated only with heavy oils in this basin. Oils similar to those illustrated are found associated with the Nordegg subcrop south-to-central basinal areas (Lac Ste. Anne area).

In southeastern Saskatchewan, oils of geochemical character similar to that of Nordegg-sourced oils have been correlated to a source facies in the Mississippian Lodgepole Formation (Brooks et al., 1987). Neither of these two source units is known to be present in southern Alberta or southwestern Saskatchewan, and yet oils of this type are quite widely distributed in Jurassic and Early Cretaceous traps. Further work on the details of secondary migration for these sulfurous oils in the foreland section will clearly be required if we are to gain a proper understanding of this part of the Mannville in the southern part of the basin.

A hypothesis currently being tested by the authors to identify mixed oils is the presence and abundance of aryl isoprenoids in Mannville conventional oils. These compounds are chlorobiaceae metabolites (Summons and Powell, 1987) and hence may be used for source difference characterization. The data illustrated in Figures 18 through 21 show that aryl isoprenoids are very abundant in the Exshaw type oils, but are present only in low quantities in Nordegg oils. Figures 22(E) and 23(E) include aryl isoprenoid m/z 133 fragmentograms of two typical lower-gravity, high-sulfur conventional Mannville oils from central basin locations. (In this context, "conventional" is used to define nonbiodegraded oil). The oil in the Manola field (Figure 23) has some Nordegg-sourced features (pristane/phytane ratio less than 1, high sulfur, low API gravity, terpane distributions) but is enriched in aryl isoprenoids. In contrast, oil from the Highvale field (Figure 22) has Exshaw-sourced characteristics (pristane/phytane ratio greater than 1, sulfur content below 1.5%) but is impoverished in aryl isoprenoids. Under the outline of the initial hypothesis, these two pools are currently considered to be examples of mixed source, conventional oil accumulations in the Mannville. The petroleum geologic aspects of the two pools would not dispute such a conclusion.

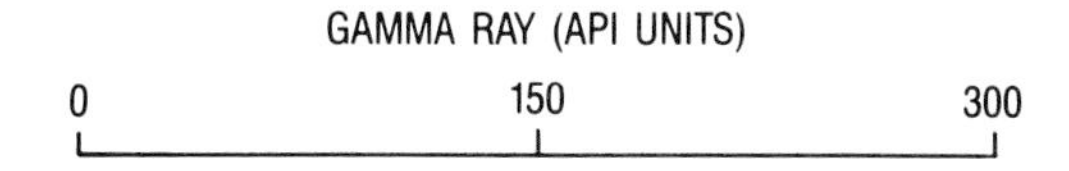

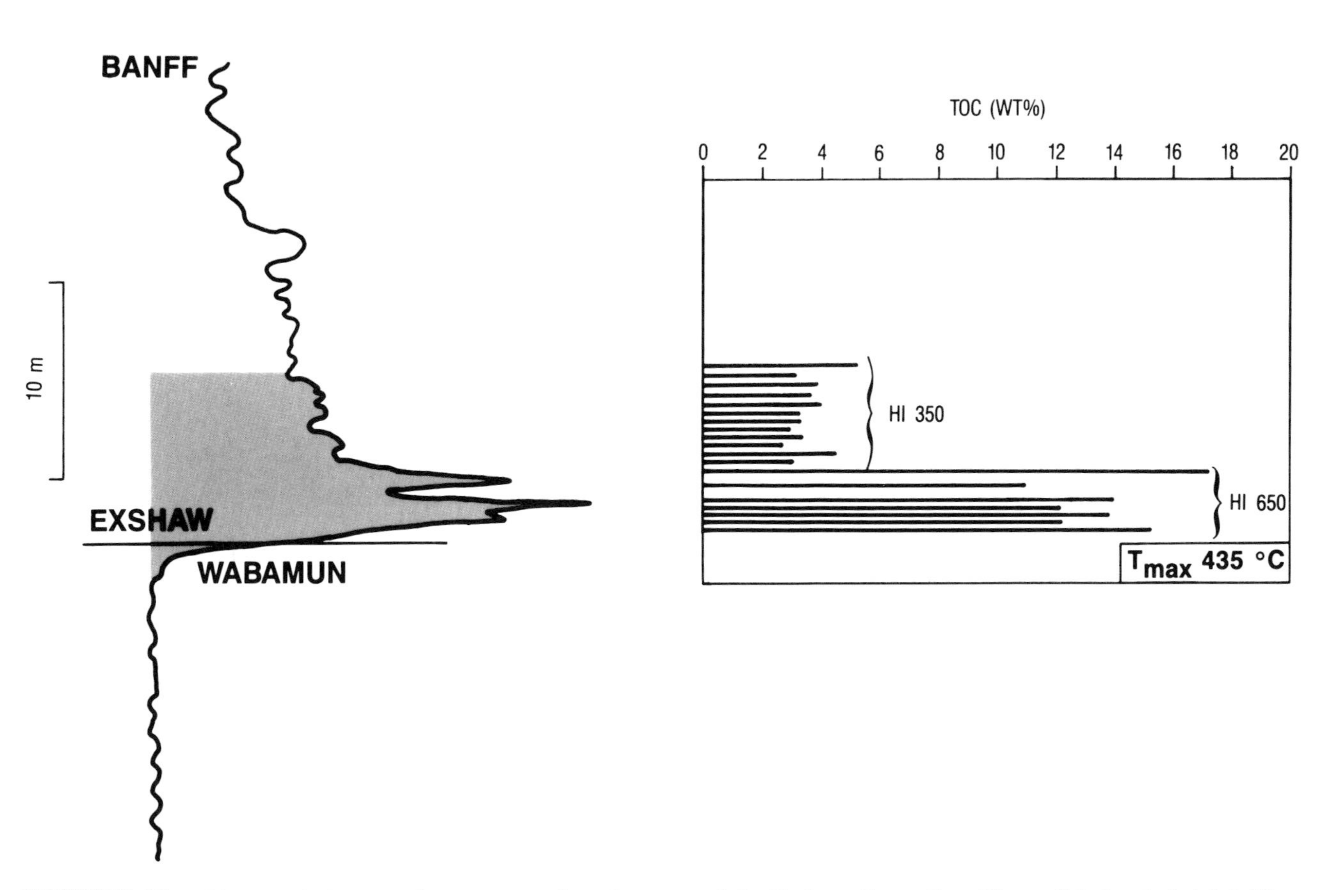

FIGURE 13. **Characteristic natural gamma-ray log signature of the Exshaw Formation. The well is located at 6-2-79-22W5. Core-measured total organic carbon, hydrogen index, and an average T_{max} value are annotated.**

Petroleum Geochemistry of Mannville Heavy Oils

The application of aryl isoprenoid distributions in oil-source correlations for the heavy oil deposits is unfortunately impossible because these compounds either were never present or have been removed by recent aerobic bacterial degradation (Milner et al., 1977). This process requires oxygenated water flow (usually meteoric) and reservoir temperatures generally less than 65°C. The impact of microbial consumption is ultimately very destructive to the quality of a crude oil, reducing high-quality, low- to medium-sulfur crude oil to heavy, high-sulfur bitumen (gravities as low as 8° API). It appears that the good reservoir quality in the basal Mannville, which allowed massive oil migration, also facilitated freshwater incursion into the basin. Figure 24 is a summary map of Mannville salinities showing the extent of this incursion with its effect on Mannville oil properties.

Commonly, sterane and terpane biomarkers are preserved in apparently unaltered to slightly altered states in the heavy oils. Biomarker analyses of numerous samples from the various accumulations (Peace River, Buffalo Head Hills, Wabasca, Athabasca, Cold Lake, and Lloydminster) all show a remarkable similarity to each other (Brooks et al., 1988). Using the relative alteration scale of Alexander et al. (1983), most of the oils would be classed as moderately altered (loss of *n*-paraffins, but partial retention of isoprenoid alkanes). Some *n*-alkanes are still present in the Peace River oils. Extremely biodegraded oil containing demethylated hopanes occurs locally in the Grosmont (Creaney and Allan, 1990) and Wabasca (Brooks et al., 1988) deposits. The preservation of some isoprenoid alkanes while aryl isoprenoids are missing circumstantially supports a strong Nordegg contribution to the deposits. Examination of heavy oil terpane biomarkers (Figure 25) shows them to be remarkably similar to those of conventional Nordegg-sourced oils (Figure 21). Given the geo-

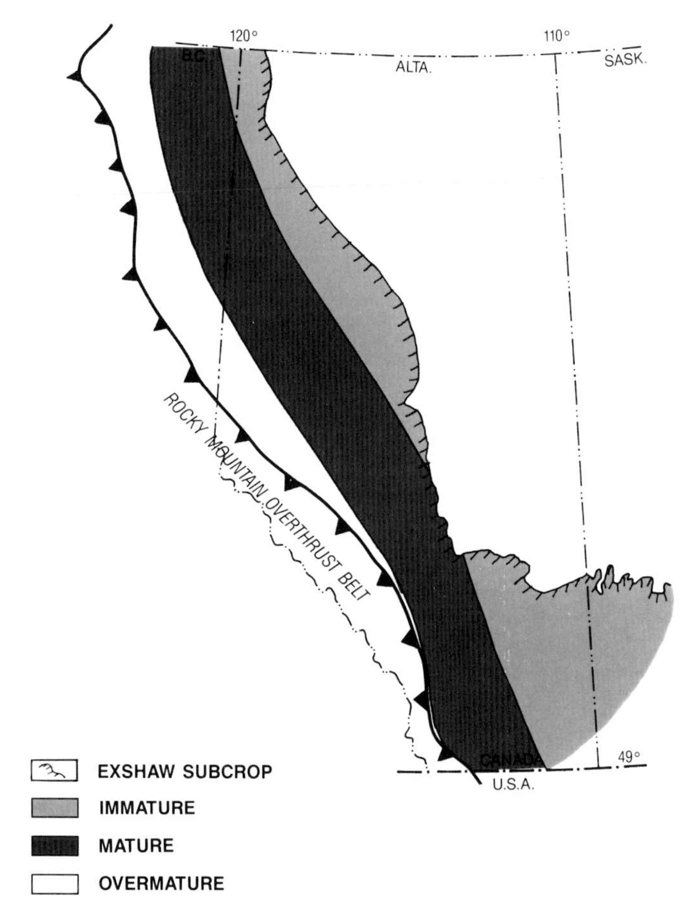

FIGURE 14. Present-day maturity map of the Exshaw Formation.

logic occurrence of this source rock and its molecular characteristics, there is a strong possibility that the Nordegg Member is the major contributor to the more northerly deposits of heavy oil (i.e., Peace River, Buffalo Head Hills, and northernmost Athabasca). Farther south (i.e., southern Athabasca, Cold Lake, and Lloydminster), a unique oil-source correlation is not possible on the basis of these compounds. The steranes (Figure 26) all show the $C_{29} > C_{27} > C_{28}$ relative abundance common to all pre-Late Cretaceous, marine oil source rocks. The C_{28}/C_{29} ratios suggest a Mesozoic age (Grantham and Wakefield, 1988), but the ratios generally fall between those seen in the Jurassic- and Mississippian-sourced type oils (Figures 21 and 19, respectively). Taken at face value, this is suggestive of mixed sources. However, abundance of diasteranes relative to regular steranes in heavy oil total sterane mixtures has been raised as an argument against a predominant Nordegg source contribution (M. Fowler, personal communication, 1990).

The origin of a particular C_{28} triterpane (28,30-bisnorhopane) that is found in all heavy oils has been mooted also as a crucial factor in the oil-source correlation question (Brooks et al., 1988). Thus far, no oil-source rock that consistently contains this compound in the extractable biomarker fractions has been located in the basin. Despite this, it is present in most Mannville-pooled oils and a variety of pre-Mannville-pooled oils including the type Nordegg oils discussed above (see Figures 20 and 21). Thus, the significance of this compound in oil-source correlation is disputable, especially because it has been found in several nonsource, non-oil-stained sediments (Figure 27).

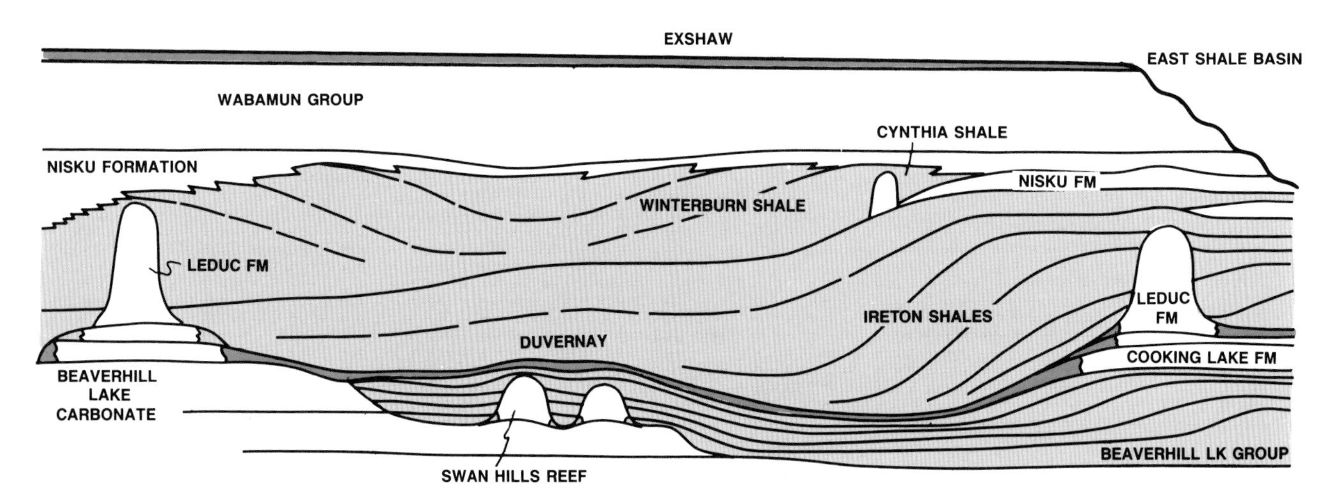

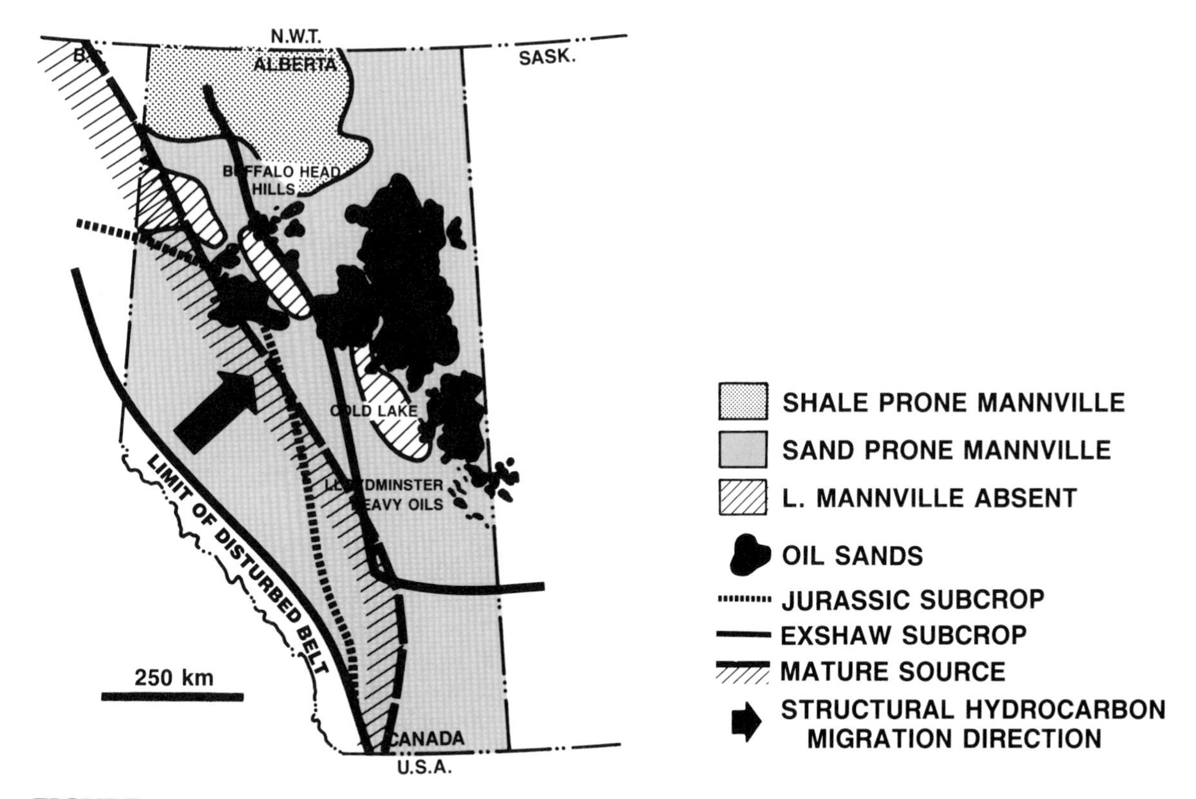

FIGURE 16. Elements of the petroleum geology of Mannville heavy oil deposits.

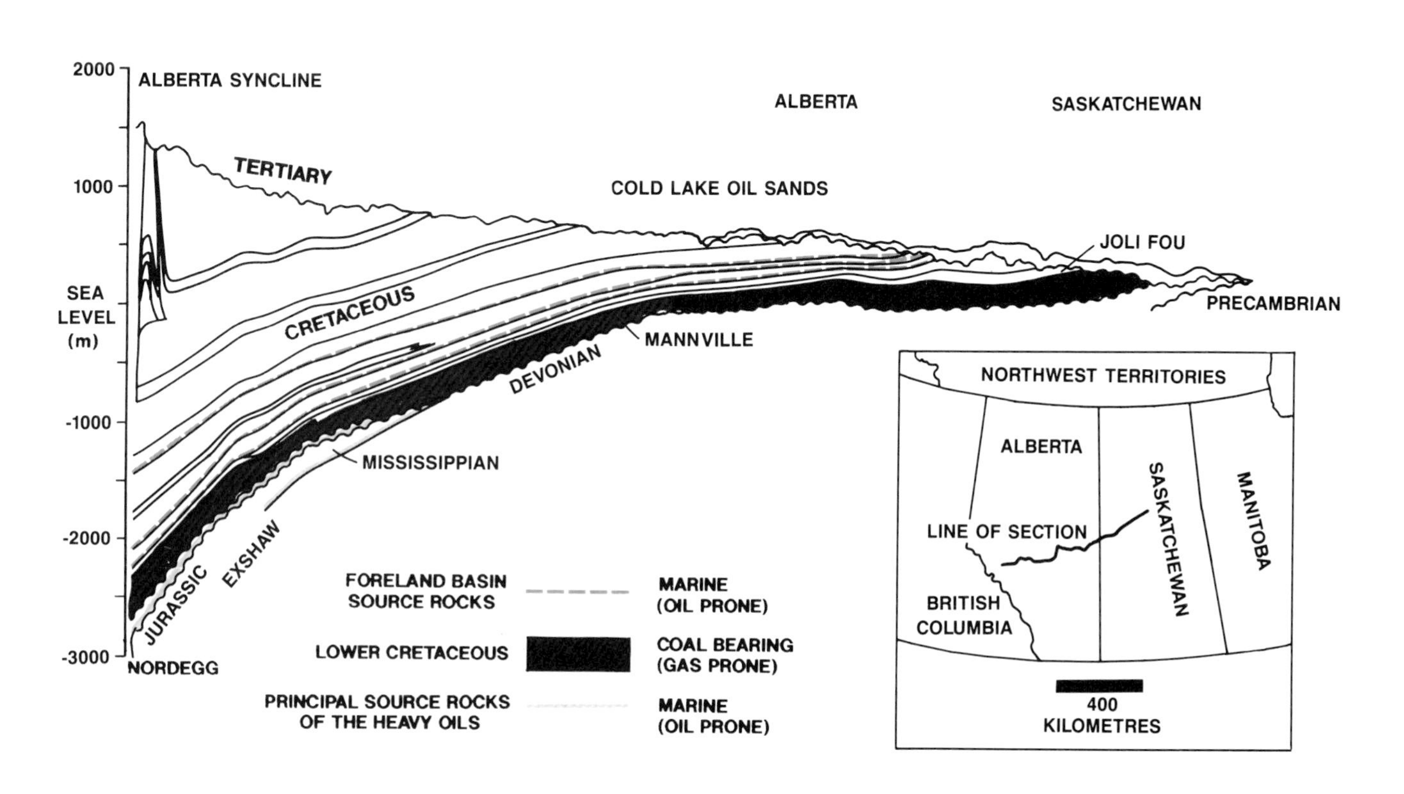

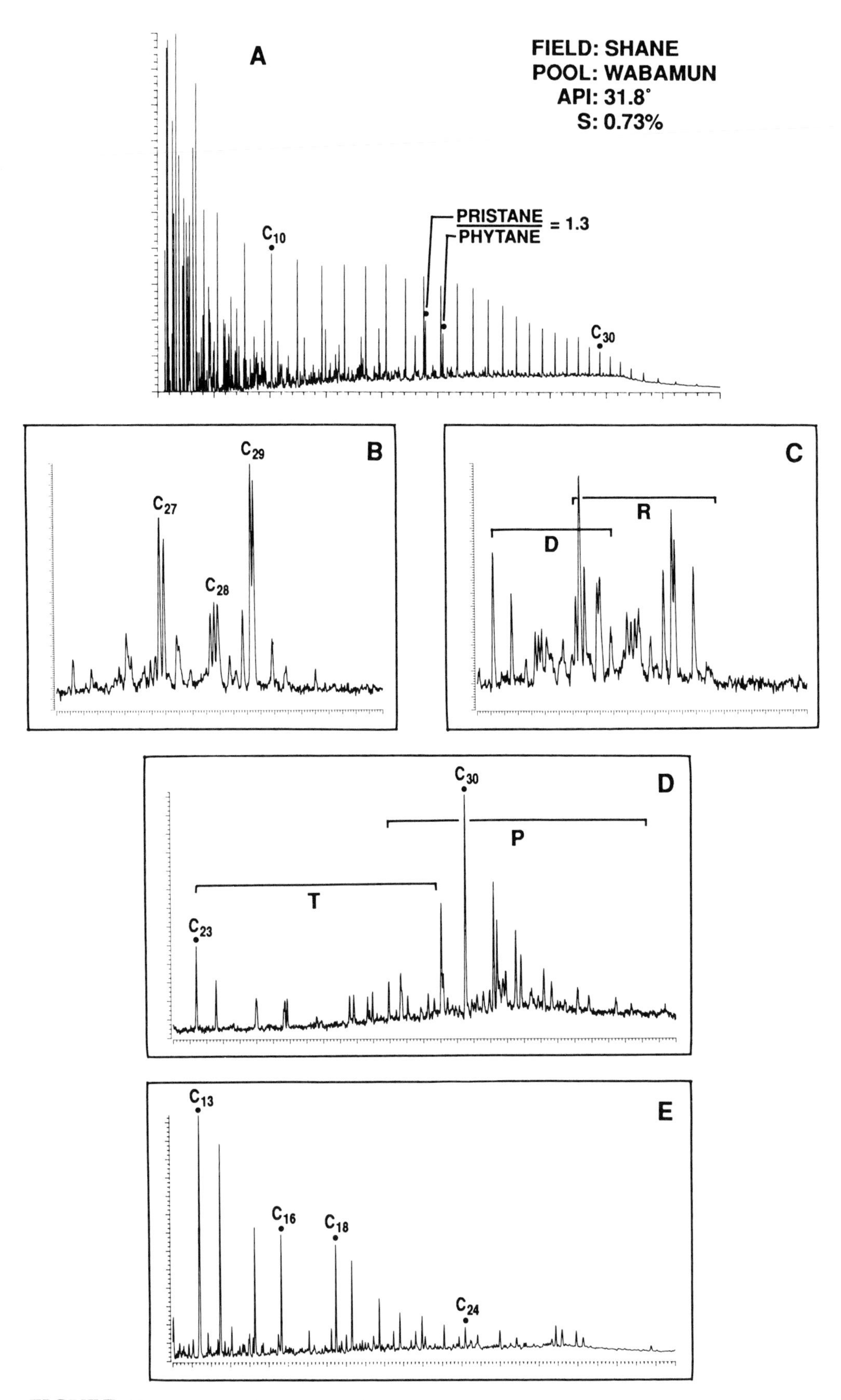

FIGURE 18. Geochemical characteristics of an Exshaw-sourced oil (Wabamun Formation, 8-1-77-2W6). See Table 1 for legend.

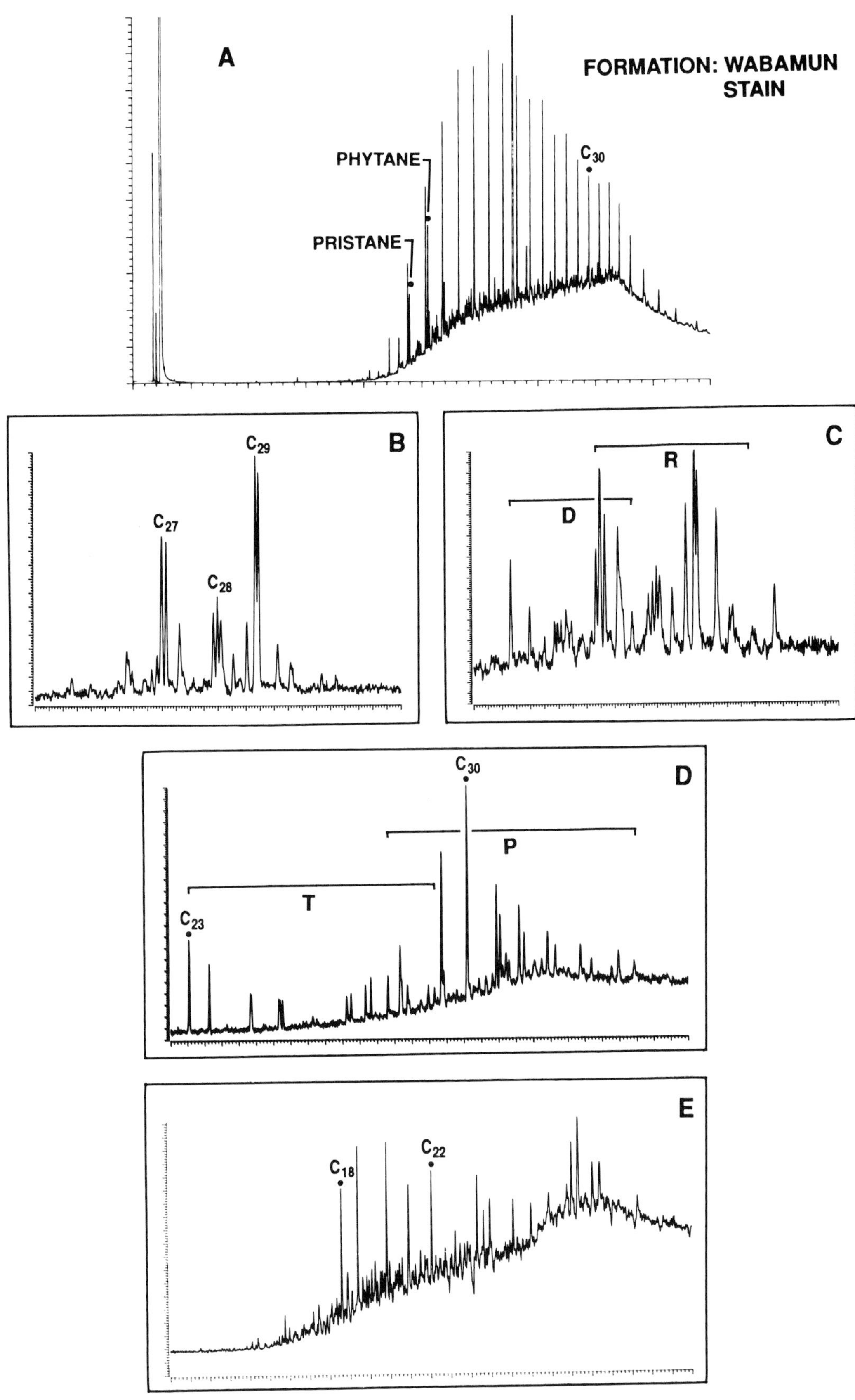

FIGURE 19. Geochemical characteristics of an Exshaw-sourced stain (core, Wabamun Formation, 1-31-13-26W4). See Table 1 for legend.

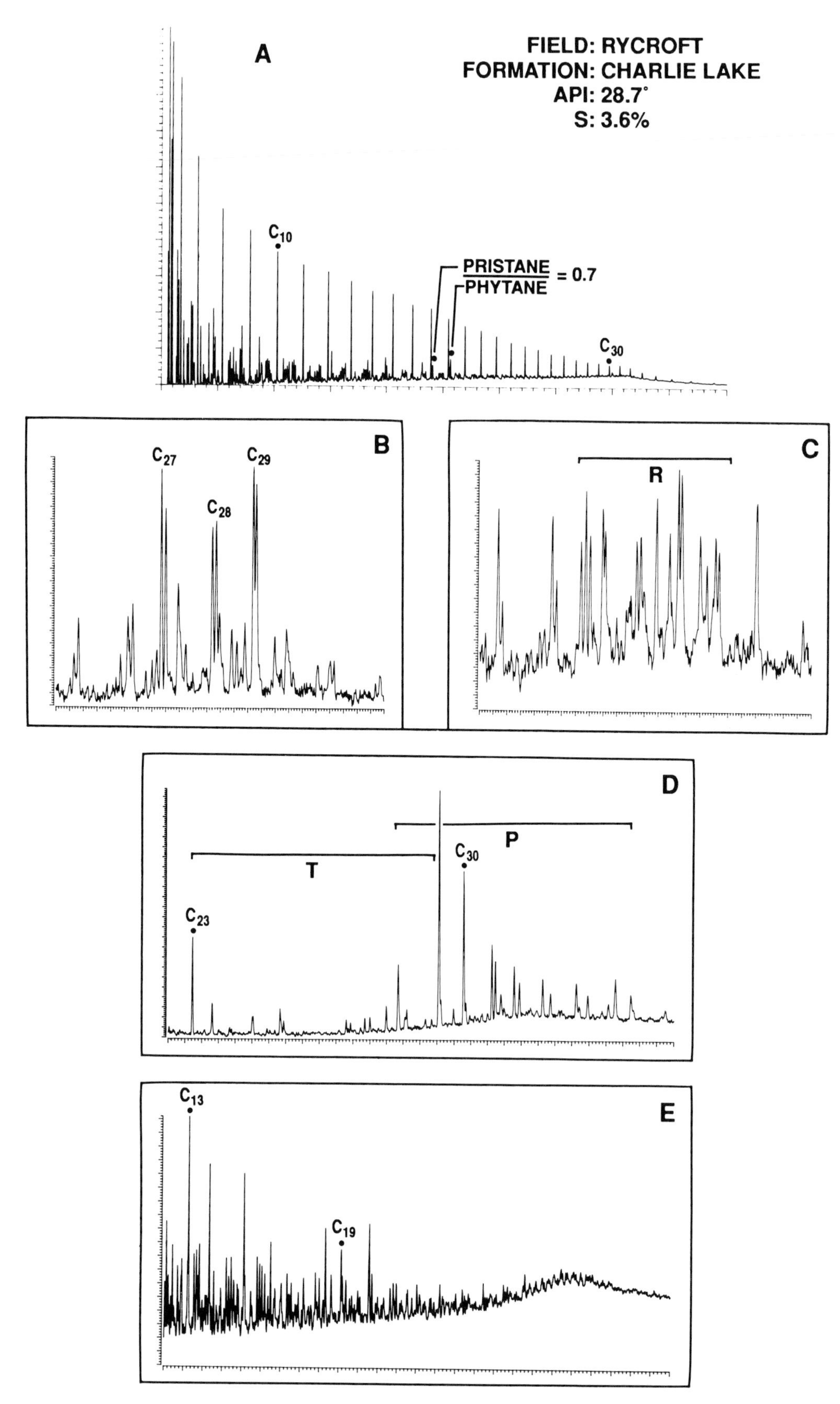

FIGURE 20. Geochemical characteristics of a Nordegg-sourced oil (Charlie Lake Formation, 8-20-77-5W6). See Table 1 for legend.

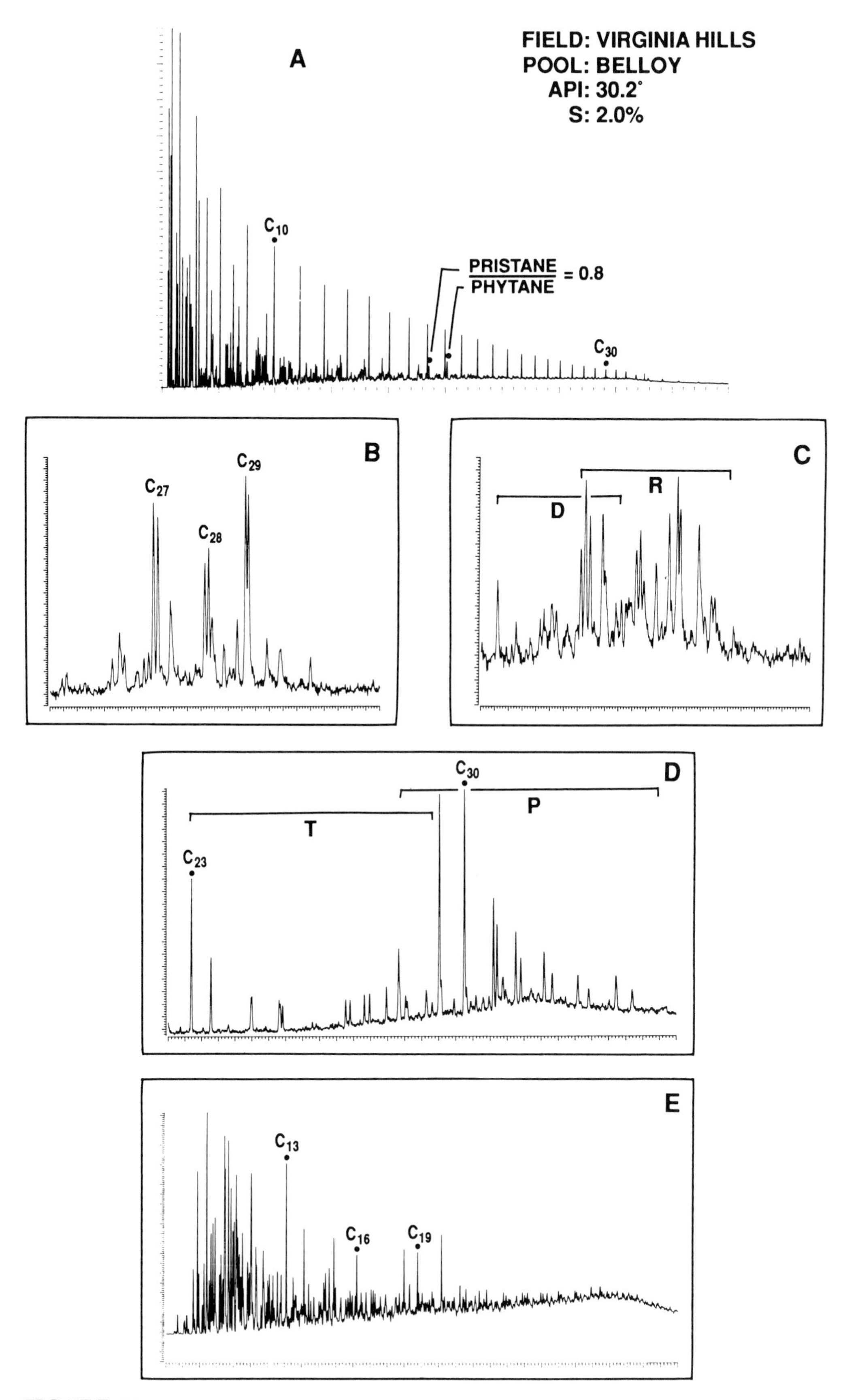

FIGURE 21. Geochemical characteristics of a Nordegg-sourced oil (Belloy Formation, 8-30-63-13W5). See Table 1 for legend.

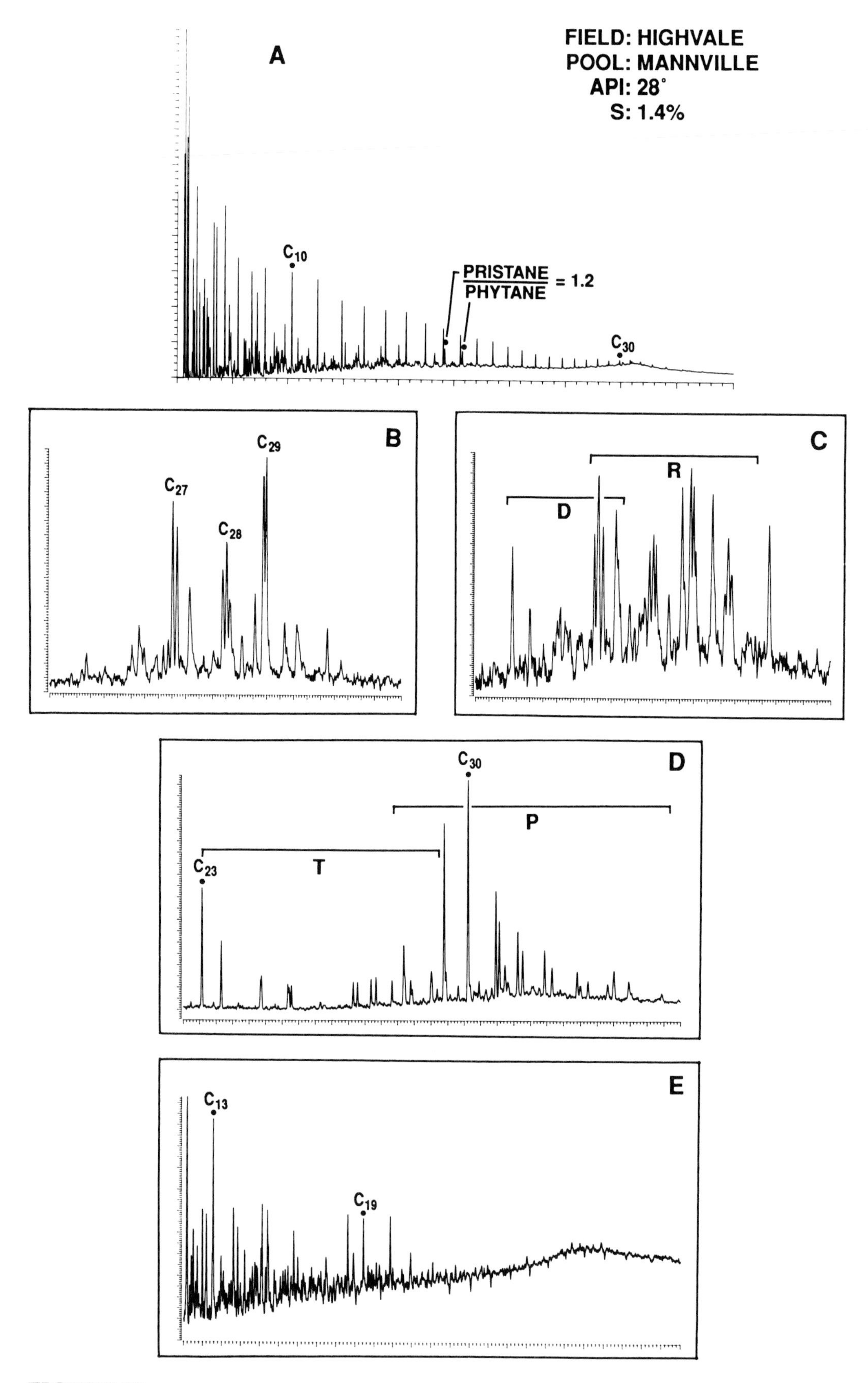

FIGURE 22. Geochemical characteristics of a conventional Mannville oil (Highvale field, 13-3-51-4W5). See Table 1 for legend.

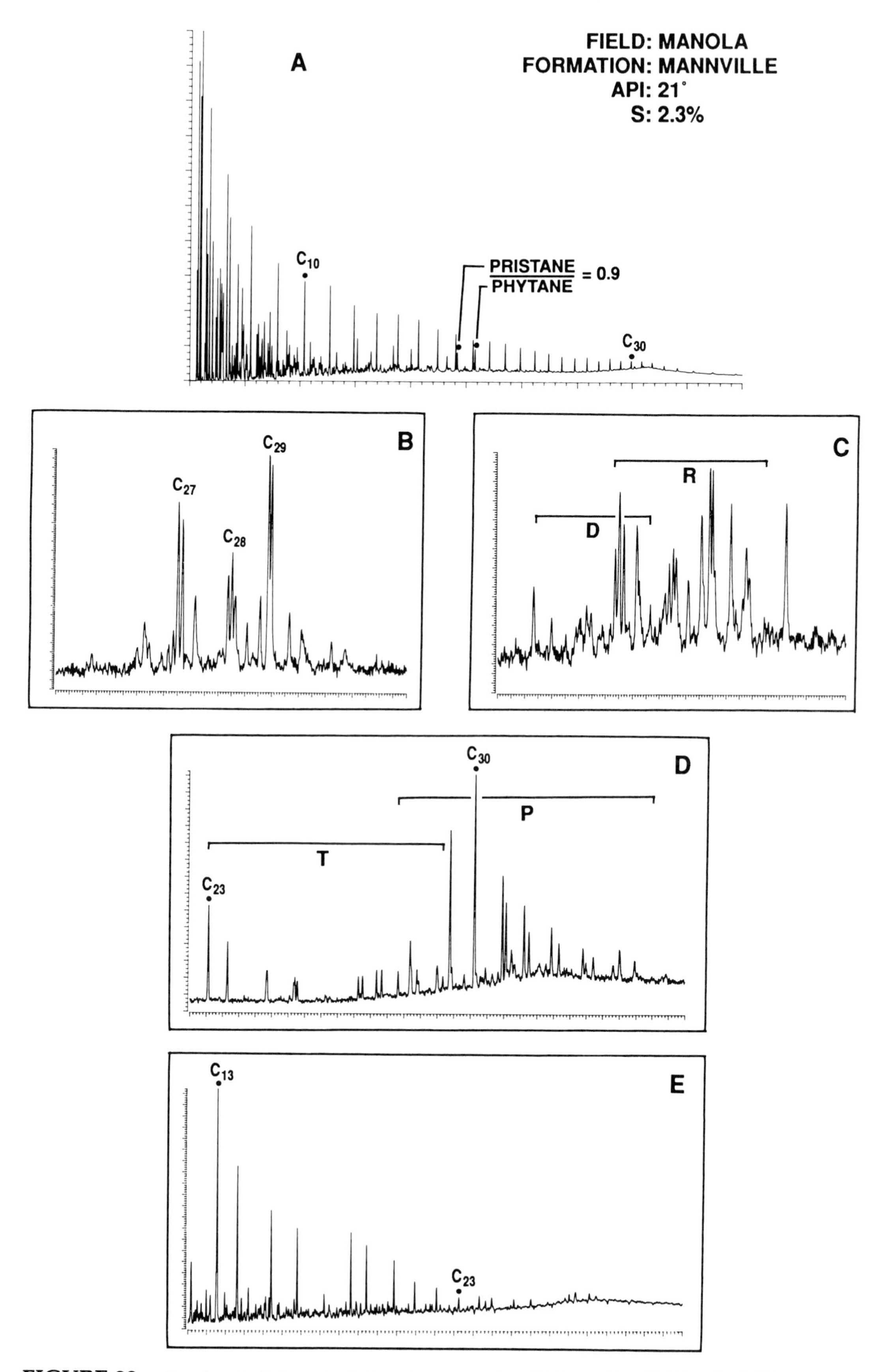

FIGURE 23. Geochemical characteristics of a conventional Mannville oil (Manola field, 6-19-59-2W5). See Table 1 for legend.

TABLE 1. Details of Geochemical Data Shown in Figures 18 to 23.

A. Gas chromatogram of whole oil (or stain, Figure 19). *n*-Paraffins C_{10} and C_{30} identified for reference.

B. Mass fragmentogram m/z 218, showing relative distribution of C_{27}–C_{29} iso- and regular steranes.

C. Mass fragmentogram m/z 217, showing relative distribution of C_{27}–C_{30} steranes. D = range of C_{27}–C_{30} diasteranes. R = range of C_{27}–C_{30} regular and isosteranes.

D. Mass fragmentogram m/z 191, showing relative distribution of tricyclic diterpanes and pentacyclic triterpanes. T = range of C_{23}–C_{31} tricyclic diterpanes. P = range of C_{27}–C_{35} pentacyclic triterpanes.

E. Mass fragmentogram m/z 133, showing relative distribution of C_{13}–C_{24} tetraalkyl benzenes.

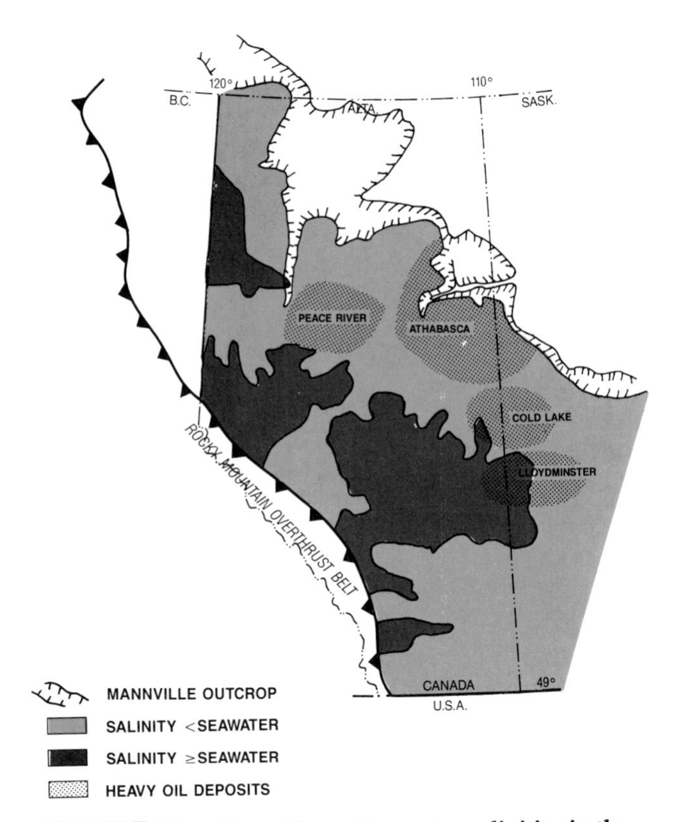

FIGURE 24. Map of formation water salinities in the Mannville group sediments. (Modified after Hitchon, 1964.)

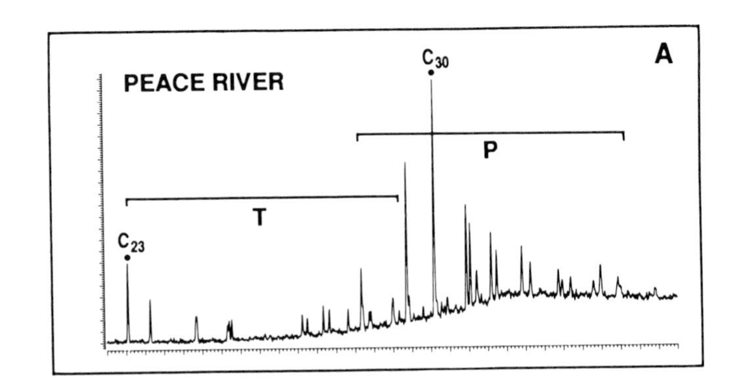

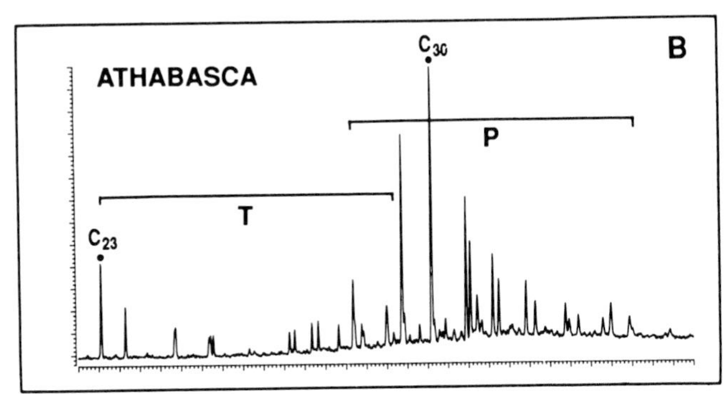

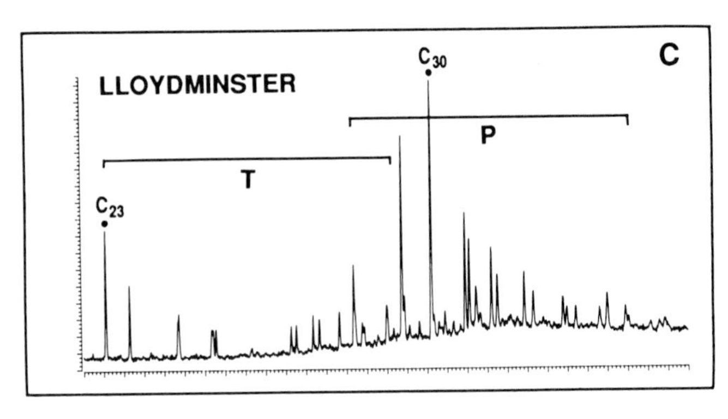

m/z 191

FIGURE 25. Characteristic triterpane distributions of a selection of heavy oil samples: (A) 4-21-85-18W5; (B) 10-21-65-13W4; (C) 12-33-49-25W3.

The above discussion has focused qualitatively on Jurassic and Mississippian source rocks as the likely sources of the heavy oil deposits. Mass balance considerations (Creaney and Allan, 1990) suggest that no single source is sufficient alone to have generated the necessary volumes of liquids. Equally, however, current geochemical practice and knowledge cannot discount minor contributions from either the Devonian Duvernay Formation or the Triassic Doig Formation, both of which lie in some or all of the drainage area of the heavy oil deposits. The geochemical characteristics of oils from both of these sources are sufficiently well known that they must be discounted as major contributors to the heavy oils, but geologic arguments can be raised to delineate areas of limited connectivity and hence limited contributions.

Colorado Group Oils

Without exception, all oils reservoired in Viking and all younger sands are sourced from the Colorado Group shales. These shales generate sweet, high-quality crude oil, with sulfur contents usually less than 0.4 wt. %. Hodg-

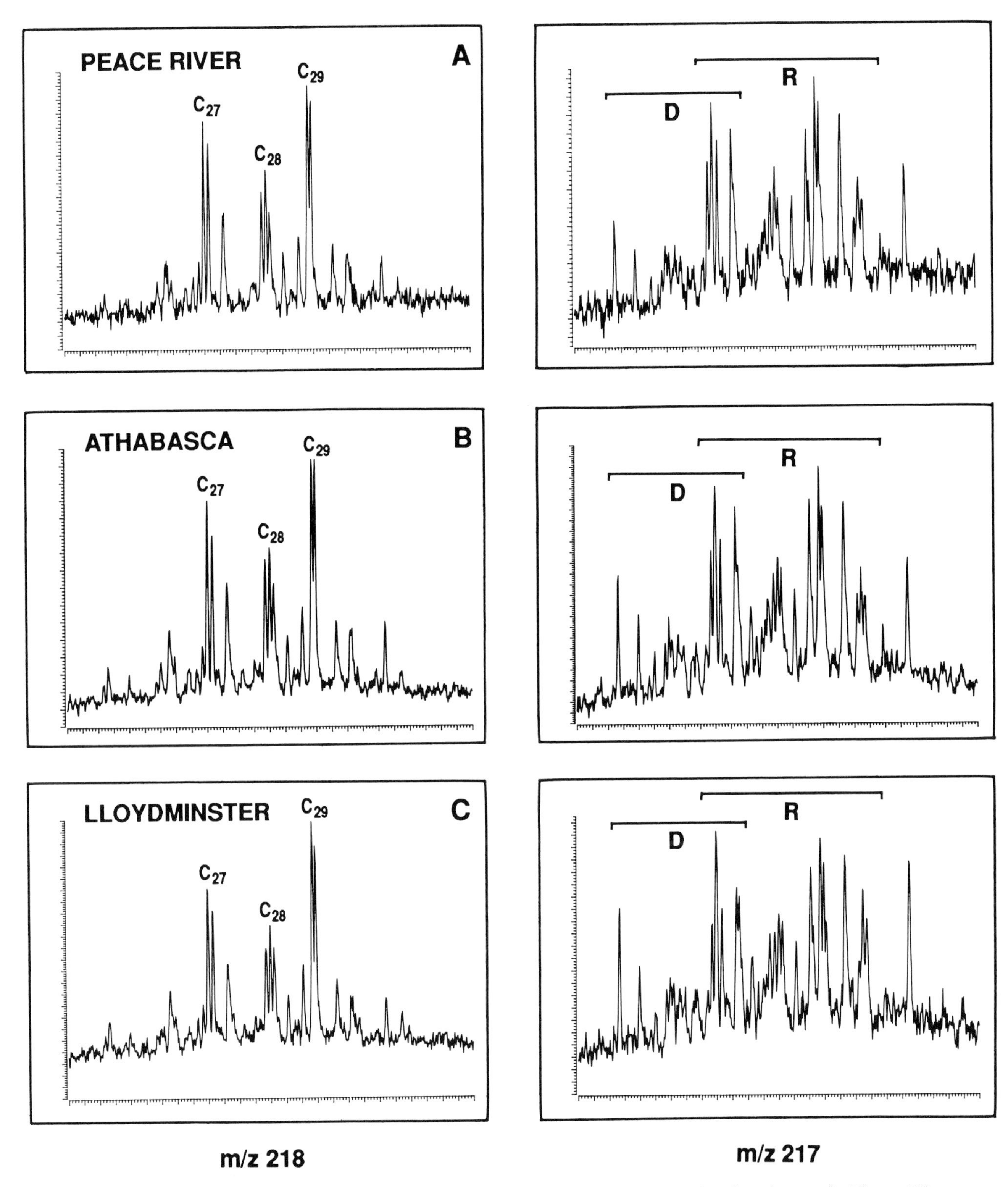

FIGURE 26. Characteristic sterane distributions of a selection of heavy oil samples (locations as in Figure 25).

son (1954) and Deroo et al. (1977) previously concluded that Late Cretaceous oils and Late Devonian (i.e., Duvernay-sourced) oils are geochemically very similar to each other, and their differentiation was a difficult task, given the knowledge and analytical tools of the day. This, in itself, is an interesting observation in that the two oil families are respective representatives of "carbonate" and "clastic" oils. Biomarker technology clearly and unequivocally differentiates the two oil families, however, and the key to the separation lies in evolutionary biochemical

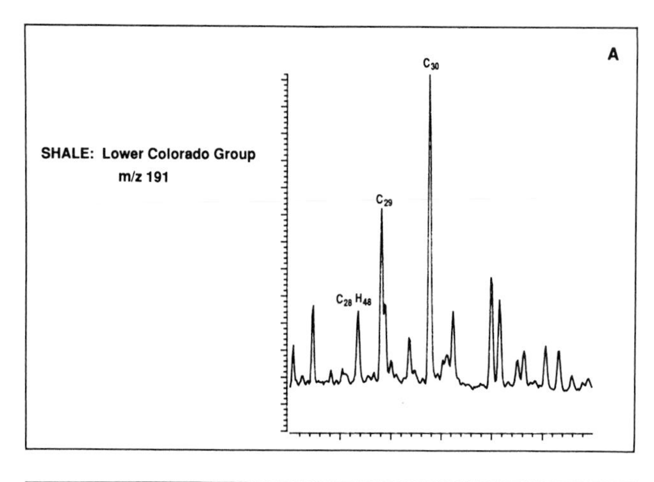

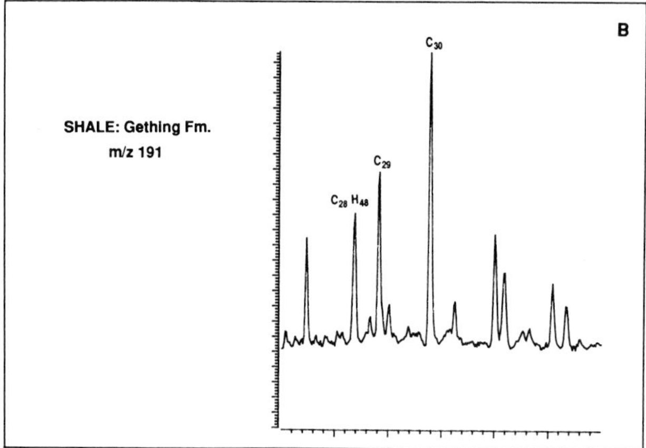

FIGURE 27. Pentacyclic triterpane distributions in unstained Mannville shale core extracts. Well locations: (A) 11-34-51-27W4; (B) 10-29-59-16W5.

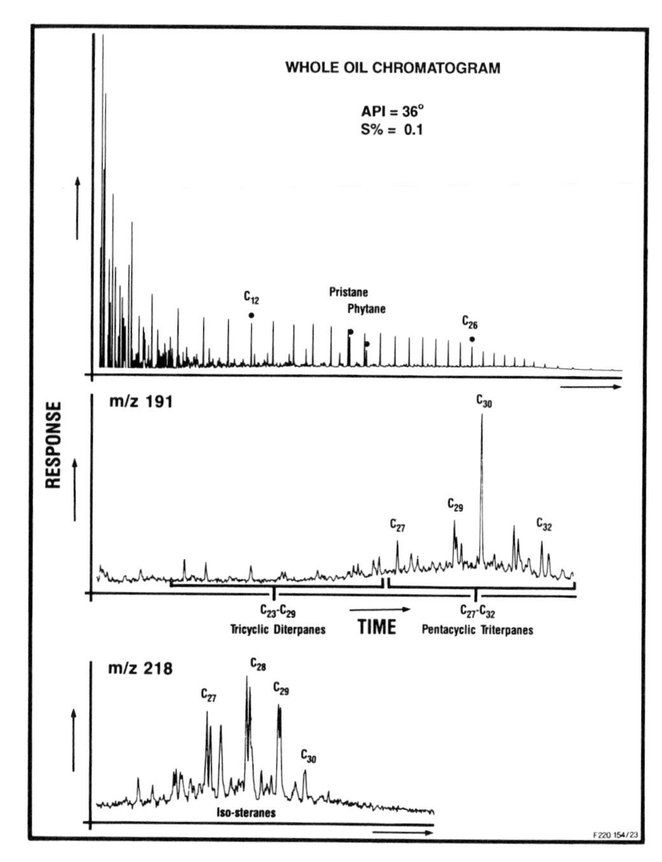

FIGURE 28. Biomarker distributions characteristic of Late Cretaceous oils (Cardium Formation, 13-2-28-5W5).

changes in oceanic planktonic organisms rather than inherent differences in carbonate and clastic environments. Steranes of Colorado Group-sourced oils consistently have a $C_{28} > C_{29} > C_{27}$ relative abundance (Figure 28), as do the steranes in organic extracts of the shales themselves. The transition from the $C_{29} > C_{27} > C_{28}$ relative pattern of all older source rocks and their oils to the Colorado Shale Group distribution occurs in Albian time and has been documented in core extracts (Figure 29). The change in sterol abundances is likely related to the emergence of coccolithophorids and silicoflagellates as prominent contributors to the steroidal budget of marine muds (Grantham and Wakefield, 1988).

Terpanes of Late Cretaceous oils (Figure 28) contain the usual range of hopanoid components, but usually only trace quantities of tricyclic compounds. Whether this indicates an environmental or an evolutionary change is not known, but Late Cretaceous source rocks are unique in that they are the only ones in the basin to be deposited in a shallow epicontinental seaway.

As previously stated, all oils found in Viking-Belly River reservoirs are sourced from the Colorado Group. Further, Colorado Group-sourced oil has not been documented in any Mannville or older sediments. Thus, there appears to be basin-wide, hydraulic disconnection between the Viking-Belly River reservoirs and Mannville and older sediments. Stratigraphically, this corresponds to the Albian Lower Shaftesbury/Joli Fou formations (Allan and Creaney, 1988; Creaney and Allan, 1990; Leckie and Smith, 1992). The lack of any correspondence in either degree of oil alteration or geographic distribution of altered oils above and below these shale formations strongly supports the concept of an oil, water, and probably gas permeability barrier at this point in the stratigraphic column.

Viking oils, which are widespread across southern Alberta and eastern Saskatchewan, have been sourced from mature Cenomanian-aged shales of the Colorado Group adjacent to the current Foothills of southern Alberta. Thus, given the complex reservoir sand facies of the Viking Formation (e.g., Reinson et al., 1988; Downing and Walker, 1988), there must exist (or have existed) continuous migration pathways from the westerly mature source zone that have distributed oil and gas into traps as far as 400 km (250 mi) from the source. Viking pools such as Provost, Eureka, and Dodsland are examples of such distant accumulations. The shallow Viking over the eastern limb of the basin is in places open to meteoric water re-

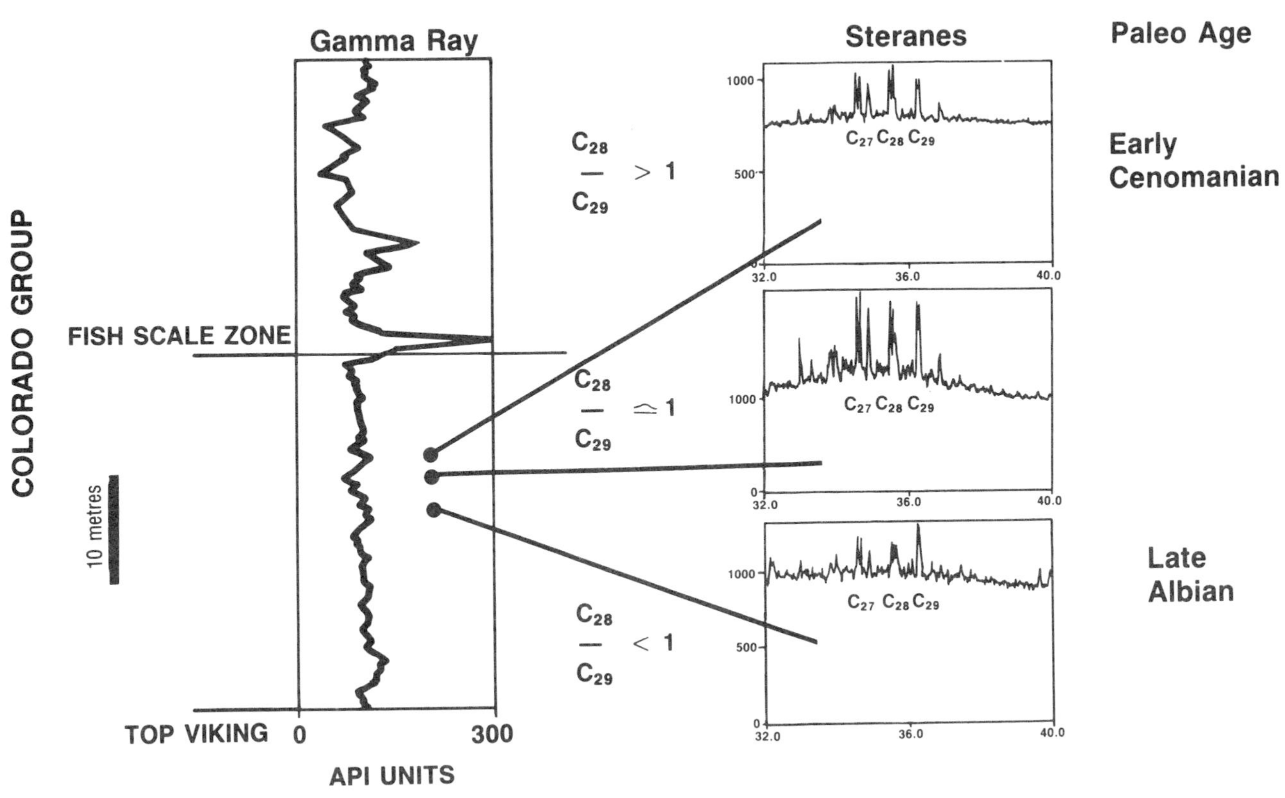

FIGURE 29. **Changes in relative concentrations of C_{27}-C_{29} steranes across the Albian/Cenomanian boundary. Well location, 6-25-42-6W5; depth interval, 2042–2108 m (6700–6917 ft).**

charge, and some minor oil biodegradation has been noted. Examples of sequential loss of *n*-alkanes in Viking oils at Hylo, Provost, and Coleville (Figure 30) attest to this freshwater incursion.

Cardium oils lie in traps that are localized at or below erosion-transgression surfaces, as described by Plint et al. (1986) and Walker (1988). The major Cardium accumulation is the giant Pembina field (Barclay and Smith, 1992; Porter, 1992), which lies in traps located below the conglomerates on the E5 surface (Plint et al., 1986). The bulk of the oil is peak-maturity liquid derived from the Second White Speckled Shale some considerable distance downdip from the reservoirs (Allan and Creaney, 1988). Similar oil is found at Cynthia/Pembina, Keystone, Carrot Creek, and Willesden Green.

Several Cardium pools occur in sandstones associated with surfaces other than the E5. Examination of their geochemical properties (maturity and biomarker profiles) suggests that they are pooled close to their sources, and are most likely derived from shales adjacent to their reservoir sands. The maturities of these oils (Kakwa, Garrington, and Crossfield fields, for example) are similar to the maturities of closely adjacent shales.

Oils in Belly River channel and shoreline sands in south-central Alberta are commonly peak-maturity liquids and are very similar to those oils in underlying Cardium pools. Quantitatively, total Belly River reserves (Porter, 1992) represent only a small fraction of the Cardium reserves, and they can be considered as a small amount of "overspill" oil from the Second White Speckled Shale that failed to access Cardium sands. Thus, a considerable cross-stratal component exists for the secondary migration pathway of Belly River oils, coupled with a significant amount of updip migration once the oil accessed channel sands.

HYDROCARBON MIGRATION

Figure 31 is a schematic cross section of the Western Canada sedimentary basin, marked with the principal migration conduits for oil and gas derived from the source rocks described in this paper.

Support for the overall hydrocarbon migration system described above is obtained from analysis of basic reserves data (Porter, 1992). Figure 32 shows conventional initial oil-in-place reserves by geologic period for Western Canada (ERCB, 1988). A total of 62% of the conventional crude oil occurs in the more or less "closed systems" of the Late Cretaceous and Devonian periods. In this context, the term "closed system" is used to describe a sedimentary package that is sourced internally, and in which structural/stratigraphic seals restrict the ability of hydrocarbons to leak out of or into the system. In contrast, the prolific source rocks of the Jurassic and Mississippian are not confined within closed systems, and nearly age-equivalent traps contain only 8.6% of the total conventional reserves. These

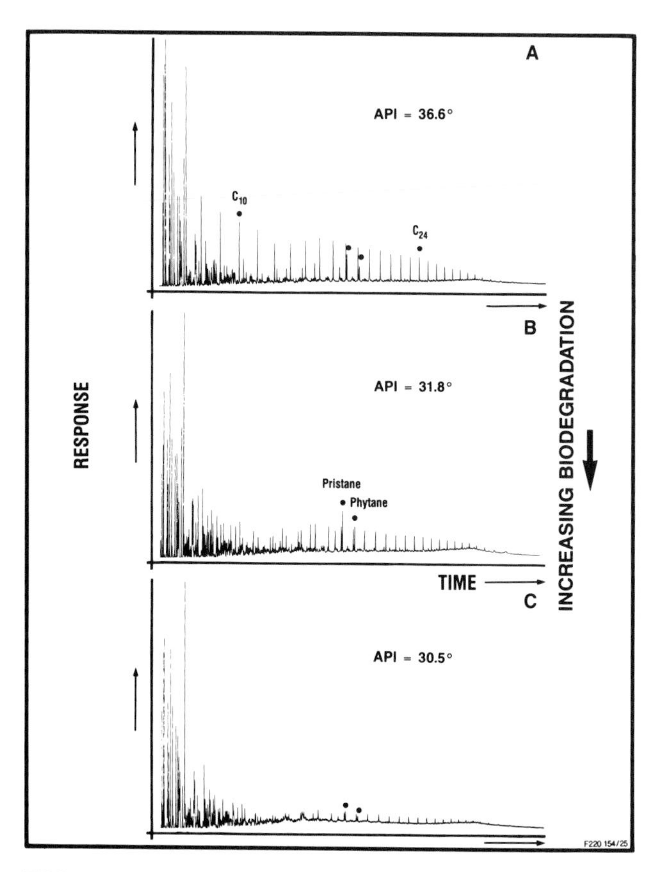

FIGURE 30. **Gas chromatograms of Viking oils demonstrating loss of *n*-alkanes through biodegradation. Well locations: (A) 12-18-24-15W4; (B) 10-13-35-7W4; (C) 14-5-31-25W3.**

sources are open to the porous and permeable sand-prone Mannville at their respective subcrops.

Conventionally, burial history curves constructed for the Western Canada basin show that the development of full maturity for any source rock did not occur until Late Cretaceous time or later (Deroo et al., 1977; Creaney and Allan, 1990). The event responsible for this is the crustal downwarping (Laramide orogeny) associated with foreland basin development (Fermor and Moffat, 1992). Previous geochemical studies have supported this dating of peak oil generation (e.g., Hacquebard in Deroo et al., 1977). An example of maturity development profiles in the southern part of the basin, currently an axial location, is shown in Figure 33. This modeling shows a 20 million year time range over which oils from Duvernay Shale to Colorado Shale oils have been generated. The maturity modeling predicts a 15 million year time gap between peak maturity of the Duvernay Formation and that of Jurassic sources, the youngest and oldest source rocks, respectively, able to contribute oils to the heavy oil deposits (the Duvernay contribution is likely to have been small). Thus, it is reasonable to suppose that oils arrived at the accumulation sites at Wabasca, Athabasca, Cold Lake, and Grosmont over roughly the same span of time. Biodegradation may have occurred after all the oils were in place (less likely), or it may have been an active process during the time over which the oils accumulated (more likely). If the former case is correct, the degree of biodegradation might be expected to be relatively uniform. If the latter case is correct, then the earliest pooled oil might well show a greater degree of biodegradation than the latest pooled oil simply because the microorganisms will have had an additional 15 million years to feed on the hydrocarbons. Variations in the degree of chemical biodegradation certainly are found (Brooks et al., 1987, 1988; Creaney and Allan, 1990), as are considerable differences in bulk physical properties such as viscosity. However, a rigorous geologic interpretation of these oil quality differences has yet to be formulated.

Within the Mannville, the following migration/pooling sequence is proposed. First, early production of biogenic methane and low-maturity coal gas sourced from Kootenay/Nikanassin and Mannville group coals and coaly shales occurred. These products migrated toward the eastern limb of the basin, producing numerous accumulations in a broad area extending from the Rainbow/Steen/Sousa pools in the northwest to the southeast. Later, as pre-Mannville source rocks matured sequentially, their expelled oils entered the Mannville and began to migrate to the eastern limb of the basin, accumulating at Peace River, Athabasca, Wabasca, Cold Lake, and Lloydminster. This oil pooled beneath the biogenic/low-maturity coal gases that locally formed gas caps within these deposits. Late-stage deep burial of Mannville coals in the west generated enormous quantities of dry gas and minor condensate which, probably as a result of inadequate reservoir quality, has been largely retained in the axial part of the basin.

Within the post-Joli Fou/Lower Shaftesbury section, the following hydrocarbon migration/pooling history is proposed. Significant quantities of bacterial methane, generated in the anoxic seafloor muds through Cenomanian–Santonian time, are preserved in immature sediments of the Colorado Group on the eastern limb of the basin, including the 8.5×10^{10} m^3 (3 tcf) Medicine Hat gas pool (Porter, 1992; Barclay and Smith, 1992). Similarly, Belly River sands have trapped early biogenic methane and low-maturity coal gases generated from the organic-rich muds and coals of the Belly River coal swamps. With the progressive development of the foredeep and rapid burial of the Colorado Group in Late Cretaceous to early Tertiary time, the marine, oil-prone source rocks within the section matured, producing oil that moved into Viking/Bow Island, Doe Creek, Dunvegan, Cardium, and Belly River sands. Variable-distance, updip migration to stratigraphic pinch-outs occurred, with small amounts of oil being retained in traps en route. Access of oil from the Second White Speckled Shale condensed interval (the most prolific Colorado Group source interval) to Cardium (E5) and Belly River conductor zones is proposed to have occurred through cross-stratal fracturing at the eastern margin of the Disturbed belt. Other Cardium sands and Dunvegan sands received oil from adjacent shales with minimal involvement of secondary migration.

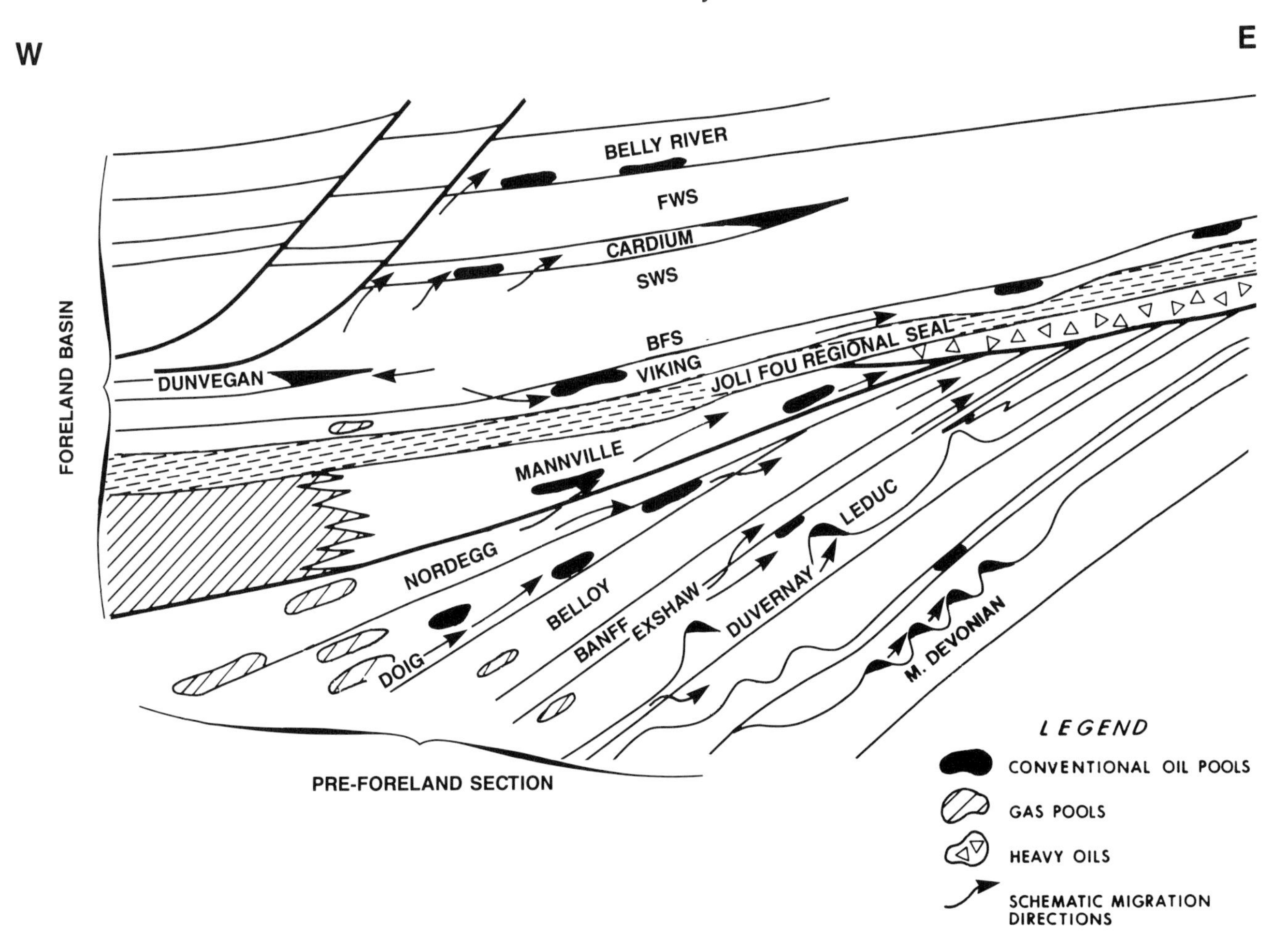

FIGURE 31. Basin cross section showing schematically the migration routes of oil from foreland basin and pre-foreland basin source rocks. General oil migration directions are indicated by arrows. (Modified after Creaney and Allan, 1990.)

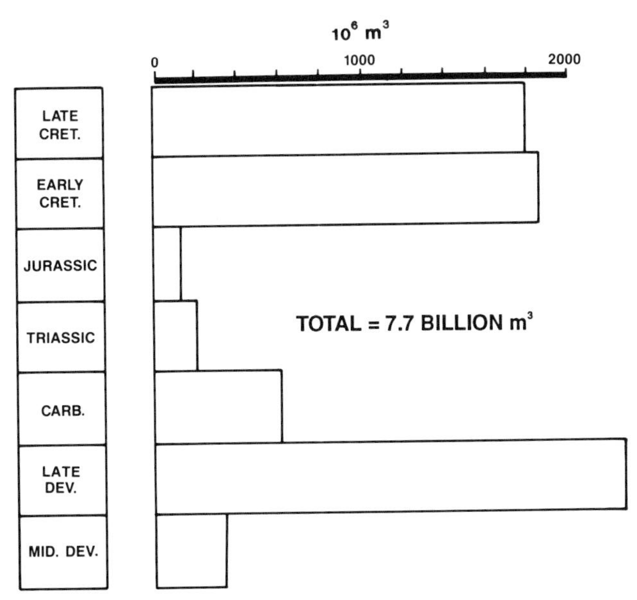

FIGURE 32. Initial oil-in-place conventional reserves (in millions of cubic meters) by reservoir age, Western Canada basin. (Data source: ERCB, 1988.)

CONCLUSIONS

Several oil and gas source rocks occur within the post-Middle Jurassic sediments of the Western Canada basin. The Late Jurassic/Early Cretaceous Nikanassin-Mannville section is dominated by continental sediments that contain significant coals and coaly shales with considerable gas potential. Rising sea level through the Cretaceous pushed shorelines landward, produced sediment starvation in more distal locations, and allowed anoxic bottom waters to develop sporadically. Organic-rich marine shales were deposited in condensed sections associated with the Basal Shaftesbury, the Fish Scale Zone (96 Ma), the Second White Speckled Shale (92 Ma), and the First White Speckled Shale (79 Ma). The Cretaceous section was completed with transgressive continental sedimentation of the Belly River and younger formations, which contain some thin coals.

Hydrocarbon migration throughout the Western Canada sedimentary basin probably began in the middle-Late Cretaceous from pre-foreland basin (pre-Middle Jurassic) oil-prone source rocks, including the Late Devonian (Frasnian) Duvernay/Majeau Lake formations, the Late Devo-

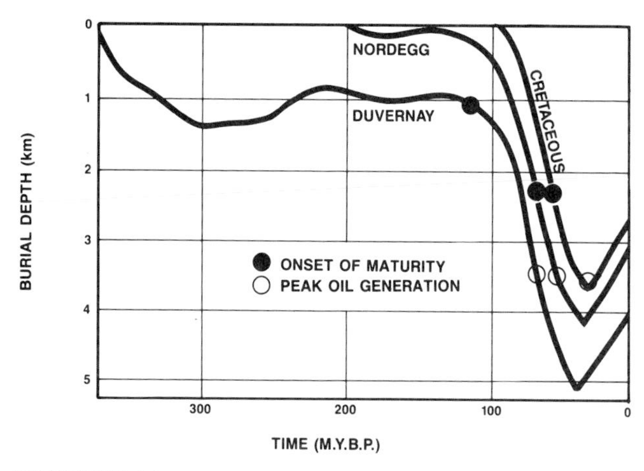

FIGURE 33. Burial history and timing of maturity for Devonian–Cretaceous source rocks, southern Alberta. (Modified after Creaney and Allan, 1990.)

nian and Early Mississippian Exshaw/Bakken Formation, the Mid-Late Triassic Doig Formation, and the Early Jurassic (Sinemurian) Nordegg and (Toarcian) "Poker Chip" units. This oil finally accumulated in a huge, probably salt-solution-controlled, low-relief anticline on the Alberta-Saskatchewan border (Athabasca/Cold Lake area). Oil and gas generated from the Fish Scale Zone, the Second White Speckled Shale, and possibly the First White Speckled Shale, within the foreland sediments, accessed Viking through Belly River sands, both directly and via thrust faults and fractures. Migration was then simply updip to stratigraphic trapping configurations. In no instance has this oil accessed the Lower Cretaceous below the Viking, and conversely no oil has leaked through the Mannville from older sources into the Viking or Late Cretaceous sands. Late stage gas generation from Mannville coals produced the huge gas deposits of the Deep basin, which are trapped in low-porosity, low-productivity-rate sands.

With extensive Tertiary erosion and exhumation of the eastern limb of the basin, freshwater incursion and biodegradation converted the Peace River and Athabasca/Cold Lake deposits to 1.7 trillion barrels of low-quality heavy oil. Mannville oils are progressively less degraded downdip. Biodegradation is only rarely encountered in the Viking and has not been recorded at all in any Late Cretaceous reservoirs.

ACKNOWLEDGMENTS

The authors would like to thank Esso Resources Canada Limited for fostering the research required to produce this paper and permission to publish it. Further, we would like to thank the members of the Western Canada exploration teams at Esso Resources Canada Limited (past and present), without whom this paper would not have been possible. D. L. Layton, S. D. Triffo, R. B. Scott, S. Hjelmeland, and P. M. D. Riggins are thanked for technical assistance, and L. J. Hasiuk is thanked for coordinating the typing of the manuscript, which benefited greatly from the reviews of P. J. R. Nederlof and an anonymous person. The Exploration Drafting and Writing Resources groups are thanked for assistance with the preparation of this paper.

REFERENCES CITED

Alexander, R., R. I. Kagi, G. W. Woodhouse, and J. K. Volkman, 1983, The geochemistry of some biograded Australian oils: Australian Petroleum Exploration Association Journal, v. 23, part 1, p. 53–63.

Allan, J., and S. Creaney, 1988, Sequence stratigraphic control of source rocks: Viking-Belly River system, *in* D. P. James and D. A. Leckie, eds., Sequences, stratigraphy, sedimentology: surface and subsurface: Canadian Society of Petroleum Geologists Memoir 15, p. 575.

Allan, J., and S. Creaney, 1991, Oil families of the Western Canada basin: Canadian Society of Petroleum Geologists Bulletin, v. 39, n. 2, p. 107–122.

Bailey, N. J. L., C. R. Evans, and C. W. D. Milner, 1974, Applying petroleum geochemistry to search for oil: examples from Western Canada basin: AAPG Bulletin, v. 58, p. 2284–2294.

Barclay, J. E., and D. G. Smith, 1992, Western Canada foreland basin oil and gas plays, this volume.

Brooks, P. W., M. G. Fowler, and R. W. Macqueen, 1988, Biological marker and conventional organic geochemistry of oil sands/heavy oils, Western Canada basin: Organic Geochemistry, v. 12, p. 519–538.

Brooks, P. W., M. G. Fowler, and R. W. Macqueen, 1989, Use of biomarkers, including aromatic steroids, to indicate relationships between oil sands/heavy oils/bitumens and conventional oils, Western Canada basin, *in* D. J. Lazar, ed., Proceedings, 1989 Eastern Oil Shale Symposium, November 15-17, 1989: Institute for Mining and Minerals Research, University of Kentucky, Lexington, Kentucky, p. 104–111.

Brooks, P. W., L. R. Snowdon, and K. G. Osadetz, 1987, Families of oils in southeastern Saskatchewan, *in* C. G. Carlson and J. E. Christopher, eds., Proceedings of the 5th International Williston Basin Symposium: Saskatchewan Geological Society Special Publication No. 9, p. 253–264.

Cameron, A., 1972, Petrography of Kootenay coals in the upper Elk River and Crowsnest areas, British Columbia and Alberta: Alberta Research Council, Information Series, n. 60, p. 31–46.

Creaney, S., and J. Allan, 1990, Hydrocarbon generation and migration in the Western Canada basin, *in* J. Brooks, ed., Classic petroleum provinces: Blackwell, London, Special Publication Number 50 of the Geological Society of London, p. 189–202.

Demaison, G. J., and G. T. Moore, 1980, Anoxic environments and oil source bed genesis: AAPG Bulletin, v. 64, p. 1179–1209.

Deroo, G., T. G. Powell, B. Tissot, and R. G. McCrossan, 1977, The origin and migration of petroleum in the Western Canadian sedimentary basin, Alberta: a geochemical and thermal maturation study: Geological Survey of Canada Bulletin 262, 136 p.

Downing, K. P., and R. G. Walker, 1988, Viking Formation, Joffre field, Alberta: shoreface origin of long, narrow

sand body encased in marine mudstones: AAPG Bulletin, v. 72, p. 1212–1228.
du Rouchet, J., 1984, Migration in fracture networks: an alternative interpretation of the supply of the "giant" tar accumulations in Alberta, Canada, I: Journal of Petroleum Geology, v. 7, p. 381–402.
du Rouchet, J., 1985, The origin and migration paths of hydrocarbons accumulated in the Lower Cretaceous "giant" tar accumulation of Alberta, Canada, II: Journal of Petroleum Geology, v. 8, p. 101–114.
Ells, S. C., 1923, Cretaceous shales of Manitoba and Saskatchewan as a possible source of crude petroleum: Canada Department of Mines Summary Report of Investigations, 1921, Report 588, p. 34–41.
ERCB, 1988, Alberta's reserves of crude oil, oil sands, gas, natural gas liquids and sulphur: ERCB ST89-18, ERCB Series Annual Statistical Series, v.p.
Fermor and Moffat, 1992, Tectonics and structure of the Western Canada foreland basin, this volume.
Fowler, M. G., and P. W. Brooks, 1987, Organic geochemistry of Western Canada basin tar sands and heavy oils: 2, Correlation of tar sands using hydrous pyrolysis of asphaltenes: Journal of Energy and Fuels, v. 1, p. 459–467.
Fowler, M. G., P. W. Brooks, and R. W. Macqueen, 1989, A comparison between the biomarker geochemistry of some samples from the Lower Jurassic Nordegg Member and Western Canada basin oil sands and heavy oils, *in* Current Research, Part D: Geological Survey of Canada, Paper 89-1D, p. 19–24.
George, A. E., R. C. Banerjee, G. T. Smiley, and H. Sawatzky, 1977, Simulated geothermal maturation of Athabasca bitumen: Bulletin of Canadian Petroleum Geology, v. 25, p. 1085–1096.
Grantham, P. J., and L. L. Wakefield, 1988, Variations in the sterane carbon number distributions of marine source rock derived crude oils through geological time: Organic Geochemistry, v. 12, p. 61–73.
Gussow, W. C., 1954, Differential entrapment of oil and gas: a fundamental principle: AAPG Bulletin, v. 38, n. 5, p. 816–853.
Gussow, W. C., 1955, Discussion of "in-situ origin of McMurray oil": AAPG Bulletin, v. 39, n. 8, p. 1625–1631.
Hall, R. L., 1984, Lithostratigraphy and biostratigraphy of the Fernie Formation (Jurassic) in the southern Canadian Rocky Mountains, *in* D. F. Stott and D. J. Glass, eds., The Mesozoic of middle North America: Canadian Society of Petroleum Geologists Memoir 9, p. 233–247.
Hitchon, B., 1963a, Geochemical studies of natural gas, Part I: Journal of Canadian Petroleum Technology, v. 2, p. 60–76.
Hitchon, B., 1963b, Geochemical studies of natural gas, Part II: Journal of Canadian Petroleum Technology, v. 2, p. 100–116.
Hitchon, B., 1963c, Geochemical studies of natural gas, Part III: Journal of Canadian Petroleum Technology, Winter, v. 2, p. 165–174.
Hitchon, B., 1964, Formation fluids, *in* R. G. McCrossan and R. P. Glaister, eds., The Geological history of Western Canada: Alberta Society of Petroleum Geologists, p. 201–217.
Hitchon, B., and R. H. Filby, 1984, Use of trace elements for classification of crude oils into families: example from Alberta, Canada: AAPG Bulletin, v. 68, p. 838–849.
Hodgson, G. W., 1954, Vanadium, nickel, and iron trace metals in crude oils of Western Canada: AAPG Bulletin, v. 38, p. 2537–2554.
Hoffmann, C. F., and O. P. Strausz, 1986, Bitumen accumulation in Grosmont platform complex, Upper Devonian, Alberta, Canada: AAPG Bulletin, v. 70, p. 1113–1128.
James, A. T., 1983, Correlation of natural gas by use of carbon isotopic distribution between hydrocarbon components: AAPG Bulletin, v. 67, p. 1176–1191.
Jervey, M. T., 1988, Siliciclastic sequence development in foreland basins—a numerical approach, *in* D. P. James and D. A. Leckie, eds., Sequences, stratigraphy, sedimentology: surface and subsurface: Canadian Society of Petroleum Geologists Memoir 15, p. 579.
Jervey, M. T., 1992, Siliciclastic sequence development in foreland basins, with examples from the Western Canada foreland basin, this volume.
Kalkreuth, W., 1982, Rank and petrographic composition of selected Jurassic–Lower Cretaceous coals from British Columbia, Canada: Bulletin of Canadian Petroleum Geology, v. 30, p. 112–139.
Kalkreuth, W., and M. McMechan, 1988, Burial history and thermal maturity, Rocky Mountain Front Ranges, Foothills, and foreland, east-central British Columbia and adjacent Alberta, Canada: AAPG Bulletin, v. 72, p. 1395–1410.
Leckie, D., and D. Smith, 1992, Regional setting, evolution, and depositional cycles of the Western Canada foreland basin, this volume.
Leckie, D. A., C. Singh, F. Goodarzi, and J. W. Wall, 1990, Organic-rich, radioactive marine shale: a case study of a shallow-water condensed section, Cretaceous Shaftesbury Formation, Alberta, Canada: Journal of Sedimentary Petrology, v. 60, p. 101–117.
Leenheer, M. J., 1984, Mississippian Bakken and equivalent formations as source rocks in the Western Canadian basin: Organic Geochemistry, v. 6, p. 521–532.
Macauley, G., 1984a, Cretaceous oil shale potential of the prairie provinces, Canada: Geological Survey of Canada Open File Report OF-977, 61 p.
Macauley, G., 1984b, Cretaceous oil shale potential in Saskatchewan, *in* J. A. Lorsong and M. A. Wilson, eds., Oil and gas in Saskatchewan: Special Publication of the Saskatchewan Geological Society, 7, p. 255–269.
Macauley, G., L. R. Snowdon, and F. D. Ball, 1985, Geochemistry and geological factors governing exploitation of selected Canadian oil shale deposits: Geological Survey of Canada Paper 85-13, 64 p.
Masters, J. A., 1979, Deep basin gas trap, Western Canada: AAPG Bulletin, v. 63, p. 152–181.
McInnes, W., 1913, The basins of Nelson and Churchill rivers: Geological Survey of Canada Memoir 30, p. 68.
Milner, C. W. D., M. A. Rogers, and C. R. Evans, 1977, Petroleum transformations in reservoirs: Journal of Geochemical Exploration, v. 7, p. 101–153.
Montgomery, D. S., D. M. Clugston, A. E. George, G. T. Smiley, and H. Sawatzky, 1974, Geochemical investigation of oils in the Western Canada tar belt, *in* L. V. Hills, ed., Oil sands: fuel of the future: Canadian Society of Petroleum Geologists Memoir 3, p. 168–183.
Moshier, S. O., and D. W. Waples, 1985, Quantitative evalua-

tion of Lower Cretaceous Mannville Group as source rock for Alberta's oil sands: AAPG Bulletin, v. 69, p 161–172.

Passey, Q. R., S. Creaney, J. B. Kulla, F. Moretti, and J. D. Stroud, 1989, Well log evaluation of organic-rich rocks: European Association of Organic Geochemists Annual Meeting, Paris, Sept. 1989.

Passey, Q. R., S. Creaney, J. B. Kulla, F. Moretti, and J. D. Stroud, 1990, A practical model for organic richness from porosity and resistivity logs: AAPG Bulletin, v. 74, p. 1777–1794.

Plint, A. G., R. G. Walker, and K. M. Bergman, 1986, Cardium Formation 6: Stratigraphy framework of the Cardium in subsurface: Bulletin of Canadian Petroleum Geology, v. 34, p. 213–225.

Porter, J., 1992, Conventional hydrocarbon reserves of the Western Canada foreland basin, this volume.

Poulton, T. P., 1984, The Jurassic of the Canadian Western Interior from 49 degrees north latitude to Beaufort Sea, *in* D. F. Stott and D. S. Glass, eds., The Mesozoic of middle North America: Canadian Society of Petroleum Geologists Memoir 9, p. 15–41.

Powell, T. G., and L. R. Snowdon, 1980, Geochemical controls on hydrocarbon generation in Canadian sedimentary basins, *in* A. D. Miall, ed., Facts and principles of world petroleum occurrence: Canadian Society of Petroleum Geologists Memoir 6, p. 421–444.

Reinson, G., J. E. Clark, and A. E. Foscolos, 1988, Reservoir geology of Crystal Viking field, lower Cretaceous estuarine tidal channel-bay complex, south central Alberta: AAPG Bulletin, v. 72, p. 1270–1294.

Rubinstein, I., O. P. Strausz, C. Spyckerelle, R. J. Crawford, and D. W. S. Westlake, 1977, The origin of the oil sand bitumens of Alberta: a chemical and microbiological simulation study: Geochimica et Cosmochimica Acta, v. 41, p. 1241–1353.

Stach, E., 1975, The fundamentals of coal petrology, *in* Stach's Textbook of Coal Petrology: Gebruder Borntraeger, p. 5–174.

Stoakes, F. A., and S. Creaney, 1984, Sedimentology of a carbonate source rock: Duvernay Formation of central Alberta, *in* L. Eliuk, J. Kaldi, N. Watts, and G. Harrison, eds., Carbonates in subsurface and outcrop: Proceedings of the 1984 Canadian Society of Petroleum Geologists Core Conference, Calgary, p. 132–147.

Stoakes, F. A., and J. C. Wendte, 1987, The Woodbend Group, *in* F. F. Krause and O. G. Burrowes, eds., Devonian lithofacies and reservoir styles in Alberta: Canadian Society of Petroleum Geologists 13th Annual Core Conference proceedings, p. 153–170.

Stronach, N. J., 1984, Depositional environments and cycles in the Jurassic Fernie Formation, Southern Canadian Rocky Mountains, *in* D. F. Stott and D. J. Glass, eds., The Mesozoic of middle North America: Canadian Society of Petroleum Geologists Memoir 9, p. 43–67.

Summons, R. E., and T. G. Powell, 1987, Identification of aryl isoprenoids in source rocks and crude oils: Biological markers for the green sulphur bacteria: Geochemica et Cosmochimica Acta, v. 51, p. 557–566.

Tyrrell, J. B., 1892, Report on northwestern Manitoba with portions of the districts of Assiniboia and Saskatchewan: Geological Survey of Canada Annual Report 5, 1890-1891, v. 5, part 1, report E, v.p.

Vigrass, L. W., 1968, Geology of Canadian heavy oil sands: AAPG Bulletin, v. 52, p. 1984–1999.

Walker, R. G., 1988, The origin and scale of sequence and erosional bounding surfaces in the Cardium Formation, *in* D. P. James and D. A. Leckie, eds., Sequences, stratigraphy, sedimentology: surface and subsurface: Canadian Society of Petroleum Geologists Memoir 15, p. 573.

Welte, D. H., R. G. Schaefer, W. Stoessinger, and M. Radke, 1984, Gas generation and migration in the Deep basin of Western Canada, *in* J. A. Masters, ed., Elmworth: case study of a Deep basin gas field: AAPG Memoir 38, p. 35–47.

Chapter 11

Petroleum in the Zagros Basin: A Late Tertiary Foreland Basin Overprinted onto the Outer Edge of a Vast Hydrocarbon-Rich Paleozoic–Mesozoic Passive-Margin Shelf

Z. R. Beydoun

*American University of Beirut, Lebanon and Marathon International Petroleum (GB) Ltd.
London, U.K.*

M. W. Hughes Clarke

Manningtree, Essex, U.K.

R. Stoneley

*Imperial College, University of London
London, U.K.*

INTRODUCTION: THE ZAGROS OROGENE AND ITS FORELAND BASIN

The Zagros Mountains lie within the Middle East as a part of the world girdle of Alpine mountain ranges.

It is now well documented, although perhaps not yet fully assimilated by all those working in the oil industry elsewhere, that, even though incompletely explored, the Middle East contains two-thirds of the world's proven reserves of oil and one-third of the world's reserves of gas (Beydoun, 1991). Some 98% of these vast volumes are lodged in the fields lying in a broad belt (approximately 2500 by 500 to 700 km, or 1555 by 310 to 435 mi) running from northern Iraq to Oman, including southwestern Iran, eastern Saudi Arabia, and the Gulf States, in an area constituting the pre-Neogene northeastern continental shelf of the Arabian plate (Figure 1).

This belt of prolific oil and gas fields (Figure 2; see Table 1 for field/discovery names) is bounded on the northeast by the high Zagros mountain range (the fields lie in the foothills of this range and across the Arabian plains areas), and on the south and west by the outcropping crystalline basement of the Arabian shield. The southeastern end lies at the Arabian Sea margin of Oman, now bounded by Indian Ocean crust, and the northwestern end is formed by the convergence of the High Zagros with the prominent north-south structural feature of the Ha'il-Ga'ara arch (Figure 1), west of which a more restless geologic history with important stratigraphic gaps has been much less favorable to hydrocarbon generation/retention.

The Zagros mountain range is an orogene created by collision of the continental Arabian plate with continental segments of the Eurasian margin; oceanic Arabian plate crust was subducted northward beneath Eurasia such that the continent-to-continent collision began locally in the late Eocene, and convergence continues today. A consequence of this collision was to create a depositional basin (the Zagros foreland basin) across the Arabian shelf southwest of the Zagros suture. Continuing convergence led to deformation of that basin area by southwest-verging reverse faults and impressive whaleback folds, essentially paralleling the suture and dominating the surface topography (Figure 3).

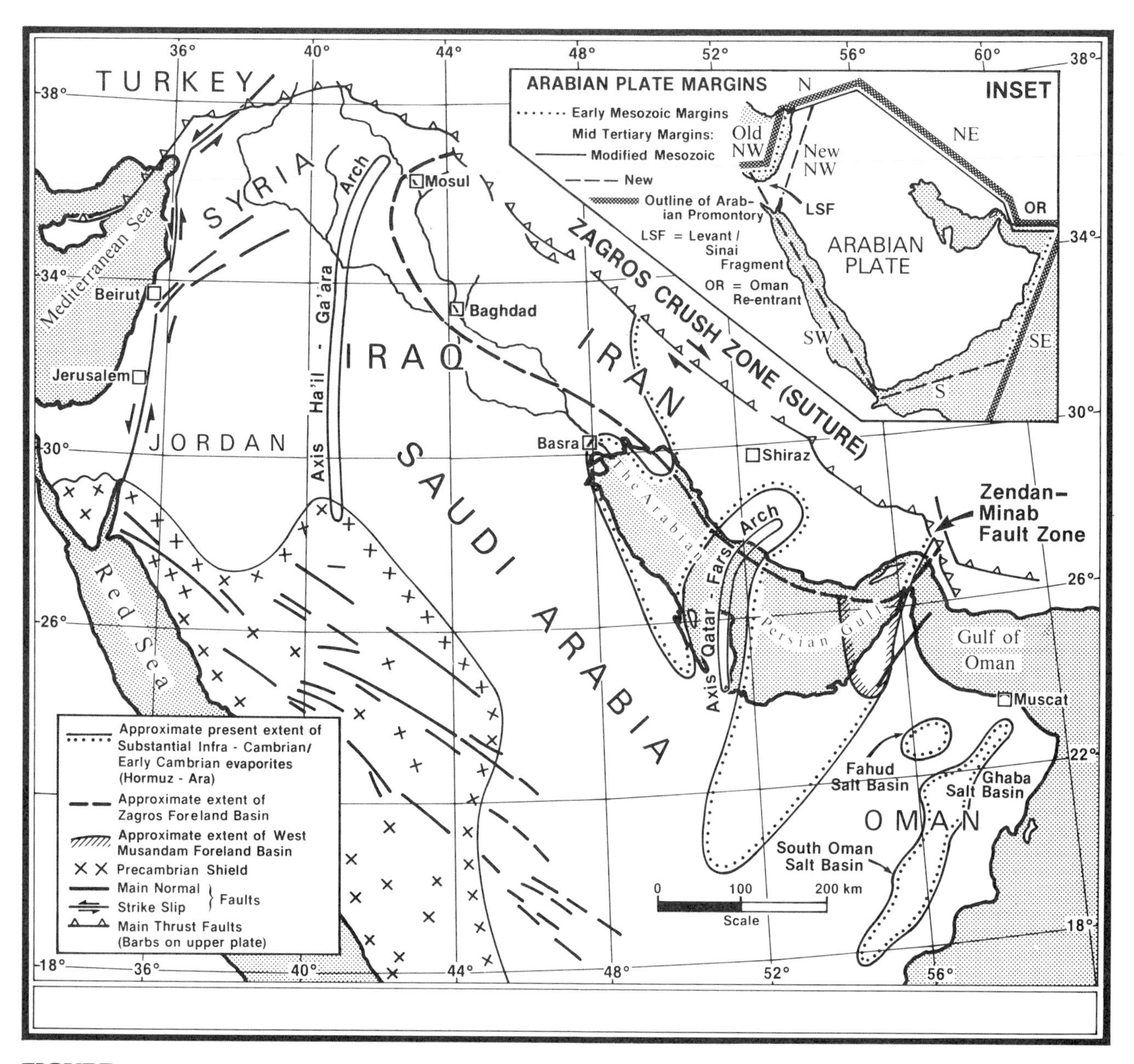

FIGURE 1. Orogenic setting of the Zagros foreland basin.

The Zagros foreland basin, the prime concern of this article, is thus defined as follows. It comprises the post-Zagros collision succession (upper Eocene to Holocene), the sediments of which were in part cannibalized from the underlying beds near the former shelf margin. This section has been involved in the subsequent folding in southwestern Iran and northeastern Iraq. For present purposes, the southwestern limit of the basin is taken arbitrarily at or close to the front of the folding. Strictly speaking, of course, the depositional "basin" does persist some distance to the southwest toward an undefinable boundary, containing, for instance, the Asmari (lower Miocene) accumulation at Abouzar (formerly Ardeshir) in offshore Iran. It is impossible to consider solely the hydrocarbons in the younger section, because pre-Eocene accumulations also have common sources and are found within common structures; in some cases even the pools are common and interconnected. The concept of the Zagros foreland basin is, therefore, inevitably somewhat loose and arbitrary.

This Zagros foreland basin contains many supergiant fields in high-amplitude structures created by the Neogene convergence, presently totaling almost one-quarter of the quoted oil reserves of the whole of the northeastern Arabian shelf. However, studies of the hydrocarbon habitats of the fields in the foreland basin have shown that all these oil (and presumably gas) reserves have been sourced from stratigraphic levels predating the late Eocene onset of collision. Hence, the understanding of the origin and appreciation of the volume of hydrocarbons now trapped in the Zagros foreland basin requires synthesis of the full history of the northeastern shelf of the Arabian craton, an essentially passive margin history that began in the late Proterozoic. Only then can the signifi-

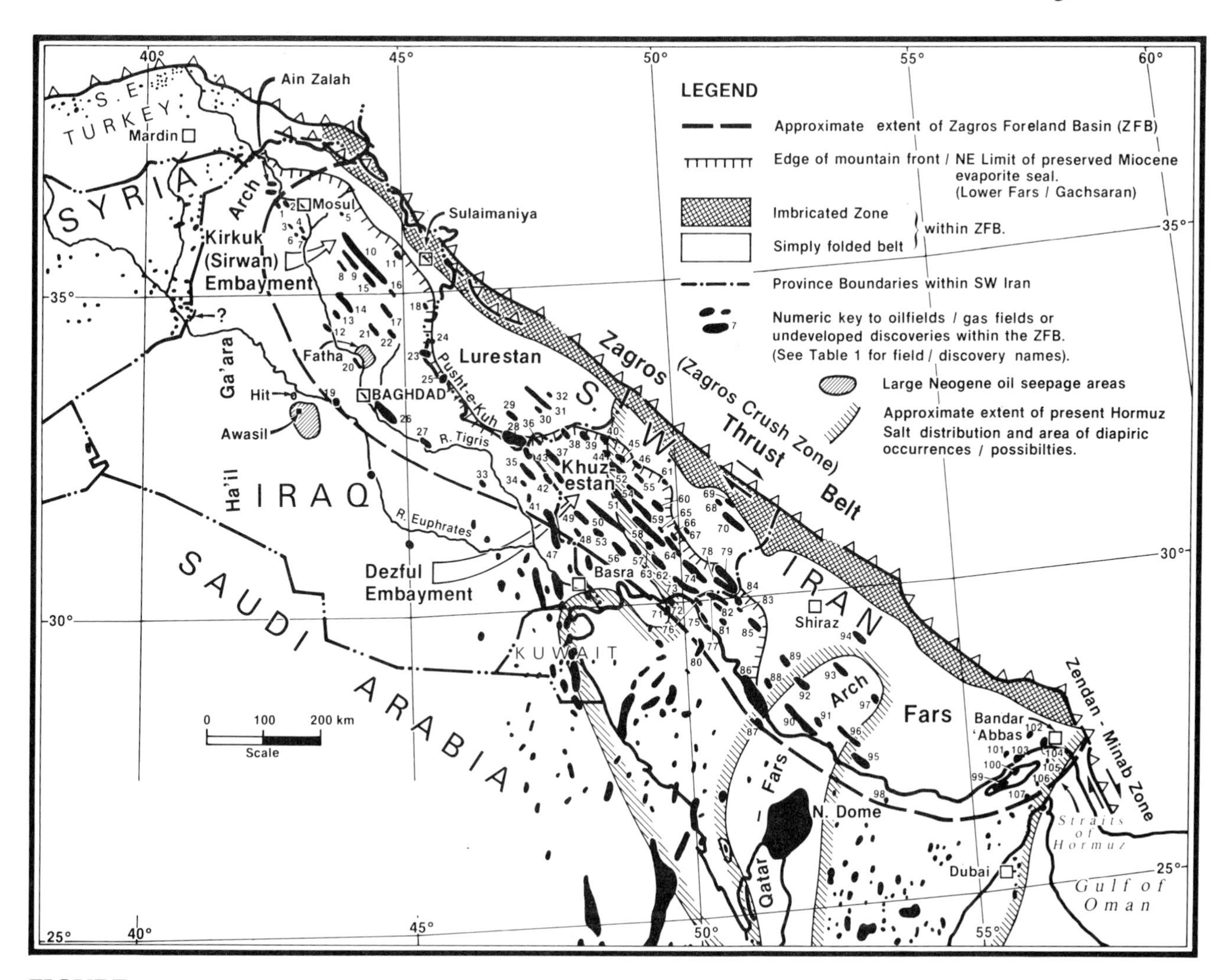

FIGURE 2. Regional oil and gas map showing locations, provinces, and key structural elements in the Zagros foreland basin.

cance of overprinting Neogene foreland deposition and deformation on this enormous hydrocarbon-rich, multilayer Paleozoic–Mesozoic sandwich be properly brought out. In the context of this article, aimed at documenting a very large part of this vast oil-rich area in the frame of a foreland basin, it is necessary to emphasize that the inherent complexity and variability must be greatly simplified. This must be done not only for the purposes of description and comparison with the type basin and other basins discussed in this volume, but also as a consequence of a surprisingly sparse cover of published petroleum geologic information. This last factor is itself a measure of the relatively immature level of exploration in the region as a whole and emphasizes that the already very large reserves must increase considerably with more intensive exploration (Grunau, 1985).

The geologic history from late Proterozoic to mid-Tertiary is of remarkable stability, punctuated by episodes of extension in the Permian–Triassic and brief convergence in the Late Cretaceous. The mid-Tertiary continental collision leading to the Zagros orogene imposed a dominant northwest-southeast fold trend that contrasts with the dominant north-south or centric pattern developed in the pre-Neogene history; this difference is even apparent in the outlines of the fields, which also emphasize the transitional change from one trend to the other (Figure 2). In discussing the foreland basin, we include all the transitional areas. The Oman segment occupies a reentrant at the southeast end of the northeastern Arabian shelf and hence largely escaped this northwest-southeast overprint. The southeast end of the Zagros foreland basin, therefore, terminates north of Oman at the NNW-SSE Zendan-Minab fault zone (Figure 1). However, a separate north-south foreland basin developed west of the northern end of the Late Cretaceous Oman overthrust belt with a complex Upper Cretaceous and Tertiary fill and structural pattern mainly dominated by west-verging thrust tectonics. This West Musandam basin (Figure 1) is quite distinct and is not included in the Zagros foreland basin. The Zagros foreland basin, as defined here, is thus about 1800 km (1120 mi) in length (Ha'il-Ga'ara arch to Zendan-Minab fault zone) and some

TABLE 1. Numeric Key to Oil Fields/Gas Fields or Undeveloped Discoveries in the Zagros Foreland Basin. See Figure 2 for locations.

No.	Field	No.	Field	No.	Field
1	Sasan	37	Cheshmeh Khush	73	Rag-e-Safid
2	Alan	38	Kabud	74	Pazanan
3	Jawan	39	Qaleh Nar	75	Binak
4	Qasab	40	Lab-e-Safid	76	Siah Makan
5	Demir Dagh	41	Al Halfayah	77	Bidi Hakimeh
6	Najmah	42	Jabal Fauqui	78	Garangan
7	Qaiyarah	43	Peydar	79	Gachsaran
8	Khabbaz	44	Palangan	80	Douroud/Kharg
9	Bai Hassan	45	Lali	81	Gulkhari
10	Kirkuk	46	Karun	82	Kilurkarim
11	Chemchemal	47	Majnoun	83	Sulabedar
12	Tikrit	48	Jufeyr	84	Chillingar
13	Saddam	49	Susangerd	85	Nargesi
14	Hamrin	50	Ahwaz	86	Kuh-e-Mand
15	Jambur	51	Ramin	87	Pars
16	Kor Mor	52	Zeloi	88	Kuh-e-Kaki
17	Pulkhana	53	Ab Teymour	89	Bushgan
18	Chiah Surkh	54	Naft Safid	90	Kangan
19	Fallujah	55	Masjid-i-Suleiman	91	Nar
12	Balad	56	Mansuri	92	Dalan
21	Injana	57	Shadegan	93	Aghar
22	Gilabat	58	Marun	94	Sarvestan
23	Naft Khaneh/Naft-i-Shahr	59	Kupal	95	Lamard
24	Emmam Hassan	60	Haft Kel	96	Varavi
25	Tang-e-Bijar	61	Par-e-Siah	97	Bandubast
26	East Baghdad	62	Ramshir	98	"T"
27	Ahdab	63	Ramshir (gas)	99	Salakh
28	Delhuran	64	Agha Jari	100	Gavarzin
29	Samand	65	Khavizi	101	S. Gashu
30	Halush	66	Karanj	102	West Namak
31	Veyzenhar/Malah Kuh	67	Paris	103	Suru
32	Sarkan	67	Doudrou	104	Sarkhun
33	Dujailah	69	Kuh-i-Rig	105	HD
34	Nur	70	Shurom	106	Hanquan
35	Buzurgan	71	Bahrgansar	107	Henjam
36	Danan	72	Hendijan		

250 to 300 km (155 to 185 mi) in width (southwest limit of the dominant northwest-southeast structural grain as outlined by surface and subsurface structures and fields trending in that direction) (Figure 2).

OIL AND GAS FIELDS IN THE REGION

Liquid oil and solid and semisolid bitumen have been obtained from the northeastern part of the Arabian plate region since antiquity—particularly from Iraq and southwest Iran. It was, however, not until 1908 that the first major oil accumulation was discovered at Masjid-i-Suleiman (this field recently became depleted) in the Zagros fold belt of southwest Iran by the D'Arcy Exploration Co. (later Anglo-Persian, then Anglo-Iranian, today BP) in an Oligocene-Miocene Asmari carbonate reservoir. Commercial oil was not discovered outside Iran until 1927 at Kirkuk in northeastern Iraq, in the northwest prolongation of the Zagros fold belt, by the Turkish, later Iraq Petroleum Co. Kirkuk is still in production and is one of the supergiant fields of the world with more than 9.54×10^8 m^3 (6×10^9 bbl) produced by 1972, the year of nationalization by Iraq. Oil was subsequently discovered outside the Tertiary Zagros belt in Bahrain in 1932 at Awali, which is still producing about 6680 m^3/day (42,000 BOPD) from Cretaceous sandstone and subordinate carbonate reservoirs. Further discoveries were made in 1938 in Saudi Arabia (Dammam field, still producing from Upper Jurassic carbonate reservoirs) and in Kuwait (Burgan field, still in production from Cretaceous sandstone reservoirs and, as part of the Greater Burgan field, probably the second largest oil field in the world after Ghawar in Saudi Arabia), and in 1939–1940 in Qatar (Dukhan, still a major producer from Upper Jurassic carbonate reservoirs). The period following World War II saw an expansion in exploration, with new discoveries being

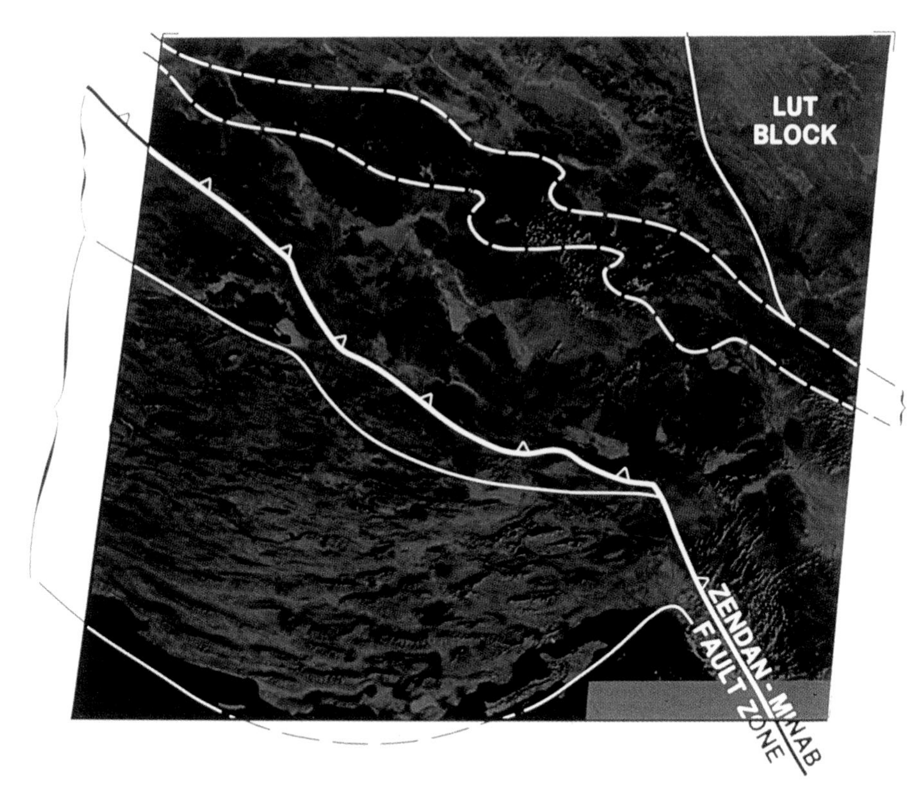

FIGURE 3. Landsat-1 mosaic covering the southeastern end of the Zagros fold belt and foreland basin and its termination at the Zendan-Minab fault zone. The striking pattern of the whaleback anticlinal mountains in the "simply folded belt" is broken by the circular dark debris masses capping the diapirs of Infracambrian/Early Cambrian Hormuz salt. Annotated from Short et al. (1976, NASA).

made in the Saudi-Kuwaiti Neutral Zone in 1953, in northeastern Syria in 1956, in the United Arab Emirates in 1958, and in northern Oman in 1964. For further details on early exploration in the region, see Beydoun (1988).

At the start of 1991, the reported recoverable oil and gas reserves of the whole Arabian plate region were given as 1.05×10^{11} m^3 (663.2×10^9 bbl) and 3.75×10^{13} m^3 (1325.4 tcf), respectively (Oil and Gas Journal, 31 Dec. 1990). Some 98.7% of the recoverable oil reserves and 97.2% of the recoverable gas reserves are located along the northeastern Arabian shelf, which extends from northwest Iraq through southwestern Iran, eastern Arabia, and the Gulf, to northern Oman. Cumulative oil production in this belt at the start of 1990 had totaled approximately 2.62×10^{10} m^3 (165×10^9 bbl), implying ultimate recovery of well over 1.27×10^{11} m^3 (800×10^9 bbl).

The recoverable oil and gas reserves of the Zagros foreland basin, as outlined above, are difficult to assess with any accuracy in light of a dearth of published figures for individual field production and remaining reserves over nearly the last two decades. However, based on the approximate volumetrics in Dunnington (1967), we estimate them at around 2.39×10^{10} m^3 (150×10^9 bbl) and some 1.7×10^{13} to 1.84×10^{13} m^3 (600 to 650 tcf), respectively, of which perhaps 50 to 60% or more are housed as substantially remigrated hydrocarbons in Oligocene-Miocene and older Tertiary Asmari and "Main Limestone" carbonate reservoirs. These reserves constitute some 23% of the total oil reserves and between 46 and 50% of the total gas reserves of the whole northeastern Arabian plate margin region. We also roughly estimate the total oil so far produced from the Zagros foreland basin fields (including that from pre-Tertiary reservoirs), using field production figures available into the middle 1970s and projected to the present, as 8.43×10^9 m^3 (53×10^9 bbl), implying an ultimate recovery from the Zagros foreland basin comfortably exceeding 3.18×10^{10} m^3 (200×10^9 bbl).

By comparison, at the beginning of 1990, the type basin discussed in this volume, the Western Canada foreland and cratonic platform basin, had estimated remaining reserves of 2.85×10^9 m^3 (17.9×10^9 bbl) of oil, 7.31×10^8 m^3 (4.6×10^9 bbl) of natural gas liquids, and some 3.76×10^{12} m^3 (132.9 tcf) of gas, of which the Western Canada foreland basin's share was 8.9×10^8 m^3 (5.6×10^9 bbl) of oil, 2.39×10^8 m^3 (1.5×10^9 bbl) of natural gas liquids, and 1.9×10^{12} m^3 (67 tcf) of gas (Porter, 1992).

To reemphasize the large size of the reserves in the Middle East, it is also of interest to make comparison with the figures for the whole of the continental United States, including Alaska, where the remaining recoverable oil reserves at the start of 1991 were estimated at 4.17×10^9 m^3 (26.2×10^9 bbl) of oil and 4.7×10^{12} m^3 (166 tcf) of gas (Oil and Gas Journal, 31 Dec. 1990). Cumulative oil production from the various continental U.S. oil fields at the end of 1989 totaled 1.24×10^{10} m^3 (78.1×10^9 bbl) (International Petroleum Encyclopedia, 1990).

STRUCTURAL AND SEDIMENTARY HISTORY OF THE NORTHEASTERN ARABIAN SHELF, AND OVERPRINTING OF THE ZAGROS FORELAND BASIN

Because the oils in the fields of the Zagros foreland basin have been shown by geochemical typing and correlation to be sourced entirely from sequences older than the initiation of the basin itself, the petroleum geologic history of the entire northeastern Arabian shelf from the shield to the Zagros suture must be summarized in order to understand the habitat of the Zagros foreland basin.

The history of this vast area falls into several recognizably distinct phases from consolidation of the Arabian basement in the late Proterozoic to the start of the Eocene convergence and collision that initiated the Zagros foreland basin and imprinted it onto the northeastern Arabian margin.

The Arabian Basement

Evolution of the Arabian basement and consolidation of the Arabian shield occurred by coalescence of several island-arc terranes that took place over a long time span commencing about 950 Ma (Gass, 1981). The neocraton formed by this Proterozoic accretion of arcs culminated at about 640 Ma. Each closure and arc collision resulted in deformation and in ophiolite obduction preserved as cryptic sutures segmenting the shield in approximately north-south units, which implies east-west compression. This gave rise to a pronounced north-south structural grain in the basement that often controlled hydrocarbon-related structural alignment in Phanerozoic sediments through periodic rejuvenation.

Much of the north-south grain was subsequently offset by the northwest-southeast-trending, latest Proterozoic-earliest Paleozoic strike-slip Najd fault system and a subordinate conjugate northeast-southwest fracture system. This was probably linked to both rift-related and collision-related events spanning the period 640 to 550 Ma, which changed the sense, initially dextral, of horizontal motion on the Najd system to sinistral, and fractured the now thick shield in several great shears. These trends are evident in other parts of the Phanerozoic-covered Arabian plate where they have been subjected to later rejuvenation in consequence of subsequent plate-margin evolution; indeed, the northwest-southeast "Zagros trend" may have resulted from a late Paleozoic plate-margin development on a "Najd trend" fracture. For further details, the reader is referred to several fine overall summary accounts dealing with this early phase of crustal evolution, such as Stoesser and Camp (1985), Brown et al. (1989), Stacey and Agar (1985) and Husseini (1988, 1989).

Infracambrian-Carboniferous

Events associated with the Najd faulting and conjugate trends had far-reaching repercussions both in the shield area and in the recently stabilized bordering craton, being responsible for the creation of several rift-related basins and crustal sags in the Arabian region bordering the shield. Greenwood et al. (1980) suggested that the Infracambrian-Lower Cambrian oil-producing salt basins of Oman are examples of such structures. Husseini (1988, 1989) extended this structuring to cover many parts of the region. These basins represent the first intraplate structuration of this newly stable shelf in this part of Gondwana. Sediments spanning the Precambrian-Phanerozoic boundary are known from outcrops in the shield, in Oman, and in parts of central Iran and southeastern Turkey, and have been reported in the subsurface where sporadic deep exploration boreholes have penetrated to this level. In the Oman outcrops and intensively drilled South Oman salt basin, they include thick clastic-carbonate cyclical sediments (the Huqf Group), with several levels of extensive organic-rich source rocks and a terminal evaporitic unit (Ara, Figure 4). These evaporites manifest themselves as piercement domes throughout the Gulf region and much of southwestern Iran (Hormuz, Figure 4), where they have also provided the detachment level for the late Neogene compressional folding (Kent, 1979; Stocklin, 1986). The subsurface extents of these clastic-carbonate-evaporitic deposits presently can only be conjectural. Gorin et al. (1982) extended them under much of Arabia and the then adjacent portions of India and central Iran where the Paleo-Tethyan border of Gondwana was located.

From the Infracambrian until the late Paleozoic, the present Arabian plate, together with the Indian, Afghan, central and northwestern Iran, and Turkish plates, collectively formed part of the long and very wide and stable northern passive margin of Gondwana bordering the Paleo-Tethys ocean. Much of the region was covered by shallow epeiric seas bordering lowlands whose areal extent altered in response to succeeding transgressions and regressions and which periodically gave rise to extensive intrashelf basins. The width of this epeiric shelf was at least 2000 km (1245 mi) and may have reached 3000 km (1865 mi), and its setting was principally in temperate latitudes of the southern hemisphere with occasional incursions into tropical regions.

The Cambrian-Ordovician sequence of the northeastern margin shelf, the region of immediate concern to the much later formed Zagros foreland basin, was principally clastic (temperate belt), although there were short episodes of warmer-water carbonate deposition during the Middle to Late Cambrian. In the Late Ordovician to earliest Silurian, central Arabia was at the northern periphery of the major Saharan glaciation (McClure, 1978; Vaslet, 1989; Combaz, 1986). Deglaciation led to major sea level rise and transgression with anoxia in intraplatformal basins and widespread deposition of Silurian organic-rich shales over the greater part of the Arabian region; these were subsequently partly removed, in an uncertainly defined pattern, by later uplift and erosion, but are represented by the Qusaiba, Safiq, and Gahkum shales of Saudi Arabia, Oman, and southwestern Iran, respectively (Figure 4).

The general absence of Devonian deposits except in northwest Arabia, parts of northeastern Iraq (Jauf and Kaista/Ora, with carbonate deposition heralding a short

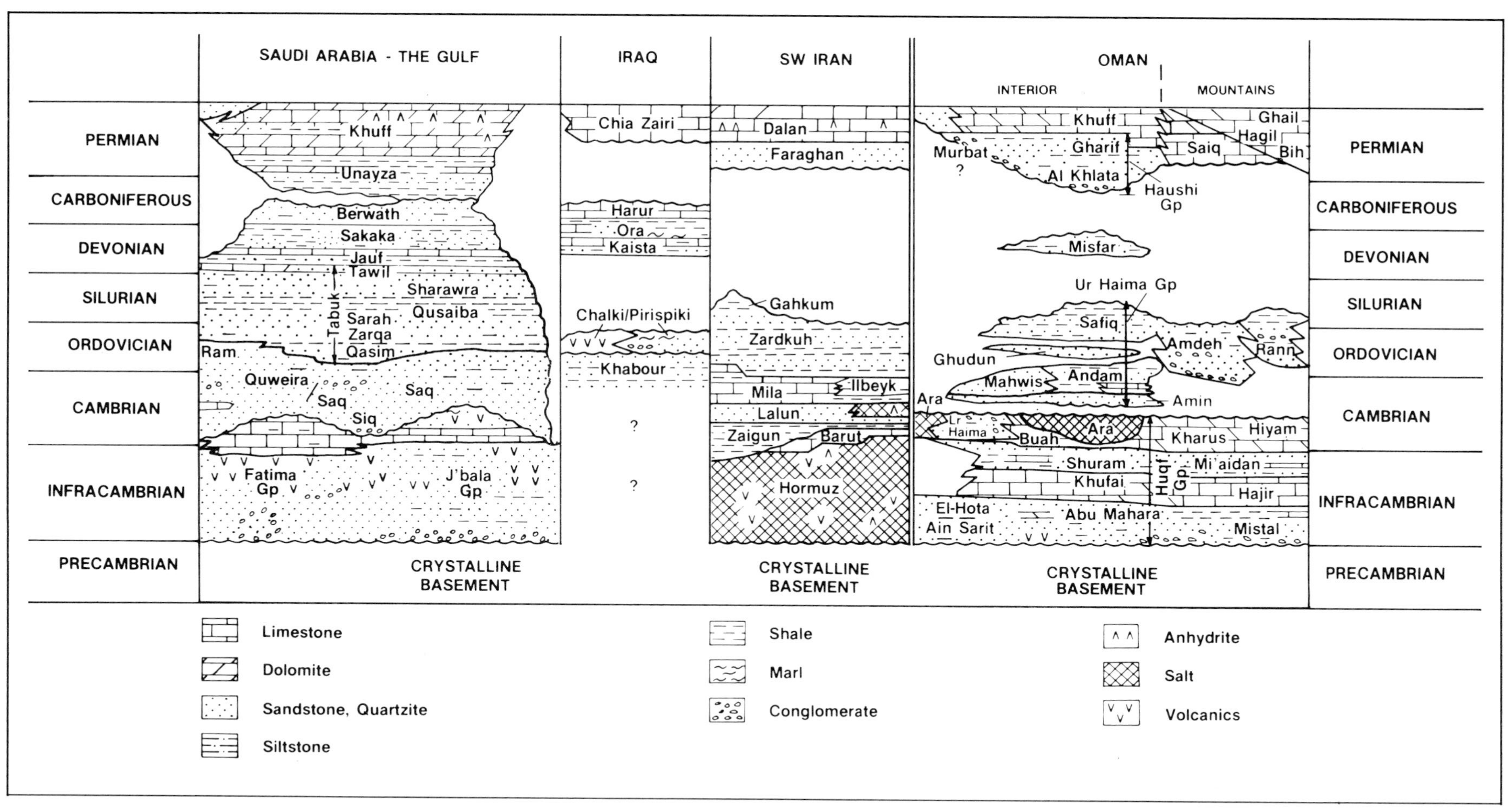

FIGURE 4. Northeastern Arabian shelf: Infracambrian–Paleozoic formations correlation chart. Modified from Beydoun (1991).

return to warmer latitudes), and the Isfahan basin of Iran (Cherven, 1986), together with the reduction of Silurian sequences or their removal from much of southwestern Iran and from parts of the Ha'il-Ga'ara arch, may perhaps reflect mid-Paleozoic epeirogenesis associated with "Caledonian" orogenic pulses elsewhere. Carboniferous exposure over much of the northeastern margin (while southern Arabia was undergoing continental, possibly montane, glaciation of the Late Carboniferous–Early Permian Gondwanan episode) reflects regional emergence, nondeposition, or erosion, perhaps together with the first updoming along the upper Paleozoic rift belts that began to break up Gondwana, in the late Paleozoic.

Permian–Triassic

At the time of the Late Carboniferous–Early Permian glacial episode in the southern Arabian sector, it has been argued that southward-directed subduction of the Paleo-Tethys began under the leading edge of northern Gondwana in central Iran. This would have transferred the Gondwanan plate edge from a passive to an active margin, with the Sanandaj-Sirjan block of Iran being identified by Sengör (1990) as an ensialic magmatic arc. Evidence of "Hercynian"-related metamorphism from the leading edge has been reported from the Oman Mountains and central Iran (Lovelock et al., 1981; Berberian and King, 1981).

A widespread marine transgression in the Late Permian resulted from several factors. Gondwanan deglaciation was accompanied by a rapid plate movement to warmer latitudes. Evidence from the Oman allochthon (Hawasina nappes) indicates crustal stretching, rifting, and basaltic volcanicity, presumably associated with crustal thinning that eventually led to Triassic separation of the Iranian blocks (central Iran, Sanandaj-Sirjan, and northwest Iran) from the northeastern Arabian margin along the Zagros line. This engendered the opening by full spreading of the Neo-Tethys, whose initial opening may be considered as a back-arc basin (Sengör, 1990). During the same period, the southeastern and northwestern Arabian margins were formed by separation of formerly adjacent parts of the Afghan/India block and the Turkish blocks, probably along transforms.

These combined events led to widespread deposition of Upper Permian carbonates (the Khuff and equivalent Dalan carbonates) over the newly forming northeastern Arabian passive-margin shelf, with deeper Permian facies in the developing marginal rifts, now seen in the Hawasina units in the Late Cretaceous allochthons. Away from the plate edge, anhydrites are interbedded, often cyclically, with the shallow-water carbonates constituting good reservoir-seal pairs; these facies continue into the Lower Triassic (Figure 4).

Triassic–Late Cretaceous

The northeastern Arabian shelf remained amazingly stable throughout much of this interval with only slow but steady subsidence that allowed widespread sheetlike deposition of shallow marine sediments on an extraordinarily wide epeiric shelf more than 2000 km (1245 mi) long (northwest to southeast) and almost 2000 km wide from shelf break to shield. Why this period of stable gentle subsidence was so long (over 150 m.y.) is one of the greatest unanswered questions in the geology of the region. The vast areal width of the shelf (for which no modern analog exists) engendered remarkably broad uniformity of deposition but, nevertheless, with slight differential subsidence and sedimentation, allowing the development of large, but in general shallow, intrashelf basins. The scale of the shelf prevented efficient water agitation and flushing from the ocean so that even the shallow basins repeatedly became stratified and anoxic except for their well-lit, organically productive surface waters. The anoxia in the basins gave rise to large areas of potential source rocks at several levels from the Triassic to the early Tertiary (Murris, 1980; Stoneley, 1987, 1990; see Figures 5 and 10), and the distribution allowed optimum juxtaposition of source, reservoir, and seal facies over wide areas, permitting very efficient horizontal migration from areally extensive source kitchens. With the onset of maturity, the areal setting allowed large hydrocarbon kitchens to drain into extraordinarily large but typically gentle structural closures, ensuring minimal loss of seal efficiency and large trap volumes (e.g., Grunau, 1977). Throughout this interval, differential sedimentation/compaction and drape over generally north–south-trending basement features, or slow salt pillowing, provided the principal mechanisms for formation of these potential traps (Sugden, 1962). Facies were predominantly carbonate with lesser intervals of fine clastics (marls, shales, argillaceous carbonates) and periodic evaporites. Continental clastics from the Arabian shield tongued out through deltas into the inner shelf areas during the Early-middle Cretaceous, interfingering to the east with marine clastics and carbonates (Figure 5); they provide the main reservoirs for the very large oil fields of Kuwait and southern Iraq.

Murris (1980, 1981) developed a persuasive model of source–reservoir–seal rock association over this remarkably wide and almost flat shelf. During transgressive periods, differentiated carbonate deposition gave rise to stratified sediment-starved basins in which potential source rocks accumulated, separated by higher-energy margins from carbonate-evaporite shelves: marginal mounds, oolite bars and shoals, rudistid banks, and regressive carbonate sands form the main reservoirs, and supratidal evaporites and thin shales form the seals.

Tectonic deformation of this very long and wide margin began in association with the onset of events leading to ophiolite obduction around the Arabian shelf commencing in the late Turonian; these formed the peri-Arabian ophiolitic crescent extending from the northwest Arabian margin to the southeast margin in Oman (Ricou, 1971).

Late Cretaceous Obduction

Oceanic sedimentary prisms and ophiolite masses were overthrust or glided onto the edge of the Arabian shelf

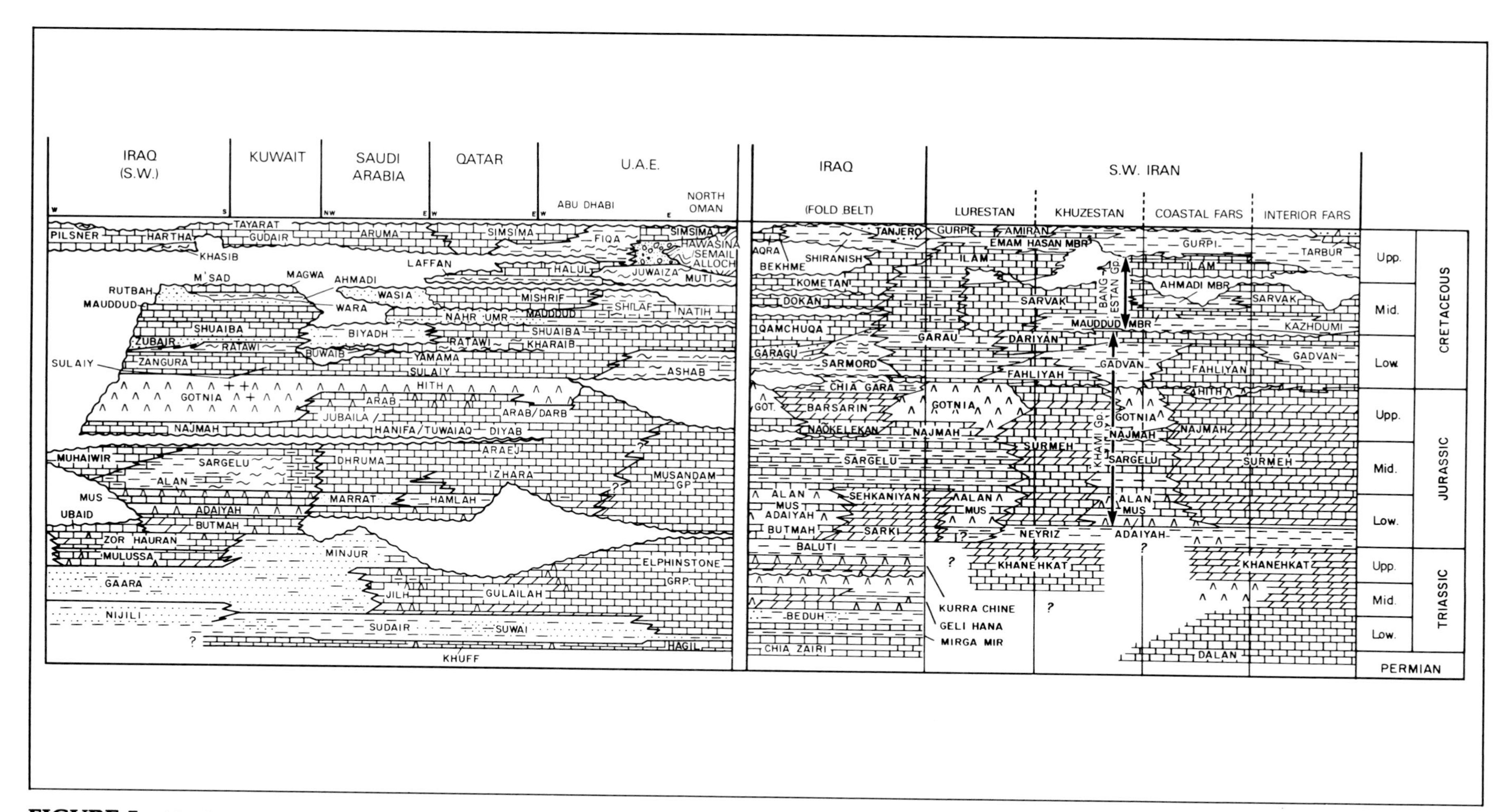

FIGURE 5. Northeastern Arabian shelf: Mesozoic formations correlation chart. Modified from Beydoun (1991). For legend, see Figure 4.

from northwestern Syria to southeastern Oman in late Turonian to about end Campanian-early Maastrichtian time. The nappes are best seen in the Oman Mountains (Glennie et al., 1973), where they extend up to 100 km (62 mi) onto the shelf from the shelf break, and a downwarped foredeep to the nappes further extends about 20 km (12 mi). In the northeastern margin shelf, intrashelf structural elements became enhanced and ancient north-south structural trends were rejuvenated right across this northeastern region (Koop and Stoneley, 1982). A belt of linear "flysch" foredeeps developed to the southwest (and to the west in Oman) of the obduction zone, and these foredeeps were filled with detritus derived from the oceanic rocks (Tanjero, Kolosh, Gercus, Amiran, Kashkan, and Juweiza formations; Figures 5 and 6). Farther to the southwest, adjacent to the Arabian carbonate platform, pelagic deeper-water conditions prevailed, extending across the shelf beyond the limits of the later Zagros foreland basin, partly reflecting Late Cretaceous sea level rise. There is little evidence from either Oman or the northeastern shelf margin, or from northwest Syria and Hatay, that the nappes formed any substantial high relief, because later shallow marine Maastrichtian-Paleogene sediments show only local detritus and commonly lie across the nappes; structures directly associated with the thrusts are generally limited to 20 to 30 km (12 to 19 mi) from the nappe front.

The extent and type of plate margin interaction and the type and degree of collision (if any) that gave rise to this Late Cretaceous deformation and obduction/thrusting are subject to debate; the various proposed models were recently reviewed by Beydoun (1988, 1991). Whatever the details of the different aspects of the various proposed models, the events were short-lived, no full-scale orogeny occurred, and a return to stable shelf conditions soon followed. Although these events were relatively short-lived, their consequences for the hydrocarbon habitat of the northeastern margin prior to the evolution of the Zagros foreland basin were important and included the following. (a) The stable shelf after long quiescence was disturbed. (b) The deeper sediments of the outer margin were "overcooked" as a consequence of the obduction load. (c) Older structures on the inner shelf were reactivated or rejuvenated—e.g., elongate north-south growth structures on basement highs and domal features over Hormuz (Infracambrian/Cambrian) salt, many of which show Late Cretaceous accentuation (Sugden, 1962); in the southeastern part of the United Arab Emirates, the Lekhwair high and other more linear folds on a unique northeast-southwest trend formed at the end of the Cretaceous and exhibit truncation of middle and Upper Cretaceous sequences under the Paleogene (Loutfi and El Bishlawy, 1986). (d) Cycles of deeper-water facies including shales, many being bentonitic (?island-arc tuff-derived), were spread across the shelf and form good seals. (e) A wide regression occurred after ophiolite emplacement as the downwarped plate margin became uplifted. This may have led to local or regional activation or reversal of hydrodynamic gradients.

In the Iraq-Iran sector, the overburden from nappe and flysch load and deformation were apparently sufficient to mature, and/or allow the escape of, oil to the surface to be recorded as bitumen pebbles in Upper Cretaceous-Paleocene clastics (Dunnington, 1958; Kent et al., 1951). In eastern areas of the United Arab Emirates, where the north-south Oman foredeep forms the western front to the northern Oman Mountains, western-verging structures from the North Oman thrusts form traps in foredeep fill of the Upper Cretaceous Juwaiza (Blinten and Wahid, 1983; Patton and O'Connor, 1986). The Upper Cretaceous and Paleogene fill of this foredeep (running into several thousand meters in the axial area) must have added maturity to Mesozoic sources and brought forward the timing of hydrocarbon expulsion (Burruss et al., 1985).

Maastrichtian to Middle Eocene

A return to stable shelf conditions commenced soon after the start of the Maastrichtian, with marine sediments being deposited over most of the shelf and adjacent areas of the northeastern margin, indicating that little or no emergent relief remained after the Late Cretaceous events. Sequences across the northeastern shelf pass from platform carbonates (and evaporites in the Eocene) to pelagic marls, which include organic-rich sediments formed in outer intrashelf and shelf-edge stratified basins; some were created by differential sedimentation, as in the Mesozoic (Pabdeh, Germav, Aaliji/Jaddala, and others; Figure 6). Along the Iran-Iraq outer shelf-margin, earlier Paleogene flysch deposits are topped by shallow shelf carbonates with local evaporites (e.g., Sachun, Kashkan, Gercus; Figure 6), probably indicating a remnant post-obduction submerged shelf-margin ridge. Deep Infracambrian/Cambrian Hormuz salt movements continued to influence sedimentation throughout this period (Koop and Stoneley, 1982).

Late Eocene to Present: Formation of the Zagros Orogene and Its Foreland Basin

The onset of continent-to-continent collision, and the additional stresses from the two-stage opening of the Red Sea and Gulf of Aden, initiated the Zagros orogeny and creation of the foreland basin.

The Zagros orogene is one segment of the Alpine mountain ranges that resulted from the closure of the Neo-Tethys. The history of ocean closure, and the complex movement of the many crustal plates involved in the Mediterranean-to-India sector, have been the subjects of numerous studies over the past decade (notably Sengör, 1984; Dercourt et al., 1986), which set the larger scene for the more specific setting of the Zagros orogene. The step-by-step development of the continental collision between northeastern Arabia and the opposed segment of Eurasia is most succinctly outlined by Hempton (1987), whose analysis is summarized here.

Prior to the Eocene, the Arabian continental block constituted a northeasterly projecting rectangular spur of the African continent (Figure 1 inset, the "Arabian Promontory"; Beydoun, 1991). With consumption of ocean crust in the adjacent part of the Neo-Tethys, mainly by

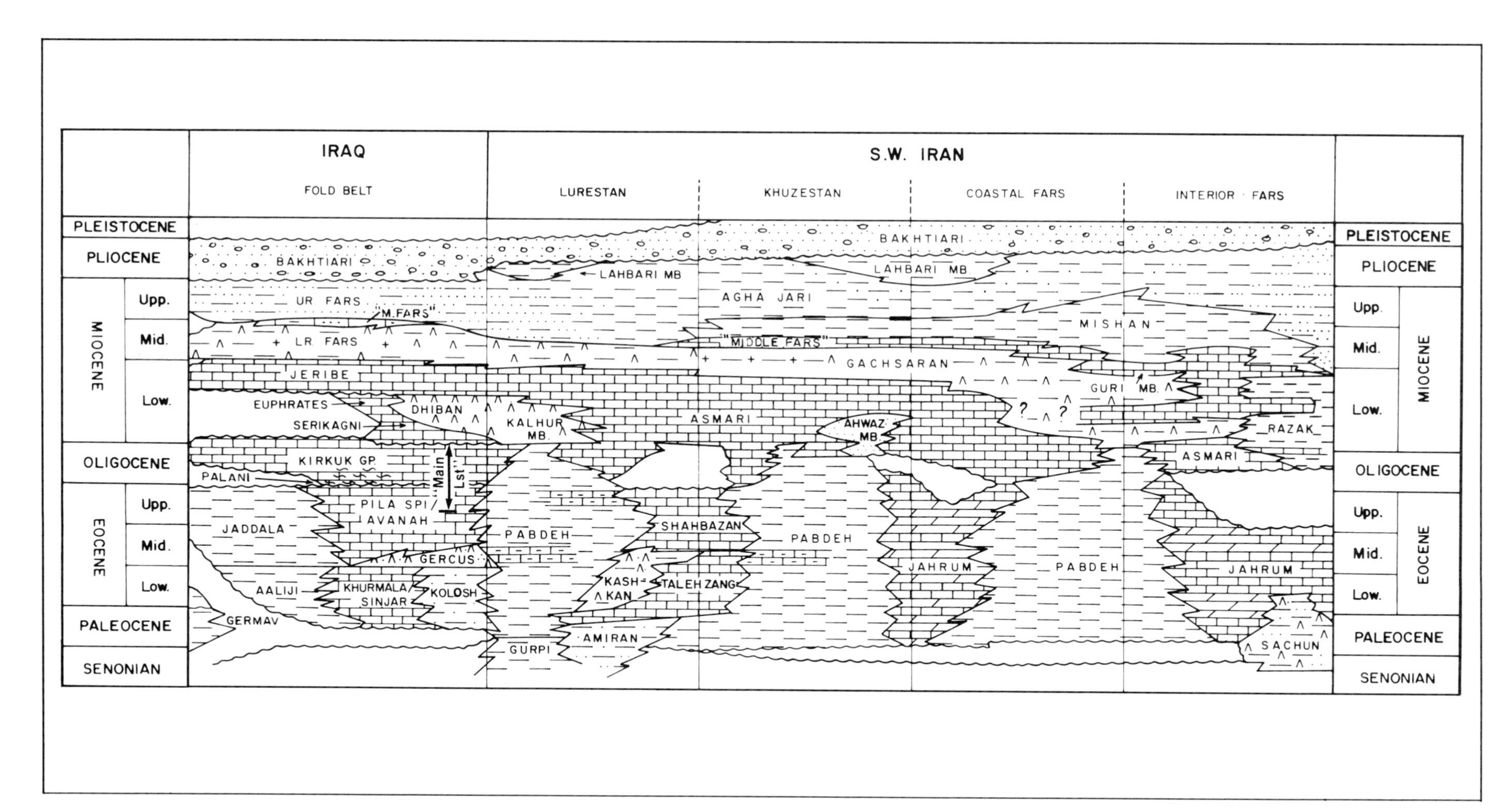

FIGURE 6. Zagros foreland basin: Cenozoic formations correlation chart. Slightly modified from Beydoun (1991). For legend, see Figure 4.

subduction beneath the active southern margin of the opposite sector of continental Eurasia (the pertinent opposite sector being the Sanandaj-Sirjan belt, together with several Turkish fragments; Berberian and King, 1981; Sengör and Yilmaz, 1981), the Arabian Promontory converged into the subduction zone and into continent-to-continent collision. The first continental suturing occurred at the northwest corner of the promontory (with the Turkish block) and spread along the northeast margin in the mid- to late Eocene. Hempton (1987) infers that the frontal edge of the Arabian plate was a stretched and thinned continental crust (Figure 7), developed in the early stages of passive rifting (Triassic); this is also envisaged from seismic and structural reasoning by Snyder and Barazangi (1986) and Ni and Barazangi (1986). This condition apparently gave rise to the extremely broad Mesozoic continental shelf and margin of Arabia, as also envisaged by Murris (1980, 1981). By continuing convergence after suturing, the thinned crust in this margin became restacked and thickened by movement on the original passive-margin extensional faults, which were reactivated in reverse in the manner described by Jackson (1980) and Jackson et al. (1981). By the middle Miocene, the thickening had produced rapid surface uplift and shedding of clastics. Analysis of the predominantly shallow-depth Zagros earthquakes (approx. 10 km, or 6 mi) supports the interpretation that convergence is now accommodated by crustal shortening and thickening rather than by subduction of continental crust (Ni and Barazangi, 1986). A single intermediate-depth earthquake in central Iran may evidence the last trace of a subducted slab of oceanic Arabian plate (Kadinski-Cade and Barazangi, 1982). Hempton (1987) also linked the continuance of continental convergence in the Zagros with the opening phases (most probably two-stage phases) of the Red Sea and Gulf of Aden rifts, documenting the maintenance of stresses in the Zagros suture and the translation of convergent movement into dextral strike slip.

An oceanic melange with Maastrichtian and Paleocene limestone exotics in a Miocene matrix evidences the persistence of a deep marine seaway along the Zagros suture until at least the early Miocene (Stoneley, 1981).

The continental collision that gave definition to a separate foreland basin on the northeast Arabian plate initially produced a downwarping of the outer shelf and uplift of the inner shelf, which restricted the depositional area of upper Eocene and lower Oligocene sediments. This warping was accentuated in the Oligocene so as to maintain the foreland basin as a depositional area throughout the global sea level fall of the mid-Oligocene (Vail et al., 1977), and the depositional basin is essentially outlined by the distribution of the Oligocene-lower Miocene marine sediments (figure 11 in Koop and Stoneley, 1982). The basin at this time was a fairly open and continuous shelf basin, but from mid-Miocene times, progressive compressive folding of the basin substrata began to break it up into many, essentially "synclinal," subbasins. The upper Eocene to lower Miocene deposition (Figure 6) was typical tropical-shelf sedimentation dominated by carbonates (Henson, 1950b; Dunnington, 1958; James and Wynd, 1965). The shallower parts of the shelf—the inner shelf to the southwest and, on the northeast, the part over the shelf-margin ridge remnant from the Late Cretaceous

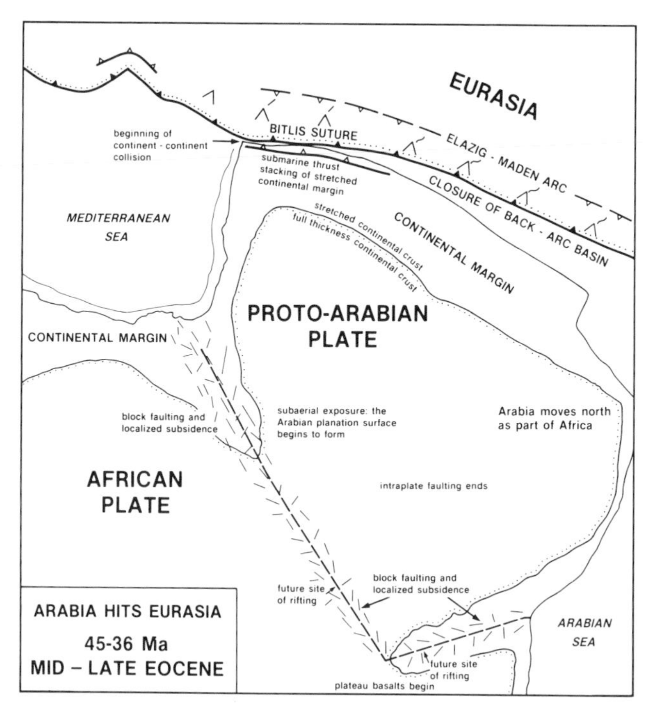

FIGURE 7. Tectonic reconstruction for the mid-late Eocene, illustrating the initial suturing between Arabia and Eurasia. From Hempton (1987). Copyright American Geophysical Union, reproduced with permission.

obduction event—developed shallow marine limestones and dolomites (Asmari and Kirkuk Main Limestones). Similar carbonate/evaporite units were deposited over the juxtaposed Eurasian continental blocks of central Iran (Qom Formation). The intervening deeper areas of the foreland basin accumulated pelagic marls (Upper Pabdeh; Palani to Serikagni in Iraq). Shallow carbonate deposition enabled closed intrashelf lows to develop in which evaporites were deposited (Kalhur, Dhiban) and one significant pulse of quartzose clastics from the craton occurred in the early Miocene (Ghar sands of Kuwait and Ahwaz sands of southwestern Iran and adjacent Iraq).

The mid-Oligocene global sea level drop, and subsequent sedimentation, constricted marine circulation from the Mediterranean to the Indian Ocean, and from the early Miocene this began to be reflected in sediment type, fossil diversity, and other factors (Ricou et al., 1986). By mid-Miocene, as the shelf seas began to be further broken up by the rising folds into a maze of subbasins, the poor circulation allowed intense evaporite production in the growing synclinal lows (Gach Saran; Lower Fars of Iraq), with thick salt from the Dezful embayment (Figure 2) to northern Iraq. In the West Musandam basin, cojoining the southeast end of the Zagros foreland basin (Figure 1), a separate thick salt-bearing evaporite sequence was developed at this time (Qeshm Formation of Kashfi, 1980). The first signs of clastic deposits from the uplifting suture zone also began to appear in the southeast (Razak), and limestones indicate the passage into more marine environments (reefal Guri) closer to the Indian Ocean in the middle Miocene.

The widespread deposition of evaporites was halted by a marine influx from the southeast, depositing marine sediments (Mishan, Upper Guri; Middle Fars of Iraq) that thin and die out to the northwest; this transgression was probably a response to a global rise in sea level (Vail et al., 1977), but the presently available dating (middle Miocene) is not sufficiently accurate to confirm this.

Subsequent deposition in the Zagros foreland basin is dominated by the southwesterly spread of clastics from the rising orogene to the northeast. The clastics are at first fine-grained and contain marginal marine fossils (Agha Jari; Upper Fars of Iraq) but grade up into a continental molasse of coarse sands, gravels, and conglomerates (Bakhtiari) that is laterally extremely variable. Both the Agha Jari and Bakhtiari have local depocenters related to the contemporaneously rising structures.

The extent of the Zagros foreland basin can also be assessed on the basis of structuration. Falcon (1967, 1969, 1974), followed by Colman-Sadd (1978), divided the Zagros into three structural zones (Figure 2). On the northeast the narrow "Thrust belt" defines the collision suture proper with intense fault shattering. Southwestward lies an "Imbricated zone" in which recognizable stratal sequences are juxtaposed by reverse faults and thrusts. The widest zone is the "Simply folded belt" and is made up of huge anticlines and synclines, reflected in the topography as anticlinal mountains and synclinal valleys as a demonstration of the youth of the folding. In the folded belt, the amplitude of the folds decreases only a little to the southwest, but the folds change in general from more southwest-verging, closely spaced asymmetric folds to more open, symmetric concentric folds (Figure 8). Most authors draw the final fold of this zone in the Gulf close to the Iranian shore, but Kassler (1973) argues that the bathymetric ridge of the Great Pearl Bank, extending from Abu Dhabi on through Qatar to join the Bahrain ridge closing the Gulf of Salwa, is the most distal Zagros "fold" and was sufficiently active to affect sedimentation in the Quaternary.

That the Zagros tectonic activity has been continuous from historical times to the present is widely evidenced. Lees and Falcon (1952) documented the growth of one anticline in the Dezful area that had uplifted the course of a Sassanian canal by approximately 20 m (66 ft) in 1700 years. Furthermore, many of the buried structures under the Mesopotamian plains are neotectonically active and cause Pleistocene–Holocene divergences in wadi drainage patterns of the Kirkuk plains of Iraq as well as in the course of the river Tigris (e.g., Balad, Tekrit, and other recently discovered fields: Al Sakini, 1975, and personal communication, 1986). The pattern of present seismicity in the Zagros shows the continuing movement of both faults and folds (Jackson and McKenzie, 1984).

The style of deformation is controlled by the reaction of the stratal sequence to convergence. The critical feature is the presence of upper Proterozoic-Cambrian "Hormuz" salt closely above crystalline basement. All of the Paleozoic to mid-Tertiary strata above the Hormuz salt act together as a "Competent Group" (Colman-Sadd, 1978) but are freed from conforming to the basement deformation by décollement in the salt. In the Imbricated zone, only a minimum degree of shortening—about 20%—can be estimated (Falcon, 1974), and there must be an unknown loss in underthrust material beneath the uplifted crystalline basement of the Sanandaj-Sirjan massif. For the folded belt, Colman-Sadd (1978) has demonstrated that the folds are essentially "parallel" (or "concentric") and hence show a contrast between the stretched outer strata and the pinched inner strata in each fold. Because the level of erosion in the young Zagros folded belt exposes only the top of the Competent Group, Colman-Sadd (1978) validly argued that this can explain why the surface impression is of broad anticlines and narrow, pinched synclines. He exemplified this in an illustration (his figure 10) showing the development of an axial reverse fault in the syncline as part of the volumetric requirement of deformation inside the curvature of the neutral surface of the synclinal fold. However, unpublished oil industry seismic surveys in the inner folded belt show evidence to suggest that such reversed faults can flatten parallel to the northeasterly flank of the southwesterly bounding anticline (Figure 9) and run beneath the northeasterly whaleback, implying that the narrow synclines may have resulted, at least partly, from thrust packing of wider-spaced anticlines. The shortening in the 250-km (155-mi) wide folded belt has been thought to be relatively small—Falcon (1974) gives a fair minimum average of 30 km (18.6 mi) for the Imbricated zone and the Simply folded belt—if the Competent Group is continuous, but thrust packing could markedly increase this. The lost 300 to 500 km (185 to 310 mi) of outer shelf implied in the simple plate reconstructions by Hempton (1987) may not be unreasonable when the possible shortening in the three belts is combined.

The concentric folding of the 6-km (3.7-mi) thick "Competent Group" requires considerable additional fault adjustment of the beds in the folds below the center of curvature. These faults preferentially occur in the shale and evaporite horizons. The folded belt has marked changes in level, the most apparent being that termed the "mountain front," which marks the most southwesterly outcrops of pre-Fars rock units. The mountain front is a major geoflexure on which the normal Zagros folds are superimposed and whose amplitude has been estimated to be about 12 km (40,000 ft) (Falcon, 1958, 1974). Ameen (1991) refers to this change in levels as a megascale geowarp that separates a region with a greater amount of crustal shortening of some 10 to 33% in the high mountain zone, as compared to the downwarped foothills zone, where 3 to 17% has been calculated, implying the involvement of a change in the basement structure to permit this greater shortening and thickening.

The mountain front is not linear, but swings to form two major northeasterly reentrants or embayments in which the main Iraqi and Iranian fields occur (the Kirkuk and Dezful embayments; Figure 2).

THE GENERATION HISTORY OF THE NORTHEASTERN ARABIAN SHELF HYDROCARBONS

Source Rocks

At least a dozen good source rock formations of significant thickness and/or of wide extent (and, in most cases,

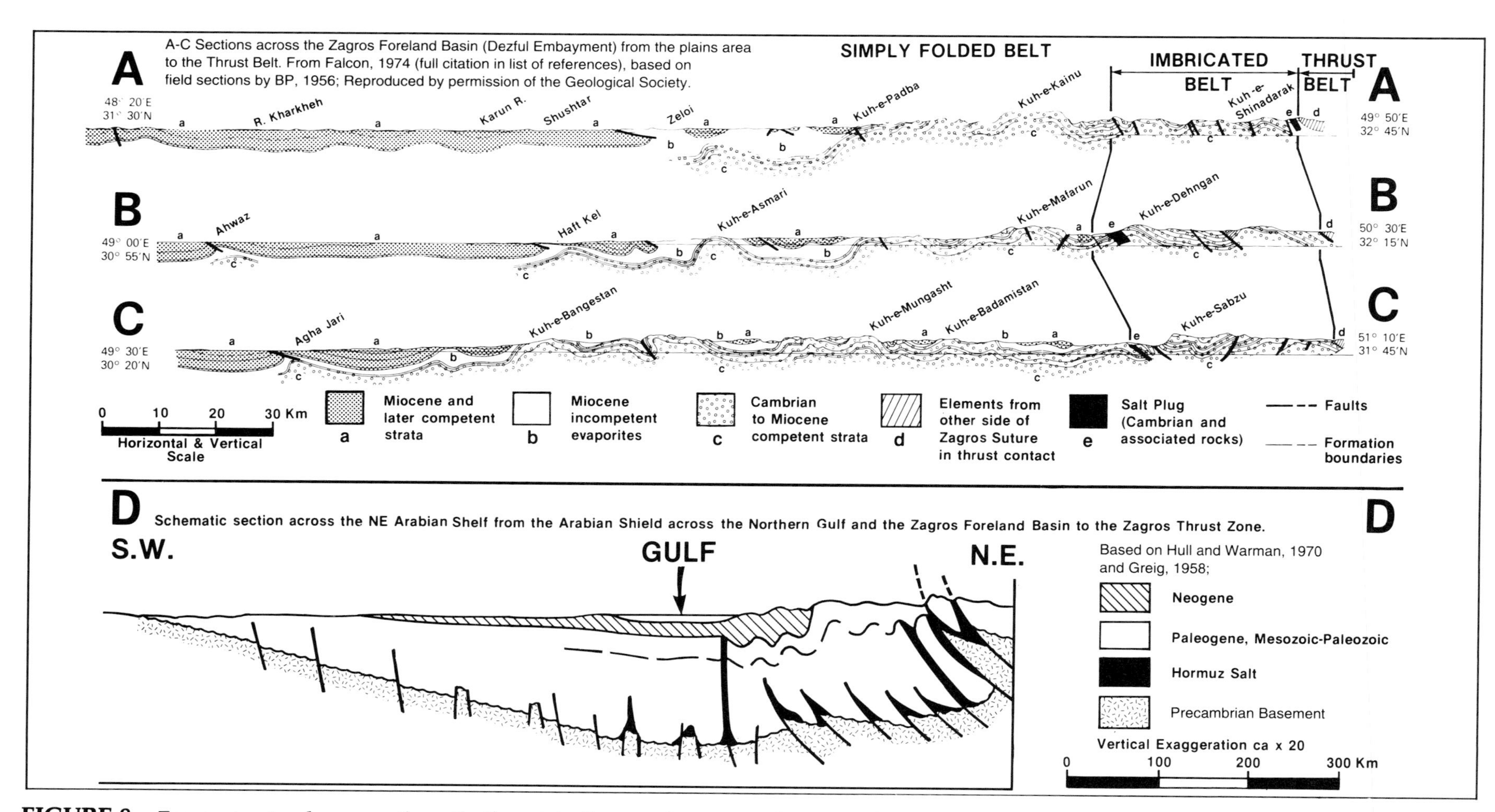

FIGURE 8. Zagros structural cross sections. Sections A to C are sections across the Zagros foreland basin (Dezful embayment) from the plains area to the thrust belt. Section D is a schematic section across the northeastern Arabian shelf from the Arabian shield across the northern Gulf and the Zagros foreland basin to the Zagros thrust zone. Sections A to C from Falcon (1974), based on field sections by BP (1956). Reproduced by permission of The Geological Society (London). Section D is based on Hull and Warman (1970) and Greig (1958).

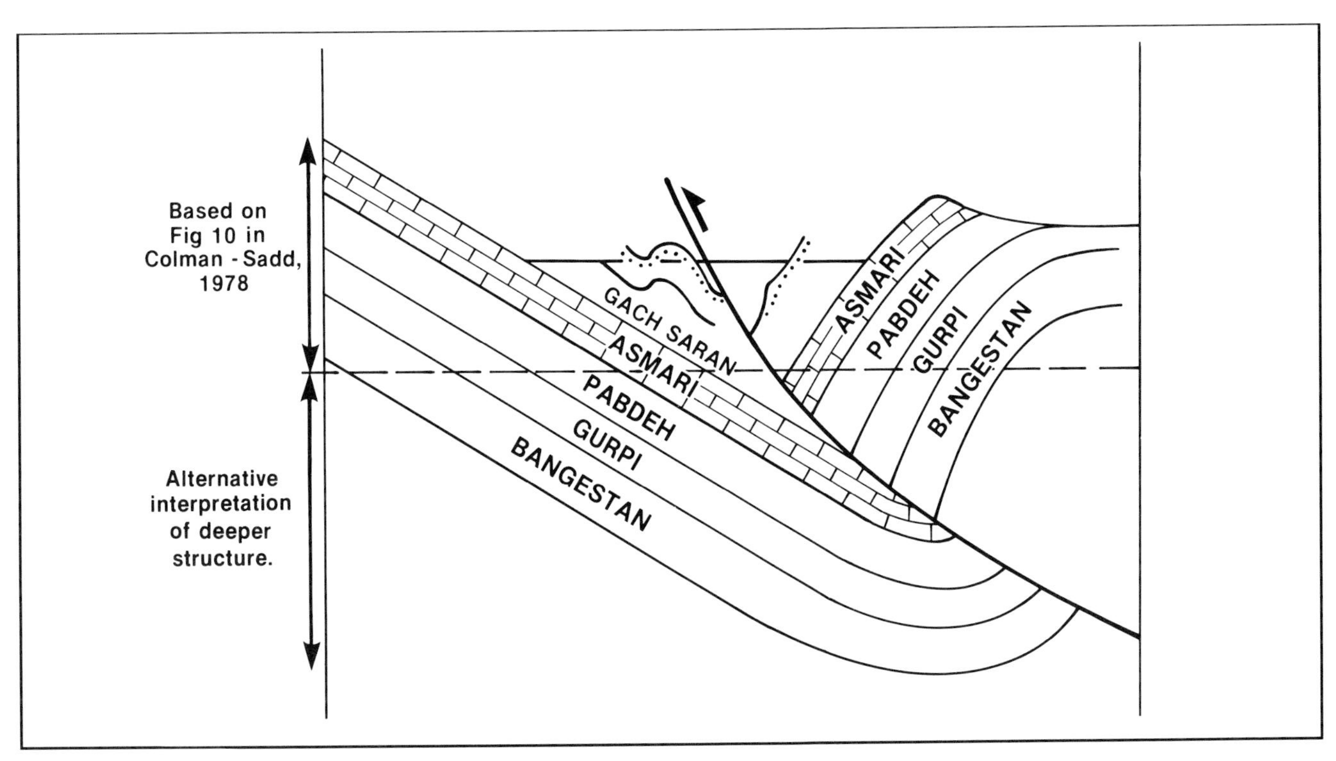

FIGURE 9. Possible thrust packing involved in narrow synclines: Zagros fold belt. This structural style implies greater shortening than the continuous folds.

oils proven by organic geochemical matching to have been derived from them) are present within the northeastern Arabian shelf (Figure 10). The source rocks occupy separate levels and geographical settings from upper Proterozoic to Tertiary. They are all marine sapropels with Type II kerogen, and hence the oils are also similar in many features. The lack of input of significant amounts of terrestrial organic material (none available pre-Devonian, and land areas since then seemingly too distant or arid) not only has reduced possible differences in oil composition but also has provided only limited levels with vitrinite in the sequence to assist maturity estimates; the Albian and Eocene are important exceptions.

Excellent thick (hundreds of meters in south Oman) source rocks in the Proterozoic Huqf Group (Hughes Clarke, 1988) have been documented by geochemical typing as having sourced over 80% of the oil reserves in Oman's oil fields in reservoirs from Proterozoic to middle Cretaceous ages (Grantham et al., 1988; Sykes and Abu Risheh, 1989; see also Al Marjeby and Nash, 1986; Mattes and Conway Morris, 1990). The overburden history within Oman has varied such that some Huqf oils began to be generated within the early Paleozoic, whereas elsewhere this 600 Ma source rock is still immature. The extent to which source rocks of this age were deposited over the whole northeastern Arabian shelf area is unproven and rather speculative. However, their association with the stratified basins producing the major Huqf/Hormuz evaporite deposits is genetically linked (Sonnenfeld, 1985; Momenzadeh, 1990); there is every possibility that Huqf-type source rocks occur in all areas where Huqf/Hormuz salt diapirs are present (Figures 1 and 2). This is probably also the case where diapirs are absent, but seismic evidence indicates anomalous stratal disturbance in sedimentary basins at or above this stratigraphic level, suggesting the former presence in depth of (mobile or dissolved) evaporites. Although documentation is poor, such basins may well be widespread and much more extensive than is currently thought (Beydoun, 1989, 1991).

Also in Oman, geochemical typing matches oil in two fields (Sahmah, Hasirah) with a Lower Silurian source rock level (Safiq Formation; Grantham et al., 1988), and large recent discoveries in Saudi Arabia are also reported to be geochemically matched to the correlative of this basal Silurian level (Qusaiba Formation; McGillavray and Husseini, 1992). This source rock is found at outcrop in southern Iran (Gakhum Formation), and the Silurian anoxic event is related to global transgression after Late Ordovician glaciation (Combaz, 1986; Bordenave and Burwood, 1990).

Over much of the northeastern parts of the northeastern Arabian shelf, any Huqf-level source rocks must now be very deeply buried and organically metamorphosed, but the evidence from Oman indicates that they could have generated ample hydrocarbons into the system at earlier times through the Paleozoic and Mesozoic. If any of these liquids survived above the oil floor, with possible remigrations through intermediate traps, their very characteristic chemical signature (carbon-isotopically light, sterane imbalance, etc.; Grantham et al., 1988) should make them identifiable. None has yet been recorded outside Oman, but some oil accumulations have been recorded in Saudi Arabia in basal Paleozoic levels (Saq Formation; McGillavray and Husseini, 1992), which, without

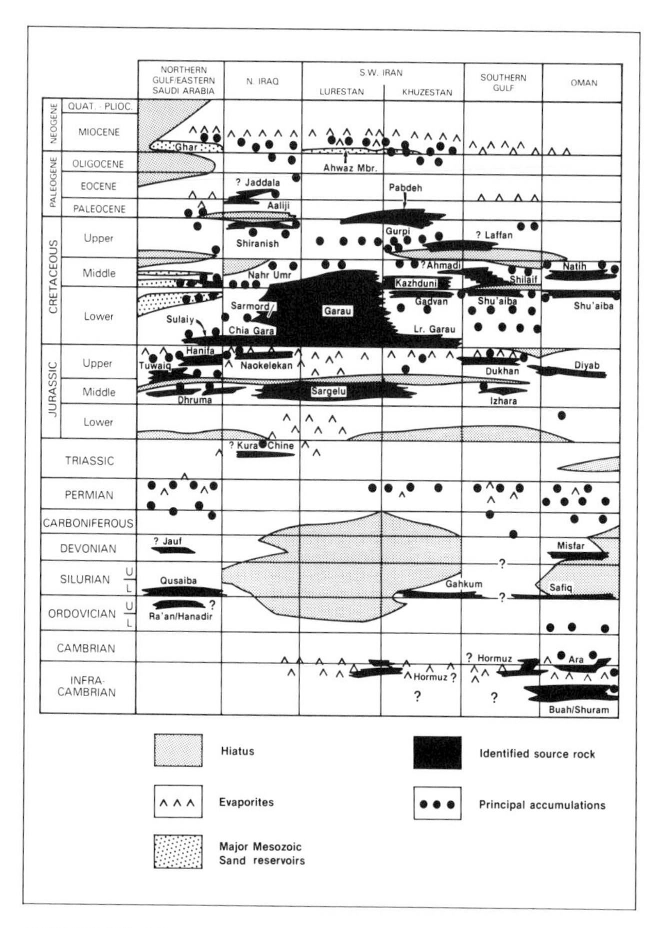

FIGURE 10. Stratigraphic position of the identified source rocks in the northeastern Arabian shelf region and Oman. Expanded and modified from Stoneley (1990). Reproduced by permission of The Geological Society (London).

sequence-disturbing tectonics, would seemingly imply charge from a stratigraphically lower level. The Silurian-sourced oils also seem to be chemically distinctive (Grantham et al., 1988), and any later remigration of these oils to stratigraphically much higher reservoirs should be recognizable unless they are very diluted in mixtures. Dry gas, however, cannot be matched chemically with a source rock (as yet), and it follows that identification of sources for the many dry gas accumulations in the northeastern Arabian shelf must presently rely on determination of the probable geometrically closest deep underlying source. This applies particularly to the many gas fields (some very large) in the Permian Khuff carbonate reservoir, which could be charged from either Huqf or Silurian Safiq/Qusaiba source rocks.

It has been inferred that the enormous Qatar North Dome Khuff gas field and surrounding fields on the Qatar-Fars arch (Figure 2) were sourced from the nearest underlying level, the Silurian (Ala et al., 1980; Bordenave and Burwood, 1990). However, the Silurian has not been penetrated there by wells and is absent by unconformity in the nearest surface outcrops in Fars province (Kuh-e-Surmeh, Stocklin and Setudehnia, 1972); at Kuh-e-Gahkum, farther away in southeast Fars, over 100 m of black silty shales of Silurian age have TOC values varying between 1 and 4.3%, which is high for an overmature source rock that has reached the graphite stage (Bordenave and Burwood, 1990). El Bishlawy (1985) and Ali and Silwadi (1989) ascribed the Khuff (and pre-Khuff) gas accumulation in the southern Gulf region to a Silurian source. Sourcing from the Huqf/Hormuz levels that provide the salt domes to the east and west of the Qatar arch remains equally possible (Figures 1 and 2); a further but less likely possibility is that the gas could have been sourced from the Khuff itself (El Bishlawy, 1985). Grunau (1985) assigned a significant role to the Silurian shales as a source of hydrocarbons not only in the main petroleum-producing basin of the Middle East (the northeastern Arabian shelf) but also elsewhere such as in the Levant region.

Several levels of source rocks occur in the Mesozoic strata of the northeastern Arabian shelf. By contrast with the older levels, however, most of these seem to be much more distinctly limited in initial depositional extent, although still covering large areas. The Mesozoic source rocks appear to be marine deposits in irregular anoxic shallow intrashelf basins, some formed mainly by differential shelf sedimentation (Murris, 1980, 1981), in some cases coeval with documented global anoxic events. The common occurrence of these in the interval from Middle Jurassic to mid-Cretaceous accounts for the bulk of the recorded source rocks that are most widely considered to have given the region its enormous oil charge (Stoneley, 1987).

Middle and Upper Jurassic source rocks are recognized in the Sargelu and Neokelekan formations of Iraq; these source rocks were deposited in the Gotnia/Garau intrashelf basin and were probably tectonically controlled (Figure 11). These source levels are capped by persistent top Jurassic evaporites (Gotnia Formation), but further source rock facies were deposited in the basin almost continuously throughout the Early and middle Cretaceous (Garau Formation) These source-rich facies marginally extend into the Dezful area of Iran (Figures 11 and 2), although one basal Cretaceous finger extends as far as Kharg island in the Gulf waters and is believed to have sourced the Lower Cretaceous accumulations there and at the Gach Saran field.

By contrast with the vertically very continuous source rock facies in the Gotnia/Garau basin, thin but excellent source rocks are also developed in the Middle and Upper Jurassic in an area of irregular intrashelf basins stretching from the Gotnia/Garau basin across the Qatar positive axis (Murris, 1981). These source rocks (Tuwaiq Mountain, Hanifa, and Diyab formations) have received more thorough geochemical study and are considered to have sourced most of the oils in the Jurassic reservoirs of Saudi Arabia, and also in both Jurassic and Lower Cretaceous reservoirs of the United Arab Emirates and Qatar (Ayres et al., 1982; Loutfi and El Bishlawy, 1986; Murris, 1981). The accumulation of these oils is primarily controlled by the presence of the top Jurassic regional evaporites (Hith/Arab/Gotnia formations), which limit vertical migration except around the evaporite edge (northern offshore Saudi Arabian fields, Ayres et al., 1982; some Abu Dhabi fields, Loutfi and El Bishlawy, 1986) or

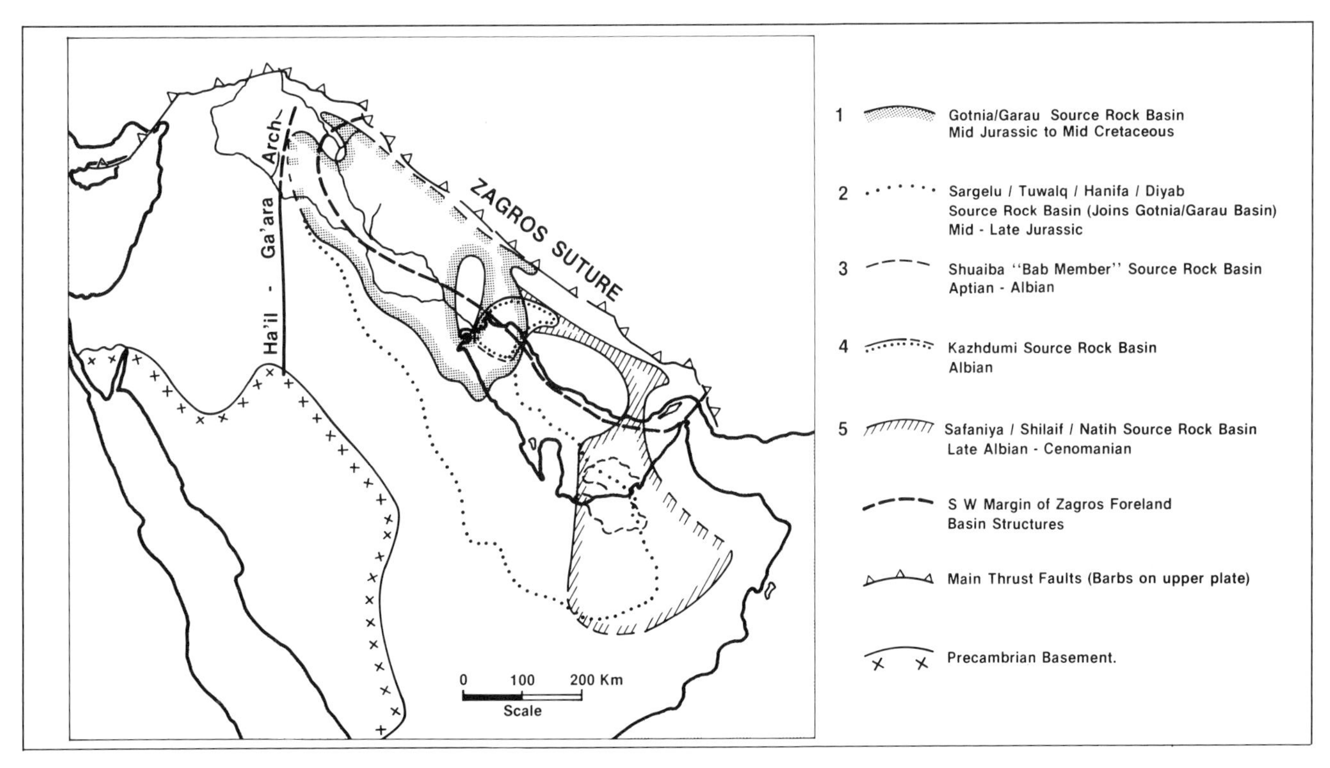

FIGURE 11. Main Mesozoic intrashelf source rock basins on the northeastern Arabian shelf. Numbers 1, 2, 3, and 5 generalized from Murris (1980); number 4 generalized from Bordenave and Burwood (1990), reproduced by permission of Pergamon Press.

conceptually where the seal is broken tectonically or by salt diapirs (Murris, 1981; Wells, 1987). The Jurassic source rocks of Saudi Arabia, Qatar, and the United Arab Emirates are known from numerous well penetrations, and, although of good quality, are not unusually thick or rich. The conclusions by Ayres et al. (1982) and Loutfi and El Bishlawy (1986) that these source rocks have generated most of the oil in the Mesozoic reservoirs in these countries has not been accepted everywhere (Wilson, 1982; Brennan, 1985; Barker and Dickey, 1984), and contributions from deeper sources in the section are suspected from volumetric calculations.

Across the middle Cretaceous, several global anoxic events appear to coincide with good source rock deposition in areas of the northeastern Arabian shelf. In the late Aptian and Albian, two intrashelf basins developed new source rock sequences, whereas in the Gotnia/Garau basin, organic-rich sedimentation continued (Figures 10 and 11). In the southern Gulf, a deeper facies (Bab Member; Alsharhan, 1985) of the shallow-shelf Shuaiba rudist-bearing carbonates was deposited in a basin lying mainly within the United Arab Emirates (Figure 11). This unit was initially thought to be the likely source for oils in Lower Cretaceous and some middle Cretaceous reservoirs of the United Arab Emirates (Clarke, 1975; Murris, 1981), but it is now inferred that these oils were sourced mainly from the Upper Jurassic (Diyab Formation) levels around the edge of the intervening Hith anhydrite seal (Hassan and Azer, 1985; Loutfi and El Bishlawy, 1986). The Aptian also saw the development of an intrashelf basin in the Dezful area of Iran in which Aptian-Albian anoxia induced deposition of rich source rocks (Gadvan-Kazhdumi formations), the second of which is shown by careful geochemical comparison to have sourced most of the oils in the large Dezful fields (Bordenave and Burwood, 1990), as discussed in the next section.

Two further levels in the middle Cretaceous over much of the northeastern Arabian shelf are found to be variably organic rich and in some areas sufficiently so to be classed as potential source rocks. A lower level lies close to the Albian-Cenomanian boundary and is considered to be a potential source in southern and eastern Arabia ("Safaniya source rock," Newell and Hennington, 1983; Shilaif, Loutfi and El Bishlawy, 1986; Natih "e," Grantham et al., 1988), and an upper level occurs close to the Cenomanian-Turonian boundary (lower Mishrif, Newell and Hennington, 1983; Natih "b," Grantham et al., 1988). In Abu Dhabi, the oils in the middle and Upper Cretaceous Mishrif and Simsima reservoirs are geochemically typed to the Shilaif source rock (Loutfi and El Bishlawy, 1986). The oils in some northern Oman fields are matched to the Natih source rocks; no chemical difference has yet been established, however, between the two Natih source levels (Grantham et al., 1988). From available age dating, they do seem to coincide reasonably well with periods of global oceanic anoxia at the highest stands of mid-Cretaceous sea level (Newell and Hennington, 1983; Arthur and Schlanger, 1979).

Prior to the development of good geochemical typing, the deeper marine facies in the Upper Cretaceous and

Paleogene were argued (mainly based on proximity) by many authors (e.g., Weeks, 1950) to be source rocks for oils in Cretaceous and Tertiary reservoirs. Chemical study, however, has established that source potential in those levels is not widespread and that they are generally immature. The best documented exception is the Paleogene Pabdeh pelagic marls of the northern Dezful area, which have been shown to have sourced geochemically characteristic oils (Bordenave and Burwood, 1990); they contribute to some of the accumulations in this area.

Maturity

The timing of maturity of the various source rock levels is as yet poorly constrained. In the dominant marine carbonate facies of the sequence from the late Paleozoic, vitrinite is very rare and maturity must be estimated by calculation using Lopatin time-temperature methods, or from less exact geochemical factors. Only the Upper Cretaceous and Tertiary sections can be calibrated using sufficient observed vitrinite reflectance measurements. Present geothermal gradients over the northeastern Arabian shelf are at or below the continental average (approx. 30°C/km; Clarke, 1975), with the lowest being in the areas of thick young sediments along the Zagros foreland. However, it is uncertain how these gradients varied in the past, particularly in the Paleozoic and early Mesozoic. Examples of the scale of the differences that occur locally as a result of great differential sediment loads are given as calibrated burial graphs (Figure 12) for the old source rocks in Oman (Sykes and Abu Risheh, 1989). These authors also state that oil was generated in southern Oman before the end of the Paleozoic and that fission track data show that later uplift froze any further maturation until the Tertiary. Murris (1981) included a calibrated burial graph (Figure 13) for the main Upper Jurassic source rock (Hanifa) of the Gulf area. Ibrahim (1983, 1984) gave geothermal gradient and time temperature maturation models for some wells in southern Iraq (Figure 14), together with TTI regional maps for selected Upper Jurassic–Lower Cretaceous intervals. Bordenave and Burwood (1990) provided burial graphs (Figure 15) for the on-structure Mesozoic–Tertiary of the Dezful area, for which they have sufficient observed vitrinite reflectance measurements to obtain confidence.

The consensus of the available maturity information suggests that source rocks in the Upper Jurassic and lowest Cretaceous (and certainly any older levels) had passed into or through the oil window by the early Tertiary prior to the onset of the Zagros orogeny. Mid-Cretaceous and younger source levels required significant Neogene overburden to reach full maturity, although Aptian–Albian source levels were arguably within the oil window in the early Miocene and perhaps as early as the Eocene (Ala et al., 1980).

This general conclusion is supported by observations of free oil in the geologic history. A Turonian fossil seep associated with a salt diapir in Iran was reported by Kent (1979). Analysis of fluid inclusions from cements in the eastern United Arab Emirates shows the presence of free oil in the interval from Turonian to Campanian (Burruss et al., 1985). Other examples are dealt with in some detail in the next section.

HYDROCARBONS IN THE ZAGROS FORELAND BASIN

Following the above summary, it is now possible to home in on the hydrocarbons found specifically within the Zagros foreland basin as defined above.

Source Rocks

Almost 60 years ago, in considering the source(s) of the prolific Oligocene–Miocene Asmari-reservoired oils in the Zagros foreland basin fold belt of southwestern Iran, Lees (1933a) conjectured that they were a mixture from various sources from the Jurassic upward. This view was further amplified by Lees (1950) and Henson (1950a, 1951), although largely based on stratigraphic relationships without geochemical backup. Weeks (1950, 1958) refuted such possibilities. Mounting evidence for substantial vertical migration was documented by Dunnington (1958) and Falcon (1958).

Dunnington (1958, 1960, 1967, 1974) progressively provided cogent arguments supported by field evidence (oil gravities, chemical composition, pressure data) in favor of common pre-Tertiary (Lower and middle Cretaceous) origin for the oil found in the Tertiary reservoirs of the Zagros basin. The onset of the Zagros orogeny prompted hydrocarbon movement within the limits of the Mesozoic reservoirs. As folding intensified, fractures developed in the overlying tight, competent, typically argillaceous carbonates, promoting vertical migration across the bedding via poor, tight but fractured, Upper Cretaceous reservoirs and into Tertiary (Eocene and Oligocene–Miocene) porous and fractured carbonate reservoirs capped by the widespread and very efficient Lower Fars (Gachsaran) evaporite cover. Where this cover was breached or developed in a nonsealing facies in the eastern part of the fold belt, the volume of oil lost could well have been prodigious.

Kent and Warman (1972) argued that numerous uncovered Asmari anticlines in the easternmost part of the Zagros fold belt have never been oil fields and that migration was so late that vertical hydrocarbon movement led straight to escape at the surface. Changes in hydrodynamic regimes as the inner Zagros rose in the late orogenic phase, and erosion removed or thinned the cover, would enhance this movement. Nevertheless, the Zagros orogeny must have destroyed a substantial number of preexisting accumulations.

Evidence of considerable hydrocarbon loss at the surface is provided by numerous indications in both northeastern Iraq and southwestern Iran that date back to middle and Late Cretaceous, Paleocene, and middle Miocene times as well as by some going on until the present. These losses are not all related to the creation of the Zagros foreland basin (as a consequence of Neogene collision and formation of the Zagrosides), but date from the

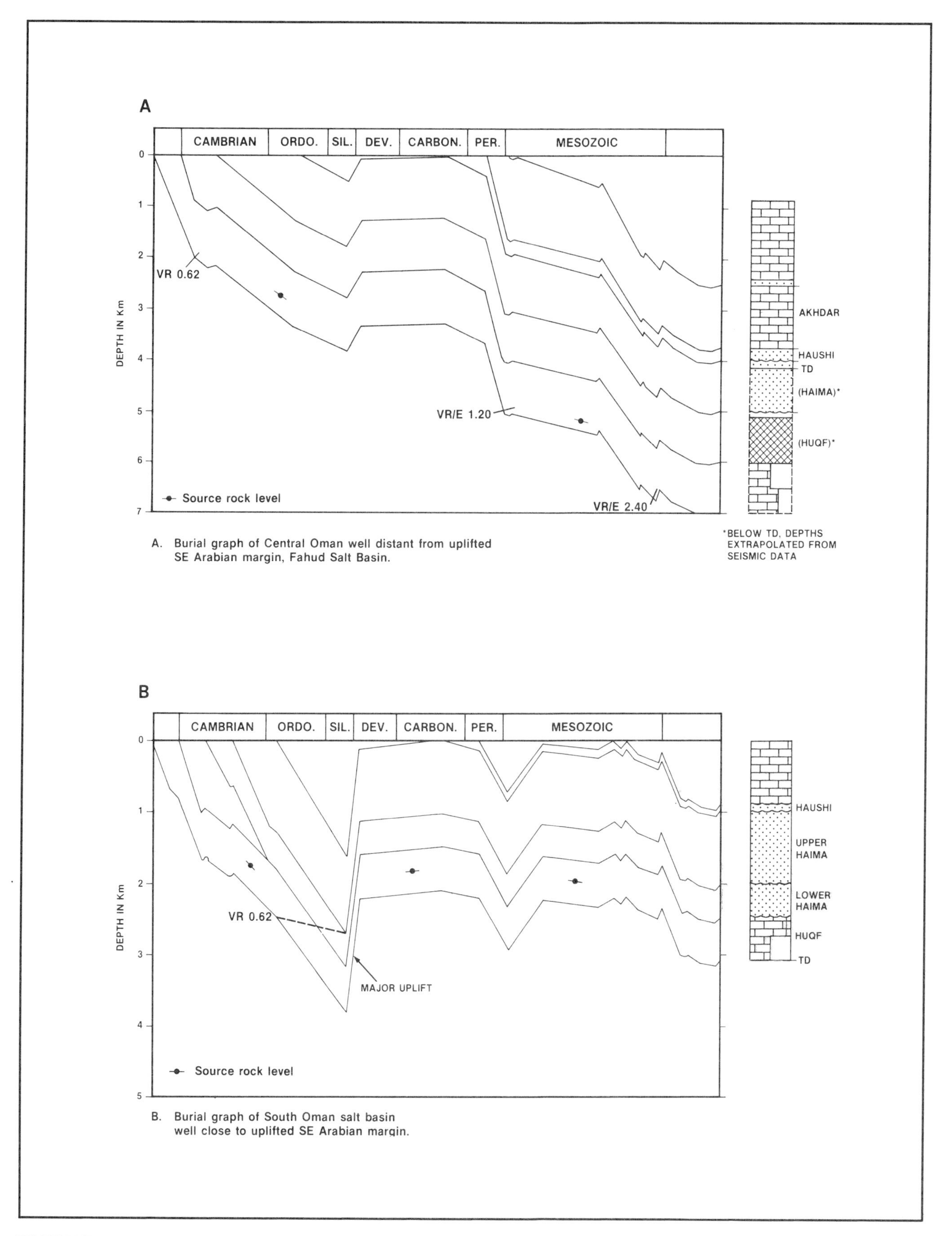

FIGURE 12. Contrasting burial history in Oman wells. From Sykes and Abu Risheh (1989). Reproduced by permission of OAPEC.

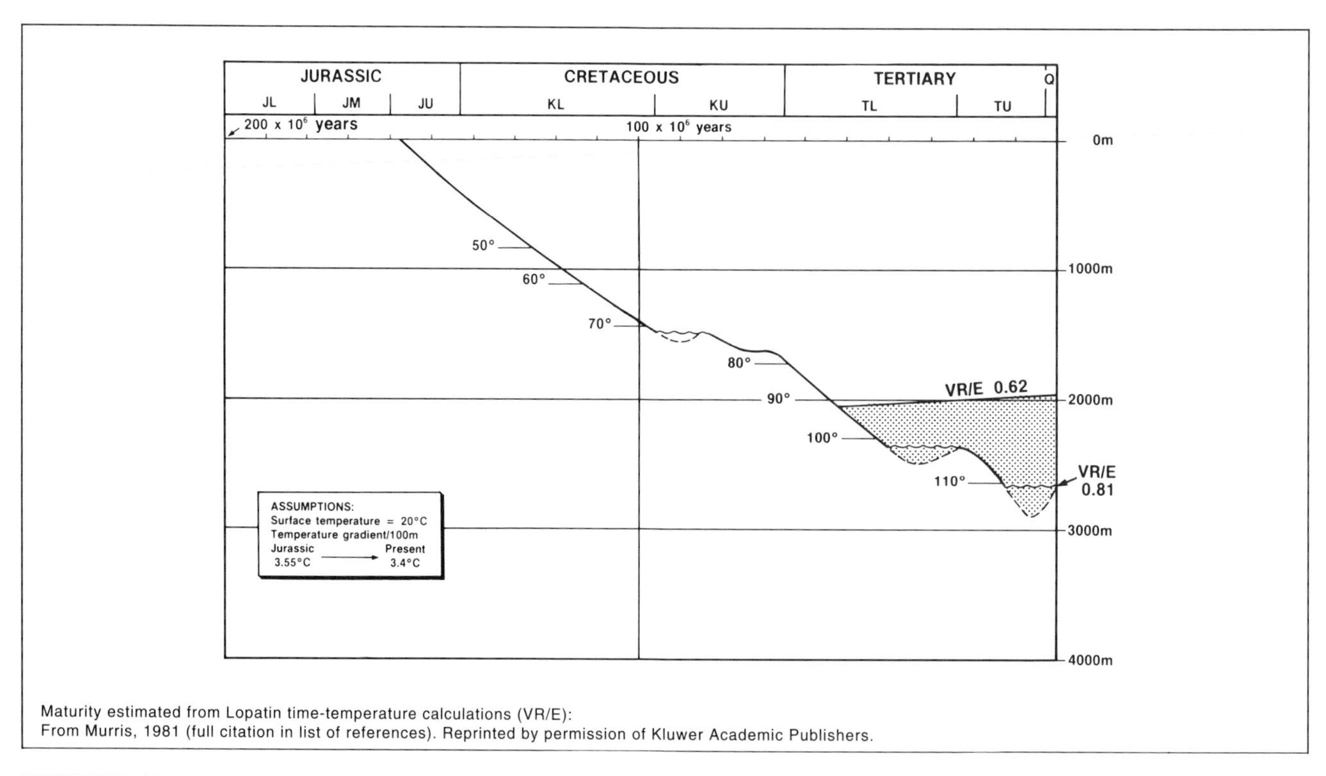

FIGURE 13. Burial graph for a central Gulf well indicating maturity onset for the Upper Jurassic Hanifa source rock. From Murris (1981). Reproduced by permission of Kluwer Academic Publishers.

earlier Late Cretaceous (locally to Paleocene) phase of collision and ophiolite obduction along the whole of the Tethyan Arabian passive margin, or from movement of Hormuz salt. Because volumetric estimates of quantities generated require many assumptions, involving unknown source quality, distribution, and maturity, it is virtually impossible to speculate on the volume of the dissipated hydrocarbons, but the loss could have run into many billions of barrels.

The indications of pre-Zagros basin hydrocarbon loss include impressive Upper Cretaceous bitumen impregnations of outcropping reef limestones in the Kurdish mountains of northeastern Iraq. These impregnations may have been generated locally from adjacent Upper Cretaceous basinal sediments but more likely were seeps fed by deeper vertical migration (Dunnington, 1958). They also include water-borne bitumen pebbles in profusion in some Paleocene–lower Eocene conglomerates in northeastern Iraq along the Iranian border (Pila Spi area near Sulaymaniyah), indicating that large oil accumulations were undergoing dissipation in that region at the time of deposition of these beds (Dunnington, 1958). Similar basal Paleocene pebble beds with detrital bitumen have been reported from many localities in southwestern Iran, indicating erosion and drainage of older oil reservoirs (Kent et al., 1951).

With the onset of the Zagros orogeny, vast quantities of sedimentary bitumen were spread over large areas of the Tigris basin of northwestern Iraq (Fatha gorge, south of the Qaiyarah region) and beyond to the southwest and south in the Hit-Awasil area of the Euphrates basin (Figure 2). This bitumen impregnated basal lower Fars sediments, shortly after deposition of the underlying lower Miocene Jeribe/Euphrates Limestone formations, by leakage along rejuvenated faults and into the shallow saline Lower Fars sea (the potential seal, as yet undeveloped). This oil is heavy and sulfurous but demonstrates that ample free oil has been in the system since at least the onset of Zagros folding; it is considered to have escaped from an Upper Jurassic (Middle Jurassic sourced) Najmah carbonate reservoir (Dunnington, 1958). Detrital bitumen in the Pliocene Bakhtiari sequence of the mountainous region of Iraqi Kurdistan and fossil seeps in the Agha Jari of Iran suggest that middle Cretaceous or Asmari limestones were traps for large accumulations of migrant oils before Pliocene and later erosion or fracturing destroyed their seals. During the Pleistocene erosional cycle, important seepages developed in the crestal regions of large anticlines that contain (or have contained) Asmari-type oil accumulations, and included much gas (the historic "eternal fires" of Kirkuk and many Iranian structures) related to surficial thrust-fault breaching of the incompetent Lower Fars seals. Secondary mineralization of gypsum into aragonite, and extensive amounts of "gach-i-tarush" (sour "earth") resulting from alteration of calcareous marls (Lees and Richardson, 1940), are present over much of the Iraq-Iran fold belt (Dunnington, 1958). The east-west-trending Ain Zalah field in northwest Iraq (although arguably just beyond the margin of the Zagros foreland basin, lying on the Ha'il-Ga'ara trend) demonstrates beautifully that vertical migration continues to the present day, with the middle Cretaceous reservoir slowly feeding and building up depleted pressure in the tight but fractured Upper Cretaceous "reservoir" that once was

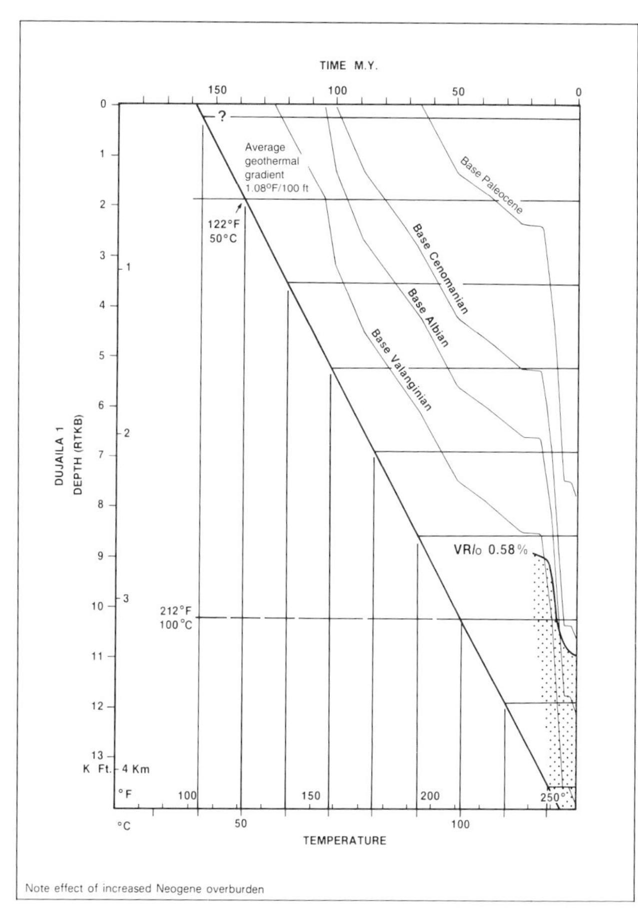

FIGURE 14. Burial graph for a southern Iraqi plains well (Dujaila 1) on the outer margin of the Zagros foreland basin. Simplified from Ibrahim (1983).

greatly depleted by short bursts of production (Daniel, 1954; Dunnington, 1958 et seq.).

Independent support for vertical migration as the origin of Tertiary-reservoired oil in the fold belt of the Zagros foreland basin, based on geochemical typing and correlation, has come from several sources. For the northeastern Iraq region within the so-called Kirkuk embayment (or Sirwan embayment; Lees, 1950), Thode and Monster (1970) examined oils from Tertiary and Upper, middle, and Lower Cretaceous reservoirs of the Zagros fold belt fields using sulfur isotope abundance. They confirmed that there are essentially no significant differences in sulfur isotope composition in any specific vertical sequence of oils, suggesting a common origin. There are, however, significant variations in composition of oils in contemporaneous reservoirs in different fields; this may reflect differences in conditions at various locations in the basin during source rock deposition or during periods of oil formation, and suggests that little or no horizontal migration or mixing occurred either before or after vertical migration took place. Comparison of the δ S34 values of these differently reservoired oils and of contemporaneous Jurassic, Cretaceous, and Tertiary evaporites in Northern Iraq led to the favoring of a Jurassic origin (despite the lack of a Jurassic oil sample for analysis). On the basis of the sulfur isotope data for Kirkuk and the surrounding producing fields, however, a Lower Cretaceous source rock is considered the most likely common origin for the oils, with a Tertiary source the least likely. The "basinal" Lower-middle Cretaceous Chia Gara and more neritic low-energy Sarmord Formation marls (essentially equivalent to the Cretaceous Garau Formation of southwestern Iran) are well situated to have generated the oils found in the Lower-middle Cretaceous Qamchuqa (and other) reservoirs of northeastern Iraq, as indicated by Dunnington (1958, 1974).

Al Shaharistani and Al Atyia (1972) provided further support for vertical migration in northeastern Iraq. Using vanadium and nickel concentrations and V/Ni ratios in oil samples from the various producing formations of the fields in the region, they concluded that these oils have a common origin and originated from similar source environments. They argued that the path of vertical oil migration from middle Cretaceous to Upper Cretaceous reservoirs and then to Tertiary reservoirs is clearly reflected by the decreasing vanadium and nickel concentrations.

Young et al. (1977) investigated the dating of oils based on maturation changes in specific hydrocarbon structures and applied this technique, among others, to some Middle East oils. They concluded that oils from four large Dezful embayment fields had ages ranging from 102 to 109 m.y. (close to the Albian Kazhdumi source rock) whereas the Lurestan Cheshmeh Khush oil was older than 120 m.y. (the Garau source rock or older levels). The Saudi Arabian and United Arab Emirates oils studied were given ages of 146 to 174 m.y. in keeping with the accepted Middle-Upper Jurassic sources.

For southwestern Iran, Ala et al. (1980) published results of geochemical work on five potential source rocks ranging in age from Silurian to Paleocene-Eocene (Pabdeh); the organic matter in all these formations is almost exclusively of marine algal origin (Type II kerogen). They concluded that the Albian Kazhdumi Formation is the major source of the oil in both the middle Cretaceous Sarvak carbonate reservoir and the Oligocene-Miocene Asmari, whereas the Paleocene-Eocene Pabdeh and the uppermost Cretaceous Gurpi potential source rocks were generally immature. Extracts from these source rocks showed little similarity to the nearby crude oils.

More recently, Khosravi Said (1987) published further results of geochemical work, which indicated that the Kazhdumi Formation was confirmed as the major source unit for Bangestan Group (middle and Upper Cretaceous) and Asmari oil in the Dezful embayment of southeastern Iran, with hydrocarbon generation and migration being contemporaneous with the Zagros orogeny. In adjacent Lurestan (to the northwest) and in northeastern Dezful, hydrocarbons were sourced from Garau and from the Pabdeh and Gurpi formations, respectively.

Bordenave and Burwood (1990) went into much more detail with regard to the overall results of the geochemical project carried out by the Iranian Oil Operating Companies in 1967-1978, intending to establish the identity (and distribution) of the source rock(s) responsible for the generation of the majority of Iranian Zagros basin oil reserves. Correlation techniques revealed that the majority of the oils variously housed in Oligocene-Miocene

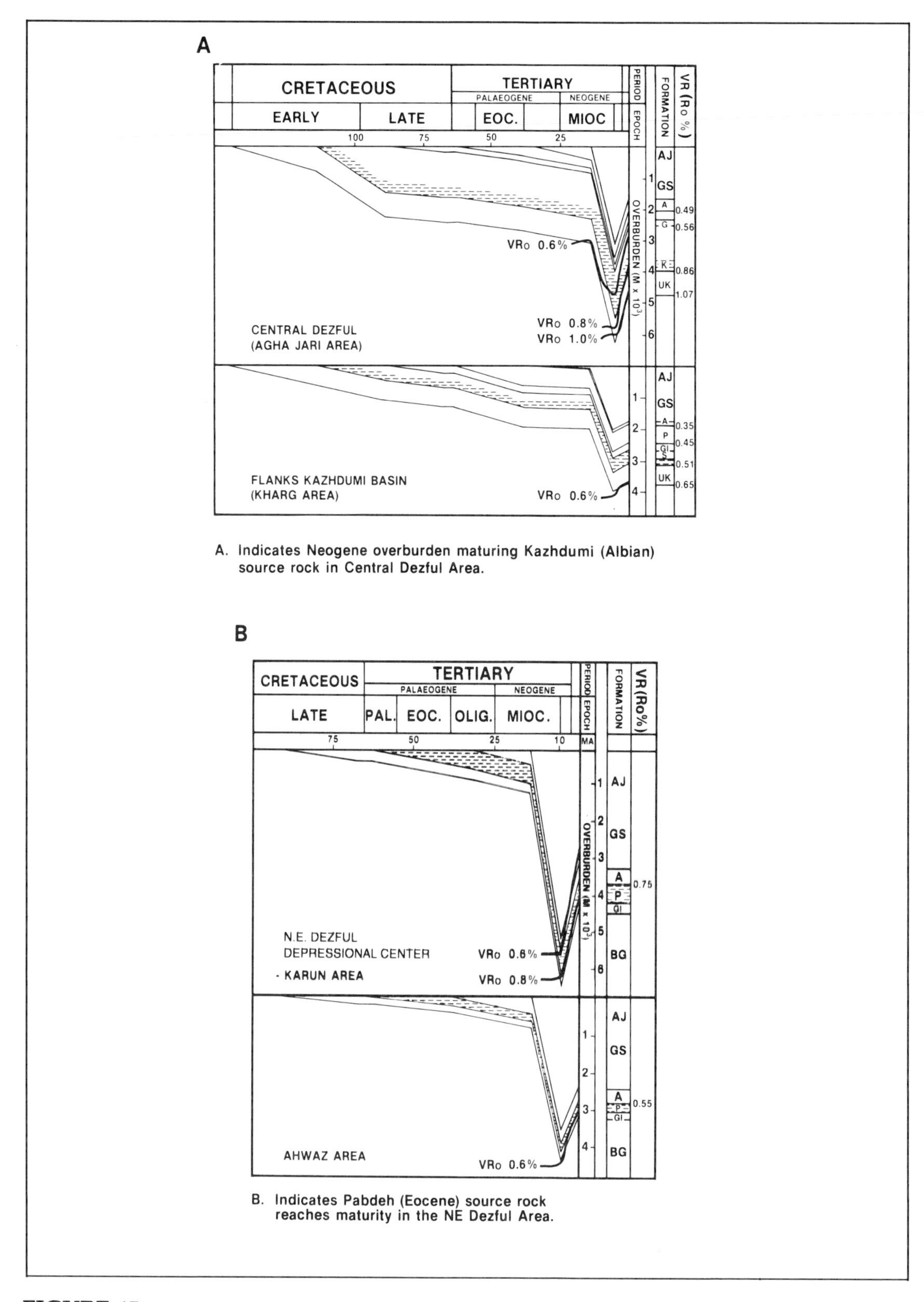

FIGURE 15. Burial graphs from the Iranian sector of the Zagros foreland basin. Modified from Bordenave and Burwood (1990). Reproduced by permission of Pergamon Press.

Asmari and middle to Upper Cretaceous Sarvak and Ilam carbonate reservoirs in the Dezful embayment show a common provenance, which focuses on the Albian Kazhdumi shales and argillaceous sediments as the principal source rocks (3.1 to 12% TOC). The Lower Cretaceous potential source rocks (Gadvan and Garau; 0.8 and 1.5 to 10% TOC, respectively), are considered to be subordinate, whereas the Sargelu (Middle Jurassic; 3.1 to 4.4% TOC), on the basis of distribution and maturity, may have been depleted prior to effective Neogene trap formation.

At the northeastern edge of the Dezful embayment on the edge of the mountain front, a second compositional family of oils is matched to the Paleocene–Eocene (locally to Oligocene) Pabdeh source unit (1.0 to 12.0% TOC), based on biomarker analysis. Another group of fields centered on Masjid-i-Suleiman has mixed-provenance oils derived from both Kazhdumi and Pabdeh sources (Bordenave and Burwood, 1990). The Garau (Lower Cretaceous Aptian and older) shows no correlation with the northeastern Dezful oils and can be eliminated as a contributory source. Other Upper Cretaceous potential sources are dismissed as significant contributory sources on the basis of kerogen type and/or thickness (e.g., Gurpi and Surgah/Laffan, 0.5 to 2.0% and 3.5% TOC, Types II/III and III, respectively). In the Lurestan province, where the mountain-front bulge of Pusht-i-Kuh separates the Dezful and Kirkuk embayments (Figure 2), the Garau facies (Lower Cretaceous) appears to have potential as a source rock, which agrees with the suggested Lower Cretaceous–middle Cretaceous Chia Gara and Sarmord marly facies as the sources of the northeastern Iraq oil, as discussed above, and with the >120 m.y. age of the Cheshmeh Khush oil given by Young et al. (1977).

The regional geothermal gradient in the Zagros region is average to below average (e.g., Clarke, 1975). Subsidence profiles calibrated by vitrinite reflectance indicate that hydrocarbon generation from the candidate sources in southwestern Iran (Kazhdumi and Pabdeh) has been largely controlled by the Zagros (Neogene) orogeny. Maturation and structure formation are both very recent, with the critical events having occurred in the last 15 m.y. Rapid accumulation of foreland basin–related lower to middle Miocene evaporites (Lower Fars-Gachsaran) occurred, followed by accumulation of thick upper Miocene–Pliocene molasse-type sediments (Upper Fars/Agha Jari and Bakhtiari), totaling between 4000 and 7000 m, synchronous with the formation of large anticlinal traps. Their rapid accumulation ensured rapid maturity of the Kazhdumi source, which had been marginally mature at the onset of the orogeny, and the maturation of the Pabdeh in the foredeep area. This occurred particularly in the synclines between the large anticlines, where considerable thicknesses (>5000 m; 3 mi) of overburden accumulated and triggered the generation and expulsion of large amounts of hydrocarbons, with fluid communication between source and reservoirs facilitated by fracturing developed over the folds during the main Zagros orogenic phase (Bordenave and Burwood, 1990; McQuillan, 1973, 1974). Early-generated oil may have been housed initially in Cretaceous traps, only to be redistributed into the Asmari reservoirs when Late Cretaceous (and younger) seals failed because of the intensified fracturing. Much of this occurred over the interval from 8 to 5 Ma but continues today in the flanking areas. Because several of the Pabdeh-sourced Asmari fields of the northeastern Dezful region also have oil columns in middle-Upper Cretaceous Bangestan Group reservoirs, it is very probable that Mesozoic sourcing is also involved there.

Kent and Warman (1972) suggested a two-phase history of vertical migration for the Asmari-reservoired oil, in view of the markedly bimodal gravity distributions. The earlier phase, associated with the primary folding, resulted in the oil losing some of its light ends.

In view of the considerable and overwhelming evidence reviewed above regarding the principally pre-Neogene (essentially pre-Tertiary) sources of the Tertiary-reservoired oils in the fold belt of the Zagros foreland basin, no valid justification remains for postulating oil sources indigenous to the reservoirs and for not accepting that virtually all the prolific accumulations in the Tertiary reservoirs of the basin (possibly barring very minor occurrences; e.g., see Kashfi, 1984) are the result of vertical migration and remigration.

Reservoirs

There is a wide stratigraphic range of reservoirs in the Zagros foreland basin region (Figure 10). They are predominantly carbonates deposited before the main continental collision and the onset of folding: their original fabric and reservoir properties have thus been strongly modified by the subsequent tectonism and now bear little relation to depositional facies. Indeed, reservoir capacity and performance depend so strongly on fold-related fracturing that an understanding of fracture distribution is the prime requirement of the production engineer (McQuillan, 1973, 1974). Formation strength therefore governs the stratigraphic distribution of potential reservoirs.

The Asmari Reservoir

The Oligocene–lower Miocene carbonates of the Asmari Group and its Iraqi partial time-equivalent reservoir, the "Main Limestone" of the Kirkuk Group (Henson, 1950b; van Bellen, 1956), form by far the most important reservoir of both the Zagros foreland basin (sensu stricto) and the entire succession in the area. It is estimated to provide at least 90% of the oil in the oil fields area of southwestern Iran (vicinity of the Dezful embayment; see Figure 2) and a lesser but still considerable percentage of the oil fields in the foothills belt of Iraq (Kirkuk embayment). An almost perfect seal is provided by the evaporites of the middle Miocene Gach Saran/Lower Fars formations of the region, which preserved their integrity throughout the Zagros folding.

The Asmari of southwestern Iran varies in thickness from some 100 m (330 ft) to more than 800 m (2625 ft) in parts of the Dezful embayment (Koop and Stoneley, 1982), and may be augmented where the directly underlying Eocene is in shallow marine carbonate facies (Jahrum). Except in thin intervals where matrix porosities may exceed 15% (the maximum recorded is 22%) and permeabilities may be a few millidarcys (Lees, 1933b), average values are seldom more than 5% and 1 millidarcy. Facies variations do provide some control on the distributions of these low values, but they appear to be significant only in the large Kirkuk field of Iraq: porous Eocene and Oligocene reefal limestone facies trend and prograde obliquely across the structure (Figure 16), providing a complex reservoir (Daniel, 1954).

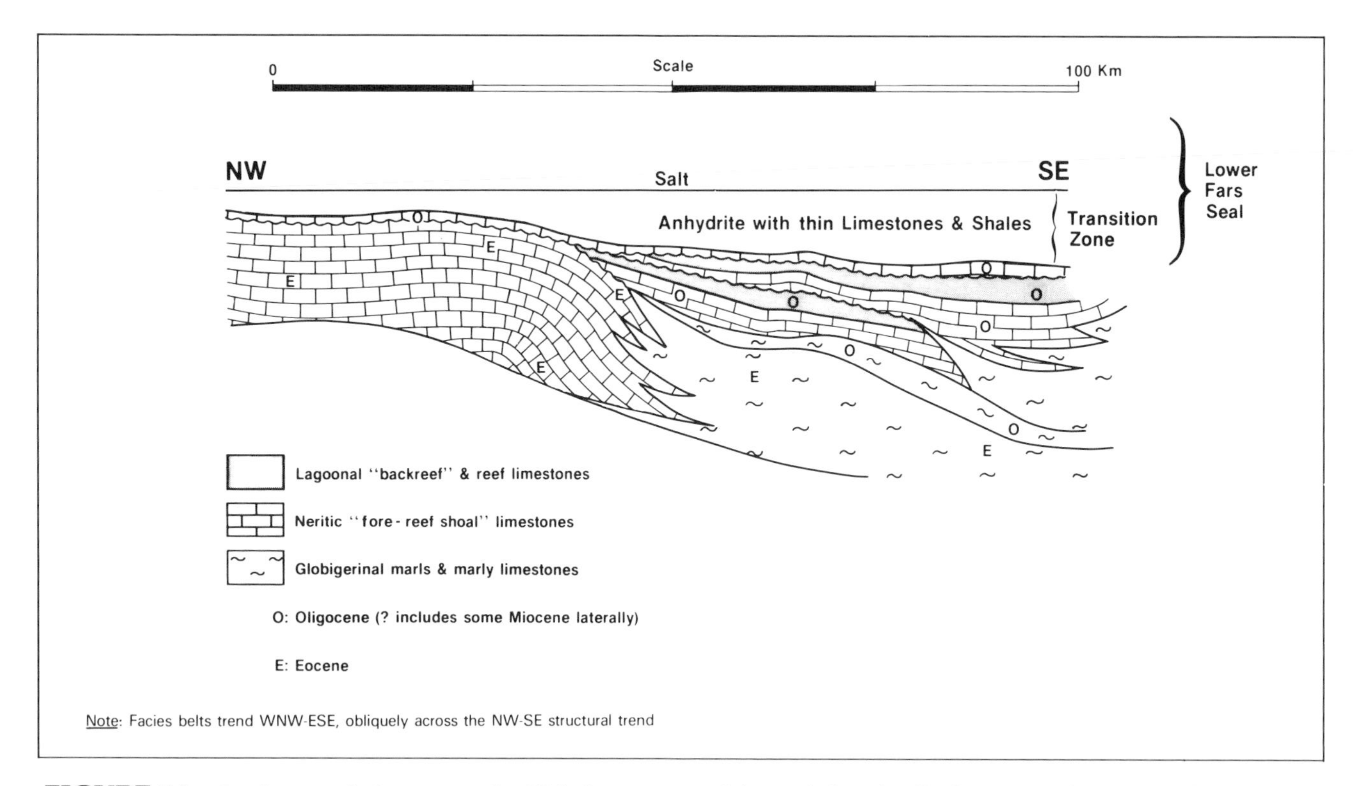

FIGURE 16. Reef progradation across the Kirkuk structure. Schematic longitudinal cross section; vertical exaggeration × 40. Simplified from Daniel (1954) and Dunnington (1958).

The "Main Limestone" of Iraq includes the Avanah Formation (Eocene) and the overlying Oligocene Kirkuk Group formations that provide the first and main oil pay zones of the Kirkuk structure and some of the adjacent structures in northeastern Iraq; along the southeast part of the Kirkuk structure, a lower Miocene back-reef limestone facies is present and forms part of the reservoir (Daniel, 1954). Thus, the Kirkuk Group formations are equivalent to the Asmari, and the Avanah Formation is equivalent to the Jahrum of southwestern Iran. "Main Limestone" reservoir thickness varies substantially up to a maximum of approximately 300 m (985 ft); average porosities and permeabilities vary considerably depending on facies and can range from 0 to over 36% and from 0 to 1000 millidarcys, respectively, with higher values in cavernous fore reef sections (Daniel, 1954).

It is clear that production from such rocks as the Asmari and "Main Limestone" is derived through fractures: wells have been tested at up to 19,080 m^3/day (120,000 BOPD) and sustained at 12,720 m^3/day (80,000 BOPD), with an estimated average for producing wells of 3180 to 3975 m^3/day (20,000 to 25,000 BOPD). Where wells have evidently failed to penetrate fractures, they are essentially nonproductive. A part of the gross porosity must be provided by the fractures themselves, but it would appear that there is slow subsurface drainage from the block matrix porosity into them; their productivity depends on the very large surface area of the fracture in contact with the oil-bearing matrix.

In parts of the Dezful embayment, the quality of the Asmari reservoir is considerably enhanced by interbedded distal tongues of a spread of quartz sand (Ahwaz sands) eastward from the vicinity of Kuwait (Ghar sands). They die out eastward across the Iranian oil fields area, but, in fields such as Ahwaz and Marun, where they are believed to have been deposited in nearshore, shoreline, and possibly eolian environments, they form a significant proportion of the reservoir. The sands are clean and poorly cemented, and have porosities of 20 to 30% and permeabilities in the darcy range.

The Bangestan/Qamchuqa Reservoir

The essentially Cenomanian–Turonian carbonates of the Bangestan Group of southwestern Iran form an important second reservoir, where they are developed in neritic facies. This is primarily in the Dezful embayment, where they reach a thickness on the order of 1000 m (3280 ft). Regionally, there is considerable facies variation: in Lurestan, for example, virtually the whole of the sequence is pelagic with little or no reservoir potential.

The approximate equivalents of this reservoir in fields of the Kirkuk embayment of northeastern Iraq are the neritic carbonates of the Qamchuqa Formation, essentially Albian–Cenomanian in age but extending down into the Aptian; reservoir thickness reaches over 200 m (655 ft) and sediments grade into more "basinal" facies to the northeast. Oil reserves in this reservoir have been established in several structures of the Kirkuk region (Dunnington, 1967).

The Bangestan carbonates of southwestern Iran have undergone a complex diagenetic history, and the secondary porosities are very low and patchy. Again, it is the fractures that provide the reservoir. Indeed, in several Iranian fields including Naft Safid, Agha Jari, and Gach Saran (NIOC-IOOC, 1965), the fractures extend upward through the argillaceous Upper Cretaceous and lower Tertiary to the Asmari, to provide communication between the two reservoirs. They can thus contribute significantly to field reserves. A similar situation is present in the Kirkuk embayment of northeastern Iraq, where relatively minor accumulations have also been left behind in tight, argillaceous but fractured, Upper Cretaceous Shiranish carbonates during vertical migration from the Qamchuqa to the "Main Limestone" in Kirkuk, Bai Hassan, and one or two other fields of the Kirkuk embayment (Dunnington, 1958, 1967).

Other Reservoirs

Several additional proven and potential reservoir formations are present in the region covered by the Zagros foreland basin, but as yet they are poorly explored and virtually undeveloped. The most important are the gas-bearing Neogene formations in the Bandar Abbas area in the far southeast, and the Permian. They are listed below stratigraphically.

The dolomites and limestones of the Permian Khuff-Dalan formations (extending up into the Lower Triassic Kangan Formation in southern Iran) contain vast reserves of gas in the Gulf and beyond. The only appraised or partly appraised discoveries within the fold belt are close to the coast at Kangan and Mand; the nearby offshore Pars accumulation in a halokinetically controlled structure beyond the folding is being developed at present.

In much of southern Iran, the Jurassic consists of a thick, massive limestone. It is little known in the subsurface and is generally tight, but it could provide a reservoir where fractured. At Masjid-i-Suleiman, sour gas, derived from a carbonate unit underlying Upper Jurassic evaporites and believed to be representative of the Najmah Formation, caused a spectacular subsurface "blowout" in the early 1960s (Koch and Taylor, 1968).

Neocomian carbonates are the reservoir in the Kharg Island-Doroud (formerly Darius) field just outside the fold belt. Locally, in both Iran and Iraq, they may form a common reservoir with Aptian-Albian limestones that have tested minor amounts of oil and gas (often sour) in a few wells in Iraq, in the Dezful embayment, and near Bandar Abbas. This Khami Group offers exploration potential for the future.

The final significant reservoir is the Guri limestone of the middle Miocene Mishan Formation, which contains undeveloped gas in several discoveries near Bandar Abbas, as also do the underlying Razak siliciclastics. The Guri, at the southeast end of the Zagros foreland basin, varies in thickness from zero to almost 1200 m (3935 ft), but no information is available on its reservoir properties. In the region where it is present, it is the prime exploration objective, the Asmari being thin and impersistent there.

Traps

Traps in the Zagros fold belt, including the Zagros foreland basin as defined in this paper, are dominated by the Neogene-Quaternary compressive folding. All of the petroleum discovered to date is in these structures.

The anticlines vary considerably in size. Some, such as those in the deeper parts of the Dezful embayment, are small, low-amplitude structures; others are enormous. The largest, Kabir Kuh, which exposes the Lower Cretaceous in the core and has tested nonhydrocarbon gases from the lower Paleozoic, is some 190 km (120 mi) in length. Amplitudes have been estimated to reach 6 to 10 km (3.7 to 6.2 mi) (Hull and Warman, 1970).

In parts of the region—for example, the Dezful embayment and much of the Iraqi sector—the folds are close to concentric in style. This is believed to reflect both the presence of several thick and rigid carbonate formations in the folded sequence and also detachment from basement on the underlying Hormuz salt. In other areas—Lurestan, for instance—the folds become generally smaller, tighter, and more acute, possibly because of the absence of the Hormuz salt and a more argillaceous succession.

The concentric style of the great anticlines containing the majority of the oil fields carries the implication that, at deeper levels, the sizes of potential traps will progressively decrease: smaller prospects can be anticipated. There will also be a common center of curvature, below which the anticline cannot persist and complex thrust cores can be expected. Deep wells on some anticlines have indeed encountered such complexities. The level of the center of curvature will vary depending on the size of the fold, but in most of the Iranian structures it probably lies at horizons between the Triassic and the Cretaceous: where the sedimentary sequence is thinner, as over much of the Fars platform, the Permian is still within the "simply folded" section. There is thus a limit to the depths at which sizable prospects can remain, a factor to be borne in mind when considering deeper exploration.

There is a tendency for the anticlines to be arranged in trends, although often offset somewhat from each other en echelon. The implication is that the trend may be underlain by an essentially common thrust detaching from the basal décollement but that, in the areas of overlap, relations may become extremely complex. This inhibits precise definition of spillpoints and clearly separates accumulations with different oil-water contact levels; this difference amounts to some 1500 m (4920 ft) between the co-joined en echelon Marun and Agha Jari fields (Hull and Warman, 1970). Initial oil columns were in keeping with the scale of the structures, with that in Gach Saran being over 1825 m (6000 ft) (Hull and Warman, 1970).

It has been noted earlier that some structuration and potential traps developed prior to the Zagros folding, during the Mesozoic and early Tertiary, as a result of two causes: drape-compaction, sometimes accompanied by normal faulting, above essentially north-south-trending basement features; and salt doming. Just outside the fold belt, the former is exemplified by the Kharg-Doroud oil field, and the latter by the Pars gas field. Although in some cases the north-south trends were active earlier, they became particularly pronounced during the Late Cretaceous, at which time they were able to receive early-generated oil from Jurassic-basal Cretaceous source rocks. The salt structures mostly date from the Early Cretaceous, but some were initiated earlier. Pars, for example, was active during the Triassic.

An important question is whether or not these pre-Zagros traps, and possibly some hydrocarbon accumulations, are still preserved within the fold belt. No discovery appears to have been made that is recognizably contained in an earlier structure, and it would appear that any such traps have been overwhelmed by the later folding. Furthermore, there appears to be no relationship between the distribution of oil in the Zagros anticlines and the positions of the earlier highs as mapped from thickness changes. This, of course, is consistent with the concept that most of the oil in Iran was generated from the Albian Kazhdumi source as the Neogene folds were developing; in northeastern Iraq, similar generation, but from slightly earlier sources, probably migrated from former basin-edge Late Cretaceous traps within the same carrier beds into the traps formed by the Neogene compression (Dunnington, 1958, 1967).

Purely stratigraphically trapped accumulations are unknown in the Zagros foreland basin. The rather widespread nature of the principal reservoir formations and the dominant role played by the folding might suggest that, other than small accumulations localized on the flanks of some anticlines, stratigraphic traps are unlikely to contribute to reserves. Varied reservoir facies in a few fields, such as Kirkuk (see above; Dunnington, 1958), may nevertheless imply some stratigraphic element in defining the movable oil pool. Southwest of the fold front, on the other hand, the presence of stratigraphic traps is quite conceivable, although no exploration appears to have been directed toward them as yet.

Similarly, hydrodynamically trapped oil seems unlikely. Clearly there is a water flow toward the Gulf through the Asmari reservoir in at least parts of the Dezful embayment, as shown by lower oil-water contacts on the southwest flanks of some fields as compared with the northeast flanks; a flow in the opposite direction is believed to trap oil beneath the Gulf (Wells, 1987). Comparable opposing two-way flow relationships exist in the Iraq sector, but here it has resulted in surface seepages, asphalt, and saltwater lakes (Figure 2) of the Awasil-Hit area (Al Mashadani, 1986). The scale of the anticlines in the fold belt is such that structural control is likely to be predominant, and exploration for hydrodynamic traps cannot be envisaged in the foreseeable future.

Future Potential

Because of the late overprinting of the Zagros foreland basin onto the outer edge of an older passive-margin shelf, as described and discussed in this chapter, accurate estimates of as-yet-undrilled future oil and gas reserves in the Zagros basin cannot be assessed separately from estimates for the northeastern Arabian shelf as a whole. Estimates for the northeastern Arabian shelf region, however, cannot be assessed with any high degree of probability because of the lack of sufficient published data; moreover, the immaturity of exploration dictates that any new data acquired will markedly revise any current estimates.

Nevertheless, a speculative broad estimate for the northeastern Arabian shelf is attempted here for the sake of completeness. This can be used as a rough indicator of the Zagros foreland basin's potential share provided that the rough percentages of Zagros basin oil and gas volumetrics given earlier in this chapter are kept in mind; these are approximately 25% of the reported oil and 50% of the reported gas reserves of the total reported reserves for the whole northeastern Arabian shelf region.

Grunau (1985) estimated the undrilled future oil potential for the Arabian plate Arab countries (excluding Iran) as roughly 2.39 to 3.18×10^{10} m^3 (150 to 200×10^9 bbl) and the gas potential as 1.13 to 1.56×10^{13} m^3 (400 to 550 tcf) (over 98.7% of the recoverable oil reserves and 97.2% of the recoverable gas reserves of the Arabian plate are located along the northeastern Arabian shelf including Iran; see earlier). These maxima had already been approached or exceeded as of the end of 1990. Current reserve figures for the northeastern Arabian region could easily increase by 50% in another decade or two of intensive exploration, or conceivably even double (Beydoun, 1991). The Zagros foreland basin's share of these increases could well be at least 20 to 25% of the oil (housed in Tertiary and older Mesozoic reservoirs) and at least 50% (but probably more) of the gas. This latter estimate results from the fact that many of the structures in the southeastern end of the basin in the Fars province of Iran are still undrilled for the deeper gas prospects and virtually none of the structures of northeastern Iraq have been tested down to Paleozoic levels.

SUMMARY OF THE CONSEQUENCES OF ZAGROS FORELAND BASIN FORMATION ON THE HYDROCARBON HABITAT

1. The extra overburden of late Tertiary sediments in the newly formed Zagros foreland basin brought mid-Cretaceous source rocks (Kazhdumi, Garau, and their Iraqi equivalents) from early maturity to full maturity.

2. The extra overburden brought limited immature Eocene source rocks (Pabdeh) into the oil window.

3. The extra overburden increased maturity of all Lower Cretaceous and older source rocks, increasing relative gas generation.

4. Zagros fold belt high-amplitude structures form immense traps, with fracturing producing well-connected reservoirs and synfolding evaporites constituting an ideal seal.

5. Zagros folding breached Mesozoic traps and allowed upward migration across the bedding to the sealed mid-Tertiary reservoirs or escape to the surface.

6. The trap-fill of the Zagros foreland basin potentially could have come from numerous source rocks. Available geochemical typing of oils is insufficient to quantify relative source contribution, although older oils (Paleozoic-Infracambrian), if involved, should be chemically characteristic.

7. Distribution of hydrocarbons in traps of the Zagros foreland basin may perhaps be significantly controlled by positions of pre-foreland basin paleotraps; few of these, however, have as yet been outlined, and known hydrocarbon distribution does not seem to be related to them.

8. The emplacement of ophiolitic thrust sheets and associated flysch prisms on the outer shelf edge in the Late Cretaceous was not part of the Zagros orogeny, but did induce hydrocarbon generation and loss to the surface.

9. Sedimentary evidence shows that much hydrocarbon was lost to the surface during development of the Zagros fold belt from the Miocene to the present. This loss cannot be quantified, but it is moot whether the loss was compensated by the extra trapped hydrocarbons induced by the factors enumerated in points 1 to 4 above.

CONCLUSIONS

The Zagros orogene is a conspicuous element of the Middle East hydrocarbon province. The Iranian and Iraqi oil fields in the Zagros foreland basin contain around one-quarter of the vast oil reserves of the northeastern Arabian shelf. Therefore, at first sight it is not surprising that this basin should qualify for inclusion in this volume of the AAPG World Petroleum Basin Project series. However, when viewed in greater depth, it becomes apparent that the hydrocarbons in the Zagros basin are part and parcel of those developed over the whole of the northeastern Arabian shelf, whose enormous hydrocarbon-sourcing potential is linked to its geologically long and remarkably stable history as a passive or divergent margin through most of the period from late Proterozoic to Late Cretaceous, and again briefly to early Tertiary.

The continent-to-continent convergence and collision that affected this plate margin from late Eocene to the present have overprinted the outer passive-margin shelf with a depositional basin and a structural grid of high-amplitude whaleback folds. The Tertiary deposition has added another carbonate reservoir level capped by an excellent evaporite seal. The folding has formed large anticlinal traps and provided fracture routes to fill them. However, although more sources were matured by the overburden, evidence of great surface hydrocarbon loss leaves the outstanding question of whether the Zagros foreland basin has not lost more hydrocarbons than it has retained. A principal factor in answering this question, and in creating the habitat frame to direct much of the future exploration in the northeastern Arabian shelf, will be a major regional effort to relate all oils (and gases) geochemically to source rocks.

ACKNOWLEDGMENTS

The authors thank the Exploration Management of Marathon International Petroleum (GB) Ltd. in London for the considerable support provided in both time and facilities during the preparation of this chapter. Sincere thanks are due to the Marathon Drawing Office, London, for preparation of the illustrations, and to Miss Susan Hambrook for patiently typing several drafts of the manuscript. The authors also thank K. W. Glennie and H. R. Grunau for reviewing the manuscript and for their helpful comments and suggestions.

REFERENCES CITED

Ala, M. A., R. R. F. Kinghorn, and M. Rahman, 1980, Organic geochemistry and source rock characteristics of the Zagros petroleum province, SW Iran: Journal of Petroleum Geology, v. 3, p. 61–89.

Ali, A. R., and S. J. Silwadi, 1989, Hydrocarbon potential of Paleozoic pre-Khuff clastics in Abu Dhabi, United Arab Emirates: Proceedings, 6th Society of Petroleum Engineers Middle East Oil Show, March 1989, Bahrain, p. 819–832.

Alsharhan, A. S., 1985, Depositional environment, reservoir unit evaluation and hydrocarbon habitat of Shuaiba Formation, Lower Cretaceous, Abu Dhabi, United Arab Emirates: AAPG Bulletin, v. 69, p. 899–912.

Ameen, M. S., 1991, Alpine geowarpings in the Zagros-Taurus range: influence on hydrocarbon generation, migration and accumulation: Journal of Petroleum Geology, v. 14, p. 417–428.

Arthur, M. A., and S. O. Schlanger, 1979, Cretaceous "oceanic anoxic events" as causal factors in development of reef-reservoired giant oil fields: AAPG Bulletin, v. 63, p. 870–885.

Ayres, M. G., M. Bilal, R. W. Jones, L. Stentz, M. Tartir, and A. O. Wilson, 1982, Hydrocarbon habitat in main producing areas, Saudi Arabia: AAPG Bulletin, v. 66, p. 1–9.

Barker, C., and P. E. Dickey, 1984, Hydrocarbon habitat in main producing areas, Saudi Arabia: discussion: AAPG Bulletin, v. 68, p. 108–109.

Bellen, R. C. van, 1956, The stratigraphy of the "Main Limestone" of the Kirkuk, Bai Hassan and Qarah Chauq Dagh structures in north Iraq: Journal of the Institute of Petroleum, v. 42, p. 233–263.

Berberian, M., and G. C. P. King, 1981, Towards a paleogeography and tectonic evolution of Iran: Canadian Journal of Earth Sciences, v. 18, p. 210–265.

Beydoun, Z. R., 1988, The Middle East: regional geology and petroleum resources: Beaconsfield, U.K., Scientific Press, 292 p.

Beydoun, Z. R., 1989, Hydrocarbon potential of the deep (pre-Mesozoic) formations in the Middle East Arab countries, *in* Hydrocarbon potential and exploration techniques: OAPEC/ADNOC Seminar on Deep Formations in the Arab Countries: Abu Dhabi, Oct. 1989, Proceedings Technical Papers, p. 31–84.

Beydoun, Z. R., 1991, Arabian plate hydrocarbon geology and potential—a plate tectonic approach: AAPG Studies in Geology, n. 33, 77 p.

Bishlawy, S. H. El-, 1985, Geology and hydrocarbon occurrences of the Khuff Formation in Abu Dhabi, United Arab Emirates: Proceedings, 4th Society of Petroleum Engineers Middle East Oil Show, March 1985, Bahrain, p. 9-24.

Blinten, J. S., and I. A. Wahid, 1983, A review of Sajaa field development, Sharjah, United Arab Emirates: Proceedings, 3rd Society of Petroleum Engineers Middle East Oil Show, Bahrain, March 1983, p. 601-606.

Bordenave, M. L., and R. Burwood, 1990, Source rock distribution and maturation in the Zagros orogenic belt: provenance of the Asmari and Bangestan reservoir oil accumulations: Organic Geochemistry, v. 16, p. 369-387.

Brennan, P., 1985, Middle Cretaceous carbonate reservoirs, Fahud field and northwestern Oman: discussion: AAPG Bulletin, v. 69, p. 809-812.

Brown, G. F., D. L. Schmidt, and A. C. Huffman, Jr., 1989, Geology of the Arabian Peninsula: shield area of western Saudi Arabia: USGS Professional Paper 560-A.

Burruss, R. C., K. R. Cercone, and P. M. Harris, 1985, Timing of hydrocarbon migration evidenced from fluid inclusions in calcite cements, tectonics and burial history, *in* N. Schneiderman and P. M. Harris, eds., Carbonate cements: SEPM Special Publication 36, p. 277-289.

Cherven, V. B., 1986, Tethys-marginal sedimentary basins in western Iran: GSA Bulletin, v. 97, p. 516-522.

Clarke, R. H., 1975, Petroleum formation and accumulation in Abu Dhabi: Proceedings, 9th Arab Petroleum Congress, Dubai, 120(B-3), 20 p.

Colman-Sadd, S. P., 1978, Fold development in Zagros Simply folded belt, southwest Iran: AAPG Bulletin, v. 62, p. 984-1003.

Combaz, A., 1986, The Silurian Gamma zone of Saharian areas: organic content and condition of sedimentation: Document BRGM 10, Greco 52-CNRS, France.

Daniel E. J., 1954, Fractured reservoirs of the Middle East: AAPG Bulletin, v. 38, p. 774-815.

Dercourt, J., L. P. Zonenshain, L. E. Ricou, V. G. Kazmin, X. Le Pichon, A. L. Knipper, C. Grandjaquet, I. M. Sbortchikov, J. Geyssant, C. Lepvrier, D. H. Pechersky, J. Boulin, J. C. Sibuet, L. A. Savostin, O. Sorokhtin, M. Westphal, M. L. Bazhenov, J. P. Lauer, and B. Biju-Duval, 1986, Geologic evolution of the Tethys belt from the Atlantic to the Pamirs since the Lias: Tectonophysics, v. 123, p. 241-315.

Dunnington, H. V., 1958, Generation, migration, accumulation and dissipation of oil in northern Iraq, *in* L. G. Weeks, ed., Habitat of oil, a symposium: Tulsa, Oklahoma, AAPG, p. 1194-1251.

Dunnington, H. V., 1960, Some problems of stratigraphy, structure and oil migration affecting the search for oil in Iraq: 2nd Arab Petroleum Congress Proceedings, Beirut, 18(B-2), 25 p.

Dunnington, H. V., 1967, Stratigraphical distribution of oilfields in the Iran-Iraq-Arabia basin: Journal of the Institute of Petroleum, v. 53, p. 129-161.

Dunnington, H. V., 1974, Aspects of Middle East oil geology: Proceedings, Geological Principles of World Oil Occurrences, Banff, Alberta, p. 89-156.

Falcon, N. L., 1958, Position of oilfields in southwest Iran with respect to relevant sedimentary basins, *in* L. G. Weeks, ed., Habitat of oil, a symposium: Tulsa, Oklahoma, AAPG, p. 1252-1278.

Falcon, N. L., 1967, The geology of the northeast margin of the Arabian basement shield: Advancement of Science, v. 24, p. 31-42.

Falcon, N. L., 1969, Problems of the relationship between surface structure and deep displacements illustrated by the Zagros Range, *in* P. E. Kent, G. E. Satterthwaite, and A. M. Spencer, eds., Time and place in orogeny: The Geological Society (London) Special Publication 3, p. 9-22.

Falcon, N. L., 1974, Southern Iran—Zagros Mountains, *in* A. M. Spencer, ed., Mesozoic-Cenozoic orogenic belts: The Geological Society (London) Special Publication 4, p. 199-211.

Gass, I. G., 1981, Pan African (Upper Proterozoic) plate tectonics of the Arabian-Nubian shield, *in* A. Kroner, ed., Precambrian plate tectonics: Amsterdam, Elsevier, p. 387-405.

Glennie, K. W., M. G. A. Boeuf, M. W. Hughes Clarke, M. Moody-Stuart, W. F. N. Pilaar, and B. M. Reinhardt, 1973, Late Cretaceous nappes in Oman Mountains and their geologic evolution: AAPG Bulletin, v. 57, p. 5-27.

Gorin, G. E., L. G. Racz, and M. R. Walter, 1982, Late Precambrian-Cambrian sediments of Huqf Group, Sultanate of Oman: AAPG Bulletin, v. 66, p. 2609-2627.

Grantham, P. J., G. W. M. Lijmbach, J. Posthuma, M. W. Hughes Clarke, and R. J. Willink, 1988, Origin of crude oils in Oman: Journal of Petroleum Geology, v. 11, p. 61-80.

Greenwood, W. R., R. E. Anderson, R. J. Fleck, and D. C. Schmidt, 1980, Precambrian geologic history and plate tectonic evolution of the Arabian shield: Saudi Arabia Ministry of Petroleum and Mineral Resources, Mineral Resources Bulletin 24, 35 p.

Greig, D. A., 1958, Oil horizons in the Middle East, *in* L. G. Weeks, ed., Habitat of oil, a symposium: Tulsa, Oklahoma, AAPG, p. 1182-1193.

Grunau, H. R., 1977, Generation, migration, entrapment and retention of hydrocarbons in the Middle East: Petroleum Times (June 10), p. 33-43.

Grunau, H. R., 1985, Future of hydrocarbon exploration in the Arab World, *in* Source and habitat of petroleum in the Arab countries: Proceedings, OAPEC/ADNOC Seminar, Oct. 1984, Kuwait, p. 451-499.

Hassan, T. H., and S. Azer, 1985, The occurrence and origin of oil in offshore Abu Dhabi, Proceedings, 4th Society of Petroleum Engineers Middle East Oil Show, March 1985, Bahrain, p. 143-155.

Hempton, M. R., 1987, Constraints on Arabian plate motion and extensional history of the Red Sea: Tectonics, v. 6, p. 687-705.

Henson, F. R. S., 1950a, Discussion of papers by G. M. Lees, F. R. S. Henson and others: Report of the 18th International Geological Congress, Great Britain, 1948, v. 6, p. 68-73.

Henson, F. R. S., 1950b, Cretaceous and Tertiary reef formations and associated sediments in Middle East: AAPG Bulletin, v. 34, p. 215-238.

Henson, F. R. S., 1951, Observations on the geology and

petroleum occurrences of the Middle East: Proceedings, 3rd World Petroleum Congress, The Hague, sec. 1, p. 118-140.

Hughes Clark, M. W., 1988, Stratigraphy and rock unit nomenclature in the oil-producing area of interior Oman: Journal of Petroleum Geology, v. 11, p. 5-59.

Hull, C. E., and H. R. Warman, 1970, Asmari oil fields of Iran, *in* M. T. Halbouty, ed., Geology of giant petroleum fields: AAPG Memoir 14, p. 428-437.

Husseini, M. I., 1988, The Arabian Infracambrian extensional system: Tectonophysics, v. 148, p. 93-103.

Husseini, M. I., 1989, Tectonic and depositional model of late Precambrian-Cambrian Arabian and adjoining plates: AAPG Bulletin, v. 73, p. 1117-1131.

Ibrahim, M. W., 1983, Petroleum geology of southern Iraq: AAPG Bulletin, v. 67, p. 97-130.

Ibrahim, M. W., 1984, Geothermal gradients and geothermal oil generation in southern Iraq: a preliminary investigation: Journal of Petroleum Geology, v. 7, p. 77-85.

International Petroleum Encyclopedia, 1990 (v. 23): Tulsa, Oklahoma, Penn Well Publishing Co.

Jackson, J. A., 1980, Reactivation of basement faults and crustal shortening in orogenic belts: Nature, v. 283, p. 343-346.

Jackson, J. A., T. J. Fitch, and D. P. McKenzie, 1981, Active thrusting and the evolution of the Zagros fold belt, *in* K. R. McClay and N. J. Price, eds., Thrust and nappe tectonics: The Geological Society (London) Special Publication 9, p. 371-379.

Jackson, J. A., and D. P. McKenzie, 1984, Active tectonics of the Alpine-Himalayan Belt between western Turkey and Pakistan: Geophysical Journal of the Royal Astronomical Society, v. 77, p. 185-264.

James, G. A., and J. G. Wynd, 1965, Stratigraphical nomenclature of Iranian Oil Consortium Agreement Area: AAPG Bulletin, v. 49, p. 2182-2245.

Kadinski-Cade, K., and M. Barazangi, 1982, Seismotectonics of southern Iran: the Oman Line: Tectonics, v. 1, p. 389-412.

Kashfi, M. S., 1980, Stratigraphic and environmental sedimentology of Lower Fars Group (Miocene), south-southwest Iran: AAPG Bulletin, v. 64, p. 2095-2107.

Kashfi, M. S., 1984, A source bed study of the Oligo-Miocene Asmari Limestone in SW Iran: Journal of Petroleum Geology, v. 7, p. 419-428.

Kassler, P., 1973, The structural and geomorphic evolution of the Persian Gulf, *in* B. H. Purser, ed., The Persian Gulf: New York, Springer-Verlag, p. 11-32.

Kent, P. E., 1979, Emergent Hormuz salt plugs of southern Iran: Journal of Petroleum Geology, v. 2, p. 117-144.

Kent, P. E., and H. R. Warman, 1972, An environmental review of the world's richest oil-bearing region: the Middle East: Proceedings, 24th International Geological Congress, Montreal, sec. 5, p. 142-152.

Kent, P. E., F. C. Slinger, and A. N. Thomas, 1951, Stratigraphical exploration surveys in southwest Persia: Proceedings, 3rd World Petroleum Congress, The Hague, sec. 1, p. 141-161.

Khosravi Said, A., 1987, Geochemical concepts on origin, migration and entrapment of oil in southwest Iran, *in* R. B. Kumar, P. D. Wivedi, V. Banerjie, and V. Gupta, eds., Petroleum geochemistry and exploration in the Afro-Asian region: Rotterdam, Balkema, p. 531-539.

Koch, R. D., and G. D. Taylor, 1968, How Iran Consortium drillers finally and successfully killed M.I.S. gas leak: Oil and Gas International, v. 8, p. 94-104.

Koop, W. J., and R. Stoneley, 1982, Subsidence history of the Middle East Zagros basin, Permian to Recent: Philosophical Transactions of the Royal Society of London, Series A, v. 305, p. 149-168.

Lees, G. M., 1933a, The source rocks of Persian oil: Proceedings, 1st World Petroleum Congress, London, v. 1, p. 3-6.

Lees, G. M., 1933b, Reservoir rocks of Persian oilfields: AAPG Bulletin, v. 17, p. 229-240.

Lees, G. M., 1950, Some structural and stratigraphical aspects of the oilfields of the Middle East: Report of the 18th International Geological Congress, Great Britain, 1948, v. 6, p. 26-33.

Lees, G. M., and F. D. S. Richardson, 1940, The geology of the oil-field belt of S.W. Iran and Iraq: Geological Magazine, v. 77, p. 227-252.

Lees, G. M., and N. L. Falcon, 1952, The geographical history of the Mesopotamian Plains: Geographical Journal, v. 118, p. 24-39.

Loutfi, G., and S. El Bishlawy, 1986, Habitat of hydrocarbon in Abu Dhabi, United Arab Emirates, *in* OAPEC seminar, the hydrocarbon potential of intense thrust zones, Abu Dhabi, Dec. 1986: Proceedings Technical Papers and Case Studies, v. II, p. 63-124.

Lovelock, P. E. R., T. L. Potter, E. B. Walsworth-Bell, and W. M. Weimer, 1981, Ordovician rocks in the Oman Mountains: the Amdeh Formation: Geologie en Mijnbouw, v. 60, p. 487-495.

Marjeby, A. Al-, and D. Nash, 1986, A summary of the geology and oil habitat of the Eastern Flank hydrocarbon province of south Oman: Marine and Petroleum Geology, v. 3, p. 306-314.

Mashadani, A. Al-, 1986, Hydrodynamic framework of the petroleum reservoirs and cap rocks of the Mesopotamian basin of Iraq: Journal of Petroleum Geology, v. 9, p. 89-110.

Mattes, B. W., and S. Conway Morris, 1990, Carbonate/evaporite deposition in the late Precambrian-Early Cambrian Ara Formation of southern Oman, *in* A. H. F. Robertson, M. P. Searle, and A. C. Ries, eds., The geology and tectonics of the Oman region: The Geological Society (London) Special Publication 49, p. 617-637.

McClure, H. A., 1978, Early Paleozoic glaciation in Arabia: Palaeogeography, Palaeoclimatology, Palaeoecology, v. 25, p. 315-326.

McGillavray, J. G., and M. I. Husseini, 1992, The Paleozoic petroleum geology of central Arabia: AAPG Bulletin, v. 72, n. 10.

McQuillan, H., 1973, Small-scale fracture density in Asmari Formation of southwest Iran and its relation to bed thickness and structural setting: AAPG Bulletin, v. 57, p. 2367-2385.

McQuillan, H., 1974, Fracture patterns on Kuh-e-Asmari anticline, southwest Iran: AAPG Bulletin, v. 58, p. 236-246.

Momenzadeh, M., 1990, Saline deposits and alkaline magmatism, a genetic model: Journal of Petroleum Geology, v. 13, p. 341-356.

Murris, R. J., 1980, The Middle East: stratigraphic evolution and oil habitat: AAPG Bulletin, v. 64, p. 597-618.

Murris, R. J., 1981, Middle East: stratigraphic evolution and oil habitat: Geologie en Mijnbouw, v. 60, p. 467-486.

Newell, K. D., and R. D. Hennington, 1983, Potential petroleum source-rock deposition in the middle Cretaceous Wasia Formation, Rub'Al Khali, Saudi Arabia: Proceedings, Society of Petroleum Engineers Middle East Oil Show, March 1983, Bahrain, p. 151-160.

Ni, J., and M. Barazangi, 1986, Seismotectonics of the Zagros continental collision zone and a comparison with the Himalayas: Journal of Geophysical Research, B, v. 91, p. 8205-8218.

NIOC-IOOC, 1965, Present status of natural gas in Iran: Proceedings, Seminar on Development and Utilization of Natural Gas Reserves, Tehran, Mineral Resources Development, Series 25, United Nations, New York, p. 64-82.

Patton, T. L., and S. J. O'Connor, 1986, Cretaceous flexural history of northern Oman mountains foredeep, United Arab Emirates: AAPG Bulletin, v. 72, p. 797-809.

Porter, J. W., 1992, Conventional hydrocarbon reserves of the Western Canada foreland basin, this volume.

Ricou, L. E., 1971, Le croissant ophiolitique péri-arabe, une ceinture de nappes mises en place en Cretacé superieur: Revue de Géographie Physique et du Géologie Dynamique (2e Serie), v. 13, p. 327-349.

Ricou, L. E., M. B. de Lepinay, and J. Marcoux, 1986, Evolution of the Tethyan seaways and implications for the oceanic circulation around the Eocene-Oligocene boundary, *in* C. Pomerol and I. Premoli Silva, eds., Terminal Eocene events: Amsterdam, Elsevier, p. 387-394.

Sakini, J. A. Al-, 1975, Usage of drainage characteristics in interpretation of subsurface structures in plains around Kirkuk fields: Journal of the Geological Society of Iraq Special Issue, p. 45-53.

Sengör, A. M. C., 1984, The Cimmeride Orogenic System and the tectonics of Eurasia: GSA Special Paper 195, 82 p.

Sengör, A. M. C., 1990, A new model for the late Paleozoic-Mesozoic tectonic evolution of Iran and implications for Oman, *in* A. H. F. Robertson, M. P. Searle, and A. C. Ries, eds., The geology and tectonics of the Oman region: The Geological Society (London) Special Publication 49, p. 797-831.

Sengör, A. M. C., and Y. Yilmaz, 1981, Tethyan evolution of Turkey: a plate tectonic approach: Tectonophysics, v. 75, p. 181-241.

Shaharistani, H. Al-, and M. J. Al Atyia, 1972, Vertical migration of oil in Iraq oilfields: evidence based on vanadium and nickel concentrations: Geochemica Cosmochimica Acta, v. 36, p. 929-938.

Short, N. M., P. D. Lowman, Jr., and S. C. Freden, 1976, Mission to earth: Landsat views the world: United States National Aeronautics and Space Administration Special Publication 360, 459 p.

Snyder, D. B., and M. Barazangi, 1986, Deep crustal structure and flexure of the Arabian plate beneath Zagros collisional mountain belt as inferred from gravity observations: Tectonics, v. 5, p. 361-373.

Sonnenfeld, P., 1985, Evaporites as oil and gas source rocks: Journal of Petroleum Geology, v. 8, p. 253-271.

Stacey, J. C., and R. A. Agar, 1985, U-Pb isotopic evidence for the accretion of a continental microplate in the Zalm region of the Saudi Arabian shield: Journal of the Geological Society of London, v. 142, p. 1189-1203.

Stocklin, J., 1986, The Vendian-Lower Cambrian salt basins of Iran, Oman and Pakistan: stratigraphy, correlations, paleogeography: Sciences de la Terre Memoir 47, p. 329-345.

Stocklin, J., and A. O. Setudehnia, 1972, Iran, *in* L. Dubertret, ed., Asie: Lexique Stratigraphique International, v. III, CNRS Paris, fasc. 9b, 376 p.

Stoesser, D. B., and V. E. Camp, 1985, Pan-African microplate accretion of the Arabian shield: GSA Bulletin, v. 96, p. 817-826.

Stoneley, R., 1981, The geology of the Kuh-e Dalneshin area of southern Iran and its bearing on the evolution of southern Tethys: Journal of the Geological Society of London, v. 138, p. 509-526.

Stoneley, R., 1987, A review of petroleum source rocks in parts of the Middle East, *in* J. Brooks and A. J. Fleet, eds., Marine petroleum source rocks: The Geological Society (London) Special Publication 26, p. 263-269.

Stoneley, R., 1990, The Middle East basin: a summary overview, *in* J. Brooks, ed., Classic petroleum provinces: The Geological Society (London) Special Publication 50, p. 293-298.

Sugden, W., 1962, Structural analysis and geometrical prediction of change of form with depth of some Arabian plains-type folds: AAPG Bulletin, v. 46, p. 2213-2228.

Sykes, R. M., and A. K. Abu Risheh, 1989, Exploration of deep Paleozoic and Pre-Cambrian plays in the Sultanate of Oman, *in* Hydrocarbon potential and exploration techniques: OAPEC/ADNOC Seminar on Deep Formations in the Arab Countries, Abu Dhabi, Oct. 1989, Proceedings Country Reports and Case Studies, p. E71-E113.

Thode, H. G., and J. Monster, 1970, Sulfur isotope abundance and genetic relations of oil accumulations in Middle East basin: AAPG Bulletin, v. 54, p. 627-637.

Vail P. R., R. M. Mitchum, Jr., and S. Thompson III, 1977, Seismic stratigraphy and global changes of sea level, part 4: global cycles of relative changes of sea level, *in* C. E. Payton, ed., Seismic stratigraphy—applications to hydrocarbon exploration: AAPG Memoir 26, p. 83-98.

Vaslet, D., 1989, Late Ordovician glacial deposits in Saudi Arabia: a lithostratigraphic revision of the early Paleozoic succession: Saudi Arabian Deputy Ministry for Mineral Resources Professional Paper 3, p. 13-44.

Weeks, L., 1950, Discussion of papers by G. M. Lees, F. R. S. Henson and others: Report of the 18th International Geological Congress, Great Britain, 1948, v. 6, p. 68-73.

Weeks, L. G., 1958, Habitat of oil and some factors that control it, *in* L. G. Weeks, ed., Habitat of oil, a symposium: Tulsa, Oklahoma, AAPG, p. 1-61.

Wells, P. R. A., 1987, Hydrodynamic trapping of oil and gas in the Cretaceous Nahr Umr Lower Sand of the North Area, offshore Qatar: Proceedings, 5th Society of Petroleum Engineers Middle East Oil Show, March 1987, Bahrain, p. 17-26.

Wilson, H. H., 1982, Hydrocarbon habitat in main producing areas, Saudi Arabia: discussion: AAPG Bulletin, v. 66, p. 2688-2691.

Young, A., P. H. Monaghan, and R. T. Schweisberger, 1977, Calculation of ages of hydrocarbons in oils—physical chemistry applied to petroleum geochemistry 1: AAPG Bulletin, v. 61, p. 573-600.

Chapter 12

Petroleum Geology of the Eastern Venezuela Foreland Basin

R. N. Erlich and S. F. Barrett

Amoco Production Company
Houston, Texas, U.S.A.

ABSTRACT

The Eastern Venezuela foreland basin is a structurally and stratigraphically complex foreland basin that contains the single largest oil accumulation in the world, as well as several other supergiant oil and gas fields. The development of these large hydrocarbon resources was a result of the widespread deposition of rich Cretaceous source rocks, multiple thick sandstone reservoirs, and long-term structural deformation.

This paper reviews the general tectonostratigraphic development of the Eastern Venezuela foreland basin in light of recent petroleum exploration and production activities. This recent work, when combined with detailed stratigraphic data compiled during the past four decades, has helped to constrain and refine models of the geohistory of northeastern Venezuela.

Data accumulated during nearly 100 years of oil exploration suggest that, despite the relative exploration maturity and drilling density in the basin, substantial undiscovered hydrocarbon resources may still be found along the present trend of giant fields.

INTRODUCTION

The Eastern Venezuela foreland basin is the collective name given to two foreland subbasins in the northeastern part of Venezuela: the Guarico subbasin in the west and the Maturin subbasin in the east (Figures 1, 2). The boundaries of the Eastern Venezuela basin are the Precambrian rocks of the Guayana shield to the south, the El Baul arch to the west, oceanic crust of the equatorial Atlantic to the east, and the Cordillera de la Costa/Villa de Cura and Araya/Paria igneous and metamorphic belts to the north (Mascle and Letouzey, 1990). The boundary that separates the Guarico and Maturin subbasins is the Urica arch (Figure 2), which was intermittently active during the Miocene to Holocene (Young et al., 1956; Dallmus, 1965; Gonzalez de Juana et al., 1980). Total area of the basin is about 165,000 km^2 (64,000 mi^2) (CEPET, 1989; Aymard et al., 1990).

The sedimentary sequence south of the metamorphic belts generally is considered to be an autochthonous part of the South American plate (e.g., Bellizzia, 1972; Erlich and Barrett, 1990; Pindell and Dewey, 1991) and consists of mainly Cretaceous and Tertiary passive margin and foreland basin rocks that unconformably overlie lower Paleo-

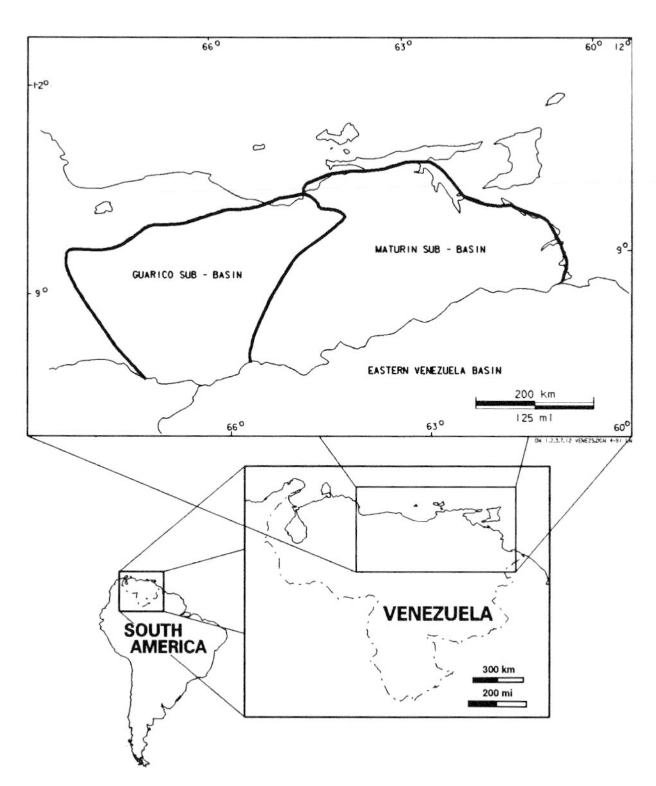

FIGURE 1. Regional map of the Eastern Venezuela basin.

zoic sediments and Paleozoic igneous and Precambrian metamorphic basement (Feo Codecido et al., 1984) (Figures 3, 4). In contrast, the origin of the igneous and metamorphic rocks that form the northern margin of the basin is less clear. Some workers consider these sections to be allochthonous to their present locations, having been transported from a few tens of kilometers to as much as 1000 km (620 mi) prior to their emplacement (Bellizzia, 1972; Stephan, 1977; Skerlec and Hargraves, 1980; Benjamini et al., 1987; Erlich and Barrett, 1990). Others consider these rocks to be part of the autochthonous passive margin section, thrust up from great depths by oblique convergence between the Caribbean and South American plates during the Tertiary (Speed, 1985; Pindell et al., 1991).

Two belts of folded sedimentary rocks are found between the Eastern Venezuela foreland basin and the igneous and metamorphic terrains to the north. These fold belts have fundamentally different origins and surface expressions (Figures 2, 3, 4). Low-angle folds and thrusts produced during emplacement of the Cordillera de la Costa/Villa de Cura allochthon are rooted to shallow detachments in the relatively thin sedimentary cover (Blanco and Sanchez, 1990; Orihuela and Franklin, 1990). In contrast, the high-amplitude folds and thrusts of the Serrania del Interior have been interpreted as having thin-skinned, upper crustal roots, but may involve tens of kilometers of crustal shortening (Lander et al., 1990; Algar et al., 1991).

Several northwest-southeast-trending right-lateral strike-slip faults (Urica, San Francisco, Los Bajos) cut the subbasins at their northern edges, but appear to have had only local influence on the formation of structures (Figure 2) (Martin and Espinoza, 1990; Fiume and Graterol, 1990). East-west faults such as the El Pilar have relatively minor offsets (40–125 km; 25–78 miles) and do not currently appear to be major components of the plate-boundary system (Schubert, 1984; Erlich and Barrett, 1990). The same is true for the northwest-southeast-trending Urica, San Francisco, and Los Bajos faults (displacements of 35 km/22 mi, 25 km/16 mi, and 10.5 km/6.5 mi, respectively), although motion on the faults generally is youngest to the east. Motion on the Bohordal fault appears to be mostly vertical (down to the east), with a minimum of 1520 m (5000 ft) of offset (Feo Codecido et al., 1984).

LATE JURASSIC–HOLOCENE TECTONIC HISTORY

The post-Paleozoic tectonic history of the basin has three main phases: Late Jurassic rifting, Cretaceous–Paleogene passive margin, and Paleogene–Quaternary strike-slip, compression/transpression, and foreland basin development. The Eastern Venezuela basin was superimposed on an area that was the interior of the supercontinent of Pangea; the sparse Paleozoic sedimentary record and the lack of marine Paleozoic rocks (Stover, 1967; Dirección de Geología, 1970; Gonzalez de Juana et al., 1980; Feo Codecido et al., 1984) suggest that, prior to rifting, the area was mostly in a continental to marginal marine setting.

Phase 1: Rifting

The rifting that affected the northern margin of South America was part of the opening of the central North Atlantic, when North America separated from Gondwana. Opening of the central North Atlantic was diachronous, being oldest in the north, and youngest in the south between North and South America (Pindell and Dewey, 1982). Two separate branches of the sea floor-spreading system formed between North America and Yucatan (Gulf of Mexico branch), and between Yucatan and northern South America (Pindell, 1985). The sedimentary record suggests that the rifting along the eastern Venezuela part of northern South America was at least pre-Barremian, whereas radiometric dates on igneous rocks indicate a Late Triassic–Early Jurassic age (Macdonald and Opdyke, 1974; Feo Codecido et al., 1984; Moticska, 1985). Triassic and Jurassic La Quinta Formation red beds occur in Colombia and northwestern Venezuela, and appear to be present as Late Jurassic basalts and red beds in the Espino graben of the Eastern Venezuela basin (Gonzalez de Juana et al., 1980; Feo Codecido et al., 1984; Moticska, 1985) (Figure 2).

The conjugate margins of the postulated rift system (Yucatan and northern South America) are similar and show evidence of minor crustal stretching during rifting. The resultant subsidence history shows no evidence of major initial rapid subsidence, therefore suggesting a somewhat different geohistory than that of a typical rifted passive margin. The apparent lack of extensive crustal stretching suggests that the initial crustal rupture was primarily by shearing resulting from strike-slip or transform faulting, rather than primarily by extension (Pindell, 1985).

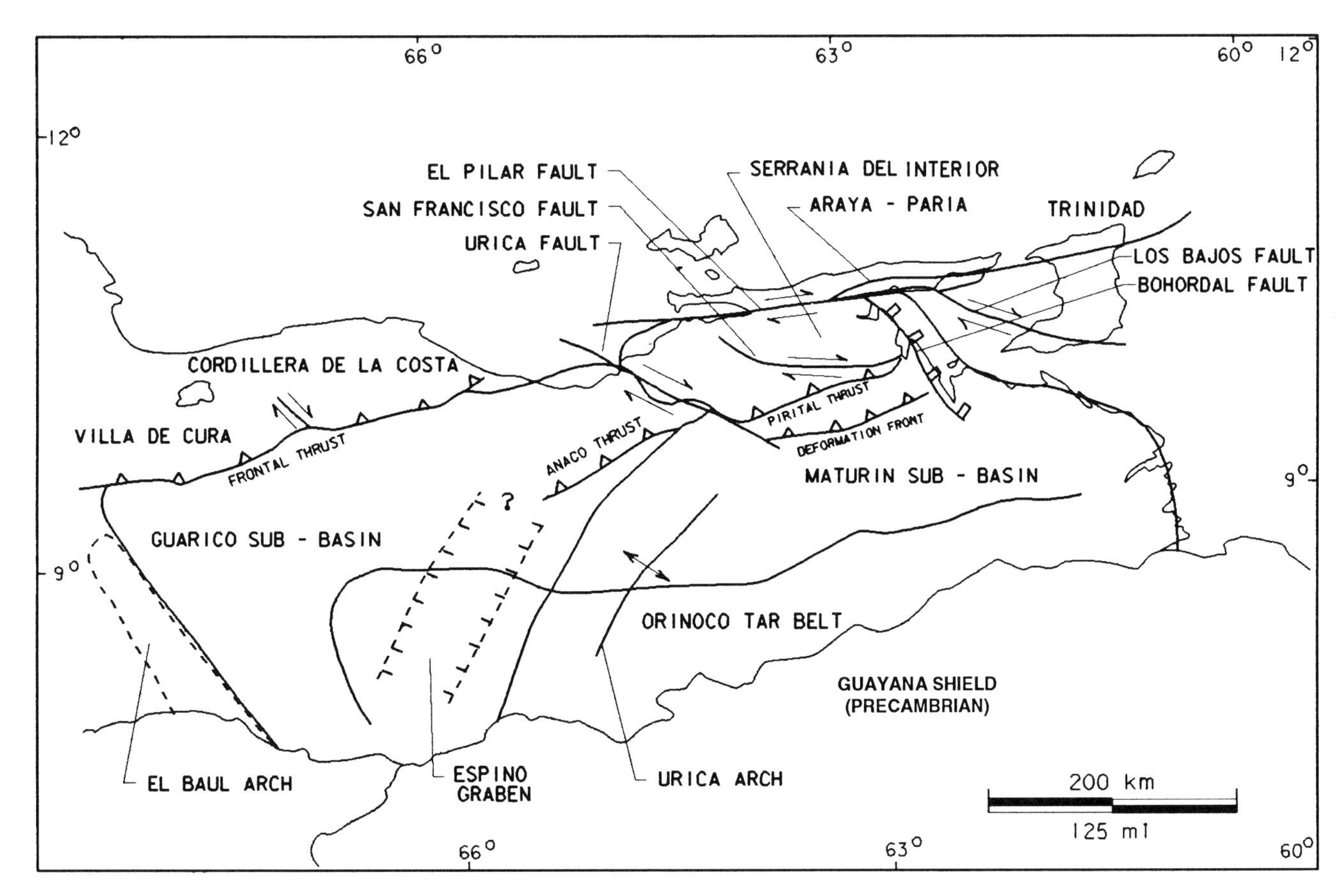

FIGURE 2. Major structural features of the Eastern Venezuela basin.

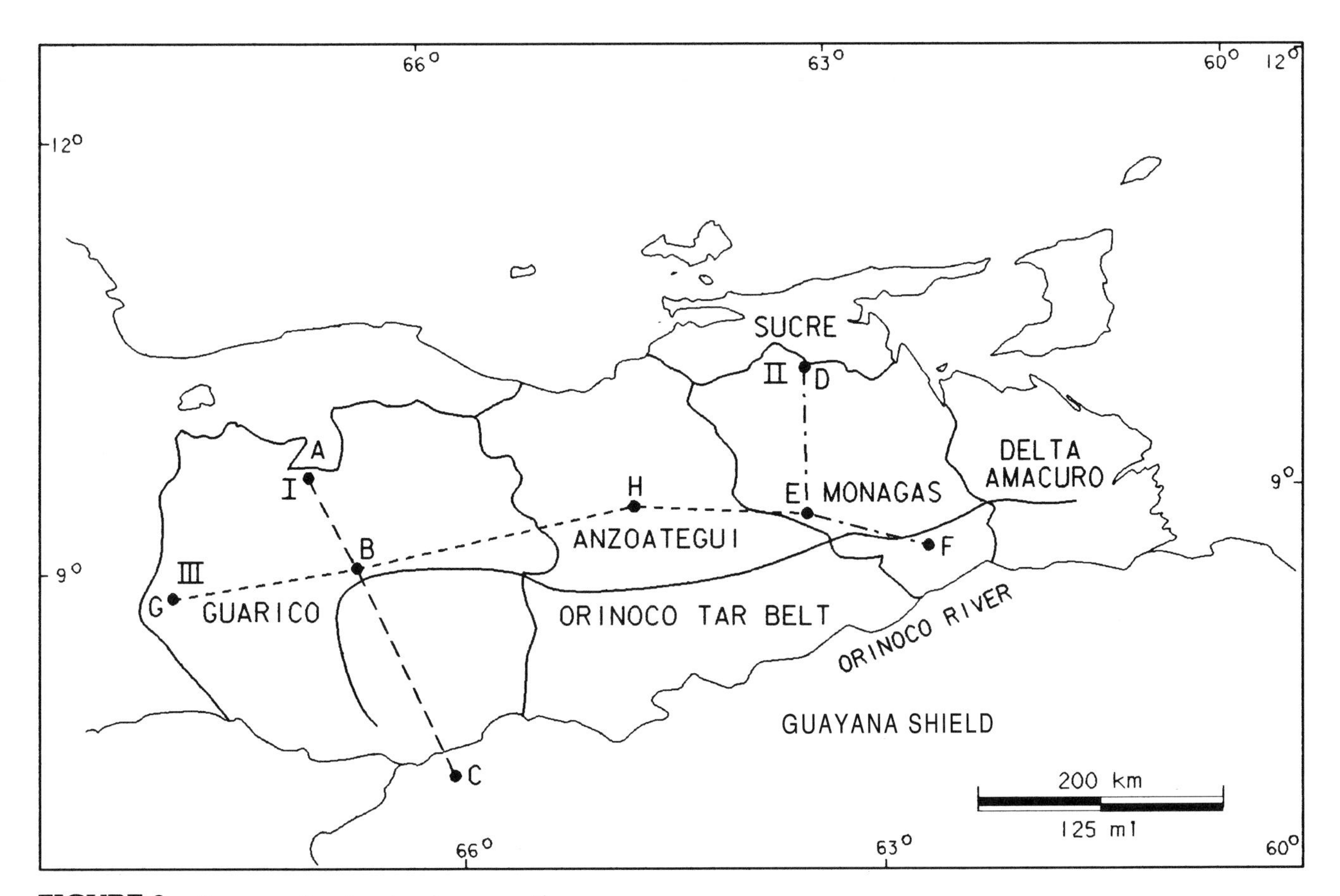

FIGURE 3. Base map for structural cross sections.

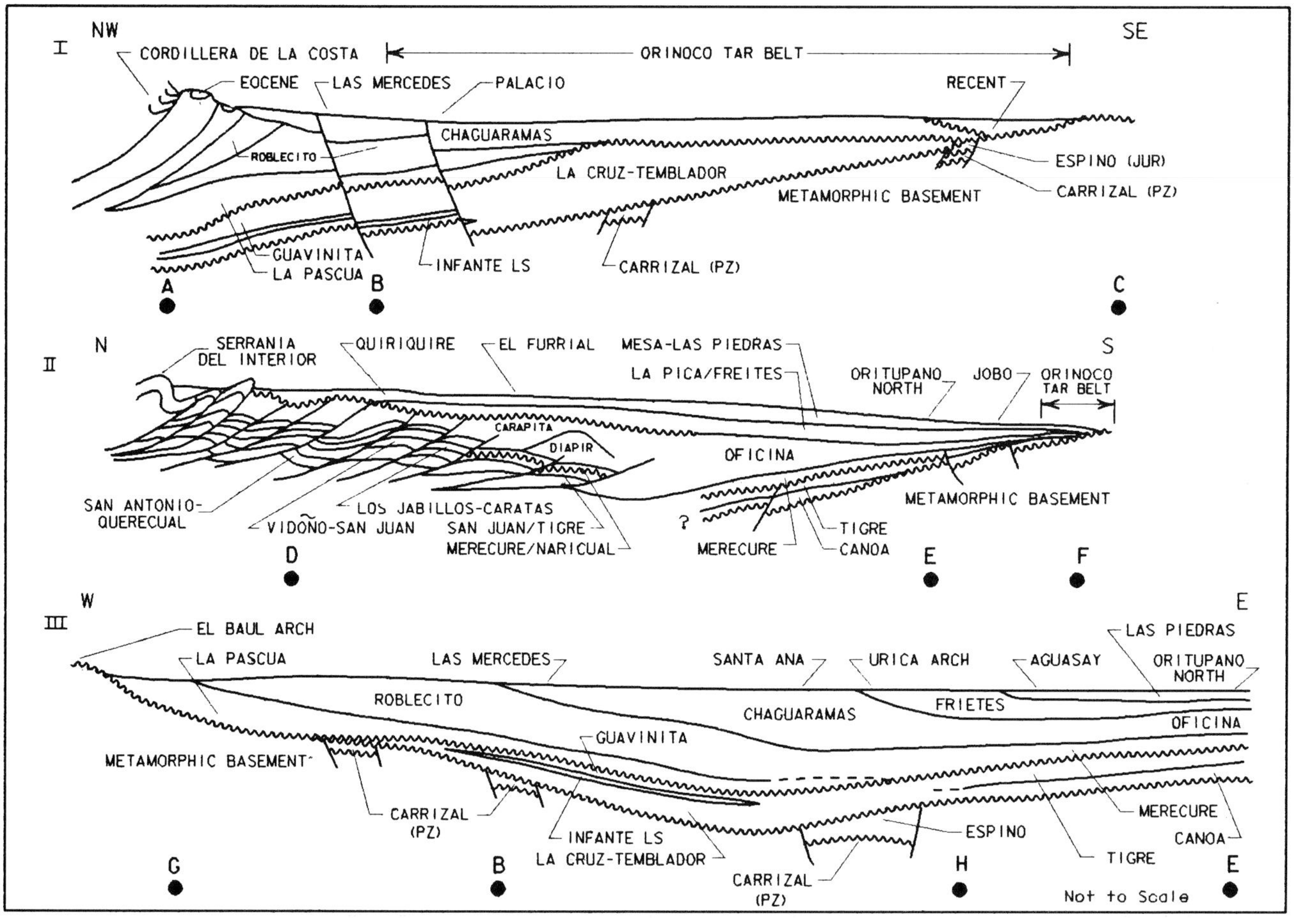

FIGURE 4. Cross sections through the Guarico and Maturin subbasins (see Figure 3 for base map), not to scale. Dots represent stratigraphic section reference points (see Figure 3) for each cross section (Figures 6A–6D). Note that cross sections have been modified from Figures 6A–6D because of complex stratigraphy and structure, and may not match reference sections exactly. Adapted from Patterson and Wilson (1953), Young et al. (1956), Gonzalez de Juana et al. (1980), Carnevali (1988), Subieta et al. (1988), CEPET (1989), Lander et al. (1990).

Phase 2: Passive Margin

This phase spans the end of rifting in the Late Jurassic or earliest Cretaceous to the onset of active tectonism in the Eocene. The northern margin of South America subsided enough to permit the accumulation of 3 to 4 km (about 12,000 ft) of predominantly marine clastic rocks, although the Early Cretaceous section also contains several thick carbonate units.

Phase 3: Strike-Slip and Compression/ Transpression

The passive margin phase of basin development apparently ended diachronously along the northern margin of South America. The onset of strike-slip and compression/ transpression is older in western Venezuela and northern Colombia (early-middle Eocene) than in the east (late Oligocene-middle Miocene in eastern Venezuela and Trinidad). Deformation throughout northern Venezuela has continued to the present, probably as a result of the eastward motion of the Caribbean relative to South America (Erlich and Barrett, 1990; Pindell and Barrett, 1990; Pindell and Dewey, 1991). During this phase of foreland basin development, the Guarico and Maturin subbasins were superimposed on the older passive margin sequences.

REGIONAL STRATIGRAPHY

The style and timing of plate boundary processes were critical factors in determining the amounts and types of sedimentary fill within the Guarico and Maturin subbasins, and the preserved record of these sediments varies greatly between the basins. For example, the present thickness of the Guarico subbasin sedimentary section is about 5 km (16,400 ft) at its maximum, although the thickness averages about 3 to 4 km (9800 to 13,000 ft) over most of the basin (Patterson and Wilson, 1953; Gonzalez de Juana et al., 1980; Orihuela and Franklin, 1990). This does not include more than 3 km (9800 ft) of sedimentary section eroded since the Oligocene (Figures 3, 4) (Patterson and Wilson, 1953; CEPET, 1989).

In contrast, the maximum thickness of the Maturin subbasin has been estimated from seismic, gravity, and magnetics data at 8.5 to 14 km (28,000 to 46,000 ft), with 1.0 to 6.1 km (3280 to 20,000 ft) of section eroded from the uplifted passive margin sequence in the Serrania del Interior (Figures 3, 4) (Hedberg, 1950; Subieta et al., 1988; Talukdar et al., 1988; CEPET, 1989; Aymard et al., 1990; Fernandez and Passalacqua, 1990; Lander et al., 1990).

Pre-Cretaceous

Pre-Cretaceous sedimentary rocks are found in isolated areas of the southern Guarico subbasin (Figures 3, 4, 5A) and are restricted to Cambrian sandstones of the Hato Viejo Formation and Cambro-Ordovician (to Carboniferous?) sandstones and shales of the Carrizal Formation (Patterson and Wilson, 1953; Gonzalez de Juana et al., 1980; Feo Codecido et al., 1984; Moticska, 1985; Fernandez and Passalacqua, 1990). Well data show that the Hato Viejo is at least 90 m (300 ft) thick and the Carrizal is at least 640 m (2100 ft) thick (Hedberg, 1950; Stover, 1967; Gonzalez de Juana et al., 1980; Feo Codecido et al., 1984; Moticska, 1985); they are separated from the overlying Cretaceous Temblador Group or Late Jurassic red beds by an angular unconformity.

Rocks of pre-Cretaceous age may have originally been more widespread prior to initial rifting and erosion in the Middle to Late Jurassic or Neocomian, although Triassic and Jurassic sedimentary rocks (other than the Espino/Ipire red beds of Moticska, 1985) do not occur south of the El Pilar fault and Cordillera de la Costa frontal thrust (Erlich and Barrett, 1990). Rift-related Jurassic igneous rocks do occur, however, in the El Baul area (Guacamayas Group) and in parts of the Guayana Precambrian shield (associated with rifting of the Takutu graben in Guyana) (Feo Codecido et al., 1984; Moticska, 1985).

Recent work (Algar et al., 1991) suggests that some of the metamorphic terrains of the Araya/Paria Peninsula and Northern Range of Trinidad are southward-thrusted sections of an autochthonous South American passive margin. One obvious difficulty with this latter idea is that thick sections of Carrizal and Hato Viejo-age rocks are preserved on the craton, but they do not exist in the Araya/Paria and Northern Range tectonic belts. In addition, rocks of similar age and lithology to those in the tectonic belts do not occur anywhere in the Eastern Venezuela basin.

Cretaceous

The thickness and facies distribution of Cretaceous rocks within the Guarico and Maturin subbasins is variable but predictable. The thickness of Cretaceous rocks varies mostly in a depositional dip (north-south) direction; in the south, the section is erosionally truncated against Precambrian rocks of the Guayana shield, but thickens rapidly northward to nearly 5 km (16,400 ft) (Hedberg, 1950; Patterson and Wilson, 1953; Gonzalez de Juana et al., 1980; Aymard et al., 1990).

The El Baul arch and the Guayana shield appear to have been the only significant structural influences on Cretaceous depositional patterns (Figure 2). Cretaceous and Paleogene sedimentary rocks thin across, or are erosionally truncated against, both features, and facies patterns reflect this influence (Figures 3, 4) (Patterson and Wilson, 1953; Gonzalez de Juana et al., 1980; CEPET, 1989).

The lithologies of Cretaceous sedimentary rocks in each subbasin reflect typical passive margin depositional cycles; continental/fluvial/deltaic clastics of Neocomian (Barremian) age rest unconformably on Precambrian basement or early Paleozoic rocks, and grade vertically and laterally into conformable sequences of shallow marine clastics and carbonates of Aptian to Cenomanian age (Figures 5A, 5B, 6A-6D).

Drowning of the Early Cretaceous carbonate platform in the late Albian was followed by deposition of organic shales, limestones, and cherts throughout the Eastern Venezuela basin (Figure 7). This sequence represents the maximum marine transgression during the passive margin phase of deposition. Later Cretaceous and Tertiary units reflect the gradual infilling and shallowing of the basin, first by deep-water and finally by shallow-water clastics (Hedberg, 1950; Rod and Maync, 1954; Gonzalez de Juana et al., 1980; CEPET, 1989; Fasola and Paredes, 1991; Pindell and Dewey, 1991).

Tertiary

Passive margin, shallow-marine clastic deposition continued across the Cretaceous/Tertiary boundary over much of the Eastern Venezuela basin through the Paleogene (Figure 5B, C). Overthrusting of the northern part of the Guarico subbasin caused flexural loading and the formation of the present foreland basin to the south; nearly 3 km (9800 ft) of Eocene-Oligocene clastics were subsequently deposited in the basin (Patterson and Wilson, 1953; Blanco and Sanchez, 1990; Orihuela and Franklin, 1990). The last phases of overthrusting provided favorable conditions for the development of small, environmentally stressed carbonate platforms on isolated uplifts east of the major clastic depocenter (Figure 5C).

Miocene-Pliocene sediments are thickest in the Maturin subbasin. Estimates range from 4 to 9 km (13,000 to 29,500 ft) (Hedberg, 1950; Lander et al., 1990) in the axis of the basin. The deposition of these units was strongly influenced by the coeval uplift and flexural loading of the crust by the Serrania del Interior to the north (Figures 3, 4) (Erlich and Barrett, 1990). The Miocene-Pliocene section thins by deposition and erosion to the north and south, and by erosion to the west in the Guarico subbasin. The section thickens rapidly to the east, and along with Pleistocene clastics may make up much of the 14-km (46,000-ft) thick sedimentary sequence southeast of Trinidad (Erlich and Barrett, 1990).

DEPOSITIONAL SYSTEMS

Depositional systems of the Eastern Venezuela basin can be clearly divided into two types: those influenced by

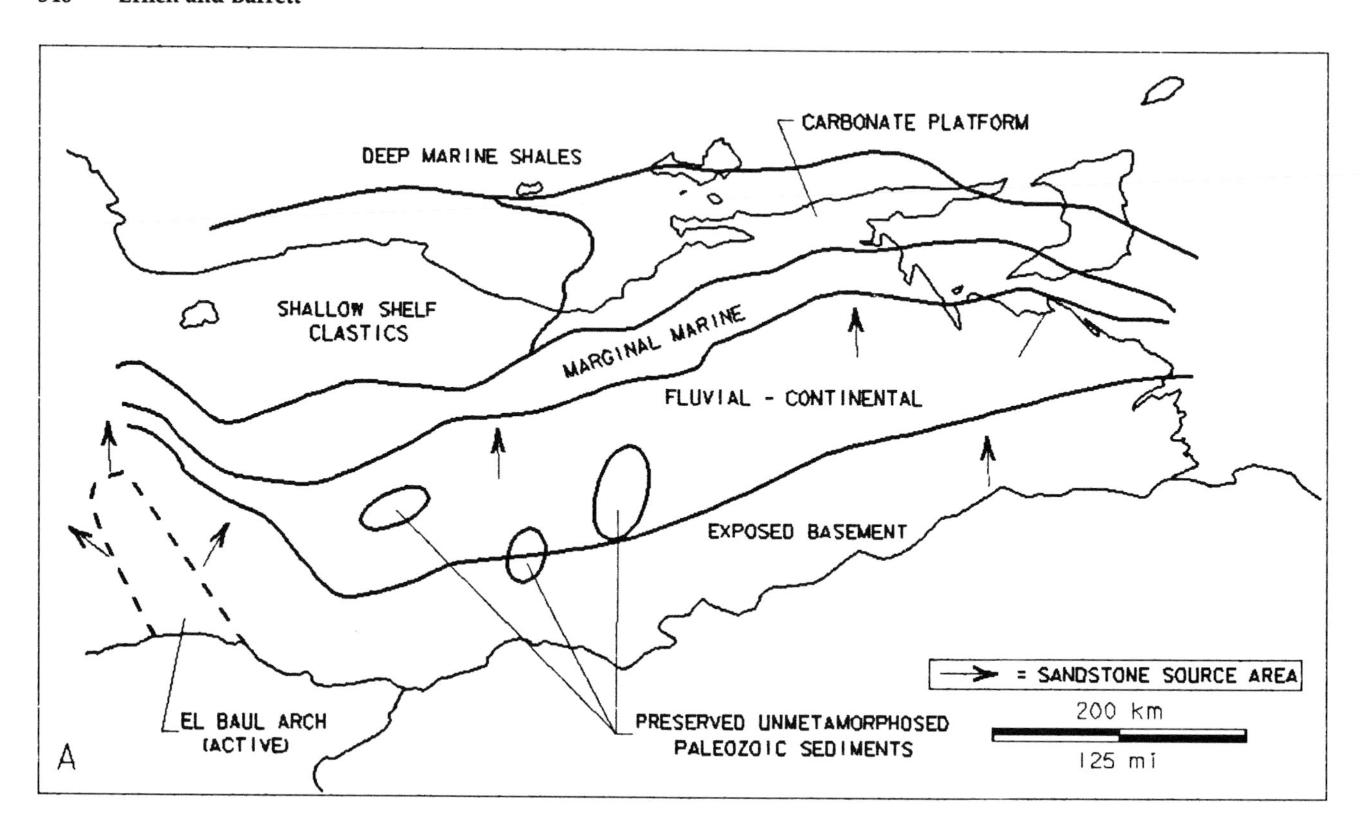

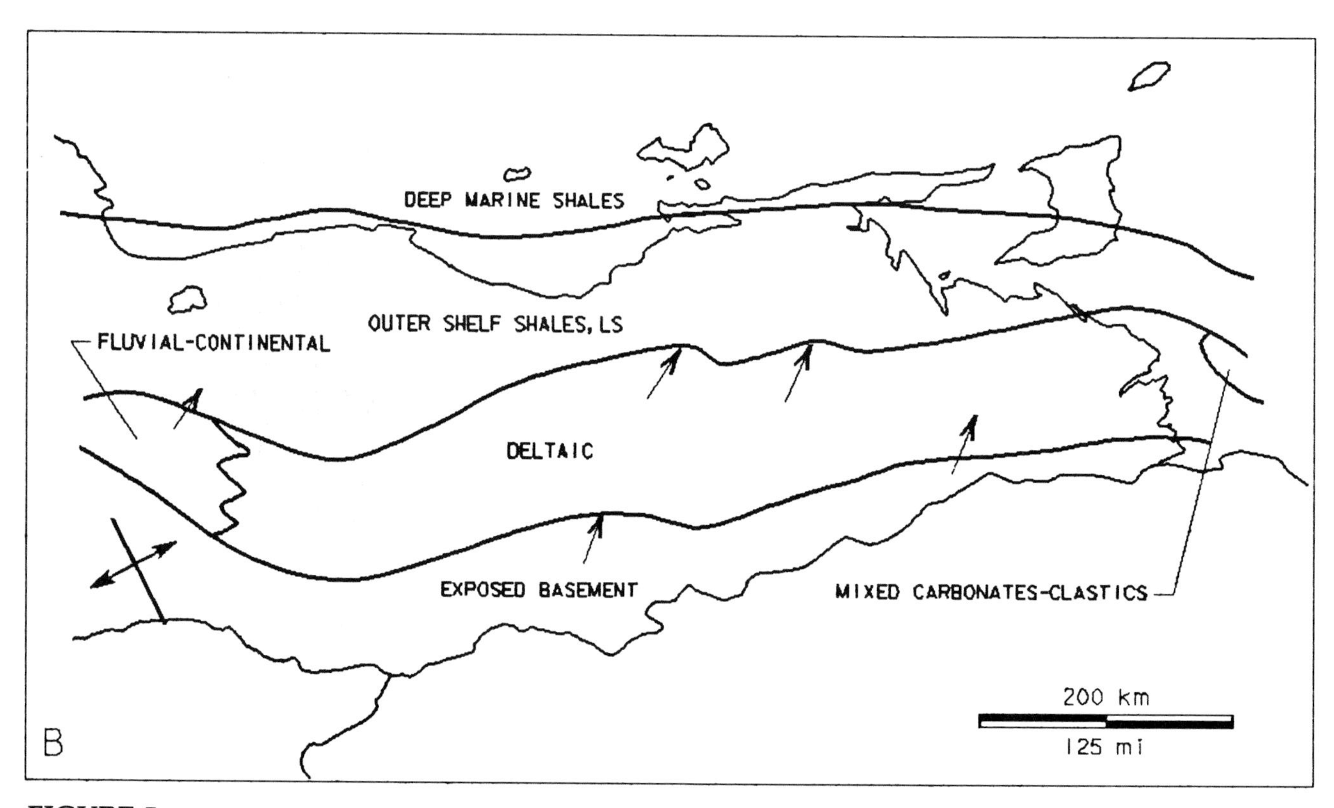

FIGURE 5. Paleogeography of northeastern Venezuela. (A) Barremian–middle Cenomanian. (B) Maastrichtian–Paleocene. (C) Eocene. Adapted from Gonzalez de Juana et al. (1980), Galea-Alvarez (1985), CEPET (1989).

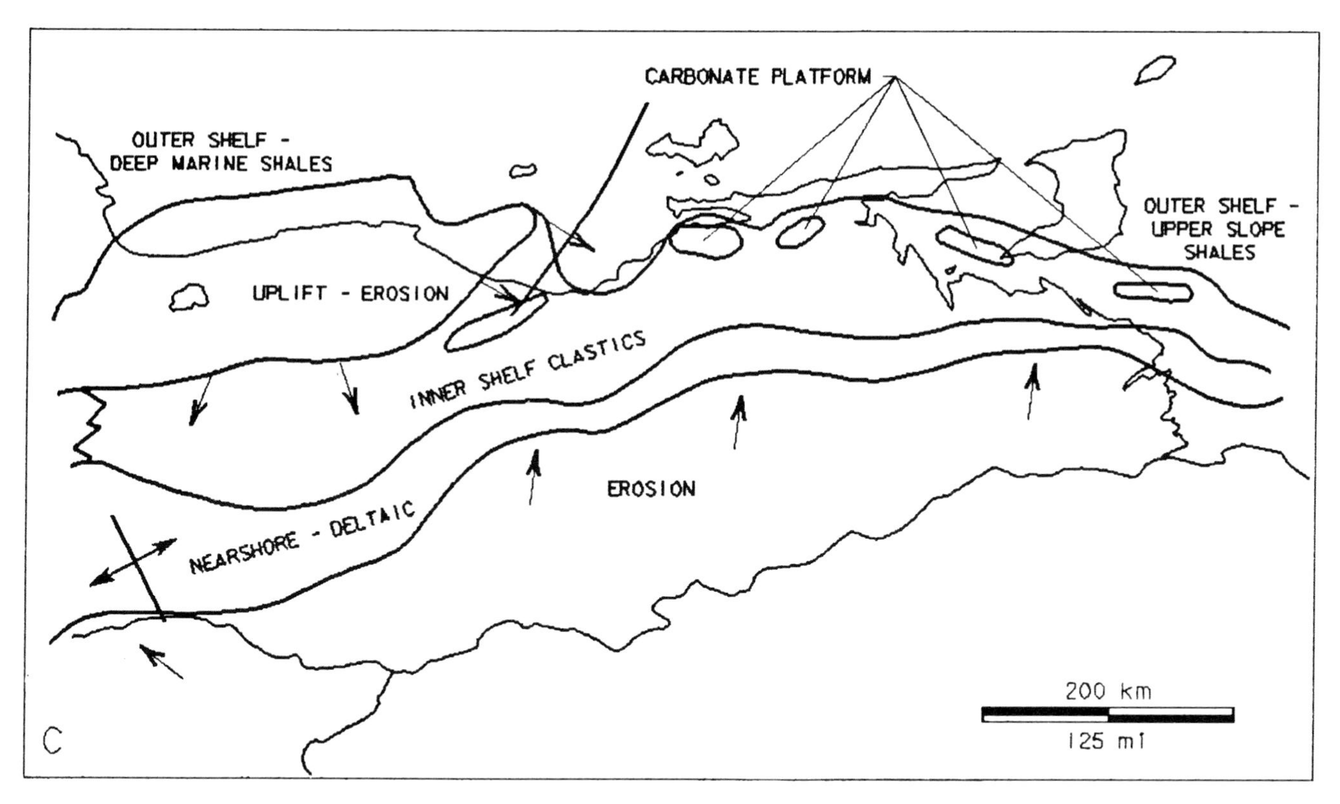

FIGURE 5. (Continued)

passive margin processes, and those influenced by foreland basin development.

Passive Margin Systems

Based on the sedimentary record of the Guarico and Maturin subbasins, passive margin deposition appears to have begun during the late Neocomian (latest Hauterivian-Barremian) with continental/fluvial/deltaic sandstones and shales of the Barranquin Formation (Figure 5A) (Hedberg, 1950; Osten, 1957; Gonzalez de Juana et al., 1980; Rossi et al., 1987). Maximum northward extent of the Barranquin shoreline is not known, although deltaic sandstones and coals occur on outcrop in the extreme northwest corner of the Maturin subbasin, in the Serrania del Interior (Barcelona area), and on islands southwest of the Araya Peninsula (Osten, 1957; Guillaume et al., 1972). Later overthrusting and folding of this area suggest that these sections were probably deposited farther to the north than their present locations.

The upper parts of the Barranquin clearly show the early influence of subsidence and marine transgression on the continental margin. Shallow-water limestones and marginal marine clastics extend southward through the Serrania, and may have been eroded from areas surrounding the Guayana shield (Figure 6B) (Rod and Maync, 1954; Osten, 1957). These rocks grade conformably up-section into shallow-water marine shales and lime wackestones and packstones of the El Cantil Formation in the northern Maturin subbasin (Figure 5A).

In the Guarico subbasin and the southern part of the Maturin subbasin, the first sedimentary rocks preserved are nearshore/marginal marine clastics of the early Aptian-Santonian Temblador Group. In the Maturin subbasin, the Temblador has been divided into upper glauconitic and lower mottled units—the Tigre and Canoa formations, respectively. The Temblador is primarily nonmarine at the base (lower-middle Canoa), grading up-section into marginal marine sandstones (upper Canoa-Tigre) (Hedberg, 1950; Gonzalez de Juana et al., 1980).

The Tigre Formation is subdivided into three members in the Guarico subbasin: the La Cruz at the base, the Infante limestone, and the Guavinita at the top. The formation contacts are generally gradational and conformable, ranging from mostly nonmarine sandstones and shales in the La Cruz, to lagoonal limestones of the Infante, and outer-shelf shales of the Guavinita (Patterson and Wilson, 1953; Gonzalez de Juana et al., 1980). The presence of lagoonal limestones of the Turonian-Coniacian Infante Member suggests that the Guarico subbasin did not initially subside as rapidly as the Maturin subbasin. One possible explanation for this may be that the El Baul arch acted as a structural "support," especially in the western and southern parts of the basin.

The deposition of shallow-water carbonate sediments in the Maturin subbasin reached a maximum during the late Albian (Figure 5A). El Cantil Formation lagoonal limestones in the northern Maturin subbasin interfingered with lagoonal limestones and nearshore shales of the Chimana and upper Canoa formations (Rossi et al., 1987).

Evidence of a large rudist barrier reef at the El Cantil shelf margin is rare, although displaced blocks of rudist

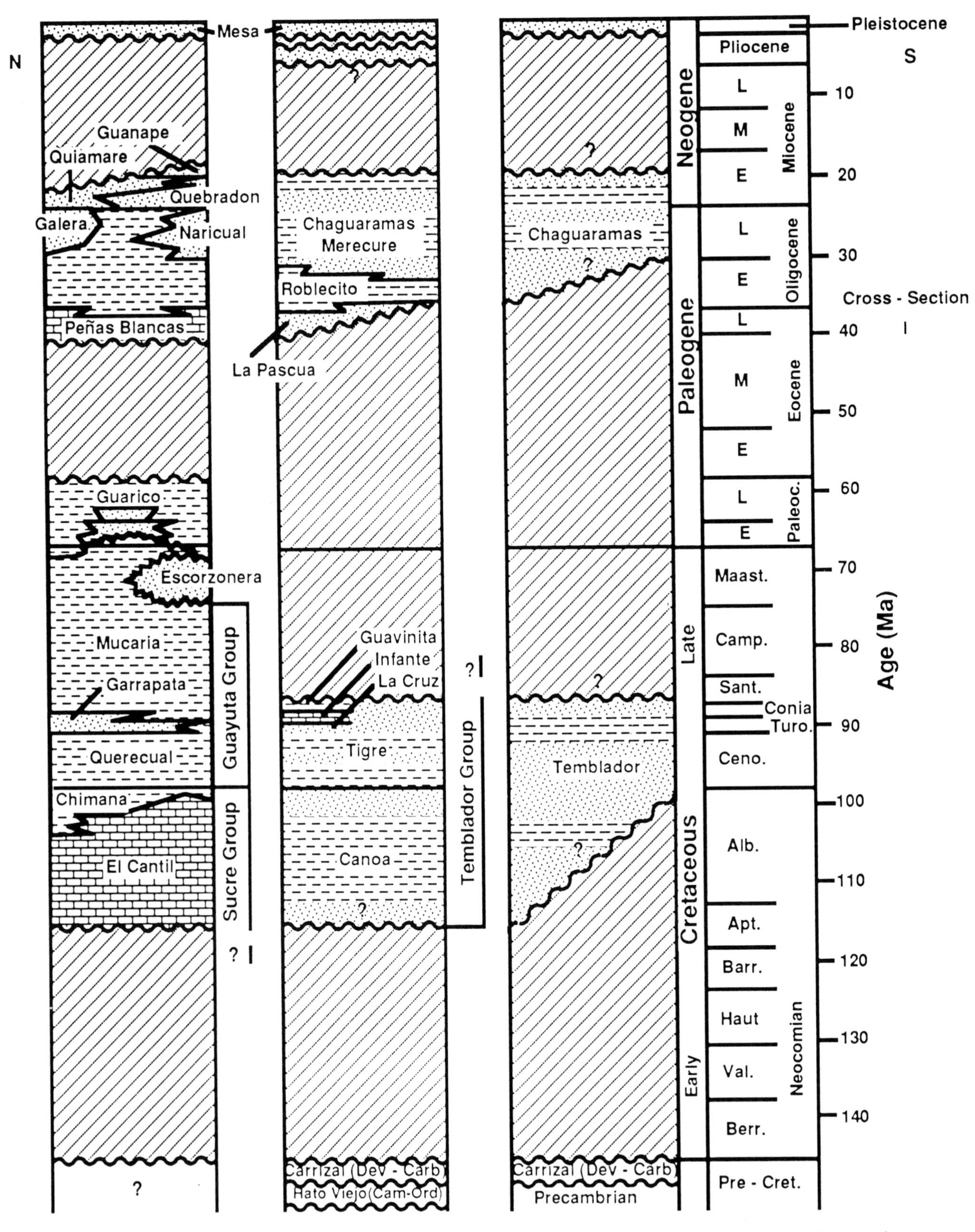

FIGURE 6A–6D. Lithostratigraphic charts for cross sections I, II, and III (see Figures 3 and 4). Differences between similar sections represent local stratigraphic changes within the reference area.

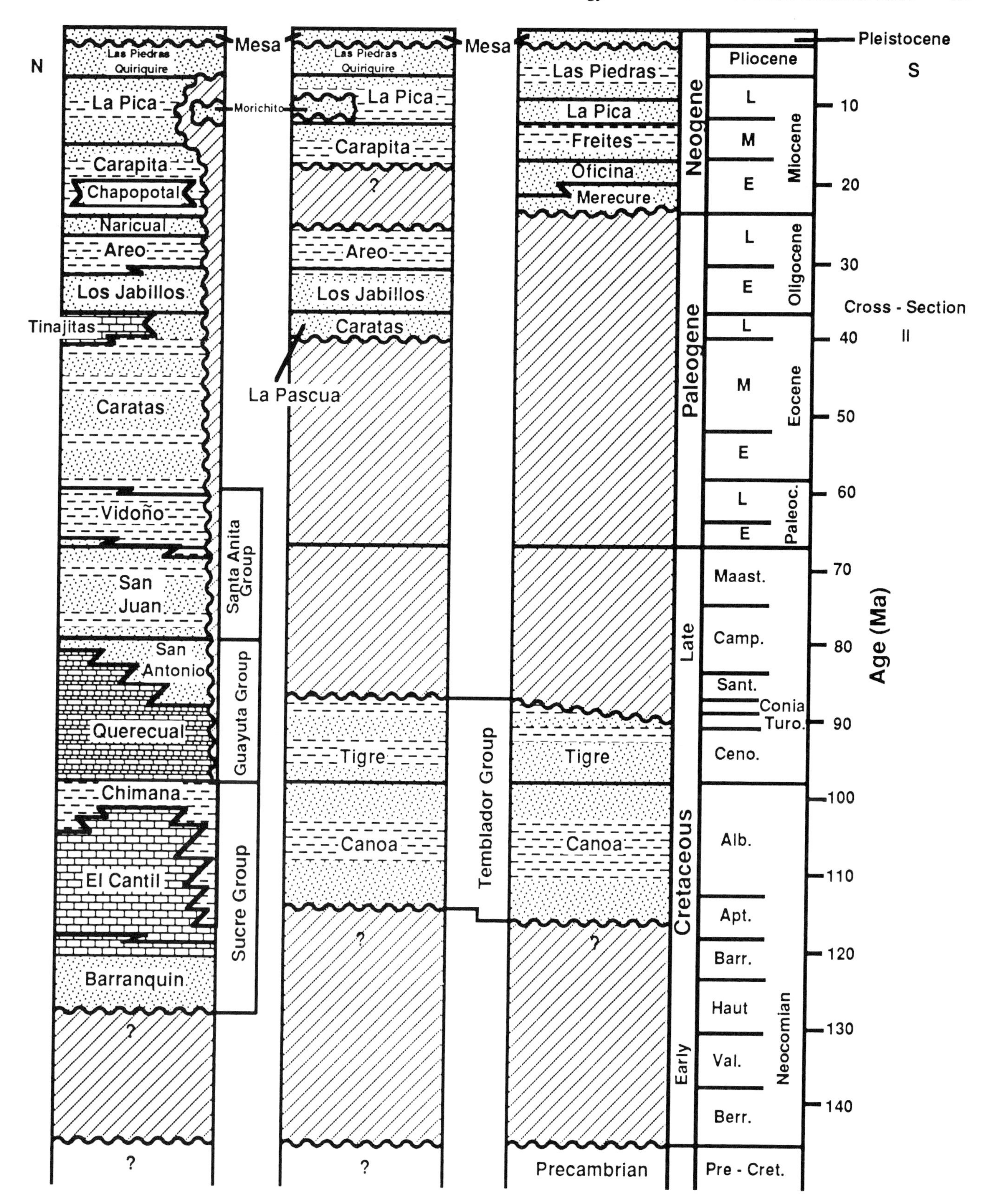

FIGURE 6B. See caption under Figure 6A.

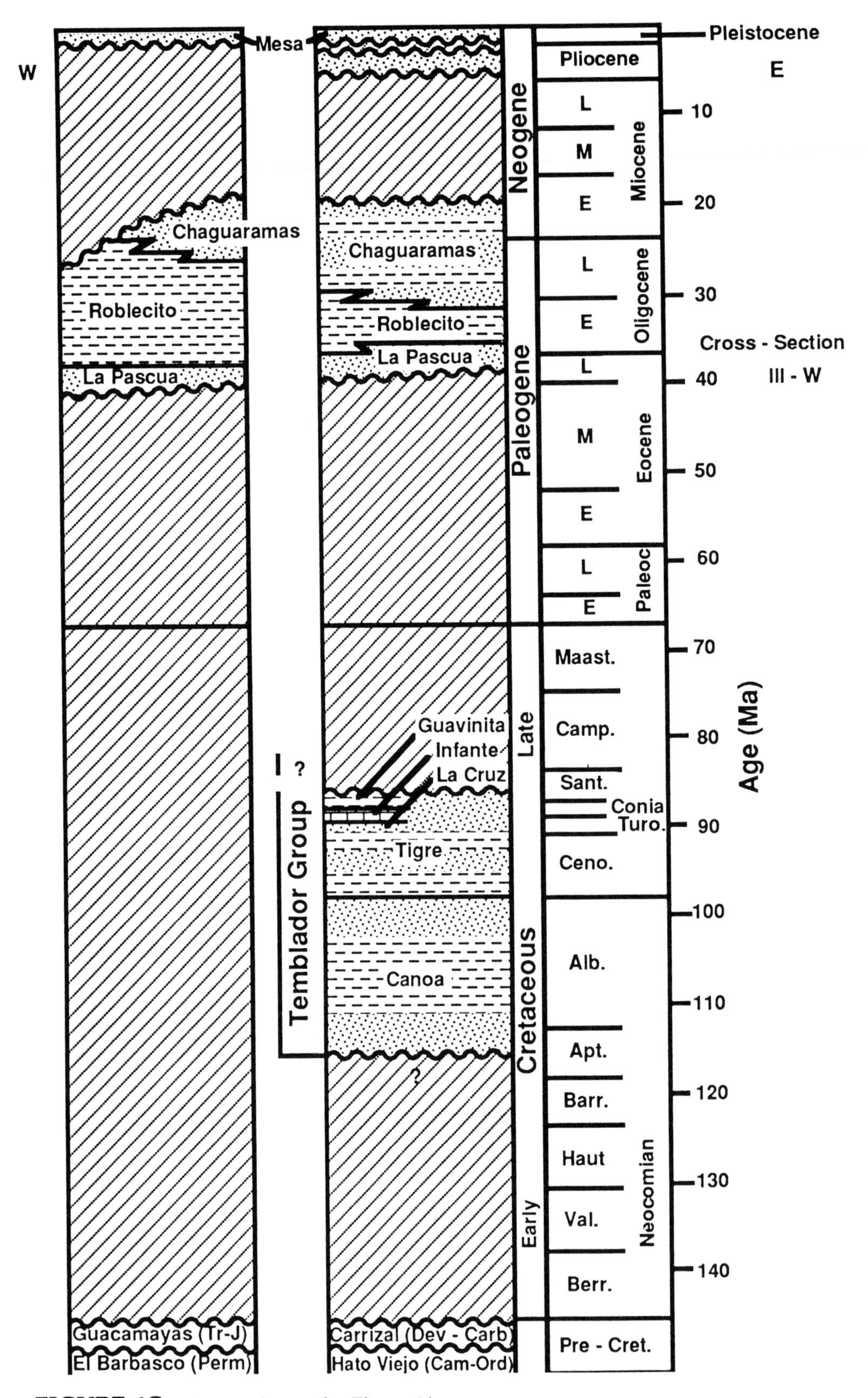

FIGURE 6C. See caption under Figure 6A.

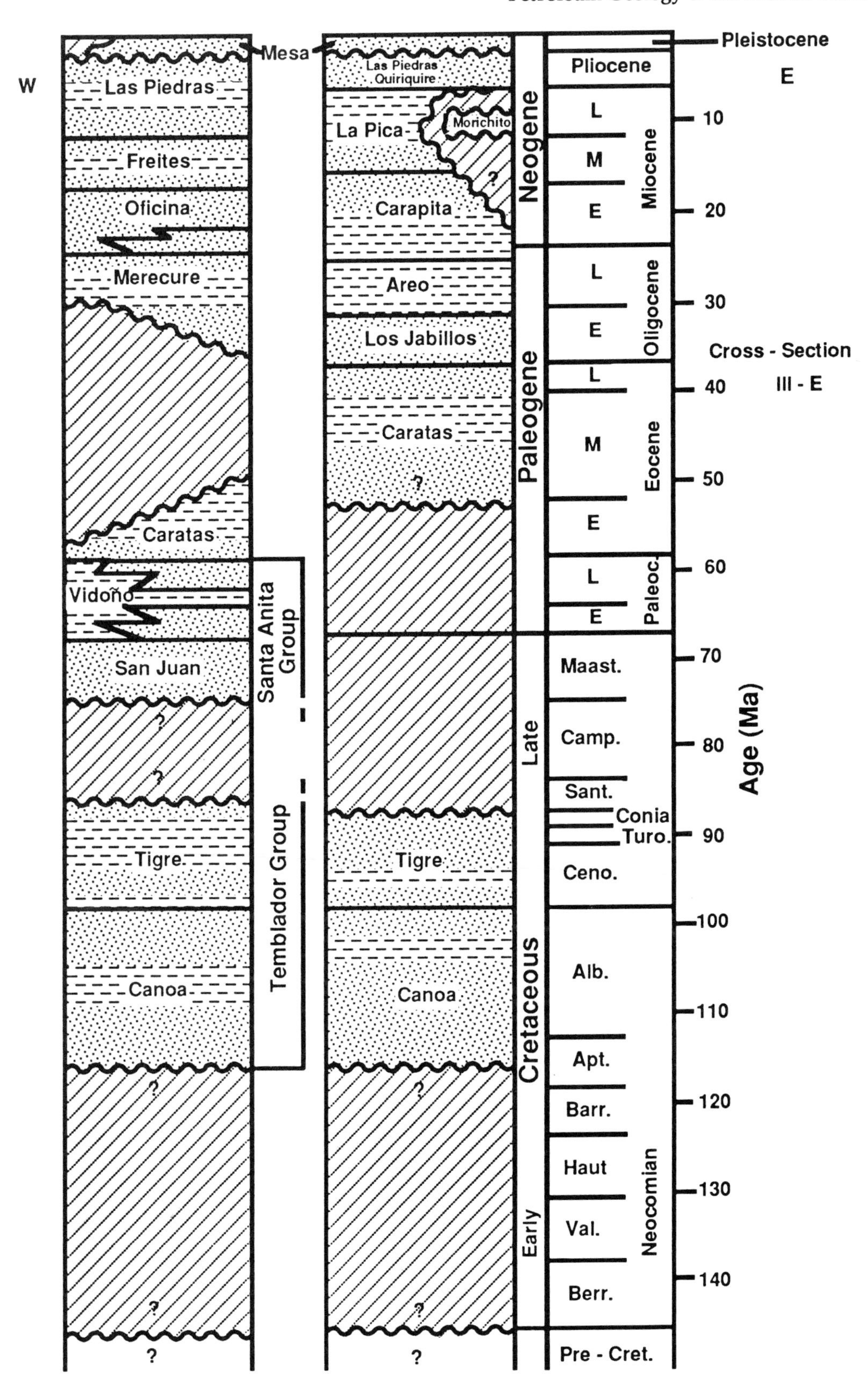

FIGURE 6D. See caption under Figure 6A.

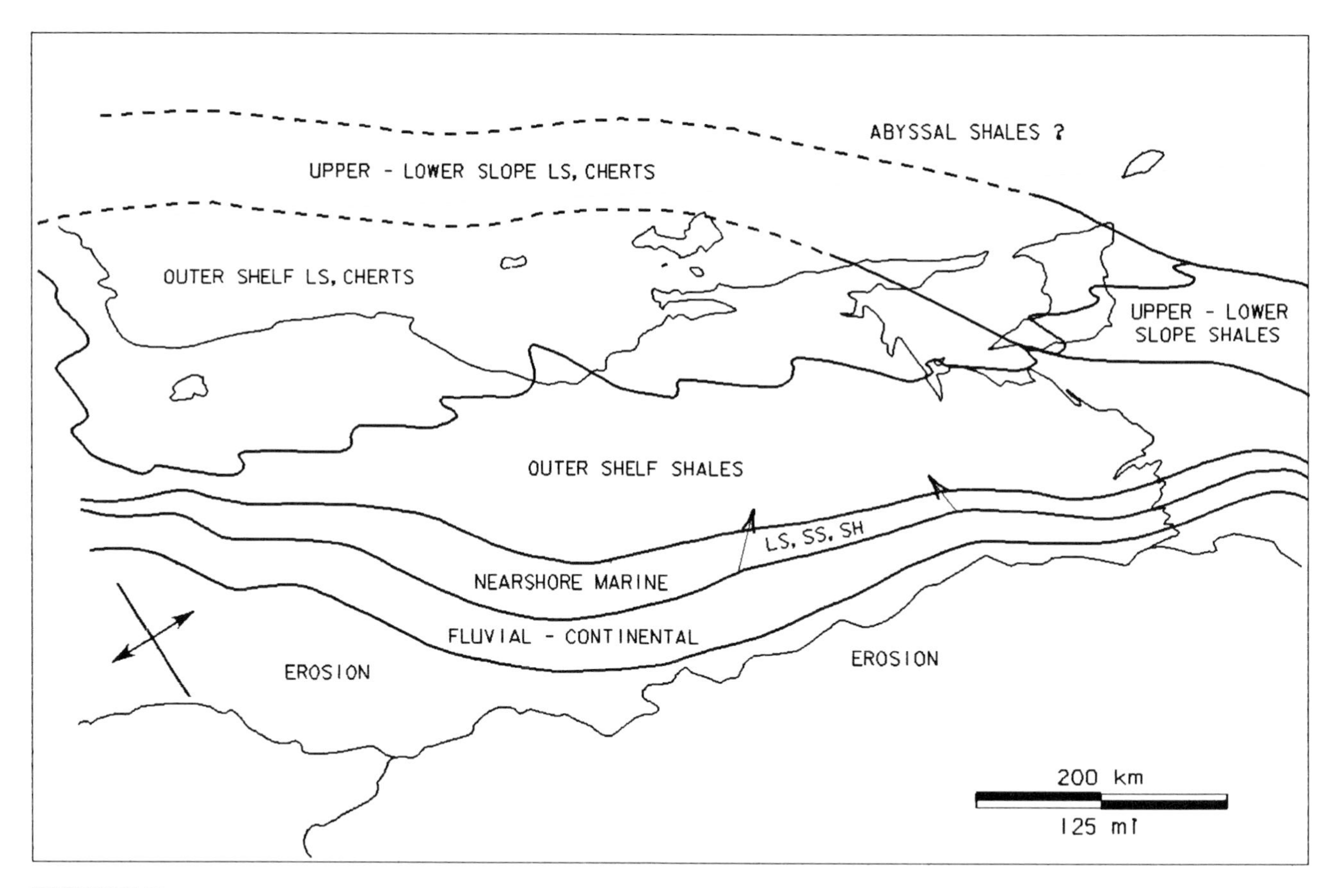

FIGURE 7. Paleogeography of northeastern Venezuela, middle Cenomanian to Campanian. Adapted from Gonzalez de Juana et al. (1980), Rossi et al. (1987), CEPET (1989).

boundstone have been found in the Casanay area of the northern Serrania (Gonzalez de Juana et al., 1980). Exotic blocks of rudist boundstone have also been found along the El Pilar fault in Trinidad (Erlich, 1987), and in Tertiary shales in central Trinidad (Wells, 1948; Kugler and Bolli, 1967).

Shallow-water carbonate sedimentation ceased over most of the Eastern Venezuela basin during the Albian. Glauconitic outer-shelf shales of the Chimana Formation conformably buried El Cantil shallow-water limestones, and grade conformably up-section into outer-shelf/slope siliceous shales and limestones of the Querecual Formation (Figure 7) (Hedberg, 1937, 1950; Rod and Maync, 1954). It should be noted, however, that this transition from shallow- to deep-water deposition may be represented in time by a condensed section. Guillaume et al. (1972) and Bolli et al. (1991) even interpreted a minor unconformity between the top of the El Cantil Formation and the overlying beds.

In other areas, this type of gradational sequence indicates drowning of a carbonate platform (Erlich et al., 1990). This suggests that El Cantil carbonate sedimentation was terminated by a combination of relative sea level rise flooding the shallow shelf and the accompanying release of excess nutrients from the inundated coastal areas (Hallock and Schlager, 1986; Erlich et al., 1991).

The Querecual Formation grades southward into shales and nearshore-marine sandstones of the Tigre Formation and westward into shales of the Guavinita Member of the Tigre Formation (Figure 7). The Querecual marks the maximum southward and westward marine transgression during the Cretaceous, extending into the Las Mercedes area of the central Guarico subbasin. In the northern parts of the Maturin and Guarico subbasins, the Querecual grades conformably into the overlying San Antonio and Mucaria formations (Hedberg, 1937, 1950; Rod and Maync, 1954).

Sandstones are generally absent from the shaly Mucaria Formation, except in the southern parts of the basin, where it grades shoreward (south) into the Escorzonera Formation (Figure 6A). The boundary between the San Antonio and the Querecual is also transitional, although it is often considered to occur at the first appearance of massive, gray, fine-grained sandstones (Hedberg, 1937, 1950). The depositional environment of the sandstones in the San Antonio is problematic, although interbedded black siliceous shales suggest at least an outer-shelf setting (e.g., Rossi et al., 1987).

Eocene to Oligocene uplift caused the erosion of Late Cretaceous and Paleocene strata from most of the southern parts of the Guarico and Maturin subbasins (Figures 5C, 8A). Data from outcrops in the Cordillera de la Costa and Serrania del Interior and wells in the northern parts of the subbasins show that formations are conformable across the Cretaceous/Tertiary boundary over much of the area (Rossi et al., 1987; Carnevali, 1988).

In the northern Maturin subbasin, the San Antonio is overlain by massive sandstones of the San Juan Formation (Figure 6B). Upper San Juan gray-olive shales grade up-

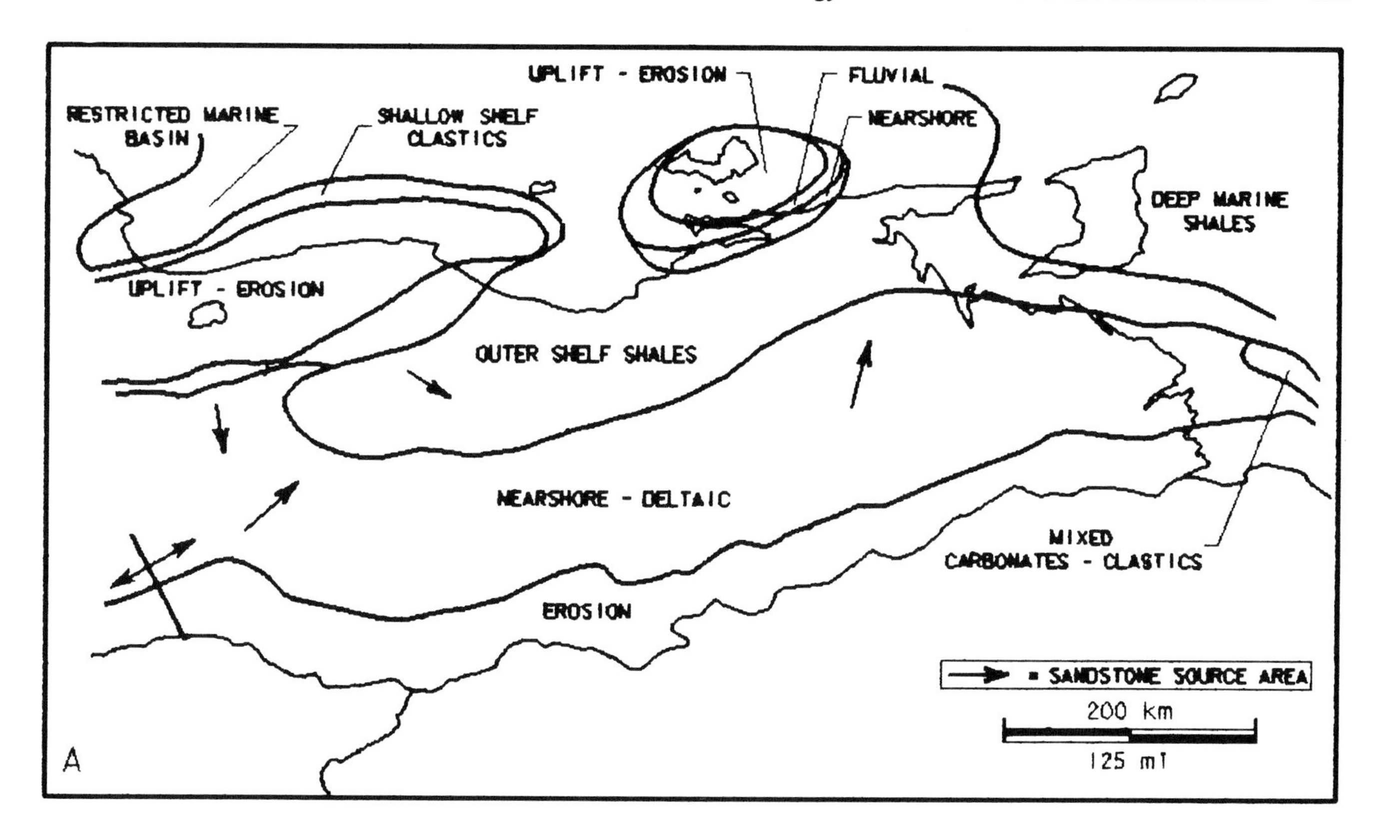

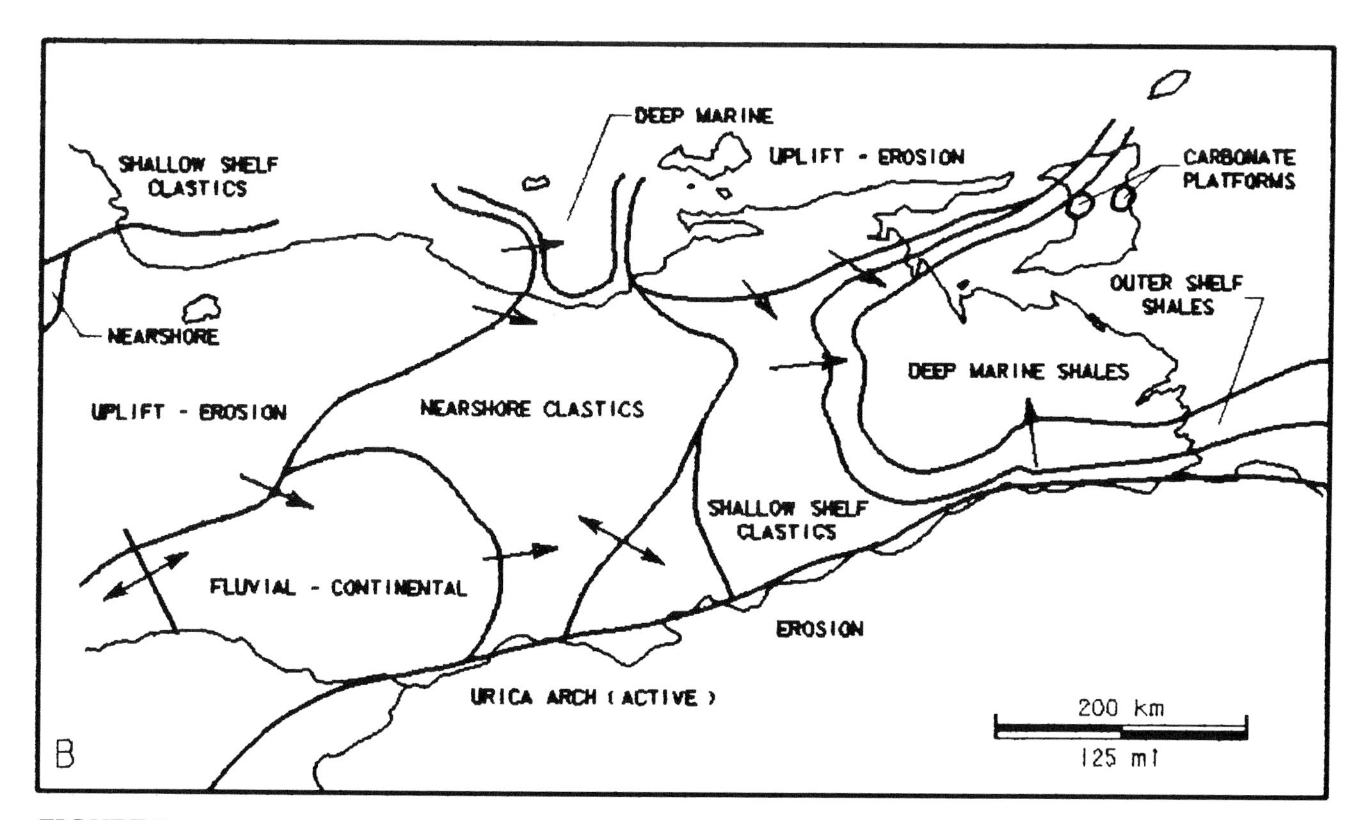

FIGURE 8. Paleogeography of northeastern Venezuela. (A) Oligocene–lower Miocene. (B) Middle–upper Miocene. (C) Pliocene–Holocene. Adapted from Gonzalez de Juana et al. (1980), Rossi et al. (1987), CEPET (1989).

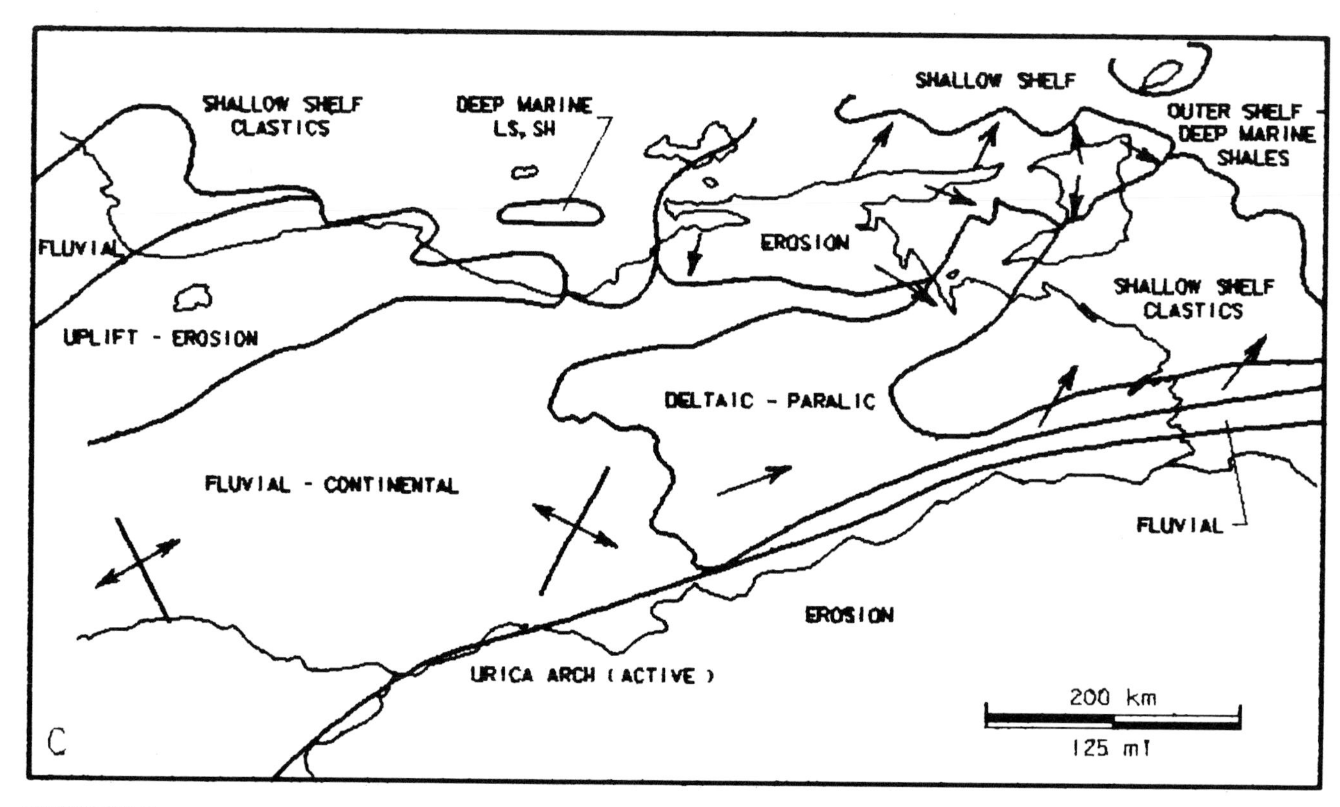

FIGURE 8. (Continued)

section into black shales of the Maastrichtian-late Paleocene Vidoño Formation (Bolli et al., 1991). The San Juan and Vidoño appear to represent a shallowing and deltaic infilling of the basin from the south and west, although the San Juan may have been deposited in an outer-shelf setting. Mucaria shales of the northern Guarico subbasin may conformably cross the Cretaceous/Tertiary boundary and appear to be conformable with the overlying Guarico Formation.

Foreland Basin Systems

In the northern and western parts of the Guarico foreland subbasin, the upper part of the Paleocene Guarico Formation is missing; it is unconformably overlain by late Eocene sandstones and shales of the La Pascua Formation and clastics and limestones of the Peñas Blancas Formation (Figures 6A, 6C) (Patterson and Wilson, 1953; Young et al., 1956; Galea-Alvarez, 1985). The eroded late Paleocene-early Eocene section is the first stratigraphic evidence of the collision and overthrusting of the Cordillera de la Costa/Villa de Cura allochthon onto the passive margin of South America. Based on regional stratigraphic relationships, Erlich and Barrett (1990) suggested a late Paleocene-middle Eocene age for uplift and erosion of this area (Figure 5C).

In the western and northern Maturin foreland subbasin, outer-shelf shales of the Vidoño are conformably overlain by shelf and prodelta shales of the Caratas Formation (Figures 6B, 6D). The Caratas thinned or was not deposited to the south, suggesting that flexural upwarping of areas immediately north of the Guayana shield may have been coincident with this first phase of compression/transpression and foreland basin development in the Guarico subbasin (Figure 5C).

The lower part of the Caratas is mostly a deep marine shale in the northern parts of the Maturin subbasin, but rapidly changes facies vertically into massive delta front and nearshore-marine sandstones (Galea-Alvarez, 1985). In the northwestern part of the Maturin subbasin and the northeastern part of the Guarico subbasin, late Eocene shallow shelf limestones of the Tinajitas Member of the Peñas Blancas Formation mark the early uplift of the area and the seaward limit of Caratas deltaic sandstones (Figure 5C). These algal/foraminiferal rocks are characteristic of high nutrient-stressed carbonate platforms (Hallock and Schlager, 1986; Erlich et al., 1991), and suggest that deltaic deposition exerted a major influence on sedimentary patterns of the area. A middle to late Eocene uplift age is also supported by radiometric and stratigraphic data from Margarita Island (just north of Araya/Paria) (Chevalier et al., 1988).

The eastward and southward migration of Caribbean-South American plate interactions is reflected by the progressive erosion of younger stratigraphic sequences in the Maturin subbasin (Figures 6B, 6D) (Hedberg, 1950; Rossi et al., 1987; Erlich and Barrett, 1990). In the west-central part of the Maturin subbasin, the upper part of the Caratas (late Eocene) and the lower part of the Merecure (early Oligocene) formations were eroded (Figures 5C, 8A). The Merecure unconformably overlies

Cretaceous Temblador Group rocks in the southeastern part of the Maturin subbasin, where flexural uplift and erosion continued into the earliest Miocene (Figure 8B).

In the north-central part of the Maturin subbasin, Oligocene Naricual Formation deltaic to marginal marine sandstones and shales unconformably overlie the Late Cretaceous San Juan Formation, indicating an intermediate stratigraphic position of the erosional unconformity (with the Paleocene and Eocene missing) (Carnevali, 1988; Fasola and Paredes, 1991).

Foreland basin sedimentation began in the Guarico subbasin after initial uplift and overthrusting in the early-middle Eocene (Figure 5C). Oligocene Roblecito Formation deltaic/marine shales conformably overlie La Pascua Formation deltaic sandstones and shales (Figure 8A). These are in turn conformably overlain by Oligocene-Miocene Chaguaramas Formation deltaic sandstones and shales. These units grade eastward into the western part of the Maturin subbasin, where Oligocene-lower Miocene deltaic sandstones and shales of the Merecure and lower Naricual formations change facies to deltaic/nearshore marine clastics of the early Oligocene Los Jabillos Formation, and inner-outer shelf shales of the Areo Formation (Figures 6B, 6D, 8A, 8B).

The onset of overthrusting and uplift in the Maturin subbasin can be timed by the late Oligocene eastward and northward progradation of the inner shelf/nearshore marine sandstones of the upper Naricual Formation (Figures 5A, 8B). Subsequent Miocene and younger sandstones also reflect multiple provenance areas, indicating uplift and erosional reworking of some of the Cretaceous-Eocene formations to the north (Hedberg, 1950; Chevalier et al., 1988).

Early Miocene compression/transpression between the Caribbean and South American plates caused overthrusting and uplift of the Serrania del Interior, downwarping of the central Maturin subbasin, and flexural or isostatic uplift of the Guarico subbasin east to the Urica arch (Figure 8B) (Erlich and Barrett, 1990; Pindell et al., 1991). Prior to uplift in the northern and western parts of the Guarico subbasin, fluvial/deltaic Chaguaramas and equivalent early Miocene Quebradon, Quiamare, and Guanape formation sandstones were deposited in a rapidly infilling basin. These units were subsequently eroded during the Miocene and Pliocene, and were buried during the Pleistocene by Mesa Formation fluvial/continental deposits (Figure 8C). In the eastern and central parts of the Guarico subbasin, some Pliocene section is still preserved (Patterson and Wilson, 1953; Young et al., 1956; Gonzalez de Juana et al., 1980).

Rapid deepening of the Maturin foreland basin produced a deep reentrant in the early Miocene shelf (Figure 8B, C). Erosion of Oligocene and Cretaceous strata in the Guarico subbasin caused initial progradation of lower Carapita Formation shales and sandstones from the west and southwest. Incipient folding and deformation of the northern part of the subbasin caused local sourcing of sandstones such as the Chapopotal and Morichito, which were deposited in narrow troughs parallel to subparallel to the fold axes (see, for example, Carnevali, 1988). The deposition of more widespread formations such as the Carapita and La Pica was also strongly controlled by contemporaneous structural deformation in the northern Maturin subbasin.

The upper Carapita prograded into the turbidite basin from the north, west, and south, causing rapid infilling and shallowing to more proximal shelf and nearshore marine environments (Figure 8B, C). Flexural and early lithostatic downwarping of the crust allowed a conformable sequence of fluvial/deltaic/nearshore marine sandstones and shales to be deposited south of the basin axis. Upper Merecure clastics are conformably overlain by Oficina, Freites, and La Pica formation rocks in the southwestern and south-central parts of the Maturin subbasin (Hedberg, 1950; Patterson and Wilson, 1953; Bolli et al., 1991).

The late Miocene-Pliocene Las Piedras Formation fluvial/deltaic section conformably overlies older rocks in the south, but is more localized in the northern part of the subbasin (Figure 6B). Diapirism and intense deformation of the lower Carapita shales occurred as southward thrusting of the Serrania del Interior continued into the Pliocene. As a result, the Las Piedras is often subdivided in these areas with a sandier lower sequence called the Quiriquire Formation (Carnevali, 1988). Minor uplift at the end of the Pliocene caused a short episode of erosion of the Las Piedras, which can be found as an angular unconformity with the overlying Mesa Formation rocks.

The northeastern extension of the filling of the Eastern Venezuela basin, which began in the late Oligocene, is reflected in continued flexural and lithostatic subsidence southeast of Trinidad (Erlich and Barrett, 1990). The Miocene stratigraphy and depositional patterns of Trinidad are the subject of another study (Farfan et al., in preparation), and will not be discussed here.

OIL AND GAS SYSTEMS

Overview

The oil and gas resources of the Eastern Venezuela basin are not distributed uniformly throughout the basin. The bulk of the hydrocarbon resource is concentrated in the Orinoco Tar belt, at the southern margins of the Maturin and Guarico subbasins (Figure 9). In-place oil reserves for the Orinoco Tar belt have been estimated by Roadifer (1986), CEPET (1989), and the Oil and Gas Journal (1991) at 1.2 trillion barrels, making it the single largest oil accumulation in the world. About 31.8 to 42.9×10^9 m^3 (200 to 270×10^9 bbl) of oil may be recoverable using conventional and unconventional means. Krause and James (1989) estimated the gas reserves of the Eastern Venezuela basin at 1.87×10^{12} m^3 (66 tcf).

Conventionally recoverable light, medium, and heavy oil reserves are found in all parts of the Eastern Venezuela basin, although over 96% (including most of the Orinoco Tar belt) occur in the Maturin subbasin (data from Carnevali, 1988; Oil and Gas Journal, 1990). At least five major producing trends (excluding the Orinoco Tar belt) can be identified within the Maturin and Guarico subbasins (Figure 9), containing a well-defined range of API oil gravities.

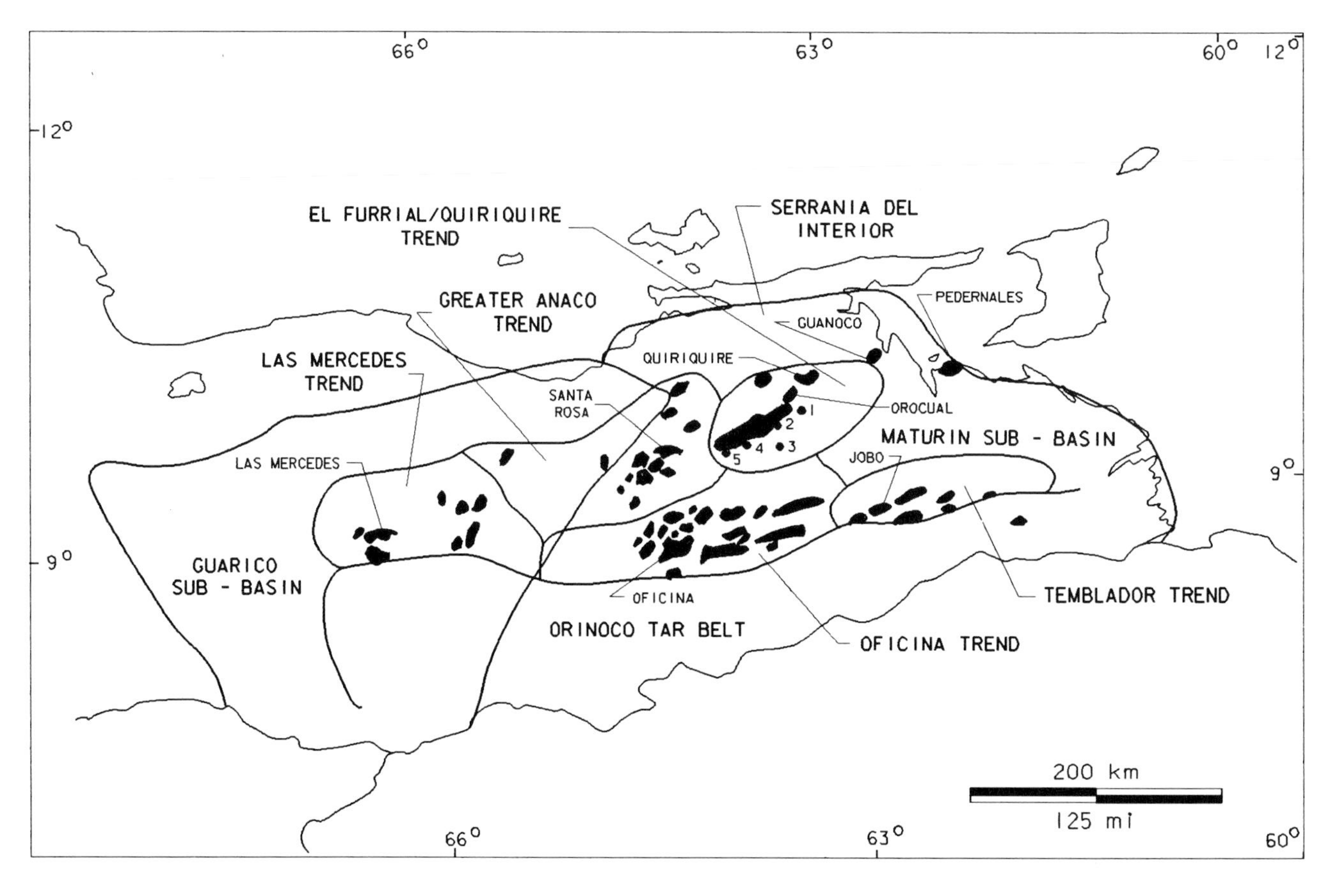

FIGURE 9. Major hydrocarbon provinces of the Eastern Venezuela basin. Numbered dots in the Maturin subbasin show the locations of recent important wells: 1 = Boqueron-3, 2 = El Furrial-1, 3 = Amarillis-1, 4 = El Carito-1X, 5 = Bosque/El Tejero-2E.

The main producing trend within the Guarico subbasin is known as the Las Mercedes trend (Figure 9), which has produced small amounts of medium-gravity (25–35° API) crude oils, but is now mostly inactive. The largest field in the trend is Las Mercedes field, which produced nearly 14.6×10^6 m^3 (92×10^6 bbl) of oil (it should be noted at this point that the cumulative production figures reported in this study do not consider the effects of government regulations, which may have restricted the ultimate potential of each field).

At the western edge of the Maturin subbasin in the southern part of Anzoategui State, the Greater Anaco trend (Figure 9) has been the site of several small light oil (35–50° API), gas, and condensate discoveries and one large field (Santa Rosa: 46.3×10^6 m^3/291×10^6 bbl of oil), but is not currently a major producing part of the basin. The rest of the Maturin subbasin contains three major producing trends. The Greater Oficina and Greater Temblador trends in the southern part of the basin produce mostly heavy- (10–25° API) and medium-gravity oil, although light oil, condensate, and gas are found in the northern part of the Greater Oficina trend. The largest field in the Greater Oficina trend is Oficina field (over 57.2 $\times 10^6$ m^3/360×10^6 bbl of oil), whereas the largest field in the Greater Temblador trend is Jobo field (about 46.1×10^6 m^3/290×10^6 bbl of oil).

The El Furrial/Quiriquire trend (previously known as the Greater Santa Barbara/Jusepin trend) in the northern part of the Maturin subbasin is actually a combination of traditional shallow (La Pica Formation) and newer deep (Carapita, San Juan formations) oil plays (Figure 9). The shallow reservoirs and the north and central parts of the trend generally produce medium-gravity crude oil, whereas the deeper reservoirs and the southern and western parts of the trend often produce a mixture of medium and light oils. The trend also contains three known supergiants ($> 0.16 \times 10^9$ m^3/10^9 bbl of oil) and one giant (>79.5 $\times 10^6$ m^3/500×10^6 bbl) oil field: El Furrial/Musipan (0.19–0.35 $\times 10^9$ m^3/1.2–2.2 $\times 10^9$ bbl), El Carito (0.39–0.43 $\times 10^9$ m^3/2.45–2.7 $\times 10^9$ bbl) (Carnevali, 1988, 1989; Aymard et al., 1990; Fasola and Paredes, 1991), Bosque/El Tejero ($>0.16 \times$ 10^9 m^3/10^9 bbl) (data from Lander et al., 1990; Mijares and Lopez, 1990), and Quiriquire (121×10^6 m^3/760×10^6 bbl) (Oil and Gas Journal, 1990). Gas reserves within the trend are also large; Carnevali (1989) estimated the combined reserves of El Carito and El Furrial/Musipan at 0.21×10^{12} m^3 (7.3 tcf).

Reservoirs, Trap Style, and Trap Timing

The main producing reservoirs in the Las Mercedes trend are sandstones of the Oligocene–Miocene Chaguaramas, Oligocene Roblecito, and La Pascua formations (La Pascua is the most significant by volume), and the Cretaceous La Cruz Formation (Patterson and Wilson, 1953; Young et al.,

1956; Gonzalez de Juana et al., 1980). Production is from normal and reverse fault closures (Figure 10).

Oil reservoirs of the Greater Oficina and Greater Anaco trends are sandstones within the Oligocene Oficina Formation and the Eocene-Oligocene Merecure Formation (Young et al., 1956; Gonzalez de Juana et al., 1980; CEPET, 1989), and are productive from normal and reverse fault closures (Figure 10). However, production in Tacat field in the northwestern part of Monagas State (Figure 7) appears to be related in part to right-lateral wrench faulting along the Urica fault (Figure 2) (Reistroffer, 1991). Oil reservoirs in the trend are also controlled by the stratigraphic and areal distribution of individual sandstone units.

Greater Temblador trend reservoirs are productive from Oligocene Oficina Formation and Cretaceous Temblador Group sandstones (Young et al., 1956; Gonzalez de Juana et al., 1980), which are also trapped in normal and reverse fault closures (Figure 10).

Oil reservoirs within the El Furrial/Quiriquire trend are wide ranging in age and trap style. The Pliocene Quiriquire Formation, Miocene-Pliocene La Pica Formation, Miocene Carapita Formation, Oligocene Naricual Formation, Eocene-Oligocene Los Jabillos Formation, Eocene Caratas Formation, and Upper Cretaceous San Juan Formation are all productive at various depths throughout the trend (Young et al., 1956; Gonzalez de Juana et al., 1980; Carnevali, 1988; Aymard et al., 1990; Fasola and Paredes, 1991). Productive sands are found at depths from less than 600 m (1970 ft) in the northern part of the trend to nearly 5950 m (about 19,500 ft) in the southern and eastern parts of the trend. Traps are in normal and reverse fault closures and in three- and four-way anticlines associated with overthrusting (Figure 10).

The timing of trap formation in the Las Mercedes trend appears to have been during the Oligocene-Holocene (Figure 10), with at least three phases of faulting noted by some workers (Patterson and Wilson, 1953; Gonzalez de Juana et al., 1980; CEPET, 1989). Northeast-southwest-oriented normal and reverse faults of Oligocene to early Miocene age probably reflect flexural uplift of the basement as foreland basin development migrated from west to east across northern Venezuela (Pindell and Dewey, 1991). Later northwest-southeast-oriented normal faulting may have been related to reactivation of southward-directed plate boundary stresses, or isostatic relaxation as those stresses migrated to the east.

Greater Oficina and Greater Temblador trend hydrocarbon traps began forming in the Oligocene (Figure 10), possibly as a result of eastward migration of the peripheral bulge during rapid downwarping of the passive margin sequence in the Maturin subbasin (Pindell and Dewey, 1991). Thrust-related anticlines and fault traps in the Greater Anaco and El Furrial/Quiriquire trends underwent several episodes of development (Subieta et al., 1988;

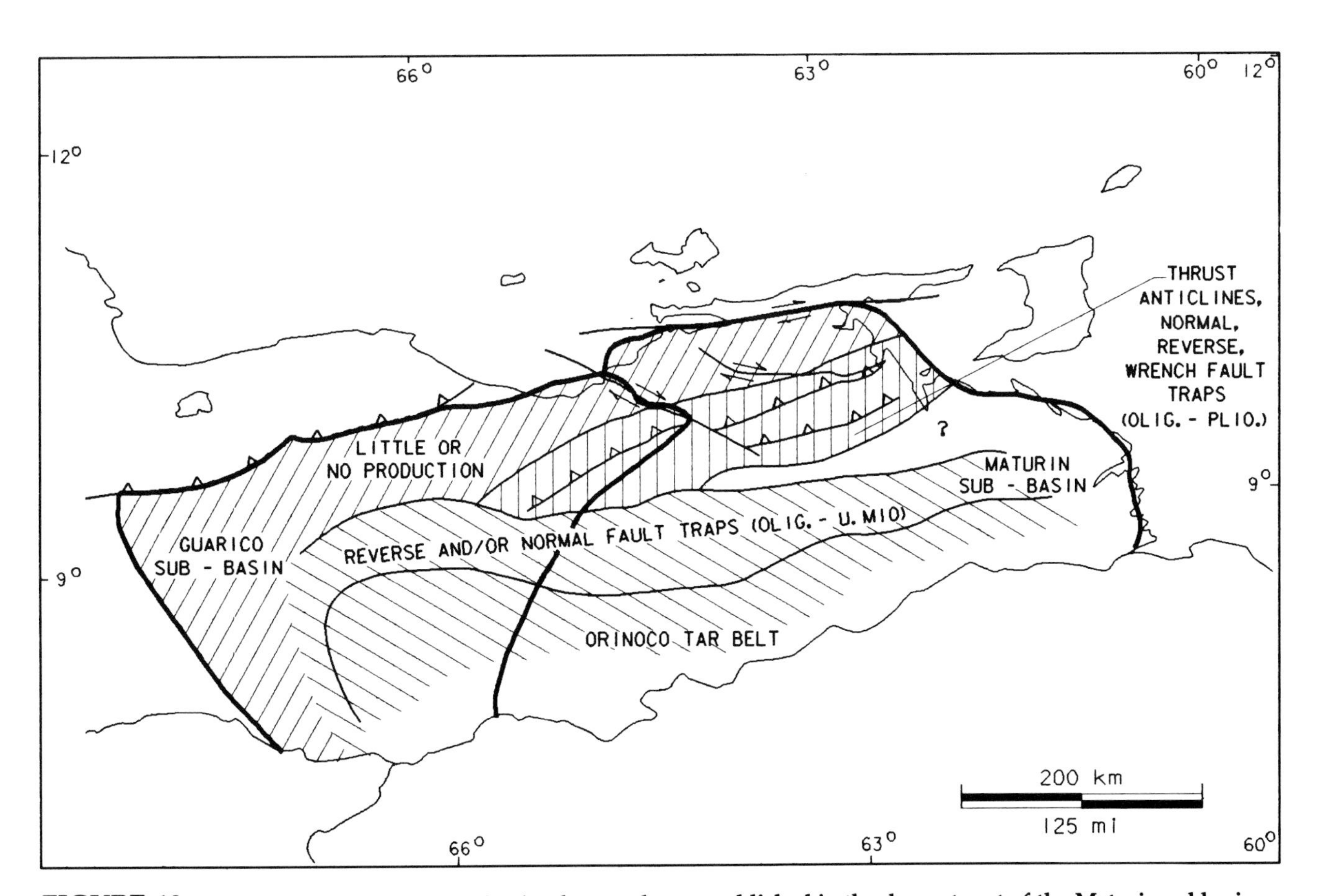

FIGURE 10. Trap style and timing. Production has not been established in the deepest part of the Maturin subbasin (area with ?), so trap style and timing for that area are unknown.

Talukdar et al., 1988; Talukdar, 1991). Thrust-related traps began forming during the late Oligocene, and continued to be affected and reoriented by overthrusting until the Pliocene (Figure 10). The giant fields of the El Furrial/ Quiriquire trend were all formed during this period.

Accompanying diapirism of Carapita Formation shales affected depositional and structural patterns in shallow Pliocene reservoirs. Continued movements on minor strike-slip faults such as the Urica fault reoriented reverse fault traps in some Greater Anaco trend fields (Reistroffer, 1991), but these relatively recent wrench-related traps are not common.

Source Rocks

Published data on source rocks of the Eastern Venezuela basin are scarce at best. The generation and migration of oil in the Maturin subbasin were studied by Talukdar et al. (1988) and Talukdar (1991), but little direct information is available from the Guarico subbasin.

Arnstein et al. (1982) and Krause and James (1989) suggested that the highly paraffinic oils of the Greater Oficina and Las Mercedes trends were derived from Oligocene Merecure and Roblecito, and Miocene Oficina and Chaguaramas formations, but presented little data in support. CEPET (1989) quoted an internal study by Talukdar et al. that acknowledged that these formations had contributed only minor amounts of oil to reservoirs in the area.

Talukdar et al. (1988) showed data that supported their conclusion that the Upper Cretaceous Querecual and San Antonio formations were the most important (by volume) source rocks for oil found in the Maturin subbasin. Based on kerogen typing, they subdivided each of the formations into two facies that corresponded to Type II or Type II to Type III organic matter using a Rock-Eval classification. Total organic carbon (TOC) for most of the Guayuta Group rocks studied ranged from 0.25 to 6.6%. They also estimated that 50 to 55% of the 610 to 1021-m (2000 to 3350-ft) thickness of the Guayuta Group could be considered potential source rocks. The analyzed hydrocarbon yields range from 15 to 454 mg HC/g TOC. Talukdar et al. (1988) estimated the average hydrogen index for immature Type II facies at 700 mg HC/g TOC, and that for immature Type II to Type III facies at 400 mg HC/g TOC.

The variability of hydrocarbon yield of the Querecual and San Antonio may be governed largely by their depositional environments. The black cherts and limestones of the Querecual were deposited in an euxinic environment across most of the Eastern Venezuela basin, becoming shaly and sandy in the south and in the west (Hedberg, 1937, 1950; Patterson and Wilson, 1953; Krause and James, 1989). Most workers suggest a pelagic marine depositional environment for the Querecual and the overlying San Antonio (Figure 7). The great thickness and areal extent of these units, combined with the long period of deposition (Figure 6B), suggest that low-oxygen bottom waters or high biologic productivity cannot sufficiently explain the vast amounts of organic carbon trapped in the section. This problem is currently the focus of another study (Erlich et al., in preparation).

Hydrocarbon Maturation and Migration

Arnstein et al. (1982), Talukdar et al. (1988), and Talukdar (1991) concluded that Cretaceous rocks of the Eastern Venezuela basin passed through the oil window progressively from north to south. Maturation and early migration may have begun in the Oligocene–early Miocene just after overthrusting in the northern part of the Guarico subbasin, and in the latest Oligocene–middle Miocene in the northern part of the Maturin subbasin (Figure 11). Talukdar et al. (1988) and Talukdar (1991) calculated that the Querecual may now be reaching maturity in the deeper parts of the Maturin subbasin. Based on a geothermal gradient of 2.4°C/100 m (1.3°F/100 ft) and a surface temperature of 23°C (74°F), they place peak oil generation of the subthrust Querecual and San Antonio formations at 7000 to 8000 m (23,000 to 26,000 ft) subsea.

Migration of oil from the Cretaceous source rocks proceeded from north to south in the Eastern Venezuela basin (Figure 11). Long-distance migration of 150 to 325 km (93 to 202 mi) probably occurred during the middle-late Miocene, before laterally continuous reservoirs and pathways were disrupted by faulting and folding (Talukdar et al., 1988; Talukdar, 1991). The Orinoco Tar belt accumulation, as well as the Greater Temblador and much of the Greater Oficina and Las Mercedes trend oils, probably were trapped during this time. Fields of the El Furrial/ Quiriquire trend and many light oil fields in the northwestern Greater Oficina and Greater Anaco trends probably were not charged until the Pliocene or Pleistocene, as Cretaceous and Tertiary source rocks subsided into the oil window as a result of loading in the axial part of the Maturin foreland subbasin (Figure 11). It should be noted, however, that elevated geothermal gradients (>3.6°C/100 m, or 2.0°F/100 ft) in the eastern part of the Guarico subbasin (overlying the Espino graben, Figure 2) may have caused Tertiary source rocks in that area to mature earlier, during the late Miocene–early Pliocene (E. Murany, 1991 personal communication).

Exploration History

The exploration history of the Eastern Venezuela basin can be divided into three phases based on the type of technical investigations performed (data from CEPET, 1989): Phase 1 (1909–1949)—surface seeps, shallow or surface structures, early geophysical methods; Phase 2 (1950–1958)—subsurface geology and early seismic methods; and Phase 3 (1959–present)—advanced subsurface geology and geophysical methods.

First Exploration Phase (1909–1949)

Although exploitation of asphalt from the Guanoco asphalt seep in the Pedernales area began in 1890 (Figure 9), the groundwork for exploration in the Eastern Venezuela basin began in 1909, when the Governor General of Venezuela granted exploration leases to the Venezuelan Development Company, Ltd. in the states of Anzoategui, Sucre, Monagas, and the Orinoco delta (see Figure 3). This led to

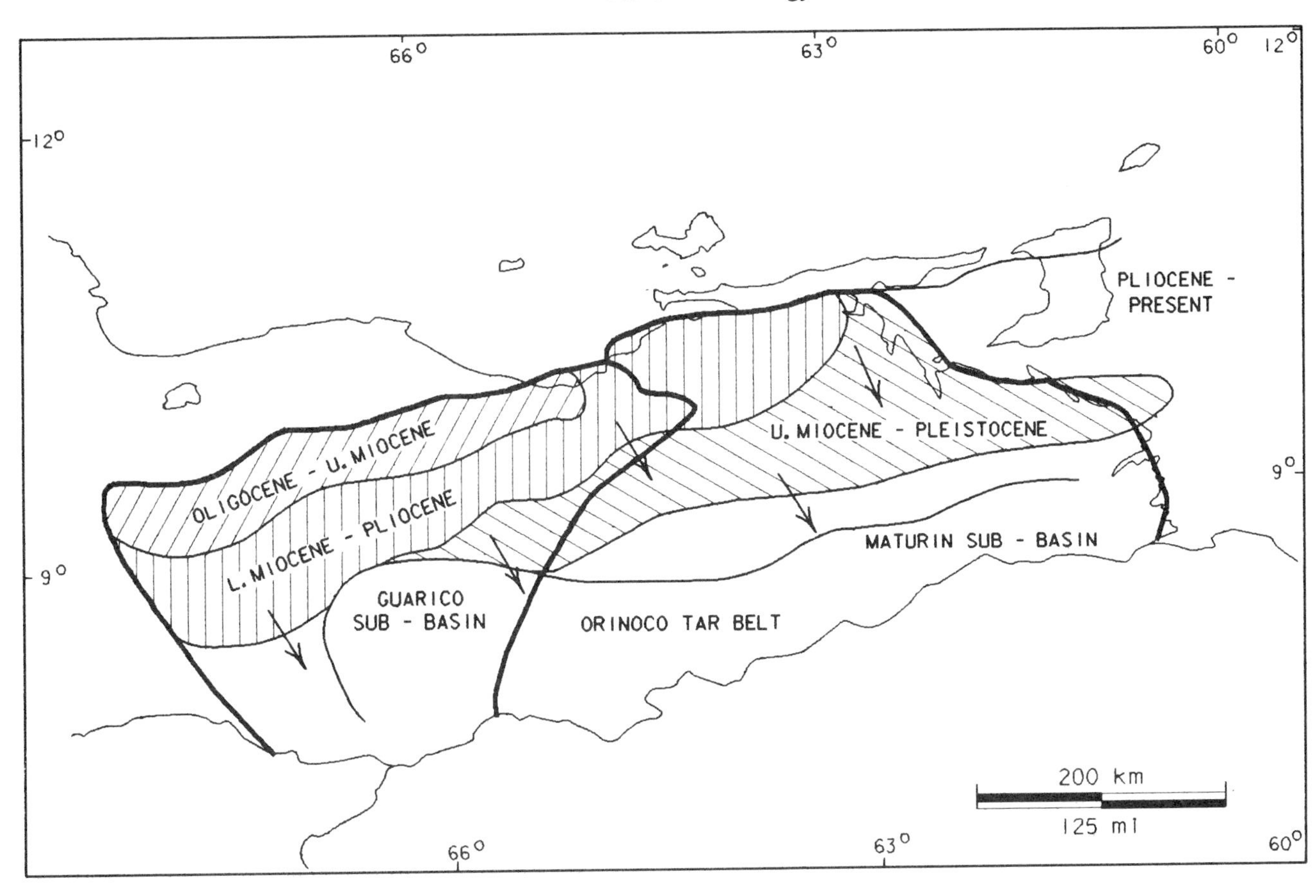

FIGURE 11. Oil "kitchens" of the Guarico and Maturin subbasins. Time indicated represents the interval during which the Cretaceous and early Tertiary source rocks were within the oil window (rocks within the Orinoco Tar belt are immature; very early stage of maturity at present). Arrows show dominant hydrocarbon migration direction. Adapted from Talukdar et al. (1988) and Talukdar (1991).

the drilling of a well in the Araya metamorphic belt by the Caribbean Petroleum Company, which later became part of the Royal Dutch Shell Group.

The first successful exploratory drilling in the Eastern Venezuela basin was done by the Bermudez Company in 1912–1913 in the Guanoco asphalt/heavy oil field, with 14 discoveries out of 24 wells drilled (Figure 9). One of these wells (Bababui-1) is considered to be the first true exploration well drilled in Venezuela (CEPET, 1989). Exploration in the basin was generally unsuccessful until Stanolind (Amoco) and Standard Oil of Venezuela (Exxon) discovered the giant Quiriquire field in Monagas State in 1928. This was followed by Gulf Oil's discovery of the Oficina field in Anzoategui State in 1934.

The use of early gravity, magnetic, and surface gamma-ray surveys led directly to the discovery of several shallow reservoirs in the Maturin and Guarico subbasins from 1934 to 1949. Exploration within and north of the Orinoco Tar belt in 1936 resulted in the discovery of Temblador field, which opened successful exploration in the heavy oil trend (Figure 9). Santa Rosa field, in the Greater Anaco trend, was discovered in 1941 using geophysical methods, as was Las Mercedes field.

Small to medium-sized fields (7.9–39.7 $\times$ 10^6 m^3/50–250 $\times$ 10^6 bbl) continued to be found until the end of the first exploration phase in 1949, although the discovery rate declined (CEPET, 1989). Over 0.5 $\times$ 10^9 m^3 (3 $\times$ 10^9 bbl) was discovered during this exploration phase (Oil and Gas Journal, 1990).

Second Exploration Phase (1950–1958)

With the advent of common-depth-point (CDP) seismic techniques in 1950, exploration for deep reservoirs within and outside of existing fields increased (CEPET, 1989). The improved technology aided in the discovery of Eocene reservoirs in Quiriquire field and the geophysically defined targets of Orocual (El Furrial/Quiriquire trend) and Jobo (Greater Temblador trend) fields (Figure 9). Mostly small to medium-sized discoveries continued to be made in the Maturin subbasin through the end of the second phase of exploration in 1958, although the total amount of oil discovered was significant (0.3 x 10^9 m^3/2 $\times$ 10^9 bbl) (Oil and Gas Journal, 1990).

Third Exploration Phase (1959–present)

The 1960 decision of the Venezuelan government not to grant new concessions caused a country-wide slump in exploratory drilling, especially in the Eastern Venezuela basin (CEPET, 1989). Between 1960 and 1975, only six marginal to medium-sized fields (0.8–31.8 × 10^6 m^3/5–200 × 10^6 bbl) were discovered (<79.5 × 10^6 m^3/500 × 10^6 bbl) (Oil and Gas Journal, 1990), as opposed to 17 discoveries in the previous nine years.

After nationalization of the oil industry on December 31, 1975, exploration in the Eastern Venezuela basin increased substantially. Five national oil companies, composed of remnants of the foreign concession holders, were formed in 1978 to consolidate the holdings into geographic operating areas. Of these, Corpoven and Lagoven emerged as the most important in the eastern part of the country.

Rapid advances in the acquisition and processing of seismic data led to several small discoveries from 1976–1981. Increased seismic surveys by Corpoven and Lagoven in the Maturin subbasin during 1981-1985 eventually revealed the existence of deeper structures underlying the shallow reservoirs (Santa Barbara/Jusepin fields) of the El Furrial/Quiriquire trend. Obscured from view by shale diapirism within the Carapita Formation, the low-amplitude thrust anticlines drilled by El Furrial-1 and Orocual (ORS)-52 proved the existence of a new trend of supergiant fields below the existing old production (Carnevali, 1988, 1989; Aymard et al., 1990). Since 1986, significant discoveries in the El Furrial/Quiriquire trend have been made on the El Carito, Bosque/El Tejero, Boqueron, and Amarillas structures (Figure 9).

FUTURE EXPLORATION POTENTIAL

Based on the theory of early Miocene-Pleistocene maturation, generation, and migration of hydrocarbons in the Eastern Venezuela basin (Talukdar et al., 1988; Talukdar, 1991), exploratory drilling will probably continue in deeper parts of the basin. Affiliates of Petroleos de Venezuela (PDVSA) plan to acquire about 50,000 km (31,000 mi) of seismic data and drill 130 exploratory wells in Venezuela in the 1991-1996 period (Oil and Gas Journal, 1991). Much of this work will be conducted in the El Furrial/Quiriquire trend (Figure 9).

Drilling in the Las Mercedes, Greater Oficina, and Greater Temblador trends probably will be limited to exploitation of existing light, medium, and heavy oil resources, a major goal of PDVSA through the year 2000 (Oil and Gas Journal, 1991). Although the discovery of large fields in these areas is still possible, most of the future major discoveries will most likely be made in the El Furrial/Quiriquire trend (Aymard et al., 1990).

ACKNOWLEDGMENTS

The authors wish to thank the following people for their help in producing this report: P. Farfan, D. Felio, W. Hale-Erlich, C. Kazmer, S. Nederbragt, J. Pindell, W. Schollnberger, J. Tunnel, and C. Williams. This report benefited from the constructive comments of AAPG reviewers E. Murany and S. Talukdar. We also wish to acknowledge the many workers, past and present, who contributed so much to our understanding of the geology of Venezuela. Their work stands as a testament to all those involved in research and study in Venezuela since the time of von Humboldt. Amoco Production Company granted permission to publish this report. H. Van Nguyen and A. Curtis drafted the illustrations.

REFERENCES CITED

Algar, S. T., J. E. Erikson, and J. L. Pindell, 1991, Geological studies in eastern Venezuela and Trinidad: from Cretaceous passive margin to Neogene transpressional thrust belt: AAPG Annual Meeting, Dallas, 7-10 April, p. 69-70.

Arnstein, R., C. Betoret, E. Molina, L. Mompart, J. Ortega, F. Russomanno, and H. Sanchez, 1982, Geología petrolera cuenca de Venezuela Oriental: XLV Reunión a Nivel de Expertos de ARPEL, Mexico, 17-21 Mayo, 1982, p. 229-252.

Aymard, R., L. Pimentel, P. Eitz, A. Chaouch, J. Navarro, J. Mijares, and J. G. Pereira, 1990, Geological integration and evaluation of Northern Monagas, Eastern Venezuelan basin, *in* J. Brooks, ed., Classic petroleum provinces: Geological Society of London Special Publication No. 50, p. 37-53.

Bellizzia, A., 1972, Is the entire Caribbean Mountain belt of northern Venezuela allochthonous?, *in* R. Shagam, ed., Studies in earth and space sciences: GSA Memoir 132, p. 363-368.

Benjamini, C., R. Shagam, and A. Mendez V., 1987, (Late?) Paleozoic age for the "Cretaceous" Tucutunemo Formation, northern Venezuela: stratigraphic and tectonic implications: Geology, v. 15, p. 922-926.

Blanco, B., and J. H. Sanchez, 1990, Prospectividad en el frente de montañas entre Acarigua y Boca de Uchire: Reunión de V Congreso Venezolano de Geofísica, Caracas, 21-25 Octubre, p. 210-218.

Bolli, H. M., J. P. Beckmann, and J. B. Saunders, 1991, Benthic foraminiferal biostratigraphy of the southern Caribbean: Cambridge, Cambridge University Press.

Carnevali, J. O., 1988, Venezuela nor-oriental: exploracion del frente de montaña, *in* A. Bellizzia, A. Leslie Escoffery, and I. Bass, eds., III Simposio Bolivariano: Boletín de la Sociedad Venezolana de Geólogos, v. 1, p. 69-89.

Carnevali, J. O., 1989, Geology of new giant oil fields in mountain front of northeastern Venezuela: Proceedings, 28th International Geological Congress, Washington, 9-19 July, v. 1, p. 1.240-1.241.

CEPET, 1989, La industria Venezolana de los hidrocarburos: Caracas, Venezuela, El Centro de Formacion y Adiestramiento de Petróleos de Venezuela, v. 1, 754 p.

Chevalier, Y., J. F. Stephan, J. R. Darboux, M. Gravelle, H. Bellon, A. Bellizzia, and R. Blanchet, 1988, Obduction et collision pre-Tertiaire dans les zones internes de la Caine Caraibe Venezuelienne, sur le transect Ile de Maragarita-Peninsule d'Araya: C. R. Academie Science Paris, t. 307, Serie II, p. 1925-1932.

Dallmus, K. F., 1965, The geology and oil accumulations of the Eastern Venezuela basins: Boletín Informativo, Asociación Venezolana de Geología, Minería y Petróleo, v. 8, p. 5-32.

Dirección de Geología, 1970, Lexico estratigráfico de Venezuela: Boletín de Geología Publicación Especial 4, 756 p.

Erlich, R. N., 1987, The stratigraphic and structural evolution of the northeast Venezuela-Trinidad area: Amoco International Geological Report R87-251.

Erlich, R. N., and S. F. Barrett, 1990, Cenozoic plate tectonic history of the northern Venezuela-Trinidad area: Tectonics, v. 9, p. 161-184.

Erlich, R. N., S. F. Barrett, and B. J. Guo, 1990, Seismic and geologic characteristics of drowning events on carbonate platforms: AAPG Bulletin, v. 74, p. 1523-1537.

Erlich, R. N., S. F. Barrett, and B. J. Guo, 1991, Drowning events on carbonate platforms: a key to hydrocarbon entrapment?: Proceedings, 23rd Annual Offshore Technology Conference, Houston, 6-9 May, p. 101-112.

Fasola, A., and I. Paredes, 1991, Late Cretaceous and mid-Tertiary palynological assemblages from the "El Furrial" area wells, Venezuela: Proceedings, AAPG Annual Meeting, Dallas, 7-10 April, p. 107.

Feo Codecido, G., F. D. Smith, Jr., N. Aboud, and E. de Di Giacomo, 1984, Basement and Paleozoic rocks of the Venezuelan Llanos basins, *in* W. E. Bonini, R. B. Hargraves, and R. Shagam, eds., The Caribbean-South American plate boundary and regional tectonics: GSA Memoir 162, p. 175-187.

Fernandez, F., and H. Passalacqua, 1990, Procesamiento y interpretación de datos gravimétricos y magnéticos en la Cuenca Oriental de Venezuela: Reunión de V Congreso Venezolano de Geofísico, Caracas, 21-25 Octubre, p. 86-93.

Fiume, G., and V. Graterol, 1990, Procesamiento y interpretación de datos aerogravimétricos y aeromagnéticos en areas de interés exploratorio en la región Pedernale-Quiriquire: Reunión de V Congreso Venezolano de Geofísico, Caracas, 21-25 Octubre, p. 458-465.

Galea-Alvarez, F. A., 1985, Biostratigraphy and depositional environment of the Upper Cretaceous-Eocene Santa Anita Group (northeastern Venezuela): Ph.D. dissertation, Free University Press, Amsterdam, 115 p.

Gonzalez de Juana, C., J. A. Arozena, and X. Picard Cadillat, 1980, Geología de Venezuela y sus cuencas petrolíferas: Caracas, Venezuela, Ediciones Foninves, 1031 p.

Guillaume, H. A., H. M. Bolli, and J. P. Beckmann, 1972, Estratigrafia del Cretaceo Inferior en la Serrania del Interior, oriente de Venezuela, *in* Memoria IV Congreso Geológico Venezolano, Tomo III: Ministerio de Minas y Hidrocarburos Boletín de Geología Publicación Especial 5, p. 1619-1658.

Hallock, P., and W. Schlager, 1986, Nutrient excess and the demise of coral reefs and platforms: Palaios, v. 1, p. 389-398.

Hedberg, H. D., 1937, Stratigraphy of the Rio Querecual section of northeastern Venezuela: GSA Bulletin, v. 48, p. 1971-2024.

Hedberg, H. D., 1950, Geology of the eastern Venezuela basin (Anzoategui-Monagas-Sucre-eastern Guarico portion): AAPG Bulletin, v. 61, p. 1173-1215.

Krause, H. H., and K. H. James, 1989, Hydrocarbon resources of Venezuela—their source rocks and structural habitat, *in* G. E. Ericksen, M. T. Canas Pinochet, and J. A. Reinemund, eds., Geology of the Andes and its relation to hydrocarbon and mineral resources: Houston, Texas, Circum-Pacific Council for Energy and Mineral Resources Earth Science Series, v. 11, chap. 29, p. 405-414.

Kugler, H. G., and H. M. Bolli, 1967, Cretaceous biostratigraphy in Trinidad: Boletín Informativo, Asociacion Venezolana de Geología, Minería y Petróleo, v. 10, p. 209-236.

Lander, R., V. Hernandez, J. Fuentes, and S. Dos Santos, 1990, Metodología integrada para la interpretación estructural de areas geológicas complejas: Bosque, norte de Monagas, cuenca oriental de Venezuela: Reunión de V Congreso Venezolano de Geofísico, Caracas, 21-25 Octubre, p. 110-117.

Macdonald, W. D., and N. D. Opdyke, 1974, Triassic paleomagnetism of northern South America: AAPG Bulletin, v. 58, p. 208-251.

Mascle, A., and P. Letouzey, 1990, Geological map of the Caribbean, 1:2,500,000, 2 sheets: Paris, France, Editions Technip.

Martin, N., and E. Espinoza, 1990, Integración de la información gravimétrica del flanco noreste de la cuenca oriental de Venezuela: Reunión de V Congreso Venezolano de Geofísico, Caracas, 21-25 Octubre, p. 78-85.

Mijares, J., and P. Lopez, 1990, La technología TOMEX en la predicción de la profundidad de reflectores de interés y la delineación estructural: Reunión de V Congreso Venezolano de Geofísico, Caracas, 21-25 Octubre, p. 166-172.

Moticska N., P., 1985, Volcanismo Mesozoico en el subsuelo de la Faja Petrolifera del Orinoco, Estado Guarico, Venezuela, *in* A. Espejo C., J. H. Rios F., N. Pimental de Bellizzia, and A. S. de Pardo, eds., Petrologia, geoquimica y geochronologia: Memoria Congresso Geológico Venezolano, VI Congreso Geológico Venezolano, Caracas, 29 Septiembre-6 Octubre, v. 6, p. 1929-1943.

Oil and Gas Journal, 1990, Worldwide production—Venezuela: Oil and Gas Journal, 31 Dec., v. 88, p. 80-81.

Oil and Gas Journal, 1991, PDVSA maps ambitious Venezuelan oil plan: Oil and Gas Journal, 14 Jan., v. 89, p. 35-38.

Orihuela, N., and R. Franklin, 1990, Modelaje gravimétrico de un perfil comprendido entre los poblados de Altagracia de Orituco, Edo. Guarico y Caraballeda, Dtto. Federal, Venezula: Reunión de V Congreso Venezolano de Geofísico, Caracas, 21-25 Octubre, p. 465-473.

Osten, E. von der, 1957, Lower Cretaceous Barranquin Formation of northeastern Venezuela: AAPG Bulletin, v. 41, p. 679-708.

Patterson, J. M., and J. G. Wilson, 1953, Oil fields of Mercedes region, Venezuela: AAPG Bulletin, v. 37, p. 2705-2733.

Pindell, J. L., 1985, Alleghenian reconstruction and the subsequent evolution of the Gulf of Mexico, Bahamas and proto-Caribbean Sea: Tectonics, v. 4, p. 1-39.

Pindell, J. L. and S. F. Barrett, 1990, Geological evolution of the Caribbean region: a plate-tectonic perspective, *in* G. Dengo and J. E. Case, eds., The Caribbean region: Boulder, Colorado, GSA, The Geology of North America, v. H, chap. 16, p. 405-432.

Pindell, J. L., and J. F. Dewey, 1982, Permo-Triassic reconstruction of western Pangea and the evolution of the Gulf of Mexico/Caribbean region: Tectonics, v. 1, p. 179-211.

Pindell, J. L. and J. F. Dewey, 1991, Cenozoic transpressional model for the tectonic and basinal development of Venezuela and Trinidad: Annual Meeting of the AAPG, 7-10 April, Dallas, p. 189-190.

Pindell, J. L., C. L. Drake, and W. C. Pitman, 1991, Preliminary assessment of a Cretaceous-Paleogene Atlantic passive margin, Serrania del Interior and Central Ranges, Venezuela/Trinidad: Annual Meeting of the AAPG, Dallas, 7-10 April, p. 190.

Reistroffer, J., 1991, Detection of tectonic and subcompacted geopressures in the western Monagas-Anzoategui overthrust belt: a structural solution: Annual Meeting of the AAPG, Dallas, 7-10 April, p. 195.

Roadifer, R. E., 1986, Size and distribution of world's largest known oil, tar accumulations: Oil and Gas Journal, 24 Feb., v. 84, p. 93-100.

Rod, E., and W. Maync, 1954, Revision of the Lower Cretaceous stratigraphy of Venezuela: AAPG Bulletin, v. 38, p. 193-283.

Rossi, T., J. F. Stephan, R. Blanchet, and G. Hernandez, 1987, Etude geologique de la Serrania del interior oriental (Venezuela) sur le transect Cariaco-Maturin: Revue de L'Institut Français du Pétrole, v. 42, p. 3-30.

Schubert, C., 1984, Basin formation along the Bocono-Moron-El Pilar fault system: Journal of Geophysical Research, v. 89, p. 5711-5718.

Skerlec, G. M., and R. B. Hargraves, 1980, Tectonic significance of paleomagnetic data from northern Venezuela: Journal of Geophysical Research, v. 85, p. 5303-5315.

Speed, R. C., 1985, Cenozoic collision of the Lesser Antilles arc and continental South America and origin of El Pilar fault: Tectonics, v. 4, p. 41-69.

Stephan, J. F., 1977, Una interpretación de los complejos con bloques asociados a los flysch Paleoceno-Eoceno de la cadena Caribe Venezolana: el emplazamiento submarino de la Napa de Lara, *in* GUA Stichting, ed., Proceedings of the Eighth Caribbean Geological Conference: Amsterdam, University of Amsterdam, p. 197-198.

Stover, L. E., 1967, Palynological dating of the Carrizal Formation of eastern Venezuela: Boletín Informativo, Asociación Venezolana de Geología, Minería y Petróleo, v. 10, p. 288-290.

Subieta, T., J. O. Carnevali, and V. Hunter, 1988, Evolución tectonoestratigráfica de la Serrania del Interior y la subcuenca de Maturin, *in* A. Bellizzia, A. L. Escoffery, and I. Bass, eds., III Simposio Bolivariano: Sociedad Venezolana de Geólogos, Caracas, Venezuela, v. 2, p. 549-578.

Talukdar, S. C., 1991, Petroleum systems of the eastern Venezuelan basin: Annual Meeting of the AAPG, Dallas, 7-10 April, p. 215.

Talukdar, S., O. Gallango, and A. Ruggiero, 1988, Generation and migration of oil in the Maturin sub-basin, eastern Venezuelan basin: Organic Geochemistry, v. 13, p. 537-547.

Wells, J. W., 1948, Lower Cretaceous corals from Trinidad, British West Indies: Journal of Paleontology, v. 22, p. 608-616.

Young, G. A., A. Bellizzia, F. W. Johnson, H. H. Renz, R. H. Robie, and J. Mas Vall, 1956, Geología de las cuencas sedimentarias de Venezuela y de sus campos petrolíferos: Boletín de Geología, Publicación Especial Dirección de Geología 2, 140 p.

Chapter 13

The North Slope Foreland Basin, Alaska

Kenneth J. Bird

U.S. Geological Survey
Menlo Park, CA, U.S.A.

Cornelius M. Molenaar

U.S. Geological Survey
Denver, CO, U.S.A.

INTRODUCTION

The North Slope foreland basin, also called the Colville basin or trough, is a late Mesozoic and Cenozoic basin that spans the entire width of the North Slope of Alaska (Figure 1). It is bounded on the south by the Brooks Range, a thrust-faulted orogenic mountain belt that is an extension of the Rocky Mountains of Canada. The northern boundary of the basin approximately coincides with the north shore of Alaska, the Beaufort Sea, where a rift shoulder that is now a subsurface passive high, the Barrow arch, separates the foreland basin to the south from the Canada basin to the north. To the west, the North Slope basin extends offshore under the Chukchi Sea as far as the northwestward-trending Herald arch and the northward-trending Chukchi platform located along the U.S.-U.S.S.R. boundary (Figure 2). On the far east, the basin narrows; along the Alaska-Canada border, the Brooks Range orogene extends almost to the coastline. Beyond that area and offshore to the east and northeast, the sedimentary fill merges with the Cretaceous and Tertiary passive margin deposits of the Mackenzie delta and Canada basin. As thus defined, the North Slope basin is about 1000 km (600 mi) long and 50 to 350 km (30 to 215 mi) wide, and covers an area of about 240,000 km^2 (93,000 mi^2).

Because of the hostile arctic environment and the distance from market outlets, hydrocarbon exploration in the North Slope basin occurred much later than that in the foreland basins of the lower 48 states and Canada. Since the discovery of the giant Prudhoe Bay field in 1967, however, greatly increased drilling and seismic exploration have resulted in additional discoveries. The commercial oil fields produce mostly from pre-foreland basin strata, including Triassic and Lower Cretaceous clastic as well as Mississippian and Pennsylvanian clastic and carbonate reservoirs along the Barrow arch in and adjacent to the greater Prudhoe Bay area. Foreland basin strata provide reservoirs for huge deposits of heavy oil in uppermost Cretaceous and lowermost Tertiary deltaic sandstones in the same general area, and noncommercial oil and gas accumulations are present in mid-Cretaceous deltaic sandstone reservoirs in anticlinal traps in the fold belt (Figure 2). (By North Slope standards, "noncommercial" could be a large commercial field in the lower 48 states.)

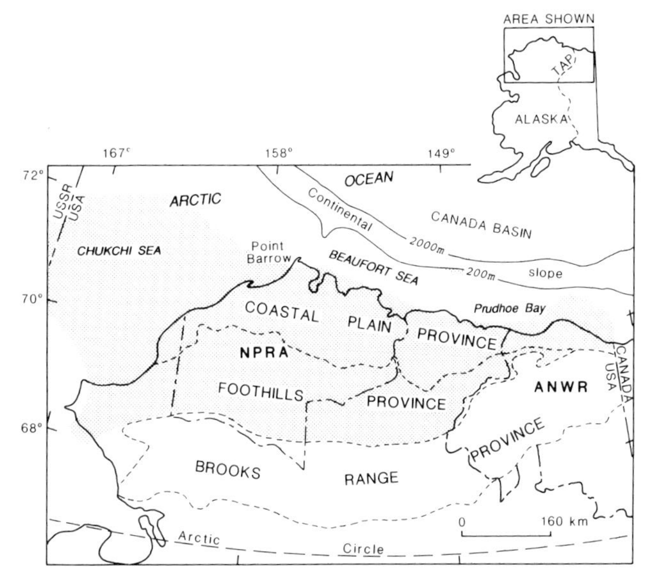

FIGURE 1. Index map showing the locations of the onshore and offshore North Slope foreland basin (shaded area) and its physiographic provinces. NPRA, National Petroleum Reserve in Alaska; ANWR, Arctic National Wildlife Refuge; TAP, Trans-Alaska pipeline.

REGIONAL SETTING

The North Slope foreland basin developed on a continental fragment, the Arctic Alaska plate, that encompasses all

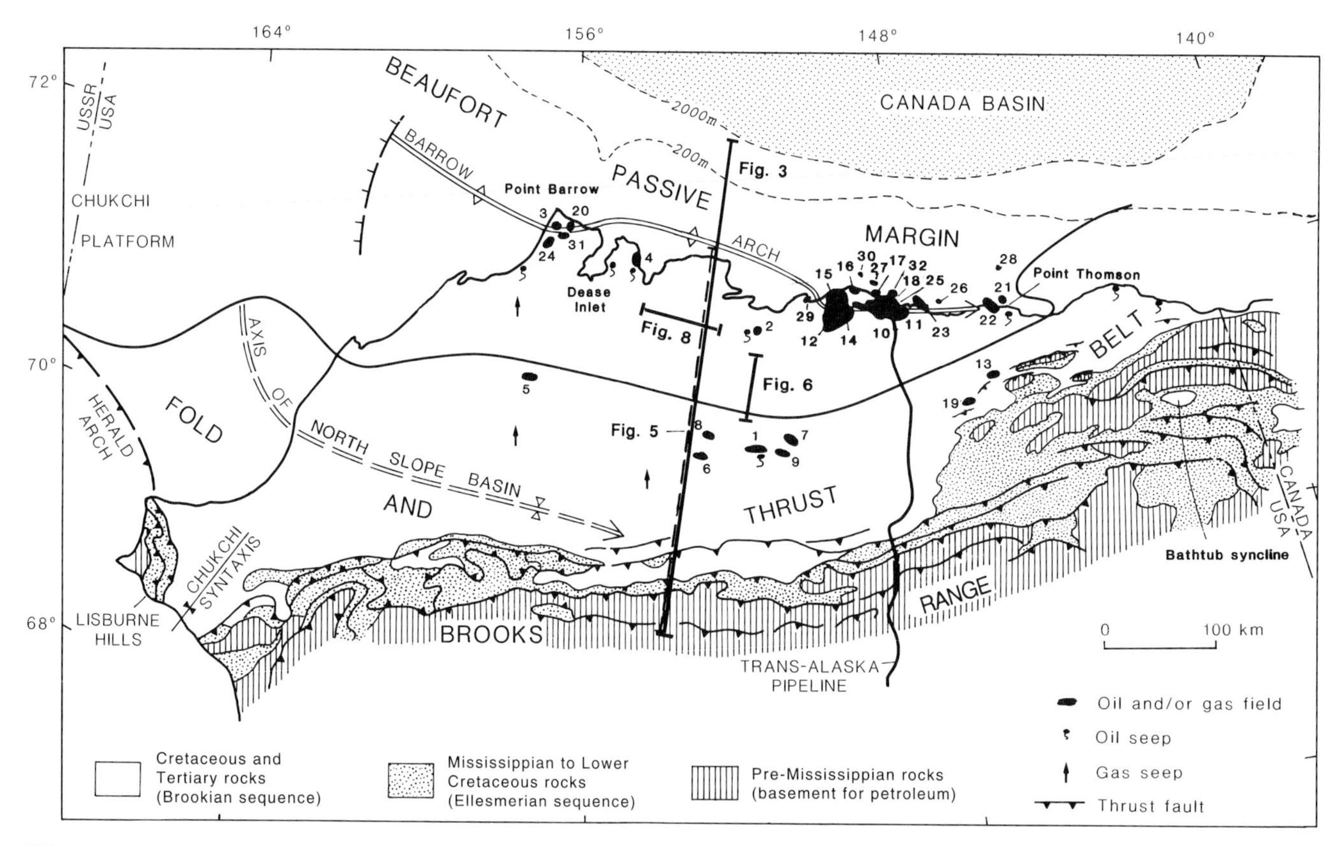

FIGURE 2. Generalized geologic map of the North Slope, showing locations of oil and gas fields, hydrocarbon seepages, major tectonic elements, and cross sections of Figures 3, 5, 6, and 8. Modified from Bird (1991). The northern edge of the Brooks Range generally coincides with the northern limit of pre-Cretaceous rocks, and the foothills–coastal plain boundary coincides with the northern edge of the fold and thrust belt. Map numbers refer to fields listed in Table 1.

of northern Alaska and adjacent parts of northwestern Canada and northeastern Siberia. In Paleozoic and early Mesozoic time, this plate was part of a passive continental margin—probably part of the North American continent north of the Canadian Arctic Islands. During Jurassic and Cretaceous time, rifting occurred along this margin, severing the continental connection and creating a separate plate. Drift and rotation of the plate away from North America produced the Canada basin and Beaufort passive margin. Concurrent with rifting, on the opposite side of the Arctic Alaska plate, collision with an oceanic island arc produced the Brooks Range orogene and the North Slope foreland basin. Because of the apparent changing orientation of the Arctic Alaska plate through time, all references to directions in this paper are in terms of the present-day orientation of northern Alaska.

The continental crust of the Arctic Alaska plate at drillable depths consists mostly of weakly metamorphosed, mostly pre-Upper Devonian argillite and greenschist facies sedimentary rocks. These basement crustal rocks are overlain by a 1000 to 2600-m (3280 to 8530-ft) thick platform sequence of dominantly Mississippian and Pennsylvanian carbonate rocks and shelf to basinal, dominantly clastic Permian to Jurassic rocks. These pre-foreland basin Paleozoic and Mesozoic rocks were deposited on a southward-facing passive continental margin in which more basinal or distal facies are to the south toward the area of the present Brooks Range.

The foreland basin-fill ranges in age from Middle(?) Jurassic through Tertiary or Quaternary and consists initially of orogenic deposits dumped into a foredeep flanking the ancestral Brooks Range orogene followed by thick northeastward-prograding basinal, basin-slope, and shallow marine and nonmarine shelf deposits of mudstone, sandstone, and conglomerate derived both from the ancestral Brooks Range orogenic belt to the south and southwest and from a source area farther to the west, now under the Chukchi Sea. In general, the depocenter of the basin migrated northeastward. In the foredeep bordering the Brooks Range in the western half of the basin, Albian and older Cretaceous rocks (younger strata are not preserved in this area) are in excess of 6000 m (19,685 ft) thick (Figure 3). Rocks of this age thin by both deep-water onlap and downlap to 100 to 200 m (328 to 656 ft) along the flank of the Barrow arch to the north and northeast. In the Point Thomson area in the northeastern part of the basin, about 4000 m (13,000 ft) of Tertiary rocks overlie a very thin, condensed Cretaceous section of basement rocks. Farther east, Tertiary rocks are much thicker in parts of the structurally complex Arctic National Wildlife Refuge (ANWR).

The southern half of the North Slope basin is a fold and thrust belt characterized by rootless, thrust-faulted,

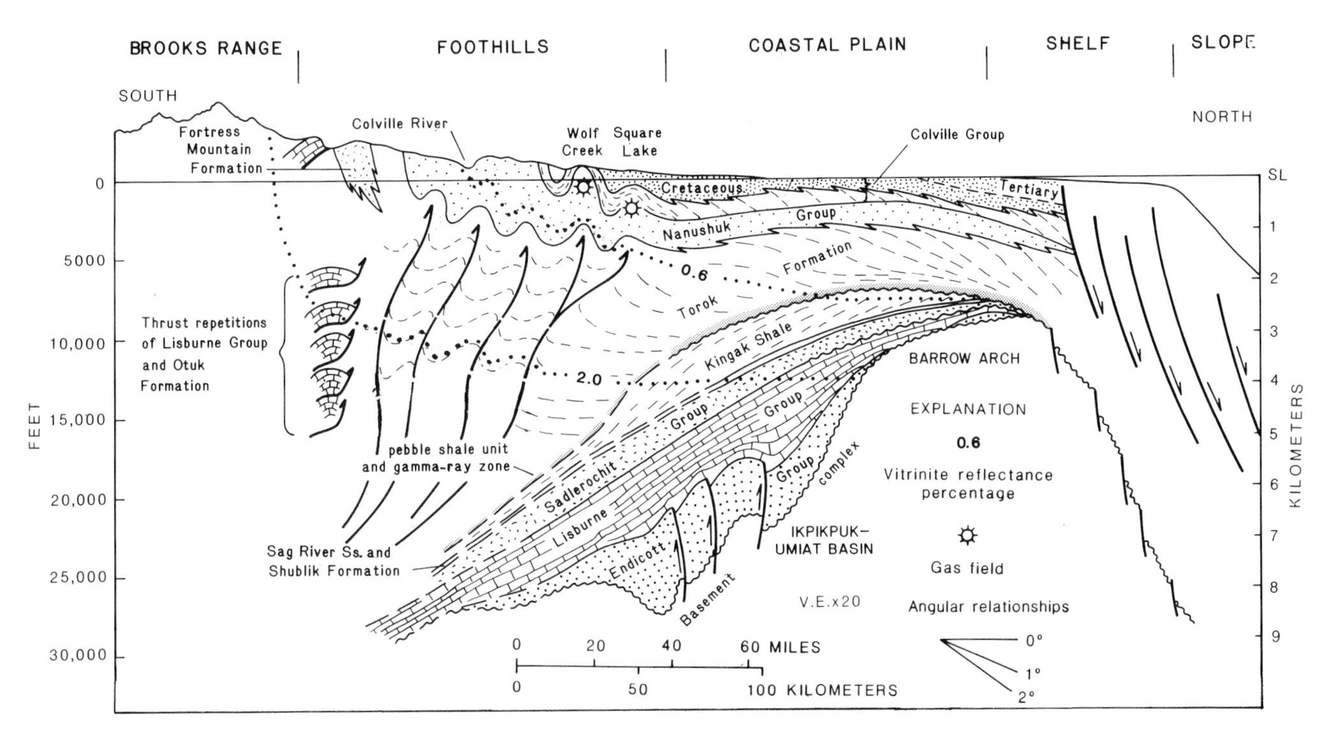

FIGURE 3. Schematic structural and stratigraphic cross section across the central North Slope (from Bird, 1987). Shaded zone between Torok Formation and Kingak Shale is the pebble shale unit and gamma-ray zone. See Figure 2 for location of section.

elongate, en echelon anticlinal folds that attenuate to the north into generally structureless strata of the northern half (Figure 3). In the ANWR, however, the fold belt, as well as the Brooks Range orogene, extends farther north, the fold belt extending under the Beaufort Sea (Figure 2).

With notable exceptions, the overall structural and depositional patterns of the North Slope basin are similar to those of the Alberta basin of Canada and the Rocky Mountain foreland basin of the Western Interior of the United States. The exceptions are that in the North Slope basin, (1) compressive tectonic activity of the orogenic belt continued in the eastern part of the basin (ANWR) into the late Tertiary, (2) the northern or outboard structural margin of the basin was part of a rifted margin of an oceanic basin (early subsidence—i.e., Hauterivian—preserved the features of this margin, which are documented by well and seismic data), (3) the thickness of the basin-fill was much greater and much of it was deposited as turbidites in deeper-water environments, and (4) a sediment source area west or southwest of the basin and eastward tilting of the basin caused the basin depocenter to migrate through time to the east or northeast, subparallel to the Brooks Range.

STRATIGRAPHY

The North Slope foreland basin-fill ranges in age from Middle(?) Jurassic to late Tertiary. Older parts of the fill (Jurassic and earliest Cretaceous) are relatively thin and consist of remnants of deep marine, mainly turbidite deposits, whereas younger parts (late Early Cretaceous and younger) of the fill consist of thick northeastward-prograding basinal, basin-slope, and shallow marine shelf and nonmarine deposits of mudstone, sandstone, and conglomerate. The sources of these deposits were the ancestral Brooks Range to the south and an area far to the west or southwest under the present Chukchi Sea. These clastic rocks are generally lithic rich in composition and are called the Brookian sequence (Lerand, 1973). Prior to and during early formation of the basin, sediments were still being derived from positive areas north of the basin. These northern-sourced sediments, which include rocks of Mississippian to Early Cretaceous (Barremian) age, are mineralogically mature and are called the Ellesmerian sequence (Lerand, 1973). Many investigators limit the Ellesmerian sequence by placing the Jurassic and lowermost Cretaceous part in a separate sequence called the Barrovian sequence (Carman and Hardwick, 1983; Noonan, 1987) or Beaufortian sequence (Hubbard et al., 1987) on the basis of the association of these rocks with early or pre-rifting tectonics offshore to the north and northwest. For the purpose of this report, however, we are using the Ellesmerian in its broader definition. As we use the terms, the Brookian and Ellesmerian sequences overlap in time and probably intertongue, in part, in the deeper parts of the basin. Recognition and differentiation of the Brookian and Ellesmerian sequences is important in understanding the depositional history and development of the North Slope foreland basin.

The following is a brief description of the stratigraphy of the pre-foreland basin Ellesmerian sequence followed by the stratigraphy of Jurassic and younger rocks, which are the strata related to the foreland basin fill. A chronostratigraphic section across the North Slope basin is shown in Figure 4.

Mississippian to Triassic Rocks of the Ellesmerian Sequence

The Ellesmerian sequence records a major northward advance of the sea following the Devonian Ellesmerian orogeny. The Mississippian to Lower Permian part of the sequence constitutes a transgressive megacycle, generally about 1 km (0.6 mi) thick but as thick as 4 km (2.5 mi) in some areas (Figure 3). This sequence unconformably overlies a folded and faulted pre-Mississippian terrane of mostly weakly metamorphosed argillite and quartzite. Initial deposits consist of nonmarine coal-bearing sandstone, shale, and conglomerate (Kekiktuk Conglomerate) that are succeeded by shallow marine black shale (Kayak Shale) or, along the northern basin margin, by red and green shale (Itkilyariak Formation). These shale units grade upward and laterally into an areally extensive carbonate platform sequence of limestone and dolomite (Lisburne Group). Marine regression in Late Pennsylvanian(?) and Early Permian time caused a withdrawal of the sea and development of a regional unconformity at the top of the Lisburne.

Advance of the sea over the eroded Lisburne platform resulted in deposition of the next megacycle, about 300 to 600 m (1000 to 2000 ft) of clastic deposits of the Permian and Triassic Sadlerochit Group. The lower part of the Sadlerochit is a northward-thinning, transgressive marine sandstone and siltstone unit (Echooka Formation). The Echooka is abruptly overlain by prodelta shale and siltstone (Kavik Member of the Ivishak Formation) that grades upward into a southward-prograding clastic wedge of marine and nonmarine sandstone and conglomerate (Ledge Sandstone Member of the Ivishak Formation). The Ledge is the main reservoir of the Prudhoe Bay field; there it consists of alluvial fan and deltaic facies (Melvin and Knight, 1984) or coastal plain complex (Lawton et al., 1987). The uppermost part of the Sadlerochit Group is a transgressive upward-fining and northward-thinning marine siltstone and argillaceous sandstone unit (Fire Creek Siltstone Member of the Ivishak Formation).

The transgression that began with deposition of the Fire Creek Siltstone Member continued in Middle and Late Triassic time with deposition of as much as 100 m (328 ft) of richly fossiliferous shale, siltstone, mudstone, and limestone (Shublik Formation). In basin-margin positions, the Shublik rests unconformably on the Sadlerochit Group or older rocks. The Shublik Formation, believed to represent deposition under upwelling conditions on a broad marine shelf (Parrish, 1987), is an important petroleum source rock (Seifert et al., 1980; Sedivy et al., 1987). A thin (0–40 m, 0–130 ft), regressive marine shelf sandstone and siltstone unit (Sag River Sandstone or Karen Creek Sandstone) gradationally overlies the Shublik Formation along the northern margin of the basin and marks the end of the Shublik transgressive cycle.

Jurassic and Lowermost Cretaceous Rocks of the Ellesmerian Sequence

Northern-sourced Jurassic and lowermost Cretaceous sediments (Kingak Shale and associated rock units) prograded southward toward the newly forming foreland basin. The Kingak Shale ranges in age from Jurassic to Valanginian (Early Cretaceous) (Molenaar, 1983, p. 1070) and is as thick as 1100 m (3600 ft). However, on the north side of the basin, erosion under a northward-truncating mid-Neocomian unconformity has removed parts or all of the Kingak (Molenaar et al., 1987) (Figure 4). In addition, intraformational disconformities based on missing faunal zones may be present within the formation (Detterman et al., 1975). The Kingak consists dominantly of dark-gray to black marine shale and is an important petroleum source rock (Magoon and Claypool, 1984; Seifert et al., 1980; Sedivy et al., 1987).

Southward-extending tongues of sandstone occur in the northern part of the Kingak. Two tongues of Jurassic age, known as the Simpson and Barrow sandstones of local usage, occur in the northern part of the National Petroleum Reserve in Alaska (NPRA). In the Kuparuk River oil field area west of Prudhoe Bay, two or three southward- or southeastward-extending tongues of sandstone occur in the Lower Cretaceous part of the section. Carman and Hardwick (1983) included these sandstone tongues and the shale and siltstone intervals separating them in their Kuparuk Formation, a unit about 85 to 115 m (280 to 375 ft) thick. The sandstone tongues are interpreted to have been deposited in a shallow marine shelf environment and are the producing reservoirs in the giant Kuparuk River field (Carman and Hardwick, 1983; Masterson and Paris, 1987; Gaynor and Scheihing, 1988). Carman and Hardwick (1983) included 91 to 152 m (300 to 500 ft) of underlying lowermost Cretaceous mudstone in their Miluveach Formation, which unconformably overlies the Jurassic part of the Kingak Shale (Figure 4). This unconformity is thought to be related to early or pre-rift structural movements to the north (Carman and Hardwick, 1983, p. 1030). On the basis of this tectonic relationship, Carman and Hardwick (1983) and Noonan (1987) placed the Miluveach and Kuparuk formations, as well as the underlying Kingak Shale, in a separate sequence that they called the Barrovian sequence. Hubbard et al. (1987) referred to this sequence as the Beaufortian sequence.

In the eastern part of the NPRA, the Kingak Shale, including its Lower Cretaceous part, prograded in a southerly direction as indicated by southward-dipping clinoforms on seismic lines (Figures 5 and 6). Relief of the clinoforms suggests water depths of at least 400 to 900 m (1312 to 2953 ft) for basinal or condensed Kingak bottomset beds (Molenaar, 1981, p. 6; 1988, p. 605). In the Prudhoe Bay area, Noonan (1987, p. 464–465) showed Kingak clinoform beds on detailed well-log correlations prograding to the southeast.

The Jurassic and lowermost Cretaceous shale that crops out along the eastern Brooks Range front and, in one area, well back in the range is included in the Kingak Shale and is considered part of the Ellesmerian sequence (Molenaar, 1983; Molenaar et al., 1987; Bird and Molenaar, 1987). The eastern part of the Brooks Range bulges farther

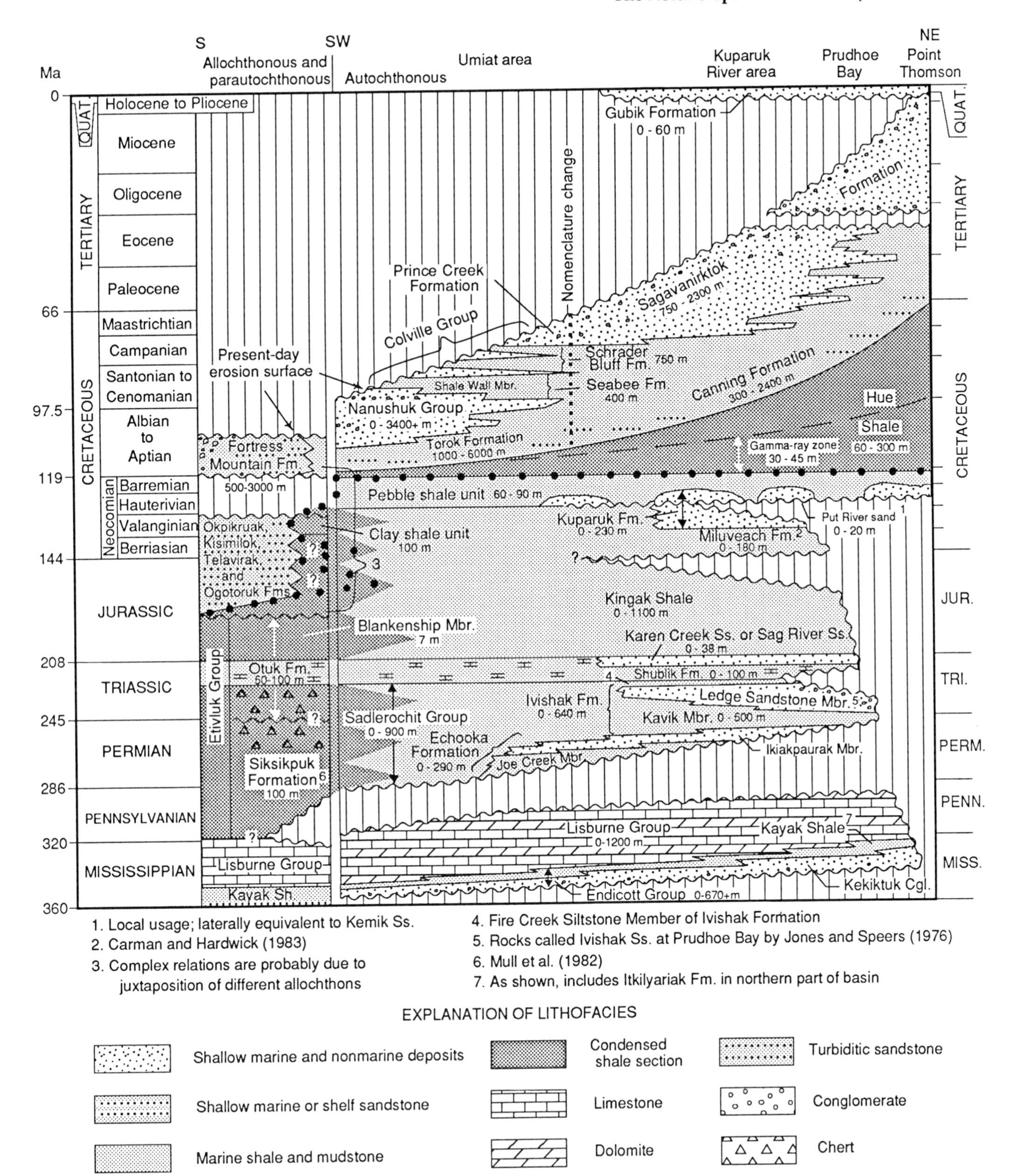

FIGURE 4. Chronostratigraphic chart of post-Devonian rocks across the North Slope basin. Heavy dotted line at top of pebble shale unit and extending down section into the Jurassic in allochthonous rocks separates Brookian sequence (foreland basin) rocks above from Ellesmerian sequence rocks below. Absolute time scale modified from Palmer (1983). Vertical time scale changes by a factor of 2 at 144 Ma.

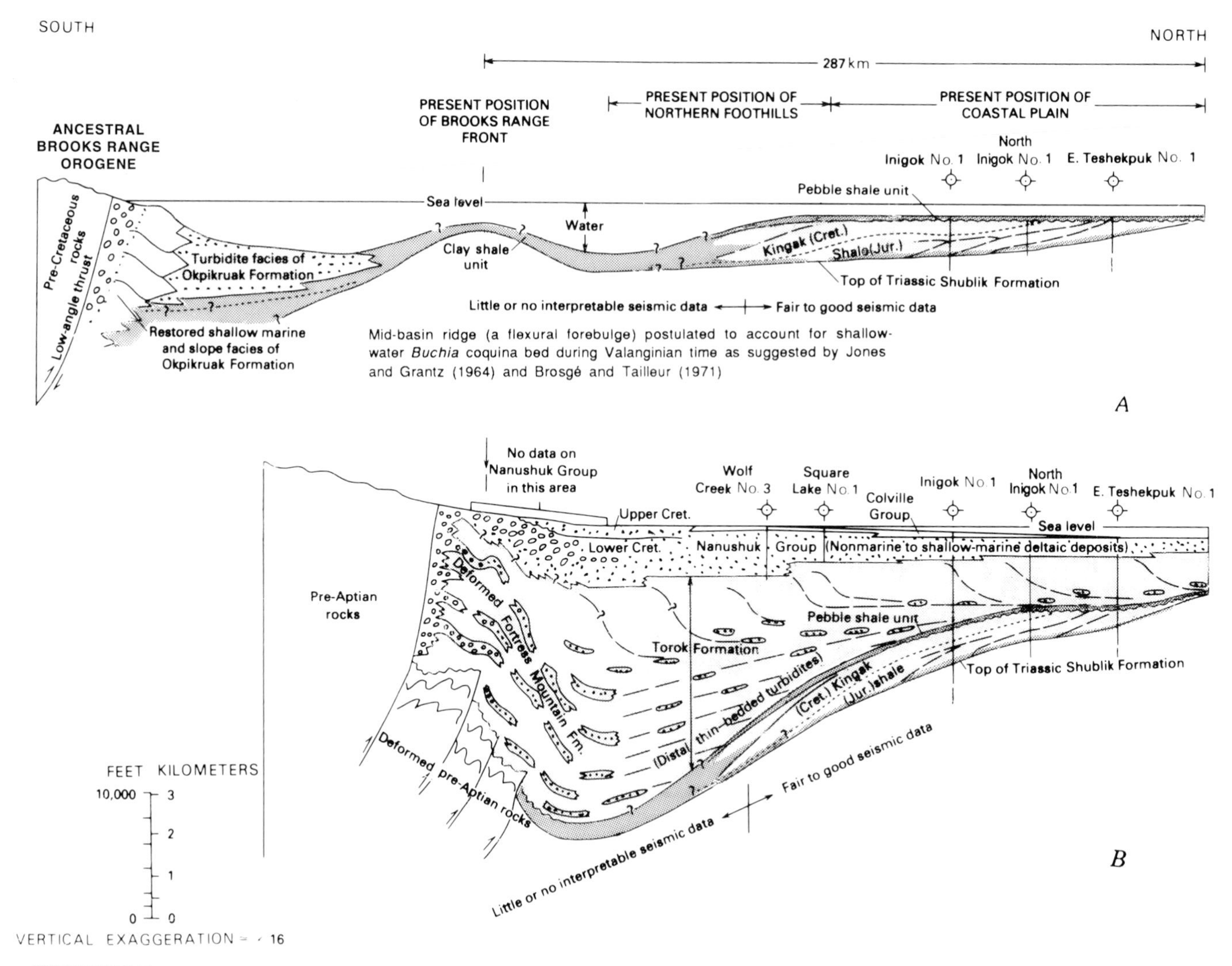

FIGURE 5. Restored cross sections across the North Slope basin, showing Lower Cretaceous rocks at two different times in the Cretaceous. Modified from Molenaar (1981, 1988). Dotted line approximates Jurassic–Cretaceous boundary; dashed lines represent bedding traces as indicated by seismic data. Southern half of section is diagrammatic. See Figure 2 for location of section and Figure 4 for explanation of lithofacies patterns. (A) End of pebble shale unit (Neocomian) deposition. (B) End of Nanushuk Group (Cenomanian) deposition. Turbidites in lower part of Torok Formation shown diagrammatically to illustrate relations with Fortress Mountain Formation. Torok Formation probably has been tectonically thickened in southern half of section.

north than the western and central parts of the range (Figure 2) owing to later thrusting that extended into the late Tertiary in that area (Leiggi, 1987; Wallace and Hanks, 1990). As a result, rocks cropping out along the range front in that area are relatively much farther north in the foreland basin than those along the range front in the central and western Brooks Range, and relate more to the northern-sourced Ellesmerian sequence than to the Brookian rocks (Molenaar, 1983; Molenaar et al., 1987). At Bathtub Ridge, a synclinal remnant of Lower Cretaceous rocks 60 km (40 mi) back into the Brooks Range and 50 to 80 km (30 to 50 mi) west of the U.S.-Canada border (Figure 2), the lower 100 to 260 m (330 to 855 ft) of the Kongakut Formation is a clay shale interval containing *Buchia* coquina beds of Valanginian age (Detterman et al., 1975, p. 22). This interval is considered a distal equivalent of the Kingak Shale (Molenaar, 1983; Molenaar et al., 1987).

In the Yukon Territory of Canada to the east, the Kingak or age-equivalent units generally had a southeastern source (Poulton, 1982). However, considerable differential plate movement (strike slip or rotation) has been postulated between northeastern Alaska and northwestern Canada since Kingak time. Although direct evidence is lacking, the area of separation of the present-day northern-sourced Kingak from the southeastern-sourced Kingak is thought to be east of the Canada-Alaska border (Poulton, 1982).

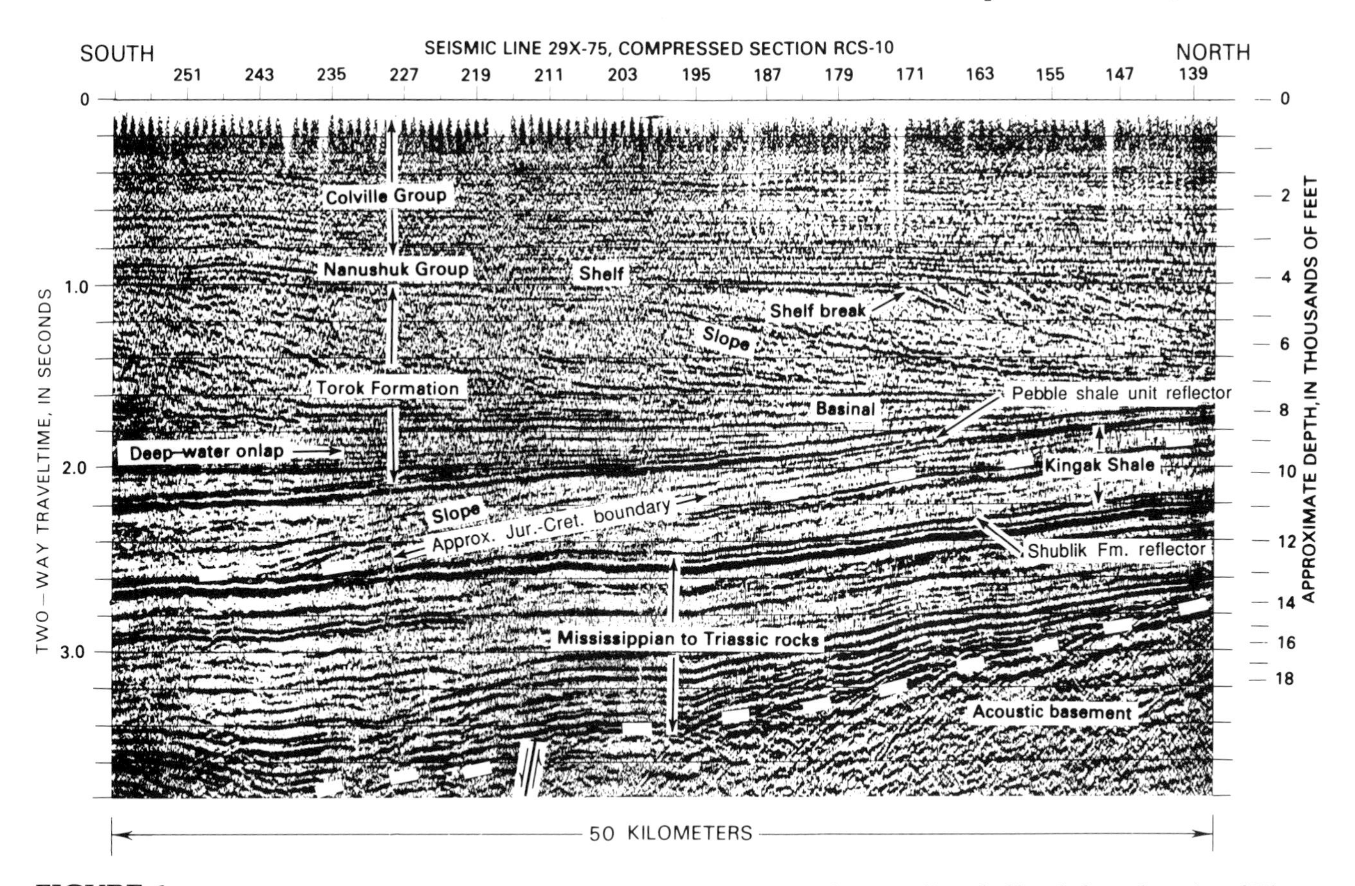

FIGURE 6. Seismic section in eastern part of NPRA, showing southward-prograding shelf and slope deposits of Kingak Shale; approximate Jurassic–Cretaceous boundary within Kingak Shale; lower part of Torok Formation onlapping southward-dipping pebble shale unit (and Hue Shale lithic equivalents); and relation of shelf, slope, and, basinal deposits of Torok-Nanushuk interval (from Molenaar, 1981, 1988). Numbers above section are shot-point numbers. See Figure 2 for location of section.

Jurassic and Lowermost Cretaceous (Berriasian and Valanginian) Rocks of the Brookian Sequence

Jurassic and lowermost Cretaceous (Berriasian and Valanginian) rocks of the Brookian sequence are known only in and along the front of the Brooks Range where they are involved in Brooks Range thrusting. Evidently these rocks, which are a deep-water facies, were deposited in a foredeep flanking the Brooks Range orogene, onlapping the gentler north flank of the basin. During this time, coeval rocks of the Ellesmerian sequence were prograding into the basin from the north (Figures 4 and 5).

Known or dated Jurassic rocks of Brookian character crop out in only a few areas, primarily along the western half of the range front. On the far west end of the Brooks Range southeast of the Lisburne Hills, Campbell (1967) reported a thick sequence of unfossiliferous mudstone (Ogotoruk Formation) and thin-bedded turbidites (Telavirak Formation). Based on his mapping, he estimated this sequence to be about 3000 m (9842 ft) thick. The inferred Jurassic age is based on an overlying section of turbidites (Kisimilok Formation) containing an Early Cretaceous (Berriasian and Valanginian) *Buchia* fauna. The sandstones are feldspathic or arkosic wackes in composition, and hence are considered to be early indicators of the Brooks Range orogeny and the beginning of the foredeep (and foreland basin) development.

In the central part of the Brooks Range front, Middle Jurassic (Bajocian) fossils were reported from tuffaceous graywacke beds in a complexly faulted area (Jones and Grantz, 1964, p. 1468). These are the oldest dated Brookian rocks and probably indicate the initiation of foreland basin development. In other places along the central and western Brooks Range front (and probably on a different allochthon), a thin black clay shale interval of middle Early Jurassic to early Middle Jurassic age is the upper member of the mostly Upper Triassic Otuk Formation and is called the Blankenship Member (Mull et al., 1982). This member may be a distal condensed section that is correlative with the lower part of the Kingak Shale, a northerly derived Ellesmerian unit (Bodnar, 1989).

Overlying the Jurassic rocks, probably conformably, is a continuation of the flysch sequence called the Okpikruak Formation (Gryc et al., 1951, p. 159). This unit is exposed in several allochthons along the Brooks Range and is at least 1000 m (3280 ft) thick (Mayfield et al., 1988). It consists of interbedded graywacke sandstone and mudstone, and much of it is considered an outer fan or distal turbidite facies (Molenaar, 1988, p. 605). In some areas, however, olistostromes and conglomerates are present, indicating proximal positions (Mull, 1985, p. 15; Crane, 1987). *Buchia* bivalves indicate a Berriasian and Valanginian age for the Okpikruak Formation (Jones and Grantz, 1964, p. 1464), although Mayfield et al. (1988) suggest that the basal part of the unit may range down into the Jurassic.

A thin clay shale interval containing *Buchia* coquinoid limestone beds of Valanginian age is present in many localities along the Brooks Range front. This interval, which is no more than 100 m thick (Brosgé and Tailleur, 1971, p. 84), is included in the Ipewik unit of Crane and Wiggins (1976) in the western part of the range front. In some places it rests on Triassic rocks, and in other places it rests on a thin Jurassic shale section (Blankenship Member of the Otuk Formation; Bodnar, 1989) with no intervening Okpikruak Formation (Jones and Grantz, 1964, p. 1471). This may be explained as being on an allochthon separated from the allochthon(s) on which the Okpikruak is present. In the western Brooks Range, however, Mayfield et al. (1988) show this clay shale interval (which they referred to as the Ipewik unit) as unconformably underlying the Okpikruak. They explain the age discrepancy of the Okpikruak Formation as a result of its basal part being diachronous, becoming younger to the north.

Because the coquinoid limestone beds were probably deposited in relatively shallow water, certainly much shallower than that in which the Okpikruak turbidites were deposited, the clay shale interval is interpreted to have been deposited on a mid-basin high, as suggested by Jones and Grantz (1964) and Brosgé and Tailleur (1971) (Figure 5). A mid-basin high of such great regional extent (900 km; 560 mi), and apparently parallel to the trend of the Brooks Range, suggests that this feature may be a flexural forebulge.

Hauterivian and Barremian Rocks of the Ellesmerian Sequence

Erosion of an uplifted segment of the rifted margin of the Arctic Alaska plate during mid-Neocomian time resulted in removal of much of the Ellesmerian sequence in northern Alaska, especially northeastern Alaska (Figures 3 and 4). Subsequent subsidence of this northern uplift resulted in northward transgression of the sea in late Neocomian time and deposition of a discontinuous transgressive sandstone unit, the Kemik Sandstone, and an associated shale unit informally named the pebble shale unit. The unconformity at the bases of these units is considered to be a breakup unconformity (Grantz and May, 1983, p. 96). Although separated from underlying Ellesmerian rocks by this major unconformity, the Kemik Sandstone and pebble shale unit are included in the Ellesmerian sequence because they were derived from the north as the sea transgressed northward.

The Kemik Sandstone crops out and is present in several wells in the eastern part of the basin. It is a fine- to very fine-grained, locally conglomeratic, quartzose, shallow marine sandstone and is as thick as 30 m (98 ft) (Mull, 1987). It directly overlies the unconformity, but to the south the unconformity dies out and the Kemik is conformable on the underlying Kingak Shale before the sandstone pinches out (Molenaar, 1983). The Kemik is laterally equivalent to the Put River sand of local usage in the Prudhoe Bay area (Jamison et al., 1980), the upper sandstone of the Kuparuk Formation of Carman and Hardwick (1983, figures 7 and 8), the Walakpa sandstone of local usage in the NPRA south of Barrow (Schindler, 1988, p. 59), and the Thomson sand of local usage in the Point Thomson area along the coast near the ANWR (Bird and Molenaar, 1987, p. 51; Gautier, 1987, p. 118).

The pebble shale unit is a transgressive shale unit, about 60 to 90 m (195 to 295 ft) thick, that crops out in the northeastern part of the basin and is present in the subsurface along the entire north flank of the basin. Where the Kemik Sandstone or equivalent sandstone units are absent, the pebble shale unit unconformably overlies the Kingak Shale and older rocks, but to the south, where the unconformity dies out, it is conformable with underlying rocks. The pebble shale unit has not been reported on the south flank of the basin and may be represented by an unconformity related to Brooks Range thrusting. In addition, Hauterivian and Barremian fossils in Brookian sequence rocks have not been reported anywhere in northern Alaska. In the eastern Brooks Range, the pebble shale unit is present in the Bathtub syncline (Figure 2). It consists of dark-gray to black, noncalcareous clayey to silty shale and contains minor scattered rounded and frosted quartz grains. Common to rare, matrix-supported chert and quartzite pebbles or granules and very rare cobbles occur throughout the unit. It is considered to be a good oil source rock (Morgridge and Smith, 1972; Magoon and Bird, 1985; Keal and Dow, 1985). Some investigators include an overlying highly radioactive shale interval, about 30 to 45 m (98 to 148 ft) thick, in the pebble shale unit, but Molenaar et al. (1987, p. 520) included this shale interval in the lower part of the Hue Shale, which they considered to represent distal Brookian deposits (Figure 4).

Transgression of the late Neocomian seaway to the north resulted in the final inundation of the northern source terrain and thus terminated deposition of the Ellesmerian sequence in what would later become the northern margin of the North Slope foreland basin.

Aptian(?) and Albian Rocks of the Brookian Sequence

Aptian(?) and Albian rocks are grouped together because they form one depositional sequence. Aptian megafossils have not been recognized in northern Alaska, but Aptian determinations have been made based on foraminifers and palynomorphs. However, some of these beds, which are the deep-water part of the clinoforms, can be traced on seismic data updip to shallower-water beds determined to be Albian in age (Molenaar, 1988, p. 611). Aptian rocks undoubtedly are present, especially in the deep part of the basin, but the exact placement of the Aptian–Albian boundary is uncertain. Hence, for this discussion, rocks of these ages are grouped together.

The great thickness of Aptian(?) and Albian rocks along and adjacent to the foredeep along the Brooks Range front, especially in the western half, indicates major downwarping of the basin (Figures 3 and 5). In addition to sediment loading, this was probably related to thrust loading by the orogene to the south. The ancestral Brooks Range was an obvious provenance; however, progradation patterns of prodelta sediments (Figure 7) indicate a major source area to the far west.

The oldest Aptian(?)–Albian rocks exposed in the foothills along the Brooks Range front are included in the Fortress Mountain Formation along most of the range front and in the upper two-thirds of the Kongakut Formation and overlying Bathtub Graywacke in the Bathtub syncline in the far east (Figure 2). The Fortress Mountain Formation is 1000 to 3000 m (3280 to 9840 ft) thick and consists of sandstone, conglomerate, mudstone, and occasional coal beds deposited in environments ranging from alluvial fan deltas, submarine canyons, and slope channels to proximal and distal turbidite fans (Molenaar et al., 1981, 1988; Crowder, 1987, 1989). The coarser facies were clearly derived from many point sources along the ancestral mountain front. To the north, the Fortress Mountain grades into deep-water mudstones of the Torok Formation (Figure 5). Molenaar et al. (1988) recommended that Fortress Mountain terminology be restricted to the outcrop belt inasmuch as the unit is dominantly mudstone in the subsurface a short distance to the north and is lithologically indistinguishable from the Torok Formation.

In the Bathtub syncline, the Brookian part of the Kongakut Formation, 300 to 700 m (984 to 2296 ft) thick,

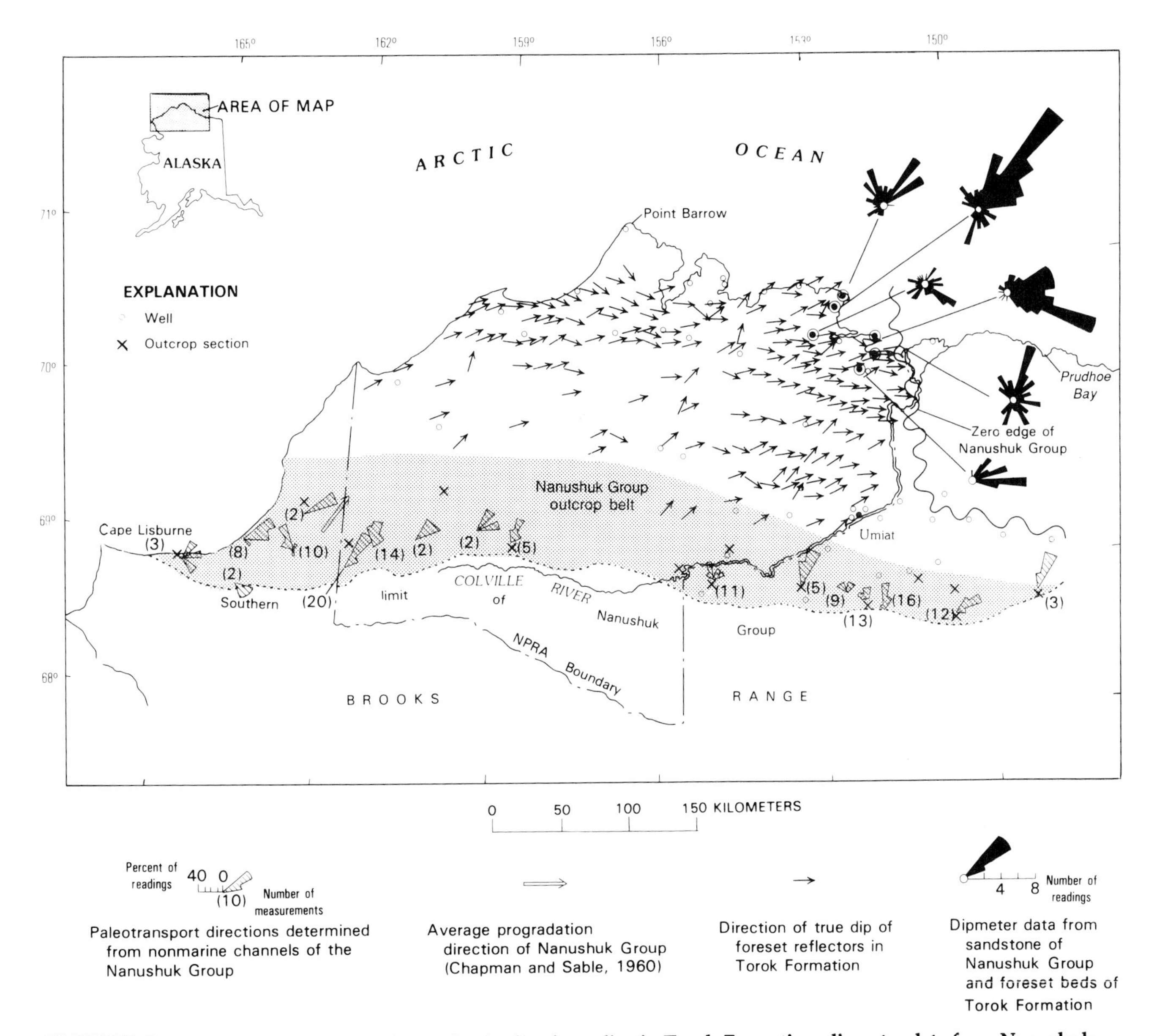

FIGURE 7. Plot of directional data from seismic clinoform dips in Torok Formation, dipmeter data from Nanushuk Group and Torok Formation, direction of Nanushuk progradation from outcrop studies (Chapman and Sable, 1960), and direction of nonmarine channel trends as measured by Ahlbrandt et al. (1979) and Huffman et al. (1985). Modified from Bird and Andrews (1979).

is interpreted to be a deep-water mudstone unit. This is overlain by 750 m (2460 ft) (top not exposed) of proximal turbidites of the Bathtub Graywacke (Detterman et al., 1975; Molenaar, 1983). Although these units are poorly dated, all but possibly the lower part are considered to be Aptian(?) and Albian in age. In the foothills along the range front in the central part of the ANWR, distal turbidites referred to as the Arctic Creek unit (Molenaar et al., 1987, p. 521) contain fossils of Albian age (Detterman et al., 1975, p. 29).

During the Aptian(?) and Albian and ranging into the Cenomanian, a thick deltaic, slope-shale, and basinal sequence prograded across the western half of the foreland basin from the west-southwest (Figure 8). In the eastern half of the foreland basin, the sequence thins dramatically, being represented in the northeastern part of the basin by a thin condensed section. The Nanushuk Group comprises the deltaic part of this sequence. It crops out in the foothills and is present in the subsurface (Figure 7). It ranges in thickness from 3444 m (11,300 ft) along the Chukchi Sea coast on the west (Smiley, 1969) to a pinch-out edge in the area of the present Colville River delta on the east (Molenaar, 1985) (Figure 9). The lower part of the Nanushuk consists of a thick section of intertonguing shallow marine sandstone and neritic shale and siltstone (Kukpowruk, Grandstand, and Tuktu formations) that grades seaward (east and northeast) into mudstone of the Torok Formation. The upper part of the Nanushuk consists dominantly of nonmarine facies of paludal shale, fluvial sandstone, and coal (Corwin and Chandler formations) that grade into the marine facies. Conglomerate is present in the southern part of the outcrop belt, which is closer to the ancestral Brooks Range source (Ahlbrandt et al., 1979; Molenaar, 1985, 1988).

The Torok Formation comprises the prodelta shelf, slope, and basinal deposits that are coeval lateral equivalents of the Nanushuk deltaic system. Except for limited areas in the foothills province (Figure 1), where the upper part of the Torok is exposed, the Torok is a subsurface formation in the northern foothills and coastal plain provinces of the western two-thirds of the North Slope. It ranges in thickness from less than 900 m (2952 ft) on the Barrow arch to about 6000 m (19,685 ft) on the south near the eastward-trending part of the Colville River (Molenaar, 1985, 1988) and consists dominantly of mudstone containing thin-bedded, fine- to very fine-grained turbiditic sandstone in its lower part.

Depositional patterns of the Nanushuk-Torok interval are well displayed on many of the seismic lines in the NPRA (Figures 5, 6, and 8). Large clinoforms dipping as steeply as 6° (but mostly 3 to 5°) show the progradation directions to be east and northeast (Figure 7). Based on the relief of the presently compacted clinoforms, the water depth in which the basinal beds were deposited is estimated to range from 700 to 1400 m (2300 to 4600 ft). It was probably greater to the south in the foothills belt, but seismic data are not clear enough to interpret the geometry of the clinoforms. Large-scale submarine slumps are apparent in the northeastern part of the NPRA (Molenaar, 1985; Weimer, 1987). Southeast of Point Barrow, where the Nanushuk overtopped the Barrow arch and prograded into the Canada basin to the north, a large submarine canyon, known as Simpson canyon (Figure 9), was cut to a

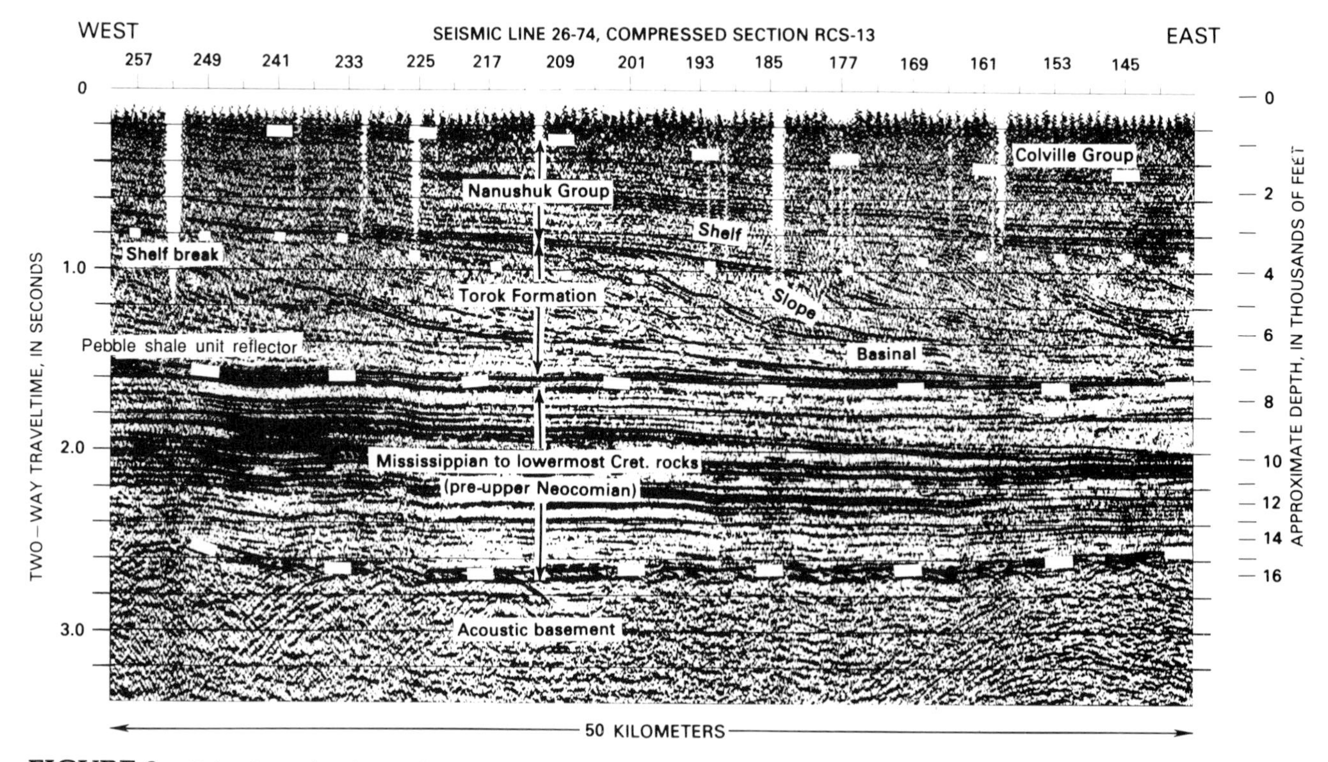

FIGURE 8. Seismic section in northeastern part of NPRA, showing stratigraphic relations of shelf, slope, and basinal deposits of Torok-Nanushuk interval (from Molenaar, 1981, 1988). Dots represent position of shelf break. Note stratigraphic rise from west to east. Numbers above section are shot-point numbers. See Figure 2 for location of section.

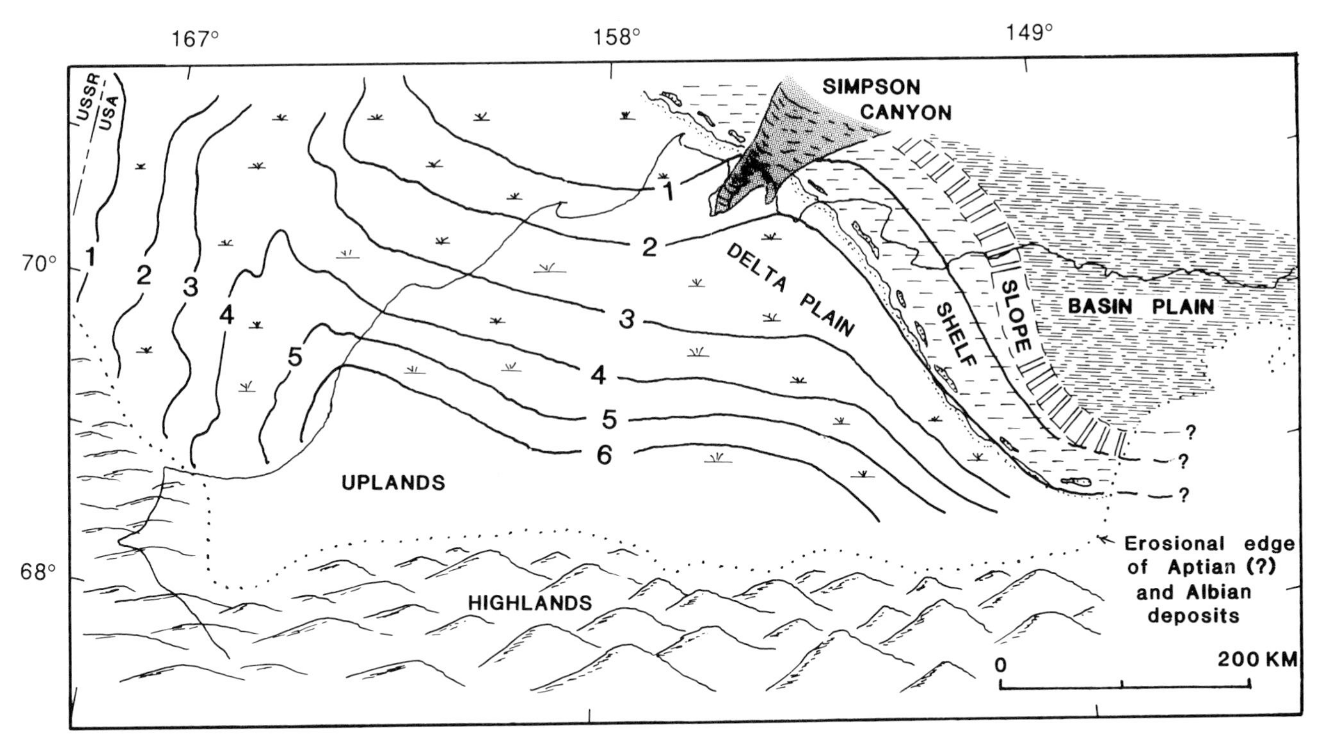

FIGURE 9. Isopach map of Torok-Nanushuk interval and paleogeographic map at time of maximum progradation of Nanushuk Group. Contours beneath Chukchi Sea are modified from Thurston and Theiss (1987, p. 50). Areas where thickness of mapped interval is greater than 4 km include tectonically thickened section. In these same areas, uplift and erosion of several kilometers of section are indicated by vitrinite reflectance values of 0.5% and greater. Simpson Canyon is a submarine canyon that was cut into the Nanushuk Group and Torok Formation in Cenomanian(?) time and filled with marine shale in Turonian time. Isopachs are in kilometers.

depth of about 1500 m (4920 ft) through the deltaic deposits of the Nanushuk and into slope shales of the Torok Formation (Molenaar, 1981, 1988). This canyon is filled with Upper Cretaceous shale and the subsequent differential compaction is probably evident today in the form of Dease Inlet, the large inlet southeast of Point Barrow (Figure 2).

Seismic data show that the basinal Torok beds downlap toward the east onto a prominent reflector (Figure 8) and onlap it in a northerly direction (Figure 6). This clearly shows that the north flank of the basin had a gentle southward-dipping slope (the flank of the Barrow arch) and that older parts of the Torok (and Fortress Mountain correlatives) onlapped this slope in a deep-water environment. This was probably the result of tectonically induced subsidence prior to sediment loading. Well data indicate that the prominent reflector underlying the Torok is a highly organic-rich shale interval known as the gamma-ray zone (GRZ) or highly radioactive zone (HRZ). This zone is about 30 to 45 m (100 to 150 ft) thick and was formerly included in the underlying transgressive pebble shale unit. Based on its Aptian and Albian age (Carman and Hardwick, 1983, p. 1017) and paleogeographic considerations, Molenaar et al. (1987) and Bird and Molenaar (1987) assigned the GRZ to the basal part of their newly named Hue Shale in the area east of the NPRA. In this area it is considered to be a condensed section of the total Aptian(?) and Albian succession, which is in excess of 6000 m (19,685 ft) thick south of Umiat (Figure 9).

Although the ancestral Brooks Range was a prominent highland that contributed sediments northward into the foredeep, a large source area to the west or southwest of the North Slope basin, probably west of the Herald arch (Figure 2), may have been the dominant source area that provided sediment for the eastward- and northeastward-prograding deltaic system. East of 149° W longitude, the deltaic Nanushuk Group is limited to a relatively narrow, eastward-trending band (Figure 9). In this area, much of the Aptian and Albian sediment from the Brooks Range on the south was trapped in the frontal foredeep, which limited the north or northeastward progradation.

Upper Cretaceous to Quaternary Rocks of the Brookian Sequence

The Nanushuk Group regression was terminated by a significant transgressive event in Cenomanian-Turonian time, which was followed by the continuation of the northeasterly progradation of thick deltaic and prodelta deposition. This progradational sequence, punctuated by several smaller transgressions and at least one unconformity in the middle or late Eocene, continued through the

Late Cretaceous and into the late Tertiary or Quaternary (Figures 4 and 10). One difference between the Lower Cretaceous section and the Upper Cretaceous and Tertiary section is that the latter contains more bentonite.

Upper Cretaceous and Tertiary rocks are present only in the eastern half of the North Slope, except for a thin veneer of Upper Cretaceous rocks preserved in the western part. Undoubtedly, however, a thick section of at least Upper Cretaceous rocks must have been deposited over the western part. The organic maturity of the upper part of the Nanushuk Group in the western areas (vitrinite reflectance values of 0.5% R_o; Magoon and Bird, 1988, p. 407) suggests that at least 2500 m (8200 ft) of strata were removed by erosion. Eastward tilting of the basin and consequent eastward shifting of depocenters probably resulted in cannibalization of older units to source younger Tertiary sediments to the east.

A 0 to 350-m (0 to 1148-ft) thick marine sandstone and shale unit (Ninuluk Formation) of Cenomanian age, which is associated with, and occurs at the base of, the transgression south of the Umiat area, is included in the Nanushuk Group (Detterman, 1956; Detterman et al., 1963; Chapman et al., 1964). The overlying Turonian shale unit, known as the Shale Wall Member of the Seabee Formation, correlates with similar transgressions throughout the Western Interior and other areas of the world and probably is the result of a eustatic sea level rise that peaked in the early Turonian (Hancock and Kauffman, 1979; Haq et al., 1987).

Many mappable units were recognized and named within the intertonguing marine and nonmarine Upper Cretaceous rocks exposed in the greater Umiat area (Gryc et al., 1951, 1956; Detterman, 1956; Whittington, 1956; Brosgé and Whittington, 1966) (Figure 4). Most of these units grade into marine shale in the subsurface a short distance to the north and northeast. Where most of these facies changes occur, which is about at the seaward pinchout of the Nanushuk Group, a different nomenclature was proposed by Molenaar et al. (1987). The thick prodelta shale or mudstone interval, including the thin-bedded turbidites in the lower part, is called the Canning Formation, and the overlying thick marine and nonmarine deltaic and coastal-plain rocks are included in the Sagavanirktok Formation (Figure 4). The basal part of the Canning Formation downlaps and partly grades into distal condensed deposits of the Hue Shale. All these units are highly diachronous.

The Hue Shale is a distal condensed section ranging in thickness from about 180 m (590 ft) in the Colville River delta area to as much as 300 m (984 ft) in the Point Thomson area. It consists of fissile, organic-rich, black clay shale and bentonite (Molenaar et al., 1986; 1987, p. 520). The basal 30 to 45 m (98 to 148 ft) is a highly radioactive zone (HRZ or GRZ) that is easily identified on gamma-ray logs. As previously discussed, this part of the Hue Shale is considered to be of Aptian and Albian age and coeval with the thick Nanushuk-Torok sequence in the west half of the North Slope basin. The GRZ contains a few thin bentonite beds, but the remaining part of the Hue Shale is much more bentonitic. The lower half of the Hue Shale contains the richest oil source rocks, some as high as 12% organic carbon (Magoon et al., 1987). The Hue Shale ranges in age from Aptian and Albian to Maastrichtian and possibly Paleocene, its uppermost part being diachronous and becoming younger from west to east (Figure 4). Because it is a condensed section, there probably are depositional hiatuses within the unit.

The Canning Formation is a thick prodelta shale unit composed of shelf and slope deposits that downlap onto the underlying Hue Shale and is overlain by deltaic deposits of the Sagavanirktok Formation (Figure 10). The Canning ranges in thickness from 1200 to 1800 m (3937 to 5905 ft) and consists of silty, bentonitic shale containing minor thin-bedded, very fine- to fine-grained, turbiditic sandstone in its lower part (Molenaar et al., 1987). Based on the relief of the clinoforms in the Canning Formation, the lower part of the formation is estimated to have been deposited in water depths ranging from 600 to 1200 m (1968 to 3937 ft). The Canning ranges in age from as old as Albian on the west to as young as Oligocene(?) on the east (Figure 4).

The Sagavanirktok Formation is a thick, shallow marine and nonmarine unit that overlies the Canning Formation (Figure 4). It ranges in thickness from 600 to 2300 m (1968 to 7545 ft), but its upper surface is a late Pliocene to Holocene erosion surface (Molenaar et al., 1987). It consists dominantly of fine- to medium-grained sandstone, generally bentonitic shale, and lesser amounts of conglomerate and coal. The Sagavanirktok gradationally overlies and intertongues with the Canning Formation in most areas, but in the Point Thomson area a conglomeratic sandstone rests directly on marine shale with no angular discordance. This disconformity, which occurs in the upper part of the Eocene section (Figure 4), may be the result of a relative drop in sea level and may correlate with the 39.5-Ma lowstand indicated by Haq et al. (1987). Seismic data in the northwestern part of the ANWR indicate that the unconformity dies out to the east or northeast (or passes into strata considered to be below wave base) as the deltaic sequence continues to rise stratigraphically to the northeast (Bird and Molenaar, 1987). The Sagavanirktok Formation ranges in age from about Campanian in the west to Pliocene in the east (Figure 4).

The Jago River Formation of Buckingham (1987) is a thick (3000 m; 9842 ft) section of predominantly nonmarine rocks exposed in a relatively small area near the Jago River in the northern part of the ANWR. This formation, of Late Cretaceous and early Tertiary age, consists of thick units of sandstone, conglomerate, and mudstone with thin beds of coal. These thick nonmarine rocks with current dispersal directions to the north and northwest are anomalous in their geographic position in the foreland basin with respect to thinner, at least partly coeval turbidite sandstone facies of the Canning Formation observed about 70 km (44 mi) to the west. Bird and Molenaar (1987, p. 59) interpret the Jago River Formation as an erosional remnant of a basin displaced northward on a thrust sheet.

The Gubik Formation, of late Pliocene and Pleistocene age (Nelson and Carter, 1985), forms a veneer across most of the coastal plain, unconformably overlying Cretaceous and Tertiary rocks (Figure 4). It ranges in thickness from a few meters to as much as 50 m (164 ft) and consists of marine and nonmarine loosely consolidated sand, conglomeratic sand, conglomerate or gravel, and silt (Brosgé and Whittington, 1966, p. 574).

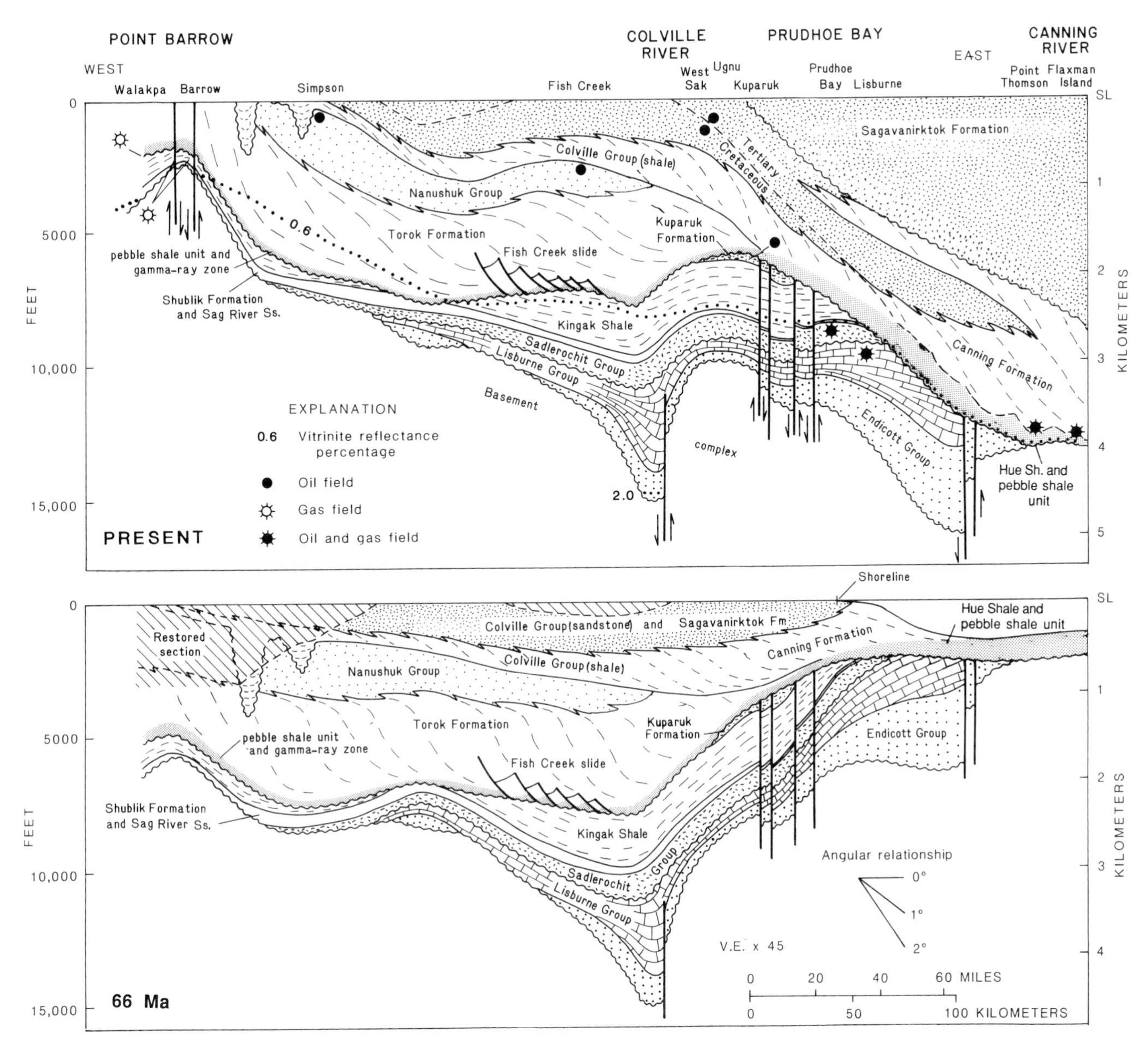

FIGURE 10. Reconstruction of west-to-east structural and stratigraphic cross section along or paralleling the Barrow arch at the end of the Cretaceous (66 Ma) and at present time. Modified from Bird (1987). See Figure 11 for location of section.

TECTONICS AND STRUCTURE

Tectonic elements that form the bounding margins of the North Slope basin controlled both the deposition and structural development of the basin. Along the south margin, the ancestral Brooks Range was certainly the dominating structural element affecting basin development. Not only was it a significant provenance, but it also provided the tectonic load necessary to form the foredeep along the south side of the basin (Figure 3). Burial depths to pre-Upper Devonian basement exceed 10 km (6.2 mi) beneath the axis of the foredeep; minimum depths as shallow as 1 km (0.6 mi) occur along the Barrow arch and on the Chukchi platform (Figures 10 and 11).

The bounding structure on the north side of the basin, prior to the inception of the North Slope foreland basin, was part of a south-facing passive margin that, early in the development of the basin, was a rift margin of the Canada basin that was forming to the north. This feature, known as the Barrow arch or Beaufort sill (Mull, 1989) or Barrow inflection (Ehm and Tailleur, 1987) (Figure 2), was a positive area during Jurassic and earliest Cretaceous time. Normal faulting associated with rifting is interpreted from seismic-sequence analysis to have been active from Middle Jurassic (175 Ma) to Early Cretaceous (113 Ma) time (Hubbard et al., 1987, p. 812). Faulting, erosional truncation, and subsequent marine transgression along this feature were responsible for the large,

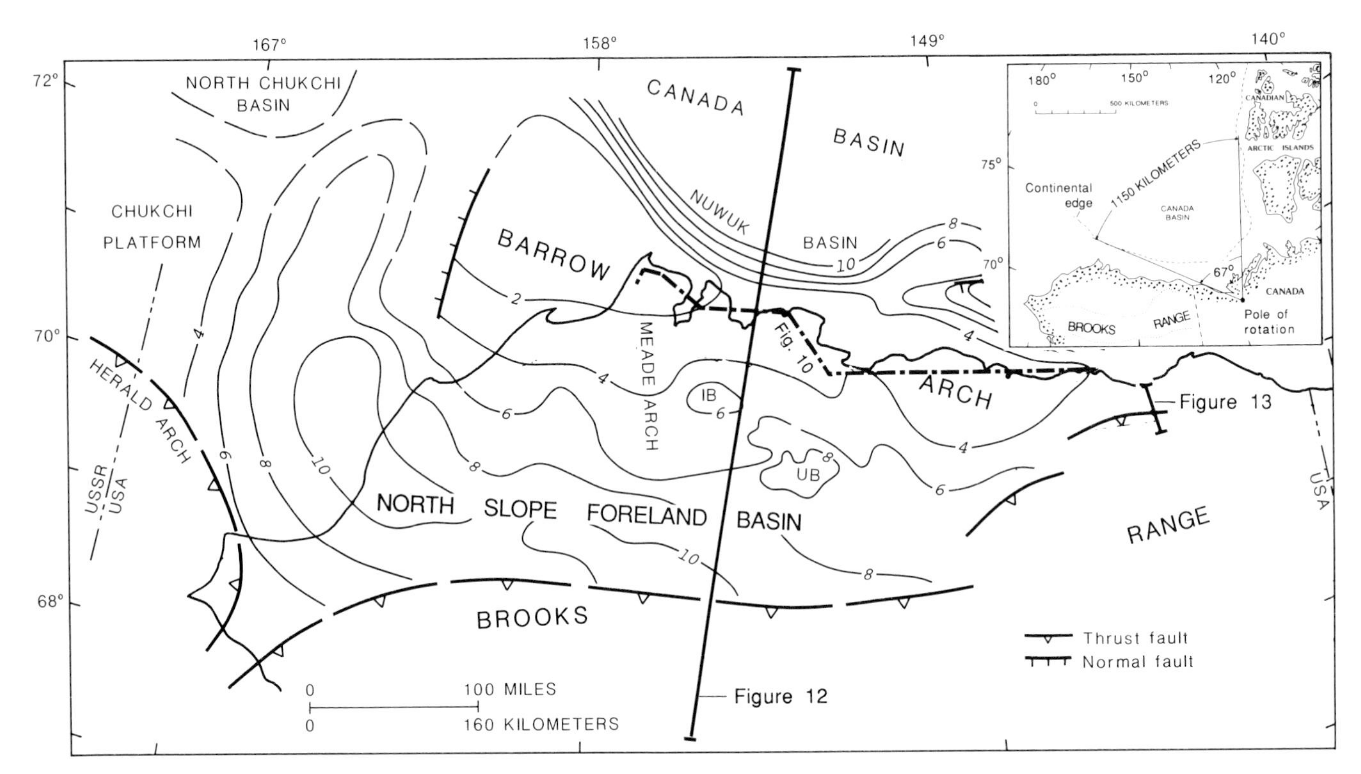

FIGURE 11. Structure contour map of the North Slope basin showing subsea depth (in km) to pre-Upper Devonian basement, lines of cross sections of Figures 10, 12, and 13, and diagram (inset) illustrating rotational hypothesis for the origin of the Arctic Alaska plate (modified from Mayfield et al., 1988). Offshore contours generated from Grantz and May (1988) and Grantz et al. (1988). IB (Ikpikpuk basin) and UB (Umiat basin) are pre-foreland basin features. (See Figure 3 for their expression.)

combination structural-stratigraphic hydrocarbon traps that lie along the north margin of the North Slope basin (Figure 10).

Brooks Range Structure

The Brooks Range and the deformed rocks of the foothills province compose a fold and thrust belt (Figure 2). The Brooks Range is a continental (A-type) subduction orogenic belt (Bally and Snelson, 1980) more than 1000 km (621 mi) long and as much as 300 km (186 mi) wide. In the western part of the North Slope, this orogenic belt makes a sharp bend (the Chukchi syntaxis), almost doubling back on itself, and continues northward through the Lisburne Hills and northwestward offshore along the Herald arch (Figure 2). Except for the area of the syntaxis, where structural trends almost oppose each other, the orogene is characterized by eastward-trending, northward-verging structures.

Important differences in structural style and time of deformation occur along the northern margin of the Brooks Range. West of 148° W longitude (the central and western Brooks Range), the range and range front are made up of as many as seven major thrust sheets or allochthons (Mayfield et al., 1988). Because of a regional westward plunge of the Brooks Range, more allochthons are present and are better preserved in the western part of the range. Each allochthon has a recognizable distinctive stratigraphy and, except for the two oldest (and structurally highest), are made up of Neocomian and older sedimentary rocks (Mull, 1982, 1989; Mayfield et al., 1988). The younger of the two older allochthons is made up dominantly of pillow basalt containing chert and mudstone interbeds of Triassic and Jurassic(?) age. The older allochthon is comprised of gabbro, peridotite, and dunite of Jurassic age and is considered to be an ophiolite suite that was obducted from oceanic crust to the south onto the Arctic Alaska plate in the early phase of Brooks Range thrusting (Roeder and Mull, 1978; Mayfield et al., 1988, p. 164). On the basis of detailed mapping in the central part of the Brooks Range front of the youngest (structurally lowest) and most extensive of the allochthons, Kelley and Bohn (1988) found that the allochthon itself was cut by several thrust décollements. Figure 12 shows schematically the structure across the Brooks Range, the foreland basin, and the Barrow arch to the Canada basin.

East of 148° W longitude, the part of the range that bulges northward from the east-west trend of the main range is a younger structural feature than the central and western parts of the Brooks Range. It exhibits a different structural style, although still one of compression. The four east-west-trending mountain ranges in the western part of the ANWR comprise anticlinoria, and the intervening valleys comprise synclinoria (Leiggi, 1987) (Figure 13). The broad anticlines are north verging and are interpreted as fault-bend folds in a duplex with a floor thrust in pre-Mississippian rocks and a roof thrust in the Mississippian

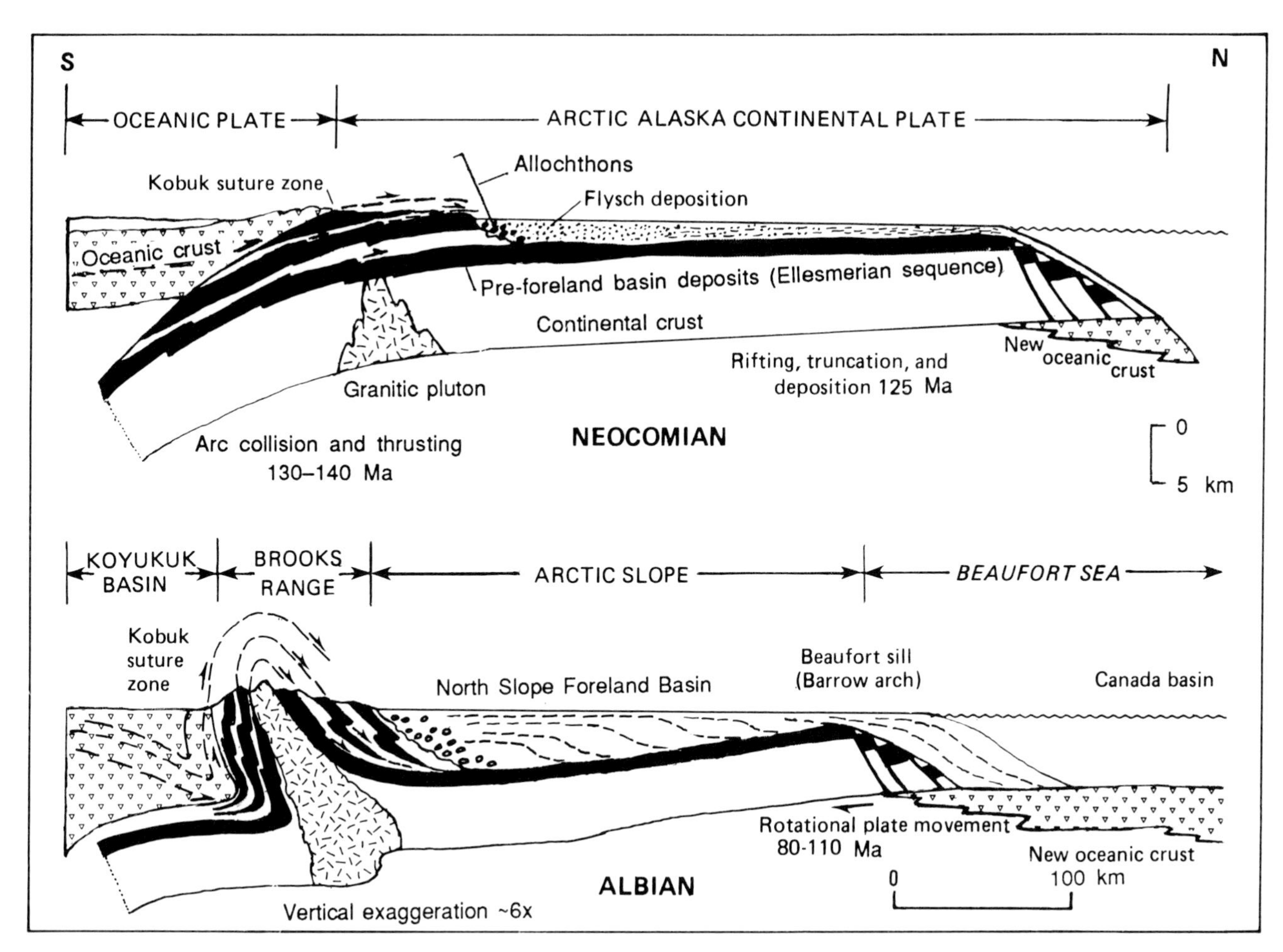

FIGURE 12. Diagrammatic cross section across the central part of the Brooks Range and North Slope basin, illustrating multiple allochthons in the Brooks Range. Modified from Mull (1989, figure 39). See Figure 11 for location of section.

Kayak Shale (Namson and Wallace, 1987; Wallace and Hanks, 1990). Concentric folding above décollement surfaces is common in some of the area. Formerly, the magnitude of thrusting in the eastern Brooks Range was considered insignificant compared with that in the western Brooks Range (Mull, 1982, p. 42), but recent studies indicate crustal shortening on the order of 400 km (250 mi), an amount comparable to the central and western parts of the range (Rattey, 1985). Except for the northwest corner of the ANWR, thrusting and folding within the Cretaceous and Tertiary rocks extended across the entire coastal plain (Figure 13) (Kelley and Foland, 1987; Bruns et al., 1987). Little work has been reported farther south in the eastern Brooks Range, but the structural style is considered to be a continuation of those of the central part of the range (Namson and Wallace, 1987; Wallace and Hanks, 1990).

At least two phases of thrusting are recognized in the Brooks Range. The early and main phase of displacement occurred during Middle or Late Jurassic through Neocomian time (Mayfield et al., 1988, p. 167). This phase is generally thought to be genetically related to the opening of the Canada basin, but most investigators interpret the opening of the Canada basin as slightly later, between the late Neocomian and Cenomanian (Lawver and Scotese, 1990). Embry and Dixon (1990) interpret the opening to be even later, between the Cenomanian and latest Cretaceous. Counterclockwise rotation of about 67° from a pole in the Mackenzie Delta area is thought to have rotated the Arctic Alaska plate away from the Canadian Arctic Islands (Figure 11). This tectonic model would provide the space for the Canada basin and the compressional stresses for the thrusting (underthrusting) in the Brooks Range (Rickwood, 1970; Tailleur, 1973; and many others). There are several other tectonic models for the evolution of the Canada basin, some of which involve a system of transform faulting; Lawver and Scotese (1990) summarize the different ideas or models. The rotational model, which currently is the most accepted model, would require about 1150 km (715 mi) of crustal shortening in the western Brooks Range (Mayfield et al., 1988, p. 170) and lesser amounts to the east. Rattey (1985) estimated that the crustal shortening across the eastern Brooks Range was more than 400 km (250 mi). This amount of shortening is consistent with the rotational model shown by Mayfield

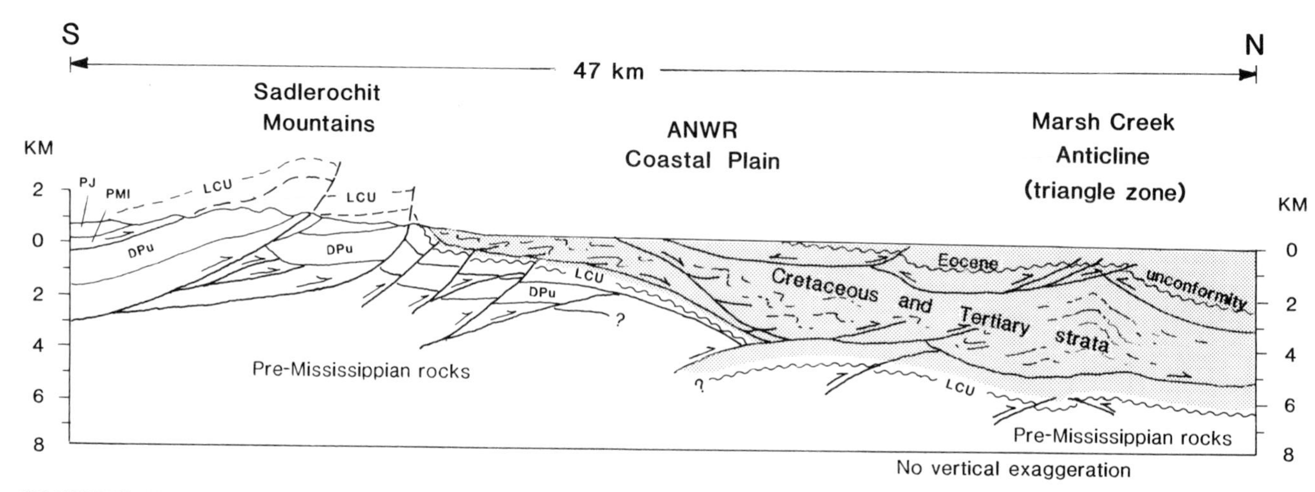

FIGURE 13. Cross section of the northern part of the ANWR. Left third modified from Leiggi (1987); remainder modified from Kelley and Foland (1987). Rock units: DPu, Proterozoic to Devonian Nanook Limestone and Katakturuk Dolomite, undivided; PMl, Mississippian and Pennsylvanian Lisburne Group; PJ, Permian to Jurassic rocks; LCU, Lower Cretaceous unconformity. Arrows on faults indicate direction of relative movement. See Figure 11 for location of section.

et al. (1988, p. 171), considering that some of the thrusting was later than that associated with the opening of the Canada basin.

The second phase of Brooks Range thrusting occurred during Late Cretaceous and Tertiary time. Evidence for this phase of thrusting is most readily observed in the development of a fold belt in the foreland basin. Patterns of development and filling of the foreland basin suggest that deformation was diachronous, becoming progressively younger to the east. In the ANWR, compressional tectonics extended into the late Tertiary (Leiggi, 1987, p. 755; Kelley and Foland, 1987). Earthquake epicenters and warped terrace deposits suggest that deformation continues even today in the northern part of the ANWR and offshore in the eastern Beaufort Sea (Grantz et al., 1988).

Foreland Basin Structure

Following the early phase of Brooks Range thrusting, major subsidence and filling of the foreland basin occurred. This phase of development is attributed to isostatic rebound of the Brooks Range followed by later, and relatively minor, compression (Mull, 1982, p. 43; Mayfield et al., 1988, p. 167). Alternatively, Howell et al. (1992) propose more or less continuous compression with a change from a thin-skinned style of deformation to a thick-skinned style, a change that produced a thickening of the orogene, resulting in a greater crustal load with major subsidence of the foredeep and uplift in the hinterland source areas.

The foreland basin fold belt is as much as 160 km (100 mi) wide (Figure 2) and is divided into two parts, the southern and northern foothills provinces. Structures in the southern foothills province are developed in the Neocomian rocks of the Okpikruak Formation and the Aptian(?) and Albian Fortress Mountain Formation and are very complex. The northern foothills province is structurally more orderly and consists of numerous, rootless en echelon folds developed in the Aptian(?) and Albian Torok Formation, the Nanushuk Group, and younger rocks. The folds consist of broad, open synclines and thrust-cored, northward-verging anticlines, some of which are as long as 170 km (106 mi) (Lathram, 1965). Few of the structures have been mapped in detail, but the relief from synclinal to anticlinal axes in the Umiat area, where more detailed mapping was conducted, ranges from a few hundred to as much as 1200 m (3937 ft) (Brosgé and Whittington, 1966; Detterman et al., 1963). The intensity of folding decreases to the north across the fold belt to a low-dipping homocline on the coastal plain (Figure 3).

Except for most of the ANWR coastal plain, the base of the foreland basin fill—that is, the top of the pebble shale unit or the Hue Shale—is not involved in the compressional folding (Figures 3 and 12). Seismic data indicate that this surface dips gently to the south at an angle ranging from about 1° in the coastal plain area to about 3° in the southern part of the northern foothills (Bird, 1987, p. 132). Under most of the coastal plain of the ANWR, the pre-foreland basin, as well as the foreland basin rocks, are involved in the thrusting (Figure 13).

Along the Barrow arch, on the north margin of the foreland basin, extensional tectonics associated with the rifted margin of the Canada basin mostly affected pre-foreland basin rocks. However, eastward tilting of the basin during the Tertiary resulted in the present high point along the Barrow arch being near Point Barrow and a subsidiary high west of Prudhoe Bay (Figure 10) (Bird, 1987, p. 139).

BASIN EVOLUTION

The North Slope foreland basin records about 170 m.y. of earth history, extending from the Middle Jurassic to the Holocene. The basin developed on a broad, south-facing continental margin of late Paleozoic and early Mesozoic age when this margin and its adjacent ocean basin collided with an intra-oceanic volcanic arc (Box, 1985; Patton and Box, 1989)—one of numerous crustal fragments that make up the collage of terranes that forms Alaska south of the Brooks Range (Plafker, 1990). The simultaneous occurrence of an extensional (rifted) northern margin and a compressional southern margin makes the North Slope foreland basin unique among North American Cordilleran foreland basins.

The evolution of the North Slope foreland basin may be thought of in two parts: an early part and a later part. The early part (Jurassic and Neocomian) is known only from fragments of the basin preserved in the Brooks Range orogene, which records hundreds of kilometers of crustal shortening. The later part (Aptian? to Holocene), which constitutes the basin as we know it today, apparently records a relatively small amount of crustal shortening, but developed as a result of much greater crustal loading by the adjacent orogene than occurred during the earlier stages of basin development. A pronounced northeastward-directed longitudinal filling of the basin may reflect shifting sites of deformation.

Subsidence History

The subsidence history of the North Slope foreland basin is variable, depending on position in the basin. Little is known of the earliest stages of basin formation—the time of subduction of the Arctic Alaska plate and northward migration of the orogene and foreland basin. Foreland basin deposits that formed during this time (Okpikruak Formation and other flysch deposits) seldom exceed 1 km (0.6 mi) in thickness. These deposits were apparently bounded on the north by a mid-basin topographic high and northward beyond that by the southward-prograding shelf and slope deposits of the upper part of the Ellesmerian sequence (Kingak Shale) (Figure 5A).

The North Slope foreland basin as we know it today began to form in Aptian(?) to Albian time when major subsidence, probably indicating a change from thin-skinned to thick-skinned deformation, depressed the south-facing shelf and slope deposits of the upper part of the Ellesmerian sequence. The basin filled, initially, with deep-water deposits onlapping and/or downlapping the drowned shelf (Figures 3, 5, and 6), and later with northeastward-prograding deltaic and shelf deposits (Figures 5B, 6, 7, and 8). Depocenters shifted from the western part of the basin in mid-Cretaceous time (Figure 9), to the central part of the basin in Late Cretaceous time, and to the eastern part of the basin (and offshore) in Tertiary time. During this time, the Barrow arch was generally a slowly subsiding region, separating thick foreland basin deposits to the south from equally thick passive margin deposits to the north. Uplift and erosion of as much as 1000 m (3280 ft) of foreland basin strata occurred on the Barrow arch in the region of Point Barrow during Late Cretaceous or Tertiary time; the eastern end of the arch subsided during Tertiary time, possibly because of tectonic loading by the Brooks Range orogene in the northern ANWR (Figure 10).

Deformation History

Two phases of compressional deformation are postulated for the development of the Brooks Range. The early phase, Middle(?) or Late Jurassic to Neocomian, resulted in hundreds of kilometers of crustal shortening and development of numerous allochthons. The later phase, Late Cretaceous and Tertiary, produced deformation in the Brooks Range and in the adjacent foreland basin. The timing of deformation in the foreland basin is poorly constrained because deformation generally occurred after basin-filling was complete. Deformation may have been diachronous, being older in the west and younger in the east. Only in northeastern Alaska, directly along the front of the central Brooks Range, and possibly in the foothills of the central part of the North Slope, is there evidence of deformation during basin-filling. In the ANWR coastal plain area (Bruns et al., 1987; Kelley and Foland, 1987) and the adjacent offshore (Grantz et al., 1987), moderately deformed Eocene and younger rocks lie over more intensely deformed older rocks. Along the front of the central Brooks Range, the Fortress Mountain Formation unconformably overlies highly deformed Neocomian and older rocks (Patton and Tailleur, 1964; Mull, 1985) and also displays local internal unconformities (Molenaar et al., 1988). In the foothills in the Umiat–Maybe Creek region (located between numbers 6 and 7 in Figure 2), Brosgé and Whittington (1966, p. 598) postulate that regional warping and local folding occurred during Late Cretaceous time based on the presence of an unconformity between the Nanushuk and Colville groups as well as textural and thickness variations in Upper Cretaceous rocks.

Apatite fission-track analysis, indicating the time of uplift and cooling of the rocks through the 100°C isotherm, gives dates of about 60 Ma for the range front along the Trans-Alaska Pipeline (Figure 2; O'Sullivan, 1990) and 45 and 32 Ma along the Canning River near the western end of the Sadlerochit Mountains (near the Kavik gas field, number 13 in Figure 2; O'Sullivan et al., 1990). Earthquake epicenters and deformed terrace deposits indicate that deformation continues in the northern ANWR and offshore in the eastern Beaufort Sea (Grantz et al., 1988).

Thermal History

Our knowledge of North Slope thermal history is based on measurements of organic metamorphism (mainly vitrinite reflectance, thermal alteration index, and conodont alteration index), present-day thermal profiles, and, most recently, apatite fission-track analysis. Depth to the top of the oil window ($R_o = 0.6\%$) in the foreland basin is shown in Figure 14. Trends of organic metamorphic thresholds for the onset of oil ($R_o = 0.6\%$) and dry gas ($R_o = 2.0\%$) generation are plotted on the north-south cross sections in Figures 3 and 15.

The organic metamorphic grade of rocks at the surface generally decreases northward across the North Slope; this is a reflection of uplift and erosion in the orogenic belt and subsidence and sedimentation in the adjacent foredeep basin and passive margin. Data from Brosgé

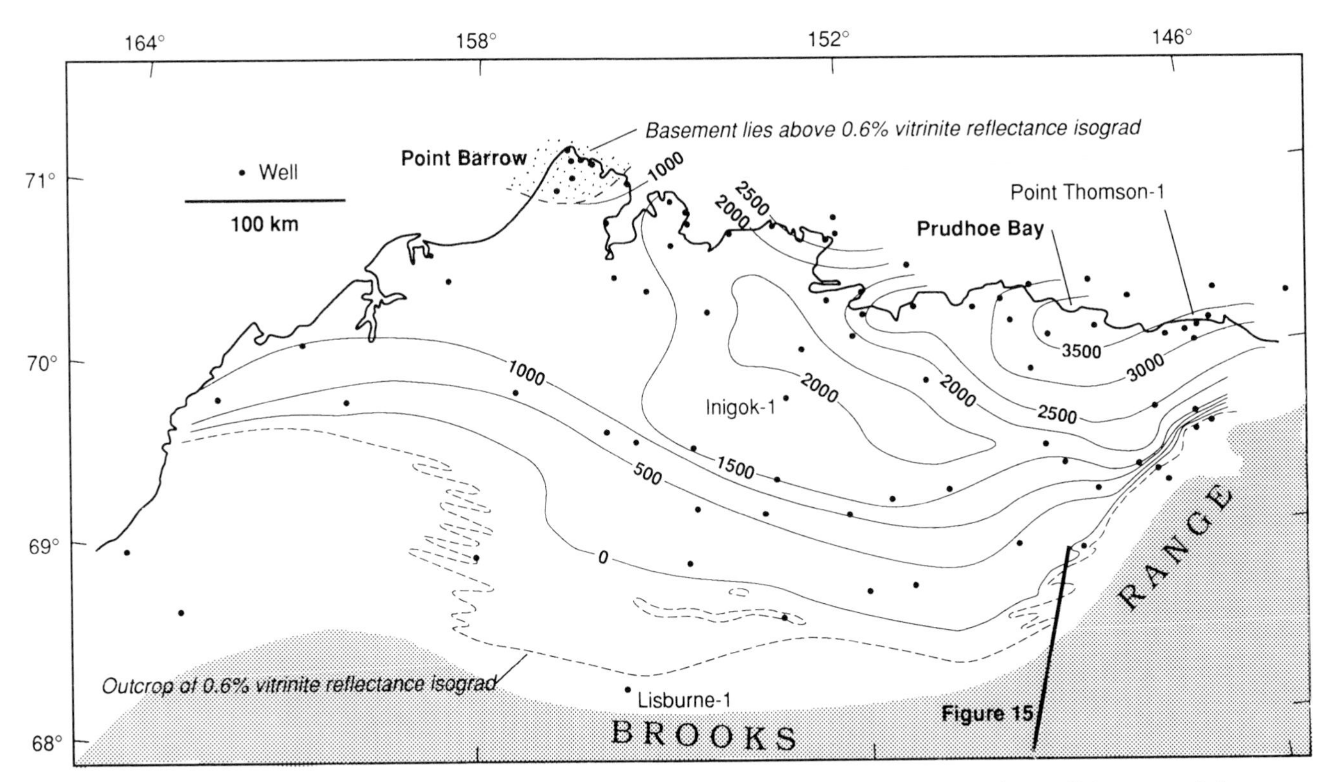

FIGURE 14. **Elevation of the top of the oil generation window (0.6% vitrinite reflectance isograd) in meters below sea level. From Johnsson et al. (1991).**

et al. (1981), Harris et al. (1987), Magoon and Bird (1985, 1988), Howell et al. (1992), as well as a large amount of new data summarized by Johnsson et al. (1991), show that the 2% R_o value is generally found at the surface within the northern margin of the Brooks Range and at depths of 4 km (2.5 mi) or more beneath the North Slope basin. The 0.6% R_o value intersects the surface in the foothills and plunges northward and eastward into the subsurface; along the coastline it occurs at a depth of about 600 m (1968 ft) near Point Barrow and about 4000 m (13,125 ft) at the northwestern corner of the ANWR (Figure 14). The vertical separation between R_o values 0.6 and 2.0% tends to increase from about 1.5 km (0.9 mi) in the north to about 4.5 km (2.8 mi) in the south (Figure 3), apparently a reflection of higher thermal gradients to the north and lower gradients to the south.

The thermal maturity data are interpreted as indicating that maximum heating of the rocks was achieved in the orogene by structural stacking and in the adjacent foreland basin by sediment burial. Within the western Brooks Range and in the Lisburne-1 well (Figure 14), thermal maturity progressively increases with depth through multiple thrust sheets (Harris et al., 1987; Magoon et al., 1988, pl. 19.38). Maturity profiles in these areas are interpreted as the result of maximum heating after structural stacking. In contrast, maturity isograds approximate stratigraphic horizons within the foreland basin. Maturity profiles in this region are interpreted as the result of maximum heating by sediment burial. Detailed sampling across foreland basin structures along the Trans-Alaska pipeline (Figure 15) shows increasing maturity with depth and evidence of warping (and perhaps fault offset?) of maturity isograds. Amplitudes of warping are nearly 3 km (1.9 mi), and at least 10 km (6.2 mi) of uplift is indicated for the core of the Brooks Range in this area (Howell et al., 1992).

The present-day North Slope thermal regime is characterized by thick permafrost and variable geothermal gradients (American Association of Petroleum Geologists and U.S. Geological Survey, 1976; Blanchard and Tailleur, 1982; Lachenbruch et al., 1988). The long-term mean surface temperature systematically varies from -12°C along the northern coast to -4°C inland near the Brooks Range. Depth to base of permafrost, the 0°C isotherm, generally ranges from 200 to 400 m (656 to 1312 ft), but in the area of the Prudhoe Bay oil field and eastward along the coast it reaches depths in excess of 600 m (1968 ft). Thermal gradients within the permafrost range from 15 to 50°C/km, whereas gradients below the permafrost range from 24 to 47°C/km. In general, higher gradients are found in coastal plain wells (area of Barrow arch and northern flank of the foreland basin) and lower gradients in foothill wells (the fold and thrust belt).

Deming et al. (1992) concluded from analysis of thermal gradients and heat flow calculations in the NPRA that the observed thermal pattern is consistent with forced convection by a topographically driven groundwater flow system. They postulate that this system transports heat by advection from areas of high elevation in the Brooks Range and foothills to areas of low elevation on the coastal plain. Furthermore, the similarity in patterns between present-day heat flow and organic maturation data suggests that the groundwater flow system has persisted for

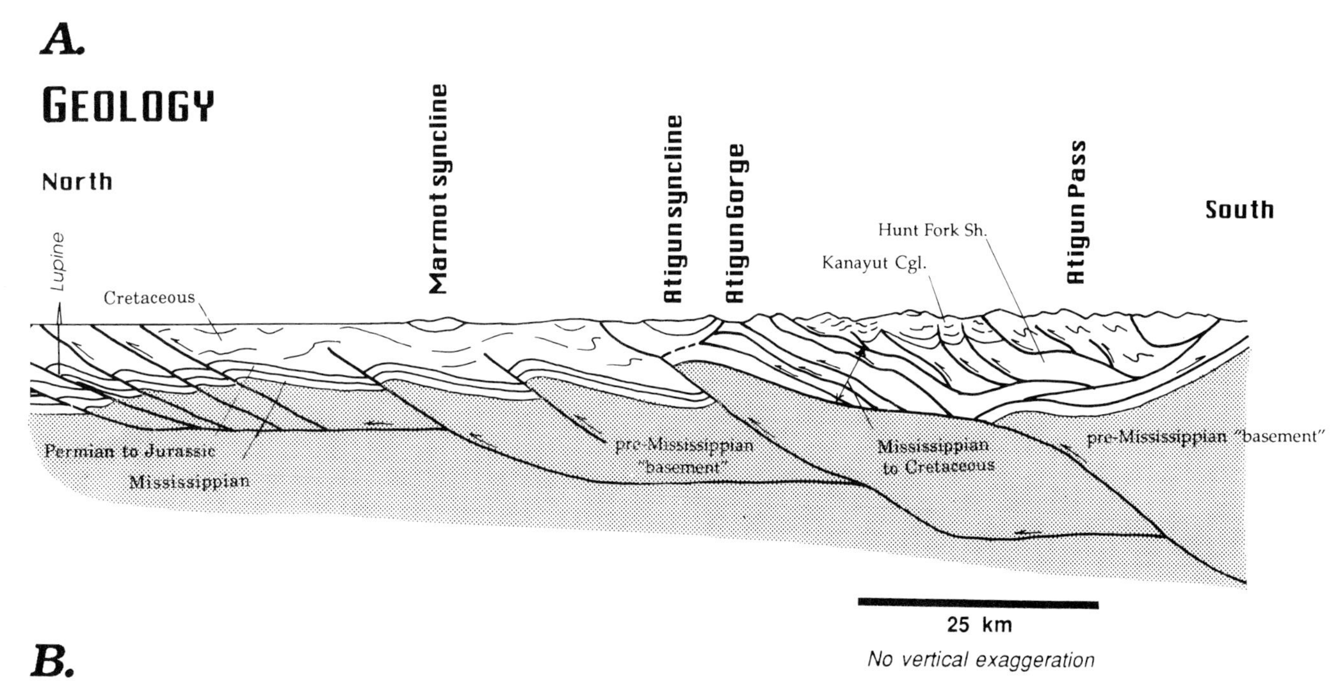

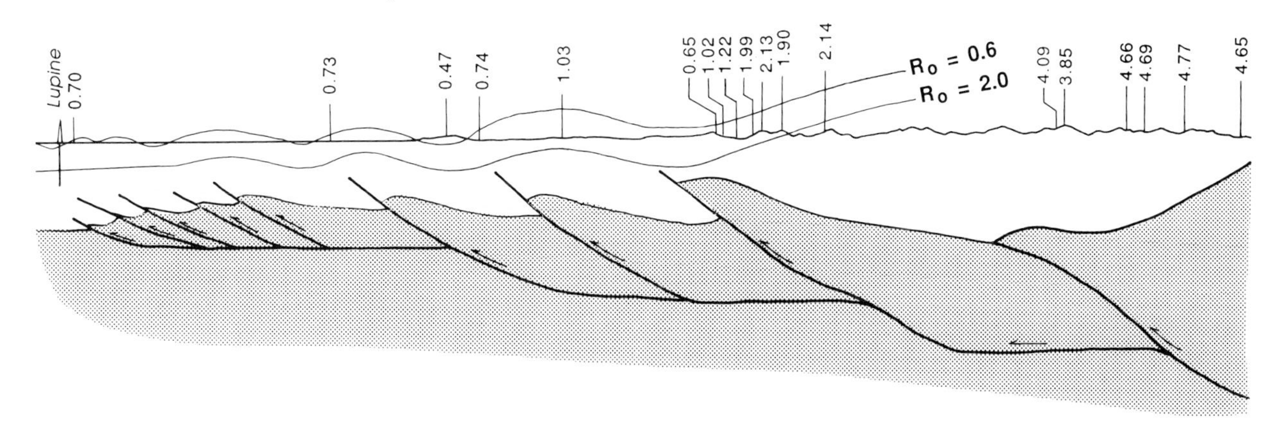

FIGURE 15. Structure section and paleothermal isograds across northern Brooks Range and southern part of North Slope basin (modified from Howell et al., 1992). (A) Diagrammatic structure section. (B) Simplified section showing vitrinite reflectance data and warping of straight-line projections of vitrinite isograds in area of Atigun and Marmot synclines. Arrows on faults indicate direction of relative movement. See Figure 14 for location of section.

tens of millions of years. Thus, basin-scale groundwater flow has probably been a feature of the foreland basin throughout its history and may have been an important factor in the timing of hydrocarbon generation, migration, and preservation.

The present-day North Slope thermal regime provides conditions suitable for the occurrence of natural gas hydrates (solids composed of light gases caged in an ice crystal lattice). Collett (1991) showed that multiple hydrate-bearing reservoirs occur in the Sagavanirktok Formation overlying the western part of the Prudhoe Bay oil field and the eastern part of the Kuparuk River oil field (Figure 2). The amount of gas contained in these hydrates is estimated at 37 to 44 $\times 10^{12}$ ft^3 (105 to 125 $\times 10^{10}$ m^3), or about the same volume of gas as contained in all conventional North Slope oil and gas fields combined.

OIL AND GAS SYSTEMS

At least three oil and gas systems are recognized on the North Slope according to Magoon (in press). These are the Ellesmerian system, the Hue-Sagavanirktok system, and the Torok-Nanushuk system. Volumetrically, the Ellesmerian system is most important, being responsible for all presently commercial North Slope oil. Oil-to-source-rock comparisons indicate that the source of the oil is in pre-foreland basin strata (primarily the Shublik Formation and

Kingak Shale). Although most oil production comes from pre-foreland basin reservoirs, a huge volume of heavy oil, more than half of the known in-place oil, occurs in foreland basin reservoirs of the Sagavanirktok Formation. In and adjacent to the ANWR, extracts of the Hue Shale compare favorably to surface oil seeps and oil stains from the Sagavanirktok Formation. More tenuous are the geochemical correlations of rock extracts from the Torok Formation with the shallow Nanushuk oil accumulations, such as those found in the presently noncommercial Umiat oil field (11 × 10^6 m^3, or 70 × 10^6 bbl, of recoverable oil) and as seeps in the Simpson area of the NPRA (number 4 in Figure 2). The light, low-sulfur oils expelled from the source rocks of these two systems differ significantly from the heavier, high-sulfur oils of the Ellesmerian system.

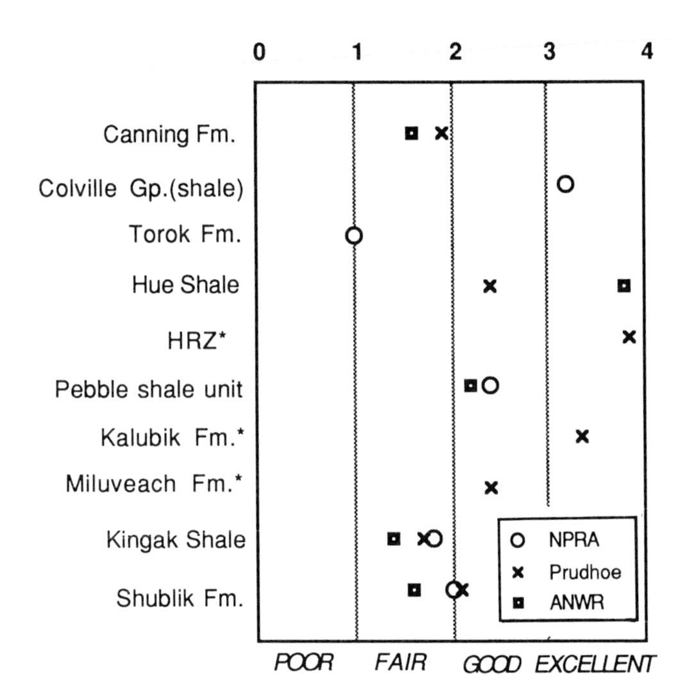

FIGURE 16. Average organic carbon content of North Slope marine shale units. NPRA area values modified from Magoon and Bird (1988); Prudhoe Bay area values from Sedivy et al. (1987); ANWR area values from Magoon et al. (1987). *Asterisks indicate rock units of Carman and Hardwick (1983). Molenaar et al. (1987) include the HRZ in the Hue Shale, the Kalubik Formation in the pebble shale unit, and the Miluveach Formation in the upper part of the Kingak Shale.

Reservoir Rocks and Diagenesis

North Slope foreland basin reservoir rocks are sandstones deposited in a variety of environments ranging from deep marine to nonmarine. The most recent or only references providing petrographic/diagenetic details of foreland basin sandstone units are Siok (1989) and Wilbur et al. (1987) for the Okpikruak Formation, Molenaar et al. (1988) for the Fortress Mountain Formation, Bartsch-Winkler and Huffman (1988) for the Nanushuk Group and Torok Formation, Bird et al. (1987) and Gautier (1987) for the Sagavanirktok and Canning formations, and McLean (1987) and Buckingham (1987) for Buckingham's Jago River Formation. Smosna (1989) formulated a burial-compaction law for Nanushuk Group sandstone.

North Slope foreland basin reservoir rocks are generally subquartzose ($<$ 75%) lithic sandstone that is distinct from the mature, quartz-chert sandstone that characterizes the pre-foreland basin reservoir rocks. Observed ranges of composition of the foreland basin sandstone units are quartz (5 to 85%), feldspar (0 to 22%), and rock fragments (15 to 85%). Quartz generally becomes more abundant in younger sandstone units. Rock fragments include sedimentary, igneous, and metamorphic types, and, in formations along the outcrop belt, their abundances often show significant areal variations that suggest local provenance. Primary matrix material is pseudomatrix, resulting from the destruction of softer clastic grains. Observed authigenic cements include calcite, silica, kaolinite, and chlorite. No zeolite cement is reported. A common theme in the analysis of these sandstone units is that porosity loss is dominated by ductile-grain deformation and to a lesser degree by cementation. Sandstone with better porosity commonly shows evidence of leached grains and cements.

Source Rock Characteristics

The organic carbon contents of most North Slope rock units exceed the threshold value of 0.5 wt. % of potential petroleum source rocks (Figure 16), although their hydrogen contents, which are indicators of propensity to generate oil or gas, vary considerably. Within the foreland basin, deltaic and prodeltaic units (Nanushuk Group, Torok Formation, Colville Group, Sagavanirktok Formation, and Canning Formation) generally have relatively high organic carbon contents but low hydrogen contents; they are considered gas-prone source rocks. The more distal marine shale units, such as the Hue Shale and perhaps parts of the Torok Formation, which have both high organic carbon and hydrogen contents, are considered oil-prone source rocks. The richest source rock for oil in the foreland basin is the Hue Shale with a total organic carbon content that averages nearly 4% (Figure 16). This value is nearly double the value of the Shublik Formation and Kingak Shale, which are identified as the primary source rocks for the Prudhoe Bay area oils. At Prudhoe Bay, the Hue Shale is immature.

Chemical analyses of North Slope oils and source rocks indicate that there are multiple oil types that have been generated by multiple source rocks. Prior to the discovery of the Prudhoe Bay field, investigations of North Slope oils from wells and seeps (McKinney et al., 1959) focused on the quality of products that could be derived from these oils rather than on their relations to source rocks or to other oils. This emphasis changed, however, with the discovery of multiple oil-bearing reservoirs in the Prudhoe area. Analysis indicated that oils from widely separated Prudhoe area reservoirs—the Sadlerochit Group, Kuparuk Formation, and Sagavanirktok Formation—were similar and thus commonly sourced (Jones and Speers, 1976).

The most comprehensive study aimed at discovering related types of North Slope oils is that of Magoon and Claypool (1981). From their analyses of 40 samples collected from seeps and wells all across the North Slope, two groups of oil were identified: (1) the Barrow-Prudhoe (from the Barrow and Prudhoe Bay fields) and (2) the Simpson-Umiat (from the Simpson seeps and Umiat field). The Barrow-Prudhoe Group, volumetrically the predominant North Slope oil, occurs in reservoirs of Mississippian to Tertiary age. It is characterized by high sulfur content, medium API gravity, light isotopic composition, and a pristane-phytane ratio less than 1.5. The Simpson-Umiat Group occurs in Cretaceous and Tertiary reservoirs and, surprisingly, is the only oil found in seeps. This group is characterized by low sulfur content, high API gravity, heavy isotopic composition, and a pristane-phytane ratio greater than 1.5.

Other studies of the same and newly discovered oils (e.g., Seifert et al., 1980; Magoon and Claypool, 1985, 1988; Curiale, 1987; Sedivy et al., 1987; Anders et al., 1987) reveal various oil types within the two groups. Many investigators now regard the Simpson and Umiat oils as being derived from different terrigenous source-rock facies. Additional oil types identified within the Simpson-Umiat group include the Manning (from the Manning Point seep in the ANWR), the Jago (from oil-stained rocks along the Jago River in the ANWR), the Kavik (from oil-stained rocks near the Kavik field, number 13 in Figure 2) in the ANWR area (Anders et al., 1987), and the pebble shale unit in the Barrow area (Magoon and Claypool, 1988). The Kingak oil in the Prudhoe area, a type within the Barrow-Prudhoe oil group, is locally derived from the marine Kingak Shale (Seifert et al., 1980).

Considerable effort has been devoted to matching North Slope oils with specific source rocks. The earliest efforts (Morgridge and Smith, 1972) identified Cretaceous shale above the Lower Cretaceous unconformity and the Kingak Shale as the most probable source rocks for the Prudhoe Bay oils, based on geologic relations and bulk geochemical characteristics of the proposed source rocks. Later, analysis of biomarker compounds from rocks and oils (Seifert et al., 1980) suggested that these oils were sourced from an assemblage of rocks including the Shublik Formation, the Kingak Shale, and deeply buried shale of the GRZ (Hue Shale). Isotopic correlations of Sedivy et al. (1987, figure 10) complement the biomarker results by showing an excellent correlation between source rocks (Shublik Formation and Kingak Shale) and the Prudhoe oils. The USGS-sponsored oil-rock correlation study (Magoon and Claypool, 1985) was a multilaboratory effort that focused on a common set of oil and candidate source-rock samples, mostly from outside of the Prudhoe Bay area. The majority (17 of 30 laboratories) agreed that the Shublik Formation and, to a lesser extent (eight laboratories), the Kingak Shale are source rocks for the Prudhoe oils. There was also general agreement (14 laboratories) that the pebble shale unit and, to a lesser extent (seven laboratories), the Torok Formation are source rocks for the Umiat oil type (Claypool and Magoon, 1985).

Maturation History

Thermal history relative to hydrocarbon maturation at the 6127-m (20,102-ft) deep Inigok well in the eastern part of the NPRA is illustrated in Figure 17A. Burial history curves for this well show almost continuous subsidence and sedimentation from Mississippian to middle Tertiary time, followed by an estimated 300 m (984 ft) of uplift and erosion. Lopatin's method of integrating time and temperature (Waples, 1980) shows that thermal conditions favorable for oil generation (TTI value of 10) in the Endicott Group and the lower part of the Lisburne Group occurred here as early as Triassic time. As subsidence and burial continued, the zone of oil generation (TTI, 10 to 1000) effectively moved upward through the stratigraphic section to its present position, encompassing rocks of Jurassic to mid-Cretaceous age between the depths of 1980 and 3690 m (6496 and 12,106 ft). The zone of oil generation is expected to have migrated northeastward across the North Slope in concert with the filling of the foreland basin. Verification of this scenario is provided by the burial history diagram of the Point Thomson Unit-1 well (Figure 17B). Analysis of this well, located on the eastern, downplunge extent of the Barrow arch (Figure 2), shows that the section consists mostly of foreland basin strata of Tertiary age, that the zone of oil generation is limited to Paleocene and older strata below a depth of 3500 m (11,483 ft), and that the onset of oil generation for the Hue Shale occurred about 5 Ma. In the general area of this well, the zone of oil generation lies within a geopressured section (Gautier et al., 1987).

Migration and Entrapment

Analysis of the thermal maturity data indicates that source rocks within the Brooks Range (and directly adjacent foothills) became mature only after thrust loading, whereas source rocks within and beneath the foreland basin became mature during and after sedimentary filling, but prior to deformation, except for the ANWR coastal plain. Deformation in the ANWR coastal plain appears to have been generally synchronous with Tertiary sedimentation. Shifting depocenters as the foreland basin filled resulted in shifting sites of maturation.

At the time of maximum burial at any particular site within the foreland basin, the directions of migration would have been generally northward (updip toward the Barrow arch) for hydrocarbons generated in pre-foreland basin strata and in the lower part of the foreland basin fill. In the upper part of the foreland basin fill, hydrocarbons may have migrated to the southwest up along clinoform bedding into shelf and deltaic sandstone reservoirs.

Deformation of the foreland basin-fill occurred after maturity was achieved, as indicated by warping of the thermal isograds (Figure 15). The creation of folds and

A

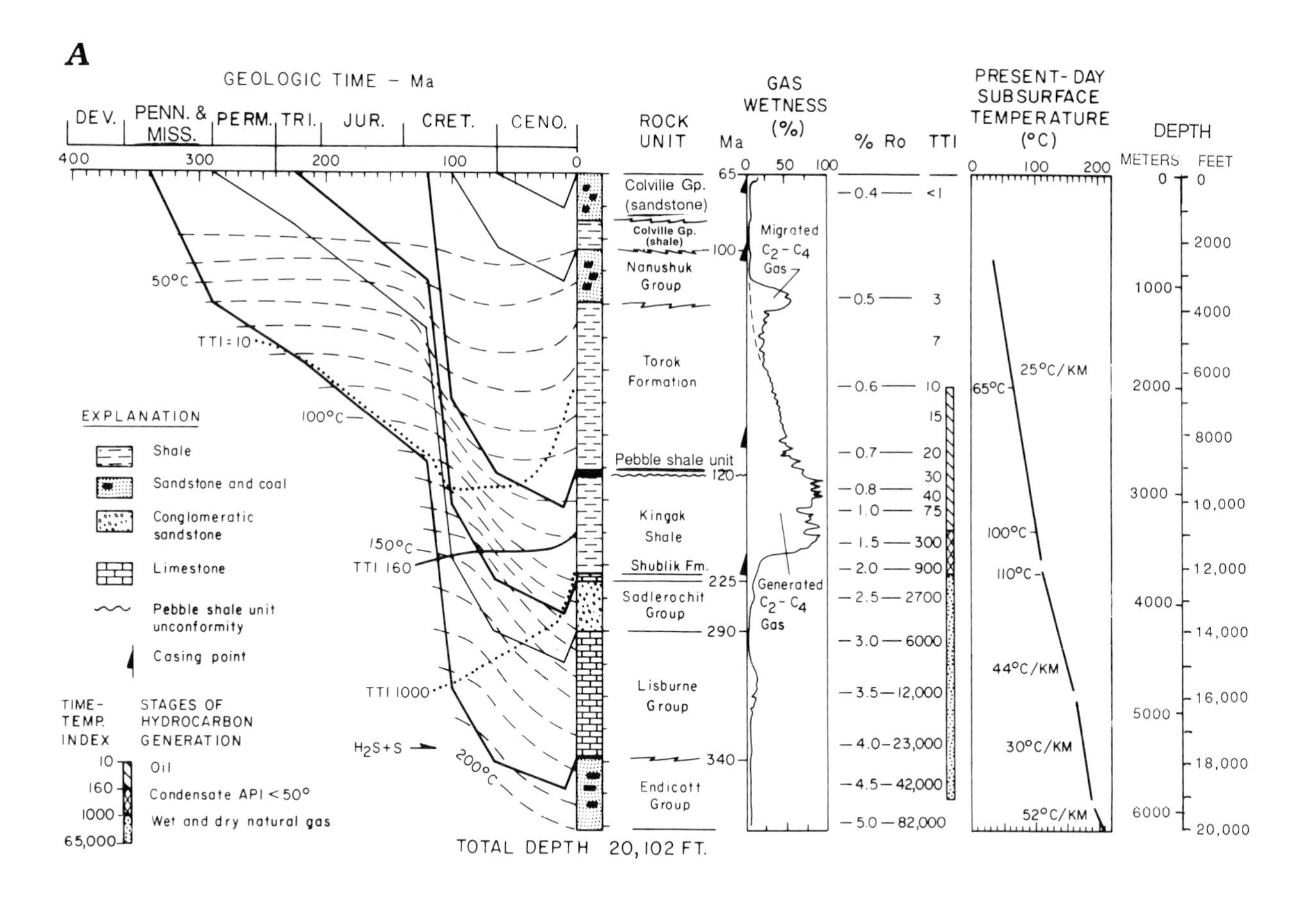

B

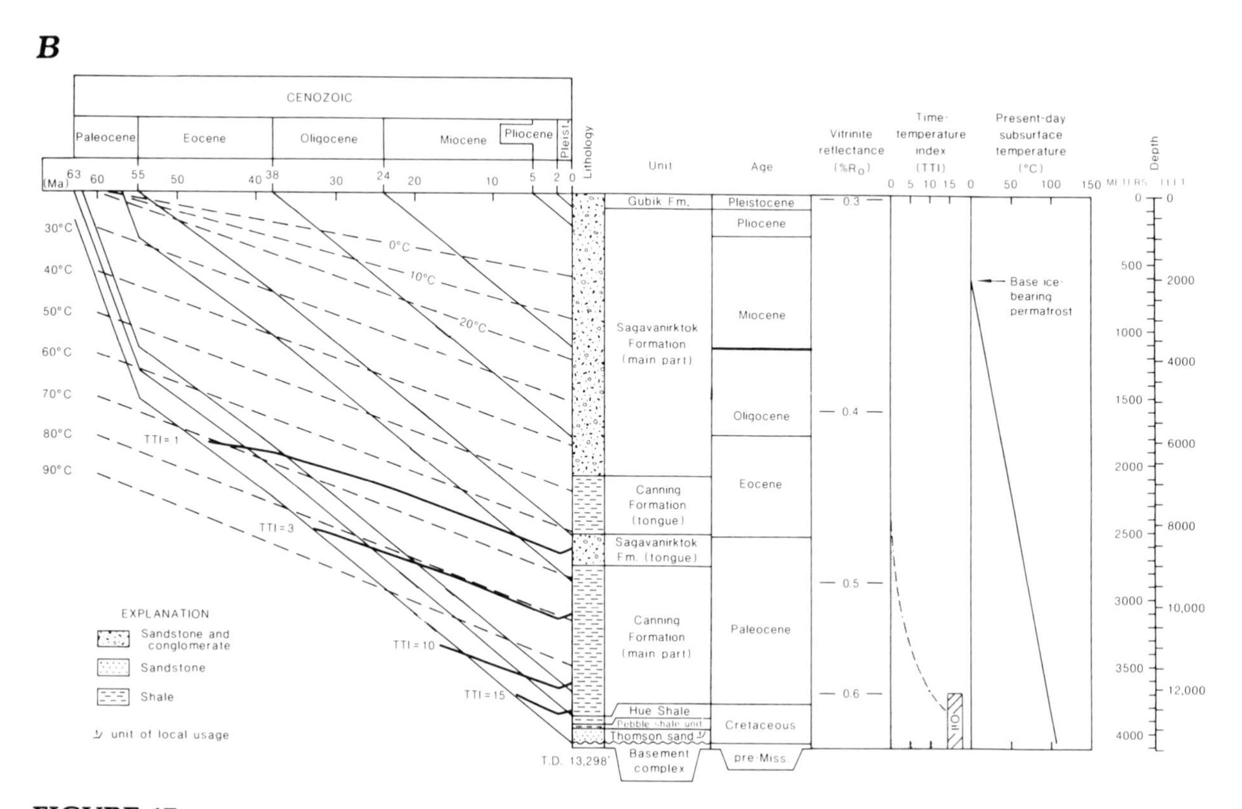

FIGURE 17. **Burial history diagrams illustrating time of maturation of source rocks on the North Slope. (A) Inigok-1 well (after Magoon and Claypool, 1983). (B) Point Thomson-1 well (after Magoon et al., 1987). See Figure 14 for well locations.**

faults at this time would be expected to have modified or destroyed the early-formed oil and gas accumulations. At the same time, faults may have provided new migration pathways for late-stage hydrocarbons.

Our preliminary conclusions from this analysis are that oil and gas exploration in the fold and thrust belt is largely a matter of looking for remigrated hydrocarbons. The best chances for discovering oil and gas in this region are (1) in the most distal (least-deformed) structures, (2) in those structures with the thickest, most-intact seals, and (3) in the youngest structures (those with the shortest time available for leakage of hydrocarbons).

Play Types

In the recently completed assessment of United States undiscovered conventional oil and gas resources (Mast et al., 1989), 12 plays were identified and assessed for the North Slope. Individual play descriptions, maps, and estimated resources for each of these plays are provided by Bird (1991). Seven of the 12 plays involve foreland basin strata, whereas the remaining five plays are limited to pre-foreland basin strata.

In the fold and thrust belt, four structural plays consisting of faulted anticlinal traps are distinguished, and all involve foreland basin strata to some degree. Two plays are identified in the area west of the Trans-Alaska pipeline (the Western Fold Belt and Western Thrust Belt plays) and two plays to the east of the pipeline (the Eastern Fold Belt and Eastern Thrust Belt plays). The Western and Eastern Thrust Belt plays encompass that area along the southern part of the foothills where thrust anticlines of pre-foreland basin carbonate and clastic reservoir rocks are present. Pre-foreland basin and foreland basin shales may serve as source rocks and seals in these plays. The plays are separated on the basis of differences in structural style, reservoir rocks, and source rocks. The Western and Eastern Fold Belt plays consist of faulted anticlines having Cretaceous and (or) Tertiary sandstone reservoirs. Foreland basin shales are expected to serve as both source rocks and seals. The plays are separated on the basis of differences in reservoir rocks and source rocks. A total of about 50 wells have been drilled in these plays resulting in the discovery of eight hydrocarbon accumulations (seven gas and one oil); all are noncommercial. Two of the gas accumulations (numbers 13 and 19 in Figure 2) occur in pre-foreland basin reservoirs.

In the coastal plain province, where stratigraphic or combination structural-stratigraphic traps provide the trapping mechanism for oil and gas, three plays are distinguished on the basis of reservoir character and source rock. The Topset play consists of deltaic sandstone reservoirs of the Nanushuk Group and Sagavanirktok Formation with prodelta marine shales as potential source rocks. The Western and Eastern Turbidite plays consist of deep marine turbidite sandstone reservoirs in the Torok and Canning formations, respectively. Potential source rocks include the gas-prone Torok and Canning formations and the oil-prone Hue Shale. Oil- and gas-prone source rocks are also present in the underlying pre-foreland basin strata. Several hundred exploratory wells and thousands of development wells have penetrated the rocks comprising these plays, but few of the wells are believed to have been drilled for prospects within these plays. Six oil accumulations are known in these plays. These include the noncommercial Fish Creek, Simpson, and Flaxman Island fields, as well as the enormous, multibillion-barrel heavy oil accumulations West Sak and Ugnu, and the commercial Schrader Bluff pool of the Milne Point field (Table 1).

Discovery, Exploration History, and Reserves

Intermittent petroleum exploration of the North Slope can be documented for 50 of the last 70 years, and the role of government in these efforts makes this region unique. Details of North Slope petroleum exploration history provided by Reed (1958), Gryc (1970, 1988), Jamison et al. (1980), Bird (1981), and Tailleur and Weimer (1987) show that government exploration was conducted intermittently beginning in 1923 and that industry exploration has been continuous since 1958.

Government exploration of the North Slope was conducted in three phases. The initial phase, following World War I, consisted of four seasons of geologic and topographic studies (Smith and Mertie, 1930) in the newly created Naval Petroleum Reserve No. 4 (now known as the National Petroleum Reserve in Alaska; NPRA in Figure 1). The second phase of exploration began during World War II (1944) and lasted ten years (Reed, 1958). This was a full-scale exploration program that included geologic and geophysical surveys. A total of 36 test wells and 45 shallow core tests were drilled, resulting in the discovery of three oil fields and five gas fields (Table 1, Figure 2). All petroleum discoveries were noncommercial, although accumulation sizes are only approximately known. Most discoveries were made on anticlinal structures in foreland basin (Cretaceous) sandstone reservoirs, although two of the three oil accumulations (Fish Creek and Simpson, numbers 2 and 4 in Figure 2) were discovered by drilling near surface oil seepages in the absence of anticlinal structures. The third and final phase of government exploration (1974–1982) was stimulated by the discovery of the Prudhoe Bay oil field (1967) and by the Arab oil embargo (1973). This program, limited to the NPRA, was a comprehensive exploration effort designed to test as many types of plays as possible. A total of 28 test wells were drilled, primarily for pre-foreland basin objectives similar to those developed at Prudhoe Bay. Two noncommercial gas accumulations were discovered (East Barrow and Walakpa, numbers 20 and 24 in Figure 2), and indications of oil and/or gas were found in nearly all the test wells. A summary of this program and results of studies may be found in Gryc (1988) and in various contractor reports that are available through the National Geophysical Data Center, NOAA, Boulder, Colorado 80303.

Industry exploration of the North Slope dates from 1958, when the government lifted a land freeze and offered acreage for lease. Until 1979, when the first offshore lease sale was held, industry exploration was restricted by land availability to the area between the NPRA and the ANWR (Figure 1). Since then, additional sales have been held, both offshore and onshore, including four sales in the NPRA and, most recently, sales in the Chukchi Sea.

TABLE 1. Summary of Selected Details for North Slope Oil and Gas Accumulations. (For additional details see table in Magoon, in press.)

Map No.[1]	Field or Accumulation Name	Discovery Date	Reservoir[2]	In Place[3] Oil (BBO)	In Place[3] Gas (tcf)	Cumulative Production[4] Oil (MBO)	Cumulative Production[4] Gas (bcf)	Reserves[5] Oil (MBO)	Reserves[5] Gas (bcf)	Data Source[6]
			Foreland Basin Accumulations							
1	Umiat	1946	Nanushuk	<1	<<1	-	-	70	?	1
2	Fish Creek	1949	Nanushuk	<<1	-	-	-	?	?	2
4	Simpson	1950	Nanushuk	<<1	<<1	-	-	12	?	2
5	Meade	1950	Nanushuk	-	<<1	-	-	-	20	2
6	Wolf Creek	1951	Nanushuk	-	<<1	-	-	-	?	2
7	Gubik	1951	Colville	-	<1	-	-	-	295	2
8	Square Lake	1952	Colville	-	<<1	-	-	-	58	2
9	East Umiat	1963	Nanushuk	-	<1	-	-	-	?	4
14	West Sak	1969	Sagavanirktok	20*	<<1	1	-	-	-	9
15	Ugnu	1969	Sagavanirktok	15*	-	-	-	-	-	9
16	Milne Point	1969	Schrader Bluff	<1	<<1	-	-	-	?	8
21	Flaxman Island	1975	Canning	?	?	-	-	?	?	4
28	Hammerhead	1985	Sagavanirktok	?	?	-	-	?	?	15
			Pre-Foreland Accumulations							
3	South Barrow	1949	Barrow	-	<<1	-	20	-	5	3
10	Prudhoe Bay	1967	Ivishak	23	27	7,026	11,951	2,700	23,441	5
11	Lisburne	1967	Lisburne	3	3	64	382	101	406	6
12	Kuparuk River	1969	Kuparuk	~4	~2	723	814	780	634	7
13	Kavik	1969	Ivishak	-	<1	-	-	-	?	8
16	Milne Point	1969	Kuparuk	<1	<<1	16	6	84	?	8
17	Gwydyr Bay	1969	Ivishak	<1	<<1	-	-	60	?	10
18	North Prudhoe	1970	Ivishak	<1	<<1	-	-	75	?	10
19	Kemik	1972	Shublik	-	<1	-	-	-	?	8
20	East Barrow	1974	Barrow	-	<<1	-	6	-	6	3
22	Point Thomson	1977	Thomson	<1	6	-	-	350	5,000	11
23	Endicott	1978	Kekiktuk	1	<2	118	127	272	907	12
24	Walakpa	1980	Walakpa	-	<<1	-	-	-	?	13

[1]Map numbers correspond to locations shown in Figure 2. [2]Main hydrocarbon-bearing rock unit; see Figure 4 for age and lithology. [3]Reported values or rounded estimates based on assumed recovery factors of about 30% (oil) and 80% (gas); BBO = 10^9 bbl = 159×10^6 m^3 (1 bbl = 42 U.S. gallons = 0.159 m^3); tcf = 10^{12} ft^3 = 283×10^8 m^3; dash indicates resource not present; asterisk indicates midpoint of range; question mark indicates amount of resource unknown. [4]Through 12-31-90; MBO = 10^6 bbl = 159×10^3 m^3; bcf = 10^9 ft^3 = 283×10^5 m^3; dash indicates resource not present. [5]Remaining reserves as of 12-31-90; question mark indicates amount of resource unknown; dash indicates resource not present. [6]Data sources: 1. Molenaar, 1982; 2. Collins and Robinson, 1967; Reed, 1958; 3. Lantz, 1981; 4. Well file; 5. State of Alaska, 1977; 6. State of Alaska, 1984; 7. Carman and Hardwick, 1983; van Poollen and Associates and State of Alaska, 1978; 8. State of Alaska, 1985; 9. Werner, 1987; 10. Van Dyke, 1980; 11. Oil and Gas Journal, 1984; Bird and Magoon, 1987; Craig et al., 1985; 12. Woidneck et al., 1987; Harris, 1987; 13. Gryc, 1988; 14. Harris, 1988; 15. Minerals Management Service, 1988; 16. Alaska Report, 1989; 17. Williams, 1989.

TABLE 1. (Continued)

				In Place[3]		Cumulative Production[4]		Reserves[5]		
Map No.[1]	Field or Accumulation Name	Discovery Date	Reservoir[2]	Oil (BBO)	Gas (tcf)	Oil (MBO)	Gas (bcf)	Oil (MBO)	Gas (bcf)	Data Source[6]
25	Niakuk	1981	Kuparuk	<1	<<1	-	-	58	30	14
26	Tern Island	1982	Kekiktuk	?	?	-	-	?	?	15
27	Seal Island	1984	Ivishak	<1	<1	-	-	150	?	16
29	Colville Delta	1985	Kuparuk?	?	?	-	-	?	?	16
30	Sandpiper	1986	Ivishak	?	?	-	-	?	?	15
31	Sikulik	1988	Barrow	-	<<1	-	-	-	?	4
32	Point McIntyre	1988	Kuparuk	1	?	-	-	~300	?	17
TOTALS				~69	~40	7,948	13,270	5,012	30,787	

Limited exploration, including seismic surveys and surface geologic studies, was allowed on a part of the ANWR coastal plain in 1984 and 1985. Results of these studies, described in Bird and Magoon (1987), indicate that this is the most promising onshore area for petroleum exploration remaining on the North Slope in both foreland basin and pre-foreland basin strata.

Industry drilling began in the foothills, testing foreland basin anticlinal objectives similar to those tested by the Navy in NPR-4; eight wells were drilled from 1964 to 1967. Although only one noncommercial gas field (East Umiat, number 9 in Figure 2) was discovered, indications of oil and gas were encountered in every well. Exploration activity then shifted northward to the coastal plain. The third coastal plain well was drilled in 1967, when exploration activity on the North Slope was winding down toward a virtual standstill. This well, the eleventh industry attempt on the North Slope, resulted in the discovery of the Prudhoe Bay field, the largest commercial oil accumulation in North America. In the year or so following the Prudhoe Bay discovery, North Slope exploration activity blossomed, and the rate of oil and gas discovery shot upward. Since then, the rate has tapered off to one discovery every two years (Figure 18).

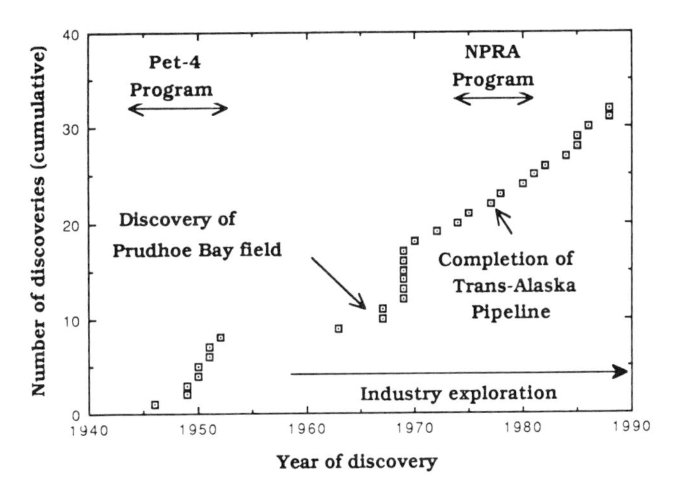

FIGURE 18. Summary of North Slope petroleum exploration history in terms of oil and gas discoveries, 1944 through 1989. Each square represents a single oil or gas accumulation. Most North Slope oil and gas accumulations are presently noncommercial. Data from Table 1. Pet-4, Naval Petroleum Reserve No. 4; NPRA, National Petroleum Reserve in Alaska.

Exploratory drilling by government and industry has been conducted for 45 years on the North Slope, resulting in the discovery of 32 oil and gas accumulations, which encompass both onshore and offshore areas. The rate of discovery is plotted in Figure 18, based on the information in Table 1. The oil and gas accumulations are located in Figure 2. Although most of the oil and gas accumulations are noncommercial by current North Slope standards, five are now producing. Their total ultimate recovery (reserves plus produced) is estimated to be nearly 1.9×10^9 m^3 (12×10^9 bbl) of oil. Even more indicative of the richness of this province are the total in-place resources shown in Table 1: nearly 11×10^9 m^3 (70×10^9 bbl) of oil and 1.13×10^{12} m^3 (40 tcf) of gas, more than 95% of which is concentrated within a 65-km (40-mi) radius of Prudhoe Bay.

Hydrocarbon production on the North Slope began in 1949, when the government produced natural gas for local use at Barrow. Commercial oil production began in 1977 from the Prudhoe Bay field when the Trans-Alaska pipeline was completed—almost ten years after the field was discovered. Production from the Prudhoe Bay and Kuparuk River fields, currently the two leading U.S. oil producers, combined with that from three other North Slope fields (Endicott, Lisburne, and Milne Point), supplied about 3.2×10^5 m^3 (2×10^6 bbl) of oil per day in 1990, or about 25% of total U.S. production. The first commercial oil production from a foreland basin reservoir on the North Slope began in 1991 when the Schrader Bluff pool of the Milne Point field began production. North Slope production rates, on a per-well basis as of 1984, averaged nearly 318 m^3 (2000 bbl) per day per well, about two orders of magnitude greater than per-well rates of any other state (Gerhard et al., 1988). Alaska's rates reflect the remarkable characteristics of the Sadlerochit reservoir at the Prudhoe Bay field—some wells produce nearly 3180 m^3 (20,000 bbl) per day (Jamison et al., 1980)—and the economics of Arctic petroleum development. At least in part for lack of a transportation system, natural gas is not presently economical, even though large volumes of gas have been discovered. Some gas, however, is produced for use in the production facilities at Prudhoe Bay and nearby oil fields.

Future Potential

The future petroleum potential of the North Slope province is considerable, as indicated by the results of the latest assessment of undiscovered recoverable oil and gas resources (Mast et al., 1989). Between 0.40 and 5.37×10^9 m^3 (2.5 and 33.8×10^9 bbl) of crude oil and between 0.297 and 4.146×10^{12} m^3 (10.5 and 146.4 tcf) of natural gas are estimated to underlie the North Slope. The range estimates for each commodity above correspond to 95 and 5% probabilities of there being more than these amounts, respectively. The mean estimates are 2.00×10^9 m^3 (12.6×10^9 bbl) of oil and 1.53×10^{12} m^3 (54.1 tcf) of gas. The majority of future potential resources (83% of the oil and 73% of the gas at the mean value) is estimated to occur in the seven plays involving foreland basin strata. By far the richest area is expected to be the eastern part of the North Slope, the area that includes the ANWR coastal plain.

ACKNOWLEDGMENTS

We gratefully acknowledge the conscientious and timely reviews by W. P. Brosgé, J. Dixon, D. G. Howell, and L. B. Magoon. We have followed many of their suggestions and think that the manuscript has thereby been improved.

REFERENCES CITED

Ahlbrandt, T. S., A. C. Huffman, J. E. Fox, and I. Pasternack, 1979, Depositional framework and reservoir-quality studies of selected Nanushuk Group outcrops, North Slope, Alaska, *in* T. S. Ahlbrandt, ed., Preliminary geologic, petrologic, and paleontologic results of the study of Nanushuk Group rocks, North Slope, Alaska: USGS Circular 794, p. 14–31.

Alaska Report, 1989, Consortium seeks extension of Beaufort Sea leases: Alaska Report, v. 35, n. 42, p. 1.

American Association of Petroleum Geologists and U.S. Geological Survey, 1976, Geothermal gradient map of North America: Tulsa, Oklahoma, AAPG Geothermal Survey of North America, scale 1:5,000,000, 2 sheets.

Anders, D. E., L. B. Magoon, and S. C. Lubeck, 1987, Geochemistry of surface oil shows and potential source rocks, *in* K. J. Bird and L. B. Magoon, eds., Petroleum geology of the northern part of the Arctic National Wildlife Refuge, northeastern Alaska: USGS Bulletin 1778, p. 181–198.

Bally, A. W., and S. Snelson, 1980, Realms of subsidence, *in* A. D. Miall, ed., Facts and principles of world petroleum occurrence: Canadian Society of Petroleum Geologists Memoir 6, p. 9–94.

Bartsch-Winkler, S., and A. C. Huffman, Jr., 1988, Sandstone petrography of the Nanushuk Group and Torok Formation, *in* G. Gryc, ed., Geology and exploration of the National Petroleum Reserve in Alaska, 1974 to 1982: USGS Professional Paper 1399, p. 801–831.

Bird, K. J., 1981, Petroleum exploration of the North Slope, Alaska, *in* J. F. Mason, ed., Petroleum geology in China: Tulsa, Oklahoma, Pennwell Publishing Company, p. 233–248.

Bird, K. J., 1987, The framework geology of the North Slope of Alaska as related to oil-source rock correlations, *in* I. Tailleur and P. Weimer, eds., Alaskan North Slope geology: Bakersfield, California, SEPM, Pacific Section, Book 50, v. 1, p. 121–143.

Bird, K. J., 1991, Geology, play descriptions, and petroleum resources of the Alaskan North Slope (petroleum provinces 58-60): USGS Open-File Report 88-450Y, 52 p.

Bird, K. J., and J. Andrews, 1979, Subsurface studies of the Nanushuk Group, North Slope, Alaska, *in* T. S. Ahlbrandt, ed., Preliminary geologic, petrologic, and paleontologic results of the study of Nanushuk Group rocks, North Slope, Alaska: USGS Circular 794, p. 32–41.

Bird, K. J., and L. B. Magoon, eds., 1987, Petroleum geology of the northern part of the Arctic National Wildlife Refuge, northeastern Alaska: USGS Bulletin 1778, 329 p.

Bird, K. J., and C. M. Molenaar, 1987, Stratigraphy, *in* K. J. Bird and L. B. Magoon, eds., Petroleum geology of the northern part of the Arctic National Wildlife Refuge, northeastern Alaska: USGS Bulletin 1778, p. 37–39.

Bird, K. J., S. B. Griscom, S. Bartsch-Winkler, and D. M. Giovannetti, 1987, Petroleum reservoir rocks, *in* K. J. Bird and L. B. Magoon, eds., Petroleum geology of the northern part of the Arctic National Wildlife Refuge, northeastern Alaska: USGS Bulletin 1778, p. 79–99.

Blanchard, D. C., and I. L. Tailleur, 1982, Preliminary geothermal isograd map, NPRA, *in* W. L. Coonrad, ed., The United States Geological Survey in Alaska—accomplishments during 1980: USGS Circular 844, p. 47–48.

Bodnar, D. A., 1989, Stratigraphy of the Otuk Formation and a Cretaceous coquinoid limestone and shale unit, *in* C. G. Mull and K. E. Adams, eds., Dalton Highway, Yukon River to Prudhoe Bay, Alaska—bedrock geology of the eastern Koyukuk basin, central Brooks Range, and eastcentral Arctic Slope: Fairbanks, Alaska, Division of Geological and Geophysical Surveys Guidebook 7, v. 2, p. 277–284.

Box, S. E., 1985, Early Cretaceous orogenic belt in northwestern Alaska: internal organization, lateral extent, and tectonic interpretation, *in* D. G. Howell, ed., Tectonostratigraphic terranes of the circum-Pacific region: Circum-Pacific Council for Energy and Mineral Resources (AAPG), Earth Science Series, n. 1, p. 137-145.

Brosgé, W. P., and I. L. Tailleur, 1971, Northern Alaska petroleum province, *in* I. H. Cram, ed., Future petroleum provinces of the United States: their geology and potential: AAPG Memoir 15, v. 1, p. 68-99.

Brosgé, W. P., and C. L. Whittington, 1966, Geology of the Umiat-Maybe Creek region, Alaska: USGS Professional Paper 303-H, p. 501-638.

Brosgé, W. P., H. N. Reiser, J. T. Dutro, Jr., and R. L. Detterman, 1981, Organic geochemical data for Mesozoic and Paleozoic shales, central and eastern Brooks Range, Alaska: USGS Open-File Report 81-551, 17 p.

Bruns, T. R., M. A. Fisher, W. J. Leinbach, Jr., and J. J. Miller, 1987, *in* K. J. Bird and L. B. Magoon, eds., Petroleum geology of the northern part of the Arctic National Wildlife Refuge, northeastern Alaska: USGS Bulletin 1778, p. 249-254.

Buckingham, M. L., 1987, Fluviodeltaic sedimentation patterns of the Upper Cretaceous to lower Tertiary Jago River Formation, Arctic National Wildlife Refuge, northeastern Alaska, *in* I. Tailleur and P. Weimer, eds., Alaskan North Slope geology: Bakersfield, California, SEPM, Pacific Section, v. 1, p. 529-540.

Campbell, R. H., 1967, Areal geology in the vicinity of the Chariot site, Lisburne Peninsula, northwestern Alaska: USGS Professional Paper 395, 71 p.

Carman, G. J., and P. Hardwick, 1983, Geology and regional setting of the Kuparuk oil field, Alaska: AAPG Bulletin, v. 67, n. 6, p. 1014-1031.

Chapman, R. M., and E. G. Sable, 1960, Geology of the Utukok-Corwin region, northwestern Alaska: USGS Professional Paper 303-C, p. 47-174.

Chapman, R. M., R. L. Detterman, and M. D. Mangus, 1964, Geology of the Killik-Etivluk Rivers region, Alaska: USGS Professional Paper 303-F, p. 325-407.

Claypool, G. E., and L. B. Magoon, 1985, Comparison of oil-source rock correlation data for Alaskan North Slope—techniques, results, and conclusions, *in* L. B. Magoon and G. E. Claypool, eds., Alaska North Slope oil/rock correlation study: AAPG Studies in Geology No. 20, p. 49-81.

Collett, T. S., 1991, Natural gas hydrates on the North Slope of Alaska: DOE/MC/20422-2968, Oak Ridge, Tennessee, Office of Scientific and Technical Information, 32 p.

Collins, F. R., and F. M. Robinson, 1967, Subsurface stratigraphic, structural, and economic geology, northern Alaska: USGS Open-File Report 287, 252 p.

Craig, J. D., K. W. Sherwood, and P. P. Johnson, 1985, Geologic report for the Beaufort Sea planning area, Alaska—regional geology, petroleum geology, environmental geology: Minerals Management Service OCS Report MMS 85-0111, 192 p.

Crane, R. C., 1987, Cretaceous olistostrome model, Brooks Range, Alaska, *in* I. Tailleur and P. Weimer, eds., Alaskan North Slope geology: Bakersfield, California, SEPM, Pacific Section, Book 50, v. 1, p. 433-440.

Crane, R. C., and V. D. Wiggins, 1976, Ipewick Formation, significant Jurassic-Neocomian map unit in northern Brooks Range fold belt (abs.): AAPG Bulletin, v. 60, n. 12, p. 2177.

Crowder, R. K., 1987, Cretaceous basin to shelf transition in northern Alaska: Deposition of the Fortress Mountain Formation, *in* I. Tailleur and P. Weimer, eds., Alaskan North Slope geology: Bakersfield, California, SEPM, Pacific Section, Book 50, v. 1, p. 449-458.

Crowder, R. K., 1989, Deposition of the Fortress Mountain Formation, *in* C. G. Mull and K. E. Adams, eds., Dalton Highway, Yukon River to Prudhoe Bay, Alaska—bedrock geology of the eastern Koyukuk basin, central Brooks Range, and eastcentral Arctic Slope: Fairbanks, Alaska, Division of Geological and Geophysical Surveys Guidebook 7, v. 2, p. 293-301.

Curiale, J. A., 1987, Crude oil chemistry and classification, Alaska North Slope, *in* I. Tailleur and P. Weimer, eds., Alaskan North Slope geology: Bakersfield, California, SEPM, Pacific Section, Book 50, p. 161-167.

Deming, D., J. H. Sass, A. H. Lachenbruch, and R. F. De Rito, 1992, Heat flow and subsurface temperature as evidence for basin-scale groundwater flow, North Slope of Alaska: GSA Bulletin, v. 104, n. 6, p. 528-542.

Detterman, R. L., 1956, New and redefined nomenclature of Nanushuk Group, *in* G. Gryc et al., Mesozoic sequence in Colville River region, northern Alaska: AAPG Bulletin, v. 40, n. 2, p. 233-244.

Detterman, R. L., R. S. Bickel, and G. Gryc, 1963, Geology of the Chandler River region, Alaska: USGS Professional Paper 303-E, p. 223-324.

Detterman, R. L., H. N. Reiser, W. P. Brosgé, and J. T. Dutro, Jr., 1975, Post-Carboniferous stratigraphy, northeastern Alaska: USGS Professional Paper 886, 46 p.

Ehm, A., and I. L. Tailleur, 1987, Refined names for Brookian age elements in northern Alaska (abs.), *in* I. Tailleur and P. Weimer, eds., Alaskan North Slope geology: Bakersfield, California, SEPM, Pacific Section, Book 50, p. 432.

Embry, A. F., and J. Dixon, 1990, The breakup unconformity of the Amerasia basin, Arctic Ocean—evidence from Arctic Canada: GSA Bulletin, v. 102, p. 1526-1534.

Gautier, D. L., 1987, Petrology of Cretaceous and Tertiary reservoir sandstones in the Point Thomson area, *in* K. J. Bird and L. B. Magoon, eds., Petroleum geology of the northern part of the Arctic National Wildlife Refuge, northeastern Alaska: USGS Bulletin 1778, p. 117-122.

Gautier, D. L., K. J. Bird, and V. A. Colten-Bradley, 1987, Relationship of clay mineralogy, thermal maturity, and geopressure in wells of the Point Thomson area, *in* K. J. Bird and L. B. Magoon, eds., Petroleum geology of the northern part of the Arctic National Wildlife Refuge, northeastern Alaska: USGS Bulletin 1778, p. 199-207.

Gaynor, G. C., and M. H. Scheihing, 1988, Shelf depositional environments and reservoir characteristics of the Kuparuk River Formation (Lower Cretaceous), Kuparuk field, North Slope, Alaska, *in* A. J. Lomando and P. M. Harris, eds., Giant oil and gas fields, a core workshop: Tulsa, Oklahoma, SEPM Core Workshop No. 12, p. 333-389.

Gerhard, L. C., L. A. Graber, and E. A. Brostuen, 1988, A look at the status of U.S. petroleum: Oil and Gas Journal, 20 June, 1988, p. 73-78.

Grantz, A., and S. D. May, 1983, Rifting history and structural development of the continental margin north of Alaska, *in* J. S. Watkins and C. L. Drake, eds., Studies in continental margin geology: AAPG Memoir 34, p. 77-100.

Grantz, A., and S. D. May, 1988, Regional geology and petroleum potential of the United States Chukchi shelf north of Point Hope, *in* G. Gryc, ed., Geology and exploration of the National Petroleum Reserve in Alaska, 1974 to 1982: USGS Professional Paper 1399, p. 209-229.

Grantz, A., D. A. Dinter, and R. C. Culotta, 1987, Structure of the continental shelf north of the Arctic National Wildlife Refuge, *in* K. J. Bird and L. B. Magoon, eds., Petroleum geology of the northern part of the Arctic National Wildlife Refuge, northeastern Alaska: USGS Bulletin 1778, p. 271-276.

Grantz, A., S. D. May, and D. A. Dinter, 1988, Geologic framework, petroleum potential, and environmental geology of the United States Beaufort and northeasternmost Chukchi Seas, *in* G. Gryc, ed., Geology and exploration of the National Petroleum Reserve in Alaska, 1974 to 1982: USGS Professional Paper 1399, p. 231-255.

Gryc, G., 1970, History of petroleum exploration in Alaska, *in* W. L. Adkison and M. M. Brosgé, eds., Proceedings of the geological seminar on the North Slope of Alaska: Los Angeles, California, AAPG, Pacific Section, p. C1-C8.

Gryc, G., ed., 1988, Geology and exploration of the National Petroleum Reserve in Alaska, 1974 to 1982: USGS Professional Paper 1399, 940 p.

Gryc, G., H. R. Bergquist, R. L. Detterman, W. W. Patton, Jr., F. M. Robinson, F. P. Rucker, and C. L. Whittington, 1956, Mesozoic sequence in Colville River region, northern Alaska: AAPG Bulletin, v. 40, n. 2, p. 209-254.

Gryc, G., W. W. Patton, Jr., and T. G. Payne, 1951, Present Cretaceous stratigraphic nomenclature of northern Alaska: Washington Academy of Science Journal, v. 41, n. 5, p. 159-167.

Hancock, J. M., and E. G. Kauffman, 1979, The great transgressions of the Late Cretaceous: Journal of the Geological Society of London, v. 136, p. 175-186.

Haq, B. U., J. Hardenbol, and P. R. Vail, 1987, Chronology of fluctuating sea levels since the Triassic: Science, v. 235, p. 1156-1167.

Harris, A. G., H. R. Lane, I. L. Tailleur, and I. Ellersieck, 1987, Conodont thermal maturation patterns in Paleozoic and Triassic rocks, northern Alaska—geologic and exploration implications, *in* I. Tailleur and P. Weimer, eds., Alaskan North Slope geology: Bakersfield, California, SEPM, Pacific Section, Book 50, p. 181-191.

Harris, M., 1987, Endicott benefits from lessons learned: Alaska Construction & Oil, October, 1987, p. 15-16.

Harris, M., 1988, Beaufort causeways: Alaska Construction, v. 29, n. 12, p. 12-16.

Howell, D. G., K. J. Bird, L. Huafu, and M. J. Johnsson, 1992, Tectonics and petroleum potential of the Brooks Range fold and thrust belt—a progress report, *in* D. C. Bradley and A. B. Ford, eds., Geologic studies in Alaska by the U.S. Geological Survey during 1990: USGS Bulletin 1999, p. 112-126.

Hubbard, R. J., S. P. Edrich, and R. P. Rattey, 1987, Geologic evolution and hydrocarbon habitat of the 'Arctic Alaska microplate,' *in* I. Tailleur and P. Weimer, eds., Alaskan North Slope geology: Bakersfield, California, SEPM, Pacific Section, Book 50, v. 2, p. 797-830.

Huffman, A. C., Jr., T. S. Ahlbrandt, I. Pasternack, G. D. Stricker, and J. E. Fox, 1985, Depositional and sedimentologic factors affecting the reservoir potential of the Cretaceous Nanushuk Group, central North Slope, Alaska, *in* A. C. Huffman, Jr., ed., Geology of the Nanushuk Group and related rocks, North Slope, Alaska: USGS Bulletin 1614, p. 61-74.

Jamison, H. C., L. D. Brockett, and R. A. McIntosh, 1980, Prudhoe Bay—a 10-year perspective, *in* M. T. Halbouty, ed., Giant oil fields of the decade 1968-1978: AAPG Memoir 30, p. 289-314.

Johnsson, M. J., K. J. Bird, D. G. Howell, L. B. Magoon, R. G. Stanley, Z. C. Valin, A. G. Harris, and M. J. Pawlewicz, 1991, Preliminary map showing thermal maturity of sedimentary rocks in Alaska (abs.): AAPG Bulletin, v. 75, n. 3, p. 603.

Jones, D. L., and A. Grantz, 1964, Stratigraphic and structural significance of Cretaceous fossils from Tiglukpuk Formation, northern Alaska: AAPG Bulletin, v. 48, n. 9, p. 1462-1474.

Jones, H. P., and R. G. Speers, 1976, Permo-Triassic reservoirs of Prudhoe Bay field, North Slope, Alaska, *in* J. Braunstein, ed., North American oil and gas fields: AAPG Memoir 24, p. 23-50.

Keal, J. E., and W. G. Dow, 1985, Alaska North Slope oils and source rocks, *in* L. B. Magoon and G. E. Claypool, eds., Alaska North Slope oil-rock correlation study—analysis of North Slope crude: AAPG Studies in Geology No. 20, p. 85-94.

Kelley, J. S., and D. Bohn, 1988, Décollements in the Endicott Mountains allochthon, north-central Brooks Range, *in* J. P. Galloway and T. D. Hamilton, eds., Geologic studies in Alaska by the U.S. Geological Survey during 1987: USGS Circular 1016, p. 44-47.

Kelley, J. S., and R. L. Foland, 1987, Structural style and framework geology of the coastal plain and adjacent Brooks Range, *in* K. J. Bird and L. B. Magoon, eds., Petroleum geology of the northern part of the Arctic National Wildlife Refuge, northeastern Alaska: USGS Bulletin 1778, p. 255-270.

Lachenbruch, A. H., J. H. Sass, L. A. Lawver, M. C. Brewer, B. V. Marshall, R. J. Munroe, J. P. Kennelly, Jr., S. P. Galanis, Jr., and T. H. Moses, Jr., 1988, Temperature and depth of permafrost on the Arctic Slope of Alaska, *in* G. George, ed., Geology and exploration of the National Petroleum Reserve in Alaska, 1974 to 1982: USGS Professional Paper 1399, p. 645-656.

Lantz, R. J., 1981, Barrow gas fields—N. Slope, Alaska: Oil and Gas Journal, v. 79, n. 13, p. 197-200.

Lathram, E. H., 1965, Preliminary geologic map of northern Alaska: USGS Open File Report (65-96), 2 sheets, scale 1:1,000,000.

Lawton, T. F., G. W. Geehan, and B. J. Voorhees, 1987, Lithofacies and depositional environments of the Ivishak Formation, Prudhoe Bay field, *in* I. Tailleur and P. Weimer, eds., Alaskan North Slope geology: Bakersfield, California, SEPM, Pacific Section, Book 50, v. 1, p. 61-76.

Lawver, L. A., and C. R. Scotese, 1990, A review of tectonic models for the evolution of the Canada basin, *in* A. Grantz, L. Johnson, and J. F. Sweeney, eds., The Arctic Ocean region: Boulder, Colorado, GSA, The Geology of North America, v. L, p. 593-618.

Leiggi, P. A., 1987, Style and age of tectonism of the Sadlerochit Mountains to Franklin Mountains, Arctic National Wildlife Refuge, Alaska, *in* I. Tailleur and P. Weimer, eds., Alaskan North Slope geology: Bakersfield, California, SEPM, Pacific Section, Book 50, v. 2, p. 749-756.

Lerand, Monti, 1973, Beaufort Sea, *in* R. G. McCrossan, ed., The future petroleum provinces of Canada—their geology and potential: Canadian Society of Petroleum Geologists Memoir 1, p. 315-386.

Magoon, L. B., in press, The geology of known oil and gas resources by petroleum system—onshore Alaska, *in* G. Plafker, D. L. Jones, and H. C. Berg, eds., The Cordilleran orogen: Alaska: Boulder, Colorado, GSA, The Geology of North America , v. G-1.

Magoon, L. B., and K. J. Bird, 1985, Alaskan North Slope petroleum geochemistry for the Shublik Formation, Kingak Shale, pebble shale unit, and Torok Formation, *in* L. B. Magoon and G. E. Claypool, eds., Alaska North Slope oil-rock correlation study—analysis of North Slope crude: AAPG Studies in Geology No. 20, p. 31-48.

Magoon, L. B., and K. J. Bird, 1988, Evaluation of petroleum source rocks in the National Petroleum Reserve in Alaska, using organic-carbon content, hydrocarbon content, visual kerogen, and vitrinite reflectance, *in* G. Gryc, ed., Geology and exploration of the National Petroleum Reserve in Alaska, 1974 to 1982: USGS Professional Paper 1399, p. 381-450.

Magoon, L. B., and G. E. Claypool, 1981, Two oil types on North Slope of Alaska—implications for exploration: AAPG Bulletin, v. 65, n. 4, p. 644-652.

Magoon, L. B., and G. E. Claypool, 1983, Petroleum geochemistry of the North Slope of Alaska—time and degree of thermal maturity, *in* M. Byorøy et al., eds., Advances in organic geochemistry 1981: Chichester, U.K., Wiley Heyden, p. 28-38.

Magoon, L. B., and G. E. Claypool, 1984, The Kingak Shale of northern Alaska—regional variations in organic geochemical properties and petroleum source rock quality: Organic Geochemistry, v. 6, p. 533-542.

Magoon, L. B., and G. E. Claypool, eds., 1985, Alaska North Slope oil-rock correlation study: AAPG Studies in Geology No. 20, 682 p.

Magoon, L. B., and G. E. Claypool, 1988, Geochemistry of oil occurrences, National Petroleum Reserve in Alaska, *in* G. Gryc, ed., Geology and exploration of the National Petroleum Reserve in Alaska, 1974 to 1982: USGS Professional Paper 1399, p. 519-549.

Magoon, L. B., K. J. Bird, G. E. Claypool, D. E. Weitzmann, and R. H. Thompson, 1988, Organic geochemistry, hydrocarbon occurrence, and stratigraphy in government-drilled wells, North Slope, Alaska, *in* G. Gryc, ed., Geology and exploration of the National Petroleum Reserve in Alaska, 1974 to 1982: USGS Professional Paper 1399, p. 483-487.

Magoon, L. B., P. V. Woodward, A. C. Banet, S. B. Griscom, and T. Daws, 1987, Thermal maturity, richness, and type of organic matter of source rocks, *in* K. J. Bird and L. B. Magoon, eds., Petroleum geology of the northern part of the Arctic National Wildlife Refuge, northeastern Alaska: USGS Bulletin 1778, p. 127-179.

Mast, R. F., G. L. Dolton, R. A. Crovelli, D. H. Root, E. D. Attanasi, P. E. Martin, L. W. Cooke, G. B. Carpenter, W. C. Pecora, and M. B. Rose, 1989, Estimates of undiscovered conventional oil and gas resources in the United States—a part of the Nation's energy endowment: U.S. Department of the Interior, 44 p.

Masterson, W. D., and C. E. Paris, 1987, Depositional history and reservoir description of the Kuparuk River Formation, North Slope, Alaska, *in* I. Tailleur and P. Weimer, eds., Alaskan North Slope geology: Bakersfield, California, SEPM, Pacific Section, Book 50, v. 1, p. 95-107.

Mayfield, C. F., I. L. Tailleur, and I. Ellersieck, 1988, Stratigraphy, structure, and palinspastic synthesis of the western Brooks Range, northwestern Alaska, *in* G. Gryc, ed., Geology and exploration of the National Petroleum Reserve in Alaska, 1974 to 1982: USGS Professional Paper 1399, p. 143-186.

McKinney, C. M., E. L. Garton, and F. G. Schwartz, 1959, Analyses of some crude oils from Alaska: U.S. Bureau of Mines Report of Investigations 5447, 19 p.

McLean, H., 1987, Petrology and reservoir potential of the Jago River Formation, *in* K. J. Bird and L. B. Magoon, eds., Petroleum geology of the northern part of the Arctic National Wildlife Refuge, northeastern Alaska: USGS Bulletin 1778, p. 123-126.

Melvin, J., and A. S. Knight, 1984, Lithofacies, diagenesis and porosity of the Ivishak Formation, Prudhoe Bay area, Alaska, *in* D. A. McDonald and R. C. Surdam, eds., Clastic diagenesis: AAPG Memoir 37, p. 347-365.

Minerals Management Service, 1988, Alaska update—January 1987-August 1988: OCS Information Report MMS 88-0073, 44 p.

Molenaar, C. M., 1981, Depositional history and seismic stratigraphy of Lower Cretaceous rocks in the National Petroleum Reserve in Alaska and adjacent areas: USGS Open-File Report 81-1084, 42 p.

Molenaar, C. M., 1982, Umiat field, an oil accumulation in a thrust-faulted anticline, North Slope, Alaska, *in* R. B. Powers, ed., Geologic studies of the Cordilleran thrust belt: Denver, Colorado, Rocky Mountain Association of Geologists, p. 537-548.

Molenaar, C. M., 1983, Depositional relations of Cretaceous and lower Tertiary rocks, northeastern Alaska: AAPG Bulletin, v. 67, n. 7, p. 1066-1080.

Molenaar, C. M., 1985, Subsurface correlations and depositional history of the Nanushuk Group and related strata, North Slope, Alaska, *in* A. C. Huffman, Jr., ed., Geology of Nanushuk Group and related rocks, North Slope, Alaska: USGS Bulletin 1614, p. 37-59.

Molenaar, C. M., 1988, Depositional history and seismic stratigraphy of Lower Cretaceous rocks in the National Petroleum Reserve in Alaska and adjacent areas, *in* G. Gryc, ed., Geology and exploration of the National Petroleum Reserve in Alaska, 1974 to 1982: USGS Professional Paper 1399, p. 593-621.

Molenaar, C. M., K. J. Bird, and T. S. Collett, 1986, Regional correlation sections across the North Slope of Alaska: USGS Miscellaneous Field Studies Map MF-1907.

Molenaar, C. M., K. J. Bird, and A. R. Kirk, 1987, Cretaceous and Tertiary stratigraphy of northeastern Alaska, *in* I. Tailleur and P. Weimer, eds., Alaskan North Slope geology: Bakersfield, California, SEPM, Pacific Section, Book 50, v. 1, p. 513-528.

Molenaar, C. M., R. M. Egbert, and L. F. Krystinik, 1981, Depositional facies, petrography, and reservoir potential of the Fortress Mountain Formation (Lower Cretaceous), central North Slope, Alaska: USGS Open-File Report 81-967, 32 p.

Molenaar, C. M., R. M. Egbert, and L. F. Krystinik, 1988, Depositional facies, petrography, and reservoir potential of the Fortress Mountain Formation (Lower Cretaceous), Central North Slope, Alaska, *in* G. Gryc, ed., Geology and exploration of the National Petroleum Reserve in Alaska, 1974 to 1982: USGS Professional Paper 1399, p. 257-280.

Morgridge, D. L., and W. B. Smith, 1972, Geology and discovery of Prudhoe Bay field, eastern Arctic Slope, *in* R. E. King, ed., Stratigraphic oil and gas fields: AAPG Memoir 16, p. 489-501.

Mull, C. G., 1982, Tectonic evolution and structural style of the Brooks Range, Alaska: an illustrated summary, *in* R. B. Powers, ed., Geologic studies of the Cordilleran thrust belt: Denver, Colorado, Rocky Mountain Association of Geologists, v. 1.

Mull, C. G., 1985, Cretaceous tectonics, depositional cycles, and the Nanushuk Group, Brooks Range and Arctic Slope, Alaska, *in* A. C. Huffman, Jr., ed., Geology of the Nanushuk Group and related rocks, North Slope, Alaska: USGS Bulletin 1614, p. 7-36.

Mull, C. G., 1987, Kemik Formation, Arctic National Wildlife Refuge, northeastern Alaska, *in* I. Tailleur and P. Weimer, eds., Alaskan North Slope geology: Bakersfield, California, SEPM, Pacific Section, Book 50, v. 1, p. 405-431.

Mull, C. G., 1989, Summary of structural style and history of Brooks Range deformation, *in* C. G. Mull and K. E. Adams, eds., Dalton Highway, Yukon River to Prudhoe Bay, Alaska—Bedrock geology of the eastern Koyukuk basin, central Brooks Range, and eastcentral Arctic Slope: Fairbanks, Alaska, Division of Geological and Geophysical Surveys Guidebook 7, v. 1, p. 47-56.

Mull, C. G., I. L. Tailleur, C. F. Mayfield, I. Ellersieck, and S. Curtis, 1982, New upper Paleozoic and lower Mesozoic stratigraphic units, central and western Brooks Range, Alaska: AAPG Bulletin, v. 66, n. 3, p. 348-362.

Namson, J. S., and W. K. Wallace, 1987, A structural transect across the northeastern Brooks Range, Alaska (abs.), *in* I. Tailleur and P. Weimer, eds., Alaskan North Slope geology: Bakersfield, California, SEPM, Pacific Section, Book 50, v. 2, p. 758.

Nelson, R. E., and L. D. Carter, 1985, Pollen analysis of a late Pliocene and early Pleistocene section from the Gubik Formation of arctic Alaska: Quaternary Research, v. 24, p. 295-306.

Noonan, W. G., 1987, Post-Ellesmerian depositional sequences of the North Slope subsurface, *in* I. Tailleur and P. Weimer, eds., Alaskan North Slope geology: Bakersfield, California, SEPM, Pacific Section, Book 50, v. 1, p. 459-477.

Oil and Gas Journal, 1984, Exxon: N. Slope gas/condensate field is a giant: Oil and Gas Journal, v. 82, n. 11, p. 30.

O'Sullivan, P. B., 1990, Preliminary results of 11 apatite fission-track analyses of samples from the Galbraith Lake-Toolik Lake region, North Slope, Alaska: Alaska Division of Geological and Geophysical Surveys Public-Data File 90-7a, 17 p.

O'Sullivan, P. B., J. Decker, S. C. Bergman, and W. K. Wallace, 1990, Constraints by apatite fission track analysis on cooling times due to uplift in the northeastern Brooks Range, Alaska (abs.): AAPG Bulletin, v. 74, n. 5, p. 733.

Palmer, A. R., 1983, The decade of North American geology 1983 geologic time scale: Geology, v. 11, p. 503-504.

Parrish, J. T., 1987, Lithology, geochemistry, and depositional environment of the Triassic Shublik Formation, northern Alaska, *in* I. Tailleur and P. Weimer, eds., Alaskan North Slope geology: Bakersfield, California, SEPM, Pacific Section, Book 50, v. 1, p. 391-396.

Patton, W. W., Jr., and S. E. Box, 1989, Tectonic setting of the Yukon-Koyukuk basin and its borderlands, western Alaska: Journal of Geophysical Research, v. 94, p. 15,807-15,820.

Patton, W. W., Jr., and I. L. Tailleur, 1964, Geology of the Killik-Itkillik region, Alaska: USGS Professional Paper 303-G, p. 409-500.

Plafker, G., 1990, Regional geology and tectonic evolution of Alaska and adjacent parts of the northeast Pacific Ocean margin: Proceedings of the Pacific Rim Congress 90, Australian Institute of Mining and Metallurgy, Queensland, Australia, p. 841-853.

Poulton, T. P., 1982, Paleogeographic and tectonic implications of the Lower and Middle Jurassic facies patterns in northern Yukon Territory and adjacent Northwest Territories, *in* A. F. Embry and H. R. Balkwill, eds., Arctic geology and geophysics: Canadian Society of Petroleum Geologists Memoir 8, p. 13-27.

Rattey, R. P., 1985, Northeastern Brooks Range, Alaska: new evidence for complex thin-skinned thrusting (abs.): AAPG Bulletin, v. 69, n. 4, p. 676-677.

Reed, J. C., 1958, Exploration of Naval Petroleum Reserve No. 4 and adjacent areas, northern Alaska, 1944-53: Part 1, History of the exploration: U.S. Geological Survey Professional Paper 301, 192 p.

Rickwood, F. K., 1970, The Prudhoe Bay field, *in* W. L. Adkison and M. M. Brosgé, eds., Proceedings of the geological seminar on the North Slope of Alaska: Los Angeles, California, AAPG, Pacific Section, p. L1-L11.

Roeder, D., and C. G. Mull, 1978, Tectonics of Brooks Range ophiolites, Alaska: AAPG Bulletin, v. 62. n. 9, p. 1696-1713.

Schindler, J. F., 1988, History of exploration in the National Petroleum Reserve in Alaska, with emphasis on the period from 1975 to 1982, *in* G. Gryc, ed., Geology and exploration of the National Petroleum Reserve in Alaska, 1974 to 1982: USGS Professional Paper 1399, p. 645-656.

Sedivy, R. A., I. E. Penfield, H. I. Halpern, R. J. Drozd, G. A. Cole, and R. Burwood, 1987, Investigation of source rock-crude oil relationships in the northern Alaska hydrocarbon habitat, *in* I. Tailleur and P. Weimer, eds., Alaskan North Slope geology: Bakersfield, California, SEPM, Pacific Section, Book 50, p. 169-179.

Seifert, W. K., J. M. Moldowan, and R. W. Jones, 1980, Application of biological marker chemistry to petroleum exploration: Proceedings of the 10th World Petroleum Congress, Bucharest, p. 425-440.

Siok, J. P., 1989, Stratigraphy and petrology of the Okpikruak Formation at Cobblestone Creek, northcentral Brooks Range, *in* C. G. Mull and K. E. Adam, eds., Dalton Highway, Yukon River to Prudhoe Bay, Alaska—Bedrock geology of the eastern Koyukuk basin, central Brooks Range, and eastcentral Arctic Slope: Fairbanks, Alaska, Division of Geological and Geophysical Surveys

Guidebook 7, v. 2, p. 285–292.

Smiley, C. J., 1969, Floral zones and correlations of Cretaceous Kukpowruk and Corwin Formations, northwestern Alaska: AAPG Bulletin, v. 53, n. 10, p. 2079–2093.

Smith, P. S., and J. B. Mertie, Jr., 1930, Geology and mineral resources of northwestern Alaska: USGS Bulletin 815, 351 p.

Smosna, R., 1989, Compaction law for Cretaceous sandstones of Alaska's North Slope: Journal of Sedimentary Petrology, v. 59, n. 4, p. 572–584.

State of Alaska, 1977, Prudhoe Bay unit operating plan: Anchorage, Oil and Gas Conservation Commission, 5 May, 1977, Conservation Hearing No. 145, Exhibit No. 8.

State of Alaska, 1984, Lisburne field rules: Anchorage, Oil and Gas Conservation Commission, Proceedings of the 29 November, 1984 public hearing, 86 p.

State of Alaska, 1985, 1984 Statistical Report: Anchorage, Oil and Gas Conservation Commission, 177 p.

Tailleur, I. L., 1973, Probable rift origin of Canada basin, Arctic Ocean, *in* M. G. Pitcher, ed., Arctic geology: AAPG Memoir 19, p. 526–535.

Tailleur, I., and P. Weimer, eds., 1987, Alaskan North Slope geology: Bakersfield, California, SEPM, Pacific Section, Book 50, 874 p.

Thurston, D. K., and L. A. Theiss, 1987, Geologic report for the Chukchi Sea planning area, Alaska: Minerals Management Service OCS Report MMS 87-0046, 193 p.

Van Dyke, W. D., 1980, Proven and probable oil and gas reserves, North Slope, Alaska: Alaska Department of Natural Resources, Division of Minerals and Energy Management, Anchorage, Alaska, 11 p.

van Poollen, H. K., and Associates, Inc., and Alaska Division of Oil and Gas, 1978, In-place hydrocarbons determination Kuparuk River Formation Prudhoe Bay area, Alaska: Anchorage, State of Alaska Department of Natural Resources, 13 p.

Wallace, W. K., and C. L. Hanks, 1990, Structural provinces of the northeastern Brooks Range, Arctic National Wildlife Refuge, Alaska: AAPG Bulletin, v. 74, n. 7, p. 1100–1118.

Waples, D. W., 1980, Time and temperature in petroleum formation—application of Lopatin's method to petroleum exploration: AAPG Bulletin, v. 64, n. 6, p. 916–926.

Weimer, P., 1987, Seismic stratigraphy of three areas of slope failure, Torok Formation, northern Alaska, *in* I. Tailleur and P. Weimer, eds., Alaskan North Slope geology: Bakersfield, California, SEPM, Pacific Section, Book 50, v. 1, p. 481–496.

Werner, M. R., 1987, West Sak and Ugnu sands; low-gravity oil zones of the Kuparuk River area, Alaskan North Slope, *in* I. Tailleur and P. Weimer, eds., Alaskan North Slope geology: Bakersfield, California, SEPM, Pacific Section, Book 50, p. 109–118.

Whittington, C. L., 1956, Revised stratigraphic nomenclature of Colville Group, *in* G. Gryc et al.: AAPG Bulletin, v. 40, n. 2, p. 244–253.

Wilbur, S., J. P. Siok, and C. G. Mull, 1987, A comparison of two petrographic suites of the Okpikruak Formation: a point count analysis, *in* I. Tailleur and P. Weimer, eds., Alaskan North Slope geology: Bakersfield, California, SEPM, Pacific Section, Book 50, p. 441–447.

Williams, B., 1989, Alaska tax hikes cloud latest giant's prospects: Oil and Gas Journal, v. 87, n. 33, p. 26.

Woidneck, K., P. Behrman, C. Soule, and J. Wu, 1987, Reservoir description of the Endicott field, North Slope, Alaska, *in* I. Tailleur and P. Weimer, eds., Alaskan North Slope geology: Bakersfield, California, SEPM, Pacific Section, p. 43–59.

Chapter 14

Structural and Stratigraphic Evolution and Hydrocarbon Distribution, Rocky Mountain Foreland

Robbie Gries

Consultant
Denver, Colorado, U.S.A.

J. C. Dolson

Amoco Production Co.
Denver, Colorado, U.S.A.

R. G. H. Raynolds

Amoco Production Co.
Denver, Colorado, U.S.A.

INTRODUCTION AND REGIONAL SETTING

The United States Rocky Mountain foreland basin (Figure 1) was superposed on the western North American passive margin, a region that had been relatively stable since Pennsylvanian time. During much of the Paleozoic and early Mesozoic, the North American craton presented a low-relief landscape, intermittently flooded by regional transgressions from the west, north, and south. As the Atlantic Ocean opened during the Jurassic, the western North American craton evolved from a passive margin to an active subduction/collision margin (Figure 2). At the onset of the resultant orogenic activity, the polarity of continental drainage networks changed from westerly to more easterly. Large Jurassic lakes and low-gradient fluvial systems dominated the protoforedeep. Lack of accommodation space resulted in the accumulation of thin sedimentary sequences. Paleosols, evaporites, dune fields, and dinosaur trails are characteristic of the Jurassic in the western United States.

In Early Cretaceous (Aptian) time, relief developed at the western collision margin, and foreland subsidence began adjacent to the Cordilleran Sevier orogenic belt (Cross, 1986) (Figure 1). Transverse streams flowed from the growing highlands, joining longitudinal drainage networks that flowed north to the boreal sea and, elsewhere, south to the ancestral Gulf Coast seaway, leaving a series of fluvial/alluvial pebble-bearing sedimentary packages including the Lakota, Lytle, Cedar Mountain, and Burro Canyon formations (Heller and Paola, 1989), which heralded foredeep sedimentation.

The first through-going connection of the Western Interior seaway along the subsiding axis of the foredeep occurred in late Aptian time, coincident with a worldwide eustatic sea level rise. In the Rocky Mountain foreland, widespread and thick marine Skull Creek/Thermopolis shales (and associated transgressive deposits) are preserved in the foredeep.

This basin architecture persisted (see also Leckie and Smith, 1992) laterally throughout the U.S. Rocky Mountain and Western Canada (Alberta) basins (Figure 1) until the onset of Laramide style deformation in the Late Cretaceous. The Western Canada and U.S. Rocky Mountain foreland basins share an early history of numerous base-level changes that resulted in widespread transgressions and regressions within an asymmetric foredeep. Thick, coarse clastics characterize the western margin, whereas thinner clastics grading to carbonates (Niobrara and Greenhorn formations) onlap the flanks of the stable craton to the east (McNeil, 1984).

Three major clastic wedges, equivalent to the Muddy, Frontier, and Mesaverde formations, were formed during these relative base-level shifts. In contrast to the Western Canada basin, withdrawal of the Cretaceous seaway com-

FIGURE 1. Map of North America, showing the Cordilleran thrust belt and the Western Canada (Alberta) and Rocky Mountain foreland basins.

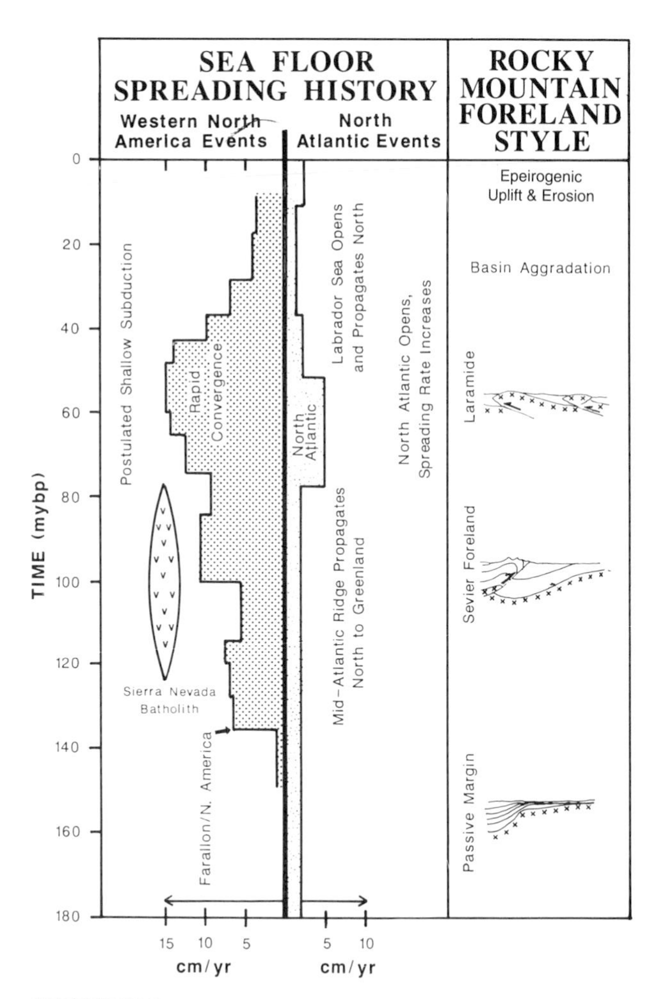

FIGURE 2. Time line and schematic summary of North American evolution. Passive margin tectonics dominated until about 120 Ma, when the Sevier orogenic episode began. Approximately 81 Ma, Laramide tectonics were superposed on the Sevier foreland basin. At 81 Ma, the North Atlantic opened and the spreading rate increased. At 75 Ma, the rate of Farallon and North American convergence significantly increased, causing flat-plate subduction beneath the Rocky Mountain foreland and subsequent breakup of basement blocks (Figure 3). The rates for North Atlantic seafloor spreading are from Pitman and Talwani (1972); the rates for Farallon convergence are from Engebretson et al. (1984).

menced in the U.S. Rocky Mountain foreland with the Laramide orogeny during the Campanian and Maastrichtian (Dickinson et al., 1988). Broad crustal warps of the foreland were initiated, possibly as a result of a change from steep to shallow plate subduction (Figures 2 and 3).

A comparable situation has been described in South America (Jordan and Allmendinger, 1986), where the northern Argentinian foreland displays a similar tectonic style, including a breakup of the Andean foredeep into basement-cored ranges and deep adjacent basins. Northern Argentina has several differences, however. One is the magnitude of basement displacement, which is much more pronounced in the U.S. Rocky Mountain foreland. This may be a result of timing and duration, because the causal flat-plate subduction in the Andes of northern Argentina is of relatively recent origin. The breakup of the Rocky Mountains started robustly in the Late Cretaceous (about 80 Ma) and lasted, although episodically, until about 38 Ma. The modern Argentinian flat-plate subduction that formed the Pampaenas Ranges started about 10 Ma (Jordan et al., 1983). Another difference is the direction of relative plate movement (and compressional force), which in Argentina has been constant, whereas there is evidence that compression was episodic and changed direction during the 42 m.y. of Laramide deformation (Gries, 1990). Figure 2 is a schematic history of the Rocky Mountain foreland plate tectonic history.

SEQUENCE STRATIGRAPHY, RESERVOIRS, AND SOURCE ROCKS

Hydrocarbon reservoirs in the Rocky Mountain foreland can be subdivided into three major time-stratigraphic groups (Figure 4) that were deposited successively on a passive margin (I), within a foreland basin (II), and in partitioned basins of Tertiary age (III) (Figure 5). The sedimentary assemblages formed in three distinct regional tectonic systems related to the larger-scale plate movements discussed above (Figure 2). These packages contain

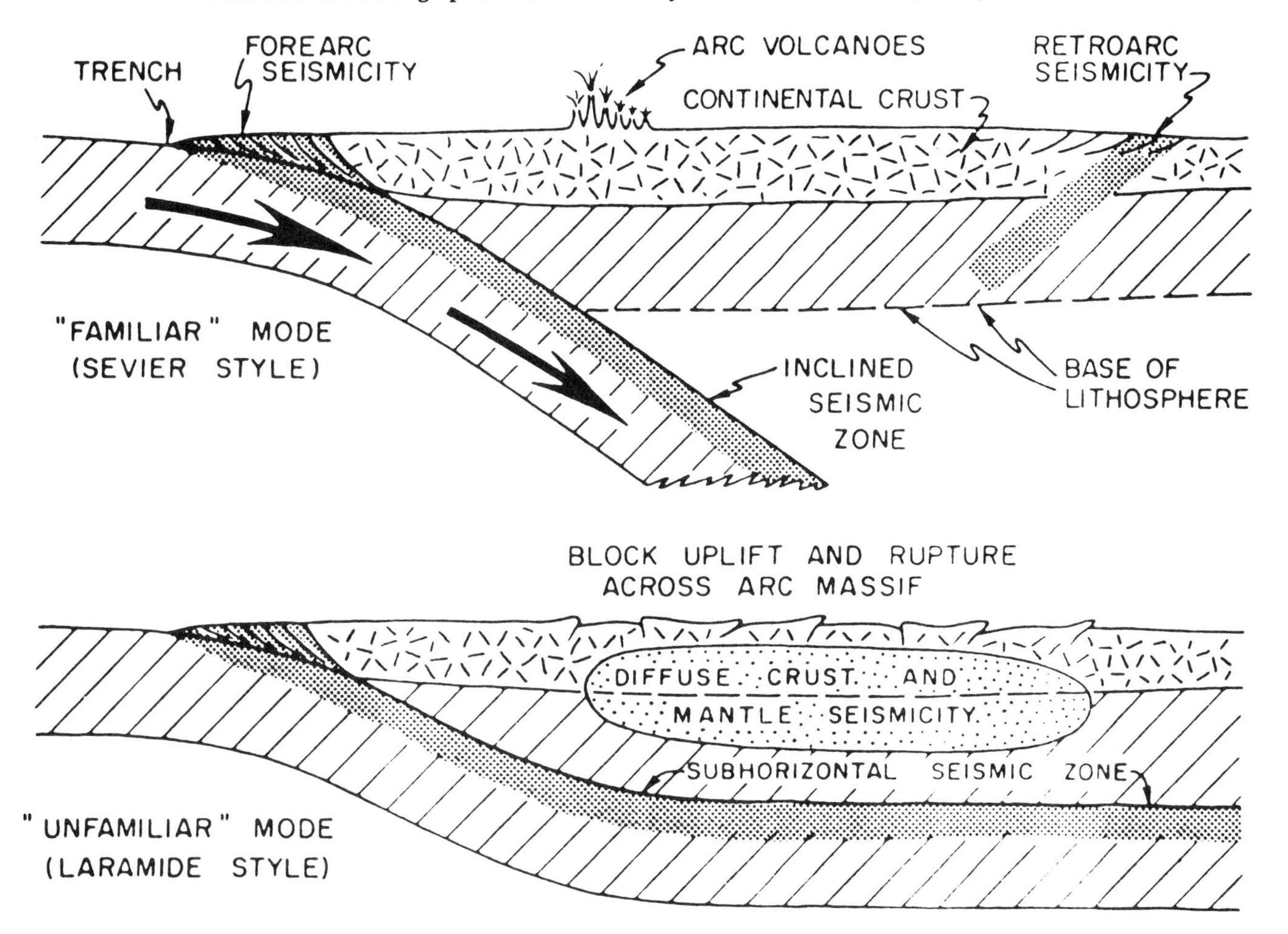

FIGURE 3. Schematic cross sections contrasting normal versus flat-plate subduction. Flat-plate subduction is proposed as the mechanism for creating the basement-involved structures in the Rocky Mountain foreland during the Laramide orogeny, 80 to 36 Ma (after Dickinson and Snyder, 1978).

various amounts of carbonate and clastic reservoirs (Table 1), all of which produce hydrocarbons as a result of Laramide maturation of source strata in the deep foreland basins.

Group I—Pre-Middle Jurassic Passive Margin Strata

Late Precambrian through Mid-Jurassic aged reservoirs formed largely in a passive margin setting. Precambrian sedimentary packages preserved in the Rocky Mountain region accumulated in aulacogens and rift settings. Potential source strata include the Chuar, Greyson, and Red Pine shales, which contain Type I kerogens.

Paleozoic source rocks formed primarily in marine condensed sections or lagoonal shales that accumulated during sea level highstands or as anoxic shales and evaporites associated with basin starvation during periods of relative sea level lowstands. These strata include the Devonian Bakken Formation and equivalent shales (Williston basin), the Mississippian Manning Canyon Formation (Utah), the Heath Shales (Montana), the Lower Pennsylvanian Tyler Formation (Montana), the Pennsylvanian Desert Creek Shales (Paradox basin), the Pennsylvanian Belden Shales (Eagle basin in Colorado), and the Pennsylvanian Leo, Virgil, and other formations (Powder River and eastern Denver basin). The Permian Phosphoria shales, which are perhaps the most significant source beds, formed in the passive margin setting and are the sources for much of the Permian and Pennsylvanian production discussed below.

Mid-Pennsylvanian Tectonic Overprint on the Passive Margin

Relative sea level fluctuations and Late Mississippian and Mid-Pennsylvanian tectonism created thick packages of mixed carbonate and clastic deposits that locally exceed 4600 m (15,000 ft) in thickness, as for example in the Paradox basin of Utah. The Mid-Pennsylvanian orogeny resulted from continental collision along the Marathon fold and thrust belt in Texas. This event translated stresses northward into Colorado and Utah, creating the Eagle,

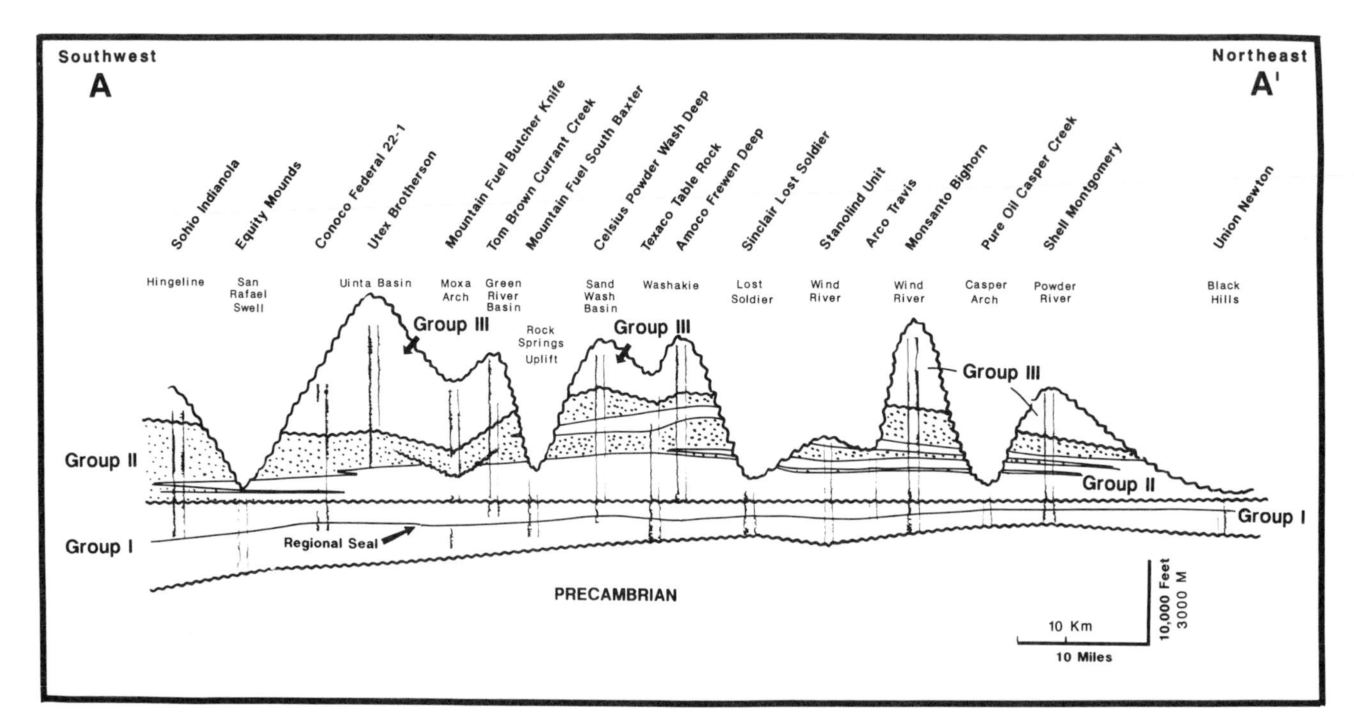

FIGURE 4. Regional cross section A–A′ (Figure 5) from the Uinta basin to the Powder River basin, showing three sedimentary packages. Group I is the passive margin package (pre-Jurassic Morrison strata) showing thin shelf sequences bounded by widespread regional unconformities. The regional seal shown is below Permian or Triassic red beds and separates Paleozoic oils from Cretaceous oils. Pennsylvanian orogenic episodes have been added to the Group I passive margin sequences because of their relatively short duration and limited areal extent within the area of interest (see text for discussion). Group II is the foredeep package (Morrison through Upper Cretaceous strata) showing eastward thinning away from the Sevier orogenic belt. Group III is the Tertiary sediment that accumulated in varying thicknesses within individual basins.

Paradox, and San Juan Paleozoic basins (Kluth and Coney, 1981). These basins were intermittently linked to open marine conditions to the south along the foredeep of the Anadarko basin. Repeated transgressions from the south through the Powder River and Denver basins also helped deposit mixed continental and marine strata. Uplifts such as the Ancestral Rockies (Colorado Front Range) and Uncompaghre (western Colorado) shed clastics into adjacent basins. By Permian time, this orogenic event had ended and the basins were largely buried beneath continental and shallow marine deposits such as the Phosphoria and Goose Egg formations.

With the exception of these Pennsylvanian basins, however, the passive margin section on the craton is thin when compared with the superposed foreland clastic wedges (Figure 4).

By Triassic time, increasingly arid climatic conditions had created the eolian sandstone and redbed shale assemblage of the Nugget and Chugwater formations. The Triassic redbeds contain little or no petroleum source beds and act primarily as vertical seals to hydrocarbons generated in underlying strata (Figure 4).

Group II—Foreland Basin Setting, Jurassic and Cretaceous Strata

Group II sediments formed during a period of tectonic readjustment as foreland thrust belt highlands and uplifts formed in response to westward plate movement. In Early Jurassic time, arid climates and shallow transgressions advancing from the north and west (Sundance and Gypsum Springs transgressions) left marine deposits rich in coquina-bearing limestones, anhydrites, and fossiliferous sandstones encased in red and green shales. Along the subsiding foredeep to the west, thick salt packages (Pruess and Arapien formations) accumulated in a restricted environment spatially associated with the early Sevier orogenic disruption of the passive margin. The latter strata provide excellent top seals to eolian Jurassic reservoirs (Nugget Formation) in the Overthrust belt. The absence of organic-rich shales in these zones has prevented Jurassic reservoirs from being important hydrocarbon targets when not juxtaposed structurally against Cretaceous source rocks.

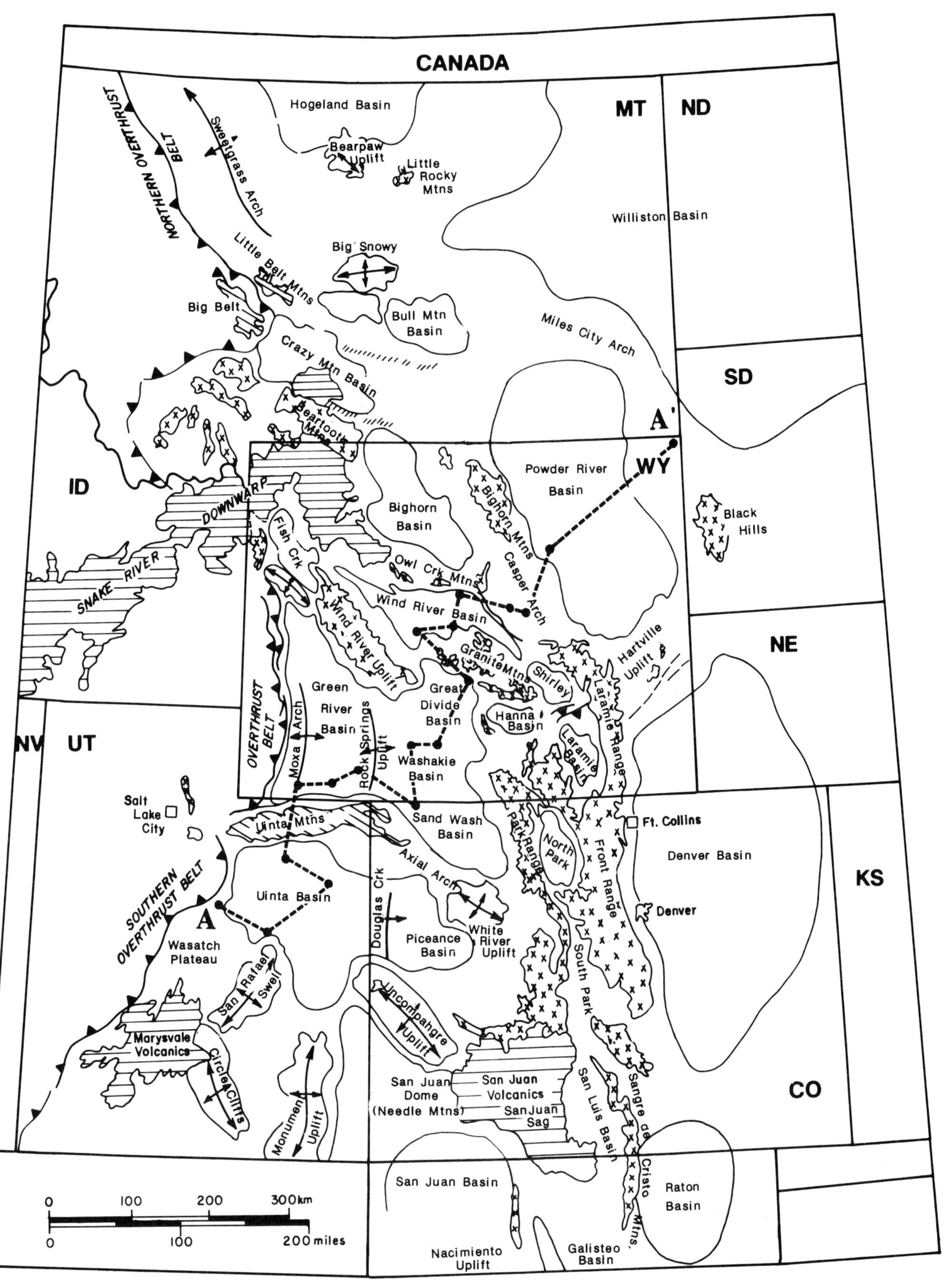

FIGURE 5. Map showing Rocky Mountain foreland basins and uplifts. Line of regional cross section shown as A–A′. X′s denote Precambrian crystalline rocks, diagonal hachures are Precambrian sedimentary rocks, and horizontal hachures are major Tertiary volcanics.

TABLE 1. Source Rock Packages and Kerogen Types for the Major Intervals of the Rocky Mountain Foreland (from Woodward et al., 1984; and Gries, personal communication).

Rock Unit	Total Organic Carbon (TOC)	Kerogen Type
Cretaceous Transgressive/Highstand Deposits		
Mowry	1.0–2.7%	II, III
Niobrara	1.5–10.0%	II, III
Mancos	0.5–1.7%	II, III
Lewis	0.1–4.0%	I, II
Laramide Lacustrine and Paludal Deposits		
Green River	1.8–21%	I, II

By Mid-Jurassic (Morrison) time, clastics were being shed from the west into flat-lying lacustrine and continental strata, a pattern that became accentuated in the Cretaceous, creating largely western-derived clastic wedges characteristic of Group II strata (Figure 4).

This simple pattern is complicated by at least 15 regional unconformities that are developed within Jurassic and Cretaceous strata. These surfaces are both submarine and subaerial in origin and, with their associated facies tracts, exerted a strong control on stratigraphic trapping of hydrocarbons. In addition, highstand deposits associated with each major marine transgression created anoxic condensed sections and source rock packages (Table 1).

The unconformities were caused by combinations of tectonic and eustatically induced relative sea level shifts (Weimer, 1983; Pipiringos and O'Sullivan, 1978). These surfaces have not yet been convincingly linked to the Exxon worldwide eustatic curves (e.g., Vail et al., 1977), and the fit may be complicated by foreland basin tectonic overprinting (cf. Jervey, 1992). Mixed provenance areas from the east, west, and southwest contributed clastic debris during these drops in relative sea level.

Cretaceous transgressions, commencing with the Skull Creek event, deposited thick successions of marine shales containing Type I and II kerogens averaging 1 to 2% TOC within the overall interval, but reaching 3 to 4% TOC in individual condensed sections, which probably accumulated at peak highstands (Table 1). Maximum water depths were reached during the Greenhorn cyclothem (Kauffman, 1977), where strata bearing Type II kerogens contain up to 10% TOC in the Niobrara Formation. Lowstands following each sea level rise formed erosional unconformities and related valley networks or low-gradient alluvial plains, many of which have given rise to regional trends of stratigraphically trapped hydrocarbons. Examples are the Lower Cretaceous Cutbank and Fall River sandstones and the Muddy Formation. During ensuing transgressions, onlapping shorelines created additional favorable trapping geometries, particularly in barrier island systems. Examples are found within the Lower Cretaceous Muddy Formation (Dolson et al., 1991), the Upper Cretaceous Almond Formation (Weimer, 1966), and the Upper Cretaceous Fruitland Formation.

Sharp-based shoreface/shelf sandstones such as the Tocito Formation in the San Juan basin, New Mexico (McCubbin, 1982) produce hydrocarbons from strike-elongated stratigraphic traps, as do the shelf sandstones of the Upper Cretaceous Sussex and Shannon formations (Spearing, 1976). In rare cases, breached highstand systems tracts have formed clastic "buried hill" traps that have trapped significant reserves. Examples include 28×10^9 m^3 (1 tcf) of gas in the Wattenberg field (Muddy Formation) (Dolson et al., 1991; Weimer et al., 1986) and 20.7×10^6 m^3 (130×10^6 BOE) from Bell Creek field (Muddy Formation) (Weimer et al., 1988).

Lowstand systems tract submarine fans have provided relatively small reserves to date. An example of a turbidite reservoir is the Hay Reservoir field, which produces from the Upper Cretaceous Lewis Formation (Grilley, 1979).

Upper Cretaceous subsidence rates were significantly higher than those in the Lower Cretaceous and Jurassic systems, leading to extensive coal deposits and thick clastic wedges of nonmarine strata, particularly to the southwest. Thick coal seams and Type III kerogen-rich shales juxtaposed against reservoir packages have created significant deep basin and coal methane reserves. The Type III kerogen packages grade eastward into Type I and Type II shales in the more marine settings of the Maastrichtian clastic wedges. The Mancos shales (lower Mesaverde Group) have TOC contents varying from 0.5 to 2%. The younger Lewis and Pierre shales are dominated by Type III kerogens (although Types II and III mixed locally) and generally do not exceed 1.5% TOC. These relatively lean shales may have generated significant volumes of hydrocarbons (Gries and Clayton, unpublished data) because of their thickness of 1200 to 1800 m (4000 to 6000 ft).

Upper Cretaceous clastic wedges (Mesaverde Group and equivalents) have produced several stacked reservoir zones, including the Sussex/Shannon, Parkman, Teapot, and Teckla formations (Powder River basin), Almond Formation (Green River basin), and Fruitland Formation (San Juan basin).

The onset of the Laramide orogeny is marked by the stratigraphic change from marine (Lewis Formation) to nonmarine deposition (Lance, Hell Canyon, and Meeteetse formations) in the Rocky Mountain foreland. Many localized unconformities formed on rising structures during the late Campanian and/or early Maastrichtian (Figures 6 and 7). Examples of reservoirs include the latest Cretaceous Almond and Ericson formations on the Moxa arch.

Group III—Tertiary Basin Partitioning

A regional unconformity marks the boundary between the Paleocene (Fort Union Formation) and the sediments of the Late Cretaceous regression (Figures 6 and 8). This unconformity is very pronounced on or adjacent to early Laramide structures. Widespread erosional surfaces developed between the Paleocene and Eocene (Figures 6 and 9) and occur locally on and near uplifts.

Traps associated with these unconformity surfaces have been economically insignificant, and most Laramide-associated production comes from structural traps located adjacent to thermally mature deep basin centers. The producing reservoirs are nonmarine fluvial and lacustrine sandstones and shales, such as the gas-bearing Paleocene Fort Union Formation at both the Madden (Brown and Shannon, 1989) and Pavillion (Mueller and Brown, 1989) fields in the Wind River basin. Lacustrine source strata concentrated in Eocene basins have provided potential source and seal in Type II and Type III kerogen-rich packages with TOC contents varying from 1 to 21%. Where clastic units intertongue with these source rocks in thermally mature settings, stratigraphic traps occur, such as the 15.9×10^6 m^3 (100×10^6 BOE) Altamont-Bluebell complex in the Uinta basin and the Haybarn field (Anders and Gerrild, 1984) in the Wind River basin.

TECTONICS AND STRUCTURE

The U.S. Rocky Mountain foreland basin was formed concurrently with the Western Canadian foreland basin as the North American plate moved westward over the Pacific plates during the opening of the Atlantic ocean (Figure 2). A thrust belt formed along the west margin of the North American plate about 140 Ma and as it shed clastics eastward, the foredeep developed (Figure 1).

About 81 Ma, as the North Atlantic began to open, the rate of westward plate movement increased (Pitman and Talwani, 1972), and about 75 Ma the Farallon and North American plate convergence rate greatly increased (Engebretson et al., 1984) (Figure 2). At this time the U.S. Rocky Mountain foreland began to break up into distinct basement-involved compressional blocks. The nature of the earliest Laramide orogeny (70–80 Ma) was the formation of folds and faulted folds essentially parallel to the Sevier belt (Figure 7) or north-south trending. The tops of these folds were eroded down through upper Cretaceous rocks, stripping them of up to 1800 m (about 6000 ft) of sediments (Figure 6). The Moxa arch, Targhee uplift, San Rafael swell/Douglas Creek arch, Rock Springs uplift, Wind River–Washakie–Beartooth trend, Front Range/Laramie Range/Sangre de Cristo system (Figures 5 and 6), and possibly the Big Horn Mountains are examples of places where this early folding and erosion occurred. Most Mesozoic foredeep sediments were eroded during this episode. For several of these uplifts, the maximum vertical displacement occurred during this time and they were relatively unaffected by younger Laramide movement (Figure 6). Large west/northwest strike-slip zones, as described by Stone (1969), probably were active at the same time.

During the Paleocene, another episode of compression is recorded in the synorogenic sediments of the separate foreland basins in the Rocky Mountain foreland; it appears that many of the earlier-formed highlands were further compressed, uplifted, and denuded. This time the compression occurred in a more northeast-southwesterly sense (Figure 8). Uplifts that exhibit maximum vertical displacement at this time include the Big Horn uplift, the Beartooth uplift, the Casper arch, the San Juan dome, and possibly the Front Range/Sangre de Cristo system (Figures 6 and 8).

Laramide orogenic activity during the Eocene primarily involved north–south-directed compression (Gries, 1983, 1990) that caused northward and southward thrusting on the flanks of east-west-trending uplifts and gave rise to their maximum vertical displacement in the northern Rocky Mountain region (Figure 9). Chapin and Cather (1981, 1983) document this change in compression direction for the southern Rocky Mountain or Colorado Plateau region and have determined from paleostress studies using dike and fracture orientations that sigma 1 was oriented east-northeast in early Laramide and changed to north-northeast in late Laramide time.

Late Laramide uplifts include the Uinta Mountains, which were uplifted 7600 to 9000 m (25,000 to 30,000 ft), and the nearby White River uplift, which was displaced vertically by 8200 m (27,000 ft). The Granite Mountains rose vertically 4500 m (15,000 ft) as did the north end of the Laramie Range; there was 3000 to 4500 m (10,000 to 15,000 ft) of movement on the Owl Creek uplift. Thrusting also occurred on the north or south margins of earlier Laramide uplifts during the Eocene, such as on the north flank of the Beartooth uplift (Dutcher et al., 1986), on the south flank of the Wind River uplift (Gries, 1983), on the north flank of the Laramie Range (Blackstone, 1988), and on the north and south flanks of the Shirley Mountains (Bergh and Snoke, 1992) (Figures 6 and 9). Eocene synorogenic deposition was extremely thick (Chapin and Cather, 1983) and reflected the large basement offsets characteristic of this part of the Laramide orogeny (Figure 9). Coincident with this change in compressional direction in the Rockies was the cessation of eastward overthrusting in the Overthrust belt, where the western Uinta Mountain thrust overrides the Hogback thrust (Lamerson, 1982). Additionally, north-trending Eocene strike/slip faults and axial basins formed on or near earlier Laramide uplifts (Gries, 1990; Chapin and Cather, 1981).

The episodic nature of the Laramide orogeny in the Rocky Mountain region is documented by the regional unconformities between each episode and the accompanying synorogenic sedimentary packages adjacent to specific foreland uplifts (Figure 6). The earliest Laramide episodes produced fine-grained synorogenic sediments, in contrast with thick conglomerates that accumulated during the later Paleocene and Eocene events. The striking angular unconformities associated with the earliest pulses, however, attest to their magnitude. The absence of thick conglomerates is mostly attributed to the fact that fine-grained Mesozoic sediments were being eroded from

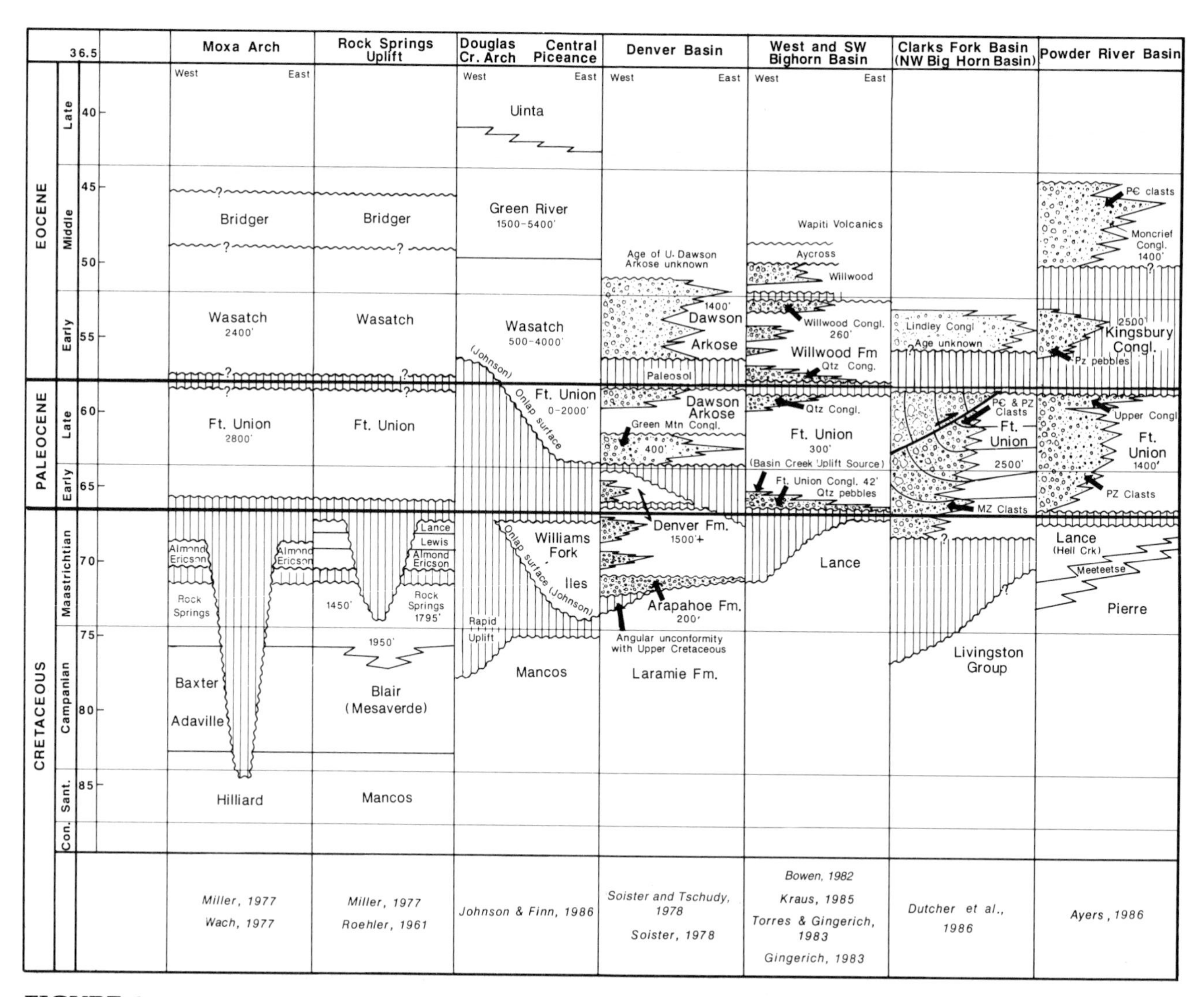

FIGURE 6. Stratigraphic correlation chart illustrating Laramide orogenic pulses preserved in the sedimentary record of basins in the Rocky Mountain foreland. Early deformation in Late Cretaceous removed several thousand feet of upper Cretaceous rocks from north-trending structures (see, for example, the Moxa arch and Rock Springs uplift). Paleocene synorogenic deposits are thick in basins adjacent to rising Paleocene uplifts such as the Big Horn, Bear Tooth, and Front Range uplifts. Late Laramide Eocene orogeny created thick conglomerates in basins such as the Great Divide, Green River, and Wind River.

these highlands. During the Paleocene and Eocene episodes, Paleozoic and Precambrian rocks were being eroded.

The uniqueness of the Laramide orogeny in breaking up the U.S. Rocky Mountain portion (mainly Colorado and Wyoming) of the North American foredeep (Figure 8) has been attributed to flat-plate subduction (Figure 3) of the Farallon plate (Lowell, 1974; Dickinson and Snyder, 1978; Livaccari et al., 1981; and Engebretson et al., 1984). According to this model, the Western Canada basin was not affected by subducted Farallon crustal "debris" and consequently did not undergo an episode of basement-involved deformation. Changes in the direction of motion of the North American craton through time have been documented by Pitman and Talwani (1972) and Engebretson et al. (1984) (Figure 10). These changes are coincident with many of the changes in the direction of compression within the Rocky Mountain foreland (Gries, 1983, 1990; Chapin and Cather, 1981, 1983) over the course of the 42-m.y. Laramide orogeny.

The role of the Colorado Plateau in foreland tectonics is significant. This thicker and/or more rigid crustal block was relatively high throughout Sevier deformation and foredeep sedimentation. Consequently, most of the Cretaceous sediments are nonmarine on the Plateau with lesser amounts of Upper Cretaceous and Tertiary rocks preserved. Basement-involved uplifts are usually bounded by high-angle reverse faults, and have less vertical and horizontal displacement than blocks north of the plateau.

The fold-thrust nature of Laramide uplifts (Figure 11) with Precambrian rocks displaced vertically from 4500 to 9000 m (15,000 to 30,000 ft) has been a source of discussion

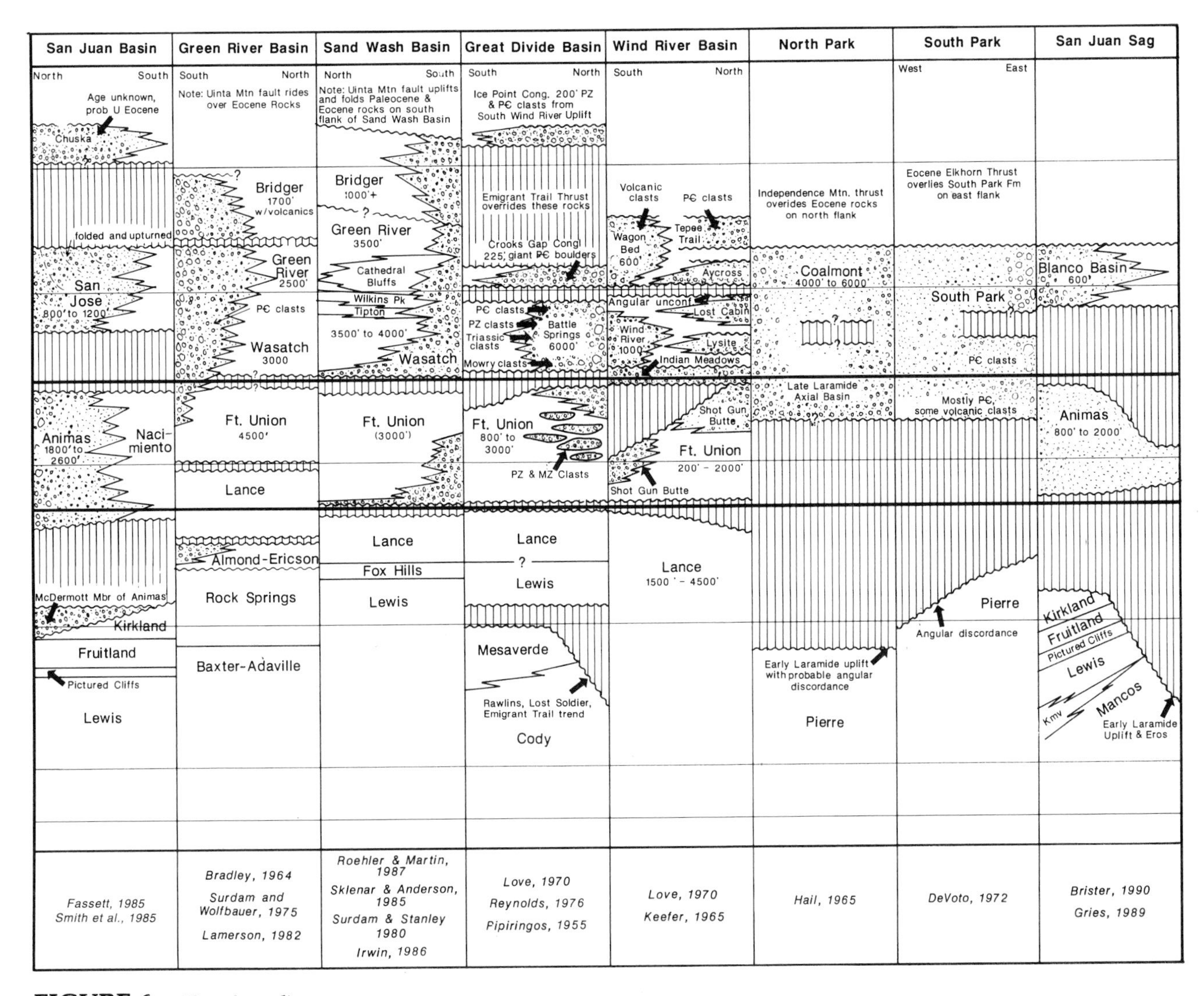

FIGURE 6. **(Continued)**

for structural geologists worldwide, especially when large gas reserves (potential 14.2 $\times$ 10^9 m^3, or 500 bcf) were discovered beneath thrusted Precambrian basement along the southwest flank of the Casper arch. Figure 12 shows a seismic line across the northeast margin of the Wind River basin onto the southwest flank of the Casper arch, where 2703 m (8865 ft) of hanging wall (including 813 m or 2665 ft of Precambrian crystalline rocks) was drilled prior to penetrating subthrust Cretaceous rocks. Large Precambrian-cored structures were uplifted along fault planes breaking roughly at a 30° angle (Stone, in press). These flank faults have been documented over the last 30 years by drilling and seismic data on virtually every Laramide mountain front in the Rocky Mountain foreland. Figure 11 is a current structural contour map on basement of the numerous Rocky Mountain foreland basins and uplifts. These structures are complex, particularly in the poorly understood subthrust section, primarily as a result of repeated changes in compression direction. This complexity is in contrast to earlier-stage foreland basins such as the Pampeanas in Argentina, which has a simpler system of basement-cored structures parallel to the adjacent thrust belt.

MATURATION HISTORY

Maturation of source rocks in Rocky Mountain foreland basins occurred with the deposition of thick, late Laramide Eocene sediments that caused Cretaceous and older source rocks to descend into the hydrocarbon-generating window (Figure 13). In certain places, such as the Piceance basin, the Uinta basin, and parts of the Green River basins, even the lower and middle Eocene lacustrine source rocks were buried deeply enough to enter the oil window. Passive-margin sediments (Group I) were buried sufficiently to enter the dry gas window in the deeper Laramide basins.

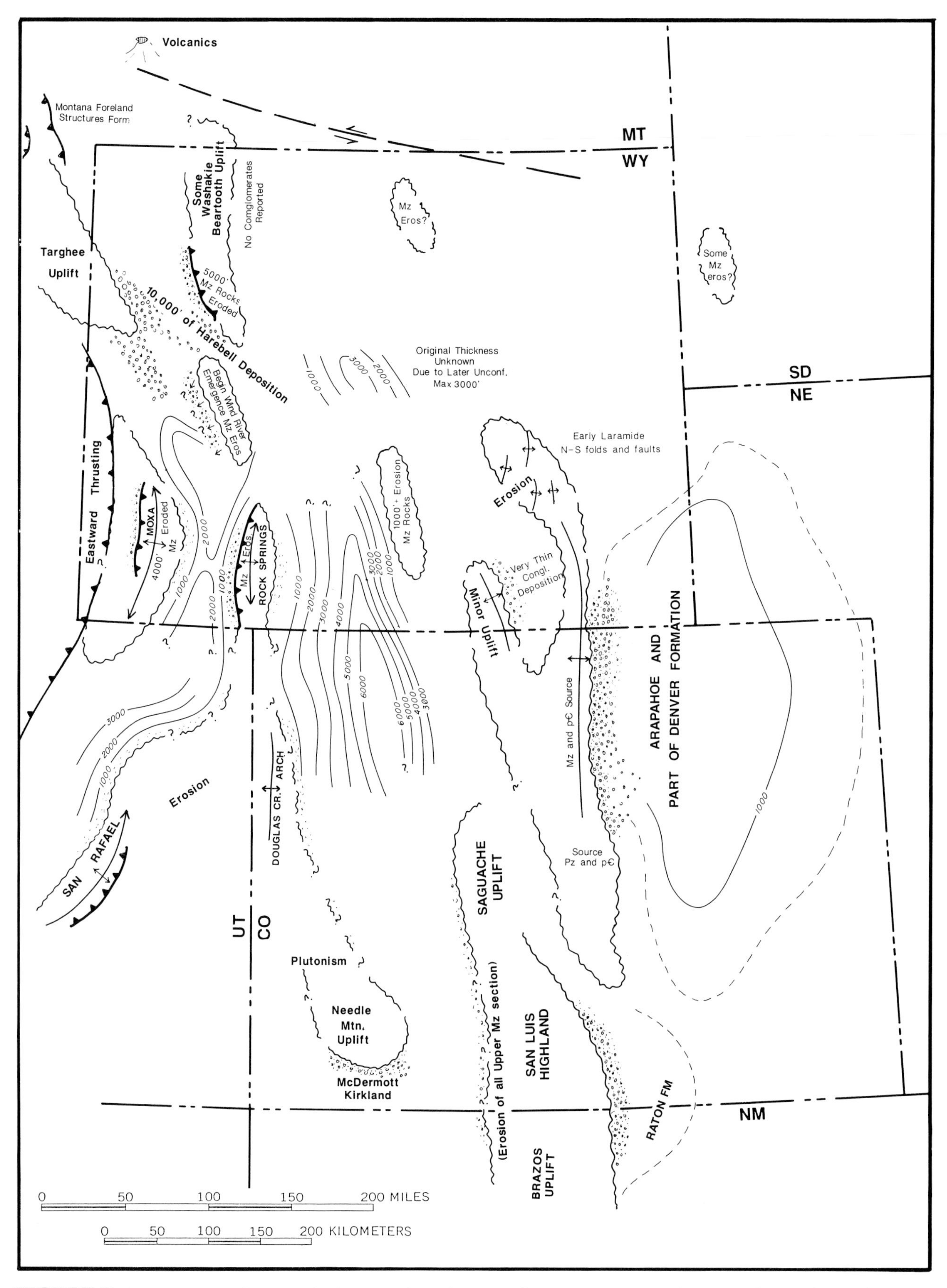

FIGURE 7. Isopach map showing thickness and areal extent of Late Cretaceous sediments. The north-south-trending thins illustrate the effects of an early Laramide orogenic pulse. Structures were mostly parallel to the Sevier overthrust belt and probably resembled structural grains in the modern Argentinian basement-involved foreland as illustrated in the Bermejo basin by Beer and Jordan (1989). Isopach interval is 305 m (1000 ft).

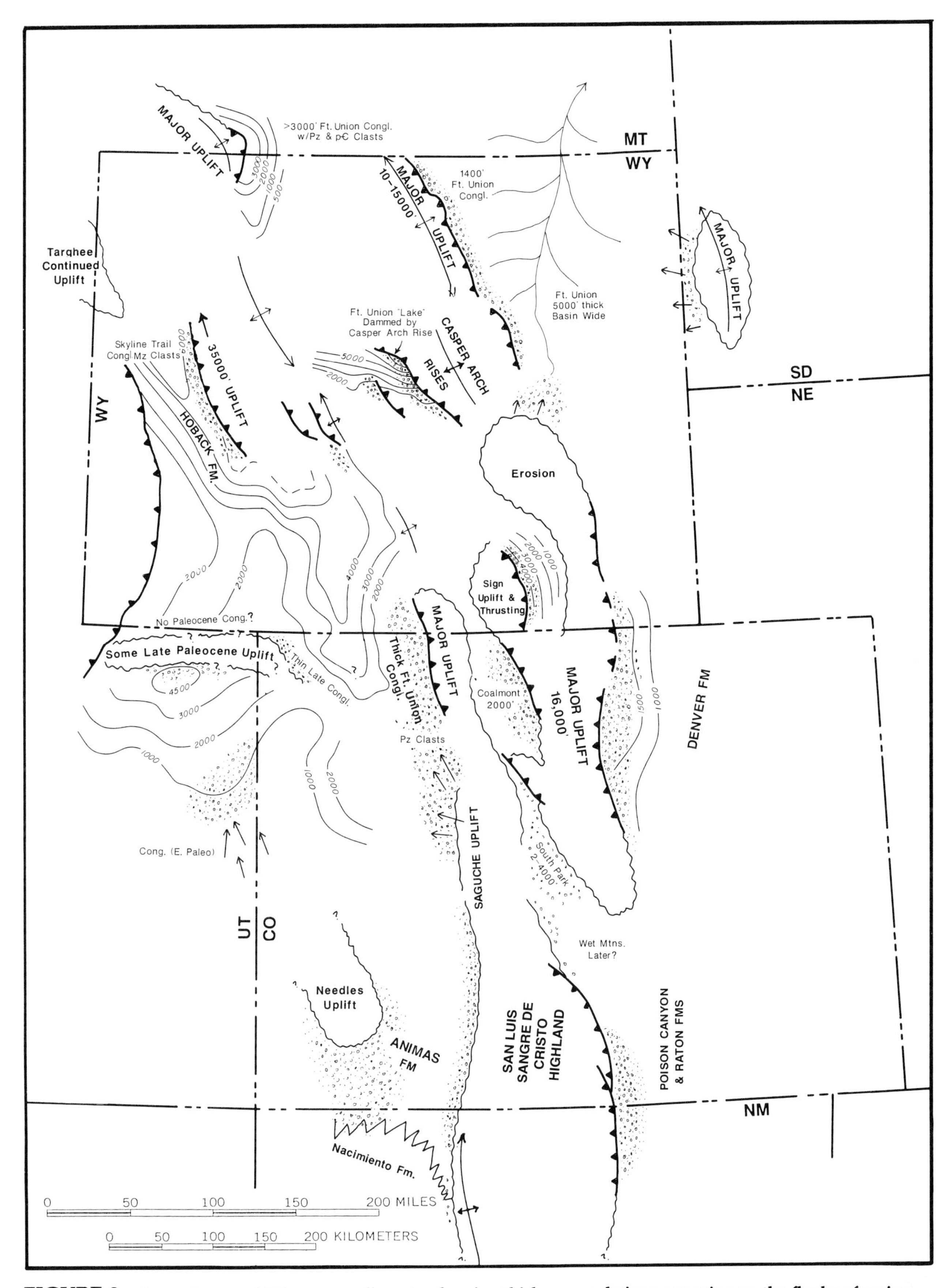

FIGURE 8. Isopach map of Paleocene sediments, showing thick accumulations occurring on the flanks of major uplifts. Large Paleocene vertical displacements occurred on the Sangre de Cristo, Front Range, Saguache, Medicine Bow, Park, Big Horn and Casper arch, Black Hills, Beartooth, and Wind River uplifts with associated thick conglomerates shed into the adjacent basins. Contour interval is 305 m (1000 ft).

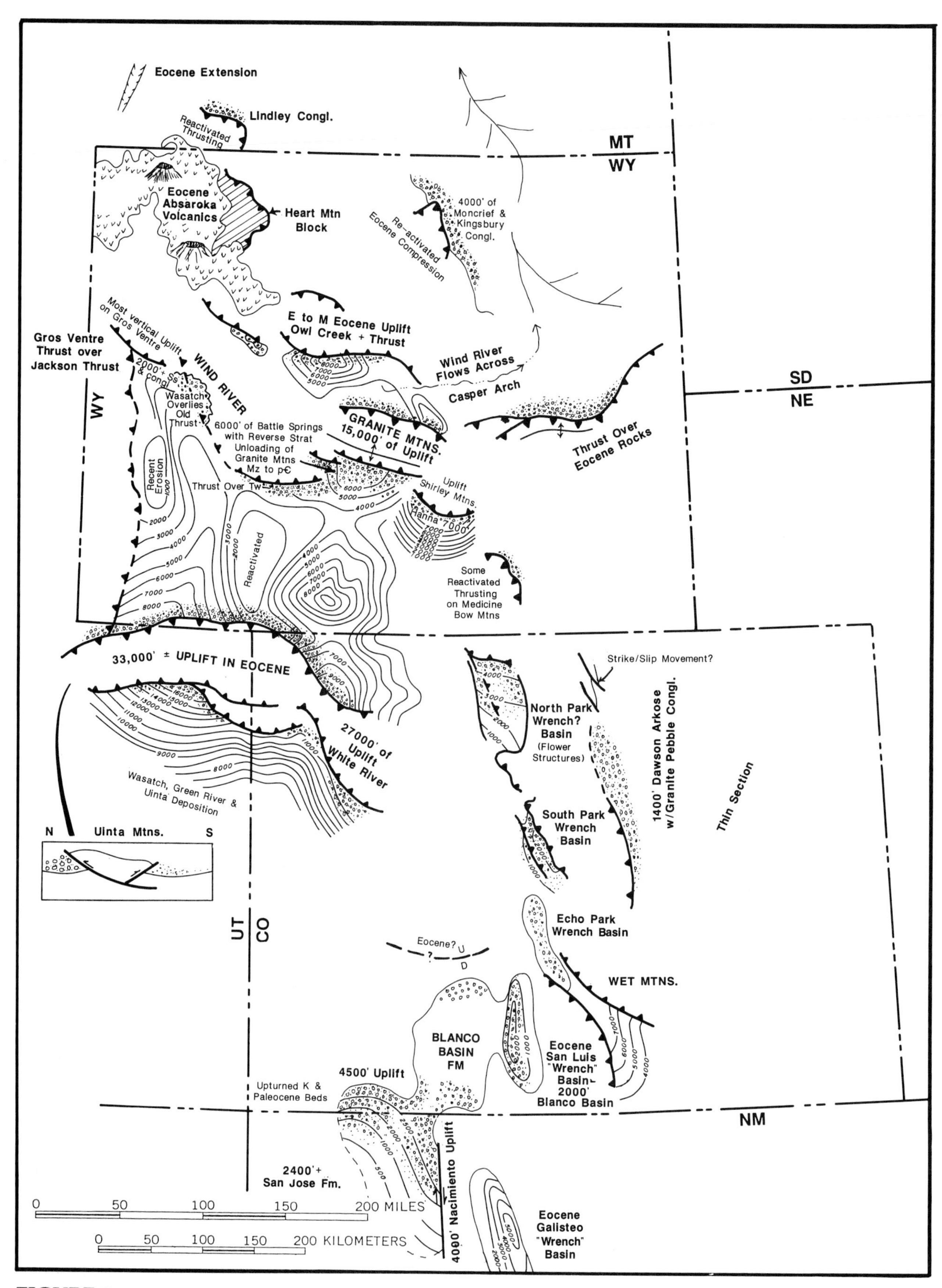

FIGURE 9. Isopach map of Eocene sediments, showing well-defined thicks in partitioned basins. Eocene compressive forces caused uplift of the Uinta–White River system (8200–10,000 m; 27,000–33,000 ft), Granite Mountains (4500 m; 15,000 ft), and Owl Creek Range (3000–4500 m; 9840–15,000 ft). Thrusts were reactivated on the north or south flanks of older uplifts and, simultaneously, wrench basins formed along the east side of the Colorado Plateau (North Park, South Park, Echo Park, San Luis, and Galisteo basins).

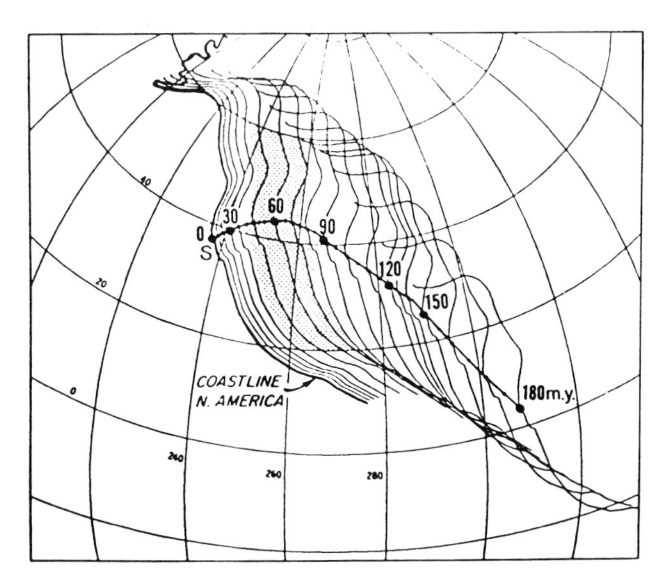

FIGURE 10. Map of the western coastline of North America through time (from Engebretson et al., 1984). A change in the direction of motion of the North American plate between 60 and 30 Ma is illustrated, consistent with the late Laramide change in compression direction.

Shallow basin platforms such as the Absaroka basin and the San Juan sag were not buried deeply enough to reach maturity during the Laramide, but became mature enough to generate hydrocarbons in Tertiary time following the deposition of several thousand feet of late Eocene or early Oligocene volcanics.

OIL AND GAS SYSTEMS

Play Types

Hydrocarbon distribution in the Rocky Mountain foreland can be considered in terms of the three major trap types: structural, stratigraphic, and combination (Figures 14 through 19). Reservoir lithologies also vary by tectonic history (Table 2), with carbonate strata confined largely to passive-margin Paleozoic reservoirs (Group I).

Structural closures and fault traps for the entire section were formed primarily by Laramide deformation and occur mostly on the margins of Laramide basins. Combination traps occur throughout all basins and are common as a result of the large number of low-relief structural noses and flank-position traps (Figure 14).

Economically significant stratigraphic traps occur mostly on gently dipping basin flanks and in basin centers. Most Rocky Mountain stratigraphic traps occur in reservoirs varying from 6 to 15 m (20 to 50 ft) thick, thus requiring large areal extents or significant column heights for large reserves. The absence of significant stratigraphic traps on the steeper basin margins is probably due to a general lack of effective lateral and bottom seals required to preserve long hydrocarbon columns where tectonically induced fractures and faults are present. An exception to this rule is the Cottonwood Creek field in the Big Horn basin of Wyoming, where tight dolomites, red beds, and anhydrites form effective seals trapping a hydrocarbon column in excess of 1800 m (6000 ft). Another exception is in areas where the steep flank has fractured tight reservoirs that contain significant oil reserves (Puerto Chiquito area—2.7×10^6 m^3, or 17 x 10^6 bbl, eastern San Juan Basin).

The largest stratigraphically trapped reserves occur in the breached 1.0×10^9 m^3 (6.5×10^9 BOE) tar deposit of the (Permian) White Rim sandstone in central Utah (Huntoon and Chan, 1987), also located on a gentle homoclinal dip.

Most of the hydrocarbons discovered in the Rocky Mountain foreland are contained in Group II foreland basin strata (Figures 20, 21). The repeated transgressions, regressions, and lowstand fluvial incision events have resulted in a rich variety of sandstone reservoir settings cast in the midst of a thick marine source-prone section. In addition, thick coal seams in upper Cretaceous strata have generated large volumes of gas, which help contribute to the large reserves shown in Figure 21.

Stratigraphic traps resulting from facies juxtapositions created by regional unconformities discussed earlier are similar to traps in related sequences in the Western Canada basin, which shared a similar early tectonic and stratigraphic history. The valley-fill traps involving the Viking Formation (Alberta) and Muddy Formation are examples (Dolson et al., 1991).

The Laramide partitioning of the Rocky Mountain foreland basin broke up many of the stratigraphic trends once shared jointly with the Western Canada basin. Sediments in intermontane lacustrine and fluvial packages (Group III) have accumulated only minor amounts of liquid hydrocarbons. Lacustrine oil shales accumulated in the deep basin centers, but absence of thermally mature source rocks is responsible for the poor production to date from Tertiary strata (Figure 21).

Thermally mature Tertiary strata are known in the Uinta and Wind River basins and production has been established. The Hanna, Piceance, and greater Green River basins are largely unproductive although their sediments are buried deeply enough to have generated hydrocarbons.

Discovery and Exploration History

The earliest visitors to the Rocky Mountain foreland were aware of the numerous seeps and oil-stained sandstone outcrops. Early medicinal use gave way to wagon-axle lubricants and, in 1862, just three years after the United States' first successful well in Pennsylvania, drilling began in the Rocky Mountains. Early drilling efforts were guided by surface seeps. As the anticlinal theory of oil accumulation gained credence, many of the obvious Laramide structures were drilled. Notable successes during this era were the Salt Creek and Elk Basin fields of Wyoming.

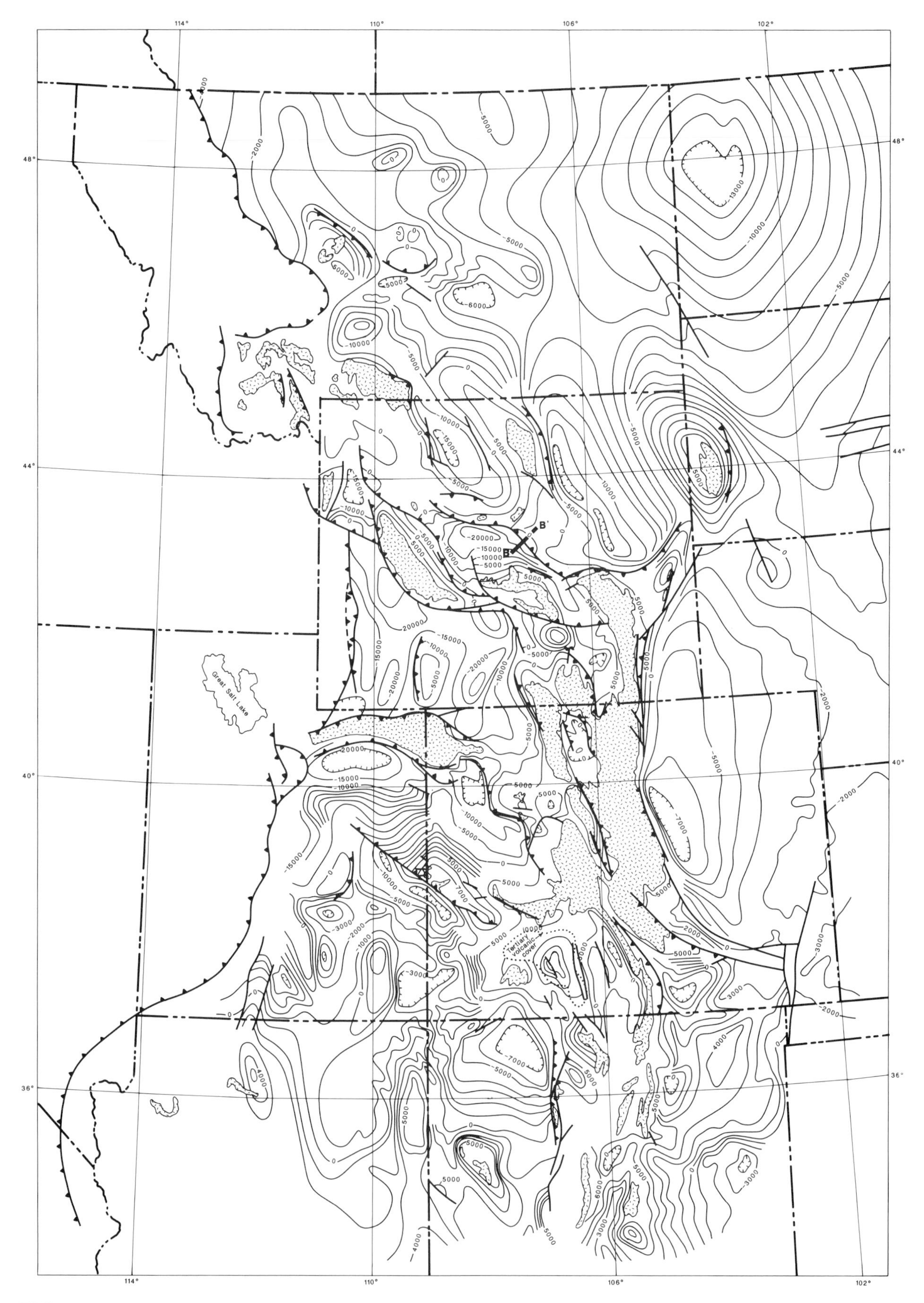

FIGURE 11. Structure contour map, top of Precambrian, showing the Rocky Mountain foreland structure today. Seismic line B–B′ is a representative area of basement-involved compressional style (see Figure 12).

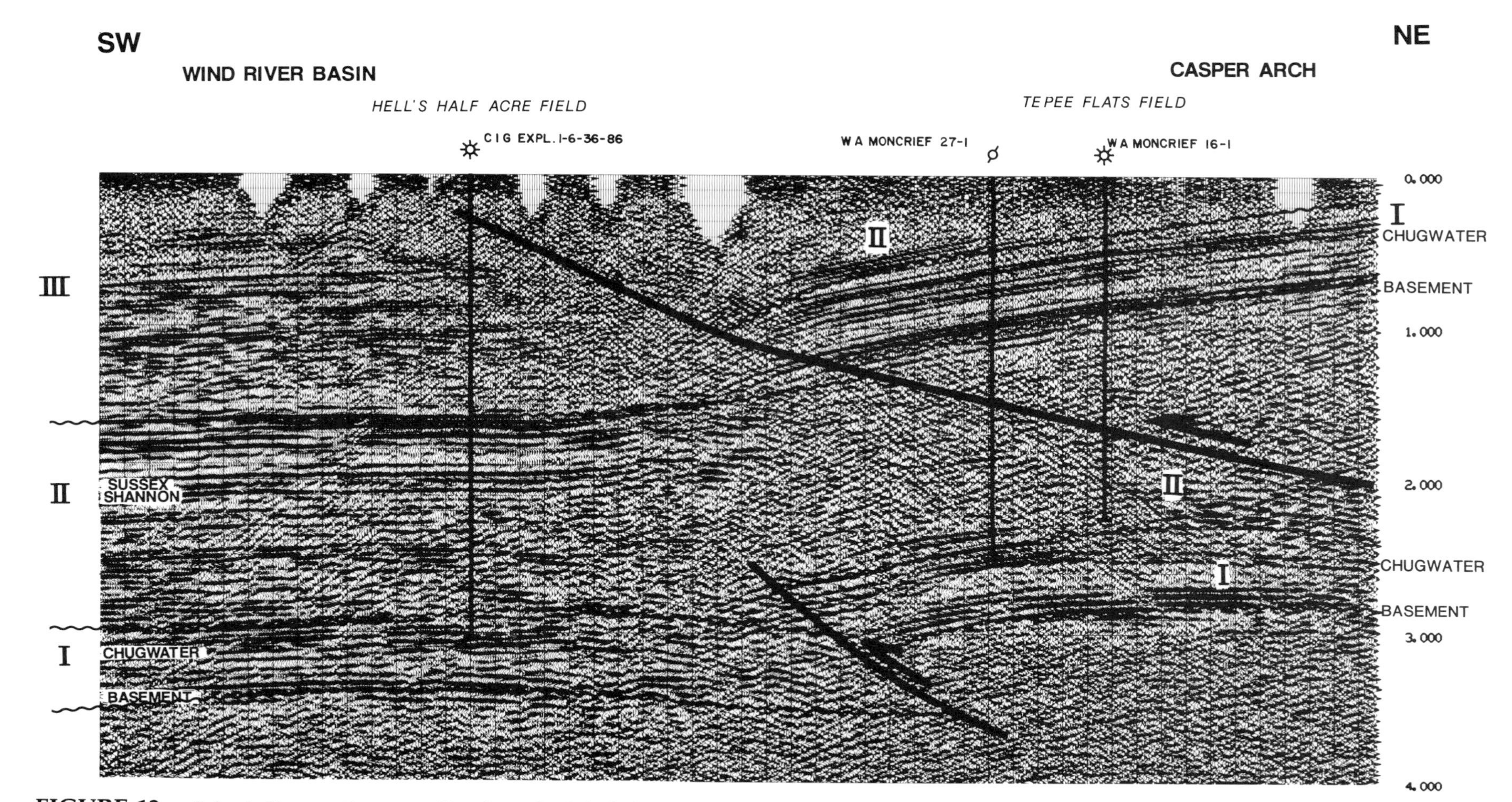

FIGURE 12. Seismic time section extending from the Wind River Basin on the southwest to the Casper arch on the northeast. The three sedimentary packages characteristic of the Rocky Mountain foreland are labeled: I, passive margin; II, foredeep; and III, Laramide. The apparent footwall structure shown is exaggerated by velocity pull-up beneath high-velocity Precambrian crystalline rocks (see Skeen and Ray, 1983).

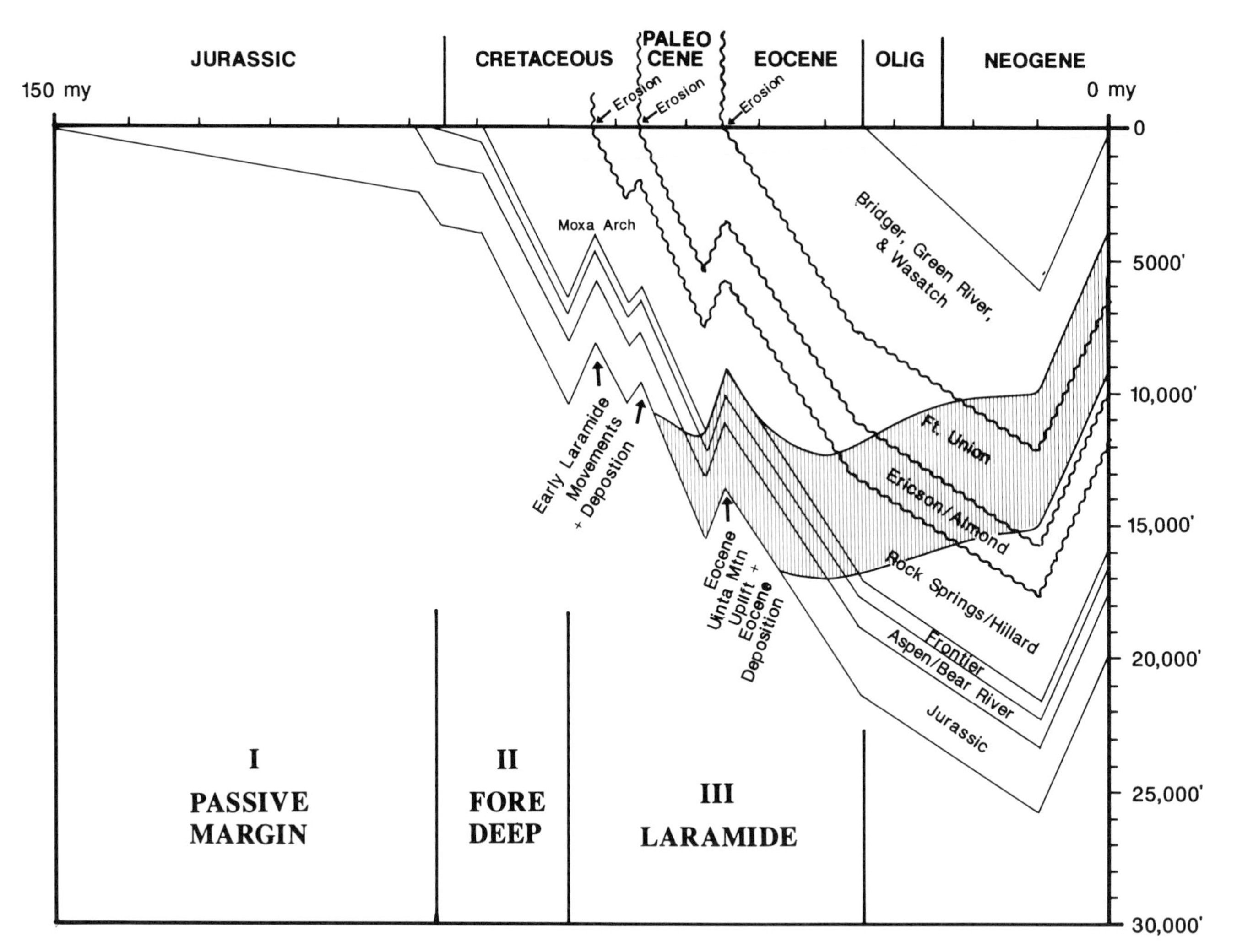

FIGURE 13. **Chart showing subsidence history through time for the Greater Green River basin, which we interpret as typical of other Rocky Mountain foreland basin subsidence histories. From 150 Ma to 80 Ma, the thin sedimentary layers are part of the Passive Margin package (I). From Late Jurassic/Early Cretaceous, the foredeep shows increased sedimentation (II). Finally, the Laramide Tertiary basin episode (III) involves burial to depths sufficient to move Upper Cretaceous source rocks into the oil window (shaded). The Green River basin model shows the relative insignificance of early Laramide episodes compared to the Eocene episodes. However, some basins, such as the San Juan sag and Absaroka basin, were so shallow that insignificant maturation occurred during the entire Laramide. Maturation was only reached after burial beneath late Eocene or Oligocene volcanic rocks.**

Developments in the oil industry found ready application in the foreland. Figures 22, 23, and 24 illustrate the succession of discoveries through time. A general acceleration of successful drilling is evident in the first quarter of the 20th century propelled by the advent of new drilling technology (rotary bits) and seismic techniques. A pause in activity is associated with the difficult economic times of the early 1930s, which was coupled with an excess of oil supply. The war effort of the 1940s saw greatly increased demand for oil, and the Rocky Mountain foreland supplied a significant portion of the United States' fuel needs during the war. Also in the 1940s, the area witnessed some of the earliest efforts at improved recovery through unitization and gas injection.

Continuing to the present, pipeline access has been a critical factor in the economics of the Rocky Mountain oil and gas fields, and there have been repeated pulses of exploration activity prompted by new pipeline construction.

In the past decades, exploration success has come from improved seismic imaging of complex structural traps not only in the Overthrust belt, but also along thrusted Precambrian-cored foreland structures. Increased resolution and modeling have been successfully applied to interpretation of subtle stratigraphic traps such as the

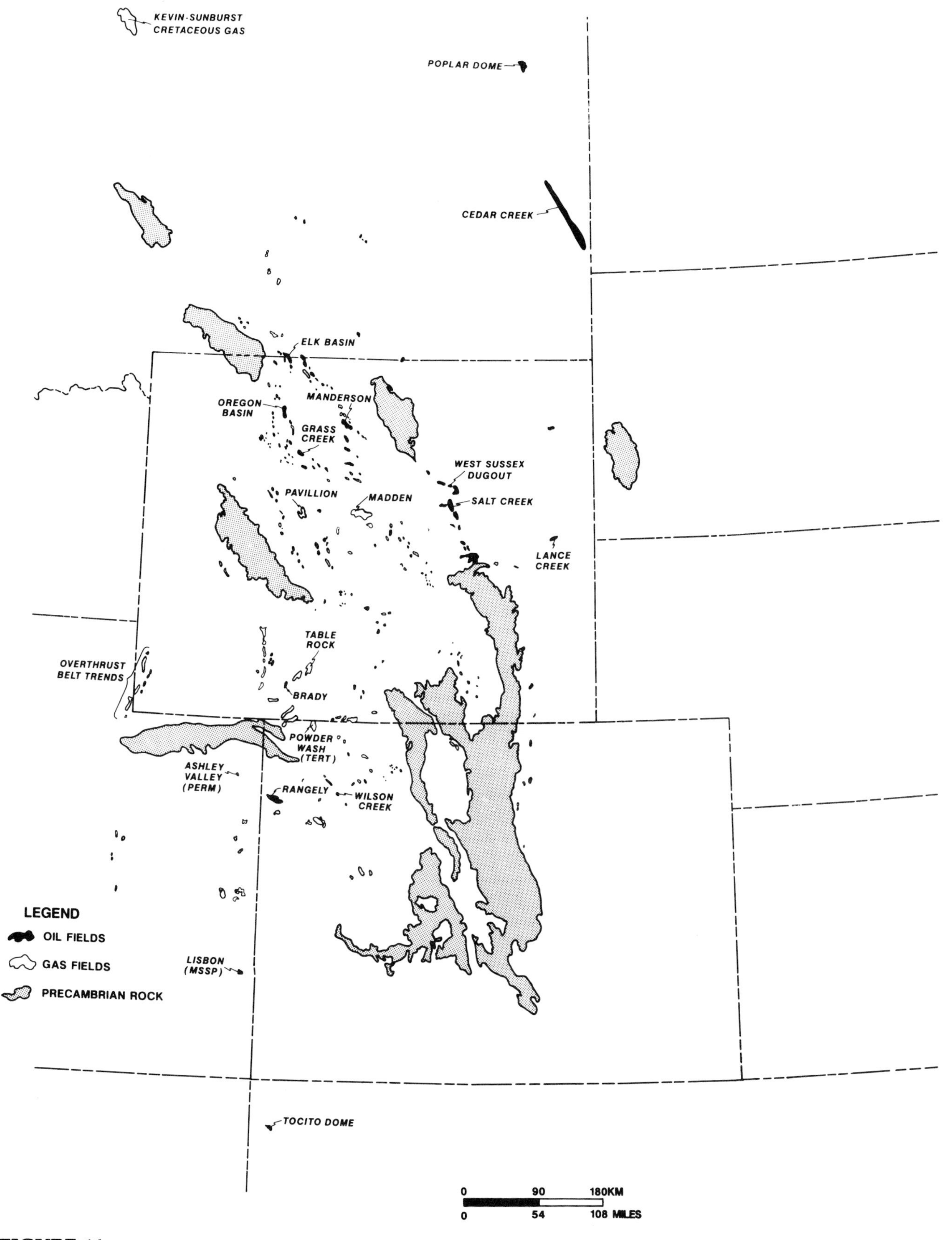

FIGURE 14. Map showing location of major structurally trapped hydrocarbons. Most structural traps are located along Tertiary basin margins, particularly in western Wyoming. See Figure 15 for a typical example. Structural traps are defined in this text as traps where the bulk of the hydrocarbons are confined by four-way or fault closure.

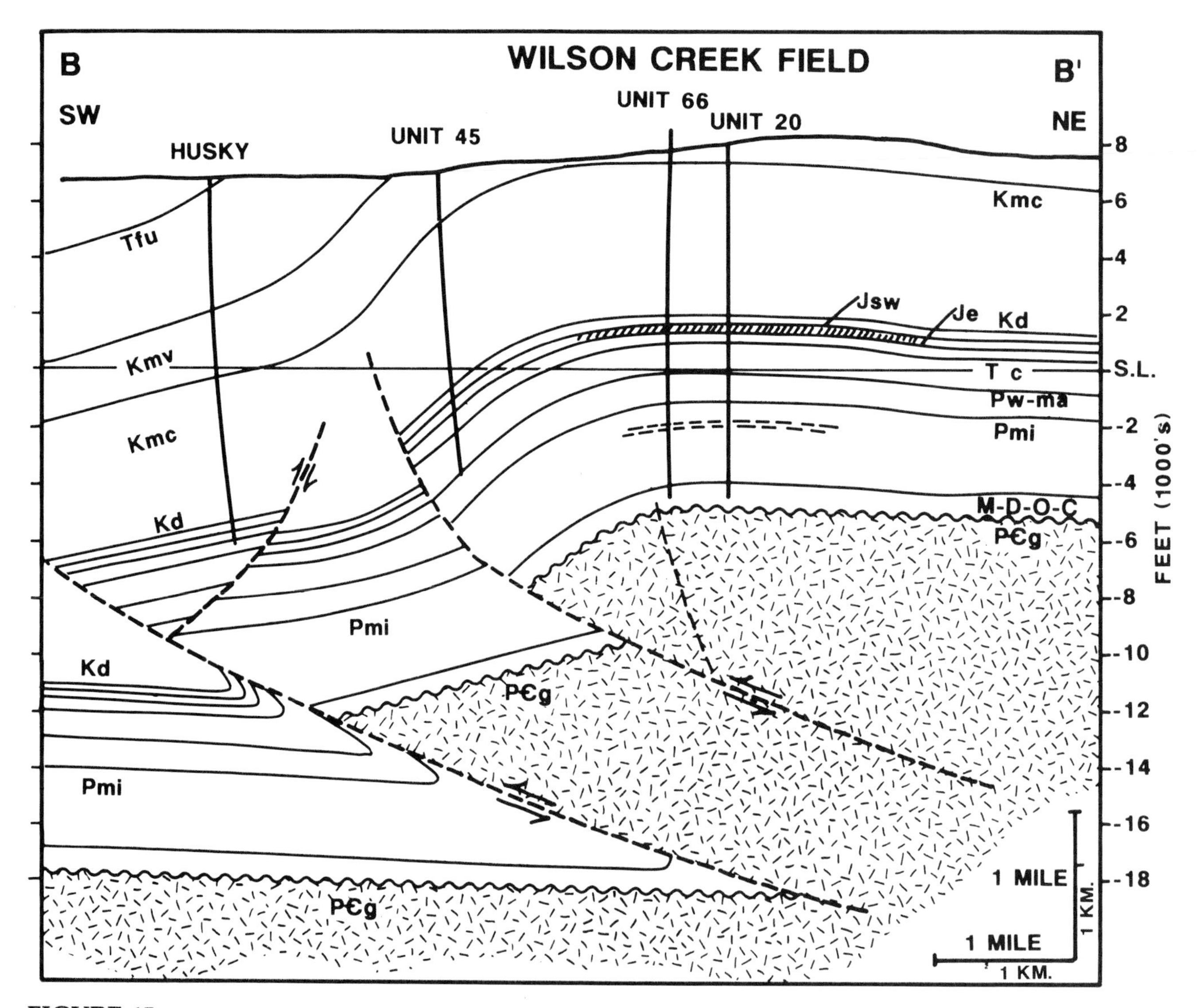

FIGURE 15. Typical example of a basement-involved foreland structural accumulation (modified from Stone, 1986).

incised valley systems of the Cretaceous Muddy and Pennsylvanian Morrow sandstones and Permian Minnelusa dunes and paleohills. Basin-centered gas accumulations and the coal-gas plays of the San Juan, Piceance, and Powder River basins are adding additional reserves.

Future Directions

Horizontal Drilling

The technological advances made with horizontal drilling are particularly applicable to the numerous Rocky Mountain foreland basins that have rich, mature source rocks highly fractured on their structural flanks (Schmoker, in press). Where these rocks are accompanied by thin, interbedded, "tight" sands, production potential will be greater. Some areas where horizontal drilling is being tested on flanks of Laramide structures include the Axial arch of Colorado, the steep east flank of the San Juan basin, the steep flank of the Denver basin, the Casper arch, and along the axis of the San Rafael swell.

Subvolcanic Plays

Laramide structures underlying the thick Eocene and Oligocene cover in the Rocky Mountain region have only recently been identified (Gries, 1989; Sundell, 1986). Seismic technological advances are primarily responsible for actualizing this potential, and initial drilling in both the San Juan and Absaroka volcanic fields has revealed the presence of reservoir rocks, Laramide structures, and mature source rocks.

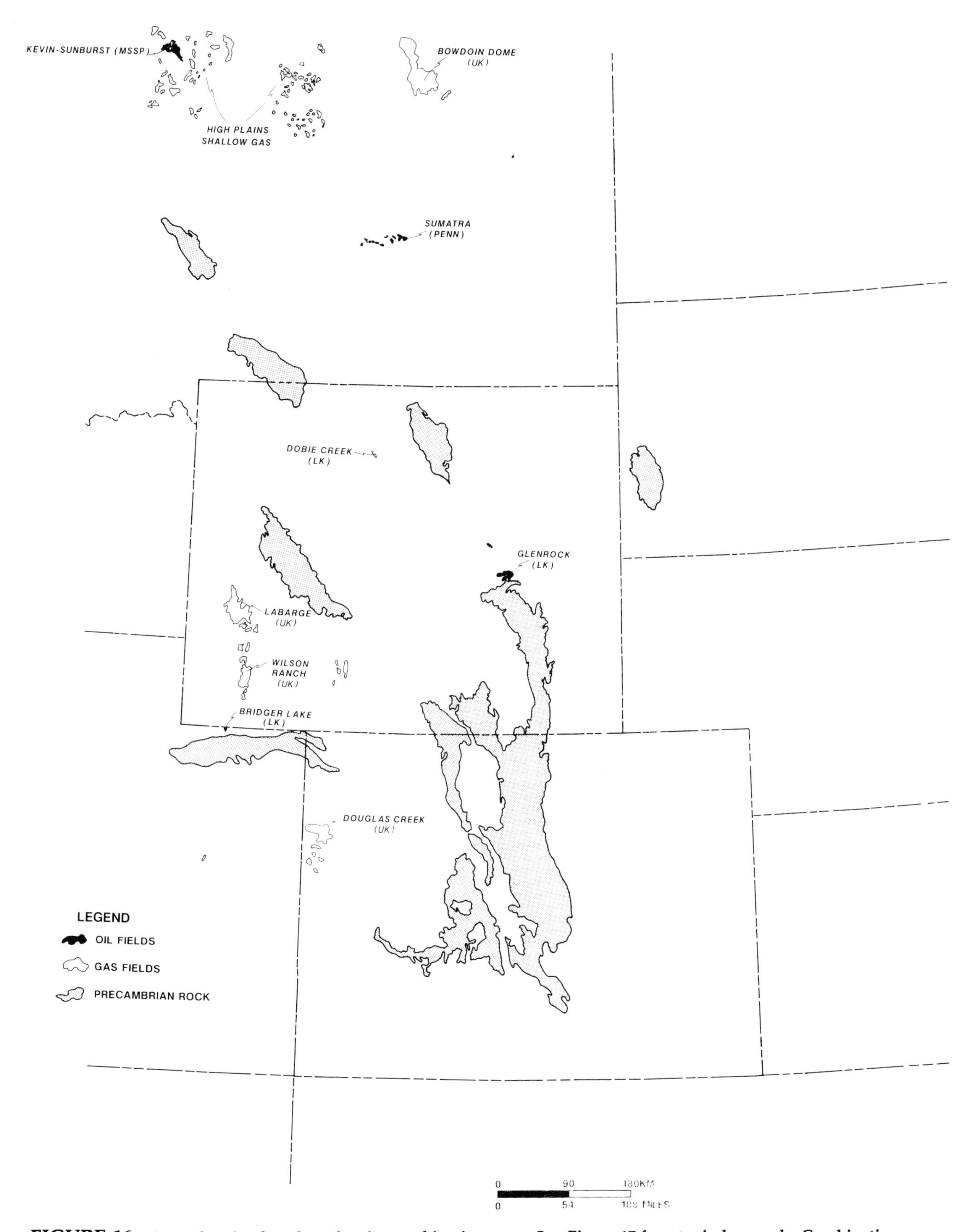

FIGURE 16. Map showing location of major combination traps. See Figure 17 for a typical example. Combination traps in this study are defined as stratigraphic traps in flank structural positions involving associated closures, fault seals or noses that help influence the hydrocarbon accumulation.

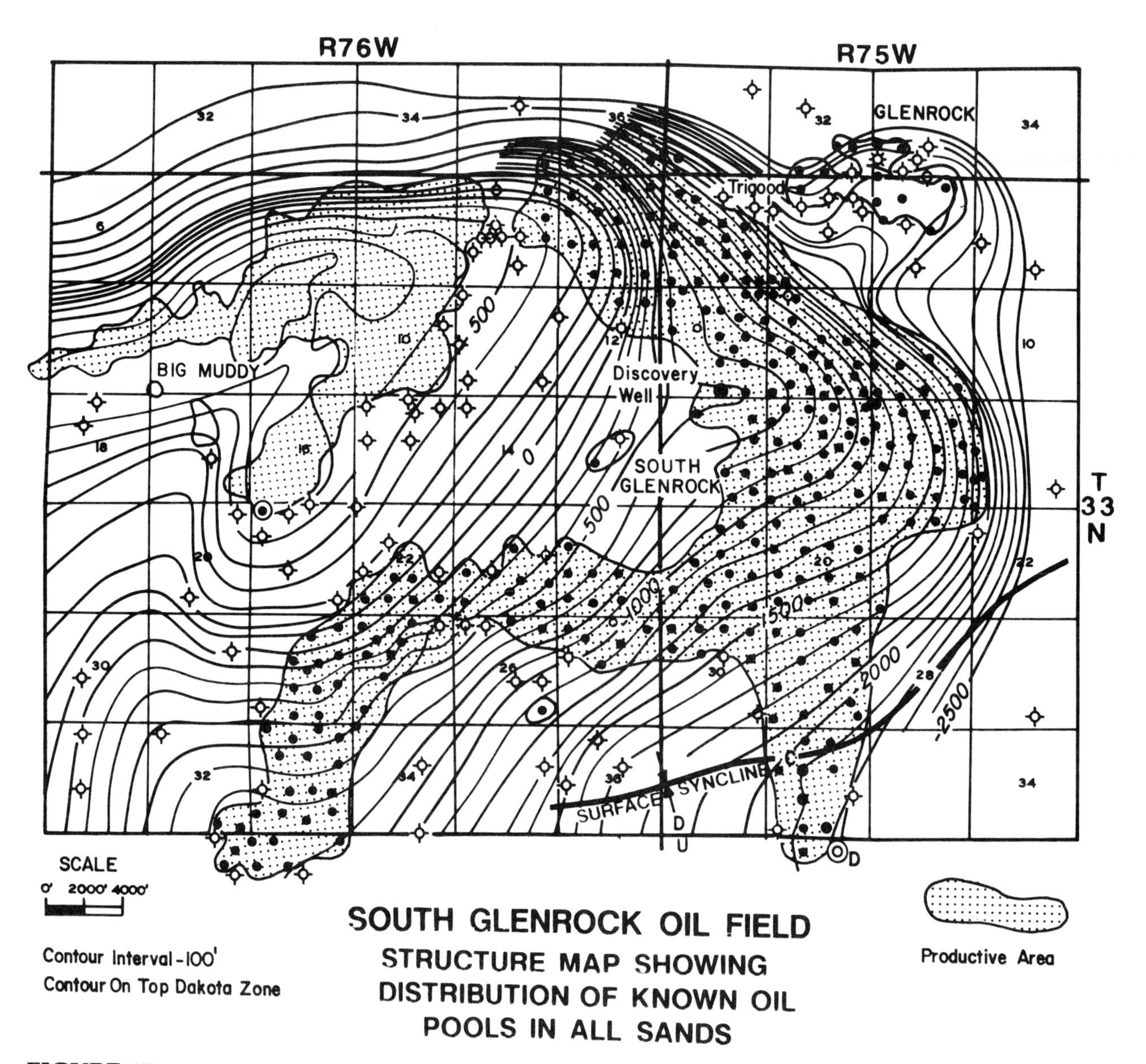

FIGURE 17. Typical example of a combination trap, Dakota Sandstone and Muddy Formation (Lower Cretaceous), South Glenrock oil field (from Curry and Curry, 1954).

Coalbed Methane

Although the upper Cretaceous coals of the San Juan basin have provided the most economic coal-bed play to date, increased gas prices and improved production technology are expected to expand this play many fold in several other Laramide basins (Schwochow et al., 1991). These will include both Upper Cretaceous coals and Paleocene coals in the Raton, Powder River, Hanna, Piceance, Wind River, Sand Wash, and many other basins.

Basin-Centered Gas

The deep gas-prone basins of western Wyoming, Utah, New Mexico, and Colorado will continue to yield significant new gas resources. This play is driven by developments in deep drilling and completion technology, because some targets are deeper than 6000 m (19,000 ft). Porosity and pore-throat modeling and prediction offer a new research frontier that could also enhance new exploration. Oil and gas reservoirs in excess of 1 darcy permeability and 18% porosity have recently been developed on the southern plunge of the Moxa arch at 4500 m (15,000 ft). Similar resources may be discovered in other deep basin settings.

Applications of Sequence Stratigraphy

New paleogeographic models are currently being developed for foreland strata based on a more thorough under-

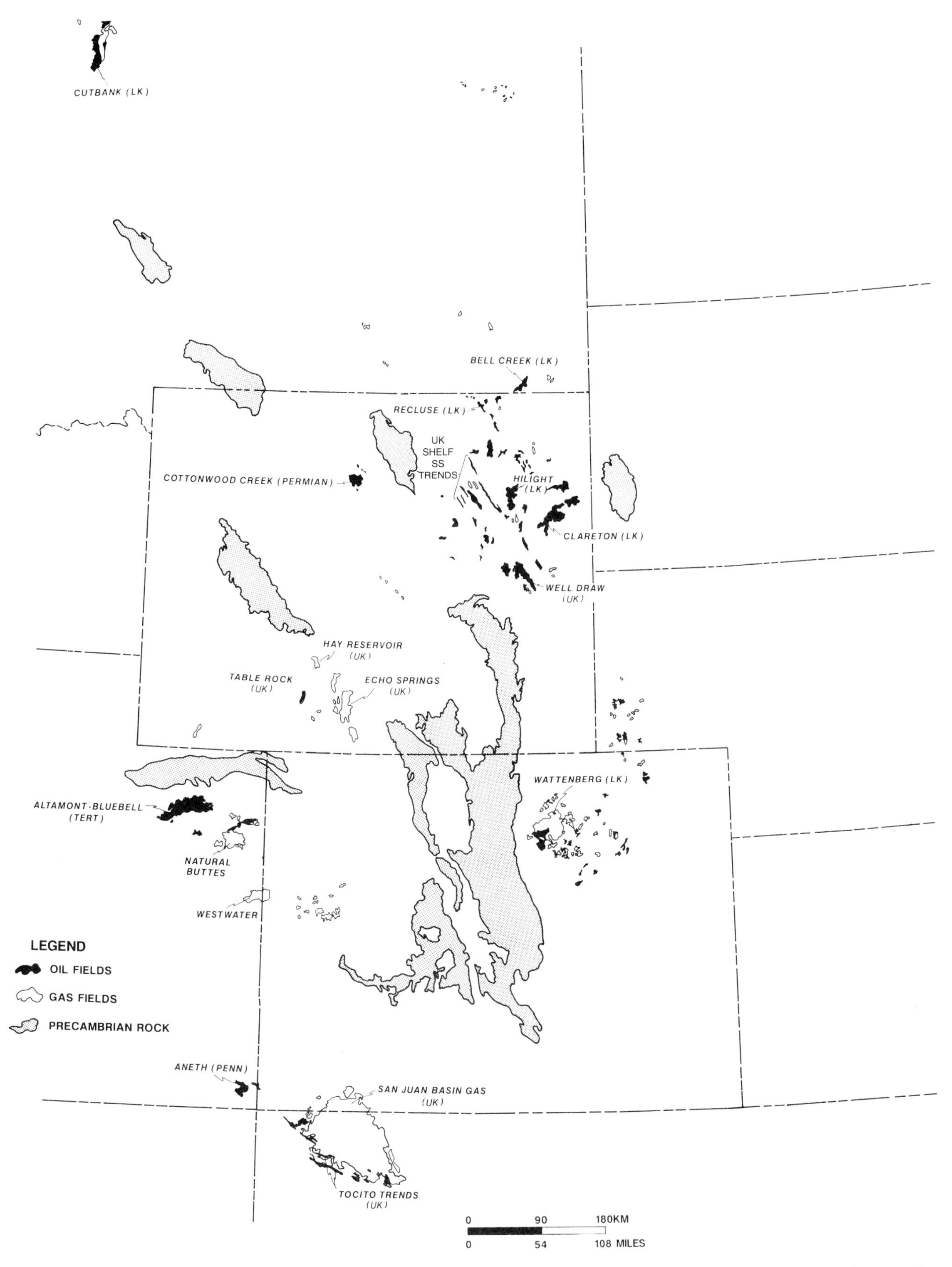

FIGURE 18. Map showing the locations of major stratigraphic traps. See Figure 19 for a typical example. Stratigraphically trapped hydrocarbons in this text are defined as traps where significant structural nosing, sealing faults, or closure is absent. The abundance of stratigraphic traps in areas of gentle homoclinal dip and within structurally simple basin centers results primarily from the presence of effective lateral and bottom seals capable of holding large column heights. See text for discussion.

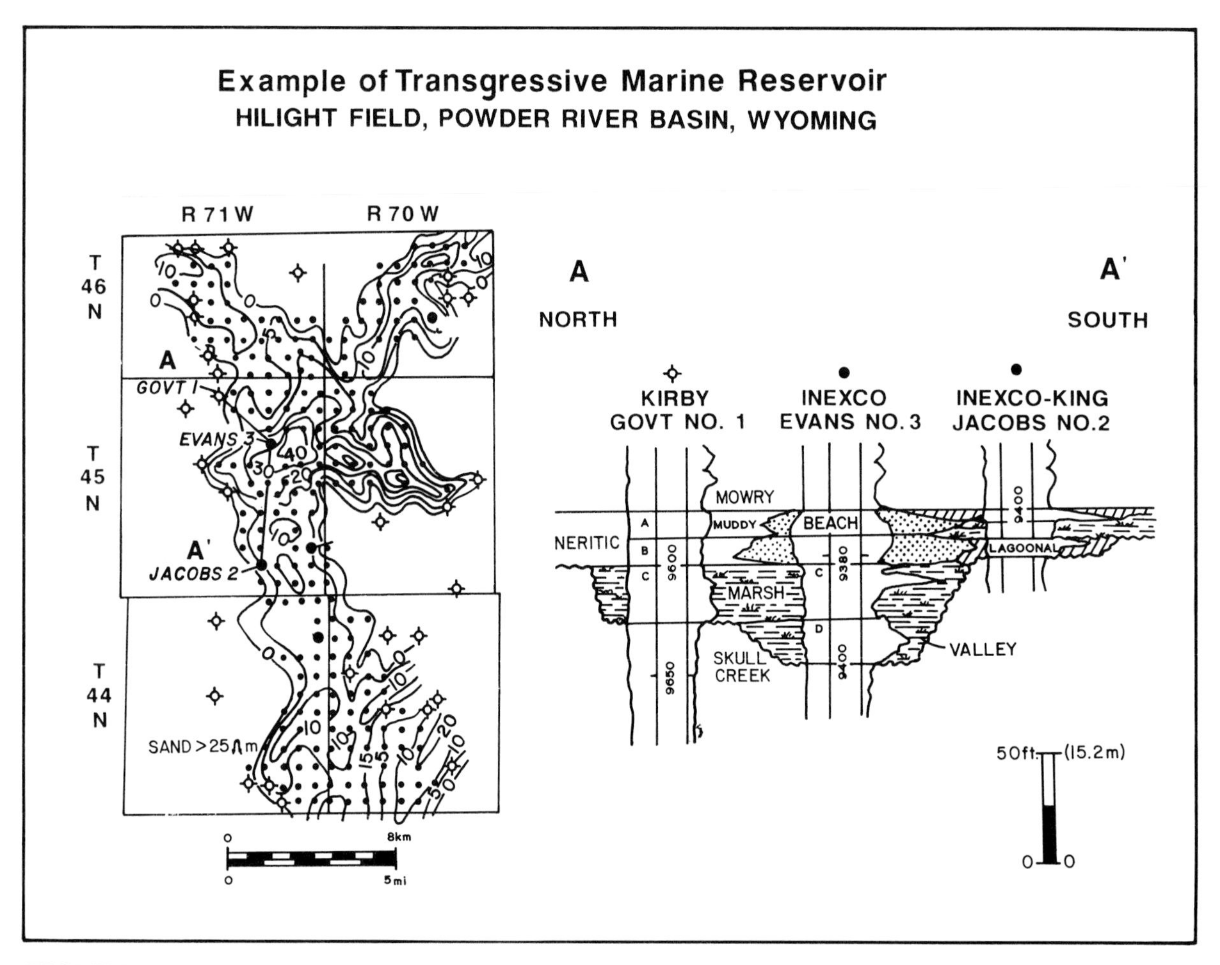

FIGURE 19. Isopach map and cross section of Hilight field, Powder River basin, Wyoming, an example of a stratigraphically trapped hydrocarbon accumulation (from Berg, 1976). Both valley-fill and spit-barrier production occur within this field, which is in an area of regional homoclinal dip. The term "transgressive marine" is used to describe reservoirs formed as progradational pulses during an overall relative sea level rise (Dolson et al., 1991). Contour interval (net) equals 3.3 m (10 ft) of sandstone with greater than 25 ohm-meters of resistivity.

standing of regional unconformities, particularly in the Cretaceous section. Ongoing research such as the Global Sedimentary Project (GSP) and the Western Interior Cretaceous (WIK) efforts will continue to shed new light on old interpretations. High-resolution application of sequence stratigraphic concepts should lead to improved secondary recovery in existing fields and should help locate new stratigraphically trapped reserves in other areas. Concurrent improvements in seismic imaging of these more subtle traps will also assist these efforts.

3-D Seismic and 2-D Seismic

Complex structural oil and gas fields are ideal targets for 3-D seismic exploration and exploitation. Many older fields in secondary recovery phases of development and injection of water or CO_2 have revealed surprising reservoir discontinuities. Much of this discontinuity is the result of fault complexities. Development geologists working in the overthrust belt of western Wyoming have successfully used 3-D seismic to enhance recovery, and we predict that these techniques will soon be successfully applied to mature foreland structural fields.

Stratigraphic targets in the Paradox basin algal mounds and the Powder River basin sand dunes (Minnelusa Formation) have also been successfully exploited with 3-D seismic technology.

Swath 2-D seismic has been used in the Rockies for years as an effective structural and stratigraphic geophysical tool that is cost effective for small company exploration and exploitation in areas with small- to medium-sized reserves.

TABLE 2. Ultimate (Known Recoverable) Hydrocarbon Reserves of Producing Horizons Grouped by Reservoir Age. See Figures 19 and 20 for production comparisons. Data courtesy of R. Nehring.

No. of Fields	Producing Formation	Dominant Lithology	Known Recoverable, × 10^6 BOE	Age
		Group I—Passive Margin Reservoirs		
1	Flathead	Ss	20.07	Cambrian
4	Big Horn	Carb	43.9	Ordovician
2	Ordovician-Silurian	Carb	8.81	
38	Red River	Carb	105.99	
4	Winnipegosis	Carb	5.09	
5	Interlake	Carb	6.78	Silurian
1	Silurian-Devonian	Carb	2.12	
3	Duperow	Carb	11.62	Devonian
2	Gunton	Carb	3.03	
10	Nisku	Carb	32.05	
3	Charles	Carb	54.25	Mississippian
2	Darwin	Ss	10.78	
3	Kibbey	Ss	10.28	
2	Lower Tyler	Ss	7.7	
32	Madison/Leadville	Carb	794.29	
5	Mission Canyon	Carb	392.07	
3	Mississippian	Carb	13.64	
1	Osagian	Carb	0.72	
3	Ratcliffe	Carb	36.8	
2	Spergen	Carb	4.28	
1	St. Louis	Carb	1.76	
5	Sun River	Carb	50.59	
13	Warsaw	Carb	477.85	
5	Amsden	Mixed Carb/Ss	14.56	Pennsylvanian
2	Barker Creek	Carb	30.11	
1	Cane Creek	Carb	1.13	
1	Cherokee	Carb	1.16	
1	Cutler	Ss	3.66	
5	Desert Creek	Carb	529.23	
1	Hermosa	Ss	0.72	
2	Hiawatha	Ss	45.66	
5	Ismay	Carb	17.97	
1	Keyes	Ss	1.73	
1	Lansing	Carb	2.2	
2	Leo	Ss	76.04	
1	Leo 1	Ss	6.84	

TABLE 2. (Continued)

No. of Fields	Producing Formation	Dominant Lithology	Known Recoverable, × 10^6 BOE	Age
2	Lower Ismay	Carb	17.66	
1	Marmaton	Carb	0.27	
1	Mississippian-Pennsylvanian	Mixed Carb/Ss	5.93	
1	Morgan	Ss	1.16	
17	Morrowan	Ss	77.53	
4	Paradox	Carb	39.35	
11	Paradox Salt	Shale	55.38	
62	Tensleep	Ss	1802.65	
7	Topeka	Carb	19.46	
15	Tyler	Ss	92.95	
2	Upper Ismay	Carb	3.52	
3	Casper	Ss	8.86	Permo-Penn.
1	Coconino	Ss	5.32	
7	Permo-Pennsylvanian	Mixed Carb/Ss	592.89	
8	Weber	Ss	1192.29	
1	Ervay	Carb	64.32	Permian
2	Kaibab	Carb	25.77	
4	Lyons	Ss	24.05	
82	Minnelusa	Ss	394.6	
36	Phosphoria	Carb	513.47	
25	Upper Minnelusa	Ss	76.34	
1	Chugwater	Ss	9.3	Triassic
4	Crow Mountain	Ss	67.36	
1	Jelm	Ss	1.83	
1	Shinarump	Ss	1	
		Group II—Foreland Reservoirs		
13	Entrada	Ss	64.09	Jurassic
14	Morrison	Ss	85.61	
25	Nugget	Ss	1176	
4	Sawtooth	Ss	20.35	
1	Stump	Carb	2.25	
12	Sundance	Ss	128.55	
4	Swift	Ss	11.86	
5	Twin Creek	Carb	37.54	
1	Bear River	Ss	93.2	
3	Bow Island	Ss	7.81	
4	Cat Creek	Ss	23.45	

TABLE 2. (Continued)

No. of Fields	Producing Formation	Dominant Lithology	Known Recoverable, × 10^6 BOE	Age
2	Cedar Mountain	Ss	4.03	
7	Cloverly	Ss	34.81	Lower Cretaceous
3	Cut Bank	Ss	5.45	
258	Dakota	Ss	3345.76 (Includes some U. Cret.)	
3	Greybull	Ss	10.51	
9	Lakota	Ss	49.87	
1	Carb	Ss	39.74	
114	Muddy	Ss	1021.73	
2	Sunburst	Ss	1.53	
11	Almond	Ss	362.6	
2	Big Elk	Ss	8	
1	Blackleaf	Ss	3.21	
4	Castlegate	Ss	4.38	
4	Chacra	Ss	64.31	
2	Cliff House	Ss	6.33	
5	Codell	Ss	92.61	
3	Cody	Ss	69.67	
1	Colorado	Ss	97.5	
2	Cozzette	Ss	2.96	
9	Eagle	Ss	87.75	
2	Ferron	Ss	25.07	
1	Fox Hills	Ss	2.25	
58	Frontier	Ss	1176.77	
7	Fruitland	Ss	39.8	
2	Fruitland Coal	Ss	209.02	
18	Gallup	Ss	153.43	
1	Hilliard	Ss	2.08	
7	Lance	Ss	39.47	
13	Lewis	Ss	127.09	
3	Lower Gallup	Ss	36.06	Upper Cretaceous
1	Lower Mesaverde	Ss	1.2	
10	Mancos	Ss	238.21	
32	Mesaverde	Ss	3454.71	
1	Montana	Ss	6.3	
1	Morapos	Ss	0.52	
2	Mowry	Ss	2.05	
15	Niobrara	Ss	108.97	
11	Parkman	Ss	33.18	
3	Peay	Ss	1.87	
12	Pictured Cliffs	Ss	1018.56	
2	Pierre	Shale	17.9	

TABLE 2. (Continued)

No. of Fields	Producing Formation	Dominant Lithology	Known Recoverable, × 10⁶ BOE	Age
2	Smoky Hill	Carb	15.5	
1	Steele	Ss	7.98	
20	Susex/Shannon	Ss	464.09	
3	Teapot	Ss	26.2	
2	Teck A	Ss	9.71	
4	Tocito	Ss	17.47	
1	Torchlight	Ss	0.71	
9	Turner	Ss	34.6	
5	Wall Creek	Ss	774.45	
		Group III—Reservoirs		
5	Almy	Ss	111.5	
1	Green River	Shale	0.22	
1	Douglas Creek	Ss	9.75	
14	Fort Union	Ss	138.45	
12	Green River	Shale	263.69	Tertiary
1	Lower Green River	Ss	31.75	
1	Uinta	Ss	2.89	
1	Wind River	Ss	16.5	
		Total All Groups	23502.17	

CONCLUSIONS

The Rocky Mountain foreland has had a unique and complex history that involved a shift from a largely Paleozoic passive margin setting to a Cretaceous foreland basin setting that was followed by a deep intermontane basin phase.

The Western Canada foreland basin is similar tectonically and stratigraphically to the Rocky Mountain foreland in its shared passive margin and foreland basin stages. The Laramide partitioning of the Rocky Mountain foreland distinguishes this area from most other foreland basin settings, particularly in terms of the extreme structural relief that developed.

A complex history of flat-plate subduction and shifting regional tectonic overprinting in Tertiary time has created numerous basins and hydrocarbon traps. Eocene intermontane basin deposition brought most of the stratigraphic section into or through the oil window along basin axes, and completely into the dry gas window in the deepest basins. Basin-flank settings and areas such as the Colorado Plateau remain largely immature.

This late phase of deformation and associated subsidence thermally matured most source strata, creating most of the 3.7×10^9 m³ (23×10^9 BOE) of resources discovered to date. Because of its complex stratigraphic and structural evolution, the Rocky Mountain foreland is an unusually fertile environment for the generation and entrapment of hydrocarbons in many different trap types. Successful exploration continues today, 130 years after the first productive drilling.

ACKNOWLEDGMENTS

This manuscript benefited from review by J. S. Bell, Steven Cather, Chuck Chapin, and Dale Leckie. Mike Ambrosia, Shari Foos, and Ann Priestman drafted the illustrations. Historical production data was supplied by R. Nehring.

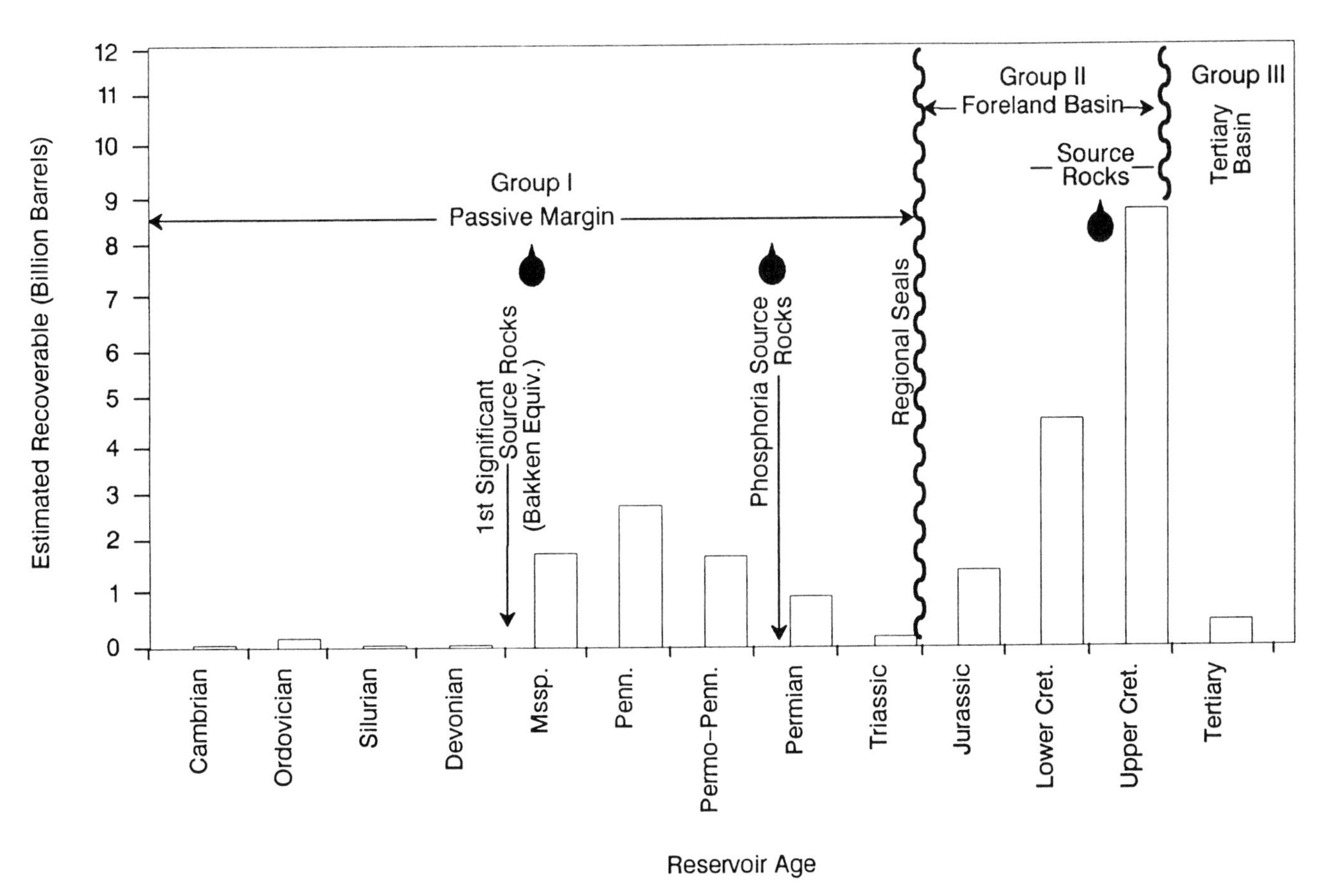

FIGURE 20. **Histogram showing estimated ultimate recovery by age of reservoir. Group I passive margin strata are sourced primarily from Pennsylvanian and Permian source rocks, accounting for most of the production in Mississippian through Triassic strata, primarily on structural closures. Triassic production is low because of a lack of source rock in a system that contains primarily sealing strata (Figure 4). Group II foreland basin strata are the largest producers, owing primarily to thick Cretaceous shales and coal that have generated large reserves. The small amount of Tertiary production is primarily a result of limited areal extent of thermally mature source rock. Data courtesy of R. Nehring.**

Dolson and Raynolds thank Amoco Production Co. for permission to publish this manuscript.

REFERENCES CITED

Anders, D. E., and P. M. Gerrild, 1984, Hydrocarbon generation in lacustrine rocks of Tertiary age, Uinta basin, Utah—organic carbon, pyrolysis yield, and light hydrocarbons, *in* J. Woodward, F. F. Meissner, and J. L. Clayton, eds., Hydrocarbon source rocks of the Greater Rocky Mountain Region: Rocky Mountain Association of Geologists, p. 513–556.

Ayers, W. B., Jr., 1986, Lacustrine and fluvial-deltaic depositional systems, Fort Union Formation (Paleocene), Powder River basin, Wyoming and Montana: AAPG Bulletin, v. 70, p. 1651–1673.

Beer, J. A., and T. E. Jordan, 1989, The effects of Neogene thrusting on deposition in the Bermejo basin: Journal of Sedimentary Petrology, v. 59, p. 330–345.

Berg, R. R., 1976, Hilight Muddy field—Lower Cretaceous transgressive deposits of the Powder River basin, Wyoming: The Mountain Geologist, v. 13, n. 2, p. 33–45.

Bergh, S. G., and A. W. Snoke, 1992, Polyphase Laramide deformation in the Shirley Mountains, south central Wyoming foreland: The Mountain Geologist, v. 29, n. 3, p. 85–100.

Blackstone, D. L., 1988, Thrust faulting: southern margin Powder River basin, Wyoming: Wyoming Geological Association Guidebook, 39th Annual Field Conference, Eastern Powder River Basin—Black Hills, p. 35–44.

Bowen, T. W., 1982, Geology, paleontology, and correlation of Eocene volcaniclastic rocks, southeast Absaroka Range, Hot Springs County, Wyoming: USGS Professional Paper 1201-A, 75 p.

Bradley, W. H., 1964, Geology of Green River Formation and associated Eocene rocks in southwestern Wyoming and adjacent parts of Colorado and Utah: USGS Professional Paper 496-A, 85 p.

Brister, B. S., 1990, Tertiary sedimentation and tectonics:

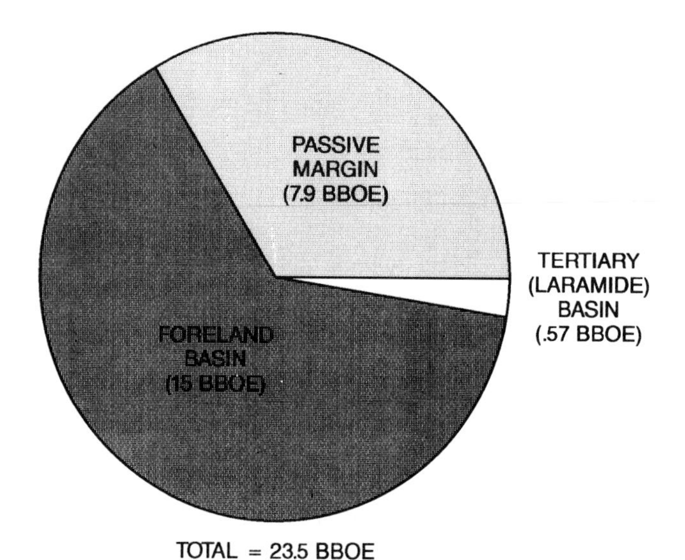

FIGURE 21. Pie diagram showing estimated ultimate recovery by reservoir package (I, II, or III). Foreland basin clastic wedges contain the bulk of the hydrocarbons produced, primarily owing to the high gas volumes generated in the deep basins of western Wyoming, Utah, and Colorado. Data courtesy of R. Nehring.

San Juan sag–San Luis basin region, Colorado and New Mexico: unpublished Ph.D. dissertation, New Mexico Institute of Mining and Technology, 267 p.

Brown, R. G., and L. T. Shannon, 1989, Madden field, *in* D. F. Cardinal, ed., Wyoming Oil and Gas Fields Symposium, Bighorn and Wind River Basins, p. 293–295.

Chapin, C. E., and S. M. Cather, 1981, Eocene tectonics and sedimentation in the Colorado Plateau—Rocky Mountain area, *in* W. R. Dickinson and M. D. Payne, eds., Relations of tectonics to ore deposits in the southern Cordillera: Arizona Geological Digest, v. 14, p. 173–198.

Chapin, C. E., and S. M. Cather, 1983, Eocene tectonics and sedimentation in the Colorado Plateau—Rocky Mountain area, *in* J. D. Lowell, ed., Rocky Mountain foreland basins and uplifts: Rocky Mountain Association of Geologists, p. 9–32.

Cross, T. A., 1986, Tectonic controls of foreland basin subsidence and Laramide style deformation, western United States, *in* P. A. Allen and P. Homewood, eds., Foreland basins: International Association of Sedimentologists Special Publication 8, p. 15–39.

Curry, W. H., Jr., and W. H. Curry III, 1954, South Glenrock, a Wyoming stratigraphic oil field: AAPG Bulletin, v. 38, p. 2119–2156.

DeVoto, R. H., 1972, Cenozoic geologic history of South Park: Mountain Geologist, v. 9, p. 211–221.

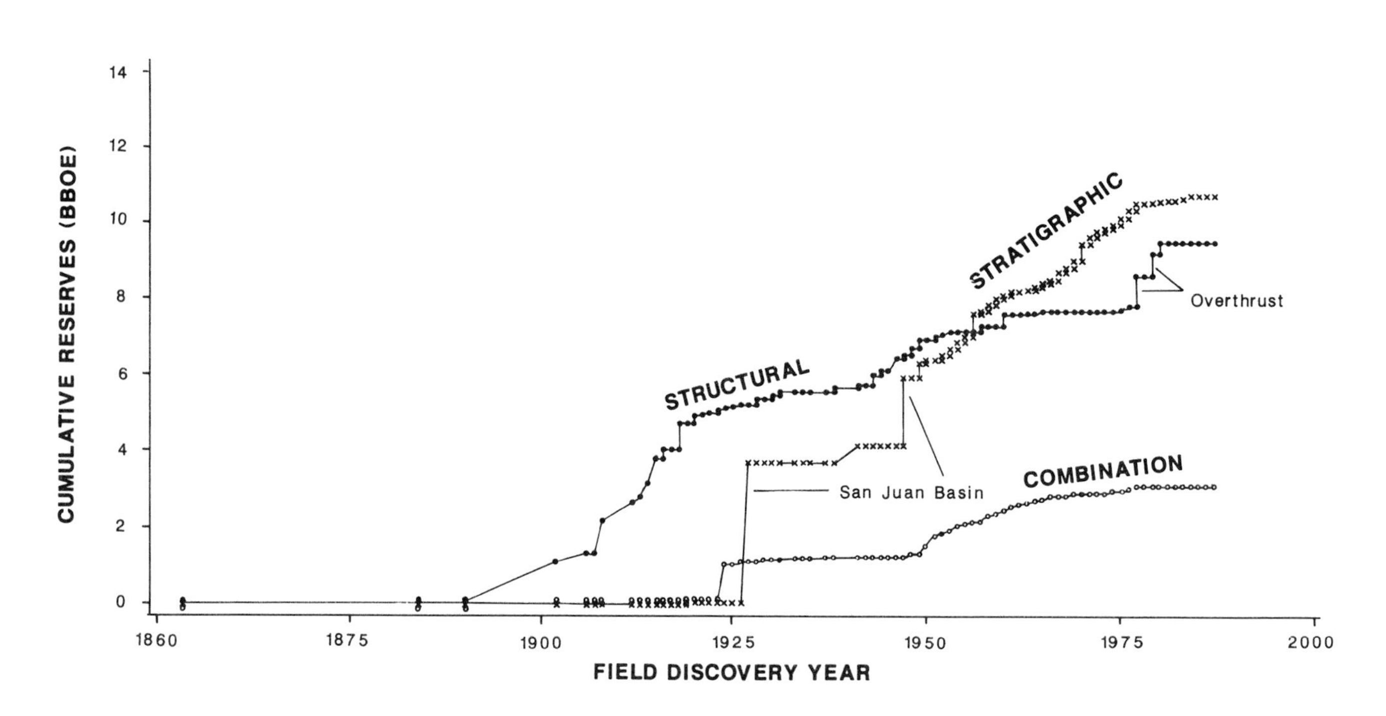

FIGURE 22. Cumulative discovery plot of reserve additions through time by trap type. Basin-centered gas reserves in the San Juan basin contain the most significant stratigraphically trapped hydrocarbon volumes. Data modified from and courtesy of R. Nehring.

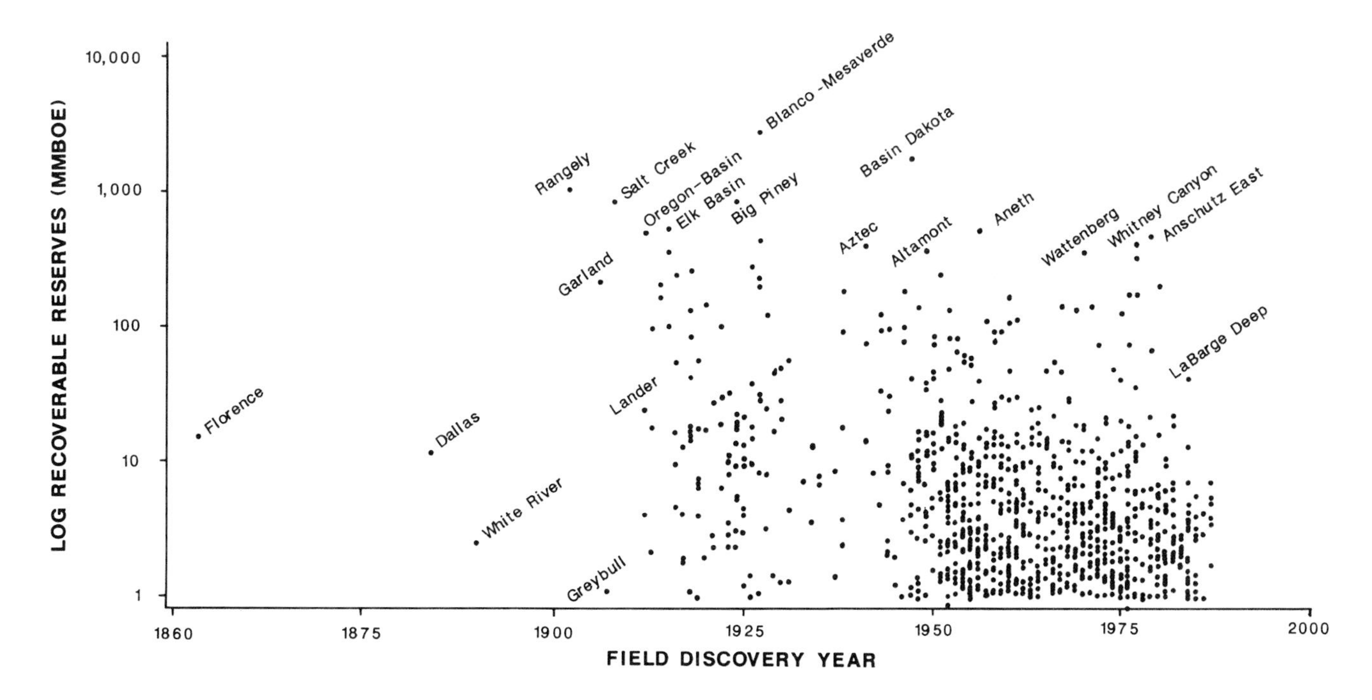

FIGURE 23. Plot of historical field discoveries and recoverable reserves per field vs. time. Note the lack of significant field size decrease through time. This is a result of new technology and deep gas exploitation (see text for explanation). Data courtesy of R. Nehring.

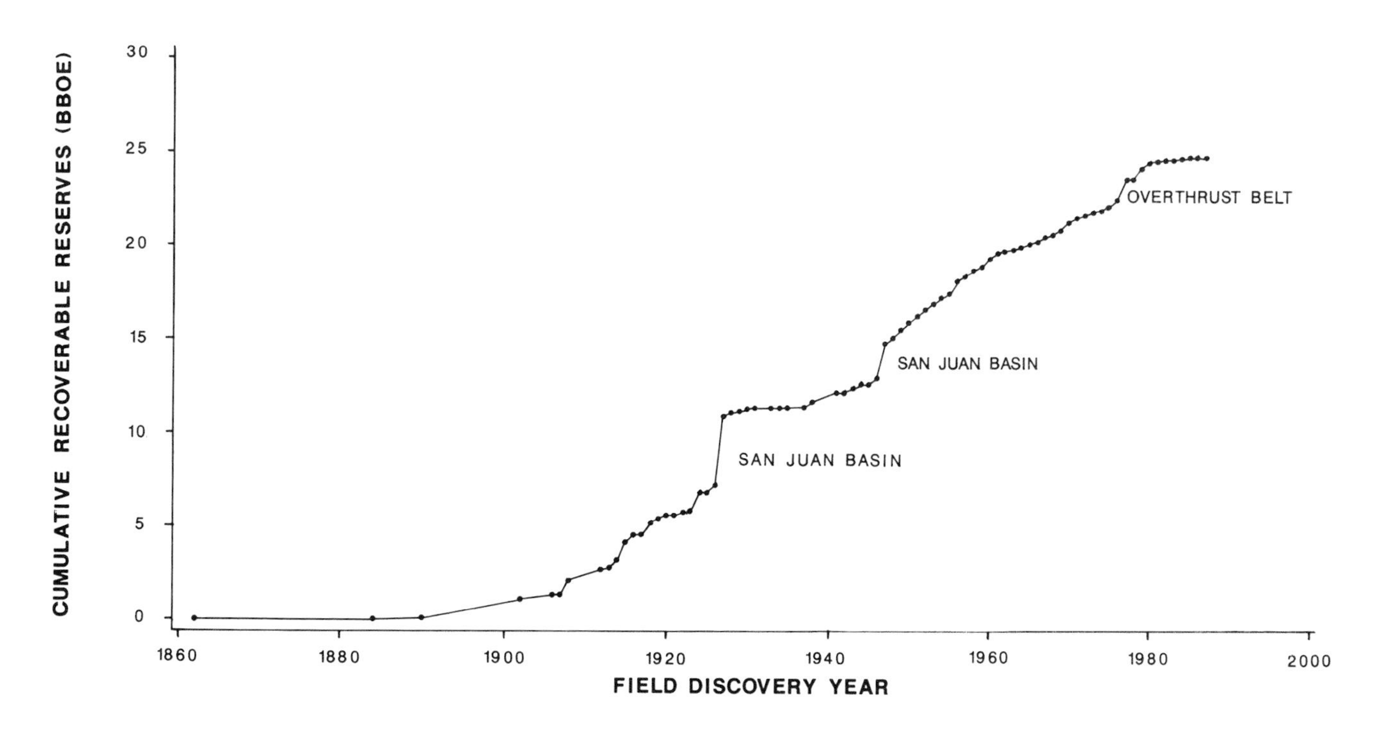

FIGURE 24. Historical discovery plot of cumulative oil and gas field discoveries by year. Presently, reserve additions are being made by developing technology and deep gas exploitation. See text for explanation. Data courtesy of R. Nehring.

Dickinson, W. R., and W. S. Snyder, 1978, Plate tectonics of the Laramide orogeny, *in* V. Mathews III, ed., Laramide folding associated with basement block faulting in the western United States: GSA Memoir 151, p. 355-366.

Dickinson, W. R., M. A. Klute, M. J. Hayes, S. U. Janecke, E. R. Lundin, M. A. McKittrick, and M. D. Olivares, 1988, Paleogeographic and paleotectonic setting of Laramide sedimentary basins in the central Rocky Mountain region: GSA Bulletin, v. 100, p. 1023-1039.

Dolson, J. C., D. S. Muller, M. J. Evetts, and J. A. Stein, 1991, Regional paleotopographic trends and production, Muddy Sandstone (Lower Cretaceous), central and northern Rocky Mountains: AAPG Bulletin, v. 75, n. 3, p. 409-435.

Dutcher, L. A. F., J. L. Jobling, and R. R. Dutcher, 1986, Stratigraphy, sedimentology, and structural geology of Laramide synorogenic sediments marginal to the Beartooth Mountains, Montana and Wyoming, *in* P. B. Garrison, Geology of the Beartooth uplift and adjacent basins: Montana Geological Society and Yellowstone Bighorn Research Association Joint Field Conference, p. 33-52.

Engebretson, D. C., A. Cox, and G. A. Thompson, 1984, Correlation of plate motions with continental tectonics: Laramide to basin and range: Tectonics, v. 3, p. 115-119.

Fassett, J. E., 1985, Early Tertiary paleogeography and paleotectonics of the San Juan basin area, New Mexico and Colorado, *in* R. M. Flores and S. S. Kaplan, eds., Cenozoic paleogeography of west-central United States: SEPM, p. 317-334.

Gingerich, P. D., 1983, Paleocene-Eocene faunal zones and a preliminary analysis of Laramide structural deformation in the Clark's Fork basin, Wyoming: Wyoming Geological Association Guidebook, 34th Annual Field Conference, Geology of the Big Horn Basin, p. 185-196.

Gries, R., 1983, North-south compression of Rocky Mountain foreland structures, *in* J. D. Lowell, ed., Rocky Mountain foreland basins and uplifts: Rocky Mountain Association of Geologists, p. 9-32.

Gries, R., 1989, San Juan sag: oil and gas exploration in a newly discovered basin beneath the San Juan volcanic field, *in* J. C. Lorenz and S. G. Lucas, eds., Energy frontiers in the Rockies: Albuquerque Geological Society, New Mexico, p. 69-78.

Gries, R., 1990, Rocky Mountain foreland structures: changes in compression direction through time, *in* J. Letouzey, ed., Petroleum and tectonics in mobile belts: IFP Exploration and Production Research Conferences, Editions Technip, p. 129-147.

Grilley, N., 1979, Hay reservoir, *in* D. F. Cardinal and W. W. Stewart, eds., Wyoming Oil and Gas Fields Symposium, Greater Green River Basin, p. 182-183.

Hail, W. J., Jr., 1965, Geology of northwestern North Park, Colorado: USGS Bulletin 1188, 133 p.

Heller, P. L., and C. Paola, 1989, The paradox of Lower Cretaceous gravels and the initiation of thrusting in the Sevier orogenic belt, United States Western Interior: GSA Bulletin, v. 101, p. 864-875.

Huntoon, J. E., and M. Z. Chan, 1987, Marine origin of paleotopographic relief on eolian White Rim Sandstone (Permian), Elaterite basin, Utah: AAPG Bulletin, v. 71, p. 1035-1045.

Irwin, C. D., 1986, Upper Cretaceous and Tertiary cross sections, Moffat County, Colorado, *in* D. S. Stone, ed., New interpretations of northwest Colorado geology: Rocky Mountain Association of Geologists, p. 151-156.

Jervey, M., 1992, Siliciclastic sequence development in foreland basins, with examples from the Western Canada foreland basin, this volume.

Johnson, R. C., and T. M. Finn, 1986, Cretaceous through Holocene history of the Douglas Creek arch, Colorado and Utah, *in* D. S. Stone, ed., New interpretations of northwest Colorado geology: Rocky Mountain Association of Geologists, p. 77-96.

Jordan, T. E., and R. W. Allmendinger, 1986, The Sierras Pampeanas of Argentina: a modern analogue of Rocky Mountain deformation: American Journal of Science, v. 286, p. 737-764.

Jordan, T. E., B. L. Isaacs, R. W. Allmendinger, J. A. Brewer, V. A. Ramos, and C. J. Ando, 1983, Andean tectonics related to geometry of subducted Nazca plate: GSA Bulletin, v. 94, p. 341-361.

Kauffman, E. G., 1977, Upper Cretaceous cyclothems, biotas and environments, Rock Canyon anticline, Pueblo, Colorado: The Mountain Geologist, v. 14, p. 129-152.

Keefer, W. R., 1965, Stratigraphy and geologic history of the uppermost Cretaceous, Paleocene and lower Eocene rocks in the Wind River basin, Wyoming: USGS Professional Paper 495-A, 77 p.

Kluth, C. F., and P. J. Coney, 1981, Plate tectonics of the Ancestral Rocky Mountains: Geology, v. 9, p. 10-15.

Kraus, M. J., 1985, Early Tertiary quartzite conglomerates of the Big Horn basin and their significance for paleogeographic reconstruction of northwest Wyoming, *in* R. M. Flores and S. S. Kaplan, eds., Cenozoic paleogeography of west-central United States: SEPM, p. 71-92.

Lamerson, P. R., 1982, The Fossil basin area and its relationship to the Absaroka thrust fault systems, *in* R. B. Powers, ed., Geologic studies of the Cordilleran thrust belt: Rocky Mountain Association of Geologists, v. II, p. 296-340.

Leckie, D. A., and D. G. Smith, 1992, Regional setting, evolution, and depositional cycles of the Western Canada foreland basin, this volume.

Livaccari, R. F., K. Burke, and A. M. C. Sengor, 1981, Was the Laramide orogeny related to subduction of an oceanic plateau?: Nature, v. 289, p. 276-278.

Love, J. D., 1970, Cenozoic geology of the Granite Mountains area, central Wyoming: USGS Professional Paper 495-C, 154 p.

Lowell, J. D., 1974, Plate tectonics and foreland basement deformation: Geology, v. 2, p. 275-278.

McCubbin, D. B., 1982, Barrier-island and strand-plain facies, *in* P. A. Scholle and D. Spearing, eds., Sandstone depositional environments: AAPG Memoir 31, p. 247-279.

McNeil, D. H., 1984, The eastern facies of the Cretaceous system in the Canadian Western Interior, *in* F. Scott and D. J. Glass, eds., The Mesozoic of Middle North America: Canadian Society of Petroleum Geologists Memoir 9, p. 145-171.

Miller, F. X., 1977, Biostratigraphic correlation of the Mesaverde Group in southwestern Wyoming and northwestern Colorado, *in* H. K. Veal, ed., Exploration frontiers of the central and southern Rockies: Rocky Mountain Association of Geologists, p. 117-138.

Mueller, C., and T. Brown, 1989, Pavillion field, *in* D. F. Cardinal, ed., Wyoming Oil and Gas Fields Symposium,

Bighorn and Wind River basins, p. 356-358.

Pipiringos, G. N., 1955, Tertiary rocks in the central part of the Great Divide basin, Sweetwater County, Wyoming: Wyoming Geological Association Guidebook, 10th Annual Field Conference, p. 100-104.

Pipiringos, G. N., and R. B. O'Sullivan, 1978, Principal unconformities in Triassic and Jurassic rocks, Western Interior United States—a preliminary survey: USGS Professional Paper 1035A.

Pitman, W. C., and M. Talwani, 1972, Sea-floor spreading in the North Atlantic: GSA Bulletin, v. 83, p. 619-646.

Reynolds, M. W., 1976, Influence of recurrent Laramide structural growth on sedimentation and petroleum accumulation, Lost Soldier area, Wyoming: AAPG Bulletin, v. 60, p. 12-33.

Roehler, H. W., 1961, The Late Cretaceous-Tertiary boundary in the Rock Springs uplift, Sweetwater County, Wyoming: Wyoming Geological Association Guidebook and Symposium on Late Cretaceous rocks of Wyoming, p. 96-100.

Roehler, H. W., and P. L. Martin, 1987, Geological investigations of the Vermillion Creek Coal Bed in the Eocene Niland Tongue of the Wasatch Formation, Sweetwater County, Wyoming: USGS Professional Paper 1314-L, 202 p.

Schmoker, J. W., in press, Geological studies relevant to horizontal drilling: examples from western North America: Rocky Mountain Association of Geologists 1992 Symposium Volume.

Schwochow, S. D., D. K. Murray, and M. F. Fahy, 1991, Coalbed methane of western North America: Rocky Mountain Association of Geologists Guidebook, 336 p.

Skeen, R. C., and R. R. Ray, 1983, Seismic models and interpretation of the Casper arch thrust: Application to the Rocky Mountain foreland structure, *in* J. D. Lowell, ed., Rocky Mountain foreland basins and uplifts: Rocky Mountain Association of Geologists, p. 99-124.

Sklenar, S. E., and D. W. Anderson, 1985, Origin and early evolution of an Eocene lake system within the Washakie basin of southwestern Wyoming, *in* R. M. Flores and S. S. Kaplan, eds., Cenozoic paleogeography of west-central United States: SEPM, p. 231-246.

Smith, L. N., S. G. Lucas, and W. E. Elston, 1985, Paleocene stratigraphy, sedimentation and volcanism of New Mexico, *in* R. M. Flores and S. S. Kaplan, eds., Cenozoic paleogeography of west-central United States: SEPM, p. 293-316.

Soister, P. E., 1978, Stratigraphy of latest Cretaceous and early Tertiary rocks of the Denver basin, *in* J. D. Pruitt and P. E. Coffin, Energy resources of the Denver basin: Rocky Mountain Association of Geologists, p. 223-230.

Soister, P. E., and R. H. Tschudy, 1978, Eocene rocks in the Denver basin, *in* J. D. Pruitt and P. E. Coffin, Energy resources of the Denver basin: Rocky Mountain Association of Geologists, p. 231-236.

Spearing, D. R., 1976, Upper Cretaceous Shannon Sandstone: an offshore shallow marine sand body: Wyoming Geological Association 28th Annual Guidebook, p. 65-72.

Stone, D. S., 1969, Wrench faulting and Rocky Mountain tectonics: The Mountain Geologist, v. 6, p. 67-79.

Stone, D. S., 1986, Geology of the Wilson Creek field, Rio Blanco County, Colorado, *in* D. S. Stone, ed., New interpretations of northwest Colorado geology: Rocky Mountain Association of Geologists, p. 229-246.

Stone, D. S., in press, Basement-involved thrust generated folds as seismically imaged in the subsurface of the Central Rocky Mountain foreland, *in* C. J. Schmidt, E. A. Erslev, and R. B. Chase, eds., Laramide basement deformation in the Rocky Mountain foreland of the western United States: GSA Special Paper.

Sundell, K. A., 1986, Petroleum exploration in the Absaroka basin of northwestern Wyoming (Abs.): AAPG Bulletin, v. 70, p. 1058.

Surdam, R. C., and K. O. Stanley, 1980, The stratigraphic and sedimentologic framework of the Green River Formation, Wyoming, *in* Stratigraphy of Wyoming: Wyoming Geological Association Guidebook, p. 205-221.

Surdam, R. C., and C. A. Wolfbauer, 1975, Green River Formation, Wyoming: a playa-lake complex: GSA Bulletin, v. 86, p. 335-345.

Torres, V., and P. D. Gingerich, 1983, Summary of Eocene stratigraphy at the base of Jim Mountain, North Fork of the Shoshone River, northwestern Wyoming, *in* W. W. Boberg, ed., Geology of the Bighorn basin: Wyoming Geological Association, p. 205-208.

Vail, P. R., R. M. Mitchum, and S. Thompson III, 1977, Seismic stratigraphy and global changes of sea level, part 3: relative changes of sea level from coastal onlap, *in* C. W. Payton, ed., Seismic stratigraphy applications to hydrocarbon exploration: AAPG Memoir 36, p. 129-144.

Wach, P. H., 1977, The Moxa arch, an overthrust model?: Wyoming Geological Association Guidebook, 29th Annual Field Conference, p. 651-654.

Weimer, R. J., 1966, Time-stratigraphic analysis and petroleum accumulations, Patrick Draw field, Sweetwater County, Wyoming: AAPG Bulletin, v. 50, p. 2150-2175.

Weimer, R. J., 1983, Relation of unconformities, tectonics and sea level changes, Cretaceous of the Denver basin and adjacent areas, *in* J. W. Reynolds and E. D. Dolly, eds., Mesozoic paleogeography of west-central United States: SEPM, Rocky Mountain Section, Rocky Mountain Paleogeography Symposium 2, p. 359-376.

Weimer, R. J., C. A. Rebne, and T. L. Davis, 1988, Geologic and seismic models, Muddy Sandstone, Lower Cretaceous, Bell Creek-Rocky Point area, Powder River basin, Montana and Wyoming, *in* R. P. Diedrich, M. A. Dyka, and W. R. Miller, eds., Wyoming Geologic Association 39th Annual Field Conference Guidebook, Eastern Powder River Basin-Black Hills, p. 161-177.

Weimer, R. J., S. A. Sonnenberg, and G. B. Young, 1986, Wattenberg field, Denver basin, Colorado, *in* C. W. Spencer and R. F. Mast, eds., Geology of tight gas reservoirs: AAPG Studies in Geology 24, p. 143-164.

Woodward, J., F. F. Meissner, and J. L. Clayton, eds., 1984, Hydrocarbon source rocks of the greater Rocky Mountain region: Rocky Mountain Association of Geologists, 557 p.

Chapter 15

Ouachita Foredeep Basins: Regional Paleogeography and Habitat of Hydrocarbons

Lawrence D. Meckel, Jr.

L. D. Meckel and Company
Houston, Texas, U.S.A.

David G. Smith

Canadian Hunter Exploration Limited
Calgary, Alberta, Canada

Leon A. Wells

Exploration Consultant
Houston, Texas, U.S.A.

INTRODUCTION

Seven present-day structure basins occur along the leading edge of the Ouachita thrust belt, an east-west Paleozoic overthrust trend in the southern United States that extends 2250 km (1400 mi) from Alabama to west Texas, where it continues into Mexico. These basins, from east to west, are: Black Warrior, Arkoma, Sherman, Fort Worth, Kerr, Val Verde, and Marfa (Figure 1). Now separated by pronounced basement uplifts or subtle arches, these basins were once part of a widespread late Paleozoic foredeep that formed in front of the Ouachita orogene as a result of tectonic loading associated with the collision of the Afro-South American and North American continents.

During early and mid-Paleozoic time, the southern boundary of North America was a passive margin. Carbonate shelf deposition was widespread. With the onset of the collision of the Afro-South American plate with North America, the southern margin of the North American craton was depressed, carbonate deposition terminated, and emergent highlands developed along the thrust belt. Clastics began to fill the foredeep trough in front of the Ouachita thrusts. The plate collision was oblique, moving from east to west with time. Consequently, major structural downwarp was time transgressive across the entire Ouachita foredeep trend. Downwarp started in the Black Warrior basin during Early Pennsylvanian time. In the Val Verde and Marfa basins, the subsidence developed during Late Pennsylvanian/Early Permian time (Figure 2). This change is also indicated by the systematic change in age of the switch from shelf deposition (dominantly carbonates) to clastic deposition (basin-fill) from east to west along the trend (see Figure 6).

In addition to creating a foredeep depression, the plate collision produced several major positive structural features along the cratonic border. Some of these, such as the Wichita Mountains, became important contributors of sediment to the foredeep in addition to the older cratonic source and the new orogenic source forming to the south and southeast along the Ouachita thrust belt.

The Ouachita foredeep trough filled during Pennsylvanian and Permian time with up to 4950 m (15,000 ft) of sediment. Environments of deposition varied from alluvial to deep marine. To date, more than 7.95×10^7 m^3 (500 million bbl) of oil and 2.83×10^{11} m^3 (10 tcf) of gas have been found in these Pennsylvanian and Permian clastics. The oil is found for the most part in conventional structural traps. The vast majority of the gas reserves are found not in conventional structural or stratigraphic traps but rather in large gas-pervasive, subnormally pressured fields called deep-basin traps (Masters, 1979).

The purposes of this chapter are (1) to document the regional paleogeography and facies of clastic reservoirs in this foredeep trough and (2) to describe the habitat of existing hydrocarbons and comment on future exploration potential.

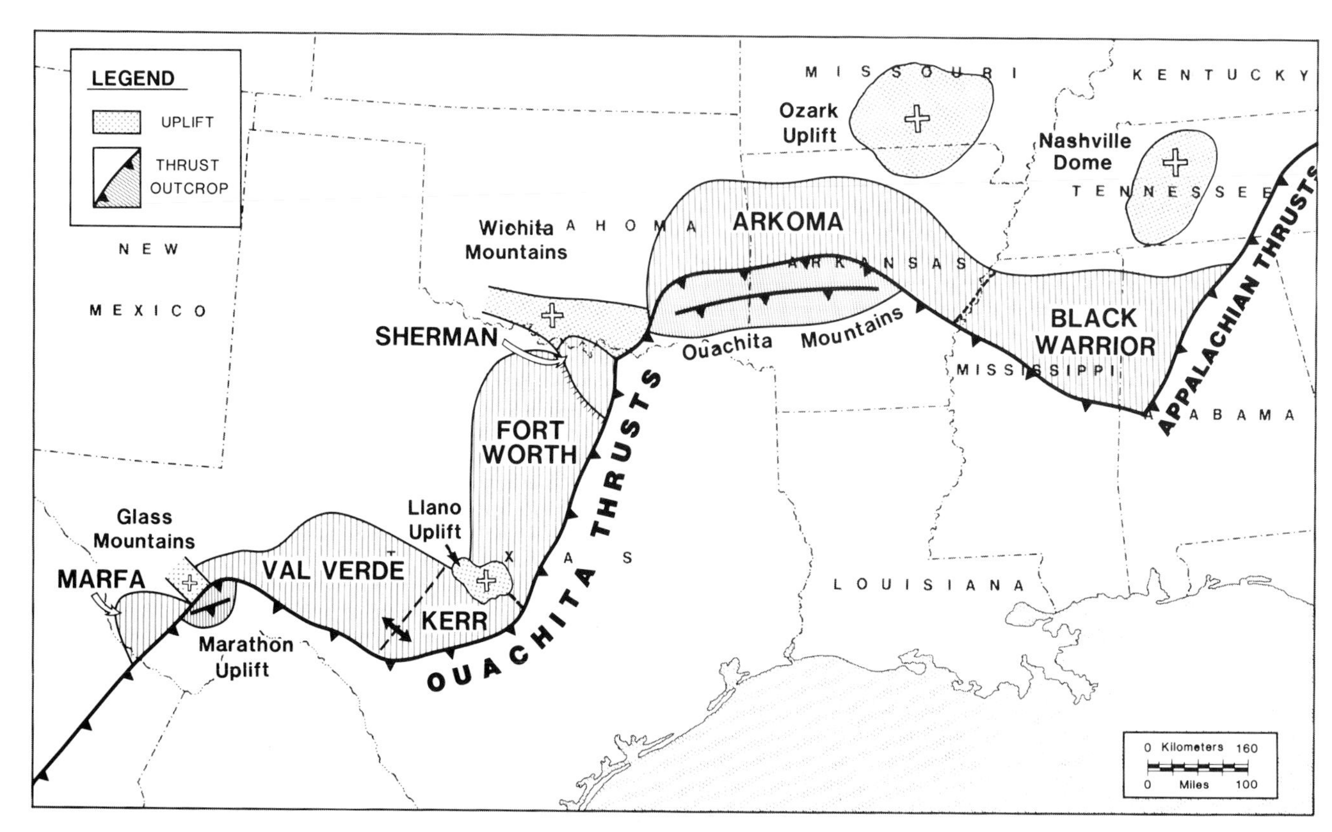

FIGURE 1. Location map showing the Ouachita foredeep basins and the adjacent structural features along the southern margin of the North American continent.

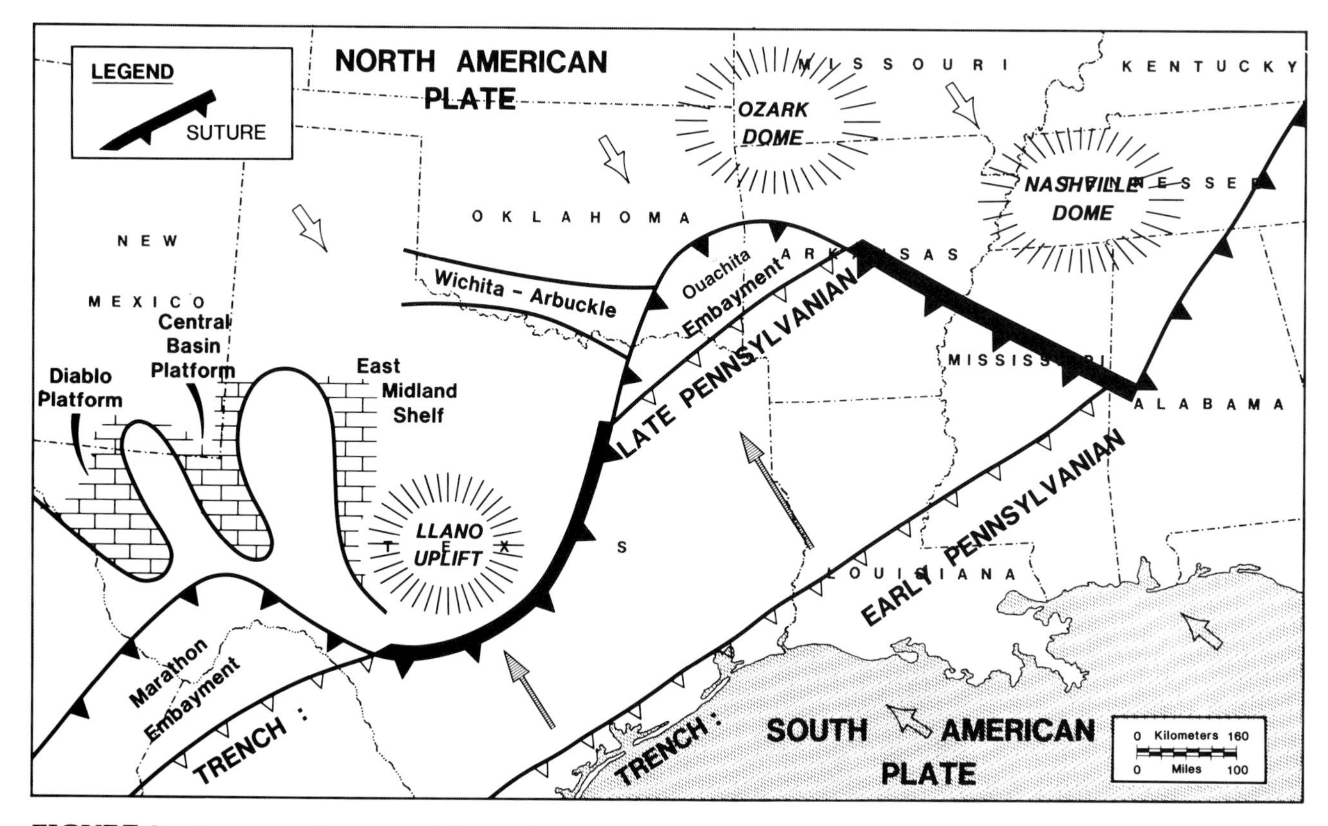

FIGURE 2. Plate tectonic context of the southern margin of the United States during Pennsylvanian time. Modified from Walper and Miller (1985).

TECTONICS AND STRUCTURE

The Ouachita thrusts, which define the southern margin of the foredeep trough, outcrop in only two areas: (1) the Ouachita Mountains in Arkansas and Oklahoma along the southern margin of the Arkoma basin and (2) the much smaller Marathon uplift in West Texas (Figure 1). Both outcrops occur in major embayments or recesses along the Ouachita trend (Figure 2). The allochthonous rocks outcropping in these two areas range from Cambrian/Ordovician to Pennsylvanian and are mostly deep-water facies composed of turbidites, black shales, and chert. These are the exposed parts of older accretionary prisms associated with the convergent southern margin of the North American continent. These allochthonous units have been studied and described in detail by many authors, including King (1978), Briggs (1974), Morris (1974a), McBride, (1978), Flores (1978), and Flawn et al. (1961).

Along the rest of the trend, the Ouachita thrusts are covered by Mesozoic units and can only be mapped in the subsurface using seismic or well control. The thrust margin and outcrops of the allochthonous thrust facies are shown in Figure 1.

In addition to the thrusts, the collision also produced several other major structural features along the southern margin of the North American platform. The largest of these is the Arbuckle-Wichita Mountains, which trend at near right angles to the Ouachitas and break the Ouachita foredeep trend into two segments. This uplift also became an important source of sediment to the foredeep. The collision also produced several other uplifts, including the Devil's River uplift (Val Verde basin), the Waco uplift (Fort Worth basin), the Muenster arch (separating the Fort Worth and Sherman basins), and the Central Mississippi uplift (Black Warrior basin); all are subsurface features (Nicholas and Rozendal, 1975).

The cratonic side of the foredeep trough is flanked by three large basement highs (Nashville dome, Ozark dome, and Llano uplift) and two stable late Paleozoic carbonate platforms (Central basin and Diablo) that occur in west Texas, the last area affected by the oblique plate collision (Figure 2).

Each structural basin along the trend is essentially a homocline dipping away from the craton and extending beneath the Ouachita thrusts (Figure 3). In several basins, a synclinal axis can be established with seismic data just in front of or beneath the leading thrust. The typical structural style is illustrated by a dip seismic line from the Black Warrior basin in Figure 4. On this line the Pennsylvanian strata dip southward from the Nashville uplift into and under the leading edge of the thrust. The synclinal axis of the basin is directly in front of the leading thrust. Both the Paleozoic basin-fill and the Ouachita thrusts are truncated and covered by Mesozoic clastics that dip gently southward into the Gulf basin.

The amount of faulting and folding within the foredeep trough varies from basin to basin along the Ouachita trend (Figure 5). The Arkoma basin shows the most structural deformation and is characterized by large normal faults (down-to-the-basin), broad synclines, and narrow anticlines, all oriented parallel to strike. Faulting was contemporaneous with Atoka deposition, as documented by the extreme basinward thickening of Atoka section across these features (Houseknecht, 1986). The Black Warrior basin lacks the large anticlines but does contain major down-to-the-basin normal faults and smaller antithetic faults that dip back into the major faults (Figure 5). Other basins, such as the Fort Worth, Kerr, and Val Verde, have very little structural deformation and are essentially ramps dipping into and under the Ouachita thrusts (Figure 5).

PALEOGEOGRAPHY AND FACIES

This study focuses on the clastics that filled the Ouachita foredeep trough. The section studied varies from Lower Pennsylvanian (Pottsville) in the east to Upper Pennsylvanian (Cisco) and Lower Permian (Wolfcamp) in the west (Figure 6). This reflects the changing time of development of the Ouachita foredeep associated with the onset of plate collision in the east and progressive westward shift of orogenic activity with time.

The depositional environments range from alluvial (channel sandstones, flood plain shales and coals) to deep-water turbidites (Figure 7). Where major basin subsidence was time-coincident with a high rate of sediment supply (Fort Worth, Black Warrior), deposition was able to keep up with subsidence and thick alluvial, coastal, and shallow marine facies developed. In those areas where subsidence preceded the arrival of a major clastic system, a deep-water basin developed directly above the underlying shelf section (Arkoma and Val Verde basins). These deep-water portions of the foredeep trough first filled with turbidites and then, as the basin shoaled, coastal and alluvial sediments. The thickest turbidite sections occur in the two major tectonic embayments (reentrants), the Arkoma basin and the Val Verde basin.

The paleogeography and distribution of reservoir facies in the Ouachita foredeep are a primary focus of this study, with emphasis on the deeper, less well-explored parts of each basin. Sufficient work was done on the shallow updip flanks of each basin to relate the various facies patterns across the basin. Detailed structure maps, cross sections, and reservoir isopach maps were constructed for the main reservoir units in the Black Warrior, Arkoma, Fort Worth, Kerr, and Val Verde basins. Less detailed work was done on the smaller Sherman and Marfa basins.

The paleogeography, depositional environments, and reservoir distribution for this thick wedge of clastics are discussed separately below for the Lower Pennsylvanian and the Upper Pennsylvanian.

Lower Pennsylvanian

The Lower Pennsylvanian, as used here, includes rocks of Pottsville age in the Black Warrior basin and rocks of Atoka (Bend) age across the Arkoma, Sherman, Fort Worth, and Kerr basins. The Lower Pennsylvanian paleogeography and facies patterns are summarized in Figure 8. Two major clastic dispersal systems, each containing sediments ranging from alluvial to deep marine, can be documented. One developed east of the Wichita-Arbuckle Mountains and

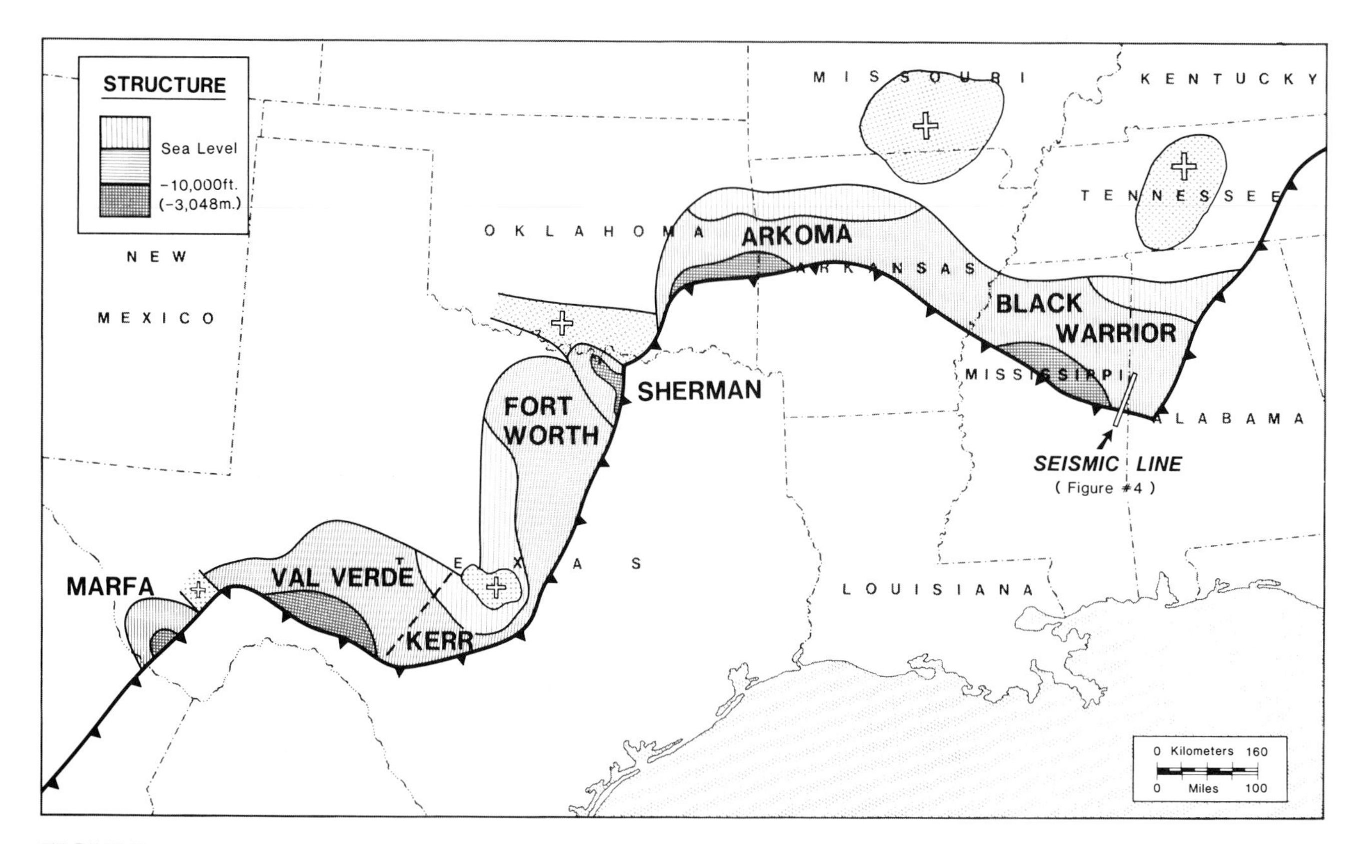

FIGURE 3. **General structure configuration of upper Paleozoic units in the Ouachita foredeep basins.**

supplied sediment to the Arkoma and Black Warrior basins. The other was located in Texas south of the Wichita-Arbuckle Mountains, a major source of sediment for the Fort Worth, Kerr, and Val Verde basins.

The eastern clastic system in the Black Warrior basin had its principal source in the newly elevated southern Appalachian orogene to the east and southeast (Figures 9 and 10). The alluvial facies found close to the source can best be documented in Pottsville outcrops in Alabama, where numerous coals, flood plain shales, and fluvial channels occur. Cross-beds in outcropping channel sands indicate a westward transport direction (Breland, 1972). The scatter of cross-bedding readings suggests a low-gradient, probably meandering, river system. This continental facies can be correlated (Figure 9) into the adjacent subsurface, where it is transitional to a well-developed coastal bar facies that trends northwest-southeast through the Black Warrior basin in eastern Mississippi and southwestern Alabama (Figure 10). The axis of this bar system shifted slightly back and forth during lower, middle, and upper Pottsville time, suggesting variations in sediment supply and/or rates of basin subsidence. Southwest of the stacked coastal-bar facies is a thick marine shale section that contains only a few thin shelf sands.

To the northwest, across the Mississippi Embayment, this coastal facies changes to a broad deltaic system (Atoka Formation) that covers the northern shelf area of the Arkoma basin (Figures 11 and 12). The deltaic clastics here are very quartzose and were derived from the craton to the north and northeast via the Illinois basin (Potter and Glass, 1958). Basinward of the east-west-trending shelf edge, the Atoka abruptly changes to turbidites (Figure 11) (Morris, 1974b; Houseknecht, 1986). Paleocurrent directions measured in the Ouachita Mountains indicate a pronounced westward transport direction longitudinally along the axis of the basin (Moiola and Shanmugam, 1984). These deep-water clastics have three possible origins: (1) the deltaic clastic systems to the north, (2) the shelf margin to the east in the Black Warrior basin, and (3) the rising orogenic belt to the south and southeast. Whichever the source (and it may be all three), the basin deepened westward along its axis so that the fill was dominantly longitudinal from east to west. These units in the Arkoma basin are well documented by Houseknecht (1986).

The second Lower Pennsylvanian clastic dispersal system originated in the Wichita Mountains near the Texas-Oklahoma border. Clastics were transported southeastward into the Fort Worth basin (Figures 13 and 14). Alluvial deposition (Bend Formation) occurs in the northwestern rim of the basin close to the source area. The alluvial facies is transitional basinward to a deltaic facies (also the Bend Formation) and then to two separate offshore bar systems (Hood and Grant formations, Figure 13). The shelf sands have a pronounced northeast-southwest trend paralleling the Ouachita trend; the upper sandstone (Grant) is the more extensive of the two (Figure 14). Marine shales occur farther basinward of these shallow marine bars.

These paralic facies flank the northwest margin of the Llano uplift and extend as far south as the northern part

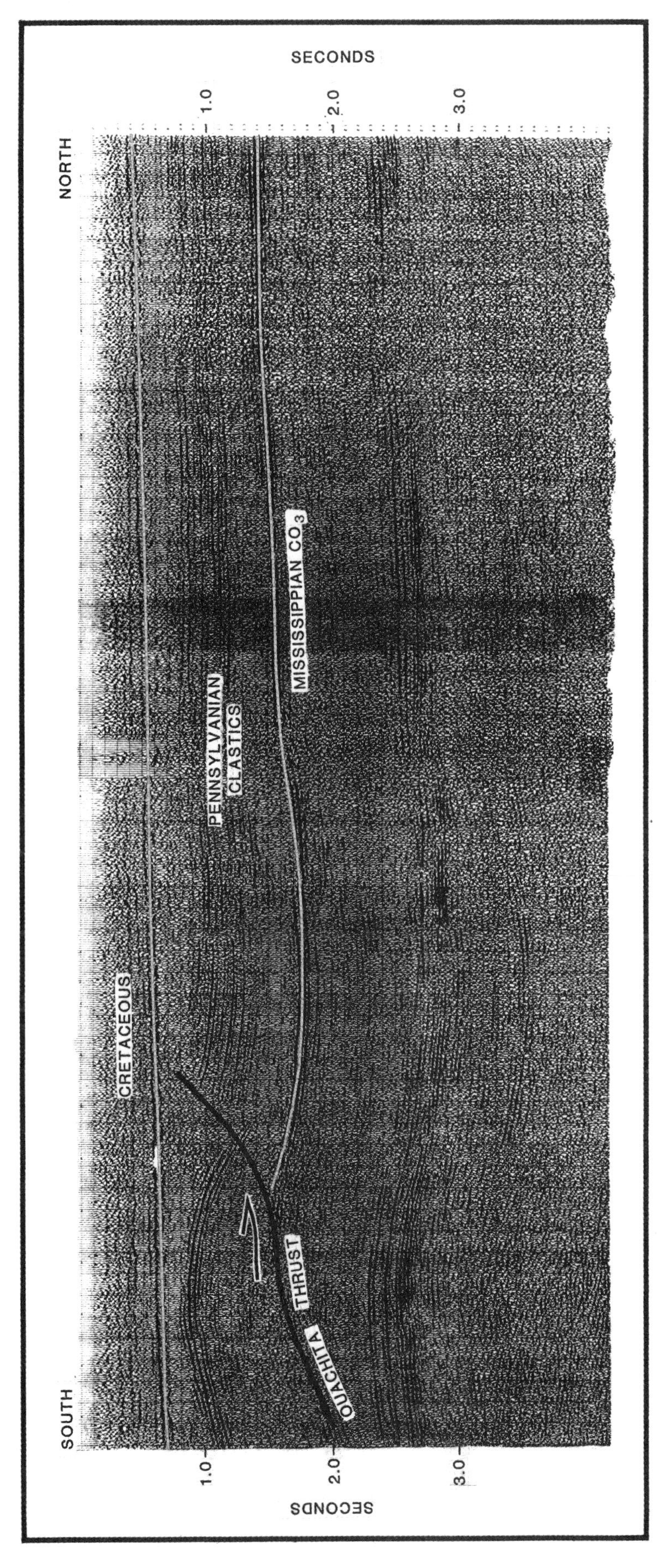

FIGURE 4. A north-south seismic line across the southern part of the Black Warrior basin, showing the structural features typical of these foredeep basins. Depths shown are two-way travel times. Line is courtesy of Western Geophysical, Houston.

A

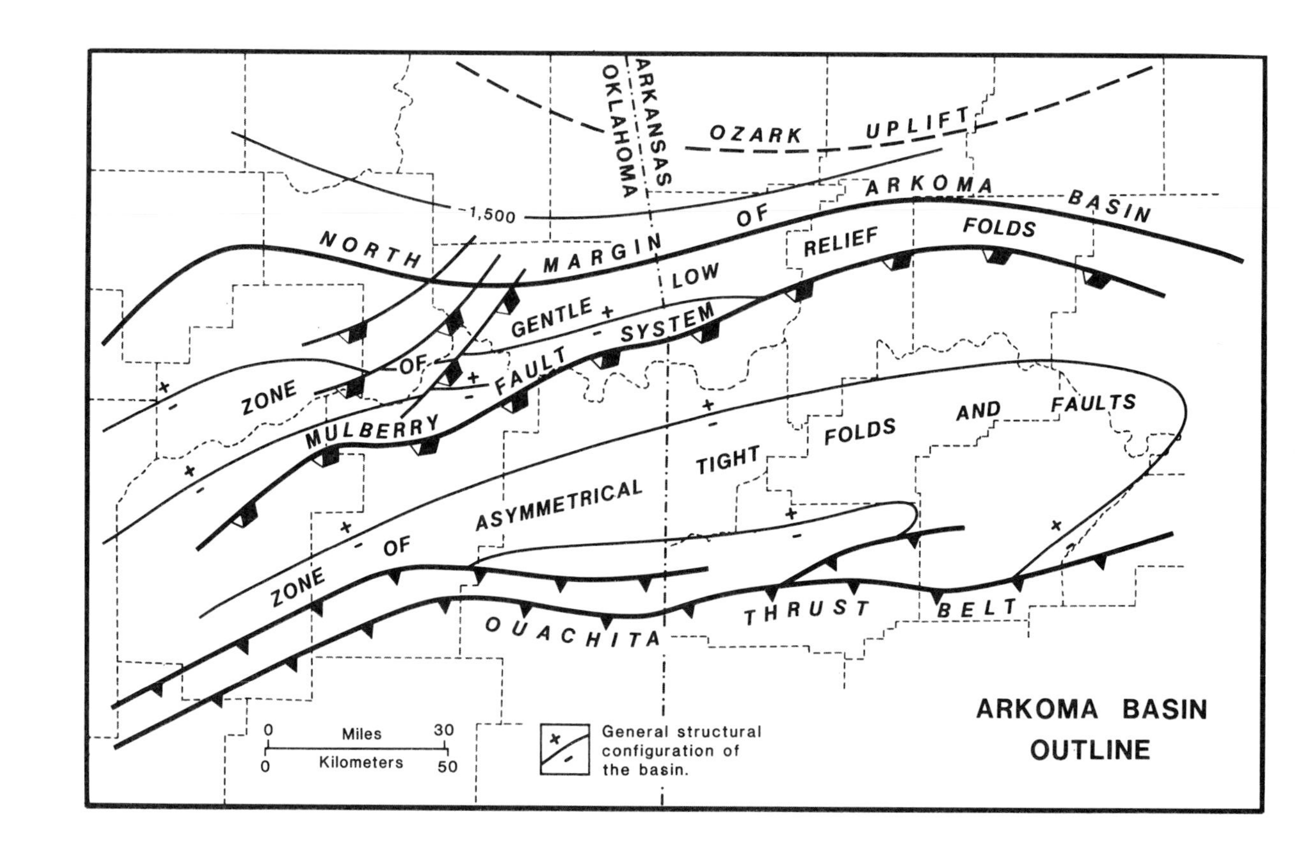

B

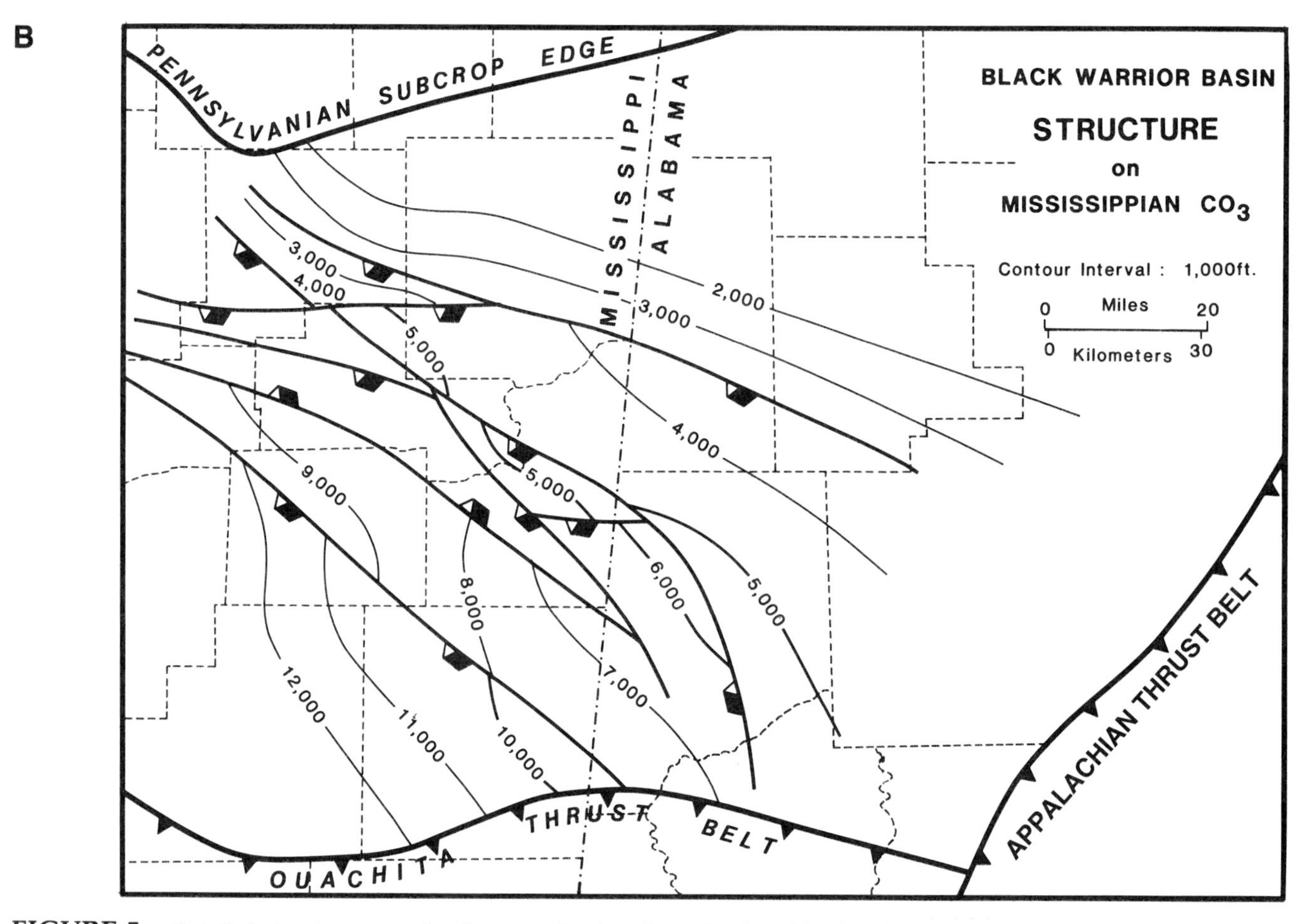

FIGURE 5. Detailed structure maps for the major basins along the Ouachita foredeep trend.

C

of the Kerr basin. Here they represent the distal facies of clastics transported southward from the Wichita Mountains around the west side of the Llano uplift.

Basinward of the shelf edge in the northern Kerr basin (Figure 8), the Lower Pennsylvanian clastics grade laterally to deep-water turbidites that outcrop on the southeastern side of the Llano uplift as the Smithwick Formation. Paleocurrent measurements in the Smithwick indicate a southwesterly transport direction (McBride and Kimberly, 1963), also suggesting longitudinal fill of the foredeep trough. From the paleocurrents, it is not clear whether these deep-water sands were derived entirely from the coastal-shelf system to the north or from the orogenic source to the southeast, or both. However, based on petrography, McBride and Kimberly (1963) concluded that these sands had an orogenic source.

Upper Pennsylvanian

The Upper Pennsylvanian, as used in this paper, includes units of Strawn, Canyon, and Cisco age found in north-central and west Texas. The Upper Pennsylvanian paleogeography for the clastics is summarized in Figure 15. There are two documented sources for the clastics entering the basin. The Wichita source, active in the Early Pennsylvanian, persisted into Late Pennsylvanian time and continued to disperse clastics southward across the Fort Worth basin (Strawn Formation) (Figure 16). The second source was the Ouachita orogene, which by Late

D

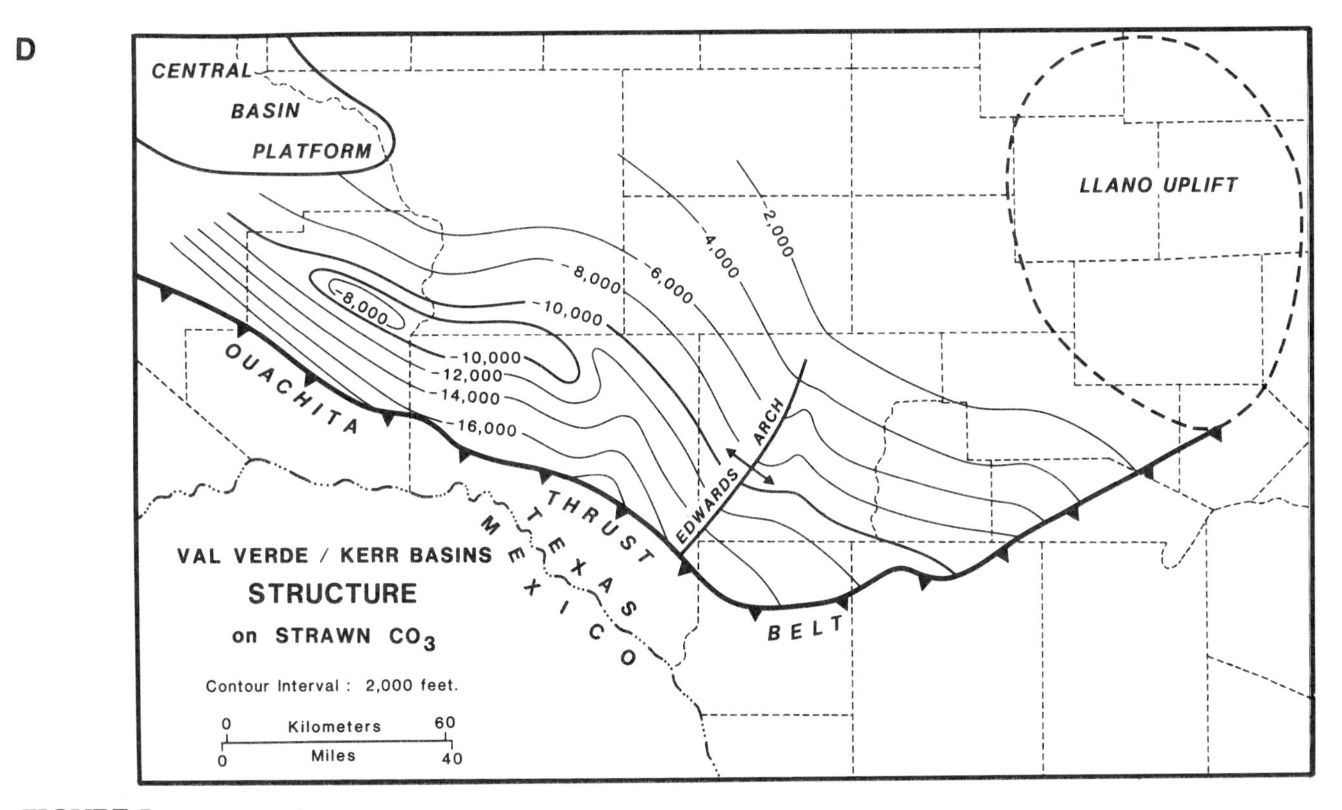

FIGURE 5. (Continued)

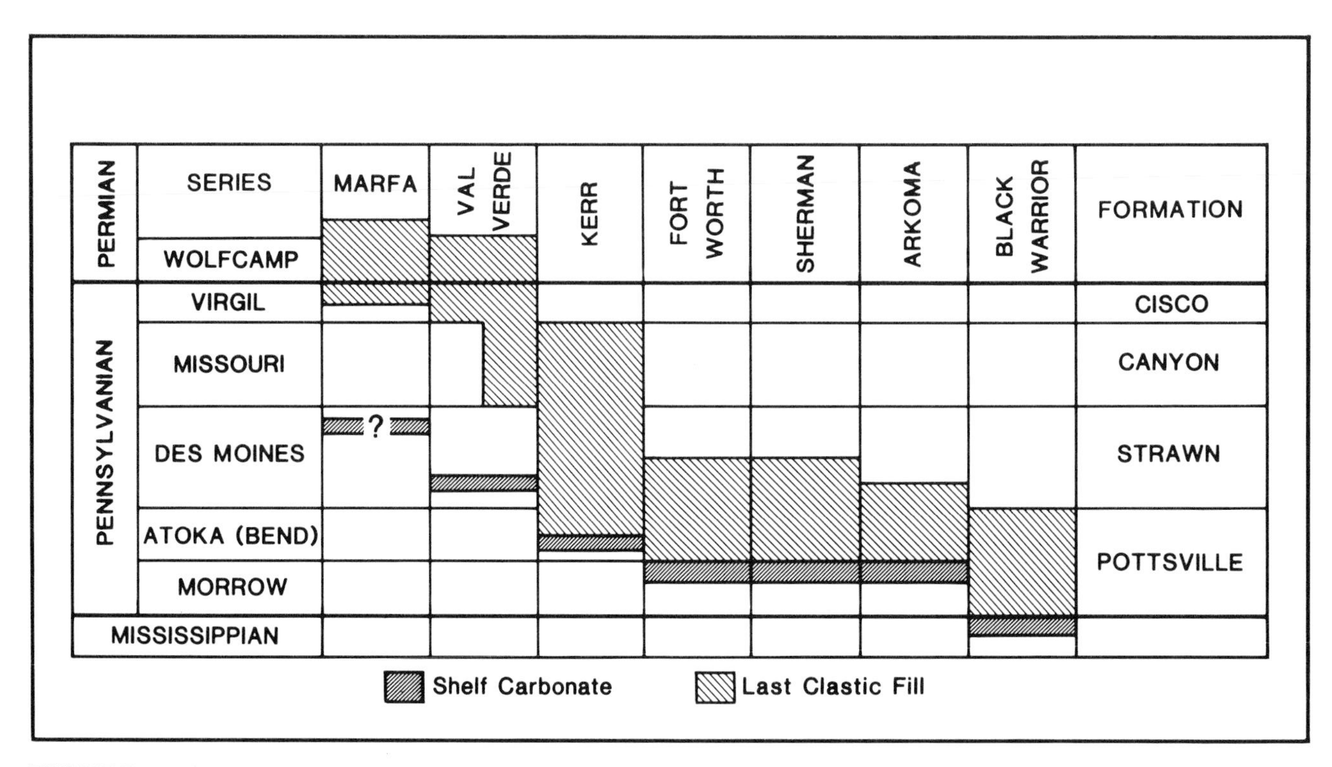

FIGURE 6. Stratigraphic column showing age of the clastic interval studied in this paper. Basins are positioned from east (right) to west (left) along the trend.

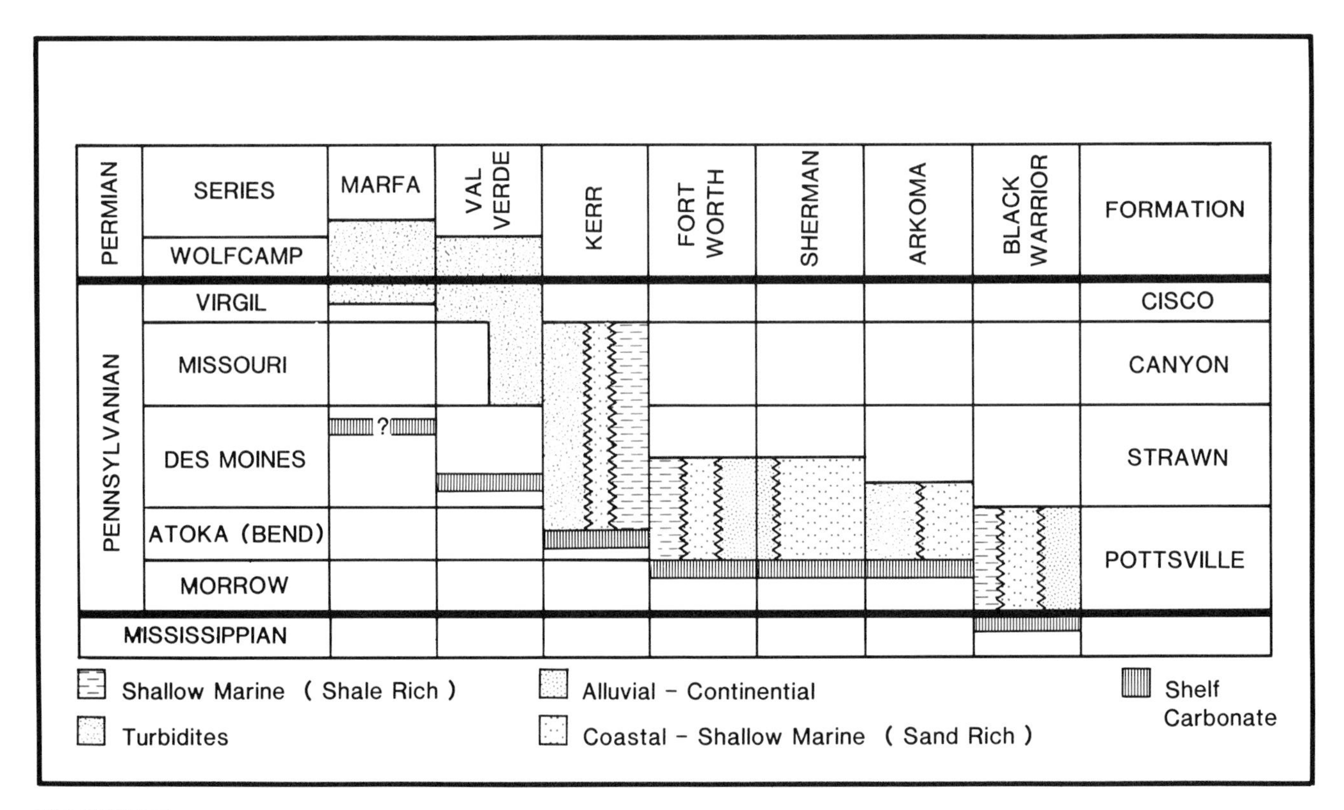

FIGURE 7. Summary of depositional environments for the units studied in each basin.

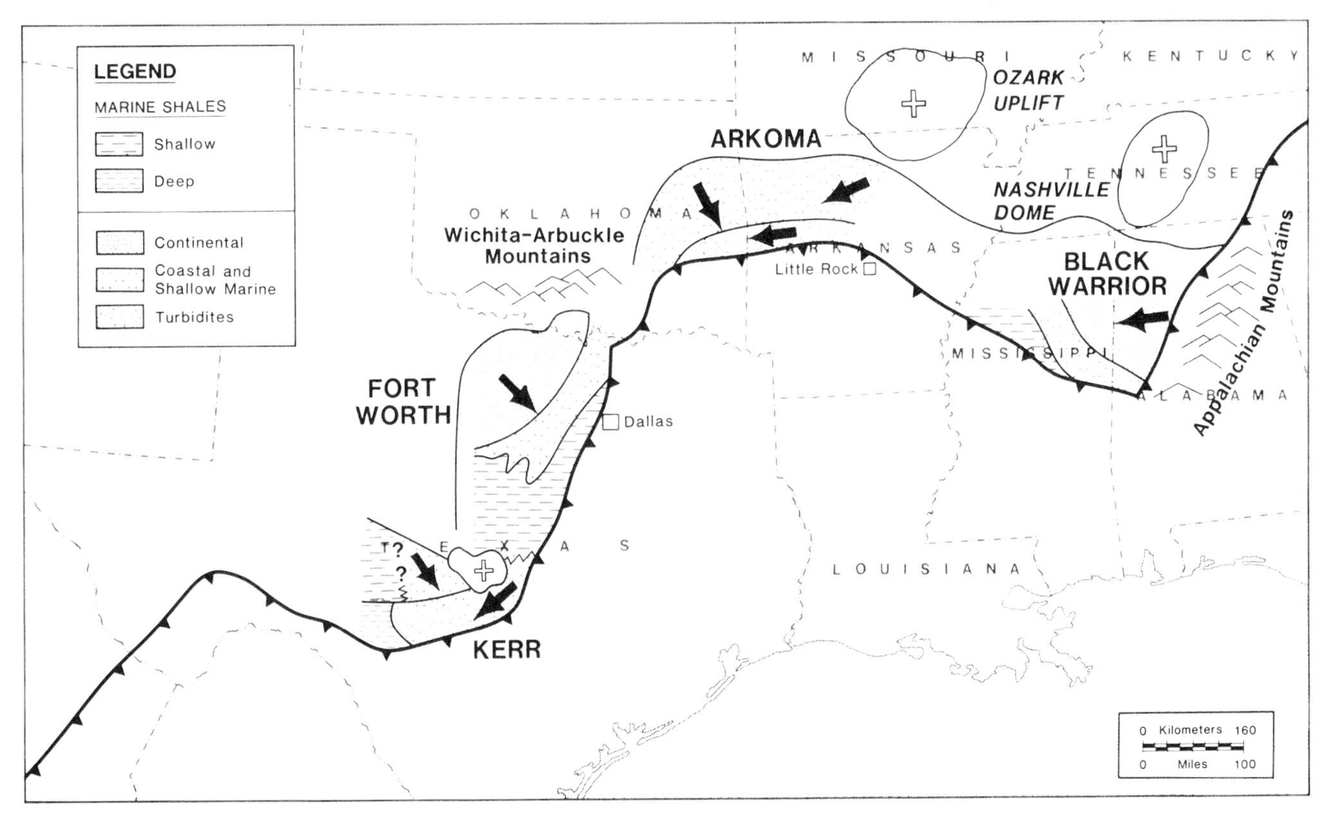

FIGURE 8. Lower Pennsylvanian paleogeography and facies distribution. This map is for Pottsville and Atoka age units.

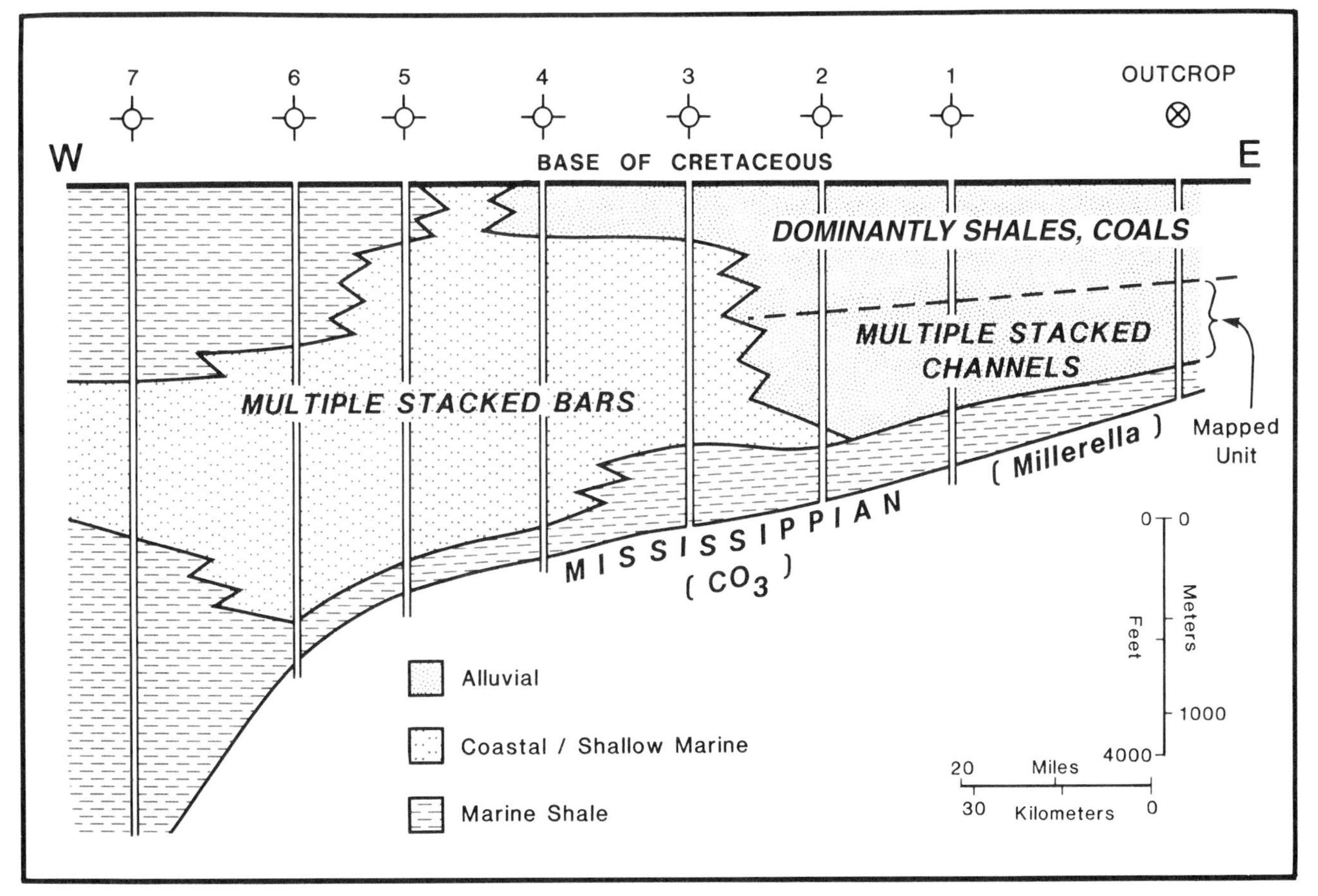

FIGURE 9. Cross section showing Pottsville facies relations in the Black Warrior basin. Location of section is indexed in Figure 10.

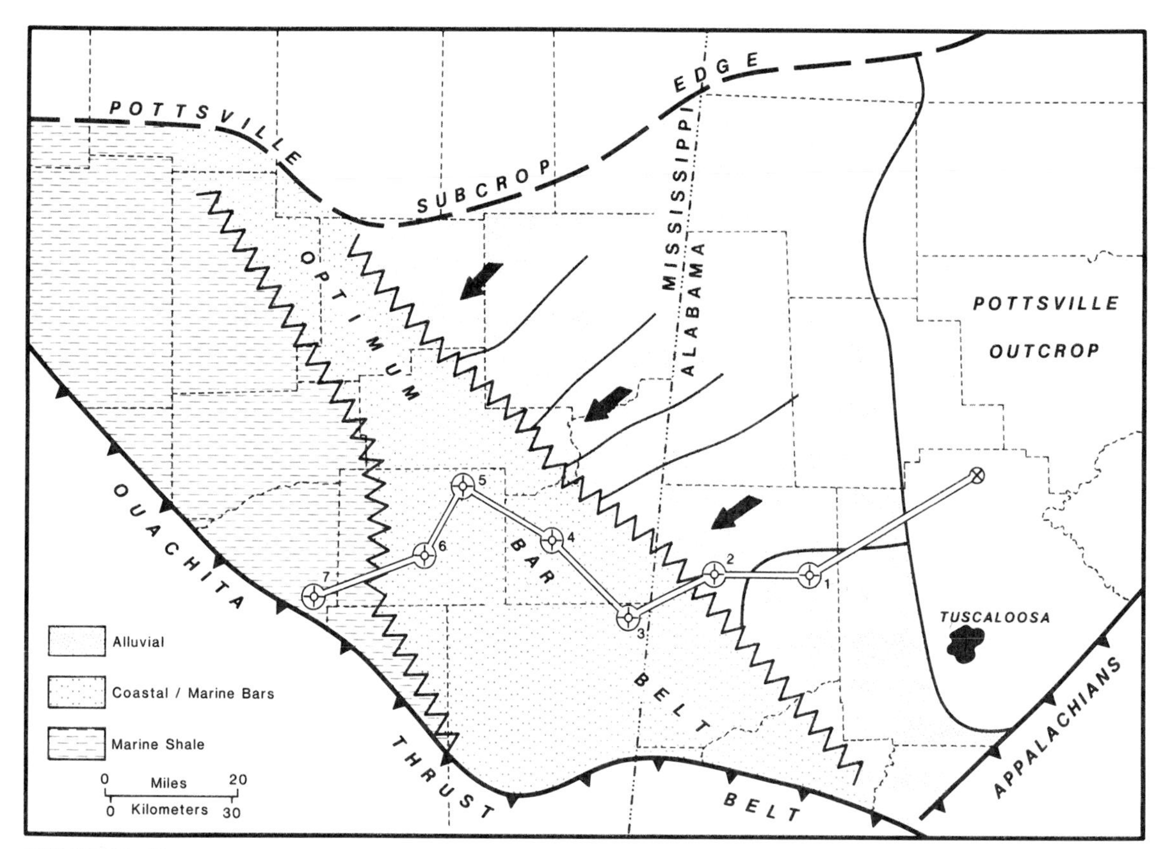

FIGURE 10. Lower Pottsville facies distribution in the Black Warrior basin. Unit mapped is indexed in Figure 9.

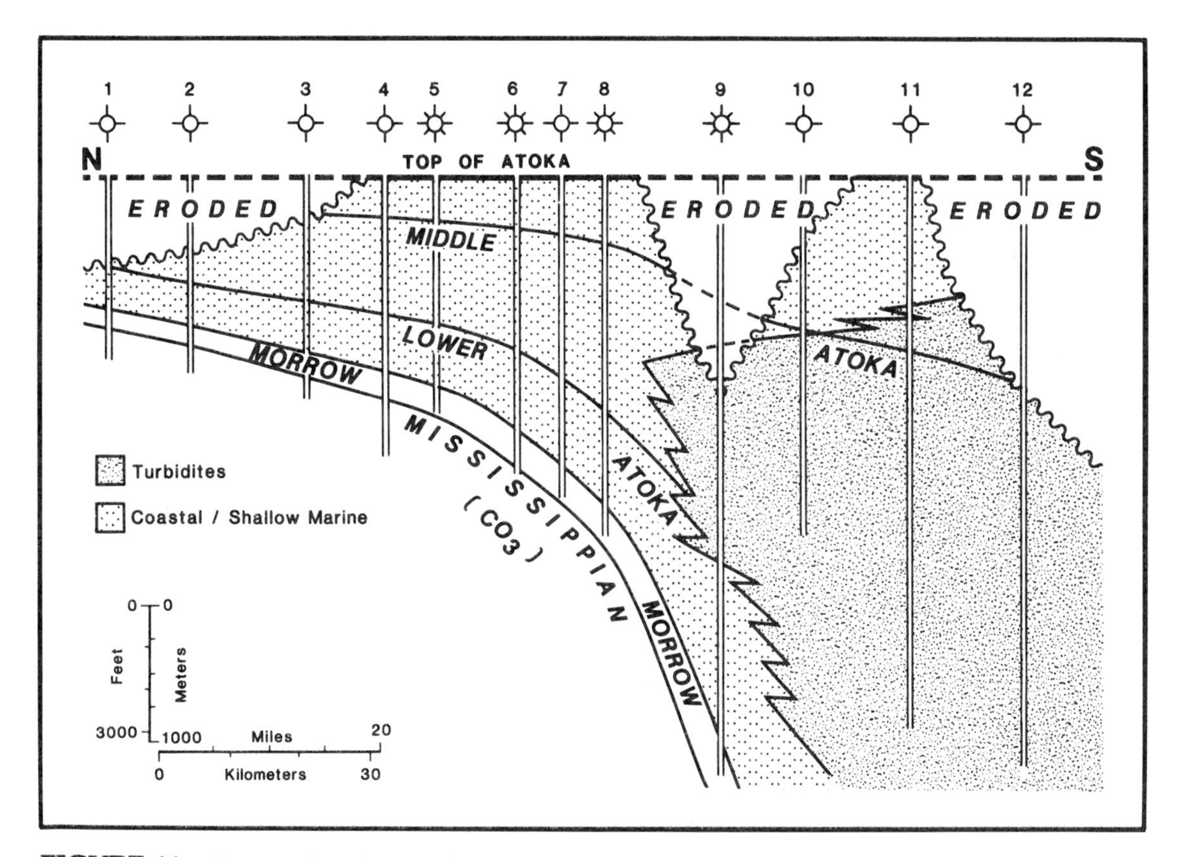

FIGURE 11. Cross section showing facies relations of the Atoka in the Arkoma basin. Location of section is shown in Figure 12. Modified from Haley (1984).

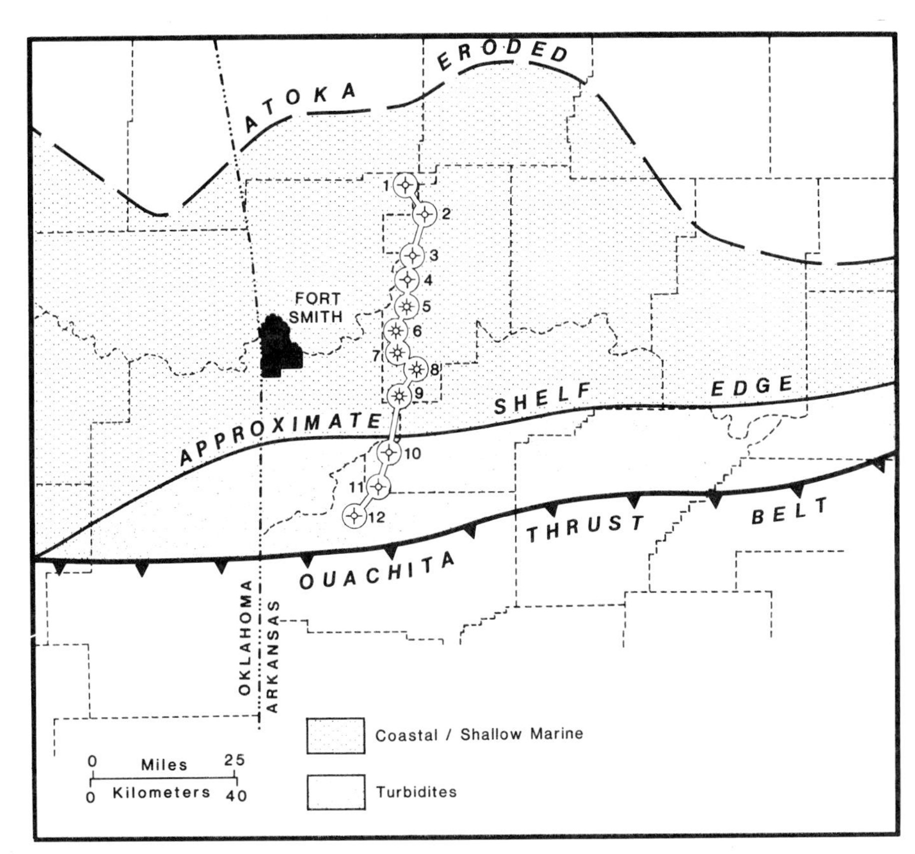

FIGURE 12. Facies distribution for the lower and middle Atoka in the Arkoma basin.

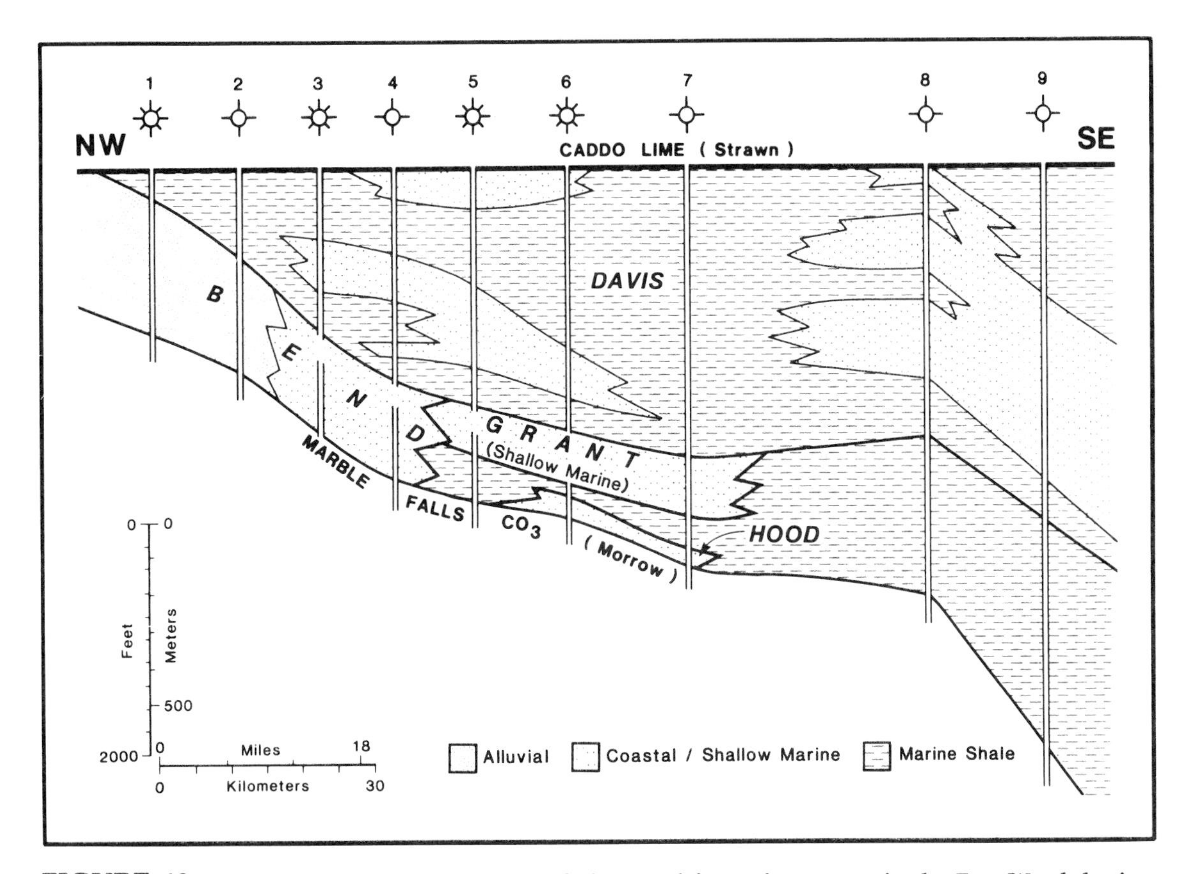

FIGURE 13. Cross section showing facies relations and formation names in the Fort Worth basin. Location of section is shown in Figure 14.

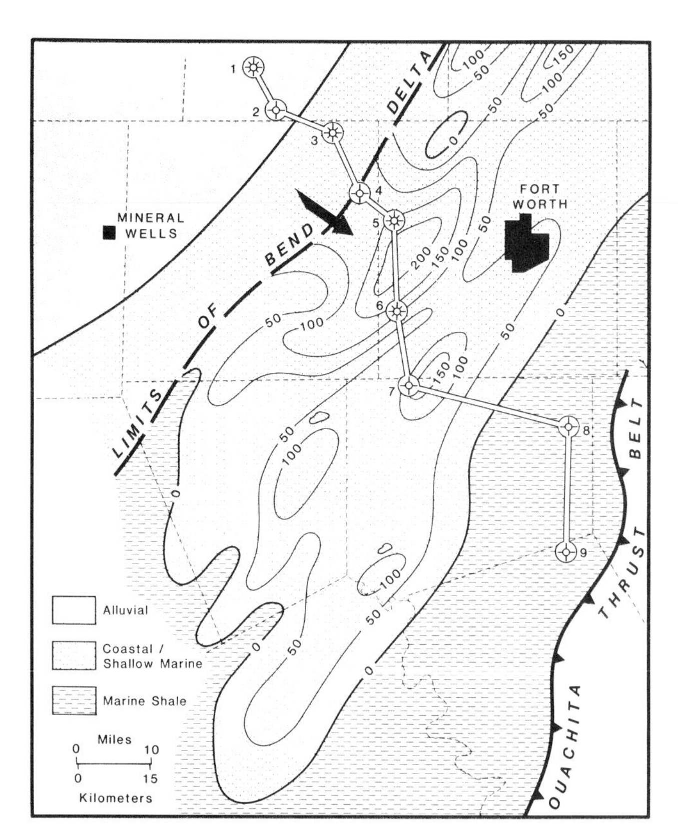

FIGURE 14. Facies distribution for the Bend and Grant sands in the Fort Worth basin. Net sand isopach is in feet.

Pennsylvanian time was emergent to the east and was dispersing clastics (Davis Sand and Strawn Formation) westward into the Fort Worth basin (Figure 16). The clastics from both sources commingled and continued westward and southwestward across the preexisting carbonate shelf of the eastern Midland basin.

These clastics reached the shelf edge in two places where they spilled into deep water, forming thick turbidite sequences. The first depocenter developed in the northwest corner of the Midland basin (Galloway and Brown, 1972; Brown et al., 1973). The second turbidite depocenter (Canyon Formation) occurs in the Kerr and Val Verde basins at the southern termination of the eastern Midland carbonate shelf (Mitchell, 1975). Large individual fans (or lobes) can be mapped in the subsurface in both the Kerr basin (Figures 17 and 18) and the Val Verde basin (Figures 19 and 20). It is not clear if these deep-water clastics were entirely derived from the north or whether part of the material came from the orogenic source to the south and east.

During the Upper Pennsylvanian, two large carbonate platforms continued to exist to the northwest and west of the Val Verde basin. These are the Central Basin platform and the Diablo platform. Deep-water Upper Pennsylvanian clastics buttress against these platforms similar to the shelf-basin relationship observed in offshore southern Belize today (James and Ginsburg, 1979).

Lower Permian

Late Pennsylvanian turbidite deposition in the Val Verde basin continued uninterrupted into Early Permian time

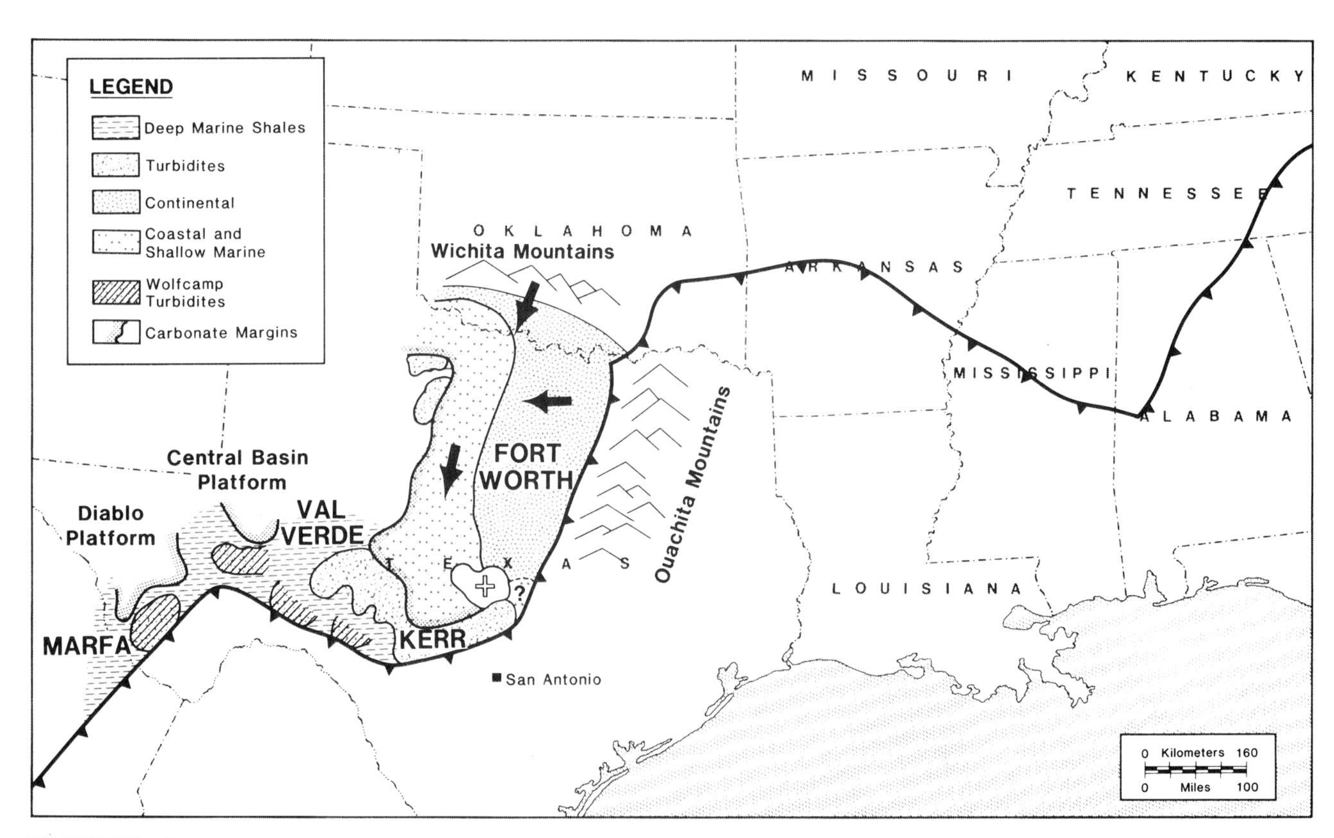

FIGURE 15. Upper Pennsylvanian paleogeography and facies distribution. This map is for Strawn, Canyon, and Cisco age units. The major Wolfcamp (Lower Permian) turbidite centers are also shown.

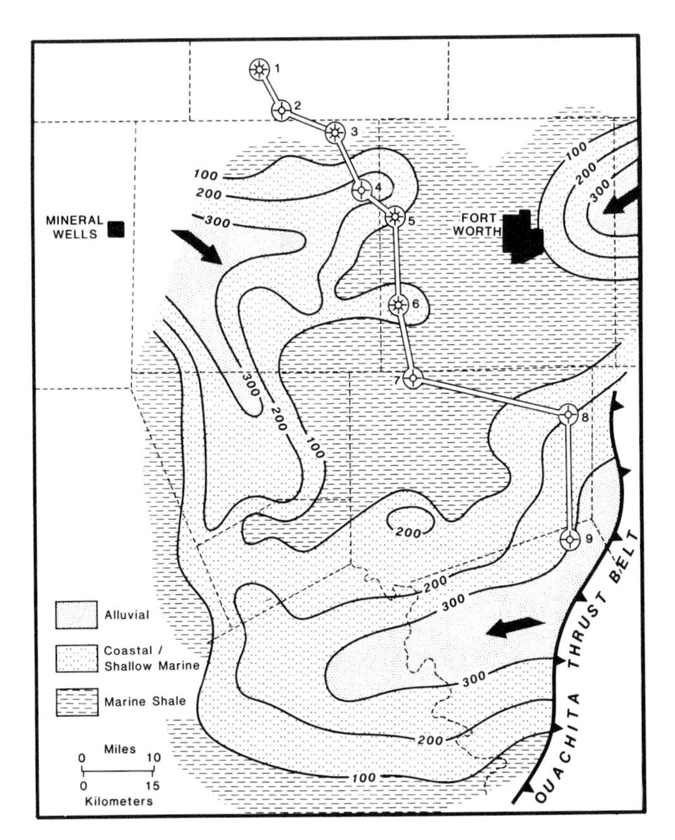

FIGURE 16. Facies distribution of the Davis sands in the Fort Worth basin. Refer to Figure 13 for the stratigraphic position of the Davis. Net sand isopach is in feet.

(Wolfcamp). Permian turbidites are also found farther west in the deep part of the Marfa basin. In the Val Verde basin, the Permian turbidite depocenters offlap the Upper Pennsylvanian fans, indicating that the system was progradational (Figures 19 and 20). Farther west, Wolfcamp turbidites buttress against the lower slopes of the prominent Central Basin and Diablo carbonate platforms. The exact transport direction and source for these Wolfcamp clastics were not obvious from our mapping. It is speculated that they are a continuation of the longitudinal fill of the Ouachita foredeep trough and that they are derived from both the cratonic shelf margin of the Val Verde basin and the newly emergent Ouachita orogenic source.

Summary

In summary, it appears that most of the Lower and Upper Pennsylvanian clastic material is derived from southern Appalachian or cratonic sources. One major entry of clastic material from a Ouachita source can be documented in the Fort Worth basin just south of Dallas. The provenance of the deep-water facies of the Kerr, Val Verde, and Marfa basins is uncertain. An isopach of the various sand systems does not permit a unique interpretation, and a definitive petrographic analysis of the sands was not part of the study. It is the authors' opinion that these clastics were transported longitudinally down the Ouachita foredeep trough from east to west and probably had both a cratonic and an orogenic source as the trough received sediment from both sides.

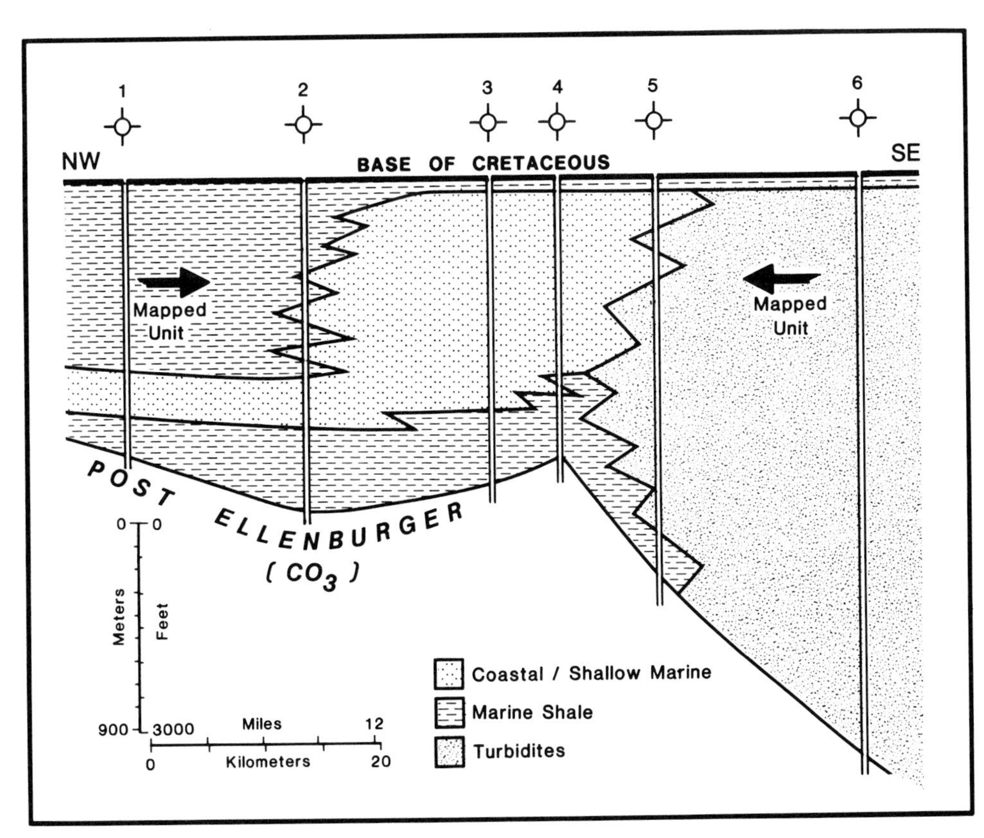

FIGURE 17. Cross section showing facies relations in the Kerr basin. Location of section is shown in Figure 18.

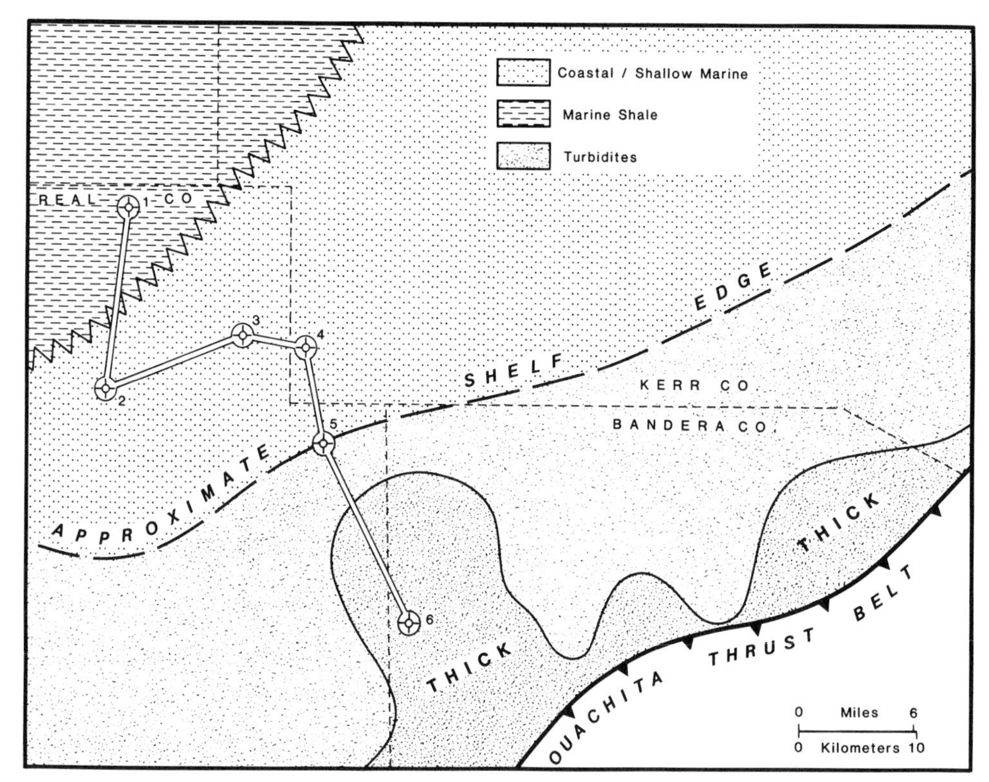

FIGURE 18. Facies distribution of the upper Canyon units in the Kerr basin. Mapped unit is indicated in Figure 17.

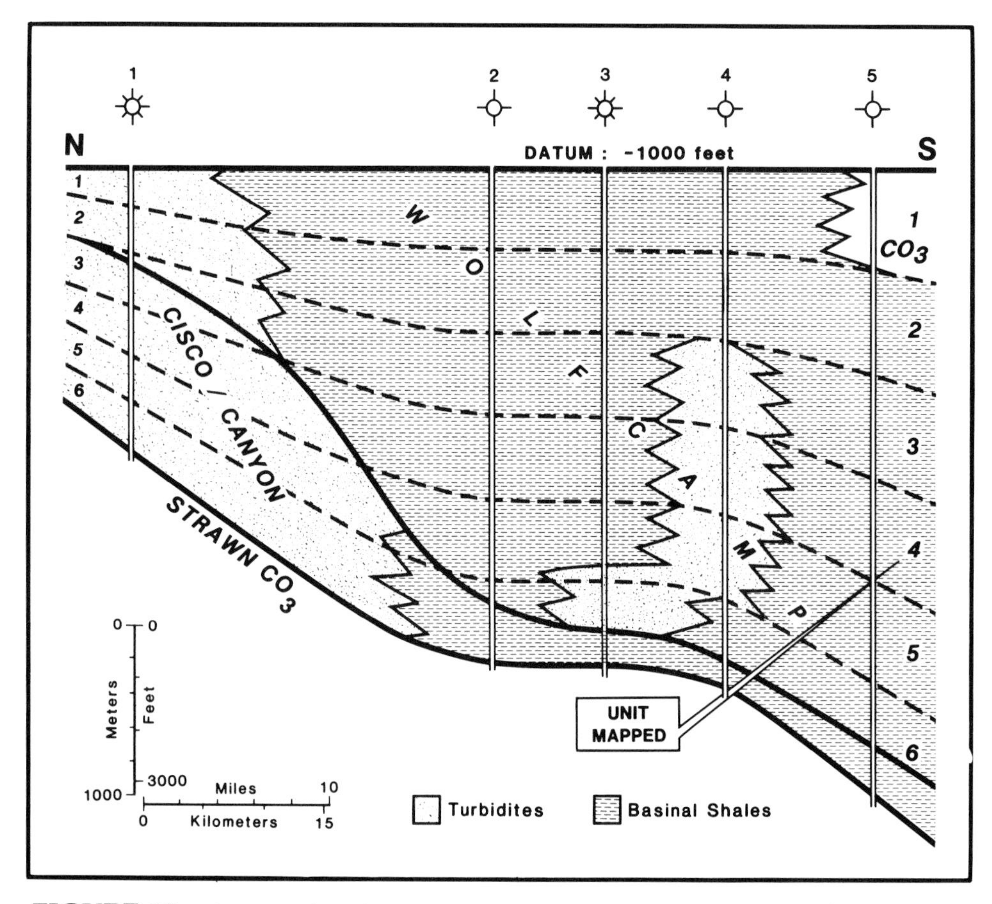

FIGURE 19. Cross section showing facies relations in the Val Verde basin. Location of section is shown in Figure 20.

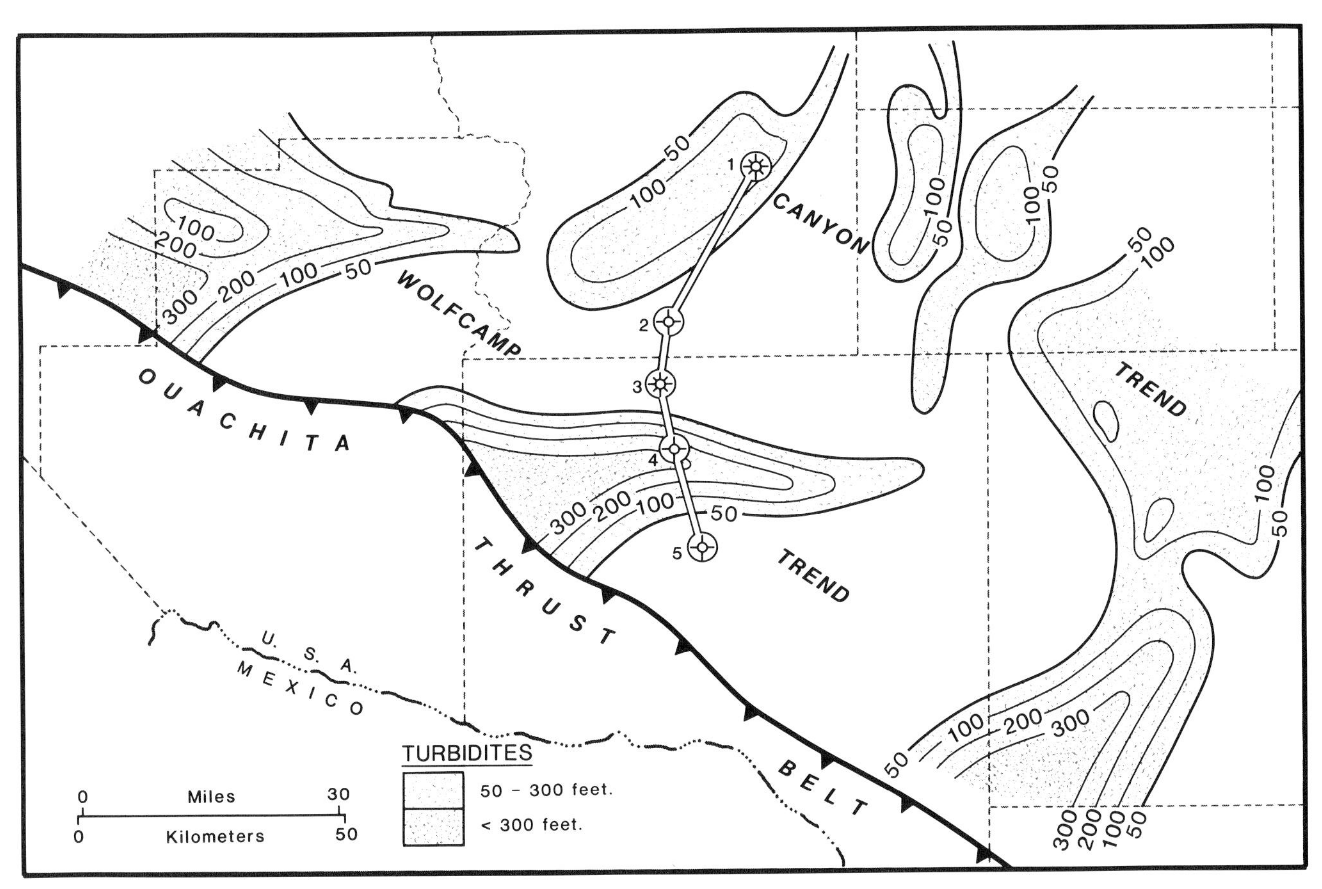

FIGURE 20. **Facies distribution of the Canyon and Wolfcamp units in the Val Verde basin. This map is for unit 4 shown in Figure 19.**

HABITAT OF HYDROCARBONS

In excess of 7.95×10^7 m^3 (500 million bbl) of oil and more than 2.83×10^{11} m^3 (10 tcf) of gas have been found in Pennsylvanian and Permian clastics in the Ouachita foredeep basins. The oil occurs principally in conventional structural and structural-stratigraphic traps. However, the vast majority of the gas occurs in large, pervasive gas traps that can best be described as deep-basin types of accumulations (as defined by Masters, 1979). This unique trap is discussed below.

The distribution of hydrocarbons discovered to date is highly unequal among the Ouachita foredeep basins (Figure 21). There is no production from these clastics in the Kerr and Marfa basins. There is only a small amount of gas (1.4×10^9 m^3; 50 bcf) in the Black Warrior basin in conventional structural traps. The Sherman basin is the only oil-prone basin in the Ouachita trend with the oil occurring in conventional structural traps (Bradfield, 1957a, b). Three basins contain the bulk of the gas reserves: the Arkoma (1.4×10^{11} m^3; 5 tcf), Val Verde (8.5×10^{10} m^3; 3 tcf), and Fort Worth (7.1×10^{10} m^3; 2.5 tcf) basins. These reserves are only estimates, because gas production records for these areas are not complete. In each of these basins, the porous units in the shallow updip parts of the basin contain small conventional structural or stratigraphic gas fields. Downdip, however, the entire stratigraphic section becomes totally gas saturated, resulting in widespread pervasive gas accumulations wherever there is sufficient reservoir rock.

It is interesting to speculate on the reasons for this variation in hydrocarbon type (oil vs. gas) and amount (lean vs. rich) from basin to basin along the trend. It is the authors' opinion that the gas-prone nature of most of these Ouachita foredeep basins probably relates to the type and distribution of the two principal source rocks: (1) the lower Paleozoic sapropelic (oil-prone) source rocks, which include the well-known Woodford/Chattanooga shales, and (2) the widespread black Pennsylvanian shales, which are very humic (gas-prone). The lower Paleozoic oil-prone source rocks are generally absent because of unconformities in most of the Ouachita foredeep basins (Branan, 1968; Pike, 1968; Blanchard et al., 1968), with the exception of the Sherman basin, where oil production is common. In contrast, Pennsylvanian humic-rich black shales (of variable age) occur in all basins and appear to be the major, mature source rocks. Of interest, these shales are very thin (almost absent) in the Black Warrior basin, where coarse clastics infill (Pottsville Formation) started very early; this basin contains significantly less gas reserves than those farther to the west along the trend, where these shales are much thicker.

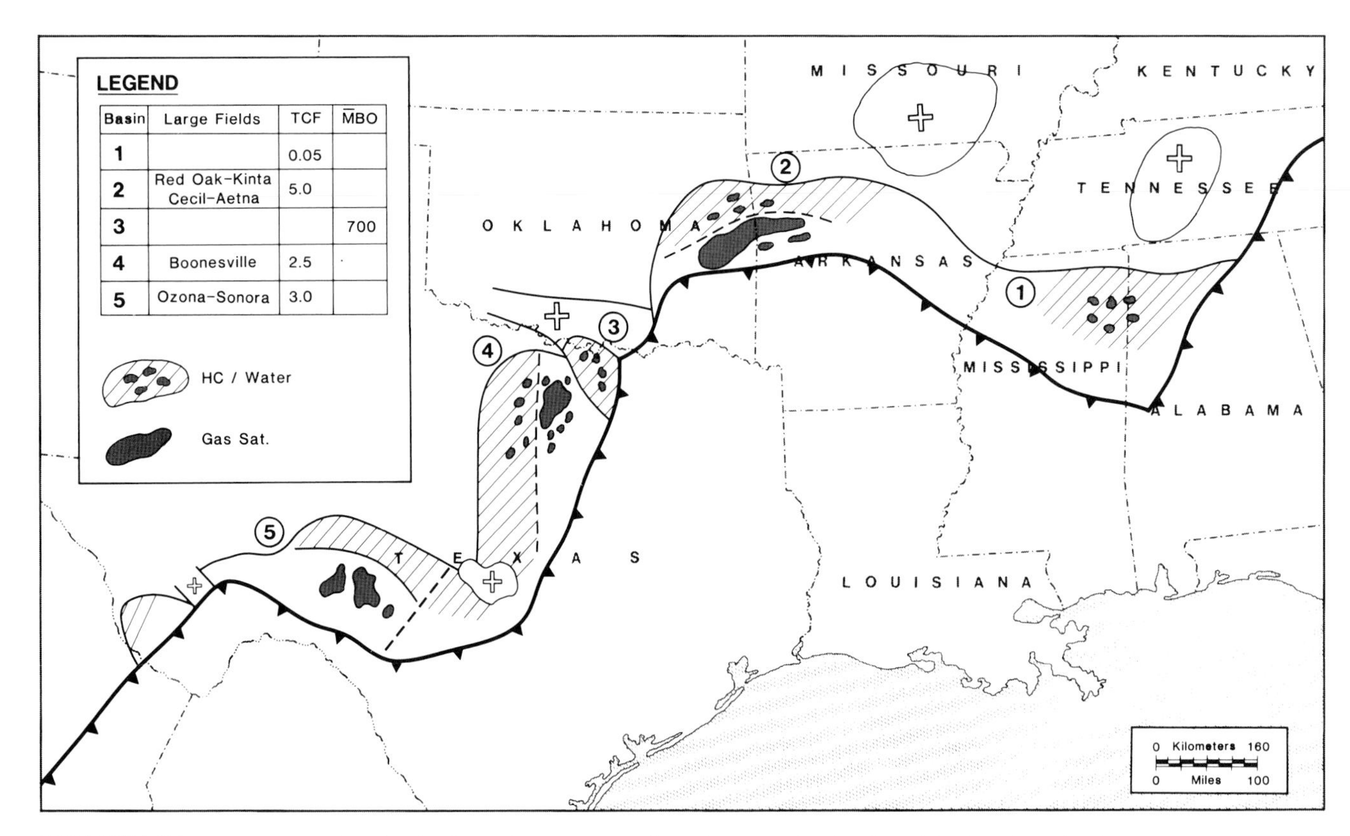

FIGURE 21. Summary of major hydrocarbon reserves found to date in the Ouachita foredeep basins.

The very large gas fields are characterized by several common attributes:

1. The section is totally gas saturated.
2. Very little or no water is produced. The major water system in these reservoirs occurs updip.
3. The reservoirs are tight, having both low porosity and low permeability. Typical reservoir properties are summarized in Figure 22.
4. The reservoirs are underpressured. The significance of this characteristic will be discussed below.

Pressure vs. depth graphs are an important tool for recognizing these large unconventional traps. Such graphs were made for individual stratigraphic units within each basin. Depths are plotted relative to sea level. Typical pressure-depth plots based on drill-stem tests (DST) or production tests are shown in Figure 23. This diagram shows such a plot for the Bend Formation in the Fort Worth basin. The solid bars show the normal water gradient in the shallower part of the basin. The broken bars represent the gas gradient, again taken from test data. The resulting graph demonstrates that (1) the Boonesville field is subnormally pressured and (2) the top of the gas section is at approximately -1219 m (-4000 ft) elevation. The Bend reservoir system is totally gas saturated below this structural elevation in the basin. Gas distribution in the Boonesville field appears to be totally independent of environment of deposition, occurring in both the alluvial and coastal facies of the Bend Formation.

The pressure-depth plot for the Atoka Formation in the Arkoma basin shows an interesting variation on this theme (Figure 23). The test data indicate that there are several major gas columns within the pervasive gas system, all underpressured. The top of the gas-saturated section varies, depending on the various gas columns, but occurs at about 1067 m (3500 ft). It is suggested that the separate gas columns reflect separate and unconnected reservoirs in the various Atoka shelf and turbidite sands. The test data for the prolific Canyon turbidites in the Val Verde basin also indicate that these reservoirs are underpressured.

The geologic boundary that separates the updip water system and the downdip gas-saturated section varies from basin to basin. In the Arkoma basin, the gas-saturated section seems to relate to the major Mulberry fault (Figure 5), with gas-saturated Atoka rocks occurring downdip (south) of this major fault. In the Val Verde basin, the shallow shelf sands are wet; deep-water sands are gas saturated, and the change appears to relate to the facies change where deep-basin sands downdip change to basin slope shales updip. In the Fort Worth basin, the change from downdip gas to updip water occurs within the alluvial facies of the Bend Formation and appears to relate to changing reservoir permeability. The low-permeability reservoirs occur downdip and are gas saturated.

TYPICAL RESERVOIR PARAMETERS				
	VAL VERDE	FORT WORTH		ARKOMA
GIANT FIELDS	OZONA SONORA	BOONESVILLE		RED OAK - KINTA CECIL-AETNA
Average Ø (%)	10 - 13	10 - 12	8 - 12	9
Average K (md)	0.2 - 0.3	4 - 10	0.3 - 0.6	<1

FIGURE 22. Summary of typical reservoir properties in the three large gas fields.

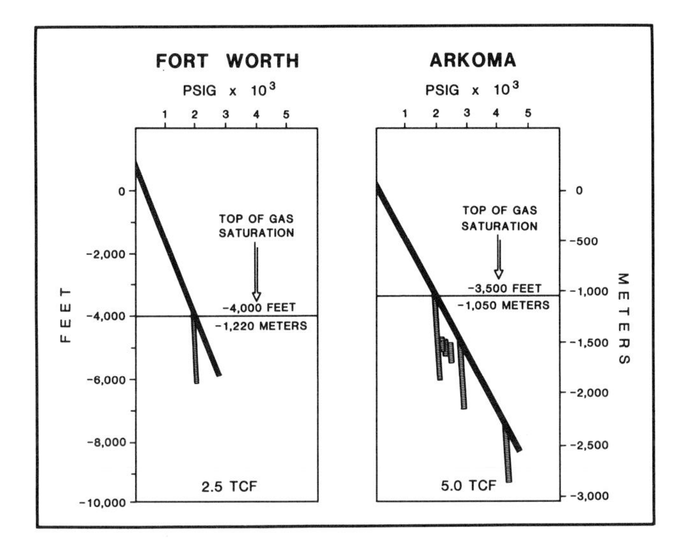

FIGURE 23. Pressure profiles for Bend reservoirs in the Fort Worth basin and the Atoka reservoirs in the Arkoma basin. Figure courtesy of Canadian Hunter Exploration Ltd.

Future exploration opportunities within the Ouachita foredeep basins can be divided into three categories, based on present knowledge (Figure 24).

First, additional conventional traps will undoubtedly be discovered in those basins where such traps have been found to date (i.e., Sherman basin). Most will be subtle stratigraphic or structural-stratigraphic traps, because most of the obvious structures have been drilled.

Second, opportunities exist in those basins and parts of basins that have been sparsely drilled. The Marfa and Kerr basins are essentially unexplored. In the Marfa basin, this is probably a result of the fact that seismic is difficult to obtain because of extensive volcanic cover. In the Kerr basin and the deep part of the Black Warrior basin, little structure is evident from seismic; consequently, there has been very little drilling.

Most wells drilled in the shallow part of the Black Warrior basin have the more prolific Mississippian shelf clastics as their target. Basinward of the Mississippian shelf edge in the deep Black Warrior basin, there are few

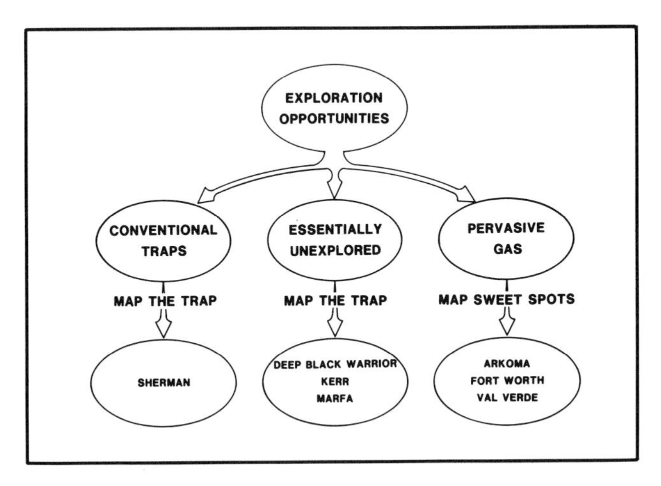

FIGURE 24. Chart depicting remaining exploration opportunities in the Ouachita foredeep trend.

Paleozoic tests. Consequently, the Pennsylvanian Pottsville bar trend found there (Figure 10) remains relatively untested.

In the sparsely drilled Kerr basin, few wells have tested the entire Pennsylvanian clastic section. The wells drilled to date have encountered numerous oil and gas shows. We can think of no other basin in the United States that has so many shows where no significant field has yet been found. This basin is clearly ripe for reexamination.

A very high temperature gradient exists in the Marfa basin. This is likely to have destroyed reservoir quality at depth. This factor seems to have discouraged exploration to date.

Third, reservoir mapping documents that an enormous amount of gas in place still remains untapped in "deep basin," gas-pervasive areas of the Arkoma, Fort Worth, and Val Verde basins. For example, the Davis Sand alone in the Fort Worth basin could conservatively contain an additional 5.7×10^{11} m^3 (20 tcf) of gas in place. Such reserve numbers are staggering in terms of future potential.

However, there are three current problems: costs are high, the reservoirs are "tight," and gas prices are depressed.

The solution to the problem centers on improved technology in drilling and completion practices (better flow rates) and more accurate geologic predictions. In these tight gas sands, the challenge for the geologist is to predict those trends (laterally or vertically) of better porosity and permeability within the overall gas-saturated tight section, commonly referred to as "sweet spots" or high-deliverability areas. Sweet spots in the existing fields relate to at least three geologic factors: (1) fractures, (2) lithology or petrology changes, and (3) environments of deposition. The role of geologists is to map and predict these high-productive fairways.

CONCLUDING REMARK

To find additional hydrocarbon reserves in the future, explorationists will typically (1) explore deeper in existing basins, (2) complete in tighter reservoirs, and (3) move to frontier areas. The Ouachita foredeep trough is a 2250-km (1400-mi) trend that offers all three of these opportunities. What's more, it is in the heart of the great American "oil patch."

REFERENCES CITED

Blanchard, K. S., O. Denman, and A. S. Knight, 1968, Natural gas in Atoka (Bend) section of northern Fort Worth basin, *in* Natural gases of North America: AAPG Memoir 9, v. 1, p. 1446-1454.

Bradfield, H. H., 1957a, The petroleum geology of Grayson County, Texas, *in* The geology and geophysics of Cooke and Grayson Counties, Texas: Dallas Geological Society Publications, p. 15-49.

Bradfield, H. H., 1957b, Subsurface geology of Cooke County, Texas, *in* The geology and geophysics of Cooke and Grayson Counties, Texas: Dallas Geological Society Publications, p. 75-98.

Branan, C. B., 1968, Natural gas in Arkoma basin of Oklahoma and Arkansas, *in* Natural gases of North America: AAPG Memoir 9, p. 1616-1635.

Breland, F. C., 1972, A petrologic and paleocurrent analysis of the Lower Pennsylvanian Pottsville Formation of the Warrior basin in Alabama: Masters thesis, University of Mississippi, 55 p.

Briggs, G., 1974, Carboniferous depositional environments in the Ouachita Mountains-Arkoma basin area of southeastern Oklahoma, *in* G. Briggs, ed., GSA Special Paper 148, p. 225-239.

Brown, L. F., A. W. Cleaves, II, and A. S. Erxleben, 1973, Pennsylvanian depositional systems in north-central Texas, a guide for interpreting terrigenous clastic facies in a cratonic basin: Bureau of Economic Geology, Guidebook No. 14, 122 p.

Flawn, P. T., A. J. Goldstein, P. B. King, and C. E. Weaver, 1961, The Ouachita system: University of Texas Publication No. 6120, 401 p.

Flores, R. M., 1978, Braided fan-delta deposits of the Pennsylvanian Upper Haymond Formation in the northeastern Marathon basin, Texas: SEPM Permian Basin Section 1978 Field Conference Guidebook, p. 149-160.

Galloway, W. E., and L. F. Brown, 1972, Depositional systems and shelf-slope relationships in Upper Pennsylvanian rocks, north-central Texas: Bureau of Economic Geology, Report of Investigations No. 75, 62 p.

Haley, B. R., 1984, Structural cross section of the eastern Ouachita Mountains: Arkansas Geol. Comm. Guidebook 84-2, p. 31.

Houseknecht, D. W., 1986, Evolution from passive margin to foreland basin: the Atoka Formation of the Arkoma basin, south-central U.S.A., *in* P. A. Allen and P. Homewood, eds., Foreland basins: International Association of Sedimentologists Special Publication 8, p. 327-346.

James, N. P., and R. N. Ginsburg, 1979, The seaward margin of the Belize barrier and atoll reefs: Special Publication of the International Association of Sedimentologists, 191 p.

King, P. B., 1978, Tectonics and sedimentation of Paleozoic rocks in the Marathon region, West Texas: SEPM Permian Basin Section 1978 Field Conference Guidebook, p. 5-37.

Masters, J. A., 1979, Deep basin gas trap, Western Canada: AAPG Bulletin, v. 63, p. 152-181.

McBride, E. F., 1978, The Ouachita trough sequence: Marathon region and Ouachita Mountains: SEPM Permian Basin Section 1978 Field Conference Guidebook, p. 39-50.

McBride, E. F., and J. E. Kimberly, 1963, Sedimentology of Smithwick shale (Pennsylvanian), eastern Llano region, Texas: AAPG Bulletin, v. 47, p. 1840-1854.

Mitchell, M. H., 1975, Depositional environment and facies relationships of the Canyon sandstone, Val Verde basin, Texas: M. A. thesis, Texas A & M University, 225 p.

Moiola, R. J., and G. Shanmugam, 1984, Submarine fan sedimentation, Ouachita Mountains, Arkansas and Oklahoma: Trans Gulf Coast Association of Geological Society, v. 34, p. 175-182.

Morris, R. C., 1974a, Sedimentary and tectonic history of the Ouachita Mountains, *in* W. R. Dickinson, ed., Tectonics and sedimentation: SEPM Special Publication 22, p. 120-142.

Morris, R. C., 1974b, Carboniferous rocks of the Ouachita Mountains, Arkansas: a study of facies patterns along the unstable slope and axis of a flysch trough: G. Briggs, ed., GSA Special Paper 148, p. 241-279.

Nicholas, R. L., and R. A. Rozendal, 1975, Subsurface positive elements within Ouachita foldbelt in Texas and their relation to Paleozoic cratonic margin: AAPG Bulletin, v. 59, p. 193-216.

Pike, S. J., 1968, Black Warrior basin, northeast Mississippi and northwest Alabama, *in* Natural gases of North America: AAPG Memoir 9, p. 1693-1701.

Potter, P. E., and H. D. Glass, 1958, Petrology and sedimentation of the Pennsylvanian sedimentation in southern Illinois—a vertical profile: Illinois Geological Survey Report of Investigations No. 24, 60 p.

Walper, J. L., and R. E. Miller, 1985, Tectonic evolution of Gulf Coast basins, *in* Habitat of oil and gas in the Gulf Coast: SEPM Fourth Annual Research Conference Proceedings, p. 25-42.

Chapter 16

Summary and Conclusions

Roger W. Macqueen and Dale A. Leckie

Geological Survey of Canada
Institute of Sedimentary and Petroleum Geology
Calgary, Alberta, Canada

INTRODUCTION

The papers in this volume describe six foreland basins and fold belts in terms of their regional setting, stratigraphy, tectonics, and structure, and their oil and gas systems. All of the basins show general similarities, but each differs significantly in detail from the others, posing something of a problem in terms of arriving at a "typical" foreland basin and fold belt. Some are major hydrocarbon producers; others are not. The major characteristics of the six foreland basins and fold belts are summarized in Tables 1 through 5, which provide a convenient means of comparing and contrasting these basins and their hydrocarbon resources.

The Western Canada foreland basin and fold belt serves as the type example for several reasons. These include: its setting and clear relationship to a major orogene of Mesozoic–Cenozoic age; the fact that it is uncomplicated by later overprinting, segmentation, or cover rocks (unlike the Ouachita, Eastern Venezuela, and U.S. Rocky Mountain foreland basins and fold belts); the fact that there is a large volume of publicly available data on the basin and an active exploration and research community; and the fact that it has reasonable oil and gas reserves in a well-defined stratigraphic framework. Perhaps more than any other basin included here, the stratigraphic framework demonstrates the complex interplay between tectonics and sea level changes, with details and understanding still evolving, and the place within this framework occupied by source facies, reservoirs, and seals. The Western Canada foreland basin and fold belt also demonstrates the interplay between the underlying passive-margin succession and the evolution of the foreland basin. This is a theme that is illustrated to varying degrees for all of the foreland basins considered here.

REGIONAL SETTING

The plate tectonic settings of foreland basins, and in particular plate-margin interactions, appear to be more closely related to deformational patterns in the case of some foreland basins and fold belts such as the Eastern Venezuela and Zagros basins, than in others such as the Western Canada and U.S. Rocky Mountain foreland basins. The length of time during which foreland basin sediments accumulated also is variable. Foreland basin fill in some basins accumulated in a short period of time (e.g., Eastern Venezuela foreland basin and Zagros basin), whereas others are relatively long-lived (e.g., Alaska North Slope, Western Canada, and U.S. Rocky Mountain foreland basins; Table 1).

With the exception of the late Paleozoic Ouachita foreland basin, which is relatively poorly known because of cover by younger rocks, all foreland basins discussed here are Mesozoic and Cenozoic features, and are associated with an existing mountain belt. Some of these belts, such as the Zagros Mountains associated with the Zagros foreland basin or the Cordillera de la Costa associated with the Eastern Venezuela foreland basin, are presently undergoing compressional deformation and overthrusting, whereas other mountain belts, such as the Western Canada and U.S. Rocky Mountains, are now inactive. It is probably significant that the two most richly hydrocarbon-endowed foreland basins, the Zagros and Eastern Venezuela basins, were located at low paleolatitudes during their development. Other factors being equal (e.g., low terrigenous input), these low-latitude locations appear to have been characterized by high rates of organic productivity with associated high preservation potential. It is the close association of type II marine organic matter, fractured carbonate reservoir rocks, and a widespread evaporite seal, combined with a very large horizontal scale for reservoirs, that make the Zagros setting as productive as it is.

The basins are generally a few hundred kilometers wide and 1000 to 2000 km long. The outstanding exception to this generalization is the composite North American Cordilleran foreland basin, including the Western Canada and U.S. Rocky Mountain basins: this composite foreland basin is more than 6000 km long, extending from the Gulf of Mexico to the Arctic Ocean. Most basins are elongate, with the longest dimension parallel to the associated orogenic belt. Only the Maturin and Guarico subbasins of the Eastern Venezuela foreland basin are roughly equidimensional. Foreland basins also vary in size, ranging from about 2.4×10^5 km^2 (Alaska North Slope) to about 1.9×10^6 km^2 (Western Canada foreland basin). Some of the

TABLE 1. Regional Setting.

Basin	Age	Location	Size	Reserves*
Western Canada Basin	• Middle Jurassic to Cretaceous	• Manitoba, Saskatchewan, Alberta, British Columbia, and Northwest Territories, Canada	• 1600 km wide east-west by 1200 km long north-south • Continuous to the Gulf of Mexico; total length greater than 6000 km • $\sim 1.9 \times 10^6$ km^2	• 5.6 BBO • 1.5 billion barrels natural gas liquids • 67 tcf gas • 1.7 trillion barrels of heavy oil • 17 tcf coal-bed methane
Zagros basin	• Late Eocene to present	• Middle East: northeast Iraq and eastern Gulf waters	• 1800 km long × 250–300 km wide • $\sim 450\text{–}540 \times 10^3$ km^2	• 150 BBO oil • 600–650 tcf gas
Eastern Venezuela basin: • Guarico subbasin • Maturin subbasin	• Early Eocene to present	• Eastern Venezuela	• 210,000 km^2 • Guarico subbasin is 480 by 260 km • Maturin subbasin is 520 by 170 km	• 20 BBO • 66 tcf gas • 1.2 trillion barrels of heavy oil
Alaska North Slope	• Middle(?) Jurassic to late Tertiary	• North Slope of Alaska	• 240,000 km^2 • 1000 km by 50 to 350 km	• 100 MBO • 400 bcf gas • multi-BB of heavy oil
U.S. Rocky Mountain basin	• Middle Jurassic to late Tertiary	• Western United States	• 1000 by 1400 km • $\sim 1.4 \times 10^6$ km^2	• 15.9 BBO
Ouachitas	• Early Pennsylvanian to Early Permian	• Alabama to Mexico in eastern North America	• 2250 km long	• 500 MBO • 10 tcf gas

*BBO, billion barrels of oil; MBO, million barrels of oil; tcf, trillion cubic feet; bcf, billion cubic feet.

foreland basins, such as the Ouachita and Eastern Venezuela foreland basins, were segmented by structural highs, whereas most are continuous.

It is in estimates of the reserves (Table 1) that we see perhaps the most outstanding variability between foreland basins and fold belts. It would be most instructive for hydrocarbon exploration if one or more of the above characteristics—age, location, tectonic setting, thickness, and size—could be clearly related to hydrocarbon reserves, but this is clearly not so. Considering conventional oil only, the range in proven reserves (Table 1) is from the relatively paltry 15.9×10^6 m^3 (100×10^6 bbl) of the Alaska North Slope foreland basin to the staggering abundance of the Zagros foreland basin at about 23.8×10^9 m^3 (150×10^9 bbl). Estimates of proven natural gas reserves follow similar patterns. As in the case of conventional oil, the Zagros basin natural gas reserves *dwarf* those of other basins, being about an order of magnitude greater than the reserves of the next richest basins, the Western Canada and Eastern Venezuela basins.

If the reserves of oil sands, heavy oils, and bitumens are taken into account, both the Western Canada basin and the Eastern Venezuela basin take on new significance in terms of their past oil-generating performance—these two basins are the world's *supergiants* in terms of unconventional heavy oil and tar sands deposits (Table 1). We will return to the question of reserves and their most favorable controls and settings below.

STRATIGRAPHY

The foreland basins reviewed herein typically overlie passive continental margin deposits, commonly marine carbonates and clastics of Paleozoic and Mesozoic age (Table 2). The precise age and facies of underlying rocks is variable, however. For example, the two richest basins in terms of hydrocarbon reserves (Table 1) differ considerably in the nature of their underlying, pre-foreland basin rocks. The Zagros foreland basin is underlain by carbonate-rich passive-margin deposits with a large number of rich and widespread source facies, whereas the Eastern Venezuela foreland basin is underlain by a diverse mixture of fluvial and deltaic clastics with only limited source facies. What appears to be most important with respect to hydrocarbons trapped in the overlying foreland successions is *not* the precise nature of the pre-foreland deposits, but rather the nature, distribution, and richness of hydrocarbon source facies in these underlying rocks, and the extent to which these source facies were thermally matured through burial by the overlying foreland basin succession. This factor is most dramatically demon-

TABLE 2. Stratigraphy.

	Pre-Foreland Basin		Foreland Basin		
Basin	**Age**	**Facies**	**Age**	**Thickness**	**Facies**
Western Canada basin	• Mississippian to Devonian • Triassic to Early Jurassic • Cambro-Ordovician and Silurian	• Passive margin carbonates, lesser clastics	• Middle Jurassic to early Tertiary	• 5000 m thick in the west	• Shallow marine and nonmarine mudstone, sandstone, and conglomerate
	• Proterozoic	• Passive and rifted margin clastics			
Zagros basin	• Cambrian to mid-Eocene	• Passive margin shelf carbonate in Mesozoic • Clastics in Paleozoic	• Late Eocene to present	• Up to 12,000 m; 4000–7000 m synorogenic	• Preorogenic: shallow-water carbonates and evaporites; some pelagic marls Miocene pulse of quartzose sand • Synorogenic: Plio-Quaternary molasse
Eastern Venezuela basin: • Guarico sub-basin • Maturin sub-basin	• Late Jurassic to Eocene	• Fluvial/deltaic sandstones and shale; shallow-water marine shale and lime mudstone, shallow- to deep-water carbonate	• Early Eocene to present	• Guarico subbasin is 5 km maximum thickness, averaging 3–4 km • Maturin subbasin is 8.5–14 km thick	• Deep marine shale, turbidites, deltaic and shallow marine sandstone
• Alaska North Slope	• Pre-Upper Devonian	• Weakly metamorphosed sedimentary rocks	• Middle (?) Jurassic to late Tertiary	• 400 to 7000 m	• Deep marine basinal, basin slope, shallow marine, and nonmarine mudstone, sandstone, and conglomerate
	• Mississippian to Pennsylvanian	• Passive margin carbonates			
	• Permian to Jurassic	• Passive margin clastics			
• U.S. Rocky Mountain basin	• Late Precambrian to Mid-Jurassic	• Passive margin mixed carbonates and clastics	• Middle Jurassic to early Tertiary		• Shallow marine and nonmarine mudstone, sandstone; lacustrine sediments
Ouachitas	• Early and mid-Paleozoic	• Passive margin carbonates	• Early Pennsylvanian to Early Permian		• Alluvial to deep marine clastics

strated by the Zagros and Western Canada foreland basins, both of which have significant volumes of oil derived from the maturation of source facies in their pre-foreland successions.

Sediment thicknesses in most foreland basins discussed herein range up to about 7000 m; the Maturin subbasin of the Eastern Venezuela foreland basin stands out because it reaches 14 km in thickness. The nature of sediment fill of foreland basin successions is in part related to proximity to the equator at the time of sedimentation (Table 2). For example, the fill of the low-latitude Zagros basin contains considerable thicknesses of shallow-water shelf carbonates. The other foreland basins lack significant volumes of carbonate sediments—they were the sites of proportionately high volumes of clastic input, in part because of their higher-latitude paleogeographic settings, which do not favor carbonate sedimentation.

Of the foreland basins considered here, the Zagros foreland basin is the richest in carbonates, including shallow marine limestones and dolomites with evaporites in intrashelf lows. An episode of quartzose clastic sedimentation derived from the Arabian craton took place during early Miocene time in Kuwait, southwestern Iran, and adjacent Iraq. The fill of the Eastern Venezuela basin was predominantly siliciclastic, with less common occurrences of carbonate platformal sedimentation.

All of the basins contain an initial component of fine-grained fill. The initial fill in the Alaska North Slope, Western Canada, Eastern Venezuela, and Ouachita basins consists of relatively deep-water marine shale and turbidite deposits. These deposits give way upward to repetitive, cyclic sequences governed by interactions between tectonic pulses in the orogenic belt and eustatically determined changes in sea level. These controls are major determinants of the nature, quality, and extent of source and reservoir facies within the foreland stratigraphic record. It is thus extremely advantageous to understand these and related factors so as to comprehend fully the source and distribution of discovered and undiscovered hydrocarbon resources.

STRUCTURE

Foreland basins have asymmetrical cross-sectional profiles, being thickest adjacent to their disturbed belt and thinning distally. At a finer scale, however, each of the foreland basins described in this volume is characterized by different structural styles (Table 3).

Structural complexities of the Western Canada foreland basin are relatively minor, with variations in sediment thickness resulting from basin flexure, salt dissolution, and what appear to be reflections of minor Precambrian-basement movements. These structures do not play a significant role in the formation of most hydrocarbon traps in the undeformed part of the Western Canada foreland basin. Hydrocarbon traps in the main part of the Western Canada foreland basin (outside the Foothills belt) are primarily stratigraphic traps. Structural traps are confined essentially to the Foothills belt, where there are large reserves of natural gas with more limited reserves of oil to date.

In contrast to the Western Canada setting, the Zagros foreland basin and fold belt contains large-amplitude, concentric anticlines that provide some of the world's largest and most effective traps. The largest anticline, Kabir Kuh, is 190 km long and has an amplitude of 6 to 10 km. The large and concentric nature of these folds results from the presence of thick, internal, relatively rigid carbonate layers that were detached from the underlying Hormuz salt during deformation. Where thicknesses of argillaceous sediment are greater and where the Hormuz salt is not present, folds are smaller, tighter, and contain much smaller volumes of hydrocarbons. Fractures, created during folding, provided conduits to a rich variety of source beds located at depth within the underlying passive margin deposits, and play a significant role in reservoir capacity and performance.

The Oligocene–Miocene hydrocarbon traps of the Eastern Venezuela foreland basin are productive from normal and reverse fault closures, without which reserves would be much more limited to nonexistent. The conventional oil reserves of this basin are more than three times those of the Western Canada basin, owing in large part to the presence of fault-bounded traps.

The Alaska North Slope foreland basin is unique in that during its development there existed a northern extensional (rifted) margin and a southern compressional margin. The Brooks Range and adjacent Foothills, which bound the Alaska North Slope foreland basin, are 1000 km long and up to 300 km wide. Structures are largely the result of Brooks Range compressional tectonics. Plays consist of faulted anticlines, classified on the basis of differences in structural style, reservoir rocks, and source rocks. Without the structural component, most traps would not exist.

United States Rocky Mountain foreland basins contain structural and combined structural-stratigraphic traps formed primarily by Laramide deformation associated with the segmentation of this foreland basin system. This is in contrast to the Western Canada foreland basin, which shows little structural influence on the formation of traps.

The late Paleozoic Ouachita foredeep in southwestern North America consists of seven subbasins—Black Warrior, Arkoma, Sherman, Fort Worth, Kerr, Val Verde, and Marfa. Each subbasin is a homocline that dips away from the craton, extends to the Ouachita thrusts, and is separated from other basins by major basement uplifts or subtle arches. Some of these uplifts, such as the Wichita Mountains, subsequently became suppliers of sediment. The structure is more directly related to basement movements than is the case in the Zagros or Eastern Venezuela basin.

SOURCE ROCKS

Characteristics of source rocks—including type, richness, thickness, distribution, and maturity level—are key factors in the hydrocarbon productivity of *any* basin. The characteristics of known and potential source rocks of the six foreland basins are summarized in Table 4. Known source rocks have been verified by organic geochemical studies used to establish oil-source correlations via gross chemical compositions, or saturate fraction biomarker geochemis-

TABLE 3. Structure.

Basin	Structure
Western Canada basin	• One asymmetric basin, thick to the west; several low-relief basement arches and basins were present; significant Paleozoic salt solution in eastern Alberta and Saskatchewan • Two main orogenic phases: Middle Jurassic to Lower Cretaceous Columbian orogeny and Upper Cretaceous to lower Tertiary Laramide orogeny, which are the result of two "superterrane" collisions • Thrust and sediment-loading subsidence • Gently to steeply dipping thrust sheets in orogene that progressively stepped into and incorporated the foredeep
Zagros basin	• Large amplitude, elongate, concentric folds with hydrocarbons trapped in whaleback anticlines of large areal extent
Eastern Venezuela basin	• Two subbasins superimposed on the older passive margin sequences
Alaska North Slope	• Structures developed from Brooks Range compressional tectonics; folding in foothills decreases northward to a low-dipping homocline on the coastal plain; north margin of basin is an eastward-plunging subsurface high, the Barrow arch, which is a buried rift shoulder that forms the southern margin of the Canada basin to the north
U.S. Rocky Mountain basin	• Basement-involved compressional blocks with vertical movement of up to 9000 m, thin sediment cover • High-angle thrust faults • Hinge zone associated with normal fault blocks
Ouachitas	• Seven subbasins within a larger foreland basin. Each basin is a homocline dipping away from craton and extending to the Ouachita thrusts. Collision also caused positive structural features such as the Arbuckle Wichita Mountains, Devils River, Central Mississippi and Waco uplifts, and the Muenster arch.

try. Potential or speculative source rocks are promising in terms of distribution, maturation level, organic matter type, etc., but conclusive oil-source correlations have not yet been established. Prime source rocks generally include type II marine organic matter and contain from about 3 to 30% total organic carbon.

A comparison of the ages of the foreland basins (Table 1) with the ages of their known/potential source rocks (Table 4) indicates that *more than half of the foreland basin source rock-bearing formations are located in pre-foreland basin sediments*. Thus, *unlike* most other types of basins (active margin basins, interior cratonic basins, lacustrine basins, etc.), the hydrocarbons are derived from underlying, predominantly passive continental margin deposits, as well as from the foreland basin successions themselves.

The richest source rocks of the foreland basin successions of the Western Canada, Alaska North Slope, and Ouachita basins are condensed sections of marine shales deposited under anoxic conditions, either within the foreland succession or in the underlying pre-foreland succession (Table 4). A major foreland basin source facies in Western Canada is the Cenomanian-Turonian Second White Speckled Shales. In the Alaska North Slope basin, oil is derived from the Shublik Formation and the Kingak Shale, which are pre-foreland basin, and from the Hue Shale and possibly from the Torok Formation, which are within the foreland basin succession.

The Zagros setting is *the most extreme case* of a hydrocarbon charge system that is entirely dependent on source rocks developed within the pre-foreland succes-

TABLE 4. Source Rocks.

Basin	Age	Type/Quality
Western Canada basin	• Devonian Duvernay Fm. • Mississippian Exshaw Fm. • Triassic Doig Fm. • Jurassic Nordegg Fm. • Cretaceous coal • Cretaceous shale, Second White Speckled Shale	• Type II, to 17% TOC • Type II, to 20% TOC • Type II, to 23% TOC • Type II, to 33% TOC • Coal-bed methane • Type II/III, to 13% TOC
Zagros basin	• Cambrian • Silurian • Jurassic-mid-Cretaceous (six levels) • Paleocene-Eocene (minor)	• Type II, variable TOC levels • Type II, variable TOC levels • Type II, variable TOC levels • Type II, variable TOC levels
Eastern Venezuela basin	• Upper Cretaceous Querecual and San Antonio Fm. • Oligocene (possible) • Miocene (possible)	• Type II, to 6.6% TOC
Alaska North Slope	• Triassic Shublik Fm. • Jurassic-lowermost Cretaceous Kingak Shale and equivalents • Hauterivian-Barremian pebble shale unit • Aptian(?)-Albian Torok Fm. • Aptian-Maastrichtian Hue Shale • Albian-Oligocene(?) Canning Fm.	• Type II/III, TOC to 7%, ~2% avg. • Type II/III, TOC to 6%, ~1.8% avg. • Type II/III, TOC to 6%, ~2.5% avg. • Type II/III, TOC to 6%, ~1% avg. • Type II, TOC to 12%, ~4% avg. • Type III, TOC to 6%, ~1.5% avg.
U.S. Rocky Mountain basin	• Turonian Greenhorn cyclothem • Upper Cretaceous coal deposits • Upper Cretaceous Mancos, Lewis, and Pierre shales • Eocene lacustrine sediment • Permian Phosphoria • Pennsylvanian and Mississippian shales	• To 10%, Type II • Type III, coal-bed methane • 0.5-2%, Type II & III • To 21%, Type II and III • To 12%, Type II • To 10%, Type II
Ouachitas	• Lower Paleozoic oil- and gas-prone Woodford and Chattanooga marine shales • Pennsylvanian marine shale	• Marine shale

sion. At least *12 formations* may have generated the hydrocarbons; confirmation by organic geochemical oil-source correlations has not been made for all source facies. The regionally extensive Middle Jurassic to mid-Cretaceous shales charged most Middle East reservoirs, including those of the Zagros foreland basin and fold belt. The Proterozoic Huqf Formation, with thicknesses up to hundreds of meters, and Silurian anoxic shales were major sources of gas and oil. Mesozoic source shales were deposited within irregular intrashelf basins. Mid-Cretaceous global anoxic events (late Aptian and Albian, Albian-Cenomanian, and Cenomanian-Turonian) played a significant role in the formation of source facies in the Persian Gulf, as elsewhere (Western Canada and U.S. Rocky Mountain foreland basins, and the Alaska North Slope foreland basin).

Of all the factors that govern the existence, size, distribution, and quality of hydrocarbon reserves in foreland basins, it seems clear that source facies, as part of the hydrocarbon system, is among the most important. Without superior or at least adequate source facies, it matters little whether reservoirs are present. Edwards and Santogrossi (1990, p. 243), in their summary of divergent/passive margin basins, made a similar observation, stating ". . . . We must emphasize the overriding importance of source rocks in achieving exploration success . . . ." The second aspect of great importance is that the pre-foreland basin source facies must be matured to an oil-expulsion level at just the correct time for migration and trapping to occur in the overlying foreland basin succession. For most basins, this must be a tall order—but the richness of the Zagros Basin attests to the favorable coincidence of all these factors.

RESERVOIRS AND TRAPS

Table 5 summarizes the range of traps, reservoir facies, and seals encountered in the six foreland basins outlined here. Western Canada foreland basin reservoirs are mostly stratigraphic traps with the exception of the Foothills belt structural traps. In their discussion of the various plays, Barclay and Smith (1992) note that stratigraphic traps, nearly all of which have sandstones as reservoirs, consist of depositional pinch-outs and unconventional deep basin hydrodynamic traps, with the geometry, stratigraphic setting, and distribution governed by the complex inter-

TABLE 5. Reservoir Rocks.

Basin	Traps	Facies	Seals
Western Canada basin	• Depositional pinch-out traps in basin	• Shoreline-related incised valley, shallow offshore bars (fluvial)	• Shale
	• Thrust faulted and folded anticlinal traps in fold/thrust belt	• Passive-margin fractured carbonates	• Shale, tight carbonates
	• Unconventional deep-basin hydrodynamic traps	• Fluvial shoreline, shallow marine conglomerates and sandstone with underpressured gas	• Updip normal pressured water ± updip shale
Zagros basin	• Large elongate, concentric folds	• Fractured carbonates, minor clastics	• Evaporites (regional), shales/marl for lower reservoirs
Eastern Venezuela basin	• Normal and reverse fault closures; right lateral wrench faults; stratigraphic traps; anticlines associated with thrusting	• Sandstones	• Shale
Alaska North Slope	• Faulted anticlinal traps of Cretaceous and/or Tertiary reservoirs in the Brooks Range Foothills • Structural-stratigraphic traps	• Deep marine to nonmarine	• Shale
U.S. Rocky Mountain basin	• Structural closures and fault traps on the margins of gently dipping basin flanks • Stratigraphic traps on gently dipping basin flanks and basin centers • Regional unconformities creating stratigraphic traps • Basin-centered gas and coal-gas traps		
Oachitas	• Conventional structural and structural-stratigraphic traps; unconventional deep-basin gas traps	• Alluvial to deep marine clastics	

play among transgressions, regressions, lowstand fluvial incision events, and tectonic pulses controlled by events in the developing Cordilleran orogene. There is an increasing level of interest in the development of "unconventional" stratigraphic traps—source facies within shales at an intermediate maturity level but within which the bulk of the matured oil is still resident. Using horizontal drilling, these source facies, especially when fractured, can be treated as reservoirs; developments in this field are promising. Foothills belt structural traps are thrust-faulted and folded anticlinal traps rich in natural gas but with only limited oil reserves.

All of the reservoirs within the Zagros foreland basin and fold belt occur in broad anticlines that originated during Neogene-Quaternary compressive folding as noted above. The largest anticline, Kabir Kuh (190 km long, 6–10 km amplitude), is a truly gigantic structure in any basin. Folding-induced fractures also created well-connected reservoirs within thick carbonate successions, and this pervasive fracturing has drastically modified depositional facies. Very effective seals are created by overlying evaporites. The vast scale of the carbonate reservoirs, their very effective porosity/permeability, and their efficient evaporite seals combine to make these reservoirs among the most prolific of any basin in the world.

For the Eastern Venezuela foreland basin, sandstone reservoir character and distribution were determined initially by depositional and tectonic factors, but the size, character, and significance of these reservoirs are controlled by the nature, distribution, and extent of normal and reverse fault closures. Without these structural factors, many reservoirs would be either nonexistent or much more limited in reserves.

Alaska North Slope foreland basin reservoirs are also sandstones, and are of relatively deep marine to nonmarine origin. Reservoir quality is governed in part by leached grains and by the presence or absence of diagenetic cements, as well as by porosity loss determined by ductile grain deformation.

The United States Rocky Mountain foreland basin has a rich variety of sandstone reservoirs, with their distribution, like those in Canada, governed by transgressions, regressions, and lowstand fluvial incision events. Unlike Canada, reservoirs are bounded by Laramide structures including basement uplifts of up to 9000 meters in the Uinta Mountains, events that are part of the Laramide influence in this region.

The Ouachita foreland basin is also a region of clastic reservoirs, with stratigraphic traps dominant and hydrodynamic deep basin traps also present.

RESERVES

Hydrocarbon reserves within the foreland basins described in this volume are immense (Table 1). With more than 3.02×10^{10} m^3 (190×10^9 bbl) of proven conventional oil (Table 1), the foreland basins discussed here contain nearly 20% of what was formerly known as the "free" world's proven conventional oil reserves (i.e., excluding the reserves of the former communist block countries), as well as the vast majority of the world's known reserves of heavy oil, oil sands, and bitumen at more than 4.6×10^{11} m^3 (2.9×10^{12} bbl). It is more difficult to rank the foreland basin contribution to world reserves of natural gas, but at 2.36×10^{13} m^3 (833 tcf) it is obvious that these are highly significant.

Reserves of conventional oil within the foreland basins are very unevenly distributed, however. One foreland basin and fold belt setting, the Zagros basin, dwarfs all others in terms of conventional oil and natural gas reserves. By comparison, the type basin, the Western Canada foreland basin, has originally proven reserves of only 1.1×10^9 m^3 (7.1×10^9 bbl) of conventional oil—well below 10% of those of the Zagros basin. Even the Eastern Venezuela foreland basin, with 3.2×10^9 m^3 (20×10^9 bbl) of conventional oil, is not in the same league as the Zagros basin. What are the unique factors that account for this vast single basin hydrocarbon endowment? For more than 50 years, geoscientists have speculated on the reasons underlying this enormous richness, as the staggering abundance became clearly known through exploration and development drilling. New knowledge of the stratigraphic framework of the Middle East basins, their source facies, thermal and burial history, and organic geochemistry helped to clarify the reasons responsible for this rich endowment compared with all other basins, foreland or otherwise. Conditions for hydrocarbon maturation, generation, migration, and trapping were optimized to a degree unknown elsewhere in the world. In the opinion of Murris (1984a, b), however, one factor stands out—the factor of *horizontal scale*. For the Middle East basin as a whole, because the pre-erosional and predeformational Mesozoic platform was 2000 to 3000 km wide and at least twice as long, and because subbasin formation was minimal and thus source rocks, reservoirs, and seals have very wide extent, the reservoir closures are of *vast* horizontal scale. Murris (1984b, p. 371) further noted that the very gentle structures have two effects: minimal loss of seal efficiency through fracturing, and large trap volumes associated with limited vertical closures.

Nonconventional oil sands and bitumen deposits of the Western Canada sedimentary basin and the Eastern Venezuela basin comprise the great bulk of the world's known reserves of these deposits (e.g., Meyer and Duford, 1989). The Athabasca-Wabasca oil sands/bitumen deposit is the world's largest known natural occurrence of hydrocarbons. For the Western Canada foreland basin oil sands and bitumen, organic geochemical characteristics point to a conventional oil source, with the present state of these materials being determined by the amounts of biodegradation, water-washing, and oxidation that have taken place (Brooks et al., 1988, 1989). This points to the enormous generative potential of this basin in the past. The critical missing factor is lack of traps and seals to preserve the enormous volumes of oil from alteration to heavy oil and bitumen. Given the huge volume of oil sands and bitumen, the Western Canada basin may have been as prolific in the past in generating oil as the Zagros foreland basin and fold belt, but lacked the preservation potential.

FUTURE HYDROCARBON POTENTIAL OF FORELAND BASINS

It should be no surprise to learn that the outstanding potential for further foreland basin reserves on a global scale is resident in the Zagros foreland basin and fold belt. Beydoun et al. (1992), noting that there is a paucity of published information on the region, advance the view that the reserves of the Arabian shelf region may increase by half again *or even double* over the next several decades, depending on the level of exploration. In terms of the Zagros foreland basin and fold belt, this translates into additional reserves in excess of about 1.59×10^{10} m^3 (100×10^9 bbl) of oil and about 5.66×10^{12} m^3 (200 tcf) of natural gas. The area of greatest potential for future discoveries is estimated to be the southeastern end of the basin in the Fars province of Iran, and in the structures of northeastern Iraq, where drilling has been limited to date.

A more rigorous estimation procedure is possible for the Western Canada foreland basin, given the large amount of publicly available data, the relatively mature exploration stage, and the presence of an active research community. Podruski et al. (1988) estimated that the Jurassic and Cretaceous portion of the Western Canada basin contains about 19% (about 108×10^6 m^3) of the undiscovered conventional oil potential. Their results suggest that there are many small pools (about 795,000 m^3 or 5×10^6 bbl, or less) remaining to be discovered, but that there are few or no remaining large pools to be discovered. For the Western Canada basin as a whole, the Devonian has about 55% of undiscovered potential (Podruski et al., 1988). Remaining pools are likely to be discovered by better mapping of stratigraphic traps, especially by recent innovations in seismic reflection acquisition and interpretation (e.g., Nielsen and Porter, 1984), and by the improving level of understanding of source and reservoir relationships within the evolving sequence stratigraphic framework (Jervey, 1992; Leckie and Smith, 1992).

Play analysis and statistical techniques applied to the Alaska North Slope foreland basin (Bird and Molenaar, 1992) suggest that 2.00×10^9 m^3 (12.6×10^9 bbl) of conventional oil and 1.53×10^{12} m^3 (54.1 tcf) of natural gas remain to be found. The majority of the future potential is estimated to occur in the seven plays of the foreland basin, particularly in the eastern part of the North Slope, including the Arctic National Wildlife Refuge coastal plain.

Erlich and Barrett (1992) suggest that there is significant potential for future conventional oil and gas reserves in the deeper parts of the Eastern Venezuela basin, particularly in the El Furrial/Quiriquire trend of the Maturin subbasin. Although no specific estimates of undiscovered potential are given, the proven conventional oil reserves to date of about 3.18×10^9 m^3 (20×10^9 bbl) of oil and 1.87×10^{12} m^3 (66 tcf) of gas suggest that this basin may have a promising exploration future.

For the U.S. Rocky Mountain belt, Gries et al. (1992) indicate that the most promising future potential lies with new and novel exploration approaches including subvolcanic plays, horizontal drilling, coal-bed methane, and basin-centered gas plays resembling the hydrodynamically trapped Elmworth field of the Western Canada setting and the deep gas resources of the Ouachita foreland basin.

Exploration potential of the Ouachita basin lies with the discovery of new conventional traps, particularly in sparsely drilled areas and at greater drilling depths. There is also potential for basin-centered natural gas resources (Meckel et al., 1992).

Two main points can be drawn on future potential of the foreland basins and fold belts. The first is that the amount of basic geologic and hydrocarbon occurrence information required to make reasonable estimates of undiscovered potential varies considerably from basin to basin. The second is that no matter how precise or accurate estimates for the various foreland basins and fold belts may become in the future, the Zagros foreland basin and fold belt will continue to stand out, in terms of proven reserves and future potential.

GENERAL PROBLEMS AND CHALLENGES OF FORELAND BASINS AND FOLD BELTS

Problems and challenges may be approached on at least two levels. The first is a *global or basin scale level,* where concerns center around increasing our understanding of fundamental questions about the origin and evolution of foreland basins and fold belts. These kinds of questions and approaches include the following.

• How has the plate tectonic setting influenced the development of a particular foreland basin, in terms of tectonic pulses of sedimentary wedges, subsidence and loading history, sequence of deformational events, terrane-docking events, and geometry and style of folding and faulting? In some foreland basins and fold belts, plate margin processes exerted a strong control on many of these factors, leading to such events as the local obduction of ophiolites as in the Zagros basin and close control of the nature and style of sedimentary fill as in the Eastern Venezuela setting. In others, such as the Western Canada foreland basin and fold belt, plate tectonic effects are more subtle. The model of correlation of clastic wedges of the foreland basin with terrane-docking events provides a framework that has yet to be substantiated. Flat-plate subduction has been offered as a reasonable explanation for Laramide segmentation of the U.S. Rocky Mountain foreland basin, which is otherwise much like the Western Canada foreland basin.

• All sedimentary basins are lithospheric features. Accordingly, any complete understanding of the origin and evolution of sedimentary basins, including foreland basins and fold belts, has to consider the role of the lithosphere, including stress, variable mechanical and chemical properties, depth-dependent and time-dependent rheologies, thermal history, and other factors. Deep reflection and refraction seismic profiling along with various potential field techniques (gravity studies, electromagnetic field studies, magnetotelluric measurements, etc.) continue to shed light on the present state of the lithosphere under sedimentary basins. Future studies may be characterized by better approximations of properties at depth, particularly in attempts to model the origin and behavior of sedimentary basins by consideration of lithospheric properties and behavior through time.

• What was the role and influence (and contribution) of the crustal succession underlying a particular foreland basin in the development of that foreland basin? In terms of hydrocarbon charge histories, underlying successions have been important to basins such as the Zagros basin and to a lesser extent the Western Canada foreland basin and fold belt. What about such factors as subsidence history, nature and location of faults, etc.? For the Western Canada foreland basin, there is mounting evidence that the tectonic history and structure developed in the underlying Precambrian craton may have influenced the location and timing of faults within the foreland basin, and possibly even the location of shorelines within the clastic wedges. These and other factors are being tested within the Canadian LITHOPROBE project, a multidisciplinary research program involving deep seismic reflection and refraction lines, along with supporting geoscience studies, designed to identify deep structure and to test relationships between the craton and the overlying Phanerozoic sedimentary succession. The concepts involved are applicable to any foreland basin; all have similar questions awaiting answers.

• What are the lateral links, beyond the foreland basin, to tectonic or depositional events within the associated orogene, and in adjacent geologic settings? In the case of the Ouachita foreland basin, there is uncertainty regarding the source of clastics within the Ouachita basin—three different source locations may have been involved. This is a difficult problem to solve in this case, because much of the evidence is obscured by younger cover rocks.

• Within the developing framework of sequence stratigraphy of foreland basins, what are the relationships between global sea level changes and (relatively) local tectonic pulses? Seismic stratigraphy fostered the concept of sequence stratigraphy, and still has much to offer in identifying, correlating, and refining the detailed lithostratigraphy. For the Western Canada foreland basin and fold belt, we have seen that tectonic pulses seem to be dominant in explaining the timing, distribution, and magnitude of coarse clastic wedges. Nevertheless, the geometry and distribution of certain clastic wedges such as the Viking and Cardium sands suggest that these may be more closely related to sea level fluctuations, probably globally controlled. It is perhaps discouraging for those who see this as a major control of foreland successions that the global record of sea level changes (Posamentier et al., 1988) shows no strong correlation with major clastic wedge development in this basin. What is the situation elsewhere? There is much to be learned on this topic, from which should stem a better understanding of the stratigraphic framework and location and form of source and reservoir facies. One of the key aspects of this research will be to obtain better temporal understanding of sequence boundaries, through the use of biostratigraphy, radiometric dates, and magnetic reversal chronologies.

• What are the plate tectonic or regional tectonic controls on foreland basin development that cause the migration of depocenters through time? Marked shifts in accumulation patterns were noticed for the Eastern Venezuela, Western Canada, Ouachita, and Alaska North Slope foreland basins, and it seems clear that these can be understood and factored into exploration programs only by obtaining a better knowledge of global or regional events and their timing.

• What can be learned of the patterns of subsidence, thermal history, and fluid flow within foreland basins? For the better-known foreland basins at least, including the Western Canada, U.S. Rocky Mountain, and Alaska North Slope foreland basins, there is scope for further maturation and fluid-flow studies, including fission track work to better constrain thermal and subsidence histories. There are continuing disagreements on the amount of heat transfer by topography-driven fluid flow within present-day foreland basins. Although this factor may *seem* to be of academic interest only, it affects the framework within which hydrocarbon maturation and migration have occurred, and therefore our understanding of these processes.

A second level at which foreland basins may be approached involves problems *specific to the generation and occurrence of hydrocarbons.* A major purpose of this volume is to provide analog models of foreland basins and their hydrocarbon resources and potentials. Foreland basins and fold belts reviewed here account for almost 20% of the "free" world's known conventional oil resources. One foreland basin, the Zagros, accounts for more than three-quarters of these reserves and for much of the future potential. Only one Middle East type of setting is known, however. Finding optimum conditions that *closely* parallel this setting elsewhere seems unlikely, and thus the question of other factors prominent in pooling large hydrocarbon reserves arises.

Even if the Middle East setting is unlikely to be duplicated elsewhere, there remains much scope for exploration in other foreland basins of the world. It would seem fruitful to identify those factors, in this case for foreland basins, that control the pooling of large volumes of oil and gas. What are the unknowns for these basins, and what kinds of exploration and research programs are needed to reduce exploration risks in these basins? Reducing the level of risk is a major goal of hydrocarbon exploration. How can this be accomplished for foreland basins and fold belts? Some factors to consider include the following.

• Carmalt and St. John (1986), in their tabulation of 509 giant oil and gas fields (defined as fields having 79.5×10^6 m^3, or 500×10^6 bbl, of recoverable oil or equivalent gas), assigned 211 of these fields to their category "foredeep and underlying platform sediments," which includes foreland basins. Carmalt and St. John (1986, p. 14) considered that the most prolific oil and gas provinces are those located "on . . . continental crust, but in a mobile zone associated with collision." These giant fields seem to represent situations in which most or all factors that govern conventional oil (or gas) occurrences are at a truly optimum level. What factors govern these occurrences, and are these factors duplicated elsewhere, yet unrecognized?

• The Western Canada sedimentary basin has nine giant fields (Carmalt and St. John, 1986), but only two, the Pembina conventional oil field and the Elmworth deep basin gas field, are located within the foreland basin succession. The giant Pembina field of the Alberta portion of the Western Canada foreland basin is Canada's largest oil field. Some 0.29×10^9 m^3 (1.81×10^9 bbl) of conventional oil were pooled at Pembina, in 66 separate reservoirs—most of them in the Cardium Formation (Nielsen and Porter, 1984). The area of the Pembina field, 2330 km^2 (900 mi^2), makes it second in area only to the Ghawar field in Saudi Arabia, the largest field in the world (Carmalt and St. John, 1986). Pembina was discovered in 1953 *by accident*—the target of the discovery well was a seismic anomaly mapped at the Devonian level and believed to be a Leduc reef. The optimum factors for this trap include: updip termination of reservoir sands by "shale-out," providing an effective seal; a subtle structural hinge related to synorogenic loading of the foreland basin by the overthrust belt, providing a large, gently flexed reservoir somewhat similar to those of the Persian Gulf in areal extent and closure but developed in clastics; many levels of reservoirs (66) with moderate porosities and permeabilities in coarse-grained reservoir rocks; the presence of a diagenetic seal developed following entrapment of oil; an overpressured gas solution-drive mechanism; and no known biodegradation of the crude oil by meteoric water invasion. Geologic play analysis combined with discovery process modeling suggests that Pembina may be unique (Podruski et al., 1988). Nevertheless, the concepts that govern the occurrence of Pembina and similar giant fields are instructive, particularly in terms of hydrocarbon charge systems, oil-source correlations, reservoir dynamics, and the ability to map the nature and extent of Cardium and similar sands using modern seismic reflection techniques.

• Accordingly, we suggest that renewed examination of the controls on giant fields, within the high levels of geologic, geophysical, and organic geochemical understanding that characterize the 1990s, should be instructive and potentially of exploration significance.

• In an attempt to understand the evolution of sedimentary basins and their hydrocarbon resources, there is no substitute for systematic study of source facies—known, potential, and newly discovered. Such work has been carried on with great success in the intracratonic Williston basin of Saskatchewan, Manitoba, and North Dakota, yielding previously unknown source facies, new and unexpected or more precise oil-source correlations, and new exploration concepts indicative of significant undiscovered oil resources (Brooks et al., 1987). Such systematic study is invaluable, and can form the basis of renewed exploration. It need not be carried out by the petroleum industry alone, but by research partnerships among the petroleum industry, government, and academic research communities.

• Resource evaluation studies such as those conducted for the Western Canada sedimentary basin (Podruski et al., 1988) or the Alaska North Slope basin (Mast et al., 1989) provide publicly available geologic play analyses and estimates of undiscovered potential. Whether one agrees with the statistical estimates presented in such studies or with the estimation process(es) used, the data gathering required to assemble actual and conceptual geologic plays is a prime source of exploration information and stimulus.

• The systematic analysis of a foreland basin via sequence stratigraphy or via hydrocarbon charge models serves the very useful purpose of integrating a great deal of sometimes disparate data. This is equally true of fluid-migration studies, backstripping computer-based models, thermal/subsidence history models, fission-track studies, reservoir diagenesis studies, source-seal relationships, etc.

• For the less well-known basins, it seems essential to accumulate more fundamental knowledge of all aspects of these basins, ideally in a publicly available format for all to use. The Western Canada, U.S. Rocky Mountain, and Alaska foreland basins provide good examples.

• For specific problems such as the origin of the huge reserves of oil sands, heavy oils, and bitumens in the Western Canada sedimentary basin or the Orinoco tar belt of the Eastern Venezuela basin, questions center around the distribution, origin, and chemical state of these materials, and their relationship to known conventional reserves. There is still no satisfactory answer for the source of the vast Canadian oil sands, but the problem has been clarified by organic geochemical studies. The level of understanding of the Orinoco tar belt seems to be at an earlier stage, but the questions are similar.

FORELAND BASIN OPPORTUNITIES

The Western Canada, Ouachita, and U.S. Rocky Mountain foreland basins can be considered as mature exploration basins. Large fields are not expected to be discovered. Much of their future potential lies in the development of new exploration and production technology to enhance secondary recovery from existing fields and to find more subtle seismic traps. In the Ouachita basin, the Kerr and Marfa basins have only been sparsely drilled. The Marfa basin, like some of the U.S. Rocky Mountain basins, is covered by volcanic strata that, prior to recent technological advances, had been seismically opaque.

In contrast, exploration in the Alaska North Slope foreland basin and in the northeastern Arabian shelf, including the Zagros basin, is immature. To date, many of the hydrocarbon accumulations in the Alaska North Slope are noncommercial, even though they would be considered as large fields in much of the rest of North America. The future potential of this basin is considerable. Mast et al. (1989) predicted a future potential of 1.65×10^9 m^3 (10.4×10^9 bbl) of oil and 1.12×10^{12} m^3 (39.5 tcf) of gas within deposits of the North Slope foreland basin. In the northeast Arabian shelf, accurate hydrocarbon assessments are difficult to carry out because of the overall immaturity of exploration. Beydoun et al. (1992) conservatively estimate the future potential of the Zagros foreland basin to be 7.95×10^9 m^3 (50×10^9 bbl) of oil and 7.79×10^{12} m^3 (275 tcf) of gas. Few of the structures in Fars province, Iran, and none in northeast Iraq, have had deep tests to date.

In the Venezuelan basin, future potential appears to lie in deeper exploration targets.

Future exploration targets within foreland basin successions will continue to be organic-rich marine shales that were deposited as condensed sections. Where thermally mature, in the appropriate basin setting, and highly fractured, these shales have the potential to act as oil source rocks *and* reservoirs. Examples include the Coniacian-Santonian Niobrara Shale in the United States and the Cenomanian-Turonian Second White Speckled Shale in Canada. The exploration potential is increased where the shale is interbedded with thin, tightly cemented sandstones. Current exploration activity is occurring on the flanks of Laramide structures in the United States and adjacent to the Rocky Mountain Foothills in Canada. These shales provide ideal horizontal drilling targets.

Coal-bed methane and minor amounts of condensate are being explored for cautiously in nonmarine coal-bearing sediments in the United States and Canada. As technology improves, coal gas plays are expected to become important in other basins.

REFERENCES CITED

Barclay, J. E., and D. G. Smith, 1992, Western Canada foreland basin oil and gas plays, this volume.

Beydoun, Z. R., M. W. Hughes Clarke, and R. Stonely, 1992, Petroleum in the Zagros basin: a late Tertiary foreland basin overprinted onto the outer edge of a vast hydrocarbon-rich Paleozoic-Mesozoic passive-margin shelf, this volume.

Bird, K. J., and C. M. Molenaar, 1992, The North Slope foreland basin, Alaska, this volume.

Brooks, P. W., M. G. Fowler, and R. W. Macqueen, 1988, Biological marker and conventional organic geochemistry of oil sands/heavy oils, Western Canada basin: Organic Geochemistry, v. 12, n. 6, p. 519–538.

Brooks, P. W., M. G. Fowler, and R. W. Macqueen, 1989, Biomarker geochemistry of Cretaceous oil sands, heavy oil and Paleozoic carbonate trend bitumens, Western Canada basin, *in* R. F. Meyer and E. J. Wiggins, eds., Proceedings, Fourth UNITAR/UNDP International Conference on Heavy Crude and Tar Sands, published by AOSTRA, Edmonton, Alberta, v. 2, Geology, Chemistry, p. 594-606.

Brooks, P. W., L. R. Snowdon, and K. G. Osadetz, 1987, Families of oils in Southeastern Saskatchewan, *in* J. E. Christopher and C. G. Carlson, eds., International Williston Basin Symposium North Dakota, p. 253-264.

Carmalt, S. W., and B. St. John, 1986, Giant oil and gas fields, *in* M. T. Halbouty, ed., Future petroleum provinces of the world: AAPG Memoir 40, p. 11-53

Edwards, J. D., and P. A. Santogrossi, eds., 1990, Divergent/passive margin basins: AAPG Memoir 48, 252 p.

Erlich, R. N., and S. F. Barrett, 1992, Petroleum geology of the Eastern Venezuela foreland basin, this volume.

Gries, R., J. C. Dolson, and R. G. H. Raynolds, 1992, Structural and stratigraphic evolution and hydrocarbon distribution, Rocky Mountain foreland, this volume.

Jervey, M. T., 1992, Siliciclastic sequence development in foreland basins, with examples from the Western Canada foreland basin, this volume.

Leckie, D. A., and D. G. Smith, 1992, Regional setting, evolution, and depositional cycles of the Western Canada foreland basin, this volume.

Mast, R. F., G. L. Dolton, R. A. Crovelli, D. H. Root, E. D. Attanasi, P. E. Martin, L. W. Cooke, G. B. Carpenter, W. C. Pecora, and M. B. Rose, 1989, Estimates of undiscovered conventional oil and gas resources in the United States—a part of the Nation's energy endowment: U.S. Department of the Interior, 44 p.

Meckel, L. D., D. G. Smith, and L. A. Wells, 1992, Ouachita foredeep basins: regional paleogeography and habitat of hydrocarbons, this volume.

Meyer, R. F., and J. M. Duford, 1989, Resources of heavy oil and natural bitumen worldwide, *in* R. F. Meyer and E. J. Wiggins, eds., Proceedings, Fourth UNITAR/UNDP International Conference on Heavy Crude and Tar Sands, published by AOSTRA, Edmonton, Alberta, v. 2, Geology, Chemistry, p. 277-307.

Murris, R. J, 1984a, Introduction, *in* G. Demaison and R. J. Murris, eds., Petroleum geochemistry and basin evaluation: AAPG Memoir 35, p. x-xii.

Murris, R. J., 1984b, Middle East: stratigraphic evolution and oil habitat, *in* G. Demaison and R. J. Murris, eds., Petroleum geochemistry and basin evaluation: AAPG Memoir 35, p. 353-372.

Nielsen, A. R., and J. W. Porter, 1984, Pembina oil field—in retrospect, *in* D. F. Stott and D. J. Glass, eds., The Mesozoic of middle North America: Canadian Society of Petroleum Geologists Memoir 9, p. 1-13.

Podruski, J. A, J. E. Barclay, A. P. Hamblin, P. J. Lee, K. G. Osadetz, R. M. Procter, G. C. Taylor, R. F. Conn, and J. A. Christie, 1988, Conventional oil resources of Western Canada (light and medium): Geological Survey of Canada Paper 87-26, 149 p.

Posamentier, H. W., M. T. Jervey, and P. R. Vail, 1988, Eustatic controls on clastic deposition I—conceptual framework, *in* C. K. Wilgus, B. S. Hastings, C. G. St. G. Kendall, H. W. Posamentier, C. A. Ross, and J. C. Van Wagoner, eds., Sea level change—an integrated approach: SEPM Special Publication 42, p. 110-124.

Index